Key

84	Alamor	93	Loja-Zamora road
7	Alto Tambo	52	Loreto road
10	Ángel, Páramo de	80	Macara
53	Archidona	9	Maldonado road
82	Arenillas	28	Machalilla
1	Atacames	73	Manglares-Churute Ecological Reserve
67	Ayampe, Río	74	Manta Real
48	Baeza	31	Maquipucuna
33	Bellavista Lodge	57	Maxus road
16	Bermejo oil-field area	32	Mindo
3	Bilsa	11	Mongus, Cerro
71	Blanco, Cerro	2	Muisne
95	Bombuscaro, Río		
5	Borbón, Río	79	Oña
83	Buenaventura		
87	Cajanuma	60	Pacuyacu, Río
26	Bahía de Caráquez	45	Papallacta pass
86	Catacocha	42	Pasochoa
85	Celica	81	Pitahaya, Puerto
39	Chinapinza	4	Playa de Oro
41	Chiriboga road	96	Quebrada Honda
56	Coca	62	Quevedo
64	Colta, Laguna de	40	Quito
76	Cuenca	43	Río Palenque Scientific Station
19	Cuyabeno, Laguna de		
46	Cuyuja	90	Sabiango
69	Ecuasal lagoons	21	Sacha Lodge
75	El Cajas National Recreation Area	49	San Isidro, Cabañas
47	El Chaco	14	San Pablo, Laguna de
6	El Placer	44	San Rafael Falls
25	Filo de Monos	29	San Sebastián, Cerro
78	Girón	99	Santiago
36	Guaillabamba, Río	37	Santo Domingo de los Colorados
77	Gualaceo-Limón road	65	Sarayacu
12	Guandera Biological Reserve	50	SierrAzul
72	Guayaquil	30	Simón Bolívar road (= Pedro Vicente Maldonado road)
51	Huacamayos, Cordillera de	91	Sozoranga
24	Imuyacocha	100	Taisha
55	Jatun Sacha	34	Tandayapa
63	Jauneche	54	Tena
70	José Velasco Ibarra, Represa	18	Tigre Playa
66	Kapawi Lodge	38	Tinalandia
15	La Bonita road	59	Tiputini Biodiversity Center
23	Lagarto, Río	92	Utuana
17	Lago Agrio	88	Vilcabamba
68	La Libertad	13	Yaguarcocha, Laguna de
27	La Plata, Isla de	35	Yanacocha
22	La Selva	58	Yuturi Lodge
97	Las Lagunillas, Cordillera	94	Zamora
20	Limoncocha	61	Zancudococha
8	Lita	89	Zapotillo
		98	Zumba

THE BIRDS OF ECUADOR

VOLUME I

THE BIRDS OF
Ecuador VOLUME I

Status, Distribution, and Taxonomy

Robert S. Ridgely and Paul J. Greenfield

With the collaboration of
Mark B. Robbins and Paul Coopmans

In association with
The Academy of Natural Sciences

COMSTOCK PUBLISHING ASSOCIATES a division of

CORNELL UNIVERSITY PRESS | Ithaca, New York

First published 2001 by Cornell University Press
First printing, Cornell Paperbacks, 2001

Printed in Hong Kong
Color plates printed in Hong Kong

Library of Congress Cataloging-in-Publication Data

Ridgely, Robert S., 1946–
   The birds of Ecuador / by Robert S. Ridgely and Paul J. Greenfield;
with the collaboration of Mark B. Robbins and Paul Coopmans.
      v.   cm.
   Includes bibliographical references (p.   ) and indexes.
   Contents: vol.   1. Status, distribution, and taxonomy—vol.   2.
Field guide.
   ISBN 0-8014-8720-X (v. 1 : pbk.)—ISBN 0-8014-8721-8 (v. 2 : pbk.)—
ISBN 0-8014-8722-6 (set : pbk.)
   1. Birds–Ecuador.   2. Birds–Ecuador–Identification.   I. Greenfield,
Paul J.   II. Title.
   QL689.E2 R53 2001
   598'.09866—dc21   00-052292

Cornell University Press strives to use environmentally responsible suppliers and materials to the fullest extent possible in the publishing of its books. Such materials include vegetable-based, low-VOC inks and acid-free papers that are recycled, totally chlorine-free, or partly composed of nonwood fibers. Books that bear the logo of the FSC (Forest Stewardship Council) use paper taken from forests that have been inspected and certified as meeting the highest standards for environmental and social responsibility. For further information, visit our website at www.cornellpress.cornell.edu.

Paperback printing   10   9   8   7   6   5   4   3   2   1

Contents

Foreword

For many decades world tourism has treated the tiny Andean country of Ecuador as a staging site for the ultimate ecotourism destination, the Galápagos Islands, some 800 km off the Ecuador coast in the Pacific Ocean. Yes, the wild remote beauty and tame wildlife in a natural laboratory of evolution guaranteed a remarkable experience. After all, the finches and mockingbirds, giant tortoises and marine iguanas, as well as other creatures had inspired Charles Darwin, and all those who were privileged to follow him to the Enchanted Isles.

Mainland Ecuador, however, awaited serious world attention. Astride the equator, this amazing little country embraces diverse habitats from the dry Pacific coast to the towering volcanoes of the Andes to the great forests of the Amazon basin. In just 283,000 sq km, hundreds of bird species beckoned, most of them little known, perhaps some even yet to be discovered. We all could manage the fascinating, but minimal, avifauna of the Galápagos. But the birds of mainland Ecuador posed a daunting challenge indeed.

More than twenty years ago, Bob Ridgely and Paul Greenfield began the process of modern exploration and discovery on the Ecuadorian mainland. Frank Chapman and others had cracked the shell and peeked inside. But it was Ridgely, Greenfield, and their colleagues who mastered the new avifauna, secret after secret. They found new birds in accessible sites thought to be well known. Year after year they documented poorly known bird faunas in the remote corners of Ecuador, corners that were gradually becoming more accessible (some of them). Their work, supported by the Academy of Natural Sciences in Philadelphia (where Ridgely worked with me in the Ornithology Department), the Museo de Ciencias Naturales in Quito, and the John D. and Catherine T. MacArthur Foundation, had several goals, one of the most important of which was to produce a well-illustrated field guide to Ecuador's birds. The hope—and I think it is justified—was that this would attract the attention of world birders and ecotourists to the untapped riches found in mainland Ecuador. The ultimate purpose was broader and grander: the conservation of the extraordinary diversity of Ecuador and beyond.

In 1984 I had the privilege of joining Bob Ridgely and Paul Greenfield on a punishing expedition to the biologically unknown Cordillera de Cutucú in southeastern Ecuador. Ridgely and Greenfield alone knew bird vocalizations well enough to pick out that one that was slightly different, and whose author would usually turn out to be a rarity. Day after day, there and elsewhere, they—together with others, notably Mark Robbins, then also at the Academy (and, justifiably, a collaborator on Volume I of this book)—made ornithological history with the rediscovery of long-lost or hardly known birds. Elsewhere they even found some totally new ones. *The Birds of Ecuador* is the grand result of this expedition, and of many others. The two volumes are landmark works in that they graduate our scientific knowledge to new heights. They are also landmarks in that at last the door to full public appreciation of Ecuador's avifauna is open.

Finally, this wonderful new book challenges us to protect what we know. No more excuses. Conservation is born first of discovery and wonder, then of understanding, and finally of action. Thank you, Bob Ridgely and Paul Greenfield, for persevering and sharing with us your profound knowledge of and—it shows—love for the birds of Ecuador. Now we have the tool we need to ensure their future.

Frank B. Gill
National Audubon Society

Preface

More than twenty years ago, in July 1978, RSR—together with a friend from Yale University, David Wilcove—was conducting research in Ecuador on the status of certain rare parrot species. We had assistance during that trip from several people in the U.S. State Department and Peace Corps, and one of them chanced to mention that a young American artist, interested in birds and married to a Guayaquileña, had just moved to Quito. People interested in birds were a rarity in those days, so I called him, and shortly we were out looking for birds in the mountains west of the city and then at Tinalandia. The artist was, of course, PJG. And from our very first meeting, it seems we were discussing the possibility of together "doing" a book on the birds of Ecuador.

Naive were we; little did we know what we were getting ourselves into! But you are holding in your hands the result of our collaborative effort—which was aided, of course, by the efforts of many others. At the outset, it is fair now to admit, we did not comprehend just how little we knew. In part it was that realization, and that we would have to make extraordinary efforts over the years to fill in the gaps of knowledge, that delayed completion of *The Birds of Ecuador*. Some despaired of our ever finishing, and indeed at times it seemed to us too that we would never reach that point, that it would continue to recede into the future as RSR continued to revise and update text and PJG to fine-tune his paintings. By 1998, however, it was clear that we were approaching the end, and a superhuman effort that extended over many months put us over the top.

The Birds of Ecuador represents the culmination of our two-decade research effort investigating Ecuador's avifauna. Of course there is much yet to be learned—and avifaunas are never static anyway—so it is our fervent desire that this ongoing period of active field work will continue. In particular we hope that the gradually increasing number of Ecuadorian nationals conducting research or simply going out to observe and enjoy birds will continue to increase. We hope *The Birds of Ecuador* itself will contribute to this burgeoning interest.

It is our even more fervent hope that our book's release will stimulate efforts to protect Ecuador's birdlife. Both of us feel incredibly fortunate to have seen as much as we have; it is still a wonderful world out there! Nothing would please us more than to know that through the active participation of both Ecuadorians and interested parties from abroad, Ecuador's avifauna continued to prosper. To this end, we encourage all readers and users of *The Birds of Ecuador* to support the activities of the numerous nongovernmental conservation organizations that have sprung up over the past few decades. Most of these—notable among them Fundación Jatun Sacha, Fundación Maquipucuna, and Arcoiris—are broadly based, but we would be remiss not to mention the two that explicitly focus on birds and the threats they face: Fundación Ornitológica del Ecuador (CECIA) and Fundación Jocotoco. We actively work with and support both organizations, and we strongly encourage anyone with an interest in Ecuador's birds to do so as well. Both of us look forward to working with you.

ROBERT S. RIDGELY
Ornithology Department
Academy of Natural Sciences
1900 Benjamin Franklin Pkwy
Philadelphia, PA 19103
USA

PAUL J. GREENFIELD
Urbanización El Bosque
2da Etapa, Calle 6, Lote 161
Edificio El Parque
Quito
Ecuador

CECIA can be contacted at:

La Tierra 203 y los Shyris
P.O. Box 17-17-906
Quito
Ecuador

Fundación Jocotoco can be contacted through:

Dr. Robert S. Ridgely   or
(address above)

Dr. Nigel Simpson
Honeysuckle Cottage
Tidenham Chase
Nr. Chepstow, Gwent
NP6 7JW United Kingdom

Acknowledgments

Any book such as ours inevitably is the result of the efforts of many individuals and organizations. Our desire here is to express our heartfelt gratitude to everyone who has been involved in what has doubtless at times seemed a never-ending process.

In particular the debt we owe to our two collaborators is immeasurable. Mark B. Robbins, formerly collection manager of the Department of Ornithology at the Academy of Natural Sciences of Philadelphia (ANSP), is surely the finest expedition organizer anywhere, and we all benefited from his logistical know-how, but it was for his skill in preparing field specimens, and enthusiasm for teaching others the minutiae of that art, that he will perhaps be longest remembered. His assistance in the preparation of the initial drafts of Volume I's text was substantial, and especially involved making correct subspecific diagnoses. Paul Coopmans is a relative newcomer to the Ecuadorian ornithological scene, but in the mid- and late 1990s he probably spent more time in the field than anyone else. In the past decade Paul has developed a particular interest in bird vocalizations, to the point where the breadth of his knowledge on the subject is perhaps unrivaled. Paul has directed his careful eye to the texts of both volumes, in Volume I sharing his many personal observations and helping to keep RSR abreast of ornithological developments in Ecuador, and in Volume II concentrating on accurately transcribing a representative range of vocalizations for each species.

Financial assistance came from various quarters, and we are grateful to all. First and foremost, we thank the John D. and Catherine T. MacArthur Foundation for its generous support of our field investigations throughout Ecuador in the early 1990s. The MacArthur funding permitted us to vastly accelerate our efforts to learn as much as we could about Ecuador's rich avifauna. Little did the foundation realize—nor did we—that almost another decade would be needed for our efforts to reach fruition, and we can only hope that the wait was worth it. Those funds were channeled through the ANSP, a marvelous institution where RSR has been privileged to be on staff for more than fifteen years; a more congenial "home" could not be imagined, and PJG too enjoyed its hospitality during his frequent working visits. We are also grateful to our host institution in Ecuador, the Museo Ecuatoriano de Ciencias Naturales (MECN) in Quito, whose staffing and infrastructure improved markedly during our period of most extensive collaboration and which was always of great assistance in permitting and other logistical details during our work in Ecuador.

Various individuals also helped keep us (particularly PJG) financially afloat through some lean years. PJG acknowledges his special gratitude to Evelyn Fowles who volunteered her time, energy, and enthusiasm and, together with Christine Bush and RSR, organized a fund-raising drive on PJG's behalf through the ANSP. Numerous persons, too many to mention every one, participated in this campaign, and we are grateful to all of them for the confidence they extended to us and hope they enjoy what they now have in front of them. The following individuals made especially important contributions to this fund: Wallace Dayton, Daisy Ford, Evelyn Fowles, William Greenfield, Kit Hansen and Stephen Greenfield, Dr. Norman Mellor, in memory of Cdr. Gregory Peirce, and Dr. and Mrs. Howard E. Wilson. We would like to single out Tom Butler, who was an early supporter of our project and made several significant financial contributions, and Jim and Jean Macaleer, who participated in a major way toward the end.

The only way to become really familiar with an avifauna is to study it intensively in the field as well as in the lab or studio. The recent upsurge in birding tours to Ecuador, a development in which both of us played a major role, has provided us with a previously unparalleled opportunity to get into the field on a regular basis. Both of us have benefited substantially from this frequent and repeated exposure to Ecuador's birds, having led numerous trips to various parts of the country for more than twenty years, especially under the auspices of Victor Emanuel Nature Tours. RSR has also worked with Wings and PJG with several other companies, among them Wilderness Adventure, Voyagers, Ornitholidays, Princeton Nature Tours, New Jersey Audubon Society, and Neblina Forest. To all these companies, and to the individuals who participated in the trips themselves, we can only extend a humble thanks and express the hope that we will see you again in Ecuador.

A few of our trips, especially those of RSR, were privately financed by individuals willing to endure the rigors of serious expedition travel. These individuals include Ken Berlin, Hugh Braswell, Nigel Simpson, Rod Thompson, Marc Weinberger, and Minturn Wright. The Catherwood Foundation underwrote the ANSP's initial exploratory trip to Ecuador, our arduous trek to the previously unexplored upper slopes of the Cordillera de Cutucú in 1984. We are grateful to one and all and trust that the stories you have been able to relate have more than compensated for any of the difficulties we encountered.

Other field work has been sponsored by commercial companies in Ecuador, including Ecuambiente (with Maria Eugenia Puente and Jorge Aguilera) and Transturi and Metropolitan Touring (and Raul Garcia and the late Pedro Proaño), and we are grateful for the opportunities provided. The owners of several private properties, notably SierrAzul (= Hacienda Aragon) in Napo and Hacienda La Libertad in Cañar, also encouraged our research visits to the areas they hold dear.

A major thrust of the ANSP's work in Ecuador has been to assemble a comprehensive working collection of modern, full-data study skins of Ecuadorian birds, the collection that both of us ultimately depended on for our work on *The Birds of Ecuador*. Assembling such a collection is not a simple task, and many individuals played crucial roles in its success. Apart from the aforementioned Mark Robbins, foremost among these was Francisco Sornoza, who was for many years our in-country project coordinator. Young and inexperienced at the start, "Pancho" quickly matured as he took on increasingly heavy loads of responsibility, gradually learned Ecuador's birds, and also learned a little English; we remain immensely grateful to him for his herculean efforts. Throughout this period we worked in close concert with the MECN in Quito. That institution, and in particular its directors, Miguel Moreno and Fausto Sarmiento, always proved a willing partner and was especially helpful in working through the intricacies of the governmental permitting process. MECN staff members participated in all of our expeditions, and we greatly appreciate the efforts of Marco Jácome, Juan Carlos Matheus, and the late Juan José Espinosa. Col. Paul Scharf of the U.S. Marine Corps enabled us to visit several remote Ecuadorian military camps and participated in several trips. William Phillips facilitated our visits to SierrAzul. Other individuals who participated in trips and expeditions, and all of whom made valuable contributions to their success, include David Agro, David Brewer, Christine Bush, Angelo Capparella, Tristan Davis, Peter English, Frank Gill, Skip Glenn, Steve Greenfield, Steve Holt, Rick Huber, Niels Krabbe, Jackson Loomis, Todd Mark, Terry Maxwell, Bob Peck, Tracy Pedersen, Gary Rosenberg, Tom Schulenberg, Fred and Jodie Sheldon, Beth Slikas, Fernando Sornoza, Doug Wechsler, and Andrew Whittaker.

Much of the actual work on *The Birds of Ecuador* was accomplished at the ANSP, and a more stimulating place to write and paint can hardly be imagined. Here we benefitted not only from the Academy's marvelous collection of study skins but also its superb library and the most comprehensive collection of bird photographs ever assembled (Visual Resources for Ornithology [VIREO]). The following individuals materially abetted our efforts in one way or another at the Academy: David Agro, Louis Bevier, Christine Bush, Sally Conyne, Frank Gill, and Mark Robbins. Over the past several years PJG's brother, Steve Greenfield, has been especially helpful in work on the distribution maps, as has Sally Conyne. Through it all we remain especially grateful to Frank Gill, for much of this period chair of the Ornithology Department and who for many years abetted and encouraged RSR's ongoing research on Ecuadorian birds, quietly enduring the many delays in our completion of the task.

Both of us, but especially RSR, have spent substantial periods of time at other institutions, especially the American Museum of Natural History in New York, or have communicated with individuals at those museums concerning Ecuadorian material in their care. At the American Museum of Natural History, John Bull, Bud Lanyon, Mary LeCroy, François Vuilleumier, and the late John Farrand were especially helpful. We also thank Steve Cardiff, John O'Neill, and Van Remsen at the Louisiana State University Museum of Zoology in Baton

Rouge; Lloyd Kiff and Manuel Marín at the Western Foundation for Vertebrate Zoology in California; Ken Parkes at the Carnegie Museum in Pittsburgh; Raymond Paynter, Jr., at the Museum of Comparative Zoology in Cambridge, Massachusetts; and Tom Schulenberg and Melvin Traylor at the Field Museum of Natural History in Chicago.

During the last several decades many other individuals have investigated one or more aspects of Ecuadorian ornithology. As word got out that we were preparing *The Birds of Ecuador*, all freely shared their questions and newly acquired information with us, thus creating what became truly a communal effort. Although our gratitude goes out to all, we here name the more important of those individuals, mainly those who have not been mentioned previously in some capacity and each of whom has had significant input into this book: John Arvin, Dusty Becker, Bob Behrstock, Karl Berg, Brinley Best, Rob Bleiweiss, Bonnie Bochan, Greg Budney, Chris Canaday, Juan Manuel Carrión, Mario Cohn-Haft, Peter English, Jon Fjeldså, Ben Haase, Steve Hilty, Steve Howell, Olaf John, Ralph Jones, Lou Jost, Guy Kirwan, Raymond Lévêque, Bernabé López-Lanús, Jane Lyons, Mitch Lysinger, Felix Man Ging, Manuel Marín, John Moore, Lelis Navarrete, Jonas Nilsson, Fernando Ortiz-Crespo, David Pearson, David Peerman, Michael Poulsen, Rick Prum, Carsten Rahbek, Jan Rasmussen, Giovanni Rivadeneira, Gary Rosenberg, John Rowlett, Rose Ann Rowlett, Paul Salaman, Paul Scharf, Fred Sibley, Francisco Sornoza, Dan and Erika Tallman, David Wege, Bret Whitney, Andrew Whittaker, Rob Williams, David Wolf, and the late Tom Davis and Ted Parker. We should especially like to single out Niels Krabbe, who put at our disposal his vast knowledge of Andean birds, and Paul Van Gasse, whose meticulous critical reading and proofing of Volume I were invaluable. Lastly, the all-important copyediting phase of the work was ably executed by Elizabeth Pierson, whose meticulous eye managed to pick up inconsistencies that crept into the text, not to mention some outright slipups; RSR's gratitude for her patient assistance is substantial indeed. Despite our—and our helpers'—best efforts, however, inevitably a few inconsistencies and misstatements will remain, errors of omission and commission for which we take full responsibility (but please tell us about them!).

As with all other aspects of this opus, the illustrations have not been produced in a vacuum. Although the brush—actually hundreds of brushes!—was ultimately wielded only by PJG over these twenty years, the plates could not have been painted without much help and consultation. The two of us worked together closely during their planning and execution, and PJG good-naturedly undertook all the corrections that RSR could spot (no matter how minor or subtle). The late Gustavo Orcés shared his knowledge with PJG when he first arrived in Quito, and both Bill Davis and later Paul Scharf were PJG's most frequent field companions for several years during this "learning process." Many individuals looked over the plates while they were in various stages of production and offered helpful comments; among them were Jon Fjeldså, Niels Krabbe, Mark Robbins, John Rowlett, Rose Ann Rowlett, Tom Schulenberg, and Guy

Tudor. Louis Bevier shared his knowledge of plumage and posture details for certain boreal migrants and pelagics. Bill Clark, John Dunning, Arthur Panzer, Doug Wechsler, and the entire VIREO team made their photographs available. Tracy Pedersen, a talented artist in her own right, graciously agreed to paint a species that was inadvertently left off the plates (we won't reveal which it was). PJG's brother Steve was always a help, and PJG's wife, Martha, assisted in cleaning up the plates.

PJG's deepest gratitude extends to Eva and Felix Hirschberg and especially Lenore Greenfield for their encouragement, help, and support; their presence was always an inspiration. Of special importance were Jorge and Martha de Kalil, whose hospitality, generosity, and love helped make PJG's life in Ecuador so special; they are missed. To William Greenfield—who will never see the fruits of this labor—PJG expresses his love and appreciation for his endless and all-encompassing positivity.

RSR was introduced to the natural world by his parents, Beverly and Barbara Ridgely, and he remains truly grateful for their forbearance, and encouragement, as he commenced his career as an ornithologist and conservationist. It may not have been financially rewarding, but the many other, less tangible rewards have more than made up for it.

Lastly, we both express our gratitude to and love for our wives, Martha Greenfield and Peg Ridgely, whose forbearance during this process has been nothing short of a marvel. Martha unselfishly awaited PJG's return from innumerable field and work trips and put up with all the years of "bird talk." Peg has been the very model of a wife and partner to RSR and has learned to put up with (and even enjoy!) all those late dinners. Neither of us could have done this without you, and our gratitude will never end.

Abbreviations

Abbreviations of the following museums and institutions housing material of significance to Ecuadorian ornithology are used repeatedly throughout Volume I. Others that are mentioned only once or infrequently are named fully.

AMNH   American Museum of Natural History
ANSP   Academy of Natural Sciences of Philadelphia
BMNH   British Museum of Natural History
FMNH   Field Museum of Natural History
LNS   Library of Natural Sounds (Cornell Laboratory of Ornithology)
LSUMZ   Louisiana State University Museum of Zoology
MECN   Museo Ecuatoriano de Ciencias Naturales
MCZ   Museum of Comparative Zoology
UKMNH   University of Kansas Museum of Natural History
USNM   United States National Museum (Smithsonian Institution)
VIREO   Visual Resources for Ornithology (at the ANSP)
WFVZ   Western Foundation for Vertebrate Zoology
ZMCOP   Zoological Museum of the University of Copenhagen

THE BIRDS OF ECUADOR

VOLUME I

Plan of the Book

An all-too-frequent complaint directed at the authors and illustrators of Neotropical bird books is that "the book is too big—I'd have to hire a porter to carry it around." Essentially these critics want it both ways. They want everything illustrated, and they want detail and distribution maps necessary to enable them to sort through the superabundance of birds. At the same time they want the book to be a manageable size—and to be cheap too! Something has to give: there simply are too many birds in most Neotropical countries to make this possible, and the complexities are too vast for simplification to be an option.

We wrestled with this conundrum for years before devising a simple but relatively novel solution: write two books! And that is what we have done, with one volume intended primarily for field use and the other for your library (or hotel room or even car). The books are designed to be used together, following precisely the same family and species sequence.

Volume I includes information on the distribution and status of each species in Ecuador, thoroughly referenced and placed in a historical context when necessary. If a species is deemed at risk, its conservation status in Ecuador is discussed. Then follows a taxonomic discussion in which species-level and subspecific systematics issues are presented, together with possible alternate treatments and our rationale for adopting the approach we did. Subspecies are evaluated in terms of their validity, and each taxon's distribution in Ecuador is given. In addition, we address the contentious issue of English names. Lastly, we give the species' worldwide, extra-Ecuador distribution.

This leaves for Volume II all matters explicitly pertaining to identification: the 96 plates, each with a facing page, and following these the individual species accounts. Each account includes a description of the species; sections on similar species, habits, and voice; and a map of that species' distribution in Ecuador.

We recognize that presenting information in such a way inevitably means repeating some things and that this takes up valuable space. We have attempted to minimize this, however, and hope that the result has more advantages than disadvantages. We also recognize that neither volume is "small" in the sense that a book dealing with the birds of a relatively species-poor area in the temperate zone can be. But really the glory of the tropics *is* its diversity: revel in it, marvel at it, and have fun trying to figure it all out. We find few other activities to be as thoroughly engrossing.

Volume I: Status, Distribution, and Taxonomy

This volume is devoted to a detailed discussion of the distribution, abundance, status, taxonomy, and overall range of each of Ecuador's nearly 1600 bird species. Information on each species is presented in much the same way, to facilitate interspecific comparison, but inevitably some accounts—for species on which one or more topics are either complex or controversial—are longer than others. An overview of how Volume I is organized and an explanation of the terms used follow.

Taxonomy

Since the early 1990s—subsequent mainly to the publication of two stimulating and vastly important synthetic general works on the birds of the world (Sibley and Ahlquist 1990; Sibley and Monroe 1990)—it has been proposed that the higher systematics of birds, all of which are classified in the Class Aves, undergo a radical transformation. This transformation, often referred to as the "Sibley sequence," is based mainly on various groups' perceived relationships as judged by the results of various biochemical techniques (especially studies of DNA-DNA hybridization). Following these changes results in a drastically altered sequence of families from the more familiar "Peters sequence" (derived from the *Birds of the World* series) employed, for instance, in the most recent American Ornithologists' Union (AOU) Check-lists (AOU 1983, 1998). After consultation with ornithologists conversant with the several proposals that have been advanced, we opted not to follow the "Sibley sequence" but rather the much less drastically changed sequence and family arrangement of Gill (1994). This arrangement retains the more or less traditional sequence of orders (whose names end in "formes") and of families (whose names end in "idae") within the nonpasserine orders. The nonpasserine orders are the ones changed most dramatically, and controversially, in the "Sibley sequence." However, Gill (1994) did adopt Sibley's more generally accepted changes of families within the order Passeriformes. As thus arranged, a total of 22 avian orders and 82 avian families occurs in Ecuador.

Not only has there been recent "turmoil" in the higher taxonomic realm of the Class Aves, of late there also has arisen a good deal of controversy concerning what actually constitutes a species, and it is this issue that of course most affects—and confuses—many people.

Each species' scientific name consists of at least two parts: a genus name, given first, followed by a species name. Both are normally written in italics (whereas the family and order names are not). A genus comprises one or more species that are judged to be more closely related to each other than they are to species in another genus. If a species is judged sufficiently distinct, it is placed in a monotypic genus (in which there is only that one species). If a species shows geographic variation in plumage, size, or some other characteristic, the forms that bear these usually minor variations are named as subspecies (or "races"—

the two terms are used interchangeably). By definition, two subspecies cannot occur together (i.e., "in sympatry," or "sympatrically").

For many decades what has been termed the "biological species concept" (BSC) has been accepted by most taxonomists (a taxonomist being a scientist trained in evolutionary biology and systematics and who investigates how organisms are related to each other). Under it, a species is defined as a population, or group of populations, of actually or potentially interbreeding individuals isolated reproductively from all other such populations. This reproductive isolation is usually reflected in the species' different appearances (in plumage, size, etc.), different vocalizations (such that species similar in appearance can recognize each other), different behaviors (e.g., courtship displays) or habitats, or even different timing in their breeding cycles. So far so good, especially where populations occur together (i.e., are "sympatric") or in very close proximity (i.e., are "parapatric") with little or no indication of interbreeding (that is, where species make mistakes). But what about populations that are not in contact (i.e., are "allopatric"), populations that, for instance, exist on two islands or two mountaintops, with unsuitable terrain in between? In such cases—and they are frequent—one enters into a more subjective situation, for there is no way to actually test for or observe the populations' reproductive isolation; they never come into contact, so you cannot directly ascertain whether they are capable of interbreeding. In such instances, the BSC forces one to make a judgement: Are the different characters the two populations exhibit different enough that the populations would not interbreed were they to come into contact? Inevitably opinions will vary, and this is the crux of the ongoing "splitting" versus "lumping" debate.

Partly in response to the difficulties posed by allopatric populations, an alternate species definition has been proposed, the "phylogenetic species concept" (PSC), in which a species is defined as "the smallest diagnosable cluster of organisms showing a parental pattern of ancestry" (J. Cracraft. 1981. Toward a phylogenetic classification of the recent birds of the world [Class Aves]. *Auk* 98[4]: 681–714). By "diagnosable" one means that the population shares a character that differs from other, similar populations. Again, so far so good, but what of populations that vary gradually across a wide range (i.e., are "clinal") and that therefore are not, strictly speaking, "diagnosable" except in a broad sense? Under the BSC, in such instances the species in question would be subdivided into subspecies, geographically defined taxonomic groups of organisms that differ subtly from other such groups, so subtly that the differences do not represent a barrier to reproduction between them. Under the PSC, no subspecies are allowed; if it is diagnosable, it is a species. Further, just how minor can a character be for it to characterize a "diagnosable unit"? For some adherents of the PSC, it can be very trivial indeed.

All this poses problems for the practicing taxonomist. As birds are more and more intensively studied, and as subtle differences and their actual or potential effects on behavior are better appreciated, the recent tendency has been for more and more similar populations to be accorded the rank of species; they are "split." But many such determinations have been and continue to be at least somewhat subjective, and all too often they are based on limited information. We have attempted to steer a course to the proverbial "middle ground" of this ongoing taxonomic debate on species-level relationships. Our basic approach has been to follow the BSC but to place somewhat more emphasis on population-level differences than similarities. The effective result has been to split some species when the evidence appears to support the likelihood that the populations would not interbreed were they to come into contact.

English Names

The subject of English names is another hotly debated topic. We have tried to steer a moderate course and have followed the main suggestions put forward in previously published works such as Hilty and Brown (1986), Ridgely and Tudor (1989, 1994), Sibley and Monroe (1990), Ridgely et al. (1998), and the 1998 AOU Check-list. Taxonomic revisions have forced a few changes, and there have been some—usually minor—alterations and clarifications. Very few totally new names have been proposed. Throughout we have attempted to follow the ongoing debate and decisions, insofar as these are available, of the Neotropical Subcommittee of the International Ornithological Congress's (IOC) Committee on English Bird Names (chaired by Frank B. Gill); RSR serves as chair of the subcommittee. We recognize that we cannot please everyone, but we have striven to be evenhanded in what will always be a contentious issue. Throughout we have indicated widely employed names that are alternatives to the one we use in *The Birds of Ecuador*.

Spanish Names

A Spanish vernacular name is suggested for each species occurring in Ecuador. PJG created the original list of names in conjunction with Fernando Ortiz-Crespo (see Ortiz-Crespo et al. 1990), and in conjunction with his son, Ilán Greenfield, PJG has further developed and updated these names over the years. Bird names are even less standardized in Spanish than they are in English. Our selections were made after due consideration of what often were several options, both Iberian and from various parts of Latin America. We have striven to make the names as broad-based as possible and have avoided names used only locally. We look forward to ongoing discussion with other interested parties on this issue, which in our view needs to be addressed on a hemispheric basis.

Area Included

After much thought and discussion, we opted not to include the avifauna of the Galápagos Islands in *The Birds of Ecuador*. We of course recognize that the Enchanted Isles are very much a legal part of Ecuador, and that they are much visited by ecotourists from around the world. The archipelago's avifauna, however, although obviously related to that of the South American continent, is so distinct as to require separate treatment, a task that we possibly envision in our future.

The actual limits of the Republic of Ecuador have been the subject of some debate in recent decades, the boundaries imposed by the Rio de Janeiro Protocol of 1942 having not been universally accepted by all parties involved until very recently. The Cordillera del Cóndor region on Ecuador's southeastern border has proven to be especially difficult. In the elfin woodlands of this region, precisely where Ecuador ends and Peru begins has often been hard to ascertain. We have personal experience in only one part of this region, but there have been a few recent expeditions to other sites where it also has not always been clear which country one is in (Schulenberg and Awbrey 1997). We have tended to be conservative, not admitting to the Ecuador list a few species that have been found just over the border in Peru, though all seem virtually certain to range into Ecuador as well and thus are mentioned parenthetically. The 1998 accord between Ecuador and Peru seems to have more definitively established the countries' respective borders.

Our oceanic limits have been to some degree imposed by practicality. Rather little pelagic work on birds has been undertaken in mainland Ecuadorian waters (principally by B. Haase and during our ANSP surveys), and what has been done has been largely restricted to within 20 to 30 km offshore, thus out to just beyond the sight of land. In a practical sense it is only these waters that are presently accessible to venturesome ornithologists, and a trip even there must be undertaken only with considerable care and with a responsible boat captain. No one, to our knowledge, has routinely been able to get out farther onto the continental shelf, much less to the "blue water" of the open ocean (the sole exceptions being on transits en route to the Galápagos). Thus although we feel satisfied that we know a reasonable amount concerning the birds of Ecuador's inshore waters, we admit to knowing next to nothing about what may occur farther offshore. We hope some interested (and intrepid) individuals will soon be able to accomplish what we have not.

Abundance Terms

Although recognizing their inherent limitations, we have employed the abundance terminology now more or less standard in works of this kind. Most birds are relatively "common" in their appropriate range and habitat, but absolute abundances of course vary on a per-area basis; thus the "common" Roadside Hawk (*Buteo magnirostris*) is much less numerous in an absolute sense than the "common" Thrush-like Wren (*Campylorhynchus turdinus*). The various terms employed must, then, be used only in comparison within the respective groups (e.g., between diurnal raptors). Recognize too that birds tend to be most numerous in the heart of their range and less so toward their periphery (both in the elevational sense and the overall geographical sense).

Even more important, recognize that some cryptic species (e.g., tinamous, rails, many antbirds and flycatchers, and others) are much more apt to be heard than seen, and that in virtually all Ecuadorian habitats—all but the most open ones—many more birds will be heard than seen. The abundance categories we use are tied to the frequency with which a particular species is recorded, whether by sight, voice, or sometimes mist-netting. Individuals not familiar with bird vocalizations clearly will be at a disadvantage and will record fewer birds; they may even come to assume that our abundance ratings are wildly inflated! *We cannot overemphasize the importance of learning bird songs and calls in Ecuador* and indeed anywhere in the Neotropics. To the uninitiated it may seem magical, but with diligence the vocalizations of many common birds can be learned after even only short exposure. You will be amply rewarded, and will in due course find more rare birds too!

These caveats aside, abundance categories do convey at least some information on a species' relative numbers. The categories we use are (roughly) defined as follows.

Abundant. In proper habitat and season, recorded daily in large numbers. Birds so termed are usually birds of open terrain or the water.

Very common. In proper habitat and season, usually recorded daily in small to moderate numbers.

Common. In proper habitat and season, recorded on most (at least 75 percent) of day trips in small to moderate numbers.

Fairly common. In proper habitat and season, recorded on about half of day trips, usually in only small numbers.

Uncommon. In proper habitat and season, recorded on fewer than half of day trips, in at most small numbers.

Rare. Even in proper habitat and season, recorded on only a small number of day trips (less than 25 percent), in at most small numbers.

Very rare. Records are extremely few, and not to be expected even in proper habitat and season; a resident occurring at only extremely low densities, or a migrant occurring at the periphery of its expected range.

Casual. Records are very few, and not to be expected even in proper habitat and season; typically a vagrant migrant or wanderer whose normal range does not include Ecuador.

Accidental. Usually only one record for Ecuador, highly unexpected and perhaps not likely to be found again; almost invariably these involve very lost migrants.

Hypothetical. A species for which there is no incontrovertible "hard" evidence that it has occurred in Ecuador. Species accepted as occurring in Ecuador should ideally be documented by some corroborating evidence. This evidence can take the form of a *specimen*, housed in a museum setting and available for future study; a *photograph*, archived in some photographic collection such as VIREO at the ANSP; or a *tape-recording*, again archived at some repository such as the Library of Natural Sounds (LNS) at the Cornell Laboratory of Ornithology. With the rapid increase in birding activity in Ecuador over the past few decades, inevitably some species have now been recorded only on the basis of sight observations, there being no documentary specimen, photograph, or tape-recording. We have accepted such species onto the Ecuador list if at least three such observations have been made by qualified observers. Nevertheless, in our judgement it is still desirable that hard evidence for such species be obtained in the future. If, however, there are fewer than three observations, or if there is still ongoing uncertainty regarding the observations, we have only provisionally accepted such species onto the Ecuador list. These species are considered to be hypothetical for the country, and their names are placed in brackets in both volumes of this work. A list of the 59 hypothetical species, as of July 1999, follows. This list also includes a few species for which specimens are available but for which there exists some uncertainty as to their actual provinance or identity. These include such species as the Common Snipe, Crimson Topaz, Plain-breasted Piculet, and Cinereous Becard.

HYPOTHETICAL SPECIES

Southern Fulmar	*Fulmarus glacialoides*
White-chinned Petrel	*Procellaria aequinoctialis*
Audubon's Shearwater	*Puffinus lherminieri*
Band-rumped Storm-Petrel	*Oceanodroma castro*
Ashy Storm-Petrel	*Oceanodroma homochroa*
Great Frigatebird	*Fregata minor*
Brown Booby	*Sula leucogaster*
Northern Shoveler	*Anas clypeata*
Stripe-backed Bittern	*Ixobrychus involucris*
Scarlet Ibis	*Eudocimus ruber*
Glossy Ibis	*Plegadis falcinellus*
Rufous-thighed Kite	*Harpagus diodon*
Mississippi Kite	*Ictinia mississippiensis*
Yellow-breasted Crake	*Porzana flaviventer*
Dunlin	*Calidris alpina*
Common Snipe	*Gallinago gallinago*
Puna Snipe	*Gallinago andina*
Pacific Golden-Plover	*Pluvialis fulva*
Chilean Skua	*Catharacta chilensis*
South Polar Skua	*Catharacta maccormicki*
Long-tailed Jaeger	*Stercorarius longicaudus*
Band-tailed Gull	*Larus belcheri*
Herring Gull	*Larus argentatus*
California Gull	*Larus californicus*
Caspian Tern	*Sterna caspia*
South American Tern	*Sterna hirundinacea*
Least Tern	*Sterna antillarum*
Marañón Pigeon	*Columba oenops*
Painted Parakeet	*Pyrrhura picta*

Tui Parakeet	*Brotogeris sanctithomae*
Saffron-headed Parrot	*Pionopsitta pyrilia*
Rusty-faced Parrot	*Hapalopsittaca amazonina*
Red-billed Ground-Cuckoo	*Neomorphus pucheranii*
Chapman's Swift	*Chaetura chapmani*
Rufous-crested Coquette	*Lophornis delattrei*
Sapphire-spangled Emerald	*Amazilia lactea*
Crimson Topaz	*Topaza pella*
Blue-tufted Starthroat	*Heliomaster furcifer*
Peruvian Sheartail	*Thaumastura cora*
Belted Kingfisher	*Megaceryle alcyon*
Plain-breasted Piculet	*Picumnus castelnau*
Bay Hornero	*Furnarius torridus*
Ash-winged Antwren	*Terenura spodioptila*
Marañón Crescentchest	*Melanopareia maranonica*
Dark-faced Ground-Tyrant	*Muscisaxicola macloviana*
Pied Water-Tyrant	*Fluvicola pica*
White-headed Marsh-Tyrant	*Arundinicola leucocephala*
White-eyed Attila	*Attila bolivianus*
Gray Kingbird	*Tyrannus dominicensis*
Cinereous Becard	*Pachyramphus rufus*
Sharpbill	*Oxyruncus cristatus*
Purple Martin	*Progne subis*
Southern Martin	*Progne elegans*
Tree Swallow	*Tachycineta bicolor*
Black-throated Green Warbler	*Dendroica virens*
Chestnut-sided Warbler	*Dendroica pensylvanica*
Bicolored Conebill	*Conirostrum bicolor*
Dickcissel	*Spiza americana*
Cinereous Finch	*Piezorhina cinerea*

Recent Species

A large number of bird species have only recently been recorded in Ecuador. "Recent" we define as being subsequent to the publication of Meyer de Schauensee's seminal *A Guide to the Birds of South America* (1970), a book that revolutionized ornithological activity on the continent, for it was only then that field identification of most South American birds began to become feasible. Meyer de Schauensee (1970) noted the bird species found in Ecuador up to that time, and as his volume marks the beginning of what can be considered to be the modern era of South American ornithology, we treat any species found or published subsequent to that year as being "recent." The names of the 229 such species (as of July 1999) are preceded by an asterisk in the text of Volume I and are also listed here. Some of these species remain hypothetical (see above) in Ecuador. Species first described to science during this period are followed by a "+" symbol in the list.

RECENT SPECIES

Bartlett's Tinamou	*Crypturellus bartletti*
Tataupa Tinamou	*Crypturellus tataupa*
Humboldt Penguin	*Spheniscus humboldti*
Black-browed Albatross	*Thalassarche melanophris*
White-chinned Petrel	*Procellaria aequinoctialis*
Buller's Shearwater	*Puffinus bulleri*
Wilson's Storm-Petrel	*Oceanites oceanicus*
Band-rumped Storm-Petrel	*Oceanodroma castro*
Markham's Storm-Petrel	*Oceanodroma markhami*
Ashy Storm-Petrel	*Oceanodroma homochroa*
Great Frigatebird	*Fregata minor*
Red-footed Booby	*Sula sula*
Peruvian Pelican	*Pelecanus thagus*
Orinoco Goose	*Neochen jubata*
Northern Shoveler	*Anas clypeata*
Chilean Flamingo	*Phoenicopterus chilensis*
Stripe-backed Bittern	*Ixobrychus involucris*
Zigzag Heron	*Zebrilus undulatus*
Great Blue Heron	*Ardea herodias*
Green Heron	*Butorides virescens*
Capped Heron	*Pilherodias pileatus*
Bare-faced Ibis	*Phimosus infuscatus*
Scarlet Ibis	*Eudocimus ruber*
Glossy Ibis	*Plegadis falcinellus*
Jabiru	*Jabiru mycteria*
Greater Yellow-headed Vulture	*Cathartes melambrotus*
White-tailed Kite	*Elanus leucurus*
Slender-billed Kite	*Rostrhamus hamatus*
Rufous-thighed Kite	*Harpagus diodon*
Mississippi Kite	*Ictinia mississippiensis*
Black-collared Hawk	*Busarellus nigricollis*
Zone-tailed Hawk	*Buteo albonotatus*
Black-and-white Hawk-Eagle	*Spizastur melanoleucus*
Mountain Caracara	*Phalcoboenus megalopterus*
Yellow-headed Caracara	*Milvago chimachima*
Slaty-backed Forest-Falcon	*Micrastur mirandollei*
Yellow-breasted Crake	*Porzana flaviventer*
Paint-billed Crake	*Neocrex erythrops*
Plumbeous Rail	*Pardirallus sanguinolentus*
Spotted Rail	*Pardirallus maculatus*
Hudsonian Godwit	*Limosa haemastica*
Red Knot	*Calidris canutus*
White-rumped Sandpiper	*Calidris fuscicollis*
Dunlin	*Calidris alpina*
Curlew Sandpiper	*Calidris ferruginea*

South American Snipe	*Gallinago paraguaiae*
Puna Snipe	*Gallinago andina*
Imperial Snipe	*Gallinago imperialis*
American Avocet	*Recurvirostra americana*
Southern Lapwing	*Vanellus chilensis*
Pacific Golden-Plover	*Pluvialis fulva*
Chilean Skua	*Catharacta chilensis*
South Polar Skua	*Catharacta maccormicki*
Long-tailed Jaeger	*Stercorarius longicaudus*
Band-tailed Gull	*Larus belcheri*
Lesser Black-backed Gull	*Larus fuscus*
Herring Gull	*Larus argentatus*
Ring-billed Gull	*Larus delawarensis*
California Gull	*Larus californicus*
Caspian Tern	*Sterna caspia*
South American Tern	*Sterna hirundinacea*
Least Tern	*Sterna antillarum*
Bridled Tern	*Sterna anaethetus*
Marañón Pigeon	*Columba oenops*
Plain-breasted Ground-Dove	*Columbina minuta*
Ruddy Ground-Dove	*Columbina talpacoti*
Violaceous Quail-Dove	*Geotrygon violacea*
El Oro Parakeet	*Pyrrhura orcesi* +
Painted Parakeet	*Pyrrhura picta*
Barred Parakeet	*Bolborhynchus lineola*
Blue-winged Parrotlet	*Forpus xanthopterygius*
Tui Parakeet	*Brotogeris sanctithomae*
Sapphire-rumped Parrotlet	*Touit purpurata*
Saffron-headed Parrot	*Pionopsitta pyrilia*
Rusty-faced Parrot	*Hapalopsittaca amazonina*
Pearly-breasted Cuckoo	*Coccyzus euleri*
Pheasant Cuckoo	*Dromococcyx phasianellus*
Red-billed Ground-Cuckoo	*Neomorphus pucheranii*
Cinnamon Screech-Owl	*Otus petersoni* +
Cloud-forest Pygmy-Owl	*Glaucidium nubicola* +
Central American Pygmy-Owl	*Glaucidium griseiceps*
Subtropical Pygmy-Owl	*Glaucidium parkeri* +
Short-tailed Nighthawk	*Lurocalis semitorquatus*
Band-tailed Nighthawk	*Nyctiprogne leucopyga*
Nacunda Nighthawk	*Podager nacunda*
White-tailed Nightjar	*Caprimulgus cayennensis*
Spot-tailed Nightjar	*Caprimulgus maculicaudus*
Rufous Nightjar	*Caprimulgus rufus*
Spot-fronted Swift	*Cypseloides cherriei*
White-chested Swift	*Cypseloides lemosi*
Chimney Swift	*Chaetura pelagica*
Chapman's Swift	*Chaetura chapmani*

Peruvian Antpitta	*Grallaricula peruviana*
Marañón Crescentchest	*Melanopareia maranonica*
Chocó Tapaculo	*Scytalopus chocoensis* +
El Oro Tapaculo	*Scytalopus robbinsi* +
Chusquea Tapaculo	*Scytalopus parkeri* +
Slender-footed Tyrannulet	*Zimmerius gracilipes*
Mouse-colored Tyrannulet	*Phaeomyias murina*
Yellow-crowned Elaenia	*Myiopagis flavivertex*
Large Elaenia	*Elaenia spectabilis*
Highland Elaenia	*Elaenia obscura*
Amazonian Scrub-Flycatcher	*Sublegatus obscurior*
Sulphur-bellied Tyrannulet	*Mecocerculus minor*
River Tyrannulet	*Serpophaga hypoleuca*
Lesser Wagtail-Tyrant	*Stigmatura napensis*
Black-crested Tit-Tyrant	*Anairetes nigrocristatus*
Rufous-browed Tyrannulet	*Phylloscartes superciliaris*
Short-tailed Pygmy-Tyrant	*Myiornis ecaudatus*
Johannes's Tody-Tyrant	*Hemitriccus iohannis*
Buff-throated Tody-Tyrant	*Hemitriccus rufigularis*
Cinnamon-breasted Tody-Tyrant	*Hemitriccus cinnamomeipectus* +
Yellow-browed Tody-Flycatcher	*Todirostrum chrysocrotaphum*
Spotted Tody-Flycatcher	*Todirostrum maculatum*
Rufous-tailed Flatbill	*Ramphotrigon ruficauda*
Large-headed Flatbill	*Ramphotrigon megacephala*
Orange-eyed Flatbill	*Tolmomyias traylori* +
Cinnamon-crested Spadebill	*Platyrinchus saturatus*
Roraiman Flycatcher	*Myiophobus roraimae*
Fuscous Flycatcher	*Cnemotriccus fuscatus*
White-browed Chat-Tyrant	*Ochthoeca leucophrys*
Cliff Flycatcher	*Hirundinea ferruginea*
Little Ground-Tyrant	*Muscisaxicola fluviatilis*
Dark-faced Ground-Tyrant	*Muscisaxicola macloviana*
Riverside Tyrant	*Knipolegus orenocensis*
Amazonian Black-Tyrant	*Knipolegus poecilocercus*
Pied Water-Tyrant	*Fluvicola pica*
White-headed Marsh-Tyrant	*Arundinicola leucocephala*
White-eyed Attila	*Attila bolivianus*
Western Sirystes	*Sirystes albogriseus*
Swainson's Flycatcher	*Myiarchus swainsoni*
Great Crested Flycatcher	*Myiarchus crinitus*
Yellow-throated Flycatcher	*Conopias parva*
Three-striped Flycatcher	*Conopias trivirgata*
White-throated Kingbird	*Tyrannus albogularis*
Gray Kingbird	*Tyrannus dominicensis*
Crested Becard	*Platypsaris validus*
Sharpbill	*Oxyruncus cristatus*
Chestnut-crested Cotinga	*Ampelion rufaxilla*
Chestnut-bellied Cotinga	*Doliornis remseni* +

Jet Manakin	*Chloropipo unicolor*
Várzea Schiffornis	*Schiffornis major*
Lemon-chested Greenlet	*Hylophilus thoracicus*
Marañón Thrush	*Turdus maranonicus*
Tropical Mockingbird	*Mimus gilvus*
Purple Martin	*Progne subis*
Southern Martin	*Progne elegans*
Tumbes Swallow	*Tachycineta stolzmanni*
Tree Swallow	*Tachycineta bicolor*
Pale-footed Swallow	*Notiochelidon flavipes*
Bar-winged Wood-Wren	*Henicorhina leucoptera* +
Collared Gnatwren	*Microbates collaris*
Golden-winged Warbler	*Vermivora chrysoptera*
Tennessee Warbler	*Vermivora peregrina*
Bay-breasted Warbler	*Dendroica castanea*
Black-throated Green Warbler	*Dendroica virens*
Chestnut-sided Warbler	*Dendroica pensylvanica*
Ovenbird	*Seiurus auricapillus*
Connecticut Warbler	*Oporornis agilis*
Short-billed Honeycreeper	*Cyanerpes nitidus*
Bicolored Conebill	*Conirostrum bicolor*
Tit-like Dacnis	*Xenodacnis parina*
Buff-bellied Tanager	*Thlypopsis inornata*
Purple-throated Euphonia	*Euphonia chlorotica*
Emerald Tanager	*Tangara florida*
Straw-backed Tanager	*Tangara argyrofenges*
Orange-throated Tanager	*Wetmorethraupis sterrhopteron*
Olive Tanager	*Chlorothraupis frenata*
Piura Hemispingus	*Hemispingus piurae*
Black-faced Tanager	*Schistochlamys melanopis*
Blue Grosbeak	*Guiraca caerulea*
Dickcissel	*Spiza americana*
Yellow-faced Grassquit	*Tiaris olivacea*
Black-billed Seed-Finch	*Oryzoborus atrirostris*
Drab Seedeater	*Sporophila simplex*
Cinereous Finch	*Piezorhina cinerea*
Red Pileated-Finch	*Coryphospingus cucullatus*
Pale-eyed Blackbird	*Agelaius xanthophthalmus*
Baltimore Oriole	*Icterus galbula*
Red-breasted Blackbird	*Sturnella militaris*
Andean Siskin	*Carduelis spinescens*
House Sparrow	*Passer domesticus*

Species Accounts

Although there is some variation in the species accounts, depending mainly on the complexity of the issues involved, a minimum of three paragraphs is included for each species, with additional paragraphs for many.

The first paragraph describes the species' distribution in Ecuador. For a few species this is straightforward; they may be of widespread occurrence in a discrete area. For many others, however, distribution details may be complex, in a few cases so much so that for clarity's sake the discussion is broken up into multiple paragraphs. In the case of recently recorded species (those marked with an asterisk), details concerning the first Ecuadorian record(s) are given. Precise distributional limits are provided where needed, together with supporting details. In the case of sight observations, generally the observers are listed by their first initial and last name; if that name is preceded by the word "fide" (from the Latin, "on the faith of"), it indicates that records may have been reported to that person by other observers. Observations attributed to "v.o." refer to "various observers." In the case of most important specimen records, we state where that specimen is housed. In the case of long-distance migrants present in Ecuador for only part of the year, we give the spread of months during which records indicate that the species has been found in the country. Finally, one or more sentences give the species' elevational spread in Ecuador (note that in the vast majority of instances, records from outside Ecuador have not been incorporated into these elevation statements).

An additional paragraph contains, where it is available, information on international recoveries of banded birds. We must admit that on the whole there are disappointingly few of these. The assistance of David Agro, Ann Demers, and Lucie Métras (the last two of the Canadian Wildlife Service Bird Banding Office) in obtaining these data is gratefully acknowledged.

In cases where a species is considered to be at risk, the next paragraph outlines its conservation status in Ecuador. Information on how the various categories are defined can be found in the Conservation chapter.

The next paragraph contains, where necessary, comments on alternate English names. A few minor issues of spelling—such as our inclusion of diacritical marks on English names, their inclusion having been deemed advisable by the IOC Committee on English Bird Names—are sometimes not directly addressed, being clear in context. In a few cases, for clarity's sake it has been necessary to defer these comments to the next paragraph.

The next paragraph contains information on the species' taxonomy. This includes comments on its generic placement as well as its status as a separate species (where this has been open to debate in recent years). We name the subspecies occurring in Ecuador, together—in some instances—with comments on the validity of these taxa. The assistance of our collaborator Mark B. Robbins in making many of these subspecific determinations during his tenure as collection manager at the ANSP should be singled out here.

A final paragraph (preceded by "**Range:**") presents the species' overall world range. This information is of necessity derived from the literature generally, though all statements insofar as they deal with South America have been carefully vetted against the site-by-site maps for every South American bird species that are maintained and updated by RSR (after having been set up originally by the late William Brown in the early and mid-1980s). Note that a semicolon in the stated range denotes a range disjunction, at least based on presently available information.

A few species (e.g., the Bluish-fronted Jacamar [*Galbula cyanescens*] and the Narrow-tailed Emerald [*Chlorostilbon stenura*]) that have been published as occurring in Ecuador in the past, but which for various reasons are considered unlikely actually to be found there, are mentioned in a supplemental paragraph following the range statement for its closest relative. The reasons for such species being considered unlikely to occur are given.

In the same position as the preceding "dismissed" species we also present a few species considered to be potential additions to the Ecuadorian avifauna. These species have not yet been recorded in Ecuador, but they range very close to the border in adjacent Colombia or Peru (in habitats that also occur in adjacent Ecuador) or are long-distance migrants that have occurred close to Ecuador.

References

References are included throughout the species accounts in Volume I; they are generally not referred to, however, in Volume II. References that are cited repeatedly or that were used extensively during the preparation of Volume I are given in full in the Bibliography at the end of this volume. References that pertain only to one or a few species are cited in full at the point in the species account where they are first mentioned; thereafter in that species account they are referred to by "op. cit." after the name(s) of the author(s). For simplicity's sake, we refer only by abbreviated title and volume or part number to two frequently cited multivolume works: *Catalogue of Birds of the Americas* (*Birds of the Americas*), authored primarily by Charles E. Hellmayr (1924–1949), and *Check-list of Birds of the World* (*Birds of the World*), authored primarily by James L. Peters (1934–1986). Full citations are provided for each of these references in the Bibliography.

Key

Geography, Climate, and Vegetation

Ecuador is a relatively small country—its land area comprising only about 283,000 sq km, or roughly the same area as the state of Colorado in the United States—and in fact is one of the smallest in South America. Its basic topographical pattern is relatively simple, for the high chain of the Andes Mountains, oriented along a north-south axis, divides the country effectively into thirds. The Pacific coastal plain to the west of the Andes is here termed the *western lowland region*, and the lowland area to the east of the Andes, all of which drains ultimately into the Amazon River system, is termed the *eastern lowland region*. The montane area in between is termed the *Andean region*. A last region, distinctive but of very limited extent in Ecuador, lies within the drainage of the Río Marañón (an affluent of the Amazon, joining it in Peru).

Western Lowland Region

This region comprises a coastal plain between 100 and 200 km wide (east-west) between the Pacific Ocean to the west and the Andes to the east. Where it abuts the Andes there is a foothill zone of varying width, where low mountains and ridges rise to between 500 and 1000 m above sea level; this foothill zone is wider to the north, especially in Esmeraldas and Pichincha. A low coastal cordillera, with maximum elevations at only about 800 to 900 m, roughly parallels the coast from southwestern Esmeraldas south into western Guayas, terminating just west of the city of Guayaquil; the northern part of this range is called the Cordillera de Mache and the southern part the Cordillera de Colonche. Much of the western coastal plain is relatively flat, and human population density is high, with about 50 percent of Ecuador's total population residing here. (Guayaquil, with more than 2.5 million inhabitants, is easily Ecuador's largest city.) Soils in many areas are relatively fertile, especially when compared to the tropical norm, and this has led to the development of extensive, industrial-style agriculture with major crops of rice, bananas, palm oil, and others now being produced. Rainfall totals vary clinally from north to south, with northern areas (especially Esmeraldas) being extremely wet—some areas here are among the wettest in the world, with 3000 to 4000 mm of rain annually—and southern areas much less so.

As a result of the high human population densities and the prevalence of agriculture, the forests that once covered this region have been almost entirely eliminated. It is estimated that south of Esmeraldas less than 5 percent of the original forest cover remains. Most of what does remain consists of patches (themselves often degraded) that persist locally on Andean foothill slopes to the east and on the slopes of the coastal cordillera. Between the coastal cordillera and the Andes lies the extensive floodplain of two sluggish rivers, the Ríos Guayas and Daule. Formerly there were extensive marshes here, but the area is now largely under rice cultivation. These two large rivers empty into the shallow Gulf of Guayaquil, whose fringes still support extensive mangrove forests (the largest in Ecuador, though they continue to be reduced for shrimp-pond development). South of the Río Guayas estuary the coastal plain becomes much narrower, and the climate quickly becomes drier, approaching desertlike conditions in southern coastal El Oro close to the Peruvian border. Comparably arid conditions are also found in an isolated area near the western tip of the Santa Elena Peninsula which juts out into the Pacific Ocean in western Guayas. Isla Puná, a large island in the Gulf of Guayaquil, and Isla de la Plata off southwestern Manabí are also quite arid.

The climate of the western lowland region is strongly seasonal, with a pronounced rainy season during the first four to five months of the year, this being caused by a slight warming of the Pacific Ocean waters off southwestern Ecuador. The remainder of the year is often persistently cloudy, but there is little sustained rain. During the rainy season, rainy periods typically alternate with sunny intervals, resulting in temperatures that are several degrees higher than they are the remainder of the year. Rainfall totals can vary significantly between years, with greatest amounts during El Niño events, when the Pacific Ocean waters off Ecuador warm dramatically. Rains can then be torrential, causing significant infrastructural damage. As noted above, rainfall totals increase northward. In northwestern Ecuador the rains tend also to be less seasonal (though they are still heaviest during the first half of the year), and they are more consistent, showing relatively little between-year variation. The persistent low cloud cover that occurs during the dry season has the effect of creating "cloud forest" conditions at relatively low elevations, where peaks and ridges of the coastal cordillera and the Andean foothills are frequently enshrouded by clouds; it does not actually rain much, but conditions remain relatively moist.

Major Habitats

Following are the six major "primary" natural habitats found in the lowlands of western Ecuador; these terms are used in both volumes of *The Birds of Ecuador*. Bear in mind that a continuum exists from one habitat to the next and that the distinctions drawn are generalizations, useful for organization and sorting purposes but not absolutes. Further note that each, as defined, is a primary habitat, meaning it is more or less free from human interference. As discussed above, such areas are now quite rare in overpopulated western Ecuador, and thus in practice most areas are at least to some degree "secondary"; i.e., they are at some stage of regrowth back toward a primary state. Often the term "secondary woodland" is used to denote such situations (whether humid or arid). Note further that various anthropogenic (human-created and -maintained) habitats are not included here; some of these—such as plantations, gardens, and pastures—are often surprisingly good for a variety of common birds. All of the habitats below are important bird habitats.

Wet forest. An evergreen forest with a canopy level of 30 to 40 m growing in areas with very high annual precipitation (3000–4000 mm, locally even more in foothill areas) in the northwestern lowlands, primarily in Esmeraldas where a good deal still persists. Sometimes called "pluvial forest."

Humid forest. A predominantly evergreen forest, similar to wet forest but growing in areas with lower annual precipitation (ca. 1500–3000 mm). This forest once grew across wide areas in the western lowland region south into central Manabí and northern Los Ríos but has been almost entirely eliminated. Southward, as annual rainfall gradually diminishes, humid forest grades into semihumid forest; it too is much diminished in extent.

Deciduous forest/woodland. A predominantly deciduous forest or woodland, fairly tall (20–25 m), in which many or most trees lose their leaves during the dry season. Rainfall totals about 500 to 1500 mm annually and is strongly seasonal. This forest originally grew in many areas of southwestern Ecuador from southern Manabí and southern Los Ríos south through El Oro and Loja, and it persists in some places, especially in hilly situations. In many areas this type of forest is quite disturbed, as evidenced by heavy viny growth and often a lack of much understory (from browsing by domestic animals).

Desert woodland and arid scrub. An open shrubby woodland with scattered and shorter trees and much lower diversity than the preceding habitats. It grows in areas with less than 500 mm annual precipitation, primarily in western Guayas (where on the western tip of the Santa Elena Peninsula precipitation levels are so low that barren, desertlike conditions prevail unless there have been recent El Niño rains), southern coastal El Oro, and southwestern Loja.

Cloud forest. At about 700 to 1100 m above sea level, cloud banks in southwestern Ecuador are so persistent that nearly continuous damp conditions prevail, despite such areas' relatively low annual precipitation levels of 1000 to 1500 mm. This habitat occurs primarily near the crest of the coastal cordillera and in southern Azuay and El Oro.

Mangroves. A depauperate but interesting forest of only a few tree species that are adapted to growing in salt or brackish, muddy water. Formerly much more extensive, mangrove forests now exist mainly in northern Esmeraldas, around the Río Guayas estuary, and in El Oro; all are much threatened by a still expanding shrimp-farming industry. An important habitat not only for certain specialized species but also for wading birds and shorebirds.

Andean Region

This region basically comprises two great parallel cordilleras that are aligned from north to south: the Western Andes (or Cordillera Occidental), rising abruptly from the Pacific coastal plain, and the Eastern Andes (or Cordillera Oriental), rising somewhat more gradually from the Amazonian lowlands to the east. Both cordilleras have numerous high peaks of 5000 m and higher in northern and central Ecuador, and several spectacular volcanoes (some of them still active) rise even higher, culminating in the 6272-m summit of Volcán Chimborazo, one of the highest mountains in the Americas. Between the Western and Eastern Andes lies an elevated intermontane valley which in the north is quite well defined, and there is often called the central valley. South of Chimborazo, Andean topography becomes more complex and includes a series of interandean valleys and intersecting, often transverse, ridges. Here peaks rarely attain elevations higher than 4000 m.

Because of their volcanic influence, soils in the central and interandean valleys are often quite rich, and thus despite its often steep slopes and high elevations, the area's agricultural productivity is high, with corn and (at higher elevations) potatoes being the primary crops grown. The area has long been inhabited, and before the Spanish conquest it was within the sphere of influence of the Incas; now virtually no area is beyond the influence of humans. Little or nothing in the way of natural vegetation remains at lower elevations and on the "inside" slopes (i.e., the east-facing slopes of the Western Andes and the west-facing slopes of the Eastern Andes), the deforestation having been so severe and complete that ecologists often cannot determine what the natural vegetation actually was. Some areas have been reforested with plantations of exotic tree species such as *Eucalyptus* and pines (*Pinus*); sadly, virtually no effort has been made to use native species such as alders (*Alnus*).

The "outside" slopes of the Andes (i.e., the west-facing slopes of the Western Andes and the east-facing slopes of the Eastern Andes) remain considerably less altered, and it is here that some of Ecuador's most spectacular landscapes are found. Until very recently these slopes were covered with nearly continuous montane forest, the actual type of forest varying with elevation and exposure (see below). Improved access has changed this to some extent, notably in the Western Andes of central Ecuador and to a lesser degree elsewhere, but many areas still support extensive forest (though because it tends to be removed from areas close to roads, it is often difficult to reach). Perhaps the most pristine montane forests of all are those found on the ridges and peaks lying east of and separated from the main ridge of the Eastern Andes. Remote and very difficult of access, there are three main such areas in Ecuador: Volcán Sumaco/Cordillera de Galeras, Cordillera de Cutucú, and Cordillera del Cóndor.

For the most part the outside slopes are uniformly humid, though both the Western and Eastern Andes are incised by a few deep river valleys in which conditions may differ. In the Western Andes this occurs especially in the drainages of the Ríos Chota, Guaillabamba, Chanchán, and Jubones valleys, portions of which are very arid. In the Eastern Andes this occurs less frequently, though there is an interesting arid area in the Río Pastaza valley near Baños in Tungurahua; for the most part, however, valleys in the Eastern Andes are wider, inhibiting the development of "rainshadow" conditions.

The Andean climate is much less strongly seasonal than that of the Pacific slope lowlands. In both the Western and Eastern Andes precipitation occurs throughout the year, occasionally in prodigous amounts and continuing for several days though more often lasting for only several hours during the afternoon or evening. There then may be several days with little or no rain at all. Somewhat less rain may fall, on average, during the months of July and August. Rainfall amounts are relatively uniform in the northwest and on the entire east slope, where annual totals of 2000 to 3000 mm are the norm, but may be higher locally in some foothill localities. The southwest tends to be somewhat drier. The central and interandean valleys are markedly drier (with annual rainfall totals of about 1000 m), and some regions, such as central Chimborazo, are quite arid. Temperature levels vary greatly, as a function of elevation. Frosts occur above about 3000 m. Treeline (the zone between the upper limit of trees and the grassy or shrubby slopes above) occurs at 3200 to 3600 m, its exact level a reflection of exposure and rainfall totals. Permanent snow commences at 5000 to 5200 m.

Major Habitats

Seven major natural primary habitats are found in the Andean region. As in the western lowland region, a continuum occurs between these montane habitats. The human hand is dominant in many areas, to the extent that in some areas natural habitats now barely exist. An additional factor on steeper slopes is the frequency of landslides, which in many areas result in there being little forest of any substantial age or stature.

Foothill forest. This very humid and floristically rich forest occurs in a zone from about 600–700 to 1200–1300 m on both outside slopes of the Andes. Annual precipitation is about 3000 mm. The canopy is about 20 to 25 m, and there is much epiphytic growth. It is alternatively called "upper tropical forest." An exceptionally important bird habitat, often of limited extent, supporting an interesting mixture of lowland and montane elements as well as numerous specialties of its own, on either side of the Andes.

Subtropical montane forest. This forest occurs in a zone from about 1200–1300 to 2300–2400 m on both outside slopes of the Andes. Annual precipitation is typically about 2000 to 2500 m, and the canopy is about 15 to 20 m. An important bird habitat, with species often replacing each other on either side of the Andes.

Temperate montane forest. This forest occurs in a zone from about 2300–2400 to 3200–3300 m on both outside slopes of the Andes and also locally on slopes above the central and interandean valleys. It doubtless used to be much more extensive on the inside slopes than it is presently. Annual precipitation is about 2000 to 2500 m, and it is substantially cooler than in the previous zone, but frosts are still infrequent. The canopy is only about 10 to 15 m with occasional emergents, and it typically presents a broken aspect; a *Chusquea* bamboo understory is often prevalent, especially in disturbed situations. An important bird habitat; more species occur widely in this zone, on both sides of the Andes, because of the greater likelihood of their being in contact.

Elfin woodland. A low-stature, ecotonal woodland that occurs at or near treeline. Its elevation varies depending on rainfall and exposure but normally is in the range of 3100 to 3400 m. In areas where paramo burning is frequent, the changeover from forest to paramo is abrupt and often occurs at lower elevations than it would normally. In areas where fire occurs rarely or not at all, the elfin woodland at treeline is more extensive and the shift from forest to paramo more gradual. Rainfall amounts are similar to those found in temperate montane forest, but conditions are even cooler (frosts can occur regularly), and evapotranspiration rates are slower. A surprising number of bird species occurs here, including several specialists, especially on the east slope.

Polylepis **woodland.** A woodland found at high elevations, typically above treeline, often in groves that are dominated by one or more species of low, gnarled, flaky-barked trees in the genus *Polylepis*. Several species of varying stature occur, some forming monospecific stands, others mixed in with upper-elevation temperate forest. There is evidence that *Polylepis* formerly was much more extensive in the Andes and that it gradually has been reduced by burning in the paramo and cutting for firewood. *Polylepis* occurs at 3500 to 4200 m and is an important habitat for a few specialized bird species.

Paramo. Open grassy areas occurring at elevations above treeline, thus typically commencing at 3200–3600 m and continuing up to well over 4000 m, with a gradual reduction in the frequency of shrubby growth as elevation increases. Frosts are of virtual nightly occurrence, and even short-lived snowfalls can occur, especially in more humid areas. In especially wet areas, to the north and on the east slope, there may be stands of *Espeletia* shrubs, whereas southward there may be *Chuquiragua* shrubs and *Puya* bromeliads. Southward the paramo tends to become drier and more resembles the puna grasslands of Peru's altiplano. Throughout there are occasional small lakes and boggy depressions. Above 4000 to 4200 m the *Stipa* grass tussocks become sparser and lower and merge onto barren rocky slopes; even higher, at about 5000 m (lower in wetter areas), one encounters perpetual snow.

Montane scrub/woodland. A relatively sparse and low-stature woodland found in montane areas where reduced precipitation levels do not permit the growth of more luxuriant montane forest. For the most part this woodland is much degraded by human activities. So far as known, it is of comparatively limited importance for birds.

Eastern Lowland Region

The vast Amazonian lowlands (often referred to locally as simply the "Oriente") consist of a more or less level area—though in some areas the relief can be rather hilly—that gradually descends eastward from the foothills of the Andes, dropping from about 400 to 200 m near most of the Peruvian border. Rivers flow from west to east, all eventually flowing into the Amazon River system. The eastern lowlands are carpeted in humid forest, a forest that even now is still mainly intact though human impacts are spreading; clearing for settlements, oil exploration and production, and intensive hunting near settled areas are all taking their toll. Human population levels remain very low—the entire region supports less than 5 percent of Ecuador's total population—and settlement is largely concentrated in a zone along the eastern base of the Andes. Only a few small towns exist eastward, almost all along navigable rivers. Aboriginal peoples still live in many areas, though they are now mainly settled in permanent villages, their hunting and gathering days having mostly ended.

The climate of the Oriente is relatively uniform. It is, of course, hot throughout the year, though on very infrequent occasions cold fronts from the south penetrate this far north during the austral winter. Rainfall continues at a more or less uniform pace every month, with July and August having perhaps the least precipitation. Although all-day, lighter rains do occur, for the most part rain occurs in the form of heavy showers that last only a few hours (sometimes less), followed by clearing. Annual rainfall is 2000 to 3000 mm, with highest amounts close to the base of the Andes.

Major Habitats

Six major primary habitats are found in the eastern lowland region. Unlike in the western lowland and Andean regions, habitats in the east tend not to blend into each other, rather standing more often as well-defined and discrete units. Of importance is that anthropogenic effects are much less evident here than they are elsewhere in Ecuador, a function of the region's low population density.

Terra firme forest. By far the most extensive forest type found in eastern Ecuador, though in most areas it is difficult to access. Terra firme forest is found away from navigable watercourses (still the primary means of transportation in most areas), and typically when a road is built into terra firme (as has happened along the eastern base of the Andes and in the western half of Sucumbíos), the forest is soon modified. Terra firme is the tall-stature forest found in upland areas; the canopy typically reaches 30 to 40 m, and often the understory is fairly open though treefalls are not that infrequent and often there is more under-growth near streams and in the small valleys found in hillier regions. Subdivisions that are to a large extent reflective of subtle shifts in soil types do occur, but from an avian perspective they tend to be minor, at least in Ecuador. Unlike sw. Amazonia, bamboo stands are scarce to nonexistent in the Ecuadorian lowlands. Terra firme forest is extremely rich in bird species (and plants and other animals as well), though many occur at notably low densities.

Várzea forest. This forest occurs in areas that are seasonally flooded, sometimes quite deeply and for protracted periods. Nevertheless the understory of várzea forest is often well developed, and the habitat is of great importance to a wide variety of bird species. Although the stature of várzea forest can be fairly high (25–30 m), in some areas it seems to remain considerably lower ("várzea wood-land"), at about 15 to 20 m.

Riparian woodland/forest. This forest eventually develops in floodplain areas along rivers and on larger river islands. It is at first dominated by *Cecropia* and *Gynerium* cane but then gradually becomes more diverse floristically. A substantial number of bird species are found here, including several habitat specialists.

River island scrub. Low-growing, early-succession scrub that grows on recently exposed river islands. *Salix* willows and *Tessaria* are often dominant plants for the first few years. Despite being relatively unprepossessing looking, this habitat supports a surprisingly distinctive avifauna, including several species found nowhere else.

Oxbow lakes. As rivers meander through their floodplains, subsequent to a major flooding event they occasionally adopt an entirely new course, abandoning their old channel. These former channels often gradually become separated from the river itself, resulting in the isolation of shallowly curved bodies of water called oxbow lakes. Gradually these lakes fill in—it takes centuries—but meanwhile a variety of specialized marshy and swampy habitats can flourish, and these provide a superb environment for a wide variety of birds. In particular there is often a fringing marsh and floating vegetation mat comprising grasses and gradually other plants, as well as swamps dominated by palms.

Moriche palm swamps. A distinctive swampy forest with more or less permanent standing water where one species of tall palm, *Mauritia flexuosa*, is dominant. Such swamps are locally distributed, being most often found around oxbow lakes or in other situations where drainage is poor.

Río Marañón Drainage

A small region in far southeastern Zamora-Chinchipe falls within the drainage of the Río Marañón. Since becoming accessible by road in the late 1980s, this area—which is isolated from the rest of the eastern lowlands by the Cordillera del Cóndor and other mountain ridges—has been found to support an interesting avifauna that is unique from an Ecuadorian perspective. Some fifteen or more bird species once thought to occur only in the upper Marañón drainage of northwestern Peru are now known to extend as far north as this still poorly known, remote region of Ecuador. The lowest-elevation areas in Ecuador are at about 650 m along the Río Mayo, and the "Marañón influence" extends upward and northward to around Valladolid. Little seems to have been recorded about the region's climate, which varies greatly depending on elevation. Around Zumba it appears to be moderate, with deciduous or semihumid forest and woodland being found (sadly, much of it quite disturbed). At higher elevations these habitats merge into subtropical montane forest.

Bird Migration in Ecuador

Despite the fact that Ecuador straddles the equator and is therefore quite distant from the source areas for most migratory birds, migrants do comprise a fairly conspicuous component of the Ecuadorian avifauna, though not to such a degree as farther north in the Neotropics. Migrants come to Ecuador from both the north and south and from various parts of the Pacific Ocean, and these different groups are segregated out in the sections that follow.

Some of these species occur in small numbers in Ecuador, and a few are truly rare; in the species lists that follow, these latter species are marked with a "V" (for "vagrant"). A small number of migratory species comprise both migrant populations and breeding populations in Ecuador, and these are marked with a "B" (for "breeder"). Species that occur mainly or entirely as transients in Ecuador, en route to or from breeding areas north or south of Ecuador, are indicated by a "T" (for "transient"). Invariably there is uncertainty regarding the migratory status of certain species, and such species are indicated with a "?".

Boreal Migrants

Boreal migrants breed in the Northern Hemisphere and migrate southward during the northern (boreal) winter. These form by far the most diverse and conspicuous group of migrants in Ecuador, with no fewer than 120 species having been recorded in the country. At least some individuals of a wide majority of these species pass some or all of the northern winter in Ecuador. However, a subset of 20 species (many of them shorebirds), occurs mainly or entirely as transients, en route to or from wintering groups farther south in South America. A few of these transients occur in Ecuador mainly or entirely on southward passage (this group includes several shorebirds whose migratory route is farther to the west during the northern autumn), whereas only one species (Yellow-billed Cuckoo) may occur primarily on northward passage. Boreal migrants occur throughout Ecuador, though the greatest number of species is found in the west (especially when the multitude of migratory shorebirds ranging mainly along the coast is taken into account.) All of the migratory Charadriiformes ("shorebirds") occurring in Ecuador breed in the Northern Hemisphere. Three

boreal migrants (Cinnamon Teal, Lesser Scaup, and Killdeer) have entirely
ceased to come to Ecuador, for reasons unknown.

BOREAL MIGRANTS (120 SPECIES)

Blue-winged Teal	*Anas discors*
Cinnamon Teal	*Anas cyanoptera* (V)
Northern Shoveler	*Anas clypeata* (V)
Lesser Scaup	*Aythya affinis* (V)
Least Bittern	*Ixobrychus exilis* (?) (B)
Great Blue Heron	*Ardea herodias* (V)
Great Egret	*Ardea alba* (?) (B)
Little Blue Heron	*Egretta florida* (?) (B)
Tricolored Heron	*Egretta tricolor* (B)
Cattle Egret	*Bubulcus ibis* (?) (B)
Green Heron	*Butorides virescens* (V)
Black-crowned Night-Heron	*Nycticorax nycticorax* (?) (B)
Glossy Ibis	*Plegadis falcinellus* (V)
Turkey Vulture	*Cathartes aura* (B)
Osprey	*Pandion haliaetus*
Swallow-tailed Kite	*Elanoides forficatus* (B)
Plumbeous Kite	*Ictinia plumbea* (?) (B)
Mississippi Kite	*Ictinia mississippiensis* (V) (T)
Broad-winged Hawk	*Buteo platypterus*
Swainson's Hawk	*Buteo swainsoni* (V) (T)
Merlin	*Falco columbarius*
Peregrine Falcon	*Falco peregrinus* (B)
Sora	*Porzana carolina*
Greater Yellowlegs	*Tringa melanoleuca*
Lesser Yellowlegs	*Tringa flavipes*
Solitary Sandpiper	*Tringa solitaria*
Willet	*Catoptrophorus semipalmatus*
Wandering Tattler	*Heteroscelus incanus*
Spotted Sandpiper	*Actitis macularia*
Upland Sandpiper	*Bartramia longicauda* (T)
Whimbrel	*Numenius phaeopus*
Hudsonian Godwit	*Limosa haemastica* (T)
Marbled Godwit	*Limosa fedoa* (V)
Ruddy Turnstone	*Arenaria interpres*
Surfbird	*Aphriza virgata*
Red Knot	*Calidris canutus* (T)
Sanderling	*Calidris alba*
Semipalmated Sandpiper	*Calidris pusilla*
Western Sandpiper	*Calidris mauri*
Least Sandpiper	*Calidris minutilla*
White-rumped Sandpiper	*Calidris fuscicollis* (T)
Baird's Sandpiper	*Calidris bairdii* (T)

Pectoral Sandpiper	*Calidris melanotos* (T)
Dunlin	*Calidris alpina* (V)
Curlew Sandpiper	*Calidris ferruginea* (V)
Stilt Sandpiper	*Micropalama himantopus*
Buff-breasted Sandpiper	*Tryngites subruficollis* (T)
Short-billed Dowitcher	*Limnodromus griseus*
Common Snipe	*Gallinago gallinago* (V)
Red Phalarope	*Phalaropus fulicaria* (T)
Red-necked Phalarope	*Phalaropus lobipes*
Wilson's Phalarope	*Phalaropus tricolor*
American Avocet	*Recurvirostra americana* (V)
Gray Plover	*Pluvialis squatarola*
American Golden-Plover	*Pluvialis dominica* (T)
Pacific Golden-Plover	*Pluvialis fulva* (V)
Semipalmated Plover	*Charadrius semipalmatus*
Killdeer	*Charadrius vociferus*
Piping Plover	*Charadrius melodus* (V)
Lesser Black-backed Gull	*Larus fuscus* (V)
Herring Gull	*Larus argentatus* (V)
Ring-billed Gull	*Larus delawarensis* (V)
California Gull	*Larus californicus* (V)
Laughing Gull	*Larus atricilla*
Franklin's Gull	*Larus pipixcan* (T)
Caspian Tern	*Sterna caspia* (V)
Royal Tern	*Sterna maxima*
Elegant Tern	*Sterna elegans* (T)
Sandwich Tern	*Sterna sandvicensis*
Common Tern	*Sterna hirundo*
Arctic Tern	*Sterna paradisaea* (T)
Least Tern	*Sterna antillarum* (V)
Black Tern	*Chlidonias niger*
Black-billed Cuckoo	*Coccyzus erythropthalmus*
Yellow-billed Cuckoo	*Coccyzus americanus* (T)
Common Nighthawk	*Chordeiles minor*
Chimney Swift	*Chaetura pelagica*
Eastern Wood-Pewee	*Contopus virens*
Western Wood-Pewee	*Contopus sordidulus*
Olive-sided Flycatcher	*Contopus cooperi*
Acadian Flycatcher	*Empidonax virescens*
Willow Flycatcher	*Empidonax traillii*
Alder Flycatcher	*Empidonax alnorum*
Great Crested Flycatcher	*Myiarchus crinitus* (V)
Sulphur-bellied Flycatcher	*Myiodynastes luteiventris*
Gray Kingbird	*Tyrannus dominicensis* (V)
Eastern Kingbird	*Tyrannus tyrannus* (T)
Red-eyed Vireo	*Vireo olivaceus* (B)
Yellow-green Vireo	*Vireo flavoviridis*

Gray-cheeked Thrush	*Catharus minimus*
Swainson's Thrush	*Catharus ustulatus*
Purple Martin	*Progne subis* (V) (T)
Tree Swallow	*Tachycineta bicolor* (V)
Sand Martin	*Riparia riparia* (T)
Barn Swallow	*Hirundo rustica*
Cliff Swallow	*Petrochelidon pyrrholeuca* (T)
Golden-winged Warbler	*Vermivora chrysoptera* (V)
Tennessee Warbler	*Vermivora peregrina*
Yellow Warbler	*Dendroica aestiva*
Cerulean Warbler	*Dendroica cerulea*
Blackpoll Warbler	*Dendroica striata*
Bay-breasted Warbler	*Dendroica castanea* (V)
Blackburnian Warbler	*Dendroica fusca*
Black-throated Green Warbler	*Dendroica virens* (V)
Chestnut-sided Warbler	*Dendroica pensylvanica* (V)
Black-and-white Warbler	*Mniotilta varia*
American Redstart	*Setophaga ruticilla*
Prothonotary Warbler	*Protonotaria citrea* (V)
Northern Waterthrush	*Seiurus noveboracensis*
Ovenbird	*Seiurus auricapillus* (V)
Mourning Warbler	*Oporornis philadelphia*
Connecticut Warbler	*Oporornis agilis* (V)
Canada Warbler	*Wilsonia canadensis*
Summer Tanager	*Piranga rubra*
Scarlet Tanager	*Piranga flava*
Rose-breasted Grosbeak	*Pheucticus ludovicianus*
Blue Grosbeak	*Guiraca caerulea* (V)
Dickcissel	*Spiza americana* (V)
Baltimore Oriole	*Icterus galbula* (V)
Bobolink	*Dolichonyx oryzivora* (T)

Austral Migrants

Austral migrants breed in the Southern Hemisphere and migrate northward during the southern (austral) winter. Only 21 species exhibit this pattern in Ecuador, and they occur in much smaller numbers than do boreal migrants. Few austral migrants attract much attention in Ecuador, the one exception being the flocks of Fork-tailed Flycatchers sometimes seen in the eastern lowlands (this species is also the sole austral migrant to occur mainly as a transient here). One species, the Red-eyed Vireo, is known to have both boreal and austral migrant populations, as well as resident populations in three different parts of Ecuador. Of particular note is that virtually none of the austral migrants occurs west of the Andes.

AUSTRAL MIGRANTS (21 SPECIES)

Pearly-breasted Cuckoo	*Coccyzus euleri* (V)
Dark-billed Cuckoo	*Coccyzus melacoryphus* (B)
Nacunda Nighthawk	*Podager nacunda* (?)
Large Elaenia	*Elaenia spectabilis*
Small-billed Elaenia	*Elaenia parvirostris*
Vermilion Flycatcher	*Pyrocephalus rubinus* (B)
White-browed Ground-Tyrant	*Muscisaxicola albilora*
Dark-faced Ground-Tyrant	*Muscisaxicola maculirostris* (V)
Swainson's Flycatcher	*Myiarchus swainsoni*
Streaked Flycatcher	*Myiodynastes maculatus* (B)
Crowned Slaty Flycatcher	*Griseotyrannus aurantioatrocristatus*
Variegated Flycatcher	*Empidonomus varius*
Tropical Kingbird	*Tyrannus melancholicus* (?) (B)
White-throated Kingbird	*Tyrannus albogularis* (V)
Fork-tailed Flycatcher	*Tyrannus savana* (T)
Red-eyed Vireo	*Vireo olivaceus* (B)
Andean Slaty-Thrush	*Turdus nigriceps* (B)
Brown-chested Martin	*Progne tapera* (B)
Southern Martin	*Progne elegans* (V)
Blue-and-white Swallow	*Notiochelidon cyanoleuca* (B)
Lined Seedeater	*Sporophila lineola*

Intratropical Migrants

Intratropical migrants breed in one area of the tropics and then migrate to another tropical area during their nonreproductive season. The distribution patterns exhibited by Ecuador's six such species vary. Two (Snowy-throated Kingbird and Crimson-breasted Finch) breed in the southwestern lowlands and disperse northward during the second half of the year (corresponding to the local dry season). Two others (Andean Slaty-Thrush and Black-and-white Tanager) breed in more montane parts of the southwest and then disperse to areas east of the Andes during the second half of the year. The Black-and-white Seedeater breeds locally in the highlands, then descends to the eastern lowlands where it occurs in flocks with other seedeaters, sometimes including the Lesson's Seedeater, whose distribution is unique; it breeds in northeastern South America and then migrates to western Amazonia. One species, the Andean Slaty-Thrush, occurs as both an intratropical and an austral migrant.

INTRATROPICAL MIGRANTS (6 SPECIES)

Snowy-throated Kingbird	*Tyrannus niveigularis*
Andean Slaty-Thrush	*Turdus nigriceps*
Black-and-white Tanager	*Conothraupis speculigera*

Crimson-breasted Finch	*Rhodospingus cruentus*
Lesson's Seedeater	*Sporophila bouvronides*
Black-and-white Seedeater	*Sporophila luctuosa*

Pelagic Visitants

Pelagic visitants occur mainly or entirely offshore in the Pacific Ocean and do not breed in continental Ecuadorian waters. Some species breed north of Ecuador and occur in Ecuadorian waters mainly during the northern winter; these are marked in the list below with an "N." Others breed south of Ecuador and occur in Ecuadorian waters primarily during the southern winter; these are marked with an "S." A few species breed in closer proximity to Ecuador, typically off the coast of Peru or on the Galápagos Islands; these are not marked at all. Note the high percentage of vagrants (14 species, almost half the total); to some extent this total may be inflated by the fact that relatively little ornithological field work has been done off the Ecuadorian coast.

PELAGIC VISITANTS (29 SPECIES)

Black-browed Albatross	*Thalassarche melanophris* (S) (V)
Southern Fulmar	*Fulmarus glacialoides* (S) (V)
Pintado Petrel	*Daption capense* (S) (V)
Dark-rumped Petrel	*Pterodroma phaeopygia*
White-chinned Petrel	*Procellaria aequinoctialis* (S) (V)
Parkinson's Petrel	*Procellaria parkinsoni* (S)
Buller's Shearwater	*Puffinus bulleri* (S) (V)
Pink-footed Shearwater	*Puffinus creatopus* (S)
Sooty Shearwater	*Puffinus griseus* (S)
Audubon's Shearwater	*Puffinus lherminieri* (V)
Wilson's Storm-Petrel	*Oceanites oceanicus* (S)
White-vented Storm-Petrel	*Oceanites gracilis*
White-faced Storm-Petrel	*Pelagodroma marina* (S) (V)
Least Storm-Petrel	*Oceanodroma microsoma* (N)
Wedge-rumped Storm-Petrel	*Oceanodroma tethys*
Band-rumped Storm-Petrel	*Oceanodroma castro* (V)
Ashy Storm-Petrel	*Oceanodroma homochroa* (N) (V)
Black Storm-Petrel	*Oceanodroma melania* (N)
Markham's Storm-Petrel	*Oceanodroma markhami*
Hornby's Storm-Petrel	*Oceanodroma hornbyi*
Great Frigatebird	*Fregata minor* (V)
Brown Booby	*Sula leucogaster* (N) (V)
Chilean Skua	*Catharacta chilensis* (S) (V)
South Polar Skua	*Catharacta maccormicki* (S) (V)
Pomarine Jaeger	*Stercorarius pomarinus* (N)

Parasitic Jaeger	*Stercorarius parasiticus* (N)
Long-tailed Jaeger	*Stercorarius longicaudus* (N) (V)
Sabine's Gull	*Xema sabini* (N)
Swallow-tailed Gull	*Creagrus furcatus*

Dispersers from Peru

Dispersers from Peru breed in coastal Peru and then disperse northward into coastal Ecuadorian waters. Some of these 11 species are known to occur regularly in Ecuador, others less so. A few seem to occur primarily or only in association with El Niño events. One of these species (Kelp Gull) has recently begun to nest in very small numbers in Ecuador, and another (Peruvian Pelican) has become much more numerous of late. It is possible that other listed species may have nested in Ecuador in the past (e.g., Peruvian Tern) or could do so in the future (e.g., South American Tern).

DISPERSERS FROM PERU (11 SPECIES)

Humboldt Penguin	*Spheniscus humboldti* (V)
Peruvian Booby	*Sula variegata*
Guanay Shag	*Phalacrocorax bougainvillii* (V)
Peruvian Pelican	*Pelecanus thagus*
Chilean Flamingo	*Phoenicopterus chilensis*
Gray Gull	*Larus modestus*
Kelp Gull	*Larus dominicanus*
Band-tailed Gull	*Larus belcheri* (V)
South American Tern	*Sterna hirundinacea* (V)
Peruvian Tern	*Sterna lorata*
Inca Tern	*Larosterna inca* (V)

Wanderers

Wanderers are species that occur in Ecuador only on an irregular basis, with their status and sometimes even their source often being uncertain. Often these are birds for which little specific information is available. Some may prove to have small resident populations in Ecuador, but in any case all appear to be rare in the country.

WANDERERS (16 SPECIES)

Stripe-backed Bittern	*Ixobrychus involucris*
Bare-faced Ibis	*Phimosus infuscatus*
Scarlet Ibis	*Eudocimus ruber*
Rufous-thighed Kite	*Harpagus diodon*

South American Snipe	*Gallinago paraguaiae*
Chapman's Swift	*Chaetura chapmani*
Rufous-crested Coquette	*Lophornis delatrii*
White-chinned Sapphire	*Hylocharis cyanus*
Sapphire-spangled Emerald	*Amazilia lactea*
Blue-tufted Starthroat	*Heliomaster furcifer*
Peruvian Sheartail	*Thaumastura cora*
Little Ground-Tyrant	*Muscisaxicola fluviatilis*
Pied Water-Tyrant	*Fluvicola pica*
White-headed Marsh-Tyrant	*Arundinicola leucocephala*
White-eyed Attila	*Attila bolivianus*
Bicolored Conebill	*Conirostrum bicolor*

Altitudinal Migrants

No definite evidence of altitudinal migration exists in Ecuador, but we suspect this may be due more to the lack of sustained careful observation at individual sites than to an actual absence. Some suggestive information indicating that some local altitudinal movements occur is already available from Jatun Sacha (fide B. Bochan). The Mindo region, with its easy access to a range of elevations, represents another promising site for such observations. Various hummingbird species (e.g., the Sparkling Violetear [*Colibri coruscans*], Green Violetear [*C. thalassinus*], and Purple-backed Thornbill [*Ramphomicron microrhynchum*]) seem likely candidates to be altitudinal migrants. Even the fragmentary evidence presently at hand would appear to indicate that some species appear out of "nowhere" when certain favored trees or shrubs (e.g., *Inga* or *Erythrina*) come into flower. A few other species (e.g., Yellow-collared Chlorophonia [*Chlorophonia flavirostris*] and Golden-rumped Euphonia [*Euphonia cyanocephala*]) appear to be erratic or seasonal in their appearances, but details of these movements await sustained, long-term study.

Ecuadorian Ornithology: Some People and a Gazetteer

Ornithological Activity in Ecuador

Chapman (1926) provides a complete review of ornithological activity in Ecuador up to the mid-1920s. What follows here is a brief summary of activities since then.

Concentrated ornithological work in Ecuador was limited during the decades following the publication of Chapman's (1926) landmark publication. A few commercial collectors—notably members of the Olalla family, C. Durán, and T. Mena—remained active in the country, procuring collections of birds and sending them to museums in Europe and North America. Much of this scattered material never was published upon, but several such collections did form the basis for a series of papers published by J. Berlioz in the late 1920s and 1930s (Berlioz 1927, 1928a, 1928b, 1932a, 1932b, 1937a, 1937b, 1937c). In 1931 J. T. Zimmer began publishing a long series of papers (eventually there would be 66) in which he discussed the systematics and distribution of Peruvian birds; considerable new information on Ecuadorian birds was also presented therein.

The pace gradually quickened during the 1940s, 1950s, and 1960s. Most notably, G. Orcés, Ecuador's first ornithologist (and also an all-round naturalist, conversant in many fields), began to assemble the first local collection of birds, first privately, then at the Colegio Mejía and Escuela Politécnica in Quito; the latter ultimately formed the basis for the collection at the MECN. A British petroleum engineer, S. Marchant, was resident on the Santa Elena Peninsula during part of the 1950s, and he produced several important articles on that area's birdlife (e.g., Marchant 1958), including the first detailed breeding accounts. An active collecting program was again initiated, this time under the auspices of the MCZ and headed by Harvard University student D. Norton. In 1964 and 1965 Norton visited various localities on the east slope of the Andes and in southern Ecuador, in the latter year accompanied by R. A. Paynter, Jr. Unfortunately this promising program was cut short by a life-threatening incident in southern Loja.

It was subsequent to the publication of Meyer de Schauensee's important synthetic works (1964, 1966, 1970) that field identification of South American birds became feasible, if difficult. In the 1970s an increasing number of ornithologists and birders began to visit Ecuador and to make their findings known. Birds were gradually becoming known as living organisms, not just as study skins. D. and E. Tallman spent more than a year at Limoncocha studying antbirds and amassing an important collection for the LSUMZ. More and more Ecuadorians began to take an interest as well, notably F. Ortiz-Crespo and a series of younger students.

The ornithological resurgence accelerated in the 1980s and 1990s. The ANSP began its scientific work in the early 1980s under the direction of RSR (who first spent time in Ecuador in 1976), M. Robbins, and F. Sornoza; ANSP personnel worked in conjunction with the MECN (where much specimen material has been deposited). Together they sponsored an extensive series of expeditions and trips that by its end in the mid-1990s had visited virtually every corner of the country. At the same time the WFVZ, under the direction of L. Kiff and M. Marín and likewise in conjunction with the MECN, also sponsored several trips to different sites. Several individuals were meanwhile conducting important sustained work in specific localities, perhaps the most notable of whom were B. Bochan at Jatun Sach and O. Jahn at Playa de Oro; others (including C. Canaday, N. Krabbe, M. Lysinger, and J. Moore) pursued their interests more widely across the country. Various groups of British and Danish university students started coming to Ecuador on expeditions that usually had a strong conservation bent. Notable among these were B. Best and his coworkers (in southwestern Ecuador) and C. Rahbek with his (in and around Podocarpus National Park).

Almost as important during these decades, a legion of birders also began to descend on Ecuador. They now number far too many for us to keep up with them all, or with everything they see. Some come under expert leadership, usually in groups organized by foreign birding tour companies (though now local companies such as Neblina Forest are also active), others independently. A growing number of Ecuadorian birders and ornithologists is also now in the field. Many groups and individuals have communicated with us—often by first contacting PJG in Quito—and if so, we have endeavored to include their more important observations and discoveries here. (If we have missed you, we apologize, and urge you to contact one of us for possible inclusion in later editions of this book.) The point is, it has been a collective effort, and it seems certain that it will continue to be.

A Gazetteer

The following localities and geographical areas are mentioned repeatedly in Volume I, and the comments here are intended to put these sites into an overall biogeographic and ornithological context. Emphasis has been given to sites in which we either have personal experience or for which we have been given information (or have access thereto); together these localities therefore provide much of the basis for the distributional data included in this book. We are aware that certain other sites about which we know little have been studied or visited, and we regret we could not include them. It should also be noted that we have usually not included older (pre-1926) sites that have not been visited or studied in the modern era (i.e., before 1960) unless new information on them has come to light.

Aguarico, Río. An important river flowing east across northeastern Ecuador, mainly through Sucumbíos; it eventually arches southward and joins the Río Napo at the Peruvian border. Island specialties do not seem to range above the Zancudococha area, thus not as far upstream as they do on the Río Napo. Much of this region remains relatively wild, particularly downstream from around Cuyabeno.

Alamor. A small agricultural town (ca. 1300 m) in the mountains of southwestern Loja. Alamor is situated northwest of Celica but lies at a considerably lower elevation. Its avifauna was first studied by AMNH personnel in the 1920s, at a time when lower montane forest was still extensive. Now forest is patchy at best and entirely lacking in many areas. Recent field work has been done mainly by B. Best and his coworkers. 4°02′S, 80°02′W.

Alto Tambo. A very small settlement in the lowlands (250–350 m) of northern Esmeraldas north of Lita along the road between Ibarra and San Lorenzo. A site situated north of town in wet forest was studied by an ANSP group in Jul.–Aug. 1990, and the region has since been visited by a few other birders and ornithologists, though poor road conditions often prevent access. 0°56′N, 78°32′W.

Angashcola, Río. A remote high valley in the Andes east of Amaluza in southeastern Loja whose avifauna was studied by groups of British students headed by R. Williams during two periods in the early 1990s. 4°35′S, 79°02′W.

Ángel, Páramo del. A fine, relatively undisturbed paramo area (3200–3600 m) in the Western Andes just south of the Colombian border in northern Carchi. It features unusually extensive and lush stands of *Espeletia*. 0°43′N, 77°58′W.

Antisana. A spectacular high mountain (5704 m) in the Eastern Andes in extreme western Napo, situated just east of the main ridge of the Andes and hence with much snow and often enveloped by clouds; occasionally it can be viewed near the pass on the Papallacta road. High paramos around its base are relatively undisturbed and support a good number of Andean Condors (*Vultur gryphus*); Laguna Micacocha harbors a population of Silvery Grebes (*Podiceps occipitalis*), and Black-faced Ibis (*Theristicus melanopis*) are present. An effort is being made to protect a vast area in this region as the Antisana Ecological Reserve. 0°30′S, 78°08′W.

Archidona. A town lying along the eastern base of the Andes along the Baeza-Tena road in western Napo (600–700 m). Nearby terrain has been at least partially deforested, but there is much second-growth and avian diversity remains high. 0°55′S, 77°48′W.

Arenillas. A town on the coastal plain of El Oro. Its surroundings must originally have been covered with rich deciduous forest and woodland, but now much has been cleared; nonetheless, birds remain numerous, and ponds and marshes of varying sizes provide habitat for a variety of waterbirds. The Pan-American Highway leads south to Huaquillas and on into Peru, passing through excellent deciduous forest, though access to most of it is limited because of the sensitive nature of this border region. 3°33′S, 80°04′W.

Atacames. A small coastal resort town west of the city of Esmeraldas in western Esmeraldas. Secondary woodland and some remnant mangrove growth can still be found. 0°52′N, 79°51′W.

Ayampe, Río. A narrow, shallow river forming the boundary between south-western Manabí and northwestern Guayas; Machalilla National Park lies just to the north. The river doubles as a "road" and normally can be driven in vehicles with reasonable clearance. Extensive deciduous and semihumid woodland and forest remains in various areas, including along the main coastal road leading south. This road passes through a range of hills (up to 150 m) before dropping back down to the coast. This hilly area, which is unprotected, is important as the only known breeding area for the endangered Esmeraldas Woodstar (*Chaetocercus berlepschi*). The recently opened Hotel Atamari is nearby. 1°40′S, 80°48′W.

Babahoyo. A hot, squalid town in the marshy, low-lying plains of southern Los Ríos along the main road between Quevedo and Guayaquil. Babahoyo is surrounded by densely inhabited rice-growing terrain. 1°49′S, 79°31′W.

Baeza. A small crossroads town situated on the east slope of the Andes (ca. 1700 m) in western Napo. It is from Baeza that the road south to Archidona and Tena cuts off from the Quito–Lago Agrio highway. Baeza was an important early collecting site, and it continues to be a useful base for exploring this area in the Andes. Montane forest in the immediate vicinity is now patchy and mostly secondary in nature, but many birds can still be found. 0°27′S, 77°53′W.

Bellavista Lodge. A recently opened ecotourism lodge near the crest of the Tandayapa Ridge on the west slope of the Andes above Mindo (2000–2200 m). It provides an excellent base from which to explore this section of the Andes. 0°15′N, 78°38′W.

Bermejo oil-field area. A petroleum-production area situated in the foothills (500–900 m) along the eastern base of the Andes north of Lumbaquí in northern Sucumbíos close to the Colombian border. Extensive primary forest remains, though some settlement is occurring. First explored by ANSP teams in 1992 and 1993; an excellent range of foothill and lower-montane birds is found here, notably the Olive Tanager (*Chlorothraupis frenata*) at its only known regular site in Ecuador. 0°07′N, 77°20′W.

Bilsa (= Bilsa Biological Station = Jatun Sacha Bilsa). A site in the coastal Cordillera de Mache (400–600 m) in southwestern Esmeraldas and the focus of an ongoing land-preservation effort by the Fundación Jatun Sacha. The region supports the last extensive area of primary wet forest south of northern Esmeraldas and is thus of international importance. In 1994 a British team headed by R. Clay found several montane bird species away from the Andes for the first time, and the area supports several rare species such as the Banded Ground-Cuckoo (*Neomorphus radiolosus*) and Long-wattled Umbrellabird (*Cephalopterus penduliger*). The region is somewhat turbulent and should not be entered without local assistance. 0°28′ N, 79°48′ W.

Blanco, Cerro. A newly established and easily accessible private reserve in the Chongón Hills west of Guayaquil in Guayas, located at the easternmost end of the coastal Cordillera de Colonche. It contains extensive deciduous forest and is most notable for its remnant population of the endangered Great Green Macaw (*Ara ambigua*), the last such in southwestern Ecuador, and now is under active protection and management by a local group, ProBosque. A good range of southwestern endemics occurs, including Ochre-bellied Dove (*Leptotila ochraceiventris*), Blackish-headed Spinetail (*Synallaxis tithys*), Henna-hooded Foliage-gleaner (*Hylocryptus erythrocephalus*), and Saffron Siskin (*Carduelis siemiradzkii*). 2°10′ S, 80°02′ W.

Bombuscaro, Río. An important river draining a large valley in the northeastern sector of Podocarpus National Park in Zamora-Chinchipe, south of the town of Zamora. One of the park's most important guard stations is situated at about 1000 m, along a trail leading from the roadhead upward to 1300 to 1400 m. The area supports extensive and rich lower montane forest, with many scarce and range-restricted species being found (notably the endemic White-breasted Parakeet [*Pyrrhura albipectus*]). The Bombuscaro area was first intensively studied by ZMCOP teams in the late 1980s and early 1990s and since has been visited by numerous birders. 4°08′ S, 79°00′ W.

Borbón, Río. A river in Esmeraldas, navigable in its lower reaches upstream from the town of the same name (1°06′ N, 78°59′ W), providing access to lower elevations of Cotacachi-Cayapas Ecological Reserve. Humid forest and secondary woodland remain, inhabited by some northwestern endemics and specialties.

Buenaventura (formerly often called "west of Piñas"). An important site in the cloud forest zone (700–1000 m) on the west slope of the Andes in western El Oro. Despite considerable deforestation, the area continues to support an extremely diverse assemblage of birds with many endemics, including most notably the newly described El Oro Parakeet (*Pyrrhura orcesi*) and El Oro Tapaculo (*Scytalopus robbinsi*). The site was first seriously studied by an ANSP team in 1985 and has since been visited by numerous birders. In 1999 some of this area was purchased by Fundación Jocotoco as a reserve; more is expected to be protected in the years to come. 3°33′ S, 79°59′ W.

Cajanuma. An important guard station in Podocarpus National Park situated on a ridge (2500–3100 m) just west of the Continental Divide in extreme eastern Loja. Little-disturbed montane forest and paramo are extensive here, and the area supports an extremely rich avifauna, first intensively studied by ZMCOP groups in the late 1980s and early 1990s. Of particular importance is the essentially undisturbed ecotone between paramo and upper montane forest. At no other site is the Imperial Snipe (*Gallinago imperialis*) so likely to be seen on the ground. 4°05′ S, 79°12′ W.

Caráquez, Bahía de. A large bay on the coast of central Manabí near the coastal city of the same name and known locally as simply "Bahía." It is still fringed to a large extent by mangroves—though their conversion to shrimp lagoons has reduced them—and extensive freshwater ponds and marshes in the drainage of the lower Río Chone lie to the east. Most surrounding terrain has been converted to agricultural use, with only small patches of deciduous forest and woodland remaining. Marshy areas to the east support some waterfowl, including the only Southern Pochards (*Netta erythrophthalma*) known to survive in Ecuador. 0°36′S, 80°23′W.

Caripero. A remote area on the west slope of the Andes in western Cotopaxi (2100–2700 m), studied by N. Krabbe and F. Sornoza in the mid- and late 1970s and notable (at least until recently) for supporting a small remnant population of Yellow-eared Parrots (*Ognorhynchus icterotis*). About 0°40′S, 79°05′W.

Catacocha. An attractive town in the highlands (1700–1900 m) of western Loja. Despite being situated in what is now mainly agricultural terrain, a sizable area of good-quality deciduous forest and woodland lies on the slopes just south of town. Various southwestern specialties such as Rufous-necked Foliage-gleaner (*Syndactyla ruficollis*), Andean Slaty-Thrush (*Turdus nigriceps*), and Black-and-white Tanager (*Conothraupis speculigera*) approach their northern limits here. 4°04′S, 79°38′W.

Catamayo. A dusty town (locally often called "La Toma") in the broad upper valley (1300 m) of the Río Catamayo west of the city of Loja in Loja. Several southwestern endemics reach their northeastern limit in this region. 3°59′S, 79°21′W.

Cayambe. A spectacular high volcano (5790 m) situated in the Eastern Andes in northeastern Pichincha, visible from Quito but nonetheless quite remote. Andean Condors (*Vultur gryphus*) remain relatively numerous. 0°03′N, 78°8′W.

Cayambe-Coca Ecological Reserve. A large protected area on the east slope of the Andes in northern Ecuador extending from paramo down at least to the lower subtropical zone; its southwestern fringe is skirted by the Lago Agrio highway near Papallacta.

Celica. A quaint old town perched high (ca. 2000 m) in the mountains of southwestern Loja. First studied ornithologically in the 1920s by AMNH personnel, the town's surroundings have once again been explored since the late 1980s by a number of birders and museum groups. Little remains of the formerly extensive subtropical forest, but numerous special birds nonetheless persist, including a number of Tumbesian endemics, the rarest of which is the endangered Gray-headed Antbird (*Myrmeciza grisc**eiceps*); the Rufous-necked Foliage-gleaner (*Syndactyla ruficollis*) is also reasonably numerous. 4°07′S, 79°59′W.

Chimborazo. The highest mountain in Ecuador (6272 m), situated in the Western Andes of extreme northwestern Chimborazo. One can drive to a climber's hut on Chimborazo's southern flank; at 4600 m, this is the highest point one can reach in a car in Ecuador. Paramo in this region is decidedly arid, and on the whole birdlife is sparse. 1°28′S, 78°48′W.

Chinapinza. A small settlement at the end (1300–1400 m) of a road that extends up the west slope of the Cordillera del Cóndor in extreme southeastern Zamora-Chinchipe. North from Chinapinza a trail extends up to a goldmining area at 1750 to 1900 m (numerous other deposits are being worked elsewhere), with minor trails continuing on to the Peruvian border along the crest of the ridge. Despite the mining activity, much good forest remains at most elevations, the major exception being at lower elevations (ca. 900–1100 m) near Paquisha where formerly luxuriant forest has been largely cut over. The region was studied by N. Krabbe and F. Sornoza in 1990 and (nearby) again in 1993 by an ANSP group; two of the three bird species recently described from the Peruvian side of the cordillera (Bar-winged Wood-Wren [*Henicorhina leucoptera*] and Cinnamon-breasted Tody-Tyrant [*Hemitriccus cinnamomeipectus*]) have been found here. An effort is presently being made to create a large binational park encompassing much of the cordillera. 4°00′S, 78°27′W.

Chindul, Cordillera de. The low coastal ridge that parallels the coast in northern Manabí and adjacent western Esmeraldas. Although once extensively forested, most has been logged off since the early 1950s. A remnant area is now protected at Bilsa.

Chiriboga road. An old road (now relatively little traveled) named for a small settlement about midway along it that descends the west slope of the Western Andes between Quito and Santo Domingo de los Colorados. The road parallels an oil pipeline. The temperate zone supports much *Chusquea* bamboo, but the foothill and subtropical zones are more cut over. A wide variety of montane birds occurs along the transect; the bridge at its lowermost point is an excellent vantage point for Torrent Ducks (*Merganetta armata*). Although a traditional birding road, Chiriboga has not been much studied by scientists; M. Marín and several WFVZ teams have worked there for short periods, their most notable find being the discovery of breeding Spot-fronted Swifts (*Cypseloides cherriei*). 0°15′S, 78°45′W.

Coca. A bustling town (also called "Puerto Francisco de Orellana") in Napo on the upper Río Napo near the confluence of the Río Coca. This is the site of an airfield with regular flights back to Quito and is the jumping-off point for many trips farther east into the Oriente. Surrounding terrain has been widely cleared, and as a result numerous bird species found in open habitats (e.g., Pearl Kite [*Gampsonyx swainsonii*], Ruddy Ground-Dove [*Columbina talpacoti*], Plain-breasted Ground-Dove [*C. minuta*], and Red-breasted Blackbird [*Sturnella militaris*]) have colonized the region, as they also have around Lago Agrio. 0°28′S, 76°58′W.

Colonche, Cordillera de. The coastal ridge that sweeps northwestward from west of Guayaquil west through Guayas into southwestern Manabí, petering out south of Bahía de Caráquez. In some areas extensive deciduous and semihumid forest remains, though much of it is disturbed to some extent. Although elevations are low (terrain above 800 m is limited), the peaks nonetheless support an interesting avifauna comprising certain foothill and montane elements, most of them only recently discovered away from the Andes. See also La Torre, Cerro; Salanguilla; and San San Sebastián, Cerro.

Colta, Laguna de. A fairly large, shallow, reed-fringed lake (3300 m) situated south of Riobamba in northern Chimborazo. Despite the nearby high human population, the lake's margins act as a magnet for migratory shorebirds. There are also resident populations of waterbirds, including Andean Ruddy-Ducks (*Oxyura ferruginea*) and Ecuadorian Rails (*Rallus aequatorialis*). 1°45′S, 78°46′W.

Cóndor, Cordillera del. A remote ridge, isolated from the main spine of the Andes, that straddles the Peruvian border in southeastern Zamora-Chinchipe; it attains elevations of 2000 to 2500 m. Clothed in montane and elfin forest, much of it remains inaccessible. See Chinapinza, the primary access point from the Ecuadorian side.

Cordoncillo, Cordillera de. A ridge on the east slope of the Andes in extreme northern Loja south of Saraguro. Although montane forest here is now relatively limited in extent, the avifauna still includes such rarities as Bearded Guan (*Penelope barbata*), Red-faced Parrot (*Hapalopsittica pyrrhops*), Crescent-faced Antpitta (*Grallaricula lineifrons*) and Orange-banded Flycatcher (*Myiophobus lintoni*). 3°40′S, 79°17′W.

Cotacachi-Cayapas Ecological Reserve. A large protected area on the west slope of the Andes in northwestern Ecuador. The reserve encompasses a broad sweep of montane habitats but remains poorly known ornithologically because of its extremely remote nature. Its western fringe is accessible from the Río Borbón (see Borbón, Río) and El Placer (which see).

Cotapino. A hacienda (apparently more properly referred to as Concepción; Paynter 1993) northeast of Tena and Archidona in western Napo. The area is covered with extensive upper tropical forest and has been a traditional Olalla collecting site. It was also visited by D. Norton for MCZ in Jun.–Jul. 1964. 0°48′S, 77°25′W.

Cotopaxi. A beautiful, high, symmetrical volcano (5897 m) in the Eastern Andes of northwestern Cotopaxi, visible from Quito and often visited by urban residents and tourists. Much of the volcano's surroundings are encompassed by a national park, though in recent decades extensive areas at lower elevations have been despoiled by pine (*Pinus*) plantations, replacing an interesting shrub habitat with an almost sterile, exotic one. Paramo here is relatively arid, though Laguna Limpiopungo lies in a wide saddle to the volcano's north; it supports some waterfowl, and Noble Snipe (*Gallinago nobilis*) occur on the "back" marshy side. 0°40′S, 78°26′W.

Cuenca. An attractive colonial city in the highlands (ca. 2550 m) of western Azuay. Most natural vegetation in its vicinity was removed long ago, but the city does provide lodging when visiting nearby areas such as El Cajas National Recreation Area and Río Mazán. 2°53′S, 78°59′W.

Curaray, mouth of the Río. A right-bank affluent of the Río Napo that joins the Napo in northeastern Peru; formerly the river's entire drainage basin lay in

Ecuador, but now only its upper reaches do. "Boca de Curaray" was an important collecting site for the Olalla family in the mid-1920s, with much material being sent to the AMNH. The usefulness of some of it is compromised, however, by the near certainty that some portion was mislabeled and not actually obtained there (but rather specimens from around the Olallas' base in the Avila/Río Suno area were mixed in with others actually taken in the Curaray region). 2°22′S, 74°05′W.

Cutucú, Cordillera de. A remote ridge, isolated from the main spine of the Andes and situated east and southeast of Macas in northwestern Morona-Santiago, not far to the north of the northern tip of the Cordillera del Cóndor (the two being separated by the Río Santiago). Like the Cóndor, the highest parts of the Cutucú reach about 2500 m; but unlike the Cóndor, there appear to be no endemic elements present. Largely unexplored until recently, the Cordillera de Cutucú was the site of a major ANSP expedition east from Logroño in Jun.–Jul. 1984 (see Robbins et al. 1987); N. Krabbe has also worked here. Only the Cutucú's western slope has so far been explored. It remains an exceptionally difficult area to penetrate, and permission to do so must be obtained from the local Shuar Indians.

Cuyabeno Faunistic (or Wildlife) Reserve. A very large protected area of lowland forest situated north of the Río Aguarico in eastern Sucumbíos. Originally restricted to an area around the blackwater lake of the same name, it was extended eastward in 1992 to the Peruvian border to encompass the Imuya-cocha/Río Lagarto area.

Cuyabeno, Laguna de and Río. A blackwater lake (0°02′N, 76°12′W) and river in the lowlands of Sucumbíos that flows southeast into the Río Aguarico. Except along its westernmost fringe (affected by oil development and settlers), the area remains relatively wild. The Universidad Católica maintains a biological station here, and recently the area has been opened up to limited ecotourism use.

Cuyuja. A small settlement on the east slope of the Andes (2400–2600 m) in western Napo along the Lago Agrio road between Papallacta and Baeza. Despite heavy traffic and numerous farms, some decent temperate forest and woodland still exist along this road, and the area remains a favored birding destination out of Quito (no other site on the east slope is so easily reached). 0°25′S, 78°01′W.

Ecuasal lagoons. Salt evaporation ponds along the south shore of the Santa Elena Peninsula west of Punta Carnero in extreme western Guayas. These privately owned lagoons provide habitat for large numbers and a wide variety of waterbirds, and they seem to be notable magnets for rarities. Numerous other water and coastal birds are resident and breeding (notably Gray-hooded Gulls [*Larus cirrocephalus*], Kelp Gulls [*L. dominicanus*], Gull-billed Terns [*Sterna nilotica*], and Snowy Plovers [*Charadrius alexandrinus*]), and usually some Chilean Flamingos (*Phoenicopterus chilensis*) are present. Surrounding terrain (some of it heavily inhabited and garbage-strewn) is almost desert-like and supports few birds; some Peruvian Thick-knees (*Burhinus superciliaris*) persist. 2°16′S, 80°54′W.

El Cajas National Recreation Area. An attractive area of relatively moist paramo and *Polylepis* groves situated on a high plateau (3500–4000 m) west of Cuenca in western Azuay. El Cajas is important as the heart of the very small range of the endemic Violet-throated Metaltail (*Metallura baroni*); Tit-like Dacnises (*Xenodacnis parina*) are much more numerous here than elsewhere in Ecuador. 2°52′W, 79°13′W.

El Chaco (NNE of). A site on the east slope of the Andes (2000–2100 m) in western Napo that was investigated by an ANSP team in Sep.–Oct. 1992. Montane forest is patchy and mostly secondary, with much *Chusquea* bamboo. 0°12′N, 77°50′W.

El Chical (SSW of). A remote site on the west slope of the Andes (1500–1600 m) near the Colombian border in Carchi which was studied by an ANSP team in Aug. 1988. Extensive subtropical forest remained as of that date, but even then deforestation was proceeding; many scarce and range-restricted birds were found here, including Dark-backed Wood-Quail (*Odontophorus melanonotus*), Cloud-forest Pygmy-Owl (*Glaucidium nubicola*), Star-chested Treerunner (*Margarornis stellatus*), and Black Solitaire (*Entomodestes coracinus*). 0°54′N, 78°12′W.

El Placer. An important ornithological site situated along the Ibarra-San Lorenzo railway in the wet foothills (500–700 m) of the Western Andes in extreme eastern Esmeraldas. First intensively studied by an ANSP team in Jul.–Aug. 1987, El Placer has since—despite the total absence of any visitor facilities—been visited by numerous intrepid birders. Despite some clearing along the tracks, forest and secondary woodland remain extensive, and the area supports an exceptionally interesting avifauna with numerous scarce and endemic elements, some difficult to see elsewhere. Especially noteworthy is the presence of the Plumbeous Forest-Falcon (*Micrastur plumbeus*), Long-wattled Umbrellabird (*Cephalopterus penduliger*), Rufous-brown Solitaire (*Cichlopsis leucogenys*), Rufous-crowned Antpitta (*Pittasoma rufopiliatum*), and Yellow-green Bush-Tanager (*Chlorospingus flavovirens*). 0°51′N, 78°34′W.

El Tambo. A controversial collecting site of the commercial collector W. Clark-Macintyre, usually listed as "El Tambo, Loja" though no town of that name is actually known to exist in Loja. Paynter (1993) suggests that the El Tambo in question is situated in Cañar and not in Loja, and we suspect he is correct. This would explain the collecting of certain west-slope species not otherwise recorded from as far south or east as Loja (e.g., Giant Antpitta [*Grallaria gigantea*], Purplish-mantled Tanager [*Iridosornis porphyrocephala*], Black-chinned Mountain-Tanager [*Anisognathus notabilis*], and the *fagani* race of Sickle-winged Guan [*Chamaepetes goudotii*]. 2°30′S, 78°54′W.

Filo de Monos. A site (400–500 m) on the coastal Cordillera de Chindul in northern Manabí that was visited by a WFVZ team in Jul. 1988. Filo de Monos then supported some remnant patchy forest, though much of that may now be gone. 0°00′, 79°55′W.

Girón. A small town in southwestern Azuay near the headwaters of the Río Girón (2100 m). The region was one of the two known sites for the critically endangered and endemic Pale-headed Brush-Finch (*Atlapetes pallidiceps*), redis-covered near here in 1998. The remnant population's woodland habitat was purchased by Fundación Jocotoco in 1999. 3°10′S, 79°08′W.

Guaillabamba, Río. A river flowing north from the Quito region through an arid valley in northern Pichincha. Most natural vegetation has been removed or severely modified, though a little woodland remains. The area is significant as it is the presumed habitat for the critically endangered Turquoise-throated Puffleg (*Eriocnemis godini*). 0°28′N, 79°25′W.

Gualaceo-Limón road. A road that leads up over the Eastern Andes from Gualaceo (2°54′S, 78°47′W) in northern Azuay east to Limón (2°58′S, 78°27′W; also known as General L. Plaza Gutiérrez) in northwestern Morona-Santiago, continuing on to Macas. It passes through fine treeline shrubbery—Masked Mountain-Tanagers (*Buthraupis wetmorei*) are regularly found—and then descends through excellent temperate forest. Forest below about 2200 m has mostly been cleared, at least near the road. The Gualaceo-Limón road has been visited by numerous birders since the early 1980s, and its avifauna now is quite well known; numerous rare species are found here. It remains unprotected.

Guandera Biological Reserve. A newly established reserve of east-slope tem-perate forest and paramo in southeastern Carchi, owned and maintained by Fundación Jatun Sacha. The area is generally similar to that of Cerro Mongus (see Mongus, Cerro), situated not far to the south; as at Cerro Mongus, the forest itself actually lies just over the Continental Divide on the west slope of the Eastern Andes. In 1988 Guandera's avifauna was surveyed by Creswell et al. (1999), and all of Cerro Mongus's more important species (e.g., Black-thighed Puffleg [*Eriocnemis derbyi*], Crescent-faced Antpitta [*Grallaricula lineifrons*], Chestnut-bellied Cotinga [*Doliornis remseni*], Masked Mountain-Tanager [*Buthraupis wetmorei*]) were recorded here. 0°36′N, 77°41′W.

Guataracu, Río. A site (usually termed "head of the Río Guataracu" and often misspelled "Guataraco") at about 1350 m on the lower slopes of Volcán Sumaco where D. Norton collected in Jul.–Aug. 1964 for the MCZ. The slightly higher Palm Peak (which see) was reached slightly later. Both sites are clothed in foothill and lower subtropical forest. Lowland species are well represented at the head of the Río Guataracu, despite its relatively high elevation. 0°46′S, 77°15′W.

Guaticocha. A small, deep lake on the southwestern slope (750 m) of Volcán Sumaco in western Napo. Guaticocha was visited by D. Norton for the MCZ in Aug. 1964. It supports extensive upper tropical forest. 0°45′S, 77°24′W.

Guayaquil. A large and teeming city at the head of the Río Guayas estuary in Guayas, Guayaquil is by far the largest city in Ecuador (with more that 2.5 million inhabitants). Despite this, natural habitats do occur quite close by, both in the Chongón Hills to the west (deciduous forest) and in the mangrove forests to the south. Gray-cheeked Parakeets (*Brotogeris pyrrhopterus*) are found in city parks, and during the rainy season Chestnut-collared Swallows (*Petroche-*

lidon rufocollaris) nest on certain buildings. Crime is a serious problem in some areas; beware. 2°10′S, 79°50′W.

Huacamayos, Cordillera de. A semi-isolated ridge (sometimes spelled "Cordillera de Guacamayos") oriented east-west on the east slope of the Andes in western Napo. It is transected by the Baeza-Tena road, on which the highest elevation achieved is about 2200 m. Good subtropical forest and woodland exist on both slopes, above Cosanga on the north side (this is the best area for the rare Bicolored Antvireo [*Dysithamnus occidentalis*]) and down to about 1600 m on the south side (below this precipitous point most forest has been cleared); slopes on the south side are precipitous. A narrow trail leading west from the crest provides access to the forest interior. Numerous scarce montane birds occur in this important bird area, including Andean Potoo (*Nyctibius maculosus*), Emerald-bellied Puffleg (*Eriocnemis alinae*), Black-billed Mountain-Toucan (*Andigena nigrirostris*), White-capped Tanager (*Sericossypha albocristata*), Masked Saltator (*Saltator cinctus*), and many others. 0°40′S, 77°50′W.

Imuyacocha. A lake in the Río Lagarto drainage, close to the Peruvian border in extreme eastern Sucumbíos and since the early 1990s the site of a remote tourist camp run by Metropolitan Tours, a local travel agency, though a fire recently forced its (temporary?) closure. Seasonally flooded blackwater várzea forest and woodland dominate; accessible terra firme is limited. Although limited collecting was done in the 1920s, this still-wild region was first intensively studied by ANSP groups in 1990 and 1991; it was since been visited by numerous birders. Imuyacocha abounds in various parrot species, with macaws and Festive Amazons (*Amazona festiva*) being especially numerous. Most notable of many local specialties is the Cocha Antshrike (*Thamnophilus praecox*), rediscovered here in 1990. 0°32′S, 75°17′W.

Inchillaqui. A Quechua settlement (550 m) near Archidona in western Napo. Accessible nearby terrain has been deforested, but there is regenerating second-growth of varying ages. Secondary woodland along the road is notable for supporting the only known Ecuadorian population of the very local Chestnut-throated Spinetail (*Synallaxis cherriei*). 0°55′S, 77°52′W.

Isimanchi, Río. A river in the drainage of the Río Marañón which descends the east slope of the Cordillera Las Lagunillas in southwestern Zamora-Chinchipe, eventually flowing into the Río Mayo north of Zumba. ANSP teams investigated an area of foothill forest at 800 to 900 m in Dec. 1990 and another region of subtropical forest at 2200 to 2300 m in Nov. 1992.

Jatun Sacha (Jatun Sacha Biological Station). A private reserve situated just south of the upper Río Napo east of Puerto Quito (ca. 400 m), owned and managed by the Fundación Jatun Sacha. It presently encompasses several thousand hectares of primary terra firme forest, with smaller areas of secondary woodland and floodplain habitat. Thanks to the efforts of B. Bochan and others, Jatun Sacha's avifauna is among the best known in Ecuador. Improved protection has enabled populations of some bird species, formerly reduced by hunting, to begin to recover. Visits by tourists and scientists are encouraged. 1°05′S, 77°40′W.

Jauneche. A small private reserve in the lowlands of northern Los Ríos, managed by the Universidad de Guayaquil. Although isolated from other forest fragments and only a few hundred hectares in extent, Jauneche is important for protecting one of the last remnants of semihumid forest in the lowlands of southwestern Ecuador. Its avifauna was studied briefly by T. Parker and P. Coopmans (others too?) in the early 1990s. 1°20′S, 79°35′W.

José Velasco Ibarra, Represa. A reservoir on the Santa Elena Peninsula of western Guayas, situated north of Punta Carnero and reached via a network of dirt roads. The reservoir's extent varies from year to year depending on local rainfall. When it is full, after heavy rains, waterfowl numbers can be impressive, with thousands of White-cheeked Pintail (*Anas bahamensis*) and Blue-winged Teal (*A. discors*) and hundreds of grebes; Southern Pochards (*Netta erythrophthalma*) were formerly present and should still be watched for. A breeding population of the Andean Coot (*Fulica ardesiaca*) occurs; this is its sole known breeding site in the Ecuadorian lowlands. 2°13′S, 80°58′W.

Kapawi Lodge. An ecotourism lodge situated near the mouth of the Río Capahuari into the Río Pastaza in far southeastern Pastaza. The 1996 opening of this fine facility presented a useful opportunity to assess the distribution and comparative abundance of birds in the Río Pastaza drainage, a region that in recent years has been much less well studied than the northeast. Habitats include extensive terra firme and várzea forest as well as numerous islands in the Río Pastaza. Various notable birds have already been found, including populations of the Orinoco Goose (*Neochen jubata*) and Red-fan Parrot (*Deroptyus accipitrinus*). 2°45′S, 76°45′W.

La Bonita road. A road that winds down the east slope of the Andes in extreme northern Ecuador in western Sucumbíos, just south of the Colombian border. Although somewhat hard to find among the maze of minor roads in the highlands south of Tulcán, the La Bonita road provides access to some temperate and upper subtropical forest and woodland, and in an Ecuadorian context it has proven to be of biogeographic significance. Its avifauna remains only cursorily studied. 0°35′N, 77°30′W.

La Ciénega. An historic, colonial hacienda, recently converted into an elegant hotel, in the central valley (3000 m) north of Latacunga near Volcán Cotopaxi. A remnant population of the Subtropical Doradito (*Pseudocolopteryx acutipennis*) occurs nearby. 0°45′S, 78°40′W.

Lagarto, Río. A river draining south into the lower Río Aguarico, forming the boundary between Peru and Ecuador. There is a small lake named Lagartococha in extreme eastern Sucumbíos upstream from the better known Imuyacocha (which see). So far as known, the area upstream has an avifauna similar to that of Imuyacocha. 0°40′S, 75°13′W.

Lago Agrio. A small city in the lowlands of northeastern Ecuador in Sucumbíos along the Río Aguarico. Lago Agrio is dominated by the oil industry (its name translates as "bitter lake" and refers to the oily ponds that predate the town's establishment), and it is the hub of a large road system. Surrounding terrain is (as around Coca) mainly deforested, and birds of semiopen areas have begun to colonize. 0°06′N, 76°53′W.

La Libertad. A fishing town on the north shore of the Santa Elena Peninsula in western Guayas. Although itself not of any particular ornithological significance, La Libertad was the embarkation point for numerous ANSP excursions into the deep waters off the Santa Elena in search of pelagic birds. 2°14′S, 80°57′W.

La Libertad, Hacienda. A privately owned hacienda situated on the east slope of the Andes (2500–3300 m) in central Cañar, La Libertad was visited by an ANSP group for a short period in Aug. 1991. Extensive forest (with much *Podocarpus*) remains above 2800 to 2900 m, but below that it has been almost entirely removed. 2°36′S, 78°42′W.

La Plata, Isla de. A precipitous island about 30 km off the coast of southwestern Manabí and now part of Machalilla National Park; visitation is regulated but can be arranged at park headquarters in Puerto López. During most years, Isla de la Plata supports large populations of nesting seabirds, including Magnificent Frigatebirds (*Fregata magnificens*), three booby (*Sula*) species, Red-billed Tropicbirds (*Phaethon aethereus*), and the only nesting population of Waved Albatrosses (*Phoebastria irrorata*) away from the Galápagos. Landbirds on the island include remarkably tame Collared Warbling-Finches (*Poospiza hispaniolensis*) and the Gray-and-white Tyrannulet (*Pseudelaenia leucospodia*). 1°16′S, 81°06′W.

La Selva. An important ecotourism lodge in the lowlands of northeastern Ecuador, a short distance north of the Río Napo. Since its opening in 1988, La Selva has become something of a mecca for birders. The lodge offers a smoothly run operation with excellent food and guides, and several of the latter have become very knowledgable about local birds. Numerous trails (often muddy) provide access to a variety of habitats, and there are several lakes. River islands with their specialized vegetation and birds can be visited, as can trails into hilly terra firme on the south bank of the Napo. More than 500 species have been recorded by the stream of birders who have visited since 1988; this total includes a number of specialties that can be found here with reasonable regularity, prizes such as the Zigzag Heron (*Zebrilus undulatus*), Ochre-striped Antpitta (*Grallaria dignissima*), and Orange-crested Manakin (*Heterocercus aurantiivertex*). 2°25′S, 76°20′W.

Las Lagunillas, Cordillera. A spectacular, remote mountain range situated in extreme southern Ecuador in southwestern Zamora-Chinchipe and adjacent southeastern Loja; it lies not far to the north of Cerro Chinguela in adjacent Peru. The cordillera is traversed by a rough road which ascends from Amaluza and Jimbura, crosses a pass at about 3500 m (nearby peaks reach 4000 m), and then drops into the Río Isimanchi valley. The east slope was studied by an ANSP team in Oct.–Nov. 1992; drier and patchier forest on the west slope was also investigated, albeit less thoroughly. In 1992, forest on the east slope and the paramo were in remarkably pristine states (as an indication, signs of Spectacled Bear [*Tremarctos ornatus*] and Woolly Mountain Tapir [*Tapirus pinchaque*] were abundant); below 2500 m some deforestation had taken place. Upper elevations on the Cordillera Las Lagunillas are noteworthy for the abundance of the Neblina Metaltail (*Metallura odomae*) and for the presence of the recently described Chestnut-bellied Cotinga (*Doliornis remseni*); this remains the only Ecuadorian site for the Andean Hillstar (*Oreotrochilus estella*) and Andean Flicker (*Colaptes rupicola*). 4°47′S, 79°22′W.

La Torre, Cerro. A peak in the coastal Cordillera de Colonche in northwestern Guayas. The area still supports some patchy semihumid and cloud forest and was briefly studied by N. Krabbe in 1992. Some of the species found previously on Cerro San Sebastián were also found here.

Limoncocha. A very important site in the lowlands of northeastern Ecuador in Napo, Limoncocha was established by the Summer Institute of Linguistics as a missionary base in the 1950s. Planes based on the lake were employed to fly to Indian groups all over eastern Ecuador; they later departed from a large grassy airfield. A large base operation gradually developed, reaching its heyday in the 1970s. In the 1980s it was taken over by the Ecuadorian military, and it since has gradually been abandoned. Originally a remote outpost, Limoncocha now can be reached by road, and deforestation in the area has been severe. Ornithologists reached it in the 1970s, when it was intensively studied by D. Pearson, D. and E. Tallman, and others; for a while Limoncocha was one of Amazonia's better known ornithological sites, and numerous interesting discoveries were made here and on the nearby Río Napo. The lake continues to be one of the best in the Oriente for birds and remains the only Ecuadorian site with a permanent population of the rare Pale-eyed Blackbird (*Agelaius xanthophthalmus*). 0°25′S, 76°38′W.

Lita. A small town (900 m) along the Ibarra-San Lorenzo railroad, and also reached by road down the arid Río Mira valley. Some forest remains nearby, but the town is best known as the jumping-off site for other destinations, such as El Placer and Alto Tambo (which see). 0°52′N, 78°28′W.

Loja. A bustling, pleasant city situated in an arid valley (2000 m) of northeastern Loja province. Now a major distribution center and a hub for roads going out in various directions. 4°00′S, 79°13′W.

Loja-Zamora road. A traditional birding road that descends the east slope of the Andes from the pass (2700 m) above Loja to the town of Zamora (ca. 1000 m). Although the region was investigated by collectors in the early 20th century and even earlier, modern ornithological work commenced in Aug.–Sep. 1965 when D. Norton collected at several elevations for the MCZ. Since then bird observations have been carried out by numerous individuals and groups. A new, high-speed paved road recently replaced the old one; certain sections of the latter remain open to traffic (now light), however, and these stretches now provide favorable opportunities for birding. The two best extend down from the pass before dead-ending at washouts at about 2400 m, and upward from above Zamora to where the old road rejoins the new at about 1500 m.

Loma Alta Ecological Reserve. A small, locally owned and managed reserve of fairly humid ridgetop forest and cloud forest along the crest of the Cordillera de Colonche in western Guayas, above a town of the same name. The avifauna of this relatively remote area was studied by Becker and López-Lanús (1997), and it harbors (presumably small) populations of several scarce Tumbesian endemics, including Esmeraldas Woodstar (*Chaetocercus berlepschi*) and Little Woodstar (*C. bombus*). 1°49′S, 80°36′W.

Loreto road. A road that cuts east from the Baeza-Tena highway at 1300 m north of Archidona and gradually descends through precipitous Andean foothill terrain toward Loreto, from there continuing on toward Coca. Constructed in 1987 as a response to the closing of the main Lago Agrio road after an earthquake, it is sometimes referred to colloquially as the "Ministerio road" because it was built by Ecuador's Ministerio de Obras Públicas. When originally built, the Loreto road passed through magnificent lower subtropical and foothill forest, but sadly—and despite government promise—habitats along the road have since deteriorated, with much settlement and clearing having taken place. Despite the forest destruction, the Loreto road continues to provide a superb ornithological experience for those willing to endure its frequent foggy conditions and periodic downpours. The variety of range-restricted and rare species found here is great and includes numerous foothill "endemics."

Los Bancos, San Miguel de. A small town in the foothills (500–600 m) of northwestern Pichincha, west of Mindo along the road terminating near Puerto Quito. Although it is mainly deforested, a few forest patches remain in its vicinity, and these support at least a vestige of the ornithological riches formerly found here. Los Bancos was first visited by PJG and W. Davis in the late 1970s. 0°02′N, 78°54′W.

Macará. A small agricultural city in extreme southern Loja, and the Ecuadorian terminus of the old Pan-American Highway (virtually all traffic now goes via the new route through the coastal lowlands). Hills nearby are still carpeted in extensive deciduous woodland and forest, and they support an assemblage of southwestern specialties, including endemics such as the Henna-hooded Foliage-gleaner (*Hylocryptus erythrocephalus*) and Saffron Siskin (*Carduelis siemiradzkii*). 4°23′S, 79°57′W.

Machalilla National Park. A large and important national park in southwestern Manabí that encompasses some of the more extensive deciduous forests left in southwestern Ecuador; Isla de la Plata and other offshore islands are now included in the park. Many people have long resided within the officially designated park area, and signs of overgrazing, wood-cutting, and outright deforestation are, sadly, widespread. Since the late 1970s the park's avifauna has been studied by many people, notably several ANSP groups and T. Parker. Machalilla, which under good management could become one of Latin America's premiere parks, continues to harbor large populations of various southwestern endemics and specialties. See further discussions under Ayampe, Río; La Plata, Isla de; and San Sebastián, Cerro. 1°42′S, 80°46′W.

Maldonado road. A remote road leading to a small town on the west slope of the Andes in extreme northwestern Ecuador in northern Carchi, just south of the Colombian border. The Maldonado road's avifauna remains only partially sampled: two sites between about 2500 and 3400 m along the road were investigated by an ANSP group in Jul.–Aug. 1988, and N. Krabbe has worked along it on several occasions, but on the whole it remains remote and undervisited. There has been some deforestation, especially close to the road. 0°54′N, 78°07′W.

Mangaurcu (E of). An Ecuadorian border outpost in extreme western Loja, not far south of Puyango. Arid woodland and scrub are extensive, though most of it is overgrazed. The area was investigated by an ANSP group in Aug. 1992. 4°10′S, 80°02′W.

Manglares-Churute Ecological Reserve. An important reserve in the lowlands of southeastern Guayas, Manglares-Churute encompasses extensive mangrove forests, a large freshwater marsh (harboring western Ecuador's only remaining Horned Screamer [*Anhima cornuta*] population), and (especially on the Churute Hills) fairly undisturbed deciduous and semihumid forest and woodland; parts of the last are controlled by the Guayaquil-based Fundación Andrade and have recently been explored biologically by groups of British students (see Pople et al. 1977). Much terrain continues to suffer from human impacts: cattle grazing continues in many areas, and numerous people live within the official reserve boundaries. 2°30′S, 79°42′W.

Manta Real. A village at the base of the Western Andes in northwestern Azuay, Manta Real was the site of intensive surveys (by ANSP and T. Parker) in the early 1990s; an ornithologically minded Peace Corps volunteer, K. Berg, had been posted to the area. The slopes above town (up to 700 m) were still mainly forested in the early 1990s. Only a few hours from Guayaquil, and the closest reasonably accessible humid forest site to that city, Manta Real has potential as an ecotourism site. 2°30′S, 79°25′W.

Maquipucuna Reserve. A private reserve situated on the west slope of the Andes in northwestern Pichincha above Nanegalito, owned and managed by the Fundación Maquipucuna. The reserve encompasses montane forest from about 1300 to over 2500 m; a small amount of secondary vegetation exists, especially at lower elevations. The reserve's avifauna has been studied by several WFVZ groups and also by numerous visitors. 0°05′N, 78°35′W.

Maxus road (S of Pompeya). A 130-km-long private road constructed in 1994–1995 by the Maxus Oil Company to service its oil-extraction activities in Block 16 in Napo, south of the Río Napo. The road, constructed with a minimum of ecological impact, traverses wildlife-rich terrain on its generally southeastward course from the south bank of the Napo near Pompeya (1°22′S, 76°38′W); it crosses three major rivers, the Tiputini, Tivacuno, and Yasuní. Public entry is prohibited, and the only colonization that has occurred has been on the part of indigenous Quechuas (north of the Río Tiputini); some Huao have settled along the road south of the Tiputini. Hunting is proscribed other than by indigenous peoples; as they gradually concentrate their settlements along the road, a slow decrease in animal populations can be expected. The area's rich avifauna has been the subject of intensive investigation by RSR and others (including C. Canaday, P. English, N. Krabbe, and F. Sornoza) since early 1994, this being part of Ecuambiente's ongoing ecological monitoring program. The recently described Yasuní Antwren (*Myrmotherula fjeldsaai*) was first found here.

Mazán, Río. A forested valley (2900–3500 m, with paramo above) east of Cuenca in the highlands of Azuay, the upper Río Mazán valley encompasses the last large forested tract remaining in the Cuenca region. The valley is protected as a water source for the city of Cuenca. Río Mazán's avifauna was first studied by a group of British students in the 1980s (see King 1989). Numerous scarce montane species (e.g., Red-faced Parrot [*Hapalopsittaca pyrrhops*], Gray-breasted Mountain-Toucan [*Andigena hypoglauca*]) have been found here, and the endemic Violet-throated Metaltail (*Metallura baroni*) also occurs. 2°49′ S, 79°13′ W.

Miazi. An Ecuadorian military camp on the upper Río Nangaritza (850–900 m) southeast of Zamora in Zamora-Chinchipe visited by T. Parker and A. Luna in Jul. 1993 (Schulenberg and Awbrey 1997) and a British team in Dec. 1994 (Balchin and Toyne 1998). It supports a diverse assemblage of foothill birds—notably the Sharpbill [*Oxyruncus cristatus*] and the Orange-throated Tanager [*Wetmorethraupis sterrhopteron*]—and numerous lowland species were here found at unusually high elevations. 4°20′ S, 78°35′ W.

Mindo. A pleasant town set in a valley (1300 m) on the lower flanks of the Western Andes in Pichincha, easily reached from Quito and frequently visited by birders. The Mindo area supports an extremely rich and diverse avifauna, and it now has a well-developed ecotourism infrastructure. Although there has been some deforestation, excellent montane forest and woodland still exist in many areas and at various elevations. Mindo was an important base for bird collectors in the 19th century, and its basic avifauna became known early on; now that birders visit regularly, many further details are becoming available (see Kirwan et al. 1996). Its most notable bird is perhaps the Giant Antpitta (*Grallaria gigantea*), but many other subtropical specialties of the Andes' west slope also occur; a highlight for many is the existence of an accessible lek of Andean Cocks-of-the-rock (*Rupicola peruviana*). 0°02′ S, 78°48′ W.

Mongus, Cerro. A site high (2800–3600 m) in the Eastern Andes of southeastern Carchi, Cerro Mongus still supports extensive montane forest and has a relatively intact treeline. First reached by F. Sornoza, its avifauna was investigated by ANSP teams in Mar. and Jun. 1992; see Robbins et al. (1994). Cerro Mongus is notable as being the site where M. Robbins and G. Rosenberg obtained the first specimens of the recently described Chestnut-bellied Cotinga (*Doliornis remseni*); Crescent-faced Antpittas (*Grallaricula lineifrons*) and Masked Mountain-Tanagers (*Buthraupis wetmorei*) are both unusually numerous. The site remains unprotected, and forest destruction is proceeding. 0°27′ N, 77°52′ W.

Muisne. A small coastal town in western Esmeraldas. There are some good freshwater marshes nearby, and limited areas of mangroves persist, but Muisne is best known as the jumping-off point to remaining humid forests that extend to the southeast. These are accessible only through logging roads, often impassable in (frequent) wet weather. These unprotected and rapidly diminishing lowland forests constitute the last remaining so far south in western Ecuador. 3°35′N, 79°59′W.

Nangaritza, Río. An affluent of the Río Zamora in eastern Zamora-Chinchipe. The headwaters of the Río Nangaritza lie in the remote southeastern part of Podocarpus National Park and flow east and then north, draining the west flank of the Cordillera del Cóndor. Various military posts along the river have been visited by ornthologists in recent years, and its avifauna was summarized by Balchin and Toyne (1998).

Napo, Río. A very important river that flows eastward from the Andes across the lowlands of northeastern Ecuador, continuing into northeastern Peru and entering the Amazon just east of Iquitos. The Napo is navigable downstream from around Puerto Napo (south of Tena). Until recently the Napo and its myriad tributaries provided the only means of local transportation; it still is the sole access to ecotourism lodges such as La Selva, Sacha Lodge, and Yuturi Lodge. The Napo is wide and shallow, with a shifting mosaic of sandbars and islands that support a diverse assemblage of island specialties; many of these species (e.g., Castelnau's Antshrike [*Thamnophilus cryptoleucus*], Black-and-white Antbird [*Myrmochanes hemileucus*], Lesser Wagtail-Tyrant [*Stigmatura napensis*], and Riverside Tyrant [*Knipolegus orenocensis*]) have only recently been first found to occur in Ecuador.

Nono-Mindo road. A well-known birding road that descends the west slope of the Andes in western Pichincha from northwest of Quito, the "old" Nono-Mindo road has recently been largely superseded by a new paved route that has diverted all but local traffic. As a result of the lighter traffic, the Nono-Mindo road provides ideal access to the rich temperate and subtropical forests that remain along some sections.

Oña. A small town in an arid interior valley (1900 m) of extreme southern Azuay. Almost all natural vegetation has been removed, though some valleys and ravines retain scrub and woodland. The area is important as one of the last known for the critically endangered Pale-headed Brush-Finch (*Atlapetes pallidiceps*), and Ecuador's only known population of the White-browed Chat-Tyrant (*Ochthoeca leucophrys*) occurs near here. 3°28′S, 79°10′W.

Pacuyacu, Río. A minor right-bank, blackwater tributary of the lower Río Aguarico in extreme eastern Napo. In the early 1990s Metropolitan Tours (a local travel agency) established a boardwalk and small camp here; the boardwalk traverses a good blackwater várzea swamp. 0°32′S, 75°38′W.

Palm Peak. A site on the lower slopes (1500 m) of Volcán Sumaco that marks the highest elevation attained by D. Norton during his explorations of the Sumaco region for the MCZ. Palm Peak was visited in Aug. 1964. 0°39′S, 77°36′W.

Panguri. A remote agricultural outpost (1500–1600 m) near the southern border of Podocarpus National Park in southern Zamora-Chinchipe. Panguri's interesting subtropical forests were studied by T. J. Davis and F. Sornoza for the ANSP on an arduous expedition in Jul.–Aug. 1992. 4°37′S, 78°58′W.

Papallacta (pass). The high point (4000 m) on the Quito-Lago Agrio highway, lying on the Pichincha/Napo border. Mountainous country extends in all directions from this point; relatively wet paramo predominates, but there are numerous patches of *Polylepis*-dominated woodland as well. A side road to the north leads even higher, reaching some microwave repeating towers at 4200 m. The area has long been a favored destination for Quito-based ornithologists and birders, and its avifauna is now well known and, despite some disturbance, remains relatively intact. Specialties to watch for include the Red-rumped Bush-Tyrant (*Cnemarchus erythropygius*), Giant Conebill (*Oreomanes fraseri*), and Black-backed Bush-Tanager (*Urothraupis stolzmanni*). 0°22′S, 78°13′W.

Pasochoa. A private reserve (2800–4000 m) managed by Fundación Natura (of Quito) as an ecotourism and environmental education center. Pasochoa protects some of the last remaining natural forests in the central valley and is readily accessible off the main road south of Quito. Andean Guans (*Penelope montagnii*) and Ocellated Tapaculos (*Acropternis orthonyx*) are unusually easy to see here. 0°28′S, 78°29′W.

Pastaza, Río. An important river that flows east across the lowlands of southeastern Ecuador before curving south and joining the Río Marañón in northern Peru. Various villages in the upper drainage of the Pastaza (notably Sarayacu), were important 19th-century collecting sites, and professional collectors continued to work at various localities into the mid 20th century. In recent decades, however, little ornithological field work has been done here other than at Kapawi Lodge (which see).

Payamino, Río. A tributary of the Río Napo that joins the Napo at Coca and drains the northeast flank of Volcán Sumaco. The general Río Payamino area has long been a favored collecting ground of the Olalla family. An excellent forest area on the south bank of the lower Payamino can be accessed by road off the main Coca-Archidona highway; some small *Guadua* bamboo patches occur here.

Pichincha, Volcán. An imposing volcano (4794 m) which provides a backdrop for the city of Quito. Even the summit of Volcán Pichincha rarely has snow for long, but it does support extensive paramo and (on its north and particularly west flanks) montane forest. Such forest extends up to an unusually high elevation (3500–3800 m) on Pichincha's northwest flank. 0°10′S, 78°33′W.

Pitahaya, Puerto. A small fishing settlement at the edge of mangrove forest along the El Oro coast north of Arenillas. Mangroves remain extensive here, though as usual some areas have been converted to shrimp ponds. Rufous-necked Wood-Rails (*Aramides axillaris*) are relatively numerous in the mangroves; there are several important heron nesting colonies. 3°25′S, 80°07′W.

Playa de Oro. A small community in the lowlands of northern Esmeraldas (adjacent to the lowermost foothills). The avifauna of Playa de Oro was intensively studied by O. Jahn and his coworkers in the late 1990s. Except along the rivers, the area is still covered mainly with wet forest, and it remains relatively little disturbed. 0°52′N, 78°48′W.

Podocarpus National Park. A large and internationally important park on the east slope of the Andes in western Zamora-Chinchipe and adjacent eastern

Loja, first intensively studied by ZMCOP teams in 1989–1990. The park protects a magnificent cross-section of montane habitats, from the foothills at 900 to 1000 m up into wet paramo and treeline vegetation at 3100 to 3300 m. Stands of the *Podocarpus* trees for which the park is named—and which elsewhere in Ecuador have usually been cut for timber—can be visited around Romerillos. Most of Podocarpus National Park is remote, with its two primary access points being at Río Bombuscaro and Cajanuma (which see). Forests are being encroached upon along the park's western border, and gold mining is affecting certain drainages, but on the whole the park is well protected. Approximately 4°10′S, 79°10′W.

Quebrada Honda. A deep valley on the east slope of the Andes (ca. 1600–2400 m) north of Valladolid in western Zamora-Chinchipe, Quebrada Honda is situated near the southwestern border of Podocarpus National Park. The valley itself has been settled for several decades and has mostly been deforested. The ornithological significance of Quebrada Honda derives from the trail to it that departs from the Loja-Zumba road a few kilometers south of the crest of the Cordillera Sabanilla. This trail cuts through patchy and disturbed temperate and subtropical forest and was the site of the Nov. 1997 discovery of the spectacular Jocotoco Antpitta (*Grallaria ridgelyi*) by RSR, L. Navarrete, and J. Moore. A wealth of other scarce east-slope birds has also been found here; see also Tapichalaca Biological Reserve. 4°30′S, 79°10′W.

Quevedo. A crowded small city in the lowlands (50 m) of northern Los Ríos. Judging from specimens taken a half-century ago, Quevedo must then have been surrounded by extensive lowland forest. Sadly, however, such forests are now but a fading memory, and the entire region is presently devoted to intensive agriculture. 1°02′S, 79°29′W.

Quito. Ecuador's capital and second-largest city, set in a broad valley at about 2800 m, beneath the east flank of Volcán Pichincha. Quito has grown rapidly in recent decades; as recently as the 1970s, much terrain toward the airport remained semiopen with scattered grazing cattle, but now much or all of that area is heavily built up. As late as the 1920s much of what is now Parque La Carolina was marshy terrain in which waterfowl were numerous. Gardens and the few parks continue to be suitable for some birds, but diversity is low and the city now is primarily a base from which to set out for more productive locales. 0°13′S, 78°30′W.

Río Palenque (Río Palenque Scientific Station). A small private reserve in the lowlands (100 m) of extreme southern Pinchincha. Owned by the Dodson family, Río Palenque has been operated as a scientific research facility since the 1970s, and its diverse avifauna is well known. Only about 100 ha are still in humid forest, and this forest is now entirely isolated from other such forest fragments by extensive agriculture; as a result of this isolation, many larger and scarcer bird species are no longer extant here. 0°35′S, 79°25′W.

Sabiango. A small farming town (800–1000 m) in southern Loja, situated along the road between Macará and Sozoranga. Most terrain in the region has been converted to agricultural use, but patches of secondary woodland and scrub hold some species of interest, notably large numbers of Black-and-white Tanagers (*Conothraupis speculigera*) in the rainy season. First studied by R. A. Paynter, Jr., and D. Norton in Oct. 1965, more recently the Sabiango area has been visited by several British and ANSP groups. 4°24′S, 79°52′W.

Sacha Lodge. An ecotourism lodge opened in 1992 and situated only a short distance upstream on the Río Napo from the slightly older La Selva lodge (which see). The habitats present are similar to those at La Selva, and the facilities are likewise excellent; there is, in fact, little to choose between the two lodges, though Sacha is a bit closer to Coca, thus reducing the transfer time. 0°27′S, 76°23′W.

Salanguilla (above). A site in the coastal Cordillera de Colonche in northwestern Guayas, above the town by that name. The area still supports some patches of semihumid and cloud forest and was briefly studied by N. Krabbe in 1992. Some of the species found previously on Cerro San Sebastián were also located here. 1°58′S, 80°34′W.

Sangay National Park. A large, wild area highlighted by the volcano (5320 m) of the same name, situated in the Eastern Andes of northwestern Morona-Santiago. Volcán Sangay represents the highest point in a major national park that encompasses much wild terrain on the east slope of the Andes down to about 1000 m northwest of Macas. Sangay is the most active volcano in Ecuador, and it nearly always rumbles and smokes ominously (however, rarely does much more happen). Most of the park is difficult to access, and it remains poorly known ornithologically. In Oct. 1976 J. O'Neill and RSR explored the upper Río Palora valley, and in Aug. 1979 RSR and R. A. Rowlett visited the Río Abanico valley west of Macas. 2°S, 78°20′W.

San Isidro, Cabañas. A small ecotourism lodge on the east slope of the Andes (1900–2200 m) in western Napo near Cosanga, just off the Baeza-Tena highway. Several trails leading from the lodge permit access into surrounding rich subtropical zone forest; doubtless the rarest bird that has been seen in the area is the Bicolored Antvireo (*Dysithamnus occidentalis*), but there are also many others (including the Giant Antpitta [*Grallaria gigantea*]). 1°35′S, 77°55′W.

San Miguel, Río. A river situated in the eastern lowlands of extreme northern Ecuador, the San Miguel joins the Río Putumayo at Puerto El Carmen de Putumayo on the Colombian border. The avifauna of this still remote—and now rather dangerous—region remains little known, and much remains to be learned (see Tigre Playa).

San Pablo, Laguna de. A beautiful deep lake, reed-fringed in part, in the highlands of Imbabura (2570 m) near Otavalo. At one time waterbirds probably were numerous on this lake, but now its shorelines are apparently too densely inhabited for many resident birds to breed successfully. Some migrants do still occur. 0°13′N, 78°12′W.

San Rafael Falls. An impressive waterfall (formerly often called "Coca Falls") of the Río Quijos on the east slope of the Andes (1250 m) in western Napo, but very close to the border with Sucumbíos. Easily accessible just off the Lago Agrio highway, San Rafael Falls has long been known as a good birding area, having first been visited by PJG and W. Davis in the 1970s. Primitive accommodations are available. Considerable lower-subtropical forest persists in the vicinity of the falls; tanagers are notably numerous. A plan to dam the gorge for electricity may eventually be realized. 0°05′S, 77°35′W.

San Sebastián, Cerro. A peak (ca. 800 m) in the coastal Cordillera de Colonche situated in Machalilla National Park in southwestern Manabí, usually reached by climbing up from the town of Aguas Blancas to the north. Elevations above about 700 m support cloud forest (though, little by little, this continues to be cut); areas below are covered by semihumid and then deciduous forest and woodland, with elevations below 300 m having arid scrub, much of it heavily grazed (despite its park status). The area is rich in endemic bird species, some of them threatened, with notably large populations of Ochre-bellied Doves (*Leptotila ochraceiventris*) and Henna-hooded Foliage-gleaners (*Hylocryptus erythrocephalus*). In early 1991 this site was briefly studied by T. Parker and a Conservation International group, and later that year more intensively by an ANSP expedition. Despite the diverse avifauna, apparently only a few birders have returned since. 1°35′S, 80°40′W.

Santa Elena Peninsula. A low, arid peninsula jutting into the Pacific Ocean and forming the westernmost part of Guayas (and Ecuador). Beaches on its north shore are protected and relatively warm watered, and this area has been developed as a major resort for Guayaquil vacationers. On the south shore the beaches are much more exposed and the water considerably colder. Oil production has been important for a half-century; the petroleum engineer S. Marchant conducted a major study of the Santa Elena's avifauna in the 1950s, and various parts of the peninsula have since been visited by ornithologists and birders. As a result of increasing human population density and disturbance, various bird species appear to have declined in the vicinity, and two (Tawny-throated Dotterel [*Oreopholus ruficollis*] and Least Seedsnipe [*Thinocorus rumicivorus*]) apparently have been extirpated. Approximately 2°15′S, 80°50′W.

Santiago. An Ecuadorian military camp (400 m) in southwestern Morona-Santiago on the north side of the Río Santiago. Most land near the encampment and the river is semiopen or in secondary woodland, but some primary forest remains on ridges to the north. Santiago's avifauna was investigated by an ANSP team in Jul.–Aug. 1989. Despite the proximity of the foothills, the primary affinity of Santiago's avifauna is clearly with the lowlands to the east. 3°03′S, 78°03′W.

Santo Domingo de los Colorados. A crowded and noisy small city in the lowlands of southwestern Pichincha, Santo Domingo (as it is usually called) has grown substantially in recent decades; it is now a major agricultural center. Formerly it was surrounded by humid forest, but very little of that remains. 0°15′S, 79°09′W.

Shaime. A military camp and small settlement in the valley of the upper Río Nangaritza in southeastern Zamora-Chinchipe. Using this camp as a base, two groups (one from WFVZ, the other from Conservation International) have investigated the avifauna of the lower slopes on the west side of the Cordillera del Cóndor. The most important discovery here was M. Marín's finding the rare Orange-throated Tanager (*Wetmorethraupis sterrhopteron*) in Jul.–Aug. 1989. 4°08′S, 78°40′W.

SierrAzul (Hacienda Aragón). A privately owned property with extensive subtropical and lower temperate zone montane forest (2200–2400 m) situated on the east slope of the Andes in western Napo near Cosanga and Cabañas San Isidro. 0°41′S, 77°55′W.

Simón Bolívar road. A minor side road in the foothills (450–500 m) on the west slope of the Andes in northwestern Pichincha that leads north from west of Pedro Vicente Maldonado off the highway between Mindo and Puerto Quito. Although much of the region is deforested, several extensive but nonetheless accessible areas of secondary woodland and forest persist along this road. In early 1995 various birders (first P. Coopmans and M. Lysinger) began visiting the Simón Bolívar road on a regular basis, and numerous scarce and interesting birds have now been found here, many at the southern limit of their distribution. The most notable have been two species new to Ecuador, the Double-banded Graytail (*Xenerpestes minlosi*) and Griscom's Antwren (*Myrmotherula ignota*). 0°06′N, 79°04′W.

Sozoranga. A pleasant, small town (1500–2000 m) in the Andes of southern Loja. The avifauna was studied in the early 1990s by several British groups coordinated by B. Best. Although the area is principally agricultural, some montane forest and woodland remain on the slopes above town. Fundación Arcoiris (of Loja) has recently protected a fine 200-ha forest, El Tundo Reserve, not far west of town. This area supports an important population of the endangered Gray-headed Antbird (*Myrmeciza grisceiceps*). Paynter (1993) indicates that the spelling of the town should be "Zozoranga," but on maps we have seen it is always spelled with the initial "S." 4°20′S, 79°45′W.

Sumaco, Volcán. A spectacular, symmetrical volcano (3807 m) isolated east of the main chain of the Andes in western Napo. A few groups and individuals (notably D. Norton and members of the Olalla family) have made ornithological explorations on its forested lower slopes as high as about 1500 m; because of the extremely difficult access, however, little or no work has been done above that. At least some of the volcano has recently been incorporated into the Sumaco-Galeras National Park, but we remain uncertain of this park's precise boundaries. 0°34′S, 77°38′W.

Taisha. A small town and military post in the lowlands (ca. 400 m) of southeastern Morona-Santiago. Taisha's avifauna was studied by an ANSP group in Aug. 1990. This hilly region is still almost entirely forested, with the only clearings being in a limited area around the airstrip and base. Although it supports primarily a lowland avifauna, a few foothill species were found at the lower limits of their distribution, notably the Fiery-throated Fruiteater (*Pipreola chlorolepidota*) and Blackish Pewee (*Contopus nigrescens*). 2°23′S, 77°30′W.

Tandayapa. A crossroads settlement (1600 m) on the west slope of the Andes in Pichincha where the old road from Quito and Nono bifurcates, one proceeding to Mindo and beyond and the other to Nanegal (and also Maquipucuna). The area, a favored one for birders, is still partially forested. 0°15′N, 78°38′W.

Tapichalaca Biological Reserve. A private reserve owned and managed by Fundacíon Jocotoco and situated on the east slope of the Andes (2000–3100 m) north of Valladolid in southwestern Zamora-Chinchipe. As of 2000 the Tapichalaca Reserve comprised some 2200 ha of (mainly) montane forest, and we hope it will be expanded further in the years to come, perhaps sufficiently to connect it to nearby Podocarpus National Park. The impetus for the reserve's establishment came from the discovery along the Quebrada Honda trail of the Jocotoco Antpitta (*Grallaria ridgelyi*); most of the world's known population of that species occurs on the reserve. Numerous other montane birds of course also occur here; see also Quebrada Honda. 4°30′S, 79°10′W.

Taracoa, Laguna. A small blackwater lake situated a few kilometers south of the Río Napo near Pompeya. Taracoa was visited in the 1970s and 1980s by many birders traveling on the Metropolitan Tours' ship *Flotel Orellana* when it provided the only simple access into Ecuador's eastern lowlands. To reach the lake, one crossed a várzea forest on a boardwalk, and beyond the lake itself there was superb, little-disturbed terra firme; in addition, a platform in an emergent *Ceiba* provided a chance to observe the community of forest canopy birds (the first such opportunity in western Amazonia). Sadly, however, civilization encroached on what was originally a pristine area still supporting Harpy Eagles (*Harpia harpyja*), and Metropolitan Tours abandoned it around 1990. 0°25′S, 76°43′W.

Tayuntza. A small settlement and military base in the lowlands (600 m) of eastern Morona-Santiago which was surveyed by a WFVZ team headed by M. Marín in Jul.–Aug. 1987. 2°43′S, 77°52′W.

Tena. A large town at the base of the Andes (500 m) in southwestern Napo, now often used as a base for exploring nearby areas, including the Loreto road and Jatun Sacha. Most of the region, and virtually all accessible terrain, has been deforested, but areas of secondary woodland exist in many areas. 0°59′S, 77°49′W.

Tigre Playa. A tiny settlement on the Río San Miguel in extreme northern Sucumbíos, just south of the Colombian border. This remote region is not well known ornithologically, but in Jul.–Aug. 1993 an ANSP group conducted an intensive survey at one site (250 m) in somewhat disturbed terra firme forest some 10 km north of the river. The avifauna there was rich, with several species scarce or absent elsewhere in Ecuador being numerous, among them the Stipple-throated Antwren (*Myrmotherula haematonota*) and Collared Gnatwren (*Microbates collaris*). 0°13′N, 76°17′W.

Tinalandia. A long-established hotel in the foothills (700–800 m) of the Western Andes just off the Quito-Santo Domingo highway in western Pichincha. Fairly large patches of mature forest continue to be protected here, and of late some open areas have been allowed to begin to revert to forest; birds are numerous and highly visible, and some species (e.g., Chocó Trogon [*Trogon comptus*] are more readily found here than elsewhere. Surrounding areas, however, have been largely deforested; this includes the almost mythical "moss forest," most of which was destroyed decades ago. Tinalandia has been a favored birding area since the early 1970s when its riches were first publicized by T. Butler. With good management, it should continue to attract visitors long into the future. 0°19′S, 79°02′W.

Tiputini Biodiversity Center. A biological research center situated in almost untouched terra firme forest on the south bank of the Río Tiputini in the western part of Yasuní National Park. Tiputini mainly caters to researchers but is also open for limited ecotourism. 0°43′S, 76°12′W.

Utuana. A small town in the Andes (2400–2500 m) of extreme southern Loja. Montane scrub and woodland are still fairly extensive, and a limited amount of moss forest remains. Utuana's avifauna was first studied by British teams in the early 1990s; they discovered Ecuador's first Black-crested Tit-Tyrants (*Anairetes nigrocristatus*) here, and Utuana also harbors Ecuador's primary population of the Piura Hemispingus (*Hemispingus piurae*). Most of the forest has recently been protected through the efforts of Fundación Arcoiris and Fundación Jocotoco. 4°22′S, 79°42′W.

Vilcabamba. A small and attractive town on the west slope of the Eastern Andes (1600 m) in southeastern Loja. The Vilcabamba area has several attractive hotels, and it makes an excellent base from which to explore various parts of nearby Podocarpus National Park. Ecuador's first Plumbeous Rails (*Pardirallus sanguinolentus*) were found here in the early 1990s. 4°12′S, 79°14′W.

Warientza. A small military base at the northern end of the Cordillera del Cóndor (ca. 950 m) in southwestern Morona-Santiago. Warientza's avifauna was surveyed by a WFVZ team headed by M. Marín in Jul.–Aug. 1988. 3°12′ S, 78°17′W.

Yaguarcocha, Laguna de. A large lake, shallow and with marshy reed-fringed borders, in Imbabura (2450 m). Waterbirds remain quite numerous, though the Cinnamon Teal (*Anas cyanoptera*) and American Coots (*Fulica americana*) that apparently were present here now have disappeared. Although surrounded by (of all things!) an automobile racetrack, the track usually is not in use. 0°22′ N, 78°06′W.

Yanacocha. A hacienda high (3200–3600 m) on the northwestern slope of Volcán Pichincha, located northwest of Quito and accessible off the road to Nono. Fine temperate forest remains, with sites nearby being the stronghold of the critically endangered Black-breasted Puffleg (*Eriocnemis nigrivestis*). 0°10′ S, 78°35′W.

Yasuní National Park. A large national park in the lowlands (ca. 150–300 m) of northeastern Ecuador south of the Río Napo in eastern Napo. The park's jurisdiction and some of its boundaries remain diffuse and somewhat subject to change. Virtually all of the park is remote, and its avifauna is little studied (though there seems to be little reason to expect that it differs substantially from surrounding areas; see, for example, the Maxus road). Visitation remains difficult to arrange except at the Tiputini Biodiversity Center (which see).

Yaupi. A small settlement in the lowlands (600 m) of central Morona-

Santiago near the eastern base of the Cordillera de Cutucú. Yaupi is the eastern terminus of the Logroño-Yaupi trail that traverses that cordillera (and was walked by the ANSP expedition to the Cutucú in 1984). It was also visited by a WFVZ group in 1987. 2°50′S, 77°55′W.

Yunguilla Reserve. A small Fundación Jocotoco reserve southwest of Girón in western Azuay (1650–1800 m) whose remnant montane scrub supports the last known population of Pale-headed Brush-Finches (*Atlapetes pallidiceps*), rediscovered in 1998 (A. Agreda, N. Krabbe, and O. Rodriguez, *Cotinga* 11: 50–54, 1999). 3°15′S, 79°15′W.

Yuturi Lodge. An ecotourism lodge on the south bank of the Río Napo near the mouth of the Río Yuturi. Its avifauna remains less well known than that of many of the other lodges in eastern Ecuador; information provided by S. Howell would indicate that both várzea and terra firme forest are extensive and in good condition. 0°33′S, 76°05′W.

Zamora. A small town at the base of the east slope of the Andes (1000 m) in western Zamora-Chinchipe. Zamora is an important agriculture and transportation center, and much surrounding terrain has been deforested, though patches of forest and woodland remain, especially on steeper slopes. Collectors visited it as early as a century ago, and the town continues to be a convenient base for birders wishing to sample the rich avifauna still found in its vicinity. The Río Bombuscaro sector of Podocarpus National Park is nearby. 4°04′S, 78°05′W.

Zancudococha. A large lake southwest of the lower Río Aguarico in extreme eastern Napo. In the early 1990s Metropolitan Tours (a local travel agency) developed an ecotourism camp (Iriparí) on the shores of Zancudococha. The lake is surrounded by undisturbed terra firme and várzea forest; birdlife is abundant and varied. Zancudococha's avifauna became known through a long visit by RSR in Sep. 1976 and by an ANSP expedition in Mar.–Apr. 1991. The site has since been visited by numerous birders. 0°32′S, 75°30′W.

Zapotillo. A dusty town in the hot lowlands (150 m) of extreme southwestern Loja. Zapotillo is surrounded by shrubby desertlike terrain, much of it now severely overgrazed; woodland, also quite disturbed, is confined to the ridges to the north and to the margins of the few watercourses. Zapotillo was first investigated ornithologically by ANSP groups in Mar. 1992 and Apr. 1993. Their most notable discovery was finding breeding Tumbes Swallows (*Tachycineta stolzmanni*). 4°23′S, 80°13′W.

Zumba. An isolated town (1000 m; nearby areas reach as low as 650 m) in extreme southern Zamora-Chinchipe, the southernmost part of Ecuador. In an Ecuadorian context Zumba is of special biogeographic importance because of its location in the drainage of the Río Marañón. Numerous Marañón "endemics," formerly known only from adjacent Peru, have recently been found to occur here. Certain of these (e.g., the Marañón Thrush [*Turdus maranonicus*]) have also recently been found as far north as around Valladolid. 4°52′S, 79°08′W.

Endemic Bird Areas in Ecuador

BirdLife International (formerly the International Council for Bird Preservation and often called simply BirdLife) pioneered the effort to identify the world's "endemic" bird species, bird species with "restricted ranges," and the conservation plight of these species. This effort has most recently been summarized in Stattersfield et al.'s exhaustive *A Global Directory of Endemic Bird Areas* (1997). As defined by Stattersfield et al., each "endemic bird area" would encompass an area within which are found at least two species whose total ranges more or less overlap in an area of under 50,000 sq km. Several such areas are found in Ecuador; in fact virtually the entire country lies within one or another such area.

For the most part, the endemic areas as defined by Stattersfield et al. (1997) work reasonably well in Ecuador. We see a few internal inconsistencies, however, and thus have modified their arrangement here so as to, in our view, more accurately reflect the intricacies of bird distribution in Ecuador. In particular we see an advantage in dividing species from Ecuador's two most species-rich endemic areas, the Chocó and the Tumbesian, into species with a *montane* distribution and those with a *lowland* distribution. The two groups break out quite well; their respective situations are in fact quite different, and there is now enough information available to make the distinction relatively straightforward. Thus we recognize "Chocó Lowlands" and "West Slope of Andes" centers in place of simply a "Chocó" center, and "Tumbesian Lowlands" and "Southwestern Highlands" centers in place of simply a "Tumbesian" center.

We are also troubled by BirdLife's "North Central Andes," "South Central Andes," and "Central Andean Paramo" centers, finding them to obscure more than to clarify. We prefer to more clearly segregate groups of species by their ranges, breaking them out by three regions, as follows: a group that occurs mainly or primarily on the west slope, a group that occurs mainly or entirely on interandean slopes and valleys, and a group that occurs mainly or entirely on the east slope. At least from an Ecuadorian perspective, this seems to work considerably better, and it results in groupings that more nearly reflect biogeographic reality. We suspect that these groupings will also be acceptable in Colombia and Peru.

Further, for clarity's sake we have occasionally extended BirdLife's boundaries to account for species' distributions that extend marginally beyond what

Stattersfield et al. (1997) established as their Endemic Area limits. Thus, for instance, the Lemon-spectacled Tanager (*Chlorothraupis olivacea*) is clearly a "Chocó Lowlands endemic" species even if its distribution extends north to extreme eastern Panama (slightly beyond the defined area). Likewise the Parrot-billed Seedeater (*Sporophila peruviana*) is obviously a "Tumbesian Lowlands endemic" even if its distribution extends south into west-central Peru. We have also done this with the Western Amazonian Lowlands center: we have added a few species whose distributions are clearly centered in this area but that may extend marginally outward (e.g., along the Amazon River system).

Numerous additional species of course occur within each Endemic Center, but these are not included—even if in Ecuador they occur only within an area encompassed by an Endemic Center—if elsewhere their ranges extend well beyond the Endemic Center in question. Thus, for instance, the Semiplumbeous Hawk (*Leucopternis semiplumbea*) and Plumbeous Hawk (*L. plumbea*) both occur in Ecuador only in the northwest, so they would appear to be "Chocó Lowlands endemics," but in fact their respective ranges extend north well into Central America, considerably beyond the northern limit of that center, so they do not qualify.

Further, any consideration of subspecies has had to be eliminated, so even if a subspecies is endemic to the Endemic Area under consideration, it is not listed. Thus the Tataupa Tinamou (*Crypturellus tataupa*), whose race *inops* is endemic to the Río Marañón, is not mentioned as a "Río Marañón endemic." This of course would change were *inops* to be raised to species rank, and a comparable situation has indeed occurred for several other taxa in the lists that follow. For example, the Pacific Hornero (*Furnarius cinnamomeus*) is now considered a "Tumbesian Lowlands endemic," having been raised to species rank; until recently it was considered only a race of a wide-ranging Pale-legged Hornero (*F. leucopus*), and then it would not have been mentioned.

Below we describe the nine endemic centers we recognize in Ecuador. *Species considered by us to be at risk are preceded by an asterisk.* These species are listed separately, by category, in the next chapter ("Conservation"), and each is also discussed at greater length in the relevant species account in Volume I. Not all restricted-range species are at risk, though many are inherently more vulnerable because of their small total ranges. Conversely, not all species that are at risk have small, restricted ranges; some threatened species occur naturally at low population densities over a broad area, or are being affected by other factors.

Choco Lowlands

The Chocó wet-forest belt of western Colombia and northwestern Ecuador supports the greatest concentration of restricted-range endemic species in the world. No fewer than 31 such species occur in Ecuador's Chocó lowlands, with a further 44 occurring in the highland zone immediately above it (here broken out for clarity's sake; see below). These two contiguous areas thus jointly harbor the staggering total of 75 range-restricted species, with an additional 12 or so occurring only in Colombia. The ranges of a few of these species extend north to just north of the Panama border, or south into the foothills of southwestern Ecuador. All species are essentially dependent on forest, with only a few tolerating even somewhat disturbed situations.

A considerable extent of forest in this region—a far higher percentage than is found farther south in western Ecuador—still stands. Most of it remains unprotected, however, and despite the near certainty that any attempt at permanent agriculture will ultimately fail because of the region's excessively wet climate, more forest continues to be lost each year. Formally protected areas include the remote and still (ornithologically) little known Awá Forest Reserve on the Colombian border, the lowermost sectors of Cotocachi-Cayapas Ecological Reserve, the private Bilsa reserve in southwestern Esmeraldas, and the very small and isolated Río Palenque Scientific Station. Much more needs to be protected.

SPECIES (31)

*Berlepsch's Tinamou	*Crypturellus berlepschi*
*Plumbeous Forest-Falcon	*Micrastur plumbeus*
*Baudó Guan	*Penelope ortoni*
*Brown Wood-Rail	*Aramides wolfi*
Dusky Pigeon	*Columba goodsoni*
Pallid Dove	*Leptotila pallida*
*Indigo-crowned Quail-Dove	*Geotrygon purpurata*
Rose-faced Parrot	*Pionopsitta pulchra*
*Banded Ground-Cuckoo	*Neomorphus radiolosus*
*Chocó Poorwill	*Nyctiphrynus rosenbergi*
White-whiskered Hermit	*Phaethornis yaruqui*
*Humboldt's Sapphire	*Hylocharis humboldtii*
Purple-chested Hummingbird	*Amazilia rosenbergi*
Chocó Trogon	*Trogon comptus*
Orange-fronted Barbet	*Capito squamatus*
*Five-colored Barbet	*Capito quinticolor*
Stripe-billed Araçari	*Pteroglossus sanguineus*
Pale-mandibled Araçari	*Pteroglossus erythropygius*
Chocó Toucan	*Ramphastos brevis*
*Lita Woodpecker	*Picumnus litae*
*Chocó Woodpecker	*Veniliornis chocoensis*
*Double-banded Graytail	*Xenerpestes minlosi*
Stub-tailed Antbird	*Myrmeciza berlepschi*
*Rufous-crowned Antpitta	*Pittasoma rufopileatum*
Pacific Flatbill	*Rhynchocyclus pacificus*
*Long-wattled Umbrellabird	*Cephalopterus penduliger*
*Scarlet-breasted Dacnis	*Dacnis berlepschi*

*Blue-whiskered Tanager	*Tangara johannae*
*Golden-chested Tanager	*Bangsia rothschildi*
Lemon-spectacled Tanager	*Chlorothraupis olivacea*
Scarlet-browed Tanager	*Heterospingus xanthopygius*

West Slope of Andes

As noted above, we have separated the montane Chocó endemics from their lowland cohort in order to better highlight the extraordinary diversity of species found at higher elevations. This group represents what is surely the highest concentration of endemic montane birds in the world, numbering no fewer than 44 species (with additional species found exclusively in the Colombian portion of this region). Two species (El Oro Parakeet [*Pyrrhura orcesi*] and El Oro Tapaculo [*Scytalopus robbinsi*]) endemic to the foothill zone of southwestern Ecuador could almost as easily have been placed in this category, but instead we place them with our "Southwestern Highlands" group. Five species listed here also occur very locally on the east slope of the Andes in Ecuador, and these are followed by a "(+)."

Virtually all members of this group are forest-dependent species (the sole exception is the Western Emerald), but as montane forest remains relatively extensive, none is as yet critically at risk. A reasonable amount of natural habitat in this area now receives formal protection, including areas incorporated within Cotacachi-Cayapas Ecological Reserve, the private Maquipucuna Reserve, and various privately held areas around Mindo. The greatest lack may be an area that includes foothill-zone cloud forest.

SPECIES (44)

*Dark-backed Wood-Quail	*Odontophorus melanonotus*
*Yellow-eared Parrot	*Ognorhynchus icterotis*
Cloud-forest Pygmy-Owl	*Glaucidium nubicola*
Western Emerald	*Chlorostilbon melanorhynchus*
Purple-bibbed Whitetip	*Urosticte benjamini*
*Empress Brilliant	*Heliodoxa imperatrix*
Brown Inca	*Coeligena wilsoni*
*Velvet-purple Coronet	*Boissonneaua jardini*
Gorgeted Sunangel	*Heliangelus strophianus*
*Hoary Puffleg	*Haplophaedia lugens*
Violet-tailed Sylph	*Aglaiocercus coelestis*
*Toucan Barbet	*Semnornis ramphastinus*
*Plate-billed Mountain-Toucan	*Andigena laminirostris*
*Pacific Tuftedcheek	*Pseudocolaptes johnsoni*
*Star-chested Treerunner	*Margarornis stellatus*
Uniform Treehunter	*Thripadectes ignobilis*
*Bicolored Antvireo	*Dysithamnus occidentalis* (+)

Esmeraldas Antbird	*Myrmeciza nigricauda*
*Giant Antpitta	*Grallaria gigantea* (+)
*Moustached Antpitta	*Grallaria alleni* (+)
Yellow-breasted Antpitta	*Grallaria flavotincta*
Nariño Tapaculo	*Scytalopus vicinior*
Chocó Tapaculo	*Scytalopus chocoensis*
Orange-crested Flycatcher	*Myiophobus phoenicomitra* (+)
Orange-breasted Fruiteater	*Pipreola jucunda*
Club-winged Manakin	*Machaeropterus deliciosus*
*Beautiful Jay	*Cyanolyca pulchra*
*Black Solitaire	*Entomodestes coracinus*
Chocó Warbler	*Basileuterus chlorophrys*
*Indigo Flowerpiercer	*Diglossopis indigotica*
Scarlet-and-white Tanager	*Erythrothlypis salmoni*
Yellow-collared Chlorophonia	*Chlorophonia flavirostris*
Glistening-green Tanager	*Chlorochrysa phoenicotis*
Rufous-throated Tanager	*Tangara rufigula*
Gray-and-gold Tanager	*Tangara palmeri*
*Purplish-mantled Tanager	*Iridosornis porphyrocephala*
Black-chinned Mountain-Tanager	*Anisognathus notabilis*
Moss-backed Tanager	*Bangsia edwardsi*
Ochre-breasted Tanager	*Chlorothraupis stolzmanni*
Dusky Bush-Tanager	*Chlorospingus semifuscus*
*Yellow-green Bush-Tanager	*Chlorospingus flavovirens*
Western Hemispingus	*Hemispingus ochraceus*
*White-rimmed Brush-Finch	*Atlapetes leucopis* (+)
*Tanager Finch	*Oreothraupis arremonops*

Tumbesian Lowlands

This exceptionally diverse Endemic Center centers on southwestern Ecuador (north mainly to Manabí and Los Ríos, with small patches of similar terrain occurring north to northwestern Esmeraldas) and northwestern Peru. It supports 59 restricted-range species, an amazingly high number considering that most of the birds inhabit nonhumid scrub and woodland habitats. Although not done by BirdLife, for clarity's sake we have segregated the restricted-range species found in this region that occur only in the highlands and discussed them in the next section. Note that several species found primarily in the lowlands—and therefore listed here—do range at least locally up into the highlands. Seven species listed here also occur in the Río Marañón center in Ecuador and are followed by a "(+)."

Over the past century most habitat in this area has been much modified by humans; very little extensive primary habitat remains, such that it is now important to try to protect the small remnants. Significant natural areas occur in Machalilla National Park (though this park suffers from significant inholding problems), Cerro Blanco Reserve near Guayaquil, and Manglares-Churute Ecological Reserve (and the adjacent private Andrade Reserve). But much, much more needs to be done, and could be accomplished with concerted effort, with perhaps the top priorities being the dry forests of the Macará region and the semihumid forest and woodland south of the Río Ayampe in northwestern Guayas.

SPECIES (59)

*Pale-browed Tinamou	*Crypturellus transfasciatus*
*Gray-backed Hawk	*Leucopternis occidentalis*
*Rufous-headed Chachalaca	*Ortalis erythroptera* (+)
Ecuadorian Ground-Dove	*Columbina buckleyi* (+)
*Ochre-bellied Dove	*Leptotila ochraceiventris*
*Red-masked Parakeet	*Aratinga erythrogenys*
Pacific Parrotlet	*Forpus coelestis*
*Gray-cheeked Parakeet	*Brotogeris pyrrhopterus*
West Peruvian Screech-Owl	*Otus roboratus*
Pacific Pygmy-Owl	*Glaucidium peruanum*
Anthony's Nightjar	*Caprimulgus anthonyi*
Baron's Hermit	*Phaethornis baroni*
*Emerald-bellied Woodnymph	*Thalurania hypochlora*
Tumbes Hummingbird	*Leucippus baeri*
Short-tailed Woodstar	*Myrmia micrura*
*Little Woodstar	*Chaetocercus bombus*
*Esmeraldas Woodstar	*Chaetocercus berlepschi*
Ecuadorian Trogon	*Trogon mesurus*
Ecuadorian Piculet	*Picumnus sclateri*
Scarlet-backed Woodpecker	*Veniliornis callonotus*
Guayaquil Woodpecker	*Campephilus gayaquilensis*
Pacific Hornero	*Furnarius cinnamomeus*
*Blackish-headed Spinetail	*Synallaxis tithys*
Necklaced Spinetail	*Synallaxis stictothorax*
*Henna-hooded Foliage-gleaner	*Hylocryptus erythrocephalus*
Collared Antshrike	*Sakesphorus bernardi*
Elegant Crescentchest	*Melanopareia elegans*
Tumbesian Tyrannulet	*Phaeomyias tumbezana*
Gray-and-white Tyrannulet	*Pseudelaenia leucospodia*
Pacific Elaenia	*Myiopagis subplacens*
Rufous-winged Tyrannulet	*Mecocerculus calopterus*
*Pacific Royal-Flycatcher	*Onychorhynchus occidentalis*
Tumbes Pewee	*Contopus punensis*
*Gray-breasted Flycatcher	*Lathrotriccus griseipectus* (+)
*Ochraceous Attila	*Attila torridus*
Sooty-crowned Flycatcher	*Myiarchus phaeocephalus* (+)
Baird's Flycatcher	*Myiodynastes bairdii*
Snowy-throated Kingbird	*Tyrannus niveigularis*

*Slaty Becard	*Pachyramphus spodiurus*
White-tailed Jay	*Cyanocorax mystacalis*
Plumbeous-backed Thrush	*Turdus reevei*
Ecuadorian Thrush	*Turdus maculirostris*
Tumbes Swallow	*Tachycineta stolzmanni*
Chestnut-collared Swallow	*Petrochelidon rufocollaris*
Fasciated Wren	*Campylorhynchus fasciatus*
Speckle-breasted Wren	*Thryothorus sclateri* (+)
Superciliated Wren	*Thryothorus superciliaris*
Black-lored Yellowthroat	*Geothlypis auricularis* (+)
Gray-and-gold Warbler	*Basileuterus fraseri*
Crimson-breasted Finch	*Rhodospingus cruentus*
Parrot-billed Seedeater	*Sporophila peruviana*
Cinereous Finch	*Piezorhina cinerea*
*Sulphur-throated Finch	*Sicalis taczanowskii*
White-headed Brush-Finch	*Atlapetes albiceps*
Black-capped Sparrow	*Arremon abeillei* (+)
Tumbes Sparrow	*Aimophila stolzmanni*
Collared Warbling-Finch	*Poospiza hispaniolensis*
White-edged Oriole	*Icterus graceannae*
*Saffron Siskin	*Carduelis siemiradzkii*

Southwestern Highlands

The highlands of the southwest—here defined as areas above about 1000 m—support a highly distinctive avifauna of 19 range-restricted species, some of them now very rare and among Ecuador's most threatened. A few range exclusively in cloud forests on the west slope, but most are found farther inland, in remanant patches of woodland and forest across Loja and adjacent El Oro.

Conservation needs in this region are great and the habitat remaining limited. The highest priority is to ensure the survival of a large cloud forest area in the range of the El Oro Parakeet, an area exceptionally rich for other species as well. Other montane needs are more scattered but hardly less important; the remnant forests and woodland around Utuana are of great importance, as are the still little-studied remaining forests in eastern El Oro. One of the most critically endangered bird species in all of Ecuador is the Pale-headed Brush-Finch, now confined to one small, recently protected site south of Cuenca near Girón; we hope this dangerously small population can soon begin to recover.

SPECIES (19)

*El Oro Parakeet	*Pyrrhura orcesi*
Rainbow Starfrontlet	*Coeligena iris*
Purple-throated Sunangel	*Heliangelus viola*

*Violet-throated Metaltail	*Metallura baroni*
Line-cheeked Spinetail	*Cranioleuca antisiensis*
*Rufous-necked Foliage-gleaner	*Syndactyla ruficollis*
Chapman's Antshrike	*Thamnophilus zarumae*
*Gray-headed Antbird	*Myrmeciza griseiceps*
Watkins's Antpitta	*Grallaria watkinsi*
El Oro Tapaculo	*Scytalopus robbinsi*
Loja Tyrannulet	*Zimmerius viridiflavus*
*Black-crested Tit-Tyrant	*Anairetes nigrocristatus*
Jelski's Chat-Tyrant	*Ochthoeca jelskii*
Three-banded Warbler	*Basileuterus tristriatus*
*Piura Hemispingus	*Hemispingus piurae*
Black-cowled Saltator	*Saltator nigriceps*
Drab Seedeater	*Sporophila simplex*
Bay-crowned Brush-Finch	*Atlapetes seebohmi*
*Pale-headed Brush-Finch	*Atlapetes pallidiceps*

Interandean Slopes and Valleys

This Endemic Center incorporates areas found at and above treeline on either the Western or Eastern Andes, on the inside slopes of both the Western and Eastern Andes, and the valleys and ridges intervening. The 16 species included are thus found either in paramo or temperate forest/woodland, but none is included if it ranges below treeline on either outside Andean slope. One species (Ecuadorian Rail) is a marsh inhabitant. A few species range into adjacent Colombia or Peru but not both; most occur in either northern or southern Ecuador, but the distribution of a minority extends the length of the country (e.g., Rainbow-bearded Thornbill).

Natural habitats in most of this region have been so severely affected by humans that little can now be done to reconstitute them. It is conceivable that some bird species went extinct before they even became known. Most of the endemics occurring here are relatively habitat-tolerant. At present the highest conservation priority should probably be placed on the two exceptionally rare pufflegs of the Quito region, the Turquoise-throated and Black-breasted; the first remains unknown in life, and the second apparently occurs only in a few unprotected woodlands high on the northwest flank of Volcán Pichincha. Protecting the Selva Alegre forest patches on the Cordillera de Cordoncillo would ensure the survival of what may be the largest extant populations of the Bearded Guan and Redfaced Parrot.

SPECIES (16)

Curve-billed Tinamou	*Nothoprocta curvirostris*
Carunculated Caracara	*Phalcoboenus carunculatus*
*Bearded Guan	*Penelope barbata*

Ecuadorian Rail	*Rallus aequatorialis*
*Red-faced Parrot	*Hapalopsittaca pyrrhops*
Ecuadorian Hillstar	*Oreotrochilus chimborazo*
*Black-breasted Puffleg	*Eriocnemis nigrivestis*
*Turquoise-throated Puffleg	*Eriocnemis godini*
Golden-breasted Puffleg	*Eriocnemis mosquera*
*Black-thighed Puffleg	*Eriocnemis derbyi*
Rainbow-bearded Thornbill	*Chalcostigma herrani*
Stout-billed Cinclodes	*Cinclodes excelsior*
Mouse-colored Thistletail	*Schizoeaca griseomurina*
Paramo Ground-Tyrant	*Muscisaxicola alpina*
*Tit-like Dacnis	*Xenodacnis parina*
White-winged Brush-Finch	*Atlapetes leucopterus*

East Slope of Andes

Numerous montane bird species have distributions in Ecuador that are restricted to the east slope of the Andes. This Endemic Center incorporates 36 restricted-range species whose distributions do not extend far to the north in Colombia or to the south in Peru; many thus have distributions that may be hundreds of kilometers "long" (in the north-south sense) but are only a few tens of kilometers—sometimes even less—"wide" (in the east-west sense). We include in this center species that are found from the foothill zone up to treeline on the east slope. Virtually all are restricted to a narrow elevational zone, typically within a range of 500 to 1000 m, substantially less in a few cases. Six species also have small populations on the west slope of the Andes (two of these, White-chested Swift and Yellow-headed Manakin, occurring away from the Andes' east slope only in Colombia); these are followed by a "(+)."

No fewer than four major parks and reserves (Sumaco-Galeras, Sangay, and Podocarpus National Parks and Cayambe-Coca Ecological Reserve) are situated on the east slope of the Andes, and habitats for virtually all endemic bird species, many of which are scarce and naturally occur in low densities, are incorporated in one or all of them. The foothill region at the 500- to 1000-m zone is, however, comparatively underrepresented, leading to some concern about the future of some bird species found only there. The recently discovered Jocotoco Antpitta has the smallest known range, but much of that limited area is now protected, and we hope future field work will reveal that this species' range is somewhat larger.

SPECIES (36)

*White-breasted Parakeet	*Pyrrhura albipectus*
Cinnamon Screech-Owl	*Otus petersoni*
Subtropical Pygmy-Owl	*Glaucidium parkeri*

*White-chested Swift	*Cypseloides lemosi* (+)
*Napo Sabrewing	*Campylopterus villaviscensio*
Wire-crested Thorntail	*Popelairia popelairii*
Rufous-vented Whitetip	*Urosticte ruficrissa*
*Ecuadorian Piedtail	*Phlogophilus hemileucurus*
*Pink-throated Brilliant	*Heliodoxa gularis*
Flame-throated Sunangel	*Heliangelus micraster*
Neblina Metaltail	*Metallura odomae*
Mountain Avocetbill	*Opisthoprora euryptera*
*Coppery-chested Jacamar	*Galbula pastazae*
Spectacled Prickletail	*Siptornis striaticollis*
*Equatorial Graytail	*Xenerpestes singularis*
*Bicolored Antvireo	*Dysithamnus occidentalis* (+)
*Giant Antpitta	*Grallaria gigantea* (+)
*Jocotoco Antpitta	*Grallaria ridgelyi*
Crescent-faced Antpitta	*Grallaricula lineifrons*
*Peruvian Antpitta	*Grallaricula peruviana*
Equatorial Rufous-vented Tapaculo	*Scytalopus micropterus*
Chusquea Tapaculo	*Scytalopus parkeri*
Ecuadorian Tyrannulet	*Phylloscartes gualaquizae*
Orange-crested Flycatcher	*Myiophobus phoenicomitra* (+)
Olive-chested Flycatcher	*Myiophobus cryptoxanthus*
*Orange-banded Flycatcher	*Myiophobus lintoni*
Yellow-cheeked Becard	*Pachyramphus xanthogenys*
*Chestnut-bellied Cotinga	*Doliornis remseni*
*Black-chested Fruiteater	*Pipreola lubomirskii*
Blue-rumped Manakin	*Pipra isidorei*
*Yellow-headed Manakin	*Chloropipo flavicapilla* (+)
Jet Manakin	*Chloropipo unicolor*
Olivaceous Greenlet	*Hylophilus olivaceus*
*Masked Mountain-Tanager	*Buthraupis wetmorei*
Black-backed Bush-Tanager	*Urothraupis stolzmanni*
*White-rimmed Brush-Finch	*Atlapetes leucopis* (+)

Isolated East-Andean Ridges

This Endemic Center incorporates three mountainous areas situated east of the actual Andes: Volcán Sumaco/Cordillera Galeras, Cordillera de Cutucú, and Cordillera del Cóndor. Only three "endemic" bird species are in Ecuador found exclusively on what BirdLife terms the "Andean ridgetop forests" center— which we found imperfectly named, preferring to call it "Isolated East-Andean Ridges"—but in northern Peru several others are also found and at least one of them, Royal Sunangel (*Heliangelus regalis*), almost certainly occurs in Ecuador as well. The relatively low-elevation terrain found between these ridges and the Andes leaves them isolated, and this isolation has the potential to permit the divergence and speciation of montane forms, divergent from those found in the Andes. A number of other rare birds (e.g., Buff-browed Foliage-gleaner [*Syndactyla rufosuperciliata*], Buff-throated Tody-Tyrant [*Hemitriccus rufigularis*], Roraiman Flycatcher [*Myiophobus roraimae*], and Rufous-browed Tyrannulet [*Phylloscartes superciliaris*]) are in Ecuador found only or principally on these outlying ridges.

Little habitat on these remote, outlying ridges has been affected by human activities, though gold mining on the Cordillera del Cóndor remains a potential risk. With the 1999 signing of the peace treaty between Ecuador and Peru, the establishment of a binational "peace park" has become a real possibility, very much to be hoped for.

SPECIES (3)

*Cinnamon-breasted Tody-Tyrant	*Hemitriccus cinnamomeipectus*
Bar-winged Wood-Wren	*Henicorhina leucoptera*
*Orange-throated Tanager	*Wetmorethraupis sterrhopteron*

Río Marañón

The important Río Marañón Endemic Center lies primarily within northwestern Peru, with only a small part of the upper Río Chinchipe (an affluent of the Río Marañón) drainage falling within Ecuadorian territory around and north of Zumba in southern Zamora-Chinchipe. The region is important from an Ecuadorian biogeographic context, for some 15 bird species found here have been recorded nowhere else in the country. A majority of these are wide-ranging South American species, and only six Marañón endemics have as yet been recorded, though others remain possible. Six other species, followed by a "(+)," occur here as disjunct populations separate from their principal ranges west of the Andes in the Tumbesian Lowlands center; the Marañón populations of several of these (e.g., Speckle-breasted Wren, Black-capped Sparrow) may in fact be distinct species in their own right.

None of this limited and still poorly explored area receives any protection. It should.

SPECIES (12)

*Rufous-headed Chachalaca	*Ortalis erythroptera* (+)
*Marañón Pigeon	*Columba oenops*
Marañón Spinetail	*Synallaxis maranonica*
*Marañón Slaty-Antshrike	*Thamnophilus leucogaster*
Marañón Crescentchest	*Melanopareia maranonica*
*Gray-breasted Flycatcher	*Lathrotriccus griseipectus* (+)
Sooty-crowned Flycatcher	*Myiarchus phaeocephalus* (+)
Marañón Thrush	*Turdus maranonicus*

Speckle-breasted Wren   *Thryothorus sclateri* (+)
Black-lored Yellowthroat   *Geothlypis auricularis* (+)
Buff-bellied Tanager   *Thlypopsis inornata*
Black-capped Sparrow   *Arremon abeillei* (+)

Western Amazonian Lowlands

Various western Amazonian localities are now known to support the highest avian diversity in the world. Upwards of 500 to 550 species have now been recorded at various well-watched localities in eastern Ecuador and eastern Peru, principally at the series of frequently visited tourist lodges now established in the Amazonian drainages of both countries. Nonetheless, there is surprisingly little range-restricted endemism, assuming one follows the definitions of BirdLife. Following is a list of 23 species we consider to be endemic to the lowlands of eastern Ecuador and northeastern Peru; the distributions of a few extend slightly farther but are still basically centered here. Most are forest birds, but a few are found in riparian or island habitats (though the distributions of most of this "river-island" group extend too far from this region for them to be considered "endemics" in this sense).

Habitats in the eastern Ecuador lowlands remain relatively intact, and there are few or no imminent threats across broad areas. The status of most of the restricted-range species listed here thus seems secure; large areas receive governmental protection in Yasuní National Park and Cuyabeno Faunistic Reserve, and there are several other privately or locally protected areas. In this region, direct persecution of birds, generally directed toward larger birds that often have large overall ranges, is actually a greater threat.

SPECIES (23)

*Salvin's Curassow	*Mitu salvini*
*Red-winged Wood-Rail	*Aramides calopterus*
Sapphire Quail-Dove	*Geotrygon sapphirina*
Black-throated Hermit	*Phaethornis atrimentalis*
Olive-spotted Hummingbird	*Leucippus chlorocercus*
White-chinned Jacamar	*Galbula tombacea*
Brown Nunlet	*Nonnula brunnea*
Plain-breasted Piculet	*Picumnus castelnau*
Dusky Spinetail	*Synallaxis moesta*
Cocha Antshrike	*Thamnophilus praecox*
Yasuní Antwren	*Myrmotherula fjeldsaai*
Río Suno Antwren	*Myrmotherula sunensis*
*Ancient Antwren	*Herpsilochmus gentryi*
Dugand's Antwren	*Herpsilochmus dugandi*
Slate-colored Antbird	*Schistocichla schistacea*
Lunulated Antbird	*Gymnopithys lunulata*

Ochre-striped Antpitta	*Grallaria dignissima*
White-lored Antpitta	*Hylopezus fulviventris*
Golden-winged Tody-Flycatcher	*Poecilotriccus calopterum*
Orange-eyed Flatbill	*Tolmomyias traylori*
Orange-crested Manakin	*Heterocercus aurantiivertex*
Ecuadorian Cacique	*Cacicus sclateri*
Band-tailed Oropendola	*Ocyalus latirostris*

Conservation

Mainland Ecuador's avifauna of nearly 1600 species is—or should be!—one of the wonders of the natural world. Nowhere else is such incredible avian diversity found in such a small country. Peru, Colombia, and Brazil may have more species within their borders, but each of these nations is many times larger than Ecuador. Birds are, happily, numerous in many parts of Ecuador; even the downtown parks of the big cities such as Quito and Guayaquil host their complement of birds.

Ecuador's human population continues to grow rapidly, however, and the impacts of this growth continue to widen across much of the country. Colonization of formerly wild and little-peopled areas continues apace, and regions already inhabited are forced to support more and more human activity. Something is going to have to change over the next few decades or we are going to start losing avian diversity—and everything else—rapidly. For the moment, the conservation situation in Ecuador can be viewed as reasonably positive and hopeful; there is increasing environmental awareness on the part of more and more Ecuadorian citizens, there is an increasing recognition of the importance of protecting representative habitats, and an increasing effort is being made to foster "sustainable" development such as ecotourism. But let's be realistic: if human population pressures continue, and if poverty therefore continues to spread, it seems almost inevitable that much will ultimately be lost. Our fervent hope is that this point will never be reached, that Ecuador's human population will reach some level of stability before our presence overwhelms the natural world.

Thankfully we are not at that point yet, and this chapter is designed to assess where we stand at present and what steps might be taken to further enhance and protect Ecuador's extraordinarily rich avian diversity.

Land Protection in Ecuador

What can be considered the start of the modern conservation era in Ecuador began in the mid-1970s under the impetus of an internationally funded program to identify and protect representative samples of Ecuador's basic biogeographic regions. Prior to this, the only significant land-protection effort in Ecuador had

Table 1. Ecuador's Major Government-Protected Natural Areas

Name	Approximate size (in hectares)
Antisana Ecological Reserve paramo of w. Napo	120,000
Awá Forest Reserve wet forest of n. Esmeraldas	100,000
Cayambe-Coca Ecological Reserve montane forest on east slope	405,000
Chimborazo Faunistic Reserve paramo of Chimborazo	55,000
Cotacachi-Cayapas Ecological Reserve montane forest on west slope	205,000
Cotopaxi National Park paramo of Cotopaxi	33,000
Cuyabeno Faunistic Reserve lowland forest of Sucumbíos	655,000
El Cajas National Recreation Area paramo of Azuay	28,000
Machalilla National Park deciduous forest of sw. Manabí	55,000
Manglares-Churute Ecological Reserve forest and mangroves in se. Guayas	35,000
Podocarpus National Park montane forest on east slope	145,000
Sangay National Park montane forest on east slope	500,000
Sumaco-Galeras National Park montane forest on east slope	205,000
Yasuní National Park lowland forest of Napo	980,000

been focused on the Galápagos Islands, portions of which had been declared a national park in the early 1960s.

This program, headed at the time by international consultant Allen Putney, established Ecuador's initial system of major national parks and reserves: a series of large areas of intact or relatively intact habitat that did indeed comprise a fairly comprehensive and representative cross-section of Ecuador's ecological diversity. The system was basically in place by the early 1980s, though actual protection and management on the ground usually did not commence until later, and are still not totally effective in some areas. Very few areas have been added to the system in subsequent years. Table 1 lists the system's major components.

This government-supported system of major parks and reserves, although not "perfect," represents a tremendous accomplishment, one of which all Ecuadorian citizens should be proud. No less than 12 percent of Ecuador's territory has now been incorporated into the national system. Management issues and difficulties—involving, for instance, privately held land inside a unit's ostensible boundaries, colonization along a unit's borders, or outright disregard of a unit by another, more powerful government agency—do exist in virtually every area, much as they do in many other countries around the globe. But thanks to this system, populations of most Ecuadorian bird species, including many sensitive and habitat-restricted ones, are protected.

Ecuador's national parks and reserves do not, however, adequately represent Ecuador's vast biodiversity. Note in particular the system's strong bias toward areas in the north and east; only one unit, Podocarpus National Park, exists south of Cuenca. There is a reason for this: the original system was specifically designed to protect *large* areas of relatively intact habitat. The areas that were set aside typically encompassed remote and inaccessible terrain, and most of them remain so. Such large areas simply did not exist in more densely inhabited southern and western Ecuador, not even in the 1970s (and certainly not now). As a result, these sectors of the country—even then recognized as supporting imperilled biodiversity elements—tended to be passed over in the planning process. And they continue to be badly under-represented, with the result that many of Ecuador's bird species that are most at risk hang on in small patches of usually degraded, unprotected habitat.

It likely is beyond the capability of the Ecuadorian government, with a multitude of other priorities to consider, to expand its system of protected natural areas. How, then, can such areas be protected? We would argue that it is now up to the private sector, that nonprofit organizations with an interest in conserving Ecuador's avian diversity must step forward. Happily, some have indeed begun to do so, with notable examples being Fundación Natura, Fundación Jatun Sacha, Fundación Maquipucuna, and Fundación Jocotoco. Each organization has now taken the lead in establishing small reserves for their biological values. To some extent the efforts of these organizations have not yet focused specifically on the degree of threat to important, under-represented regions or species, but this too seems to be changing. One thus can envision—we hope we are not being too optimistic!—a complementary system in which the government manages its large and often relatively remote protected areas and the private sector fills in the gaps with a series of smaller, often more accessible reserves designed to protect species and ecosystems that otherwise would be left out. A representative sample of Ecuador's privately owned reserves that are of particular importance to birds is given in Table 2.

Conservation Status of At-Risk Bird Species

Extensive field work since the early 1970s across virtually all of Ecuador has resulted in a tremendous amount of new information concerning the status of Ecuador's birds. A large part of this information is derived from the ANSP's

Table 2. Ecuador's Privately Protected Reserves of Particular Importance to Birds

Name	Approximate size (in hectares)
Bilsa Biological Station sw. Esmeraldas	3,000
Buenaventura Biological Reserve w. El Oro	500
Cabañas San Isidro w. Napo	1,200
Cerro Blanco Reserve w. Guayas	650
El Tundo Reserve sw. Loja	200
Guandera Biological Reserve se. Carachi	1,000
Jatun Sacha Biological Station w. Napo	1,800
Loma Alta Ecological Reserve w. Guayas	1,000
Maquipucuna Reserve w. Pichincha	15,000
Pasochoa Reserve cen. Pichincha	500
Río Palenque Scientific Station s. Pichincha	200
Tapichalaca Biological Reserve sw. Zamora-Chinchipe	2,200

program of field work that commenced in the country in 1984 and has been headed up by RSR, with major funding from the MacArthur Foundation and to a lesser degree from other sources, including important contributions from numerous other organizations and individuals. Comprising thousands and thousands of hours in the field, this work forms the core of what is presented in *The Birds of Ecuador*. It is our belief that Ecuador's avifauna now must rank among the best known of any Neotropical country, and we also now have the ability to much more accurately assess the status of bird species that until recently were hardly known at all.

In the species accounts in Volume I, where appropriate we include a paragraph summarizing a species' status in Ecuador, our assessment as to its degree of threat (if any), and any change in its numbers. We here collate that information, presenting each species in the various "at-risk" categories, basically as defined by BirdLife International (in particular Collar et al. 1994 and Wege and Long 1995, pp. 305–306), whose definitions we paraphrase below. These formal categories are capitalized (e.g., "Critical") when referred to in the text.

A species' status in Ecuador *only* has been considered when making the vast majority of these judgements (the exceptions concerning species that migrate or wander to Ecuador but breed elsewhere, in which case their status in their breeding range is considered paramount). One result has been that we judge certain species to be more at risk in Ecuador than they are generally in their respective ranges. For example, the Red-and-green Macaw (*Ara chloroptera*) and the Horned Screamer (*Anhima cornuta*) have both declined substantially in Ecuador, but elsewhere in their ranges they often are still numerous; thus we give the macaw a Vulnerable rating and the screamer an Endangered rating for Ecuador, but on a world level neither is considered to be of great conservation concern, so neither is listed by Collar et al. (1992, 1994). Perceived downward trends in population size (and the rapidity of those declines) have been crucial to making these assessments, and to some extent they carry as much weight— sometimes more—than a species' actual population size.

We emphasize that most "restricted-range" species, in the sense of BirdLife International (e.g., Wege and Long 1995), have been included in our lists of at-risk species unless new and positive information on them has recently come to light. But just because a species has a small range in Ecuador does not necessarily mean we consider it to be threatened; some other anthropogenic factor must be having an impact. Thus, for example, most of the "Marañón endemics" that have recently been discovered to occur in the Río Chinchipe drainage of extreme southeastern Zamora-Chinchipe around Zumba are not considered to be at risk, even though their Ecuadorian ranges are miniscule. The great majority of these birds in fact inhabit secondary, human-altered habitats and thus do not seem destined to decline in numbers (and certainly have not up until now; some may even have spread into Ecuador as a result of clearing). The few that do seem to be restricted to less-disturbed habitats, such as the Marañón Slaty-Antshrike (*Thamnophilus leucogaster*), are included in our lists of at-risk species.

Finally, species believed to occur only as wanderers to Ecuador (e.g., Scarlet Ibis [*Eudocimus ruber*], Bare-faced Ibis [*Phimosus infuscatus*]) are not given a status rating despite their evident rarity in the country. Others (e.g., Orinoco Goose [*Neochen jubata*], Jabiru [*Jabiru mycteria*]), which seem likely once to have been resident in Ecuador but now only occur as wanderers, are given a status assessment.

The formal "at-risk" categories are described below, and a list of the species in each category is presented in Table 3.

Extirpated/Extinct. A taxon for which there is no reasonable doubt that the last individual has died. "Extirpated" refers to taxa that no longer exist in a portion of the taxon's range but that still occur elsewhere. In Ecuador there are four Extirpated species but no Extinct ones.

Critical. A taxon that in Ecuador is facing an extremely high risk of local extinction in the wild in the immediate future (i.e., in the next few years).

Endangered. A taxon that in Ecuador is not quite so gravely at risk as taxa judged to be Critical but which nonetheless is facing a very high risk

Table 3. Species Considered to Be at Risk in Ecuador

Extirpated/Extinct (4 species)

Cinnamon Teal	*Anas cyanoptera borreroi* *
American Coot	*Fulica americana columbiana* *
Least Seedsnipe	*Thinocorus rumicivorus cuneicauda*
Tawny-throated Dotterel	*Oreopholus ruficollis pallidus*

Critical (10 species)

Southern Pochard	*Netta e. erythrophthalma*
Jabiru	*Jabiru mycteria*
Great Curassow	*Crax r. rubra*
Wattled Curassow	*Crax globulosa*
Great Green Macaw	*Ara ambigua*
Yellow-eared Parrot	*Ognorhynchus icterotis* *
Black-breasted Puffleg	*Eriocnemis nigrivestis* **
Turquoise-throated Puffleg	*Eriocnemis godini* **
Pale-headed Brush-Finch	*Atlapetes pallidiceps* **
Grasshopper Sparrow	*Ammodramus savanarum caucae*

Endangered (16 species)

Dark-rumped Petrel	*Pterodroma phaeopygia*
Horned Screamer	*Anhima cornuta*
Orinoco Goose	*Neochen jubata*
Black-faced Ibis	*Theristicus melanopis*
Andean Condor	*Vultur gryphus*
Gray-backed Hawk	*Leucopternis occidentalis* *
Peregrine Falcon	*Falco peregrinus*
Brown Wood-Rail	*Aramides wolfi* *
Ochre-bellied Dove	*Leptotila ochraceiventris* *
Red-faced Parrot	*Hapalopsittaca pyrrhops* *
Banded Ground-Cuckoo	*Neomorphus radiolosus* *
Esmeraldas Woodstar	*Chaetocercus berlepschi* **
Gray-headed Antbird	*Myrmeciza griseiceps* *
Bicolored Antpitta	*Grallaria rufocinerea* *
Jocotoco Antpitta	*Grallaria ridgelyi* **
Orange-throated Tanager	*Wetmorethraupis sterrhopteron* *

Vulnerable (63 species)

Silvery Grebe	*Podiceps occipitalis*
Waved Albatross	*Phoebastria irrorata*
Parkinson's Petrel	*Procellaria parkinsoni*
Pink-footed Shearwater	*Puffinus creatopus*
Muscovy Duck	*Cairina moschata*
Comb Duck	*Sarkidiornis melanotos*
Snail Kite	*Rostrhamus sociabilis*
Cinereous Harrier	*Circus cinereus*
Solitary Eagle	*Harpyhaliaetus solitarius*
Crested Eagle	*Morphnus guianensis*
Harpy Eagle	*Harpia harpyja*
Plumbeous Forest-Falcon	*Micrastur plumbeus* *
Orange-breasted Falcon	*Falco deiroleucus*
Bearded Guan	*Penelope barbata* *
Baudó Guan	*Penelope ortoni* *
Crested Guan	*Penelope purpurascens*
Wattled Guan	*Aburria aburri*
Clapper Rail	*Rallus longirostris*
Peruvian Thick-knee	*Burhinus superciliaris*

Table 3—*cont.*

Piping Plover	*Charadrius melodus*
Bridled Tern	*Sterna anaethetus*
Marañón Pigeon	*Columba oenops* *
Indigo-crowned Quail-Dove	*Geotrygon purpurata* *
Military Macaw	*Ara militaris*
Red-and-green Macaw	*Ara chloroptera*
Red-masked Parakeet	*Aratinga erythrogenys* *
Golden-plumed Parakeet	*Leptosittaca branickii*
El Oro Parakeet	*Pyrrhura orcesi* **
Red-lored Amazon	*Amazona autumnalis*
Central American Pygmy-Owl	*Glaucidium griseiceps*
White-chested Swift	*Cypseloides lemosi* *
Pink-throated Brilliant	*Heliodoxa gularis* *
Little Woodstar	*Chaetocercus bombus* *
Five-colored Barbet	*Capito quinticolor* *
Blackish-headed Spinetail	*Synallaxis tithys* *
Double-banded Graytail	*Xenerpestes minlosi* *
Pacific Tuftedcheek	*Pseudocolaptes johnsoni* *
Star-chested Treerunner	*Margarornis stellatus* *
Rufous-necked Foliage-gleaner	*Syndactyla ruficollis* *
Slaty-winged Foliage-gleaner	*Philydor fuscipennis*
Henna-hooded Foliage-gleaner	*Hylocryptus erythrocephalus* *
Marañón Slaty-Antshrike	*Thamnophilus leucogaster* *
Bicolored Antvireo	*Dysithamnus occidentalis* *
Ocellated Antbird	*Phaenostictus mcleannani*
Rufous-crowned Antpitta	*Pittasoma rufopileatum* *
Giant Antpitta	*Grallaria gigantea* *
Moustached Antpitta	*Grallaria alleni*
Cinnamon-breasted Tody-Tyrant	*Hemitriccus cinnamomeipectus* *
Pacific Royal-Flycatcher	*Onychorhynchus occidentalis* *
Gray-breasted Flycatcher	*Lathrotriccus griseipectus* *
White-tailed Shrike-Tyrant	*Agriornis andicola*
Ochraceous Attila	*Attila torridus* *
Chestnut-bellied Cotinga	*Doliornis remseni* **
Long-wattled Umbrellabird	*Cephalopterus penduliger* *
Yellow-headed Manakin	*Chloropipo flavicapilla* *
Rufous-brown Solitaire	*Cichlopsis leucogenys*
Scarlet-breasted Dacnis	*Dacnis berlepschi* *
Indigo Flowerpiercer	*Diglossopis indigotica* *
Blue-whiskered Tanager	*Tangara johannae* *
Purplish-mantled Tanager	*Iridosornis prohyrocephala* *
Masked Mountain-Tanager	*Buthraupis wetmorei* *
Yellow-green Bush-Tanager	*Chlorospingus flavovirens* *
Tanager Finch	*Oreothraupis arremonops* *

Data Deficient (25 species)

Brown Tinamou	*Crypturellus obsoletus*
White-vented Storm-Petrel	*Oceanites gracilis*
Markham's Storm-Petrel	*Oceanodroma markhami*
Hornby's Storm-Petrel	*Oceanodroma hornbyi*
Starred Wood-Quail	*Odontophorus stellatus*
Red-winged Wood-Rail	*Aramides calopterus* *
Maroon-chested Ground-Dove	*Claravis mondetoura*
Scarlet-fronted Parakeet	*Aratinga wagleri*
Saffron-headed Parrot	*Pionopsitta pyrilia*

Table 3—*cont.*

Rusty-faced Parrot	*Hapalopsittaca amazonina*
Spot-fronted Swift	*Cypseloides cherriei*
Lazuline Sabrewing	*Campylopterus falcatus*
Humboldt's Sapphire	*Hylocharis humboldtii* *
Crimson Topaz	*Topaza pella*
Emerald-bellied Puffleg	*Eriocnemis alinae*
White-faced Nunbird	*Hapaloptila castanea*
Yellow-eared Toucanet	*Selenidera spectabilis*
Greater Scythebill	*Campylorhamphus pucherani*
Ancient Antwren	*Herpsilochmus gentryi* *
Striated Antbird	*Drymophila devillei*
Wing-banded Antbird	*Myrmornis torquata*
Red-ruffed Fruitcrow	*Pyroderus scutatus*
Black-collared Jay	*Cyanolyca armillata*
Straw-backed Tanager	*Tangara argyrofenges*
Sulphur-throated Finch	*Sicalis taczanowskii* *

Near-threatened (85 species)

Gray Tinamou	*Tinamus tao*
Berlepsch's Tinamou	*Crypturellus berlepschi* *
Pale-browed Tinamou	*Crypturellus transfasciatus* *
Humboldt Penguin	*Spheniscus humboldti*
Buller's Shearwater	*Puffinus bulleri*
Wedge-rumped Storm-Petrel	*Oceanodroma tethys*
Ashy Storm-Petrel	*Oceanodroma homochroa*
Peruvian Booby	*Sula variegata*
Guanay Shag	*Phalacrocorax bougainvillii*
Pinnated Bittern	*Botaurus pinnatus*
Plumbeous Hawk	*Leucopternis plumbea*
Semiplumbeous Hawk	*Leucopternis semiplumbea*
Common Black-Hawk	*Buteogallus anthracinus*
Black-and-chestnut Eagle	*Oroaetus isidori*
Rufous-headed Chachalaca	*Ortalis erythroptera* *
Salvin's Curassow	*Mitu salvini*
Dark-backed Wood-Quail	*Odontophorus melanonotus* *
Rufous-breasted Wood-Quail	*Odontophorus speciosus*
Tawny-faced Quail	*Rhynchortyx cinctus*
Rufous-necked Wood-Rail	*Aramides axillaris*
Gray-winged Trumpeter	*Psophia crepitans*
Hudsonian Godwit	*Limosa haemastica*
Buff-breasted Sandpiper	*Tryngites subruficollis*
Pied Plover	*Hoploxypterus cayanus*
Elegant Tern	*Sterna elegans*
Peruvian Tern	*Sterna lorata*
Large-billed Tern	*Phaetusa simplex*
Black Skimmer	*Rynchops niger*
Olive-backed Quail-Dove	*Geotrygon veraguensis*
White-breasted Parakeet	*Pyrrhura albipectus* **
Gray-cheeked Parakeet	*Brotogeris pyrrhopterus* *
Blue-fronted Parrotlet	*Touit dilectissima*
Spot-winged Parrotlet	*Touit stictoptera*
Buff-fronted Owl	*Aegolius harrisii*
Chocó Poorwill	*Nyctiphrynus rosenbergi* *
Napo Sabrewing	*Campylopterus villaviscensio* *
Fiery-tailed Awlbill	*Avocettula recurvirostris*

Table 3—*cont.*

Emerald-bellied Woodnymph	*Thalurania hypochlora* **
White-vented Plumeleteer	*Chalybura buffonii*
Ecuadorian Piedtail	*Phlogophilus hemileucurus* *
Empress Brilliant	*Heliodoxa imperatrix* *
Velvet-purple Coronet	*Boissonneaua jardini* *
Black-thighed Puffleg	*Eriocnemis derbyi* *
Hoary Puffleg	*Haplophaedia lugens* *
Violet-throated Metaltail	*Metallura baroni* **
Coppery-chested Jacamar	*Galbula pastazae* *
Black-streaked Puffbird	*Malacoptila fulvogularis*
Toucan Barbet	*Semnornis ramphastinus* *
Plate-billed Mountain-Toucan	*Andigena laminirostris* *
Gray-breasted Mountain-Toucan	*Andigena hypoglauca*
Black-billed Mountain-Toucan	*Andigena nigrirostris*
Black-mandibled Toucan	*Ramphastos ambiguus*
Lita Woodpecker	*Piculus litae* *
Chocó Woodpecker	*Veniliornis chocoensis* *
Chestnut-throated Spinetail	*Synallaxis cherriei*
Equatorial Graytail	*Xenerpestes singularis*
Peruvian Antpitta	*Grallaricula peruviana* *
Black-crested Tit-Tyrant	*Anairetes nigrocristatus* *
Subtropical Doradito	*Pseudocolopteryx acutipennis*
Spectacled Bristle-Tyrant	*Pogonotriccus orbitalis*
Yellow-throated Spadebill	*Platyrinchus flavigularis*
Orange-banded Flycatcher	*Myiophobus lintoni* *
Slaty Becard	*Pachyramphus spodiurus* *
Black-chested Fruiteater	*Pipreola lubomirskii*
Fiery-throated Fruiteater	*Pipreola chlorolepidota*
Andean Laniisoma	*Laniisoma buckleyi*
Speckled Mourner	*Laniocera rufescens*
Gray-tailed Piha	*Lathria subalaris*
Black-tipped Cotinga	*Carpodectes hopkei*
Black-necked Red-Cotinga	*Phoenicircus nigricollis*
Broad-billed Sapayoa	*Sapayoa aenigma*
Beautiful Jay	*Cyanolyca pulchra* *
Black Solitaire	*Entomodestes coracinus* *
Andean Slaty-Thrush	*Turdus nigriceps*
Slate-throated Gnatcatcher	*Polioptila schistaceigula*
Giant Conebill	*Oreomanes fraseri*
Tit-like Dacnis	*Xenodacnis parini*
Golden-chested Tanager	*Bangsia rothschildi* *
Piura Hemispingus	*Hemispingus piurae* *
Masked Saltator	*Saltator cinctus*
Large-billed Seed-Finch	*Oryzoborus crassirostris*
Black-billed Seed-Finch	*Oryzoborus atrirostris*
White-rimmed Brush-Finch	*Atlapetes leucopis*
Pale-eyed Blackbird	*Agelaius xanthophthalmus*
Saffron Siskin	*Carduelis siemiradzkii* *

Note: Subspecies are included only in the Extirpated/Extinct and Critical categories.
* A species with a distribution shared by either Colombia or Peru (a "shared endemic").
** A species endemic to Ecuador.

of local extinction in the wild in the near future (i.e., in the next decade or two).

Vulnerable. A taxon that in Ecuador is not so gravely at risk as taxa judged to be Critical or Endangered, but which nonetheless is facing a high risk of local extinction in the future (i.e., in the next several decades).

Data Deficient. A taxon for which there still is insufficient information to accurately assess its risk of extinction in Ecuador. Some such species may be declining for uncertain reasons in Ecuador, or their Ecuadorian populations may be inexplicably small. Even if there are few—or even no—recent records for a species in Ecuador, if its habitat is known and is judged not to be at risk, then we do not necessarily consider the species to warrant Data Deficient (or any other) status.

Near-threatened. A taxon judged to be not seriously at risk at present but whose status gives some cause for concern, and which will require careful monitoring in the future.

A few overall considerations follow:

1. *Only four bird species appear to have been extirpated from Ecuador* (see Table 3), but 10 others are perilously close to that point, such that strong measures will be needed if they are to continue to exist within the country (these being the Critical species listed in Table 3). No confirmation that the Turquoise-throated Puffleg exists has come to light for over a century; it may already be Extinct, but a concerted effort may yet result in its rediscovery. The Southern Pochard and Yellow-eared Parrot both seem to be slipping into Ecuadorian oblivion, and the two curassows are so rare that one has to suspect that no viable population still exists here. No one knows whether the Grasshopper Sparrow still is found in Ecuador; despite the absence of recent records, elsewhere it is a nomadic species, prone to reappearing when and where conditions are right, and we harbor the hope that it survives. The Jabiru no longer appears to be resident in the country (and we are only assuming that it once was, albeit in very small numbers). The other three species (Great Green Macaw, Black-breasted Puffleg, and Pale-headed Brush-Finch), though extremely rare, give at least some reason for hope.

2. *Of Ecuador's 14 endemic bird species (see the list below), none is numerous, and all—not surprisingly—have very limited ranges.* The most poorly known is the Turquoise-throated Puffleg, which perhaps is extinct or, conversely, possibly is not even a valid species. A few century-old specimens were taken near Quito, where the species may have inhabited some specialized type of intermontane forest or woodland, long since cut over. Next rarest is the Pale-headed

Brush-Finch, confined so far as known to one patch of scrubby woodland near Girón where it was rediscovered in late 1998; that patch was purchased by the Fundación Jocotoco in 1999, and an effort to increase its tiny population has begun. Also rare is the dramatic Jocotoco Antpitta, discovered in 1997 on the Cordillera de Sabanilla of the southeastern Andes, in terrain that has also been purchased by the foundation of that name. A pair of *Pyrrhura* parakeets, the White-breasted Parakeet and the recently described El Oro Parakeet, is found in southern Ecuador, one on either side of the Andes; the White-breasted is turning out to be more numerous than had been thought (and occurs in Podocarpus National Park), but the El Oro's cloud forest habitat continues to be whittled away, though the Fundación Jocotoco has, as of August 1999, begun to purchase and protect the forests at its Buenaventura type locality. Preserving and rehabilitating the Buenaventura forests would have the additional benefit of ensuring that the also recently described El Oro Tapaculo persists into the future. Two hummingbirds, the Black-breasted Puffleg and the Esmeraldas Woodstar, are both very rare, and little is known about their habitat requirements and seasonal movements; neither is known to occur in a formally protected area (though the woodstar may occur at Loma Alta Ecological Reserve), and both need additional attention. A third hummingbird, the Violet-throated Metaltail, ranges in high montane scrub and woodland above Cuenca and occurs in El Cajas National Recreation Area; it does not appear especially at risk. Little is known about a fourth hummingbird, the Emerald-bellied Woodnymph, found locally in more humid forests of the southwest; the only protected area where it is known to occur is in the recently established Buenaventura Biological Reserve in El Oro. The Pale-mandibled Araçari ranges fairly widely in remaining forests and woodlands of the western lowlands, and though doubtless much reduced in overall numbers because of deforestation, it seems to be relatively tolerant of habitat disturbance and does not seem to be in any trouble. The recently described Chestnut-bellied Cotinga, though certainly far from numerous, seems likely to persist indefinitely in its treeline woodland on the east slope of the Andes; a population occurs in the newly established Guandera Biological Reserve, and the species may range north into Colombia as well (and thus not be an Ecuadorian endemic). And lastly, the Cocha Antshrike (only recently rediscovered) and the Yellow-crested Manakin of the northeastern lowlands—both of which seem likely to occur in adjacent Colombia and/or Peru—do not appear to be in any difficulty, and their specialized várzea woodland habitat is found extensively in the Cuyabeno Faunistic Reserve.

ECUADOR'S 14 ENDEMIC BIRD SPECIES

El Oro Parakeet	*Pyrrhura orcesi*
White-breasted Parakeet	*Pyrrhura albipectus*
Emerald-bellied Woodnymph	*Thalurania hypochlora*

Turquoise-throated Puffleg	*Eriocnemis godini*
Black-breasted Puffleg	*Eriocnemis nigrivestis*
Violet-throated Metaltail	*Metallura baroni*
Esmeraldas Woodstar	*Chaetocercus berlepschi*
Pale-mandibled Araçari	*Pteroglossus erythropygius*
Cocha Antshrike	*Thamnophilus praecox*
Jocotoco Antpitta	*Grallaria ridgelyi*
El Oro Tapaculo	*Scytalopus robbinsi*
Orange-crested Manakin	*Heterocercus aurantiivertex*
Chestnut-bellied Cotinga	*Doliornis remseni*
Pale-headed Brush-Finch	*Atlapetes pallidiceps*

3. *Larger birds figure prominently, especially in the highest risk categories,* and indeed many areas still supporting reasonably intact natural habitat have effectively been "defaunated" (meaning that virtually all larger animals have been locally extirpated or their populations brought to very low levels). The two *Crax* curassows, for instance, have both been affected by overhunting far more than any other factor (as too has been the Salvin's Curassow, though this species' status is nowhere near as dire); suitable habitat continues to exist for both *Crax* curassows, though they have been eliminated from all or virtually all of their ranges (either could be the focus of a reintroduction effort). Several guan species are also faring poorly, though their situations are not quite so critical. Various parrots, especially the larger species, are in decline virtually wherever they occur, but numbers of most—an exception being the Red-and-green Macaw—are holding up relatively well in the eastern lowlands, though many have declined, some drastically, in the western lowlands and in the Andes (especially in the south). The Great Green Macaw and the Yellow-eared Parrot are both the focus of much-needed recovery campaigns at present, but the needs of several other parrots (e.g., Red-faced Parrot, Scarlet-fronted Parakeet, and Red-masked Parakeet) have until now gone largely unaddressed.

4. *Human impact on the central valley of northern Ecuador has been profound.* Not only has this resulted in the extirpation of two of the four Extirpated/Extinct species in Ecuador (Cinnamon Teal and American Coot), but two other species (Turquoise-throated Puffleg and Grasshopper Sparrow) once found here are so critically rare that they may not still exist; neither species has been recorded for more than a century. Habitats have been so severely disturbed that no one really knows what the original, natural woodland and scrub habitats were like (or indeed what birds they presumably supported). The ponds and marshes that still exist are but an impoverished echo of what they must have been a century ago.

5. *Species diversity may be high in the eastern lowlands, but endangerment levels are not.* Almost all the species occurring in this region that are con-

sidered to be at risk (there are not that many, all things considered) are larger birds, often gamebirds or other persecuted species. Habitat does not seem to be a problem for these birds; in fact they often are absent, or occur at unnaturally low densities, in areas where forest habitat is disturbed little if at all. Most of these species have relatively wide ranges outside Ecuador. There are only a few exceptions to this "large bird" generalization: a few hummingbirds about which little is known, too large *Oryzoborus* seed-finches and the Pale-eyed Blackbird which occur locally in marshy areas around a few lakes, and the Black-necked Red-Cotinga (males of which are much persecuted for their red feathers).

6. *In contrast, a tremendous number of at-risk species occur in what little remains of the forests of the western lowlands*, species ranging from the enigmatic and very rare Banded Ground-Cuckoo and the bizarre Long-wattled Umbrellabird to various tanagers, antbirds, and other smaller birds. These western forests have been so disastrously impacted by human activity that *any* bird occurring at a low density or restricted to the forest interior could be considered to be at risk; for practical reasons, only the less numerous and more sensitive have been. Sadly, little can now be done to restore these forests to their former glory, for virtually the entire region has been converted to agriculture. Some areas, particularly in northern Manabí, that are now used for marginally productive cattle ranching could perhaps eventually be restored to forest, and perhaps the few precious fragments that remain (at places such as Río Palenque and Jauneche) could gradually be expanded outward. Fringing foothill forests have fared somewhat better, and at about 500 m or so a good percentage of the lowland avifauna is still present, albeit often at reduced abundance levels compared to in the actual lowlands. These remnant forests, too, are precious indeed, and no effort should be spared to protect some of the best, at such sites as Pedro Vicente Maldonado in northwestern Pichincha, west of Quevedo in eastern Los Ríos, Manta Real in northwestern Azuay, and anywhere along the coastal cordillera (such as at Bilsa and Loma Alta). The still relatively extensive lowland forests of Esmeraldas are fundamentally different—wetter—than those found farther south, but they too are under increasing pressure, and all too little (aside from the poorly known Awá Forest Reserve in the extreme north) actually enjoys any form of protection.

7. Perhaps most importantly, *much remains to be accomplished in southwestern Ecuador, where—despite the need—to date relatively little conservation work has been done. The number of at-risk bird species occurring in southern Ecuador is tremendous, and population levels of many are very small.* (By way of comparison, various species occurring in the foothills and lower subtropics on the east slope of the Andes—many of them considered to be Near-threatened—have populations that are much larger, and most occur in one or several large protected areas such as Sumaco-Galeras, Sangay, or Podocarpus National Parks.) A range of habitats has been affected in the southwest, with

the current status of deciduous forest being the most serious, it having been reduced to a series of fragments, most of those small and degraded. Cloud and montane forests have also been greatly reduced in extent, and even more arid habitats such as the open desert scrub found on the Santa Elena Peninsula have been badly disturbed (two of Ecuador's four species of extinct birds, the Least Seedsnipe and the Tawny-throated Dotterel, once occurred there). Priority sites vying for conservation attention would include: the rich cloud forests of Buenaventura near Piñas, El Oro; deciduous and semihumid forests east of Portoviejo in eastern El Oro; the semihumid forests south of the Río Ayampe in northwestern Guayas; forest remnants near Utuana in the highlands of southern Loja; the superb and extensive deciduous forests along the Peruvian border west of Macará in far southern Loja (and perhaps the less accessible ones at Tambo Negro east of Macará); montane forest patches west of Celica, north of Alamor, and around El Limo; and Selva Alegre to the west of Saraguro in northern Loja.

Consider this a call to arms. We feel it should be the Ecuadorian conservation community's goal *to strive for the protection of at least one population—ideally more—of every species of bird that naturally occurs in Ecuador.*

But formally protected areas and lists of at-risk species are only part of the conservation picture. One must also be concerned about several more general aspects of the conservation situation in Ecuador, and we will conclude by raising a few such issues here.

The basic problem, as it is throughout the Neotropics (and indeed the rest of the world), is one of general habitat degradation and destruction. Hunting and other forms of direct persecution have had—and continue to have—an effect on population levels of certain birds, especially diurnal raptors, parrots, and so-called gamebirds. But it is ongoing habitat loss that presents far and away the gravest risk to the future of these birds.

In Ecuador, habitat loss almost invariably means the loss of forest and woodland cover and their conversion to agricultural activities and human settlement. Two additional issues are important in western Ecuador: the conversion of marshland and other low-lying terrain to intensive rice cultivation and the destruction of mangrove forest to make way for artificial shrimp ponds. The expansion of rice cultivation has had a devastating impact on numerous freshwater-inhabiting birds in the lowlands of southwestern Ecuador, none so gravely as the Southern Pochard. The loss of mangrove forest cover has also been severe, and though actual data are scanty, it must surely have had a major impact on various bird species whose distributions are strictly tied to mangroves. In some areas, such as southwestern Esmeraldas, mangrove forests have been entirely destroyed, and in many other areas destruction has also been severe. The most extensive surviving mangrove forests lie in northern Esmeraldas and around the Gulf of Guayaquil, but these two areas both remain vulnerable to destruction.

As noted above, other than hunting of gamebirds, direct persecution and consumption of bird populations have not had severe impacts on bird populations in Ecuador—with one major exception. For several decades Ecuador has enforced a prohibition on the export of its birds, including parrots, and though a few are smuggled into northern Peru for re-export from there (a loophole in Peruvian law makes it legal to do so), in general the Ecuadorian export ban has worked well. But that has not ended the pressure on wild parrot populations, for everyone in Ecuador still seems to want a parrot as a pet, preferably an amazon or even a macaw. Parrots are for sale, usually informally, in almost every town market—and this is perfectly legal! As no (or virtually no) parrots are bred in captivity in Ecuador, the drain on wild populations is substantial and the damage to parrot breeding potential (because of nest-site disturbance and destruction) great. *Restrictions are needed.*

One important positive development has been making significant headway in Ecuador over the last several decades. What has come to be called "ecotourism," or tourism whose aim is to explore the natural world, has become a postive force, and facilities (e.g., hotels, lodges, and reserves) in which this is emphasized are becoming increasingly prevalent across the country. Having long been an important facet of the economy of the Galápagos Islands, ecotourism now is becoming increasingly important on the mainland as well. One important component of ecotourism is, of course, "avitourism"—for birds represent perhaps the most popular group of all living organisms. We can only hope that this trend accelerates, for its potential for positive impacts cannot be overemphasized. Once a society has begun to reap some economic reward for protecting its natural features, it inevitably will hold those features in higher regard, and a greater effort to protect the natural patrimony will result.

Lastly, let us revisit the issue of specimen collecting for scientific purposes. As will be clear to readers of our two volumes, scientific collecting has been a major component of our research on Ecuadorian birds over the past several decades. We are convinced that our doing so has had zero impact on bird populations in Ecuador, indeed that with the publication of our work, bird populations should have the opportunity to expand. We feel strongly that whatever success *The Birds of Ecuador* enjoys, it will in no small measure be due to the fact that both of us had a comprehensive collection of modern bird skins at our elbows to use as reference material, PJG for helping to ensure the accuracy of his illustrations and RSR for descriptive details and for researching subspecific (sometimes even specific) identifications. These bird skins now form the core of the MECN collection in Quito (along, of course, with material obtained by others) and are a vital component of the collection at the ANSP. We must also recognize, and deplore, the fact that obtaining such material is—for a variety of reasons—now becoming more and more difficult from a procedural (= permitting) standpoint, not only in Ecuador but also elsewhere in the Neotropics and around the world (including, notably so, the United

States). We urge the appropriate authorities, in Ecuador and elsewhere, to rethink their positions on this matter; closing off the opportunity to accomplish research such as we have accomplished in Ecuador would, we are convinced, be a profound long-term mistake.

SPECIES ACCOUNTS

Tinamiformes
Tinamidae   Tinamous

Sixteen species in four genera. Tinamous are exclusively Neotropical in distribution.

Tinamus tao   Gray Tinamou   Tinamú Gris

Very rare and apparently local on the ground inside terra firme forest in the foothills and lower subtropical zone on the east slope of the Andes, locally out into adjacent eastern lowlands. There are only a few Ecuadorian records of this large tinamou, including old specimens from San José (Salvadori and Festa 1900) and "Cutucú" (*Birds of the Americas*, pt. 1, no. 1). More recently there have been a few records from Jatun Sacha (fide B. Bochan), along the Loreto road north of Archidona (N. Krabbe), in the Río Bombuscaro sector of Podocarpus National Park (first in Sep. 1991; R. Williams et al.), and in Zamora-Chinchipe at Panguri (one seen on 21 Jul. 1992; T. J. Davis) and Coangos (Schulenberg and Awbrey 1997). The Gray Tinamou has not been confirmed as occurring in the lowlands at any distance away from the Andes. Recorded from about 400 to 1600 m.

The Gray Tinamou may have been overlooked to some extent in Ecuador, but it does have a fairly distinctive primary vocalization, and present evidence indicates that it is genuinely scarce in the country. We consider the species to be Near-threatened in Ecuador; it is not, however, thought to be at risk in its overall range (Collar et al. 1994).

One race: *kleei*. All Ecuadorian specimens of *T. tao* are old, and the possibility of intergradation with *larensis* (of Colombia) exists in the north.

Range: locally from n. Colombia, n. Venezuela, and Guyana to n. Bolivia and Amazonian Brazil.

It remains possible that the rare Black Tinamou (*T. osgoodi*), a species known from only a few foothill localities in Colombia and Peru, will yet be found in the forests of the upper tropical zone along the east slope of the Andes in Ecuador.

Tinamus major   Great Tinamou   Tinamú Grande

Rare to locally fairly common on the ground inside humid forest in the low-lands of e. and nw. Ecuador (in the east mainly in terra firme), also occurring in smaller numbers up into the foothills. In w. Ecuador recorded mainly from Esmeraldas, adjacent sw. Imbabura, and Pichincha, in the last province found south to Río Palenque (where as of the mid-1990s a few birds persisted; in the late 1990s, however, none have been heard [P. Coopmans]). The Great Tinamou formerly also occurred at least in n. Manabí, and in the late 19th century the type specimen of the western subspecies was obtained as far south as Balzar in n. Guayas. Surprisingly, in Aug. 1996 the Great Tinamou was reported from as far south as the humid forested hill summits of Manglares-Churute Ecological Reserve (Pople et al. 1997); presumably this represents an isolated population. Recorded mostly below 700 m, in smaller numbers up to about 1200 m on both slopes of the Andes, and once as high as 1350 m (head of the Río Guataraco; MCZ).

The Great Tinamou is now much reduced in numbers and is very local in the west, especially in Pichincha, as a result of deforestation and hunting. In the still extensively forested drainages of the Ríos Santiago and Cayapas in n. Esmeraldas, the species was found to be widespread and fairly common in 1995–1998, but populations are markedly diminished surrounding settlements (O. Jahn et al.). Even in e. Ecuador the Great Tinamou becomes scarce near settled areas as a result of hunting, even where patches of relatively undisturbed forest persist. Even so, in our judgment it does not yet appear to be at risk in Ecuador.

Two rather different races occur in Ecuador, *peruvianus* in the east and *latifrons* in the northwest. *Latifrons* differs most notably in its rather long crest and browner and darker (not rufous) crown.

Range: s. Mexico to w. Ecuador, n. Bolivia, and Amazonian Brazil.

Tinamus guttatus   White-throated Tinamou   Tinamú Goliblanco

Uncommon to locally fairly common on the ground inside terra firme forest in the lowlands of e. Ecuador. Now that the voice of the reclusive White-throated Tinamou is becoming better known, it is evident that the species is not any-where near as rare in Ecuador as had been previously thought. Nonetheless it remains an infrequently seen bird, its presence being made known almost exclusively through its far-carrying calls. Recorded up only to about 400 m.

Monotypic.

Range: se. Colombia and s. Venezuela to n. Bolivia and Amazonian Brazil.

Nothocercus bonapartei   Highland Tinamou   Tinamú Serrano

Rare to locally uncommon on the ground inside montane forest in the sub-tropical zone on the east slope of the Andes. There are no confirmed records from the west slope of the Andes in Ecuador, but a tinamou briefly seen in Jul. 1988 as it walked on the forest floor in Carchi above Maldonado (2525 m) was believed likely to have been this species (M. B. Robbins). The secretive Highland Tinamou usually escapes detection except when it is singing, and we believe it likely that its distribution in Ecuador is more continuous than the relative paucity of records would seem to indicate; not only is it reclusive, but it also seems to vocalize relatively little. Recorded mostly between about 1600 and 2200 m, but once seen at the remarkably high elevation of 3075 m at Oyacachi (Krabbe et al. 1997).

One race: *plumbeiceps*.

Range: mountains of Costa Rica and w. Panama; mountains of n. Venezuela and Andes from w. Venezuela to extreme n. Peru.

Nothocercus julius   Tawny-breasted Tinamou   Tinamú Pechileonado

Uncommon to locally fairly common on the ground inside montane forest, occasionally emerging into adjacent clearings and fields when feeding, in the temperate zone on the east slope of the Andes. On the west slope recorded south to w. Cotopaxi (an old record from Pilaló, and recent records from the Caripero area [N. Krabbe and F. Sornoza]). The Tawny-breasted Tinamou normally goes unrecorded except when it is singing, which—at least compared to the Highland Tinamou—it appears to do frequently. Recorded mostly from about 2300 to 3400 m, infrequently down to 2100–2200 m at Cabañas San Isidro (M. Lysinger).

Monotypic, but see the discussion in Blake (1977, p. 26). If Peruvian birds are separated as a still-unnamed subspecies, then Ecuadorian birds would be referable to the nominate race.

Range: Andes from w. Venezuela to s. Peru.

Crypturellus cinereus   Cinereous Tinamou   Tinamú Cinéreo

Fairly common on the ground inside humid forest (favoring várzea and floodplain forests but also occurring in terra firme near streams and other wet places) in the lowlands of e. Ecuador. Heard far more often than it is observed, this seems to be a particularly difficult tinamou to actually see, especially compared to the often sympatric Undulated Tinamou (*C. undulatus*). Recorded mostly below 600 m, but at least locally ranges higher, perhaps mainly in the floodplains of larger rivers (e.g., to 900 m along the Río Nangaritza in Zamora-Chinchipe).

Monotypic (when the trans-Andean *C. berlepschi* [Berlepsch's Tinamou] is considered a separate species).

Range: e. Colombia, sw. Venezuela, and the Guianas to n. Bolivia and Amazonian Brazil.

Crypturellus berlepschi   Berlepsch's Tinamou   Tinamú de Berlepsch

Rare to locally uncommon on the ground inside very humid forest and mature secondary woodland in the lowlands and lower foothills of nw. Ecuador.

Recorded primarily from Esmeraldas, mainly in the north of that province, but also with a few records from Pichincha, including a specimen from "Mindo/ Milpe" (somewhere west of Mindo along the former trail to Milpe). In addition, there have been a few recent reports of what was believed to be this species from the coastal Cordillera de Mache at Bilsa (fide J. Hornbuckle), and there were also such reports from Río Palenque in the 1970s and early 1980s (R. Webster et al.). Whether any of the Río Palenque reports were fully corroborated is uncertain; in any case, it now appears that the species does not presently occur at this isolated forest tract. More recently (1995–1998), the Berlepsch's Tinamou was found to be widespread in the humid forest belt of n. Esmeraldas, where it prefers stream-rich forest sections and forest borders at the base of the Andes (O. Jahn and P. Mena Valenzuela et al.). A nest found in 1998 was located inside forest (away from any stream), but local hunters claim that the species also breeds regularly in the borders of mixed-culture plantations, as does the Little Tinamou (*C. soui*; O. Jahn et al.). Recorded mainly below 300 m (but the elevation at which the "Mindo/Milpe" specimen was taken is unknown; it almost has to be higher, as Milpe lies at 700 m; Paynter 1993).

Given the paucity of records of this species, and the deforestation that has taken place across much of its range, we assess its Ecuadorian status as Near-threatened. It has not, however, been considered to be at risk by Collar et al. (1992, 1994).

C. berlepschi is regarded as a monotypic species separate from the cis-Andean *C. cinereus* (Cinereous Tinamou), following Blake (1977) and most recent authors.

Range: w. Colombia and nw. Ecuador.

Crypturellus soui   **Little Tinamou**   **Tinamú Chico**

Fairly common to common on the ground inside secondary woodland, the borders of humid forest, and in overgrown clearings and plantations in the lowlands and foothills of e. and w. Ecuador. In the west found only in more humid regions and occurring south in small numbers to Guayas, El Oro, and sw. Loja (one heard near El Limo on 17 Apr. 1993; RSR). Although like other tinamous the Little is shy and infrequently seen, it does seem to be observed more frequently than most other tinamous, this perhaps mainly a reflection of its overall abundance. Recorded up to about 1200 m.

Two similar races are found in Ecuador, *nigriceps* in the east and *harterti* in the west. *Nigriceps* has somewhat more deeply colored underparts, especially in females.

Range: s. Mexico to sw. Ecuador, n. Bolivia, and Amazonian Brazil; e. Brazil.

Crypturellus obsoletus   **Brown Tinamou**   **Tinamú Pardo**

Uncertain. Apparently rare and local on the ground inside humid forest in the foothills along the eastern base of the Andes. The Brown Tinamou is recorded in Ecuador only from two old specimens taken in w. Napo (Río Suno and

Concepción) and from a few recent reports from the Río Bombuscaro valley in Podocarpus National Park in Zamora-Chinchipe; the species has also been reported from the Cajanuma sector of Podocarpus National Park (Bloch et al. 1991). The Podocarpus reports consist only of birds believed to have been heard, and as the voice of the Brown Tinamou in Ecuador is not certainly known, these reports must be treated with some caution. The possible record mentioned by Fjeldså and Krabbe (1990) from the Río Guaillabamba valley does not, based on what is now known of the species' distribution, seem credible. Recorded only between about 500 and 1100 m, but in Peru known from as high as 2900 m (Fjeldså and Krabbe 1990), so it possibly occurs at comparable elevations in Ecuador.

Given the lack of information on this species' status in Ecuador, its evident rarity, and the significant deforestation that has occurred in much of its presumed range, we feel we must accord it Data Deficient status in Ecuador. It was not considered to be globally at risk by Collar et al. (1992, 1994).

One race: *castaneus*.

Range: locally from n. and w. Venezuela to w. Bolivia and sw. Amazonian Brazil; se. Brazil, e. Paraguay, and ne. Argentina.

Crypturellus undulatus   Undulated Tinamou   Tinamú Ondulado

Fairly common to common on the ground inside humid forest (principally várzea), forest borders, riparian forest and woodland (including larger river islands), and adjacent clearings with heavy ground cover in the lowlands of e. Ecuador. Like other tinamous, the Undulated is shy and heard much more often than it is seen, though it is glimpsed more often than the others found in Amazonia, particularly along the water's edge when it comes to drink. Recorded up to about 600 m.

One race: *yapura*.

Range: e. Colombia, s. Venezuela, and the Guianas to n. Bolivia, n. Argentina, and s.-cen. Brazil.

Crypturellus variegatus   Variegated Tinamou   Tinamú Abigarrado

Uncommon to locally fairly common on the ground inside humid forest (principally terra firme) in the lowlands of e. Ecuador. Like other Ecuadorian tinamous, the Variegated is heard much more often than it is seen. Recorded below 400 m.

Monotypic. We follow Blake (1977) in considering the race *salvini*, to which Ecuadorian birds were often formerly ascribed (e.g., Chapman 1926), to be too variable to be worthy of recognition.

Range: se. Colombia, s. Venezuela, and the Guianas to n. Bolivia and Amazonian Brazil; se. Brazil.

**Crypturellus bartletti*   Bartlett's Tinamou   Tinamú de Bartlett

Rare to locally uncommon on the ground inside terra firme forest in the lowlands of e. Ecuador. So far known mainly from the drainages of the Ríos Napo and Aguarico, though recent reports (1996 and subsequently) from around Kapawi Lodge near the Peruvian border on the Río Pastaza make it likely that the species occurs through the southeastern lowlands as well. The first Ecuadorian records of the Bartlett's Tinamou were not made until early 1990, when various observers found it at La Selva. It has since been found at various localities recently made accessible for ecotourism in ne. Ecuador, and it now appears that it occurs wherever extensive terra firme forest is found. The only Ecuadorian specimen is a male (ANSP) taken by F. Sornoza at Zancudococha on 5 Apr. 1991. The species seems almost certain to occur in adjacent Colombia, where it has not been recorded (Hilty and Brown 1986). Recorded up to about 400 m.

Monotypic. The relationship of C. *bartletti* with C. *brevirostris* (Rusty Tinamou) remains uncertain: they may prove to be conspecific, but there appears to be overlap in parts of w. Amazonia. If merged, the name *brevirostris* has priority.

Range: e. Ecuador, e. Peru, nw. Bolivia, and w. Amazonian Brazil.

Crypturellus transfasciatus   Pale-browed Tinamou   Tinamú Cejiblanco

Fairly common on the ground inside deciduous forest and woodland, and at least locally in regenerating scrubby woodland as well (this seemingly is especially the case in coastal El Oro), in the lowlands of sw. Ecuador, apparently in two disjunct populations. The Pale-browed Tinamou ranges from cen. Manabí (north to the Chone and Bahía de Caráquez region) and Los Ríos (north in small numbers to around Quevedo, and also numerous [fide P. Coopmans] at Jauneche) south into Guayas (east to Cerro Blanco in the Chongón Hills, with a few extending to Manglares-Churute Ecological Reserve [Pople et al. 1997]). It also occurs widely in El Oro and w. Loja (east to the Macará region near Sabiango and the Sozoranga region at El Tundo Reserve). Recorded mostly below about 800 m, but has been heard up to about 1500 m along the El Empalme-Celica road (P. Coopmans) and to 1600 m at El Tundo Reserve.

The Pale-browed Tinamou's numbers seem to be holding up well in most regions, and the species appears to tolerate substantial fragmentation of its habitat, indeed often remaining quite numerous in such areas. Nonetheless, because of the widespread destruction of forest and woodland across major portions of its comparatively limited range, we consider the species to merit Near-threatened status in Ecuador. This is in accord with the assessment of Collar et al. (1994).

Monotypic.

Range: sw. Ecuador and nw. Peru.

**Crypturellus tataupa*   Tataupa Tinamou   Tinamú Tataupá

Uncommon on the ground inside deciduous woodland and in regenerating scrub and clearings in extreme se. Ecuador in the Río Marañón drainage of s. Zamora-Chinchipe near Zumba. First recorded on 12 Dec. 1991, when at least three birds were heard calling in the Río Mayo valley east of La Chonta (RSR). One was collected here by T. J. Davis on 9 Aug. 1992 (ANSP), and the species has since been found by several observers, both here and north of Zumba in the Río Isimanchi valley. Recorded from 650 to 950 m.

One race: *inops*. This subspecies was heretofore known only from a few localities in the upper Río Marañón valley of Cajamarca, Peru.

Range: locally in arid intermontane valleys of extreme s. Ecuador to cen. Peru; n. and e. Bolivia to n. Argentina and s. and e. Brazil.

Nothoprocta pentlandii   **Andean Tinamou**   **Tinamú Andino**

Uncommon and apparently local in montane scrub, grassy clearings, and adjacent agricultural fields in the highlands of s. Ecuador, where so far recorded mainly from Loja though it likely also occurs in adjacent e. El Oro. This species has also recently been found to range as far north as adjacent s. Azuay in the Río Jubones drainage (N. Krabbe). Long known in Ecuador from a single specimen (AMNH) obtained in Dec. 1920 at Punta Santa Ana, along the trail between Portovelo and Loja, and from a series of no fewer than 10 specimens (FMNH) obtained at Malacatos in e. Loja, the Andean Tinamou has in recent years been found to occur at several localities, though almost always in small numbers. Like other tinamous it is known primarily from its vocalizations. Recorded between about 1000 and 2300 m.

One race: *ambigua*.

Range: Andes from s. Ecuador to extreme n. Chile and w.-cen. Argentina.

Nothoprocta curvirostris   **Curve-billed Tinamou**   **Tinamú Piquicurvo**

Uncommon and somewhat local in shrubby paramo grassland and pastures, usually near patches of woodland, in the Andes from se. Carchi (several recent records from Guandera Biological Reserve northeast of San Gabriel; Creswell et al. 1999), s. Imbabura (specimen taken by C. G. and D. C. Schmitt west of Laguna Cuicocha on 15 Sep. 1983; LSUMZ), and Pichincha south sparingly to e. Azuay (old specimens from Sigsig and Bestión) and Zamora-Chinchipe (where known only from one sighting at Cerro Toledo in the southern part of Podocarpus National Park; N. Krabbe). There still are no confirmed records from Loja, but the species perhaps occurs there locally as it has been found just south of the border on Cerro Chinguela in extreme n. Peru (Parker et al. 1985). A good population of Curve-billed Tinamous now occurs in Cotopaxi National Park, even adjacent to the otherwise nearly sterile pine (*Pinus*) plantations, and the species is also considered to be not uncommon on the slopes of Volcán Pichincha (N. Krabbe). Recorded mostly from 3000 to 3900 m.

The Curve-billed Tinamou is generally a scarce and infrequently encountered bird in Ecuador, where it usually is found only by chance while scurrying across or along a roadside. Its numbers may have been affected by overhunting in some areas, but overall the species does not appear to be at risk.

One race occurs in Ecuador, nominate *curvirostris*. Although nearly endemic to Ecuador, this is evidently the form that was collected on Cerro Chinguela (Fjeldså and Krabbe 1990).

Range: Andes from n. Ecuador to cen. Peru.

Podicipediformes
Podicipedidae   Grebes

Three species in three genera. Grebes are wide-ranging in aquatic habitats, but they breed only on freshwater.

Tachybaptus dominicus   Least Grebe   Zambullidor Menor

Fairly common to common on ponds and lakes (sometimes quite small and even ephemeral) in the lowlands of w. Ecuador from Esmeraldas in the Atacames and Muisne areas (RSR, P. Coopmans), with also one report from Porvenir near San Lorenzo (one bird seen in Nov. 1997; K. Berg), south through coastal El Oro. Very rare and local in the lowlands of the east, whence there is only a single published record (Orcés 1944), that of specimens reportedly in the Mejía collection—though they do not seem to be there now—involving birds collected at Río Suno in w. Napo and Macas in nw. Morona-Santiago. The only recent record from the east concerns a pair taken on a pond some 30 km southeast of Coca in w. Napo on 20 Nov. 1988 (WFVZ, MECN). The number of Least Grebes present in some areas of w. Ecuador can fluctuate dramatically, apparently in response to changes in water levels. When rains are abundant, pairs evidently disperse widely, even to relatively small and ephemeral ponds, in order to breed; when water levels drop, birds gather on larger ponds and reservoirs such as the Represa José Velasco Ibarra on the Santa Elena Peninsula (where hundreds have been known to congregate) and the Represa Chongón west of Guayaquil. Unlike the Pied-billed Grebe (*Podilymbus podiceps*), the Least seems almost never to occur on saltwater ponds and lagoons, and at least in Ecuador it seems not to occur in the highlands. Recorded up to 700 m at Tinalandia.

Formerly placed in the genus *Podiceps*. Two similar races are found in Ecuador, *brachyrhynchus* in the east and the recently described *eisenmanni* in the west (see R. W. Storer and T. Getty, *Neotropical Ornithology*, Ornithol. Monograph no. 36: 31–39, 1985). *Eisenmanni* differs only in being slightly smaller; it is nearly endemic to w. Ecuador.

Range: sw. United States to nw. Peru, n. Argentina, and s. Brazil, but absent from much of Amazonia; Greater Antilles.

Podilymbus podiceps   Pied-billed Grebe   Zambullidor Piquipinto

Fairly common to common on lakes and ponds in the lowlands of w. Ecuador from w. Esmeraldas (southeast of Muisne; RSR) south through coastal El Oro.

Also found very locally on certain lakes in the central valley of n. Ecuador, where known from Yaguarcocha and Laguna de San Pablo in Imbabura, the La Carolina marshes in Quito (formerly), and Laguna Yambo on the Cotopaxi/Tungurahua border (N. Krabbe); the only report from the southern highlands is a single presumed wandering bird seen at El Cajas on 21 Feb. 1997 (P. Coopmans et al.). There are no reports at all from the e. lowlands. Recorded mostly below 200 m, but locally from 2100 to 3200 m in the highlands.

One race: *antarcticus*.

Range: North America to s. South America, but almost entirely absent from Amazonia and ne. Brazil; West Indies.

Podiceps occipitalis   Silvery Grebe   **Zambullidor Plateado**

Rare to uncommon and very local on open lakes in the central valley and in paramo, with overall numbers now much reduced. Never a numerous or widespread bird in Ecuador—or elsewhere in the n. Andes—the Silvery Grebe is now a very scarce bird here, with only a few extant populations we are aware of. Silvery Grebes are found on a few lagoons on Páramo del Ángel (J. R. Fletcher), Cuicocha in s. Imbabura, on a small lake near Papallacta pass (17 seen on 16 Nov. 1996; P. Coopmans et al.), Micacocha on the paramo of Volcán Antisana in extreme w. Napo (where 400–500 were estimated to be present in Nov. 1994 [N. Krabbe], though after water levels were artificially raised numbers have been much lower), and on paramo lakes on the Altar/Sangay massif in extreme nw. Morona-Santiago (1990 sightings by C. A. Vogt, fide N. Krabbe). Silvery Grebes formerly occurred in substantial numbers on Yaguarcocha in Imbabura, where breeding continued at least into the early 1980s, and a few individuals may still wander there irregularly; the species also used to occur on Laguna de San Pablo in Imbabura (including a Mejía specimen taken in 1939). In addition, a specimen was taken in Oct. 1896 in Azuay on Lago de Culebrillas, a lake that cannot be located but is believed (fide N. Krabbe) to lie on the Cajas plateau. Possibly a few birds are still found in the Cajas area, as indicated by sightings reported without details by King (1989), but there have been no subsequent reports. Recorded between about 2200 and 4100 m.

The future of the Silvery Grebe in Ecuador seems tenuous unless improved protection for its few remaining sites can be assured. We thus consider the species to merit Vulnerable status. Elsewhere in its wide range this relatively numerous species is not considered to be at risk (Collar et al. 1992, 1994). However, if *juninensis* is separated specifically, as seems a possibility, then it likely would have to be.

One race: *juninensis*. As noted above, this form may prove to be a monotypic species (Northern Silvery-Grebe) separate from *P. occipitalis* (Southern Silvery-Grebe) of s. South America.

Range: Andes from s. Colombia to Chile and Argentina, also in the lowlands in the last two countries.

The Great Grebe (*P. major*) has in recent years been increasing and spreading northward in coastal Peru, and in 1999 it was seen in Piura, Peru, not far to the south of the Ecuador border (P. Coopmans). It is thus not inconceivable that in the future the Great Grebe could occur in sw. Ecuador.

Sphenisciformes
Spheniscidae   Penguins

One species. Penguins are found primarily in southern oceans (the northernmost-ranging being the endemic Galápagos Penguin [*Spheniscus mendiculus*]); none breeds in mainland Ecuador.

Spheniscus humboldti   Humboldt Penguin   Pingüino de Humboldt

Apparently a casual visitant to the coast of s. Ecuador, where known from a few reports from Guayas and s. Manabí, with all records involving birds found dead or dying on beaches. A juvenile was found by a fisherman near Guayaquil in Oct. 1983; another was reportedly found near Playas in the late 1970s (fide F. Ortiz-Crespo); and one was picked up near Puerto López in early 1993 (fide M. Jácome), with its skeleton now deposited in the museum at Salango (fide N. Krabbe). The provenance of these penguins must be viewed with some skepticism, and it certainly is possible that some or all of the records pertain to ship-assisted birds (i.e., individuals captured in their normal range and transported northward, where they either escaped or were released).

The Humboldt Penguin was considered Near-threatened by Collar et al. (1992, 1994) because of persecution and disturbance on its nesting grounds, as well as the removal of the guano deposits in which it formerly burrowed and nested; the recent increased frequency and severity of El Niño events likely also are having an impact on its population size. We feel this species' status may actually be more critical, and that Vulnerable may be a more appropriate status rating, but for the present we remain in accord with Collar et al. (op. cit.).

Monotypic.

Range: coast of Peru and n. Chile, very rarely wandering to s. Ecuador.

Procellariiformes
Diomedeidae   Albatrosses

Two species in two genera. Albatrosses are wide-ranging on oceans, though they are absent from the North Atlantic.

Phoebastria irrorata   Waved Albatross   Albatros de Galápagos

A rare to occasionally uncommon visitant to offshore and (occasionally) coastal waters in s. Ecuador. Small numbers appear to range year-round on the seas off the Santa Elena Peninsula (though despite the birds' great size, they are rarely visible from shore), but the Waved Albatross is best known in mainland Ecuadorian waters from Isla de la Plata, where a small breeding colony has been present for some years. This colony was first described by O. T. Owre (*Ibis* 118[3]: 419–420, 1976), who in May 1975 found at least five nests in the Punta Machete area. The lighthouse keeper at the time claimed that the birds had only a few years previously begun to nest on the island, but their at least occasional presence on the island long before this is borne out by Murphy's (1936) having seen the carcass of an albatross that had been shot there in Nov. 1924. Since the 1970s the number of nests has fluctuated, with up to 10 having been reported in 1990 (F. I. Ortiz-Crespo and P. Agnew, *Bull. B. O. C.* 112[2]: 66–73, 1992), though only 3 were occupied in Jul. 1993 (RSR and F. Sornoza). The estimate of "10–50 pairs" given by Enticott and Tipling (1997, p. 26) is overly optimistic. With full protection of the island subsequent to its incorporation into Machalilla National Park, the albatross population will, we hope, slowly increase; the control or—ideally—elimination of exotic mammals such as goats, cats, and rats should remain a high priority.

Given the very small size of the Isla de la Plata breeding population, we consider the Waved Albatross to be Vulnerable in the Ecuadorian mainland area; were it to be declining, the species would have to be considered Critical. The species as a whole is classified only as Near-threatened by Collar et al. (1994).

The species is sometimes called the Galápagos Albatross.

The species was formerly placed in the genus *Diomedea*, but we follow G. B. Nunn et al. (*Auk* 113[4]: 784–801, 1996) and the 1998 AOU Check-list in placing it in *Phoebastria*. Monotypic.

Range: breeds on Isla Española (= Hood Island) in the Galápagos Islands, and on Isla de la Plata off sw. Ecuador; nonbreeding birds disperse to e. Pacific Ocean from off Panama to off Peru.

Thalassarche melanophris   Black-browed Albatross   Albatros Ojeroso

A casual visitant to offshore and coastal waters in s. Ecuador. We are aware of only two records of the Black-browed Albatross from Ecuador. The first involves a dead adult found washed up on a beach on the Santa Elena Peninsula on 1 Nov. 1965 (M. P. Harris, *Ardea* 56: 284–285, 1968); though measured and photographed, the specimen was not saved (and the photos themselves were accidentally destroyed, fide M. P. Harris). The second record involves an adult seen resting on the water in Canal de Morro, the channel from Guayaquil out into the Gulf of Guayaquil, on 2 May 1984 (N. Krabbe). Of possible significance is the fact that the 1965 and 1984 records occurred during or subsequent to an El Niño event; assisted passage to Ecuadorian waters can also, of course, not be ruled out.

This species was formerly placed in the genus *Diomedea*, but we follow G. B. Nunn et al. (*Auk* 113[4]: 784–801, 1996) and the 1998 AOU Check-list in placing it in *Thalassarche*. One race: nominate *melanophris*.

Range: breeds on many subantarctic islands; nonbreeding birds disperse northward, in e. Pacific Ocean ranging regularly as far as Humboldt Current waters off Peru.

Procellariidae   Shearwaters and Petrels

Nine species in five genera. Shearwaters and petrels occur on all oceans but reach their highest diversity in the Southern Hemisphere. None is known to breed in mainland Ecuadorian waters.

[*Fulmarus glacialoides*   Southern Fulmar   Fulmar Sureño]

Apparently an accidental visitor to coastal and offshore waters in s. Ecuador. M. P. Harris (*Ardea* 56: 284–285, 1968) found a moribund bird that had washed up on a beach on the Santa Elena Peninsula on 1 Nov. 1965; the bird was measured and photographed, but the photos subsequently were destroyed in a fire (fide M. P. Harris). Mills (1967) records seeing two Southern Fulmars with Pintado Petrels (*Daption capense*) on 2 Oct. 1965 west of the "Mancora Banks," which we cannot locate; however, the coordinates given (3°48′ S, 81°28′ W) clearly lie within Peruvian waters, west of Tumbes.

   The species is sometimes called the Antarctic or Silver-gray Fulmar.
   Monotypic.
   Range: breeds widely on subantarctic islands and on Antarctic coasts; nonbreeding birds disperse northward, in e. Pacific ranging regularly as far as Humboldt Current waters off Peru.

Daption capense   Pintado Petrel   Petrel Pintado

An irregular and rare visitor to coastal and offshore waters of s. Ecuador, with one report from as far north as off Manta in Manabí (Murphy 1936). The unmistakable Pintado Petrel is not reported on an annual basis in Ecuadorian waters, and numbers found here are always small. Mills (1967) observed four on 2 Oct. 1965 west of the "Mancora Banks," but the coordinates given (3°40′ S, 81°21.5′ W) clearly lie in Peruvian waters, west of Tumbes. Recorded at least from Jul. to Feb., but the species could perhaps occur at any time of year.

   The species is often called the Cape Petrel.
   Presumed race: nominate *capense*.
   Range: breeds widely on subantarctic islands and on Antarctic coasts; nonbreeding birds disperse northward, ranging regularly and in substantial numbers in e. Pacific as far as Humboldt Current waters off Peru.

Pterodroma phaeopygia   Dark-rumped Petrel   Petrel Lomioscuro

A rare and perhaps irregular visitor to offshore waters off s. Ecuador, with the northernmost specific report being a sighting off Isla de la Plata on 4 Sep. 1919

(Murphy 1936); there is also a sighting from "off northern Ecuador" on 5 Dec. 1924 (Chapman 1926, p. 184). In recent years only a few individuals of this species have been seen during the series of pelagic trips made off the Santa Elena Peninsula (ANSP). These small numbers contrast strikingly with the large numbers that apparently were seen in the years between 1910 and 1930 (Murphy, op. cit.). So far recorded only from Jul. to Jan., but perhaps occurs year-round.

This apparent decrease in numbers correlates with the species' decline on its nesting grounds, at higher elevations on some of the Galápagos Islands, which has been caused by predation—mainly by rats—and nest-site destruction by introduced mammals. The species' status has recently been considered to be Critical (Collar et al. 1994). However, given the recent successful efforts to protect nesting birds at some colonies (see J. B. Cruz and F. Cruz, *Bird Conserv. Int.* 6[1]: 23–32, 1996), and given the extremely small population sizes of other Ecuadorian birds that have been accorded Critical status, we give this species Endangered status. A minimum of 10,000 pairs continues to breed on the Galápagos (Collar et al. 1994).

R. J. Tomkins and B. J. Milne (*Notornis* 38: 1–35, 1991) suggest considering the birds breeding on the Galápagos Islands as a monotypic species (*P. phaeopygia*, Galápagos Petrel) separate from those breeding on the Hawaiian Islands (*P. sandwichensis*, Hawaiian Petrel), this being based on minor differences in morphology and vocalizations. However, we continue to follow most authors, including the 1998 AOU Check-list, in considering these forms to be conspecific. One race occurs in Ecuadorian waters, nominate *phaeopygia*.

Range: breeds on Galápagos and Hawaiian Islands; nonbreeding birds disperse into e. Pacific Ocean east to off Mexico and Panama and south to off Peru.

[Procellaria aequinoctialis White-chinned Petrel Petrel Barbiblanco]

An accidental visitor to far offshore waters. Only one sighting appears to be valid, that of two birds seen well west of the Santa Elena Peninsula, en route to the Galápagos, on 25 Aug. 1983 (T. and M. Southerland et al.). Other reported sightings likely refer to the Parkinson's Petrel (*P. parkinsoni*), a species that until recently was not known to occur as regularly and frequently in Ecuadorian waters as is now known to be the case.

Monotypic.

Range: breeds widely on subantarctic islands; nonbreeding birds disperse northward, small numbers reaching as far north as Humboldt Current waters off n. Peru. Excludes Spectacled Petrel (*P. conspicillatus*) of South Atlantic Ocean (see P. G. Ryan, *Bird Conserv. Int.* 8[3]: 223–235, 1998).

Procellaria parkinsoni Parkinson's Petrel Petrel de Parkinson

A fairly common but perhaps local visitor to offshore waters around the Santa Elena Peninsula, but with relatively few reports from elsewhere (though this is perhaps merely due to confusion with other large dark petrels and shearwaters). First recorded from two unpublished specimens (AMNH) that were taken in Apr. 1941 by R. C. Murphy and J. G. Correia off n. Ecuador, one west of Cabo San Francisco in w. Esmeraldas on 2 Apr., the other west of Cabo Pasado in n. Manabí on 16 Apr. Since then, virtually all records have come from the Santa Elena Peninsula region, where the species appears to be present year-round (contra many references, e.g., Enticott and Tipling 1997) though perhaps occurring in largest numbers from May to Sep. A few additional specimens (ANSP, MECN) have been taken, in the months of Jan., May, Jul., and Sep. R. L. Pitman and L. T. Ballance (*Condor* 94[4]: 825–835, 1992) describe the Parkinson's Petrel's frequent feeding association with marine mammals (especially the Melon-headed Whale [*Peponocephala electra*] and the False Killer Whale [*Pseudorca crassidens*]) in the e. Pacific Ocean, though off the Santa Elena Peninsula we have only seen this species feeding around fishing boats.

Thanks to the removal of predators from its nesting islands, the small world population of Parkinson's Petrel appears to be slowly increasing, and this likely is the explanation for the species' apparent recent increase in Ecuadorian waters. Nonetheless, the species is still considered to be Vulnerable by Collar et al. (1994), and it is so considered here.

The species has sometimes been called the Black, or Parkinson's Black, Petrel (e.g., Sibley and Monroe 1990). However, we concur with the 1998 AOU Check-list in preferring the English name of Parkinson's Petrel for the species, which is no "blacker" than several other *Procellaria*.

Monotypic.

Range: breeds on two islands off New Zealand; nonbreeding birds disperse to tropical e. Pacific Ocean from off Mexico to off Ecuador.

Puffinus creatopus   **Pink-footed Shearwater**   Pardela Patirrosada

An uncommon visitant to offshore waters, where known from along the entire coast though noted with greatest frequency and apparently in largest numbers off the Santa Elena Peninsula; seemingly occurs year-round. The only specimens of the Pink-footed Shearwater that have been taken in Ecuadorian waters are the four birds (ANSP, MECN) obtained by J. Gerwin and F. Sornoza off the Santa Elena Peninsula on 5 May 1991.

Because of threats on its nesting islands (principally from introduced predators), the Pink-footed Shearwater is considered to merit Vulnerable status by Collar et al. (1994), and it is so treated here.

Monotypic.

Range: breeds on islands off Chile; nonbreeding birds disperse to n. Pacific Ocean.

We believe it certain that the birds reported as Wedge-tailed Shearwaters (*Puffinus pacificus*) that were seen in the Gulf of Guayaquil by R. Lévêque on 14 Mar. 1962 (*Ibis* 106: 53–54, 1964), and again by M. P. Harris on 15 Oct.

1965 (*Ardea* 56: 284–285, 1968), were actually Pink-footed Shearwaters. Lévêque's published description is inadequate to confirm the identification of the birds as Wedge-tailed Shearwaters, and Harris provided no details. There are no other reports of Wedge-tailed Shearwaters from anywhere near mainland Ecuadorian waters.

Puffinus bulleri   Buller's Shearwater   Pardela de Buller

Accidental. There is only one Ecuadorian record of the Buller's Shearwater, a bird washed up on a beach near Palmar in w. Guayas found by N. Krabbe on 1 Jan. 1992 (specimen to MECN). Presumably the species normally passes Ecuador rapidly and only very far offshore, perhaps mainly on southward passage toward its breeding grounds.

The Buller's Shearwater is considered to be Near-threatened by Collar et al. (1994) because of threats on its nesting islands (though it is apparently increasing in numbers); it is so treated here.

The species was formerly sometimes called the Gray-backed or the New Zealand Shearwater.

Monotypic.

Range: breeds on a few islands off New Zealand; nonbreeding birds disperse north into n. Pacific Ocean and also occur south in e. Pacific Ocean to off Peru and Chile.

Puffinus griseus   Sooty Shearwater   Pardela Sombría

An uncommon to fairly common visitor to offshore and, to a lesser extent, coastal waters along the entire coast, though apparently most numerous off the Santa Elena Peninsula; apparently occurs year-round, with largest numbers perhaps being found during the second half of the year. The Sooty is the shearwater most likely to be seen from land, but it is always more frequent at least a few kilometers out to sea. Marchant (1958) found that at times many dead Sooties washed ashore on beaches around the Santa Elena Peninsula.

Monotypic.

Range: breeds on many islands off s. South America and around New Zealand and se. Australia; nonbreeding birds disperse northward into n. Atlantic and Pacific Oceans.

[*Puffinus lherminieri*   Audubon's Shearwater   Pardela de Audubon]

Apparently a casual visitor to offshore and coastal waters. Meyer de Schauensee (1970, p. 16) recorded the Audubon's Shearwater from "off the Pacific coast of Ecuador" but provided no details. There are only two additional sight reports from Ecuadorian waters: a single bird seen from Salinas on an unspecified date in Aug. 1989 (B. Haase) and three reported seen near Isla de la Plata on 31 Jul. 1996 (S. Howell and R. Behrstock et al.). Given the lack of much pelagic work in the warmer waters off the coast of n. Ecuador—where the Audubon's Shearwater would be most likely to occur—it remains possible that the species may prove to be more regular than the present paucity of reports would seem to indicate.

The species was formerly sometimes called the Dusky-backed Shearwater.

The race of *P. lherminieri* occurring in mainland Ecuadorian waters is unknown, but it almost certainly is *subalaris*, the race breeding in the Galápagos Islands; this form is known to range as far east as the waters off Panama and n. Colombia.

Range: breeds locally on many smaller islands in tropical seas; nonbreeders tend not to disperse very far. More than one species is perhaps involved.

Hydrobatidae   Storm-Petrels

Ten species in three genera. Storm-petrels are wide-ranging and occur on all oceans. No storm-petrel is known to breed in mainland Ecuadorian waters.

Oceanites oceanicus   Wilson's Storm-Petrel   Paíño de Wilson

Apparently a very rare and perhaps irregular austral winter visitor to offshore waters (but perhaps overlooked); only three reports are known to us. In *Birds of the Americas* (pt. 1, no. 2) this species was listed as being "accidental" in Ecuador, but no details were given; Meyer de Schauensee (1966) repeated this, again providing no details. Recent reports are as follows: about 15 were reported seen near Isla de la Plata on 9 Jul. 1976 (F. Sibley et al.); one was seen off the Santa Elena Peninsula on 14 Sep. 1983 (B. Haase); and about 3 were seen off the Santa Elena Peninsula on 20 Jun. 1992, with one of these birds being collected (ANSP).

This species, together with all the members of the Hydrobatidae, formerly went by the group name of "petrel," and that name is occasionally still used.

The Santa Elena Peninsula specimen is referable, on the basis of its large size, to nominate *oceanicus*. The longer-winged *exasperatus* could also occur in Ecuadorian waters.

Range: breeds on many subantarctic islands and along Antarctic coast; nonbreeding birds disperse northward, especially into n. Atlantic Ocean (but are rare in North Pacific).

Oceanites gracilis   White-vented Storm-Petrel   Paíño Grácil

Apparently a rare and perhaps irregular visitant to coastal and offshore waters; more likely to occur inshore than the similar Wilson's Storm-Petrel (*O. oceanicus*). The White-vented Storm-Petrel was first recorded from a sighting of R. C. Murphy's in the "Bay of Santa Elena" on 11 Feb. 1925 (Murphy 1936, p. 759), and six were seen from shore on the Santa Elena Peninsula at Chepipe (near Salinas) on 30 Dec. 1982 (PJG). One was reported seen near Isla de la Plata on 9 Jul. 1976 (F. Sibley et al.), and the species was reportedly seen here in Oct. 1985 (J. B. Nowak, *Noticias de Galápagos* 44: 17, 1987). The only other reports are from 1992, when on 20 Jun. about 10 were seen off the Santa Elena Peninsula, with 3 of these being collected (ANSP, MECN), and on 23 Jun. when about 3 were seen off Isla de la Plata (RSR and F. Sornoza). The storm-petrels reported seen on the Santa Elena Peninsula, apparently from land, by Marchant (1958, p. 360), though assumed by him "probably" to be Wilson's Storm-Petrels, seem just as likely (perhaps more so) to have been White-venteds. There seems to be no discernible pattern to the occurrences of the White-vented Storm-Petrel in mainland Ecuadorian waters; at least one of the records was made during an El Niño event (Dec. 1982), but the others were not. Although this species is relatively numerous in the Humboldt Current and around the Galápagos Islands, only one nest has ever been found, in Chile (R. P. Schlatter and M. A. Marín, *Gerfaut* 73: 197–199, 1983); remarkably, no nest has ever been found on the Galápagos Islands, where the species is commonly seen. It is not inconceivable that the species could even nest in mainland Ecuadorian waters, though there is no real evidence that it does so.

The White-vented Storm-Petrel was considered to be a Data Deficient species by Collar et al. (1994), primarily because so little is known about its breeding; it is so considered here. There is no evidence of any actual decline in numbers.

The species is sometimes called the Elliot's Storm-Petrel.

Only nominate *gracilis* is known from mainland Ecuadorian waters.

Range: breeding grounds essentially unknown (presumed to be on islands and along coasts of Peru and n. Chile, and on Galápagos Islands); dispersal when not breeding appears to be essentially local.

Pelagodroma marina   White-faced Storm-Petrel   Paíño Cariblanco

Apparently a casual visitor to offshore waters, perhaps only occurring far off the coast. There are only two old records of this distinctive storm-petrel. One was collected (USNM) after it flew aboard a ship just west of Punta Santa Elena on 23 Jun. 1922 (A. Wetmore, *Condor* 25: 171, 1923). R. C. Murphy and S. Irving (*Am. Mus. Novitates* 1506, 1951) recorded another individual that was collected (AMNH) on 10 Jul. 1938 "off the Gulf of Guayaquil" (from the coordinates provided [2°35′ S, 81°20′ W] this would place the bird about 50 km southwest of the Santa Elena Peninsula). Despite considerable field work off the Santa Elena Peninsula in the 1990s, there have been no recent records.

The species is sometimes called the Frigate Storm-Petrel.

Murphy and Irving (op. cit.) refer the 1938 specimen to the subspecies *maoriana*, which breeds on islands off New Zealand and disperses northward into the tropical Pacific Ocean (a pattern similar to that seen in *Procellaria parkinsoni* [Parkinson's Petrel]). Birds collected in the vicinity of the Galápagos Islands have also been referred to *maoriana*.

Range: breeds locally on islands in Atlantic Ocean, and off New Zealand and Australia; nonbreeders of latter population disperse northward into Indian and tropical Pacific Oceans.

Oceanodroma microsoma   Least Storm-Petrel   Paíño Menudo

Apparently a rare to occasionally common boreal winter visitor to offshore waters, where known only from off Manabí, the Santa Elena Peninsula, and in the Gulf of Guayaquil; it presumably occurs off Esmeraldas as well. The two Feb. specimens obtained by Salvadori and Festa (1900, p. 48) at 1°30′ N "off the Ecuadorian coast" may or may not have been obtained in Ecuadorian waters (that latitude being close to the Colombian border), which means that a single bird (ANSP) taken off the Santa Elena Peninsula on 22 Jan. 1991 may be the only definite Ecuadorian specimen. Least Storm-Petrels have otherwise been seen, occasionally in large numbers, between Dec. and Feb.; B. Haase observed as many as 250 around fishing boats in the Gulf of Guayaquil in early Feb. 1988.

Formerly placed in the monotypic genus *Halocyptena*, and sometimes still so separated from *Oceanodroma* (e.g., in *Birds of the World*, vol. 1). We follow, however, the 1998 AOU Check-list and most recent authors in placing it in in the genus *Oceanodroma*. Monotypic.

Range: breeds on islands off nw. Mexico; nonbreeding birds disperse in e. Pacific Ocean south as far as off s. Ecuador and north to off California.

Oceanodroma tethys   Wedge-rumped Storm-Petrel   Paíño Danzarín

A fairly common visitor to offshore waters of s. Ecuador, where recorded north to off s. Manabí (a few records from the vicinity of Isla de la Plata). The Wedge-rumped Storm-Petrel doubtless also occurs northward, for there are numerous records of the species from the waters off Colombia and Panama. Recorded only from May to Oct., but almost certainly occurs later and probably earlier as well.

The species is considered Near-threatened by Collar et al. (1994), because so little is known of the Peruvian breeding population, and it is so treated here. The Galápagos population is relatively large and presumably secure.

The species is sometimes called the Galápagos Storm-Petrel. We concur with the 1998 AOU Check-list in calling it the Wedge-rumped Storm-Petrel.

One race: *kelsalli*, which nests off Peru. The larger nominate race, which nests in the Galápagos, might also range as far east as mainland Ecuadorian waters.

Range: breeds on a few islands off Peruvian coast and on Galápagos Islands; nonbreeding birds disperse north in e. Pacific Ocean to off Mexico and south to off Chile.

*[Oceanodroma castro   Band-rumped Storm-Petrel   Paíño Lomibandeado]

Uncertain. Apparently a casual visitor to offshore waters in the Gulf of Guayaquil southwest of the Santa Elena Peninsula, probably only well off the coast. Known in Ecuador from only two sightings, both in 1988 when B. Haase spent considerable time offshore on fishing boats: he observed one in Jan. and 17 in Feb. Of note is the fact that no Wedge-rumped Storm-Petrels (*O. tethys*) were seen on these dates, nor are there any Ecuadorian records of that species from these months.

The species is often called either the Madeiran or the Harcourt's Storm-Petrel. We follow the 1998 AOU Check-list in calling it the Band-rumped Storm-Petrel.

Monotypic.

Range: breeds on various islands in tropical Atlantic and Pacific Oceans, including Galápagos Islands; nonbreeding birds disperse widely in warmer oceanic waters.

Oceanodroma markhami   Markham's Storm-Petrel   Paíño de Markham

Apparently an uncommon visitor to offshore waters of s. Ecuador, perhaps mostly (entirely?) during the austral summer months, but status is confused because of this species' similarity to the Black Storm-Petrel (*O. melania*). The three Sep. specimens published as Markham's Storm-Petrel by Chapman (1926, p. 182) were later reidentified as Black Storm-Petrels by Murphy (1936, p. 740), and RSR examined two of these at the AMNH in Sep. 1991. The Markham's Storm-Petrel was subsequently not recorded in Ecuador until the late 1980s, when B. Haase began observing birds off the Santa Elena Peninsula, sometimes in large numbers, which he was confident were this species (and not the closely similar Black Storm-Petrel). The Markham's Storm-Petrel's presence in Ecuadorian waters was confirmed on 22 Jan. 1991 when about 25 very similar birds were seen off the Santa Elena Peninsula; 2 of these were collected (ANSP and MECN). Since then small numbers of what were identified merely as "Markham's/Black Storm-Petrels" have been seen during various months off the Santa Elena Peninsula, but no further examples have been taken; our suspicion, however, is that many or most of these birds have been Black Storm-Petrels.

The Markham's Storm-Petrel is considered to be a Data Deficient species by Collar et al. (1994) because so little is known of its breeding distribution and biology; it is so treated here.

The species was formerly often called the Sooty Storm-Petrel (e.g., Meyer de Schauensee 1966, 1970).

Monotypic.

Range: breeds very locally on coast of Peru (where first documented only in 1987; Collar and Andrew 1988) and (presumably) n. Chile, but distribution still imperfectly known; nonbreeding birds disperse north to off Ecuador, in small numbers as far as Central America and even Mexico.

Oceanodroma melania   Black Storm-Petrel   Paíño Negro

Apparently a fairly common visitor to coastal and offshore waters along the entire Ecuadorian coast but with status somewhat confused because of the presence here of the very similar Markham's Storm-Petrel (*O. markhami*). Recorded at least from Apr. to Nov., but some birds are probably present year-round.

Based on its long legs, formerly placed in the monotypic genus *Loomelania*. Monotypic.

Range: breeds on islands off s. California and nw. Mexico; nonbreeding birds disperse south in e. Pacific Ocean to n. Peru.

*[Oceanodroma homochroa   Ashy Storm-Petrel   Paíño Cinéreo]

Uncertain. Apparently a casual boreal winter visitor to offshore waters. Known in Ecuador from only two sightings. A single bird was observed about 8 km off Isla de la Plata on an unspecified date in Sep. 1980 (P. Harrison); and five birds believed to be this species were seen on an unspecified date in Jan. 1990 off the Santa Elena Peninsula (B. Haase). These represent the only reports of the Ashy Storm-Petrel from anywhere in South American waters, and they come from an area substantially south of the species' normal range off the coasts of California and Mexico. However, the Ashy Storm-Petrel is known to disperse southward when not breeding, so its presence in Ecuadorian waters is not entirely unexpected. Observers should be alert to the possibility of additional sightings; specimen confirmation would be desirable.

The Ashy Storm-Petrel is considered to be Near-threatened by Collar et al. (1994) and is so treated here. Its total population is quite small, though evidently there is little or no evidence of any recent change in numbers.

Monotypic.

Range: breeds on islands off California and nw. Mexico; nonbreeding birds disperse southward in e. Pacific Ocean to at least some extent, apparently some as far as off Ecuador.

Oceanodroma hornbyi   Hornby's Storm-Petrel   Paíño de Hornby

Apparently an irregular visitant to offshore waters in s. Ecuador, with the northernmost record being a specimen (AMNH) obtained off the coast at 1° S (placing it off s. Manabí) on 13 Sep. 1922. Marchant (1958) also records two beach-washed specimens from the Santa Elena Peninsula, but all other records are from the Gulf of Guayaquil. A few observers have found this species numerous here, most recently on 11 Jan. 1984 when a flock of 130 was observed (M. Van Beirs). More typically, however, the Hornby's Storm-Petrel is found in very small numbers or is entirely absent. Recorded only from Aug. to Jan., but perhaps present year-round. Despite the recent upsurge in pelagic activity off the Santa Elena Peninsula by ANSP personnel and B. Haase, the Hornby's Storm-Petrel has not been found there, and indeed the only recent report is the 1984 record mentioned above. Being so distinctive, the species seems unlikely to have been overlooked; one is forced to suspect that a genuine decline in its numbers has occurred.

The Hornby's Storm-Petrel is considered to be Data Deficient by Collar et al. (1994) because nothing is known of its breeding distribution and biology (no nest has ever been discovered); it is so treated here as well. As noted above, we suspect that this species, like so many others found in Humboldt Current waters, may have declined in numbers; we suspect that it could definitely be at risk.

The species is sometimes called the Ringed Storm-Petrel (e.g., Meyer de Schauensee 1966, 1970; Sibley and Monroe 1990). However, we concur with Carboneras (1992) and most other recent books on seabirds in employing the patronym "Hornby's."

Monotypic.

Range: at sea in Humboldt Current waters off Peru and Chile, but actual nesting grounds remain unknown; disperses north to off s. Ecuador, with one record from as far north as Isla Gorgona off Colombia.

Pelecaniformes
Phaethontidae   Tropicbirds

One species. Tropicbirds are wide-ranging in tropical oceans.

Phaethon aethereus   Red-billed Tropicbird   Rabijunco Piquirrojo

A fairly common breeder on Isla de la Plata off the coast of s. Manabí, where about 20 pairs were counted in Jun. 1992 and Jul. 1993 (RSR and F. Sornoza). Based on the small numbers that were recorded here until at least the 1970s, there may have been some increase in this population during recent years, perhaps the result of heightened protection efforts on the island. La Plata represents the southernmost nesting site for the species along the Pacific coast of South America. The Red-billed Tropicbird is pelagic when not nesting and is only rarely seen from the mainland; over the years B. Haase has, however, observed it on a very few occasions from various lookouts in w. Guayas, and one was seen in the bay of Puerto López on 27 Jan. 1997 (O. Jahn). There are also a few sightings from north to off Esmeraldas.

One race: *mesonauta*.

Range: breeds locally in tropical Atlantic, e. Pacific, and n. Indian Oceans; nonbreeders disperse widely across tropical seas.

Fregatidae   Frigatebirds

Two species in one genus. Frigatebirds are found widely in tropical oceans.

Fregata magnificens   Magnificent Frigatebird   Fragata Magnífica

Common and widespread on coastal (mainly) and offshore waters along the entire coast, with numbers largest in Manabí (especially in the Bahía de Caráquez and the south) and Guayas (especially around the Santa Elena Peninsula and in the Gulf of Guayaquil); numbers drop off sharply southward along the El Oro coast. Most numerous along the immediate coast, where flocks congregate around fishing trawlers, in various harbors where fishing boats are offloaded, and at roosting sites; numbers drop off rapidly on the open ocean, with only the odd individual being seen well beyond the sight of land. There is also one amazing report away from the coast, a female seen flying high above the east slope of the Andes along the Gualaceo-Limón road (1900 m) on 12 Aug. 1990 (J. Arvin and J. Rowlett et al.); another was seen flying up the Río Santiago near Playa de Oro (some 45 km inland) on 28 Nov. 1995 (O. Jahn et al.). The Magnificent Frigatebird is known to breed on both Isla de la Plata off s. Manabí and on Isla Santa Clara in the Gulf of Guayaquil; these colonies represent the species' southernmost nesting sites along the Pacific coast of South America. On Isla Santa Clara an estimated minimum of 1000 pairs was completing their nesting cycle on 5 Jun. 1993 (RSR et al.), whereas on Isla de la Plata hundreds of pairs, perhaps more, nest during most years (RSR et al.); 2598 birds were believed to be present in Jul.–Aug. 1990 (F. Ortiz-Crespo and P. Agnew, *Bull. B. O. C.* 112[2]: 69–70, 1992). Smaller numbers perhaps also nest on islets elsewhere (e.g., off the coast of s. Manabí near Puerto López), but this has never been confirmed. Roosting birds are found in many areas, often on small mangrove-covered islands, where—despite the presence of ballooning males—breeding does not take place. A large roost was found, for instance, in mangroves at the mouth of the Río Verde in n. Esmeraldas, where in 1995–1996 some 3000 frigatebirds (with about 500 pelicans) were counted; this roost was abandoned during the second half of 1997, probably because of the strong effect of El Niño that year (O. Jahn et al.).

The species is generally considered to be monotypic (see Blake 1977).

Range: breeds on islands and coasts in tropical Atlantic and e. Pacific Oceans.

*[*Fregata minor*   Great Frigatebird]   Fragata Grande

Uncertain. Apparently an accidental visitant to the coast of w. Guayas. There is only one report, an immature seen on the coast of w. Guayas at Olón on 31 May 1987 (B. Haase). The Great Frigatebird could conceivably be being overlooked among the often large numbers of Magnificent Frigatebirds (*F. magnificens*) that range along the coast, but the Great is obviously at best only a very infrequent wanderer to the Ecuadorian mainland. Its closest nesting site is on the Galápagos Islands.

Presumed race: *ridgwayi*, the subspecies occurring on the Galápagos Islands.

Range: breeds on numerous tropical islands, especially in Pacific Ocean.

Sulidae   Boobies and Gannets

Five species in one genus. Boobies and gannets are wide-ranging on oceans, the boobies at more tropical latitudes and the gannets to the north and south.

Sula nebouxii   Blue-footed Booby   Piquero Patiazul

Locally fairly common to common in offshore waters along the entire coast, with largest numbers along the coasts off sw. Manabí and w. Guayas, fewer off El Oro and Esmeraldas. Roosting birds favor rocky islands. Only two major breeding colonies of the Blue-footed Booby are known from mainland Ecuadorian waters, these being on Isla de la Plata off Manabí and Isla Santa Clara in the Gulf of Guayaquil; 1000 or more pairs are present on both islands during most seasons (but when rains are heavy and as a result vegetation luxuriant, few or no birds may nest). A few small nesting colonies on islets along the s. Manabí coast have also been reported, but whether these are only roosting sites remains unknown. Numbers apparently fluctuate seasonally, being highest during Apr.–Oct. (B. Haase; RSR); substantial numbers may starve and die during periods of relatively warm water (such as occurred during the latter part of 1997). An extraordinary inland record involves a specimen (FMNH) taken by the Olallas at Isla Silva, along the Río Babahoyo in n. Guayas near Babahoyo, in Sep. 1931.

That at least occasionally birds nesting on the Galápagos Islands range east to waters off the mainland coast of Ecuador is borne out by the recovery on 5 Aug. 1973 of a banded individual on the Esmeraldas coast (M. P. Harris, *The Ring* 7[79]: 133, 1974).

One race: nominate *nebouxii*.

Range: locally along Pacific coast from n. Mexico to cent. Peru, and on Galápagos Islands.

Sula variegata   Peruvian Booby   Piquero Peruano

An irregular visitor to coastal and offshore waters along the coast of s. Ecuador, ranging north principally to w. Guayas and s. Manabí. There are only a few reports from north of the Santa Elena Peninsula area, usually of single individuals among larger numbers of Blue-footed Boobies (*S. nebouxii*), with the northernmost site being the Atacames/Súa area in w. Esmeraldas (v.o.). The Peruvian Booby is has occasionally been numerous, particularly around the Santa Elena Peninsula and especially during and after warm-water El Niño events along the coast of Peru; in Nov. 1997, however, despite a strong El Niño event, none could be found among the unusually large numbers of Blue-footed Boobies at various sites on the Santa Elena Peninsula (RSR). The most recent major incursion of Peruvian Boobies occurred in 1983, and some were also reported seen north to the Isla de la Plata area in Jul.–Aug. 1996 (S. Howell and R. Behrstock et al.). During most years, however, no Peruvian Boobies at all are present in Ecuadorian waters, or at most there is a scatter of single birds.

Although the Peruvian Booby is not considered to be at risk globally (Collar et al. 1992, 1994), its total world population is now much reduced as a result of disturbance on nesting colonies and of human overfishing of the boobies' primary prey. Repeated and severe El Niño events in recent decades have also had a substantial effect, and numbers have not rebounded well during years that have been more normal. Although the species is not yet truly rare, we feel it should be considered Near-threatened.

Monotypic.

Range: coast of Peru and n. Chile, irregularly erupting north to Ecuador (on rare occasions even to Panama).

Sula granti   Nazca Booby   Piquero de Nazca

A common breeder on Isla de La Plata off the coast of s. Manabí, where almost 400 pairs were estimated to be nesting in Jul. 1993 (RSR and F. Sornoza). Over the past several decades this population appears to have gradually increased in numbers, this subsequent to some persecution (see O. T. Owre, *Ibis* 118: 419–420, 1976). Given the island's present protected status as part of Machalilla National Park, the population there could well continue to grow. The Nazca Booby is primarily pelagic when not nesting, dispersing solitarily or in very small groups out over the open ocean; a few are regularly to be found, for example, on the waters well off the Santa Elena Peninsula, but the species is not definitely known from Ecuadorian waters south of there. Peruvian records all appear to refer to the Masked Booby (*S. dactylatra*), now considered a separate species (see below). The Nazca Booby is only very infrequently seen from the mainland. Nonetheless there is a specimen (FMNH, fide *Birds of the Americas*, pt. 1, no. 2) labeled as having been taken at Silva Isla along the Río Babahoyo in inland Guayas; it would seem, however, surely to have been mislabeled.

S. granti is regarded as a monotypic species distinct from *S. dactylatra*, following R. L. Pitman and J. R. Jehl, Jr. (*Wilson Bull.* 110[2]: 155–170, 1998), this based especially on assortative mating on the few islands where both species nest together. *S. granti* has an orange bill, whereas that of *S. dactylatra* is greenish yellow. There do not appear to be any records of *S. dactylatra* from Ecuadorian mainland waters, though it is possible that a few may occur. The English name of Nazca Booby for *S. granti* was suggested by Pitman and Jehl (op. cit.).

Range: nests mainly on Galápagos Islands and Malpelo Island off Colombia, with smaller numbers off Ecuador (Isla de la Plata) and off w. Mexico; disperses locally in e. Pacific Ocean.

[*Sula leucogaster*   Brown Booby   Piquero Pardo]

Uncertain. Probably a very rare visitor to offshore waters along the coast of n. Ecuador in Esmeraldas, but apparently there are no definite records; very little pelagic work has been done in this oceanic sector, and this may account for the lack of definite records. There exists one published mention of the Brown Booby from Ecuador: Murphy (1936, p. 860) writes that "Cape San Francisco, Ecuador, probably marks the southern bounds of its distribution [on the Pacific coast of South America]." Unfortunately, Murphy does not mention the actual basis for this statement. In fact, he elsewhere (op. cit., p. 313) specifically states that "it has not been recorded farther southward" than west of Tumaco in s. Colombia, where he observed a single individual on 13 Mar. 1925. Chapman (1926) refers to this sighting as having taken place in Ecuadorian waters. We thus cannot be certain precisely where Murphy observed the bird. However, as the Brown Booby nests as close to Ecuador as Malpelo Island off the coast of s. Colombia, its occurrence in Ecuadorian waters remains likely. We consider that the sighting of E. L. Mills (*Ibis* 109: 535, 1967) from much farther south at Isla Pelado, just north of La Libertad on the Santa Elena Peninsula, surely refers only to juveniles of the Blue-footed Booby (*S. nebouxii*), a species numerous there.

Presumed race: *etesiaca*.

Range: breeds widely on islands in tropical oceans, dispersing mainly quite close to its nesting islands.

Sula sula   Red-footed Booby   Piquero Patirrojo

Recently found to be an uncommon breeder on Isla de la Plata off the coast of s. Manabí, where a few pairs have been nesting in a small colony since at least 1989 (fide B. Haase and C. Zambrano). In 1989 eight pairs were counted, and in Jul. 1993 eleven nests with incubating adults plus an additional two recently fledged juveniles were noted (RSR and F. Sornoza). It would appear that the Red-footed Booby has only recently colonized Isla de la Plata; at least we can only presume that such a conspicuous species would not have been missed by earlier observers. All birds present appear to be of the brown morph. Isla de la Plata represents the southernmost nesting site for the species along the Pacific coast of South America, and is presumably the source for the occasional sightings from the Santa Elena Peninsula (B. Haase) and elsewhere off Manabí (R. G. B. Brown). The Red-footed Booby is entirely pelagic when not nesting and is unlikely to be seen from the mainland.

Presumed race: *websteri*.

Range: breeds widely on islands in tropical oceans, dispersing extensively.

Phalacrocoracidae   Cormorants and Shags

Two species in one genus. Cormorants are wide-ranging along coasts and on freshwater.

Phalacrocorax brasilianus   Neotropic Cormorant   Cormorán Neotropical

Rare to uncommon and local along rivers and on certain lakes in the lowlands of e. Ecuador. In w. Ecuador more widespread and much more numerous in similar habitats, and also occurring on salt water along the coast and in lagoons.

Neotropic Cormorants are most numerous in the lower Río Guayas drainage and its estuary, where flocks in the hundreds and even in the low thousands are sometimes encountered; in some places, particularly around shrimp ponds, cormorants are even considered to be pests, but fortunately at least at present this species does not seem to be much persecuted. Breeding colonies are rarely reported in Ecuador, where we have never encountered one. Recorded mainly below about 800 m, individuals occasionally wandering somewhat higher, especially along rivers; somewhat surprisingly, there seem to be no modern reports from lakes in the central valley, though there is an old record from Yaguarcocha in Imbabura given by Lönnberg and Rendahl (1922).

The species was formerly called the Olivaceous Cormorant.

The species was formerly called *P. olivaceus*, but M. R. Browning (*Wilson Bull.* 110[1]: 101–106, 1989) determined that the name *brasilianus* was applicable, and the latter name was used in the 1998 AOU Check-list. D. Siegel-Causey (*Condor* 90[4]: 885–905, 1988) suggests separating this species and several others in the genus *Hypoleucos*. One race: nominate *brasilianus*.

Range: sw. United States to s. Chile and Argentina; Cuba.

Phalacrocorax bougainvillii　　　**Guanay Shag**　　　**Cormorán Guanero**

A highly irregular visitant to coastal waters along the coast of s. Ecuador, where it has been recorded north at least to w. Guayas (Santa Elena Peninsula), but as there are records from Colombia and Panama it doubtless also occurs at least occasionally farther north in Ecuador. On very rare occasions Guanay Shags have even moved up larger rivers such as the Río Guayas (Murphy 1936). The Guanay Shag has occasionally been briefly numerous in Ecuadorian waters, with its incursions mainly seeming to be tied to warm-water El Niño events along the coast of Peru, where the species breeds in large numbers in the usually cold waters of the Humboldt Current. In the vast majority of years, however, no Guanay Shags at all are present in Ecuadorian waters; the only report subsequent to 1987 is a single individual seen at La Libertad on 23 Nov. 1997 (RSR), during a period when a strong El Niño event was occurring.

The size of the overall Guanay Shag population is now much smaller than it was formerly as a result of human exploitation of the cormorant's primary food, a small fish called the "anchoveta." Repeated severe El Niño events have also caused breeding failures. The species is still thought to be perhaps the world's most numerous cormorant (Enticott and Tipling 1997), but given the magnitude of the decline in its numbers over recent decades, we still favor according it Near-threatened status.

The species was formerly called the Guanay Cormorant, but D. Siegel-Causey (*Condor* 90[4]: 885–905, 1988) presented evidence showing that it is more closely allied to several other cormorant species that are generally known as "shags." The species is also sometimes called simply the Guanay.

Siegel-Causey (op. cit.) suggests separating this species and several other shags in the genus *Leucocarbo*. Monotypic.

Range: breeds along coasts of Peru and n. Chile (also a small population in s. Argentina), occasionally erupting north to Ecuador and rarely even beyond.

Contra some references (e.g., Enticott and Tipling 1997), there is no evidence that the Red-legged Shag (*Phalacrocorax gaimardi*) has reached Ecuadorian waters, even during strong El Niño events.

Anhingidae   Darters

One species. Darters (as the Old World species are called) and anhingas are found widely on freshwater ponds and rivers.

Anhinga anhinga   Anhinga   Aninga

Rare to locally fairly common along rivers and around freshwater ponds and lakes in the lowlands of e. Ecuador; rare and local in the lowlands of w. Ecuador in n. Esmeraldas (along the Río Santiago from Selva Alegre downstream, possibly breeding downstream in Cayapas-Mataje Ecological Reserve; O. Jahn et al.) and from Manabí and s. Pichincha (Río Palenque) south to se. Guayas (Manglares-Churute Ecological Reserve). It is suspected that over-hunting may have reduced numbers in some eastern areas (e.g., around Jatun Sacha; B. Bochan). Seasonal movements have also been suspected, for at some sites the species does not seem to be resident (PJG). Recorded almost entirely below about 400 m, but there is also an anomalous record of one obtained at the La Carolina marshes in Quito (2800 m) on 15 May 1900 (Lönnberg and Rendahl 1992) which, if not incorrectly labeled, would represent a remarkable record.

One race: nominate *anhinga*.

Range: se. United States and Mexico to sw. Ecuador, n. Argentina, Uruguay, and s. Brazil; Cuba.

Pelecanidae   Pelicans

Two species in one genus. Pelicans are found widely along coasts as well as on lakes and ponds; Ecuador's two species are exclusively coastal.

Pelecanus occidentalis   Brown Pelican   Pelícano Pardo

Common and widespread in coastal waters, bays, and harbors along the entire coast, with numbers perhaps largest in w. Guayas on the Santa Elena Peninsula and in the Gulf of Guayaquil. Numbers decrease rapidly as one proceeds out to sea, and only rarely is the species seen more than about 10 km out;

Peruvian Pelicans (*P. thagus*) predominate in offshore waters. The Brown Pelican is rather strictly tied to salt water, though very occasionally the odd individual will wander up a river or be found on a freshwater lake near the coast. Unlike Peruvian Pelican's, Browns frequently roost in trees, often in mangroves. Most Brown Pelicans occurring in Ecuador are apparently visitants from populations that must nest elsewhere, for the sole known breeding colony continues to be found on Isla Santa Clara in the Gulf of Guayaquil, as was also the case early in the 20th century (Chapman 1926). At least 500 pairs were observed to be early in the nesting cycle on Isla Santa Clara on 5 Jun. 1993 (RSR et al.). It remains possible that pelican colonies exist elsewhere, but if this is the case they have not been reported, and confusion with what are merely roosting sites is likely; it is curious that this species does not nest on the seemingly suitable Isla de la Plata. Isla Santa Clara is the Brown Pelican's southernmost nesting colony on the Pacific coast (when the Peruvian Pelican is separated specifically).

One race: *murphyi*.

Range: coasts of s. United States to Ecuador and n. Brazil; Galápagos Islands; West Indies.

Pelecanus thagus   Peruvian Pelican   Pelícano Peruano

Recently found to be an uncommon to locally fairly common austral visitant to coastal and offshore waters along the coast of s. Ecuador, where recorded in largest numbers around the Santa Elena Peninsula, in particular at the Ecuasal lagoons, where hundreds (up to 500+) sometimes roost on inaccessible dikes and sandy islands, and in the harbor at La Libertad. They seem to be most numerous from May through Sep., with smaller numbers being in evidence during other months; from Nov. to Feb. sometimes none at all are present, though large numbers (many hundreds) were present on 23 Nov. 1997, during a potent El Niño event. Peruvian Pelicans only rarely venture north of w. Guayas in the Monteverde/Olón region, with the northernmost report being a single adult seen on Isla de la Plata on 16–17 Jul. 1993 (RSR and F. Sornoza); there are also band recoveries from as far north as near the equator. The Peruvian Pelican seems to avoid the shallow waters of the Río Guayas estuary, though small numbers are seen regularly around Playas, and about 150 (mainly subadult or younger) were found on Isla Santa Clara on 5 Jun. 1993 (RSR et al.). This species often congregates around fishing boats in offshore waters and is much more frequent at and beyond the sight of land than is the Brown Pelican (*P. occidentalis*). Although it may have been overlooked until recently (see Marchant 1958), the absence of early records may in part simply be due to the Peruvian Pelican not being distinguished from resident populations of the Brown Pelican. Aside from banding recoveries (see below), the Peruvian Pelican was first definitely recorded in Ecuador on 3 Sep. 1983 when about 35 were seen at Salinas (PJG et al.). Numbers appear to have increased in recent years, which is curious given the general decline in the numbers of most Humboldt Current seabirds.

Eighteen Peruvian Pelicans banded at their natal colonies in cen. Peru have been recovered in Ecuador, mostly between Apr. and Aug. Almost all birds recovered were less than one year old, which is in accord with the relative proportion (about 1:10) of adults to juveniles seen in the field.

P. thagus is regarded as a monotypic species separate from the much smaller *P. occidentalis*, following most recent authors.

Range: breeds on islands along coast from n. Peru to cen. Chile, dispersing north regularly to s. Ecuador.

Anseriformes
Anhimidae   Screamers

One species. Screamers are exclusively South American in distribution.

Anhima cornuta   **Horned Screamer**   **Gritador Unicornio**

Rare and local in marshes and around the margins of lakes, occasionally also on sandbars along rivers, in the lowlands of e. and sw. Ecuador. Horned Screamers now occur principally in the far east, especially in the lower Ríos Aguarico and Lagartococha region as well as near the lower Río Pastaza (where they remain, for example, reasonably numerous around Kapawi Lodge); populations may also still be found in other remote areas of the southeast. A pair remains more or less in residence at La Selva's well-protected Mandicocha, but screamers disappeared from Limoncocha around 1980 (D. Pearson) and from along the lower Río Jivino apparently by the late 1980s (P. Coopmans). In the southwest the Horned Screamer was known formerly from the floodplain of the lower Río Guayas where it remained in the Yaguachi marshes at least into the mid-1970s (RSR); there also is a 19th-century specimen from as far north as Balzar in n. Guayas. Horned Screamers presently occur in the large marsh at Manglares-Churute Ecological Reserve in se. Guayas. Here, according to reserve guards, the population was censused at 52 birds in 1988, but (fide B. Best et al.) by early 1996 it was up to 113 birds. Possibly correlated with this increase, a single calling screamer was also seen in marshes north of Santa Rosa in coastal El Oro on an unspecified date in Jul. 1994 (Green 1996). The population in sw. Ecuador was long thought to be unique on South America's Pacific slope, but screamers have also recently been found at Laguna del Trueno in w. Nariño, Colombia (Salaman 1994). Recorded below about 300 m.

The spectacular Horned Screamer is now much reduced in overall numbers and range in Ecuador, a reduction that seems to be primarily the result of hunting pressure. In the west, marshland conversion (to rice fields) may have also played a role. We conclude that the species merits Endangered status in

Ecuador; elsewhere in its broad range, however, it is, not considered to be at risk (Collar et al. 1994), though numbers have certainly been reduced locally in many other areas as well.

Monotypic.

Range: very locally in nw. Venezuela, w. Colombia, and sw. Ecuador; e. Colombia, much of Venezuela, and the Guianas to n. and e. Bolivia and s. Brazil.

Anatidae   Ducks, Geese, and Swans

Sixteen species in ten genera. The family is found widely on and around water on all continents, with the geese and swans being found primarily at temperate latitudes.

Dendrocygna bicolor   Fulvous Whistling-Duck   Pato Silbador Canelo

Locally or seasonally fairly common to common in freshwater marshes and on ponds (sometimes also in rice fields and occasionally even on coastal tidal flats) in the lowlands of w. Ecuador from w. Esmeraldas (sightings south of Atacames and Muisne in Sep. 1990 [D. Pearson; PJG]) south locally to coastal El Oro (south at least occasionally to around Arenillas and Puerto Pitahaya, e.g., small numbers seen here on 2–3 Apr. 1989; RSR and PJG et al.); numbers are highest in parts of Manabí, Los Ríos, and Guayas. Although various authors since Chapman (1926, p. 210) have claimed that the Fulvous Whistling-Duck occurred in both e. and w. Ecuador, in fact there are no records from e. Ecuador, nor indeed does the species occur regularly anywhere in Amazonia. Fulvous Whistling-Ducks clearly move about in response to local water conditions, and when water levels are generally low they may occur in very large aggregations where water persists; the amazing total of 30,000 was estimated to be present at Manglares-Churute Ecological Reserve on 12 Aug. 1990 (S. Hilty et al.). Recorded only below about 100 m.

Unlike many other larger birds that were formerly found in large numbers in w. Ecuador but now are scarce, the Fulvous Whistling-Duck does not appear to have declined appreciably in numbers.

The species was formerly often called the Fulvous Tree-Duck.

Monotypic.

Range: s. United States south locally to n. Argentina (but absent from a wide area across Amazonia and Brazil); Greater Antilles; also Africa and s. Asia.

Dendrocygna autumnalis   Black-bellied   Pato Silbador
                                Whistling-Duck   Aliblanco

Locally or seasonally fairly common to common in freshwater marshes and on ponds, also locally in mangroves (in which it sometimes roosts), in the lowlands of w. Ecuador from Manabí (Bahía de Caráquez region) south locally to coastal El Oro (south at least occasionally to around Huaquillas) and sw. Loja (4 seen southwest of Zapotillo on 13 Feb. 1999; P. Coopmans and T. Gullick). There are also two sightings from e. Ecuador, a single bird seen at Limoncocha on 19 Jan. 1972 (D. Pearson, *Condor* 77[1]: 97, 1975) and another singleton seen near La Selva on 21 Feb. 1994 (M. Lysinger et al.). In Ecuador the Black-bellied Whistling-Duck does not seem to gather in the extremely large flocks sometimes noted for the Fulvous Whistling-Duck (*D. bicolor*); occasionally they congregate together. Recorded mostly below 200 m, but there is also an old record (Lönnberg and Rendahl 1922) of a presumed wandering bird from the La Carolina marshes near Quito (2800 m).

The species was formerly often called the Black-bellied Tree-Duck.

One race: *fulgens*.

Range: sw. United States to n. Argentina, Paraguay, and s. Brazil, but essentially absent from n. and w. Amazonia.

Neochen jubata   Orinoco Goose   Ganso del Orinoco

Very rare on sandbars along major rivers in the lowlands of most of e. Ecuador, but still reasonably numerous along the Río Pastaza near the Peruvian frontier (e.g., around Kapawi Lodge). All Ecuadorian records of the Orinoco Goose are relatively recent, though it seems virtually certain that long ago the species was more numerous and widespread than it is at present. It was first formally recorded from two birds collected in se. Pastaza at Capitán Chiriboga, near the mouth of the Río Capaguary into the Río Pastaza, on 15 Dec. 1973, with another taken on 7 Nov. 1974 (Orcés 1974). They species was noted as being "not rare" at that time, and (as noted above) continues to be present in this region, though how far upstream on the Río Pastaza it occurs remains unknown; the indigenous Achuar Indians apparently consider the species sacred and therefore do not molest it (fide N. Krabbe). Despite an abundance of clearly suitable habitat in ne. Ecuador, the Orinoco Goose does not appear presently to be resident anywhere along the Ríos Napo or Aguarico, presumably because of the area's high human population density and because the species is too conspicuous and easy to hunt. The few birds that have been seen here seem to represent wandering birds. Some of the more recent records from this region involve a pair seen on the Río Napo near San Carlos on 19 Nov. 1990 (P. Coopmans et al.), with another individual being seen along the lower Río Aguarico near Zancudococha on several dates in Nov. 1991 (H. Kasteleijn and P. Hartley). Recorded below 300 m.

The Orinoco Goose was considered to be a Near-threatened species by Collar et al. (1994). As noted above, however, the species' current situation in Ecuador is considerably less secure, and we consider it as meriting Endangered status here. Its protection along the Río Pastaza should be a high priority, as should surveys to determine the size of the population there. At present no population occurs within any formally protected area.

The species has sometimes been called the Orinoco Sheldgoose.

Monotypic.

Range: locally from e. Colombia and cen. Venezuela to nw. Argentina, w. Paraguay, and Amazonian Brazil.

Cairina moschata   Muscovy Duck   Pato Real

Rare to locally uncommon along remote rivers, lakeshores, and in marshes in the lowlands of e. and sw. Ecuador. Although the Muscovy may once have been more widespread and numerous, there are records from surprisingly few specific Ecuadorian localities. In the east, Muscovies have been found mainly in the north, with recent reports coming primarily from the Imuyacocha region and Cuyabeno. There are also recent reports of it from around Kapawi Lodge on the Río Pastaza near the Peruvian border, where it appears to be more numerous than elsewhere in Ecuador; it likely also ranges elsewhere in the Río Pastaza drainage, and there is even an old record (Salvadori and Festa 1900) from the Río Bomboiza, near Gualaquiza in Morona-Santiago. Muscovies formerly occurred at Limoncocha, but the last records from there date back to the 1970s. In the west the Muscovy has been found mainly in Guayas, where there are records from the Yaguachi marshes (including a few sightings from as late as the mid-1980s; PJG) and Manglares-Churute Ecological Reserve (where it is known from only a single sighting, a surprising count of about 25 birds seen on 22 Mar. 1993; J. C. Matheus et al.). There are also early-20th-century sightings from Santa Rosa in coastal El Oro (Chapman 1926, p. 209), and even reports with no details from as far north as Río Palenque in extreme s. Pichincha. Throughout its range the possibility of confusion with domesticated Muscovy Ducks is a potential problem. Recorded only below about 300 m.

The Muscovy Duck's numbers in Ecuador have declined substantially in recent decades, and its range has contracted, almost entirely as a result of intense hunting pressure. The species continues to be much persecuted in most areas. We therefore consider its Ecuadorian status to be Vulnerable. Elsewhere in its large range the species can be locally numerous and is not considered to be at risk (Collar et al. 1994). With improved protection, Ecuadorian populations could rebound as well.

Monotypic.

Range: n. Mexico to sw. Ecuador, n. Argentina, and Uruguay.

Sarkidiornis melanotos   Comb Duck   Pato Crestudo

Apparently a rare and local resident in the lowlands of sw. Ecuador, with—surprisingly—all records being relatively recent; it is even conceivable that numbers may be increasing. The Comb Duck was apparently first recorded from Ecuador by Norton et al. (1972) on the basis of two birds (MCZ) that were shot and preserved by a hunter on the slopes of Volcán Cayambe in Feb. 1951 and Feb. 1952. This anomalous record has not been repeated, anywhere in the highlands. However, there have been several recent reports of Comb Ducks from Manabí, Guayas, and Loja lowlands; here they have principally been found along rivers and in small marshy areas with rice cultivation. They have also once been reported roosting in mangroves (a behavior unreported from anywhere else in South America) by F. Ortiz-Crespo (*Bull. B. O. C.* 108[3]: 141–144, 1988), who described seeing several flocks of birds in Manglares-Churute Ecological Reserve on 5 Nov. 1987. In Feb.–Mar. 1991, "small numbers" were seen along the Río Sabiango in extreme s. Loja west of Macará (Best et al. 1993, p. 167), and on 3 Mar. 1994 a singleton and a pair were seen in two areas near Macará (P. Coopmans). On 20–24 Apr. 1993 a group of 5 apparently young birds was located by RSR et al. north of Zapotillo, also in extreme s. Loja (ANSP, MECN); a flock of 16 was seen southwest of Zapotillo and 3 at Macará on 13 Feb. 1999 (P. Coopmans and T. Gullick); and 2 were seen along the Río Chira at Lalamor on 16 Sep. 1998 (RSR et al.). It thus seems that a small population is resident in extreme s. Loja. Lastly, two female-plumaged birds were seen at a lake surrounded by rice cultivation near Chone in Manabí on 30 Jul. 1996 (S. Howell and R. Behrstock et al.). Recorded mainly below about 500 m (though a few individuals have been known to wander higher).

Given its small range and population in Ecuador, we consider the Comb Duck to warrant Vulnerable status here. It is perhaps more threatened by hunting than by any other factor. Larger and nonthreatened populations occur elsewhere in South America, and from an overall standpoint the species has therefore not been considered to be at risk (Collar et al. 1994).

One race: *sylvicola*. The rather distinct American form *sylvicola* has sometimes been accorded specific rank separate from Old World populations (e.g., Livezey 1997). It would then be called either the American Comb-Duck or (Livezey 1997) the Black-sided Comb-Duck.

Range: locally from e. Panama to n. Argentina, Uruguay, and se. Brazil, but essentially absent from much of Amazonia; also in Africa and s. Asia.

Merganetta armata   **Torrent Duck**   Pato Torrentero

Uncommon to locally fairly common along swift-flowing, rocky rivers and larger streams from the foothills up into the temperate zone on both slopes of the Andes, and also locally (where water conditions permit) in the central and interandean valleys. On the west slope the Torrent Duck has seemingly not been reported from farther south than Cotopaxi, though there would appear to be no reason why it should not extend south of there. Recorded from 700 m— it is regular this low along the Río Toachi west of Tinalandia—up to at least 3200 m.

Numbers of the Torrent Duck appear to be holding up reasonably well in Ecuador, though localized reductions have doubtless occurred near larger towns and cities, especially in the central valley. Torrent Ducks are not persecuted to any great extent, and they can occur more or less independently of forest cover, so declines are mostly the result of pollution.

One race: *colombiana*. A recently obtained specimen (ANSP) from the Río Isimanchi in extreme s. Zamora-Chinchipe near the Peruvian border apparently extends the known range of this subspecies southward. *Colombiana* had been recorded south only to "central Ecuador" (*Birds of the World*, vol. 1), and it even had been supposed that the Peruvian race *leucogenis* occurred north into Ecuador (Madge and Burn 1988, p. 179), though this is evidently not the case.

Range: Andes from w. Venezuela to Tierra del Fuego.

Anas andium   **Andean Teal**   **Cerceta Andina**

Locally fairly common on lakes and ponds in paramo from Carchi (sightings from Laguna Verde; N. Krabbe) south to Azuay (El Cajas and Bestión); also recently found in extreme s. Loja (Cordillera Las Lagunillas, where recorded in Oct. 1992; ANSP). The Andean Teal seems likely to occur in adjacent extreme n. Peru in the Cerro Chinguela area, though it has not been found there (Parker et al. 1985). Recorded mostly from 3000 to 4000 m.

   A. andium is here regarded as a separate species from *A. flavirostris* (Speckled Teal) of n.-cen. Peru to Tierra del Fuego and s. Brazil because of the striking difference in its bill color and other plumage differences. A teal with "an obviously yellow bill" was described as having been seen on "paramo lakes" above Mazán on 26 Oct. 1987 (King 1989, p. 143). Whether this individual was a wandering *A. flavirostris* is uncertain; to our knowledge, no intermediates between *A. a. andium* and *A. flavirostris oxyptera* (the northernmost-ranging race of *A. flavirostris*) are known. One race: nominate *andium*.

Range: Andes from w. Venezuela to s. Ecuador.

Anas bahamensis   **White-cheeked Pintail**   **Anade Cariblanco**

Locally fairly common to common on ponds and lagoons near the coast, particularly on artificial ponds created for salt production and shrimp cultivation, also on freshwater ponds and lakes, in the lowlands of w. Ecuador from w. Esmeraldas (sightings since at least 1990 in the Atacames area; v.o.) south locally to El Oro (numerous sightings since at least 1979 from Santa Rosa south to Huaquillas; v.o.). The White-cheeked Pintail was first found in Ecuador on the Santa Elena Peninsula as recently as the 1950s (Marchant 1958), and its range and population have increased substantially since then, an increase that has doubtless been greatly abetted by the recent construction of so many artificial ponds. Numbers are still always highest on the Santa Elena Peninsula, where flocks of up to 2000–4000 birds have been seen at various times on the Ecuasal lagoons and on the Represa José Velasco Ibarra (RSR). Recorded only below about 50 m.

   The species was formerly often called the Bahama Pintail.

   One race: *rubrirostris*.

Range: along Caribbean coast from n. Colombia to the Guianas; along Pacific coast from nw. Ecuador to extreme n. Chile, and in cen. Chile; locally from s. Bolivia and w. Paraguay to cen. Argentina (wandering farther south), Uruguay, and extreme se. Brazil; locally in ne. Brazil; Galápagos Islands and locally in West Indies.

Anas georgica   **Yellow-billed Pintail**   **Anade Piquiamarillo**

Locally uncommon to fairly common on lakes and ponds in paramo and the central valley from Carchi and Imbabura south to Azuay (El Cajas); there also is a single record (FMNH, fide *Birds of the Americas*, pt. 1, no. 2) from considerably farther south, at "Nudo de Sabanilla" on the Loja/ Zamora-Chinchipe border. The Yellow-billed Pintail now occurs mainly in paramo and is very local in the central valley because of drainage and excessive disturbance at most sites; formerly it likely was more widespread and much more numerous in highland valleys. It does, however, remain relatively numerous at Laguna Yaguarcocha in Imbabura. Recorded mostly from 2200 to 4000 m.

One race: *spinicauda. Spinicauda* of the South American mainland is sometimes regarded as a species, *A. spinicauda*, separate from the population isolated on South Georgia Island in the South Atlantic, which is then called *A. georgica*, South Georgia Pintail. *A. spinicauda* of the mainland would include the apparently extinct *niceforoi* of Colombia as a race. *A. georgica* and *A. spinicauda* were treated as separate species as recently as 1948 in *Birds of the Americas* (pt. 1, no. 2), but in subsequent works (e.g., Meyer de Schauensee 1966, 1970) were considered to be conspecific. We are not aware of any published justification for this merger, but in the absence of any new information we continue to lump the two forms. *Georgica* is smaller, darker, shorter tailed, and females have dark bills.

Range: locally in Andes of Colombia and Ecuador, and from cen. Peru south to Tierra del Fuego, spreading east across Argentinian lowlands to Uruguay and extreme se. Brazil (a few wanderers northward).

Anas discors   **Blue-winged Teal**   **Cerceta Aliazul**

Locally a fairly common to common boreal winter visitant to lakes, ponds, and marshes—though always unusual on salt water (generally only as a temporary transient)—in the lowlands of w. Ecuador, and to lakes and marshy areas in the central valley and paramo of n. and cen. Ecuador (south to Chimborazo, with an old record from as far south as Azuay at Laguna de Kingora, near Sígsig); there also are a few reports from the lowlands of ne. Ecuador in Sucumbíos and Napo. In the west recorded south only to Guayas, but a few likely occur south through coastal El Oro as well. Recorded mostly from Oct. to Apr., but there are also several reports, almost always of single individuals, from the northern summer months, these probably involving sick or wounded birds that can no longer migrate. Blue-winged Teal numbers are greatest in the western lowlands, where congregations of several thousand birds are occasionally encountered; the largest numbers found in recent years have been on the Represa José Velasco Ibarra on the Santa Elena Peninsula. Recorded up to at least 3200 m.

One race: nominate (though *orphna* could also occur). The species is, however, often considered to be monotypic.

Range: breeds in North America, wintering from s. United States south to Ecuador and the Guianas (also in West Indies), a few straggling much farther south (the odd individual as far as s. Argentina).

Anas cyanoptera   Cinnamon Teal   Cerceta Colorada

Uncertain, with two subspecies having been recorded but neither of them occurring at present: the resident *borreroi* and the boreal migrant *septentrionalium*. *Borreroi* occurred very locally on lakes in the central valley of n. Ecuador. *Septentrionalium* occurred locally on lakes in the central valley of n. Ecuador in Imbabura, Pichincha, and Cotopaxi. Dated records attributable to boreal migrants range from Dec. to Apr. *Borreroi* is found primarily in the Colombian Andes and is known in Ecuador only from a series of six specimens with no data (now in ZMCOP) which were obtained very early in the 20th century (J. Fjeldså, *Neotropical Ornithology*, Ornithol. Monogr. no. 36, 1985); these were believed to have been obtained in the Ibarra area of Imbabura (J. Fjeldså, pers. comm.). Why northern migrant *septentrionalium*—which remains numerous on its breeding grounds in w. North America—should have apparently ceased to migrate as far south as Ecuador (as well as Colombia and even Panama) remains an enigma. Migrant Cinnamon Teal might as likely occur in the western lowlands as in the highlands, for numbers of Blue-winged Teal (*A. discors*) are now usually greater there than on highland lakes. Recorded only from about 2500 to 2800 m.

The Cinnamon Teal has not been recorded in Ecuador since 1938, when a male and a female *septentrionalium* (both in the Mejía collection, where they were examined by us in 1993) were obtained on Laguna de San Pablo. *Borreroi* must be presumed to be Extirpated from Ecuador and indeed now is very scarce and local in Colombia as well (Hilty and Brown 1986).

Male *borreroi* have some black spotting on their dark chestnut breast and belly; in the smaller *septentrionalium* the breast is slightly paler and black spots are lacking. Females of the two taxa would be indistinguishable in the field. Some early specimens (see Chapman 1926) were not separated subspecifically, the subspecies *borreroi* not having been described until 1951. It is also conceivable that *orinomus* (which is sometimes merged into nominate *cyanoptera*) might wander north from Peru to the coastal lowlands of the southwest, most likely to El Oro. Although *orinomus*/nominate *cyanoptera* is not known from within the present borders of Ecuador, there is a 19th-century record (which we presume to refer to this taxon and not to *septentrionalium*) from Santa Luzia, a town now situated just over the border in Peru but which was formerly in Ecuador. There are no recent reports from Tumbes, but the Cinnamon Teal has been found not far to the south (see T. S. Schulenberg and T. A. Parker III, *Condor* 83[3]: 210–211, 1981). Conceivably this could explain the published report of *A. cyanoptera* from the Río Llaviuco at El Cajas in Azuay (King 1989), but no details on this sighting were published.

Range: breeds in w. North America and n. Mexico, wintering from Mexico to Costa Rica, small numbers (at least formerly) to n. Ecuador; resident very locally in Andes of Colombia and (formerly) n. Ecuador; nw. Peru to cen. Chile and cen. Argentina.

[Anas clypeata   Northern Shoveler   Pato Cuchara Norteño]

An accidental boreal winter visitant to lakes and lagoons near the coast of sw. Ecuador. Known in Ecuador from only a single sighting, that of a male and a female seen in w. Guayas at the Ecuasal lagoons on 29 Feb. 1988 (P. Shepherd and A. Marshall). This represents the southernmost American record for the species.

Monotypic.

Range: breeds in North America, wintering from s. United States south to nw. South America, also in West Indies; also in Eurasia.

Netta erythrophthalma   Southern Pochard   Porrón Sureño

Very rare and local on freshwater ponds, lakes, and marshes in the lowlands of sw. Ecuador in Manabí, s. Los Ríos, and Guayas and also in the highlands of n. Ecuador in Imbabura and Pichincha. All recent records are from the lowlands. In the highlands the Southern Pochard is known only from a few early-20th-century specimens taken in the La Carolina marshes (2800 m) in what is now Quito, and at Laguna San Pablo (2600 m), the last being a female in the Mejía collection obtained in Nov. 1939 (Orcés 1944). The last highland report is a male seen on the large, nearly barren lake above Papallacta on 1 Jul. 1976 (F. Sibley et al.). The Southern Pochard was first recorded in the Ecuadorian lowlands on the Santa Elena Peninsula in w. Guayas by Marchant (1958), who in the 1950s found it to be a "visitor . . . sometimes in hundreds" to a freshwater impoundment (we presume that this is what is now called the Represa José Velasco Ibarra). Somewhat surprisingly, and despite much searching, there have been no recent reports from this region. The next sighting involves a group of seven seen at a pond south of Babahoyo on 5 Aug. 1988 (RSR and PJG et al.). This area was even then being converted to rice cultivation, and that process is now essentially complete; there have been no further reports of the Southern Pochard from there. All subsequent reports come from two marsh and pond complexes in w.-cen. Manabí, west of Chone between San Antonio and Tosagua, and just west of Rocafuerte. The species was first discovered in this region on 21 Feb. 1991 when a pair was seen (P. Coopmans); the largest number found has been a flock of (only) 8 seen on 24 Jun. 1992 (RSR et al.). To 3200 m (at least formerly); recent records only from below 100 m.

Our judgement is that the Southern Pochard merits Critical status in Ecuador. No Ecuadorian site where it is known to occur—even sporadically—receives any protection at present, and all such areas appear to be under imminent threat of drainage or conversion for agricultural purposes. Hunting has also likely had an effect on the species, for the numbers surviving seem very small even in the largest areas of remaining suitable habitat. The Southern Pochard is rare, local, and declining throughout much of its South American range, though it seems to be somewhat more numerous in e. Brazil than elsewhere. Perhaps because of the latter, the Southern Pochard has been deemed not to be at risk as a whole (Collar et al. 1992, 1994). Another subspecies (*brunnea*) remains relatively numerous in Africa.

The American population, nominate *erythrophthalma*, has sometimes been considered as a species distinct from the African population, in which case a monotypic *N. erythrophthalma* would be called the South American Pochard.

Range: locally from n. Colombia and nw. Venezuela to nw. Argentina; locally in e. Brazil; also in Africa.

Aythya affinis   Lesser Scaup   Porrón Menor

An accidental boreal winter visitant to lakes and marshes in the highlands of n. Ecuador. There is only one Ecuadorian record: a bird collected in the La Carolina marshes (2800 m) in what is now Quito, on the seemingly unlikely date of 12 May (Lönnberg and Rendahl 1922); perhaps the specimen was misdated, or a previously injured bird was taken on that date. This represents the Lesser Scaup's southernmost record.

Monotypic.

Range: breeds in n. North America, wintering south in small numbers to n. South America.

Oxyura ferruginea   Andean Ruddy-Duck   Pato Rojizo Andino

Locally uncommon to fairly common on relatively deep lakes in the central valley and in paramo from Carchi (sightings on Loma El Voladero; N. Krabbe) south locally to Azuay (El Cajas). Although numbers of the Andean Ruddy-Duck have doubtless declined in many accessible localities, it does remain quite numerous on some lakes, notably on Lago Yambo near the Cotopaxi-Tungurahua border and on Laguna de Colta in Chimborazo, on both of which numbers present total at least in the low hundreds. Recorded from about 2100 to 4000 m.

O. ferruginea of the Andes is considered a species distinct from *O. jamaicensis* (Northern Ruddy-Duck) of North America, following the evidence presented by W. R. Siegfried (*Auk* 93[3]: 560–570, 1976) and B. Livezey (*Wilson Bull.* 107[2]: 214–234, 1995). These authors conclude that *O. ferruginea* is more closely related to *O. vittata* (Lake Duck) of s. South America than it is to *O. jamaicensis*. The 1998 AOU Check-list concurs. One race occurs in Ecuador, nominate *ferruginea* (*andina* of the Colombian Andes is considered a subspecies of *O. ferruginea*).

The species is sometimes called simply the Andean Duck.

Range: Andes from n. Colombia to Chile and Argentina, locally down to sea level in Peru.

Nomonyx dominicus   Masked Duck   Pato Enmascarado

Very local and secretive (thus likely often overlooked), at times not uncommon, in marshy freshwater ponds heavily overgrown with aquatic vegetation in the lowlands of w. Ecuador from w. Esmeraldas (south of Muisne; ANSP) south through Manabí and Guayas to coastal El Oro (a few recent sightings near Arenillas; P. Coopmans). There are also a few records from the eastern lowlands, principally from Limoncocha (where a few birds are present at least periodically; v.o.) and at the Lago Agrio airport (where a few are also present at irregular intervals; PJG); in addition, a bizarre sighting involved five apparently dispersing female-plumaged birds seen to alight on the swift-flowing Río Napo near Hacienda Primavera soon after dawn on 22 Feb. 1985 (RSR and T. S. Schulenberg et al.). Mostly found below 300 m, but there are also two highland records. These involve an early-20th-century specimen record from the La Carolina marshes in Quito, and a sighting of two birds at Laguna de Colta (3500 m) on 7–8 Aug. 1976 (F. Sibley et al.). These highland records are not as bizarre as they might appear, for the Masked Duck is also known from highland lakes in Colombia (e.g., on the Bogotá savanna, where it is syntopic with *andina* Andean Ruddy-Ducks [*Oxyura ferruginea andina*]); however, only Andean Ruddy-Ducks have been seen on Laguna de Colta subsequently. The Masked Duck is at least to some extent nomadic, moving in response to local water conditions, and is sporadic in its occurrences at many localities.

Formerly placed in the genus *Oxyura* with the other stifftail ducks, but we follow B. Livezey (*Wilson Bull.* 107[2]: 214–234, 1995) in placing it in the monotypic genus *Nomonyx*. The 1998 AOU Check-list concurs. Monotypic.

Range: Texas and Mexico to nw. Peru, n. Argentina, and Uruguay; West Indies.

Phoenicopteriformes
Phoenicopteridae   Flamingos

One species. Flamingos are found locally along coasts and on lakes (mainly saline) on most tropical continents.

*Phoenicopterus chilensis   Chilean Flamingo   Flamenco Chileno

An irregular and apparently nomadic visitant to saltwater lagoons and tidal flats along the Pacific coast, where thus far recorded only from various points in Guayas, most regularly and in largest numbers at the Ecuasal lagoons on the Santa Elena Peninsula (but even here there are periods when no birds are to be found). The Chilean Flamingo was first recorded in Ecuador in the 1950s by Marchant (1958), who then observed birds on the Santa Elena Peninsula. Since then the number seen at the Ecuasal saltworks has varied tremendously, with up to about 200 individuals having been seen on occasion, though normally no more than a few dozen are present. Flamingos have also been seen on rare occasions at other coastal sites, with the northernmost being one seen flying low over the ocean off Olón in nw. Guayas on an unspecified date in Apr. 1990 (B. Haase). A few have also been seen in the Playas area (PJG), and two were seen on tidal flats at Manglares-Churute Ecological Reserve on 25 Jan. 1991 (RSR and F. Sornoza), and four were seen here on 9 Feb. 1997 (P. Coopmans et al.). The status of Ecuadorian birds remains uncertain: there is certainly no evidence of breeding, nor is there any evidence of a seasonality component to their occurrences, nor any indication that Ecuadorian birds are dispersing juveniles (on the contrary, most individuals seen are in adult plumage). The first and only specimen (ANSP) obtained in Ecuador involved a moribund sick bird, an adult, salvaged at Ecuasal on 13 Aug. 1991.

Monotypic.

Range: sw. Ecuador, w. and s. Peru, w. Bolivia, locally in Chile and much of Argentina south to Tierra del Fuego; southern breeders winter north to w. Paraguay and se. Brazil.

Ciconiiformes
Ardeidae   Herons, Bitterns, and Egrets

Twenty species in 13 genera. The family is found widely near water (both fresh and salt) on all continents.

Botaurus pinnatus   **Pinnated Bittern**   **Mirasol Neotropical**

Rare and local in extensive freshwater marshes with abundant and lush vegetation, wet pastures with rank and tall grass, and adjacent rice fields in the lowlands of sw. Ecuador. Long known in Ecuador only from a 19th-century specimen at Vinces in s. Los Ríos, but there has been a spate of reports since the late 1980s from three disjunct regions, and it now appears clear that the reclusive Pinnated Bittern was until recently being overlooked; it actually may range wherever there is extensive suitable habitat. Since 1989 it has been found in cen. Manabí from the Bahía de Caráquez area south to around Rocafuerte (numerous sightings; also ANSP), in se. Guayas at Manglares-Churute Ecological Reserve, and at several localities in coastal El Oro near Santa Rosa and Arenillas (numerous sightings; also ANSP and MECN). Recorded only below 50 m.

The recent clarification of the Pinnated Bittern's distribution and numbers in Ecuador has resulted in the species no longer being considered as rare as it was formerly. Given the rapidity with which many west-Ecuadorian marshy areas are being drained and converted to agricultural uses, however, the species' future status will certainly need to be monitored. We believe it merits Near-threatened status in the country. The species is not considered to be generally at risk (Collar et al. 1994).

Sometimes called the South American Bittern (e.g., Martinez-Vilalta and Motis 1992).

One race: nominate *pinnatus*.

Range: locally from s. Mexico to Nicaragua and Costa Rica, from Colombia to the Guianas and extreme n. Brazil, in sw. Ecuador, and from e. and s. Brazil to n. Argentina and Uruguay.

Ixobrychus exilis   **Least Bittern**   **Mirasol Menor**

Status complex and still imperfectly known. The Least Bittern has recently been found to be rare and very local in freshwater marshes and along the edge of oxbow lakes in the lowlands of ne. Ecuador, where thus far it is known only from four sites, all of them in Napo. In recent years it has also been found to occur very locally in freshwater marshes in the lowlands of sw. Ecuador, to date only from Manabí and El Oro. Lastly, two specimens have been recovered in the Quito area. We discuss the three areas separately.

The Least Bittern was first recorded in Ecuador from two specimens (MCZ) that were obtained at Limoncocha on 6 Jul. 1963; these were described as the race *limoncochae* (Norton 1965). A fledgling was subsequently collected at Río Arajuno, confirming that the *limoncochae* Least Bitterns are local breeding residents in e. Ecuador (Norton et al. 1972). The species continues to be seen at Limoncocha, as well as at La Selva and Sacha Lodge. Presumably it is more widespread in e. Ecuador than the present paucity of known sites indicates, but that it is indeed local can be inferred from the species' apparent absence at well-worked sites such as Imuyacocha.

The first Least Bittern sighting from the w. Ecuador lowlands involved a single bird seen at the large marsh complex north of Santa Rosa in coastal El Oro in Sep. 1988 (P. Coopmans and R. Jones). There have been several subsequent sightings from this area (Bloch et al. 1991; RSR et al.), as well as in similar habitat west of Chone in cen. Manabí (where first found by G. H. Rosenberg and R. Behrstock et al. on 25 Jan. 1993). Curiously, however, Least Bitterns appear to be absent during the northern summer months, with records so far extending from only Sep. to Apr. A concerted effort to locate the species at Santa Rosa in Jun. 1993, for instance, did not reveal any to be present, though in this area several had been seen the previous Apr. This raises the possibility that western birds may be boreal migrants, though nominate *exilis* has not definitely been recorded south of Panama. It perhaps is more likely that they may simply be engaging in local movements, or moving to and from coastal Peru. Norton (1965) discusses a specimen he examined in the Colegio San Bolívar in Ambato that presumably was obtained in Ecuador (it bears no data) and which he tentatively refers to the race resident on the Peruvian coast, *peruvianus*. *Peruvianus* is not otherwise known from north of Lambayeque (T. S. Schulenberg and

T. A. Parker III, *Condor* 83[3]: 210, 1981). Whether the west-Ecuadorian birds represent resident *peruvianus* or the boreal migrant nominate race thus remains to be determined.

Further complicating the situation, two Least Bittern specimens (MECN) have recently been obtained from residential areas within the city limits of Quito (J. M. Carrión). Both are juveniles and were picked up in a weakened condition; both subsequently expired. The first, a juvenile female, was found on 5 Jul. 1994; the second, a slightly older juvenile male, on 22 Jul. 1996. These two specimens need to be critically compared in a comprehensive collection; we have only been able to examine color photographs. *Peruvianus* would appear to be eliminated on the basis of the specimens' relatively small size. On the basis of the male's rufous cheeks, we believe these birds likely are referable to the race *erythromelas*, though *bogotensis* is also conceivable. Where these birds came from remains unknown; we presume that both were dispersing from a breeding population somewhere not too distant, perhaps as close as the extensive marshes around lakes north of Quito (suitable marshes no longer being extant near Quito itself).

Some of the described subspecies of *I. exilis* are not well diagnosed, and the paucity of material precludes definitive statements. In the field the only characters likely to be of much value are the color of the cheeks and wing-coverts. The cheeks are dark (tawny to chestnut) in *limoncochae* and *erythromelas* but paler ochraceous in nominate *exilis*, *peruvianus*, and *bogotensis*. *Bogotensis* and *peruvianus* have the wing-coverts more ochraceous (less tawny) than in *erythromelas*, *limoncochae*, and nominate *exilis*. *Peruvianus* and *limoncochae* have longer wing and bill measurements. Perhaps further complicating the situation, striking differences in the vocalizations of some of these taxa have recently been described by R. Behrstock (*Cotinga* 5: 55–61, 1996); it is even possible that more than one species may be involved.

Range: United States south locally to sw. Peru, n. Bolivia, Paraguay, ne. Argentina, and se. Brazil; West Indies; boreal migrants are recorded south to Panama.

**[Ixobrychus involucris*   Stripe-backed Bittern   Mirasol Dorsiestriado]

Uncertain. Apparently an accidental vagrant to floating vegetation around the edge of oxbow lakes in the lowlands of ne. Ecuador. Known from only one record, a bird seen on at least three occasions between 16 and 19 Mar. 1988 on Mandicocha at La Selva (H. and P. Brodkin et al.). Only Least Bitterns (*I. exilis*) have been found subsequently at this site. The Stripe-backed Bittern is found primarily in s. and e. South America, with an uncertain status northward in many areas (long-distance migration or wandering perhaps being involved). It is known from only one site in e. Colombia (Hilty and Brown 1986) and from only one record in Peru, the latter a vagrant found at Manu National Park in the same habitat as the La Selva bird (J. W. Terborgh, J. W. Fitzpatrick, and L. Emmons, *Fieldiana, Zool.*, new series 21: 1–29, 1984).

Monotypic.

Range: very locally from n. Colombia, n. Venezuela, and the Guianas to n. Bolivia and ne. Brazil; more widely in se. Brazil, Paraguay, and n. and cen. Argentina, and in cen. Chile.

*Zebrilus undulatus   Zigzag Heron   Garcilla Cebra

Rare to locally uncommon in tangled undergrowth along sluggish forested streams and around the edge of oxbow lakes in várzea forest in the lowlands of e. Ecuador. Still known from only a few Ecuadorian localities, the Zigzag Heron was first recorded from a specimen (in the Escuela Politécnica collection) obtained by T. Mena at Chicherota in se. Pastaza on 25 Nov. 1971 (Orcés 1974). The species since has been found at a few sites near the Ríos Napo and lower Aguarico, including Taracoa (sightings in the early and mid-1980s; PJG), La Selva (where it can be found reasonably easily, with repeated sightings since 1988), and Imuyacocha (sightings since 1990, and one specimen [ANSP]). The nest of the still poorly known Zigzag Heron was first described from five found at La Selva (P. English, *Wilson Bull.* 103[4]: 661–664, 1991). Recorded below 300 m.

The Zigzag Heron was considered a Near-threatened species by Collar et al. (1992, 1994), but given its ample range and normally undisturbed habitat, we do not feel that this status is justified and do not consider the species to be at risk. It likely remains substantially underrecorded and overlooked because of its reclusive habitats; now that its voice is known, it is being recorded much more frequently than in the past.

Monotypic.

Range: the Guianas, e. and s. Venezuela, and e. Colombia to n. Bolivia and Amazonian Brazil.

Tigrisoma lineatum   Rufescent Tiger-Heron   Garza Tigre Castaña

Rare to locally fairly common around the forested shores of oxbow lakes and along streams in humid forest (occasionally even in terra firme, though more frequent in várzea and floodplain forest) in the lowlands of e. Ecuador, occasionally wandering up along larger rivers into the lower foothills. Also very rare in comparable habitat in the lowlands of w. Ecuador, whence there are several old records from the Manabí/Los Ríos border (Río Peripa) south into Guayas (Vinces); we suspect that some or all of the old specimen records from elsewhere in w. Ecuador (e.g., Hacienda Paramba, Santo Domingo, Pallatanga) refer to the Fasciated Tiger-Heron (*T. fasciatum*). The Rufescent Tiger-Heron was found in 1989–1990 at Manglares-Churute Ecological Reserve (R. Jones), but there have been no reports since; in addition an immature believed to have been this species (and not the Fasciated Tiger-Heron) was seen at Río Palenque on 23 Oct. 1997 (P. Coopmans et al.). Recorded mostly below 500 m, but occasionally (perhaps mainly or only juveniles) wandering up along rivers somewhat higher.

One race: nominate *lineatum*.
Range: Honduras to sw. Ecuador, n. Argentina, and Uruguay.

Tigrisoma fasciatum   Fasciated Tiger-Heron   Garza Tigre Barreteada

Rare and local along rapidly flowing and rocky rivers and larger streams in the foothills and subtropical zone on both slopes of the Andes. On the west slope recorded south to w. Loja (early-20th-century specimens from Guainche and Las Piñas, and a recent report of one seen at Tierra Colorada near Alamor on 11 Feb. 1991 [Best et al. 1993]). Of great interest, and coming as a distinct surprise, was the recent sighting of an adult on the slopes of the coastal Cordillera de Colonche at Loma Alta in w. Guayas in Dec. 1996 (Becker and López-Lanús 1997); this represents the only Ecuadorian report of this heron away from the Andes. The Fasciated Tiger-Heron occurs almost entirely above the range of the similar Rufescent Tiger-Heron (*T. lineatum*), though occasional birds, especially the difficult-to-identify juveniles, do wander into the range of the other species. In 1996–1997 the Fasciated Tiger-Heron was regularly observed along the Río Santiago in the lowermost foothill zone near Playa de Oro in n. Esmeraldas, at an elevation of only about 100 m; it there appears to be resident (O. Jahn et al.). Recorded mostly from 600 to 2200 m, occasionally or locally lower.

The Fasciated Tiger-Heron was given Near-threatened status by Collar et al. (1992, 1994), but we do not feel that this is justified, at least not in Ecuador where there seems to be little or no evidence of any decline or negative habitat impact.

One race: *salmoni*.
Range: locally in highlands of Costa Rica to mountains of n. Venezuela and Andes from w. Venezuela to nw. Argentina; locally in highlands of cen. and se. Brazil and ne. Argentina.

Ardea cocoi   Cocoi Heron   Garzón Cocoi

Uncommon to locally fairly common in a variety of freshwater habitats and in mangroves and on tidal flats in the lowlands of both e. and w. Ecuador. In the east the Cocoi Heron usually occurs solitarily along rivers and along the margins of oxbow lakes. In the west it is known only from cen. Manabí (Bahía de Caráquez area) south locally through Guayas and coastal El Oro, with a few recent reports from extreme s. Loja in the Macará and Zapotillo region (Best et al. 1993; RSR). The old record from Mindo (Lönnberg and Rendahl 1922) seems likely to represent a mislabeled specimen as the species has in Ecuador otherwise never been found so far north in w. Ecuador, or at such a high elevation. In Ecuador the Cocoi Heron is most numerous in and near mangroves along the south coast (e.g., at Manglares-Churute Ecological Reserve); numbers elsewhere may have decreased as a result of disturbance and hunting. Recorded up to about 400 m.

The species was formerly often called the White-necked Heron, but we favor

using the English name of Cocoi Heron. This is in accord with Ridgely and Gwynne (1989), Sibley and Monroe (1990), and the 1998 AOU Check-list.
Monotypic.
Range: Panama to cen. Argentina and s. Chile, though very local in coastal Peru and n. Chile, and only wanders to higher elevations in Andes.

Ardea herodias   Great Blue Heron   Garzón Azulado

Apparently a casual northern winter visitant to coastal lagoons, freshwater marshes, and lakes in w. Ecuador, whence there are four recent sightings from three sites. The first involves an adult seen at the Ecuasal lagoons in w. Guayas on 13 Jul. 1978 (RSR and D. Wilcove); it or another bird was again there on 12–13 Feb. 1980 (RSR and D. Finch et al.; see Ridgely 1980). Another adult was seen near Guayaquil on 16 Jan. 1988 (R. A. Rowlett et al.), and another was noted at Yaguarcocha in Imbabura on 1 Mar. 1988 (PJG). The Ecuador reports represent the southernmost reports of the species. We suspect that a few uncorroborated reports from the eastern lowlands refer to immature Cocoi Herons (*A. cocoi*). To 2300 m.
Presumed race: nominate *herodias*. As noted by Ridgely (1980), however, it is conceivable that *cognatus* of the Galápagos Islands could be involved, particularly with the Ecuasal birds.
Range: breeds in North America, Mexico, locally in Greater Antilles, and on Galápagos Islands; winters south sparingly to nw. South America.

Ardea alba   Great Egret   Garceta Grande

Uncommon to locally very common in the lowlands of both e. and w. Ecuador, occupying a variety of fresh- and saltwater habitats—locally even occurring on ocean beaches—but with numbers greatest in the southwest from Manabí and Los Ríos south through coastal El Oro. In the eastern lowlands, Great Egrets are usually seen as scattered single birds or small groups standing or feeding on sandbars (less often logs or rocks) along major rivers; numbers here seem somewhat higher during the boreal winter, when some migrants from the north may occur. Breeding occurs at least locally along the shores of the Río Guayas estuary and in coastal El Oro. Small numbers of Great Egrets also occasionally wander (some may be boreal migrants?) to lakes in the central valley up to 2800 m, particularly in n. Ecuador (e.g., at Yaguarcocha, Laguna San Pablo, La Carolina marshes near Quito). There even is an old record from as high as Laguna de Colta (3500 m).
The species is sometimes called the Great White Egret, or even—in the Old World—the Great White Heron.
This species was formerly placed in the monotypic genus *Casmerodius*. Some authors, however, have placed the species in the genus *Egretta*. F. Sheldon (*Auk* 104[1]: 97–108, 1987) demonstrated that genetically it is at least as close to the genus *Ardea*, and we follow the 1998 AOU Check-list in placing the species in *Ardea*. One race: *egretta*.

Range: United States to Chile and s. Argentina; West Indies; also widely in much of Old World, especially at tropical latitudes. More than one species is perhaps involved.

Egretta thula   Snowy Egret   Garceta Nívea

Uncommon to locally very common in the lowlands of both e. and w. Ecuador, occupying a variety of fresh- and saltwater habitats (locally occurring even on ocean beaches); by a substantial margin numbers are greatest in the southwest, especially in the lower Río Guayas estuary and along the shores of the Gulf of Guayaquil. Small numbers also occur occasionally around lakes in the central valley, with the highest count being 16 birds seen at Yaguarcocha in Imbabura on 7 Aug. 1981 (PJG). In the eastern lowlands the Snowy Egret is usually seen as scattered solitary individuals (at most small groups) feeding on sandbars along major rivers; singletons occasionally occur elsewhere. Breeding occurs locally in mangroves along the Río Guayas estuary and the coast of El Oro (possibly elsewhere?). As yet there is no definite evidence of migration from North America, though this remains possible. Ranges mainly below about 500 m, though individuals on highland lakes range up to 2600 m.

One race: nominate *thula*.

Range: United States south to cen. Chile and cen. Argentina; West Indies.

Egretta caerulea   Little Blue Heron   Garceta Azul

Fairly common to common in a variety of fresh- and saltwater habitats (most numerous in mangroves and on tidal flats) in the lowlands of sw. Ecuador in the Río Guayas estuary and along the El Oro coast, where it breeds locally; decidedly less numerous and more local northward in the lowlands of w. Ecuador; also one sighting of an adult along the Río Chira in extreme s. Loja at Lalamor on 15–16 Sep. 1998 (RSR et al.). A rare to occasionally uncommon boreal winter visitant to the shores of lakes and rivers in the lowlands of ne. Ecuador, where thus far recorded only from the drainages of the Ríos Napo and Aguarico, and to lakes in the central valley of the highlands, mainly in Imbabura (e.g., at Yaguarcocha and Laguna San Pablo); likely some migrants occur in the west as well. As a migrant recorded mainly from Oct. to Mar., with one May sighting; these birds cannot be certainly ascribed to northern populations (there are no band recoveries, etc.), but the recorded dates are strongly suggestive. Recorded mainly below 600 m, though individuals on highland lakes range up to 2800 m.

Formerly placed in the monotypic genus *Florida*. Monotypic.

Range: e. United States and Mexico to sw. Peru, n. and e. Bolivia, and se. Brazil, though absent from wide areas in Amazonia and interior Brazil; West Indies.

Egretta tricolor   Tricolored Heron   Garceta Tricolor

Rare to locally fairly common in mangroves and on tidal flats in the lowlands of sw. Ecuador in the Río Guayas estuary and along the El Oro coast; perhaps most numerous near Puerto Pitahaya in s. El Oro, where in Apr. 1993 a total of perhaps 100 pairs were found in each of four heron colonies located (RSR and F. Sornoza). Very small numbers are also found northward locally along the coast to Esmeraldas, but in general the Tricolored remains one of the least numerous Ecuadorian herons. It has never been found any distance inland or in the eastern lowlands; aside from the banding recovery mentioned below, there is no evidence that boreal migrants occur.

A Tricolored Heron banded as a nestling in Virginia was recovered somewhere in Ecuador (reported as "Quito," though this locality seems unlikely), the only definite evidence of long-distance migration to Ecuador for any heron species.

The species was formerly called the Louisiana Heron.

Formerly placed in the monotypic genus *Hydranassa*. One race: *ruficollis*.

Range: e. United States and Mexico south, mainly coastally, to sw. Peru and ne. Brazil; Greater Antilles.

Bubulcus ibis   Cattle Egret   Garceta Bueyera

Status complex, with a large population now resident in open agricultural areas in the western lowlands, smaller numbers occurring elsewhere; boreal migrants seem almost certainly to be present seasonally. The Cattle Egret, which until relatively recently was found only in the Old World, apparently invaded South America during the late 19th century and has since undergone a remarkable range expansion through much of the New World; it now ranges nearly anywhere pastures or savannas with large numbers of grazing mammals are found, with populations being greatest in the tropical zone. The species was first recorded in Ecuador on 10 Nov. 1958 when a specimen (Escuela Politécnica) was obtained by M. Olalla at Nueva Rocafuerte along the Río Napo in ne. Ecuador. By the early 1960s, small numbers of Cattle Egrets were being seen in the Guayaquil region (R. Lévêque, *Ibis* 106[1]: 55, 1964). Populations have since increased dramatically in the west, to the point where the species is now common to very common in pastures and other agricultural land nearly throughout w. Ecuador. Numbers are highest from cen. Manabí and Los Ríos south through much of Guayas and coastal El Oro, though the species remains relatively scarce on the arid Santa Elena Peninsula; in much of this region it is, together with the Black Vulture (*Coragyps atratus*), one of the most frequently seen roadside birds, and it probably is now the most numerous Ecuadorian heron. Elsewhere in w. Ecuador the species is less numerous and more local, but it is recorded north to Esmeraldas and south to the Peruvian border in sw. Loja.

The Cattle Egret is also uncommon to fairly common on pastures and in agricultural land in the lowlands of e. Ecuador and in the central valley of n. Ecuador in Imbabura and Pichincha. There is, however, as yet no evidence that the species is breeding in these regions, and as numbers here are much higher from Oct. to Mar. (with only a few records of scattered single birds from the rest of the year) it is presumed to be occurring here mainly as a boreal migrant. Its already large numbers in the western lowlands are also apparently augmented by boreal migrants. The Cattle Egret's migratory behavior is further indicated by B. Haase's seawatch observations of flocks flying southward over the ocean along the coast of w. Guayas in Jul. and Aug., and flying northward in Feb. and Mar.; in addition, what appear to be arriving flocks have been seen at several sites in the eastern lowlands in Oct. and Nov. (RSR; B. Bochan). Surprisingly, however, there are only a few reports from the southeastern lowlands, recent sightings from Kapawi Lodge on the Río Pastaza near the Peruvian border; one was also seen along the Loja-Zamora road on 2 Jan. 1995 (D. Wolf). Recorded up to about 2800 m in the Quito region, exceptionally (migrants only?) as high as 3300 m (e.g., one seen at the lake above Papallacta on 4 Oct. 1997 [P. Coopmans], also a few prior sightings from there [RSR; PJG]).

The monotypic genus *Bubulcus* is sometimes merged into *Egretta*, but F. Sheldon (*Auk* 104[1]: 97–108, 1987) suggested that it is at least as close to the genus *Ardea*. We follow Sibley and Monroe (1990) and the 1998 AOU Checklist in maintaining it as a separate genus. One race: nominate *ibis*.

Range: United States south to cen. Chile and cen. Argentina; West Indies; Galápagos Islands; also widely in warmer parts of Old World (its original range). More than one species is perhaps involved.

Butorides striatus Striated Heron Garcilla Estriada

Uncommon to locally common and widespread in a variety of fresh- and saltwater habitats ranging from the borders of lakes and rivers to mangroves in the lowlands of both e. and w. Ecuador, ranging up in much smaller numbers to the central valley, especially in n. Ecuador. The Striated Heron is particularly numerous along streams through várzea forest in the east, and in areas primarily devoted to rice cultivation in the west. It is least numerous, predictably, in very arid regions, and is scarce on the actual coast. Recorded mostly below 500 m, in decreasing numbers up to 2500–2800 m in the central valley.

This and *B. virescens* (Green Heron) (together with *B. sundevalii* [Lava Heron] of the Galápagos Islands) are sometimes placed in the genus *Ardeola*, but we follow Sibley and Monroe (1990) and most recent authors in retaining them in the genus *Butorides*. One race: nominate *striatus*.

Range: Panama to w. Peru, n. Argentina, and Uruguay; also warmer parts of Old World.

**Butorides virescens* Green Heron Garcilla Verde

A very rare boreal winter visitant to the forested borders of streams and smaller rivers and oxbow lakes, and in mangroves, in the lowlands of n. Ecuador. Has been found both east of the Andes in Napo and Sucumbíos, and west of them in Esmeraldas. Recorded between Dec. and Mar. but probably occurs earlier as well. The Green Heron was first recorded from Ecuador from two immature specimens discussed by Norton et al. (1972): one (Universidad Nacional) was collected at Concepción on 25 Dec. 1929, the other (MCZ) at Edén on 7 Dec.

1964. There have been a few subsequent sight records of the easily recognized adults, with the farthest south being one seen at Jatun Sacha on 10 Jan. 1992 (B. Bochan). The only report from west of the Andes concerns an adult seen at Súa in w. Esmeraldas on 10 Mar. 1994 (R. P. Clay et al.). It is possible that the Green Heron may be more numerous in Ecuador than the few records summarized above would seem to indicate, for immatures are probably not distinguishable in the field from immature Striated Herons (*B. striatus*) and thus are likely often passed over or left unidentified. These represent the southernmost records for the species. Recorded only from below about 400 m, but could also occur around highland lakes.

B. virescens has sometimes been regarded as conspecific with *B. striatus* of South America and many tropical areas in the Old World; the expanded species is usually called the Green-backed Heron. However, evidence recently presented by B. L. Monroe, Jr., and M. R. Browning (*Bull. B. O. C.* 112[2]: 81–85, 1992) demonstrates that the two are best considered as allospecies, and this treatment was followed by Sibley and Monroe (1990) and the 1998 AOU Check-list. The racial identification of the two Ecuadorian specimens—both immatures, as noted above—was left as uncertain by Norton et al. (1972), but almost certainly it is nominate *virescens* that is involved; the other races of *B. virescens* are not known to migrate anywhere near as far south as South America.

Range: breeds from United States to Panama, and in West Indies south to islands off northern coast of South America; northern birds winter south in small numbers to n. South America.

Agamia agami   Agami Heron   Garza Agamí

Rare and inconspicuous (though doubtless under-recorded) along forested streams, around the forested margins of oxbow lakes, and in várzea forest in the lowlands of ne. Ecuador, where thus far known only from the Ríos Napo and Aguarico drainages. Although there would seem to be no reason why the Agami Heron should not occur in the Río Pastaza drainage as well, apparently there are no records of it from there. There also is a single 19th-century specimen record from Pambilár in coastal Esmeraldas (E. Hartert, *Nov. Zoöl.* 9: 599–617, 1902), which, surprisingly, remains the only record from the Chocó region of w. Colombia and w. Ecuador. Recorded only below about 300 m.

The Agami Heron was accorded Near-threatened status by Collar et al. (1992, 1994), but given its vast range and largely undisturbed habitat, we cannot agree that this is justified. We do not consider the species to be at risk.

The species was formerly often called the Chestnut-bellied Heron, but we favor using the name of Agami Heron. This is in accord with Ridgely and Gwynne (1989), Sibley and Monroe (1990), and the 1998 AOU Check-list.

Monotypic.

Range: s. Mexico to n. Bolivia and Amazonian Brazil, with one old record from nw. Ecuador.

Pilherodius pileatus   Capped Heron   Garza Pileada

Uncommon along larger rivers (where it favors secluded backwater channels and sandbars), in swampy forest, and in nearby damp pastures in the lowlands of e. Ecuador; recorded mainly from Napo and Sucumbíos, though reports from Kapawi Lodge along the Río Pastaza near the Peruvian frontier indicate that the species likely is found equally widely in the southeast. The first Ecuadorian record involves a hitherto unpublished specimen (ANSP) taken by "Olalla y Hijos" at the mouth of the Río Cotapino on 22 Jul. 1925; this specimen apparently was overlooked by Meyer de Schauensee (1966). Not being aware of this record, both Norton (1965) and Tallman and Tallman (1977) claimed supposed first Ecuadorian records for the Capped Heron, these involving birds collected and seen at Limoncocha. Ridgely (1980, p. 243), likewise unaware of the unpublished ANSP specimen, discussed the species' "apparent increase" in Ecuador, though it now seems likely that the Capped Heron has been a low-density resident all along. In recent decades there have been numerous sightings of this attractive heron from several sites in both the Río Napo and the Río Aguarico drainages, where it appears to be relatively widespread. Recorded only below 400 m.

    The species has sometimes been placed in the genus *Nycticorax*. Monotypic.
Range: Panama to n. and e. Bolivia, n. Paraguay, and s. Brazil.

Nycticorax nycticorax   Black-crowned   Garza Nocturna
                             Night-Heron   Coroninegra

Rare to locally fairly common in mangroves and on tidal flats, freshwater marshes, and around the forested margins of lakes and ponds in the lowlands of w. and ne. Ecuador and in the northern highlands; numbers are highest in sw. Ecuador from Manabí (Bahía de Caráquez and Rocafuerte region) south to El Oro, but even in this region the species is decidedly local. We are not aware that Black-crowned Night-Heron nests have ever been found in Ecuador but are confident that the species must be breeding at least locally along mangrove-lined shorelines of the Río Guayas estuary and south into El Oro. It was also presumably once resident around reed-fringed lakes in the highlands of n. Ecuador (e.g., at Lagunas Yaguarcocha and San Pablo in Imbabura and in the La Carolina marshes in Quito), but numbers now are very much reduced in the highlands because of excessive disturbance; a few may persist in Imbabura. An anomalous report concerns an adult seen at Laguna Llaviuco in El Cajas National Recreation Area on 27 Feb. 1999 (L. Navarrete et al.); this may pertain to an austral migrant. Some of the relatively few records from the eastern lowlands may also pertain to boreal migrants, as may some of the birds found in w. Ecuador; however, small numbers were found roosting at Sacha Lodge in Jul. 1994 (L. Jost), and a few now appear to be roosting year-round at La Selva (P. Coopmans). Recorded up to 2800–3000 m, but in recent years mainly below 400 m.

The Black-crowned Night-Heron appears to have declined in the Jatun Sacha area as a result of hunting (B. Bochan), but as a whole this wide-ranging species does not appear to be at risk in Ecuador.

One race: the American *hoactli* (which is sometimes merged into nominate *nycticorax* of the Old World).

Range: s. Canada south to Tierra del Fuego, though absent from much of Amazonia; West Indies; also in much of Old World except Australia.

Nyctanassa violacea	Yellow-crowned Night-Heron	Garza Nocturna Cangrejera

Locally fairly common to common (but often inconspicuous) in mangroves and on tidal flats along the coast of sw. Ecuador from cen. Manabí (Bahía de Caráquez) south locally to El Oro; also occurs locally in Esmeraldas, where there is an old record from Vaquería and a Sep. 1994 sighting from Muisne (P. Coopmans). Breeds at least locally where habitat is suitable, sometimes in quite large numbers (e.g., in Apr. 1993 near Puerto Pitahaya in El Oro; RSR). Away from the coast the Yellow-crowned Night-Heron is very rare; there is an old specimen from Babahoyo, and a single sighting from Río Palenque (6 immature-plumaged birds seen on 30 Aug. 1975; R. Webster). Recorded below 200 m.

Sometimes placed in the genus *Nycticorax*, but we follow F. Sheldon (*Auk* 104[1]: 97–108, 1987) and Sibley and Monroe (1990) in maintaining *Nyctanassa* as a distinct, monotypic genus. One race, *caliginis*, occurs in Ecuador.

Range: e. United States and Mexico south, mainly along coasts, to extreme nw. Peru and se. Brazil; Galápagos Islands; West Indies.

Cochlearius cochlearius	Boat-billed Heron	Garza Cucharón

Rare to locally fairly common around the margins of freshwater ponds and lakes, also feeding at pools in swampy forest and along streams and rivers, in the lowlands of e. Ecuador. Recorded primarily from the northeast in the Ríos Napo and Aguarico drainages, but there also is a single 19th-century record from Sarayacu and recent reports from Kapawi Lodge on the Río Pastaza near the Peruvian border, suggesting the species may also occur widely in the southeast. The Boat-billed Heron is also known from old records from w. Ecuador, though there are no modern reports from west of the Andes; curiously, it appears to be absent from the Pacific lowlands of sw. Colombia (Hilty and Brown 1986). *Birds of the Americas* (pt. 1, no. 2) records a specimen from Salidero in Esmeraldas; two specimens were collected at "Río Blanco, below Mindo" (Lönnberg and Rendahl 1922), seemingly an unlikely site for this lowland bird; two additional specimens (FMNH) were taken by the Olallas at Isla Silva on the Río Babahoyo in n. Guayas on 17 and 19 Aug. 1931. Recorded up only to about 400 m.

The genus *Cochlearius* was formerly placed in a monotypic family, Cochleari-idae, but almost all recent authors have favored placing it with the other herons in the Ardeidae. One race, nominate *cochlearius*, occurs in Ecuador.

Range: n. Mexico to n. and e. Bolivia, Paraguay, ne. Argentina, and s. Brazil.

Threskiornithidae   Ibises and Spoonbills

Seven species in six genera. Ibises and spoonbills are found widely near water (both fresh and salt) on all continents, with the greatest diversity being found in tropical regions.

Theristicus melanopis   Black-faced Ibis   Bandurria Carinegra

Rare and very local in paramo and around lakes in the vicinity of Volcán Anti-sana in extreme w. Napo and of Volcán Cotopaxi in e. Cotopaxi and adjacent w. Napo. Ecuador represents the northernmost limit of this species' range, and the population here is well isolated from the remainder of the range of the species, which otherwise is not known north of Ancash in the Peruvian Andes or north of Lambayeque on the Peruvian coast (T. S. Schulenberg and T. A. Parker III, *Condor* 83[3]: 210, 1981). Recorded between about 3800 and 4300 m.

The range and numbers of the Black-faced Ibis have apparently always been very limited in Ecuador, and at least at present birds are seen only in very remote areas; the total Ecuadorian population must be very small, likely under 100 individuals. Local landowners on Antisana have begun (fide N. Krabbe) to institute protective measures, however, leading one to be guardedly more optimistic regarding the species' future at least in the Antisana Ecologi-cal Reserve. Both principal areas where Black-faced Ibises at present occur, the Antisana Ecological Reserve and Cotopaxi National Park, are under at least nominal protection. Nonetheless, given the species' small population size and its inherently vulnerable situation, we consider it to be Endangered in Ecuador. The Black-faced Ibis becomes more numerous farther south in its range, and the species as a whole is not considered at risk (Collar et al. 1992, 1994).

T. melanopis (with *branickii*) of the Andes and far s. South America is here regarded as a species separate from *T. caudatus* (Buff-necked Ibis), the latter being found in the more open lowlands of n. and e. South America; this treat-ment is in accord with *Birds of the World* (vol. 1) and Fjeldså and Krabbe (1990). One race occurs in Ecuador, *branickii*. Some authors (e.g., Sibley and Monroe 1990) further separate the high-Andean form *branickii* as a monotypic species (Andean Ibis).

Range: Andes of n. Ecuador, and from cen. Peru to w. Bolivia and extreme n. Chile, and locally in coastal w. Peru; cen. Chile and s. Argentina south to Tierra del Fuego, wintering north in Argentina.

Mesembrinibis cayennensis   Green Ibis   Ibis Verde

Uncommon to locally fairly common (but inconspicuous and probably often overlooked) in várzea forest and along the forested or wooded margins of rivers and lakes in the lowlands of ne. Ecuador. There seems to be only one old record from the southeast, from the Río Copotaza (*Birds of the Americas*, pt. 1, no. 2); it likely, however, is more widespread there. The Green Ibis appears to be particularly numerous at Limoncocha. Recorded below about 300 m.

Monotypic.

Range: Costa Rica to n. and e. Bolivia, e. Paraguay, ne. Argentina, and se. Brazil.

*Phimosus infuscatus   Bare-faced Ibis   Ibis Caripelado

A very rare visitant to sandbars and adjacent grassy areas along major rivers in the lowlands of ne. Ecuador. There are only a few Ecuadorian records of the Bare-faced Ibis, all of them recent. The first two involve birds that were collected from along the Río Napo: one (MCZ) at Edén on 17 Dec. 1964 (Norton et al. 1972) and another (LSUMZ) near Limoncocha on 1 May 1976 (Tallman and Tallman 1977). Since then there have been a few sightings from along the Río Napo, in the lower Río Aguarico/Lagartococha region, and near Lago Agrio; no seasonality pattern is evident. We suspect that the scatter of records indicates that the Bare-faced Ibis is only an irregular wanderer to Ecuador from populations in e. Colombia, where the species is quite numerous, especially in the northeast. Recorded below 300 m.

The species has been called the Whispering Ibis.

One race: *berlepschi*.

Range: n. and e. Colombia and n. and w. Venezuela, wandering to ne. Ecuador; n. and e. Bolivia and cen. Brazil south to n. Argentina and Uruguay.

Eudocimus albus   White Ibis   Ibis Blanco

Locally uncommon to common in mangroves, tidal flats, freshwater ponds, marshes, and drying rice fields near the coast of w. Ecuador. The White Ibis is numerous in Guayas and El Oro, with an additional population occurring in cen. Manabí from the Bahía de Caráquez region south to around Rocafuerte, and there are a few in w. Esmeraldas (Atacames/Muisne area); there is also a century-old record from the coast of extreme n. Esmeraldas (Vaquería) near the Colombian border. Inland there is only one sighting, of a flock of about 20 birds seen at Río Palenque on 30 Oct. 1986 (PJG et al.). Numbers of the White Ibis are highest in the lower Río Guayas estuary (an estimated 750 birds were seen during one afternoon at Manglares-Churute Ecological Reserve on 25 Jan. 1991; RSR and F. Sornoza), where the species is resident and certainly breeds. A small nesting colony was also seen in mangroves near Puerto Pitahaya in El Oro on 16 Apr. 1993 (RSR and F. Sornoza). Presumably the species breeds locally in Manabí and Esmeraldas as well, as there is no evidence of any surge in numbers when migrants would be expected to be arriving from the north. Recorded mostly near sea level.

The species has sometimes been called the American White Ibis.

Monotypic.

Range: se. United States and Mexico south locally (mainly along coasts) to extreme nw. Peru and n. Colombia and n. Venezuela, with small numbers in llanos of the last two countries; occasional wanderers occur east to the Guianas; Greater Antilles.

*[*Eudocimus ruber*   Scarlet Ibis   Ibis Escarlata]

A casual visitor to lakeshores, and presumably the banks of larger rivers, in the lowlands of ne. Ecuador. There are only two reports, a subadult ("with a mosaic of gray and scarlet patches") seen at Limoncocha on 7 May 1964 (H. M. Stevenson, *Wilson Bull.* 84[1]: 99, 1972) and an immature seen at Cuyabeno on 27–28 Mar. 1995 (J. Hendriks, fide P. Coopmans). In Colombia, undoubtedly the source of the Ecuadorian birds, the Scarlet Ibis occurs regularly south to Meta in the northeast, occasionally straggling slightly farther south (Hilty and Brown 1986). Recorded below 300 m.

It has been suggested (e.g., C. Ramo and B. Busto, *Doñana Acta Vertebr.* 9: 404–408, 1982) that *albus* and *ruber* might be conspecific under the name of *E. ruber* (which name has priority). Interbreeding in the wild is relatively infrequent in mixed colonies, however, and we prefer to continue to treat them as distinct species, following most recent authors (e.g., Sibley and Monroe 1990, AOU 1998). Monotypic.

Range: n. Colombia east along coast to n. Brazil, also inland in llanos of Venezuela and ne. Colombia, wandering to ne. Ecuador; coastal se. Brazil.

*[*Plegadis falcinellus*   Glossy Ibis   Ibis Morito]

Apparently a casual visitant to freshwater marshes and rice fields in the lowlands of sw. Ecuador in Manabí and Guayas; status uncertain. There are only three Ecuadorian reports: one seen along the road between Guayaquil and Babahoyo on 5 Jul. 1974 (C. Leck, *Am. Birds* 34[3]: 312, 1980); one found at Manglares-Churute Ecological Reserve on 25 Mar. 1993 (J. C. Matheus and B. Sage et al.); and several flocks seen between Oct. 1996 and Apr. 1997 at Segua Marsh at the head of the Bahía de Caráquez west of Chone in cen. Manabí (B. López-Lanús and P. Gastezzi, *Cotinga* 13: 60, 2000). Recorded below 50 m.

Monotypic.

Range: e. and s. United States and Greater Antilles, also locally in n. Venezuela and s. Mexico, northern-breeding birds wintering south in small numbers to Panama, n. Colombia, and Venezuela, casually to sw. Ecuador; also warmer parts of Old World.

Ajaia ajaja   Roseate Spoonbill   Espátula Rosada

Rare to locally fairly common in mangroves, tidal flats, and adjacent fresh-water marshes in the coastal lowlands of sw. Ecuador, primarily ranging from w. Guayas (north to the lower Río Guayas) south through coastal El Oro, but also with several recent sightings from cen. Manabí from the Bahía de Caráquez area south to around Rocafuerte. Very small numbers occur along the Río Napo upriver (at least rarely) to above Coca, but especially in the Pañacocha area (where one was shot [specimen to ANSP] on 13 Jan. 1991). There also have been a few recent sightings along the lower Río Aguarico, and there is one record from the upper Río Pastaza (Orcés 1944). The Roseate Spoonbill's largest Ecuadorian population is found in the lower Río Guayas estuary; the species is, for example, numerous in the mangroves and tidal portions of Manglares-Churute Ecological Reserve. In the west recorded at or near sea level, in the east below 300 m.

Sometimes placed in the genus *Platalea* (e.g., *Birds of the World*, vol. 1). However, we follow the 1998 AOU Check-list in continuing to place it in the monotypic genus *Ajaia*. Monotypic.

Range: se. United States and Mexico south locally to nw. Peru (wandering farther south), n. Argentina, and Uruguay; Greater Antilles; absent from a broad area of Amazonia and n. Brazil.

Ciconiidae   Storks

Two species in two genera. Storks are found widely near water on all continents, reaching their greatest diversity in the tropics.

Mycteria americana   Wood Stork   Cigueña Americana

Locally uncommon to fairly common—but notably nomadic and erratic in its occurrences in many places (often just seen in flight overhead)—in freshwater marshes, around ponds and shallow lakes, and along rivers in the lowlands of sw. Ecuador from cen. Manabí (Bahía de Caráquez area) and s. Pichincha (Río Palenque) south through coastal El Oro (where regular); there are also two sightings from extreme s. Loja (5 birds south of Zapotillo along the Río Chira on 21 Apr. 1993 [RSR and F. Sornoza] and 2 seen along the Río Chira at Lalamor on 13 Sep. 1998 [RSR et al.]). A flock of 50–60 individuals seen flying northward on 11 Nov. 1995 near Maldonado on the Río Santiago is seemingly the only report for Esmeraldas (O. Jahn et al.). Although Wood Storks must do so occasionally at least locally, there is no confirmed record of nesting in Ecuador. Very rare and seemingly erratic in the eastern lowlands, where the species does not appear to be of regular occurrence anywhere (there are recent reports known to us from Limoncocha, La Selva, and Kapawi Lodge), with birds primarily being seen in high soaring flight, presumably on passage elsewhere. Wood Storks are prone to wander, and apparently they even cross the Andes, at least occasionally. There is, for example, an old report from the La Carolina marshes (2800 m) in what is now Quito, and a flock of eight was seen flying west over Papallacta (3000 m) in Feb. 1987 (P. Coopmans), whereas three others were seen soaring westward over the road to Loreto (1200 m) north of Tena on 17 Mar. 1990 (RSR and PJG et al.); another was seen flying over Panguri (1600 m) in Zamora-Chinchipe on 26 Jul. 1992 (T. J. Davis), and a flock of 24 was seen near Sabanilla (1600 m) on the Loja-Zamora road on 12 Mar. 1995 (P. Coopmans et al.). Wood Storks are most numerous and regular in the southwest; however, the largest number ever reported in Ecuador is a remarkable total of 575 birds seen flying from a roost near the eastern end of the Bahía de Caráquez in Manabí on 14 Sep. 1992 (RSR and F. Sornoza), an area where Wood Storks are often not present at all.

Monotypic.

Range: s. United States and Mexico south locally to nw. Peru (wandering farther south), n. Argentina, and Uruguay; Cuba.

*Jabiru mycteria   Jabiru   Jabirú

Apparently a casual wanderer to sandbars and river margins in the lowlands of ne. Ecuador in Napo and Sucumbíos. There are only three Ecuadorian reports: what was believed likely to be the same individual was seen occasionally, mainly in flight, at Limoncocha in 1971–1972 (D. Pearson); three flying birds were seen from an airplane over the lower Río Yasuní in Sep. 1976 (D. Pearson); and one was seen on a sandbar in the Río San Miguel near Tigre Playa on 11 Aug. 1993 (F. Sornoza). The unmistakable and conspicuous Jabiru is not known to be resident closer to Ecuador than the llanos of ne. Colombia and very locally in swampy areas of ne. Peru. Recorded only below 300 m.

Possibly in the distant past a small, resident population of Jabirus existed in e. Ecuador, but if so it was surely extirpated long ago. It now seems implausible to hope for the Jabiru's ever being more than a casual vagrant to Ecuador. Given the possibility of continued vagrancy to Ecuador, we opt to give it Critical status for the country. The Jabiru is not considered to be at risk generally (Collar et al. 1992, 1994), though it is known to be declining in certain parts of its range, notably in Middle America.

Monotypic.

Range: locally from s. Mexico to n. Argentina and Uruguay; now rare in many areas, extirpated in some.

Cathartidae   American Vultures

Five species in four genera. American Vultures are found widely in the New World, though none ranges in far n. North America. The family was long considered to be part of the Falconiformes, but most—though not all (see C. Griffiths, *Auk* 111[4]: 787–805, 1994)—recent work (e.g., Sibley and Ahlquist

1990; AOU 1998) has supported a closer relationship with the Ciconiidae, and we follow this here. Gill (1994) retained the Cathartidae in the Falconiformes.

Vultur gryphus   Andean Condor   Cóndor Andino

Rare to locally uncommon in paramo and sparsely inhabited agricultural terrain from Carchi (Páramo del Ángel; N. Krabbe) south locally into Loja (where the species is now very scarce, though it was seen in Aug. 1990 at Loma Angash-cola; R. Williams). Increased human population levels in the Quito region have resulted in the condor's being rarely seen there, even in flight overhead, though in the late 1990s several individuals apparently were still occasionally being seen on Volcán Pichincha. Small populations persist on the slopes of Volcán Cayambe and Volcán Antisana in Pichincha, as well as at El Cajas in Azuay. Recorded mostly between 2000 and 4000 m, occasionally as low as 1700 m (Río Chota valley in Imbabura; RSR) or as high as 4500 m.

Numbers of the Andean Condor in Ecuador are gradually declining, a decrease probably caused entirely by persecution, especially poisoning. Having seen them feeding on dead domestic animals, or even afterbirths, some country residents persist in believing that condors are responsible for killing stock. This belief has resulted in their elimination from various areas that appear to be ideal for them (e.g., from most of Loja). As a consequence, we conclude that the Andean Condor merits Endangered status in Ecuador. A reintroduction campaign utilizing captive-bred birds—the species is quite easily bred in captivity—and modeled after the successful efforts in Venezuela and Colombia should be considered for Ecuador. Also helpful might be the establishment of "condor feeding stations" at which carcasses are laid out on a regular basis, much as has been done for California Condors (*Gymnogyps californianus*) in the United States and with several Old World vulture species at various sites in Europe and Africa; these could become important tourist attractions as well. The species as a whole is not believed to be at risk (Collar et al. 1992, 1994).

Monotypic.

Range: Andes from w. Venezuela to Tierra del Fuego, but very local from Ecuador northward; also locally down to Pacific coast from Peru southward, and in Patagonia; Sierra de Perijá and Santa Marta Mountains.

Sarcoramphus papa   King Vulture   Gallinazo Rey

Uncommon to locally fairly common in humid forest (both terra firme and várzea) and adjacent cleared areas in the lowlands of e. Ecuador. Rare and now very local in and near humid and deciduous forest in the lowlands of w. Ecuador from Esmeraldas (recent sightings from the Río Santiago and Río Cayapas drainages, southeast of Muisne, at Bilsa, and northwest of Alto Tambo) south through Guayas to w. Loja (recent sightings from the Zapotillo, Macará, Alto Tambo, and Sozoranga regions [Best et al. 1993; RSR]). In the west the King

Vulture seems to avoid areas with very heavy rainfall and persistent low cloud cover, not being able to forage regularly enough in such areas; in n. Esmeraldas it was seen primarily during less humid months (O. Jahn). Recorded mostly below 500 m; Best et al. (1993) report sightings from as high as 2000 m in Loja, but this is exceptional.

The spectacular King Vulture was once widespread and numerous across much of w. Ecuador—it was considered "common" by Chapman (1926, p. 216)—but has declined greatly over the past 50 or more years as a result of deforestation and agricultural development. Even in comparatively wild areas (e.g., Machalilla National Park in sw. Manabí, and in s. Loja) it is now seen only infrequently, and numbers seen are usually very small, apparently because of the loss or diminution in numbers of its natural prey base in such areas. Probably the largest population persists in the extensively forested hills west of Macará in extreme s. Loja where, for instance, five birds (including 2 immatures) were seen simultaneously on 17 Sep. 1998 (RSR et al.). Numbers of King Vultures appear to be holding up relatively well in the Oriente where upwards of six to eight or more can be seen in a day when flying conditions are good.

Monotypic.

Range: s. Mexico to n. Argentina and se. Brazil; w. Ecuador and extreme nw. Peru.

Coragyps atratus   **Black Vulture**   **Gallinazo Negro**

Common to locally abundant in semiopen areas, particularly numerous around towns and cities (where large numbers gather at garbage dumps, also prospering from the all-too-frequent practice of dropping refuse along roadsides), in the lowlands of both e. and w. Ecuador; ranges up in diminished numbers onto Andean slopes, in small numbers even reaching the central valley (e.g., around Quito). Ubiquitous in settled areas, the Black Vulture actually is quite rare in regions that are still primarily forested, where it is mainly restricted to river-edge situations. With clearing and increased human activity (especially the advent of garbage dumps), Black Vulture numbers increase quickly, as was witnessed along the Maxus road in 1994–1995 (RSR et al.). Recorded mainly below 2000 m, in smaller numbers to 3000 m or slightly higher.

The species has sometimes been called the American Black Vulture.

Coragyps atratus is now usually considered to be monotypic (e.g., *Birds of the World*, vol. 1; *Birds of the Americas*, pt. 1, no. 4), though some authors continue to recognize several races. If races are recognized, western birds in Ecuador would be referable to *foetens*, whereas those of the east would be assignable to *brasiliensis*. We consider the species to be monotypic.

Range: s. United States to s. Chile and cen. Argentina; vagrants to West Indies.

Cathartes aura   **Turkey Vulture**   **Gallinazo Cabecirrojo**

Fairly common to common and widespread in a variety of forested, wooded, and open habitats (including coastlines, especially where rocky) in the lowlands of w. Ecuador, ranging up in smaller numbers into the foothills and subtropical zone on the west slope of the Andes, occasionally as high as the central valley (e.g., around Quito). Also resident on various offshore islands, such as Isla de la Plata off s. Manabí and Isla Santa Clara in the Gulf of Guayaquil. The Turkey Vulture is considerably less numerous and more local in e. Ecuador, occurring mainly fairly close to the Andes though it is found in small numbers along much of the Río Napo and has been recorded recently from around Kapawi Lodge on the Río Pastaza near the Peruvian frontier; in addition, Turkey Vultures are regular in parts of the southeast (e.g., around Zamora and in the Bombuscaro and Romerillos sectors of Podocarpus National Park). In the east the species favors partially cleared and riparian areas and is uncommon in the foothills and lower subtropical zone. Although boreal migrants quite possibly occur in Ecuador, as far as we can determine there are no specific records; they should especially be watched for in the eastern lowlands, where at least on occasion they might even occur in flocks. Recorded mostly below 2000 m, in small numbers up to about 3000 m.

Racial allocations in *C. aura* are difficult to make, being hindered by the inadequacy of accurately labeled specimen material. The general pattern is as follows, but observations of living birds are still needed and will likely result in clarifications. The race *falklandica* is resident on offshore islands. *Falklandica's* bare skin on its head is bright reddish in living birds. The similar race *jota* is resident in w. Ecuador, including locally in the highlands; it differs from the slightly smaller *falklandica* in having less intensely colored bare red skin on the head and in usually lacking *falklandica's* often extensive grayish or brownish edging on its wing feathers (particularly the coverts). The race *ruficollis* is resident locally in the eastern lowlands and on the east slope of the Andes; the bare skin on its head is also basically red, but it also has an area of pale yellowish or grayish stripes, often quite prominent, across its nape. Several birds observed in Sep. 1998 in the Buenaventura area of El Oro clearly showed pale striping on the nape; their racial allocation must remain uncertain. In addition to these resident forms, a migratory northern race, *meridionalis*, was recorded from Ecuador without details by Blake (1977); we have not, however, been able to determine the basis for this, though it seems likely that boreal migrants do occur in Ecuador.

Range: North America south widely to Tierra del Fuego, though tending to avoid heavily forested tropical areas; breeders from w. North America winter south into South America (large numbers pass as far south as Panama), but details of this migration remain unknown; Greater Antilles.

---

*Cathartes melambrotus Greater Yellow- Gallinazo Cabeciamarillo
headed Vulture Mayor

Uncommon to locally fairly common in humid forest (both terra firme and várzea) in the lowlands of e. Ecuador. Although the species is now recognized to be numerous and widespread in the Oriente, the first specimens taken in

Ecuador were reportedly not obtained until 1989 (Marín et al. 1992). These authors, however, overlooked the old Gualaquiza specimen (Salvadori and Festa 1900) labeled as "*C. burrovianus.*" At that time, that was the only recognized species of "Yellow-headed Vulture," *C. melambrotus* of Amazonia not having been described until 1964 (A. Wetmore. A revision of the American vultures of the genus *Cathartes. Smith. Misc. Coll.* 146: 1–48, 1964). The Gualaquiza specimen seems almost certain to have been a Greater Yellow-headed Vulture, but it has apparently not been critically reexamined. Recorded mostly below 800 m, in smaller numbers and seemingly locally up to 1200–1300 m.

The species has sometimes been called the Forest Vulture.

The correct spelling (Sibley and Monroe 1990) of the species name is *melambrotus* and not *melambrotos*, contra Meyer de Schauensee (1966, 1970). Monotypic.

Range: se. Colombia, s. and e. Venezuela, and the Guianas to n. Bolivia and Amazonian Brazil.

We should note that we cannot accept the published sighting (Tallman and Tallman 1977) of Lesser Yellow-headed Vulture (*C. burrovianus*) from Limoncocha, believing it far more likely that the report involved only individuals of the Greater Yellow-headed Vulture. Despite careful checking by various observers subsequently, even in areas with habitat much more suitable for the Lesser than is found at Limoncocha (such as on islands in the Río Napo farther downstream, e.g., near the mouth of the Río Aguarico), no additional sightings of the species have been made. The closest known site for the Lesser Yellow-headed Vulture in Amazonia remains the Iquitos region of ne. Peru.

Falconiformes
Accipitridae   Kites, Eagles, Hawks, and Osprey

Forty-eight species in 24 genera. The family is worldwide in distribution. The Osprey (*Pandion haliaetus*) was formerly separated in a monotypic family, the Pandionidae, but it is now generally placed as a subfamily within the Accipitridae (Sibley and Ahlquist 1990; AOU 1998).

Pandion haliaetus   Osprey   **Aguila Pescadora**

Uncommon to locally fairly common and conspicuous near open areas of water, both fresh and salt, virtually throughout, though most numerous in the lowlands; in the Andes, where it may occur primarily as a transient, Ospreys are usually found around highland lakes, especially in the central valley. Although the species occurs year-round, numbers are greatest between Oct. and Mar. The Osprey does not breed in Ecuador, nor does it do so anywhere in South America, though why this is the case remains a matter of conjecture; there is one 19th-century report of supposed breeding in nw. Peru (Berlepsch and Taczanowski 1884). Presumably it is prebreeders that are found in Ecuador during the northern summer months (ca. Apr. to Sep.); Ospreys generally do not begin to nest until they are at least three years of age. Recorded up to about 3000 m, but a few occur even higher while on passage (e.g., one was at Papallacta lake on 16 Nov. 1996; P. Coopmans et al.), when they also may overfly areas that are far from water.

A total of 35 Ospreys banded in various parts of North America have been recovered in Ecuador, almost all of them (regrettably) after they were shot. Most had been killed between Sep. and Mar.; of the few that were recovered during the boreal summer, almost all were only one or two years old.

One race: *carolinensis*.

Range: nearly cosmopolitan, breeding on every continent except South America and Middle America south of Belize, where North American breeders do winter widely.

Leptodon cayanensis   **Gray-headed Kite**   **Elanio Cabecigris**

Rare to uncommon in the canopy and borders of humid forest (both terra firme and várzea) in the lowlands of both e. and w. Ecuador; slightly more numerous in the east than the west, where numbers have doubtless been reduced by deforestation. In w. Ecuador recorded south to sw. Manabí (a sighting from Machalilla National Park in Jan. 1990; Parker and Carr 1992), Los Ríos (a 19th-century record from Babahoyo, but seemingly no recent reports), and El Oro (sightings from Buenaventura and near Balsas; v.o.). Recorded mostly below 700 m, occasionally or locally up to 900 m (in El Oro).

One race: nominate *cayanensis*.

Range: s. Mexico to sw. Ecuador, n. and e. Bolivia, e. Paraguay, ne. Argentina, and se. Brazil.

Chondrohierax uncinatus   **Hook-billed Kite**   **Elanio Piquiganchudo**

Rare and local in the canopy and borders of humid, semihumid, and montane forest and woodland in the lowlands of both e. and w. Ecuador, and in the subtropical zone on the west slope of the Andes (where there is also one sighting from the east slope, a bird seen at Cabañas San Isidro on 18 Apr. 1998; M. Lysinger). In w. Ecuador recorded from Esmeraldas south very locally to cen. Manabí, Guayas (an old specimen [FMNH] from Samborondón [*Birds of the Americas*, pt. 1, no. 4]; also recent reports from Cerro Blanco [Berg 1994] and in the Naranjal area [RSR and PJG]), El Oro (Buenaventura), and w. Loja (Tierra Adentro near Alamor); in the last two provinces recorded only from sightings of Best et al. (1993) in Feb.–Mar. 1991. In the west recorded regularly up to about 1800 m, occasionally as high as 2300 m above Mindo (M. B. Robbins); in the east known principally from below about 800 m, but on rare occasions as high as 2000 m.

We suspect that the Hook-billed Kite was formerly more numerous in w. Ecuador than in the east—at least considerably more specimens are available from there. Extensive deforestation in the west during the past half-century has apparently resulted in a major decline in overall numbers in that area. This wide-ranging raptor is not, however, considered to be at risk (Collar et al. 1992, 1994).

One race: nominate *uncinatus*.

Range: s. Texas and Mexico to n. Argentina, Paraguay, and se. Brazil (but seemingly absent from much of n. Amazonia and e. Brazil); w. Ecuador and nw. Peru; Cuba and Grenada. Cuban birds are sometimes regarded as a separate species (Cuban Kite, *C. wilsonii*).

Elanoides forficatus   Swallow-tailed Kite   Elanio Tijereta

Uncommon to locally common (and conspicuous and, where numerous, often gregarious) in humid and montane forest and woodland and their borders in the lowlands of both e. and w. Ecuador, and in the subtropical zone on both slopes of the Andes; favors hilly or mountainous areas, and usually most numerous in the foothills (e.g., at Tinalandia). In w. Ecuador found south locally to El Oro and w. Loja, where it occurs locally even in relatively decid-uous forest (e.g., at Alto Tambo; Best et al. 1993). The lovely Swallow-tailed Kite seems to persist well in quite fragmented and degraded forest patches, more so than most other forest-based raptors. Although this species is known in Ecuador primarily from presumably resident populations, transients and non-breeding visitants from populations nesting to the north are suspected to occur, though there is only one definite record (see below). Birds breeding toward the southern end of the range in South America also are known to depart that area during the austral winter, and these could also occur as far north as Ecuador, but as the same subspecies is involved these would be essentially impossible—barring the recovery of a bird that had been banded to the south—to differentiate from resident birds. Recorded mostly below about 1500 m, but smaller numbers regularly range higher, reaching 2000 m or rarely even more, and presumably transient birds can be seen virtually anywhere, even over paramo at 3500 m or more (as seen by PJG east of Salcedo in Cotopaxi on 30 Dec. 1981).

The species has sometimes been called the American Swallow-tailed Kite.

Two races are known to occur in Ecuador, *yetapa* and nominate *forficatus*. The former is the widespread form breeding in Ecuador and has its back and upperwing-coverts glossed with green, at least in fresh plumage. Nominate *forficatus*, which has its back and upperwing-coverts glossed with blue when in fresh plumage, and which breeds in the s. United States, is known in Ecuador only from an old specimen (AMNH) taken in Dec. at Bucay in e. Guayas. As the migratory form could so easily be overlooked among resident birds—only a specimen record could be regarded as definite—we suspect that boreal migrants may be more regular than this single record would seem to indicate.

Range: se. United States and Mexico to extreme nw. Peru, n. Argentina, e. Paraguay, and se. Brazil; birds breeding from Panama north withdraw south-ward into South America during boreal winter (but their exact wintering range is still imperfectly known), and likewise southern breeders are known to withdraw northward during austral winter.

Gampsonyx swainsonii   **Pearl Kite**   **Elanio Perla**

Uncommon but conspicuous in arid scrub, light deciduous woodland, and agricultural terrain in the lowlands of sw. Ecuador from Manabí (Bahía de Caráquez area) south through El Oro and extreme w. Loja. Since the mid-1970s the Pearl Kite has also been gradually moving into cleared areas in the lowlands of ne. Ecuador, where to date it has been found only in Sucumbíos and Napo, so far primarily in the Lago Agrio/Coca/Tena region, but we expect that it will continue to spread and increase, colonizing newly cleared areas. The first sighting in the east dates from 27 Sep. 1976 at Tiputini in far e. Napo; it and several subsequent early reports are summarized by Ridgely (1980). In e. Ecuador recently recorded as high as 800 m north of Archidona (P. Coopmans and M. Van Beirs et al.); in the west not known above about 300 m.

To date no specimens are available from e. Ecuador, where presumably it is nominate *swainsonii* that occurs; this is similar to *magnus* of the southwest but is slightly smaller and not so richly colored below, especially on the flanks.

Range: w. Nicaragua; Panama, n. Colombia, much of Venezuela, and the Guianas to ne. Ecuador, extreme ne. and se. Peru, n. and e. Bolivia, n. Argentina, and s. Brazil; sw. Ecuador and nw. Peru.

Elanus leucurus*   **White-tailed Kite   **Elanio Coliblanco**

Recently found to be rare and very local in pastures and semiopen agricultural terrain in the lowlands (once in the lower subtropical zone) of both ne. and nw. Ecuador, where so far known only from Carchi, w. Sucumbíos and w. Napo, and w. Esmeraldas, adjacent Manabí, and w. Pichincha. The White-tailed Kite was first found in Ecuador on 7 Apr. 1984 when a single bird was observed at the Lago Agrio airport (PJG et al.). One has been seen in that vicinity irregularly ever since, but the species has so far shown little indication of spreading in e. Ecuador, though it may eventually do so (as it has in many other parts of the Neotropics). The only other reports from the east slope involve probable wandering birds seen about 10 km west of San Rafael Falls (1200 m) on 11 Aug. 1993 (F. Sornoza) and at Cabañas San Isidro on 18 Nov. 1998 (M. Lysinger). In the northwest the White-tailed Kite is known from a few sightings in the Same/Atacames area in w. Esmeraldas, where first noted in Aug. 1991 (H. Kasteleijn), and also from Sep. 1995 sightings from Finca Paraiso de los Papagaios north of Quinindé (J. R. Fletcher et al.); from a family group of five birds found west of El Carmen in extreme n. Manabí on 14 Sep. 1992 (1 of which was collected [ANSP], the only specimen taken in Ecuador); and from one seen south of Santo Domingo de los Colorados in Pichincha on 25 Aug. 1994 (P. Coopmans and R. Still et al.) and another seen south of La Concordia on 7 Sep. 1995 (J. R. Fletcher and M. Swan). More recently a single bird was seen at Páramo del Ángel (3300 m) on 26 Dec. 1997 (J. Hendriks), a most surprising site for this species. Recorded mostly below 300 m, though what are presumably wanderers or dispersing birds have been seen higher.

E. leucurus has occasionally been considered conspecific with *E. caeruleus* (Black-shouldered Kite) of the Old World (e.g., AOU 1983), but the case for considering them as full species was convincingly made by W. S. Clark and R. C. Banks (*Wilson Bull.* 104[4]: 571–579, 1992); this treatment was followed in the 1998 AOU Check-list. One race: nominate *leucurus*.

Range: s. United States to cen. Chile and cen. Argentina, though absent from virtually all of Amazonia and n. and ne. Brazil. The species continues to spread and increase in many regions.

Rostrhamus sociabilis   Snail Kite   Elanio Caracolero

Locally fairly common and conspicuous in freshwater marshes and adjacent rice fields and canals in the lowlands of sw. Ecuador, where recorded mainly from Manabí (in the Chone and Rocafuerte area), Los Ríos, and e. Guayas (south—erratically and in small numbers—to Manglares-Churute Ecological Reserve), with scattered sightings of presumed wandering birds from w. Esmeraldas at Same on 24 Aug. 1994 (P. Coopmans), extreme s. Pichincha at Río Palenque, and El Oro north of Santa Rosa (3 seen on 13 Apr. 1996; P. Coopmans). There is also a small, apparently resident, population of Snail Kites at Limoncocha in the lowlands of ne. Ecuador, and in 1993–1994 a few occurred regularly at Sacha Lodge (L. Jost), but otherwise the Snail Kite is nearly unknown from the eastern lowlands aside from a single sighting of a wandering bird at Jatun Sacha (B. Bochan); there also is a 19th-century "Napo" specimen. In the west recorded only below about 50 m; in the east at 250–400 m.

Although it seems to have gone largely unnoticed, a massive Snail Kite decline has occurred since at least the late 1970s across most or all of the species' west Ecuadorian range. In the mid-1970s flocks and roosts of hundreds and even thousands of birds were routine in some areas, notably at several sites in the Yaguachi marshes east of Guayaquil; observers from as long ago as the 1920s (Chapman 1926) commented on the high numbers there. Now a count of a few dozen birds in any single area is a high one. This decline seems attributable to marsh drainage and the widespread conversion of such areas to rice cultivation. Pesticide residues may also be having an effect, depressing reproductive output. As little of its optimal habitat is formally protected—indeed a great deal has already been converted to agricultural pursuits—it is possible that the Snail Kite, once such a conspicuous feature of the southwestern marshes, could continue gradually to decline toward local extirpation. Numbers occurring in the east have always been very small, and there seems to have been little or no change in their status there. Given the rapidity of the Snail Kite's decline in w. Ecuador, however, we believe the species should be given an Ecuadorian status of Vulnerable. As a whole this wide-ranging species is not, however, considered to be at risk (Collar et al. 1992, 1994).

Until recently the species was often called the Everglade Kite, an evocative name—far more so than "Snail Kite," especially given that several other

Neotropical kites have a diet comprised of snails—but one that seems unduly parochial given that only a very marginal population exists in the Everglades of Florida.

One race: nominate *sociabilis*.

Range: Florida and s. Mexico to n. Argentina and Uruguay, though absent from much of n. Amazonia; sw. Ecuador; Cuba.

*Rostrhamus hamatus   Slender-billed Kite   Elanio Piquigarfio

Rare to uncommon and local in várzea and swampy forest and along the margins of ponds and rivers in the lowlands of ne. Ecuador, where recorded almost exclusively from near the Río Napo. We suspect that reports from elsewhere in e. Ecuador may be the result of confusion with the similar Slate-colored Hawk (*Leucopternis schistacea*). The Slender-billed Kite was first recorded from Ecuador at Limoncocha (D. L. Pearson, *Condor* 77[1]: 97, 1975) and is still known primarily from that vicinity; a small population continues to reside around the edge of the lake. A few are also noted regularly around Sacha Lodge (first by L. Jost in 1993–1994), and a few were seen in riparian forest near the south bank of the Río Napo along the Maxus road southeast of Pompeya in 1994–1995 (RSR); there also have been some sightings downstream to around La Selva and the Río Pañayacu as well as upstream as far as the vicinity of Coca (a few possible sightings also exist from Jatun Sacha, fide B. Bochan). The Slender-billed Kite appears, however, to be absent from the Río Aguarico drainage where—despite seemingly suitable habitat—snail-eating birds in general are very scarce or, more often, absent. There also are no confirmed records from the southeastern lowlands. Recorded below 300 m.

This species was formerly placed in the monotypic genus *Helicolestes*, but recent authors agree that there seems to be no justification for maintaining it as a separate genus from the similar *Rostrhamus*. Monotypic.

Range: locally in e. Panama, n. Colombia, n. and e. Venezuela, the Guianas, Amazonian Brazil, n. Bolivia, e. Peru, extreme se. Colombia, and ne. Ecuador.

Harpagus bidentatus   Double-toothed Kite   Elanio Bidentado

Uncommon to fairly common in the canopy and borders of humid forest (both terra firme and várzea) in the lowlands of both e. and w. Ecuador, and rare to uncommon in montane forest in the subtropical zone on both slopes of the Andes. In w. Ecuador recorded south, at least formerly, to w. Chimborazo (Chimbo), but in recent years not found south of s. Pichincha (Río Palenque) and w. Cotopaxi (one seen east of Quevedo, Los Ríos, on 28 Jul. 1991; RSR and G. H. Rosenberg). There is also a recent report (with no details) from Cerro Blanco in the Chongón Hills of Guayas (Berg 1994). Recorded mainly below 1500 m, but locally ranges considerably higher, in Pichincha regularly to 1800 m and occasionally even as high as 2200 m.

Two rather different races are found in Ecuador, nominate *bidentatus* in the east and *fasciatus* in the west. Both sexes of the nominate race are quite uniform and rich rufous below with only a little (or even no) barring, the barring being especially sparse in females. Both sexes of *fasciatus* are, by contrast, more noticeably barred below and males are predominately gray on the underparts, showing only a little rufous, mainly as a wash on the sides of the breast.

Range: s. Mexico to w. Ecuador, n. and e. Bolivia, and Amazonian Brazil; e. Brazil.

*[*Harpagus diodon*   **Rufous-thighed Kite**   **Elanio Muslirrufo**]

Uncertain. Apparently an accidental visitant to the lowlands of e. Ecuador. Known from an adult observed at close range by RSR and R. A. Rowlett et al. in várzea forest at Limoncocha on 11 Jul. 1979 (Ridgely 1980), and from an evident immature seen there on 23 Oct. 1979 (D. Pearson and PJG, the identity of this bird having been regarded as uncertain until recently). Ridgely (op. cit.) speculated that the individual seen in Ecuador might have been an austral migrant, and recent observations from Paraguay, where the species appears to be absent during the austral winter (F. E. Hayes, P. A. Scharf, and R. S. Ridgely, *Condor* 96[1]: 85–86, 1994), support this possibility. There have been, however, no additional reports from w. Amazonia, so the Limoncocha birds must be presumed to have been extreme vagrants. Recorded at around 300 m.

Monotypic.

Range: locally from e. Bolivia to n. Argentina and e. Brazil, but with a scatter of records from Amazonian Brazil and the Guianas; ne. Ecuador.

Ictinia plumbea   **Plumbeous Kite**   **Elanio Plomizo**

Uncommon to locally fairly common and conspicuous in the canopy of humid forest, forest borders, and adjacent clearings with scattered large trees in the lowlands and foothills of both e. and w. Ecuador. Except in Esmeraldas, the Plumbeous Kite is rather local and not very numerous in w. Ecuador, where it has been recorded south locally to se. Guayas (a few recent sightings from Manglares-Churute Ecological Reserve) and El Oro (an old record from Santa Rosa, but only two recent reports: 2 seen west of Piñas on 16 Mar. 1996, and 5 or more there on 9 Feb. 1998 [P. Coopmans et al.]); the Plumbeous Kite is also present, and presumably breeds, at Cerro Blanco in the Chongón Hills. Like the Swallow-tailed Kite (*Elanoides forficatus*), the Plumbeous is known in Ecuador primarily from presumed resident populations; however, transients and nonbreeding visitants from populations nesting to Ecuador's north and south also almost certainly occur, for both regions are known to be vacated during their respective nonbreeding seasons. The single bird that died on Isla de la Plata on an unspecified date in 1990 (F. I. Ortiz-Crespo and P. Agnew, *Bull. B. O. C.* 112[2]: 70, 1992) may have been derived from one of these migratory populations. Recorded mostly below about 1000 m, in smaller numbers somewhat higher (and doubtless might occur considerably higher on migration).

Monotypic.

Range: ne. Mexico to sw. Ecuador, n. Argentina, and se. Brazil; breeders toward either end of range withdraw toward equator during their respective winters.

[Ictinia mississippiensis   Mississippi Kite   Elanio de Mississippi]

Apparently a casual transient. Known in Ecuador from only a single definite sighting: G. H. Rosenberg et al. observed three individuals in flight over the Loreto road (1000 m) north of Archidona in w. Napo on 14 Apr. 1992. Other small flocks of apparently migrating *Ictinia* kites seen flying high over Zancudococha in early Apr. 1991 (M. B. Robbins) were also believed to have been Mississippi Kites, but the observer could not be entirely certain of their identity. Apparently the usual migratory routes of the Mississippi Kite pass east of Ecuador, though as with the Swainson's Hawk (*Buteo swainsoni*) it is possible that some kites do regularly pass over the country, perhaps usually remaining so high above the ground as to go unobserved.

Monotypic.

Range: breeds in s. United States, wintering in interior s.-cen. South America; migratory routes and exact wintering range still imperfectly known.

Circus cinereus   Cinereous Harrier   Aguilucho Cenizo

Rare to locally uncommon in paramo and adjacent open, grassy, agricultural terrain from Carchi (Páramo del Ángel) south locally to Cañar (a 19th-century specimen from Cañar); there are, however, no recent reports from south of Chimborazo. The absence of the Cinereous Harrier from s. Ecuador (and from n. Peru, for that matter) seems inexplicable but appears to be genuine. Recorded mostly from 3000 to 4000 m.

The Cinereous Harrier seems always to have been a scarce bird in Ecuador, but our suspicion is that its numbers have declined markedly over the past several decades. The only area where it now is seen at all regularly is on the Páramo del Ángel in Carchi. We thus consider the species to be Vulnerable in Ecuador. Elsewhere in its extensive range the Cinereous Harrier remains relatively numerous (especially southward), and the species has therefore not been viewed as being generally at risk (Collar et al. 1992, 1994).

Monotypic.

Range: locally in Andes of Colombia and Ecuador, and from cen. Peru south to Tierra del Fuego, spreading east across Argentina to Uruguay, e. Paraguay, and extreme se. Brazil.

A sighting (E. L. Mills, *Ibis* 109: 536, 1967) of a female-plumaged harrier on the Santa Elena Peninsula east of La Libertad on 6 May 1966 is—presuming the bird was correctly identified to genus—inexplicable. The record could conceivably refer to an out-of-range austral migrant Cinereous Harrier, or to a wanderer of that species down from the highlands (though the species is not otherwise known to wander in tropical South America). That it was a Northern, or Hen, Harrier (*C. cyaneus*), a holarctic species known occasionally to reach sw. Colombia as a migrant during the northern winter months (Hilty and Brown 1986), is also possible.

Accipiter ventralis   Plain-breasted Hawk   Azor Pechillano

Uncommon in montane forest and woodland and their borders (also sometimes coming out into adjacent agricultural terrain) in the subtropical and temperate zones on both slopes of the Andes, also locally on slopes above the central and interandean valleys where at least patchy wooded habitat remains. On the west slope recorded south to w. Loja (numerous recent sightings from the Celica region, Sozoranga, Utuana, and elsewhere; v.o.). Recorded mostly from 1700 to 3500 m, occasionally somewhat lower (perhaps especially in the southwest, as it has been seen in Tumbes National Forest in adjacent Peru; Parker et al. 1995). Two old specimen records exist from the Guayas lowlands (Bucay, Naranjo), but there are no recent reports from there.

We treat *A. ventralis* as an allospecies separate from *A. striatus* (Sharp-shinned Hawk) of North America (which breeds south to Mexico) and the Greater Antilles, with *A. chionogaster* (White-breasted Hawk) of Middle America and *A. erythronemius* (Rufous-thighed Hawk) of s. South America also as allospecies, as was also done in Sibley and Monroe (1990) and Bierregaard (1994). Monotypic.

Range: mountains of n. Venezuela, and Andes from w. Venezuela to w. Bolivia; Sierra de Perijá and Santa Marta Mountains; locally in tepuis of s. Venezuela.

Accipiter collaris   Semicollared Hawk   Azor Semicollarejo

Very rare to rare and seemingly very local (but perhaps mostly overlooked) in montane forest, forest borders, and adjacent clearings with large trees in the subtropical zone on both slopes of the Andes. On the west slope known only from Pichincha (old specimens from Nanegal and Gualea, and recent sightings from the Mindo/Tandayapa region; v.o.). On the east slope known only from w. Napo (a few recent sightings from the Baeza region, and 1 specimen [ANSP] obtained northeast of El Chaco on 3 Oct. 1992), Morona-Santiago (a few sightings along the Gualaceo-Limón road), and Zamora-Chinchipe (sightings along the Loja-Zamora road; J. Arvin et al.; J. Rowlett and R. A. Rowlett et al.; Rasmussen et al. 1994). Recorded from about 1500 to 2200 m.

The Semicollared Hawk is considered to be Near-threatened by Collar et al. (1992, 1994). However, we do not believe there is sufficient evidence to indicate that this species is at risk, at least not in Ecuador where it seems always to occur at very low densities, and perhaps even favors areas where montane forest cover has been fragmented.

Monotypic.

Range: locally in Andes of w. Venezuela, Colombia, and Ecuador, with one record from se. Peru; Santa Marta Mountains.

Accipiter superciliosus   Tiny Hawk   Azor Chico

Rare in humid forest (principally terra firme), forest borders, and adjacent clearings with large trees in the lowlands of both e. and w. Ecuador; more numerous in w. Ecuador, and decidedly rare and local in the east. In w. Ecuador recorded south to nw. Guayas (one seen along the Río Ayampe on 7 Jul. 1978; RSR and D. Wilcove), nw. Azuay (Aug. 1991 sightings from Manta Real; G. H. Rosenberg and RSR), and El Oro (one seen at Buenaventura on 19 Sep. 1998; RSR et al.). Details regarding three specimens recently collected in w. Ecuador are given by Marín et al. (1992). Of note is the rarity or absence of the Tiny Hawk from much of e. Ecuador; at several well-studied sites in the upper Río Napo drainage—such as Jatun Sacha, Limoncocha, the Maxus road southeast of Pompeya, and La Selva—the species is virtually unknown. Recorded mostly below 900 m; an old specimen supposedly procured at Mindo (1300 m) may not have actually been obtained there (the species has not been found recently in the Mindo region).

Two similar races are found in Ecuador, nominate *superciliosus* in the east and *fontanieri* in the west. *Fontanieri* has somewhat wider and blacker ventral barring.

Range: Nicaragua to sw. Ecuador, n. Bolivia, and Amazonian Brazil; se. Brazil, e. Paraguay, and ne. Argentina.

Accipiter bicolor   Bicolored Hawk   Azor Bicolor

Rare—but secretive and doubtless often overlooked—in humid, deciduous, and montane forest and woodland, their borders, and adjacent clearings in the lowlands of both e. and w. Ecuador (in the east particularly in terra firme) and in the lower subtropical zone on both slopes of the Andes. In the west recorded south through w. Loja in the Alamor area and at Sozoranga. Recorded mostly below 1800 m, but ranges at least occasionally or locally up to 2200–2700 m (e.g., an ANSP specimen taken northeast of El Chaco in w. Napo; old specimens from El Pun and Alunguncho).

One race: nominate *bicolor*.

Range: e. Mexico to nw. Peru, n. Argentina, and Uruguay. Excludes Chilean Hawk (*A. chilensis*) of s. Chile and s. Argentina.

Accipiter poliogaster   Gray-bellied Hawk   Azor Ventrigris

Very rare in humid forest and forest borders (both in terra firme and várzea) in the lowlands of e. Ecuador. There are only a few Ecuadorian records of this little-known hawk. It has been suggested (e.g., Hilty and Brown 1986) that the Gray-bellied Hawk might occur in n. South America as an austral migrant. This remains possible, but several Ecuadorian dates do not fit such a pattern. There are only two Ecuadorian specimens, the first taken in the 19th century at Sarayacu (M. A. Jenkinson and M. D. Tuttle, *Auk* 93[1]: 188–189, 1976) and the second, an immature in the *pectoralis*-type plumage (see below), obtained on 24 Apr. 1923 at Río Suno (AMNH). There have been a few subsequent sightings: an adult seen along the lower Río Yasuní on 18 Sep. 1976 (RSR), another adult at Imuyacocha on 23 Mar. 1991 (RSR and F. Sornoza), an immature along the Maxus road southeast of Pompeya on 20 Sep. 1995 (F. Sornoza), an immature at La Selva on 25 Dec. 1997 (P. Coopmans et al.), adults seen on two occasions at Kapawi Lodge in Jan. 1998 (PJG and D. Wolf), and an adult seen at Tiputini Biodiversity Center in Feb. 1999 (M. Lysinger). Recorded below about 400 m.

The Gray-bellied Hawk was treated as a Near-threatened species by Collar et al. (1992, 1994). From an Ecuadorian perspective, however, we are not aware of any factors that can be interpreted as affecting it adversely, and we therefore do not consider it to be at risk.

The species has sometimes been called the Gray-bellied Goshawk.

This large *Accipiter* is notable for its striking immature plumage, a plumage so strongly reminiscent of the adult *Spizaetus ornatus* (Ornate Hawk-Eagle) that these birds were long considered to represent a separate species, *A. pectoralis*. See W. H. Partridge (*Condor* 63[6]: 505–506, 1961) for an explanation as to why they are now considered to be conspecific; A. Whittaker and D. C. Oren (*Bull B. O. C.* 119[4]: 235–260, 1999) provide photos of a subadult plumage, heretofore unknown. Monotypic.

Range: locally from n. Colombia, s. Venezuela, and the Guianas to n. Bolivia and Amazonian Brazil; se. Brazil, se. Paraguay, and ne. Argentina.

Geranospiza caerulescens   Crane Hawk   Gavilán Zancón

Rare to locally uncommon in both humid and deciduous forest and woodland and their borders in the lowlands of both e. and w. Ecuador. In the east perhaps slightly more numerous in várzea than in other habitats. In the west, where it favors deciduous forest and borders, the Crane Hawk is recorded from Manabí (an old specimen from Río Peripa [Salvadori and Festa 1900]; recent reports from Machalilla National Park and north of Jipijapa) and s. Pichincha (Río Palenque) south locally through coastal El Oro and extreme w. Loja (east of Mangaurco; ANSP). Recorded only below about 400 m.

Two similar races are found in Ecuador, nominate *caerulescens* in the east and *balzarensis* in the west. *Balzarensis* is slightly darker overall.

Range: n. Mexico to n. Argentina and Uruguay; sw. Ecuador and nw. Peru.

Leucopternis schistacea   Slate-colored Hawk   Gavilán Pizarroso

Rare to locally uncommon in humid forest (both terra firme and várzea, but especially near water) in the lowlands of e. Ecuador, where recorded mainly from the northeast in the Ríos Napo and Aguarico drainages. There are many fewer records from the southeast, though this doubtless is mainly a reflection of much less complete observer coverage there; the species has, for instance, been seen around Kapawi Lodge along the Río Pastaza near the Peruvian border. Recorded up only to about 400 m.

Monotypic.

Range: e. Colombia and sw. Venezuela to n. Bolivia, Amazonian Brazil, and French Guiana.

Leucopternis plumbea   **Plumbeous Hawk**   Gavilán Plomizo

Rare to locally uncommon (but easily overlooked) in very humid to humid and montane forest in the lowlands and—perhaps especially—the lower subtropical zone of nw. Ecuador, where recorded from Esmeraldas, adjacent sw. Imbabura (a pair seen and tape-recorded north of the Salto del Tigre bridge over the Río Guaillabamba on 15 Jul. 1997; P. Coopmans et al.), Pichincha, n. Manabí (an old specimen from Río de Oro; Chapman 1926), and n. Los Ríos (where known only from a 1950 specimen [ANSP] taken at Quevedo). In the wet forest belt of n. Esmeraldas, the Plumbeous Hawk is a forest-interior species, and unlike the Semiplumbeous Hawk (*L. semiplumbea*) it usually does not occur at forest borders (O. Jahn and P. Mena Valenzuela et al.). A single 19th-century record from Tumbes, Peru, leads one to suspect that the Plumbeous Hawk might also occur in sw. Ecuador. Despite a tremendous amount of field work in this region in recent years, however, there are no confirmed records supporting this supposition, though the species was listed (Berg 1994)—without details—as occurring at Cerro Blanco in the Chongón Hills, an area with seemingly unsuitable habitat. The Plumbeous Hawk still ranges in the patches of forest at Tinalandia and is also present at Río Palenque (with a record as recently as Dec. 1996; P. Coopmans et al.). Recorded up to about 1700 m.

The Plumbeous Hawk is considered to be a Near-threatened species by Collar et al. (1992, 1994), an assessment that seems appropriate from a strictly Ecuadorian perspective and with which we concur.

Monotypic.

Range: Panama to nw. Ecuador, possibly to extreme nw. Peru.

Leucopternis semiplumbea   **Semiplumbeous**   **Gavilán**
                              Hawk   **Semiplomizo**

Rare to locally uncommon (but easily overlooked) in very humid to humid forest and forest borders in the lowlands and lower foothills of nw. Ecuador in Esmeraldas and nw. Pichincha. The Semiplumbeous Hawk is known from relatively few records in Ecuador. In Esmeraldas it is known from old specimens from San Javier and Río Cayapas and from recent reports from the Ríos Santiago and Cayapas drainages, where it is considered to be of widespread occurrence (P. Mena Valenzuela and O. Jahn et al.). In Pichincha it has been recorded from Río Palenque (where it has apparently not been found since 1986; PJG) and the Río Guaycuyacu (one seen on 26 Aug. 1991; N. Krabbe). It also has been seen near and north of Pedro Vicente Maldonado, and a single specimen (MECN) was taken in this area on 8 Jun. 1992. Recorded up to about 600 m; a few recent sightings from the Mindo region—at an elevation that would be exceptionally high for this species—are regarded as uncertain.

The Semiplumbeous Hawk is considered to be a Near-threatened species by Collar et al. (1992, 1994), an assessment that seems appropriate from a purely Ecuadorian perspective and with which we concur.

Monotypic.

Range: Honduras to nw. Ecuador.

Leucopternis melanops   Black-faced Hawk   Gavilán Carinegro

Rare and seemingly somewhat local (though it may merely occur at low densities, and is certainly inconspicuous) in humid forest in the lowlands of e. Ecuador from Sucumbíos south into Morona-Santiago (a 1990 sighting at Taisha; M. B. Robbins). The attractive Black-faced Hawk appears to favor terra firme but is occasionally also noted in várzea and floodplain forests. Recorded mostly below 500 m, but very locally as high as 900 m (sightings in the Bermejo oil-field area of Sucumbíos [M. B. Robbins; A. Capparella]).

Monotypic.

Range: se. Colombia, ne. Ecuador, ne. Peru, n. Amazonian Brazil, s. Venezuela, and the Guianas.

Leucopternis albicollis   White Hawk   Gavilán Blanco

Rare to locally uncommon in humid forest and forest borders (almost entirely in terra firme) in the lowlands and foothills of e. Ecuador. The White Hawk seems to be rather locally distributed in Ecuador. With the notable exception of the Maxus oil-field area southeast of Pompeya (where in 1994–1995 it was found to be not infrequent; RSR et al.), the species is usually distinctly scarce at any distance east of the Andes. Recorded regularly up to about 1100 m, with a single record from as high as 1500 m (MCZ specimen at Palm Peak on the lower slopes of Volcán Sumaco taken on 8 Aug. 1964).

One race: nominate *albicollis*.

Range: s. Mexico to w. Colombia, n. Bolivia, and Amazonian Brazil.

Leucopternis occidentalis   Gray-backed Hawk   Gavilán Dorsigris

Uncommon to fairly common but now quite local in humid and montane woodland and forest, even where surprisingly fragmented, in the lowlands, foothills, and subtropical zone of w. Ecuador from s. Esmeraldas (southeast of Muisne and Cerro Mutiles, Bilsa, and south of Chontaduro on the west bank of the Río Verde [ANSP]) south through El Oro and w. Loja (south to the Alamor/Celica region and near Sozoranga). Recorded up to about 1800 m near Celica and Sozoranga, though found mostly below 1300 m.

The handsome Gray-backed Hawk, which is virtually endemic to Ecuador, has recently been considered to merit Endangered status (Collar et al. 1992,

1994). We concur, for its overall numbers have certainly declined significantly over the past half-century, almost entirely as a result of the massive deforestation that has taken place over a large part of its former range. Despite the fact that this species appears to persist well in fragmented habitat—it even often seems to remain tolerably numerous there—we fear that populations in such situations may not be viable over long periods, a concern also voiced by Best et al. (1993). Also worrisome is the fact that only three Ecuadorian populations of the Gray-backed Hawk occur in areas presently receiving protection: Machalilla National Park (where good numbers also survive in areas just outside the park, particularly south of the Río Ayampe), Cerro Blanco Reserve, and the El Tundo Reserve in s. Loja.

It has been suggested (e.g., *Birds of the Americas*, pt. 1, no. 4) that *occidentalis* might be only a well-marked race of *L. albicollis* (White Hawk), but most authors have not so considered it, and we do not believe that they are that closely related. Monotypic.

Range: w. Ecuador and extreme nw. Peru.

Leucopternis princeps   **Barred Hawk**   **Gavilán Barreteado**

Rare to locally uncommon (but often very conspicuous, soaring high and regularly calling loudly and drawing attention to itself) in montane forest and forest borders in the foothills and subtropical zone on both slopes of the Andes. Also recently found locally on the coastal Cordillera de Mache in w. Esmeraldas at Bilsa (R. P. Clay et al.) and on the Cordillera de Colonche in w. Guayas on Cerro La Torre (N. Krabbe). On the west slope of the Andes recorded south locally to w. Loja (sightings in Feb. 1991 at Tierra Colorada near Alamor; Best et al. 1993), on the east slope south locally to Zamora-Chinchipe (a sighting of one bird in the Río Bombuscaro valley of Podocarpus National Park on 2 Jul. 1992 [P. Coopmans] and a sighting of 3 birds near Chinapinza on the Cordillera del Cóndor on 13 Jun. 1993 [RSR and F. Sornoza]). We now suspect that the record of the Gray-backed Hawk (*L. occidentalis*) from the Cordillera del Cóndor (see Robbins et al. 1987, p. 249) was the result of confusion with the Barred Hawk, a species much more likely on biogeographic grounds, and which is now known to occur south to the Cordillera del Cóndor; the Barred Hawk is still not formally recorded from Peru (Parker et al. 1982), but it was seen in Jul. 1994 on the Peruvian side of the Cordillera del Cóndor (T. S. Schulenberg; Schulenberg and Awbrey 1997). Recorded mostly from 700 to 2200 m, locally down to nearly sea level in n. Esmeraldas in the Ríos Santiago and Cayapas drainages (O. Jahn and P. Mena Valenzuela et al.); there also is an old specimen taken at Río Cayapas.

The species is sometimes called the Black-chested, or the Prince, Hawk.

Monotypic. We follow Blake (1977) in not recognizing the subspecies *zimmeri* for South American birds.

Range: highlands of Costa Rica and Panama; Andes from Colombia to extreme n. Peru.

Buteogallus meridionalis   Savanna Hawk   Gavilán Sabanero

Uncommon to fairly common in open and semiopen areas with scattered large trees and on the edges of deciduous and semihumid forest and woodland in the lowlands and foothills of sw. Ecuador from cen. Manabí (Bahía de Caráquez area) and s. Pichincha (at least one sighting at Río Palenque) southward through El Oro and w. Loja. The Savanna Hawk occurs regularly in agricultural terrain, especially favoring areas given over primarily to cattle grazing, and it likely has increased in recent years; nonetheless it seems to have shown little propensity to spread northward into what appears to be suitable habitat in nw. Ecuador (e.g., in n. Manabí and w. Esmeraldas). Recorded mostly below 1000 m, but occurs widely up to about 2000 m in cleared areas of Loja; an anomalous specimen (FMNH, in *Birds of the Americas*, pt. 1, no. 4) from the northern highlands at Cayambe seems likely to have been mislabeled.

Formerly placed in the monotypic genus *Heterospizias*, based mainly on the species' longer legs and wings; however, we follow most recent authorities (e.g., *Birds of the World*, vol 1; AOU 1983, 1998) in merging this into *Buteogallus*. One race: nominate *meridionalis*.

Range: Panama; n. Colombia to n. Argentina and Uruguay (though absent from all of n. and w. Amazonia); sw. Ecuador and nw. Peru.

Buteogallus anthracinus   Common Black-Hawk   Gavilán Cangrejero

Rare to uncommon in mangroves (occasionally in adjacent areas, but seems never to range very far from mangroves) along the coast of sw. Ecuador from the Río Guayas estuary in e. Guayas south locally along the El Oro coast to the Peruvian border. Despite being known from the coast of w. Colombia, there is no definite evidence to substantiate the presence of the Common Black-Hawk from northward along the Ecuadorian coast; if it occurs elsewhere, it would appear to be most likely in n. Esmeraldas, where mangrove forests are still extensive.

The Common Black-Hawk has doubtless declined significantly in Ecuador as a result of the clearing of much mangrove forest, mainly in order to create artificial shrimp ponds. This could account for the species no longer being present in some parts of its range where mangrove forests have been reduced to but a small fraction of their former extent (e.g., around the Bahía de Caráquez). A population does inhabit Manglares-Churute Ecological Reserve in se. Guayas. We consider the species to merit Near-threatened status in Ecuador, though it is not considered to be at risk in its range generally (Collar et al. 1992, 1994).

There is continuing taxonomic dispute as to whether the black-hawks occurring along the Pacific coast from s. Mexico to nw. Peru represent a separate species, *B. subtilis* (Mangrove Black-Hawk) or are representatives of the more wide-ranging *B. anthracinus*. We do not find the evidence for considering *B. subtilis* as a separate species persuasive and thus opt to consider it conspecific with *B. anthracinus*. Stiles and Skutch (1989) and Howell and Webb (1995) provide some of the rationale for continuing to treat the forms as conspecific; however, the 1983 and 1998 AOU Check-lists and *Birds of the World* (vol. 1) still consider the two forms as separate species. One race of *B. anthracinus* occurs in Ecuador, *subtilis*.

Range: sw. United States to n. and e. Venezuela and Guyana, and along Pacific coast to extreme nw. Peru.

Buteogallus urubitinga   Great Black-Hawk   Gavilán Negro Mayor

Rare to uncommon and local in humid forest (both várzea and along streams in terra firme) and along rivers in the lowlands of e. Ecuador, where thus far known mainly from the drainages of the Ríos Napo and Aguarico though there are also recent reports from Kapawi Lodge on the Río Pastaza near the Peruvian border; it likely ranges farther upstream in the Río Pastaza drainage as well. Rare in both humid and deciduous forest and their borders in the lowlands of w. Ecuador, where known mainly from Manabí and Los Ríos south locally through El Oro and w. Loja; there also are old records from n. Esmeraldas (Pulún and Río Bogotá) and w. Imbabura (Achotal, fide *Birds of the Americas*, pt. 1, no. 4). In the east not recorded above about 400 m (Jatun Sacha), but ranges higher in the west, regularly to 900–1400 m in El Oro and w. Loja; there is one sighting from as high as 2000 m (west of Portovelo in El Oro; J. Arvin and R. A. Rowlett et al.).

One race: nominate *urubitinga*.

Range: n. Mexico to n. Argentina and Uruguay; locally in sw. Colombia; and in sw. Ecuador and nw. Peru.

Harpyhaliaetus solitarius   Solitary Eagle   Aguila Solitaria

Rare and local in montane forest and forest borders, favoring expansive valleys, in the foothills and subtropical zone on both slopes of the Andes, but ranging mainly in se. Ecuador. On the west slope known only from a few recent sightings at Buenaventura in El Oro, where a single adult was seen on several occasions in Jun. 1985, again on 1 Apr. 1989 (Robbins and Ridgely 1990), and as recently as 13 Mar. 1994 (P. Coopmans et al.); also from two sightings in w. Loja, first a soaring bird near Sozoranga on 22 Sep. 1989 (Best et al. 1993) and then a pair along the road between El Empalme and Macará on 10 Apr. 1994 (P. Coopmans and M. Lysinger). It seems surprising that this eagle, which normally requires extensive forest, could have continued to survive in this region of such highly fragmented forest patches for so long; likely the observed birds were last-surviving—and perhaps no longer reproducing—individuals from a formerly larger population. On the east slope the Solitary Eagle is some-what more numerous and widespread and is recorded north to Tungurahua (a pair seen west of Shell-Mera on 21 Aug. 1987; RSR, PJG, and P. Scharf et al.) and w. Napo on the Cordillera de Huacamayos (several sightings from both its north and south slopes; v.o.). It likely also ranges northward along the east slope of the Andes into Colombia. Solitary Eagles can occasionally be seen along both the Loja-Zamora and the Gualaceo-Limón roads, especially when conditions are suitable for soaring. We should emphasize that there are no confirmed records of the Solitary Eagle from the eastern lowlands; the several reported sightings are all believed to have been in error for the Great Black-Hawk (*Buteogallus urubitinga*). Recorded between about 900 and 1800 m.

The Solitary Eagle is considered to be a Near-threatened species by Collar et al. (1992, 1994), but we judge it to merit Vulnerable status in Ecuador. Its total population in the country seems unlikely to constitute more than a few dozen pairs.

The species has sometimes been called the Black Solitary Eagle.

One race: nominate *solitarius*.

Range: very locally in highlands from nw. Mexico to Panama, mountains of n. Venezuela, and Andes from Colombia to extreme nw. Argentina.

The specimen recorded as Crowned Eagle (*H. coronatus*) from the Río Topo (an affluent of the Río Pastaza) mentioned by Taczanowski and Berlepsch (1885) must surely refer to the similar Solitary Eagle. These two species were long confused, but the Crowned Eagle is now known to occur only in s.-cen. South America.

Parabuteo unicinctus   Harris's Hawk   Gavilán Alicastaño

Uncommon to locally fairly common in agricultural terrain and deciduous scrub and woodland in the more arid lowlands of w. Ecuador from w. Esmeraldas (one seen south of Atacames on 27 Jan. 1993; RSR and F. Sornoza) south through much of Manabí to El Oro and w. Loja, southward ranging up regularly on Andean slopes to nearly 2000 m (e.g., below Celica), with old records from Chimborazo (Pallatanga) and s. Azuay at Gima (fide *Birds of the Americas*, pt. 1, no. 4). Also found in arid intermontane valleys of n. Ecuador from Carchi south through Pichincha (including small numbers persisting north of Quito in the Río Guaillabamba valley). The only record from anywhere in e. Ecuador involves a single wandering bird seen near Zamora in Zamora-Chinchipe on 23 Jul. 1978 (RSR and D. Wilcove). The Harris's Hawk is perhaps spreading north in w. Ecuador (e.g., into w. Esmeraldas) as a result of deforestation, but in general it seems to avoid humid regions; it has never, for instance, been reported from even such a well-watched area as Río Palenque. In the Andes recorded between about 1500 and 2900 m; in the southwest ranges up to 1700–1800 m, sparingly up to about 2000 m.

The species has sometimes been called the Bay-winged Hawk, but we concur with the 1998 AOU Check-list in calling it the Harris's Hawk.

One race: *harrisi*.

Range: sw. United States south locally to n. Venezuela; sw. Colombia and n. Ecuador; w. Ecuador, w. Peru, and n. Chile; n. and e. Bolivia and s. Brazil to n. Argentina and Uruguay; cen. Chile.

*Busarellus nigricollis   **Black-collared Hawk**   Gavilán de Ciénega

Uncommon and very local in semiopen, marshy terrain around the edges of lakes and along rivers in the lowlands of extreme ne. Ecuador in e. Sucumbíos. The Black-collared Hawk was first reported in 1986 from the Cuyabeno area (fide R. Sierra), where it has been seen on several occasions and where we understand it was photographed (we have not seen these photos, however). Since 1990 a few pairs have also been seen regularly along the Río Lagarto, especially near Imuyacocha. As yet there are no specimens of this unmistakable raptor from Ecuador. Recorded below 250 m.

One race: nominate *nigricollis*.

Range: locally from s. Mexico to n. Argentina, Uruguay, and sw. Brazil, but largely absent from n. Amazonia.

Geranoaetus melanoleucus   **Black-chested**   Aguila Pechinegra
                                Buzzard-Eagle

Uncommon to locally fairly common in open and semiopen terrain, favoring areas where there are rocky cliffs, in paramo and the arid temperate zone (including the central valley) from Carchi south through Loja; occurs in smaller numbers on the outside slopes of the Andes, most regularly in areas that have been extensively cleared. Recorded mostly from 2000 to 3600 m, at least occasionally wandering higher (up to at least 4000 m) or lower; there is even a single anomalous sighting of an adult near sea level just east of Guayaquil on 16 Jun. 1985 (RSR and K. Berlin).

One race: *australis*. We follow Blake (1977) and *Birds of the World* (vol. 1) in synonymizing *meridensis*, in which Ecuadorian birds were often formerly placed.

Range: Andes from w. Venezuela to Tierra del Fuego, also in lowlands from w. Peru and s. Bolivia southward, and ranging locally north in highlands of e. Brazil.

Buteo nitidus   **Gray Hawk**   Gavilán Gris

Rare to locally uncommon in deciduous woodland and forest, borders, and adjacent clearings with scattered large trees in the lowlands of w. Ecuador, where recorded from sw. Manabí (Machalilla National Park area) and Pichincha (regular at Río Palenque, with at least one Mar. 1985 sighting from as far north as Santo Domingo de los Colorados [RSR and T. Schulenberg et al.]) south locally to El Oro (seen east of El Empalme in Jun. 1985; RSR and M. B. Robbins); there is also one report from farther north, an immature seen and tape-recorded in sw. Imbabura near the Salto del Tigre bridge over the Río Guaillabamba on 14 Jul. 1997 (P. Coopmans and F. Sornoza). The Gray Hawk is also known from a few records from along the base of the Andes in se. Ecuador, whence there are old specimens from Zamora-Chinchipe (Zamora) and Morona-Santiago (Gualaquiza), with a single sighting from as far north as Logroño, Morona-Santiago, on 12 Jun. 1984 (RSR and M. B. Robbins); there is also a single possible sighting from Jatun Sacha (fide B. Bochan). In the west recorded only below about 450 m; in the southeast known from 600 to 1000 m.

Some authorities (e.g., D. Amadon, *Am. Mus. Novitates* 2741, 1982; Amadon and Bull 1988; 1998 AOU Check-list) recognize the genus *Asturina* for this species, which has long been placed in *Buteo*. We prefer, however, to maintain it in *Buteo*, not having seen any compelling evidence for separating it generically from *Buteo* (if it is separated, then other species now in *Buteo* also probably deserve to be separated with it). When it is placed in *Asturina*, the species name is spelled *nitida*. Sibley and Monroe (1990) recognize Middle American birds (ranging south to Costa Rica) as a separate species, *B. plagiatus* (Gray Hawk), but we follow the 1998 AOU Check-list in uniting that form with *nitidus*. If two species are recognized, then the population ranging from Costa Rica southward (*B. nitidus*) is best called the Gray-lined Hawk; the Middle American *B. plagiatus* would then retain the English name of Gray Hawk. One race occurs in Ecuador, nominate *nitidus*. We follow Blake (1977) in synonymizing *costaricensis*, the name that has sometimes been given to trans-Andean birds north to s. Costa Rica.

Range: sw. United States to se. Peru, n. and e. Bolivia, n. Argentina, Paraguay, and se. Brazil; w. Ecuador and locally in se. Ecuador.

Buteo magnirostris   **Roadside Hawk**   **Gavilán Campestre**

Uncommon to locally common in clearings and agricultural areas with scattered trees, secondary woodland, and forest and river borders in the lowlands of both e. and w. Ecuador, and in the subtropical zone on both slopes of the Andes. Often most numerous in the foothill zone. Conspicuous and noisy, in many areas one of the most frequently seen raptors, the Roadside Hawk avoids extensively forested regions, favoring areas where forest cover has become more fragmentary, a situation now all too prevalent as a result of the clearing of forest; as a result, and unlike most other Ecuadorian birds of prey, this species has doubtless increased in recent decades. In Ecuador the Roadside Hawk also avoids more arid regions, and thus it is decidedly local in the southwest, where even in areas of seemingly suitable humid habitat it is distinctly scarce. Ranges mostly below 1600 m, locally and usually in small numbers up to about 2500 m, on the east slope exceptionally as high as 3000 m (at Oyacachi and Anatenorio; Krabbe et al. 1997).

Sometimes placed in the genus *Asturina* or *Rupornis*. One race occurs in Ecuador, nominate *magnirostris*. We follow *Birds of the World* (vol. 1) in not recognizing the very similar *ecuadoriensis*, to which the birds of w. Ecuador are sometimes (e.g., Blake 1977) ascribed.

Range: n. Mexico to nw. Peru, n. Argentina, and Uruguay.

Buteo leucorrhous   White-rumped Hawk   Gavilán Lomiblanco

Rare to uncommon in montane forest, forest borders, and adjacent clearings with scattered trees in the upper subtropical and temperate zones on both slopes of the Andes, ranging at least locally or occasionally to slopes above the central valley (e.g., one was seen at Pasochoa on 4 Apr. 1982; PJG). On the west slope recorded south locally to El Oro and w. Loja at Utuana (Best et al. 1993). Recorded mostly from 2000 to 3200 m, rarely as high as 3550 m (west slope of Cerro Parcato in nw. Cotopaxi; specimen [MECN] taken by N. Krabbe on 4 Jan. 1991).

Monotypic.

Range: mountains of n. Venezuela, and Andes from w. Venezuela to nw. Argentina; se. Brazil, e. Paraguay, and ne. Argentina.

Buteo platypterus   Broad-winged Hawk   Gavilán Aludo

An uncommon (though usually conspicuous) boreal winter resident in the borders of montane forest and woodland, adjacent clearings, and plantations with scattered large trees in the foothills and subtropical zone on both slopes of the Andes, with a few found up into the central valley (e.g., regularly noted in some of the city parks of Quito) and in the eastern lowlands. On the west slope recorded south only to e. Guayas, though occasional birds likely wander farther south as there are a few records from w. Peru. The Broad-winged Hawk is almost unrecorded from the lowlands in Ecuador, with most birds that have been encountered there apparently pertaining to transients; there is one sighting from Limoncocha (16 Oct.), one from Imuyacocha (1 Apr.), and one from Naranjal in se. Guayas (20 Oct.). In Nov. 1994, however, a substantial number of Broad-winged Hawks (at least 12) appeared to be establishing wintering territories in terra firme forest along the Maxus oil-field road southeast of Pompeya, and small numbers were still there in Jan.–Feb. 1995 (RSR et al.); a few were again noted there in Nov. 1995. Recorded mostly from Oct. to early Apr.; some older records labeled as from Jun. to Aug. are considered almost certainly misdated. A large percentage of the individuals occurring in Ecuador are in adult plumage. Found mostly between about 800 and 2400 m, a few as high as 2800 m in the central valley and on the east slope of the Andes; a few occur in the eastern lowlands as low as 200 m.

One race: nominate *platypterus*.

Range: breeds in North America, wintering from s. United States and Mexico south to w. Bolivia, in South America mainly on Andean slopes with increasingly smaller numbers east across Amazonia to the Guianas; also resident in West Indies.

Buteo brachyurus   Short-tailed Hawk   Gavilán Colicorto

Uncommon around the borders of humid and deciduous forest and woodland, and in partially cleared areas, in the lowlands of both e. and w. Ecuador, also ranging up into the subtropical zone on both slopes of the Andes. In the west recorded south to El Oro and w. Loja (numerous recent sightings from various localities). The Short-tailed Hawk avoids both very arid and extensively forested regions. Its numbers may be increasing as a result of clearing, though because it is such a difficult bird to collect—for such a numerous, widespread raptor it is remarkably scarce in collections—one cannot easily assess its former abundance. Recorded mainly below about 1600 m, in small numbers as high as 2200 m.

One race: nominate *brachyurus*.

Range: Florida; n. Mexico to nw. Peru, n. Argentina, Paraguay, and s. Brazil.

Buteo albigula   White-throated Hawk   Gavilán Goliblanco

Rare at borders of montane forest and woodland and in adjacent cleared areas in the upper subtropical and temperate zones on both slopes of the Andes, also locally in the central and interandean valleys (where recorded mainly from the Quito area, on the slopes of Volcán Pichincha immediately above the city; N. Krabbe has also seen it above Calacali and at Pasochoa). E. F. Pavez (*J. Raptor Res.* 34[2]: 143–147, 2000) suggests that the White-throated Hawk may occur in the n. Andes only as an austral migrant, and that the species may breed exclusively in Chile and Argentina. Most or all known dates of occurrence in Ecuador appear to support this contention. Recorded mostly from 2200 to 3200 m.

Albigula has occasionally been considered a race of *B. brachyurus* (e.g., *Birds of the World*, vol. 1; Blake 1977), but there now exists ample evidence to support its treatment as a separate species, and it has been so regarded by all recent authors. Monotypic.

Range: locally in Andes from w. Venezuela to Argentina and Chile.

Buteo swainsoni   Swainson's Hawk   Gavilán de Swainson

A very rare transient, possibly overlooked, en route between its North American breeding grounds and Argentinian wintering grounds. Only a few Ecuadorian records of the Swainson's Hawk are known: an old specimen taken at Zambiza in the Quito area, supposedly on 15 Jan.; one seen soaring with other raptors at Limoncocha on 31 Mar. 1979 (PJG and W. Davis); a light-morph adult seen heading northwest low over the crest of the Cordillera de Huacamayos on 10 Apr. 1993 (RSR and F. Sornoza); and an immature seen west of Santa Elena in w. Guayas on 23 Nov. 1997 (RSR et al.; a few other Swainson's Hawks have been reported seen in the Santa Elena area, but details are unavailable). A sighting from Río Palenque lacks details. Single individuals or small groups of Swainson's Hawks might appear virtually anywhere; there are also a few reports from nw. Peru (T. Schulenberg and T. A. Parker III, *Condor* 83[3]: 214–215, 1981). The species would always be most likely during its main migratory periods (Oct.–Nov. and Mar.–Apr.), but it now seems evident that the its normal migratory route lies east of Ecuador (conceivably some birds regularly overfly the country but at heights so high as to be invisible from the ground). To about 2500 m.

Monotypic.

Range: breeds in w. North America, wintering mainly in Argentina; migratory routes in South America still imperfectly understood.

Buteo albonotatus Zone-tailed Hawk Gavilán Colifajeado

Rare to locally uncommon at the borders of forest and woodland (especially deciduous) and in adjacent cleared areas in the lowlands and lower subtropical zone of sw. Ecuador from cen. Manabí (Bahía de Caráquez area) and s. Pichincha (Río Palenque) south through El Oro and w. Loja. Also known from the lowlands of e. Ecuador adjacent to the Andes, where it has been found very locally from w. Napo (primarily in the Lago Agrio and Coca regions) south locally to Morona-Santiago (Santiago; Jul. 1990 sightings by M. B. Robbins and H. Braswell) and Zamora-Chinchipe (Apr. 1994 sighting near Zumba; P. Coopmans and M. Lysinger). Despite the frequency of recent sightings, there still are no specimens of the Zone-tailed Hawk from Ecuador. See Ridgely (1980) for a discussion of the first Ecuadorian reports, which were made as recently as the 1970s. Recorded up to about 1500 m (sightings near Pozul in Loja).

Monotypic.

Range: locally from sw. United States to the Guianas, nw. Peru, nw. Argentina, and w. Paraguay; ne. Brazil.

Buteo polyosoma Variable Hawk Gavilán Variable

Uncommon to locally fairly common in open and semiopen, arid habitats in intermontane valleys, Andean slopes (even spreading locally onto outside slopes where cleared conditions permit, e.g., a few reports from along the Nono-Mindo and Santo Domingo roads), and paramo (where it favors cliffs and rocky ledges) from Carchi south locally through El Oro and Loja. Also fairly common in the lowlands of sw. Ecuador, principally on the Santa Elena Peninsula of w. Guayas, with a single sighting from sw. Manabí near Jipijapa on 31 Jul. 1991 (RSR) and another sighting of a vagrant subadult on Isla de la Plata on 23 Jun. 1992 (RSR). Habitats occupied by the Variable Hawk in Ecuador thus range from nearly barren desert in the Santa Elena Peninsula region through a variety of types of agricultural terrain, scrub, and light woodland in intermontane valleys to high-elevation paramo. Arid terrain is, however, always favored, and the species is absent or very scarce in more humid regions, where it is nearly or entirely restricted to cleared areas. In n. and cen. Ecuador it occurs primarily from 2000 to 4400 m; in El Oro and Loja it ranges from about 700 to 3500 m; in Guayas it is recorded only below about 200 m.

Despite persistent doubts registered by some workers—who were recognizing the difficulty in separating certain examples of the two—*B. polyosoma* (Red-backed Hawk) and *B. poecilochrous* (Puna Hawk) were long considered to be distinct species by most authors. However, C. C. Farquhar (*Condor* 100[1]: 27–43, 1998) has presented evidence showing overlap in measurements and supposed plumage characters, and he concluded that they cannot be considered separate species. Farquhar (op. cit) furthermore concludes that only one race, nominate *polyosoma*, should be recognized as occurring on the mainland of South America; an additional form, *exsul*, occurs on the Juan Fernández Islands off Chile. The coastal population found in Guayas has sometimes been recognized as subspecifically distinct from highland birds (as *peruviensis*, the sometimes used *ecuadoriensis* being preoccupied), but Farquhar (op. cit) suggests that it too is best no longer recognized.

Farquhar (op. cit.) proposes calling the expanded *B. polyosoma* the Variable Hawk. Considering the plumage variation shown by the species, we agree that this is a suitable English name. The species is also notably "variable" in the range of habitats occupied. The high-elevation populations formerly called *B. poecilochrous* also sometimes went by the name of Variable Hawk.

Range: Andes from s. Colombia to Tierra del Fuego, also in Pacific slope lowlands southward from sw. Ecuador, as well as in lowlands of Argentina.

Early records (see discussion in Chapman 1926) of the White-tailed Hawk (*Buteo albicaudatus*) from the highlands of n. Ecuador are now known to refer to *B. polyosoma*.

Morphnus guianensis   Crested Eagle   Aguila Crestada

Very rare and local in humid forest (mainly terra firme) and forest borders in the lowlands of e. Ecuador, where thus far known definitely from two old specimens from relatively close to the base of the Andes (Sarayacu and "below San José"), from a light-morph adult seen in the Maxus oil-field area southeast of Pompeya on 14 Nov. 1994 (P. English and S. O'Donnell), and from a light-morph adult seen repeatedly at Sacha Lodge in 1994–1995 (L. Jost, fide PJG); there have also been sightings from the La Selva Lodge vicinity and from the Tiputini Biodiversity Center. In addition, a captive adult seen in Zamora in Dec. 1991 (RSR et al.) supposedly was obtained locally. In nw. Ecuador the Crested Eagle is known from several sightings in the Tandayapa/Mindo area of Pichincha, with almost certainly the same subadult individual being involved in some of the reports; it was seen above Tandayapa on 23 Feb. 1985 (T. Schulenberg and RSR et al.), above Mindo on 9 Jul. 1985 (R. Rowlett et al.), and near Tandapi on an unspecified date in 1986 (PJG). More recently, an immature was again seen above Mindo on 23 Dec. 1997 (P. Coopmans et al.), and two soaring birds were observed above Bellavista in Mar. 1998 (M. Lysinger). The Crested Eagle is mainly found below 300 m, though in Pichincha it ranges unusually high, about 1800–2200 m.

The Crested Eagle was considered to be a Near-threatened species by Collar et al. (1992, 1994). Our judgement is that in Ecuador it should be viewed as meriting at least Vulnerable status. This is based on its apparently small population size in the country (where it seems always to have been "outnumbered" by the Harpy Eagle [*Harpia harpyja*]), extensive habitat destruction in many areas, and continuing persecution in some areas where suitable habitat yet remains. A population of uncertain, but possibly sustantial, size occurs in Yasuní National Park.

The species has sometimes been called the Guiana Crested Eagle.
Monotypic.

Range: locally from Guatemala and Belize to nw. Ecuador, n. Bolivia, Amazonian and e. Brazil, and ne. Argentina.

Harpia harpyja   Harpy Eagle   Aguila Harpía

Rare and local in humid forest (both terra firme and várzea) and forest borders in the lowlands of both e. and nw. Ecuador. Relatively widespread in the east where extensive forest remains, but even here its territory size is large and numbers seen are always very small. In the northwest almost all records are of birds that were shot or that were wounded and then brought into captivity. The Harpy Eagle was first formally recorded from w. Ecuador only in the 1970s (see Orcés 1974), but based on the number of examples by then obtained it must, at least until recently, have been comparatively numerous there. We are not aware of any records in the west subsequent to PJG et al.'s observation of two birds near Tinalandia on 30 Jul. 1983, but there are rumors of the species' continued existence in the Machalilla National Park region of sw. Manabí (which seems to us surprising), and it may persist in the still-extensive forests east of Muisne and on the Cabeceras de Bilsa of w. Esmeraldas. Harpies are encountered comparatively regularly in the lower Río Aguarico region (an eyrie was occupied in 1992–1993 near the Río Pacucyacu), though even here actually seeing one is always a major event; a pair also seems to reside in the La Selva/Sacha Lodge region. There are, however, strangely few records from the Río Pastaza drainage; we presume this is due to underrecording, for certainly there still is a great deal of suitable habitat for it there. Recorded below 400 m apart from the Tinalandia record, at 700 m.

Never numerous even in uninhabited areas, the Harpy Eagle is now much reduced in numbers except in remote parts of the Oriente. This rarity is due in part to habitat destruction but mainly to the species so often being shot as a nearly irresistible trophy or target. Harpy Eagles tend not to fear humans, and thus perched birds can often be easily approached. The species is considered as being only Near-threatened by Collar et al. (1992, 1994). In Ecuador, however, based on the total collapse of the western population and the species' apparent rarity in the east, we judge it to warrant Vulnerable status. A population of uncertain, but possibly sustantial, size does occur in Yasuní National Park.

Monotypic.

Range: locally from s. Mexico to nw. Ecuador, nw. Argentina, and Amazonian Brazil; se. Brazil, e. Paraguay, and ne. Argentina.

**Spizastur melanoleucus*   Black-and-white   Aguila Azor Blanquinegro
                                 Hawk-Eagle

Apparently rare and local in humid forest and forest borders in the lowlands and lower subtropical zone in nw. and e. Ecuador. This splendid eagle is still known in Ecuador only from sight reports, all of them relatively recent. Black-and-white Hawk-Eagles were reportedly seen "regularly" during the late 1970s in Pichincha at Tinalandia, Río Palenque and around Santo Domingo de los Colorados (C. Leck, *Am. Birds* 34[3]: 313, 1980), but no details were provided and none have been seen in this area since. The only other reports from the northwest involve birds seen near Mindo in Oct. 1986 (PJG) and at Bilsa in w. Esmeraldas in Feb. 1994 (R. P. Clay et al.). The species has also been seen recently in ne. Ecuador in Sucumbíos and Napo, both along the east slope of the Andes and in the lowlands; these reports include one seen east of San Rafael Falls on 10 Aug. 1983 (PJG), several sightings from north of Archidona along the Loreto road (P. Coopmans and M. Van Beirs; PJG), one seen at Cuyabeno on 25 Mar. 1991 (R. ter Ellen), and one seen along the Maxus road southeast of Pompeya on 18 Jul. 1994 (S. Howell and RSR). In the southeast it is known from only two recent Zamora-Chinchipe reports, birds seen in the Bombuscaro sector of Podocarpus National Park on 3 Oct. 1996 (D. Wolf) and at Coangos on the Cordillera del Cóndor in Jul. 1993 (Schulenberg and Awbrey 1997). Recorded up to 1400 m.

The Black-and-white Hawk-Eagle was considered to be a Near-threatened species by Collar et al. (1992, 1994). In Ecuador, however, and despite the fact that it is scarce here, we cannot see that any factors are adversely affecting it and therefore do not consider it to be at risk.

Monotypic.

Range: locally from e. Mexico to nw. Ecuador, n. Argentina, and se. Brazil.

Spizaetus tyrannus   **Black Hawk-Eagle**   **Aguila Azor Negro**

Rare to locally uncommon in humid forest in the lowlands of e. Ecuador, in its frequent soaring flight sometimes ranging out over nearby partially deforested terrain; ranges up at least locally into the foothills and lower subtropical zone (e.g., north of Archidona along the road to Loreto). In w. Ecuador, from which it was first recorded only in the 1970s (see Ridgely 1980), the Black Hawk-Eagle seems largely confined to montane forest in the foothills and even occasionally up into the lower subtropical zone. Only in w. and n. Esmeraldas does the Black Hawk-Eagle presently occur in the lowlands of the west. Despite the absence of actual records, we suspect that the Black Hawk-Eagle formerly occurred more widely in the western lowlands, where it seems likely to have been eliminated by the deforestation that has occurred during the last half-century. There are a few recent reports from south into w. and s. Loja (e.g., near El Limo [RSR] and at Sozoranga [Best et al. 1993]), and it was once (Jan. 1991) reported seen on the coastal Cordillera de Colonche of sw. Manabí at Cerro San Sebastián (Parker and Carr 1922) and in w. Guayas at Loma Alta in Dec. 1996 (Becker and López-Lanús 1997); there are also a few recent sightings from Cerro Blanco in the Chongón Hills (Berg 1994; R. Clay). Black Hawk-Eagles do seem to be capable of persisting in somewhat patchy forest (e.g., a pair still appeared to be resident in the late 1990s in the Buenaventura area west of Piñas in El Oro). In the east recorded up to about 1200 m (along the Loreto road), in the west to 1700 m above Mindo.

One race: *serus*.

Range: Mexico to extreme n. Colombia and nw. Venezuela; w. Ecuador and extreme nw. Peru; se. Colombia, e. and s. Venezuela, and the Guianas to n. Bolivia and Amazonian Brazil; e. Brazil, e. Paraguay, and ne. Argentina.

Spizaetus ornatus   Ornate Hawk-Eagle   Aguila Azor Adornado

Rare to locally uncommon in humid forest in the lowlands of e. Ecuador. In w. Ecuador found very locally in humid and montane forest in the lowlands and foothills south to El Oro (Buenaventura; see Robbins and Ridgely 1990). The Ornate Hawk-Eagle was once (Jan. 1991) seen on the coastal Cordillera de Colonche of sw. Manabí at Cerro San Sebastián (Parker and Carr 1992), and it was also reported seen in Dec. 1996 at Loma Alta in w. Guayas (Becker and López-Lanús 1997). In the east not recorded above about 500 m; in the west found up to 1000 m.

Like the Black Hawk-Eagle (*S. tyrannus*), the Ornate was doubtless formerly much more numerous and widespread in w. Ecuador than it is presently, having declined greatly there as a result of deforestation. Nonetheless, in part because of the relatively large population that remains in the east, we do not judge that the species as a whole is at risk in the country.

Two similar races are found in Ecuador, nominate *ornatus* in the east and *vicarius* in the west. The latter has the sides of the head and neck somewhat duller and browner (not such a bright rufous).

Range: n. Mexico to n. Colombia and extreme nw. Peru; locally in n. Venezuela; e. Colombia, e. and s. Venezuela, and the Guianas to nw. Argentina and Amazonian Brazil; e. Brazil, e. Paraguay, and ne. Argentina.

Oroaetus isidori   Black-and-chestnut Eagle   Aguila Andina

Rare to locally uncommon in montane forest and, to a lesser extent, adjacent clearings in the subtropical and temperate zones on both slopes of the Andes. On the west slope the only report from south of Pichincha (where the species is noted regularly in the Tandayapa/Mindo region) is a sighting of a pair seen on 3–6 Mar. 1991 at Sural in Azuay (N. Krabbe); it has also been found in the Caripero area of w. Cotopaxi (N. Krabbe and F. Sornoza). Recorded mostly from 1500 to 3000 m, but at least locally it also occurs somewhat lower (in Jun. 1984 a pair was seen on the Cordillera de Cutucú at 1100 m; Robbins et al. 1987).

The magnificent Black-and-chestnut Eagle requires a large home range, and though it soars regularly when conditions permit, in general it is not very often encountered. It was considered to be a Near-threatened species by Collar et al. (1992, 1994). From an Ecuadorian perspective this assessment would appear to be appropriate. The total population in the country cannot be very large, probably under 100 pairs, and it is possible that the species may need to be downgraded in the future. Small populations do exist in several montane national parks and reserves.

Monotypic.

Range: locally in mountains of n. Venezuela, and Andes from w. Venezuela to nw. Argentina; Sierra de Perijá and Santa Marta Mountains.

Falconidae   Falcons and Caracaras

Nineteen species in eight genera. Falcons are found widely on all continents, but the caracaras are strictly New World (and mainly tropical) in distribution.

Daptrius ater   Black Caracara   Caracara Negro

Fairly common to common and generally conspicuous in open areas (especially prevalent along rivers), clearings, and at the edge of humid forest (but in large areas of continuous forest found mainly in swampy places, with a predilection for *Mauritia* palm swamps) in the lowlands (especially) and foothills of e. Ecuador. Numbers of the Black Caracara in Ecuador are likely increasing as a result of ongoing forest destruction, and the species regularly moves in along recently constructed roads. Ranges up in diminishing numbers onto the lower slopes of the Andes, regularly up to 1200–1350 m (e.g., along the Loreto road and in the Zamora region), occasionally even higher (once as high as 1550 m on the south slope of the Cordillera de Huacamayos; P. Coopmans et al.), apparently following clearings upward.
    Monotypic.
    Range: e. Colombia, nw., s., and e. Venezuela, and the Guianas to n. Bolivia and Amazonian Brazil.

Ibycter americanus   Red-throated Caracara   Caracara Ventriblanco

Uncommon to locally fairly common—and because of its incredible vocalizations, conspicuous wherever it occurs—in the canopy and borders of humid forest (mainly terra firme but also várzea) in the lowlands of e. Ecuador. Also known from the lowlands of w. Ecuador, where found in semihumid as well as very humid to humid forest. Old specimens were taken as far south as n. Manabí (Pato de Pájaro and Río de Oro), Los Ríos (a 1939 specimen from Vinces is in the Mejía collection), and Guayas (Samborondón and Guayaquil, fide *Birds of the Americas*, pt. 1, no. 4). The species perhaps persists in Esmeraldas though there seems to have been only one recent report from there, an Aug. 1987 sighting at El Placer (RSR and G. S. Glenn). Red-throated Caracaras were not found in the early 1990s in the largest remaining forested areas of w. Esmeraldas southeast of Muisne (Parker and Carr 1992; RSR) or at Bilsa (v.o.), nor were they found in 1995–1998 in the still-extensive forests of n. Esmeraldas in the Ríos Santiago and Cayapas drainages (O. Jahn and P. Mena Valenzuela et al.). The species seems likely to have disappeared from this region. Thus the subsequent discovery (Berg 1994) in the early and mid-1990s of an extant Red-throated Caracara population in the semihumid forests of the Chongón Hills at Cerro Blanco and nearby in Guayas came as a distinct surprise; the species has been judged to be "relatively common" at that site (R. Clay). Recorded up to about 800 m on the east slope of the Andes (seen in Mar. 1987 west of Archidona in w. Napo; PJG and P. Scharf), on the west slope up to about 650 m at El Placer.

Before massive deforestation swept across much of w. Ecuador, the Red-throated Caracara was surely much more widespread and numerous there; it has also undergone a decline in most of Middle America (see Ridgely and Gwynne 1989). Fortunately the species seems to have decreased relatively little across its vast Amazonian range, and it therefore has not been judged to be generally at risk (Collar et al. 1992, 1994). It seems not to have declined in e. Ecuador.

The species has usually been placed in the genus *Daptrius*, but C. S. Griffiths (*Condor* 96[1]: 127–140, 1994) demonstrated that based on syringeal evidence it and *D. ater* (Black Caracara) should be separated generically; she proposed resurrecting the genus *Ibycter* for *americanus*. The species is now usually considered to be monotypic (e.g., Blake 1977).

Range: very locally from s. Mexico to nw. Venezuela and w. Ecuador (declining or extirpated in much of this area); more widely from e. Colombia, s. and e. Venezuela, and the Guianas to n. Bolivia and interior se. Brazil.

Caracara cheriway	Northern Crested-Caracara	Caracara Crestado Norteño

Uncommon to fairly common in open country (including predominantly agricultural terrain), arid scrub, and deciduous woodland in the lowlands of sw. Ecuador from cen. Manabí (Bahía de Caráquez area) and Los Ríos southward. The Northern Crested-Caracara also ranges regularly up in arid intermontane valleys to 2000 m or higher, now particularly in the south, with occasional individuals sometimes wandering up into the central valley (to 3000 m or even higher), though whether any resident populations persist in the highlands is uncertain. In the past the species was likely more numerous here.

Formerly often placed in the genus *Polyborus*, but this name has been shown to be antedated by the name *Caracara* (see R. C. Banks and C. J. Dove, *Proc. Biol. Soc. Wash.* 105: 420–425, 1992). The generic name *Caracara* was adopted in the 1998 AOU Check-list. C. J. Dove and R. C. Banks (*Wilson Bull.* 111[3]: 330–339, 1999) concluded that the genus was better considered to consist of three biological species, with Ecuadorian birds belonging to the more northerly population, *C. cheriway*. This leaves southern birds as belonging to *C. plancus* (Southern Crested-Caracara). *C. cheriway* is monotypic.

Range: Florida and sw. United States to nw. Peru, s. Venezuela, the Guianas, and along lower and middle Amazon River in n. Brazil; Cuba.

Phalcoboenus carunculatus	Carunculated Caracara	Caracara Curiquingue

Rare to locally fairly common in paramo, agricultural fields, and cliffs in n. and cen. Ecuador south to Azuay (e.g., at El Cajas), El Oro (an old specimen from Taraguacocha; Chapman 1926), and n. Loja, where the southernmost record is of a single bird seen at Uritusinga in Mar. 1989 (Bloch et al. 1991). The Carunculated Caracara is most numerous in and near the Antisana Ecological Reserve. Recorded mostly from 3000 to 4200 m, but when foraging occasionally wanders somewhat lower or higher (e.g., to the mountaineering refugios on Ecuador's larger volcanos).

Monotypic.

Range: Andes of s. Colombia and Ecuador.

*Phalcoboenus megalopterus   Mountain   Caracara Montañero
                               Caracara

Uncommon and local in paramo and adjacent temperate-zone woodland near treeline in the Andes of s. Ecuador in e. Loja and adjacent Zamora-Chinchipe. Recorded by Fjeldså and Krabbe (1990, p. 108) from "paramos on the Peru/Ecuador border," the basis for this being a specimen obtained at "Warinja Lake" (= Lagunas Las Huarinjas) in 1917 (O. Bangs and G. K. Noble, *Auk* 35: 443, 1918). The first recent reports we are aware of were made in 1990, when a pair was seen by G. S. Glenn and F. Sornoza on the Cordillera Las Lagunillas on 20 Apr.; the species was also recorded later that year east of Amaluza on the Cordillera de Sabanilla by R. Williams et al. Mountain Caracaras were again found on the Cordillera Las Lagunillas in Dec. 1991 and Oct. 1992, and the first Ecuadorian specimen (ANSP) was obtained here on 25 Oct. 1992. Also in 1992, small numbers of the Mountain Caracara were found considerably farther north in Loja, at both Uritusinga (just south of Loja [city]) and at Cerro Acanamá south of Saraguro (B. O. Poulsen, *Wilson Bull.* 105[4]: 688–691, 1993); of particular interest is the fact that the Carunculated Caracara (*P. carunculatus*) has also been reported seen at these two localities (Bloch et al. 1991), apparently establishing sympatry between these two species. Recorded from 2900 to 3500 m.

Poulsen (op. cit.) discussed plumage variation in these two species of *Phalcoboenus* caracaras and concluded that there is at present no evidence of intergradation between *P. megalopterus* and *P. carunculatus*. He demonstrated that the two central Peruvian specimens of *P. megalopterus* believed by J. T. Zimmer (*Field Mus. Nat. Hist. Publ.* 282, 1930) to represent "hybrids" are actually just young adult *P. megalopterus*. Further, the ANSP specimen of *P. megalopterus* from the Cordillera Las Lagunillas, the northernmost specimen known, shows no evidence of any intergradation and appears to be phenotypically pure *P. megalopterus*. Monotypic.

Range: Andes from extreme s. Ecuador to cen. Chile and w.-cen. Argentina.

*Milvago chimachima   Yellow-headed Caracara   Caracara Bayo

Recently found to be uncommon in the lowlands of ne. Ecuador in Napo and Sucumbíos, where found mainly on sandbars and in adjacent riparian vegetation along larger rivers (especially along the Ríos Napo and Aguarico, upriver along the former to around Coca). Surprisingly, there appear to be no reports from the Río Pastaza drainage, though it seems quite possible that the species could spread there. The Yellow-headed Caracara may have begun to move into cleared areas away from the vicinity of rivers, though in Ecuador it certainly remains far less numerous in such habitats than it is elsewhere in its range. The species has only recently moved into Ecuador, the first report being a sighting by E. W. Stiles at Limoncocha in Jun. 1972 (D. Pearson, pers. comm.). A reported sighting from the paramo of Azuay at Río Mazán (King 1989) seems highly improbable and surely refers to a juvenile Carunculated Caracara (*Phalcoboenus carunculatus*), the plumage of which is not dissimilar. Recorded mainly below about 300 m, though birds have also been reported seen at the Hotel Auca near Tena (500 m; Green 1996).

As best as we can determine, no specimen of *M. chimachima* has been taken in Ecuador. Thus the subspecies found here remains uncertain; likely it is nominate *chimachima* that occurs.

Range: Costa Rica to n. Argentina and Uruguay, but local or absent from much of w. Amazonia.

Micrastur ruficollis   Barred Forest-Falcon   Halcón Montés Barreteado

Uncommon to locally fairly common (recorded mainly through its vocalizations) inside both humid and montane forest and secondary woodland in the lowlands of both e. and w. Ecuador, and in the subtropical zone on both slopes of the Andes. In the eastern lowlands occurs locally in both terra firme and várzea forest. In w. Ecuador recorded south to Guayas, El Oro, and w. Loja (a few recent reports from the Alamor vicinity at Tierra Colorada [Best et al. 1993] and near El Limo [RSR]). Ranges mainly up to about 2000 m on both slopes of the Andes, but occurring at least locally as high as 2900 m in Imbabura (Loma Taminanga on the Otavalo-Selva Alegre road [N. Krabbe; MECN]).

Two similar races are found in Ecuador, *interstes* in the west and what is believed to be *zonothorax* on the east slope of the Andes (2 specimens in ANSP). Apparently no specimens of *M. ruficollis* have been taken in the lowlands of the east; this population may prove to be referable to *concentricus*, which is known from the lowlands of se. Colombia (Hilty and Brown 1986; also see P. Schwartz, *Condor* 74: 399–415, 1972).

Range: e. Mexico to extreme nw. Peru; n. Colombia and n. Venezuela; e. Colombia, s. Venezuela, and the Guianas to n. Argentina, e. Paraguay, and se. Brazil.

Micrastur gilvicollis   Lined Forest-Falcon   Halcón Montés Lineado

Rare to uncommon (recorded mainly through its vocalizations) inside humid forest (principally terra firme) in the lowlands of e. Ecuador. This species and the similar Barred Forest-Falcon (*M. ruficollis*) are now known to occur syntopically at a number of Ecuadorian localities (e.g., at well-studied sites such as La Selva, Zancudococha, Jatun Sacha, and along the Maxus road southeast of Pompeya; also Taisha in Morona-Santiago and the Bermejo oil-field area north of Lumbaquí in w. Sucumbíos). How they segregate remains unclear, but in

Ecuador the Lined appears to be more strictly confined to terra firme forest, with the Barred ranging more widely (it is quite often heard in várzea). Recorded mostly below 700 m, but found in small (?) numbers up into the foothills and lower subtropical zone on the east slope to 1500 m (Palm Peak; MCZ).

This species was long confused with *Micrastur ruficollis*, and the two were often considered conspecific. P. Schwartz (*Condor* 74: 399–415, 1972) demonstrated that they must represent two broadly overlapping species, and this has been amply confirmed in subsequent decades. Monotypic.

Range: se. Colombia, s. Venezuela, and the Guianas to n. Bolivia and Amazonian Brazil.

| *Micrastur plumbeus* | Plumbeous Forest-Falcon | Halcón Montés Plomizo |

Rare—but probably overlooked—inside very humid forest in the lowlands and foothills of nw. Ecuador, where thus far recorded mainly from Esmeraldas though there is also one record from n. Pichincha (an only recently identified bird tape-recorded by N. Krabbe at Río Guaycuyacu on 25 Aug. 1991). There are relatively few specimens of this little-known forest-falcon. Known Ecuadorian localities include three old collecting sites on the coastal plain of n. Esmeraldas: Carondelet on the Río Bogotá (the species' type locality), Pulún (= Bulún), and the Río Zapallo Grande (= Río Sapayo). The first recent record came from the foothills of n. Esmeraldas at El Placer where in Aug. 1987 the Plumbeous Forest-Falcon was found to be rare, with two specimens being taken (ANSP, MECN). The species also occurs at Bilsa in sw. Esmeraldas, where in Feb. 1998 K. Berg tape-recorded it. Recorded up to about 775 m; recent records from as high as 1400 m in Nariño, Colombia (Salaman 1994), suggest that the species may occur this high in Ecuador as well.

The Plumbeous Forest-Falcon was considered to be an Endangered species by Collar et al. (1994). Our assessment is that, despite the paucity of records, it likely is not that gravely at risk. Considerable suitable habitat remains and seems likely to continue to, and the species surely has been underrecorded. We thus suggest that according it Vulnerable status is more appropriate.

Subsequent to *M. gilvicollis* (Lined Forest-Falcon) being recognized as a separate species from *M. ruficollis* (Barred Forest-Falcon), *plumbeus* has occasionally (e.g., *Birds of the World*, vol. 1) been treated as a trans-Andean subspecies of *M. gilvicollis*. In our view, even with *M. gilvicollis* being considered a full species, we see no compelling reason for not regarding *M. plumbeus* as a monotypic species in its own right.

Range: locally in sw. Colombia and nw. Ecuador.

| **Micrastur mirandollei* | Slaty-backed Forest-Falcon | Halcón Montés Dorsigrís |

Rare inside humid forest, principally if not entirely in terra firme, in the lowlands of e. and extreme nw. Ecuador. There are only a few Ecuadorian records of this *Micrastur*, all of them relatively recent. Two specimens obtained by M. Olalla at Sarayacu in Jul. 1972 (Orcés 1974), in the Escuela Politécnica collection, were apparently the first to be taken in Ecuador; we have not seen these skins, and in view of the frequency of specimen misidentification in the genus, they should be critically examined. A bird was reported seen at Limoncocha on 21 Oct. 1975 (Tallman and Tallman 1977), and there have been a few sightings (some supported by tape-recordings) from a few other sites (e.g., La Selva, Sacha Lodge, near Coca, Jatun Sacha, Yuturi, Tiputini Biodiversity Center, and Kapawi Lodge), but on the whole this remains a scarce and elusive bird in Ecuador. The Slaty-backed Forest-Falcon seems likely to occur in the far northwest, but the only report to date involves a tape-recording considered likely of this species made near San Lorenzo in n. Esmeraldas on 23 Feb. 1997 (J. Nilsson, fide P. Coopmans). Recorded up only to about 400 m.

Monotypic.

Range: Costa Rica to sw. Colombia; se. Colombia, s. Venezuela, and the Guianas to n. Bolivia and Amazonian Brazil; se. Brazil.

Micrastur semitorquatus	Collared Forest-Falcon	Halcón Montés Collarejo

Rare to locally uncommon (recorded mainly through its vocalizations) in humid forest, secondary woodland, and their borders in the lowlands of both e. and w. Ecuador; in the west recorded south through El Oro, and probably occurs at least in extreme w. Loja though there do not appear to be any definite reports. The Collared Forest-Falcon remains an elusive and infrequently encountered bird in Ecuador, even by those who know its voice. Ranges in small numbers up into the subtropical zone on both slopes of the Andes to about 1500 m, occasionally even higher (recorded as high as 1800 m at Maquipucuna [P. Coopmans] and to 2000–2100 m at Cabañas San Isidro [v.o.]); exceptionally it ranges as high as 2450 m (Caripero in w. Cotopaxi; N. Krabbe and F. Sornoza).

Two similar races are found in Ecuador, nominate *semitorquatus* in the east and *naso* in the west. *Naso* reportedly averages slightly darker, and its immatures are more coarsely barred below (but there is much individual variation).

Range: n. Mexico to extreme nw. Peru, n. Argentina, Paraguay, and se. Brazil.

Micrastur buckleyi	Buckley's Forest-Falcon	Halcón Montés de Buckley

Rare—though likely overlooked—in humid forest, secondary woodland, and forest borders in the lowlands of e. Ecuador, so far mainly in the northeast, and perhaps primarily from várzea forest along river or lake edges though there are also some reports from terra firme. The Buckley's remains an elusive and little-known *Micrastur* throughout its relatively limited range in nw. Amazonia, though now that its voice is known it is being reported more frequently. There are only five known Ecuadorian specimens (though there may be a few more misidentified as immature Barred Forest-Falcons [*M. ruficollis*] or the very similar Collared Forest-Falcon [*M. semitorquatus*], with which the Buckley's was long confused). These are the type from Sarayacu (in BMNH) and two examples (in AMNH) taken in w. Napo (one at Río Suno above Avila on 17 Feb. 1923, the other at San José de Sumaco on 30 Mar. 1923). In addition, two specimens (also in BMNH) were reportedly taken by L. Gómez at 1800 m on the Cordillera de Cutucú in Nov. 1938 (Robbins et al. 1987; Collar et al. 1992), but in light of what is now known concerning the species' range and preferred habitats in the lowlands, we are dubious regarding the actual provenance of these two birds. Recent reports include single birds seen east of Lago Agrio on 26 Jul. 1976 (F. Sibley et al.), at Lagartococha on 6 Dec. 1990 (PJG, RSR, and S. Greenfield), near the mouth of the Río Aguarico into the Napo on 20 Sep. 1992 (RSR and T. J. Davis), along the Río Pacuyacu on 20 Nov. 1993 (P. Coopmans et al.), near the Río Napo at Sacha Lodge (first in 1993–1994; L. Jost et al.), La Selva (first in Oct. 1996; P. Coopmans), Tiputini Biodiversity Center (first in Aug. 1995; J. Arvin et al.), and Kapawi Lodge on the Río Pastaza (v.o.). Recorded mostly below 300 m (possibly higher?).

The Buckley's Forest-Falcon was considered to warrant Data Deficient status by Collar et al. (1992), but more recent data have shown that it should not be viewed as seriously at risk (e.g., Wege and Long 1995). We concur with the latter assessment.

The species has sometimes been called the Lesser Collared Forest-Falcon. Monotypic.

Range: locally in e. Ecuador, e. Peru, extreme se. Colombia, and extreme w. Amazonian Brazil.

Herpetotheres cachinnans   Laughing Falcon   Halcón Reidor

Uncommon to fairly common in a variety of wooded and forested habitats (both humid and deciduous), ranging mainly in the canopy and at edge (also to some extent out into adjacent clearings), in the lowlands of both e. and w. Ecuador. Recorded mainly below about 800 m, occasionally somewhat higher.

Two similar races are found in Ecuador, nominate *cachinnans* in the east and *fulvescens* in the west. *Fulvescens* is more richly colored on its underparts and crown (more a cinnamon-buff, not merely buffy whitish).

Range: n. Mexico to nw. Peru, n. and e. Bolivia, ne. Argentina, and se. Brazil.

Falco sparverius   American Kestrel   Cernícalo Americano

Fairly common to common and conspicuous in open country of various types (even on the outskirts of towns and cities) in the highlands, where it occurs mainly in the central and interandean valleys and slopes. Especially numerous and widespread in agricultural areas in arid regions, but it also occurs in more humid regions where these have been deforested, even moving onto both outside Andean slopes where such suitable conditions exist. Kestrels are inexplicably absent from most of the western lowlands, however. Here the species occurs primarily in El Oro and w. and s. Loja, though it is perhaps now beginning to spread north into Guayas, where there are a few recent reports from around Guayaquil, the Chongón Hills at Cerro Blanco (Berg 1994; R. Clay), Loma Alta (Becker and López-Lanús 1997), and the Santa Elena Peninsula (Green 1996). Possibly the kestrel is beginning to colonize the abundance of seemingly ideal terrain for it, both natural and anthropogenic, in Guayas. We suspect that the species' being listed as "uncommon" by Parker and Carr (1992) for Machalilla National Park in s. Manabí was a mistake; no other observers have found it in this region. Recorded mostly from 1500 to 3200 m; in n. Ecuador small numbers locally occur somewhat lower, whereas in the south (as noted above) it occurs widely at low elevations, from sea level up to at least 2500 m.

Two races are found in Ecuador, *aequatorialis* south to Chimborazo and the slightly smaller *peruvianus* from El Oro and Azuay southward. Males of the two races are quite readily distinguishable, *peruvianus* being notably paler rufescent below and usually more profusely spotted; females are similar. Presumably these two taxa intergrade somewhere in the Chimborazo/Azuay region, but the precise area where this happens does not appear to be known.

Range: North America to Tierra del Fuego (boreal migrants occur south to Panama), but absent from Amazonia; West Indies.

Falco columbarius   Merlin   Esmerejón

A rare to uncommon boreal winter visitant to semiopen terrain of various types and at varying elevations, but principally occurring near water; there are no records from the eastern lowlands, and only one report exists from the east slope of the Andes. Although there are several old records from the Quito region and elsewhere in the central valley, at least in recent years the Merlin has been found mainly on or near the coast. Recent exceptions to this pattern have been single birds seen near Catamayo in Loja on 25 Oct. 1993 (PJG and D. Wolf) and at Baeza in w. Napo on 27 Oct. 1994 (A. Jaramillo). Ranges up to about 3000 m. Recorded mostly from late Oct. to Mar.; an old record from "June" was likely mislabeled.

One race: nominate *columbarius*.

Range: breeds in n. and w. North America, wintering south in small numbers to the Guianas and nw. Peru; also in Old World.

Falco femoralis   Aplomado Falcon   Halcón Aplomado

Rare to locally uncommon in paramo and open terrain of more arid regions in the central valley and on slopes above interandean valleys from Pichincha and Cotopaxi southward. Despite regularly occurring in the lowlands elsewhere, in Ecuador the Aplomado Falcon does not do so except as a vagrant; such birds have been seen near the mouth of the Río Ayampe on the Manabí-Guayas border on 4 Jan. 1991 (J. Hendriks, fide P. Coopmans) and crossing the Río Napo near Pompeya on 10 Nov. 1999 (P. Coopmans et al.). Recorded mostly from 3000 to 4100 m, very rarely wandering lower.

In Ecuador the Aplomado Falcon occurs only locally and is uncommon at best. There is no strong evidence indicating that its always small population in the country has diminished in recent decades, though we suspect that it may have done so. Its future status in the country should be monitored.

One race: *pichinchae*. Although nominate *femoralis* is listed as occurring in both e. and w. Ecuador by Blake (1977), we can find no supporting evidence for this and consider it highly unlikely.

Range: open areas from extreme sw. United States to Tierra del Fuego; absent from virtually all of Amazonia.

Falco rufigularis   Bat Falcon   Halcón Cazamurciélagos

Uncommon to fairly common at the borders of humid and deciduous forest and secondary woodland, and in adjacent clearings, in the lowlands of both e. and w. Ecuador. In the west ranges south to w. Loja (Tambo Negro and Puyango [Best et al. 1993; RSR]). The Bat Falcon is less numerous in more arid regions than elsewhere, and it thus is rather scarce and local throughout sw. Ecuador, and is entirely lacking from desert-like regions. Recorded mainly below about 1000 m, in smaller numbers as high as 1200–1500 m.

Some evidence indicates that the Bat Falcon may be declining, for reasons unknown, in at least some parts of the eastern lowlands (P. Coopmans; RSR).

Two similar races are usually recognized in Ecuador, nominate *rufigularis* east of the Andes and *petoensis* in the west. Some authors (e.g., Blake 1977), do not recognize *petoensis*, merging it into the nominate race. The color of the throat, sides of neck, and chest in western birds varies: the entire area can be whitish, or only the throat can be whitish, with remaining areas buff to brighter orange-buff. In birds from east of the Andes this area tends to be paler, with at most a buff wash on the sides of the neck and chest only.

Range: n. Mexico to nw. Peru, n. Argentina, Paraguay, and se. Brazil.

Falco deiroleucus   Orange-breasted Falcon   Halcón Pechinaranja

Rare and local at the borders of humid and montane forest, and in adjacent clearings (regularly perching on dead snags), in the lowlands of e. Ecuador and (especially) in the foothills and lower subtropical zone on the east slope of the Andes. All records come from the north, in the drainages of the Ríos Napo and Aguarico, though the species seems possible in the south as well; the Gualaquiza specimen (Salvadori and Festa 1900) proves to be referable to the Bat Falcon (*F. rufigularis*; fide *Birds of the Americas*, pt. 1, no. 4). There are no confirmed records of the Orange-breasted Falcon from nw. Ecuador, the earlier reported specimens (from Hacienda Paramba and Santo Domingo de los Colorados) belonging to the Bat Falcon; they were mistakenly listed under the Orange-breasted by Chapman (1926). A few recent sight reports from the northwest of birds identified as Orange-breasted Falcons lack, in our view, sufficient details to be fully credited; there does exist, however, a single recent sighting from w. Nariño in adjacent Colombia (Salaman 1994). The Orange-breasted Falcon is most often found near cliff faces, on which its eyries are usually placed; large dead snags and emergent trees are also occasionally used as nest sites. Recorded up to as high as 2000–2500 m on the east slope of the Andes, but perhaps ranging this high only while feeding; usually found below 1400 m.

The Orange-breasted Falcon was considered a Near-threatened species by Collar et al. (1994). From an Ecuadorian perspective, we would consider that it is better treated as Vulnerable, this because of its very small population size—certainly no more than a few dozen pairs (perhaps fewer) are resident in the country—and its inherent vulnerability when nesting. We are not aware of any evidence suggesting that this always rare falcon has actually declined in Ecuador, and suggest that the best course of action would be simply never to disturb it.

Monotypic.

Range: locally from s. Mexico to n. Argentina, Paraguay, and se. Brazil; absent from most of Amazonia.

Falco peregrinus   **Peregrine Falcon**   Halcón Peregrino

Primarily a rare to locally uncommon transient and boreal winter visitant to w. Ecuador, where most numerous and regular along the immediate coast; a few records exist from a scatter of sites in the Andes and the central valley (some of which probably refer to resident birds; see below). As far as we are aware, there are no definite reports of the Peregrine from the eastern lowlands, though the odd individual could well occur there, probably especially as a transient and along larger rivers. Austral migrants may also occur in Ecuador, though as yet there seem to be no definite records; such proof will doubtless remain elusive because of the difficulty in separating the subspecies involved. In addition to the nonbreeders, a very small population is resident very locally in the Andes, breeding during the austral summer (Dec.–Feb.). The first documented record of Peregrine nesting in Ecuador involved an eyrie situated on a cliff face north of Quito near Guaillabamba described by J. P. Jenny, F. Ortiz-Crespo, and M. D. Arnold (*Condor* 83[4]: 387, 1981); this site continues to be occupied (fide N. Hilgert). As yet there is no proof of breeding at other sites, but this remains possible. Nesting might even occur along the Ecuadorian coast; a few pairs of Peregrines are known, for example, to nest on cliffs along the Peruvian coast. The only additional evidence of local nesting is provided by two specimens (BMNH) of nestlings (long misidentified as being Orange-breasted Falcons [*F. deiroleucus*]) taken by C. Buckley in Dec. 1877 at an unlocated site named Yanayacu (C. M. White, *Condor* 91[4]: 995–997, 1989). Recorded up to about 2800 m, but may occur higher.

Four Peregrines banded in North America have been recovered in Ecuador, all of them shot: three from the highlands and one from Babahoyo in the western lowlands.

Populations of the Peregrine Falcon have for several decades been seriously at risk in North America and Europe, though subsequent to the banning of persistent pesticides most of these populations have gradually recovered (some are now more numerous than they were before the advent of pesticides). There is no evidence of any change in the extremely small number of Peregrines resident in Ecuador. Because of the resident population's minute size, at most a few pairs being involved, and because of its inherent vulnerability when nesting, we suggest that the Peregrine deserves to be considered as an Endangered species in Ecuador.

The racial allocation of Ecuadorian birds remains problematic, the material available being very limited. Only boreal migrants have actually been collected in the country, these being either *anatum* or the similar (but averaging larger) *tundrius*. The breeding population is presumably referable to the South American race *cassini*, but no specimens are available. *Cassini* resembles *anatum* and *tundrius* but is somewhat darker gray above with blacker sides of the head, and is slightly more coarsely marked with black on the underparts which tend to have a buffier ground color.

Range: virtually cosmopolitan, though nesting only locally in most regions and absent as a breeder from many; status in some areas confused because of the presence of breeders as well as boreal and austral migrants.

Galliformes
Cracidae   Curassows, Guans, and Chachalacas

Fourteen species in eight genera. The family is exclusively Neotropical in distribution.

Ortalis guttata   Speckled Chachalaca   Chachalaca Jaspeada

Uncommon to fairly common in borders of humid forest (both terra firme and várzea), secondary woodland, and riparian woodland and forest (often on river islands) in the lowlands of e. Ecuador. Also found recently in the Río Marañón drainage of s. Zamora-Chinchipe in the Zumba and Valladolid areas. Recorded up to about 1250 m below Valladolid, but elsewhere generally not found above about 1000–1100 m.

O. *guttata* is regarded as a species distinct from O. *columbiana* (Colombian Chachalaca) of the Andean slopes of Colombia, and from various other species in e. South America, following Sick (1993). One race of O. *guttata* occurs in Ecuador, the nominate race.

Range: se. Colombia, e. Ecuador, e. Peru, n. Bolivia, and w. and cen. Amazonian Brazil.

Ortalis erythroptera   Rufous-headed Chachalaca   Chachalaca Cabecirrufa

Uncommon to locally fairly common in semihumid and deciduous woodland and the borders of both humid and semihumid forest, often even where decidedly patchy, in the lowlands and foothills of w. Ecuador from Esmeraldas south through Guayas, El Oro, and w. Loja. In Loja the Rufous-headed Chachalaca occurs as far east as the Sozoranga area (Best et al. 1993) and as far south as south of Sabanilla (RSR). In Esmeraldas known mainly from the west and the south, but it recently has been reported from the coastal north as well: it was heard near Porvenir in late Nov. 1997 (K. Berg), seen and heard near Borbón in 1997, and is also now known to range along the major rivers upstream to Playa de Oro on the Río Santiago and at least to San Miguel on the Río Cayapas (O. Jahn). Given the proximity of Porvenir in particular to the Colombian border, it is indeed possible that this chachalaca does range into adjacent Nariño, as earlier suggested by Hilty and Brown (1986). The Rufous-headed Chachalaca does not range regularly into dry forest or woodland, rather seeming to favor more humid regions. Recorded mostly below about 1000 m, though ranging up as high as 1500–1800 m at several sites in w. Loja, once recorded as high as 1850 m near Sozoranga (Best et al. 1993); in Esmeraldas it occurs below 100 m.

Although numbers as a whole have unquestionably declined across broad sections of this species' original range, the nearly endemic Rufous-headed Chachalaca seems capable of persisting even where forest cover has been reduced to pitiful remnants; there it may become quite shy and perhaps even less vocal. In many areas local people assert that it is very difficult to hunt, and the species therefore appears to be under little hunting pressure at present. The species' overall conservation status was reviewed by B. J. Best and N. Krabbe (*World Pheasant Assoc. Jour.* 17 and 18: 45–56, 1992 and 1993), though these authors omitted several sites located by ANSP workers and others. Collar et al. (1994) felt that the Rufous-headed Chachalaca merited Vulnerable status. Given its tenacity in the face of major habitat destruction and its shyness in the face of hunting pressure, however, our assessment is that at least in Ecuador the species only deserves Near-threatened status.

Monotypic.

Range: w. Ecuador and extreme nw. Peru.

Penelope barbata   **Bearded Guan**   Pava Barbada

Rare to uncommon and local in montane forest and woodland in the subtropical and temperate zones of the Andes in s. Ecuador, where recorded definitely from El Oro (an early-20th-century specimen from Taraguacocha) south locally through Loja and adjacent Zamora-Chinchipe to the Peruvian border. The Bearded Guan ranges mainly on slopes above various interandean valleys, and it occurs only to a limited degree on the actual east slope of the Andes, where it spills over from westward-facing slopes. Northward it is replaced by the Andean Guan (*P. montagnii*). Sightings of birds reported to have been Bearded Guans from Azuay in the Río Mazán valley are now believed (fide N. Krabbe) to refer to the Andean Guan (certainly Andeans occur there; King 1989 and subsequent observations), though it is conceivable that both species are present at that locality. Bearded Guans have also been reported at a few other sites in

Azuay (Collar et al. 1992), though none of these reports has been confirmed. Note that the type specimen comes from Taraguacocha in e. El Oro, not from Buenaventura in w. El Oro (contra Wege and Long 1995). Recorded between 1900 and 3100 m, but most numerous below 2700 m.

The Bearded Guan is now extremely local because of deforestation across virtually all of what has always been a small range; it presumably is also hunted, though in general this relatively small cracid seems not to be under heavy hunting pressure. A population is protected in Podocarpus National Park (e.g., at Cajanuma), though greatest numbers tend to occur on the west slope at levels somewhat below actual park boundaries (Rasmussen et al. 1996). A few also occur at upper levels of the new Tapichalaca Reserve on the Cordillera de Sabanilla in se. Loja. The largest population may occur in the still unprotected remnant forest near Selva Alegre in the Cordillera de Chilla of n. Loja. Collar et al. (1994) consider the species' status to be Vulnerable; we concur, but in our view it almost deserves Endangered status.

P. barbata is considered a species distinct from *P. argyrotis* (Band-tailed Guan) of Venezuela and Colombia, following most recent authors. Monotypic.

Range: Andes of s. Ecuador and nw. Peru.

Penelope montagnii   Andean Guan   Pava Andina

Rare to locally fairly common in montane forest and woodland in the temperate zone on both slopes of the Andes, and locally on slopes above the central valley (e.g., at Pasochoa), in n. and cen. Ecuador south to Azuay at Portete (Berlioz 1932b) and Río Mazán (King 1989) and sw. Morona-Santiago (Gualaceo-Limón road; N. Krabbe). The Andean Guan is replaced southward by the Bearded Guan (*P. barbata*), with some persisting confusion as to the identity of some populations (the Río Mazán birds have been variously reported as Andeans and Beardeds, but they are now [fide N. Krabbe] believed to refer to Andeans). Andean Guans can become very tame where they are not subjected to hunting pressure; deforestation and hunting have reduced their numbers in many more accessible areas. Recorded mostly from 2500 to 3600 m (occurring up to treeline in many areas), ranging down locally to about 2200 m (e.g., near the crest of the Cordillera de Huacamayos); we suspect that specimens (MECN) labeled as having been taken at Baeza (1900 m) were actually taken on the slopes well above that town.

Two races occur in Ecuador, *brooki* on the east slope of the Andes and *atrogularis* on the west slope and above the central valley. *Brooki* is somewhat darker and less rufescent above, and the feathers of the upper back, neck, and breast are more broadly fringed with whitish (imparting a more conspicuously scaly appearance). *Atrogularis* is closely similar to the nominate race of Colombia and Venezuela.

Range: Andes from w. Venezuela to cen. Ecuador, and from n. Peru to w. Bolivia; Sierra de Perijá.

Penelope ortoni   Baudó Guan   Pava del Chocó

Rare to locally uncommon (in more remote areas) in very humid to humid forest in the lowlands and (mostly) foothills of w. Ecuador south to w. Chimborazo and ne. Guayas; there are, however, no recent confirmed records from south of Pichincha. The Baudó Guan also possibly ranges farther south than the Bucay/Chimbo region of e. Guayas, where several specimens were obtained in the early 20th century. Unconfirmed sightings (RSR, PJG, and R. A. Rowlett) in Aug. 1980 of what appeared to be small *Penelope* guans were made west of Piñas in El Oro, and a small *Penelope* seen at Buenaventura in Jan. 1990 (P. Coopmans) also appeared to be this species. C. Vaurie (*Am. Mus. Novitates* 2251, 1966) suggested that the Baudó Guan occurred sympatrically with the similarly sized Andean Guan (*P. montagnii*), but recent reports have shown that this is not the case, the supposed overlap in specimen-collection sites having doubtless been the result of the two species being taken at different elevations. However, the Baudó Guan does occur together with the Crested Guan (*P. purpurascens*), though now such occurrences are very local. In the Playa de Oro area of n. Esmeraldas the Baudó Guan is still regularly observed in the foothill zone above 200 m, though it has been extirpated close to the village itself (O. Jahn et al.). Recorded up to about 1000 m, perhaps ranging a little higher at least locally or formerly (there are old specimens from "Milpe Mindo" and "Huila Mindo," both sites apparently situated an uncertain distance west of the town of Mindo).

Collar et al. 1992 gave the Baudó Guan only Near-threatened status, but Collar et al. 1994 considered it to merit being upgraded to Vulnerable. The latter seems more appropriate, and the species is so treated here, though we suggest that it may even deserve Endangered status. No area where it occurs presently receives adequate protection, and the species is in fact not known to be numerous anywhere in Ecuador. It is considered to be unusually tame for a *Penelope* guan, more so than the Crested, this perhaps accounting for its disappearance under even light hunting pressure (O. Jahn). Presumably populations are found at lower levels of Cotacachi-Cayapas Ecological Reserve (at least it is present at nearby El Placer).

F. Vuilleumier (*Bull. Mus. Comp. Zool.* 134[1], 1965) suggested that *ortoni* might be only a race of *P. montagnii*, but given various differences in its plumage pattern and basic distribution this treatment does not seem tenable. All other recent authors have regarded it as a monotypic species, as do we.

Range: locally in w. Colombia and w. Ecuador.

Penelope jacquacu   Spix's Guan   Pava de Spix

Rare to locally uncommon in humid forest (both várzea and terra firme) in the lowlands and foothills of e. Ecuador. As a result of hunting pressure the Spix's Guan is now scarce anywhere near inhabited areas; it is usually outnumbered by the Common Piping-Guan (*Pipile pipile*), though the opposite is true at Jatun Sacha (B. Bochan). Recorded mostly below 500 m, but found locally as high as 1000 m (north of Lumbaquí in the Bermejo oil-field area; M. B. Robbins).

One race: nominate *jacquacu*.

Range: e. Colombia, s. Venezuela, and Guyana to n. Bolivia and w. and cen. Amazonian Brazil.

Penelope purpurascens   **Crested Guan**   **Pava Crestada**

Rare to locally uncommon in very humid to humid and montane forest in the foothills and lower subtropical zone of w. Ecuador south to w. Loja (recent reports from 2 areas near Alamor [Best et al. 1993], and near El Limo [RSR]). Crested Guans also persist very locally on the slopes of the coastal Cordillera de Mache at Bilsa in sw. Esmeraldas (R. P. Clay et al.) and of the Cordillera de Colonche in sw. Manabí (Cerro San Sebastián in Machalilla National Park) and w. Guayas (Cerro La Torre and above Salanguilla [N. Krabbe]; Loma Alta [Becker and López-Lanús 1997]). There are rather few records of the large Crested Guan from the lowlands, and almost none of these records is recent; this apparently reflects the relatively early destruction of humid forest at lower elevations. In the Playa de Oro area of n. Esmeraldas the Crested Guan favors the lowermost foothill zone (50–250 m), above which it decreases and is replaced by the Baudó Guan (*P. ortoni*; O. Jahn et al.). Some sightings of *Penelope* guans (e.g., from Río Palenque) are perhaps better viewed as uncertain because of possible confusion with the similar Baudó Guan; the two species are known to be sympatric at some localities (e.g., there are specimens of both from Santo Domingo de los Colorados and San Miguel de los Bancos, and recent sightings of both species from Bilsa and Playa de Oro). Recorded up to 1500 m; small numbers still occur around Mindo. We suspect that the report of this species from 2400–2500 m above Celica (Bloch et al. 1991) is in error, probably for the Bearded Guan (*P. barbata*); otherwise the Crested Guan has never been reported at anywhere near so high an elevation.

The Crested Guan's numbers in Ecuador have declined greatly during recent decades, this being the combined result of both extensive deforestation and overhunting. We thus consider the species to merit Vulnerable status in the country. In some other parts of its range it remains relatively numerous, however, and hence the species was not considered to be at risk by Collar et al. (1992, 1994).

One race: *aequatorialis*.

Range: n. Mexico to n. Venezuela and sw. Ecuador.

As noted above, the Crested Guan is not known to occur in deciduous forest in Ecuador, though this habitat is inhabited by the species elsewhere in its range. Thus the discovery near Arenillas in coastal El Oro by RSR, PJG, and R. A. Rowlett in Aug. 1980 of White-tailed Jays (*Cyanocorax mystacalis*) which were clearly imitating various vocalizations of *Penelope* guans is intriguing. We speculate that the guan involved might have been the extremely rare White-winged Guan (*P. albipennis*) of nw. Peru, which in the 19th century was recorded from nearby Tumbes (then part of Ecuador but now in Peru). Unfortunately, guans themselves were never seen, nor have they been found in this area since, nor even have jays imitating guans been heard. In Jan. 1997 unseen guans were also heard on semihumid-forested slopes north of Macará in extreme s. Loja (PJG et al.); although the possibility that these birds could have been Crested Guans cannot be ruled out, it is conceivable that the White-winged Guan was the species involved. Yet another tantalizing report involves a tape-recording of what sounds like a *Penelope* guan (again none was seen, however) made in very extensive and little-disturbed deciduous forest west of Macará in Sep. 1998 (F. Sornoza and RSR et al.). The discovery of an Ecuadorian population, however small, of this critically Endangered species—it was thought to be extinct for more than a century until its rediscovery in nw. Peru in the 1970s—would be an electrifying event indeed and is much to be hoped for. The Macará region may even be an appropriate site for an attempted introduction of birds raised in captivity in Peru.

Pipile pipile   Common Piping-Guan   Pava Silbosa Común

Uncommon to locally fairly common in humid forest (both terra firme and várzea) in the lowlands of e. Ecuador. The Common Piping-Guan is now scarce near inhabited areas as a result of hunting pressure, but it remains numerous in more remote areas (e.g., along the Maxus road southeast of Pompeya). Recorded mostly below 400 m, but ranging up in small numbers to as high as 700 m (Guaticocha; MCZ) and perhaps locally or formerly even higher (there is a specimen from the 1890s taken at "Valle de Zamora").

The genus *Pipile* is sometimes merged into the Andean genus *Aburria*. *P. pipile* is here considered as a single polytypic species, in agreement with Hilty and Brown (1986) and Stotz et al. (1996); this is based on the close similarity of all the forms placed in *P. pipile*, and on relatively frequent apparent hybrid *cumanensis* x *cujubi* observed in the piping-guan population of ne. Bolivia and sw. Brazil, where these two forms come into contact. Most recent authors (e.g., Sibley and Monroe 1990) have considered *P. cumanensis* (Blue-throated Piping-Guan) of n. and w. Amazonia, *P. pipile* (Trinidad Piping-Guan) of Trinidad, and *P. cujubi* (Red-throated Piping-Guan) of s. Amazonia to be allospecies. Further work may indeed perhaps show that they are better regarded as such. As noted above, however, we feel that the presently available evidence better supports their being treated as conspecific. One race of *P. pipile* occurs in Ecuador, *cumanensis*.

Range: the Guianas, e. and s. Venezuela, and se. Colombia to n. and e. Bolivia, n. Paraguay, and w. and cen. Amazonian Brazil.

Aburria aburri   Wattled Guan   Pava Carunculada

Rare to uncommon and rather local in montane forest and forest borders in the subtropical zone on the east slope of the Andes. There are also a few recent records from the west slope of the Andes in w. Pichincha where—after many rumors—a small population is now known to exist; given the absence of early records from this heavily collected area, this has to be viewed as rather a surprising discovery. The first western record involves a bird (now in the Universidad Católica collection) taken by a hunter near Alluriquín on 23 Dec. 1989 (Marín et al. 1992). In addition, there have been a few subsequent reports from above Mindo, first on 18 Jul. 1992 (J. Rasmussen et al.); one was also heard along the Calacali-Nanegalito road on 5 Feb. 1995 (P. Coopmans, M. Lysinger, and T. Walla). The Wattled Guan had not previously been known on the west slope of the Andes south of Cauca, Colombia. Recorded mostly from 1200 to 2100 m.

Despite its secretive nature, this comparatively large guan seems less capable of persisting in partially settled regions than are various other (smaller) montane guans such as the Sickle-winged (*Chamaepetes goudotti*), Andean (*Penelope montagnii*), and Bearded (*P. barbata*). Collar et al. (1994) gave the species Near-threatened status; from a purely Ecuadorian perspective, our assessment is that the species deserves Vulnerable status; where protected, it can be not uncommon (e.g., at Cabañas San Isidro; M. Lysinger).

Monotypic.

Range: locally in Andes from w. Venezuela to s. Peru; Sierra de Perijá.

Chamaepetes goudotii   Sickle-winged Guan   Pava Ala de Hoz

Uncommon to locally fairly common in montane forest and forest borders in the subtropical and lower temperate zones on both slopes of the Andes. On the west slope ranges south to El Oro (old specimens from Zaruma and Salvias; numerous recent reports from the Buenaventura area). C. Vaurie (*Am. Mus. Novitates* 2299, 1967, p. 11) mentions specimens from "El Tambo, Loja," the same locality from which there are several other anomalous records (e.g., the Giant Antpitta [*Grallaria gigantea*] and Purplish-mantled Tanager [*Iridosornis porphyrocephala*]); Paynter (1993) considers the only Ecuadorian collecting locality called "El Tambo" to be situated along the Guayaquil-Quito railway in Cañar, which probably is where these birds were actually obtained. There are no other records of the Sickle-winged Guan from Loja, and we consider its occurrence so far south to be unproven. Recorded between about 900 and 2600 m.

The Sickle-winged Guan persists better than other guans in partially deforested terrain, and it therefore is the most numerous cracid in many lower-montane localities, especially on the west slope.

Two distinctly different races are found in Ecuador, *tschudii* on the east slope and *fagani* on the west slope. *Tschudii* is substantially larger (weighing almost twice as much), longer tailed, and slightly darker and bronzier above; it also lacks *fagani*'s brownish gray chest, the rufous of its underparts extending up onto the throat. It is possible that two separate species are involved.

Range: Andes from n. Colombia to w. Bolivia; Santa Marta Mountains.

Nothocrax urumutum   Nocturnal Curassow   Pavón Nocturno

Uncommon to locally fairly common in humid forest (both terra firme and várzea) in the lowlands of e. Ecuador, apparently favoring areas reasonably close to rivers and less frequent in well-drained terra firme. The well-named Nocturnal Curassow is a very secretive and elusive cracid (only a very fortunate few have ever seen it during daylight); as a result it is apparently not heavily persecuted, being much too difficult to shoot in appreciable numbers. Recorded mostly below about 400 m, but ranges up in small numbers, perhaps only locally, in foothills along the eastern base of the Andes to about 900 m (Bermejo oil-field area in w. Sucumbíos; M. B. Robbins and RSR et al.).

It has occasionally been suggested that the genus *Nothocrax* be merged into *Crax*, but most authors (e.g., Sibley and Monroe 1990) maintain it as distinct. Monotypic.

Range: sw. Venezuela, extreme nw. Brazil, se. Colombia, e. Ecuador, and ne. Peru; locally in s. Amazonian Brazil.

Mitu salvini   Salvin's Curassow   Pavón de Salvin

Rare to locally fairly common in humid forest (both várzea and terra firme) in the lowlands of e. Ecuador. Populations of the spectacular Salvin's Curassow are now much reduced in most areas because of heavy hunting pressure, but in remote areas it can remain tolerably numerous (e.g., along the Maxus road southeast of Pompeya). Recorded mostly below about 400 m; there are a few, mainly old, records from as high as 700–900 m (e.g., north of Eugenio [MCZ], "valle del Zamora," "valle del Río Santiago").

The Salvin's Curassow has not been judged to be at risk overall (Collar et al. 1994), but from a purely Ecuadorian perspective we assess its status as at least Near-threatened. As noted above, in most regions—even where habitat remains extensive—this curassow remains under intense hunting pressure. A major population does now receive protection in Yasuní National Park, and the species is also present in the Cuyabeno Faunistic Reserve.

The genus *Mitu* has occasionally been merged into *Crax*, but we follow most recent authors (e.g., Sibley and Monroe 1990) in considering it worthy of recognition. Monotypic.

Range: se. Colombia, e. Ecuador, and ne. Peru.

Crax rubra   Great Curassow   Pavón Grande

Now at best very rare and extremely local in humid and deciduous forest in the lowlands of w. Ecuador south to Guayas; most reports are vague. Despite the fact that well into the 19th century the Great Curassow may have been relatively widespread and numerous in w. Ecuador, there are remarkably few definite records; these include specimens obtained at Hacienda Paramba in w. Imbabura and at Pulún in n. Esmeraldas in the 19th century, and in the Chongón Hills (at "Cerro de Bajo Verde") of Guayas in Jul. 1922. In recent years the

Great Curassow has been reported almost entirely by hunters, with most such reports being secondhand and old. It reputedly persisted in the Chongón Hills until at least the late 1970s, in the vicinity of Cerro San Sebastián in Machalilla National Park in sw. Guayas into the 1980s, and southeast of Muisne in w. Esmeraldas into at least the 1980s. The only recent (post-1990) reports come from Bilsa, west of San Miguel de los Bancos, and the still extensively forested region of the upper Río Santiago and Río Cayapas region in n. Esmeraldas. Even at the last site, in theory one of the best remaining in Ecuador for the species, it is judged to be either extinct or virtually extinct, with the last shot bird having been obtained in 1995 (O. Jahn and P. Mena Valenzuela et al.). Recorded up to about 700 m.

It was feared that the Great Curassow had been extirpated in Ecuador, but it now appears that a few may yet persist in the very few regions in w. Ecuador still remote enough that hunting pressure has not totally eliminated it. The largest Ecuadorian population may, as noted above, persist in remote lower foothill areas of n. Esmeraldas extending into the lower reaches of Cotacachi-Cayapas Ecological Reserve. Curassows were recorded as being "rare" at Bilsa on the Cordillera de Mache in w. Esmeraldas in Feb. 1994 (R. Clay et al.), but we have no details as to any specific records and some recent observers have questioned whether the species is extant in the region. In addition there is a report of a male recently shot by hunters "deep into the forests" west of San Miguel de los Bancos near San Vicente de Andoas; this bird was examined by D. Cisneros (fide P. Coopmans) on 29 Apr. 1995. The species' Ecuadorian status must obviously be viewed as Critical. Populations in some other parts of its range remain relatively robust, and for this reason the Great Curassow was not mentioned at all by Collar et al. (1992, 1994). Should vigilant protection be established at the Cerro Blanco Reserve in the Chongón Hills, or in Machalilla National Park, there seems to be no reason why the Great Curassow could not be reintroduced into these areas, and indeed this should be a high priority.

One race: nominate *rubra*.

Range: locally from n. Mexico to sw. Ecuador.

Crax globulosa   **Wattled Curassow**   **Pavón Carunculado**

Uncertain. At best very rare and local in the lowlands of far ne. Ecuador, with numbers very much reduced by heavy hunting pressure. The Wattled Curassow is known in Ecuador from a single 19th-century specimen labeled only as having been taken at "Napo" and from a single female specimen labeled as from "Río Negro" (C. Vaurie, *Am. Mus. Novitates* 2305, 1967). Neither of these localities can be precisely located, and it is quite possible that both specimens may have been obtained outside what are now the boundaries of Ecuador; several specimens (AMNH) were in fact taken in the 1920s at the mouth of the Río Curaray, which flows into the Napo not too far downstream from the Ecuador/Peru border. There are also a few recent hunters' reports of what seems likely to have been this species in the lower Río Aguarico drainage of e. Napo and e. Sucumbíos (A. Johnson and M. Hedemark; RSR), notably in the Zábalo vicinity. It is also not inconceivable that the species could occur in the Río Pastaza drainage. Based on records from elsewhere in its range, the Wattled Curassow favors várzea and riparian forest (sometimes even occurring on river islands), and the relative accessibility of this habitat—in contrast to that of much terra firme forest—likely accounts for its present rarity across virtually all of its range.

As far as we are aware, no ornithologist has ever seen *C. globulosa* in Ecuador. The species was accorded Indeterminate status by Collar et al. (1992) and Vulnerable status by Collar et al. (1994). From an Ecuadorian perspective, however, we would judge it amply to deserve a Critical rating. Even the intensive efforts of A. Johnson and M. Hedemark in the most remote sectors of the Oriente have failed to turn up verifiable evidence that this curassow persists in Ecuador. Note that the Jun. 1982 sighting from "Quebrada Papaya" ascribed to Ecuador by Collar et al. (1992, p. 164) actually refers to a site far down the Río Napo, well within the present boundaries of Peru; this doubtless was the basis for the incorrect assertion of del Hoyo (1994) that the species had not been found in Ecuador subsequent to that year.

Monotypic.

Range: locally in se. Colombia, e. Ecuador, e. Peru, n. Bolivia, and w. Amazonian Brazil.

The Black Curassow (*Crax alector*) has been found in s. Colombia as close as Tres Troncos along the Río Caquetá in e. Caquetá (Hilty and Brown 1986), not far north of the Ecuadorian border. The species could conceivably occur as far south as far ne. Ecuador, though as yet it remains unrecorded.

Odontophoridae   New World Quails

Six species in two genera. New World quails are, as the name implies, exclusively American in distribution. We follow Sibley and Ahlquist (1990) and Gill (1994) in considering them as a family separate from the pheasants, grouse, Old World quails, and turkeys (Phasianidae). The 1998 AOU Check-list concurs.

Odontophorus gujanensis   Marbled Wood-Quail   Corcovado Carirrojo

Uncommon to locally fairly common (though this and all the other wood-quail are recorded almost entirely by voice) on or near the ground inside humid forest and mature secondary woodland (principally terra firme) in the lowlands of e. Ecuador, small numbers ranging up into the foothills. As wood-quail are so difficult to see, there is still uncertainty as to the upper elevational limit of this species and whether it ever overlaps with the Rufous-breasted Wood-Quail (*O. erythrops*) of the foothills and east-slope subtropics. Wood-quail have, for example, been heard repeatedly in the Río Marañón drainage of s. Zamora-

Chinchipe around and north of Zumba but have never been seen and therefore could not be specifically identified. Recorded up to about 900 m.

One race: *buckleyi.*

Range: Costa Rica to n. Colombia, n. Bolivia, and Amazonian Brazil.

Odontophorus erythrops	Rufous-fronted Wood-Quail	Corcovado Frenticolorado

Uncommon to locally fairly common on or near the ground inside humid and montane forest and mature secondary woodland in the lowlands and foothills of w. Ecuador south to sw. Manabí (Cerro San Sebastián in Machalilla National Park), Guayas in the west (Cerro La Torre and above Salanguilla [N. Krabbe]; Loma Alta [Becker and López-Lanús 1997]) and north (Hacienda Pacaritambo; A. Brosset, *Oiseau* 34: 1–24, 1964), and along Andean foothills and adjacent lowlands south to El Oro and w. Loja (Alamor area). There are also unsubstantiated reports of this species from the Chongón Hills at Cerro Blanco (Berg 1994), but this is an area that does not seem to support habitat suitable for a wood-quail. Recorded up locally to about 1600 m in w. Loja, but mainly below 900 m.

Although overall numbers of the Rufous-fronted Wood-Quail in Ecuador have doubtless declined as a result of deforestation, at least in some areas the species seems capable of persisting where forest cover has been reduced to relatively small patches (e.g., at Tinalandia). Given that it is not much hunted, we do not consider the species yet to be at risk in the country, but we recognize that its future status will require monitoring.

Two similar races are found in Ecuador, *parambae* south to w. and n. Guayas (though no specimens are available from south of Manabí) and nominate *erythrops* from w. Chimborazo southward; these two taxa probably intergrade in the intervening area, whence no specimen material is available. The nominate race differs from *parambae* in being a paler, more olivaceous brown above.

Range: w. Colombia and w. Ecuador. Black-eared Wood-Quail (*O. melanotis*), found from Honduras to e. Panama and formerly often considered conspecific with Rufous-fronted, is now usually treated as a separate species (e.g., Ridgely and Gwynne 1989, AOU 1998).

Odontophorus melanonotus	Dark-backed Wood-Quail	Corcovado Dorsioscuro

Uncommon to locally fairly common on or near the ground inside montane forest in the subtropical zone of nw. Ecuador from Carchi south to w. Cotopaxi in the Caripero area (N. Krabbe and F. Sornoza). The Dark-backed Wood-Quail, which is nearly endemic to nw. Ecuador, remains tolerably numerous in its limited range, though it usually is very difficult to see. Coveys, however, are regularly encountered along the forest trails at the Maquipucuna Reserve (P. Coopmans) and can also be seen in some areas around Mindo (PJG). Recorded mostly from 1200 to 2700 m (Caripero).

Collar et al. (1992, 1994) judged the Dark-backed Wood-Quail to be a Near-threatened species. We concur in this assessment, though noting that significant populations do occur in three protected areas in Ecuador (which is more than can be said for many other potentially threatened species): Maquipucuna Reserve, Mindo-Nambillo Ecological Reserve, and Cotacachi-Cayapas Ecological Reserve.

Monotypic.

Range: west slope of Andes of extreme sw. Colombia and nw. Ecuador.

| *Odontophorus speciosus* | Rufous-breasted | Corcovado |
| | Wood-Quail | Pechirrufo |

Rare to uncommon and seemingly local on or near the ground inside montane forest in the foothills and subtropical zone on the east slope of the Andes from w. Napo (north to the Volcán Sumaco region) southward. Contrary to many published sources (e.g., Meyer de Schauensee 1966, 1970; Blake 1977), the Rufous-breasted Wood-Quail does not occur in the lowlands of Amazonia. Recorded mostly from 800 m (e.g., around Zumba) to 1800–2000 m; the highest site on record is Cabañas San Isidro, where it was heard on 22 Jul. 1999; M. Lysinger).

We consider the Rufous-breasted Wood-Quail to merit Near-threatened status in Ecuador because of its overall rarity, limited range, and the substantial habitat destruction that has occurred within its range. A population of unknown size does exist in Podocarpus National Park. Collar et al. (1992, 1994) did not consider the species to be at risk.

One race: *soederstroemii*.

Range: locally on east slope of Andes from n. Ecuador to w. Bolivia.

| *Odontophorus stellatus* | Starred Wood-Quail | Corcovado Estrellado |

Uncertain. Known from a few records from the lowlands of e. Ecuador, mainly (perhaps entirely) in the southeast. Elsewhere the Starred Wood-Quail occurs inside humid forest, principally in terra firme but also in transitional and floodplain forest. There is an old specimen from "Río Napo" (BMNH), and others (FMNH) are recorded as having been taken at Sarayacu and "Valladolid, Loja" (*Birds of the Americas*, pt. 1, no. 4, p. 279); the last locality, however, being in the lower subtropical zone, seems highly questionable. In addition, MECN has three specimens of this species, two of them with limited data (a female from "Oriente" and a male from "Región del Bobonaza"); a third is labeled as having been obtained at Río Bufeo on the lower Río Bobonaza in Jan. 1963 by R. Olalla. As the handwriting on all three tags is similar, it seems likely that all were obtained by Olalla on the same expedition. The only recent reports involve a single individual believed seen at Jatun Sacha on 29 Jan. 1988 (B. Bochan) and sightings from Yuturi Lodge (J. Hendriks, fide P. Coopmans) and Kapawi Lodge on the Río Pastaza near the Peruvian border. The species' ecological relationship with the much more numerous Marbled Wood-Quail (*O. gujanensis*) remains to be elucidated, as does whether they ever occur together in Ecuador. Recorded below 400 m.

Given this species' evident rarity in Ecuador, we accord it Data Deficient status. The species was not considered to be generally at risk by Collar et al. (1992, 1994).

Monotypic.

Range: locally in e. Ecuador; more widely in e. Peru, n. Bolivia, and w. Amazonian Brazil.

Rhynchortyx cinctus   **Tawny-faced Quail**   Codorniz Carirrufa

Uncommon to locally fairly common (but easily overlooked unless its vocalizations are known) on or near the ground inside very humid to humid forest in the lowlands and foothills of nw. Ecuador in n. Esmeraldas. There are recent records of this inconspicuous small quail from near Alto Tambo and El Placer (ANSP). It also was found to be widespread and not uncommon in the very wet forest belt of the Río Santiago and Río Cayapas drainages, occurring mainly inside mature forest and less often in secondary forest (O. Jahn and P. Mena et al.); in this region, the Tawny-faced Quail appears to be more numerous than the Rufous-fronted Wood-Quail (*Odontophorus erythrops*). Tawny-faced Quail are also reported to be numerous east of San Lorenzo at Río Mataje (M. Lysinger). Recorded up to about 600 m.

We consider the Tawny-faced Quail to merit Near-threatened status in Ecuador because of its relative rarity and very limited range, and the substantial amount of habitat destruction taking place there. A population does exist in the lower-lying portions of Cotacachi-Cayapas Ecological Reserve. The species as a whole is not considered to be at risk (Collar et al. 1992, 1994).

One race: *australis*.

Range: Honduras to nw. Ecuador.

Gruiformes
Rallidae   Rails, Gallinules, and Coots

Twenty-five species in 11 genera. The family is worldwide in distribution.

Laterallus exilis   **Gray-breasted Crake**   Polluela Pechigris

Uncommon to locally fairly common (easily overlooked except through its oft-given calls) in areas with dense tall grass and other shrubby vegetation, usually but by no means always near water, in the lowlands of e. and nw. Ecuador. In the east the Gray-breasted Crake is necessarily local, but its population is probably increasing as a result of deforestation, and its frequent calling can reveal it to be numerous where conditions are optimal. It is often most common in the tall grass bordering airstrips, and also occurs on the floating grass mats surrounding oxbow lakes (though often outnumbered by Rufous-sided Crakes [*L. melanophaius*] in such situations); it also can be numerous on younger river islands. In the northwest the Gray-breasted Crake is known only from w. Esmeraldas south to s. Pichincha (there is at least one record from Río Palenque), but it may be more wide-ranging. Recorded mostly below 500 m, but locally up to as high as 850 m at Miazi in the Río Nangaritza valley of s. Zamora-Chinchipe (Schulenberg and Awbrey 1997).

Monotypic.

Range: locally from Belize to n. Bolivia, Paraguay, and s. Brazil; nw. Ecuador.

Laterallus albigularis   **White-throated Crake**   **Polluela Goliblanca**

Fairly common to common (easily overlooked, except through its frequently given calls) in freshwater marshes with abundant emergent vegetation, and in damp grassy areas such as pastures (usually where the grass is tall, but often occurring far from any open water) in the lowlands and foothills of w. Ecuador south to El Oro (south as far as the Arenillas area and near Balzas). Recorded mostly below about 1000 m, but ranging up regularly to 1300–1400 m in the Mindo area and once heard as high as 1700 m above Mindo (P. Coopmans).

L. *albigularis* is sometimes considered conspecific with the cis-Andean *L. melanophaius* (Rufous-sided Crake). The two are indubitably closely related and at least occasionally respond to each other's vocalizations (though it should be noted that even *L. exilis* [Gray-breasted Crake] will sometimes respond to the calls of *L. melanophaius*). One race occurs in Ecuador, nominate *albigularis*.

Range: Honduras to sw. Ecuador.

Laterallus melanophaius   **Rufous-sided Crake**   **Polluela Flanquirrufa**

Uncommon to locally fairly common (easily overlooked, except through its frequently given calls) in floating mats of vegetation surrounding oxbow lakes, and in damp grassy areas such as pastures (where the grass is tall) in the lowlands and foothills of e. Ecuador. Recorded mostly below 700 m, but locally ranging as high as 1350 m in w. Sucumbíos (near Reventador in Aug. 1993; RSR and F. Sornoza).

Does not include trans-Andean *L. albigularis* (White-throated Crake). One race: *oenops*.

Range: locally in n. Venezuela and the Guianas; se. Colombia to n. Argentina and Uruguay.

Laterallus fasciatus   **Black-banded Crake**   **Polluela Negrilineada**

Rare to locally fairly common (easily overlooked except through its frequently given calls) in dense shrubby and grassy vegetation in clearings and at forest and woodland edges, almost always where damp and usually near streams or at least seepage zones, and on river islands, in the lowlands and foothills of ne. Ecuador. The Black-banded Crake has been recorded primarily from Napo and Sucumbíos. The only known records from the Río Pastaza drainage come from Kapawi Lodge in 1996 (J. Moore), but there seems to be no reason why the species should not occur more widely. Black-banded Crakes do seem to be most numerous close to the base of the Andes (e.g., in the Jatun Sacha region); the species has, however, been seen at Cuyabeno (C. Canaday) and has also been recorded along the Maxus road southeast of Pompeya (RSR). Recorded mostly below 600 m, rarely as high as 1100 m (north of Archidona along the road to Loreto; RSR and PJG et al.).

S. L. Olson (*Wilson Bull.* 85[4]: 381–416, 1973) suggested that this and the following species should be transferred to the genus *Anurolimnas*. However, information obtained subsequently concerning their vocalizations appears clearly to place both in the genus *Laterallus*, where they traditionally have been classified (e.g., Meyer de Schauensee 1966, 1970). Monotypic.

Range: se. Colombia, e. Ecuador, e. Peru, and w. Amazonian Brazil.

Laterallus viridis   **Russet-crowned Crake**   **Polluela Coronirrojiza**

Locally rare to uncommon—but easily overlooked, except through its frequently given calls—in dense second-growth thickets in e. Ecuador, apparently mostly in the foothills along the eastern base of the Andes. Seems less tied to the presence of water than the Black-banded Crake (*L. fasciatus*). The only published Ecuadorian record of the Russet-crowned Crake involves an FMNH specimen taken at "Quijos" (*Birds of the Americas*, pt. 1, no. 4, p. 385), a locality that cannot be precisely located but presumably lies somewhere in the upper Río Napo drainage on or near the Río Quijos. The only recent confirmed reports come from the Río Marañón drainage of extreme s. Zamora-Chinchipe in the Zumba area, first on 1 May 1989 when several were heard and one tape-recorded (P. Coopmans and P. Kaestner); there have been a few subsequent reports from here as well, but still no specimen material. A sighting from Jatun Sacha (fide B. Bochan) is regarded as uncertain. Recorded from about 650 to 1200 m.

Sick (1993) suggested that this species may merit generic separation in the monotypic genus *Rufirallus*, and indeed it does seem distinct from other *Laterallus* species. As noted under *L. fasciatus*, that species and *L. viridis* have also been placed in the genus *Anurolimnas*. One race occurs in Ecuador, nominate *viridis*.

Range: e. Colombia, s. Venezuela, and the Guianas to e. and s. Amazonian Brazil, n. Bolivia, and e. Peru; e. Brazil; locally in e. Ecuador and n. Colombia.

Anurolimnas castaneiceps   **Chestnut-headed Crake**   **Polla Cabecicastaña**

Uncommon to fairly common (though easily overlooked, except through its frequently given calls) on or near the ground in dense secondary woodland and the borders of humid forest in the lowlands of e. Ecuador. Although often near streams or other wet places (swampy depressions, etc.), the Chestnut-headed Crake is not particularly associated with the presence of surface water. Recorded mostly below 700 m, locally as high as 1000 m (e.g., north of Archidona [P. Coopmans] and in the Río Zamora and Río Nangaritza valleys in Zamora-Chinchipe).

Two races are found in Ecuador, *coccineipes* in the northeast and nominate *castaneiceps* southward. Where these taxa replace each other remains poorly understood, though it may lie near the Río Napo; birds from the Sumaco region can be referred to the nominate race. The two taxa differ strikingly in leg color, being red or reddish in *coccineipes* and dusky or gray in the nominate race; other supposed characters seem not to hold up in recently obtained material.

Range: se. Colombia, e. Ecuador, e. Peru, and extreme nw. Bolivia.

Amaurolimnas concolor   Uniform Crake   Rascón Unicolor

Very rare to rare and local (but doubtless overlooked, and can be locally more numerous) inside swampy forest, secondary woodland, and overgrown plantations, and along sluggish forest-interior streams in the lowlands of e. and nw. Ecuador. In the northwest recorded from old specimens taken in Esmeraldas at San Javier and San Mateo (the latter site cannot be located; Paynter 1993); the only recent reports of this secretive species come from Playa de Oro, where in 1995–1998 it was found (mainly by voice) to be fairly common in young successional vegetation of overgrown plantations and clearings mainly in the floodplain of the Río Santiago (O. Jahn et al.). In the east known from old specimens taken in Napo at "Raya Yacu" (thought to be near Loreto; Paynter 1993) and in Pastaza at Sarayacu. The only recent reports from the east are from Kapawi Lodge (listed as having been recorded there, but we are unaware of details), and a single bird that was seen and tape-recorded at Limoncocha on 12 Dec. 1998 (G. Rivadeneira and C. Canaday). Recorded only below 300 m.

Two similar races are found in Ecuador, *guatemalensis* in the northwest and *castaneus* in the east. *Guatemalensis* has sometimes been synonymized into *castaneus* (e.g., *Birds of the Americas*, pt. 1, no. 4), or its range has been restricted to the area from Panama northward (Blake 1977). We follow Taylor and van Perlo (1998 and pers. comm.) in placing birds from w. Ecuador—which we have not seen—in *guatemalensis*. *Castaneus* is slightly larger and more rufescent below than *guatemalensis*.

Range: locally from s. Mexico to nw. Ecuador; e. Ecuador, e. Peru, n. Bolivia, and w. Amazonian Brazil; the Guianas and lower Amazonian Brazil; e. Brazil; Jamaica (where perhaps extinct).

Porzana carolina   Sora   Sora

A rare to locally (and perhaps only temporarily) uncommon boreal winter visitant to freshwater marshes, damp grassy areas, and rice fields in the lowlands of w. Ecuador south to Guayas, and around a few lakes in the highlands of n. Ecuador south to Chimborazo (an old record from Cayandeled), once as far as Azuay (a single bird seen at El Cajas on 28 Mar. 1996; P. Coopmans et al.). The Sora may also sparingly occur farther south in the western lowlands, as there is an old record from just across the border in Tumbes, Peru (at Santa Lucía). There is also a single specimen record (MECN) from the eastern lowlands, taken by M. Olalla at Sarayacu in Pastaza on 15 Mar. 1973. The one site in Ecuador where the Sora appears to be regular is around Yaguarcocha in Imbabura. Recorded from Dec. to Mar., but undoubtedly a few arrive earlier and linger later. Ranges up to about 2800 m.

The Sora seems now to occur less often in Ecuador than it did formerly; at least there are rather few recent reports.

The species has sometimes been called the Sora Crake.

Monotypic.

Range: breeds in North America, wintering south to n. South America, exceptionally as far south as Peru.

*[Porzana flaviventer   Yellow-breasted Crake   Polluela Pechiamarilla]

Uncertain. Known in Ecuador only from a few recent sightings from grassy marshes in the lowlands of w. Ecuador in n. Manabí near the head of the Bahía de Caráquez west of Chone. Individual Yellow-breasted Crakes were seen walking on floating vegetation here, at Segua Marsh, on 23 Nov. 1996 and 27 Feb. 1997 (B. López-Lanús and P. Gastezzi, *Cotinga* 13: 60, 2000). Presumably a small population is resident in this area, but obviously more information on its status is needed. Resident populations of the Yellow-breasted Crake are known from as close to Ecuador as sw. Colombia at the Laguna de Sonso marshes of Valle (Hilty and Brown 1986). Recorded at slightly above sea level.

Monotypic.

Range: locally from s. Mexico to n. Argentina and s. Brazil, but absent from wide areas including Amazonia; Greater Antilles.

*Neocrex erythrops   Paint-billed Crake   Polluela Piquipinta

Rare and local (but apparently overlooked, and perhaps more numerous at least locally) in moist areas with tall grass, the muddy margins of ponds and ditches, and rice fields in the lowlands of sw. Ecuador. A male was collected (ANSP) by F. Sornoza and RSR west of San Antonio in the Bahía de Caráquez area of Manabí on 14 Sep. 1992, and others were heard there on that date; one and probably several others were seen near Chone on 30 Jul. 1996 (S. Howell and R. Behrstock et al.). In addition, two locally captured captive birds were seen and photographed at Macará in extreme s. Loja in Jul. 1991 (R. Williams); and one was seen at what is now Manglares-Churute Ecological Reserve on 28 Jun. 1985 (RSR and K. Berlin). Recorded below about 300 m.

Somewhat surprisingly, the one Ecuadorian specimen seems to be referable to the cis-Andean *olivascens* and not to nominate *erythrops* of nw. Peru and the Galápagos Islands.

Range: very locally from Panama to nw. Argentina, Paraguay, and e. Brazil; Galápagos Islands.

Neocrex colombianus   Colombian Crake   Polluela Colombiana

Rare to locally fairly common (but likely overlooked) in damp grassy areas and the muddy margins of ponds and ditches in the lowlands of nw. Ecuador from Esmeraldas south to Pichincha (Santo Domingo de los Colorados) and along the base of the Andes to w. Chimborazo (Puente de Chimbo). Most of the few Ecuadorian records of this species are old, but it was considered to be not uncommon during intensive survey work in 1995–1998 at Playa de Oro in n. Esmeraldas, with several birds mist-netted or trapped, photographed, and released (O. Jahn et al.). In addition, two were seen at a marsh near Same, west of Esmeraldas (city), on 28–29 Aug. 1991 (H. Kasteleijn, fide P. Coopmans). There also is a specimen (MECN) of this species from considerably farther south taken by J. Yépez N. near Alamor ("1 km antes de El Derrumbe") in w. Loja on 3 Jan. 1992. Recorded mostly below 500 m, but the Alamor specimen was taken considerably higher, at about 1100 m, and in Colombia the species has been found as high as 2100 m (Hilty and Brown 1986).

Collar et al. (1992, 1994) concluded that the poorly known Colombian Crake warranted Near-threatened status. We are not, however, aware of any factor that would be expected to affect it adversely and thus do not consider it to be at risk, at least not in Ecuador.

N. colombianus has sometimes been considered conspecific with *N. erythrops* (Paint-billed Crake), but the two are now known to occur in near sympatry in w. Ecuador, and they are also apparently locally sympatric in Panama (Ridgely and Gwynne 1989). One race occurs in Ecuador, nominate *colombianus*.

Range: Panama to sw. Ecuador.

Rallus longirostris   Clapper Rail   Rascón Manglero

Very rare and local (doubtless often overlooked; we have never heard it vocalize in Ecuador) in mangroves along the coast. The Clapper Rail is known from only two Ecuadorian specimens, one taken at Vaquería in n. Esmeraldas and the other merely labeled "Guayas," the latter being an undated MCZ specimen without further locality information. Although in the 1970s PJG observed the species on a few occasions in mangroves just west of Guayaquil, it has gone essentially unrecorded in recent years and must be viewed as inexplicably scarce in Ecuador. A sighting of 3 birds in Feb. 1993 in a freshwater situation west of Chone in cen. Manabí (B. López-Lanús and P. Gastezzi, *Cotinga* 13: 60, 2000) is considered unconfirmed. As there are records from adjacent Tumbes in Peru, conceivably it also occurs along the El Oro coast. Found at sea level.

Because of the infrequency of recent records, and because of the massive destruction of mangrove forests that has taken place in many areas, we consider the status of the Clapper Rail in Ecuador to be Vulnerable. Given the extent of mangrove forests in Manglares-Churute Ecological Reserve, the species seems likely to occur there, but this supposition has never been confirmed. Elsewhere this wide-ranging bird cannot be considered at risk (Collar et al. 1992, 1994), though a few other localized populations, particularly some of those found in the tropics (such as the one in Ecuador), may be.

Following most recent authors, including the AOU (1998) and Taylor and van Perlo (1998), we consider the closely related King Rail (*R. elegans*) of the e. United States and Cuba to be a distinct species. One race of *R. longirostris* occurs in Ecuador, *cypereti*.

Range: coastal e. United States, and locally in sw. United States and Mexico and south (mainly coastally) to extreme nw. Peru and e. Brazil, and West Indies.

Rallus aequatorialis   Ecuadorian Rail   Rascón Ecuatoriano

Rare to locally fairly common (often overlooked) in reedbeds and adjacent wet grassy areas in the highlands of n. and cen. Ecuador from Carchi (Páramo del Ángel; MECN) south locally to n. Loja (San Lucas; J. Fjeldså). The Ecuadorian Rail is quite numerous in the reedbeds at Laguna de Colta in Chimborazo, and it may be equally numerous around a few other highland lakes (e.g., Lago San Pablo in Imbabura), but in general it is scarce and infrequently recorded in Ecuador. We can find no basis for the suggestion (Fjeldså and Krabbe 1990) that the Ecuadorian Rail occurs in the western lowlands. Recorded mostly from 2200 to 3500 m, occurring locally up to 3800 m (Laguna Limpiopungo in Cotopaxi National Park; v.o.); one specimen apparently was procured much lower (1200 m; see below).

Given that its voice rather differs from that of North American birds, and in view of its extreme range disjunction, we consider *R. aequatorialis* to be a species separate from *R. limicola* (Virginia Rail) of North America and Mexico. In recent years *aequatorialis* has generally been considered as a Neotropical subspecies of *R. limicola*. One race occurs in Ecuador, the nearly endemic nominate race; it also ranges in Nariño, Colombia. J. Fjeldså (*Steenstrupia* 16[7]: 109–116, 1990) identified a specimen (BMNH) from Intag (1200 m) in Imbabura as belonging to his newly described subspecies *meyerdeschauenseei* (which ranges primarily in coastal Peru); this form differs from *aequatorialis* in being somewhat more cinnamon below with more contrasting black barring on flanks.

The species has sometimes (e.g., Meyer de Schauensee 1966) been called the Lesser Rail (though that name implies that there are only 2 rail species). We prefer to highlight the fact that the species is mainly found in Ecuador.

Range: Andes of extreme s. Colombia and Ecuador; coastal Peru.

Salvadori and Festa (1900) record a specimen that they identified as Bogotá Rail (*Rallus semiplumbeus*), but it is actually a misidentified juvenile Ecuadorian Rail (S. D. Ripley, pers. comm.).

Pardirallus nigricans   Blackish Rail   Rascón Negruzco

Uncommon to locally fairly common—though regularly vocalizing, probably often overlooked—in small marshes, around shallow ponds, and in damp places with tall grass in the lowlands and foothills of e. Ecuador along the base of the Andes from w. Sucumbíos (north to around Lumbaquí and Reventador; ANSP, MECN) south to Zamora-Chinchipe (in the Zamora area, on the lower slopes of the Cordillera del Cóndor, and at Valladolid). Apparently it does not occur at any distance east away from the Andes. The Blackish Rail seems almost certain to occur along the base of the Andes north into adjacent Colombia, from whence it is unrecorded (Hilty and Brown 1986). Recorded between about 400 and 1650 m.

This and the following two species have sometimes been separated in the genus *Ortygonax*. This and the following two species have also been placed in the genus *Rallus*, but we follow Sibley and Monroe (1990) in placing them in *Pardirallus*. One race occurs in Ecuador, nominate *nigricans*.

Range: locally in w. Colombia; e. Ecuador, e. Peru, n. Bolivia, and sw. Amazonian Brazil; e. Brazil, e. Paraguay, and ne. Argentina.

Pardirallus sanguinolentus   Plumbeous Rail   Rascón Plomizo

Locally rare to uncommon in wet grassy areas (and at least occasionally in nearby sugarcane fields) and near streams in the highlands of s. Loja. Discovered in Ecuador by H. Bloch and R. Tapia on 26 Nov. and 4 Dec. 1991, when two birds were seen at Vilcabamba (Rasmussen et al. 1996); two specimens (ANSP, MECN) were collected here by F. Sornoza and RSR on 10 Dec. 1991, and the species continues to be found in this region, where it is perhaps increasing. N. Krabbe had previously tape-recorded what turned out to be this species somewhat earlier (on 7 Apr. 1991) at Cariamanga in s. Loja. We suspect that more intensive searching, particularly with prerecorded tapes, will reveal the Plumbeous Rail to be more widespread in the s. Loja highlands. More recently, one was reported seen as far north as north of Loja (city) on 2 Jul. 1995 (O. Janni). Recorded between about 1500 and 1900 m.

One race: *simonsi*.

Range: extreme s. Ecuador and w. Peru to Chile, Argentina, and se. Brazil.

Pardirallus maculatus   Spotted Rail   Rascón Moteado

Rare and local—doubtless much overlooked—in reedbeds in the coastal lowlands of w. Ecuador. First recorded from a bird found by RSR and K. Berlin near San Vicente in coastal El Oro on 24 Jun. 1984, with one being collected the next day (ANSP). There have been several subsequent sightings from El Oro in the Arenillas and Santa Rosa areas, and a few have also been noted at the large freshwater marsh at Manglares-Churute Ecological Reserve in se. Guayas. The only reports from elsewhere involve sightings of single birds at three sites in coastal w. Esmeraldas, first south of Atacames on 10 Jan. 1991 (R. ter Ellen), then south of Esmeraldas (city) on 8 Feb. 1993 (J. and V. Strasser, fide P. Coopmans), and finally at Same on 24 and 26 Sep. 1994 (P. Coopmans and G. Muñoz). The species was also reported at Jauneche in Los Ríos (Parker and Carr 1992), seemingly a peculiar place for it given the absence or limited extent of suitable habitat for it there. We expect that additional field work will show that Spotted Rails occur more widely in w. Ecuador; the marshes in the Bahía de Caráquez region of Manabí, for instance, would appear to provide acceptable habitat. Recorded only below 100 m.

One race: nominate *maculatus*.

Range: locally from s. Mexico to n. Argentina and Uruguay, but absent from much of Amazonia and interior Brazil; Greater Antilles.

Aramides axillaris   **Rufous-necked Wood-Rail**   Rascón Montés Cuellirrufo

Rare to locally fairly common (but easy to overlook, as it does not seem to vocalize very much) in mangroves along the coast, where mainly recorded from sw. Ecuador in Guayas (the lower Río Guayas estuary) and El Oro (south to Puerto Pitahaya north of Arenillas); in addition, there is a single sighting from the northwest, of three birds seen in mangroves at Muisne on 25 Sep. 1994 (P. Coopmans and G. Muñoz). The Rufous-necked Wood-Rail also ranges, at least locally, into deciduous forest and woodland; in this habitat it is recorded only from the Chongón Hills in Guayas, from an old specimen taken at La Chonta in El Oro, and from a sighting of one bird at Sozoranga in s. Loja on 11 Mar. 1991 (Best et al. 1993). Despite generalized statements in other references indicating that the species is widespread along the Ecuadorian coast, the only record from north of the Río Guayas estuarine region in Ecuador is the Esmeraldas sighting mentioned above; populations may also exist elsewhere along the coast of Esmeraldas, in at least some parts of which mangrove forests remain extensive. Although in Ecuador found primarily at sea level, the Rufous-necked Wood-Rail has been recorded locally as high as 1400 m at Sozoranga (Best et al. 1993).

Prior to the destruction of so much mangrove forest, this wood-rail's range seems likely to have been considerably more extensive. We therefore consider it to merit Near-threatened status in Ecuador. Elsewhere the Rufous-necked Wood-Rail has not been judged to be at risk (Collar et al. 1994).

Monotypic.

Range: w. and se. Mexico south locally (mainly near coasts) to extreme nw. Peru and the Guianas.

Aramides wolfi   **Brown Wood-Rail**   Rascón Montés Moreno

Rare and very local—though doubtless overlooked—along streams and in swampy areas inside humid forest and secondary woodland and in mangroves in the lowlands and foothills of w. Ecuador from Esmeraldas south through El Oro. All indications are that the Brown Wood-Rail has decreased substantially during the past several decades, almost certainly as a result of the deforestation that has occurred across much of its former range. Recently recorded only north of Quinindé in Esmeraldas (birds tape-recorded on 8 Sep. 1994; P. Coopmans), at Bilsa in w. Esmeraldas (sightings by R. P. Clay et al.), at Río Palenque (a few reports continuing to come from there, the latest being a pair heard in Nov. 1998; P. Coopmans), and at Manglares-Churute Ecological Reserve (2 individuals seen in mangroves on 28 Dec. 1989; PJG and R. Jones). The last is the only indication that the species occurs in mangroves in Ecuador. It was not recorded at all during intensive survey work in 1995–1998 in the Río Santiago and Río Cayapas drainages of n. Esmeraldas (O. Jahn and P. Mena Valenzuela et al.). Recorded up to about 1300 m near Mindo (ANSP).

Although Collar et al. (1992) gave the Brown Wood-Rail only Near-threatened status, Collar et al. (1994) judged it to deserve Vulnerable status. From a purely Ecuadorian standpoint it better deserves an Endangered rating. An apparently large population has recently been discovered in Colombia (B. Porteous and C. Acevedo, *Cotinga* 6: 31–32, 1996), but the species was only recently first recorded from northernmost Peru (G. R. Graves, *Gerfaut* 72: 237–238, 1982) and presumably is rare there. Why the Brown Wood-Rail should apparently be so much more sensitive to habitat disturbance than the seemingly closely related Gray-necked Wood-Rail (*A. cajanea*) remains unknown. Unknown too is its ecological relationship with the Rufous-necked Wood-Rail (*A. axillaris*); though this species and the Brown Wood-Rail have been found to be locally sympatric in mangroves, the Brown appears much more often to occur away from the coast.

The species has sometimes been called the Brown-backed Wood-Rail.

Monotypic.

Range: nw. Colombia to extreme nw. Peru.

Aramides cajanea   Gray-necked Wood-Rail   Rascón Montés Cuelligris

Uncommon to fairly common (except by voice, easy to overlook; this species seems to vocalize more frequently than other *Aramides*) along streams and around ponds and marshy areas in várzea and riparian forest and woodland in the lowlands of e. Ecuador. The Gray-necked Wood-Rail is known primarily from the drainages of the Ríos Napo and Aguarico. In the southeast it appears to have been recorded only from recent records at Kapawi Lodge on the Río Pastaza near the Peruvian border, though it seems likely to occur widely there as well. Recorded up to about 400 m.

One race: nominate *cajanea*.

Range: s. Mexico to w. Colombia, n. Argentina, and Uruguay. More than one species is perhaps involved.

Aramides calopterus   Red-winged Wood-Rail   Rascón Montés Alirrojizo

Very rare and seemingly local—but presumably much overlooked—along streams inside humid forest in the lowlands and foothills of e. Ecuador. Although still poorly known, the Red-winged Wood-Rail apparently favors areas with some relief, and it seems not to be found in the flatter, swampy terrain where the Gray-necked Wood-Rail (*A. cajanea*) may replace it. Further, the Red-necked seems not to have been reliably reported well to the east of the Andes. The only sites in Ecuador from which there are confirmed recent reports are the following: Montalvo in Pastaza and Guaticocha in w. Napo (Norton 1965);

Tayuntza in Morona-Santiago (one taken on 1 Aug. 1987; WFVZ); and along the lower part of the Loreto road north of Archidona (a pair seen in Mar. 1990; B. Whitney). Recorded between about 250 and 900 m.

Given the species' evident rarity and local distribution, we accord it a Data Deficient rating in Ecuador. Elsewhere it has not been considered to be at risk (Collar et al. 1992, 1994). The Red-winged Wood-Rail is not known to occur in any area presently receiving formal protection, but as it has been recorded from along the Loreto road, it likely occurs in the nearby Sumaco-Galeras National Park.

Monotypic.

Range: locally in e. Ecuador, ne. Peru, and w. Amazonian Brazil.

Porphyrula martinica   **Purple Gallinule**   Gallareta Púrpura

Uncommon to locally common in freshwater marshes, around the margins of ponds and lakes, and along sluggish streams in the lowlands of both ne. and w. Ecuador; more numerous and widespread in the west. In the east the Purple Gallinule seems very local and, though there is no apparent reason why it could not occur southward as well, it apparently has been recorded only from the drainages of the Ríos Napo and Aguarico. In the west recorded from w. Esmeraldas south to coastal El Oro in the Arenillas region. The Purple Gallinule can be locally very numerous in w. Ecuador (e.g., in the freshwater complex of ponds and marshes at Manglares-Churute Ecological Reserve). It is, however, prone to wandering, and numbers vary seasonally at many sites. Several early-20th-century records exist from the highlands near Quito and at Latacunga in Cotopaxi; more recently a few have been seen or picked up in weakened condition in the Quito area (PJG), one of which (found on 28 Jan. 1992) was prepared as a specimen (MECN). A few are occasionally present at Yaguarcocha (2200 m) in Imbabura, and the species has also been seen at Laguna Yambo (N. Krabbe). Recorded mostly below 400 m, but wanderers occur up to about 2800 m in the highlands.

The species has sometimes been called the American Purple Gallinule.

This and the next species are sometimes placed in the Old World genus *Porphyrio* (e.g., Sibley and Monroe 1990), with this species then being called *Porphyrio martinicus*, but we follow the 1983 and 1998 AOU Check-lists in continuing to recognize the genus *Porphyrula* for them. Monotypic.

Range: se. United States and Mexico to sw. Ecuador, n. Argentina, and Uruguay; West Indies.

Porphyrula flavirostris   **Azure Gallinule**   Gallareta Azulada

Uncommon and local around the margins of certain oxbow lakes, favoring floating mats of vegetation, in the lowlands of ne. Ecuador. First recorded from Limoncocha in 1963 (Norton 1965), the Azure Gallinule has subsequently been found at several other localities in the Río Napo and lower Río Aguarico drainages. The only record from elsewhere remains the single male taken at

Conambo on an unspecified date (Orcés 1974). In accord with the conclusions of J. V. Remsen, Jr., and T. A. Parker III (*Wilson Bull.* 102[3]: 380–399, 1990), Ecuadorian populations of the Azure Gallinule appear to be resident; the apparent variation in numbers present at certain localities is probably more due to local fluctuations in water level, the species being harder to see when water levels are high, than it is to long-distance migration. A recently fledged bird was seen at Limoncocha on 20 Feb. 1985 (RSR and T. S. Schulenberg et al.). Recorded only below 300 m.

Monotypic.

Range: locally from e. Colombia, much of Venezuela, and the Guianas to Paraguay, n. Argentina, and se. Brazil.

Gallinula chloropus   Common Gallinule   Gallareta Común

Locally common on shallow, reed-fringed, marshy lakes and ponds in the lowlands of w. Ecuador from w. Esmeraldas (Same/Muisne area) south through coastal El Oro in the Arenillas region; a substantial population also occurs along the Río Chira in the Lalamor area of extreme s. Loja (Sep. 1998 observations; RSR et al.), and one was also seen near Sabiango on 14 Sep. 1998 (RSR et al.). Common Gallinules also are found on certain lakes in the highlands of n. Ecuador, now mainly in Imbabura (e.g., at Yaguarcocha), but at least formerly they were found south to Chimborazo (a 1937 specimen from "Ciénega de Chimborazo"; FMNH). The species also occurs very locally in the lowlands of ne. Ecuador, where thus far known only from a small but apparently resident population at Limoncocha. Recorded mostly below 300 m, but the highland birds range (or ranged) between about 2200 and 2800 m.

Although the species long went by the English name of Common Gallinule in the Americas (e.g., Meyer de Schauensee 1966, 1970), the conversion of this species' English name to the British group name of "Moorhen" began in the 1970s, this despite what we believe is its inappropriateness in a global context. We thus agree with Hilty and Brown (1986) and Fjeldså and Krabbe (1990), both of whom retained "gallinule," and favor continuing to employ that more suitable name.

Subspecific determinations in this species are difficult, but *pauxilla* is apparently the race found in w. Ecuador and in the highlands; no specimens have been taken in the eastern lowlands, but this population would presumably be referable to *galeata*.

Range: United States to extreme n. Chile, n. Argentina, and Uruguay; West Indies; also widely in Old World.

Fulica ardesiaca   Andean Coot   Focha Andina

Uncommon to locally fairly common on reed-fringed lakes in the highlands and paramo zone of n. and cen. Ecuador, where recorded from Carchi (Loma El Voladero; N. Krabbe) south locally to Chimborazo (where a large population, probably the largest in Ecuador, presently inhabits Laguna de Colta) and Azuay (where found recently only at Laguna Sorrocucho at El Cajas, where a small breeding population was first noted on 9 Mar. 1994; P. Coopmans at al.). A population of Andean Coots also occurs at a single locality near the coast in the southwestern lowlands, the Represa José Velasco Ibarra on the Santa Elena Peninsula south of Santa Elena in w. Guayas. Although not recorded by Marchant (1958), coots were seen here on 11 Aug. 1991 (RSR), when a total of about 40 birds was estimated; they have been present here since. At least 200 individuals, including many young birds, were observed on 14–15 Jul. 1993; one immature was collected on the latter date (ANSP). The report from the "Guayaquil area" referred to by Fjeldså and Krabbe (1990, p. 153) is based on a single sighting of 150 coots seen at an unspecified locality in the Guayaquil area on 23 Oct. 1974 by J. O. Byskof et al. (J. Fjeldså, pers. comm.). No coots have otherwise been reported from this part of Guayas. Recorded near sea level, and between 2200 and 3900 m.

The species formerly was usually called the Slate-colored Coot (e.g., Meyer de Schauensee 1966, 1970). Although the species' range is not entirely confined to the Andes, and though there are additional coot species ranging exclusively in the (southern) Andes, we favor following Fjeldså and Krabbe (1990) in employing the English name of Andean Coot for *F. ardesiaca*. All coots are "slate-colored."

The systematics and species limits of the coots found in the Andes are complex and still not completely understood. What is now called *F. ardesiaca* exhibits remarkable variation in bill and frontal shield color; for a review, see J. Fjeldså (*Steenstrupia* 8: 1–21, 1982). One race occurs in Ecuador, *atrura* (see J. Fjeldså, *Bull. B. O. C.* 103[1]: 18–22, 1983).

Range: Andes in extreme s. Colombia and Ecuador, and from cen. Peru to n. Chile and nw. Argentina, also locally along coast in sw. Ecuador and Peru.

Fulica americana   **American Coot**   Focha Americana

Apparently extirpated in Ecuador, where the species is known only from a single male collected at Yaguarcocha in Imbabura on 12 Mar. 1925 (AMNH). For at least the last twenty years all the coots present on this lake have been Andeans (*F. ardesiaca*). The American Coot has apparently also disappeared from its former range in s. Colombia (J. Fjeldså, *Steenstrupia* 9, 1983), and it now persists only locally in a few areas farther north in Colombia (though remaining numerous at some sites). Recorded at about 2200 m.

Although apparently Extirpated in Ecuador, the wide-ranging American Coot is certainly not at risk as a species, though the subspecies resident in the Colombian Andes is far from numerous. Why this species of coot should have disappeared, whereas the Andean Coot continues to persist in relatively large numbers, is unknown; competitive interactions between the two species may have put the Common at some disadvantage, but human activities could have also played a role. As with the formerly resident race of the Cinnamon Teal

(*Anas cyanoptera borreroi*), it seems doubtful whether the American Coot will ever again occur in Ecuador.

One race: *columbiana.*

Range: North America to Costa Rica, wintering to Panama; locally in Andes of Colombia and (formerly) extreme n. Ecuador; Greater Antilles. The species was formerly thought to range farther south in Andes, but these birds are now considered to refer to the Andean Coot.

Eurypygidae   Sunbittern

One species. The Sunbittern is Neotropical in distribution.

Eurypyga helias   **Sunbittern**   **Garceta Sol**

Rare to uncommon and somewhat local (or just unobtrusive?) along forested streams and rivers in the lowlands and foothills of e. and w. Ecuador; in the east occurs also around the margins of oxbow lakes and in várzea forest. In the east the Sunbittern is recorded mainly from the drainages of the Ríos Napo and Aguarico, with the only reports from farther south being a few recent sightings from the Río Bombuscaro sector of Podocarpus National Park. Despite the apparent lack of records, there seems to be no reason why the species should not also occur throughout the southeastern lowlands. In the west recorded mainly or entirely in the foothills (and not in the lowlands at any distance from the Andes) south as far as nw. Azuay (where seen by J. C. Matheus at San Luis in Jan. 1991). In the east found up to 1000 m; in the west recorded mostly between 500 and 1500 m.

Two rather different races occur in Ecuador, nominate *helias* in the east and *major* in the west. *Major* is slightly larger and has a heavier bill, but the two races differ more in the patterning of their upperparts, the nominate race being browner above with narrower buff banding, *major* being grayer above with wider blackish banding.

Range: Guatemala to w. Ecuador, n. and e. Bolivia, and Amazonian Brazil.

Heliornithidae   Finfoots

One species. Finfoots are pantropical in distribution.

Heliornis fulica   **Sungrebe**   **Ave Sol Americano**

Rare to uncommon and quite local (but easily overlooked) on sluggish, forest-fringed streams and small rivers and along the edge of shallow marshy lakes in the lowlands of e. and w. Ecuador. In the east recorded mainly from the drainages of the Ríos Napo and Aguarico in the northeast, though there are a few (mainly old) specimens from the Río Pastaza drainage and recent reports from Kapawi Lodge on the Río Pastaza near the Peruvian border. The Sungrebe seems particularly numerous at Limoncocha. In the west it is recorded south in more humid areas to Guayas (west of Guayaquil, where a pair is sometimes present on the pond at the Cemento Nacional property). Recorded below 400 m.

The species has sometimes been called the American Finfoot.

Monotypic.

Range: e. Mexico to sw. Ecuador, n. and e. Bolivia, e. Paraguay, ne. Argentina, and se. Brazil.

Aramidae   Limpkin

One species. The Limpkin is American in distribution.

Aramus guarauna   Limpkin   Carrao

Uncommon and local in marshy areas and the edges of rice fields in the lowlands of sw. Ecuador from Manabí (Bahía de Caráquez area; ANSP) and Los Ríos south to El Oro (Santa Rosa), and in the lowlands of ne. Ecuador. In the east not recorded south of the Río Napo drainage, and absent from many localities which, though appearing suitable, apparently do not support adequate populations of snails. Recorded mostly below 300 m, occasionally wandering to 400 m (Jatun Sacha).

The Limpkin has undergone a substantial population reduction in w. Ecuador because of the conversion of so much marshy and grassy country to intensive rice cultivation. As yet, however, the species remains relatively numerous, and it appears to be ecologically tolerant enough not to be considered at risk.

One race: nominate *guarauna*.

Range: Florida; s. Mexico to sw. Ecuador, n. Argentina, and Uruguay; Greater Antilles.

Psophiidae   Trumpeters

One species. Trumpeters are exclusively South American in distribution.

Psophia crepitans   Gray-winged Trumpeter   Trompetero Aligris

Rare to uncommon and local on or near the ground inside humid forest (mainly terra firme) in the lowlands and (in small numbers) up into the foothills of e. Ecuador. Even light hunting pressure reduces trumpeter numbers quickly, and birds become wary. Ranges mostly below 700 m, but has been recorded as high as 1350 at the "head of the Río Guataracu" in the Sumaco region of w. Napo (MCZ).

In overall status this wide-ranging species is not judged to be at risk (Collar et al. 1994), but from an Ecuadorian perspective we consider the Gray-winged Trumpeter to be Near-threatened, and it perhaps should even be classed as Vulnerable. Populations of unknown size do receive protection in Yasuní National Park and the Cuyabeno Faunistic Reserve.

One race: *napensis.*

Range: the Guianas, e. and s. Venezuela, se. Colombia, e. Ecuador, ne. Peru, and n. Amazonian Brazil.

Charadriiformes
Jacanidae   Jacanas

One species. Jacanas are pantropical in distribution.

Jacana jacana   Wattled Jacana   Jacana Carunculada

Common and conspicuous in marshes, around shallow lakes, and in rice fields in the lowlands of w. Ecuador from w. Esmeraldas and Pichincha south to coastal El Oro; uncommon to locally fairly common in marshes and on the floating vegetation fringing certain lakes in the lowlands of e. Ecuador south at least to e. Pastaza (around Kapawi Lodge on the Río Pastaza near the Peruvian border); curiously, jacanas seem to be unknown from Morona-Santiago or Zamora-Chinchipe. Recorded up occasionally and in small numbers to about 700 m (e.g., wandering birds at Tinalandia), but most numerous below 300 m.

Unlike so many other marsh-inhabiting birds in w. Ecuador, the Wattled Jacana has adapted readily to the conversion of much of its former natural habitat into rice fields. It may even be more numerous and widespread than it was formerly.

Two very distinct races are found in Ecuador, *scapularis* in the west and *intermedia* in the east. *Scapularis* differs in having a markedly paler mantle (rufous rather than chestnut), black on its scapulars, some chestnut on its lower underparts (underparts are all black in *intermedia*), and very pale (almost white) outermost two primaries. *Scapularis* is nearly endemic to w. Ecuador; it barely ranges into Tumbes in extreme nw. Peru, where it must be rare because of a lack of much adequate habitat.

Range: Panama to n. Argentina and Uruguay; w. Ecuador and extreme nw. Peru.

Scolopacidae   Sandpipers, Snipes, and Phalaropes

Thirty-four species in 15 genera. The family is worldwide in distribution, but a majority of its species breed in the Holarctic.

Tringa melanoleuca   Greater Yellowlegs   Patiamarillo Mayor

An uncommon to locally (and perhaps only briefly) common transient and boreal winter visitor to the muddy or sandy margins of shallow ponds, marshes, wet fields, and rivers virtually throughout, though in largest numbers near the coast in the southwestern lowlands. In general favors freshwater situations in Ecuador. The Greater Yellowlegs occurs throughout the year in Ecuador, but numbers are highest from Aug. to Oct. (800+ were counted on flooded fields north of Churute in s. Guayas on 5 Aug. 1988; RSR and PJG et al.) and are very low from May to Jul. (presumably representing only first-year prebreeders). Small numbers are regularly found around certain paramo lakes up to at least 3800 m (e.g., at Cotopaxi National Park).
Monotypic.
Range: breeds in n. North America, wintering from s. United States to s. South America.

Tringa flavipes   Lesser Yellowlegs   Patiamarillo Menor

An uncommon to locally common transient and boreal winter visitor to the shallow ponds and lagoons, marshes, and wet fields virtually throughout, though in largest numbers near the coast in the southwest. The largest concentrations of Lesser Yellowlegs in Ecuador are consistently to be found at the Ecuasal lagoons in w. Guayas, where counts regularly range from 500 to 1000. Recorded throughout the year, though only very small numbers (presumed prebreeders) occur during the northern summer months (May–Jul.). As with the Greater Yellowlegs (*T. melanoleuca*), a few Lessers occur around certain paramo lakes up to at least 3800 m (e.g., at Cotopaxi National Park).
Monotypic.
Range: breeds in n. North America, wintering from s. United States to s. South America.

Tringa solitaria   Solitary Sandpiper   Andarríos Solitario

An uncommon to fairly common and widespread transient and boreal winter visitor to the shallow margins of ponds, streams and rivers, and marshes virtually throughout. Numbers are much larger in the lowlands (especially in the west); at higher elevations (e.g., in the central valley) the species occurs mainly as a transient. Solitary Sandpipers occur exclusively on freshwater. True to its name, the species is often solitary (though sometimes occurring in groups of up to 3–5 birds, especially on migration), and it regularly occurs in situations not frequented by shorebirds other than the ubiquituous Spotted Sandpiper (*Actitis macularia*). Recorded mainly from Aug. to early Apr., exceptionally as early as late Jul. Contra the assertion of Blake (1977), we can find no evidence that any individuals pass the entire northern summer in Ecuador (nor, for that matter, do they appear to do so anywhere in the Neotropics). Recorded mostly below 3000 m.

Both of the two very similar races of this species, nominate *solitaria* and *cinnamomea*, have been recorded in Ecuador.

Range: breeds in n. North America, wintering in Middle and South America.

Catoptrophorus semipalmatus   Willet   Vadeador Aliblanco

A fairly common transient and boreal winter visitant along the coast of sw. Ecuador from the Bahía de Caráquez area in Manabí southward; also occurs northward along the coast, though so far reported only in relatively small numbers (because of less extensive suitable habitat). Willets are found exclusively on or very close to salt water, where they favor lagoons, mudflats, and beaches. The only Ecuadorian report of the Willet away from the immediate coast is a specimen (FMNH) taken by the Olallas at Isla Silva, in n. Guayas along the Río Babahoyo, on 9 Sep. 1931. The largest concentrations in Ecuador apparently occur in the Río Guayas estuary, where counts of 250+ individuals have been made. Recorded throughout the year, perhaps in largest numbers during the northern winter (especially Nov.–Feb.); only very small numbers occur during the northern summer (especially May–Jul.).

Inornatus, the western (inland-breeding) subspecies, is by far the more numerous of the two races occurring in Ecuador. The only record of the nominate race is the Isla Silva specimen mentioned above; indeed this appears to be its only record on the Pacific coast of South America. The two races are closely similar when in nonbreeding plumage, though *inornatus* averages larger.

Range: breeds in North America and locally in West Indies, wintering along coasts from s. United States to n. and w. South America.

Heteroscelus incanus   Wandering Tattler   Playero Vagabundo

A rare to uncommon and local transient and boreal winter visitant to rocky areas along the coast, where to date recorded only from Manabí (north to the Bahía de Caráquez region) and w. Guayas (east to Playas). Wandering Tattlers seem likely also to occur along rocky portions of the Esmeraldas coast, but no records are known to us. Small numbers occur throughout the year with little or no discernible seasonal fluctuation in numbers; the species is perhaps most numerous around the Santa Elena Peninsula, but even there it is exceptional to see more than three to five birds at a time.

Monotypic.

Range: breeds in Alaska and nw. Canada, wintering on rocky coasts south to w. South America and on Pacific Ocean islands.

Actitis macularia   Spotted Sandpiper   Andarríos Coleador

A fairly common to very common and widespread transient and boreal winter visitant to the rocky or sandy margins of rivers, larger streams, lakes, ponds, and coastal lagoons, mudflats, and mangroves; most common along the coast, seemingly especially in s. Guayas and El Oro. The Spotted Sandpiper shows a special predilection for the artificial shrimp ponds now so prevalent in many coastal regions; in such situations it is by far the most numerous shorebird, and often is the only one present. Recorded mainly from Aug. to Apr., with a few birds arriving as early as late Jul. or lingering into May; there is no confirmed evidence that any pass the entire northern summer in Ecuador. Although the species is most numerous in the lowlands, small numbers are also noted up to rather high elevations in the Andes, likely occurring there especially as transients; occasional birds have been seen up to almost 4000 m.

Monotypic (*rava* now usually being considered a synonym; see Monroe 1968).

Range: breeds in North America, wintering mainly in Middle and South America.

Bartramia longicauda   **Upland Sandpiper**   **Pradero Colilargo**

A rare to uncommon transient to open fields and paramo in the highlands, and in pastures and on sandbars along major rivers in the eastern lowlands. There is apparently only one record from the western lowlands: a bird collected (FMNH) at Río Blanco in s. Esmeraldas on 14 Sep. 1931. Recorded mostly during southward passage from Aug. to Oct., but there is at least one record during northward passage as well (Mar.). Upland Sandpipers have occasionally been noted in flocks—PJG has seen as many as 75 birds together—but more often they are seen singly or in small groups. Even more than other shorebirds, Upland Sandpipers sometimes drop down in seemingly improbable places; one, for instance, was parading around on the small lawn at the Instituto Geográfico Militar in downtown Quito on 20 Oct. 1992 (A. Whittaker et al.). Recorded up to almost 4000 m.

Monotypic.

Range: breeds in North America, wintering in s. South America.

Numenius phaeopus   **Whimbrel**   **Zarapito Trinador**

An uncommon to locally common transient and boreal winter visitant along the entire coast; widespread, occurring on sandy beaches, rocky areas, mudflats, and mangroves. Whimbrels are most numerous around the estuary of the Río Guayas (e.g., at Manglares-Churute Ecological Reserve) where hundreds of birds can be seen daily. On rare occasions the species also can occur inland, probably mainly as a transient; for example, one was seen on the Río Cayapas at Borbón in n. Esmeraldas on 17 Aug. 1995 (O. Jahn), and one was recorded as having been collected at the La Carolina marshes in what is now Quito (Lönnberg and Rendahl 1922). There is also a single remarkable record from the lowlands of e. Ecuador, involving a specimen (LSUMZ) obtained along the Río Napo near Limoncocha on 28 Mar. 1976 (Tallman and Tallman 1977); this represents the first record from anywhere in Amazonia. Recorded throughout the year, with lowest numbers occurring during the northern summer months (May–Aug.), though even then first-year prebreeders are regularly found.

One race: *hudsonicus*. It has been suggested (e.g., Sibley 1996) that American populations might better be separated as a distinct species, *N. hudsonicus* (Hudsonian Curlew or Hudsonian Whimbrel). However, we continue to follow the 1998 AOU Check-list in considering all forms in the complex as conspecific.

Range: breeds in arctic North America and Eurasia, wintering along coasts of Middle and South America, Africa, and Australasia.

**Limosa haemastica*   Hudsonian Godwit   Aguja Hudsoniana

A very rare transient and boreal winter visitant to mudflats and the shallow margins of ponds and lakes, with records most frequent during southward passage. The Hudsonian Godwit seemingly can occur almost anywhere, though most Ecuadorian reports come from the Ecuasal lagoons in w. Guayas. All Ecuadorian records refer to single birds, and all of them are recent, but the species' occurrences in Ecuador may be becoming more frequent, as its population levels slowly recover from excessive gunning in the late 19th and early 20th centuries. The first record involved a bird collected (LSUMZ) on the Río Napo near Limoncocha on 2 Nov. 1975 (Tallman and Tallman 1977), one of the few records from anywhere in Amazonia (Stotz et al. 1992). A second specimen (ANSP) was obtained at the Ecuasal lagoons in w. Guayas on 19 Jan. 1991, and there are several additional sightings from there (from Apr., May, Jul., and Aug.) as well as one from the El Oro coast (a single bird seen on 29 Mar. 1989 near Puerto Bolívar; Bloch et al. 1991). The only report from the highlands is a single bird seen at Laguna Limpiopungo in Cotopaxi National Park on 2 Nov. 1992 (P. Coopmans et al.); what was presumably the same individual had also been seen there in Sep. 1992 (N. Krabbe). Recorded to 3800 m.

The Hudsonian Godwit was considered Near-threatened by Collar et al. (1992, 1994) and is so treated here. Numbers have increased after reaching a critically low point early in the 20th century, and to our knowledge there is no evidence to indicate that the species is not stable or continuing to increase.

Monotypic.

Range: breeds in arctic North America, wintering primarily along coasts of s. South America.

Limosa fedoa   Marbled Godwit   Aguja Canela

A casual boreal winter visitant to the coast of sw. Ecuador, favoring mudflats, ponds, and lagoons. There are only two Ecuadorian records of the Marbled Godwit: a late-19th-century specimen obtained at Santa Rosa in El Oro (Chapman 1926) and a sighting of a single bird made at the Ecuasal lagoons in w. Guayas on 8 Feb. 1980 (D. Finch; Ridgely 1980). The town of Santa Lucía, where the species was also recorded in the late 19th century (see Chapman 1926, p. 194), is now situated within the boundaries of Tumbes, Peru.

One race: nominate *fedoa*.

Range: breeds in w. North America, nonbreeding birds occurring south regularly to Panama, casually on west coast of South America as far south as n. Chile.

Arenaria interpres   Ruddy Turnstone   Vuelvepiedras Rojizo

A fairly common to common transient and boreal winter visitant to the entire coast, favoring rocky areas but also at times feeding on sandy or muddy substrates, rarely even among mangroves or on short grass. Most numerous around the Santa Elena Peninsula in w. Guayas. There is also one report from the lowlands of e. Ecuador, involving two individuals seen along the Río Napo at Nueva Rocafuerte on 11 Sep. 1977 (D. Pearson), one of the few reports of the Ruddy Turnstone from anywhere in Amazonia (Stotz et al. 1992). Recorded throughout the year in Ecuador, with birds occurring during the northern summer months (May–Jul.) being mostly first-year prebreeders, though some apparent adults (then in full alternate plumage) are present at that time as well.

One race: *morinella*.

Range: breeds in high-arctic North America and Eurasia, wintering along coasts from United States south to South America, and in Africa and Australasia.

Aphriza virgata   Surfbird   Rompientero

A rare to locally uncommon transient and boreal winter visitant mainly to rocky areas along the entire coast, often consorting with larger numbers of Ruddy Turnstones (*Arenaria interpres*). Surfbirds are most numerous around the Santa Elena Peninsula in w. Guayas. Recorded from Aug. to Apr., with no reports of birds passing the entire northern summer in Ecuador (though this seems possible). Perhaps most numerous during southward passage (e.g., ca. 60 apparently recently arrived transients were noted on mudflats at the Ecuasal lagoons on 13 Aug. 1991; RSR et al.).

Monotypic.

Range: breeds in mountains of Alaska and adjacent Canada, wintering along Pacific coast south to Chile.

*Calidris canutus   Red Knot   Playero Rojo

A rare transient and boreal winter visitant to mudflats, ponds, and lagoons along the coast of sw. Ecuador. Thus far recorded only from w. Guayas, principally at the Ecuasal lagoons, but presumably occurs more widely and must surely be present locally on the extensive mudflats of the Gulf of Guayaquil. The first Ecuadorian report was made at Monteverde on the Guayas coast, where six birds were seen with dowitchers (*Limnodromus*) on 9 Jan. 1976 (R. Levêque, *Alauda* 45[1]: 125–126, 1977). Subsequent reports, all of them from Ecuasal, include one bird in basic plumage seen on 3 Aug. 1980 (RSR and R. A. Rowlett et al.); one in basic plumage seen on 19 Jan. 1991 (RSR); one mist-netted on 10 Apr. 1991 (D. Liley); and one in basic plumage collected (ANSP) on 11 Aug. 1991, with a second bird molting out of alternate plumage seen on 13 Aug. An unusually large number of Red Knots was present at Ecuasal in Dec. 1995, with 7 being seen on 1 Dec. (P. Coopmans et al.) and then up to 20 on 9 Dec. (PJG et al.). The town of Santa Lucía, where the Red Knot was recorded in the 19th century (see Chapman 1926, p. 199), is now situated within the boundaries of Tumbes, Peru.

In Britain and elsewhere in the Old World, the species is usually known simply as the Knot.

The sole Ecuadorian specimen is a basic-plumaged adult and thus cannot definitely be assigned to subspecies. The American-breeding race *rufa* would seem most likely, but it has been suggested that *rogersi* (which breeds in ne. Siberia and perhaps also in w. Alaska) might—though it winters mainly on coasts of Australia and New Zealand—also occur on the Pacific coast of South America. In alternate plumage, *rogersi* might be distinguishable on the basis of its slightly more deeply colored underparts and reduced white on lower belly.

Range: breeds in high-arctic North America and Siberia, wintering south to coasts of South America, Africa, and Australasia.

Calidris alba   Sanderling   Playero Arenero

A fairly common to common transient and boreal winter visitant to sandy beaches and nearby lagoons and rocky areas along the entire coast. There is also a single report from the lowlands of e. Ecuador, in Oct.–Nov. 1975 when the Sanderling was recorded as "uncommon" on sandbars along the Río Napo near Limoncocha (Tallman and Tallman 1977); specimens (LSUMZ) were taken on 7 Oct. and 12 Nov. and represent the basis for one of the few Sanderling records from anywhere in Amazonia (see Stotz et al. 1992). The species is also listed, without details, as having been seen along the Río Napo at Sacha Lodge. Otherwise there are no reports of the Sanderling from e. Ecuador; indeed there are no other reports from anywhere away from the immediate coast. Sanderling numbers are greatest, with flocks of several hundred or more being routine (especially during migratory periods), in w. Guayas from the Monteverde region south to the Ecuasal lagoons on the south side of the Santa Elena Peninsula. Recorded throughout the year, but numbers are very low during the northern summer months (May–Jul.), when most birds present are probably prebreeders (which remain in basic plumage).

Monotypic.

Range: breeds in high-arctic North America and Eurasia, wintering along coasts south to South America, Africa, and Australia.

Calidris pusilla   Semipalmated Sandpiper   Playero Semipalmeado

A fairly common to locally very common transient and boreal winter visitant to mudflats and the margins of lagoons and ponds along the entire coast, but most common in Guayas and El Oro. There are only two inland records. Five specimens were collected (FMNH) by the Olallas at Río San Antonio in Los Ríos on 22 Sep. 1931, and a basic-plumaged adult was seen along the Río Chira at Lalamor in extreme s. Loja on 16 Sep. 1998 (RSR et al.). Numbers of Semipalmated Sandpipers are apparently highest around the Río Guayas estuary (e.g., at Manglares-Churute Ecological Reserve), where many thousands (perhaps tens of thousands) pass the northern winter. It should be noted that to some extent estimates of numbers of the Semipalmated Sandpiper are confused by the species' extreme similarity, when in basic plumage, to the Western Sandpiper (*C. mauri*); except at very close range, these two species of small sandpipers are best left as "Semi/Westerns." Our overall impression is that the Semipalmated is slightly outnumbered by the Western in Ecuador. Recorded throughout the year, but with only a few presumed prebreeders occurring during the northern summer months (May–Jul.).

Monotypic.

Range: breeds in arctic North America, wintering mainly along coasts of Middle America and n. South America.

Calidris mauri   **Western Sandpiper**   **Playero Occidental**

A fairly common to locally very common transient and boreal winter visitant to mudflats and the margins of lagoons and ponds along the entire coast, most common in Guayas and El Oro. There is also a single report from the highlands, of two birds seen at Laguna de Colta (3300 m) in Chimborazo on 7–8 Aug. 1976 (F. Sibley et al.); this represents one of the few records of the Western Sandpiper away from the immediate coast in all of South America and may be the only one (at least we are not aware of any other). Western Sandpiper numbers are apparently greatest around the Río Guayas estuary (e.g., at Manglares-Churute Ecological Reserve), where many thousands (likely tens of thousands) pass the northern winter. Recorded throughout the year, but with only a few presumed prebreeders occurring during the northern summer months (May–Jul.).

Monotypic.

Range: breeds mainly in arctic Alaska and ne. Siberia, wintering along coasts from s. United States south to n. South America.

Calidris minutilla   **Least Sandpiper**   **Playero Menudo**

A fairly common to common transient and boreal winter visitant to mudflats, margins of pools and lagoons, marshes, and wet grassy fields along the entire coast; found in vegetated areas much more often than are the Semipalmated Sandpiper (*C. pusilla*) and Western Sandpiper (*C. mauri*), and more often found slightly inland. Elsewhere the Least Sandpiper occurs mainly as a rare to uncommon transient in similar situations, including the marshy margins around highland lakes and on sandbars along larger rivers in the eastern lowlands; a few may overwinter locally in such situations. Least Sandpipers appear to be most numerous at the Ecuasal lagoons in w. Guayas, where counts of up to about

500 birds are regular during the northern winter. Recorded mostly from late Jul. to Apr., with numbers extremely low during the northern summer months; usually none at all are to be found at that time, even at the most favorable localities for the species. Recorded up to at least 3800 m (Cotopaxi National Park).

Monotypic.

Range: breeds in n. North America, wintering from s. United States to n. South America.

*Calidris fuscicollis   **White-rumped Sandpiper**   Playero Lomiblanco

A very rare southbound transient around the margins of ponds and lakes and on damp short-grass fields in the Andes; also occurs in the lowlands of e. Ecuador, where it feeds on damp short-grass fields and has also been found on sandbars along larger rivers. First recorded from Ecuador on 16 Aug. 1975 when two birds were seen at Laguna Limpiopungo in Cotopaxi National Park (T. Davis et al.). Subsequent highland records include two birds seen at Laguna de Colta in Chimborazo on 7–8 Aug. 1976 (F. Sibley et al.); two more seen in Cotopaxi National Park on 9–10 Oct. 1983 (N. Krabbe); and 10 seen near Yaguarcocha in Imbabura on 27 Oct. 1988 (PJG). The only record from the eastern lowlands is of birds that were found at Limoncocha in Oct. 1975 (Tallman and Tallman 1977); one specimen (LSUMZ), still unique for Ecuador, was obtained on 27 Oct. 1975. Recorded only from Aug. to Oct.

Monotypic.

Range: breeds in high-arctic North America, wintering mainly along coasts of Argentina and s. Chile.

Calidris bairdii   **Baird's Sandpiper**   Playero de Baird

An uncommon to locally fairly common transient in the highlands around the margins of ponds and lakes, in marshes and on wet short-grass fields, and on paramo (where it sometimes occurs far from open water, though usually where conditions are at least damp). Rare in the lowlands of both e. and w. Ecuador, the records from the west coming primarily from the southwestern coast, those from the east from sandbars along the upper Río Napo downriver to about La Selva. The last sightings represent some of the few reports of the Baird's Sandpiper from anywhere in Amazonia; M. Pearman (*Bull. B. O. C.* 113[2]: 67, 1993) mentions a bird seen under similar circumstances near the Amazon River in Colombia. Baird's Sandpipers are most numerous around certain highland lakes, with numbers perhaps being greatest at Laguna de Colta in Chimborazo; the species' migration route, down the length of the Andes, is unusual for a shorebird. Recorded mainly from late Jul. (once as early as 11 Jul.) to early Nov., and again in Mar., with only a few lingering through the northern winter months. Recorded up to at least 4000 m.

Monotypic.

Range: breeds in arctic North America and ne. Siberia, wintering mainly in w. South America from Peru south to Tierra del Fuego.

Calidris melanotos   Pectoral Sandpiper   Playero Pectoral

An uncommon to locally fairly common southbound transient to wet short-grass fields, marshes, and margins of ponds, lakes, and rivers in the highlands, and in the lowlands of both e. and w. Ecuador (in the east occurring also on sandbars along major rivers). Recorded mainly from Aug. to Nov., with a few arriving as early as late Jul. or lingering into Dec. As yet there are no reports from the northern winter months, and there are only a few involving north-bound transients, these solely from along the coast in the southwest; Marchant (1958) recorded Pectoral Sandpipers on the Santa Elena Peninsula in Apr. and (surprisingly) Jun., and the species was seen north of Santa Rosa in El Oro on 30 Mar. 1989 (Bloch et al. 1991). Recorded up to about 3500 m.

Monotypic.

Range: breeds in arctic North America and Siberia, wintering mainly in s. South America (a few in Australia and New Zealand).

*[Calidris alpina   Dunlin   Playero Ventrinegro]

A casual boreal winter visitant to the coast of w. Guayas, favoring coastal lagoons and mudflats. There are several recent sightings, all of them of basic-plumaged individuals at the Ecuasal lagoons. Two were seen on 14–15 Feb. 1989 (B. Harrington et al.); one on 5 Apr. 1990 (RSR and M. Letzer et al.); and one on 19 Jan. 1991 (RSR). More surprisingly, a loose group of at least 12 birds was found here, in association with Stilt Sandpipers (*Micropalama himantopus*) and various smaller "peeps" on 9 Dec. 1995 (PJG et al.). The only other report of the Dunlin from the Pacific coast of South America comes from Peru (W. R. Petersen, P. K. Donahue, and N. Atkins, *Am. Birds* 35[3]: 342–343, 1981).

With no specimen being available, the subspecies involved in the Ecuador reports remains unknown. *Pacifica*, which winters on the Pacific coast from Canada south at least to Guatemala, is the most likely, but *hudsonia* (wintering mainly on the Atlantic coast from the e. United States to Mexico) is also possible. Both are relatively long-billed races.

Range: breeds in arctic North America and n. Eurasia, wintering mainly in North America, Europe, and s. Asia, very small numbers wandering farther south.

*Calidris ferruginea   Curlew Sandpiper   Playero Zarapito

An accidental visitant to the coast of w. Guayas. An adult only just beginning to molt out of alternate plumage was discovered at the Ecuasal lagoons on 25 Jul. 1991 (B. Haase); it was still present on 14 Aug., when it was photographed (photos to VIREO [D. and M. Zimmerman and R. Behrstock]). What seems likely to have been the same bird, by then in basic plumage, was relocated at Ecuasal on 21 Jan. 1992 (G. H. Rosenberg and R. Behrstock et al.). This represents only the second fully corroborated record of the Curlew Sandpiper from

South America, the first having involved a bird obtained on the Peruvian coast in Jun. 1976 (G. R. Graves and M. Plenge, *Condor* 80[4]: 455, 1978). Monotypic.

Range: breeds in arctic Asia, wintering in Africa, s. Asia, and Australia; small numbers are now found annually in North America, and there are a few records from Middle and South America.

Micropalama himantopus   Stilt Sandpiper   Playero Tarsilargo

A common but very local transient and boreal winter visitant to lagoons and mudflats along the coast of sw. Ecuador. In recent years the Stilt Sandpiper has been found almost exclusively in w. Guayas, where substantial numbers regularly occur at the salt-evaporation ponds south of Monteverde and at the Ecuasal lagoons; 1000 or even more have sometimes been present at the latter site. There also are 19th-century records from Babahoyo and Vinces in Los Ríos. In addition, there have been two sightings from the highlands of what evidently were transient flocks of Stilt Sandpipers: 20 were found at Laguna de Colta in Chimborazo on 7–8 Aug. 1976 (F. Sibley et al.), and 4 were seen at Laguna Limpiopungo in Cotopaxi National Park on 7 Sep. 1993 (P. Coopmans et al.). Finally, there is even a single report from the eastern lowlands, of a bird seen at Kapawi Lodge on 24–26 Aug. 1999 (M. Foster, P. Coopmans, and P. Snetsinger); this represents one of the few records from anywhere in Amazonia. In recent years recorded only from Aug. to Apr., but Marchant (1958) found it on the Santa Elena Peninsula during the intervening months in the 1950s.

The monotypic genus *Micropalama* is sometimes merged into *Calidris* (e.g., 1998 AOU Check-list), but we follow Sibley and Monroe (1990) in maintaining it. Monotypic.

Range: breeds in arctic North America, wintering locally (mainly coastally) from s. United States to s. South America.

Tryngites subruficollis   Buff-breasted Sandpiper   Praderito Canelo

A rare to uncommon transient to open fields with short grass, paramo, and (in the Oriente) sandbars along major rivers; occurs mainly in the highlands and eastern lowlands. So far known only from n. Ecuador, with no records from south of Chimborazo. The only record from west of the Andes involves a late-19th-century record (Chapman 1926) from Hacienda Paramba in Imbabura. Recorded mainly during southward passage from Jul. to Oct., with only three reports during northward passage, one from Mar. and two from Apr., one of the latter being an observation of a surprising 50 individuals—by far the largest number ever seen together in Ecuador—at Laguna de Colta in Chimborazo on 23 Apr. 1960 (R. Levêque, *Ibis* 106[1]: 58, 1964).

As in the Hudsonian Godwit (*Limosa haemastica*), numbers of the Buff-breasted Sandpiper were brought low by market gunning in the 19th and early 20th centuries. The species has since recovered to some extent, but its total population (which perhaps never was very large) remains quite small, any increase perhaps being impeded by extensive drainage and disturbance on the wet pampas of Argentina and Uruguay where it spends the boreal winter. We suspect that there are not any more Buff-breasted Sandpipers than there are Hudsonian Godwits, and as the godwit was accorded Near-threatened status by Collar et al. (1994), which we follow, we think the Buff-breasted Sandpiper should be too.

Monotypic.

Range: breeds in high-arctic North America, wintering mainly in e. Argentina and Uruguay.

The Ruff (*Philomachus pugnax*), a distinctive shorebird ranging mainly in the Old World, is now recorded annually in small numbers in North America, and there are a few reports of it from Middle and South America as well. It seems likely to occur in Ecuador, and indeed one was reported (to P. Coopmans) as having been seen (and photographed?) somewhere in the highlands in 1990, but no details of this record are presently available.

Limnodromus griseus   Short-billed Dowitcher   Agujeta Piquicorta

A common but very local transient and boreal winter visitant to lagoons and mudflats along the coast, mainly occurring in sw. Ecuador and in largest numbers in w. Guayas (particularly at the Ecuasal lagoons). Curiously, however, the Short-billed Dowitcher was not recorded from this region by Marchant (1958), despite his having found virtually all of the other shorebirds regularly occurring there. There is also a single late-19th-century record (Chapman 1926) from the highlands in Cañar, one of the few records of the Short-billed Dow-itcher from anywhere in South America away from the immediate coast. Dow-itchers are found in Ecuador throughout the year, though only small numbers occur during the northern summer months (May–Jul.), when most or all birds present remain in basic plumage and are presumed to be first-year prebreeders. Early arriving adults still in alternate plumage regularly appear as early as the first few days of Aug., a few even in late Jul.

Racial diagnoses in *L. griseus* are notoriously difficult to make, and none have been possible given the limited specimen material at hand, which is mainly in basic plumage. Some recently arrived alternate-plumaged birds have been seen that appeared to show characters of *hendersoni* (with extensively rufous underparts, encompassing the belly, and boldly marked upperparts). Others have appeared to be *caurinus*, and it is not impossible that nominate *griseus* might also occur (though it is mainly an Atlantic-coast form). Unfortunately, none could be collected.

Range: breeds in n. North America, wintering along coasts from s. United States to n. South America.

There are no confirmed records of the Long-billed Dowitcher (*L. scolopaceus*) from Ecuador, and indeed very few from anywhere in South America. Early records attributed to this species were doubtless the result of confusion with the very similar Short-billed Dowitcher; the two were not confirmed as separate species until relatively recently.

[*Gallinago gallinago*   **Common Snipe**   Becasina Común]

Apparently a casual boreal winter visitant to marshes, wet fields, and pond margins. There are only a few records of the Common Snipe from Ecuador, and no specimen from the country is known to be extant. A female was reported collected near Mapoto (1200 m) in Tungurahua on 19 Oct. 1939 (Orcés 1944). This specimen—which was identified by Orcés (op. cit.) as of the migratory race *delicata*—was apparently placed in the Mejía collection but was reportedly in poor condition, and we fear it may have been discarded; in any case, its present whereabouts are unknown. On the basis of date and range, we consider it likely that the specimen was correctly identified, but at present there is of course no way to verify this. The only additional Ecuadorian report of the Common Snipe concerns one or two individuals seen on a wet field at Mindo from 21 Dec. 1997 through at least 17 Jan. 1998 (v.o.; first seen by P. Coopmans). On the basis of range and dates, these birds are presumed to have almost certainly been Common Snipes, there being no indication that the South American Snipe (*G. paraguaiae*) occurs west of the Andes. Other sightings of snipes in the north-eastern lowlands cannot be assigned to species; we suspect that most or all of them actually refer to the South American Snipe. The Common Snipe occurs regularly as far south as w. Colombia. Recorded at about 1200–1300 m.

One race: *delicata*. It has been suggested (e.g., Sibley and Monroe 1990) that *G. delicata* might better be considered as a monotypic species (Wilson's Snipe) distinct from Old World birds, but for the present they continue to be considered conspecific (e.g., in the 1998 AOU Check-list). They differ in certain morphological and plumage details, and their flight displays also differ.

Range: breeds in North America and Eurasia, wintering as far south as Middle America, n. Africa, and s. Asia, with small numbers reaching n. South America.

Gallinago paraguaiae*   **South American Snipe   Becasina Sudamericana

Uncertain. Known from a few records from the lowlands of ne. Ecuador. A female was collected (ANSP) out of two seen on soaked fields with short grass at Zancudococha in e. Napo on 6 Apr. 1991; one had been seen there first on 4 Apr. and one remained until at least 15 Apr. At the time we believed this to be the first confirmed Ecuadorian record of the South American Snipe. At RSR's request, however, J. V. Remsen and S. W. Cardiff (pers. comm.) reexamined the snipe obtained by D. and E. Tallman at Limoncocha on 20 Jan. 1976 (LSUMZ) that had originally been identified as a Common Snipe (*G. g. delicata*), and they concluded that the specimen is actually referable to the South American Snipe. The only additional report refers to several snipes seen at Cuyabeno in Jan. 1988 (PJG and P. Scharf), but whether these were South American (as seems most likely) or Common was impossible to determine. There have also been a few subsequent sightings from this area, but no evidence of breeding. Recorded below 300 m.

The species is sometimes called the Magellan, or Paraguayan, Snipe.

G. *paraguaiae* is now usually (e.g., Sibley and Monroe 1990) regarded as a species separate from G. *gallinago*, though for many decades they were regarded as conspecific (e.g., Meyer de Schauensee 1966, 1970). Although very similar morphologically, their flight displays differ. One race occurs in Ecuador, nominate *paraguaiae*.

Range: e. Colombia, much of Venezuela, and the Guianas to s. Chile and s. Argentina (but absent or virtually so from much of Amazonia).

[Gallinago andina   Puna Snipe   Becasina de Puna]

Uncertain. Known only from a single sighting from extreme s. Ecuador, a bird flushed and well seen in wet paramo on the east side of the Cordillera Las Lagunillas (3300 m) in extreme sw. Zamora-Chinchipe on 27 Oct. 1992 (M. B. Robbins). The Puna Snipe is known from adjacent n. Peru (Parker et al. 1985).

Andina has been considered (e.g., *Birds of the Americas*, pt. 1, no. 4) as a small, Andean subspecies of the widespread G. *paraguaiae* (South American Snipe), but given its morphological and habitat differences this seems unlikely. Presumed race: nominate *andina*.

Range: Andes from extreme s. Ecuador and n. Peru to n. Chile and nw. Argentina.

Gallinago nobilis   Noble Snipe   Becasina Noble

Uncommon and somewhat local in marshes, bogs, and adjacent wet paramo in the Andes from Carchi (Páramo del Ángel [N. Krabbe; specimen MECN]) south to Azuay (few records), w. Morona-Santiago (Sangay National Park and along the Gualaceo-Limón road), n. Loja (Cordillera de Cordoncillo, fide P. Coopmans), and n. Zamora-Chinchipe (a few sightings in the Cajanuma sector of Podocarpus National Park; Bloch et al. 1991). A substantial population of this snipe occurs on the "back" side of Laguna Limpiopungo in Cotopaxi National Park. Recorded from 2900 to 4100 m.

The Noble Snipe seems to be declining in many accessible areas, perhaps because of overhunting, but as yet it does not appear to merit formal threatened status.

The species has been called the Paramo Snipe.

Monotypic.

Range: Andes from extreme sw. Venezuela to s. Ecuador.

Gallinago jamesoni   Andean Snipe   Becasina Andina

Uncommon to locally fairly common in paramo, pastures, and adjacent shrubby areas and the edges of *Polylepis* woodland in the Andes. Andean Snipe concentrate in damp areas when actually feeding, but they often are found well away from water when at rest; the species appears to be much less tied to water than is the Noble Snipe (G. *nobilis*). It is most conspicuous when roding high in the sky before dawn and just after dusk, often also when fully dark. Recorded from 3100 to 4400 m.

This species and the next are sometimes placed in the genus *Chubbia*. *G. jamesoni* is regarded as a species distinct from *G. stricklandii* (Fuegian Snipe) of far southern South America, following most recent authors. When these are considered conspecific, the species is usually called the Cordilleran Snipe. Monotypic.

Range: Andes from w. Venezuela to w. Bolivia; Santa Marta Mountains.

*Gallinago imperialis   Imperial Snipe   Becasina Imperial

Rare and local in montane forest and forest borders in the upper temperate zone just below treeline, mainly on the east slope of the Andes. N. Krabbe initially discovered this rare and local snipe in Ecuador, finding it at Loma Yanayacu on the northwestern slopes of Volcán Pichincha on 17 Dec. 1990 (Krabbe 1992). One was eventually collected there on 8 Mar. 1992 (MECN), and this remains the only Ecuadorian specimen. Since then the Imperial Snipe has also been recorded in the Cajanuma sector of Podocarpus National Park (Rasmussen et al. 1996), and in Oct.–Nov. 1992 it was also found on the east slope of the Cordillera Las Lagunillas in extreme sw. Zamora-Chinchipe (RSR and M. B. Robbins et al.; N. Krabbe). In May 1992 the Imperial Snipe was also recorded in w. Napo on the Cordillera de los Llanganates, and near the Morona-Santiago/Azuay border on the Páramos de Matanga and south of Saraguro near Acanamá in Loja (N. Krabbe); on 27 Mar. 1996 it was heard along the Gualaceo-Limón road (P. Coopmans et al.). Although only identified years later, a displaying bird was tape-recorded along the upper Loja-Zamora road on 30 Aug. 1988 (P. Coopmans). It thus appears that the Imperial Snipe may be locally distributed all along the east slope of the Andes where appropriate habitat exists; it was doubtless long overlooked because of its extremely retiring habits (almost never being noted except when roding high in the sky around dawn and dusk). Recorded from 2700 to 3800 m.

The Imperial Snipe was considered to be a Near-threatened species by Collar et al. (1992, 1994). Although agreeing that the species' population size is doubtless small, we suspect that it probably always has been; nonetheless we concur with their assessment.

The species is sometimes called the Banded Snipe. Monotypic.

Range: locally in Andes from Colombia to s. Peru.

Phalaropus fulicaria   Red Phalarope   Falaropo Rojo

A rare transient and boreal winter visitant to offshore waters, only occasionally coming within sight of land but also rarely found on salt lagoons such as Ecuasal in w. Guayas (though there outnumbered by Red-necked Phalaropes [*P. lobatus*] and, by a huge margin, Wilson's Phalaropes [*P. tricolor*]). E. L. Mills (*Ibis* 109: 536, 1967) recorded the Red Phalarope as "numerous" on the Mancora Banks in the Gulf of Guayaquil on 2 Oct. 1965, and in recent years B. Haase and ANSP personnel have found it in small numbers off the Santa

Elena Peninsula in w. Guayas at various times of the year, collecting a male in basic plumage at Ecuasal on 21 Jun. 1992 (ANSP).

In Britain and elsewhere in the Old World, the species is still often called the Gray Phalarope.

This and the next two species were formerly placed in their own family, the Phalaropodidae. K. C. Parkes (*Bull. B. O. C.* 102: 84–85, 1982) established that the correct spelling of the species name should be *fulicaria* and not *fulicarius*. Monotypic.

Range: breeds in high-arctic North America and Asia, wintering south in Pacific Ocean to off Chile and in South Atlantic Ocean.

Phalaropus lobatus   **Red-necked Phalarope**   **Falaropo Picofino**

An uncommon transient and boreal winter resident to offshore waters, occasionally coming within sight of land or occurring in small numbers on salt lagoons such as Ecuasal in w. Guayas. Recorded mainly from Aug. to Mar., but some prebreeders may occur throughout the northern summer months.

In American literature this species was formerly known as the Northern Phalarope.

Formerly placed in the monotypic genus *Lobipes*. Monotypic.

Range: breeds in arctic North America, wintering south in Pacific Ocean to off Chile, also some in s. Atlantic Ocean; also in Old World.

Phalaropus tricolor   **Wilson's Phalarope**   **Falaropo Tricolor**

An uncommon to locally very common transient and boreal winter visitant to ponds and lagoons along the coast (especially in w. Guayas); also occurs in small numbers on lakes in the highlands, and one was seen along the Río Chira at Lalamor in extreme s. Loja on 16 Sep. 1998 (RSR et al.). The Wilson's is much less pelagic than the other two phalaropes, rarely or never being seen on the ocean; it also is less aquatic, often scampering about actively on muddy margins. The numbers of Wilson's Phalaropes recorded in Ecuador seem to vary greatly from year to year. At times staggering numbers occur at the Ecuasal lagoons in w. Guayas: for example, on 10–11 Sep. 1992, B. Haase and RSR independently estimated that about 20,000 birds were present. In other years the number present, though still substantial, is smaller, with only a few hundred to several thousand being found. Recorded throughout the year, with females in full breeding plumage sometimes being seen in Jun. and Jul.; southbound birds appear in numbers in Aug.

This species was long separated in the monotypic genus *Steganopus*. Osteological data support placing the three phalaropes together in *Phalaropus*, but genetic data (e.g., D. L. Dittman, R. M. Zink, and J. A. Gerwin, *Auk* 106[2]: 326–331, 1989) suggest that *tricolor* is distinct from the other two phalarope species. Sibley and Monroe (1990) agree with the latter, maintaining *Steganopus*; we opt, however, to follow the 1998 AOU Check-list in placing *tricolor* in *Phalaropus*. Monotypic.

Range: breeds in w. North America, wintering mainly in w. and s. South America.

Thinocoridae	Seedsnipes

Two species in two genera. Seedsnipes are exclusively South American in distribution.

Attagis gayi	**Rufous-bellied Seedsnipe**	**Agachona Ventrirrufa**

Rare to uncommon, but inconspicuous and easily overlooked (though often tame once located), on very high and barren slopes with sparse low vegetation in the high Andes from Pichincha (Volcán Cayambe, Volcán Pichincha, and Corazón) and extreme w. Napo (Volcán Antisana; Papallacta pass) south locally to Chimborazo (Volcán Chimborazo) and nw. Morona-Santiago (Volcán Sangay, fide N. Krabbe); there also are recent reports from El Cajas in Azuay (J. Hendriks, fide P. Coopmans). Small numbers can be seen along the roads leading up to the mountaineering refugios on Volcán Cotopaxi and Volcán Chimborazo, and near the antenna towers above the Papallacta pass. Recorded mostly from 4000 to 4600 m.

One race: *latreillii*. This endemic and distinctive form may even have reached full species status (Ecuadorian Seedsnipe).

Range: Andes of Ecuador, and from cen. Peru to s. Chile and s. Argentina.

Thinocorus rumicivorus	**Least Seedsnipe**	**Agachona Chica**

Uncertain; apparently Extirpated in Ecuador. The Least Seedsnipe is known only from a series of 10 specimens obtained on the Santa Elena Peninsula in w. Guayas in Jan.–Feb. 1898 (Salvadori and Festa 1900), from a "possible" sighting of about eight birds near Salinas in Apr. 1956 (Marchant 1958), and from a sighting of two near Salinas on 11 Jul. 1974 (T. Davis et al.). Despite much recent field work in this area, there have been no subsequent reports. As early as the early 20th century the species appears to have been rare on the Santa Elena Peninsula, for Chapman (1926, p. 190) commented that "special search for this bird at Santa Elena by two of our expeditions has been unavailing." Although Marchant (1958, echoing the comments in Chapman 1926 and in *Birds of the Americas*, pt. 1, no 3) suggested that the Least Seedsnipe was perhaps only an austral migrant to Ecuador, the fact that the specimens (two of them specifically noted as "immatures") were obtained during the middle of the southern summer suggests that in fact it was a breeding resident. In Peru there is a resident population that inhabits the coastal plain north locally to at least Lambayeque.

Why the Least Seedsnipe should have disappeared from Ecuador remains uncertain; its favored barren habitat—sandy or gravel plains with sparse vegetation—though now disturbed in many areas, remains largely intact. This disappearance is particularly odd given that the seemingly more "sensitive" Peruvian Thick-knee (*Burhinus superciliaris*) continues to persist in reasonable numbers in several regions. The Least Seedsnipe is not considered to be at risk in the rest of its wide range (Collar et al. 1992, 1994).

One race: *cuneicauda*. The Santa Elena birds were described as the race *pallidus*, but we accept the diagnosis in *Birds of the Americas* (pt. 1, no 3., p. 235) that *pallidus* is "inseparable" from *cuneicauda*.

Range: w. Peru to Chile and Argentina; formerly north to sw. Ecuador.

Burhinidae   Thick-knees

One species. Thick-knees occur worldwide except in North America.

Burhinus superciliaris   Peruvian Thick-knee   Alcaraván Peruano

Rare and local—and inconspicuous—in open grassy scrub and on dry open fields in the lowlands of sw. Ecuador, where known only from w. Guayas (Santa Elena Peninsula area east, at least formerly, to the Playas area), s. coastal El Oro (near Huaquillas; sightings by RSR, PJG, and R. A. Rowlett in Aug. 1980), and extreme sw. Loja (near Zapotillo; a sighting of a single bird by F. Sornoza and M. Jácome on 8 Apr. 1992). Also seems likely to occur on Isla Puná. A small population of Peruvian Thick-knees persists on the Santa Elena Peninsula, especially from the Ecuasal lagoons east toward Anconcito; a specimen (ANSP) was obtained here, near the Represa José Velasco Ibarra, on 21 Jun. 1992. Recorded up to about 150 m in s. Loja.

Populations of this fine large shorebird have declined in Ecuador—where it seems never to have been numerous—because of the recent local upsurge in human numbers in its range. We therefore suggest that the Peruvian Thick-knee deserves formal Vulnerable status in Ecuador. The species is not, however, considered to be at risk elsewhere in its range (Collar et al. 1992, 1994), though in our view this assessment may be overly optimistic.

Monotypic.

Range: sw. Ecuador to extreme n. Chile.

Haematopodidae   Oystercatchers

One species. Oystercatchers are worldwide in distribution, occurring principally along coasts.

Haematopus palliatus   American Oystercatcher   Ostrero Americano

Rare to uncommon and local along the coast of sw. Ecuador, breeding locally and favoring more or less deserted beaches and lagoons, in Ecuador not seeming to forage with any regularity in rocky areas. The American Oystercatcher seems to have been recorded only from Guayas (north to around Monteverde) and El Oro; there still are no records from farther north along the Ecuadorian coast, though we are at a loss to explain the species' apparent absence there, especially from the seemingly ideal rocky coastlines such as are found in sw. Manabí.

One race: *pitanay*.

Range: coasts from e. United States and Mexico south locally to s. South America; West Indies, Galápagos Islands.

Recurvirostridae   Stilts and Avocets

Two species in two genera. Stilts and avocets are worldwide in distribution.

Himantopus mexicanus   **Black-necked Stilt**   Cigüeñuela Cuellinegra

Locally fairly common to common in marshes, rice fields, and around the margins of shallow lagoons and ponds in the lowlands of w. Ecuador from w. Esmeraldas (Atacames/Muisne area) and Los Ríos south through coastal El Oro; about six were also seen along the Río Chira at Lalamor in extreme s. Loja on 16 Sep. 1998 (RSR et al.). The number of Black-necked Stilts found at most sites varies substantially, apparently in response to fluctuations in water levels. At least in most years, large numbers breed at the Ecuasal lagoons and around the nearby Represa José Velasco Ibarra. Recorded only below about 100 m.

H. *mexicanus* is regarded as a species separate from *H. himantopus* (Black-winged Stilt) of the Old World and (at least tentatively) from *H. melanurus* (White-backed Stilt) of s. South America, following most recent authorities (e.g., Sibley and Monroe 1990). One race occurs in Ecuador, nominate *mexicanus*.

Range: s. United States south to Peru and ne. Brazil.

*Recurvirostra americana   **American Avocet**   Avoceta Americana

An accidental vagrant, known from two records from coastal salt lagoons. Four birds in basic plumage, accompanying Black-necked Stilts (*Himantopus mexicanus*), were seen near Atacames in w. Esmeraldas on 10 Jan. 1991 (R. ter Ellen and L. Steijn, *Dutch Birding* 16[2]: 59–60, 1994), and one in alternate plumage was found at the Ecuasal lagoons in w. Guayas on 26–27 Jul. 1992 (T. Kemp and E. Durbin); what was presumably the latter bird was also seen by G. Speight and M. Coverdale on 26 Aug. 1992 (R. ter Ellen and L. Steijn, op. cit.) and again on 11 Sep. 1992 (ANSP). One suspects that the Ecuasal bird was one of the four that turned up at Atacames. These represent the first records of the American Avocet from the South American mainland; there are a few records from islands (Bonaire, Tobago) off the continent's north coast.

Monotypic.

Range: breeds in w. North America, wintering south mainly in s. United States and n. Middle America, only a few straggling farther south.

Charadriidae   Plovers and Lapwings

Thirteen species in five genera. Plovers and lapwings are worldwide in distribution.

*Vanellus chilensis   **Southern Lapwing**   Avefría Sureña

Recently found to be uncommon and local on large grassy fields and on sandbars along larger rivers in the lowlands of e. Ecuador. The Southern Lapwing was first recorded (Norton et al. 1972) on the somewhat unsatisfactory basis of two unlabeled specimens, presumed to have been obtained in Ecuador, in collections in Quito (Collegio Mejía) and Ambato (Colegio Bolívar). A male (now at MECN) was subsequently obtained at the headwaters of the Río Bobonaza in Pastaza on 26 Jan. 1969 (Orcés 1974). Since then Southern Lapwing numbers have increased considerably, and its distribution has expanded, in particular since the 1980s. It seems inconceivable that this very conspicuous shorebird had previously been overlooked, which makes it clear that a genuine increase in numbers has taken place, though the cause (deforestation?) remains obscure. Breeding appears likely but apparently has not yet been confirmed in Ecuador, though what appear to be territorial pairs have been observed at several sites. Small numbers of Southern Lapwings can now be found regularly through much of ne. Ecuador south to the Puyo area and east to Zancudococha (where an additional specimen was taken on 6 Apr. 1991; ANSP) and to near the mouth of the Río Aguarico into the Río Napo. Groups of as many as 12 or more individuals have been seen on sandbars along the Río Napo (e.g., above La Selva on 21 Feb. 1991; D. Wolf and PJG et al.). It might be noted that a pair of Southern Lapwings was seen just over the Peruvian border near Pantoja on 21 Sep. 1992 (RSR); the species appears not to have been previously recorded from Peru (Parker et al. 1982). With continuing forest destruction, the Southern Lapwing seems destined to continue to increase in Ecuador. Recorded mainly below about 400 m, apparently occasionally wandering higher.

One race: *cayennensis.*

Range: Panama to Chile and Argentina, but absent from much of n. and w. Amazonia and from Pacific slope south of Colombia except in Chile. More than one species is perhaps involved.

Vanellus resplendens   **Andean Lapwing**   Avefría Andina

Fairly common and conspicuous, but somewhat local, in paramo and open areas and pastures with short grass, almost always near water, in the Andes from

Carchi (a few pairs resident at Hacienda Intihuasi on Páramo del Ángel; N. Krabbe), Imbabura (an old record from Ibarra, and seen at least once on Cerro Imbabura near the Quito-Otavalo highway [N. Krabbe]), Pichincha (north to Volcán Cayambe; ANSP), and w. Napo (Antisana) south locally to Azuay and El Oro (Taraguacocha); seemingly absent from Loja. There is also a single extraordinary record of a pair that frequented the airstrip at Limoncocha in the eastern lowlands for four months between Feb. and Jun. 1963, when both birds were collected (MCZ; Norton 1965); this latter represents the only record of the Andean Lapwing from Amazonia. In Peru the species is known to wander down to the coast during the austral winter (D. L. Pearson and M. A. Plenge, *Auk* 91[3]: 628, 1974), and this could conceivably occur in Ecuador as well. Recorded mostly from 3500 to 4400 m, in small numbers occasionally down to 2900 m, once at 300 m (Limoncocha).

Monotypic.

Range: Andes from s. Colombia to n. Chile and nw. Argentina.

Pluvialis squatarola   **Gray Plover**   **Chorlo Gris**

An uncommon to fairly common transient and boreal winter visitant along the entire coast, favoring sandy beaches, coastal mudflats, and shores of ponds and lagoons; most numerous in w. Guayas. Elsewhere known only from a single record of a bird collected (LSUMZ) on the Río Napo near Limoncocha on 20 Dec. 1975 (Tallman and Tallman 1977); this represents one of the few records of the Gray Plover from anywhere in the Amazon basin (Stotz et al. 1992). Recorded throughout the year, though from late Apr. to Jul. only very small numbers are found, all or most of these doubtless being prebreeding first-year birds; numbers gradually increase in Aug., with arriving adults still in full alternate plumage regularly being seen in Aug. and Sep.

The species is called the Black-bellied Plover in American literature, but we concur with the arguments presented by Beaman (1995) in favor of using the modifier Gray, long employed in Britain and the Old World. This and all the golden-plovers have similar black bellies when in alternate plumage.

Monotypic.

Range: breeds in arctic North America and Asia, wintering south to coasts of South America, Africa, and Australia.

Pluvialis dominica   **American Golden-Plover**   **Chorlo Dorado Americano**

A rare transient on open grassy or sandy areas, usually but not always near water; the American Golden-Plover can occur virtually anywhere in Ecuador but is never numerous. There exists a notable geographical scatter of records, with some from the coast in the southwest (especially on the Santa Elena Peninsula in w. Guayas), the highlands, and the eastern lowlands. Recorded mostly from Aug. to Nov. and again in Mar. and Apr. The only known record from the northern winter months comes from Laguna Limpiopungo in Cotopaxi National Park when a single basic-plumaged bird was photographed (photos to

VIREO) on 31 Jan. 1991 (R. Behrstock). The only possible Ecuadorian specimen was obtained at Andoas in the southeastern lowlands on 15 Oct. 1936
(Berlioz 1937c), but as Andoas is very close to the present Ecuador-Peru boundary, this bird may not actually have been taken in Ecuador.

The species is sometimes called the Lesser Golden-Plover, especially when the
Pacific Golden-Plover (*P. fulva*) is not separated as a full species.

The species name is sometimes incorrectly spelled *dominicus*. Monotypic.

Range: breeds in arctic North America, wintering mainly in s. South America.

[Pluvialis fulva   Pacific Golden-Plover   Chorlo Dorado del Pacífico]

An accidental vagrant to the coast of w. Guayas. Known from only a single
sighting of a well-documented bird in nonbreeding plumage seen at the Ecuasal
lagoons on 26 Apr. 1991 (D. Liley); it could not be relocated the next day. This
is one of the few records of the species from South America; Hayman et al.
(1986) mention that vagrants have occurred in Chile and on the Galápagos
Islands.

P. fulva is treated as a species separate from *P. dominica* (American
Golden-Plover), following most recent authors, including the AOU 1998.
Monotypic.

Range: breeds in arctic Siberia and Alaska, wintering on Pacific Ocean islands
and in Australasia, with a few records of transients from coast of North America
and islands off Mexico.

Hoploxypterus cayanus   Pied Plover   Chorlo Pinto

Uncommon to locally fairly common on sandbars along major rivers in the lowlands of e. Ecuador, where principally known from along the Río Napo upriver
regularly to around Pompeya and Limoncocha, in small numbers as far as Jatun
Sacha (B. Bochan) and once up to the Río Misahualí near Tena (where one was
seen on 4 Mar. 1992; PJG and D. Wolf) and the lower Río Aguarico upriver to
around the Río Pacuyacu. The species also ranges along the Río Pastaza upriver
at least to Sarayacu; more unusually, a single individual was seen along the Río
Upano near Macas in Morona-Santiago on 12 Dec. 1996 (P. Coopmans). In
addition, the Pied Plover is known from a few 19th-century records from sw.
Ecuador in s. Los Ríos (Babahoyo) and Guayas (Balzar); there are also two
recent reports from the west, where the total population must be very small.
Single birds were seen at Río Palenque on 23 Jul. 1992 (P. Coopmans and
M. Van Beirs) and along the Río Calabí north of Zapotal in n. Los Ríos on 25
Aug. 1994 (P. Coopmans et al.). Numbers of this dapper plover fluctuate
markedly even in ideal habitat, perhaps in response to changes in water levels.
Recorded mostly below about 300 m, occasionally wandering as high as 500 m
near Tena and even to 850 m (near Macas).

The numbers of the Pied Plover occurring in Ecuador seem to have undergone a general decline in recent decades, a decline that presumably is due to
disturbance, especially when nesting. The species is now especially rare in the

west. We thus consider it to warrant Near-threatened status. The species has a wide range across much of tropical South America and is not globally at risk.

The species is sometimes placed in the genus *Vanellus* and then is often called the Pied Lapwing. However, J. G. Strauch, Jr. (*Trans. Zool. Soc. Lond.* 34: 263–345, 1978), and J. Fjeldså and B. Nielson (*Rec. S. Aust. Mus.* 23[1]: 69–72, 1989), suggest that the species is best maintained in its monotypic genus, and that it is more closely related to *Charadrius* than to *Vanellus* (despite its having a carpal spur). Monotypic.

Range: e. Colombia, Venezuela, and the Guianas to n. and e. Bolivia, Paraguay, and se. Brazil; w. Ecuador.

Charadrius semipalmatus   Semipalmated Plover   Chorlo Semipalmeado

A fairly common to common transient and boreal winter visitant to the entire coast, favoring mudflats and margins of ponds and lagoons; most numerous in sw. Ecuador, particularly in w. Guayas and the Río Guayas estuary though a few birds are also recorded inland in the Río Guayas/Daule drainage (e.g., 4 specimens [FMNH] taken at Isla Silva and Río San Antonio in Sep. 1931). The Semipalmated Plover is also known from a few records in the highlands, where it has most often been noted at Laguna Limpiopungo in Cotopaxi National Park, though the number of reports from here may reflect mainly that area's being more frequently checked for migrating shorebirds than are other highland sites. There are even a few records from the eastern lowlands, the first involving a bird collected [LSUMZ] on the Río Napo near Limoncocha on 6 Nov. 1975 (Tallman and Tallman 1977), the second a remarkable group of at least 8 individuals seen with a flock of 30 or more Collared Plovers (*C. collaris*) near Coca on 5 Mar. 1992 (D. Wolf and PJG et al.), and the last a single bird seen along the Río Upano near Macas in Morona-Santiago on 12 Dec. 1996 (P. Coopmans). The species has also been reported seen at Sacha Lodge (no details). These eastern-lowland records represent some of the very few for the Semipalmated Plover from anywhere in the Amazon basin (see Stotz et al. 1992). Recorded throughout the year in Ecuador, though numbers during the northern summer months (May–Jul.) are much smaller, probably mainly involving prebreeding birds.

Monotypic.

Range: breeds in n. North America, wintering on coasts (mainly) from s. United States to South America.

Charadrius wilsonia   Wilson's Plover   Chorlo de Wilson

Uncommon to locally fairly common along the coast of Ecuador from Esmeraldas (beach at Camarones; MECN) and Manabí (in the Bahía de Caráquez area) southward, breeding locally. There seem to be no records from the coast of n. Manabí, but the species seems likely to occur there as well. Wilson's Plovers favor the muddy margins of coastal lagoons and estuaries, and they appear to readily adapt to the artificial shrimp ponds that have become all too prevalent in many areas; however, they seem generally to avoid open sandy beaches. That Wilson's Plovers are breeding residents on the Pacific coast of South America seems often to have been overlooked, despite the fact that as early as the 1950s Marchant (1958) had found them nesting on the Santa Elena Peninsula. The sight reports in 1960 of small flocks of Wilson's Plovers resting on airstrips at Tena and Puyo in the eastern lowlands, as reported by Levêque (1964), have never been repeated and seem improbable for so strictly coastal a bird; we believe it virtually certain that it was the Collared Plover (*C. collaris*), which regularly does rest on airstrips, that was actually involved.

The species is sometimes called the Thick-billed Plover.

One race: *beldingi*. In addition to the resident population, it is possible that migrants from farther north may also occur in Ecuador; some authors (e.g., Blake 1977) have presumed that this was the case. However, there actually is no indication of any increase in Wilson's Plover numbers in Ecuador during the appropriate months for boreal migrants.

Range: coasts from se. United States and Mexico to n. South America.

Charadrius melodus   Piping Plover   Chorlo Silbador

An accidental vagrant to the coast of w. Guayas. Only one record, a female collected (BMNH) "on a sandy shore at the water's edge" at Salinas on 15 Oct. 1955 (S. Marchant, *Ibis* 98: 533–534, 1956). The identity of this remarkable vagrant has recently been reconfirmed by M. Walters and P. Hayman (pers. comm.). This represents the only record of the Piping Plover from mainland South America and is by far the southernmost one known; the only other South American record involved a sighting from Bonaire (K. Voous, *Birds of The Netherlands Antilles*, Curaçao, De Walburg Press, 1983).

Collar et al. (1992, 1994) accorded the Piping Plover Vulnerable status because of beach development and disturbance on its nesting grounds, and we concur with that assessment.

Monotypic (*circumcinctus* now usually being synonymized).

Range: breeds locally in n. North America, wintering south to Mexico and Greater Antilles.

Charadrius alexandrinus   Snowy Plover   Chorlo Níveo

Fairly common to common but very local along the coast of w. Guayas from the Monteverde area south to the Ecuasal lagoons, where particularly numerous; the southernmost known locality is Playas, where the species was seen in the early 1960s (Levêque 1964). The Snowy Plover seems rather strictly confined to the margins and immediate vicinity of salt-evaporation ponds (less often shrimp ponds), and it rarely or never is seen away from them. So far the Snowy Plover has not been found elsewhere along the Ecuadorian coast, though its occurrence seems possible, especially in El Oro. The Snowy Plover is perhaps increasing in its limited Ecuadorian range. Marchant (1958) stated that he was unable to obtain breeding evidence in the 1950s and thought it possible that the birds he obtained were only migrants. In recent years, however, pairs with young have been repeatedly noted at several sites on the Santa Elena Peninsula, with the largest population (consisting of dozens of pairs at least) at the Ecuasal lagoons.

English names present a problem in this widespread species. The name Kentish Plover is generally used for Old World populations, but it seems singularly inappropriate in a global context, and we follow the 1998 AOU Check-list in employing the name Snowy Plover, at least for the American populations.

American birds are perhaps better treated as a distinct species (*C. nivosus*) separate from those of the Old World; their calls appear to differ. One race occurs in Ecuador, *occidentalis*. It closely resembles *nivosus* (of s. North America, Mexico, and the West Indies), but is slightly larger, white-lored, and apparently exhibits less sexual dimorphism.

Range: locally along coasts and inland in s. United States and Mexico, West Indies and islands off n. South America, sw. Ecuador and w. Peru; also in Old World. North American breeders are known to winter south to Panama (G. Castro and J. P. Myers, *Am. Birds* 42[3]: 374, 1988).

Charadrius collaris   Collared Plover   Chorlo Collarejo

An uncommon to locally fairly common resident—though local movements are likely—along larger rivers in the lowlands of both e. and w. Ecuador. More local in the west, where it ranges south through coastal El Oro; also found along the Río Chira near Lalamor in extreme s. Loja (12 or more seen on 15–16 Sep. 1998; RSR et al.). The Collared Plover favors open sandy riverine islands in the east, whereas in the west it is also found along rivers but is more numerous around lagoons and estuaries along the coast, and is also capable of adapting to artificial shrimp ponds. Small flocks of Collared Plovers also occasionally appear on damp grassy fields, often far from any extensive open water; this may occur especially when wet weather is causing flooding in the species' normal riverine habitat. They breed locally, both along rivers and along or near the coast, with known coastal nesting sites now including various areas in Manabí, w. Guayas, and El Oro. Recorded up to about 500 m in e. Ecuador (e.g., near Tena).

Monotypic. We follow most recent authors (e.g., Blake 1977) in not recognizing *gracilis*, described on the basis of its supposedly shorter wings; if accepted as valid, then Ecuadorian populations would be assigned to it.

Range: n. Mexico to nw. Peru, Argentina, and cen. Chile.

Charadrius vociferus   Killdeer   Chorlo Tildío

Status complex. Originally recorded in Ecuador as a rare boreal winter visitant to fields and lakeshores in the highlands of n. Ecuador south to Pichincha. Rather inexplicably, there are no recent reports of boreal migrants; these are the southernmost records of this population of the Killdeer (it is rare even considerably north of Ecuador, e.g., in Panama; see Ridgely and Gwynne 1989), and they perhaps pertain only to vagrant individuals. More recently the Killdeer has been found to be an uncommon to locally fairly common resident around the margins of marshy ponds, rice fields, and wet pastures in the lowlands of sw. Ecuador from Manabí (north to the Chone/Rocafuerte area) south through s. Los Ríos, Guayas, and El Oro into w. Loja. There is also a recent report of a displaying bird seen considerably farther north, near Salinas in the Río Mira valley of Carchi on 25 May 1996 (N. Krabbe), an indication that the species is apparently colonizing this area as well. There are no early records of the Killdeer from this region; even as late as the 1950s, Marchant (1958) failed to find it on the Santa Elena Peninsula, an area where it is now locally numerous. The first southwestern reports date from the early 1980s, when PJG and others began seeing a few birds in the vicinity of Guayaquil and Daule. Specimens (ANSP, MECN) have now been obtained in Loja and Manabí, the first having been taken south of Sabanilla in s. Loja on 18 Aug. 1989. Resident birds are recorded up to about 800 m in Loja and 1200 m in Carchi, migrants between about 2200 and 2800 m.

Despite its being so widely disjunct in distribution, the resident race *peruvianus* (which ranges also in coastal w. Peru) closely resembles the slightly larger and migratory nominate *vociferus*. In fresh plumage, *peruvianus* reportedly shows somewhat wider rufous margins on its mantle feathers.

Range: breeds in North America, Mexico, and West Indies, wintering south in small numbers as far south as n. South America; resident in w. Ecuador, w. Peru, and n. Chile.

| *Oreopholus ruficollis* | Tawny-throated Dotterel | Chorlo Cabezón Cuellicanelo |

Uncertain; apparently Extirpated in Ecuador. Known only from a pair of specimens obtained on the Santa Elena Peninsula in w. Guayas in Jan. or Feb 1898 (Salvadori and Festa 1900). The Tawny-throated Dotterel was not recorded by Marchant (1958) during his long period of residence on the Santa Elena Peninsula in the 1950s, and it has also not been found subsequently. Nonetheless, the species should still be watched for, perhaps especially in coastal El Oro; a flock of 25 was seen just across the border in Tumbes, Peru, on 4 Aug. 1976 (RSR). The species favors fields and plains that are nearly barren or have only short and sparse grass cover. An additional remarkable record involves a single individual photographed on Hood Island in the Galápagos Islands on 23 Jun. 1991 (R. Harshaw); this would seem to indicate that the species' occurrence on the Ecuadorian mainland, at least as a vagrant or austral migrant, remains a possibility.

The Tawny-throated Dotterel is not considered to be at risk elsewhere in its range (Collar et al. 1992, 1994).

This plover is usually placed in the monotypic genus *Oreopholus*, though a few favor subsuming this into *Eudromias*. The subspecies found in Ecuador remains uncertain; we have not examined the Ecuadorian specimens. A population (*pallidus*) of *O. ruficollis* is resident locally along the coast of Peru north at least to Lambayeque (T. S. Schulenberg and T. A. Parker III, *Condor* 83[3]: 211, 1981), but austral migration is also known in the species, probably involving only the (southern-breeding) nominate race. We assume that the Santa Elena birds were taken from a resident population (presumably *pallidus*) because they were obtained during the austral summer, when birds from a migratory population would not be expected to be present in Ecuador; further, one of the birds was noted as an "immature." In other references (e.g., Blake 1977; Wiersma 1996) they have been assumed to refer to austral migrants of the nominate race, but as far as we are aware no one has examined them recently. Whether the birds RSR observed in Tumbes were (resident) *pallidus* or the (austral migrant) nominate subspecies is unknown. *Pallidus* is smaller than the nominate race and is paler and grayer generally.

Range: nw. Peru south to Chile and Argentina; formerly north to sw. Ecuador.

Stercorariidae   Skuas and Jaegers

Five species in two genera. Skuas and jaegers breed in polar regions (the jaegers only in the Arctic), but they disperse widely on all oceans when not nesting. We follow Gill (1994) in recognizing the skuas and jaegers as a full family; its members are sometimes subsumed into the Laridae.

*[Catharacta chilensis   Chilean Skua   Salteador Chileno]

Apparently an accidental visitant during the austral winter to the coastal (and doubtless offshore) waters of w. Guayas. There is only one report, of a single bird seen between Apr. and Jun. 1987 at Salinas (B. Haase). The specific identity of this individual was not positively determined, but the distinct "rufous tinge" noted on the underparts would indicate that the Chilean Skua was almost certainly the taxon involved.

The genus *Catharacta* is sometimes merged into *Stercorarius*, but we follow the 1998 AOU Check-list in maintaining it. The various austral-breeding forms of *Catharacta* are here treated as species separate from *C. skua* (Great Skua) of the North Atlantic Ocean, following most recent authorities (e.g., Sibley and Monroe 1990). Monotypic.

Range: breeds along coast of s. Chile and Tierra del Fuego, some birds wintering north in Humboldt Current waters to cen. and n. Peru, accidentally to s. Ecuador.

*[Catharacta maccormicki   South Polar Skua   Salteador del Polo Sur]

Apparently an accidental austral winter visitant to coastal (and doubtless offshore) waters of w. Guayas. There is only one report, of a single bird seen in La Libertad harbor on 30 Sep. 1989 (D. Roberson).

It has been suggested (e.g., W. R. P. Bourne and W. F. Curtis, *Brit. Birds* 87[6]: 289, 1994) that the old English name of McCormick's Skua might be more appropriate for this species, which in fact hardly ever reaches the South Pole. The species is now widely known as the South Polar Skua, however, and it is the southernmost breeding member of the genus; we therefore are reluctant to adopt this proposed change and retain South Polar as the modifier.

As with *C. chilensis* (Chilean Skua), *C. maccormicki* is now usually considered to be a species distinct from *C. skua* (Great Skua) of the North Atlantic Ocean. Monotypic.

Range: breeds in Antarctica and on subantarctic islands, some birds (mainly juveniles?) wintering north into North Pacific and North Atlantic Oceans.

Stercorarius pomarinus   Pomarine Jaeger   Págalo Pomarino

A rare to uncommon transient and boreal winter visitant to offshore waters, only rarely coming within sight of land. Most reports are from off the Santa Elena Peninsula in w. Guayas, though there are also a few reports from Manabí and Esmeraldas. The sole Ecuadorian specimen (ANSP) was obtained off the Santa Elena Peninsula on 22 Jan. 1991. Recorded at least from Nov. to May, with occasional birds perhaps occurring year-round.

In Britain and elsewhere in the Old World, the three *Stercorarius* species are generally called skuas.

Monotypic.

Range: breeds in high-arctic North America and Asia, wintering southward, mainly on high seas.

Stercorarius parasiticus   Parasitic Jaeger   Págalo Parásito

An uncommon transient and boreal winter visitant to offshore and coastal waters; seen more often from land than the Pomarine Jaeger (*S. pomarinus*) but even so, seeing one from shore is an infrequent event. Most reports are from off and along the Santa Elena Peninsula in w. Guayas; there is also at least one sighting from Esmeraldas, but none that we are aware of from elsewhere. Parasitic Jaegers occur in Ecuadorian waters throughout the year, though numbers present vary seasonally, being highest during southward passage in Oct. and Nov. (B. Haase). No specimens appear to have been taken in Ecuador.

In Britain and elsewhere in the Old World, the species is generally called the Arctic Skua, though in fact its breeding range extends farther south than that of the other two jaegers.

Monotypic.

Range: breeds in arctic North America and Eurasia, wintering southward, mainly on high seas though somewhat less pelagic than other jaegers.

*[*Stercorarius longicaudus*   Long-tailed Jaeger   Págalo Colilargo]

Apparently an accidental transient to coastal (and doubtless offshore) waters. There is only one report, of a single bird identified at Salinas in w. Guayas on 7 Nov. 1987 (B. Haase). The Long-tailed, the most pelagic of the jaegers, probably migrates past the Ecuadorian coast on a regular basis, but it likely almost always occurs very far offshore.

Race unknown, no specimens having been taken, but birds occurring in Ecuador would presumably be referable to the American-breeding race *pallescens*.

Range: breeds in high-arctic North America and Eurasia, wintering southward on high seas.

Laridae   Gulls and Terns

Twenty-eight species in seven genera. Gulls and terns are worldwide in distribution. The two groups are sometimes separated at the family level.

Larus modestus   Gray Gull   Gaviota Gris

An uncommon to locally (or seasonally?) common visitor to sandy beaches, harbors, lagoons, and nearshore coastal waters of w. Guayas, especially on the Santa Elena Peninsula; there are a few reports from as far north as s. Manabí (Manta, Machalilla, and Isla de la Plata) but none from farther north. One was also seen on Isla de los Muertos in the Gulf of Guayaquil on 5 Jun. 1993 (RSR et al.), but there are only a few records from the El Oro coast, including two Jul. 1922 specimens (AMNH) taken on Isla Jambelí; this may merely reflect the inaccessibility of much of that province's outer coast. Remarkably, four specimens (FMNH) have also been procured well inland, all in Sep. 1931, by the Olallas at Isla Silva and Río San Antonio in n. Guayas and adjacent Los Ríos near Babahoyo; there are no other inland records. Despite the suggestion of Chapman (1926) that the Gray Gull might breed in Ecuador, it clearly does not, nesting being restricted to the extreme deserts of n. Chile. Numbers of the Gray Gull in Ecuador are usually highest during the austral winter (ca. May–Aug.), though some individuals (mainly first- and second-year birds) do occur year-round; the largest numbers RSR has ever observed occurred in Nov. 1997 in the midst of a major El Niño event.

Monotypic.

Range: breeds in deserts of n. Chile (possibly in Peru as well?), dispersing northward and southward along Pacific coast when not nesting.

*[*Larus belcheri*   Band-tailed Gull   Gaviota Colifajeada]

A casual visitor to the El Oro coast. The Band-tailed Gull is known in Ecuador from only a single sighting, of an apparent adult seen at Puerto Bolívar near

Machala on 21 Jan. 1999 (B. Bartrug and C. Hood). In fact it seems surprising that there have not been more reports of this gull in Ecuador, for it is regular along the Peruvian coast within a few hundred kilometers south of the Ecuadorian border. Blake (1977, p. 624) refers to this species as occurring "on the coast of southern Ecuador (*fide* Olrog)." We can only assume that this refers to C. C. Olrog (*Condor* 69[1]: 42–48, 1967), in which Olrog discusses the breeding and distribution of the Band-tailed Gull and its close relative of the Argentinian coast, *L. atlanticus* (now known as Olrog's Gull and generally considered to be a separate species). In fact, however, there is no mention of the species' occurring in Ecuador in that article; the same error was repeated in Burger and Gochfeld (1996, p. 601).

The species is sometimes called the Belcher's Gull.

Monotypic.

Range: coast of Peru and n. Chile, casually dispersing north to Panama and even farther.

Larus dominicanus   Kelp Gull   Gaviota Dominicana

A rare to uncommon visitor and very local breeder on lagoons, sandy beaches, and coastal waters of w. Guayas, principally on the Santa Elena Peninsula though with at least one report from the Río Guayas estuary. Numbers occurring in recent years have usually been small, until 1993 more than six never having been seen together, with birds in adult plumage predominating. In the 1920s, however, the Kelp Gull was recorded as "abundant" in the Gulf of Guayaquil, being "especially common in the harbor of Guayaquil" (Chapman 1926, p. 189); perhaps a misidentification was involved. Given the small numbers of Kelp Gulls that were occurring in Ecuador, it came as a distinct surprise that on 29 Jan. 1994 a small nesting colony with at least 17 nests was found on an open island in the Ecuasal lagoons, with about 10 young in the process of being successfully raised (B. Haase). This represents the northernmost breeding site on the Pacific coast for the Kelp Gull; it had not formerly been known to nest north of n. Peru.

The species is also known as the Dominican or the Southern Black-backed Gull.

Apparently monotypic. Ecuadorian birds would belong to nominate *dominicanus* should separate races in South Africa and Antarctica be recognized.

Range: along Pacific coast from s. Ecuador south to Tierra del Fuego, and along Atlantic coast from Tierra del Fuego north to s. Brazil; also s. Africa and Australia/New Zealand area. There are also a few recent sight reports from Gulf of Mexico region, and breeding has even been reported; records are, however, considered uncorroborated in 1998 AOU Check-list.

**Larus fuscus*   Lesser Black-backed Gull   Gaviota Dorsinegra Menor

An accidental visitor from the north to the coast of w. Guayas. An individual in adult plumage was observed and photographed (photo to VIREO) at the

Ecuasal lagoons in w. Guayas on 26 Jan. 1996 (G. H. Rosenberg et al.). This represents the first record of this Eurasian gull from the Pacific coast of South America. For the last several decades the Lesser Black-backed Gull has been known to be increasing in North America, and a bird wintered in Panama for several consecutive years (N. G. Smith, *Am. Birds* 36[3]: 336–337, 1982).

Race unknown, but in all likelihood the Ecuadorian bird would be referable to European-breeding *graellsii*, which is by far the most frequently recorded subspecies of *L. fuscus* in North America.

Range: breeds in n. Eurasia, wintering south to Africa and s. Asia; increasing in North America, where small numbers are now regular in east.

[Larus argentatus   Herring Gull   Gaviota Argéntea]

An accidental visitor from the north, known from a single sighting. An individual in adult plumage was seen and well described along the Río Napo upstream from La Selva on 13 Feb. 1991 (A. Chartier). This represents the first report from Amazonia, and is one of the few from anywhere in South America.

Race unknown, but presumably the American *smithsonianus*.

Range: breeds in North America, small numbers wintering south regularly to Middle America (casually to n. South America); also in Old World. More than one species is likely involved.

Larus delawarensis   Ring-billed Gull   Gaviota Piquianillada

An accidental visitor from the north to the coast of w. Guayas. The Ring-billed Gull is known in Ecuador from only one record: an adult in basic plumage seen and photographed on a beach at La Libertad in w. Guayas from at least 30 Apr. to 7 May 1991 (D. Liley and B. Haase; photo by the latter to VIREO). What likely was the same individual was again seen and photographed at La Libertad in Oct. 1991 (B. Haase). This represents the species' southernmost known record.

Monotypic.

Range: breeds in n. North America, small numbers wintering south regularly to Middle America, a few straggling to n. South America.

[Larus californicus   California Gull   Gaviota de California]

An accidental visitor from the north to the coast of w. Guayas. The California Gull is known in Ecuador from only one sighting: an apparent subadult carefully studied at the Ecuasal lagoons in w. Guayas on 8 Feb. 1980 (D. Finch and RSR et al.); for additional details, see Ridgely (1980). This represents the only South American report.

Race unknown.

Range: breeds in w. North America, wintering south to Mexico, accidentally to Ecuador.

Larus cirrocephalus   Gray-hooded Gull   Gaviota Cabecigris

Locally uncommon to fairly common in salt lagoons and along the coast (especially in harbors with fishing activity) of sw. Ecuador; does not range very far offshore. Recorded north in small numbers to Manabí (Manta), with a single report of an immature from as far north as Súa in w. Esmeraldas on 11 Mar. 1994 (R. P. Clay et al.). Small numbers also occur locally in the Río Guayas estuary and along the El Oro coast, but in Ecuador the Gray-hooded Gull is most numerous at certain sites on the Santa Elena Peninsula in w. Guayas. Here several hundred or more pairs now breed south of La Libertad at the Ecuasal lagoons (in association with Gull-billed Terns [*Sterna nilotica*]), having first been found nesting here as recently as 1978 (R. S. Ridgely and D. Wilcove, *Condor* 83: 438–439, 1981); their numbers here slowly continue to increase, despite a certain amount of disturbance by local residents. As of 1993, smaller numbers also breed at the salt lagoons and shrimp ponds south of Monteverde.

Numbers of the Gray-hooded Gull in Ecuador have apparently increased substantially in the 20th century. It is noteworthy that the species was not recorded from the country by either Chapman (1926) or Murphy (1936).

The species is sometimes called the Gray-headed Gull.

One race: nominate *cirrocephalus*.

Range: breeds locally along coast of s. Ecuador and Peru, and in e. Argentina (where it also occurs inland), wintering north to Paraguay and s. Brazil; also in Africa.

Larus serranus   Andean Gull   Gaviota Andina

Uncommon to locally fairly common around lakes and ponds in the temperate zone and paramo, often nesting on islets in surprisingly small bodies of water and also occasionally seen overflying areas far from water. Known from Imbabura (Laguna San Pablo) and Pichincha south locally through Loja and adjacent Zamora-Chinchipe; quite local and generally scarce southward. Numbers are perhaps greatest at Laguna de Colta in Chimborazo. Surprisingly, there is also at least one sighting from the eastern lowlands (a group of 5 was seen along the Río Napo near the Jaguar Inn on 26 Aug. 1980; H. Brokaw). Contrary to some assertions, there are no known reports of Andean Gulls from the Pacific coast or western lowlands of Ecuador, though as the species is found at least occasionally on the Peruvian coast such occurrences are not out of the question. Recorded mostly from 3000 to 4200 m, rarely wandering lower (to about 2500 m) or higher (up to 4500 m; R. Levêque, *Ibis* 106[1]: 58, 1964).

Monotypic.

Range: Andes from extreme s. Colombia (Nariño; see R. Strewe, *Bull. B. O. C.* 120[3]: 190, 2000) to cen. Chile and nw. Argentina, wandering down to coast in Peru and Chile.

Larus atricilla   Laughing Gull   Gaviota Reidora

A locally common to abundant boreal winter resident along the entire coast, where it occurs in virtually all saltwater habitats though in largest numbers in lagoons (such as Ecuasal) and estuaries (such as the Río Guayas). Rarely ranges very far offshore (almost never beyond sight of land); only relatively small numbers occur around freshwater ponds and rivers, then usually rather close to the coast (though one was seen by R. Webster at Río Palenque on 7 Aug. 1975). There also are two recent highland records, an adult in nonbreeding plumage seen and photographed with Andean Gulls (*L. serranus*) near Laguna de Colta (3300 m) on 20 Nov. 1991 (W. Hoogendoorn, *Bull. B. O. C.* 114[3]: 206–207, 1994) and another seen above Papallacta (3500 m) on 26 Nov. 1997 (O. Janni). Even more surprising, there are also several recent records from the eastern low-lands: a specimen (LSUMZ) was taken along the Río Napo near Limoncocha on 2 Dec. 1975 (Tallman and Tallman 1977); an adult in basic plumage was seen on the Río Napo just downstream from Coca on 8–10 Jan. 1994 (R. Behrstock and G. H. Rosenberg et al.); six birds in basic plumage were seen in flight at Lago Agrio on 14 Dec. 1994 (L. Navarrete et al.); an adult in basic plumage was again near Coca on 24 and 31 Dec. 1997 (P. Coopmans et al.); and another was near Coca in Jan. 1999 (G. H. Rosenberg et al.). These reports represent almost the only records of the Laughing Gull from anywhere in Amazonia, though possibly the species will prove to be more regular than had been realized. Laughing Gulls occur throughout the year along the coast, at least in favorable localities, though numbers are always small from May to Sep.

As of 1994, 35 banded Laughing Gulls had been recovered along the coast of Ecuador, all but one of them originally captured as nestlings on the east coast of the United States (the exception being one banded in Texas). Most were recovered in the Nov.–Mar. period; birds recovered in the May–Aug. period were all second-year prebreeders.

Monotypic.

Range: breeds locally along coasts of e. United States, w. Mexico, West Indies, and n. South America, wintering south to Peru and n. Brazil, occasionally strag-gling farther south.

Larus pipixcan   Franklin's Gull   **Gaviota de Franklin**

An uncommon to locally fairly common transient (primarily) and boreal winter visitor along the entire coast, also occasionally inland over freshwater. Numbers are typically greatest in Guayas (e.g., on the Santa Elena Peninsula, where the species seems to occur principally as a transient) and around Guayaquil, where sometimes substantial numbers can be found hawking insects over rice fields. A few have also been seen north into Esmeraldas and south into El Oro, and one was seen inland over the Río Santiago at Playa de Oro in Esmeraldas on 2 Dec. 1997 (O. Jahn). There are even a few recent reports of Franklin's Gulls well away from the Pacific coast: a single adult in nonbreed-ing plumage was seen and photographed while consorting with a flock of Andean Gulls (*L. serranus*) south of Cañar (city) in Cañar on 21 Nov. 1991 (W. Hoogendoorn, *Bull. B. O. C.* 114[3]: 206–207, 1994), a second-year bird was seen at Yaguarcocha in Imbabura in Mar. 1993 (B. Haase), and individuals have also been seen on Laguna de Colta (N. Krabbe). Recorded mainly from Oct. to May, with only a few stragglers being reported from the northern summer months.

Monotypic.

Range: breeds in w. North America, wintering primarily along coast of Peru and n. Chile.

Xema sabini   Sabine's Gull   Gaviota de Sabine

A rare to occasionally uncommon transient and boreal winter visitor to off-shore waters along the entire coast. Only rarely seen from land; during the course of his many years of seawatching, B. Haase has observed it only once from shore. First recorded from Ecuadorian waters by S. E. Chapman (*Ibis* 111[4]: 615–617, 1969), who had four sightings of up to 16 birds at different localities from off Esmeraldas south to the Gulf of Guayaquil. Since then the Sabine's Gull has apparently been recorded only off the Santa Elena Peninsula in w. Guayas. Apparently no specimen has ever been taken in Ecuador. Recorded mostly from Sep. to Mar., with a few sightings (RSR) from as early as late Jul.

Monotypic.

Range: breeds in high-arctic North America and Asia, wintering south in Pacific Ocean to off Chile, also in South Atlantic Ocean.

Creagrus furcatus   Swallow-tailed Gull   Gaviota Tijereta

An uncommon visitor to offshore waters of sw. Ecuador, where recorded principally from off the Santa Elena Peninsula and in the Gulf of Guayaquil, with the northernmost known reports being from off Manabí (though the species seems possible farther north, as there is a small nesting colony off Colombia). Only very rarely can Swallow-tailed Gulls be seen from shore. They can occur sporadically at virtually any time of year, though numbers seem highest from May to Jul. The colonization of Isla de la Plata off Manabí would seem possible, though as yet no birds have been reported lingering there.

Monotypic.

Range: breeds principally on Galápagos Islands, also on Malpelo Island off s. Colombia; at least some nonbreeders disperse east into Humboldt Current waters off s. Ecuador, Peru, and n. Chile.

Sterna nilotica   Gull-billed Tern   Gaviotín Piquigrueso

Uncommon to locally common along and near the coast of sw. Ecuador north to cen. Manabí (Bahía de Caráquez), with several old specimen records from coastal Esmeraldas at Vaquería as well. Gull-billed Terns favor artificial shrimp or salt-evaporation lagoons and mudflats, though they occasionally feed over inshore ocean waters or somewhat inland on fresh water. Most numerous in Guayas on the Santa Elena Peninsula and in the Río Guayas estuary. The Gull-billed Tern breeds locally in w. Guayas, with known colonies at the Ecuasal lagoons (where at least several hundred pairs nest, consorting with a comparable number of Gray-hooded Gulls [*Larus cirrocephalus*]) and on salt-evaporation ponds south of Monteverde; breeding, probably in substantial numbers, likely also occurs in the Río Guayas estuary (though as yet the only evidence of nesting there comes from Isla Puná; B. Haase), and perhaps also along the El Oro coast. These birds in sw. Ecuador represent the only known breeding population on the Pacific coast of South America, though nesting also seems possible in nw. Peru. It is conceivable that boreal migrants may also occur in Ecuador, though we are not aware of any evidence that this is the case (contra Gochfeld and Burger 1996); in fact, there seems to be no evidence of an increase in numbers during the northern winter months.

The species has sometimes been placed in the monotypic genus *Gelochelidon*, but we follow the 1998 AOU Check-list in placing it in *Sterna*. The subspecific identity of Ecuadorian birds remains unresolved. Measurements (bill, wing chord) of recently obtained examples (ANSP) of breeding birds at Ecuasal are closer to the slightly smaller race *aranea* (which breeds from the e. United States to the Greater Antilles and the Yucatán Peninsula) than to the slightly larger *vanrossemi* (which breeds principally along the coasts of nw. Mexico). The possibility—though we are not aware of its basis—of their being assignable to *groenvoldi* of the Atlantic coast of Brazil and Argentina has also been suggested (Gochfeld and Burger 1996).

Range: locally along and near coasts from s. North America to w. Peru and ne. Argentina (but nesting only very locally); also in Old World.

[Sterna caspia   Caspian Tern   Gaviotín Piquirrojo]

An accidental visitor to near the coast of cen. Manabí. There is only one Ecuadorian report, of a lone individual seen flying over a lake southwest of Chone on 30 Jul. 1996 (S. Howell et al.). This represents the first report from the Pacific coast of South America; numbers of Caspian Terns being seen in Panama have been increasing in recent years (fide D. Engleman).

The species is sometimes placed in the monotypic genus *Hydroprogne* (e.g., Gochfeld and Burger 1996). We follow the 1998 AOU Check-list in placing it in *Sterna*. Monotypic.

Range: breeds locally in North America and nw. Mexico, wintering south in small numbers to Caribbean coast of n. Colombia and nw. Venezuela; also in Old World.

Sterna maxima   Royal Tern   Gaviotín Real

A fairly common to common transient and nonbreeding visitor along the entire coast, favoring lagoons and sandy coastlines, occurring in largest numbers around the Santa Elena Peninsula in w. Guayas; the odd individual may forage well offshore (farther from land than most other terns regularly do), but most remain within sight of land. Although Royal Terns are present year-round, numbers seem highest from Oct. to Apr. Birds in alternate plumage are seen occasionally from Apr. to May, but there is no evidence of nesting in Ecuador, and indeed breeding is unknown anywhere along the Pacific coast of South America. Juveniles hatched in the United States, however, have been seen in Peru when still being fed by their banded, presumptive parents (N. P. Ashmole and H. Tovar S., *Auk* 85[1]: 90–100, 1968).

A total of 61 banded Royal Terns have been recovered in Ecuador during all months of the year except Jun. and Jul., with peaks in Sep. and from Dec. to Mar. All birds had been banded at nesting colonies in the se. United States. The few birds recovered in Apr. and May were all second-year prebreeders; adult birds were recovered as early as Aug.

This and the following two species are sometimes placed in the genus *Thalasseus* (e.g., Gochfeld and Burger 1996), but we follow the 1998 AOU Check-list in retaining them in *Sterna*. One race occurs in Ecuador, nominate *maxima*.

Range: breeds locally along coast of s. North America, off Middle America and West Indies, on islands off n. South America, and in s. Brazil and Argentina (also in w. Africa); North American breeders winter as far south as Peru.

Sterna elegans   Elegant Tern   Gaviotín Elegante

An uncommon to locally fairly common transient and boreal winter visitant along the entire coast, like the Royal Tern (*S. maxima*), with which it often consorts, favoring lagoons and sandy coastlines and occurring in largest numbers around the Santa Elena Peninsula in w. Guayas. The Elegant Tern seems somewhat erratic in Ecuador, in some years appearing in numbers as early as Jun., in others not arriving until Oct. and remaining rather scarce; records exist from throughout the year, though numbers seem lowest from Dec. to Apr. Early arriving birds in Aug. and Sep. often are still in full alternate plumage and may even engage in apparent courtship feeding chases and displays.

The Elegant Tern is considered to be a Near-threatened species by Collar et al. (1994), presumably because of its inherent vulnerability on its nesting islands. We see no reason to differ with this assessment.

Monotypic.

Range: breeds locally on islands off coast of nw. Mexico and s. California, wintering primarily along coast of Peru and n. Chile.

Sterna sandvicensis   Sandwich Tern   Gaviotín de Sandwich

An uncommon to locally fairly common transient and boreal winter visitant along the entire coast, like the Royal Tern (*S. maxima*) and Elegant Tern (*S. elegans*), with which it regularly consorts, favoring lagoons and sandy coastlines; like the Elegant Tern, the Sandwich does not tend to forage very far offshore. Most numerous around the Santa Elena Peninsula in w. Guayas. Although a few are found regularly throughout the year, numbers are greatest from Dec. to Mar. The Sandwich Tern seems to have increased in recent decades in Ecuador; though nothing like the count of 1350 birds at the Ecuasal lagoons on 13 Feb. 1980 (Ridgely 1980) has been witnessed since, the species remains far from "casual" (as considered by Meyer de Schauensee 1970, p. 88).

Two Sandwich Terns banded as nestlings in North Carolina have been recovered on the coast of Ecuador.

One race: *acuflavida*.

Range: breeds locally along coasts of se. United States and in West Indies, wintering south to Peru and the Guianas, a few wandering as far south as Chile and Brazil; also in Old World. This arrangement excludes the yellow-billed Cayenne Tern (*S. eurygnatha*), of islands in s. Caribbean Sea and locally on coast of n. and e. South America, which is sometimes considered conspecific.

Sterna hirundo   Common Tern   Gaviotín Común

An uncommon to locally common transient and boreal winter visitant along the entire coast, favoring lagoons and sandy coastlines. The Common Tern is most numerous around the Santa Elena Peninsula, where numbers are usually greatest from Jun. to Aug. At that time almost all birds present are first-year prebreeders, and these may congregate and roost in large numbers at the Ecuasal lagoons (upwards of 2000 have at times been counted [B. Haase; RSR]); adults in the process of losing their alternate plumage gradually arrive thereafter. The only records of the Common Tern away from the coast are from the lowlands of the northeast, where it is but a highly unusual vagrant; the first involves a bird collected (LSUMZ) along the Río Napo near Limoncocha on 27 Dec. 1975 (Tallman and Tallman 1977), the second a bird in alternate plumage seen at Zancudococha on 9 May 1993 (L. Navarrete et al.). These represent some of the few records of the Common Tern from anywhere in Amazonia.

One race: nominate *hirundo*.

Range: breeds in North America and locally in West Indies, wintering south, mainly coastally, to n. Chile and Argentina; also in Old World.

Sterna paradisaea   Arctic Tern   Gaviotín del Artico

An uncommon transient to offshore waters, usually with only small numbers appearing along the actual coast; recorded mainly in Apr. and May, and again from Aug. to Nov. Although the Arctic Tern probably occurs all along the coast, at least well offshore, most if not all available reports come from the Santa Elena Peninsula in w. Guayas. An apparently migrating group of 800+ birds roosted during the night of 30 Apr. 1990 at the Ecuasal lagoons, evidently having flown in at dusk from the ocean (B. Haase); this represents by far the highest number of Arctic Terns ever recorded in Ecuador, at most 20 or 30 birds in a day being more typically recorded. Adults in full alternate plumage are regularly noted during northward passage and again during southward passage in Aug. and Sep.; later during the latter period mostly first-year birds are seen. There is also a single report from e. Ecuador, of an adult in alternate plumage seen on the Río Napo upriver from La Selva on 8 May 1993 (D. Fisher et al.). While seemingly astonishing for such a basically oceanic species, in fact such reports of migrating Arctic Terns in Amazonia are not entirely without precedent (we are aware of sightings from Peru and Bolivia, and there may be others).

Monotypic.

Range: breeds in n. North America, wintering from off Chile and Argentina to pack-ice zone off Antarctica; also in Old World.

*[*Sterna hirundinacea*   South American Tern   Gaviotín Sudamericano]

Apparently a casual visitor to sandy beaches and lagoons along the coast of the Santa Elena Peninsula in w. Guayas. The only documented record we know of is the sighting of four birds on the beach south of the Ecuasal lagoons from late Jul. into Aug. 1997 (B. Haase); these birds were paired up and appeared to be preparing to breed—courtship displays were observed—but they then disappeared. B. Haase (pers. comm.) indicated that he had seen a few on earlier occasions as well. We are not aware of the basis for the statement of Gochfeld and Burger (1996) that the species was known previously from Ecuadorian waters.

Monotypic.

Range: coasts of s. South America from Peru and Brazil southward, with a few wandering occasionally as far north as s. Ecuador.

Sterna superciliaris   Yellow-billed Tern   Gaviotín Amazónico

Fairly common along larger rivers in the lowlands of e. Ecuador, where recorded primarily from drainages of the Ríos Napo and Aguarico though also occurring along the Río Pastaza (e.g., around Kapawi Lodge). The Yellow-billed Tern is the most numerous of the three river-haunting larids in e. Ecuador (the others being the Large-billed Tern [*Phaetusa simplex*] and the Black Skimmer [*Rynchops niger*]), and it is by a considerable margin also the most widespread, ranging farther upriver than the other species. There is no evidence of any major seasonal fluctuation in numbers. Although found primarily along rivers, resting (and nesting locally) on sandbars, it also regularly forages over nearby oxbow lakes. Recorded up to about 400 m along the Río Napo upriver to near Misahuallí.

The species has been called the Amazon Tern.

Monotypic.

Range: n. Colombia; e. Colombia, much of Venezuela, and the Guianas to n. Bolivia, ne. Argentina, and Uruguay.

*[*Sterna antillarum*   Least Tern   Gaviotín Menor]

Apparently a casual visitor to sandy beaches and inshore waters along the coast. Known from only two sightings of single individuals: one at Olón in w. Guayas on 14 Aug. 1989 and another at Muisne in w. Esmeraldas on 7 Sep. 1989 (B. Haase). The Least Tern perhaps occurs more regularly in Ecuadorian waters, for it has been recorded as far south as the coast of s. Peru (T. S. Schulenberg, *Gerfaut* 77: 271–273, 1987; R. A. Hughes, *Bull. B. O. C.* 108[1]: 39, 1988).

S. antillarum is regarded as a species separate from *S. albifrons* (Little Tern) of the Old World, following most recent authors (e.g., B. W. Massey [*Auk* 93(4): 760–773, 1976] and AOU 1983, 1998). The subspecies occurring in Ecuador remains unknown, and in any case specimens in nonbreeding plumage would be difficult to assign to race (see M. A. Patten and R. A. Erickson, *Condor* 98[4]: 888–889, 1996).

Range: breeds locally in North and Middle America, and in West Indies and islands off n. South America, wintering south primarily along Atlantic coast of the Guianas and ne. Brazil, with only very small numbers along Pacific coast south to Peru.

Sterna lorata   **Peruvian Tern**   **Gaviotín Peruano**

A rare to uncommon austral visitant to sandy beaches and coastal waters along the coast of sw. Ecuador. Ranges north to Manabí, where a few were seen at Manta in Jul. 1976 (F. Sibley et al.) and in the Puerto López area in Jul. 1978 (RSR and D. Wilcove). The Peruvian Tern is seemingly erratic, with none being seen in some years, small to even substantial numbers in others. In recent decades, however, the species has never been termed "abundant," as it was described by Chapman (1926, p. 187) off Isla Santa Clara in the Gulf of Guayaquil in Jul. 1922. In fact, there appears to be only one report of the Peruvian Tern from all of the 1990s: a sighting of three to four birds in immature or nonbreeding plumage at La Libertad on 2 Aug. 1996 (S. Howell and R. Behrstock et al.). Recorded only from Jul. to Oct. Despite the fact that some individuals seen have been in alternate plumage, there is no evidence of breeding in Ecuador, though this remains conceivable.

In recent years rather few—usually no—Peruvian Terns have been found in Ecuador, an absence that we suspect may reflect a decline in their numbers in Peru. We believe the species merits Near-threatened status, though it is not mentioned at all by Collar et al. (1992, 1994).

Monotypic.

Range: breeds locally along coast of Peru and n. Chile; small numbers disperse north to sw. Ecuador.

Sterna anaethetus*   **Bridled Tern   **Gaviotín Embridado**

An uncommon and local breeding visitor to offshore waters along the coast of w. Guayas. Since 1987 a few Bridled Terns have been seen off the coast near Olón (B. Haase), with one sighting in Jul. 1989 from as far south as off Salinas. On 6 May 1991, J. Gerwin and F. Sornoza discovered that small numbers were nesting on Isla Pelado off Ayangua (specimens ANSP, MECN); the precise number involved could not be ascertained but certainly numbered no more than a few dozen pairs. This colony is the southernmost known along the Pacific coast of South America; we suspect that careful searching may reveal that Bridled Terns also nest on islets off s. Manabí. The only report outside Guayas, however, involves a single bird seen en route to Isla de la Plata on 31 Jul. 1996 (S. Howell and R. Behrstock et al.); presumably Bridleds also occur in the ornithologically less well known waters to the north. To date recorded only from May to Sep.

Because of the very small numbers involved, from an Ecuadorian perspective we consider the Bridled Tern to merit Vulnerable status. Elsewhere in its wide range the species is not at risk (Collar et al. 1992, 1994).

The racial determination of Ecuadorian birds remains problematic, but they likely can be referred to *nelsoni*.

Range: locally in tropical oceans, though absent from all but the eastern sector of Pacific; mainly pelagic when not nesting.

With the confirmation of Bridled Terns nesting on Isla Pelado, it seems certain that previous sightings (Mills 1967) of terns identified as Sooty Terns (*S. fuscata*) from the same site are invalid. We conclude that the presence of the Sooty Tern in mainland Ecuadorian waters has yet to be established.

Phaetusa simplex   **Large-billed Tern**   **Gaviotín Picudo**

Rare to locally fairly common along the larger rivers in the lowlands of e. Ecuador, resting (and nesting very locally, often with larger numbers of Yellow-billed Terns [*Sterna superciliaris*]) on sandbars, sometimes feeding over nearby oxbow lakes. Most numerous in far e. Ecuador, particularly along the lower Río Aguarico. There also are a few old records of the Large-billed Tern from west of the Andes on the lower Río Guayas, with four having been obtained in Sep. 1931 by the Olallas as far upriver as Isla Silva near Babahoyo. There are no recent reports from this now heavily populated region, however, and the species seems almost certain to have been extirpated from the lowlands of w. Ecuador. Occasional wandering Large-billed Terns have also turned up on high-land lakes, at least in Chimborazo, whence there is a 19th-century record from Riobamba and a sighting of one bird at Laguna de Colta in Apr. 1974 (PJG). Mainly recorded below about 250 m.

Like the Black Skimmer (*Rynchops niger*), the Large-billed Tern is declining because of disturbance (especially egg-robbing) at its conspicuous nesting colonies. As a result of the marked decrease in numbers, we judge that the Large-billed Tern deserves Near-threatened status in Ecuador. Elsewhere in its extensive range the species is not judged to be at risk (Collar et al. 1992, 1994).

Monotypic; we follow Blake (1977) in not recognizing *chloropoda* as a valid taxon.

Range: e. Colombia, much of Venezuela, and the Guianas to n. Argentina and Uruguay; locally in n. Colombia and sw. Ecuador, rarely wandering to Panama.

Chlidonias niger   Black Tern   Gaviotín Negro

An uncommon and seemingly erratic transient and boreal winter visitor near the coast of sw. Ecuador, also sometimes occurring offshore and in the Gulf of Guayaquil. In recent years recorded mostly from Aug. to Apr., though S. Marchant's only records (1958, p. 371), on the Represa José Velasco Ibarra near Santa Elena, were from Apr. to Jul. 1957. Up to the present the Black Tern has apparently been recorded only from Manabí and Guayas, though it almost surely also occurs elsewhere along the coast, at least briefly as a transient. In Ecuador the species feeds mainly over salt water (on lagoons, estuaries, and open ocean), though occasionally it may also forage over nearby freshwater ponds and lakes. There also exist two remarkable reports from along the upper Río Napo, where the species was recorded as "rare" near Limoncocha in Dec. 1975 by Tallman and Tallman (1977); these observers (pers. comm. to RSR) observed "small flocks of about six individuals on two or three occasions." In addition, a single basic-plumaged bird was seen near Coca on 13 Nov. 1998 (P. Coopmans). There are few if any other reports of the Black Tern from anywhere in Amazonia.

One race: *surinamensis*.

Range: breeds in North America, wintering locally from Middle America south to n. South America; also in Old World.

Larosterna inca   Inca Tern   Gaviotín Inca

A very rare and erratic visitor to the coast of w. Guayas, favoring rocky coastlines and harbors and usually not occurring too far offshore. During most years no Inca Terns are present in Ecuadorian waters, at most a very few; evidently water temperatures here are too high to sustain populations, as they are for other Humboldt Current birds. Substantial numbers of the Inca Tern occasionally invade, however; it was reported as "common" around Playas and Salinas in 1972 (T. Butler). These incursions may have been correlated with major El Niño events, yet no Inca Terns were reported as having been seen during such events in 1982–1983, 1993–1994, and 1997–1998. The Inca Tern seems never to have been collected in Ecuador.

Monotypic.

Range: breeds along coasts from n. Peru (Piura) south to cen. Chile, in some years dispersing north to s. Ecuador (accidentally even to Panama).

Rynchopidae   Skimmers

One species. Skimmers are pantropical in distribution, with the sole American species also occurring north and south into the temperate zone. We follow Gill (1994) in recognizing the skimmers as a full family; it is sometimes considered a subfamily within an expanded Laridae.

Rynchops niger   Black Skimmer   Rayador Negro

Uncommon along larger rivers in the lowlands of e. Ecuador, with records principally from various points in the Río Napo and lower Río Aguarico drainages though it presumably also ranges in the Río Pastaza drainage. Skimmers also occur along the coast and larger rivers in the lowlands of sw. Ecuador, though they are now very rare there. Although skimmers were reported as "tolerably common" and "familiar" around the shores of the Gulf of Guayaquil in the early 20th century (Murphy 1936, pp. 1173–1174), in recent years there have been only a few reports, the northernmost being a single bird observed flying south along the coast at Olón in w. Guayas on 28 Jun. 1987 (B. Haase). In the 19th and early 20th centuries Black Skimmers also ranged up the Río Babahoyo as far as near Babahoyo, where one specimen (FMNH)—apparently the sole example from w. Ecuador—was obtained by the Olallas at Río San Antonio on 16 Sep. 1931. In the east the species occurs at least occasionally (and mainly formerly) up to about 300 m along the Río Napo near Coca.

Skimmers have declined in the east as a result of persecution at their nesting colonies, and they are numerous only in comparatively remote sectors of the far east. We suspect that the species may no longer be resident in the southwest, and furthermore, contrary to the assertion of Zusi (1996), there is no evidence—merely a presumption—that skimmers ever nested in sw. Ecuador. As a result of the marked decline, we must conclude that the Black Skimmer deserves Near-threatened status here. Elsewhere in its range the species is not judged to be at risk (Collar et al. 1992, 1994).

One race: *cinerascens*. We follow Blake (1977) in not recognizing *intermedia*, to which the Pacific coast population has often been assigned. Specimen material is scanty, however, and the two races may ultimately prove to deserve separation. Pacific coast birds tend to have the secondaries more widely tipped white; in birds from Amazonia these tips are, if present at all, typically very narrow. The underwing is comparatively dark, in contrast to the mainly white underwing of nominate birds from North America.

Range: breeds locally along coast of e. and s. United States, and in Mexico and Guatemala, a few wintering south to Panama (accidentally to n. South America); also resident (breeding locally, and prone to wandering) from Colombia, Venezuela, and the Guianas to n. Argentina and cen. Chile, but rare (breeding status unclear) on Pacific coast.

Columbiformes
Columbidae   Pigeons and Doves

Twenty-eight species in seven genera. Pigeons and doves are worldwide in distribution.

Columba livia   Rock Pigeon   Paloma Doméstica

As with the House Sparrow (*Passer domesticus*)—also introduced—feral populations of the Rock Pigeon exist in many Ecuadorian cities and towns, especially in the highlands and western lowlands. As far as we are aware, none exists in areas away from human habitation.

Often called the Rock Dove or Feral Pigeon.

Range: native to Eurasia; introduced into America in 19th century or earlier.

Columba fasciata   **Band-tailed Pigeon**   Paloma Collareja

Fairly common to common and widespread in the canopy and borders of montane forest, secondary woodland, and clearings with large trees in the subtropical and temperate zones on both slopes of the Andes, including the slopes above interandean valleys; sometimes even occurs up into woodland patches above treeline. The Band-tailed Pigeon regularly occurs in surprisingly deforested areas, where despite its size it does not seem to be directly persecuted to any great extent. Recorded mostly between about 1500 and 3000 m, occasionally as low as 700–800 m in Esmeraldas and El Oro or as high as 3500–4000 m; N. Krabbe (pers. comm.) has once even seen a single bird passing the antennas above the Papallacta pass at 4300 m.

One race: *albilinea*.

Range: w. North America to w. Panama; mountains of n. Venezuela, and Andes from w. Venezuela to nw. Argentina; Sierra de Perijá and Santa Marta Mountains; tepuis of s. Venezuela.

Columba speciosa   **Scaled Pigeon**   Paloma Escamosa

Rare to uncommon and rather local and perhaps seasonal in the canopy and borders of humid forest and secondary woodland in the lowlands and foothills of both e. and w. Ecuador. In the east not found any distance away from the Andes, with no records from farther east than the Jatun Sacha area (where it is only a rare seasonal visitor, mainly between Jun. and Aug.; B. Bochan). In the west recorded from Esmeraldas south locally to s. Manabí and Guayas (a few reports from hills south of Río Ayampe, and at Cerro Blanco in the Chongón Hills). In n. Esmeraldas the Scaled Pigeon has recently been found to be widespread in secondary forest and borders in the lowlands around Playa de Oro, but there it usually avoids the canopy of continuous forest (O. Jahn et al.). There is also a single recent sighting of at least one bird near Puyango in w. Loja on 19 Apr. 1993 (RSR) but, surprisingly, there are no reports at all from El Oro. Scaled Pigeons were seen in nearby Tumbes National Forest on a few occasions in Feb. 1986, and Parker et al. (1995) suggest that the Scaled Pigeon may occur this far south only during the rainy season; the Loja record and at least some of the Guayas reports mentioned above would fit this pattern. Recorded mainly below about 1100–1200 m, but occasionally wandering considerably higher (at least once as high as 1700 m, at Sabanilla in Zamora-Chinchipe; ANSP).

Monotypic.

Range: s. Mexico to extreme nw. Peru, n. Bolivia, and Amazonian Brazil (but absent from much of w. Amazonia); e. Brazil, e. Paraguay, and ne. Argentina.

Columba cayennensis   **Pale-vented Pigeon**   Paloma Ventripálida

Uncommon to locally common in the canopy and borders of forest, secondary woodland, clearings with scattered large trees, and at least locally in mangroves, in the lowlands of both e. and w. Ecuador. In the east the Pale-vented Pigeon is rather local, but it can be numerous around lake margins and along larger rivers, somewhat less so in cleared areas; it is absent from extensively forested areas. In the west somewhat less numerous and recorded south to se. Guayas in the Naranjal area. In addition, small numbers of Pale-vented Pigeons were seen in mid-Apr. 1993 in coastal El Oro in mangroves at Puerto Pitahaya and around Puyango in adjacent sw. Loja (RSR); this was at the end of a particularly wet rainy season, and whether the species regularly ranges so far south is uncertain (it has not otherwise been reported from El Oro or Loja). In the lowlands of n. Esmeraldas the Pale-vented Piegon is replaced or at least outnumbered by the Scaled Pigeon (*C. speciosa*; O. Jahn and P. Mena Valenzuela et al.). Recorded mostly below about 500 m, locally occurring higher along river valleys in the east (e.g., at 850 m along the Río Upano near Macas in Morona-Santiago; P. Coopmans) and wandering higher in the west (e.g., occasional birds at Tinalandia). We assume that the 19th-century specimens labeled as having been taken higher on the west slope of the Andes (e.g., at Pallatanga and Nanegal) were actually procured at much lower elevations.

Two races are found in Ecuador, nominate *cayennensis* in the east and *occidentalis* in the west. Although similar, *occidentalis* differs in being slightly paler gray on its lower belly.

Range: s. Mexico to sw. Ecuador, n. Argentina, and Uruguay.

*[Columba oenops   **Marañón Pigeon**   Paloma del Marañón]

Apparently rare in the canopy and borders of deciduous and semihumid forest and woodland in extreme se. Ecuador in the Río Marañón drainage of s. Zamora-Chinchipe near Zumba. The only confirmed report involves a pair seen near the Río Mayo east of La Chonta on 6 Apr. 1994 (P. Coopmans and M. Lysinger), but there were several prior, more uncertain sightings. Recorded from about 650 to 750 m.

The Marañón Pigeon was accorded Vulnerable status by Collar et al. (1994). As noted above, there is no information regarding its status in Ecuador. We therefore agree with this general assessment, though given the species' very limited range in Ecuador, it could be even more gravely at risk in Ecuador than it is in Peru.

The species is usually called the Peruvian Pigeon (e.g., Meyer de Schauensee 1966, 1970), though as its range is confined to the upper Río Marañón drainage, and as so many other pigeons occur more widely in Peru, it seems much more helpful to give it a more restricted geographic epithet.

Monotypic.

Range: upper Río Marañón drainage in nw. Peru and extreme s. Ecuador.

Columba subvinacea   **Ruddy Pigeon**   **Paloma Rojiza**

Uncommon to fairly common in the canopy and borders of humid forest in the lowlands of both e. and w. Ecuador (in the east both in terra firme and várzea) and in montane forest in the lower subtropical zone on both slopes of the Andes. In the west recorded south to n. Manabí, e. Guayas, El Oro, and w. Loja (Alamor/El Limo region); in addition there is an apparently isolated population on the slopes of the coastal Cordillera de Colonche in sw. Manabí (Cerro San Sebastián in Machalilla National Park) and w. Guayas (Cerro La Torre [N. Krabbe] and Loma Alta [Becker and López-Lanús 1997]). The ecological relationship of this species and the Plumbeous Pigeon (*C. plumbea*) remains to be fully determined. The two species are syntopic at many localities, especially in the lowlands of e. Ecuador; at La Selva and in the lower Río Aguarico drainage, for example, the two species are about equally numerous, whereas at Taisha in Morona-Santiago the Ruddy is considerably more common, and along the Maxus road southeast of Pompeya the Plumbeous is more numerous. Recorded up to 1700 m on both slopes of the Andes, in small numbers as high as about 2000 m.

Two races are found in Ecuador, *ogilviegranti* in the east and *berlepschi* in the west; this subspecific arrangement follows that of *Birds of the Americas* (pt. 1, no. 1). *Ogilviegranti* is somewhat more vinaceous (less brownish) overall with darker and more contrasting bronzy wings (the wings of *berlepschi* are a more uniform vinaceous brown).

Range: Costa Rica to sw. Ecuador; e. Colombia, much of Venezuela, and the Guianas to n. Bolivia and Amazonian Brazil.

Columba plumbea   **Plumbeous Pigeon**   **Paloma Plomiza**

Fairly common in the canopy and borders of humid forest (both terra firme and várzea) in the lowlands and foothills of e. Ecuador, though essentially absent from higher on the east slope of the Andes (a few birds have been heard as high as 1400–1700 m on the south slope of the Cordillera de Huacamayos [P. Coopmans], and there is an old specimen from as high as Baeza). Also fairly common in montane forest in the foothills and subtropical zone on the west slope of the Andes south to El Oro (Buenaventura; ANSP and many sightings), but—unlike the *berlepschi* race of the Ruddy Pigeon (*C. subvinacea*)—the Plumbeous Pigeon appears to be absent from the western lowlands. In the east recorded mainly below about 1000 m (a few ranging higher, at least locally; see above); on the west slope recorded only between about 600 and 2000 m.

Three races are apparently found in Ecuador: *chapmani* on the west slope, *bogotensis* in most of the eastern lowlands, and *pallescens* in the southeast; *pallescens* is known from old specimens from the Río Tigre and "Raya Yacu," and a recent one (ANSP) from Taisha. This subspecific arrangement follows that of *Birds of the Americas* (pt. 1, no. 1). *Chapmani* is somewhat darker and more vinaceous (not so gray) on the head, neck, and underparts than is *bogotensis*; *pallescens* is duller and more vinaceous below than *bogotensis* and is decidedly darker above.

Range: extreme e. Panama to n. Colombia and sw. Ecuador; se. Colombia, extreme se. Venezuela, and the Guianas to n. Bolivia and Amazonian Brazil; e. Brazil. More than one species is likely involved.

Columba goodsoni   Dusky Pigeon   Paloma Oscura

Uncommon to locally fairly common in the canopy and borders of humid forest in the lowlands and foothills of w. Ecuador south to w. Esmeraldas (southeast of Muisne and Bilsa), n. Manabí (Filo de Monos; WFVZ) and s. Pichincha (Río Palenque, where it appears to be declining and may well soon disappear; P. Coopmans). Reports of the Dusky Pigeon from the foothills south along the western base of the Andes to El Oro (Buenaventura) remain uncorroborated, and we suspect that confusion with the songs of the *chapmani* race of the Plumbeous Pigeon (*C. plumbea*) is the explanation for the Dusky having been reported from there. The Dusky is the most numerous *Columba* pigeon in the forest patches near Pedro Vicente Maldonado in the foothills of nw. Pichincha (PJG). Recorded up to about 800 m. An old specimen (Lönnberg and Rendahl 1922) listed as having been taken at Gualea is regarded as likely having been obtained at a lower elevation; there are no recent confirmed records from as high as the Mindo/Gualea area.

Monotypic.

Range: extreme e. Panama to nw. Ecuador.

Zenaida auriculata   Eared Dove   Tórtola Orejuda

Fairly common to common in semiopen and agricultural areas and in gardens in towns and cities in the more arid parts of w. Ecuador (including Isla de la Plata); small numbers occur locally north near the coast as far as w. Esmeraldas (Atacames area), but the Eared Dove is most numerous in parts of the central valley (and is common in and around Quito itself). With the exception of an early (mislabeled?) specimen from Gualaquiza in Morona-Santiago, the Eared Dove is not recorded at all from e. Ecuador. Ranges up to about 3200 m in the vicinity of Quito, with a few having been seen as high as 3500 m on the slopes of Volcán Iliniza (N. Krabbe).

One race: *hypoleuca*.

Range: open areas from much of Colombia, Venezuela, and the Guianas south to Chile and Argentina, but absent from virtually all of Amazon basin; Lesser Antilles.

Zenaida meloda   West Peruvian Dove   Tórtola Melódica

Uncommon and somewhat local in riparian woodland, desert scrub, and mangroves in the more arid lowlands of sw. Ecuador; recorded only from sw. Manabí (north to the Bahía de Caráquez area; RSR and T. J. Davis et al.), w. Guayas (mainly on the Santa Elena Peninsula but also east in very small numbers—perhaps only erratically—as far as the Guayaquil region), Isla Puná, coastal El Oro (south of Arenillas, especially around Huaquillas), and sw. Loja (Río Catamayo valley south to Zapotillo, where it is particularly numerous). Found mostly near sea level, but locally recorded up to 700 m in Loja.

The species has also been called the Pacific Dove, but this seems an unsatisfactory choice of an English name given the very large number of other dove species found elsewhere on islands in the Pacific Ocean.

Z. meloda is regarded as a monotypic species separate from *Z. asiatica* (White-winged Dove) of North and Middle America south to w. Panama, this based on its very different song and highly disjunct distribution.

Range: sw. Ecuador to n. Chile.

Columbina passerina   Common Ground-Dove   Tortolita Común

Uncommon to locally fairly common in arid scrub and agricultural regions in the more arid highlands in the central valley from Carchi (La Concepción and the upper Río Mira valley) and Imbabura south locally to s. Azuay (several pairs seen in the Río Léon valley north of Oña in Dec. 1991; RSR and F. Sornoza). We prefer to regard the undocumented report (Rasmuseen et al. 1994) from near Vilcabamba in s. Loja as being unverified; only the Croaking Ground-Dove (*C. cruziana*) and Ecuadorian Ground-Dove (*C. buckleyi*) have definitely been recorded from this region. Recorded mostly between 1300 and 3000 m, locally somewhat lower (down to 900 m in the Río Mira valley of Carchi and Imbabura; N. Krabbe) and higher (at least occasionally as high as 3300 m below Volcán Cotopaxi; RSR and PJG et al.).

One race: *quitensis*.

Range: s. United States to Costa Rica, locally in Panama; n. and e. Colombia, Andean valleys of Ecuador, Venezuela, the Guianas, and n. and e. Amazonian Brazil; West Indies.

*Columbina minuta   Plain-breasted Ground-Dove   Tortolita Menuda

Uncertain; apparently colonizing the lowlands of both e. and sw. Ecuador. Recently found to be rare to uncommon and local (perhaps somewhat overlooked) in agricultural regions of sw. Ecuador, where reported only from s. Los Ríos (near Babahoyo) and Guayas (the Guayaquil/Duran area south to Manglares-Churute Ecological Reserve). Rare to locally uncommon in open grassy areas near lakes and along roads in the lowlands of ne. Ecuador, with the first record being a single specimen (LSUMZ) obtained at Limoncocha on 6 Apr. 1976 (Tallman and Tallman 1977); a second specimen (ANSP) was taken at Imuyacocha on 31 Mar. 1991. Small numbers of Plain-breasted Ground-Doves have also been seen in clearings near Jatun Sacha since 1987 (B. Bochan et al.), and a male was seen near Tena on 4 Mar. 1992 (D. Wolf). It appears that the Plain-breasted Ground-Dove is in the process of colonizing cleared areas in ne. Ecuador, presumably from populations resident in se. Colombia. Western birds also apparently represent a population that has only recently become established in Ecuador, at least there were no records prior to the 1970s; they still are not confirmed by a voucher specimen, and more detailed information on their status is needed. Mostly below 500 m.

Eastern birds can be referred to nominate *minuta*. Birds from the west probably belong to the similar *amazilia*, which otherwise ranges in w. Peru. It has been suggested that *amazilia* should be synonymized into the nominate race.

Range: s. Mexico to n. and e. Bolivia, Paraguay, ne. Argentina, and s. Brazil, but absent from virtually all of Amazonia.

Columbina talpacoti   Ruddy Ground-Dove   Tortolita Colorada

Recently found to be locally uncommon to fairly common in clearings and areas near towns, and on river islands, in the lowlands of e. Ecuador. The first report involved two pairs seen at the Tiputini army base in e. Napo on 27 Sep. 1976 (RSR); we cannot find the basis for the species being recorded from "eastern Ecuador" by Meyer de Schauensee (1966). Since then the Ruddy Ground-Dove has spread and increased through much of the Lago Agrio-Tena region of w. Napo; outlying populations were also found at Taisha, Morona-Santiago, in Aug. 1990 (RSR and M. B. Robbins), a pair was seen at Imuyacocha on 30 Mar. 1991 (RSR and F. Sornoza), and in 1996 the species was found near Kapawi Lodge on river islands in the Río Pastaza near the Peruvian border. The first specimen for Ecuador was taken on an island in the Río Napo near the mouth of the Río Aguarico on 22 Sep. 1992 (ANSP); the species was quite numerous there. With continuing deforestation the Ruddy Ground-Dove seems destined to continue to increase and expand its range in Ecuador. Recorded up to around 500 m in the Tena area.

Does not include trans-Andean *C. buckleyi* (Ecuadorian Ground-Dove), following most recent authors. One race occurs in Ecuador, nominate *talpacoti*.

Range: n. Mexico to n. Argentina, Uruguay, and se. Brazil.

Columbina buckleyi   Ecuadorian Ground-Dove   Tortolita Ecuatoriana

Fairly common in clearings, gardens, agricultural areas, and young secondary vegetation in the more humid lowlands of w. Ecuador from n. Esmeraldas (north to Borbón; Oct. 1983 specimens [LSUMZ] taken by C. G. and D. C. Schmitt) south through much of Loja (east to near Malacatos; RSR). It appears, however, still to be absent from the very wet lowlands of n. Esmeraldas in the Río Santiago and Río Cayapas drainages, where it was not found during intensive survey work in 1995–1998 (O. Jahn and P. Mena Valenzuela et al.). The Ecuadorian Ground-Dove seems likely to spread north into coastal

Nariño in adjacent Colombia, and indeed it is already present on Gorgona Island off the Colombian coast (B. Ortiz-von Halle, *Caldasia* 16: 211, 1990); there has also been at least one recent report from the Tumaco area (fide P. Coopmans). It also could easily occur in, or spread to, the Río Marañón drainage in the Zumba area, for it is not uncommon just to the south in the Río Chinchipe valley of adjacent Peru. Recorded up to about 2000 m in the Celica region.

The species is sometimes called the Buckley's Ground-Dove.

C. buckleyi is regarded as a species distinct from *C. talpacoti* (Ruddy Ground-Dove), widespread in the Neotropics though in Ecuador occurring only east of the Andes. One race is known to occur in Ecuador, nominate *buckleyi*. If the species is eventually found in the Zumba area (see above), however, that population presumably would be referable to *dorsti* (see M. Koepcke, *Beitr. Zool. Neotrop.*, Fauna 2: 295–301, 1962).

Range: extreme sw. Colombia to extreme nw. Peru, and on Gorgona Island off s. Colombia; locally in upper Río Marañón valley of nw. Peru.

Columbina cruziana   Croaking Ground-Dove   Tortolita Croante

Very common to common in arid scrub, agricultural areas, and gardens in the more arid lowlands of w. Ecuador from n. Esmeraldas (north to the Borbón area; O. Jahn) southward (including Isla de la Plata), also ranging inland in arid interandean valleys of Azuay, El Oro, and Loja. The Croaking Ground-Dove replaces the Ecuadorian Ground-Dove (*C. buckleyi*) in more arid regions, but there is a broad zone where both species are found (sometimes even in the same flock), such as around Guayaquil. It has recently been found to be increasing in the Tumaco region of sw. Nariño, Colombia (P. Coopmans and P. Salaman), having first been found in this area as early as 1976 (Hilty and Brown 1986, p. 192); these sightings seem credible to us, and in fact came from Tumaco, fide D. Engleman). As the Croaking Ground-Dove is found in the upper Río Marañón drainage of Peru, it seems likely eventually to spread into the Zumba region of extreme s. Ecuador—and indeed it seems surprising that it has not already done so. Recorded mostly below about 2200 m, but ranging locally or occasionally as high as 2600 m in intermontane valleys of Azuay (P. Coopmans).

Monotypic.

Range: extreme sw. Colombia to n. Chile.

Claravis pretiosa   Blue Ground-Dove   Tortolita Azul

Uncommon to locally fairly common on the ground and in undergrowth of secondary woodland, humid forest borders, and shrubby clearings in the lowlands of both e. and w. Ecuador. In the west ranges south through much of Guayas, El Oro, and w. Loja (Tambo Negro; Best et al. 1993). In the wet-forest lowlands of n. Esmeraldas the Blue Ground-Dove is largely restricted to second-growth and the edge of clearings and planatations, and is absent from continuously forested regions (O. Jahn and P. Mena Valenzuela et al.). Neither of the two *Claravis* ground-doves is as conspicuous or numerous as the *Columbina* ground-doves. The Blue Ground-Dove, though much the more numerous of the two *Claravis*, is quite nomadic and local, especially in the east; at least in the west, it seems capable of persisting in oil-palm plantations, one of the few forest-based species that manages to do so. Recorded mostly below about 1000 m, occasionally wandering higher (e.g., a female seen at 1300 m in Mindo on 18 Oct. 1997; J. Lyons and V. Perez).

Monotypic.

Range: n. Mexico to extreme nw. Peru, n. Argentina, Paraguay, and se. Brazil.

| *Claravis mondetoura* | Maroon-chested Ground-Dove | Tortolita Pechimarrón |

Very rare and local (erratic and/or nomadic?) in the undergrowth of montane forest on both slopes of the Andes. There are only a few records of this attractive dove from Ecuador, where it appears to be even less numerous than in most other parts of its range. There are old east-slope specimens from San Rafael in Tungurahua and Zamora in Zamora-Chinchipe, and in addition the Mejía collection has two specimens taken in w. Napo: a female from San José obtained in Oct. 1939 and a male from Loreto obtained in Sep. 1940. There appear to be only three recent east-slope sightings: a pair seen at Hacienda San Isidro in Sep. 1996 (M. Lysinger), an undated sighting from the Quebrada Honda sector of Podocarpus National Park (Rasmussen et al. 1994), and one heard at Acanamá in n. Loja on 3 Oct. 1996 (P. Coopmans et al.). On the west slope there are four records: a male (in the Escuela Politécnica collection) taken by M. Olalla at Yanacocha on 4 Nov. 1962 (Orcés 1974), with another male seen and videotaped at Tinalandia in 1997 (fide L. Navarrete), a male seen in the Caripero area of w. Cotopaxi in 1998 (N. Krabbe and F. Sornoza), and a male (MECN) taken above Manta Real in Azuay on 28 Nov. 1990 (J. C. Matheus). Recorded between about 500 and 3500 m.

Little is known regarding the status of the Maroon-chested Ground-Dove in Ecuador. We suspect that it may have been affected by the fragmentation of montane forest that has taken place over the past century, but very little information is available. We therefore suggest that the species be given Data Deficient status. The species was not listed by Collar et al. (1992, 1994).

One race: nominate *mondetoura*. The species is perhaps better regarded as monotypic which, given its vagabond nature, would not be entirely surprising.

Range: locally in highlands from s. Mexico to w. Panama; mountains of n. Venezuela, and Andes from w. Venezuela to w. Bolivia; Sierra de Perijá.

| *Metriopelia melanoptera* | Black-winged Ground-Dove | Tortolita Alinegra |

Uncommon to locally fairly common in paramo, adjacent agricultural areas, and around outlying farm buildings (though rarely or never actually in towns or villages) in the Andes from n. Pichincha (Volcán Cayambe and Volcán Antisana) south locally to Cañar and Azuay (Palmas). The Black-winged Ground-Dove perhaps also occurs farther north or south of this region; the race found in Nariño, Colombia, is the same as the Ecuadorian form. There are two specimens in the Conover collection (FMNH; *Birds of the Americas*, pt. 1, no. 1, p. 564) listed as having been taken at "Nudo de Sabanilla, Loja" on the border between Loja and Zamora-Chinchipe, a very wet area of paramo that does not appear to have suitable habitat for the species. Recorded mostly between 3300 and 4300 m.

One race: *saturatior*.

Range: Andes of extreme s. Colombia and Ecuador, and from cen. Peru to s. Chile and s. Argentina, occasionally wandering (?) down to coast in Peru and Chile.

Leptotila verreauxi   **White-tipped Dove**   **Paloma Apical**

Uncommon to fairly common on the ground and in the undergrowth of semi-humid and deciduous forest and woodland, clearings, and scrub in agricultural areas in the lowlands of w. Ecuador from Carchi and w. Esmeraldas southward, ranging up in arid intermontane valleys and in cleared areas onto the west slope of the Andes and into interandean valleys. On the east slope recorded from the Río Marañón drainage around Zumba in s. Zamora-Chinchipe, and a single bird was seen at Hacienda La Libertad in Cañar in Aug. 1991 (RSR et al.); with ongoing deforestation, the White-tipped Dove may gradually spread farther into the upper reaches of eastward-flowing rivers. In addition there have been a few recent (1993 and subsequent) reports of White-tipped Doves (G. Rivadeneira et al.; tape-recordings by J. Moore) from islands in the Río Napo near Sacha Lodge, La Selva, and Yuturi. Although the White-tipped Dove is absent from much of w. Amazonia, it has also been found recently in comparable habitat along the lower Río Napo in ne. Peru. Ranges up in small numbers to about 3000–3400 m (e.g., on the slopes of Volcán Pichincha; N. Krabbe), but most numerous below 1500 m.

The race recorded from w. Ecuador, *decolor*, is a rather variable form, with birds from higher elevations being darker generally. These birds have purplish or red eye-rings, though the eye-rings of some other races of *L. verreauxi* are grayish blue. *Decolor* may even represent a separate species ranging from w. Colombia to w. Peru (*L. decolor*, Salvin's Dove). The subspecific affinty of eastern birds is unknown, no specimens having been taken, but they would probably be referable to *decipiens*.

Range: Texas and Mexico to w. Peru, n. Argentina, and Uruguay (but not recorded from much of w. Amazonia). More than one species may be involved.

Leptotila pallida   **Pallid Dove**   **Paloma Pálida**

Uncommon to locally fairly common on the ground and in the undergrowth of humid forest, mature secondary woodland, and adjacent cleared areas (including even oil-palm plantations) in the lowlands of w. Ecuador south locally to Guayas (hills south of the Río Ayampe, above Salanguilla, Loma Alta, Cerro Blanco, and Manglares-Churute Ecological Reserve) and El Oro (an early-20th-century specimen from Santa Rosa, and Jun. 1985 sightings from above La Avanzada [RSR and K. Berlin]). The Pallid Dove replaces the White-tipped Dove (*L. verreauxi*) in more humid forest; the two species occasionally overlap in nearby clearings. In n. Esmeraldas Pallid Doves occur in river-edge forest and woodland as well as in overgrown clearings and shrubby forest borders, and do not range inside continuous forest (O. Jahn and P. Valenzuela et al.). Recorded up to about 800 m at Tinalandia.

L. *pallida* has occasionally been considered conspecific with cis-Andean *L. rufaxilla* (Gray-fronted Dove). Monotypic.

Range: sw. Colombia and w. Ecuador.

Leptotila rufaxilla   Gray-fronted Dove   Paloma Frentigris

Uncommon to locally fairly common on the ground and in the undergrowth of humid forest (mainly várzea; in terra firme ranging principally along shady streams and in swampy places) and in overgrown clearings and riparian areas in the lowlands of e. Ecuador; also occurs on river islands. Recorded up along the eastern base of the Andes to about 1100 m.

One race: *dubusi*.

Range: e. Colombia, s. and e. Venezeuela, and the Guianas to n. and e. Bolivia, e. Paraguay, ne. Argentina, and Uruguay. We consider Gray-headed Dove (*L. plumbeiceps*) of Middle America to represent a separate species, contra the 1983 and 1998 AOU Check-lists but in accord with Sibley and Monroe (1990) and Sibley (1996).

Leptotila ochraceiventris   Ochre-bellied Dove   Paloma Ventriocrácea

Rare to locally uncommon on the ground and in the undergrowth of humid and semihumid forest and mature secondary woodland in the lowlands and (especially) the foothills and lower subtropical zone on the west slope of the Andes in sw. Ecuador from cen. Manabí (old specimens from Chone) and Guayas south locally to El Oro and w. Loja. The Ochre-bellied Dove has undergone a major decline in recent years and now is very local and generally scarce. Reports since 1980 come from Machalilla National Park in sw. Manabí (whose still fairly extensive and relatively undisturbed forests and woodlands may harbor the species' largest surviving population; ANSP, MECN); Jauneche in Los Ríos (one seen and tape-recorded by P. Coopmans on 1 Sep. 1991); the Chongón Hills in Guayas (occasional records since the late 1970s [RSR; PJG], more regularly in recent years at Cerro Blanco [e.g., Berg 1994]); on Isla de la Plata (a remarkable Jul. 1995 sighting by J. Hendriks, fide P. Coopmans); Buenaventura in El Oro (various sightings and tape-recordings since the mid-1980s; RSR and PJG et al.); east of Portovelo near the El Oro/Loja border (P. Coopmans et al.); west of Celica (Aug. 1990 sightings; RSR et al.); in the El

Limo/Puyango area of Loja (Apr. 1993 sightings; RSR); and several sites in the Sozoranga region in s. Loja (Best et al. 1993). The species likely is not present throughout the year at all of these sites, with local movements or perhaps even altitudinal migration perhaps being involved. Such movements are further indicated by the report from Isla de la Plata, where the Ochre-bellied Dove is certainly not normally present—indeed even the ecologically tolerant White-tipped Dove (*L. verreauxi*) has never been found on that island. A summary of the Ochre-bellied Dove's distribution and behavior was presented by B. J. Best (*Cotinga* 1: 30–33, 1994). Recorded up to about 1700 m.

Deforestation and the degradation of undergrowth (through overgrazing) in much of what remains has caused the Ochre-bellied Dove to no longer be present at several sites where it was found early in the 20th century and before. As a result this handsome dove, nearly endemic to sw. Ecuador, was given Vulnerable status by Collar et al. (1994). Our sense is that the species' situation is even more precarious than that, and we accord it Endangered status. Because it was seemingly never numerous (or only very locally so), we consider it to be one of the two most threatened Tumbesian endemic bird species (the other being the Gray-headed Antbird [*Myrmeciza grisceiceps*]).

Monotypic.

Range: sw. Ecuador and extreme nw. Peru.

Geotrygon saphirina   Sapphire Quail-Dove   Paloma Perdiz Zafiro

Rare to uncommon on or near the ground inside terra firme forest (especially where the terrain is hilly, perhaps particularly in ravines and along streams) in the lowlands and foothills of e. Ecuador from Sucumbíos and Napo south through Morona-Santiago and locally into Zamora-Chinchipe (Jul. 1993 sightings from Miazi in the Río Nangaritza valley; Schulenberg and Awbrey 1997). The Sapphire Quail-Dove likely ranges northward into adjacent Colombia, though it has not been recorded from there (Hilty and Brown 1986). Recorded mostly below 1100 m, but recorded sparingly (and perhaps only locally) up into the lower subtropical zone on the east slope of the Andes to 1350 m (head of the Río Guataracu; MCZ).

Does not include trans-Andean *G. purpurata* (Indigo-crowned Quail-Dove), which differs in various plumage characters. One race occurs in Ecuador, nominate *saphirina*.

Range: e. Ecuador, e. Peru, and extreme w. Amazonian Brazil.

Geotrygon purpurata   Indigo-crowned   Paloma Perdiz Corona
                           Quail-Dove   Indigo

Rare to uncommon on or near the ground inside very humid to humid forest in the lowlands and (especially) foothills of nw. Ecuador, where recorded from various sites in Esmeraldas, adjacent Imbabura, and nw. Pichincha; in the last it has been found as far south as Santo Domingo de los Colorados and Río Toachi (old specimens) and from sightings at Tinalandia until about 1984, as well as along the road north of Simón Bolívar near Pedro Vicente Maldonado (first on 7 Feb. 1995; P. Coopmans and M. Lysinger). At Playa de Oro in Esmeraldas the Indigo-crowned Quail-Dove occurs in lower foothill forest above about 200 m, being replaced below that level by the Olive-backed Quail-Dove (*G. veraguensis*; O. Jahn et al.). In recent years small numbers of this beautiful quail-dove have been found at El Placer and Bilsa. Recorded mostly between about 200 and 700 m; perhaps occurs down to sea level, but at least in Ecuador not yet definitely recorded from so low.

We conclude that the scarce Indigo-crowned Quail-Dove deserves Vulnerable status on the basis of its limited range and dependence on primary forest habitat. A population does occur in the lower part of Cotacachi-Cayapas Ecological Reserve. The species was not mentioned by Collar et al. (1992, 1994), probably because of the difference in the taxonomy they employed.

G. purpurata is regarded as a species distinct from cis-Andean *G. saphirina* (Sapphire Quail-Dove) of w. Amazonia, based on several striking plumage differences and its disjunct range. Monotypic.

Range: sw. Colombia and nw. Ecuador

Geotrygon veraguensis	Olive-backed Quail-Dove	Paloma Perdiz Dorsioliva

Rare to locally fairly common (but easily overlooked) on or near the ground inside very humid to humid forest in the lowlands of nw. Ecuador in n. Esmeraldas. The only sites where the Olive-backed Quail-Dove has recently been found in Ecuador are northwest of Alto Tambo (small numbers seen in Apr. and Jul. 1990; also ANSP) and at Playa de Oro. At the latter site, during 1995–1998 survey work it was found to be widespread in the wet-forest belt of the Río Santiago and Río Cayapas drainages, occurring in mature as well as secondary forest, often in the vicinity of streams or treefalls (O. Jahn and P. Mena Valenzuela et al.). The species replaces the Indigo-crowned Quail-Dove (*G. purpurata*) at altitudes below about 200–300 m. There are also old specimens from Cachabí, Río Zapollo Grande, and "Charco Redondo." Recorded only below about 300 m.

We conclude that the Olive-backed Quail-Dove deserves Near-threatened status on the basis of its very limited Ecuadorian range and strict forest-habitat dependence. The species was not considered to be generally at risk by Collar et al. (1992, 1994).

Monotypic.

Range: Costa Rica to nw. Ecuador.

Geotrygon frenata	White-throated Quail-Dove	Paloma Perdiz Goliblanca

Uncommon to locally fairly common on or near the ground inside montane forest and woodland (occasionally coming to edge or even feeding in adjacent clearings) in the subtropical and lower temperate zones on both slopes of the Andes; at least in s. Ecuador also occurs on ridges above some interandean valleys. On the west slope recorded south to w. Loja. Recorded mostly between

1300 and 2600 m, locally and in small numbers as high as 3300 m at Cerro Mongus in e. Carchi (RSR and PJG) and down to 800 m on the west slope (e.g., at Tinalandia; G. H. Rosenberg et al.). There is also an MECN specimen labeled as having been taken as low as 500 m at Alto Tambo in e. Esmeraldas, but this seems likely to have been obtained on the slopes above that town.

Two similar races of G. *frenata* are found in Ecuador, *bourcieri* in most of the species' Ecuadorian range and *subgrisea* in w. Loja. *Subgrisea* differs in its paler forecrown and slightly paler underparts.

Range: Andes from cen. Colombia to nw. Argentina.

The Dark Quail-Dove (G. *erythropareia*), known from three old specimens (BMNH) taken on the east slope of the Andes in Ecuador and an additional example from Bolivia, is considered to be of "doubtful" specific status, following the conclusions of Chapman (1926), *Birds of the Americas* (pt. 1, no. 1), and Meyer de Schauensee (1966). It is much darker and browner on the underparts, and more rufescent above; it presumably represents a dark morph of the *bourcieri* race of the White-throated Quail-Dove.

Geotrygon montana   **Ruddy Quail-Dove**   **Paloma Perdiz Rojiza**

Uncommon to fairly common (but inconspicuous) on or near the ground inside humid forest, secondary woodland, and plantations in the lowlands of both e. and w. Ecuador; in the east occurs in both várzea and terra firme, in the latter primarily along streams and in swampy places. It also ranges locally up into montane forest in the lower subtropical zone on both slopes. In w. Ecuador recorded south locally to s. Loja virtually on the Peruvian border (Sozoranga and Tambo Negro; Best et al. 1993) and seems likely to range southward into adjacent Peru. In the wet-forest belt of n. Esmeraldas the Ruddy Quail-Dove is outnumbered by the Olive-backed Quail-Dove (G. *veraguensis*; O. Jahn et al.). Recorded up in the lower subtropical zone on both sides of the Andes to about 1300 m, locally as high as 1600 m in Loja (Sozoranga) and also in Zamora-Chinchipe (Panguri; ANSP); there is also an old specimen from Baeza (1900 m) in w. Napo (Chapman 1926).

One race: nominate *montana*.

Range: n. Mexico to sw. Ecuador, nw. Argentina, and Amazonian Brazil; se. Brazil, e. Paraguay, and ne. Argentina; West Indies.

*Geotrygon violacea   **Violaceous Quail-Dove**   **Paloma Perdiz Violácea**

Uncertain. Apparently very rare and local on or near the ground inside terra firme forest near the edge of lagoons (where the forest is seasonally or at least occasionally flooded) in the lowlands of ne. Ecuador in Sucumbíos at Cuyabeno. The only records occurred on 2–10 Aug. 1992 when three individuals were located, of which one was seen and tape-recorded (B. Whitney et al.). 250 m.

Presumed race: nominate *violacea*.

Range: locally from Nicaragua to w. Venezuela, se. Colombia, and ne. Ecuador; e. Venezuela and Suriname; Bolivia; e. Brazil, e. Paraguay, and ne. Argentina.

Psittaciformes

Psittacidae   Parrots and Macaws

Forty-six species in 17 genera. Parrots are pantropical in distribution, but macaws occur only in the Neotropics.

Ara ararauna   **Blue-and-yellow Macaw**   Guacamayo Azuliamarillo

Uncommon to locally fairly common in the canopy and borders of humid forest (especially várzea; in terra firme forest found primarily in swampy places, particularly where there are stands of *Mauritia* palms, generally only overflying other areas) in the lowlands of e. Ecuador. There are also a few records, mostly recent sightings, from the lowlands of sw. Ecuador, though whether these involve remnants of a truly wild population—and not just a few birds that have escaped from captivity—remains an open question; the species was recorded by Salvadori and Festa (1900) at Río Peripa on the Manabí/Pichincha border, but even this could represent a record of a captive bird. Blue-and-yellow Macaws have been seen in and south of the Chongón Hills in w. Guayas (though not since T. Butler observed a pair on 8 Aug. 1972) and in w. Cañar (2 seen near San José de Tambo on 9 Aug. 1986; M. B. Robbins). Recorded mainly below 500 m, with a few (mainly old) records from as high as 900 m.

The Blue-and-yellow Macaw is the most widespread and frequently seen large macaw in the Oriente; its largest populations are now found in the far east (e.g., in the Río Lagarto drainage and around Kapawi Lodge). The species has, however, declined markedly in and around settled areas such as near Lago Agrio, Coca, and Tena, and it is now most often seen only overflying such regions. It does not yet appear to be at risk, however.

This macaw has sometimes been called, especially in avicultural circles, the Blue-and-gold Macaw.

Monotypic.

Range: e. Panama and n. Colombia; se. Colombia, s. and e. Venezuela, and the Guianas to n. Bolivia and Amazonian and cen. Brazil (perhaps formerly occurring farther south).

Ara militaris   **Military Macaw**   Guacamayo Militar

Rare to uncommon and very local in the canopy and borders of montane forest in the foothills and lower subtropical zone on the east slope of the Andes, where in recent decades recorded from w. Napo (repeated sightings of up to 20 or so birds north of Archidona along the road to Loreto), nw. Morona-Santiago (sightings of up to 19 birds in the Río Upano valley in Sangay National Park in Jul.–Aug. 1979; Ridgely 1980), and Zamora-Chinchipe (flock of 14 seen in the Bombuscaro sector of Podocarpus National Park on 30 Jun. 1992 [P. Coopmans]; pair seen north of Zumba on 13–14 Dec. 1991 [RSR], and small flocks seen there in Apr. 1994 [P. Coopmans]). Recorded mostly from 800 to 1500 m, but occasionally wanders as low as

400 m in the adjacent lowlands (e.g., a pair was present at Jatun Sacha 1–6 Oct 1990; B. Bochan).

The Military Macaw was accorded Vulnerable status by Collar et al. (1994). Its range and numbers have likely always been limited in Ecuador, and there is some evidence of a decline in numbers over the past several decades. We thus also give it a Vulnerable status from an Ecuadorian perspective, with the caveat that further field work may reveal that it deserves Endangered status here; certainly numbers in Ecuador cannot be very large.

One race: nominate *militaris*.

Range: locally in Mexico, n. Venezuela, n. and w. Colombia, and south on east slope of Andes from s. Colombia to s. Peru, and in s. Bolivia and nw. Argentina.

Ara ambigua   Great Green Macaw   Guacamayo Verde Mayor

Very rare to locally rare in the canopy and borders of forest (both humid and deciduous) in the lowlands and foothills of w. Ecuador from Esmeraldas south (now very locally) to Guayas in the Chongón Hills. Prior to the era of massive deforestation that began in the late 1940s, the Great Green Macaw almost surely ranged through much of lowland w. Ecuador; there are sight reports from the early 1960s at Hacienda Pacaritambo in Los Ríos and from the 1970s at Finca Victoria in Pichincha, and apparently it once occurred in the Río Palenque area as well (fide C. Dodson). It presently occurs only in Esmeraldas and in the Chongón Hills near Guayaquil. Recorded up to about 800 m.

The present status of the Great Green Macaw in Ecuador is precarious, and we feel that it merits Critical status in the country. Collar et al. (1994) did not mention the species at all, but this was evidently because they considered it to be conspecific with the Military Macaw (*A. militaris*), a combination that, based on presently available evidence (see below), we believe to be premature. As of the early 1990s, only a few pairs of Great Green Macaws survived in the species' best-known Ecuadorian site, at and near Cerro Blanco in the Chongón Hills west of Guayaquil, with breeding still taking place there (Berg 1994); this site is now under improved protection, giving some cause for hope that the population, though extremely small, may yet survive. A few may also persist in the coastal cordillera east of Manglaralto in w. Guayas, but details remain sketchy; local residents informed us that the species disappeared from the Machalilla National Park area several decades before the first major field work in that area took place in 1990–1991. The primary Ecuadorian population—though doubtless it too is not very large—is now probably found in n. Esmeraldas and immediately adjacent Imbabura, from whence the species was recorded in the 1980s and 1990s at El Placer, northwest of Alto Tambo, Lita, and Playa de Oro. At Playa de Oro it occurs in most areas only as a feeding visitor, with breeding seemingly restricted to a narrow zone between 150 and 300 m (O. Jahn et al.); it is still locally hunted for food (P. Mena Valenzuela et al.). There are three old specimens from two other sites in Esmeraldas (Río

Zapallo Grande and Río Durango, neither of which can be precisely located), and there is one recent sighting from w. Esmeraldas (a flock of 8 birds seen on the west bank of the Río Verde ca. 30 km south of Chontaduro on 1 Jul. 1992; T. J. Davis and F. Sornoza).

The species is sometimes called the Buffon's Macaw (e.g., Forshaw 1989).

J. Fjeldså, N. Krabbe, and R. S. Ridgely (*Bull. B. O. C.* 107[1]: 28–31, 1987) discuss the west-Ecuadorian race *guayaquilensis* and propose, based on its variable characters, synonymizing it into the nominate race; this would make *A. ambigua* monotypic. Fjeldså et al. (op. cit.) suggest that *A. militaris* and *A. ambigua* may even be conspecific, but we feel that more information is needed before this possibility can be resolved.

Range: Honduras south locally to sw. Ecuador.

Ara macao   Scarlet Macaw   Guacamayo Escarlata

Uncommon to fairly common in the canopy and borders of humid forest (principally terra firme) in the lowlands of e. Ecuador. The Scarlet Macaw has declined markedly near settled areas over the past several decades and now is found principally in remote regions such as in the lower Río Aguarico drainage and along the Maxus road southeast of Pompeya. Recorded only below about 400 m.

One race: nominate *macao*. See D. A. Wiedenfeld (*Ornitol. Neotrop.* 5: 99–104, 1994) for the recent description of the race *cyanoptera* from Middle America.

Range: locally from s. Mexico to w. Panama; n. Colombia; e. Colombia, much of Venezuela, and the Guianas to n. Bolivia and Amazonian Brazil.

Ara chloroptera   Red-and-green Macaw   Guacamayo Rojo y Verde

Rare and very local in the canopy and borders of terra firme forest in the lowlands of e. Ecuador. The Red-and-green Macaw was always the least numerous of the three great macaws in Ecuador's Oriente, and though it was formerly widespread (though always seeming to favor hilly areas), it now is very local and infrequently encountered. There are recent records from the Zancudococha area (though evidently few or none since RSR sightings in the mid-1970s), the Bermejo oil-field area north of Lumbaquí (RSR), Taisha (RSR), Yuturi (S. Howell), and a few from south of the Río Napo at La Selva, but on the whole this is now a very rarely seen bird in Ecuador. Recorded up to about 500 m.

Numbers of the Red-and-green Macaw have declined substantially in Ecuador in recent decades, a decline that remains unexplained and has proceeded more rapidly than that of the other large macaw species. Even in suitable habitat and in the most remote regions, the Red-and-green is now either extremely scarce or does not occur at all. For example, there still is not a single record from along the Maxus road southeast of Pompeya (RSR et al.), despite there having been extensive field work by several experienced observers in an area where both Scarlet Macaws (*A. macao*) and Blue-and-yellow Macaws (*A. ararauna*) remain numerous. We therefore accord the species Vulnerable status within the country; as a species this wide-ranging macaw is not, however, judged to be generally at risk (Collar et al. 1994).

The species is sometimes called the Green-winged Macaw (e.g., Forshaw 1989).

Monotypic.

Range: e. Panama and nw. Colombia; n. Colombia, w. and s. Venezuela, and the Guianas to n. and e. Bolivia, e. Paraguay, extreme n. Argentina (formerly), and Amazonian and cen. Brazil.

Ara severa   **Chestnut-fronted Macaw**   Guacamayo Frenticastaño

Uncommon to locally fairly common in borders of humid forest (both terra firme and várzea), and in clearings with scattered large trees left standing, in the lowlands of both e. and w. Ecuador. In the west recorded south in small numbers to n. Manabí, se. Guayas in the Naranjal area, and nw. Azuay at Manta Real. Chestnut-fronted Macaws were not recorded at all during intensive survey work in 1995–1998 around Playa de Oro in n. Esmeraldas (O. Jahn and P. Mena Valenzuela et al.). Ranges up in small numbers onto the east slope of the Andes (mainly in partially cleared areas) as high as about 1400 m, occasionally wandering as high as 1800 m on the Cordillera de Huacamayos (N. Krabbe); on the west slope recorded up to 1500 m (an old specimen from Pallatanga in Chimborazo), but mainly recorded below 600 m.

The Chestnut-fronted Macaw is now very local in the west as a result of extensive deforestation. Unlike the larger macaws, however, the Chestnut-fronted actually appears to increase after a certain amount of clearing, and it sometimes seems most numerous in partially deforested terrain. In recent years, however, it is believed to have declined substantially in the Jatun Sacha area (B. Bochan). On balance, and despite the localized declines, the species does not seem to be at risk overall.

One race: *castaneifrons*.

Range: e. Panama to sw. Ecuador, n. Bolivia, and Amazonian Brazil.

Orthopsittaca manilata   **Red-bellied Macaw**   Guacamayo Ventrirrojo

Uncommon to locally fairly common in the canopy and borders of várzea forest, swampy areas in terra firme forest, and nearby partially cleared areas in the lowlands of e. Ecuador. Recorded primarily from the northeast in the drainages of the Río Napo and Río Aguarico, but there are also recent reports from the southeast at Kapawi Lodge on the Río Pastaza near the Peruvian border, and there is an old record from Sarayacu. Likely the Red-bellied Macaw occurs more widely in the Río Pastaza drainage; specimens were procured at the uncertainly located "Laguna de Siguin," thought to be somewhere along that river (Berlioz 1937c). The Red-bellied Macaw shows a strong predilection for

Mauritia palms, in which it often feeds or roosts, and always is most numerous where these palms occur; elsewhere it occurs almost exclusively as overflying flocks. Recorded up to about 400 m.

Formerly placed in the genus *Ara*. We follow H. Sick (*Ararajuba* 1: 111–112, 1990) in placing this distinctive macaw in the monotypic genus *Orthopsittaca*. Monotypic.

Range: e. Colombia, e. Venezuela, and the Guianas to n. Bolivia and Amazonian and cen. Brazil.

Aratinga erythrogenys   **Red-masked Parakeet**   Perico Caretirrojo

Rare to locally fairly common in the canopy and borders of deciduous forest and woodland, adjacent agricultural terrain, and desert scrub in the lowlands and lower subtropical zone of w. Ecuador, occurring from w. Esmeraldas (small groups seen southeast of Muisne, in the Same-Súa area, and north of Quinindé at Finca Paraiso de Papagaios in 1993–1994 [RSR; P. Coopmans]), Pichincha (locally and perhaps only seasonally), Manabí, Los Ríos, and w. Chimborazo (Pallatanga) south through El Oro and much of Loja (inland to around Vilcabamba and in the Sozoranga area). The Red-masked Parakeet also ranges at least locally into more humid woodland and forest, such as at Machalilla National Park of sw. Manabí, as well as in Pichincha and w. Esmeraldas. Found mostly below 1300 m, ranging occasionally or seasonally up to 1900 m, locally (and perhaps only infrequently) as high as 2500 m at Utuana in Loja (Best et al. 1993).

The Red-masked Parakeet has decreased greatly over most or all of its Ecuadorian range in the past several decades, a decline caused by a combination of capture for the pet market (especially?) and habitat destruction and deterioration; see the review by B. J. Best, N. Krabbe, C. T. Clarke, and A. L. Best (*Bird Conserv. Int.* 5[2–3]: 233–250, 1995). The decline has been especially severe in Guayas and El Oro, where the species now occurs at most in scattered remnant pairs or small groups but often not at all. In much of this region it was formerly very common, occurring in roosting flocks and nonbreeding aggregations numbering in the thousands. Fortunately it remains reasonably numerous and widespread in Loja, though here too there is some indication of a decline. It is possible, however, that ongoing deforestation toward the northern edge of the species' range in Esmeraldas and Pichincha may be facilitating a modest increase in this area. Nonetheless, in our view the Red-masked Parakeet deserves formal Vulnerable status, though it was considered only Near-threatened by Collar et al. (1994); its decline seems to have been more precipitous than that seen in the always relatively uncommon Gray-cheeked Parakeet (*Brotogeris pyrrhopterus*).

This species and all members of the genus *Aratinga* sometimes go by the English group name of conure (e.g., Forshaw 1989; Juniper and Parr 1998). Monotypic.

Range: w. Ecuador and nw. Peru.

Aratinga wagleri   Scarlet-fronted Parakeet   Perico Frentiescarlata

Very rare (possibly more numerous formerly) in the canopy and borders of montane forest and woodland and adjacent cleared and agricultural areas in the highlands of s. Ecuador, where known definitely only from Loja (perhaps also in adjacent se. El Oro). There are only a few Ecuadorian records of this species, which to some extent may have been confused with the Red-masked Parakeet (*A. erythrogenys*). These include old specimens from San Lucas and Lunamá, and more recent sightings of small flocks near Catamayo in Feb. 1980 (RSR and D. Finch), Jimbura in Apr. 1990 (G. S. Glenn and F. Sornoza) and at unspecified localities in the upper Río Puyango valley (N. Krabbe). As Scarlet-fronted Parakeets are found in some numbers in the upper Río Marañón drainage of nw. Peru, it also seems possible that at least wandering individuals could occur as far north as adjacent Ecuador in the Zumba region. Recorded from about 1000 to 2500 m.

Although there is no evidence that it was ever a numerous bird in Ecuador, we suspect that, like the smaller but similar Red-masked Parakeet, the Scarlet-fronted has declined in recent decades because of disturbance and habitat degradation. Certainly there are few records (old or recent) of the species, and it is not known to occur regularly anywhere in the country. Although relatively little is known about it, we accord it Data Deficient status. The species was not mentioned by Collar et al. (1994).

The species is sometimes called the Red-fronted Parakeet (or Conure; e.g., Forshaw 1989).

A. frontata (with *minor* of Peru) is perhaps better regarded as a full species, *A. frontata* (Cordilleran Parakeet), distinct from true *A. wagleri* of Colombia and Venezuela (which lacks red on the bend of its wing, etc.). One race of *A. wagleri* occurs in Ecuador, *frontata*.

Range: mountains of n. Venezuela; Andes of w. Venezuela and Colombia, and extreme s. Ecuador and Peru; Sierra de Perijá and Santa Marta Mountains.

Aratinga leucophthalmus   White-eyed Parakeet   Perico Ojiblanco

Uncommon to locally common in the canopy and borders of várzea forest and riparian woodland and in adjacent clearings in the lowlands and foothills of e. Ecuador. The White-eyed Parakeet is often seen in large bands flying high overhead to and from its roosting sites, which in many areas are situated on river islands; rarely or never does it actually alight in extensive terra firme forest. In many (perhaps most) regions it seems to be quite seasonal in its appearances. Found mostly below 1100 m, but ranging locally or seasonally up in partially cleared terrain on the east slope of the Andes as high as 1600–1700 m, especially in s. Ecuador.

One race: *callogenys*.

Range: e. Venezuela; se. Colombia, e. Ecuador, e. Peru, Amazonian Brazil, and the Guianas to n. Argentina and Uruguay (but absent from ne. Brazil).

Aratinga weddellii   Dusky-headed Parakeet   Perico Cabecioscuro

Fairly common in the borders of humid forest (mainly várzea and floodplain forest and woodland) and in clearings with large trees and snags in the lowlands of e. Ecuador. The Dusky-headed Parakeet is essentially absent from extensive terra firme forest, and only rarely is it even noted overflying such areas. Basically a nonforest parakeet, the Dusky-headed may—despite its local popularity as a pet—be gradually increasing in Ecuador as a result of continued deforestation. Recorded mainly below about 500 m, but small numbers have been recorded up the Río Nangaritza valley to 900 m at Shaime (Balchin and Toyne 1998).
Monotypic.
Range: se. Colombia to n. Bolivia and w. Amazonian Brazil.

Leptosittaca branickii   Golden-plumed Parakeet   Perico Cachetidorado

Uncommon to locally (or perhaps temporarily) fairly common in the canopy and borders of montane forest and woodland in the temperate zone on the east slope of the Andes, also occurring at least occasionally in forests on slopes above interandean valleys. The Golden-plumed Parakeet is found principally in s. Ecuador from Cañar (Hacienda La Libertad; seen in Aug. 1991 by RSR and F. Sornoza et al.) and Morona-Santiago (Gualaceo-Limón road) southward; certainly it is most numerous there. In n. Ecuador it is known from only two records: two Sep. 1931 specimens (FMNH) taken at Pimampiro in e. Imbabura, and a sighting of a flock on the west slope of Cerro Mongus in se. Carchi on 20 Mar. 1992 (F. Sornoza). FMNH also has specimens from adjacent Colombia. In recent years this apparently nomadic parakeet has been recorded by various observers at several localities in s. Ecuador, with the largest numbers perhaps occurring at certain sites in n. Loja (especially around Saraguro), Podocarpus National Park (especially southward, e.g., on the Cordillera de Sabanilla), and on the east slope of the Cordillera Las Lagunillas. Recorded mostly from 2400 to 3400 m.

Overall numbers of the Golden-plumed Parakeet have doubtless declined because of deforestation in many parts of the species' range, and it was thus accorded Vulnerable status by Collar et al. (1994). Our sense, from an Ecuadorian perspective, is that it may not be so gravely at risk here, but we concur with this ranking. At least in Ecuador the species seems capable of persisting in somewhat fragmented habitat, perhaps because it can overfly fairly wide expanses of deforested terrain. Populations do occur, at least seasonally, in Podocarpus National Park and the nearby Tapichalaca Biological Reserve, where it nests. It should be noted that, unlike many other Neotropical parrots, the Golden-plumed Parakeet has rarely been involved in the captive-bird trade, a rarity that perhaps more reflects its overall scarcity than any other factor.

The species is sometimes called the Golden-plumed Conure.

The monotypic genus *Leptosittaca* is sometimes merged into the genus *Aratinga*, but based on vocal and morphological differences it seems preferable to maintain it as distinct. Monotypic.

Range: locally in Andes from cen. Colombia to s. Peru.

Ognorhynchus icterotis   Yellow-eared Parrot   Loro Orejiamarillo

Very rare in the canopy and borders of montane forest in the subtropical and lower temperate zones on the west slope of the Andes in n. Ecuador. A strong association with *Ceroxylon* wax palms (both for nesting and feeding) was noted in the past, but it may not always be as strict as has been believed; in recent years birds have been seen feeding mainly on the fruits of *Sapium* and *Croton* fruits (N. Krabbe). The Yellow-eared Parrot's status in Ecuador remains shrouded in mystery, and most of the few records are imprecise. There are old specimens from (presumably above) Intag in Imbabura and from Hacienda Piganta in Pichincha. Several were shot from a flock near Guallupe (ca. 1200 m) in the 1970s in Imbabura (fide G. Orcés and T. Arndt), including one labeled as having been taken on "Via San Lorenzo Lita?" (MECN). Others have recently been shot at an undisclosed site farther south in w. Ecuador; one of these was discarded because of its poor condition, two were stolen from the Mundo Juvenil collection in 1995 (fide N. Krabbe), and the fourth is in MECN. The collecting site of these four birds was finally located by N. Krabbe in 1994 at an elevation of 2300–2400 m in the Caripero region of w. Cotopaxi, and he and F. Sornoza continue to make an effort to protect the species there. Nineteen individuals were found in Aug. 1995, but only 14 in late Nov. of that year, and few or none in more recent years. Owing to fears on the part of N. Krabbe as to the danger of capture by aviculturists (as well as the danger of illegal hunting), the precise location of the southern site remains undisclosed. To the north, a small group was believed heard calling in montane forest above (east of) Maldonado in Carchi on 3 Jan. 1982 (RSR, PJG, and S. Greenfield), but they could never be seen; a group of seven birds was seen in much the same area, above Morán (3200 m), on 6 Aug. 1991 (J. Hendriks, fide P. Coopmans). Recorded between about 1800 and 3200 m.

The Yellow-eared Parrot's situation in Ecuador remains uncertain, though obviously it is extremely precarious and may even be beyond salvation. The species was first accorded Endangered status (Collar et al. 1992) and more recently Critical status (Collar et al. 1994). In Ecuador we judge it now to be (extremely) Critical. Areas where the species has recently been found remain largely unprotected, and the populations so small as perhaps not to be salvageable. Although deforestation has doubtless played a major role in its decline, this parrot now is so rare in many areas where it used to occur and where montane forest remains little altered that one has to suspect that additional factors may be involved.

The species is sometimes called the Yellow-eared Parakeet, or Conure.

Monotypic.
Range: locally in Andes of Colombia and nw. Ecuador.

Pyrrhura melanura   Maroon-tailed Parakeet   Perico Colimarrón

Uncommon to locally common in the canopy and borders of humid forest (both terra firme and várzea) and in adjacent clearings with large trees in the lowlands of e. Ecuador south to Morona-Santiago (Taynutza [WFVZ] and Taisha [seen in Aug. 1990 by RSR and M. B. Robbins]). The Maroon-tailed Parakeet also occurs locally in montane forest in the lower subtropical zone on the east slope of the Andes in Napo and Morona-Santiago (west slope of the Cordillera de Cutucú [ANSP] and Río Upano valley [RSR]), and is fairly common in humid forest in the foothills of nw. Ecuador from Esmeraldas south to w. Cotopaxi (Caripero area; N. Krabbe and F. Sornoza); the report from Río Palenque (R. S. Ridgely and M. B. Robbins, *Wilson Bull.* 100[2]: 173–182, 1988) is in error. On the east slope ranges mostly below 1200 m, in small numbers up to 1500–1600 m in Napo and Morona-Santiago; in the northwest recorded mostly between 500 and 1400 m, locally as low as 250 m (along the Río Verde in e. Esmeraldas; F. Sornoza) and as high as 1900 m (Caripero).

This species and all members of the genus *Pyrrhura* sometimes go by the group name of conure.

Four races of *P. melanura* occur in Ecuador; their characters and distribution were clarified by Ridgely and Robbins (op. cit.). The nominate race (with very narrow pale scaling on the breast) is found in the eastern lowlands. *Souancei* (with broader breast scaling) is found along the eastern base of the Andes in n. Ecuador in Sucumbíos, Napo, Pastaza, and Morona-Santiago. *Berlepschi* is found southward along the eastern base of the Andes; it thus far has been recorded only from Morona-Santiago on the Cordillera de Cutucú (ANSP) and west of Macas in the lower Río Upano valley of Sangay National Park (RSR and R. A. Rowlett), perhaps occurring southward as well (though it appears to be "replaced" in the latter area by the White-breasted Parakeet [*P. albipectus*]). *Berlepschi* is similar to *souancei* but has even broader whitish breast scaling, such that in some individuals this area is almost solidly white. Finally, on the Pacific slope is found the very distinct *pacifica*, which resembles the nominate race (i.e., it has narrow breast scaling, though the breast is heavily suffused with a pinkish wash) but differs in its shorter tail and grayish (not white) periophthalmic ring. *Pacifica* may deserve full species status (Chocó Parakeet).

Range: se. Colombia and sw. Venezuela to ne. Peru and w. Amazonian Brazil; sw. Colombia and nw. Ecuador; locally in Andes of s. Colombia (same species?).

*Pyrrhura orcesi   El Oro Parakeet   Perico de Orcés

Uncommon to locally fairly common in the canopy and borders of montane forest in the foothills of sw. Ecuador in Azuay (north to above Manta Real, where a few were seen in Aug. 1991; Parker and Carr 1992, RSR and G. H.

Rosenberg) and El Oro (south to Buenaventura). This only recently described species (R. S. Ridgely and M. B. Robbins, *Wilson Bull.* 100[2]: 173–182, 1988) is endemic to a very limited area in sw. Ecuador. Recorded mainly between 600 and 1200 m. Apparently the El Oro Parakeet formerly occurred lower—a specimen (BMNH) was obtained by L. Gómez on 9 Sep. 1939 at Piedras (300 m) in El Oro—but all suitable forest habitat for this species has long since been removed from this region. Small numbers also occur locally, or perhaps seasonally, at somewhat higher elevations; N. Krabbe has twice seen small flocks as high as 1500–1550 m. It was suggested (Collar et al. 1992) that where forest habitat still exists the species might still range down into the lowlands, but if so, this has not been confirmed.

The El Oro Parakeet is threatened by continued clearing of what remains of its highly fragmented cloud-forest habitat. Until recently no portion of its range received formal protection, but in 1999 Fundación Jocotoco began to purchase and protect tracts of land at and around the species' type locality, at Buenaventura in El Oro. It was accorded Vulnerable status by Collar et al. (1994), an assessment with which we agree.

Monotypic.

Range: west slope of Andes in sw. Ecuador.

Pyrrhura albipectus   **White-breasted Parakeet**   Perico Pechiblanco

Uncommon to locally fairly common in the canopy and borders of montane forest in the foothills and lower subtropical zone on the east slope of the Andes in s. Ecuador in nw. Morona-Santiago (Cordillera de Cutucú; Robbins et al. 1987) and in s. Zamora-Chinchipe at various sites in Podocarpus National Park (especially the Bombuscaro and Romerillos sectors), along the lower Loja-Zamora road, on the Cordillera del Cóndor at Chinapinza (Sep. 1990 specimens [ANSP and MECN]; Krabbe and Sornoza 1994) and Pachicutza (specimen [WFVZ] taken on 7 Aug. 1989), and at Panguri (Jul. 1992 specimens [ANSP and MECN]). The White-breasted Parakeet went unrecorded for more than 50 years subsequent to its description in 1914; it was rediscovered in Feb. 1980 when several small flocks were seen along the Loja-Zamora road below Sabanilla and in the Río Jamboé valley south of Zamora (RSR and D. Finch et al.). The species seems less numerous toward the northern edge of its range, on the Cordillera de Cutucú, than it does southward. Thus far an Ecuadorian endemic, the White-breasted Parakeet likely occurs in adjacent Peru as well. Recorded mostly between 900 and 1800 m, perhaps with some seasonal elevational movements.

Until recently the White-breasted Parakeet was thought to be a rare species (e.g., Collar and Andrew 1988) and was even given Vulnerable status (Collar et al. 1992, 1994). In recent years, however substantial numbers have been found at several sites in Zamora-Chinchipe (v.o.), including various localities in and near Podocarpus National Park. Given the still extensive amount of forest remaining in its range, and the protected status of a significant fraction of it, the White-breasted Parakeet in our view should not be considered to be gravely at risk, and we therefore accord it only Near-threatened status. The extent of its forested habitat is considerably greater, for instance, than it is for the El Oro Parakeet (*P. orcesi*). E. P. Toyne, M. T. Jeffcote, and J. N. Flanagan (*Bird Conserv. Int.* 2[4]: 327–338, 1992) discuss the status, distribution, and ecology of the White-breasted Parakeet in Podocarpus National Park, which remains the only area in the species' range to receive formal protection, and in which its populations are fairly large.

The species is sometimes called the White-necked Parakeet (e.g., Forshaw 1989).

Robbins et al. (1987) discuss the possibility of limited hybridization occurring between *P. albipectus* and *P. melanura berlepschi* (Maroon-tailed Parakeet) on the Cordillera de Cutucú. There still is not enough specimen material to make a definitive statement regarding this situation; however, the series of specimens obtained in the early 1990s from several sites in Zamora-Chinchipe (ANSP, MECN, WFVZ) are all morphologically pure *P. albipectus*. Monotypic.

Range: east slope of Andes in se. Ecuador.

*[Pyrrhura picta   Painted Parakeet   Perico Pintado]

Uncertain. Known in Ecuador only from sightings of "several flocks of 10–15 individuals along the Río Macuma" in the lowlands of se. Ecuador in n. Morona-Santiago on 1 Feb. 1972 (D. L. Pearson, *Condor* 77[1]: 98, 1975). The closest localities from which the Painted Parakeet is known are in the drainages of the Río Cenepa and the Río Santiago in n. Amazonas, Peru (north of the Río Marañón and east of the Cordillera del Cóndor), where at least nine specimens (FMNH, LSUMZ) were taken in the 1960s and 1970s. Photographic or specimen confirmation of the species occurring as far north as Ecuador is needed. Note that the statement (Juniper and Parr 1998) that the Painted Parakeet occurs to 1200 m in Ecuador is in error.

Presumed race: *lucianii* (which has a dusky brown crown).

Range: s. Venezuela, the Guianas, Amazonian Brazil, extreme ne. and nw. Bolivia, and e. Peru; isolated populations in n. Colombia, nw. Venezuela; once in se. Ecuador. More than one species is almost certainly involved.

*Bolborhynchus lineola   Barred Parakeet   Perico Barreteado

Uncommon—but until recently overlooked—in montane forest and forest borders in the subtropical and temperate zones on both slopes of the Andes. On the west slope—where it may be more numerous, especially northward— there are only a few reports from south of Pichincha; these include flocks seen in w. Cotopaxi (above Pilaló on 28 Jul. 1991; RSR and G. H. Rosenberg), Bolívar (northwest of Salinas; N. Krabbe), Azuay (Sural and Chaucha; N. Krabbe), and Loja (once in 1991 west of Celica; P. Coopmans). On the east slope the only reports from south of the Cordillera de Huacamayos and

Hacienda San Isidro area in w. Napo (v.o.) are a few sightings from along the Gualaceo-Limón road in nw. Morona-Santiago and in the Cajanuma sector of Podocarpus National Park (Bloch et al. 1991; N. Krabbe); the species was also seen once at Urutisinga near the last locality (Rasmussen et al. 1996). Details of the first Ecuadorian sighting—a group of six birds seen above Mindo on 25 Jun. 1979 by RSR, PJG, and R. A. Rowlett—were given by Ridgely (1980). There still are no Ecuadorian specimens of the Barred Parakeet, but the species is now seen regularly in several areas, almost always in flocks flying high overhead; only rarely is it actually seen perched. Recorded between about 1700 and 3100 m.

Presumed race: *tigrinus*, which is the subspecies known from both north and south of Ecuador.

Range: locally in highlands from s. Mexico to w. Panama; mountains of n. Venezuela, and in Andes from w. Venezuela to s. Peru; Santa Marta Mountains.

**Forpus xanthopterygius* Blue-winged Parrotlet Periquito Aliazul

Uncommon and rather local in the canopy and borders of várzea and riparian forest and woodland, and in clearings with scattered trees, in the lowlands of e. Ecuador. Although surely present earlier, the Blue-winged Parrotlet was not first recorded in Ecuador until numerous records were made at Limoncocha and nearby areas in the mid-1970s (Tallman and Tallman 1977; RSR). It has since spread into cleared areas around Lago Agrio, Coca, Tena, and Puyo, but farther east (where forest cover remains more continuous) the species remains decidedly more local, occurring mainly along larger rivers; this is also the case in se. Ecuador. Recorded mostly below 500 m, but ranges up locally to 1000 m in the foothills on the east slope of the Andes (including a male taken at Pachicutza in Morona-Santiago on 8 Aug. 1989; WFVZ).

This species long went by the name of *Forpus xanthopterygius* (e.g., Meyer de Schauensee 1966, 1970), but *crassirostris* was proposed as the applicable name (Pinto 1978; Stotz et al. 1996). More recently, however, B. M. Whitney and J. F. Pacheco (*Bull B. O. C.* 119[4]: 211–214, 1999) demonstrated that *xanthopterygius* was, after all, the correct name for the species. One race: *crassirostris*.

Range: n. Colombia; e. Ecuador, e. Peru, extreme se. Colombia, n. Bolivia, and w. Amazonian Brazil; e. Brazil, extreme e. Bolivia, e. Paraguay, and ne. Argentina.

Forpus coelestis Pacific Parrotlet Periquito del Pacífico

Fairly common to very common and widespread in deciduous forest and woodland, arid (even desert) scrub, cleared and agricultural areas, and gardens in the more arid lowlands of w. Ecuador north mainly to n. Manabí and Pichincha. Since 1990, Pacific Parrotlets have also been seen in w. Esmeraldas in the Muisne and Atacames region, as well as around Quinindé, and in Sep. 1995 they were observed by P. Coopmans as far north as the San Mateo and Tachina area east of Esmeraldas (city); in 1997 a flock was seen as far north as the Borbón area (O. Jahn). There are no early records of the Pacific Parrotlet from Esmeraldas (Chapman 1926), and it seems likely to have only recently spread this far north, an expansion doubtless abetted by ongoing deforestation in many areas; we predict that the species may eventually even spread into adjacent Nariño in sw. Colombia. In fact, S. Hilty (pers. comm.) now suspects that the *Forpus* he reported seeing near Tumaco in sw. Nariño (Hilty and Brown 1986) may well have been the Pacific Parrotlet (and not, as reported, the Spectacled Parrotlet [*F. conspicillatus*]). The Pacific Parrotlet is also now spreading northward in n. Pichincha, in 1997 having been seen nearly as far as adjacent Imbabura (P. Coopmans). Although primarily found in drier regions, and often most numerous where conditions are arid or even desert-like (e.g., on the Santa Elena Peninsula), the Pacific Parrotlet also ranges locally into areas of more humid forest, such as that found at Río Palenque and Tinalandia; at least for the most part it occurs in such humid habitats where these are fragmented. The species has also been known since at least the 1970s from certain arid parts of the upper Río Marañón valley in nw. Peru, and it seems possible that it may eventually spread north into the Zumba region, where up to now no *Forpus* parrotlet has been found. Recorded mostly below 800 m, but in w. Loja ranges up in partially cultivated terrain to about 1500 m, rarely or locally as high as 2000 m around Celica.

Monotypic.

Range: w. Ecuador and nw. Peru.

Fjeldså and Krabbe (1990, p. 212) record birds considered as "possibly" the Spectacled Parrotlet being seen at three sites in the highlands of Loja. Although their actually being that species seems highly improbable, we are at a loss to identify the parrot that these experienced observers did see. During much recent field work by many workers in the province, the only *Forpus* seen there has been the common Pacific Parrotlet.

Forpus sclateri   **Dusky-billed Parrotlet**   **Perico Piquioscuro**

Rare to uncommon and local in the canopy and borders of humid forest (both terra firme and várzea, but especially the former), secondary woodland, and adjacent clearings in the lowlands of e. Ecuador. Ranges up at least locally to about 900–1000 m along the eastern base of the Andes (Bermejo oil-field area north of Lumbaquí [ANSP], an old specimen [AMNH] from Zamora, and recent reports from Shaime [Balchin and Toyne 1998]) but mainly found below 500 m.

The species is sometimes called the Sclater's Parrotlet (e.g., Forshaw 1989). One race occurs in Ecuador, nominate *sclateri*.

Range: locally in s. Venezuela and n. Guyana; se. Colombia, e. Ecuador, e. Peru, n. Bolivia, and w. Amazonian Brazil; locally in e. Amazonian Brazil and French Guiana.

Brotogeris pyrrhopterus   Gray-cheeked Parakeet   Perico Cachetigris

Uncommon to locally fairly common in deciduous and semihumid forest and woodland, agricultural terrain where a few large trees are left standing, and gardens and parks in the lowlands of sw. Ecuador from cen. Manabí (north to around Bahía de Caráquez and Flavio Alfaro; a Nov. 1990 specimen [ANSP] from as far north as the north side of Bahía de Caráquez near San Vicente) and Los Ríos (Quevedo area) south through El Oro and w. Loja (inland to the Sozoranga area). The Gray-cheeked Parakeet ranges locally into more humid forest (e.g., above Manta Real in nw. Azuay) but has not been found in the more humid forests of nw. Ecuador (e.g., it has never been recorded from Río Palenque). In addition, and inexplicably, the Gray-cheeked Parakeet seems to be absent from seemingly ideal terrain in sw. Manabí (e.g., at Machalilla National Park) and w. Guayas. Mainly recorded below about 1000 m, though very small numbers have been noted as high as about 1500 m in El Oro and Loja.

Although concern has been expressed regarding the status of this species (B. J. Best, N. Krabbe, C. T. Clarke, and A. L. Best, *Bird Conserv. Int.* 5[2–3]: 233–250, 1995), in our view the numbers of the Gray-cheeked Parakeet have held up relatively well; we thus agree with Collar et al. (1994) in considering it no more than Near-threatened. The Gray-cheeked Parakeet persists well in many mostly deforested agricultural areas, and pairs are even resident in some of Guayaquil's most congested city parks. Numbers of the Red-masked Parakeet (*Aratinga erythrogenys*), by contrast, have declined much more over the past several decades. Nonetheless, the Gray-cheeked Parakeet's status should continue to be monitored, particularly because large numbers apparently continue to be captured for the pet market; some of these are evidently smuggled into Peru and then re-exported from there.

Monotypic.
Range: sw. Ecuador and extreme nw. Peru.

Brotogeris cyanoptera   Cobalt-winged Parakeet   Perico Alicobáltico

Common and widespread in the canopy and borders of humid forest (both terra firme and várzea), secondary woodland, and partially cleared areas with scattered large trees left standing in the lowlands of e. Ecuador. In most or all areas in the Oriente the Cobalt-winged Parakeet is the most numerous parrot. Recorded mostly below about 600 m, in small numbers at least occasionally as high as 900–1000 m (e.g., along the Loreto road [v.o.], in the Bombuscaro sector of Podocarpus National Park [P. Coopmans], and in the Río Nangaritza valley at Shaime [Balchin and Toyne 1998] and Miazi [Schulenberg and Awbrey 1997]).

One race: nominate *cyanoptera*.
Range: e. Colombia and sw. Venezuela to n. Bolivia and w. Amazonian Brazil.

The White-winged Parakeet (*B. versicolor*; formerly called the Canarywinged Parakeet), found along the Amazon River and some of its tributaries, has often been credited to the avifauna of Ecuador (e.g., Meyer de Schauensee 1966, 1970; AOU 1998). However, the only published reference to this species' presence in the country (Goodfellow 1902) indicated that it was not found within the present boundaries of Ecuador ("unknown on the upper parts of the [Napo] River"). Although it remains possible that the species will be recorded within Ecuador's present borders, unlike many other river-island specialists it has never been found. Note that *B. chiriri* (Golden-chevroned Parakeet) of s.-cen. South America is now usually regarded as a separate species.

[**Brotogeris sanctithomae*   Tui Parakeet   Perico Tui]

Uncertain. Although in the literature sometimes credited to the avifauna of Ecuador (e.g., *Birds of the World*, vol. 3), the only published record (Goodfellow 1902) of this species actually refers to birds recorded outside the present boundaries of Ecuador (e.g., below the mouth of the Río Curaray); we suspect that the "Napo" record mentioned by Chapman (1926) also refers to a specimen obtained lower down on the river. Nonetheless a few recent sightings, all from the northeast, indicate that the Tui Parakeet does apparently occur in very small numbers (perhaps seasonally?) in várzea and riverine forest within the present borders of Ecuador; one possible explanation for the paucity of definite records is the presence in the same areas of such large numbers of the similar Cobalt-winged Parakeet (*B. cyanoptera*). There are sightings from along the Río Napo near Limoncocha (e.g., 3 were believed seen by S. Hilty et al. on 4 Aug. 1981) and at La Selva (e.g., a few were seen by V. Emanuel and P. English et al. in Feb. 1990, another by PJG and RSR et al. on 25 Mar. 1990). Photographic or specimen corroboration is still needed because of individual variation in the amount of yellow shown on the forecrowns of Cobalt-winged Parakeets. Recorded below 300 m.

Presumed race: nominate *sanctithomae*.

Range: mainly along Amazon River and its major tributaries in Brazil, extreme se. Colombia, n. Bolivia, e. Peru, and ne. Ecuador.

*Touit purpurata   Sapphire-rumped Parrotlet   Periquito Lomizafiro

Rare to uncommon in the canopy of humid forest (both terra firme and várzea) in the lowlands of e. Ecuador. The Sapphire-rumped Parrotlet is still known from only four sites in Ecuador. The first record was a male (MCZ) taken by T. Mena at Río Rutuno in Pastaza on 11 Mar. 1951 (Norton et al. 1972). Since then, small numbers have been found at Imuyacocha and Zancudococha in Dec. 1990, Mar.–Apr. 1992, and Jan. 1993 (including specimens at ANSP and MECN), and there are several reports from Cuyabeno. The species is perhaps more widespread; *Touit* parrotlets are notoriously difficult to detect and identify correctly. Recorded below 300 m.

One race: *viridiceps*.

Range: the Guianas, s. Venezuela, se. Colombia, e. Ecuador, ne. Peru, and locally in Amazonian Brazil.

Touit huetii   **Scarlet-shouldered Parrotlet**   **Periquito Hombrirrojo**

Rare to uncommon and local (perhaps mainly nomadic or erratic; locally or temporarily more numerous) in the canopy of terra firme forest in the lowlands of e. Ecuador. This small parrot, scarce throughout its wide range (but perhaps mainly just overlooked), is known from only a scatter of Ecuadorian sites. There is a century-old specimen from below Baeza (Goodfellow 1902); sightings of two flocks along the lower Río Yasuní on 18 Sep. 1976 (RSR); sightings of flocks near Macas in Morona-Santiago on 3 Aug. 1979 (RSR and R. A. Rowlett) and 30 Jun. 1984 (RSR and F. Gill); near Zancudococha on the lower Río Aguarico (flock of 15 seen on 17 Sep. 1992; RSR et al.); Hacienda Primavera (flock of 8–10 seen on 17 Oct. 1993; B. Gee et al.); La Selva; and along the Maxus road southeast of Pompeya. At La Selva the Scarlet-shouldered Parrotlet has been a regular feature at the "interior saladero" where, at least from Oct. to Apr., fairly large flocks (up to 50 or more) come to feed on mineral-impregnated soil and water. The earlier reports were discussed by Ridgely (1980). Recorded at least occasionally as high as 1400 m.

Monotypic.

Range: locally from n. Guyana and e. and s. Venezuela to se. Colombia, e. Ecuador, e. Peru, nw. Bolivia, and Amazonian Brazil.

Touit dilectissima   **Blue-fronted Parrotlet**   **Periquito Frentiazul**

Rare to locally (or perhaps seasonally) uncommon or even fairly common in the canopy and borders of montane forest in the foothills and lower subtropical zone on the west slope of the Andes south to El Oro (Buenaventura; Robbins and Ridgely 1990). The Blue-fronted Parrotlet also ranges down locally (perhaps seasonally?) into adjacent humid lowlands, where there are a few sightings from Río Palenque (none of them recent), a specimen (MECN) taken by C. Durán at Santo Domingo de los Colorados in Jan. 1967, and two specimens (ANSP) taken by T. Mena at Quevedo in Oct. 1950. It has also been recorded recently at Bilsa on the coastal Cordillera de Mache in w. Esmeraldas (R. P. Clay et al.), where a population may be resident. Recorded mostly from 500 to 1400 m, occasionally ranging down to as low as about 200 m or even lower (a nesting pair was observed at 50 m along the Mataje road in n. Esmeraldas near San Lorenzo in Aug. 1998; P. Coopmans). Exceptionally it also occurs as high as 2300–2500 m (3 specimens [MECN] obtained in nw. Cotopaxi at El Triunfo Grande on 21 Jun. 1998; F. Sornoza).

Numbers of all members of this genus are notoriously difficult to assess as they are naturally scarce and wide-ranging birds, extremely difficult to detect except in flight. We suspect that the Blue-fronted Parrotlet's numbers have declined substantially in Ecuador during recent decades as a result of deforestation and thus accord it Near-threatened status. The closely related Red-fronted Parrotlet (*T. costaricensis*) (see below) has also recently been given Near-threatened status (Collar et al. 1994).

T. costaricensis of Costa Rica and w. Panama is here treated as a separate species from *T. dilectissima*, in accord with most recent authors. If these two taxa are considered conspecific, the enlarged species is called the Red-winged Parrotlet, *T. dilectissima*. Monotypic.

Range: e. Panama to sw. Ecuador; nw. Venezuela and ne. Colombia.

Touit stictoptera   **Spot-winged Parrotlet**   **Periquito Alipunteado**

Rare to locally uncommon in the canopy and borders of montane forest in the subtropical zone on the east slope of the Andes. There are rather few records of this small parrot from Ecuador (as likewise throughout its small range), but this is probably less a reflection of its actual rarity than of the difficulty in detecting it. To date the Spot-winged Parrotlet has been found in only three regions in Ecuador: at San Rafael Falls and in the Baeza/Cordillera de Huaca-mayos area in w. Napo, on the Cordillera de Cutucú (ANSP; Robbins et al. 1987) and in the upper Río Upano valley west of Macas in w. Morona-Santiago (several small flocks seen in Aug. 1979; RSR and R. A. Rowlett), and in Zamora-Chinchipe at Panguri (sightings in Jul. 1992; T. J. Davis) and on the Cordillera del Cóndor (sightings in Jul. 1993; Schulenberg and Awbrey 1997). Presumably its distribution is actually more continuous. There is also an anom-alous record of a female collected on 25 Nov. 1930 at Aucayacu in the low-lands of far e. Ecuador close to the Peruvian frontier (E. Stresemann, *Ornith. Monatsber.* 46[4]: 115–116, 1938); as the species has not otherwise been found in the lowlands anywhere near so far away from the Andes, we suspect, however, that mislabeling may have been involved. Recorded mostly from 1100 to 1800 m.

The Spot-winged Parrotlet was accorded Vulnerable status by Collar et al. (1994). We suspect that its supposed rarity may be more apparent than real, and therefore suggest that it only merits a status of Near-threatened, at least in Ecuador.

Monotypic.

Range: locally on east slope of Andes from cen. Colombia to n. Peru.

Pionites melanocephala   **Black-headed Parrot**   **Loro Coroninegro**

Fairly common and widespread in the canopy and borders of humid forest (both terra firme and várzea) in the lowlands of e. Ecuador. Recorded up only to about 400 m.

The species is sometimes called the Black-headed Caique (e.g., Forshaw 1989).

One race: *pallida*. Some Ecuadorian examples in ANSP show an indication of intergradation with the nominate race (which ranges to the north and east), notably in their having some orange intermixed in the yellow on lower flanks and thighs; these include a specimen recently obtained at Imuyacocha (though others from that site appear to be typical of *pallida*) and one from Montalvo.

Range: the Guianas, e. and s. Venezuela, se. Colombia, e. Ecuador, ne. Peru, and n. Amazonian Brazil (north of the Amazon River).

The White-bellied Parrot (*P. leucogaster*) is sometimes credited to Ecuador's avifauna (e.g., Meyer de Schauensee 1966, 1970), but this was apparently based on a misunderstanding of its actual range; the White-bellied Parrot only occurs *south* of the Amazon River and Río Marañón.

Pionopsitta pulchra   Rose-faced Parrot   Loro Cachetirrosa

Uncommon to locally fairly common in the canopy and borders of very humid to humid forest in the lowlands and (perhaps especially) foothills of w. Ecuador south to n. Manabí, s. Pichincha (Río Palenque), and El Oro (old specimens from El Chiral; recent sightings from Buenaventura). The Rose-faced Parrot is now most numerous in Esmeraldas and adjacent Pichincha. Recorded in small numbers up to 1200–1300 m.

The Rose-faced Parrot has doubtless declined substantially because of widespread deforestation, and it no longer occurs regularly at Tinalandia, where formerly it was not uncommon. Small numbers do, however, seem capable of persisting in some regions where forest cover has become quite fragmented. In the wet-forest belt of n. Esmeraldas local residents consider the Rose-faced Parrot to be a pest because it eats bananas, one of the most important human foods in the region; it therefore is shot by the hundreds annually (P. Mena Valenzuela and O. Jahn et al.) and could soon become threatened. The species probably does not yet deserve formal threatened status in Ecuador, though it clearly merits further monitoring. It was not listed by Collar et al. (1994).

The species has sometimes been called the Beautiful Parrot (e.g., Meyer de Schauensee 1966).

P. pulchra is here considered a species separate from *P. haematotis* (Brownhooded Parrot) of nw. Colombia and Middle America, in agreement with most recent authors. Monotypic.

Range: w. Colombia and w. Ecuador.

*[Pionopsitta pyrilia   Saffron-headed Parrot   Loro Cabeciazafrán]

Uncertain. There are two recent reports of this striking parrot from the canopy and borders of humid forest in the lowlands of the northwest. A pair was reported seen at Salto de Bravo in n. Esmeraldas, near the junction of the Ríos Cayapas and San Miguel in the lower part of Cotocachi-Cayapas Ecological Reserve, in early Oct. 1992 (J. C. Matheus). A flock of about 20 birds was subsequently reported seen west of Pedro Vicente Maldonado in nw. Pichincha on 8 Feb. 1998 (F. Sornoza). These represent by far the southernmost records of the species, which heretofore was known from Valle northward in the Pacific lowlands of Colombia (Hilty and Brown 1986); possibly only wandering birds were involved. To 450 m.

Given the obviously limited Ecuadorian range of the Saffron-headed Parrot—
if indeed it occurs at all on a regular basis—and given the deforestation that is
occurring in much of northern lowland Esmeraldas, we feel we must assess the
species status here as Data Deficient. The species was not mentioned by Collar
et al. (1992, 1994).

Monotypic.

Range: e. Panama to nw. Venezuela and n. Colombia; nw. Ecuador.

Pionopsitta barrabandi   Orange-cheeked Parrot   Loro Cachetinaranja

Uncommon to fairly common in the canopy and borders of humid forest
(mainly terra firme, but locally also into várzea) in the lowlands of e.
Ecuador. Recorded mainly below 400 m, but occasionally or locally occurs
higher (e.g., to 900 m north of Lumbaquí in the Bermejo oil-field area; M. B.
Robbins).

The species is sometimes called the Barraband's Parrot (e.g., Forshaw
1989).

One race: nominate *barrabandi*. In the limited amount of specimen
material at hand in ANSP, we cannot see any indication of intergradation
with *aurantiigena*, contra Forshaw (1989) who suggested that this may occur
in Ecuador. *Aurantiigena* would appear to range only south of the Amazon
River.

Range: se. Colombia and s. Venezuela to n. Bolivia and w. Amazonian Brazil.

[Hapalopsittaca amazonina   Rusty-faced Parrot   Loro Carirrojizo]

Uncertain. Apparently very rare in the canopy and borders of montane forest
in the temperate zone on the east slope of the Andes in extreme n. Ecuador.
Known from only a single sighting: a pair of *Hapalopsittaca* parrots, presumed
by range to be this species, was seen by G. H. Rosenberg on 22 Mar. 1992 on
the west slope of Cerro Mongus in e. Carchi. The Rusty-faced Parrot had not
previously been known from farther south than the upper Río Magdalena valley
in Colombia (Hilty and Brown 1986). Recorded at about 3300 m.

The Rusty-faced Parrot was considered Vulnerable by Collar et al. (1992) and
was upgraded to Endangered by Collar et al. (1994). There is virtually no
Ecuadorian information on which to base a judgment, so we consider the species
to be Data Deficient. What is known is that the forest at Cerro Mongus continues to be reduced in extent.

Does not include *H. pyrrhops* (Red-faced Parrot). Race unknown.

Range: locally in Andes from w. Venezuela to extreme n. Ecuador.

Hapalopsittaca pyrrhops   Red-faced Parrot   Loro Carirrojo

Rare to locally uncommon, but apparently quite nomadic, in the canopy and
borders of montane forest in the temperate zone of the Andes in s. Ecuador,
especially on interandean (as opposed to outside) slopes. Recorded north to s.

Cañar (Hacienda La Libertad, where a flock of 6 was seen on 20 Aug. 1991; P. Stafford) and w. Morona-Santiago (in the upper Río Palora valley of Sangay National Park, where groups of 2–4 birds were seen on 12–14 Oct. 1976; RSR) and south to the east slope of the Cordillera Las Lagunillas (a sighting of a pair in Oct. 1992; T. J. Davis). The Red-faced Parrot seems never to have been a numerous bird, and it remains known from only a handful of specimens, the only recent ones being single individuals (ANSP) taken on the Cordillera de Cordoncillo southeast of Saraguro in n. Loja on 31 Mar. 1992 and at Cajanuma in Podocarpus National Park on 3 Aug. 1992. Recorded mostly from 2700 to 3500 m, locally (perhaps just seasonally?) as low as 2400 m (Arenales on the Azuay/Morona-Santiago border, where 4 were found on 13 Feb. 1993; N. Krabbe).

Overall numbers of the Red-faced Parrot have declined substantially as a result of deforestation, though small flocks continue to be seen at a number of localities that retain at least some forest. A population is found in west-slope forests from Cajanuma south to above Yangana in the relatively well protected Podocarpus National Park, but the species appears to be most numerous in the unprotected and increasingly deforested Cordillera de Chilla in n. Loja (Rasmussen et al. 1996). It was considered Endangered by Collar et al. (1994), an assessment with which we agree.

H. pyrrhops is regarded as a monotypic species distinct from *H. amazonina* (Rusty-faced Parrot) found farther north in the Andes, following G. R. Graves and D. U. Restrepo (*Wilson Bull.* 101[3]: 369–376, 1989).

Range: Andes of s. Ecuador and extreme n. Peru.

Graydidascalus brachyurus   Short-tailed Parrot   Loro Colicorto

Rare to uncommon and apparently somewhat local (but conspicuous and noisy) in the canopy and borders of várzea and riverine forest and around lakes in the lowlands of ne. Ecuador. The Short-tailed Parrot is known in small numbers, perhaps occurring only seasonally (?), along the Río Napo upriver to Pompeya (see Ridgely 1980), the Río Coca (Goodfellow 1902), and the Río Jivino (specimen [MECN] taken on 7 Dec. 1971 by C. Durán). Larger numbers occur along the lower Río Aguarico upriver to the Zancudococha/Imuyacocha region, and the species certainly is resident in this area. Short-tailed Parrots have not been found in the Río Pastaza drainage of the southeast. Recorded below about 300 m.

Monotypic.

Range: mainly along Amazon River and some of its major tributaries in Brazil, extreme s. Colombia, ne. Peru, and ne. Ecuador; also French Guiana.

Pionus menstruus   Blue-headed Parrot   Loro Cabeciazul

Fairly common to common and widespread in the canopy and borders of humid forest, secondary woodland, clearings with scattered large trees, sometimes in gardens or raiding fields, and locally even in mangroves in the lowlands of both

e. and w. Ecuador. In the west recorded south to Manabí, n. Guayas, and El Oro (Santa Rosa, west of La Avanzada, and Puerto Pitahaya). The Blue-headed Parrot is most numerous in the east and now is rather local in much of the western parts of its range because of deforestation and local persecution. Recorded mostly below about 1100 m, but locally or in small numbers up to 1400 m (e.g., along the lower Loja-Zamora road; E. P. Toyne and M. T. Jeffcote, *Bull. B. O. C.* 114[2]: 124–127, 1994).

Two races are found in Ecuador, nominate *menstruus* in the east and *rubrigularis* in the west. The head and neck color of *rubrigularis* is a somewhat less deep and intense blue, and it has an obscure patch of red feathers on the lower throat (lacking in the nominate race).

Range: Costa Rica to nw. Venezuela, n. and w. Colombia, and sw. Ecuador; the Guianas, e. and s. Venezuela, and e. Colombia to n. Bolivia and Amazonian Brazil; e. Brazil.

Pionus sordidus   **Red-billed Parrot**   Loro Piquirrojo

Uncommon to fairly common in the canopy and borders of montane forest and (less often) in adjacent clearings in the subtropical zone on both slopes of the Andes. On the east slope recorded from Sucumbíos in the San Rafael Falls area south through Zamora-Chinchipe. On the west slope recorded from Pichincha south locally to El Oro (sightings from the Piñas and Uzhcurrumi areas [RSR and PJG; Best et al. 1993]). Recorded mostly from 1200 to 2300 m, occasionally down to 1000 m, and locally as high as 2500–2600 m (Quebrada Honda; L. Navarrete and RSR et al.).

One race: *corallinus*. We concur with most recent authors—including Forshaw (1989)—that *mindoensis*, to which the birds of w. Ecuador have been assigned, is unworthy of recognition.

Range: mountains of n. Venezuela and n. Colombia; Andes from s. Colombia to n. Peru, and in w. Bolivia.

Pionus seniloides   **White-capped Parrot**   Loro Gorriblanco

Rare to locally (and seasonally?) fairly common in the canopy and borders of montane forest in the subtropical and temperate zones on both slopes of the Andes. On the west slope recorded south at least formerly to El Oro (old records from Zaruma), but in recent years the only reports from south of w. Cotopaxi are 1995 sightings (N. Krabbe) from Bolívar (northwest of Salinas) and Azuay (Chaucha). Recorded mostly from 1500 to 3200 m, small numbers occasionally somewhat lower and higher; rarely it even overflies paramo.

The species is sometimes called the White-headed Parrot (e.g., Forshaw 1989).

We follow Forshaw (1989) and Fjeldså and Krabbe (1990) in continuing to treat the northern *P. seniloides* as a monotypic species distinct from *P. tumultuosus* (Plum-crowned Parrot) of cen. Peru to Bolivia. J. P. O'Neill and T. A. Parker III (*Condor* 79[2]: 274, 1976) had suggested treating *seniloides* as a subspecies of *P. tumultuosus*, based on what they considered to be intergradation between the two taxa in n. Peru. If considered to be conspecific—which they may prove to be—the enlarged species would be called the Speckle-faced Parrot.

Range: Andes from w. Venezuela to n. Peru.

Pionus chalcopterus   **Bronze-winged Parrot**   **Loro Alibronceado**

Uncommon to locally common in the canopy and borders of humid and semi-humid forest and woodland and in adjacent clearings with scattered large trees in the lowlands and foothills of w. Ecuador south to Guayas and w. Loja (south to the Alamor region). The Bronze-winged Parrot tends to be most numerous in humid foothill regions. Recorded mostly below 1400 m, ranging locally up to about 1650 m (e.g., near Chical in Carchi; M. B. Robbins).

The Bronze-winged Parrot persists relatively well in fragmented forest and secondary woodland, and thus it seems to have been less affected by deforestation than some other parrots; even so, overall numbers have doubtless declined substantially over the past several decades. At least formerly large flocks occurred around Tinalandia, though there seems to have been a recent decline there (P. Coopmans). For now neither we nor Collar et al. (1994) consider it Near-threatened, but its status bears monitoring in the future.

Monotypic. We concur with Forshaw (1989) that *cyanescens*, to which Ecuadorian birds have sometimes been assigned, is not worthy of recognition.

Range: locally in Andes from w. Venezuela to extreme nw. Peru; Sierra de Perijá.

Amazona autumnalis   **Red-lored Amazon**   **Amazona Frentirroja**

Rare to locally fairly common in the canopy and borders of humid and deciduous forest and mature secondary woodland (locally also in mangroves) in the lowlands of w. Ecuador from Esmeraldas (sightings from San Lorenzo [F. Vuilleumier], El Placer [RSR and M. B. Robbins et al.], southeast of Muisne and Bilsa [v.o.], the Same-Súa area [P. Coopmans], and near Quinindé [P. Coopmans]) south very locally through Guayas to coastal El Oro at Arenillas (a 1939 specimen in the Mejía collection) and Santa Rosa (where recorded by S. Hilty et al. on 11 Aug. 1990). In most of w. Ecuador the Red-lored Amazon is outnumbered by the Mealy Amazon (*A. farinosa*), though in se. Guayas (e.g., around Naranjal) this situation is reversed, with the Red-lored being more numerous. In the wettest parts of Esmeraldas the Red-lored Amazon is usually not present at all (it was never recorded at Playa de Oro in 1995–1998 survey work; O. Jahn and P. Mena Valenzuela et al.), perhaps occurring as a scarce nonbreeding visitor in some (drier?) years. Recorded mostly below 700 m, but ranges very locally (perhaps only seasonally?) as high as 1300 m (e.g., in the Mindo region; Kirwan et al. 1996).

Overall numbers of the Red-lored Amazon in Ecuador have declined substantially as a result of deforestation and persecution. Some persist on the slopes

of the Cordillera de Colonche, but the largest surviving population apparently
now exists in se. Guayas, where it remains unprotected. Our judgement is that
the species merits Vulnerable status in Ecuador. As various large populations in
the species' large range are not at risk, the Red-lored Amazon was not mentioned by Collar et al. (1994).

This species and all members of the genus *Amazona* often go by the group
name "parrot," but we prefer to use the distinctive group name "amazon"—
long in avicultural use—for the genus; see Forshaw (1989).

One race: the endemic *lilacina*.

Range: n. Mexico to n. and w. Colombia, extreme nw. Venezuela, and sw.
Ecuador. Excludes Diademed Amazon (*A. diadema*) found in a limited area of
n. Amazonian Brazil.

Amazona festiva   Festive Amazon   Amazona Festiva

Locally uncommon to fairly common in the canopy and borders of várzea and
riparian forest and woodland in the lowlands of e. Ecuador, mainly in the
northeast in e. Napo and Sucumbíos along the Ríos Napo and Aguarico and
some of their tributaries. The first record of the Festive Amazon from Ecuador
comes from a male and a female collected in 1936 by the "Mision Flornoy" at
"Laguna de Siguin" (Berlioz 1937c); Siguin cannot be precisely located but is
thought to be somewhere along the north bank of the Río Pastaza. The first
report from the northeast involved a sighting of at least four individuals at a
large *Amazona* roost on an island in the Río Napo below San Roque on 25
Mar. 1990 (RSR and PJG et al.); subsequently birds have been seen here and
nearby on a number of occasions by various observers, but none have been
found farther upstream on the Napo. The Festive Amazon was then discovered
to be fairly numerous and nesting in woodland fringing Imuyacocha (especially)
and Zancudococha in Dec. 1990 and Mar.–Apr. 1991 (one specimen, ANSP).
It has also been seen at Cuyabeno and along the Río Pacuyacu. Recorded below
about 250 m.

One race: nominate *festiva*.

Range: locally along Amazon River and some of its larger tributaries in Brazil,
extreme se. Colombia, ne. Peru, and ne. Ecuador; locally along Orinoco River
in Venezuela and extreme ne. Venezuela.

Amazona ochrocephala   Yellow-crowned   Amazona
                           Amazon   Coroniamarilla

Uncommon to locally fairly common in the canopy and borders of humid forest
(both terra firme and várzea, though especially the former) in the lowlands of
e. Ecuador. A few 19th-century records from the lowlands of w. Ecuador in
Guayas (Naranjito and Balzar; see Chapman 1926) appear likely to refer to
cagebirds, the species having long been regarded as an excellent pet and therefore transported widely, even to areas well outside its normal range. There have
been no subsequent records from the west. Recorded up only to about 400 m.

The Yellow-crowned Amazon is apparently more sensitive to habitat disturbance than the other amazons found in e. Ecuador. Trapping for the pet market
doubtless has also played a role in its evident decline, as this species is considered to be the best "talking" Ecuadorian parrot. Despite the continued existence of considerable forest, populations of it have, for instance, declined
substantially in the Jatun Sacha area (B. Bochan). It remains most numerous in
regions where forest cover remains extensive, such as along the Maxus road
southeast of Pompeya (RSR et al.).

Both *A. oratrix* (Yellow-headed Amazon) of Mexico and *A. auropalliata*
(Yellow-naped Amazon) of Central America are here treated as separate species,
following most recent authors. Formerly (e.g., Meyer de Schauensee 1966,
1970), when all were considered conspecific under the name *A. ochrocephala*,
the English name of Yellow-headed Amazon was usually employed. One race
of *A. ochrocephala* occurs in Ecuador, *nattereri*.

Range: Panama to n. Bolivia and Amazonian Brazil.

Amazona amazonica   Orange-winged Amazon   Amazona Alinaranja

Fairly common to locally common in the canopy and borders of várzea and
riparian forest and woodland and in adjacent cleared areas with scattered large
trees in the lowlands of e. Ecuador. The Orange-winged Amazon is most numerous along larger rivers, perhaps especially the Río Napo, where it roosts on
certain islands (sometimes in large numbers) and scatters out soon after day-
break to feed (it is the first parrot to start flying after dawn, often becoming
active almost before first light). Recorded mostly below 500 m; early records
from higher can perhaps be ascribed to captive birds.

One race: nominate *amazonica*.

Range: n. Colombia; e. Colombia, much of Venezuela, and the Guianas to n.
and e. Bolivia and Amazonian and cen. Brazil (but absent from most of sw.
Amazonia); locally in e. Brazil.

Amazona mercenaria   Scaly-naped Amazon   Amazona Nuquiescamosa

Uncommon in the canopy and borders of montane forest in the subtropical
and temperate zones on both slopes of the Andes; on the west slope recorded
south locally to Azuay (sightings at Sural and Chaucha; N. Krabbe). We regard
the anomalous report from Limoncocha in the eastern lowlands (Tallman and
Tallman 1977) as uncorroborated. Recorded mostly from 1200 to 2600 m,
but occasionally ranging down to 950 m along the eastern base of the Andes
(e.g., in the Río Bombuscaro sector of Podocarpus National Park; Bloch et al.
1991) or up to 3000 m or even higher (e.g., several pairs seen repeatedly as they
overflew paramo at 3400 m in the Cordillera Las Lagunillas in Oct.–Nov. 1992
[RSR and M. B. Robbins et al.], and doing the same over the Gualaceo-Limón
road at 3300 m on 26 Mar. 1996 [P. Coopmans et al.]). In addition, ANSP
has a specimen labeled as having been taken at Loreto, which at 550 m would
be an exceptionally low elevation for the species; we presume that it was

actually obtained on the slopes of Volcán Sumaco well above the town of Loreto.

Two races of *A. mercenaria* are currently accepted (see Forshaw 1999), the nominate race of the Peruvian and Bolivian Andes and *canipalliata* of Colombia and Venezuela. Ecuadorian birds appear—though we have in fact been able to examine very little material—to be intermediate. The southern nominate race has a large red speculum on its inner secondaries, but in most examples (ANSP) from Colombia little or no red shows, and the tone of the red is duller and more maroon; in one east-slope ("Loreto") Ecuador specimen it is dull maroon. Likely there is a north-south cline toward increased red on the speculum, both in extent and brightness. Scaly-naped Amazons showing an obvious red speculum were seen above Valladolid in s. Zamora-Chinchipe on 24 Mar. 1996 (P. Coopmans) and again at nearby Quebrada Honda in Jan. 1998 (RSR et al.). Birds from this vicinity may therefore be more or less typical of nominate *mercenaria*, though as far as we are aware there is no specimen material from this region.

Range: Andes from w. Venezuela to nw. Argentina; Sierra de Perijá and Santa Marta Mountains.

Amazona farinosa   Mealy Amazon   Amazona Harinosa

Uncommon to locally common in the canopy and borders of humid forest (in the east mainly in terra firme and most numerous in areas with extensive forest) in the lowlands of both e. and w. Ecuador; considerably more numerous and widespread in the east. In the west recorded south to se. Guayas in the Naranjal area (v.o.) and at Manglares-Churute Ecological Reserve (Pople et al. 1997) and adjacent Azuay (seen at San Luis in Jan. 1991; J. C. Mattheus). Recorded mostly below about 700 m, but at least locally or occasionally ranges up to 900 m along the eastern base of the Andes north of Lumbaquí in the Bermejo oilfield area (M. B. Robbins).

Deforestation has resulted in a major decline in the Mealy Amazon's overall numbers in the west, where it is now very local in most areas. In addition, it is regularly hunted for food in n. Esmeraldas (P. Mena Valenzuela and O. Jahn et al.). Numbers are now greatest in Esmeraldas, where forest cover remains relatively extensive.

Two similar races occur in Ecuador, nominate *farinosa* in the east and *inornata* in the west. As noted by Forshaw (1989), *inornata* may not be worthy of recognition.

Range: s. Mexico to sw. Ecuador, n. Bolivia, and Amazonian Brazil; se. Brazil.

Deroptyus accipitrinus   Red-fan Parrot   Loro de Abanico

Apparently rare or at least very local in the canopy and borders of terra firme forest in the lowlands of se. Ecuador, where recorded only from s. Pastaza and adjacent n. Morona-Santiago. There are only a few records of this beautiful parrot from Ecuador. The first specimen was collected by M. Olalla at Andoas on 6 Jun. 1940 (Orcés 1944; specimen in the Mejía collection). Three more were obtained by A. Mena at Capitán Chiriboga, two on 10 Nov. 1972 and one on 18 Jul. 1974 (Orcés 1974; the two 1972 birds are in MECN, but we do not know the present whereabouts of the 1974 bird). It is believed that "Capitán Chiriboga" is situated at the mouth of the Río Capahuari into the Río Pastaza. It is possible that the "Andoas" in question refers to the site now within Peru (there are now two towns going by the name of "Andoas," one in Ecuador and one in Peru), from which the species has not otherwise been recorded. More recently (since 1996), small numbers of Red-fan Parrots have been found in the same general area around Kapawi Lodge along the Río Pastaza near the Peruvian border. Recorded at about 200 m.

The species is sometimes called the Hawk-headed Parrot (e.g., Forshaw 1989).

One race: nominate *accipitrinus.*

Range: e. Colombia, s. Venezuela, and the Guianas to cen. and e. Amazonian Brazil; se. Ecuador (adjacent Peru?).

Cuculiformes
Cuculidae   Cuckoos and Anis

Seventeen species in six genera. Cuckoos are worldwide in distribution, though the anis are found exclusively in the New World, principally in the Neotropics. We follow Gill (1994) in not separating the Cuculidae into five families as suggested by Sibley and Ahlquist (1990). In Ecuador these families would be the Coccyzidae (New World Cuckoos), the Crotophagidae (Anis), and the Neomorphidae (Ground-Cuckoos).

Coccyzus erythropthalmus   **Black-billed Cuckoo**   Cuclillo Piquinegro

A rare to perhaps briefly fairly common northbound transient in deciduous woodland and scrub and in shrubby clearings in w. Ecuador, mainly in the lowlands but also recorded as high as the central valley in the Quito region; seemingly most numerous in the southwest from Guayas south through El Oro to w. Loja. The only records from e. Ecuador, and the only ones from during southward passage, involve a specimen (MECN) taken by R. Olalla at Conambo on 26 Nov. 1960 and single birds seen at the edge of terra firme forest along the Maxus road southeast of Pompeya on 9 Nov. 1994 (RSR, K. Karlson, and F. Sornoza et al.) and again a year later on 9 Nov. 1995 (RSR, D. Agro, and F. Sornoza et al.). There also are a few records from the boreal winter, including an apparent roadkill found near the Papallacta pass on 1 Feb. 1990 (ANSP) and single birds seen (P. Coopmans et al.) at Buenaventura in El Oro on 10 Feb. 1997 and in the Río Bombuscaro sector of Podocarpus National Park on 15 Feb. 1997. An MECN specimen dated as having been taken on the improbable date of 8 Jul. 1966 is regarded as not having been reliably labeled. Being so unobtrusive, the Black-billed Cuckoo may have been over-looked in other areas and at other times, but it does appear that it may employ entirely different migratory routes on southward and northward passage. Recorded mainly from Mar. to Apr., with a few reports from Nov. and Feb. as noted above. Recorded as high as 3950 m (Papallacta pass), but normally found below 900–1000 m.

Monotypic.

Range: breeds in North America, wintering mainly in e. Peru and n. and e. Bolivia.

Coccyzus americanus   **Yellow-billed Cuckoo**   Cuclillo Piquiamarillo

Uncertain. A very rare to rare transient and (perhaps only formerly?) boreal winter resident in deciduous woodland and scrub and in shrubby clearings; so far recorded mainly from the highlands, but there are a few lowland records as well. Given the number of early specimens, most of them taken in the Quito region, the Yellow-billed Cuckoo is now an unaccountably scarce bird in Ecuador. The only recent highland record involves an injured bird captured beneath a window in Quito on 7 Mar. 1986 (PJG); it was eventually released. In the eastern lowlands the only specimen (MECN) is a female taken by R. Olalla at Conambo in Pastaza on 29 Nov. 1960; there are also two recent sightings from Jatun Sacha, one on 15 Mar. 1994 and the other on 3 May 1994 (fide B. Bochan), but we believe that it is not inconceivable that one or both could refer to the Pearly-breasted Cuckoo (*C. euleri*). Recorded Sep. to early May. Recorded up to 2800 m in the Quito region.

The racial affinity of Ecuadorian birds remains uncertain. Chapman (1926), apparently without actually seeing any, referred Ecuadorian specimens to nominate *americanus*, but there seems no reason why *occidentalis* should not also occur. A specimen mentioned by Lönnberg and Rendahl (1922) was referred to *occidentalis*. Given the drastic decline of *occidentalis* on its nesting grounds in w. North America, if indeed it is *occidentalis* that mainly occurs in Ecuador, this might account for its apparent decline here. Although the validity of *occidentalis* has been questioned (most recently by R. C. Banks, *Condor* 90[2]: 473–477, 1988 and *Condor* 92[2]: 538, 1990); K. E. Franzreb and S. A. Laymon (*Western Birds* 24[1]: 17–28, 1993) present evidence indicating that it is a valid taxon, differing from the eastern-breeding nominate race in averaging slightly larger. They also suggest that *occidentalis* may differ in bill color (being more orange) and perhaps even in its vocalizations.

Range: breeds in North America and Greater Antilles, wintering mainly in s.-cen. South America.

*Coccyzus euleri   **Pearly-breasted Cuckoo**   Cuclillo Pechiperlado

Uncertain. Apparently a casual transient through Ecuador, with only a few records. There are two specimens, both of them unpublished: a female (now in the Mejía collection) taken by C. Olalla in Mar. 1936 at Mindo and a juvenile male (ANSP) obtained by T. J. Davis on 4 Sep. 1992 in disturbed humid forest east of Muisne in w. Esmeraldas. In addition, a single bird was believed seen at Jatun Sacha on 20 Apr. 1992 (B. Bochan). The wing and tail measurements of the two specimens clearly fall within the range for the Pearly-breasted Cuckoo as given by R. C. Banks (*Bull. B. O. C.* 108[2]: 87–91, 1988). The status and even the general distribution of this scarce cuckoo remain very poorly understood. Although the Pearly-breasted Cuckoo is mainly a bird of the lowlands, the Mindo specimen was presumably taken at around 1300 m.

Banks (op. cit.) showed that the name *julieni* had priority over the long-used *euleri*, but the name *julieni* has recently been formally suppressed (I. C. Z. N., *Bull. Zool. Nomen.* 49: 178–179, 1992). Monotypic.

Range: a scatter of records from n. Colombia and n. Venezuela to ne. Argentina and se. Brazil.

Coccyzus melacoryphus   Dark-billed Cuckoo   Cuclillo Piquioscuro

Status complex and still not entirely understood. In e. Ecuador an uncommon to locally fairly common austral migrant to shrubby clearings, secondary woodland, and forest borders in the lowlands; there are also a few records of presumed austral migrants from the central valley in n. Ecuador. In this region recorded mostly from late Apr. to early Oct., exceptionally as early as Feb. and as late as Nov. (Jatun Sacha, fide B. Bochan). The status of the Dark-billed Cuckoo in w. Ecuador remains enigmatic: a few specimens and several sightings are believed to pertain to austral migrants, but there also apparently is a resident population, at least in the lowlands and foothills of the far southwest (e.g., south of Sabanilla in w. Loja, where several specimens in breeding condition were taken in Feb. 1991; WFVZ, MECN). Recorded mostly below about 1000 m, occasionally up to at least 1900 m (above Catamayo, where a bird was seen on 4 Apr. 1994; P. Coopmans and M. Lysinger) and rarely even higher (migrants?).

Monotypic. As a consequence it is impossible to distinguish austral migrants from resident birds.

Range: Colombia, Venezuela, and the Guianas south to n. Argentina; southern breeders withdraw northward during austral winter.

Coccyzus lansbergi   Gray-capped Cuckoo   Cuclillo Cabecigris

Rare to occasionally fairly common in the undergrowth of deciduous and secondary woodland, borders, and adjacent clearings in the lowlands of w. Ecuador, mainly in the southwest from Manabí and Los Ríos southward. The handsome Gray-capped Cuckoo presents a problem in that most Ecuadorian records are from Jan. to Jun. (corresponding to the rainy season), when birds breed and regularly are heard calling; numbers seem to fluctuate substantially from year to year, being higher during years with more ample rainfall. It is presumed that all or most birds either disperse or undertake long-distance migration during the remainder of the year, though at least some individuals do apparently remain; the relative paucity of records from this period may to some extent be explained by the birds not then vocalizing. During some years the Gray-capped Cuckoo ranges farther north in w. Ecuador, and it is then occasionally not uncommon, for instance, at Tinalandia. PJG has even once (in Feb. 1980) seen a vagrant on the east slope of the Andes, at Baeza in w. Napo; the species also has been recorded as a vagrant in the Galápagos Islands. Recorded mostly below 900 m, but has been seen up to 1300 m in the Mindo area, exceptionally as high as 1800 m (Baeza).

Monotypic.

Range: locally from nw. Venezuela to w. Peru; breeding distribution poorly understood.

Piaya cayana   Squirrel Cuckoo   Cuco Ardilla

Uncommon to fairly common and widespread in the canopy, subcanopy, and borders of forest, mature secondary woodland, and adjacent clearings, in the lowlands of both e. and w. Ecuador, becoming less numerous in the subtropical zone on both slopes of the Andes. In the west occurs south through El Oro and w. Loja. In the east the Squirrel Cuckoo ranges in terra firme, várzea, and riparian forest; in the west found in both humid and fairly arid regions, the requirement being groves of large trees (the species thus tends to avoid scrubby areas). Ranges up regularly to about 1500 m, in small numbers as high as 2000–2200 m and occasionally up to 2500 m in Carchi (above Maldonado; M. B. Robbins) and even to 2700 m in Loja (Cajanuma; RSR et al.).

Two quite distinct races occur in Ecuador, *mesura* in the east and *nigricrissa* in the west. *Nigricrissa* has a greenish yellow (not red) orbital area, more extensive black on the lower belly, and is a paler rufous (less rufous-chestnut) above. Vocally they appear to be similar.

Range: n. Mexico to extreme nw. Peru, n. Argentina, and Uruguay.

Piaya melanogaster   Black-bellied Cuckoo   Cuco Ventrinegro

Rare to uncommon and perhaps local in the canopy and subcanopy of terra firme forest in the lowlands of e. Ecuador. The Black-bellied is regularly sympatric with the more numerous Squirrel Cuckoo (*P. cayana*), though it is less often found in borders and openings and is more restricted to primary forest. Recorded mainly below about 400 m; in Mar. 1993 one was seen at the exceptionally high elevation of 900 m at the Bermejo oil-field area north of Lumbaquí (M. B. Robbins).

One race: nominate *melanogaster*.

Range: se. Colombia, s. Venezuela, and the Guianas to extreme nw. and ne. Bolivia and Amazonian Brazil.

Piaya minuta   Little Cuckoo   Cuco Menudo

Uncommon in the undergrowth of dense shrubby woodland, forest borders, clearings, and around oxbow lakes in the lowlands of both e. and w. Ecuador.

In the west recorded from Esmeraldas (El Placer—one seen on 16 Dec. 1992 [P. Coopmans]; Playa de Oro [O. Jahn et al.]) south to El Oro at Palmales (Taczanowski 1877), Piedras (ANSP), and Buenaventura (Robbins and Ridgely 1990); on 23 Dec. 1999 one was seen and tape-recorded farther south, near the Peruvian border east of Macará in s. Loja (P. Coopmans). Recorded mostly below about 600 m, though on the east slope a few follow river valleys up to 800–900 m; on the west slope occurs locally up to about 1500 m above Mindo and at Maquipucuna.

Two races are found in Ecuador, nominate *minuta* in the east and the endemic *gracilis* in the west. *Gracilis* differs in its paler overall coloration, both above (where it is rufous rather than rufous-chestnut) and below.

Range: Panama to n. and e. Bolivia and Amazonian Brazil; w. Ecuador.

Crotophaga major   Greater Ani   Garrapatero Mayor

Fairly common in trees and shrubbery near water in the lowlands of e. Ecuador; favors the vicinity of oxbow lakes and smaller rivers. A small population of Greater Anis also exists in sw. Ecuador, where the species is known only from s. Los Ríos (at and north of Babahoyo, and at Jauneche [where seen on 2 Sep. 1991; P. Coopmans and M. Catsis]), Guayas (south of Baquerizo Moreno [where a group of 5 was seen, and one collected (ANSP), on 22 Oct. 1992] and east of Guayaquil [2 seen on 5 Mar. 1995; P. Coopmans et al.]) and coastal El Oro (a series of 7 specimens [ANSP] taken in Oct. 1950 by T. and R. Olalla at Piedras and Santa Rosa). Recorded up to about 1000 m in e. Ecuador, ranging up along some major river valleys to the base of the Andes, but most numerous below 400 m; in the west recorded only below 50 m.

Monotypic, which is surprising given the geographic isolation of the population in sw. Ecuador. This newly discovered population does not seem to be morphologically distinguishable from cis-Andean birds (M. B. Robbins; RSR).

Range: Panama to w. Colombia, n. and e. Bolivia, and n. Argentina; sw. Ecuador; possibly (only formerly?) in ne. Mexico.

Crotophaga ani   Smooth-billed Ani   Garrapatero Piquiliso

Fairly common to common and widespread in semiopen, usually grassy or agricultural areas, and around houses in the lowlands and foothills of both e. and w. Ecuador. Necessarily rather local in the eastern lowlands; subsequent to deforestation the Smooth-billed Ani can increase and spread rapidly, being one of the first birds to colonize newly cleared areas. In undisturbed regions it is found mainly near water (and is frequent on river islands), often in the same areas as the Greater Ani (*C. major*), though the two species never seem to range in the same flock. In the west not found in more arid regions, where it is replaced by the Groove-billed Ani (*C. sulcirostris*); ranges south to the El Oro lowlands (though there is also an old record from Portovelo) and to extreme sw. Loja in the Puyango area (RSR). Recorded mainly below about 1400 m (e.g., in the Mindo area), but in recent years small numbers have begun to spread upward—

following clearings—and they now reach 1800–2000 m on the east slope of the Andes (e.g., around Baeza). In addition, a group was seen at Yaguarcocha (2200 m) north of Ibarra in Imbabura on 2 Jan. 1982 (RSR, PJG, and S. Greenfield), though the species has not been found there more recently.

Monotypic.

Range: Costa Rica to sw. Ecuador, n. Argentina, and Uruguay; Florida and West Indies.

Crotophaga sulcirostris   **Groove-billed Ani**   **Garrapatero Piquiestriado**

Fairly common to common in agricultural regions, arid scrub, and gardens in the lowlands of w. Ecuador from w. Esmeraldas and Pichincha (north to Tinalandia) southward. Also found in small numbers in the Río Marañón drainage of extreme s. Zamora-Chinchipe in the Zumba region, where only first noted in Apr. 1994 (P. Coopmans and M. Lysinger). Although there are a few recent sightings of the Groove-billed Ani from as far north as Esmeraldas and n. Pichincha, for the most part it has not spread extensively into humid regions, even where these have been cleared of their natural forest cover; however, one obviously dispersing individual was seen at Playa de Oro in n. Esmeraldas on 24 Aug. 1995 (O. Jahn), the only individual ever seen there during four years of field work. Recorded mostly below 1200 m, but in El Oro and Loja regularly ranges considerably higher, often up to about 2000–2300 m, exceptionally as high as 2750 m (a group of 3 seen on the Cordillera de Chilla on 14 Feb. 1991; N. Krabbe).

One race: nominate *sulcirostris*.

Range: extreme sw. United States to Panama; n. Colombia and much of Venezuela; extreme sw. Colombia to n. Chile.

Tapera naevia   **Striped Cuckoo**   **Cuclillo Crespín**

Uncommon to fairly common in shrubby clearings, overgrown pastures, and hedgerows in the lowlands and foothills of w. Ecuador from Esmeraldas, nw. Imbabura, and Pichincha south through El Oro and w. Loja. Aside from an old "upper Napo" record of a bird that likely was mislabeled, the Striped Cuckoo was only recently first recorded from e. Ecuador; it there remains known only from the southeast. There are recent reports from the Río Marañón drainage in extreme s. Zamora-Chinchipe around Zumba (first in Aug. 1989; RSR and M. B. Robbins) and from Morona-Santiago at Santiago (Jul. 1989; M. B. Robbins) and Taisha (Aug. 1990; RSR and M. B. Robbins); the species is probably spreading in this region as a result of deforestation. Recorded mostly below 800 m, in smaller numbers up to 1500 m, mainly in deforested areas (especially in Loja, also in Pichincha), a few even having been found as high as 2000–2300 m in Loja in the Sozoranga/Utuana area.

The species has been called the American Striped Cuckoo.

One race: nominate *naevia*.

Range: s. Mexico to n. Argentina and Uruguay (but absent from part of w. Amazonia); sw. Colombia to nw. Peru.

*Dromococcyx phasianellus   Pheasant Cuckoo   Cuco Faisán

Rare to locally uncommon in the lower and middle growth of várzea forest and secondary woodland in the lowlands of e. Ecuador. There are relatively few Ecuadorian records of the Pheasant Cuckoo, and all of them are recent; aside from its vocalizations, the Pheasant Cuckoo is an inconspicuous bird and therefore has been much overlooked until recently. An unpublished specimen (MECN) was taken by A. Mena at Capitán Chiriboga in e. Pastaza on 15 Jul. 1974; one was heard at Taisha on 3–5 Aug. 1990 (RSR and M. B. Robbins); a male (ANSP) was collected along the Río Lagarto below Imuyacocha on 4 Dec. 1990; one was mist-netted near Puerto Reira, south of the Río San Miguel in Sucumbíos, on 31 Jan. 1991 (R. ter Ellen); one was seen at Zancudococha on 23 Jan. 1995 (P. Coopmans et al.); several were recorded near Canelos in Pastaza in Sep. 1996 (N. Krabbe); and one was seen and taperecorded at La Selva on 11 Oct. 1997 (P. Coopmans et al.). The species has also recently been reported from Kapawi Lodge along the Río Pastaza near the Peruvian frontier, where (fide P. Coopmans) it appears to be more numerous than elsewhere in Ecuador. Recorded up to 700 m (at Canelos in Pastaza; N. Krabbe).

One race: nominate *phasianellus*.

Range: locally from s. Mexico to n. Colombia and n. Venezuela; e. Ecuador, e. Peru, and n. Bolivia to Amazonian (almost entirely south of Amazon River) and e. Brazil, e. Paraguay, and ne. Argentina.

Dromococcyx pavoninus   Pavonine Cuckoo   Cuco Pavonino

Uncertain. Known in Ecuador from only one definite record, a male (AMNH) taken by the Olallas on 5 Apr. 1923 at "Río Suno, above Avila" in w. Napo. The species was also reported, without details, in Jul. 1993 from the slopes of the Cordillera del Cóndor at Coangos (Schulenberg and Awbrey 1997). Elsewhere the species inhabits tangled thickets in humid forest, in some areas favoring patches of bamboo. Aside from its vocalizations—which closely resemble those of the only slightly more numerous Pheasant Cuckoo (*D. phasianellus*)—it is extremely inconspicuous and as a result is frequently overlooked. There are no other published records from anywhere near Ecuador. About 400 m.

Despite its rarity virtually everywhere, the Pavonine Cuckoo is not considered to be at risk across its wide, though patchy, South American distribution (Collar et al. 1992, 1994).

Monotypic.

Range: very locally in n. Colombia and n. Venezuela; Guyana, s. Venezuela, and n. Amazonian Brazil; se. Peru, n. Bolivia, and sw. Amazonian Brazil; se. Brazil, e. Paraguay, and ne. Argentina; one record from e. Ecuador.

Neomorphus geoffroyi Rufous-vented Cuco Hormiguero
Ground-Cuckoo Ventrirrufo

Rare and local on or near the ground in terra firme forest in the lowlands of e.
Ecuador, mainly in areas fairly close to the base of the Andes and from south
of the Río Napo; however, there is a specimen (AMNH) recorded as having
been taken at Río Lagartococha (J. Haffer, *Bonn. Zool. Beitr.* 28: 59, 1977).
The Rufous-vented Ground-Cuckoo requires extensive expanses of undisturbed
forest, and it seems to occur at very low densities. There are only a few recent
reports: it was present at least in the 1970s at Limoncocha (D. Pearson;
LSUMZ); one was seen near La Selva in Feb. 1989 (M. Hedemark and A.
Johnson), where there have also been a few subsequent sightings, mainly from
south of the Río Napo; a few recent sightings from Yuturi Lodge (fide P. Coopmans); a few sightings at Jatun Sacha (fide B. Bochan); a Jan. 1995 sighting
from along the Maxus road southeast of Pompeya (P. English); and a Jul. 1989
sighting from Santiago in Morona-Santiago (S. Holt). Recorded up to about
400 m.

Despite the species' obvious rarity in Ecuador, so much of the Rufous-vented
Ground-Cuckoo's range in the country remains remote and minimally affected
by human activity that we cannot see that it should be considered to be at risk.
One race: *aequatorialis*.
Range: Nicaragua to nw. Colombia; se. Colombia to nw. Bolivia and Ama-
zonian Brazil (south of Amazon River); e. Brazil.

*[Neomorphus pucheranii Red-billed Cuco Hormiguero
Ground-Cuckoo Piquirrojo]

Apparently very rare and local in terra firme forest in the lowlands of ne.
Ecuador. The Red-billed Ground-Cuckoo is known from only two recent sightings. A pair was seen at a swarm of army ants near Zábalo, north of the Río
Aguarico in e. Sucumbíos, on 7 Dec. 1992 (M. Hedemark); and another was
seen at Yuturi Lodge on 17 Mar. 1998 (B. López-Lanus, *Cotinga* 12: 73, 1999).
These represent the westernmost reports of this scarce and rarely observed
species, heretofore known as close to Ecuador only as the mouth of the Río
Curaray into the Río Napo (J. Haffer, *Bonn. Zool. Beitr.* 28: 64, 1977). The
locality of Yuturi is of particular note given that the Rufous-vented Ground-
Cuckoo (*N. geoffroyi*) has also been reported from there; this may be the first
reported instance of sympatry between the two *Neomorphus* ground-cuckoos.
Recorded at about 200 m.

Despite its presumed very limited known range in Ecuador, and its obvious
rarity there, we do not believe the species can be viewed as being at risk; Zábalo
lies in the Cuyabeno Faunistic Reserve, and Yuturi is a well-known ecotourism
destination.

The subspecies found in Ecuador remains uncertain, but in all likelihood it
is nominate *pucheranii*, the form occurring north of the Amazon River.
Range: ne. Ecuador, ne. Peru, and w. Amazonian Brazil.

Neomorphus radiolosus Banded Ground- Cuco Hormiguero
Cuckoo Franjeado

Very rare and local on or near the ground in humid forest in the foothills of nw. Ecuador, where mainly recorded from Esmeraldas and adjacent Imbabura; an old record from "Gualea" is the only record from Pichincha. There are only two known recent reports of this spectacular ground-cuckoo from Ecuador: a single bird seen northwest of Alto Tambo in n. Esmeraldas on 13–15 Feb. 1992 (N. Krabbe) and singletons (quite likely different individuals) seen on 6 and 15 Sep. 1996 at Bilsa (J. Hornbuckle, *Cotinga* 8: 90, 1997). In Feb. 1998 the species was observed here again on two days, both times at army antswarms (B. López-Lanús, K. S. Berg, R. Strewe, and P. G. W. Salaman, *Cotinga* 11: 42–45, 1999). As an indication of this species' great rarity, it was never encountered during four years (1995–1998) of intensive field work in the Playa de Oro area of n. Esmeraldas, and the species was not known to any at the local hunters there (O. Jahn and P. Mena Valenzuela et al.). Recorded at least in the past up to about 1500 m (Gualea, though whether the specimen in question, the type, was actually obtained so high has to be viewed as uncertain); in recent years only recorded below 500 m.

The Banded Ground-Cuckoo, which appears not to be numerous anywhere—only a handful of specimens have ever been taken—was given Endangered status by Collar et al. (1994). We concur, though noting that it is somewhat reassuring that the species has recently been found in the forests at Bilsa, which is now receiving increased protection. Presumably a population also exists in remote lower-elevation forests of Cotocachi-Cayapas Ecological Reserve.
Monotypic.
Range: sw. Colombia nw. Ecuador.

Opisthocomidae Hoatzin

One species. The Hoatzin is South American in distribution. The affinities of this bizarre bird continue to be debated; it was long placed in the Galliformes, but recent evidence (e.g., C. Sibley and J. Ahlquist, *Auk* 90[1]: 1–13, 1973; Sibley and Ahlquist 1990) points to its being closer to the cuckoos (Cuculidae). Some authors, however, continue to dispute this assignment.

Opisthocomus hoazin Hoatzin Hoazín

Locally fairly common to common and conspicuous in trees and shrubbery around lakes and along certain rivers in the lowlands of e. Ecuador. Substantial populations of the Hoatzin are found at Limoncocha, La Selva, and various sites in the lower Río Aguarico drainage. Despite its large size, for the most part the Hoatzin remains unmolested by local residents (who consider it distasteful). Recorded up to about 600 m at Tayuntza in Morona-Santiago (WFVZ), but mostly below 400 m.
Monotypic.

Range: e. Colombia, much of Venezuela, and locally in the Guianas south to n. Bolivia and Amazonian Brazil.

Strigiformes
Tytonidae   Barn Owls

One species. The family is worldwide in distribution.

Tyto alba   Barn Owl   Lechuza Campanaria

Uncommon to locally fairly common in semiopen and agricultural areas in many parts of Ecuador, ranging from the western lowlands (where sometimes numerous in oil-palm plantations) to the temperate zone of the Andes (both slopes) and the central valley. Very local in the eastern lowlands, doubtless because of the general absence of extensive open habitat; the Barn Owl is presently recorded only from Limoncocha and Lago Agrio, but with deforestation it seems likely to increase and spread. Recorded mainly below about 2000 m, occasionally higher (perhaps mainly wandering birds); one was seen at La Ciénega in Aug. 1988 (P. Coopmans).

The species is sometimes called the Common Barn-Owl.

One race: *contempta*.

Range: nearly cosmopolitan, especially in open and agricultural regions; in New World from s. Canada to Chile and Argentina, but mainly absent from Amazonia (though increasing with clearing).

Strigidae   Typical Owls

Twenty-seven species in nine genera. Owls are worldwide in distribution.

Otus roraimae   Foothill Screech-Owl   Autillo Tropandino

Rare to uncommon and local (doubtless much under-recorded) in the lower and middle growth of humid forest in the foothills and lower subtropical zone on the east slope of the Andes, also ranging out at least locally into the adjacent lowlands. There are only a few Ecuadorian records of this inconspicuous screech-owl. It is known from specimens taken in the Sumaco region (below San José, Guaticocha; AMNH, MCZ), near Canelos in w. Pastaza (a male collected [MECN] on 13 Sep. 1996; N. Krabbe), and near Chinapinza on the Cordillera del Cóndor (a male [ANSP] collected on 13 Jun. 1993). The species has also been heard and seen north of Archidona along the Loreto road, and M. B. Robbins tape-recorded it at Santiago, Morona-Santiago, in Jul. 1989. Recorded mostly between 500 and 1400 m.

Following Hardy et al. (1999), we treat O. *roraimae* as a species distinct from O. *vermiculatus* (Vermiculated Screech-Owl), based primarily on its very different voice. Hardy et al. (op. cit.) employ the English name of Tepui Screech-Owl for the species, a name that hardly seems appropriate for a species that ranges widely on lower Andean slopes. We thus suggest calling O. *roraimae* the Foothill Screech-Owl. One race of O. *roraimae* occurs in Ecuador, *napensis*. It is even possible that *napensis* will prove not to be conspecific with O. *roraimae*; in that case O. *napensis* could be called the Napo Screech-Owl and O. *roraimae* the Roraiman Screech-Owl.

Range: e. Ecuador, e. Peru, and n. Bolivia; s. Venezuela and Guyana.

Otus centralis   Chocó Screech-Owl   Autillo del Chocó

Rare to locally uncommon (but easily overlooked) in the lower and middle growth of very humid and montane forest and secondary woodland in the lowlands and foothills on the west slope of the Andes. There is apparently an old specimen from Hacienda Paramba in Imbabura (J. T. Marshall, R. A. Behrstock, and C. König, *Wilson Bull.* 103[2]: 315, 1991); we have no details regarding this otherwise unpublished specimen, but possibly it was the source for the species having been listed from w. Ecuador by Meyer de Schauensee (1970). In addition a male (FMNH) was collected at Santo Domingo de los Colorados on 23 Jun. 1967 by R. Book (fide T. Schulenberg). More recently, the Chocó Screech-Owl has been recorded from a specimen (MECN) taken at Buenaventura in El Oro by N. Krabbe on 15 Nov. 1991 (N. Krabbe). The species has also been heard and tape-recorded at El Placer, Esmeraldas, in Aug. 1987 (RSR and M. B. Robbins et al.) and on 15 Dec. 1992 (P. Coopmans), and at Tinalandia (first in May 1990 [PJG; v.o.]); it has also been found in recent years near Pedro Vicente Maldonado north of Simón Bolívar (P. Coopmans). Further, during 1995–1998 field work it was found to be widespread around Playa de Oro in n. Esmeraldas, mainly in the lower foothills but smaller numbers also in the lowlands (O. Jahn et al.). Recorded mostly between 500 and 1000 m, but in n. Esmeraldas ranging down to below 100 m.

Following Hardy et al. (1999), we treat O. *centralis* as species distinct from O. *guatemalae* (Middle American, or Guatemalan, Screech-Owl), based primarily on its distinctive voice. *Centralis* was described as a subspecies of O. *guatemalae* by G. P. Hekstra (*Bull. Zool. Mus. Univ. Amsterdam* 9[7]: 49–64, 1982). Monotypic.

Range: Panama to sw. Ecuador.

Otus choliba   Tropical Screech-Owl   Autillo Tropical

Uncommon to locally common in the borders of várzea and riparian forest, plantations, and clearings in the lowlands of e. Ecuador. The Tropical Screech-Owl is recorded mainly from the Río Napo drainage northward. Apparently the only record from south of here is a bird heard at Kapawi Lodge on 22 Aug. 1999 (P. Coopmans et al.), but the species seems likely to be more widespread, and it should increase here, as elsewhere, with ongoing forest clearance. Tropical Screech-Owls are particularly numerous in the woodlands surrounding various lakes in the Río Lagarto drainage of the extreme northeast. Recorded up to about 650 m in the Archidona area.

One race: *crucigerus*.

Range: Costa Rica to w. Colombia, n. Argentina, and Uruguay.

Otus roboratus   West Peruvian Screech-Owl   Autillo Roborado

Uncommon to locally fairly common in deciduous forest and woodland and scrub in the lowlands and foothills of sw. Ecuador from sw. Manabí and Los Ríos southward. Chapman (1926) recorded this species with some uncertainty from a specimen taken at Vinces in Los Ríos (Salvadori and Festa 1900), but this early record was overlooked (or dismissed?) by Meyer de Schauensee (1966, 1970). Although neither we nor the describers of the West Peruvian Screech-Owl (O. Bangs and G. K. Noble, *Auk* 35: 448, 1918) examined the Vinces specimen, in light of the series of recent Ecuadorian sites at which the West Peruvian Screech-Owl has been found (including one in Los Ríos), we are confident that the specimen does indeed refer to that species (the Chocó Screech-Owl would not be expected to occur so far out in the lowlands). Another 19th-century specimen (BMNH) was taken in the "Balzar Mountains" of n. Guayas in Mar. 1880 (N. K. Johnson and R. E. Jones, *Wilson Bull.* 102[2]: 199–212, 1990). The first modern Ecuadorian records of the West Peruvian Screech-Owl were made by J. Rowlett and R. A. Rowlett, and independently by P. Coopmans; they found several birds east of Celica along the road to El Empalme in early Feb. 1989. In Aug. 1989 the species was relocated both here and south of El Empalme along the road toward Macará, and the first modern Ecuadorian specimens were obtained (ANSP, MECN). West Peruvian Screech-Owls have since been found on the Santa Elena Peninsula, as far north as Machalilla National Park (Jan. 1991; Parker and Carr 1992) in Manabí and Jauneche in Los Ríos (P. Coopmans), and at several localities in s. Loja; it is now clear that the species is relatively widespread in this region, occurring virtually wherever appropriate habitat remains. Recorded mostly below about 1200 m, but ranging locally or in very small numbers as high as 1800 m at Sozoranga (Best et al. 1993).

The English name of Peruvian Screech-Owl has recently been suggested for *O. roboratus* (N. K. Johnson and R. E. Jones, op. cit.). However, given that there are two other species of *Otus* endemic to Peru (Koepcke's Screech-Owl [*O. koepckeae*] and Cloud-forest Screech-Owl [*O. marshalli*], it seems preferable to retain the long-used (e.g., Meyer de Schauensee 1966, 1970) English name of West Peruvian Screech-Owl.

Only one subspecies of *O. roboratus* is recorded from Ecuador, *pacificus*, described as a race of *O. guatemalae* (Vermiculated Screech-Owl) by G. P. Hekstra (*Bull Zool. Mus. Univ. Amsterdam* 9[7]: 49–64, 1982). However, the similar but larger nominate race likely occurs in the Río Marañón drainage of extreme s. Zamora-Chinchipe in the Zumba region; it has been recorded from Perico in nearby Cajamarca, Peru.

Range: sw. Ecuador and nw. Peru.

Otus ingens   Rufescent Screech-Owl   Autillo Rojizo

Rare to uncommon inside montane forest in the subtropical zone on both slopes of the Andes, on the east slope apparently occurring mostly on outlying ridges. On the west slope recorded south only to Pichincha; first recorded in Ecuador from a "long-misidentified" specimen from Nanegal in the Paris Museum (J. Fitzpatrick and J. P. O'Neill, *Wilson Bull.* 98[1]: 1–14, 1986). More recent west-slope records exist from Carchi (a male tape-recorded and collected [ANSP] above Maldonado on 17 Aug. 1988 by M. B. Robbins) and from Pichincha at Maquipucuna (a bird tape-recorded and collected [MECN] on 31 Mar. 1993 by N. Krabbe and P. Coopmans) and above Mindo (v.o.). Despite the relative paucity of records, on the east slope the distribution of this inconspicuous screech-owl is presumed to be more or less continuous where appropriate habitat exists. Recent reports include birds found along the Loreto road north of Archidona, on the Cordillera de Huacamayos, on the Cordillera de Cutucú (ANSP, MECN), at Río Huamboya west of Macas in Morona-Santiago (collected [MECN] in Apr. 1998; F. Sornoza), and on the Cordillera del Cóndor (ANSP). On the east slope recorded from about 1200 to 2250 m, on the west slope from 1300 to 2300 m.

Two similar races occur in Ecuador, nominate *ingens* on the east slope of the Andes and *colombianus* on the west slope. *Colombianus* was for a period often considered (e.g., by Fitzpatrick and O'Neill, op. cit.) as a full species (Colombian Screech-Owl), and still is by some (e.g., Marks et al. 1999). The situation is complex and confusing, but we opt to follow Sibley (1996) in considering *colombianus* as a west-slope subspecies of *O. ingens*; their voices are very similar.

When considered to be a full species, *O. colombianus* was accorded Near-threatened status by Collar et al. (1994), presumably on the basis of ongoing deforestation in its limited range. With *colombianus* now being treated as a race of the more widespread *O. ingens*, we do not believe that the entire species should be considered at risk.

Range: west slope of Andes in sw. Colombia and nw. Ecuador; locally in mountains of n. Venezuela and in Andes of w. Venezuela, and on east slope of Andes from extreme s. Colombia to w. Bolivia; Sierra de Perijá.

*Otus petersoni   Cinnamon Screech-Owl   Autillo Canelo

Rare to uncommon and local in montane forest in the subtropical and lower temperate zones on the east slope of the Andes, occurring mainly on outlying ridges. This recently described species (J. Fitzpatrick and J. P. O'Neill, *Wilson Bull.* 98[1]: 1–14, 1986) was first recorded in Ecuador from specimens (ANSP, MECN) obtained on the Cordillera de Cutucú in w. Morona-Santiago in Jul. 1984 (Robbins et al. 1987). It has since been found in w. Napo northeast of El Chaco (one collected on 4 Oct. 1992; ANSP), in Morona-Santiago along the Gualaceo-Limón road (one tape-recorded and seen in Mar. 1990; B. Whitney), and in Zamora-Chinchipe at Chinapinza on the Cordillera del Cóndor (3 spec-

imens obtained in Sep. 1990 [MECN, ANSP]; see Krabbe and Sornoza 1994).
The Cinnamon Screech-Owl occurs locally in sympatry with the larger and
apparently more widespread Rufescent Screech-Owl (*O. ingens*). Recorded
from about 1650 to 2225 m.

J. T. Marshall (*in* Sibley 1996) suggested that the name *petersoni* was
preoccupied by *huberi*, a 1936 name based on "Bogotá" specimens. The name
huberi, however, is now considered to be "invalid" (Marks et al. 1999, p. 179).
Monotypic.

Range: locally on east slope of Andes from Colombia ("Bogotá" specimens
only) and n. Ecuador to n. Peru. Excludes Cloud-forest Screech-Owl (*O.
marshalli*), found locally on east slope of Andes in s. Peru.

Otus watsonii   Tawny-bellied Screech-Owl   Autillo Ventrileonado

Uncommon to fairly common in the lower and middle growth of humid forest
(both terra firme and várzea, but seemingly more numerous in the former) in
the lowlands of e. Ecuador. Recorded mainly below 500 m, rarely or locally up
to 900 m.

One race: nominate *watsonii*. We regard *usta* of s. Amazonia as a southern
subspecies of *O. watsonii*, though it has been suggested (e.g., Sibley 1996)
that it represents a species (Usta, or Austral, Screech-Owl) separate from *O.
watsonii*.

Range: e. Colombia, s. Venezuela, and the Guianas to n. Bolivia and
Amazonian Brazil; nw. Venezuela.

Otus albogularis   White-throated Screech-Owl   Autillo Goliblanco

Uncommon to locally fairly common in the canopy and borders of montane
forest and woodland in the upper subtropical and temperate zones on both
slopes of the Andes, sometimes occurring up to not far below treeline. Although
there are relatively few early records, field work—facilitated by knowledge of
the species' voice—in the 1980s and 1990s has demonstrated that the White-
throated Screech-Owl is considerably more numerous and widespread in
Ecuador than had been realized. There are recent records from sites in Carchi,
w. Napo, Bolívar, Cañar, Azuay, Morona-Santiago, and Loja, and the species is
probably continuously distributed in appropriate habitat; it even persists where
forest has been somewhat fragmented and disturbed. Recorded mostly from
2500 to 3400 m, locally as low as 2100 m on the crest of the Cordillera de
Huacamayos (N. Krabbe).

Two similar races are found, *macabrum* on the west slope and nominate
albogularis on the east slope.

Range: Andes from w. Venezuela to w. Bolivia; Sierra de Perijá.

Bubo virginianus   Great Horned Owl   Búho Coronado Americano

Rare to uncommon and seemingly local in less disturbed patches of woodland
(inluding *Polylepis*) and in adjacent semiopen and less disturbed agricultural

areas and paramo in the Andes. The infrequently encountered Great Horned Owl can at least be heard regularly at El Cajas National Recreation Area in Azuay. Recorded mainly from 3200 to 4500 m (it has been found around the mountaineering refugio on Volcán Cotopaxi), perhaps occasionally or locally down to 2500 m.

One race: the sooty *nigrescens*, the darkest of all the many races of the wide-ranging *B. virginianus*.

Range: arctic North America to se. South America, but very local or (usually) absent from forested tropical lowlands, and absent from nearly all of Amazonia. More than one species is apparently involved, with Magellanic (or Lesser) Horned Owl (*Bubo magellanicus*) of s. South America having recently been shown to represent a separate species (C. König, *Stutt. Beitr. Natur.*, ser. A, no. 540, 1996), but the exact relationship of various forms remains unresolved.

Glaucidium jardinii   Andean Pygmy-Owl   Mochuelo Andino

Uncommon in montane forest and woodland and their borders in the subtropical and temperate zones on both slopes of the Andes; also found locally on slopes above the central and interandean valleys, perhaps especially in n. Ecuador (e.g., at Pasochoa). On the west slope recorded south to Azuay (several recent records by N. Krabbe). Recorded mostly from 2000 to 3500 m, locally as high as 3700 m (e.g., on Volcán Pichincha).

Monotypic.

Range: Andes from w. Venezuela to cen. Peru; Sierra de Perijá. Excludes Costa Rican Pygmy-Owl (*G. costaricanum*) of highlands of Costa Rica and w. Panama, following M. B. Robbins and F. G. Stiles (*Auk* 116[2]: 305–315, 1999), and Bolivian Pygmy-Owl (*G. bolivianum*) of Andes from s. Peru to nw. Argentina.

*Glaucidium nubicola   Cloud-forest   Mochuelo Nuboselvático
                           Pygmy-Owl

Uncommon in the canopy and borders of montane forest and adjacent cleared areas in the subtropical zone on the west slope of the Andes in Carchi and Pichincha (south to the Mindo area, but perhaps extending farther southward). This only recently described species (M. B. Robbins and F. G. Stiles, *Auk* 116[2]: 305–315, 1999) was long confused with the Andean Pygmy-Owl (*G. jardinii*), which it closely resembles in plumage. Recorded from about 1400 to 2000 m.

Monotypic.

Range: west slope of Andes in sw. Colombia and nw. Ecuador.

*Glaucidium griseiceps   Central American   Mochuelo Cabecigris
                             Pygmy-Owl

Uncommon and local (though probably overlooked) in the canopy and borders of very humid forest in the lower foothills of extreme nw. Ecuador in n.

Esmeraldas. First recorded northwest of Alto Tambo, where in Jul.–Aug. 1990 M. B. Robbins saw and tape-recorded (recordings to LNS) the vocalizations of two pairs. The species was subsequently tape-recorded in 1997 at Playa de Oro on the Río Santiago, where it is seemingly restricted to the lowermost foothills; one dispersing bird was heard at only 70 m (O. Jahn et al.). There are no specimens from Ecuador. Recorded at about 200–400 m, exceptionally lower.

Given this species' apparent rarity, specialized habitat requirements, and limited range in Ecuador, we agree with O. Jahn et al. (pers. comm.) that the Central American Pygmy-Owl deserves Vulnerable status in the country. Small populations may occur in the Awá Forest Reserve and the lower parts of Cotocachi-Cayapas Ecological Reserve, but the species' actual status and population size at both have not yet been evaluated. The species was not considered to be at risk by Collar et al. (1992, 1994), as it is rather widely distributed outside Ecuador.

G. griseiceps of s. Mexico south to nw. Ecuador is regarded as a species separate from *G. minutissimum* (East Brazilian, or Least, Pygmy-Owl) following the data presented by S. N. G. Howell and M. B. Robbins (*Wilson Bull.* 107[1]: 7–25, 1995); in this paper several other allopatric taxa are also separated from true *G. minutissimum* of the se. Brazil region. The race occurring in Ecuador is unknown.

Range: s. Mexico to nw. Colombia, and in nw. Ecuador.

Glaucidium parkeri   Subtropical Pygmy-Owl   Mochuelo Subtropical

Rare and perhaps local (doubtless often overlooked) in the canopy and borders of montane forest in the subtropical zone on the east slope of the Andes from Sucumbíos southward. This recently described species (M. B. Robbins and S. N. G. Howell, *Wilson Bull.* 107[1]: 1–6, 1995) has now been found at 6 Ecuadorian localities, all in the past 15 or so years. Although it may prove to be more widespread, most sites where it is known to occur are on outlying ridges slightly east of the main Andes chain. The first record involved birds tape-recorded but never seen on the Cordillera de Cutucú in Jun. 1984 (M. B. Robbins, tape-recordings to LNS; these were misidentified as Andean Pygmy-Owls [*G. jardinii*] in Robbins et al. 1987). Subsequently one was tape-recorded on Volcán Sumaco on 21 Jan. 1991 (B. Whitney); one was tape-recorded and collected (the type specimen, at ANSP) at Panguri in s. Zamora-Chinchipe on 22 Jul. 1992 (F. Sornoza); one was seen by T. Parker at Coangos near the northern end of the Cordillera del Cóndor on 20 Jul. 1993 (Schulenberg and Awbrey 1997); one was seen at San Rafael Falls on 24 Jul. 1994 (S. Howell); and one was seen and tape-recorded north of Archidona along the Loreto road in Dec. 1997 (M. Lysinger). Recorded between about 1100 and 2000 m.

Monotypic.

Range: locally on east slope of Andes from n. Ecuador to w. Bolivia (fide M. B. Robbins).

Glaucidium brasilianum   Ferruginous   Mochuelo Ferruginoso
                                        Pygmy-Owl

Uncommon to fairly common in the canopy and borders of várzea and riparian forest, secondary woodland, and clearings in the lowlands of e. Ecuador;
rare in the canopy and borders of terra firme forest, to which it may only be
spreading as a result of deforestation. Recorded mostly below 500m, but one
was recorded at 750m in second-growth northwest of Archidona on 8–9 Mar.
1999 (L. Navarrete et al.).

Does not include trans-Andean populations that were formerly considered
conspecific with *G. brasilianum* but are now recognized as a separate species,
G. peruanum (Pacific Pygmy-Owl). The racial allocation of the Ecuadorian
population of *G. brasilianum* is uncertain—specimen material is surprisingly
scanty—but at least some birds can be referred to *ucayalae.*

Range: extreme sw. United States and Mexico to w. Panama; n. Colombia;
e. Colombia, most of Venezuela, and Guyana to n. Argentina and Uruguay.

Glaucidium peruanum   Pacific Pygmy-Owl   Mochuelo del Pacífico

Fairly common to common in a variety of wooded habitats ranging from deciduous forest to rather sparse arid scrub, and in gardens and town plazas, in the
lowlands of sw. Ecuador from n. Manabí (north to near Flavio Alfaro; RSR),
s. Pichincha (Patricia Pilar; one heard on 17 Jul. 1997 [P. Coopmans]), and Los
Ríos southward. In s. Ecuador (El Oro and Loja) the Pacific Pygmy-Owl also
ranges up in montane scrub and woodland to considerably higher elevations,
and in Loja it occurs east to the Loja (city) and Vilcabamba region. In Loja it
occurs locally as high as about 2400m around the Sozoranga region and at
Utuana; elsewhere found mostly below 1200m though in El Oro and s. Azuay
also ranging up locally to 1500–2000m.

G. peruanum was recently specifically separated from the wide-ranging
G. brasilianum (Ferruginous Pygmy-Owl; C. König, *Okol. Vogel.* 13: 56–61,
1991), mainly on the basis of its very different, faster song. It also differs from
brasilianum in its more broadly pale-streaked crown and typically paler and
more cinnamon-rufous dorsal color (though a gray-brown morph does occur).
At higher elevations in *peruanum*'s range, the plumage is at least typically more
grayish brown (not rufous) and the crown is more dotted with white, with
streaking restricted to the forehead; these birds also show larger white spots
on their scapulars and wing-coverts. This population almost appears to represent a separate species of *Glaucidium*, but it is similar in voice to *G. peruanum*
found at lower elevations, and they are therefore still considered conspecific.
Monotypic.

König (op. cit.) suggested the English name of Peruvian Pygmy-Owl for *G.
peruanum*, but as a significant portion of the species' range lies outside Peru,
and as several other *Glaucidium* species also occur in Peru, we prefer to draw
attention to its Pacific-slope range, where it is the only *Glaucidium*.

Range: sw. Ecuador, w. Peru, and n. Chile.

Athene cunicularia   Burrowing Owl   Búho Terrestre

Uncommon to locally fairly common in semiopen, usually sandy areas in two distinctly different regions of Ecuador: arid intermontane valleys and slopes from Carchi (La Concepción) and Imbabura south locally into Loja (Valle Catamayo and Vilcabamba; MECN); and near the coast from cen. Manabí (Bahía de Caráquez area) south through arid Guayas and El Oro, thence inland into the lowlands of sw. Loja to around Zapotillo. A published report from Mindo (Kirwan et al. 1996) requires confirmation and certainly is anomalous. Montane populations are found mostly from 1500 to 3000 m, whereas those of the western lowlands range only below about 150 m.

We follow the 1998 AOU Check-list in merging the monotypic genus *Speotyto*, in which this species was formerly often placed, in *Athene*. Two races of *A. cunicularia* are found in Ecuador, *pichinchae* in the highlands and *punensis* in the lowlands of the southwest. *Punensis* is comparatively small and decidedly paler overall, with less extensive brown barring on the underparts and a less prominent whitish brow.

Range: w. and s. United States to Mexico, a few wintering south into Central America; much of Venezuela and adjacent Guyana; locally in n. and w. Colombia; w. Ecuador south to Chile and Argentina, eastward to s. Brazil; West Indies.

Lophostrix cristata   Crested Owl   Búho Penachudo

Uncommon in humid forest, mature secondary woodland, and adjacent clearings with large trees in the lowlands of both e. and w. Ecuador. The first record of the Crested Owl for w. Ecuador was obtained only in 1971, a female (Escuela Polytécnica collection) taken near Santo Domingo de los Colorados on 19 Oct. (Orcés 1974). There since have been a number of reports, and Crested Owls are now known to occur south locally to sw. Manabí (Machalilla National Park; G. H. Rosenberg), Guayas at Loma Alta (Becker and López-Lanús 1997) and Cerro Blanco (R. Clay et al.), and nw. Azuay at Manta Real (J. C. Matheus; G. H. Rosenberg and RSR et al.). Recorded up to about 800 m.

Two races are found in Ecuador. Eastern birds belong to nominate *cristata*, whereas material from the west can be assigned to the generally darker (especially above) *wedeli*, heretofore known only from n. and w. Colombia.

Range: s. Mexico to extreme nw. Venezuela and sw. Ecuador; se. Colombia, s. Venezuela, and the Guianas to n. Bolivia and Amazonian Brazil (but largely absent from nw. Amazonia).

Pulsatrix perspicillata   Spectacled Owl   Búho de Anteojos

Uncommon in humid and deciduous forest and woodland and their borders, and in adjacent clearings with large trees, in the lowlands of both e. and w. Ecuador. In the east occurs in both terra firme and várzea forest, perhaps in larger numbers in the latter; in the west recorded south through El Oro and w.

Loja. Recorded mostly below 1000 m, but at least on the west slope—in the absence of its congener the Band-bellied Owl (*P. melanota*)—it locally ranges much higher (e.g., it was heard at 1700 m above Maldonado in Carchi in Aug. 1988; M. B. Robbins). The highest elevation on the east slope at which the Spectacled Owl has been recorded is 1250 m (one tape-recorded and seen along the Loreto road north of Archidona on 28 Aug. 1996; N. Krabbe and P. Coopmans), well above elevations at which Band-bellied Owls have been found along the same road.

Two similar races are found in Ecuador, nominate *perspicillata* in the east and *chapmani* in the west. *Chapmani* differs in being slightly larger and darker overall.

Range: s. Mexico to nw. Peru, n. Argentina, and se. Brazil.

Pulsatrix melanota Band-bellied Owl Búho Ventribandeado

Uncommon to locally fairly common in montane forest, forest borders, and adjacent clearings with large trees in the foothills on the east slope of the Andes from w. Sucumbíos (San Rafael Falls) south through Zamora-Chinchipe (Cumbaratza [MCZ]; Zamora and the Río Bombuscaro sector of Podocarpus National Park). The Band-bellied Owl was until recently unrecorded from Colombia except for a single specimen without specific locality (Hilty and Brown 1986); there are recent records from w. Putumayo and e. Cauca (Salaman et al. 1999), so in Ecuador it is likely to occur north all the way to the border. The Band-bellied Owl is now known to occur in sympatry with the larger Spectacled Owl (*P. perspicillata*), at least along the Loreto road north of Archidona. Recorded from about 900 to 1500 m.

One race: nominate *melanota*. The species is possibly better considered to be monotypic.

Range: east slope of Andes and along their base from s. Colombia to w. Bolivia.

Strix nigrolineata Black-and-white Owl Búho Blanquinegro

Uncommon in humid forest, secondary woodland, and borders in the lowlands of w. Ecuador south locally to sw. Manabí (Machalilla National Park), Los Ríos (Jauneche), Guayas (a few recent reports from Cerro Blanco [R. Clay et al.] and Manglares-Churute Ecological Reserve [Pople et al. 1997]), and El Oro (a pair seen at Buenaventura on 27 Jul. 1992; P. Coopmans and M. Van Beirs et al.). Most recent Ecuadorian reports of the Black-and-white Owl come from Tinalandia and (especially) Río Palenque. Recorded mainly below 900 m, but in recent years a few have been noted regularly as high as 1300–1400 m in the Mindo area (V. Perez).

We follow Sibley and Monroe (1990) and Howell and Webb (1995) in placing this and the following three species in the genus *Strix*; they and the other Neotropical members of the genus were formerly separated in the genus *Ciccaba*, and they still are in the 1998 AOU Check-list. Trans-Andean *S. nigro-*

lineata is sometimes considered conspecific with *S. huhula* (Black-banded Owl) of Amazonia, and despite distinct plumage differences this may prove best; their vocalizations are very similar, and they regularly respond to each others' calls. Monotypic.

Range: s. Mexico to n. Venezuela and extreme nw. Peru.

Strix huhula   Black-banded Owl   Búho Negribandeado

Uncommon in humid forest (mainly terra firme) and forest borders in the lowlands of e. Ecuador. Although in general an unobtrusive owl, the beautiful Black-banded can be seen with fair regularity from the canopy observation platforms at both La Selva and Sacha Lodges. Recorded up to about 900 m (old specimens from the mouth of the Río Bombuscaro and "Río Napo"; Chapman 1926).
Monotypic.

Range: se. Colombia, s. Venezuela, and the Guianas to nw. Argentina and Amazonian Brazil; se. Brazil and ne. Argentina.

Strix virgata   Mottled Owl   Búho Moteado

Uncommon and apparently local in humid and montane forest and mature secondary woodland and in large trees of adjacent clearings in the more humid lowlands and subtropical zone on the west slope of the Andes. Recorded south to sw. Manabí (Machalilla National Park), Guayas (coastal Cordillera de Colonche at Cerro La Torre and above Sanguilla [N. Krabbe] and Loma Alta [Becker and López-Lanús 1997]; Cerro Blanco; Manglares-Churute Ecological Reserve), El Oro (numerous reports from Buenaventura, first by Robbins and Ridgely [1990]), and w. Loja (a few reports from the Alamor and Celica regions [P. Coopmans; RSR] and one even from above Sozoranga [P. Coopmans]); almost all records are recent. As it ranges so close to the Peruvian border, it seems likely that the Mottled Owl also occurs in nw. Peru, where suitable habitat does appear to exist locally. Recorded up to about 2000 m at Cabañas San Isidro (M. Lysinger) and near Celica.

In the lowlands of e. Ecuador there are very few records of the Mottled Owl—indeed it appears to be a scarce and local bird throughout w. Amazonia—with the earliest being a male (ANSP) obtained by C. Olalla at Loreto in w. Napo on 10 Apr. 1942. There have been only a few subsequent reports, most of them involving birds that were only heard, but a pair was seen on several occasions at their day roost in a dense thicket in hilly terra firme forest along the Maxus road southeast of Pompeya in Oct. and Nov. 1994 (P. English). The precise status and distribution of the Mottled Owl in the east remain to be determined; we should emphasize that vocalizations of both this species and the Black-banded Owl (*S. huhula*) are variable and quite similar, and thus reports of birds that have only been heard must be viewed with some caution (certain calls of the Spectacled Owl [*Pulsatrix perspicillata*] and even of the Long-tailed Potoo

[*Nyctibius aethereus*] are also similar, fide P. Coopmans). Recorded below about 600 m.

Two races are found in Ecuador, nominate *virgata* ranging widely in the west, whereas in the east only the one ANSP specimen is known, and because of the general paucity of available specimen material in the species, its subspecific identity remains uncertain. It differs from nominate *virgata* in being generally more rufescent (not so dusky) above and in having a quite prominent white brow contrasting with a solidly dark crown as well as large buffy whitish scapular spots. The specimen appears to be similar to *macconnellii*, though that taxon seems to have been recorded only from the geographically distant Guianas; the somewhat similar *superciliaris* (with an equally prominent white brow) is known from at least s. Amazonian Brazil and n. Bolivia.

Range: n. Mexico to sw. Ecuador, n. Bolivia, and Amazonian Brazil; se. Brazil, e. Paraguay, and ne. Argentina.

Strix albitarsis   **Rufous-banded Owl**   **Búho Rufibandeado**

Uncommon to locally fairly common (but inconspicuous and not often seen; heard much more often) in montane forest in the subtropical and temperate zones on both slopes of the Andes. On the west slope until recently not recorded south of Pichincha, but in 1995–1996 it was recorded at Salinas in Bolívar and Chaucha in Azuay (Krabbe et al. 1997). Foraging birds sometimes emerge to feed at forest borders and even out into trees in adjacent clearings. Recorded mostly between about 1900 and 3100 m, locally ranging as high as 3700 m on the north slope of Volcán Pichincha (N. Krabbe).

Monotypic. The species name is usually spelled *albitarsus*, but this is evidently incorrect (Marks et al. 1999).

Range: Andes from w. Venezuela to w. Bolivia.

Asio clamator   **Striped Owl**   **Búho Listado**

Rare and seemingly local in semiopen terrain, lighter woodland, sometimes even around houses, in the lowlands of w. Ecuador from sw. Manabí (Machalilla National Park), nw. Cotopaxi (specimen from Guasaganda taken on 7 Dec. 1985; MECN), and w. Chimborazo (Chimbo) south through El Oro to w. Loja (MCZ specimen from west of Sabiango taken on 13 Oct. 1965). Since 1994 there have also been a few reports of Striped Owls in early-successional vegetation on river islands in the Río Napo near Sacha Lodge (G. Rivadeneira); a sighting from near Primavera in Feb. 1987 (P. Coopmans) was also likely this species. Although seemingly out of place in upper Amazonia, the Striped Owl has been found in similar habitat along the lower Río Napo near Iquitos, Peru (T. Parker, pers. comm.). In the southwest recorded up to about 700 m, in the east below 300 m.

The species was formerly either separated in the monotypic genus *Rhinoptynx* or (e.g., S. L. Olson [*Bull. B. O. C.* 115(1): 35–39, 1995] and the

1998 AOU Check-list) placed in the genus *Pseudoscops*. We follow, however, Howell and Webb (1995) and Marks et al. (1999) in placing it in *Asio*, which in most respects it certainly appears to resemble. One race: nominate *clamator*.

Range: s. Mexico to n. Colombia, most of Venezuela, and the Guianas; w. Ecuador; very locally in Amazonia; extreme se. Peru, n. Bolivia, and cen. and e. Brazil to n. Argentina and Uruguay.

Asio stygius   Stygian Owl   Búho Estigio

Rare to uncommon and seemingly local in borders of montane forest and woodland (often where fragmented, sometimes decidedly so), groves of trees in agricultural areas, and sometimes even around houses in the Andes from Imbabura (Hacienda Anagumbo; ANSP) south to s. Loja (Sozoranga; Best et al. 1993). Remarkably, there seem to be no (published) records from Peru where it seems virtually certain to occur. The generally scarce Stygian Owl ranges in both arid and semihumid regions, but it appears to be mainly a nonforest owl. Recorded from 1700 to 3100 m.

One race: *robustus*.

Range: locally (mainly in highlands) from Mexico to Nicaragua; locally in Andes from nw. Venezuela to s. Ecuador, and from n. Bolivia to n. Argentina and se. Brazil; Greater Antilles.

Asio flammeus   Short-eared Owl   Búho Orejicorto

Rare to locally uncommon in paramo and adjacent open agricultural terrain in the Andes from Carchi (one seen on Páramo del Ángel on 3 Jan. 1982 [RSR, PJG, and S. Greenfield]), Imbabura (a specimen [ANSP] taken in 1950 at Hacienda Anagumba; also 3 seen near Ibarra on 10 Jul. 1988 [M. B. Robbins and G. S. Glenn]), and extreme w. Sucumbíos (a specimen [USNM] taken in Mar. 1992 at Cocha Seca) south locally in the Andes to Azuay (sightings at El Cajas [PJG] and Guagualoma [N. Krabbe]). The Short-eared Owl also seems likely to occur in far s. Ecuador as well, but as yet there are no records. Recorded mostly from 3000 to 4000 m, locally or occasionally ranging lower, including an old specimen (ANSP) supposedly taken as low as 2300 m at Zámbiza.

One race: *bogotensis*.

Range: North America to Mexico, a few wintering south into Central America; locally in Andes and savannas in lowlands from Colombia and Venezuela to Chile and Argentina (but entirely absent from Amazonia); Greater Antilles; also Eurasia. More than one species may be involved.

Aegolius harrisii   Buff-fronted Owl   Buhito Frentianteado

Very rare and seemingly local—but probably underrecorded—in montane forest and woodland in the temperate zone on both slopes of the Andes; on the west slope known only from Pichincha. The rarity of the Buff-fronted Owl in Ecuador is puzzling; perhaps it is being overlooked, but the evidence to date is that it must indeed be a very scarce and inconspicuous bird. Despite considerable field work in areas that seemed likely to have populations of it, the species has continued to be virtually unrecorded. The sole Ecuadorian specimen remains a 19th-century bird taken at Zámbiza in n. Pichincha not far from Quito. One was mist-netted below the Cajanuma station in Podocarpus National Park on 15 Nov. 1991 (Bloch et al. 1991); one was seen at its day roost being mobbed by various passerine birds in the upper Río Angashcola valley in Loja on an unspecified date in Aug. 1990 (R. Williams and J. Tobias); one was purchased at a Quito market in 1991 and kept alive for several weeks before it escaped (N. Hilgert); and a pair was seen and tape-recorded in the Río Yunguilla valley of w. Azuay in June 2000 (N. Krabbe). Recorded from about 1900 to 3100 m.

The Buff-fronted Owl was given Near-threatened status by Collar et al. (1994), presumably on the basis of its apparent rarity over much or all of its entire range (perhaps especially in the Andes). Given its (inexplicable) rarity in Ecuador, we must agree.

Whether this very distinct species is properly classified in *Aegolius* remains to be demonstrated. One race occurs in Ecuador, nominate *harrisii*.

Range: very locally in mountains of n. Venezuela, Andes from w. Venezuela to nw. Argentina, and in se. Brazil, e. Paraguay, and ne. Argentina; a few scattered records from tepuis of s. Venezuela and from ne. Brazil.

Caprimulgiformes
Steatornithidae   Oilbird

One species. The Oilbird is found almost exclusively in South America (there are a few Panama records).

Steatornis caripensis   Oilbird   Guácharo

Uncommon and local in the canopy and borders of montane forest in the foothills and subtropical zone on both slopes of the Andes. In the west recorded south mainly to Pichincha, with the only records from farther south being sightings at Caripero in w. Cotopaxi (N. Krabbe and F. Sornoza) and the remains of a dead bird found at Buenaventura in El Oro in Jan. 1990 (P. Coopmans and R. Schofield et al.). Oilbirds roost and nest very locally in colonies situated in caves and deeply shaded ravines. The largest colony known in Ecuador is at the Cueva de los Tayos in Morona-Santiago, near the southern end of the Cordillera de Cutucú. The only colony known from the west slope or from the highlands is the tiny one (now consisting of only a few pairs) situated near Puellaro in n. Pichincha (see F. Ortiz-Crespo, *Auk* 96[1]: 187–189, 1981). Occasional individuals wander far out into the lowlands and may not return by day (e.g., the single roosting bird that was seen on a tree limb at Cuyabeno on 10–24 Jul. 1991 [J. Rowlett and R. A. Rowlett et al.]). Recorded mostly between 700 and

2400 m, occasionally wandering as high as 2600 m (one was attracted to camp lights in the Río Angashcola valley east of Amaluza, Loja, in Sep. 1990; R. Williams et al.) or down into the eastern lowlands.

Monotypic.

Range: locally in mountains of n. Venezuela, and in Andes from w. Venezuela to w. Bolivia; also in tepuis of s. Venezuela, e. Colombia, and Guyana; wanders to Panama.

Nyctibiidae   Potoos

Five species in one genus. Potoos are Neotropical in distribution.

Nyctibius grandis   **Great Potoo**   **Nictibio Grande**

Uncommon to locally fairly common in the canopy and borders of humid forest (both várzea and terra firme) in the lowlands of e. Ecuador. As with the other potoos, this species' memorable and far-carrying calls are heard more often than the birds are ever seen. At night Great Potoos regularly leave the cover of the forest interior to forage from lookouts at edge and sometimes even from trees and posts in adjacent clearings. Recorded up to about 400 m.

One race: nominate *grandis*. The species may, however, better be considered to be monotypic (Cohn-Haft 1999).

Range: s. Mexico to n. Colombia, n. Bolivia, and Amazonian Brazil; se. Brazil.

Nyctibius aethereus   **Long-tailed Potoo**   **Nictibio Colilargo**

Rare to uncommon in the lower growth and mid-levels of terra firme forest in the lowlands of e. Ecuador. Until recently there were few Ecuadorian records of the Long-tailed Potoo; they included 19th-century specimens from Sarayacu and "Valle del Santiago" and two early-20th-century specimens from the Sumaco region ("below San José" and Eugenio). With its very distinctive vocalizations having recently become known, since the late 1980s there has been a spate of reports of this spectacular potoo, which is proving to be considerably more widespread than had previously been realized. It is now a well-known resident at La Selva and along the Maxus road southeast of Pompeya and has also been heard at Zancudococha (G. Budney), Pañacocha (PJG), Jatun Sacha, Iro on the lower Río Tiputini (M. Hedemark and A. Johnson), and along the Río Pacuyacu off the lower Río Aguarico (RSR and F. Sornoza). Unlike the Great Potoo (*N. grandis*) and Common Potoo (*N. griseus*), the Long-tailed rarely leaves the cover of terra firme forest, even when foraging at night. Given the recent record from w. Nariño in sw. Colombia (Salaman 1994), it is also not inconceivable that the species could be found south into nw. Ecuador. Recorded up to 700 m (Eugenio; MCZ).

One race: *longicaudatus*.

Range: w. Colombia; se. Colombia, s. Venezuela, and Guyana to se. Peru, ne. Bolivia, and Amazonian Brazil; se. Brazil, e. Paraguay, and ne. Argentina. More than one species is perhaps involved.

Nyctibius griseus   Common Potoo   Nictibio Común

Uncommon to fairly common in the canopy and edge of humid and deciduous forest and secondary woodland and in adjacent clearings in the lowlands of both e. and w. Ecuador, also ranging up in the lower subtropical zone on both slopes of the Andes. In the east quite widespread, occurring in várzea and riparian forest and woodland as well as the borders of terra firme forest. In the west recorded south locally into w. Loja (Sozoranga [Best et al. 1993] and the Macará area [v.o.]). Ranges up locally to 2100–2300 m on the west slope of the Andes in w. Pichincha (above Mindo and in the Tandayapa area), but on the east slope seemingly not found above about 1700 m (and there only locally, only in the south).

The species is sometimes called the Gray Potoo.

Racial allocations in *N. griseus* are somewhat uncertain, with characters weakly defined. We follow Cohn-Haft (1999) in placing western birds in *panamensis* and eastern birds in the nominate race.

Range: Nicaragua to extreme nw. Peru, n. Argentina, and Uruguay. Excludes Northern Potoo (*N. jamaicensis*) of n. Middle America, Jamaica, and Hispaniola.

Nyctibius maculosus   Andean Potoo   Nictibio Andino

Very rare to rare and apparently local (perhaps under-recorded) in the canopy and borders of montane forest in the subtropical zone on the east slope of the Andes. There are only a few Ecuadorian records of this potoo, a species seemingly scarce throughout its range. For many years the only record of this species in Ecuador was the 19th-century specimen labeled as from Ambato (but doubtless actually taken somewhere nearby on the east slope of the Andes). Recent reports come mainly from w. Napo on the Cordillera de Huacamayos and at nearby SierrAzul; at the latter no fewer than four individuals were observed (M. B. Robbins et al.) calling and chasing each other on 6 Mar. 1993 (ANSP, MECN). A few have also been seen recently around the nearby Cabañas San Isidro (M. Lysinger). In addition, two were heard in the Río Isimanchi valley (2250 m) in extreme sw. Zamora-Chinchipe on 7 Nov. 1992 (M. B. Robbins et al.). We believe that the several reported sightings of the Andean Potoo from the west slope of the Andes in Pichincha (e.g., Fjeldså and Krabbe 1990) require further confirmation, despite the published report (Salaman 1994) from w. Nariño in adjacent Colombia; some examples of the Common Potoo (*N. griseus*) may also show considerable white on the wing-coverts. Recorded mostly between about 1800 and 2300 m, but has been seen as high as 3200 m at Oyacachi (a pair observed by N. Krabbe on 31 Mar. 1997).

N. maculosus is regarded as a species separate from the lowland-inhabiting

N. leucopterus (White-winged Potoo) found locally in e. Amazonia and e. Brazil, following the evidence presented by T. S. Schulenberg, S. E. Allen, D. F. Stotz, and D. A. Wiedenfeld (*Gerfaut* 74[1]: 59–62, 1984) and M. Cohn-Haft (*Auk* 110[2]: 391–394, 1993). Monotypic.

Range: locally in Andes from w. Venezuela to w. Bolivia.

Nyctibius bracteatus   Rufous Potoo   Nictibio Rufo

Very rare and apparently local—but doubtless under-recorded—in seasonally flooded forest and in adjacent terra firme in the lowlands of e. Ecuador. There are only five definite records of this reclusive, little-known potoo from Ecuador. A 19th-century specimen was obtained at Sarayacu; a male in the Mejía collection was taken at Loreto on an unspecified date in 1937; one was heard nightly at Imuyacocha from 4 to 8 Dec. 1990 (RSR, PJG, and P. Coopmans et al.), with doubtless the same individual collected there on 20 Mar. 1991 (ANSP); one was seen in Oct. 1996 at its day roost near Kapawi Lodge on the Río Pastaza near the Peruvian border (G. Rivadeneira and J. Moore); and another was discovered at its day roost at Tiputini Biodiversity Center in Feb. 1999 (M. Lysinger et al.). In addition, there is a specimen of the Rufous Potoo in the MECN that lacks data but which presumably was taken somewhere in Ecuador. Recorded up to about 550 m at Loreto.

Monotypic.

Range: locally in e. Ecuador, e. Peru, nw. Bolivia, Amazonian Brazil, and Guyana.

Caprimulgidae   Nightjars and Nighthawks

Nineteen species in nine genera. Nightjars are worldwide in distribution, though nighthawks occur only in the Americas.

**Lurocalis semitorquatus*   Short-tailed Nighthawk   Añapero Colicorto

Uncommon and seemingly local (surely under-recorded, at least in the past; all Ecuadorian records are recent) in humid forest and adjacent clearings in the lowlands of w. and ne. Ecuador. In the east still recorded only from Sucumbíos and Napo, though it seems likely that the species also occurs in the southeast. In the west known locally from Esmeraldas (El Placer) south to sw. Manabí (Cerro San Sebastián in Machalilla National Park; ANSP), w. Guayas (above Salanguilla; N. Krabbe), and El Oro (Buenaventura, where 3 were seen on 26–27 Jul. 1992; P. Coopmans and M. Van Beirs et al.). Short-tailed Nighhawks are usually recorded around dusk, when birds emerge from the forest canopy to forage, sometimes well above the ground but more often just above the canopy. Only one specimen has been taken in Ecuador, a female (ANSP) of a pair seen at Cerro San Sebastián in sw. Manabí on 6 Aug. 1991; note that Meyer de Schauensee's (1966, 1970) listing of this species from e. Ecuador was based on the presence of the Rufous-bellied Nighthawk (*L. rufiventris*), then regarded as conspecific with the Short-tailed (see below). Recorded up locally to about 900 m.

The species was formerly called the Semicollared Nighthawk.

Excludes *L. rufiventris* of the Andes. The sole Ecuadorian specimen of *L. semitorquatus*, obtained on the west slope (see above), was tentatively assigned by M. B. Robbins to the race *stonei*, but specimen material in the species is so scanty that subspecific determinations remain difficult. That caveat aside, our conclusion is that the Ecuadorian example and the type of *stonei* (at ANSP, from Nicaragua) are essentially indistinguishable, and we therefore propose that the supposed Panamanian race *noctivagus* be synonymized, as Wetmore (1968) had earlier suggested might be the appropriate course. *Lurocalis* nighthawks in the lowlands of e. Ecuador—which appear to be resident (pairs have been seen and vocalizations heard)—are probably referable to *nattereri*, which is darker and more rufescent below than *stonei*; as noted above, no specimens are available.

Range: s. Mexico to n. Colombia, n. Bolivia, e. Paraguay, ne. Argentina, and se. Brazil; w. Ecuador; southern breeders are migratory.

Lurocalis rufiventris   **Rufous-bellied Nighthawk**   Añapero Ventrirrufo

Uncommon in montane forest and adjacent clearings in the subtropical and temperate zones on both slopes of the Andes. On the west slope recorded south to w. Chimborazo (an old record from Cayandeled) and w. Azuay (recent sightings at Chaucha; Krabbe et al. 1997). Most records of the Rufous-bellied Nighthawk are recent, and it appears that the species is more numerous and widespread than until recently had been realized; unless one makes a special effort for it, the species is a hard one to collect, and this accounts for the paucity of older records. Like the smaller Short-tailed Nighthawk (*L. semitorquatus*), the Rufous-bellied is usually seen at dusk, when single birds or pairs begin to forage over the forest canopy and adjacent clearings. Recorded mostly from 1500 to 2500 m, locally (or perhaps seasonally) as high as 3300 m in Carchi on the west slope of Cerro Mongus.

L. rufiventris is regarded as a species separate from the *L. semitorquatus* of the lowlands, on the basis of its marked morphological, habitat, and vocal differences. This separation had earlier been suggested by T. A. Parker III, A. Castillo U., M. Gell-Mann, and O. Rocha O. (*Bull. B. O. C.* 111[3]: 120–138, 1991). Monotypic.

Range: Andes from w. Venezuela to w. Bolivia; Sierra de Perijá.

Chordeiles acutipennis   **Lesser Nighthawk**   Añapero Menor

Uncommon to locally (or erratically?) fairly common in semiopen scrub and grassy areas, often near water, in the lowlands of w. Ecuador from coastal Esmeraldas south to El Oro and w. Loja; also known from the arid intermontane valleys of n. Ecuador in Imbabura (displaying males seen and tape-recorded near Tumbabiro in Mar. 1993; M. B. Robbins). There are also a few recent sightings of birds believed to be Lesser Nighthawks (and not Commons [*C. minor*]) from the northeastern lowlands (e.g., near Lago Agrio in Aug. 1981 [PJG et al.] and at Cuyabeno in Jul. 1991 [J. Rowlett and R. A. Rowlett et al.]). Found mainly below about 800 m, but recorded very locally as high as about 2000 m (Tumbabiro).

One race: *aequatorialis*. The birds found in the eastern lowlands would, however, presumably be referable to the similar nominate race. At present there is no evidence to indicate that boreal migrant *C. acutipennis* reach Ecuador, though it is possible that small numbers do so, the race *texensis* having been recorded as far south as Cauca in sw. Colombia (Hilty and Brown 1986).

Range: sw. United States to n. Chile, n. and e. Bolivia, and s. Brazil (but absent from much of w. and cen. Amazonia).

Chordeiles minor   Common Nighthawk   Añapero Común

An uncommon to occasionally fairly common transient and less numerous boreal winter resident in lighter woodland (often on river islands), forest borders, and clearings in the lowlands of e. Ecuador. In addition, small numbers of transients have been found on a semiregular basis in the highlands in and around Quito. There are, however, only a few records from w. Ecuador, all of transients; these include an Oct. specimen from Portovelo in El Oro, two Nov. sightings from Tinalandia and another from near Pedro Vicente Maldonado (P. Coopmans), and a May specimen from Mindo. Recorded from Sep. to May, with one exceptionally early bird having been seen at Kapawi Lodge on 17 Aug. 1999 (P. Coopmans et al.). Recorded up to at least 2800 m.

Presumably all or most of the described races pass the boreal winter months in South America, and each may occur at least as a transient in Ecuador, but the paucity of specimen material, coupled with only minor differences among some purported subspecies, precludes a more detailed analysis.

Range: breeds from North America south to Panama, wintering mainly or entirely in South America south to n. Argentina and Uruguay.

Chordeiles rupestris   Sand-colored Nighthawk   Añapero Arenizco

Rare to fairly common but very local on sandbars and islands in rivers in the lowlands of e. Ecuador. The Sand-colored Nighthawk is recorded mainly from the Río Napo (upriver to just below Coca) and the Río Aguarico (upriver to just below Chiritza) in the northeast; on very rare occasions it wanders as far up the Río Napo as Jatun Sacha (where there is but one sighting, fide B. Bochan). The only records from the Río Pastaza drainage are recent sightings from the vicinity of Kapawi Lodge near the Peruvian border. The only recent specimens (ANSP, MECN) taken in Ecuador were obtained on 15 Dec. 1990 on the Río Aguarico just below Lagartococha. Meyer de Schauensee (1966, 1970) mentions the species as occurring in e. Ecuador based on a specimen obtained on 4 Jan. 1931 at Boca Suno (E. Streseman, *Ornith. Monatsber.* 46[4]: 116, 1938). Recorded mostly below about 300 m, rarely up to about 400 m. One race: nominate *rupestris*.

Range: e. Colombia and sw. Venezuela to n. Bolivia and Amazonian Brazil.

*Nyctiprogne leucopyga   Band-tailed Nighthawk   Añapero Colibandeado

Uncommon and very local around certain blackwater lakes in the lowlands of far ne. Ecuador in e. Sucumbíos. First recorded in Ecuador on 16 Feb. 1990 at Garzacocha on the Río Lagarto, when about 30 were seen (M. Hedemark); comparable numbers were also seen here on 27 Mar. 1991 (RSR; F. Sornoza et al.), but unfortunately no examples could be collected. There have also been a few recent sightings from Cuyabeno (H. Kasteleijn; J. Rowlett and R. A. Rowlett et al.). The Band-tailed Nighthawk is almost invariably seen as it emerges from adjacent woodland to feed low over lakes and rivers around dusk, sometimes accompanied by other nighthawks. Recorded at about 200 m.

Race uncertain, but *exigua* seems the most likely.

Range: e. Colombia and cen. and sw. Venezuela to n. and e. Bolivia and Amazonian and cen. Brazil.

*Podager nacunda   Nacunda Nighthawk   Añapero Nacunda

Uncertain. Probably a very rare austral migrant to the lowlands of the northeast. There are only four known Ecuador sightings of this striking large nighthawk, mainly from the period when austral migrants might be expected to occur. One was seen over the Río Napo near Limoncocha on 29 Jun. 1981 (R. A. Rowlett et al.); nine were seen over the Río Napo at Primavera on 4 Jul. 1985 (J. Rowlett et al.); one was seen over Taracoa on 18 Aug. 1985 (PJG et al.); and one was seen along the Río Aguarico near Poca Pena on 9 Nov. 1992 (L. Navarrete et al.). The species is also listed—with no details—as having been seen at La Selva. Nacunda Nighthawks are usually seen at dusk, when they fly about conspicuously over semiopen areas and water. Recorded below 300 m.

Presumed race: nominate *nacunda*.

Range: Colombia, Venezuela, and the Guianas to n. Argentina and Uruguay; apparently nomadic, with precise breeding distribution and timing of presumed migratory movements still poorly understood.

Nyctidromus albicollis   Pauraque   Pauraque

Fairly common to common and widespread in lighter woodland, forest borders, and clearings in the lowlands and foothills of both e. and w. Ecuador. In the west recorded south through El Oro and w. Loja. The attractive Pauraque is more numerous and widespread in w. Ecuador than it is in the east, this mainly being a reflection of the far greater extent of deforestation that has occurred in that region. It is local and relatively scarce in mostly forested regions, there being mainly restricted to riparian areas, lake edges, and second-growth. Recorded up to about 1850 m near Celica and Alamor in w. Loja (ANSP), but mostly found below 1000–1200 m.

Sometimes (e.g., 1998 AOU Check-list) called the Common Pauraque.

One race: nominate *albicollis*.

Range: Texas and Mexico to extreme nw. Peru, ne. Argentina, and se. Brazil.

Nyctiphrynus ocellatus   Ocellated Poorwill   Chotacabras Ocelado

Rare to uncommon and perhaps local (though more probably just overlooked) in the lower and middle growth of terra firme forest and mature secondary woodland in the lowlands of e. Ecuador from Sucumbíos (north of Tigre Playa; ANSP) south at least locally through Morona-Santiago (Taisha; ANSP, MECN) and Pastaza (Kapawi Lodge). The first record from adjacent s. Colombia is given by M. B. Robbins and R. S. Ridgely (*Condor* 94[4]: 986, 1992). The Ocellated Poorwill remains a relatively poorly known bird in Ecuador, though small numbers have in recent years been found to occur at several localities where terra firme forest is extensive. Recorded up to about 500 m.

Does not include trans-Andean *N. rosenbergi* (Chocó Poorwill); see M. B. Robbins and R. S. Ridgely (*Condor* 94[4]: 984–987, 1992). One race: nominate *ocellatus*.

Range: Honduras to Costa Rica; extreme se. Colombia, e. Ecuador, e. Peru, n. Bolivia, and s. and cen. Amazonian Brazil; e. Brazil, e. Paraguay, and ne. Argentina.

Nyctiphrynus rosenbergi   Chocó Poorwill   Chotacabras del Chocó

Uncommon to locally fairly common in very humid forest in the lowlands and foothills of nw. Ecuador, favoring treefalls, river edges, and rivers. Recorded mainly from n. Esmeraldas, though there is one recent report from nw. Pichincha along the road north of Simón Bolívar near Pedro Vicente Maldonado (3 heard on 6 Jan. 1996; J. Nilsson, fide PJG) and another from sw. Imbabura (3 or more seen and tape-recorded near the Salto del Tigre bridge over the Río Guaillabamba on 14 Jul. 1997; P. Coopmans et al.). There are only a few Ecuadorian records of the Chocó Poorwill, but in Jul. 1990 the species was found to be relatively numerous (at least based on calling males; few were actually seen) northwest of Alto Tambo (ANSP and MECN), and it has also been found to be fairly common at Playa de Oro (O. Jahn). At the latter site it is widespread and is the only resident nightjar in extensive forest. Recorded up to about 600 m below El Placer.

The Chocó Poorwill was considered to be a Near-threatened species by Collar et al. (1994). Because of the paucity of Ecuadorian records, we feel we must concur, though we suspect that the species is more overlooked than actually rare. Several years ago, M. B. Robbins and R. S. Ridgely (*Condor* 94[4]: 984–987, 1992) were more optimistic regarding its status.

N. rosenbergi was shown to be a species distinct from *N. ocellatus* (Ocellated Poorwill) by Robbins and Ridgely (op. cit.), this being based on its very different voice, several major plumage differences, and disjunct distribution. We suspect that they may not even be sister taxa. Monotypic.

Range: w. Colombia and nw. Ecuador.

Caprimulgus longirostris Band-winged Nightjar Chotacabras Alifajeado

Uncommon to fairly common in semiopen and scrubby areas, paramo, and montane woodland borders in the Andes from Carchi and w. Sucumbíos south through Loja and adjacent Zamora-Chinchipe. Recorded mostly from 1800 to 3700 m, locally and in small numbers somewhat higher and (especially in arid regions) lower, having been reported as low as 1400 m in the Río Rircay valley in Azuay (Best et al. 1993).

One race: *ruficervix*.

Range: mountains of n. Venezuela, and Andes from w. Venezuela to Chile and Argentina; in the latter two countries and also in w. Peru also occurring in adjacent lowlands; in austral winter southern breeders withdraw northward into Paraguay and s. and e. Brazil (where the species may also breed); tepuis of s. Venezuela; Sierra de Perijá and Santa Marta Mountains. More than one species is perhaps involved.

*Caprimulgus cayennensis White-tailed Nightjar Chotacabras Coliblanco

Locally common in arid scrub of intermontane valleys in the lower subtropical zone of nw. Ecuador in Imbabura; may well also occur in adjacent Carchi. The White-tailed Nightjar was first found in Ecuador on 17 Oct. 1983 when a male was collected (LSUMZ) and several others were seen 6 km north of Salinas in the Río Palacara valley (C. G. Schmitt and D. C. Schmitt, *Bull. B. O. C.* 110[3]: 139–140, 1990). On 2–3 Mar. 1993, it was found at a site with similar habitat 3 km northeast of Tumbabiro in the Río La Chota valley; three specimens (ANSP, MECN) were taken at that time. Recorded from about 1400 to 2000 m.

One race: *apertus*.

Range: Costa Rica and Panama; n. and w. Colombia, most of Venezuela, the Guianas, and extreme n. Brazil; Martinique.

Caprimulgus anthonyi Anthony's Nightjar Chotacabras de Anthony

Uncommon to locally fairly common (perhaps seasonal or nomadic?) in grassy areas and adjacent scrub and lighter woodland in the more arid lowlands of w. Ecuador. Until recently very poorly known, in recent years the Anthony's Nightjar has been found at a number of localities: w. Esmeraldas (Vaquería; Atacames area, where numerous and breeding in Jan. 1993 [RSR and F. Sornoza]), Manabí (San Mateo; Puerto Cayo; north of Puerto López), Guayas (Santa Elena Peninsula; Chongón Hills), El Oro (Portovelo; Manglares-Churute Ecological Reserve; north of Santa Rosa; north of Huaquillas), and Loja (east of Mangaurco; Zapotillo). Numerous specimens have been taken (ANSP, MECN). Recorded up to at least 650 m.

Numbers of the Anthony's Nightjar seem to be increasing, which likely is the result of the clearance of dense scrub and woodland for pasturage.

The species is sometimes called the Scrub Nightjar.

C. anthonyi is regarded as a species separate from *C. parvulus* (Little Nightjar) of e. and cen. South America, following the initial suggestion of P. Schwartz (*Condor* 70[2]: 223–227, 1968) and then supported by recent information on living birds (see M. B. Robbins, R. S. Ridgely, and S. W. Cardiff, *Condor* 96[1]: 224–228, 1994). Not only does *C. anthonyi* have a very different tail pattern and other plumage differences from *C. parvulus*, but its voice is utterly different. Monotypic.

Range: w. Ecuador and nw. Peru.

*Caprimulgus maculicaudus	Spot-tailed Nightjar	Chotacabras Colipunteado

Uncertain. Known in Ecuador from only a single female (UKMNH) taken at Santa Cecilia west of Lago Agrio in Sucumbíos on 3 Jul. 1971. Prepared originally as an alcoholic specimen, and identified as *C. parvulus* (Little Nightjar), this specimen was first brought to our attention by T. J. Davis. Upon examining the bird, M. B. Robbins and RSR tentatively reidentified it as a Spot-tailed Nightjar, and its subsequent preparation as a dried skin confirms that this is indeed the case (fide T. J. Davis). Despite the lack of additional reports, we suspect that the Spot-tailed Nightjar may be a local resident in damp pastures and low shrubby areas in the lowlands of ne. Ecuador; elsewhere in its range the species is spreading and increasing in such anthropogenic habitats. Recorded at 340 m.

Monotypic.

Range: s. Mexico (where apparently migratory); Honduras and Nicaragua; nw. Colombia; sw. Venezuela, ne. Colombia, and (once) ne. Ecuador; e. Venezuela and the Guianas; extreme se. Peru, n. and e. Bolivia, and s. Amazonian and se. Brazil.

Caprimulgus nigrescens	Blackish Nightjar	Chotacabras Negruzco

Rare to uncommon and apparently local in or near rocky or gravelly areas with sparse or no vegetation (e.g., at landslides or roadcuts) in terra firme forest in the lowlands and foothills of e. Ecuador. The Blackish Nightjar seems to be absent from the far northeast and east, apparently because of the absence or scarcity of areas with much relief. Recorded up to about 1200 m along the Loreto road, and apparently most numerous in hilly areas, especially near the base of the Andes; there are few records in Ecuador from below 300 m.

Monotypic.

Range: se. Colombia, s. Venezuela, and the Guianas to n. Bolivia and Amazonian Brazil.

*Caprimulgus rufus   Rufous Nightjar   Chotacabras Rufo

Uncommon in deciduous woodland and borders in the Río Marañón drainage of extreme s. Zamora-Chinchipe in the Zumba region. The Rufous Nightjar was first recorded in Ecuador on 15 Aug. 1989, when one was heard singing in the predawn near Zumba (RSR and M. B. Robbins et al.). Several more were heard near the Río Mayo on 12 Dec. 1991 (RSR and F. Sornoza), and one was tape-recorded there in early Aug. 1992 (F. Sornoza), but as yet it has not proven possible to obtain a specimen. There are recent records from as close as San Martín in n. Peru (fide M. B. Robbins). Recorded between about 700 and 1100 m.

Given the absence of specimens, the racial allocation of Ecuadorian birds cannot be known.

Range: Costa Rica to n. Venezuela; locally in s. Venezuela, se. Ecuador, and e. Peru; e. Bolivia and e. Amazonian and ne. Brazil south to n. Argentina; locally in Suriname and French Guiana; St. Lucia. Includes *otiosus* (M. B. Robbins and T. A. Parker III, *Studies in Neotropical Ornithology Honoring Ted Parker*, Ornithol. Monogr. no. 48: 601–607, 1997).

It remains possible that the boreal migrant Chuck-will's-widow (*C. carolinensis*) will be found as far south as Ecuador for it has been recorded from as close as Ricaurte in adjacent Nariño, Colombia (ANSP; based on a male taken on the seemingly improbable date of 7 Jun. 1944 by K. von Sneidern).

Hydropsalis climacocerca   Ladder-tailed   Chotacabras Coliescalera
                                 Nightjar

Uncommon to fairly common in semiopen areas along rivers (where often especially numerous on islands) and along lakeshores in the lowlands of e. Ecuador. The beautiful Ladder-tailed Nightjar usually roosts on sandy islands, often amongst piles of driftwood. Found mostly below 400 m, but recorded in small numbers up to about 1000 m along the Río Upano at Macas in Morona-Santiago (Chapman 1926) and along the Río Nangaritza at Shaime in Zamora-Chinchipe (WFVZ); the species may occur along other larger rivers at comparable elevations.

One race: nominate *climacocerca*.

Range: se. Colombia, s. Venezuela, and the Guianas to n. Bolivia and Amazonian Brazil.

Uropsalis segmentata   Swallow-tailed Nightjar   Chotacabras Tijereta

Rare and apparently local near damp cliffs at the edge of montane forest and in shrubby areas (e.g., at landslides) in the temperate zone on both slopes of the Andes. On the west slope recorded only from Carchi (females collected just west of Páramo del Ángel on 2 Aug. 1988 and in Nov. 1990; ANSP, MECN), Imbabura (seen along the Apuela road in Jun. 1987; J. Fjeldså), Pichincha (a few sightings near the Tandayapa ridge [e.g., M. R. Welford, *Cotinga* 10: 41–42,

1998] and from the upper Chiriboga road [v.o.]), and Azuay (a Mar. 1991 sighting from near Sural; N. Krabbe). On the east slope known from w. Napo, e. Chimborazo, Cañar, e. Loja, and adjacent Zamora-Chinchipe. Recorded mostly between 2500 and 3200 m, but locally occurs down to 2100–2150 m near the crest of the Cordillera de Huacamayos (e.g., a male [ANSP] taken on 14 Oct. 1992)—where it occurs with larger numbers of the Lyre-tailed Nightjar (*U. lyra*)—and to 1950–2000 m above Mindo.

One race: nominate *segmentata*.

Range: Andes from n. Colombia to w. Bolivia.

Uropsalis lyra   **Lyre-tailed Nightjar**   **Chotacabras Colilira**

Uncommon and apparently decidedly local near damp cliffs and along rocky streams and rivers in mostly forested areas in the foothills and subtropical zone on both slopes of the Andes. On the west slope known only from Pichincha, but presumably it is found northward as well as the species has been recorded recently from adjacent Nariño in s. Colombia (Salaman 1994). The Lyre-tailed Nightjar has presumably been overlooked in many parts of its range in Ecuador, much of which is difficult to access; in recent years it has been seen regularly at various more accessible sites in Pichincha (above Mindo, lower Chiriboga road, etc.) and in w. Napo on the Cordillera de Huacamayos. Recorded from about 800 to 2250 m.

One race: nominate *lyra*.

Range: Andes from w. Venezuela to nw. Argentina.

Apodiformes
Apodidae   Swifts

Fifteen species in six genera. Swifts are worldwide in distribution.

Streptoprocne zonaris   **White-collared Swift**   **Vencejo Cuelliblanco**

Fairly common to common and widespread at virtually all elevations in the Andes, locally nesting or roosting even well above treeline. White-collared Swifts range out—sometimes in large, screeching, high-flying flocks of several hundred birds—to feed over the adjacent lowlands (usually in humid regions), with decreasing frequency at increased distances from the Andes. The species occasionally ranges all the way down to the coast, especially in s. Ecuador. It nests in groups behind waterfalls and on damp cliffs and the sides of caves in montane or precipituous areas. Ranges from sea level up to about 4000 m, occasionally even higher.

Two similar races are found in Ecuador, *altissima* at very high elevations (it typically nests above 3000 m) and the smaller *subtropicalis* (see K. C. Parkes, *Avocetta* 17: 95–100, 1993) at lower elevations; until the erection of the latter name, these birds were assigned to the race *albicincta* (now confined to ne.

South America). During their long feeding flights there surely is elevational overlap, and it seems likely that the two taxa even at times forage together.

Range: n. Mexico to n. Argentina, e. Paraguay, and se. Brazil (though absent or infrequent in much of Amazonia and in ne. Brazil); Greater Antilles.

Cypseloides rutilus   Chestnut-collared Swift   Vencejo Cuellicastaño

Uncommon to fairly common and widespread in the foothills and subtropical zone on both slopes of the Andes and above interandean valleys, also sometimes flying out over the adjacent lowlands. In the west seemingly occurs only in small numbers (or just locally) south of Pichincha, though there are several recent sightings from the Buenaventura area of El Oro and in the Alamor/Celica/Macará region of w. Loja. There are also two reports of the Chestnut-collared Swift from areas well away from the Andes, one on each slope: several were seen on 25 Feb. 1994 flying with other swifts over the coastal Cordillera de Mache in w. Esmeraldas at Bilsa (R. P. Clay et al.), and small numbers have also been seen along the Maxus road southeast of Pompeya (S. Howell and RSR et al.). We suspect that the usually high-flying Chestnut-collared Swift will probably be shown to occur regularly over the adjacent eastern lowlands, and it may occur regularly in the coastal cordillera as well. It often accompanies usually larger numbers of White-collared Swifts (*Streptoprocne zonaris*), also sometimes various *Chaetura* swifts. Recorded mostly between 1000 and 2700 m, but routinely drops lower while feeding over the adjacent lowlands (regularly down to 400–500 m, occasionally even lower).

M. Marín and F. G. Stiles (*Proc. West. Found. Vert. Zool.* 4, no. 5, 1992) suggested that this species be placed in the genus *Streptoprocne*, primarily on the basis of its nesting behavior; this treatment was adopted in the 1998 AOU Check-list. However, based on its vocal and structural similarity to the genus *Cypseloides*, to which the species has long been assigned, we retain it there (in accord with Howell and Webb 1995 and Chantler 1999). One race: *brunnitorques* (which closely resembles the nominate race, into which it is perhaps better synonymized).

Range: locally from n. Mexico to n. Venezuela and w. Bolivia; tepuis of s. Venezuela.

Cypseloides cryptus   White-chinned Swift   Vencejo Barbiblanco

Uncertain. Only a single Ecuadorian specimen (BMNH) of the White-chinned Swift is definitely known; unfortunately it is labeled only as being from "Ecuador" (C. T. Collins, *Bull. B. O. C.* 88[8]: 133, 1968). A 19th-century specimen taken at Gualaquiza in Morona-Santiago, and then recorded as the Sooty Swift (*C. fumigatus*)—this before the White-chinned had been specifically separated from that species—seems likely to refer to the White-chinned Swift, but it apparently has not been reexamined (J. T. Zimmer, *Auk* 62[4]: 586–592, 1945). There are also a few recent sightings of this swift, which seemingly is rare throughout its range (though its supposed rarity may be due mainly to the difficulty in collecting or identifying it with any assurance). These include three seen at Taracoa on 7 Jan. 1982 (RSR, PJG, and S. Greenfield); a small group with Band-rumped Swifts (*Chaetura spinicauda*) at El Placer on 13 Aug. 1987 (RSR); a group of about eight seen at close range with a flock of White-collared Swifts (*Streptoprocne zonaris*) and Gray-rumped Swifts (*C. cinereiventris*) in the western lowlands at Río Palenque on 30 Mar. 1989 (RSR, PJG, G. Tudor et al.); and two to three seen with various other swifts at Yuturi on 11 Aug. 1996 (S. Howell and S. Webb). The White-chinned Swift presumably nests in the foothills and subtropical zone on both slopes of the Andes, at times dispersing out into the adjacent lowlands to feed.

Monotypic.

Range: very locally from Costa Rica, and in Guyana, Venezuela, Colombia, Ecuador, and Peru.

Cypseloides cherriei   Spot-fronted Swift   Vencejo Frentipunteado

Apparently rare and local in and above montane forest in the foothills and subtropical zone on both slopes of the Andes in n. Ecuador, with the paucity of records perhaps being as much due to identification difficulties as actual rarity. The Spot-fronted Swift was first recorded in Ecuador from two specimens (WFVZ) taken by M. Marín and J. M. Carrión on 19 Jun. 1989 at a nesting site near a waterfall along a stream flowing through montane forest at Las Palmeras along the Chiriboga road in w. Pichincha; another was obtained at the same site on 26 Aug. 1990 (see M. A. Marín and F. G. Stiles, *Condor* 95[2]: 479–483, 1993). One was believed seen on the Cordillera de Huacamayos on 3 Mar. 1992 (D. Wolf and PJG et al.), and there are also two recent sightings from the Tinalandia area (a group of 14 on 17 Mar. 1995 [M. Lysinger] and a group of 20 on 3 Dec. 1996 [P. Coopmans et al.]), as well as another from near Pedro Vicente Maldonado north of Simón Bolívar (a flock of about 20 on 22 Dec. 1997; P. Coopmans). These represent the southernmost records of a species that appears to be rare and/or local throughout its limited range. Recorded from about 500 to 1900 m.

Given the Spot-fronted Swift's rarity throughout its small range, we believe it appropriate to accord the species Data Deficient status for Ecuador; it may well deserve such status throughout its range.

Monotypic.

Range: very locally in Costa Rica, Venezuela, Colombia, and Ecuador.

Cypseloides lemosi   White-chested Swift   Vencejo Pechiblanco

Known in Ecuador only from a series of recent sightings from the foothills and subtropical zone on the east slope of the Andes in w. Napo and in the lowlands farther east in Napo. The first Ecuadorian report involved at least two birds seen with other *Cypseloides* swifts north of Archidona along the road to Loreto on 31 Mar. 1990 (B. Whitney); one was then seen near a flock of *Chaetura* swifts near Archidona (600 m) on 22 Nov. 1992 (P. Coopmans and M.

Lysinger); singletons were seen with other swifts along the Maxus road south-east of Pompeya on 17 and 20 Jul. 1994 (S. Howell and RSR); and two to four birds were seen at Yuturi on 20 Jul. and 11 Aug. 1996 and from 7 to 13 Aug. 1998 (S. Howell et al.). Apparently small numbers of White-chested Swifts, accompanying larger numbers of several other swift species (primarily Chestnut-collared Swifts [*C. rutilus*] and White-collared Swifts [*Streptoprocne zonaris*]), are dispersing from the Andes and fanning out to feed over the adjacent lowlands, at least in Jul. and Aug. In addition to the above "scatter" of reports, there was a series of 1995–1996 sightings of up to 25–30 birds seen at once (usually accompanying White-collared and Chestnut-collared Swifts) at and near Cabañas San Isidro (D. Stejskal; M. Lysinger). Specimen or photographic confirmation for Ecuador of this poorly known species (E. Eisenmann and F. C. Lehmann, *Am. Mus. Novitates* 2117, 1962) is still needed. The White-chested Swift was heretofore known only from the Andes of sw. Colombia; it is not known whether the Ecuador birds represent a resident population or merely wanderers from Colombia. Recorded up to at least 2000 m.

Collar et al. (1992, 1994) suspected that this swift may have undergone a decline in recent decades, and in the latter publication it was therefore accorded Vulnerable status. The recent Ecuador reports would seem to suggest otherwise and that the species' current situation may not be as critical as had been feared. In the absence of solid evidence as to the species' status in Ecuador, however, we feel we must concur with Collar et al.'s (1994) assessment.

Monotypic.

Range: very locally in s. Colombia and e. Ecuador, mainly in Andes.

Black Swifts (*C. niger*) of the northern race *borealis* were recently recorded at a site in sw. Colombia (F. G. Stiles and A. J. Negret, *Condor* 96[4]: 1091–1094, 1994), apparently occurring as southbound boreal migrants in Sep. and Oct. The species therefore needs to be watched for in Ecuador.

Chaetura pelagica   Chimney Swift   Vencejo de Chimenea

Apparently an uncommon transient throughout, and probably also a boreal winter resident in the eastern lowlands and apparently the northwestern lowlands as well. There seems to be only one Ecuadorian specimen of this swift, and surprisingly few sightings, but it has been almost certainly under-recorded because of the difficulty of collecting *Chaetura* swifts or of definitely identifying them in the field. The specimen (FMNH; heretofore unpublished) was taken by B. L. Clauson and R. M. Timm at the mouth of the Río Cuyabeno into the Río Aguarico on 23 Oct. 1983 (fide T. S. Schulenberg). Sightings include several seen along the Río Cayapas in Esmeraldas on 5–7 Mar. 1980 (PJG and S. Greenfield), small numbers seen at Bilsa in w. Esmeraldas in the latter half of Feb. 1994 (R. Clay et al.), about 10 seen near Tena on 10 Mar. 1989 (RSR, PJG, and G. Tudor et al.), two seen with other *Chaetura* near Limoncocha on 5 Nov. 1994 (RSR), and individuals seen at Cajanuma (3200 m) in Podocarpus National Park on 9 Mar. 1989 (Bloch et al. 1991).

Monotypic.

Range: breeds in North America, with (so far as known) most birds wintering in Amazonia, some in w. Peru and n. Chile, and probably in w. Ecuador.

*[*Chaetura chapmani*   Chapman's Swift   Vencejo de Chapman]

Uncertain. Known in Ecuador only from a single sighting from the lowlands of e. Ecuador, of two birds seen at Jatun Sacha on 31 Jan. 1994 (R. Clay, J. Vincent, and S. Jack). This obscure swift, difficult to identify with certainty, would be easily overlooked among other swifts unless carefully studied under ideal circumstances. Given the difficulties involved in field identification of *Chaetura* swifts, the single Ecuador record should perhaps not even be included, but the observers were confident that the Chapman's Swift was indeed the species involved. Its status in Ecuador remains to be determined. Recorded at 400 m.

The systematics of *C. chapmani* and its allies were revised by M. Marín (*Studies in Neotropical Ornithology Honoring Ted Parker*, Ornithol. Monogr. no. 48: 431–443, 1997). Marín (op. cit.) proposed that the taxon *viridipennis*, formerly considered a subspecies of *C. chapmani*, be raised to species status (Amazonian, or Mato Grosso, Swift), with a distribution mainly in sw. Amazonia. *C. chapmani* then becomes monotypic. We tentatively follow this new arrangement, but it is not universally accepted (see Chantler 1999). It is impossible to know with certainty whether the birds observed at Jatun Sacha were *C. chapmani* or *C. viridipennis*; on the basis mainly of range, we very tentatively assume the former.

Range: locally from Panama to e. Ecuador (one sighting), n. Amazonian Brazil, and the Guianas.

Chaetura brachyura   Short-tailed Swift   Vencejo Colicorto

Fairly common over semiopen areas and forest borders in the lowlands of e. Ecuador. In extensively forested areas the Short-tailed Swift seems to favor swampy places where palms (e.g., *Mauritia* spp.), abound. Recorded mostly below about 700 m, occasionally somewhat higher in cleared areas. We consider the published report (Krabbe and Sornoza 1994) of this species from the Cordillera del Cóndor near Chinapinza (1650–1700 m) to be uncertain; the species has not otherwise been reported from anywhere near as high, and in fact it does not even seem to occur in the Río Nangaritza valley below the Cordillera del Cóndor (v.o.).

Does not include *C. ocypetes* (Tumbes Swift). There exists confusion concerning the race to which Ecuadorian birds should be assigned, nominate *brachyura* or *cinereocauda*. There is very little specimen material available. ANSP has only one Ecuadorian specimen, a recently obtained female. It appears—based on the criteria of J. T. Zimmer (*Am. Mus. Novitates* 1609, 1953) and on comparison with the type in ANSP—to be referable to *cinereocauda*, this being based on its uniform blackish throat and a wing length of 118 mm.

Range: Panama to n. and e. Bolivia and Amazonian Brazil.

Chaetura ocypetes   Tumbes Swift   Vencejo de Tumbes

Uncommon and somewhat local at the edge of deciduous forest and woodland and over adjacent cleared areas in the lowlands of sw. Ecuador from sw. Manabí (several sightings from the Jipijapa area in the early 1990s; RSR) and Guayas (vicinity of Guayaquil and at Cerro Blanco in the Chongón Hills) south locally through El Oro and w. Loja (numerous sightings from the Macará and Zapotillo areas, also near Mangaurco [ANSP], with a few having been seen as high as Pindal [P. Coopmans]). Recorded up to 1000 m.

C. *ocypetes*, originally described by J. T. Zimmer (*Am. Mus. Novitates* 1609, 1953) as a subspecies of *C. brachyura* (Short-tailed Swift), is here considered a monotypic species separate from *C. brachyura* on the basis of several distinctive morphological characters as well as its different voice and disjunct range.

Range: sw. Ecuador and nw. Peru.

Chaetura spinicauda   Band-rumped Swift   Vencejo Lomifajeado

Uncommon to locally fairly common over humid forest and adjacent clearings in the lowlands, foothills, and lower subtropical zone of nw. Ecuador from Esmeraldas (El Placer [ANSP] and Playa de Oro [O. Jahn et al.]) south to n. Manabí (Filo de Monos; WFVZ specimen and sightings) and w. Chimborazo (a 19th-century specimen from Chimbo). No recent reports exist from south of w. Cotopaxi, though there seems no particular reason to expect that the species no longer occurs southward. See the summary of some recent Ecuadorian records by Marín et al. (1992). In Ecuador the Band-rumped Swift occurs primarily in the foothills and lower subtropical zone (including locally on the coastal Cordillera de Chindul in n. Manabí). Recorded mostly between about 300 and 1500 m, but has been found lower (to 50 m) at Playa de Oro.

One race: *aetherodroma*.

Range: Costa Rica to nw. Ecuador; s. and e. Venezuela, the Guianas, and n. and e. Amazonian Brazil.

Chaetura cinereiventris   Gray-rumped Swift   Vencejo Lomigris

Uncommon to locally fairly common over humid forest and clearings in the lowlands and foothills of both e. and w. Ecuador, ranging up in smaller numbers into the lower subtropics on both slopes. In the west recorded from n. Esmeraldas (seasonally [?] as far north as Playa de Oro; O. Jahn et al.) south to nw. Guayas (hills south of the lower Río Ayampe [RSR] and at Cerro La Torre [N. Krabbe] and Loma Alta [Becker and López-Lanús 1997]), El Oro (El Chiral and Buenaventura; v.o.), and w. Loja (Las Piñas). The Gray-rumped Swift has also been reported from as far south as Tambo Negro in extreme s. Loja (Best et al. 1993), an unusually dry locality for this species and where its presence needs to be confirmed. Recorded up to about 1700 m on the east slope of the Andes; generally not above 900–1000 m on the west slope (though an old specimen from El Chiral in El Oro was taken at about 1600 m).

Two rather different races are found in Ecuador, *sclateri* in the east and *occidentalis* in the west. *Occidentalis* has uniform pure gray underparts (lacking *sclateri*'s paler throat) and its rump is rather dark and uncontrasting (whereas *sclateri*'s is more contrastingly pale gray).

Range: Nicaragua to w. Panama; w. Colombia; w. Ecuador and extreme nw. Peru; Venezuela and Guyana to n. Bolivia and Amazonian Brazil; e. Brazil, e. Paraguay, and ne. Argentina.

Chaetura egregia   Pale-rumped Swift   Vencejo Lomipálido

Rare to uncommon over humid forest and clearings in the lowlands and foothills of e. Ecuador, primarily in Morona-Santiago and Zamora-Chinchipe. First recorded from sightings and a specimen (WFVZ) obtained by M. Marín at Tayuntza in Morona-Santiago on 18 Aug. 1987. Three additional specimens (WFVZ, Los Angeles County Museum of Natural History) were obtained by M. Marín and others near Paquisha in Zamora-Chinchipe on 8 Aug. 1989, and sightings were made by the same observers before and after this at several additional sites in the Río Zamora and Nangaritza valleys (see Marín et al. 1992). In addition, one bird was seen at Taisha in Morona-Santiago on 4 Aug. 1990 (RSR), at least three were seen along the Río Lagarto above Imuyacocha in e. Napo on 8 Dec. 1990 (RSR and PJG et al.), the species was seen at La Selva first on 10 Jan. 1991 (PJG; also subsequent records), and three were seen with other swifts along the Maxus road southeast of Pompeya on 9 Nov. 1995 (RSR); there likely have been other sightings as well. It appears that until recently the Pale-rumped Swift was overlooked among the Gray-rumped Swifts (*C. cinereiventris*) with which it occurs in e. Ecuador; to date all records seem to fall within the Aug.–Dec. period, so it is not inconceivable that the species may be a long-distance migrant to Ecuador. Recorded up to about 1000 m.

Monotypic.

Range: e. Ecuador, e. Peru, n. Bolivia, and sw. Amazonian Brazil.

Aeronautes montivagus   White-tipped Swift   Vencejo Filipunteado

Rare to locally fairly common in the subtropical and lower temperate zones on both slopes of the Andes in n. Ecuador and, at least locally, in the central valley. The White-tipped Swift is best known from various arid valleys in the Río Guaillabamba drainage north and east of Quito, whence a series of specimens (WFVZ, MECN, ANSP) has been obtained, these being the first from Ecuador (see Marín et al. 1992). We do not know the basis for the species' inclusion in Ecuador's avifauna in Meyer de Schauensee (1970, but not 1966); we can find no published records from prior to that year. Flocks have also been regularly seen as far north as the Río Chota valley near the Carchi/Imbabura border. There are only a few reports known to us from s. Ecuador, including a group of six seen near Oña in s. Azuay on 25 Feb. 1980 (RSR and D. Finch et al.) and a small group seen near Catacocha in Loja in Feb. 1989 (P. Coopmans); the species was also reported seen in the Río Rircay valley of sw. Azuay and at

Buenaventura in El Oro (Best et al. 1993). More surprisingly, there is also one report of a wandering bird well studied on 25 Feb. 1994 as it flew with other swifts over the coastal Cordillera de Mache in w. Esmeraldas at Bilsa (600–700 m; R. Clay et al.). Recorded mostly from about 1300 to 2700 m, occasionally as high as 3200 m (e.g., at Cerro Mongus in e. Carchi), but most numerous between 1500 and 2000 m.

One race: nominate *montivagus*.

Range: locally in mountains of n. Venezuela, and in Andes from w. Venezuela to w. Bolivia and nw. Argentina; Sierra de Perijá and Santa Marta Mountains; tepuis of s. Venezuela, Guyana, and n. Brazil.

Panyptila cayennensis   Lesser Swallow-tailed   Vencejo Tijereta
                                 Swift   Menor

Uncommon and perhaps somewhat local in the lowlands and foothills of e. and w. Ecuador, mainly ranging over semiopen areas and forest borders (less often over more or less continuous forest). In the west recorded south very locally through Guayas, El Oro (Buenaventura), and w. and s. Loja (2 specimens [ANSP, MECN] obtained along the Río Catamayo near El Empalme in Aug. 1989; also one seen north of Guayquichuma on 11 Feb. 1997 [P. Coopmans et al.] and 2 seen over hills west of Macará on 17 Sep. 1998 [RSR et al.]). See the summary of some recent records by Marín et al. (1992). Recorded mostly below 900 m, but has been seen to 1300 m near Nanegalito in w. Pichincha (M. Marín).

One race: nominate *cayennensis*.

Range: s. Mexico to nw. Peru, n. Bolivia, and Amazonian Brazil; e. Brazil.

Tachornis squamata   Neotropical Palm-Swift   Vencejo de Morete

Fairly common but somewhat local in the lowlands of e. Ecuador, generally in semiopen terrain or in the vicinity of water, and often near *Mauritia* or *Bactris* palms (other palms as well?), nesting on the underside of their fronds. What are—surprisingly—the first specimens for Ecuador were obtained by M. Marín et al. at Pachicutza in Zamora-Chinchipe in Jul. and Aug. 1989 (WFVZ, MECN); see the summary of some recent records by Marín et al. (1992). Recorded mostly below 500 m, but at least locally ranges higher in Zamora-Chinchipe (seen at 950 m in the Zamora region in Aug. 1990 [P. Coopmans] and recorded at 1000 m at Pachicutza).

The species has generally been called the Fork-tailed Palm-Swift, but its long tail is actually rarely seen to be forked, being instead almost always held in a needlelike point. There are ecologically convergent swift species in tropical Africa (African Palm-Swift [*Cypsiurus parvus*]) and tropical Asia (Asian Palm-Swift [*C. balasiensis*]).

The species was formerly (e.g., Meyer de Schauensee 1966, 1970) usually placed in the monotypic genus *Reinarda*; we follow Sibley and Monroe (1990) in placing it in *Tachornis*. One race: *semota*.

Range: e. Colombia, most of Venezuela, and the Guianas to n. Bolivia and Amazonian and e. Brazil.

Trochilidae   Hummingbirds

One hundred thirty-two species in 57 genera. Hummingbirds are exclusively American in distribution, reaching by far their highest diversity at tropical latitudes and especially in the Andes. We should note that collecting hummingbirds was very prevalent in Ecuador in the late 19th and early 20th centuries, with much of this effort being made by so-called "native collectors." The accuracy of many of their recorded localities is questionable. In the following accounts, if an old collecting locality is strongly at variance from what is presently known to be that species' distribution, then that old locality has usually been ignored.

Glaucis hirsuta   Rufous-breasted Hermit   Ermitaño Pechicanelo

Uncommon to locally common in the undergrowth of humid forest borders (both terra firme and várzea), secondary woodland, and adjacent clearings in the lowlands of e. Ecuador. Favors *Heliconia* thickets and areas near water. Recorded up to about 1100 m.

The species has also been called the Hairy Hermit.

One race: nominate *hirsuta*.

Range: Panama to n. Colombia, n. Bolivia, and Amazonian Brazil; e. Brazil; Grenada.

Glaucis aenea   Bronzy Hermit   Ermitaño Bronceado

Uncommon to locally fairly common in the undergrowth of secondary woodland and humid forest borders in the lowlands of nw. Ecuador from Esmeraldas south to s. Pichincha (Río Palenque, where it is regular) and n. Los Ríos (Jauneche; P. Coopmans). Favors *Heliconia* thickets and areas near water, and usually does not occur inside continuous forest, where it is replaced by the Band-tailed Barbthroat (*Threnetes ruckeri*; O. Jahn). Recorded up locally to about 600 m (below El Placer in Esmeraldas; RSR and M. B. Robbins).

G. aenea has sometimes been considered conspecific with *G. hirsuta* (Rufous-breasted Hermit) though this seems unlikely given *G. aenea*'s decidedly larger size and other plumage differences; contra Meyer de Schauensee (1966, p. 158), the two species do not "occur together commonly on the Pacific slope of Colombia." *G. aenea* is often considered to be monotypic, despite its disjunct distribution, but we follow Wetmore (1968) in recognizing *columbiana* of the South American portion of the species' range.

Range: Nicaragua to w. Panama; w. Colombia and nw. Ecuador.

Threnetes niger   Pale-tailed Barbthroat   Barbita Colipálida

Uncommon to fairly common in the undergrowth of humid forest (both terra firme and várzea), secondary woodland, and borders in the lowlands of e. Ecuador. Seems less numerous or more local in far e. Ecuador than it is in areas closer to the Andes. The Pale-tailed Barbthroat favors *Heliconia* thickets and other vegetation near water. Regularly ranges up to about 1100 m, in Morona-Santiago and Zamora-Chinchipe up at least locally to as high as 1600 m (e.g., on the Cordillera de Cutucú [Hamburg collection, fide C. Hinkelmann], along the Loja-Zamora road [MCZ specimen; recent sightings], and along the road between Valladolid and Palanda).

This species long went by the name *T. leucurus*, but Schuchmann (1999) concluded that *T. niger*, formerly considered a separate species (Black Barbthroat), was actually only a localized morph of the widespread species. The name *niger* has priority, so the species name must become *T. niger*. One race occurs in Ecuador: *cervinicauda*.

Range: se. Colombia, s. Venezuela, and the Guianas to n. Bolivia and Amazonian Brazil.

Threnetes ruckeri   Band-tailed Barbthroat   Barbita Colibandeada

Uncommon to locally common in the undergrowth of humid forest, secondary woodland, and borders in the lowlands of w. Ecuador south locally to sw. Manabí (Machalilla National Park; MECN), n. Guayas (Pacaritambo), s. Los Ríos (at least formerly at Babahoyo), se. Guayas (hills in Manglares-Churute Ecological Reserve; Pople et al. 1997), and south along the base of the Andes to El Oro (Buenaventura; Robbins and Ridgely 1990). The Band-tailed Barbthroat favors *Heliconia* thickets and banana plantations near wooded areas, but in Esmeraldas and elsewhere it frequently occurs inside forest as well, especially near streams and around treefalls (O. Jahn). Recorded up to about 900 m.

One race: nominate *ruckeri*.

Range: Guatemala and Belize to w. Venezuela and sw. Ecuador.

Phaethornis yaruqui   White-whiskered Hermit   Ermitaño Bigotiblanco

Common in the undergrowth of humid forest, secondary woodland, and borders in the lowlands and foothills of w. Ecuador south to El Oro (La Chonta, Buenaventura). The White-whiskered Hermit is numerous at both Tinalandia and Río Palenque. Recorded mostly below about 1300 m, in smaller numbers as high as 1500–1750 m (e.g., above Mindo and at Maquipucuna).

Following the suggestion of C. Hinkelmann (pers. comm.), we consider *P. yaruqui* to be monotypic. Ecuadorian birds were previously assigned to nominate *yaruqui*. Colombian birds, formerly placed in *sanctijohannis*, were distinguished on characters based on juvenile and subadult birds.

Range: w. Colombia and w. Ecuador.

Phaethornis guy   Green Hermit   Ermitaño Verde

Fairly common to common in the undergrowth and borders of montane forest in the foothills and lower subtropical zone on the east slope of the Andes. Although there is some overlap, the Green tends to occur at lower elevations than the Tawny-bellied Hermit (*P. syrmatophorus*); at a few sites the two species have even been captured in the same mist-nets. Recorded mostly from about 900 to 1800 m, perhaps locally somewhat lower (e.g., there is an AMNH specimen from "2000 feet, Zamora," though this is perhaps mislabeled).

One race: *apicalis*.

Range: highlands of Costa Rica to e. Panama; mountains of ne. Venezuela, and Andes from w. Venezuela to s. Peru; Sierra de Perijá.

Phaethornis syrmatophorus   Tawny-bellied   Ermitaño
                                   Hermit   Ventrileonado

Uncommon to locally fairly common in the undergrowth of montane forest and forest borders in the subtropical zone on both slopes of the Andes. On the west slope recorded mainly south to Pichincha, but there are old records from Chimborazo (Cayandeled and Pallatanga) and recent ones from El Oro (e.g., Buenaventura; VIREO photograph) and w. Loja (Sozoranga; Best et al. 1993). Recorded from about 900 to 2200 m, but mostly found above 1300 m; one specimen (LSUMZ) was collected by C. G. and D. C. Schmitt as high as 3025 m on the Laguna Cuicocha-Apuela road. The Tawny-bellied Hermit tends to range lower on the west slope, this presumably in response to the absence there of the Green Hermit (*P. guy*).

Two similar races occur in Ecuador, *columbianus* on the east slope and nominate *syrmatophorus* on the west slope. The nominate race is comparatively uniform orange ochraceous below, whereas *columbianus* is slightly darker and browner on the sides of the throat and foreneck.

Range: Andes from n. Colombia to n. Peru.

Phaethornis baroni   Baron's Hermit   Ermitaño de Baron

Fairly common in humid and deciduous forest, woodland, and their borders in the lowlands and foothills of w. Ecuador from w. Esmeraldas (Esmeraldas [city]), w. Imbabura (an old specimen from Intag), and n. Pichincha south through El Oro to w. Loja. Recorded mostly below about 1300 m (Mindo area), locally as high as 1700 m near Amaluza in s. Loja (R. Williams et al.).

C. Hinkelmann (*Ornitol. Neotrop.* 7: 119–148, 1996) considered trans-Andean forms of what was formerly considered *P. superciliosus* (Long-tailed Hermit, or Eastern Long-tailed Hermit) as a separate species (*P. longirostris*, Long-billed Hermit, or Western Long-tailed Hermit). *P. superciliosus* is now considered to range only in n. South America. Hinkelmann (op. cit.) treated the taxon *baroni*, a near-endemic of Ecuador, as a subspecies of *P. longirostris*. However, we conclude that given *baroni*'s disjunct range, distinct plumage differences, and different vocalizations of lekking males, it is more appropriate to treat *P. baroni* as a separate, monotypic species.

Range: w. Ecuador and extreme nw. Peru.

Phaethornis malaris   Great-billed Hermit   Ermitaño Piquigrande

Fairly common to common in the undergrowth of humid forest (mostly terra firme) in the lowlands and foothills of e. Ecuador. Recorded locally up to about 1350 m (an MCZ specimen from Río Guataracu), exceptionally even higher (a specimen is labeled as having been taken at El Chaco, at about 1600 m; N. Gyldenstolpe, *Ark. Zool.*, 2nd ser.: 1–320, 1951), but mostly found below about 1000 m.

We follow C. Hinkelmann (*Ornitol. Neotrop.* 7: 119–148, 1996) in considering *moorei*, the taxon found in Ecuador, as a subspecies of an expanded *P. malaris* and not of *P. superciliosus* (Long-tailed Hermit) as it was long treated (e.g., *Birds of the Americas*, pt. 2, no. 1; Meyer de Schauensee 1966, 1970).

Range: se. Colombia, s. Venezuela, and the Guianas to n. Bolivia and Amazonian Brazil; e. Brazil.

Phaethornis hispidus   White-bearded Hermit   Ermitaño Barbiblanco

Uncommon to fairly common in the undergrowth of várzea and riparian forest and forest borders (including river islands) in the lowlands of e. Ecuador. Recorded up only to about 600 m (e.g., in the Tena area).

Monotypic.

Range: e. Colombia and s. Venezuela to n. Bolivia and w. Amazonian Brazil.

Phaethornis bourcieri   Straight-billed Hermit   Ermitaño Piquirrecto

Uncommon to locally fairly common in the undergrowth of terra firme forest in the lowlands and foothills of e. Ecuador, favoring well-drained areas along ridges and not typically found near water. Recorded mostly below 700 m, locally as high as 900 m (e.g., in the Bermejo oil-field area north of Lumbaquí; M. B. Robbins).

One race: nominate *bourcieri*.

Range: the Guianas, s. Venezuela, se. Colombia, e. Ecuador, ne. Peru, and n. Amazonian Brazil.

Phaethornis ruber   Reddish Hermit   Ermitaño Rojizo

Rare to uncommon and perhaps local in the undergrowth and borders of humid forest and secondary woodland in the lowlands of e. Ecuador, primarily in terra firme situations; sometimes also ranges out into adjacent clearings. The Reddish Hermit appears to be scarcer in Ecuador than it is in many other parts of its range, and here it seems to favor blackwater drainages though also being found in some other areas (e.g., at Jatun Sacha). Recorded only below about 400 m.

One race: *nigricinctus*.

Range: se. Colombia, s. Venezuela, and the Guianas to n. Bolivia and Amazonian Brazil; e. Brazil.

Phaethornis griseogularis   Gray-chinned Hermit   Ermitaño Barbigris

Uncommon to locally fairly common in the undergrowth of montane forest, secondary woodland, and borders in the foothills and lower subtropical zone on the east slope of the Andes from Sucumbíos (Bermejo oil-field area north of Lumbaquí; ANSP) south through Zamora-Chinchipe (including the Zumba area; ANSP). Also found locally and in small numbers in s. Loja where (fide C. Hinkelmann) an unpublished specimen (Hamburg collection) was obtained by L. Gómez in 1939 at Malacatos. Prior to the this specimen being brought to our attention, there had been several recent sightings of small hermits from other localities in Loja (these were not definitely distinguished between Stripe-throated [*P. striigularis*] and Gray-chinned, but it now seems clear that Gray-chinned was the species involved): in the Río Catamayo valley and near Cruzpamba in Aug. 1989 (RSR et al.) and in the Sozoranga area (Best et al. 1993; RSR). On the east slope recorded between 600 and 1700 m; in Loja between 900 and 2000 m.

Two races occur in Ecuador, nominate *griseogularis* on the east slope and *porcullae* in Loja; the latter was previously recorded only from nw. Peru. *Porcullae* is larger than the nominate race and is less bronzy on the crown and nape and paler ochraceous below. C. Hinklemann (pers. comm.) considers it possible that *porcullae* may be better regarded as a separate species (Porculla Hermit).

Range: locally from extreme sw. Venezuela to n. Peru; Sierra de Perijá; very locally on tepuis of s. Venezuela.

Phaethornis striigularis   Stripe-throated Hermit   Ermitaño Golirrayado

Uncommon to fairly common in the undergrowth of both humid and deciduous forest, secondary woodland, and borders in the lowlands of w. Ecuador south to Guayas (Cordillera de Colonche at Loma Alta [Becker and López-Lanús 1997] and in the Chongón Hills) and El Oro (repeated recent sightings from Buenaventura). Recorded mostly below 800 m, but locally up to 1350 m at Maquipucuna (P. Coopmans; WFVZ).

We follow Hinkelmann and Schuchmann (1997) in considering Middle American and trans-Andean *P. striigularis* as a species distinct from *P. longuemareus* (Little Hermit) of ne. South America (and also from *P. atrimentalis* [Black-throated Hermit]; see below). Sibley (1996) treats Middle American forms south to Panama as another separate species, *P. adolphi* (Boucard's Hermit), considering *P. striigularis* to range only from Colombia and nw. Venezuela to w. Ecuador. The English names employed here follow the suggestions of C. Hinkelmann (pers. comm.) and Sibley (1996). One race: *subrufescens*.

Range: s. Mexico to nw. Venezuela and sw. Ecuador.

Phaethornis atrimentalis   Black-throated Hermit   Ermitaño Golinegro

Rare to uncommon and apparently local in the undergrowth and borders of humid forest and secondary woodland (especially in swampy or riparian situations; in terra firme principally near streams) in the lowlands of e. Ecuador. Recorded mostly below 800 m, locally up to 900 m (Miazi in the Río Nangaritza valley; Schulenberg and Awbrey 1997).

We follow Hinkelmann and Schuchmann (1997) in considering *P. atrimentalis* as an upper Amazonian species distinct from the rest of the *P. longuemareus* (Little Hermit) complex. One race: nominate *atrimentalis*.

Range: se. Colombia to e. Peru.

Eutoxeres aquila   White-tipped Sicklebill   Pico de Hoz Puntiblanco

Uncommon to locally common in the undergrowth and borders of very humid and montane forest and mature secondary woodland in the foothills and lower subtropical zone on both slopes of the Andes. On the west slope ranges south to El Oro (old specimens from La Chonta and El Chiral; many recent records from Buenaventura), with small numbers also being found in adjacent humid lowlands (especially in n. Esmeraldas; small numbers also occur at Río Palenque, and there are old records from n. Manabí at Río Peripa and Río de Oro). Also recorded recently from the coastal Cordillera de Mache in w. Esmeraldas (Cabeceras de Bilsa [Parker and Carr 1992] and Bilsa [first by R. P. Clay et al.]) and n. Manabí (Filo de Monos; WFVZ). The White-tipped Sicklebill is generally an inconspicuous and infrequently seen hummingbird, though substantial numbers are often mist-netted in favorable areas, especially in and around the *Heliconia* stands in which they preferentially feed. Recorded mostly from 100 to 1600 m, lowest in Esmeraldas (elsewhere mainly above 400–500 m), and perhaps regularly somewhat higher in Zamora-Chinchipe (one was taken as high as 1850 m at Quebrada Avioneta in Podocarpus National Park [ANSP], and a few were seen at 1825 m on the Cordillera del Cóndor [RSR]).

Two similar races are found in Ecuador, nominate *aquila* on the east slope and *heterura* on the west slope. On average *heterura* has somewhat smaller white tail-tips.

Range: Costa Rica to n. Peru.

Eutoxeres condamini   Buff-tailed Sicklebill   Pico de Hoz Colihabano

Uncommon in the undergrowth and borders of humid and montane forest and mature secondary woodland in the lowlands and foothills of e. Ecuador, not ranging very far east of the Andes, with the easternmost reports coming from along the Maxus road southeast of Pompeya. The Buff-tailed Sicklebill favors ravines and vegetation near streams and rivers, and in the lowlands it seems to occur only in hilly areas (never in várzea and seemingly not even in terra firme growing on flat terrain). It is perhaps not so tied to the presence of *Heliconia* as is the White-tipped Sicklebill (*E. aquila*). Basically, the two *Eutoxeres* replace each other altitudinally, with a zone of overlap at about 600 to 1000 m; contra Norton (1965, p. 7), the Buff-tailed Sicklebill does not replace the White-tipped in "open cultivated areas," and in fact in several areas we have mist-netted both sicklebills together. Found mostly between about 300 and 800 m, but recorded up in small numbers to as high as 1300 m (e.g., at San Rafael Falls; MECN).

One race: nominate *condamini*.

Range: se. Colombia, e. Ecuador, e. Peru, and w. Bolivia.

Androdon aequatorialis	Tooth-billed Hummingbird	Colibrí Piquidentado

Rare to locally fairly common in the lower and middle growth of humid and montane forest in the lowlands and foothills of nw. Ecuador. The Tooth-billed Hummingbird is recorded in Ecuador primarily from Esmeraldas, and there it can be locally numerous; a few reports extend its range south into w. Pichincha, in the Santo Domingo area and along the lower Chiriboga road, and adjacent sw. Imbabura (one seen near the Salto del Tigre bridge on the Río Guaillabamba on 15 Jul. 1997; P. Coopmans et al.). There are also recent reports of it from the coastal Cordillera de Mache at Bilsa (J. Hornbuckle et al.). The Tooth-billed is considered to be the most numerous hummingbird inside forest in the lower foothills above Playa de Oro in n. Esmeraldas (O. Jahn et al.). Recorded mostly from about 100 to 800 m, lowest in Esmeraldas (southward not below about 400 m).

Monotypic.

Range: e. Panama to nw. Ecuador.

Doryfera johannae	Blue-fronted Lancebill	Picolanza Frentiazul

Uncommon in the lower growth of montane forest in the foothills and lower subtropical zone on the east slope of the Andes, occasionally occurring out into the adjacent lowlands (e.g., at Jatun Sacha), possibly only seasonally. The Blue-fronted Lancebill seems to favor the vicinity of water, with individuals often found near shady wet rockfaces and caves, and like the Green-fronted Lancebill (*D. ludovicae*), rarely or never in sunny, more open places. Recorded mostly between 400 and 1400 m, occasionally a little higher.

One race: nominate *johannae*.

Range: along east slope of Andes from Colombia to cen. Peru; tepuis of s. Venezuela and w. Guyana.

Doryfera ludovicae	Green-fronted Lancebill	Picolanza Frentiverde

Uncommon in the lower growth of montane forest in the subtropical zone on both slopes of the Andes; on the west slope recorded south only to Cotopaxi (one seen east of Quevedo on 28 Jul. 1991; RSR and G. H. Rosenberg). The inconspicuous Green-fronted Lancebill seems especially to favor the vicinity of shady streams. Although this species generally occurs somewhat higher than the Blue-fronted Lancebill (*D. johannae*), there is also a broad zone of overlap. Both, for instance, are regularly found north of Archidona along the road to Loreto, where they have even been seen feeding simultaneously at the same small flowering tree. Recorded mostly from 1100 to 1700 m, in small numbers (and perhaps locally?) as low as 700 m and as high as 1900 m (Baeza; MECN).

The correct spelling of the species name is *ludovicae* and not *ludoviciae* (Sibley and Monroe 1990).

One race: nominate *ludovicae*. We follow J. T. Zimmer (*Am. Mus. Novitates* 1449, 1950) in synonymizing the race *rectirostris*, in which Ecuadorian birds have often been placed. Ecuadorian examples are on average very slightly longer-billed than birds from elsewhere in the species' South American range.

Range: Costa Rica to e. Panama; Andes from w. Venezuela to w. Bolivia.

Campylopterus largipennis   Gray-breasted Sabrewing   Alasable Pechigris

Uncommon to locally fairly common in humid forest (both terra firme and várzea, favoring the vicinity of streams), forest borders, and secondary woodland in the lowlands of e. Ecuador. Most numerous in the lowlands near the Andes, and apparently not occurring at all in the far eastern lowlands, with the easternmost records being a specimen (ANSP) taken north of Tigre Playa in Sucumbíos on 8 Aug. 1993 and a single bird seen at La Selva on 25 Jan. 1995 (M. Lysinger et al.). Recorded mostly from 300 to 600 m, in small numbers up to about 750 m (Guaticocha; MCZ) and down to as low as 200–250 m; an old record from Baeza (1900 m) would seem likely to have been mislabeled.

One race: *aequatorialis*.

Range: se. Colombia, s. Venezuela, and the Guianas to n. Bolivia and Amazonian Brazil; very locally in interior se. Brazil.

Campylopterus falcatus   Lazuline Sabrewing   Alasable Lazulita

Uncertain. Apparently very rare and local in the canopy and borders of montane forest and adjacent clearings in the subtropical zone on the east slope of the Andes in w. Napo. There is a series of records of this hummingbird from the early 20th century from Baeza, "above Archidona" and "below Oyacachi." Despite the considerable amount of recent field work in this region, however, there is only a single recent report of the species that can be regarded as certain, a female seen for several days in Jun. 1999 at Cabañas San Isidro (C. Bustamente, F. Sornoza, and M. Lysinger). A sighting (with no details) from somewhere along the Loja-Zamora road in Zamora-Chinchipe (Bloch et al. 1991) seems plausible, though this site is much farther south than the species' known range; furthermore, it was omitted from those authors' more recent publication on Podocarpus National Park birds (Rasmussen et al. 1994). A published report from the Río Mazán valley near Cuenca in Azuay (King 1989) seems, in view of what is known of the species' ecological requirements, very unlikely. The elevations at which the Lazuline Sabrewing has been found in Ecuador are uncertain, mainly because of the imprecisely located collecting localities from which the species has been recorded, but probably center on about 2000 m.

Elsewhere (e.g., w. Venezuela), the Lazuline Sabrewing does not appear to be especially sensitive to habitat disturbance, and in any case there still exists a considerable amount of relatively undisturbed montane forest within its Ecuadorian range. Its apparent decline here remains an enigma, but seems to be real, and we thus give it an Ecuadorian status of Data Deficient.

Monotypic.

Range: mountains of n. Venezuela, and Andes from w. Venezuela south locally to n. Ecuador; Sierra de Perijá.

Campylopterus villaviscensio   Napo Sabrewing   Alasable del Napo

Uncommon and seemingly local in the lower growth and borders of montane forest in the foothills and lower subtropical zone on the east slope of the Andes. This near endemic—it has recently been found in adjacent Colombia and Peru—has been recorded from a limited number of localities from w. Napo (Volcán Sumaco region) south locally into Zamora-Chinchipe (Warientza and Pachicutza on the Cordillera del Cóndor [WFVZ]; also a single male seen at Río Bombuscaro on 23 Mar. 1996 [P. Coopmans]). Contra Meyer de Schauensee (1966, p. 163), there is no evidence that the Napo Sabrewing occurs in the "tropical" zone; in fact, its distribution appears to be centered almost entirely on mountain ranges isolated from the main chain of the Andes. Small numbers can be found north of Archidona along the Loreto road. Recorded between about 900 and 1700 m.

The Napo Sabrewing was accorded Near-threatened status by Collar et al. (1994), an assessment with which we agree. Little of its limited range has been afforded formal protection, though a substantial population should occur in Sumaco-Galeras National Park, and evidently there are a few in Podocarpus National Park.

Monotypic.

Range: locally on east slope of Andes in s. Colombia (Salaman and Mazariegos 1998), Ecuador, and n. Peru (T. J. Davis, *Condor* 88[1]: 50–56, 1986).

Florisuga mellivora   White-necked Jacobin   Jacobino Nuquiblanco

Uncommon to locally (or perhaps seasonally) fairly common in the canopy and borders of humid forest, second-growth woodland, and adjacent clearings and gardens in the lowlands and—to a lesser degree—foothills in both e. and w. Ecuador. In the west recorded south to Guayas (Isla Puná, Manglares-Churute Ecological Reserve, and near Naranjal). Recorded mostly below 800 m, but found quite regularly in the Mindo area, occasionally as high as 1550 m.

One race: nominate *mellivora*.

Range: s. Mexico to sw. Ecuador, n. Bolivia, and Amazonian Brazil.

Colibri delphinae   Brown Violetear   Orejivioleta Parda

Rare to uncommon and seemingly local (seasonal?) in the canopy and borders of montane forest and, perhaps especially, in adjacent clearings in the foothills and subtropical zone on both slopes of the Andes; there are also some old records from the central valley in the Quito region. In the west not known south of Pichincha (Chiriboga road), whereas in the east it has been recorded mainly from w. Napo. The Brown Violetear appears to be scarce in s. Ecuador, where most of the few records come from Zamora-Chinchipe (including one seen by P. Coopmans et al. above Zamora on 13 Apr. 1989; 3 seen by P. Coopmans et al. north of Zumba on 3 May 1989; and one seen between Valladolid and Palanda on 21 May 1989 [Rasmussen et al. 1996]); one was also seen in Morona-Santiago at San Isidro west of Macas on 27 Jul. 1979 (RSR and R. A. Rowlett). In addition, there is one report from the coastal Cordillera de Colonche in w. Guayas at Loma Alta, where a single bird was observed in Dec. 1996 (Becker and López-Lanús 1997). Recorded mostly from 1000 to 1800 m, but there are early records from as high as 2400–2700 m in the central valley around Quito (but N. Krabbe suggests that these may be the result of confusion with melanistic Sparkling Violetears [*C. coruscans*]). It has also been seen as low as 500 m (one observed near Tena on 18 Dec. 1990; P. Coopmans and N. Krabbe et al.).

Monotypic.

Range: locally in highlands from Guatemala and Belize to e. Panama; mountains of n. Venezuela, and Andes from w. Venezuela to w. Bolivia; s. Venezuela, Guyana, and Suriname; very locally in e. Brazil; Sierra de Perijá and Santa Marta Mountains.

Colibri thalassinus   Green Violetear   Orejivioleta Verde

Uncommon to locally (or seasonally?) common in the canopy and borders of montane forest, secondary woodland and scrub, and clearings and gardens in the subtropical zone on both slopes of the Andes. In the west recorded south to w. Loja (Celica and Sozoranga). On the whole the Green Violetear seems rather less numerous in Ecuador than it typically is elsewhere across its wide range. Recorded mostly from 1200 to 2300 m, thus primarily below the elevational range of the Sparkling Violetear (*C. coruscans*), though the two do occasionally occur sympatrically; small numbers occasionally (e.g., in Feb.; D. Wolf and PJG et al.) wander as low as about 500 m in the Tena area, with occurrence there perhaps being tied to the flowering of *Erythrina* trees.

One race: *cyanotus*. This form, together with other races occurring from Costa Rica southward, is sometimes separated as a distinct species, *C. cyanotus* (Montane Violetear).

Range: highlands from s. Mexico to Honduras; highlands of Costa Rica and w. Panama; mountains of n. Venezuela and Andes from w. Venezuela to w. Bolivia; Sierra de Perijá and Santa Marta Mountains.

Colibri coruscans   Sparkling Violetear   Orejivioleta Ventriazul

Fairly common to common and widespread (occurring in both humid and arid regions) in scrub, gardens, plantations (even of introduced species such as *Eucalyptus*), secondary woodland, and borders of montane forest in the subtropical and temperate zones on both slopes of the Andes, and through much of the central and interandean valleys. The Sparkling Violetear is the most numerous hummingbird in the parks and gardens of Quito. Recorded mostly from 1000 to 3500 m, with some evidence of seasonal elevational movements; there is even one sighting from as low as 600 m near Tena on 18 Dec. 1990 (P. Coopmans et al.).

One race: nominate *coruscans*.

Range: mountains of n. Venezuela, and Andes from w. Venezuela to nw. Argentina and n. Chile; tepuis of s. Venezuela; Sierra de Perijá and Santa Marta Mountains.

Anthracothorax nigricollis   Black-throated Mango   Mango Gorjinegro

Rare and apparently local in clearings and borders of humid forest (both terra firme and várzea) in the lowlands of e. Ecuador; uncommon to locally fairly common in desert scrub, semiopen areas, clearings and gardens, and borders of deciduous and semihumid woodland in the more arid lowlands of w. Ecuador. In the west ranges from w. Esmeraldas south to El Oro and w. Loja (sightings near Cruzpamba in Aug. 1989; RSR et al.), including Isla de la Plata (at least formerly; we are not aware of any recent reports). As far as we are aware, the first definite report from e. Ecuador was obtained as late as 15 Apr. 1976, when one was seen at Limoncocha (Tallman and Tallman 1977); a few have since been recorded in the lowlands of w. Napo, and a nest was found near Tena in Mar. 1992 (D. Wolf). Small numbers were also found at Imuyacocha in Dec. 1990, with a female taken there on 6 Dec. (ANSP), representing the first specimen for e. Ecuador; the species has also been seen along the nearby Río Pacuyacu (P. Coopmans). Several more examples (ANSP, MECN) have since been obtained in the vicinity of Lago Agrio, and birds have also been seen in the Andean foothills north of Archidona (P. Coopmans). This nonforest hummingbird seems destined to become more numerous in e. Ecuador because of continuing deforestation. In the east recorded mostly below 400 m, but with a few reports from as high as 900 m; in the west found mainly below 500 m, but recorded as high as 1200 m in Loja.

Two races occur in Ecuador, nominate *nigricollis* in the east and *iridescens* in the west. They are very similar, aside from the slightly longer bill of *iridescens* (greater than 23.5 mm, fide R. Banks). The affinities of *iridescens* have been debated. Long considered a race of *A. nigricollis* (e.g., Chapman 1926; *Birds of the World*, vol. 5), it was transferred to *A. prevostii* (Green-breasted Mango) by J. T. Zimmer (*Am. Mus. Novitates* 1463, 1950). Most subsequent authors have retained it there. Based on plumage characters, we—with the concurrence of R. Banks—consider *iridescens* to be better placed with *A. nigricollis*. Therefore *A. prevostii* should be deleted from the list of birds occurring in Ecuador. An alternative treatment would be to consider the geographically isolated, but nonetheless not very differentiated, *iridescens* as a full species (Ecuadorian Mango). *Iridescens* is nearly endemic to w. Ecuador, also occurring in ajacent nw. Peru (the Valle, Colombia, specimens previously identified as *iridescens* are referable to *A. nigricollis*, fide R. Banks).

Range: Panama to Bolivia, e. Paraguay, ne. Argentina, Uruguay, and s. Brazil; w. Ecuador and extreme nw. Peru.

Avocettula recurvirostris   **Fiery-tailed Awlbill**   Colibrí Piquipunzón

Uncertain. Evidently very rare (overlooked?) in the canopy and borders of humid forest (apparently both terra firme and várzea, based in part on records from outside Ecuador) in the lowlands of e. Ecuador. There are only three known Ecuadorian records of this species, which is scarce throughout its entire range. J. Berlioz (*L'Oiseau et Rev. Franç. d'Orn.* 8: 15–16, 1938) recorded the first Ecuadorian record, an immature male specimen taken near Avila by an unnamed local collector on an unspecified date. A male that at least formerly was in the Politécnica collection in Quito (its present whereabouts seem to be unknown) was taken on 6 Aug. 1959 by M. Olalla at Bovera; this locality cannot definitely be located but is believed to be north of Montalvo on the Río Bobonaza. Finally, two birds (a male and a female) were seen feeding at a flowering tree in blackwater várzea forest at Imuyacocha on 5 Dec. 1990 (RSR and PJG et al.). Recorded below about 300 m.

The Fiery-tailed Awlbill has a relatively wide range in Amazonia, but it seems to be rare and local everywhere. The species is considered to be Near-threatened by Collar et al. (1994), an assessment with which we cannot disagree. We are not aware of any factor that would actually be causing any decline in the numbers of this species.

The monotypic genus *Avocettula* has occasionally been subsumed into *Anthracothorax* (e.g., Schuchmann 1999). Monotypic.

Range: locally in e. Venezuela, the Guianas, and e. Amazonian Brazil; a few records from e. Ecuador and sw. Amazonian Brazil.

Klais guimeti   **Violet-headed Hummingbird**   Colibrí Cabecivioleta

Uncommon to locally (seasonally?) common in the lower growth and borders of montane forest and in adjacent clearings and gardens in the foothills and lower subtropical zone on the east slope of the Andes. The Violet-headed Hummingbird is frequently seen feeding at flowering *Inga* trees. Recorded mostly from 800 to 1700 m, locally as low as 600 m (Avila in w. Napo; MECN) and as high as 1900 m (Baeza; MECN).

One race: nominate *guimeti* (though birds from s. Ecuador show some intermediacy toward *pallidiventris* of e. Peru and Bolivia).

Range: Honduras to w. and n. Venezuela, and along eastern base of Andes from Colombia to w. Bolivia.

Lophornis stictolophus   Spangled Coquette   Coqueta Lentejuelada

Rare and local (perhaps seasonally uncommon) in the borders of lowland and foothill forest and (perhaps especially) in clearings and gardens along the eastern base of the Andes. Recorded from Napo (one sighting from Limoncocha [P. Coopmans] and a few from Jatun Sacha [B. Bochan], at the latter mainly in Apr. when it has been seen feeding at the flowering tree *Chimarrhis glabriflora*), Morona-Santiago (in the Macas/Logroño region), and Zamora-Chinchipe (around Zamora and at Zumba). The only report of the Spangled Coquette from well east of the Andes is a female seen feeding at *Lantana* at the Zancudococha clearing on 10 Jan. 1993 (RSR et al.). Recorded up to 1200 m.

The Spangled Coquette was considered to be Near-threatened by Collar et al. (1994). Our belief, however, is that this species is actually somewhat favored by habitat disturbance—at least this appears to be the case in Ecuador—and thus, though scarce, we cannot see that it should be considered to be at risk.

The correct spelling of the species name is *stictolophus*, not *stictolopha* (Sibley and Monroe 1990).

Monotypic.

Range: locally from n. and w. Venezuela to n. Peru.

*[Lophornis delattrei   **Rufous-crested Coquette**   Coqueta Crestirrufa]

Uncertain. Apparently very rare—perhaps only overlooked, or possibly only a seasonal visitant—at the borders of humid forest in the lowlands near the eastern base of the Andes in w. Napo. Known in Ecuador only from a sighting of two birds (a male and a female) at Jatun Sacha on 19–25 Apr. 1992 (B. Bochan). Recorded at about 400 m.

Race unknown. *Lessoni* (of Colombia northward) is smaller and males have longer crest feathers entirely rufous, whereas nominate *delattrei* (of Peru southward) has the male's crest feathers minutely tipped with green.

Range: locally in Costa Rica, Panama, and w. Colombia (east to along eastern base of Andes); locally along eastern base of Andes in Peru and w. Bolivia; one record from e. Ecuador. Excludes Short-crested Coquette (*L. brachylophus*) of Mexico.

Lophornis chalybeus   Festive Coquette   Coqueta Festiva

Very rare in the canopy and borders of humid forest (evidently mostly in terra firme) and in adjacent clearings in the lowlands of e. Ecuador; likely overlooked to a certain extent. There are very few Ecuadorian reports of the Festive Coquette, so far all of them from the upper Río Napo valley. These include old specimens from "Ecuador" and "Río Napo" and a few recent sightings (PJG et al.) from the vicinity of Tena; several were also seen along the Maxus road southeast of Pompeya in 1994–1995 (RSR et al.). Festive Coquettes have occasionally been seen feeding at flowering *Erythrina* trees (PJG). Recorded up to about 500 m.

The correct spelling of the species name is *chalybeus*, not *chalybea* (Sibley and Monroe 1990).

One race: *verreauxii*.

Range: se. Venezuela, w. and cen. Amazonian Brazil, se. Colombia, e. Ecuador, e. Peru, and n. Bolivia; locally in se. Brazil and ne. Argentina.

Popelairia popelairii   Wire-crested Thorntail   Colicerda Crestuda

Uncommon to locally (seasonally?) fairly common in the canopy and borders of humid and montane forest and in clearings and gardens in the foothills along the eastern base of the Andes, ranging in smaller numbers out into the adjacent lowlands and up into the lower subtropical zone. The Wire-crested Thorntail is seen most often when feeding at flowering trees, especially at *Inga* and *Erythrina*; several will often congregate at a single tree. Recorded mostly between about 600 and 1600 m, rarely wandering as high as about 2000 m at Cabañas San Isidro (S. Connop, fide M. Lysinger). The species perhaps also occurs (only seasonally?) out into the eastern lowlands, for J. T. Zimmer (*Am. Mus. Novitates* 1463, 1950) records two specimens taken at Andoas (200 m), adjacent to the Ecuador/Peru border near the Río Pastaza; there have, however, been no recent reports from anywhere near so low or so far from the Andes.

The members of the genus *Popelairia* are sometimes placed in the genus *Discosura*. Monotypic.

Range: locally in se. Colombia, e. Ecuador, and e. Peru (mainly along eastern base of Andes).

Popelairia langsdorffi   Black-bellied Thorntail   Colicerda Ventrinegra

Rare in the canopy and borders of humid forest (apparently mostly in terra firme) in the lowlands of ne. Ecuador. To date this scarce hummingbird—which likely has been under-recorded because of its small size and habit of generally remaining high above the ground—has been found primarily in Napo and Sucumbíos, but it likely occurs southward as well; there is one specimen (MECN) from nw. Pastaza at Mera. There are recent sightings from Puerto Napo, Jatun Sacha, La Selva, along the Maxus road southeast of Pompeya, and Zancudococha. An immature male was collected (ANSP) north of Lumbaquí in the Bermejo oil-field area on 12 Mar. 1993. Recorded mostly below 400 m, once as high as 900 m (Bermejo; ANSP).

One race: *melanosternon*.

Range: extreme sw. Venezuela, se. Colombia, e. Ecuador, e. Peru, nw. Bolivia, and w. Amazonian Brazil; se. Brazil.

Popelairia conversii   Green Thorntail   Colicerda Verde

Uncommon in the canopy and borders of humid forest, forest borders, and adjacent clearings and gardens in the lowlands and foothills of w. Ecuador, where recorded in more humid areas on the west slope of the Andes south to El Oro (an old specimen from La Chonta and several recent sightings from Buenaventura) and w. Loja (a single sighting of a female east of El Limo on 18 Apr. 1993; RSR). Green Thorntails are seen most often where and when *Inga* trees are in flower. Ranges mostly between 300 and 1000 m, but locally or seasonally ranging lower (especially in Esmeraldas, where it has been seen down to 70 m at Playa de Oro [O. Jahn et al.], and there is an old specimen from Esmeraldas [city] that must have been taken near sea level). Small numbers also range somewhat higher (perhaps seasonally? or in response to the flowering of certain trees), having occasionally been seen at 1400 m in the Mindo area and once as high as 1550 m (Pallatanga; MECN).

Monotypic.

Range: Costa Rica to sw. Ecuador.

Chlorestes notatus   **Blue-chinned Sapphire**   Zafiro Barbiazul

Rare at borders of humid forest (both terra firme and várzea) and in adjacent shrubby clearings in the lowlands of e. Ecuador. Although numerous in other parts of its range, the Blue-chinned Sapphire remains a poorly known hummingbird in Ecuador, where apparently only two specimens have ever been taken (Chapman 1926); one of those, listed as from "east Ecuador," was perhaps not obtained within the present borders of the country. There are recent sightings from Taracoa, La Selva, Jatun Sacha, and along the Maxus road southeast of Pompeya. Recorded below 300 m.

One race: *puruensis*.

Range: ne. Colombia, much of Venezuela, and the Guianas to Amazonian and e. Brazil, ne. Peru, and e. Ecuador.

Chlorostilbon mellisugus   **Blue-tailed Emerald**   Esmeralda Coliazul

Rare to locally uncommon at the borders of humid forest (perhaps mainly in várzea and floodplain forest) and in adjacent clearings in the lowlands of ne. Ecuador in Sucumbíos and Napo. Uncommon and seemingly local at borders of montane forest and scrub and in clearings on the east slope of the Andes in se. Ecuador, where recorded mainly from Zamora-Chinchipe (foothills in the Zamora area and on the Cordillera del Cóndor, up locally and perhaps only seasonally into the temperate zone in Podocarpus National Park), but also found at least once in Morona-Santiago (several seen on 29 Jul. 1979 at San Isidro near Macas; RSR and R. A. Rowlett). Northern birds are recorded up to about 750 m, whereas in the southeast the species is known from between about 900 and 2600 m.

Excludes *C. melanorhynchus* (Western Emerald), following F. G. Stiles (*Wilson Bull.* 108[1]: 1–27, 1996). We follow J. T. Zimmer (*Am. Mus. Novitates* 1474, 1950) and Stiles (op. cit.) in assigning northeast Ecuadorian birds to the race *napensis*; note that Zimmer (op. cit.) dismisses *C. "vitticeps"* (listed as an "uncertain form" in *Birds of the World*, vol. 5) as being only immature examples of *C. m. napensis*. The subspecific allocation of birds from Zamora-Chinchipe remains uncertain: they belong either to *phoeopygus*, the race recorded from adjacent Peru, or to *napensis*, which is recorded from Gualaquiza (Zimmer, op. cit.). The one specimen from there (a male [MCZ] taken by D. Norton on 8 Sep. 1965 along the Loja-Zamora road at 1600 m) is in such poor condition that we cannot determine its subspecific identity. In any case, these two taxa differ only marginally; Stiles (op. cit., p. 18) considered *phoeopygus* as "possibly indistinguishable" from *napensis*.

Range: e. Colombia, Venezuela, and the Guianas to n. Bolivia and Amazonian Brazil, but absent from much of n. Amazonia.

The Narrow-tailed Emerald (*Chlorostilbon stenura*) of Andean slopes in w. Venezuela and ne. Colombia was recorded as occurring in Ecuador on the basis of a single specimen supposedly procured at Baeza (Oberholser 1902). Given the species' presently understood distribution and the absence of any subsequent records, we consider it almost certain that the specimen in question was either misidentified or mislabeled.

Chlorostilbon melanorhynchus   Western Emerald   Esmeralda Occidental

Fairly common to locally common in scrub and gardens in arid intermontane valleys of nw. Ecuador from Carchi in the Río Chota valley south to Pichincha in the Quito and Cumbaya regions; uncommon to locally fairly common at borders of montane and humid forest and in clearings and gardens on the west slope of the Andes and in adjacent western lowlands from Pichincha (Mindo region) south to ne. Guayas and sw. Chimborazo. Recorded between about 1500 and 2700 m in the northwestern intermontane valleys; farther south, found mainly between about 600 and 1800 m, but also recorded (perhaps only seasonally?) from lowlands (e.g., at Santo Domingo de los Colorados and Río Palenque).

We follow F. G. Stiles (*Wilson Bull.* 108[1]: 1–27, 1996) in considering the *Chlorostilbon* emeralds ranging in w. Ecuador (and w. Colombia) as a species (*C. melanorhynchus*) separate from the similar *C. mellisugus* (Blue-tailed Emerald) of e. Ecuador. They were formerly all considered conspecific under the name *C. mellisugus*. Two races of *C. melanorhynchus* are found in Ecuador. Birds from the northern intermontane valleys have been referred to nominate *melanorhynchus*, whereas those from the west slope of the Andes have traditionally been assigned to *pumilus*. We can discern no significant differences between these taxa, however, and propose that *pumilus* be synonymized into *melanorhynchus*; Stiles (op. cit., p. 20) also noted that these taxa "might not be separable." Both "*pumilus*" and *melanorhynchus* have by some authors (e.g., *Birds of the World*, vol. 5) been assigned to the species *C. gibsoni* (Red-billed Emerald) of Colombia and Venezuela, but this is clearly incorrect for both have, inter alia, entirely black bills.

These *Chlorostilbon* emeralds are exceedingly similar to each other, and thus establishing appropriate English names for them is difficult. Stiles (op. cit.) suggested West Andean Emerald for *C. melanorhynchus*, but using that would require a modifier for what has long simply been called the Andean Emerald

(*Amazilia franciae*). We prefer to shorten his proposed name for *C. melanorhynchus* to Western Emerald, its range in fact being the most western of any South American *Chlorostilbon*.

Range: w. Colombia and w. Ecuador.

Thalurania furcata   **Fork-tailed Woodnymph**   Ninfa Tijereta

Uncommon to locally common and widespread in the lower growth of humid forest (principally terra firme) in the lowlands of e. Ecuador, ranging up in smaller numbers through the foothills to the lower subtropical zone on the east slope of the Andes. Recorded mainly below about 1000 m, in small numbers (or locally) as high as 1700 m (e.g., on the Cordillera de Cutucú; ANSP).

One race: *viridipectus*.

Range: e. Colombia, e. and s. Venezuela, and the Guianas to Bolivia, n. Argentina, e. Paraguay, and much of Brazil. More than one species is perhaps involved.

Thalurania fannyi   **Green-crowned Woodnymph**   Ninfa Coroniverde

Uncommon to locally fairly common in the lower and middle growth of very humid to humid forest, secondary woodland, and adjacent clearings in the lowlands and foothills of nw. Ecuador south mainly to cen. Manabí and n. Los Ríos (Quevedo area; ANSP); there are also specimens from w. Chimborazo (Puente de Chimbo and Pallatanga; MECN), and the species may occur—or once occurred—across s. Los Ríos as well. Very small numbers range also on the coastal Cordillera de Colonche in sw. Manabí (Cerro San Sebastián in Machalilla National Park; RSR) and w. Guayas (Cerro La Torre [N. Krabbe] and Loma Alta [Becker and López-Lanús 1997]). Recorded mostly below 800 m, in smaller numbers up into the lower subtropical zone as high as at least 1500 m (e.g., above Mindo [ANSP] and at Pallatanga).

Trans-Andean populations of *Thalurania* have in recent years usually been considered a single polytypic species, males with a glittering crown (either green or violet), going by the English name of Crowned Woodnymph. A few authors (e.g., Meyer de Schauensee 1966, 1970) have even considered the trans-Andean *colombica* group to be conspecific with the cis-Andean *T. furcata* (Fork-tailed Woodnymph). P. Escalante-Pliego and A. T. Peterson (*Wilson Bull.* 104[2]: 205–219, 1992) presented evidence showing that the Middle American populations of *Thalurania* are better regarded as a complex of three species. We follow this suggestion, and thus the Ecuadorian race *verticeps* is placed as a subspecies of *T. fannyi*. As *hypochlora*, endemic to sw. Ecuador, shows differences at least as pronounced from *T. fannyi* as the latter shows from the other species in the complex (*T. colombica* [Violet-crowned Woodnymph], *T. ridgwayi* [Mexican Woodnymph]), we consider *T. hypochlora* (Emerald-bellied Woodnymph) as a fourth species. Male *T. fannyi verticeps* have a glittering green crown and a glittering violet-blue belly contrasting with glittering green anterior underparts, whereas females have underparts that are strongly bicolored (pale gray on throat and breast and much darker grayish green on belly). Male *T. hypochlora* differ in having their entire underparts (including the belly) glittering green, whereas females are uniform pale grayish below (showing no light-dark contrast).

Range: e. Panama to w. Venezuela and nw. Ecuador.

Thalurania hypochlora	Emerald-bellied Woodnymph	Ninfa Ventriesmeralda

Uncommon to locally fairly common in the lower growth of humid forest, secondary woodland, and adjacent clearings in the lowlands and foothills of sw. Ecuador from ne. Guayas and sw. Chimborazo south through El Oro to w. Loja (an old record from Las Piñas). We regard the specimens of *hypochlora* supposedly taken at "Gualea" as almost certainly having been mislabeled, for all birds taken and seen in the Mindo region in recent years have clearly been referable to the *verticeps* form of the Green-crowned Woodnymph (*T. fannyi*). There appears to be an area in w. Bolívar and much of Los Ríos in which no *Thalurania* is known to occur; this apparent distribution gap requires further investigation. Recorded up to about 1100 m.

Overall numbers of the Emerald-bellied Woodnymph have doubtless declined substantially as a result of deforestation, but it remains reasonably numerous at Manta Real in nw. Azuay and Buenaventura in El Oro, though both are areas where a significant amount of forest destruction has taken place. We therefore consider the species to be Near-threatened.

T. hypochlora is here regarded as a monotypic species distinct from *T. fannyi*. See the discussion under that species.

Range: sw. Ecuador and extreme nw. Peru.

Damophila julie	Violet-bellied Hummingbird	Colibrí Ventrivioleta

Uncommon to locally fairly common in the lower growth of forest (both humid and deciduous) borders, secondary woodland, and clearings and gardens in the lowlands and foothills of w. Ecuador from n. Esmeraldas (north to Playa de Oro; O. Jahn et al.) south through Manabí, El Oro, and w. Loja (old specimens from the Alamor region at Cebollal and Las Piñas). The species' presence in n. Esmeraldas not far to the south of the Colombian border would indicate that it likely also occurs north into adjacent Nariño (where it has not been recorded [Hilty and Brown 1986; Salaman and Mazariegos 1998]). Recorded up to about 1100 m in the southwest.

One race: the nearly endemic *feliciana*.

Range: Panama and n. Colombia; w. Ecuador and extreme nw. Peru.

Hylocharis sapphirina	Rufous-throated Sapphire	Zafiro Barbirrufo

Apparently rare and local in the canopy and borders of terra firme forest in the lowlands of e. Ecuador; known from only two specimens and a few recent sightings. The first specimen is a male taken by R. Olalla in Mar. 1939 at Río

Copotaza (Orcés 1944), the Copotaza being a small affluent of the Río Pastaza in e. Pastaza; we have been unable to determine the present whereabouts of this specimen. The second specimen is a female found dead by M. Lysinger and T. Walla at La Selva on 20 Apr. 1994 (ANSP). Two individuals, a male and a female, were also seen feeding at flowering mistletoe here for several weeks in Jan. 1995 (M. Lysinger et al.), and a male was seen and tape-recorded along the Maxus road southeast of Pompeya on 29 Jan. 1995 (RSR and F. Sornoza). Elsewhere the Rufous-throated Sapphire is generally much more numerous, and in addition is found in clearings adjacent to forest. Recorded below about 250 m.

H. sapphirina is generally considered to be monotypic. However, the 1994 female specimen has the upperside of its tail bronzy green with the upper tail-coverts more coppery bronze and slightly contrasting (not the entirely coppery rufous tail found elsewhere in the species' range); the male seen along the Maxus road (see above) had a similarly colored tail and also appeared to have more violet-blue on the head than is normal in *H. sapphirina*. Ecuadorian birds thus seem likely to be separable as an as-yet-undescribed subspecies.

Range: the Guianas, s. Venezuela, se. Colombia, ne. Peru, ne. Ecuador, and e. and cen. Amazonian Brazil south to n. Bolivia; se. Brazil; e. Paraguay and ne. Argentina.

Hylocharis cyanus   **White-chinned Sapphire**   Zafiro Barbiblanco

Uncertain. Apparently very rare and local in the lowlands of se. Ecuador. Known in Ecuador only from a single specimen, a male, collected by R. Olalla in Mar. 1939 at Río Copotaza (250 m) in e. Pastaza (Orcés 1944). We have been unable to determine the present whereabouts of this specimen. Elsewhere the White-chinned Sapphire is found in the canopy and borders of forest (both humid and deciduous) and in clearings.

Despite the White-chinned Sapphire's evident rarity in Ecuador, based on the generalized (and often secondary) habitats it favors elsewhere in its range we cannot consider it to be at risk.

Presumed race: *rostrata*.

Range: the Guianas, s. Venezuela, and extreme e. Colombia to n. Bolivia and Amazonian Brazil; e. Brazil; nw. Venezuela and n. Colombia.

Hylocharis grayi   **Blue-headed Sapphire**   Zafiro Cabeciazul

Rare to locally fairly common in arid scrub and gardens in arid intermontane valleys of n. Ecuador from Carchi and Imbabura (Ríos Chota and Mira valleys) south to n. Pichincha (Río Guaillabamba valley; Mejía collection). We regard the localities in the western lowlands and western subtropical zone (e.g., Babahoyo, Vinces, Nanegal) recorded by Simon (1907 *in* Chapman 1926) as being of highly dubious provenance; all were so-called "native skins," known frequently to have incorrect locality data. There have been no subsequent records of the Blue-headed Sapphire from anywhere other than the arid inter-montane valleys of the north. In early Mar. 1993 it was found to be fairly common near Tumbabiro in Imbabura (ANSP, MECN), and it has since been found in several other areas in the Río Mira valley. Recorded between about 1200 and 2200 m.

H. *humboldtii* (Humboldt's Sapphire) of coastal nw. Ecuador north to extreme se. Panama is regarded as a monotypic species distinct from *H. grayi* based on a number of striking plumage differences in both sexes, and on their entirely disjunct distributions and different habitats. The two were regarded as separate species into the early 20th century (e.g., in *Birds of the Americas*, pt. 2, no 1.), but then were merged in *Birds of the World* (vol. 5), seemingly for no reason, though that author perhaps was influenced by the range confusion that resulted from native-collected skins with unreliable locality data. There is no evidence that the two taxa come into contact.

Range: interior sw. Colombia and n. Ecuador.

Hylocharis humboldtii	Humboldt's Sapphire	Zafiro de Humboldt

Uncertain. Rare in and adjacent to mangroves along the coast of nw. Ecuador in n. Esmeraldas. The Humboldt's Sapphire is recorded in Ecuador only from specimens from Vacquería (4 males in AMNH), Esmeraldas (city; whence recorded by Chapman [1926], though on a Mar. 1995 visit RSR could not locate any specimens from there), and "Veguina" (a site that cannot be located). The type locality, "Río Mira," if correct likely refers to a site on or near the coast in adjacent sw. Colombia. The habitat of the species in Ecuador can only be inferred and is based on information derived from Colombia and Panama (Hilty and Brown 1986; Ridgely and Gwynne 1989). The only recent Ecuadorian report is a sighting—with no details—from Cerro Mutiles (east of the lower Río Esmeraldas) in early Feb. 1991 (Parker and Carr 1992). The species was not recorded during extensive field work in 1995–1998 in n. Esmeraldas in the Río Santiago and Río Cayapas drainages (O. Jahn et al.). Recorded only near sea level.

Given the apparent rarity of the Humboldt's Sapphire in Ecuador, and the absence of information regarding its numbers, we consider the species to merit Data Deficient status here; it may even deserve Endangered status. The species was not considered to be at risk by Collar et al. (1994), probably because those authors did not consider *H. humboldtii* to represent a species separate from the more numerous *H. grayi*.

H. humboldtii is here regarded as a species separate from *H. grayi* (Blue-headed Sapphire); see the discussion under that species. We employ the English name for *H. humboldtii* suggested in *Birds of the Americas* (pt. 2, no. 1).

Range: along coast from extreme e. Panama to nw. Ecuador.

Chrysuronia oenone	Golden-tailed Sapphire	Zafiro Colidorado

Fairly common to locally (seasonally?) common in the borders of humid and montane forest, secondary woodland, and adjacent clearings and gardens in the foothills along the eastern base of the Andes and the adjacent lowlands. Decreasing numbers of the Golden-tailed Sapphire also occur in the lowlands away from the Andes, but whether the species occurs—at least with any regularity—in the far eastern lowlands remains uncertain. Golden-tailed Sapphires are often seen feeding in the semiopen at flowering trees such as *Erythrina* and *Inga*. Recorded mostly from 400 to 1200 m, in much smaller numbers down into the eastern lowlands, also in small numbers up into the subtropical zone on the east slope of the Andes as high as 1600 m (near Sabanilla on the Loja-Zamora road; MCZ).

One race: nominate *oenone*. The race *azurea*, described from w. Ecuador (where the species does not occur, so it was presumably actually obtained on the east slope), refers to subadult *oenone* (Schuchmann 1999).

Range: n. Venezuela, near eastern base of Andes in Colombia, e. Ecuador, e. Peru, w. Bolivia, and extreme w. Amazonian Brazil.

Polytmus theresiae   Green-tailed Goldenthroat   Gorjioro Coliverde

Uncertain. Apparently very rare and local in the lowlands of se. Ecuador. Known in Ecuador only from a male and a female collected in 1936 by the "Mission Flornoy" at "Laguna de Siguin" (Berlioz 1937c); Siguin cannot be precisely located but is thought to be somewhere on the north bank of the Río Pastaza in Pastaza. Elsewhere the Green-tailed Goldenthroat is found in shrubby areas and wet savannas, often at the edge of forest or woodland.

One race: *leucorrhous*.

Range: the Guianas, locally in Amazonian Brazil, sw. Venezuela, and extreme e. Colombia; locally in ne. Peru and se. Ecuador.

Taphrospilus hypostictus   Many-spotted   Colibrí Multipunteado
                                 Hummingbird

Rare to perhaps locally uncommon and apparently local at the borders of humid and montane forest and in adjacent clearings in the foothills along the eastern base of the Andes from w. Napo (old specimens from Río Suno and below San José, and numerous recent sight reports from along the Loreto road north of Archidona) south locally through Zamora-Chinchipe in the Zamora region. The Many-spotted Hummingbird may be most numerous in s. Ecuador; it has been seen on several occasions around Zamora, where it sometimes feeds at flowering *Inga* trees. Recorded mostly from 500 to 1200 m.

The monotypic genus *Taphrospilus* is sometimes (e.g., Schuchmann 1999) subsumed into the genus *Leucippus*, but we continue to follow Meyer de Schauensee (1966, 1970) and Sibley and Monroe (1990) in maintaining it. One race: nominate *hypostictus*.

Range: locally on east slope of Andes from n. Ecuador to w. Bolivia.

Leucippus chlorocercus   Olive-spotted   Colibrí Olivipunteado
                              Hummingbird

Uncommon to fairly common but very local in early-succession scrub and low woodland on river islands in the lowlands of ne. Ecuador along the Río Napo upriver to around Misahuallí. The Olive-spotted Hummingbird was first recorded in Ecuador by Norton et al. (1972) based on two specimens obtained at Misahuallí on 14 Jul. 1959; presumably Meyer de Schauensee (1970) mentioned Ecuador in the species' range based on these. It has since been found on a number of Río Napo islands, including several near La Selva and Sacha Lodge, but not (at least so far) on the Río Aguarico or the Río Pastaza. Recorded below about 400 m.

Monotypic.

Range: islands along Amazon River and its major tributaries in w. Amazonian Brazil, se. Colombia, ne. Peru, and ne. Ecuador.

*Leucippus baeri   **Tumbes Hummingbird**   **Colibrí de Tumbes**

Uncommon to locally (seasonally?) fairly common and local in arid scrub in extreme sw. Ecuador in sw. Loja. First recorded from specimens (ANSP, MECN) taken in Aug. 1989 in the Río Catamayo valley about 6 road km south of El Empalme. Small numbers of the Tumbes Hummingbird have since been seen regularly in that vicinity, and it has also been seen near Zapotillo (M. B. Robbins and G. H. Rosenberg) and Macará (v.o.). Recorded up to about 1000 m near El Empalme.

Monotypic.

Range: extreme sw. Ecuador and nw. Peru.

The Spot-throated Hummingbird (*L. taczanowskii*) is numerous in deciduous woodland and arid scrub in the Jaen/San Ignacio area of Cajamarca in n. Peru, not far south of Zumba in adjacent Ecuador. It should be watched for in the Zumba region.

Amazilia tzacatl   **Rufous-tailed Hummingbird**   **Amazilia Colirrufa**

Fairly common to common and widespread in clearings, gardens, agricultural areas, secondary woodland, and forest borders in the lowlands of w. Ecuador, and in the subtropical zone on the west slope of the Andes, south into much of Guayas (at least in more humid areas) and to w. Loja (south to the Alamor/El Limo region). The Rufous-tailed Hummingbird occurs in both humid and arid regions, though it largely avoids both the very humid forested portions of Esmeraldas (where it is less numerous than it typically is elsewhere) and very dry areas (e.g., the Santa Elena Peninsula). In some regions, such as the humid hills south of the lower Río Ayampe in extreme nw. Guayas, it occurs syntopically with the Amazilia Hummingbird (*A. amazilia*). Ranges up on the west slope of the Andes regularly to about 1500 m around Mindo and to 2000 m at Hacienda La Florida east of Apuela in Imbabura, in small numbers up into the central valley in Pichincha to almost 2500 m in the Río Guaillabamba and Tumbaco valleys north and east of Quito, a few even wandering to gardens in Quito itself (N. Krabbe).

One race: *jucunda*.

Range: s. Mexico to w. Venezuela and sw. Ecuador (seems likely to occur also in immediately adjacent Peru).

Amazilia amazilia   Amazilia Hummingbird   Amazilia Ventrirrufa

Fairly common to (seasonally?) common in desert scrub, shrubby areas, deciduous woodland, and gardens in the lowlands of w. Ecuador from w. Esmeraldas southward, also ranging up into the intermontane valleys of s. Ecuador from
s. Azuay (Valle de Yunguilla and Oña) southward. The Amazilia Hummingbird
favors arid regions and certainly is most numerous in such areas; it does,
however, also range into more humid regions, in some areas occurring with
about equal numbers of the Rufous-tailed Hummingbird (*A. tzacatl*). Chapman
(1926) records a few old specimens as having been taken on the east slope of
the Andes in Zamora-Chinchipe (at Zamora and Sabanilla), these presumably
representing wanderers from the highlands; there is one recent report from this
area, a bird seen above Zamora on 27 Jul. 1992 (D. and M. Wolf). Recorded
up to at least 2200 m.

Three races of *A. amazilia* are found in Ecuador, *dumerilii* in the western
lowlands, *alticola* in the highlands (and occasionally on the east slope) of El
Oro and Loja, and an unnamed form found in s. Azuay (MCZ and ANSP
specimens). It has recently been suggested (Schuchmann 1999) that *alticola*
be recognized as a monotypic species (Loja Hummingbird), but we hesitate to
do so, especially given the existence of the still-undescribed taxon mentioned
below. These three taxa are similar, but *alticola* has a mostly rufous tail (though
the central pair of rectrices is bronzy green above, sometimes making the tail,
when viewed from above, look green in the field) and *dumerilii* an entirely
bronzy green tail. These two races seem to vary individually in the extent of
rufous across the lower breast and belly, but both show at least some. The undescribed Azuay race has mainly white underparts (with rufous confined to a
limited amount on the flanks); it also differs from the other two Ecuadorian
races in having its upper tail-coverts mostly rufous and perhaps in having less
extensive reddish pink on the bill.

Range: w. Ecuador and w. Peru.

Amazilia franciae   Andean Emerald   Amazilia Andina

Uncommon and somewhat local at borders of humid and montane forest and
in adjacent clearings and gardens in the more humid lowlands and foothills of
w. Ecuador south through El Oro to w. Loja (Punta Santa Ana and Las Piñas),
ranging up locally into the subtropical zone. Recently also found to be uncommon to fairly common in the same habitats in the foothills on the east slope of
the Andes in s. Zamora-Chinchipe from the Zumba area north to around
Zamora. In addition, in Dec. 1996 the species was found to be uncommon in
the coastal Cordillera de Colonche of w. Guayas at Loma Alta (Becker and
López-Lanús 1997). In the west recorded principally below about 1400 m,
locally or in small numbers up to nearly 2000 m (e.g., 2 were seen at Hacienda
La Florida east of Apuela in Mar. 1991; P. Coopmans); on the east slope
recorded from 900 to 1600 m.

The species has been placed in the genus *Agyrtria* (e.g., Schuchmann 1999).
Two rather different races range in Ecuador, the nearly endemic—it also occurs

in w. Nariño, Colombia—*viridiceps* in the west and *cyanocollis* in the southeast. The latter was only recently confirmed from Ecuador—it was heretofore known only from Peru—with specimens and a nest (ANSP) taken north of Zumba on 15 Aug. 1989 and south of Palanda on 10 Jul. 1992. There also are numerous sightings from that region, and a few from around Zamora. *Cyanocollis* differs from *viridiceps* in its mainly glittering blue crown and neck (reduced and mainly on the sides of the neck in females); this area is glittering green in *viridiceps*.

Range: Andes of Colombia and w. Ecuador (where also in adjacent lowlands), and in se. Ecuador and n. Peru.

Amazilia fimbriata   Glittering-throated Emerald   Amazilia Gorjibrillante

Fairly common to common and widespread in clearings, gardens, secondary woodland (including early-succession growth and riparian woodland on river islands and along rivers), and borders of humid forest (both terra firme and várzea) in the lowlands and foothills of e. Ecuador. The Glittering-throated Emerald is considerably more numerous in inhabited areas than it is in regions still predominantly forested; the species is probably increasing rapidly in Ecuador. Ranges up on the east slope of the Andes to about 1200 m.

This and the following three species have been placed in the genus *Polyerata* (e.g., Schuchmann 1999). One race: *fluviatilis*.

Range: e. Colombia, much of Venezuela, and the Guianas to ne. Peru, n. Bolivia, and e. and s. Brazil.

[Amazilia lactea   Sapphire-spangled Emerald   Amazilia Pechizafiro]

Uncertain. Known in Ecuador from only one sighting, a bird (presumed to have been a male from its intermittent singing) seen in a semiopen swampy area in the lowlands of e. Ecuador along the Maxus oil-field road southeast of Pompeya (300 m) on 28 Jan. 1995 (RSR and F. Sornoza); attempts to relocate the bird on subsequent dates proved unsuccessful, as were attempts to photograph it. Whether this individual was merely a wanderer to Ecuador, or the species has a small population in remote sections of the lowlands of e. Ecuador (as apparently do several other hummingbird species), remains unknown. The Sapphire-spangled Emerald is known from as close to Ecuador as the lower Río Napo area of ne. Peru.

Race unknown but probably *bartletti*.

Range: se. Venezuela; e. Peru, n. Bolivia, and sw. Amazonian Brazil; se. Brazil.

Amazilia amabilis   Blue-chested Hummingbird   Amazilia Pechiazul

Uncommon in the lower growth of humid forest and secondary woodland and in adjacent clearings, plantations, and gardens in the more humid lowlands of w. Ecuador south to Manabí, e. Guayas, and nw. Azuay (Manta Real; RSR); there is also a single report from as far south as El Oro (a male seen on 10 Aug. 1990 above Zaruma; S. Hilty et al.). In the wet-forest belt of n. Esmeraldas, the

Blue-chested Hummingbird is largely outnumbered by the similar Purple-chested Hummingbird (*A. rosenbergi*), as attested to by extensive mist-netting samples around Playa de Oro (O. Jahn et al.). Small numbers occur at Río Palenque. Recorded mostly below 300 m (aside from the sole Zaruma report, which was at about 1200 m).

One race: nominate *amabilis*.

Range: Nicaragua to sw. Ecuador.

Amazilia rosenbergi	Purple-chested Hummingbird	Amazilia Pechimorada

Fairly common to (northward) common in the undergrowth and mid-levels of very humid forest and forest borders in the lowlands and foothills of nw. Ecuador, where known mainly from n. Esmeraldas, adjacent Imbabura (Lita; ANSP), and nw. Pichincha (a May 1990 specimen from Salcedo Lindo and a Jun. 1992 specimen taken 19 km north of Pedro Vicente Maldonado; both MECN); there also are numerous recent (since 1995) sightings from the last area (v.o.). There is also a unique report (PJG et al.) of several birds seen as far south as s. Pichincha at Río Palenque, on 8 Jan. 1981; otherwise only the Blue-chested Hummingbird (*A. amabilis*) has been found there. Recorded mostly below 600 m, with one specimen (ANSP) apparently having been taken at about 900 m (labeled as from Lita).

Monotypic.

Range: w. Colombia and nw. Ecuador.

The Steely-vented Hummingbird (*A. saucerrottei*) of w. Colombia and nw. Venezuela has recently been recorded in the Ipiales region of Nariño, Colombia (Salaman and Mazariegos 1998), a locality only a few kilometers north of the Ecuadorian frontier. The species could easily wander south into Ecuadorian territory, and should be watched for.

The Snowy-bellied Hummingbird (*A. edward*) was recorded from Ecuador based on two specimens supposedly procured at Santo Domingo de los Colorados (Oberholser 1902). There is, however, no other information to indicate that this Central American hummingbird ranges any farther east or south than e. Panama (Ridgely and Gwynne 1989; AOU, 1998); it is not even recorded from Colombia (Hilty and Brown 1986). Mislabeled specimens were surely involved.

Chalybura buffonii	White-vented Plumeleteer	Calzonario de Buffón

Uncommon to locally (or formerly?) fairly common in the lower and middle growth of secondary woodland, forest borders, and adjacent clearings in the foothills and lower subtropical zone of sw. Ecuador in El Oro and w. Loja (Alamor region). A few sightings from somewhat farther north are regarded as uncertain. The sighting from w. Esmeraldas at Cabeceras de Bilsa (Parker and Carr 1992) seems surely to have been in error. Other workers in this region have failed to find either species of plumeleteer; from a biogeographic perspective, the Bronze-tailed (*C. urochrysia*) is far more likely. Recorded mostly from 500 to 1750 m.

Judging from the large number of specimens taken in the 19th and early 20th centuries compared to the relative paucity of recent sightings, it appears that this taxon, nearly endemic to sw. Ecuador (having recently also been found in adjacent Tumbes, Peru; Wiedenfeld et al. 1985), has declined substantially in recent decades, likely as a result of widespread deforestation. A substantial population was located east of El Limo in w. Loja in Apr. 1993 (ANSP, MECN). We consider the species to merit Near-threatened status in Ecuador.

One race occurs in Ecuador, *intermedia*. *Intermedia*'s range is widely disjunct from those of the other races of *C. buffonii*, and it also has a different elevational range—not being found in the tropical zone as are the other races of *C. buffonii*. Despite the absence of plumage divergence (though its lower mandible is reddish and its feet pinkish gray, both being black in other races of *C. buffonii*), *intermedia* may be better considered as a separate species (Ecuadorian Plumeleteer). Alternatively, it has been suggested (Schuchmann 1999) that *intermedia* could be treated as a disjunct race of *C. urochrysia*, though we consider this unlikely.

Range: Panama to w. and n. Venezuela; sw. Ecuador and extreme nw. Peru.

Chalybura urochrysia   Bronze-tailed Plumeleteer   Calzonario Patirrojo

Fairly common to common in the lower and middle growth of very humid forest and forest borders in the foothills and adjacent lowlands of extreme nw. Ecuador in n. Esmeraldas, where apparently not recorded south of the Playa de Oro area. At Playa de Oro the Bronze-tailed Plumeleteer was found to occur exclusively in the interior of forest, and—unlike the other hummingbirds of the area—it was not found in adjacent plantations (O. Jahn et al.). Reports from farther south (e.g., Río Palenque) are regarded as uncorroborated. Mainly recorded up to about 800 m (El Placer; ANSP, MECN).

The species has been called the Red-footed Plumeleteer.

One race: nominate *urochrysia*.

Range: Nicaragua to extreme nw. Ecuador.

Adelomyia melanogenys   Speckled Hummingbird   Colibrí Jaspeado

Fairly common to common in the lower growth and borders of montane forest, secondary woodland, and shrubby clearings in the subtropical and temperate zones on both slopes of the Andes, and locally on slopes above the central and interandean valleys as well. Also recently recorded from the coastal Cordillera de Colonche in sw. Manabí (Cerro Achi and Cerro San Sebastián [Parker and Carr 1992; ANSP]) and w. Guayas (Cerro La Torre and above Salanguilla [N. Krabbe]; Loma Alta [Becker and López-Lanús 1997]). Recorded mostly from 1400 to 2800 m, with smaller numbers somewhat higher and lower (once as high as 3400 m at Yanacocha; MECN). On the coastal cordillera, however, the Speckled Hummingbird occurs notably lower, between about 600 and 800 m.

Two very similar races are found in Ecuador, nominate *melanogenys* on the east slope and *maculata* on the west slope. *Maculata* has somewhat wider buff tail-tipping.

Range: mountains of n. Venezuela, and Andes from w. Venezuela to nw. Argentina; Sierra de Perijá.

Urosticte benjamini   **Purple-bibbed Whitetip**   **Puntiblanca Pechipúrpura**

Rare to uncommon and perhaps local in the lower growth of montane forest and forest borders in the foothills and subtropical zone on the west slope of the Andes. Recorded mainly from Esmeraldas (e.g., El Placer), Imbabura, and Pichincha (numerous records from the Mindo, Gualea, and Nanegal regions, and found south at least to the lower Chiriboga road), but doubtless also occurring in Carchi. There are also a few sightings (which need to be confirmed) from much farther south: in nw. Azuay (above Manta Real, where seen in Nov. 1990 by J. C. Matheus) and in El Oro (a single sighting from Buenaventura, where one was seen on 18 Mar. 1988; R. A. Rowlett et al.). Recorded from 600 m (above Manta Real; J. C. Matheus) to 1600 m; mostly above 1000 m.

  U. ruficrissa (Rufous-vented Whitetip) of the east slope of the Andes is regarded as a separate species from *U. benjamini* of the west slope, in agreement with Hilty and Brown (1986) and most other recent authors. Apart from *benjamini*'s being larger, there are several striking plumage differences in males. One race: nominate *benjamini*.

Range: west slope of Andes in Colombia and Ecuador.

Urosticte ruficrissa   **Rufous-vented Whitetip**   **Puntiblanca Pechiverde**

Rare to uncommon and perhaps local in the lower growth of montane forest and forest borders in the subtropical zone on the east slope of the Andes. The Rufous-vented Whitetip tends to be inconspicuous and is often encountered mainly through mist-net captures. Recorded from about 1300 m (San Rafael Falls; ANSP) to 2300 m.

One race: nominate *ruficrissa*.

Range: east slope of Andes from s. Colombia to n. Peru.

Phlogophilus hemileucurus   **Ecuadorian Piedtail**   **Colipinto Ecuatoriano**

Uncommon to fairly common but seemingly local in the undergrowth of humid forest in the foothills along the eastern base of the Andes from w. Napo (old specimens from Río Suno and below San José, and numerous recent reports from along the Loreto road north of Archidona) south into Zamora-Chinchipe. Presumably the Ecuadorian Piedtail occurs in Sucumbíos as well, as it is known from adjacent Colombia (Hilty and Brown 1986). Recorded only from a narrow elevational zone between about 900 and 1300 m; a specimen (MECN) supposedly from Baños (1850 m) would, however, be from considerably higher (but perhaps it is not correctly labeled).

The Ecuadorian Piedtail is considered to be a Near-threatened species by Collar et al. (1994), an assessment with which we concur.

Monotypic.

Range: eastern base of Andes from s. Colombia to n. Peru.

Heliodoxa imperatrix   **Empress Brilliant**   **Brillante Emperatriz**

Rare to perhaps locally uncommon in the undergrowth and borders of montane forest in the subtropical zone on the west slope of the Andes from Imbabura (old specimens from Intag; recent specimens from near Chical [ANSP]) south to Pichincha (various localities near and around Mindo). An old specimen from "Santo Domingo" is regarded as probably mislabeled. Even where both species are known to occur, the Empress Brilliant is almost invariably outnumbered by its smaller congener the Green-crowned (*H. jacula*); occasionally the two species are even seen foraging together, or the Empress feeds with the Fawn-breasted Brilliant (*H. rubinoides*). Recorded mostly from 1500 to 2100 m, locally (or seasonally?) down to about 1000 m near San Miguel de los Bancos (PJG).

Because of its relative rarity and small range, we consider the Empress Brilliant to merit Near-threatened status. Somewhat surprisingly, the species was not mentioned by Collar et al. (1994).

We follow the sequence of species in the genus *Heliodoxa* suggested by J. A. Gerwin and R. M. Zink (*Wilson Bull.* 101[4]: 525–544, 1989). Monotypic.

Range: west slope of Andes in Colombia and nw. Ecuador.

Heliodoxa jacula   **Green-crowned Brilliant**   **Brillante Coroniverde**

Uncommon to fairly common in the undergrowth and borders of montane forest, mature secondary woodland, and adjacent clearings and gardens in the foothills and subtropical zone on the west slope of the Andes from Esmeraldas south to El Oro; ranges down in smaller numbers (only seasonally?) into the more humid lowlands south to Manabí and Pichincha. Also recently found to occur locally on the coastal Cordillera de Mache in w. Esmeraldas (Cabeceras de Bilsa and Bilsa [Parker and Carr 1992; R. P. Clay et al.]) and the Cordillera de Colonche in sw. Manabí (Cerro San Sebastián; RSR and G. H. Rosenberg) and w. Guayas (Loma Alta; Becker and López-Lanús 1997). Recorded mostly from 500 to 1550 m, locally down as low as 300 m.

One race: *jamesoni*. *Jamesoni* is geographically isolated from the other races of *H. jacula* and is nearly endemic to Ecuador; it was only recently recorded for the first time from adjacent Nariño, Colombia (Salaman 1994).

Range: Costa Rica to n. Colombia; extreme sw. Colombia and w. Ecuador.

Heliodoxa leadbeateri   Violet-fronted Brilliant   Brillante Frentivioleta

Uncommon to fairly common and somewhat local (also rather inconspicuous, hence not often recorded by sight though numbers mist-netted can sometimes be substantial) in the undergrowth and borders of montane forest in the foothills and subtropical zone on the east slope of the Andes; perhaps more numerous southward. Recorded mostly between about 1300 and 2100 m, locally down to 1100 m (e.g., on the Cordillera de Cutucú; ANSP).

One race: *sagitta.*

Range: mountains of n. Venezuela, and Andes (mainly on east slope) from w. Venezuela to w. Bolivia; Sierra de Perijá.

Heliodoxa gularis   Pink-throated Brilliant   Brillante Gorjirrosado

Very rare to rare and seemingly local in the undergrowth of montane and humid forest in the foothills and adjacent lowlands along the eastern base of the Andes in n. Ecuador, where recorded only from w. Sucumbíos and w. Napo; perhaps occurs mainly or entirely near the base of outlying ridges and mountains. There are only a few records of this poorly known hummingbird. At least six specimens (AMNH) were taken in the lower Sumaco region ("below San José," "lower Río Suno," and "Río Suno, above Avila") in the early 20th century: males were obtained at Avila on 10 Aug. 1935 (Berlioz 1937b) and in 1940 (Mejía collection); and four specimens (UKMNH) were taken at Santa Cecilia (350 m) in w. Sucumbíos in Jun.–Jul. 1971. More recently, R. Bleiweiss observed the species along the Río Due in 1980, RSR saw a male north of Archidona along the road to Loreto at about 950 m on 13 Nov. 1991, and a female (ANSP) was collected north of Lumbaquí in the Bermejo oil-field area (900 m) on 13 Mar. 1993. Alhough not yet found south of the Volcán Sumaco region in w. Napo, the Pink-throated Brilliant seems likely also to occur at least locally southward along the base of the Andes into s. Ecuador, for it has also been found south of the Río Marañón in San Martín, Peru. Recorded between about 350 and 950 m.

The Pink-throated Brilliant was given Near-threatened status by Collar et al. (1994). In our view, however, this brilliant is considerably rarer and more local than several other hummingbird species that sometimes occur with it (e.g., the Napo Sabrewing [*Campylopterus villaviscensio*] and the Ecuadorian Piedtail [*Phlogophilus hemileucurus*]) and that were also considered to be Near-threatened. We thus believe that this brilliant should be considered as Vulnerable. No population is known to occur in any area presently receiving formal protection.

Monotypic.

Range: very locally along eastern base of Andes in extreme s. Colombia and n. Ecuador, and in n. Peru.

Heliodoxa schreibersii   Black-throated Brilliant   Brillante Gorjinegro

Rare to uncommon and apparently local in the undergrowth of terra firme forest (also occasionally feeding up into the canopy; P. Coopmans) in the lowlands of e. Ecuador, and in montane forest in the foothills along the base of the Andes. This scarce hummingbird favors the vicinity of streams, and east of the Andes it mainly occurs in areas with some relief. In the northeast it has not been recorded east of the Maxus road southeast of Pompeya (RSR et al.) and Yuturi (fide S. Howell); there are also recent sightings from Kapawi Lodge near the Peruvian border on the Río Pastaza. Mostly recorded below 1250 m, locally to as high as 1450 m (Cordillera del Cóndor near Chinapinza; ANSP).

One race: nominate *schreibersii*.

Range: se. Colombia and extreme nw. Amazonian Brazil to e. Peru.

Heliodoxa rubinoides   Fawn-breasted Brilliant   Brillante Pechianteado

Rare to uncommon and perhaps local in the lower growth and borders of montane forest in the subtropical zone on both slopes of the Andes. On the west slope recorded from Pichincha (Nanegal and Tandayapa areas) south to El Oro and adjacent w. Loja (an old specimen from Punta Santa Ana); it apparently is not found northward to the Colombian frontier. On the east slope recorded mainly from Sucumbíos (El Guanderal on the La Bonita road [MECN]) south to w. Napo at various localities south to the Cordillera de Huacamayos. The Fawn-breasted Brilliant is, however, seemingly quite local south of there (perhaps being just under-recorded?), with a few scattered records from w. Morona-Santiago (Río Upano valley west of Macas; one mist-netted in Jul. 1979 [RSR and R. A. Rowlett]) and Zamora-Chinchipe (female collected [ANSP] in the Romerillos sector of Podocarpus National Park on 5 Jan. 1992 [Rasmussen et al. 1996]; female collected above Chinapinza on the Cordillera del Cóndor on 8 Jun. 1993 [MECN]; and one seen in the Río Isimanchi valley in Nov. 1992 [M. B. Robbins]). Recorded mostly from 1100 to 2100 m, but locally ranging to as high as 2400 m in Sucumbíos.

Two similar races are found in Ecuador, *cervinigularis* in the east and *aequatorialis* in the west. The slightly smaller *aequatorialis* differs in its mainly coppery bronze greater wing-coverts (bronzy green in *cervinigularis*).

Range: Andes from n. Colombia to s. Peru.

Heliodoxa aurescens   Gould's Jewelfront   Brillante Frentijoya

Rare to uncommon and perhaps local (certainly inconspicuous) in the lower growth and borders of terra firme forest (also at least occasionally foraging up into the subcanopy) in the lowlands of e. Ecuador from Sucumbíos south to Zamora-Chinchipe (Warientza on the Cordillera del Cóndor; WFVZ). Recorded mostly below 500 m, but occurs locally up to about 900 m (e.g., at Warientza, and north of Lumbaquí in the Bermejo oil-field area [ANSP]).

The species is sometimes simply called the Jewelfront.

This species was formerly placed in the monotypic genus *Polyplancta*, but J. A. Gerwin and R. M. Zink (*Wilson Bull.* 101[4]: 525–544, 1989) suggest that its merger into *Heliodoxa* is warranted. Monotypic.

Range: s. Venezuela and se. Colombia to n. Bolivia and w. and cen. Amazonian Brazil.

Topaza pyra   Fiery Topaz   Topacio Fuego

Rare to uncommon but very local in várzea forest and along streams and rivers in the lowlands of e. Ecuador. Occurs mainly or entirely in blackwater drainages, and most often seen along watercourses in such areas, either foraging low—sometimes perching on dead branches protruding from the water—or in the canopy nearby; rarely or never seen elsewhere. In the 1970s the Fiery Topaz was regularly found along the blackwater stream at Taracoa. It has recently been seen near Imuyacocha, Zancudococha, and Río Pacuyacu in the lower Río Aguarico drainage of the far northeast, and there are also several reports from the Tiputini Biodiversity Center and Kapawi Lodge (v.o.). Two specimens (ANSP, MECN) were taken near the Río San Miguel north of Tigre Playa on 2 Aug. 1993. Recorded up to 750 m (MCZ specimen from Guaticocha), but usually ranging below 400 m.

This species has occasionally (e.g., Schuchmann 1999) been considered conspecific with *T. pella* (Crimson Topaz). One race: *amaruni* (see D.-S. Hu, L. Joseph, and D. Agro [*Ornitol. Neotrop.* 11[2]: 123–142, 2000]).

Range: e. Ecuador, ne. Peru, w. Amaz. Brazil, and sw. Venezuela.

[*Topaza pella*   Crimson Topaz   Topacio Carmesí]

Uncertain. Supposedly very rare and local in the lowlands of ne. Ecuador in w. Napo. The Crimson Topaz is still known in Ecuador only from three specimens (USNM) taken around 1900 (Oberholser 1902). These are labeled as having been obtained at the mouth of the Río Suno (presumably into the Río Napo), a locality situated not far southwest of the town of Coca near the northwest bank of the Napo. Nothing further is known about the species in Ecuador where, despite much searching in appropriate habitat, in recent years only the similar Fiery Topaz (*T. pyra*) has been found. Recorded at about 300 m.

As noted above, the Crimson Topaz remains unknown in life in Ecuador. In light of this absence of information, we consider it to deserve Data Deficient status. The species as a whole is not considered to be at risk (Collar et al. 1992, 1994).

The Río Suno specimens mentioned above were the basis for the description of the subspecies *pamprepta* (Oberholser 1902); see also Norton (1965). D.-S. Hu, L. Joseph, and D. Agro (*Ornitol. Neotrop.* 11[2]: 123–142, 2000) recently concluded that *pamprepta* is a synonym of *T. pella smaragdula* of French Guiana and Amapá, Brazil, and that the specimens in question were not actually obtained in Ecuador. The Crimson Topaz therefore should be deleted from the Ecuadorian list, but this information was received too late to be incorporated here. Schuchmann (1999) considered *T. pyra* and *T. pella* to be conspecific; based on plumage and morphometric differences, Hu et al. (op. cit.) disagreed.

Range: se. Venezuela, the Guianas, and e. Amazonian Brazil; one record from Rondonia (Cachoeira Nazaré) in sw. Amazonian Brazil and (?) e. Ecuador.

Oreotrochilus chimborazo   Ecuadorian Hillstar   Estrella Ecuatoriana

Uncommon to fairly common in paramo in the Andes from Imbabura and Pichincha south to Azuay (El Cajas National Recreation Area, ANSP), primarily in fairly arid regions. The northernmost locality on record was formerly on Cerro Cotacachi/Cerro Yana Urco de Piñán in Imbabura, where birds were observed in 1975 and 1980 (F. Ortiz-Crespo and R. Bleiweiss, *Auk* 99[2]: 376–378, 1982), but recently small numbers have been seen in Páramo del Ángel on the southwestern slope of Volcán Chiles in Carchi, hardly 1 km south of the Colombian boundary (S. Woods, F. Ortiz-Crespo, and P. M. Ramsay, *Cotinga* 10: 37–40, 1998). There is a Jul. 1991 record from even farther north in adjacent Nariño at El Tambo, the first Colombian record (Salaman et al. 1994; photos to VIREO). Ortiz-Crespo and Bleiweiss (op. cit.) suggest that to some extent this species' local distribution may reflect anthropogenic effects, especially the result of burning and grazing. The Ecuadorian Hillstar feeds mainly, perhaps exclusively, on the orange-yellow flowers of the small shrub *Chuquiragua insignis* which occurs widely on arid slopes above treeline. Recorded mostly from 3600 to 4600 m, occasionally somewhat higher.

The species has sometimes been called the Chimborazo Hillstar, but it would seem preferable to use a broader geographical epithet as its distribution is by no means confined to Volcán Chimborazo. The species is, however, virtually endemic to Ecuador.

Three races are recorded from Ecuador: nominate *chimborazo* is confined to the slopes of Volcán Chimborazo; *soederstromi* to the slopes of Volcán Quillotoa; and *jamesonii* occurs elsewhere. Males of *jamesonii* have an entirely glittering purple hood; *soederstromi* is similar but has a scattering of green feathers on the lower throat; and in the nominate race the lower throat is contrastingly glittering green. These three taxa have by some (e.g., Meyer de Schauensee 1966, 1970) been considered subspecies of *O. estella* (Andean Hillstar). This was mostly because it appeared that the Ecuadorian taxa varied clinally from north to south, with the violet-hooded taxon (*jamesonii*) most different from *O. estella* occurring farthest to the north, and more or less "intermediate" forms southward toward the range of the green-hooded *O. estella*. However, the collection (ANSP) in Aug. 1986 of males that are clearly referable to purple-hooded *jamesonii* near El Cajas, the southernmost known locality for *O. chimborazo*, confirms that this is not the case, and demonstrates that the green-throated nominate *chimborazo* is actually found only on Volcán Chimborazo and is "surrounded" by the violet-hooded *jamesonii*. The ANSP material also refutes the doubts expressed by J. T. Zimmer (*Am. Mus. Novitates* 1513, 1951) regarding the provenance of a violet-hooded male obtained in the 19th century by O. T. Baron near Cuenca.

Range: Andes of Ecuador and extreme s. Colombia.

*Oreotrochilus estella   **Andean Hillstar**   **Estrella Andina**

Uncommon in rocky paramo with a few patches of woodland in extreme s. Ecuador in se. Loja. First found in Ecuador on 16–17 Dec. 1991, when at least two males were seen on the west slope of the Cordillera Las Lagunillas along the Jimbura-Zumba road (RSR). In Oct. 1992 several specimens (ANSP, MECN) were obtained in the same area, but the species could not be located elsewhere. Surprisingly, the Andean Hillstar has not been found in adjacent extreme n. Peru (Parker et al. 1985). Recorded only between 3250 and 3500 m.

Does not include *O. chimborazo* (Ecuadorian Hillstar) of the Ecuadorian Andes, sometimes considered conspecific with this species; see discussion under that species. One race of *O. estella* occurs in Ecuador, *stolzmanni*. It has recently been suggested (Schuchmann 1999) that *stolzmanni* itself deserves recognition as a separate species (Green-headed Hillstar).

Range: Andes from extreme s. Ecuador to n. Chile and nw. Argentina.

Urochroa bougueri   White-tailed Hillstar   Estrella Coliblanca

Uncommon to rare in the lower growth and borders of montane forest in the subtropical zone on both slopes of the Andes in n. Ecuador; seems considerably scarcer and more local in the west, where largely confined to the vicinity of rushing mountain streams. On the east slope recorded south at least to Tungurahua (Baños and San Francisco), nw. Pastaza (Cerros de Abitagua) in the Río Pastaza valley, and nw. Morona-Santiago in the Río Huamboya valley west of Macas (collected [MECN] in Apr. 1998; F. Sornoza), but it may also occur southward, for the same race has been found in San Martín, n. Peru (T. A. Parker III and S. A. Parker, *Bull. B. O. C.* 102[2]: 64, 1982). On the west slope known south only to Pichincha (lower Chiriboga road, and Tandapi [MECN]). Recorded mostly from 1100 to 2000 m.

Two very different races occur in Ecuador, *leucura* on the east slope and nominate *bougueri* on the west slope. *Leucura* differs from nominate *bougueri* in being green above (not decidedly bronze), its central rectrices are bronzy green (not black) and outer rectrices mainly white (not black), and it lacks the nominate race's prominent rufous at gape. In view of their plumage and habitat differences, separate species may be involved; west-slope birds could be called the Rufous-gaped Hillstar.

Range: Andes of s. Colombia and nw. and e. Ecuador, and at one site on their east slope in n. Peru.

Patagona gigas   Giant Hummingbird   Colibrí Gigante

Uncommon to locally (or seasonally) fairly common in shrubby and cultivated areas, mainly in arid regions (and often in largest numbers in truly xeric areas), in the Andes from Carchi (a sighting at Gruta de la Paz in the Río Chota valley on 8 Jan. 1991 [Marín et al. 1992]; more recently also seen on Páramo del Ángel on the southwestern slope of Volcán Chiles just south of the Colombian border [S. Woods, F. Ortiz-Crespo, and P. M. Ramsay, *Cotinga* 10: 37–40, 1998) south to the Oña area in s. Azuay (whence numerous sightings). There seems to be only one definite report from Loja, of a bird seen on 22 May 1989 along the Malacatos-Gonzanamá road (Bloch et al. 1991). F. Ortiz-Crespo (*Ibis* 116[3]: 347–359, 1974) suggests that because the Giant Hummingbird feeds to such a large extent on the flowers of *Agave*, and because *Agave* is not native to Ecuador—having been introduced several centuries ago from Mexico—the hummingbird may have extended its range northward from Peru only subsequent to *Agave*'s introduction. We consider this unlikely, however, if only because the species feeds on such a wide variety of other flowering plants. Ortiz-Crespo (op. cit.) also suggested that the Giant Hummingbird perhaps is only a seasonal, nonbreeding visitant in the Quito area and northward; however, our recent observations (together with those of J. C. Matheus and others), including the discovery of several nests, have shown that this definitely is not the case. Recorded mostly between 1800 and 3300 m, but has been seen (J. R. King and S. J. Holloway, *Bull. B. O. C.* 110[2]: 79–80, 1990) up to 3800 m in the Río Mazán valley near Cuenca, perhaps occurring that high only seasonally.

One race: *peruviana*.

Range: Andes from extreme s. Colombia (recently found to be a seasonal wanderer to s. Nariño; Salaman and Mazariegos 1998) to cen. Chile (where it occurs down to sea level) and w. Argentina.

Aglaeactis cupripennis   Shining Sunbeam   Rayito Brillante

Fairly common to common in shrubby areas, scrub, and borders of montane woodland (often at treeline) in the temperate and paramo zones of the Andes, including slopes above the central and interandean valleys. The Shining Sunbeam occurs both in humid and fairly arid areas. Recorded mostly between 2800 and 3600 m, occasionally or locally up to 3900–4100 m in the north (e.g., around the Papallacta pass and on Volcán Pichincha) or as low as 2400 m in the south (Utuana in Loja; MECN); there is also an old record of a single specimen (AMNH) from as low as Salvias, at about 1700 m.

Two races are found in Ecuador, nominate *cupripennis* in most of the species' Ecuadorian range, and *parvulus* in the south in Azuay, El Oro, Loja, and adjacent Zamora-Chinchipe. *Parvulus* differs in its slightly smaller size, proportionately shorter bill, more deeply colored underparts, and more extensively rufous tail (with bronzy area being restricted to about the tail's distal third; in the nominate race the tail is mainly bronzy, with rufous only toward the basal quarter).

Range: Andes from n. Colombia to s. Peru.

Lafresnaya lafresnayi   Mountain Velvetbreast   Colibrí Terciopelo

Uncommon to locally fairly common at the shrubby borders of montane forest and in low woodland (sometimes at treeline) in the temperate zone on both slopes of the Andes, and locally also on slopes above the central and interandean valleys. Recorded mostly from 2500 to 3500 m, locally down to 2200 m (e.g., on the Cordillera de Huacamayos) and up to 3700 m.

One race: *saul*.

Range: Andes from w. Venezuela to s. Peru; Sierra de Perijá and Santa Marta Mountains.

Pterophanes cyanopterus   **Great Sapphirewing**   **Alazafiro Grande**

Uncommon in shrubby forest borders and low woodland (regularly at treeline) in the temperate zone on both slopes of the Andes, and very locally on the slopes above the central valley; throughout favors more humid regions. On the west slope not recorded south of Azuay in the Cuenca region (El Cajas National Recreation Area and Río Mazán valley). Recorded mostly from 3000 to 3600 m.

One race: *peruvianus* (see J. T. Zimmer, *Am. Mus. Novitates* 1540, 1951).
Range: Andes from n. Colombia to w. Bolivia.

Coeligena coeligena   **Bronzy Inca**   **Inca Bronceado**

Uncommon to locally fairly common in the lower growth and borders of montane forest in the subtropical zone on the east slope of the Andes; at least locally it also comes out into adjacent clearings (e.g., at Cabañas San Isidro). Recorded mostly from 1400 to 2300 m.

One race: *obscura*. We follow J. T. Zimmer (*Am. Mus. Novitates* 1513: 23–51, 1951) in placing all Ecuadorian birds in *obscura*, a taxon that ranges from s. Colombia south through Peru. Some other authors (e.g., *Birds of the World*, vol. 5) have placed Ecuadorian examples in *columbiana*, which is slightly paler below.
Range: mountains of n. Venezuela, and Andes from w. Venezuela to w. Bolivia; Sierra de Perijá.

Coeligena wilsoni   **Brown Inca**   **Inca Pardo**

Uncommon to fairly common in the lower growth and borders of montane forest in the foothills and subtropical zone on the west slope of the Andes south to El Oro and w. Loja (Vicentino; Best et al. 1993). Recorded mostly from 800 to 2000 m, but found lower southward (locally or perhaps seasonally as low as 400 m in nw. Azuay).
Monotypic.
Range: west slope of Andes in sw. Colombia and w. Ecuador.

Coeligena torquata   **Collared Inca**   **Inca Collarejo**

Fairly common in the lower growth and borders of montane forest and in adjacent clearings in the upper subtropical and temperate zones on both slopes of the Andes. On the west slope recorded south at least formerly to w. Chimborazo (various localities), but in recent years not found south of Cotopaxi. Recorded mostly from 2100 to 3000 m.

Two similar races occur in Ecuador, nominate *torquata* on the east slope and *fulgidigula* on the west. Male *fulgidigula* have more green-spangled feathers on the lower underparts (this area more solidly black in the nominate race), much more green iridescence on the throat (this area quite blackish in the nominate race), and a glittering blue frontal patch (glittering violet in the nominate race). Birds from Carchi (above Maldonado; ANSP specimens) appear to be somewhat intermediate, being blacker below than typical *fulgidigula* though in other respects typical of that race.

Range: Andes from w. Venezuela to cen. Peru. Excludes Gould's Inca (*C. inca*) of Andes of s. Peru and w. Bolivia.

Coeligena lutetiae   Buff-winged Starfrontlet   Frentiestrella Alianteada

Uncommon to locally common in the lower growth and borders of montane forest and woodland in the temperate zone on both slopes of the Andes. On the west slope recorded south mainly to Pichincha (various localities on Volcán Pichincha), but also seen on 4–5 Jan. 1991 on the west slope of Volcán Parcato in w. Cotopaxi (N. Krabbe), at Caripero in w. Cotopaxi (N. Krabbe and F. Sornoza), and as far south as Salinas in Bolívar (Krabbe et al. 1997). Recorded mostly from 2700 to 3500 m, locally down to 2450 m in Podocarpus National Park (C. Rahbek et al.) and as high as 3700 m on the north side of Volcán Pichincha (N. Krabbe). Rarely or never occurs in actual paramo, ranging up only as high as treeline.

Monotypic.

Range: Andes from w. Colombia to extreme n. Peru.

Coeligena iris   Rainbow Starfrontlet   Frentiestrella Arcoiris

Fairly common to locally common in montane forest borders, secondary woodland, shrubby areas, and gardens in the upper subtropical and temperate zones of s. Ecuador from sw. Chimborazo (19th-century specimen from Ceche) and Cañar (several seen at Hacienda La Libertad in Aug. 1991; RSR) south through much of Azuay, e. El Oro, and Loja. With the exception of the Cañar site mentioned above, the Rainbow Starfrontlet does not seem to occur on the actual east slope of the Andes; we presume that an old specimen labeled as from Zamora (J. T. Zimmer, *Am. Mus. Novitates* 1513, 1951) was mislabeled. In its range this splendid large hummingbird is one of the more numerous and widespread members of its family, and it is quite tolerant of substantial habitat disturbance. The Rainbow Starfrontlet is numerous above Cuenca in Azuay, and indeed was considered to be the "commonest hummingbird" at 3100–3300 m in the Río Mazán valley (King 1989, p. 144). Recorded mostly from 2000 to 3300 m, locally as low as 1700 m in the Sozoranga region of Loja (Best et al. 1993; RSR).

Two rather different races are found in Ecuador, *hesperus* in s. Chimborazo, Cañar, and most of Azuay, and nominate *iris* from s. Azuay (Oña area) south through El Oro and Loja. Both sexes of the nominate race are more extensively rufous below (that color covering the entire belly, not just the lower belly and crissum as in *hesperus*) and are also more extensively rufous on the rump and lower back (only the uppertail-coverts in *hesperus*). Furthermore, in males of the nominate race the glittering crown feathers are mainly gold and green; in *hesperus* they are mainly coppery red.

Range: Andes of s. Ecuador and nw. Peru.

Ensifera ensifera   Sword-billed Hummingbird   Colibrí Pico Espada

Rare to locally fairly common in the canopy and borders of montane forest, secondary woodland, and adjacent shrubby areas and gardens in the temperate zone on both slopes of the Andes, locally on the slopes above the central valley, and rarely even in the valley itself (e.g., seen at La Ciénega in Aug. 1988; P. Coopmans). On the west slope recorded south to e. El Oro (Taraguacocha). Recorded mostly between 2500 and 3300 m, locally down to 2150 m in Podocarpus National Park (C. Rahbek et al.).

Monotypic. G. R. Graves (*Bull. B. O. C.* 111[3]: 139–140, 1991) presented evidence showing that *caerulescens*, known only from the unique type (and of unknown origin), is not a valid taxon.

Range: Andes from w. Venezuela to w. Bolivia.

Boissonneaua flavescens   Buff-tailed Coronet   Coronita Colianteada

Uncommon to fairly common and local (or seasonal) in the canopy and borders of montane forest and in nearby clearings in the subtropical zone on both slopes of the Andes in n. Ecuador. On the west slope recorded south to w. Cotopaxi (Caripero area; N. Krabbe and F. Sornoza). On the east slope recorded mainly from Pan de Azucar near Volcán Sumaco, where a series (WFVZ, MECN) was obtained by F. Sibley et al. on 14–20 Oct. 1989; there also is an old record from "Antisana" (which presumably was taken on the east slope), and one was also seen at the feeders at Cabañas San Isidro in late Jul. and early Aug. 1999 (M. Lysinger). Recorded mostly from 1500 to 2400 m, in small numbers (or perhaps seasonally?) somewhat lower; the Pan de Azucar birds were found, however, at 2900 m.

Two races occur in Ecuador, *tinochlora* on the west slope of the Andes and nominate *flavescens* on the east slope. The two forms are similar, but the buff in the tail of the nominate race is paler.

Range: Andes from w. Venezuela to n. Ecuador.

Boissonneaua matthewsii   Chestnut-breasted   Coronita
                             Coronet   Pechicastaña

Uncommon to fairly common in the canopy and borders of montane forest and in nearby clearings in the upper subtropical and temperate zones on both slopes of the Andes. Recorded locally along the entire east slope, and on the west slope from several localities in w. Chimborazo south locally through El Oro (El Chiral) to w. Loja (above Celica; MCZ specimens taken by D. Norton in Aug. 1965). The Chestnut-breasted Coronet is most numerous in se. Ecuador, and is especially common at higher elevations on the Cordillera del Cóndor. Ranges mostly from 1900 to 2700 m, but recorded down to 1600 m on the Cordillera del Cóndor and found locally up to 3300 m at Chaucha in Azuay (MECN).

Monotypic.

Range: Andes from s. Colombia to s. Peru.

Boissonneaua jardini   Velvet-purple Coronet   Coronita Aterciopelada

Rare to locally (or seasonally) uncommon in the lower and middle growth of montane forest and forest borders in the foothills and subtropical zone on the west slope of the Andes in nw. Ecuador south to Pichincha (various localities in the Mindo and Nanegal regions). Small numbers of this lovely hummingbird are sometimes in evidence above Mindo, but at other times they appear to be absent. Ranges mostly from 800 to 1700 m.

Because of its relative rarity and small range, we consider the Velvet-purple Coronet to merit Near-threatened status. Somewhat surprisingly, it was not mentioned by Collar et al. (1994).

Monotypic.

Range: west slope of Andes in w. Colombia and nw. Ecuador.

Heliangelus amethysticollis   Amethyst-throated   Solángel
                                  Sunangel   Gorjiamatista

Uncommon to locally fairly common in the borders of montane forest and in adjacent clearings in the subtropical and lower temperate zones on the east slope of the Andes in s. Ecuador. Thus far recorded only from Morona-Santiago (north to the Gualaceo-Limón road and [ANSP] on the Cordillera de Cutucú) and Zamora-Chinchipe. A few old records (some mentioned in Chapman 1926) are regarded as uncorroborated (but see below). Small numbers of the Amethyst-throated Sunangel can be found along both the Gualaceo-Limón and the Loja-Zamora roads, but in Ecuador it appears to be most numerous near the crests of the Cordillera de Cutucú and Cordillera del Cóndor. Recorded mostly between about 1900 and 2700 m, exceptionally as high as 3025 m (at Cordillera Las Lagunillas; Krabbe et al. 1997).

Only one race is definitely known from Ecuador, *laticlavius*. However, a male sunangel that differed strikingly from southern Ecuador birds in having a conspicuous buff (not white) pectoral collar was seen on 8 Dec. 1994 in n. Ecuador at Cerro Mongus in se. Carchi (R. Williams). Conceivably an undescribed taxon is involved; no further sightings have been reported. An additional taxon in *Heliangelus*, *violicollis*, was described in 1892 on the basis of two specimens from "Sarayacu" (which locality, being well out in the eastern lowlands, must surely be incorrect); it has not been recorded subsequently. It differs from *H. amethysticollis laticlavius* in having the gorget violet-blue (not red or rosy).

Range: east slope of Andes from s. Ecuador to w. Bolivia. Excludes Longuemare's Sunangel (*H. clarisse*) of the Andes in w. Venezuela and n. Colombia.

Heliangelus strophianus   Gorgeted Sunangel   Solángel de Gorguera

Uncommon to locally fairly common in the canopy and borders of montane forest, secondary woodland, and in shrubby clearings in the subtropical zone on the west slope of the Andes. Recorded mainly from Carchi south to Pichincha, but there are also recent records from farther south: it has been seen in the Caripero area of w. Cotopaxi (N. Krabbe and F. Sornoza), one was seen above Cochancay in sw. Cañar on 5 Aug. 1980 (RSR, PJG, and R. A. Rowlett), one was seen near Corona de Oro on the Puerto Inca-Molleturo road in Azuay on 21 Feb. 1997 (P. Coopmans et al.), and there are several reports from Buenaventura in El Oro (first on 17 Sep. 1990; P. Coopmans). Small numbers of the Gorgeted Sunangel, a virtual Ecuadorian endemic, can be found above Mindo, especially near the crest of the road between Mindo and Tandayapa in the vicinity of Bellavista Lodge. Recorded mostly from 1700 to 2300 m, but occasionally found as high as 3000 m, and the birds in Azuay and El Oro were at 900 to 1100 m.

Monotypic.

Range: west slope of Andes in extreme s. Colombia (recent records in w. Nariño; Salaman and Mazariegos 1998) and Ecuador.

Heliangelus exortis   Tourmaline Sunangel   Solángel Turmalina

Uncommon to fairly common in shrubby lower growth of montane forest borders and in adjacent clearings in the temperate zone on the east slope of the Andes in Ecuador south to nw. Morona-Santiago (specimens [LSUMZ] and sightings [RSR] from Planchas in the upper Río Palora valley in Sangay National Park in Oct. 1976). There exists striking individual variation (polychromatism) in the throat color of females, some having a contrasting white throat, others resembling males; this unusual situation was discussed in detail by R. Bleiweiss (*Am. Mus. Novitates* 2811, 1985). We are not aware of the source for this species being mentioned as occurring "very locally on northwest slope of Ecuador" (Fjeldså and Krabbe 1990, p. 270); no such records are known to us. Recorded mostly from 2200 to 3100 m, occasionally or locally as low as 1900 m (Baeza; MECN).

Monotypic. *H. micraster* (Flame-throated Sunangel) of s. Ecuador is here regarded as a species distinct from *H. exortis*.

Range: Andes from n. Colombia to s. Ecuador.

Heliangelus micraster   Flame-throated Sunangel   Solángel Gorjidorado

Uncommon in the lower growth and borders of montane forest and woodland in the temperate zone on the east slope of the Andes in s. Ecuador in Morona-Santiago (north to the Gualaceo-Limón road; specimens [ZMCOP] and numerous sightings), adjacent Azuay (Quebrada Monteverde; Berlioz 1932b), and Zamora-Chinchipe, also spilling over the Continental Divide into Loja at various sites (e.g., near Loja, at Cajanuma and Quebrada Honda in

Podocarpus National Park, and on the Cordillera Las Lagunillas). Recorded mostly from 2400 to 3100 m.

H. micraster (with *cutervensis* of n. Peru) is regarded as a species separate from *H. exortis* (Tourmaline Sunangel). Although their ranges approach each other quite closely on the east slope of the Andes in cen. Ecuador, recent specimens show no evidence of intergradation, and we find the arguments of J. T. Zimmer (*Am. Mus. Novitates* 1540, 1951), who was the first to propose treating them as conspecific, unconvincing. R. Bleiweiss (*Biol. Jour. Linn. Soc.* 45: 291–314, 1992) also treated *H. micraster* as a separate species from *H. exortis*. One race occurs in Ecuador, nominate *micraster*.

The suggested English name for *H. micraster* has usually been the Little Sunangel (e.g., *Birds of the Americas*, pt. 2, no. 1; Meyer de Schauensee 1966). Given that this species is no smaller than its congeners, however, we find this an unsatisfactory choice and prefer to highlight the flame color seen in its gorget (so strikingly different from the Tourmaline Sunangel's rosy purple gorget).

Range: east slope of Andes in s. Ecuador and n. Peru.

| *Heliangelus viola* | **Purple-throated Sunangel** | **Solángel Gorjipúrpura** |

Fairly common to locally common in shrubby areas, lighter woodland, borders of montane forest, and gardens in the subtropical and temperate zones of s. Ecuador from s. Chimborazo south through Cañar, Azuay, e. El Oro, and Loja. Apart from a large population found at Hacienda La Libertad in Cañar in Aug. 1991 (RSR and P. Stafford), the beautiful Purple-throated Sunangel has not been found on the actual east slope of the Andes; a possibly wandering bird was also seen at Cerro Toledo in s. Zamora-Chinchipe (N. Krabbe). Early records (e.g., Oberholser 1902) from as far north as Pichincha seem very doubtful. The species is capable of persisting in substantial numbers in disturbed terrain, where it can be one of the more numerous and widespread hummingbirds. Recorded mostly from 1800 to 3300 m, in small numbers slightly higher and lower.

Monotypic.

Range: Andes of s. Ecuador and n. Peru.

The Royal Sunangel (*H. regalis*), a species only recently described from n. Peru, has been found in elfin forest very close to the Ecuadorian border near the crest of the ridge of the Cordillera del Cóndor (Schulenberg and Awbrey 1997). It seems likely also to occur in Ecuadorian territory, though to date efforts to find it here have proved fruitless.

| *Eriocnemis nigrivestis* | **Black-breasted Puffleg** | **Zamarrito Pechinegro** |

Rare in montane forest, mainly along ridge crests ("nudos") in the temperate zone on the west slope of the Andes in w. Pichincha, where known only from a few sites on the north slope of Volcán Pichincha and on nearby Volcán Atacazo. Much information on the range, habitat requirements, and feeding ecology of this now very scarce, endemic hummingbird was given by R. Bleiweiss and M. Olalla P. (*Wilson Bull.* 95[4]: 656–661, 1983). There are only a few reports since the early 1980s, when a few birds were seen (PJG and RSR et al.) on several occasions in May–Jul. above Yanacocha, where the species may be only a seasonal visitant. N. Krabbe and M. B. Robbins et al. saw a male and two females, and mist-netted and photographed one of the latter (photos to VIREO), at Loma Gramalote near Yanacocha from 27 Feb. to 1 Mar. 1993 (see *Cotinga* 1: 8–9, 1994), and two were again seen at Yanacocha on 21 May 1995 (M. Lysinger). Recorded mostly in a narrow zone between 2850 and 3300 m, perhaps seasonally as high as 3500 m.

Bleiweiss and Ollala (op. cit.) considered the Black-breasted Puffleg to be endangered, based on the large number of museum specimens taken in the 19th and the first half of the 20th centuries (no fewer than 8 were collected by T. Mena in just 2 days in Jul. 1950 at Verdecocha; ANSP) compared with the species' rarity now. A significant decline does indeed appear to have taken place, presumably as a result of deforestation, but more information is needed. No Collar et al. (1994) concluded that the species merited Critical status. We agree.

Monotypic.

Range: Andes of nw. Ecuador.

The Söderstrom's Puffleg (*Eriocnemis soederstromi*), known from a single specimen taken at Nono in 1890 and long considered either a valid species or an aberrant Turquoise-throated Puffleg (*E. godini*), has been shown to represent a hybrid Black-breasted Puffleg × Sapphire-vented Puffleg (*E. luciani*; G. R. Graves, *Proc. Biol. Soc. Wash.* 109[4]: 764–769, 1996).

Eriocnemis vestitus   **Glowing Puffleg**   **Zamarrito Luciente**

Uncommon to fairly common in the lower growth of montane forest, forest borders, and shrubby clearings in the temperate zone on the entire east slope of the Andes. In Cañar (near Hacienda La Libertad; RSR) and Azuay (Río Mazán; King 1989) it also occurs westward locally on slopes above certain interandean valleys. The Glowing Puffleg is generally more numerous southward in Ecuador. Recorded mostly from 2500 to 3500 m, rarely or locally as low as 2250 m (on the east slope of the Cordillera Las Lagunillas; ANSP).

One race: *smaragdinipectus*.

Range: Andes from w. Venezuela to extreme n. Peru.

Eriocnemis godini   **Turquoise-throated Puffleg**   **Zamarrito Gorjiturquesa**

Distribution—and indeed taxonomic status—uncertain. Known from only four specimens, all obtained in the 19th century, and only one with reasonably precise locality data, that taken in 1850 in n. Pichincha at Guaillabamba south of Perucho (Collar et al. 1992), which would place it in a surprisingly low and arid region. There are no recent certain reports of the Turquoise-throated Puffleg, which may have a distribution centered on the still partially wooded quebradas in the Río Guaillabamba drainage. The species' closest relative appears to be the Glowing Puffleg (*E. vestitus*), which is found near treeline on the east slope of the Andes; it has been hypothesized that the Turquoise-throated Puffleg might occur in similar habitat on the west slope of the Andes north of Pichincha, especially as there are several "Bogotá" trade skins that presumably were taken somewhere in Colombia. There is no other evidence that this is the case, however, despite considerable recent field work at appropriate elevations in the region. The only reasonably certain elevations that are available are at about 2100–2300 m.

The Turquoise-throated Puffleg was considered Endangered or even perhaps Extinct by Collar et al. (1992); it was given Critical status by Collar et al. (1994). We concur with the latter; clearly a concerted effort needs be made to determine whether any populations still exist, and if so, to take steps to protect them.

Monotypic. Although it has been suggested that *godini* might be only a subspecies of *E. vestitus*, given its apparently very different habitat preference, we favor continuing to treat it as a full species. G. R. Graves (*Proc. Biol. Soc. Wash.* 109[4]: 764–769, 1996) suggested that *E. godini* could be of hybrid origin.

Range: apparently a few intermontane valleys in nw. Ecuador, perhaps also in adjacent Colombia (there are two old "Bogotá" specimens).

Eriocnemis luciani   Sapphire-vented Puffleg   Zamarrito Colilargo

Uncommon to locally common in montane forest, forest borders, patches of *Polylepis*-dominated woodland, and shrubby areas in the temperate zone on the west slope of the Andes and (especially) on the slopes above the central and interandean valleys; there are relatively few records from the actual east slope. Recorded from se. Carchi (a Mar. 1992 sighting on the west slope of Cerro Mongus [Robbins et al. 1994]; also an old record from Huaca, the type locality of the species), s. Imbabura (an old record from Mojanda near the Pichincha border), and Pichincha (where rather widespread) south to Azuay (El Cajas National Recreation Area, and the Río Mazán area). There are also a few recent sightings from as far south as e. Loja in Podocarpus National Park at Cajanuma and near the crest of the Loja-Zamora road (Bloch et al. 1991; Rasmussen et al. 1994), but we believe that full confirmation of the species from this far south is still needed. The beautiful Sapphire-vented Puffleg tends to be less numerous or more seasonal on the true outside slopes of the Andes; it can be numerous at Pasochoa and just west of the pass along the uppermost part of Chiriboga road, and around Nono. Recorded mostly from 2700 to 3700 m, but mostly below 3400 m.

Monotypic.

Range: Andes from s. Colombia to s. Ecuador. Excludes Coppery-naped Puffleg (*E. sapphiropygia*) of Peruvian Andes.

Eriocnemis mosquera   **Golden-breasted Puffleg**   **Zamarrito Pechidorado**

Uncommon to fairly common and somewhat local in montane forest, forest borders, and adjacent shrubby areas in the temperate zone on both slopes of the Andes, mainly in n. Ecuador. On the west slope recorded from Carchi south to Pichincha on the south slope of Volcán Pichincha. On the east slope recorded south to w. Napo and (apparently more sparsely) to Tungurahua (a 19th-century record from San Rafael), e. Chimborazo (a 1911 specimen from Chambo; ANSP), and w. Morona-Santiago (a male seen along the Gualaceo-Limón road on 15 Jul. 1991; PJG). Recent sightings from Cajanuma in Podocarpus National Park (Bloch et al. 1991) are regarded as uncorroborated, and the species was not included in their more recent bird list from the park (Rasmussen et al. 1994). Most recent reports of the Golden-breasted Puffleg come from the north slope of Volcán Pichincha (e.g., in the Yanacocha area where it is numerous). In Mar. and Jun. 1992 it was also found in substantial numbers on the west slope of Cerro Mongus in se. Carchi (Robbins et al. 1994). Recorded mostly from 3000 to 3600 m.

Monotypic (see R. Bleiweiss, *Am. Mus. Novitates* 2913, 1988).

Range: Andes from n. Colombia to n. Ecuador.

Eriocnemis alinae   **Emerald-bellied Puffleg**   **Zamarrito Pechiblanco**

Rare and apparently local in montane forest and forest borders in the subtropical zone on the east slope of the Andes. It is conceivable that this species also occurs on the west slope in the extreme north, for it is known from the Pacific slope of Nariño in adjacent Colombia (Hilty and Brown 1986; Salaman and Mazariegos 1998). There are only three known Ecuadorian localities for this hummingbird, which is poorly known and scarce throughout its range. Four were collected at Río Anguchaca (northwest of Macas in Morona-Santiago) in Oct. 1933 (Berlioz 1937b), and a male was taken on the Cordillera de Huacamayos in w. Napo on 7 Oct. 1988 (WFVZ). There have been a few subsequent sightings from the Cordillera de Huacamayos (v.o.), and on 20 Jul. 1991 a male was seen along the Loja-Zamora road (PJG). Recorded only between about 1800 and 2250 m.

Given this species' rarity in Ecuador, and the lack of any explanation for its small numbers, we consider the Emerald-bellied Puffleg to merit Data Deficient status in Ecuador. The species was not considered to be at risk by Collar et al. (1994).

One race: nominate *alinae*.

Range: very locally on east slope of Andes from n. Colombia to n. Peru.

Eriocnemis derbyi   **Black-thighed Puffleg**   **Zamarrito Muslinegro**

Uncommon to locally fairly common in montane forest, forest borders, and adjacent shrubby areas in the temperate zone on both slopes of the Andes in n. Ecuador in Carchi, Imbabura (Atuntaqui and Pimampiro), and on the Carchi-

Napo border (El Troje). The only recent Ecuadorian reports come from Páramo del Ángel (ANSP) and from the west side of Cerro Mongus in se. Carchi (Robbins et al. 1994). Recorded only from a narrow zone between 3000 and 3500 m.

The Black-thighed Puffleg was considered to be a Near-threatened species by Collar et al. (1994), an assessment with which we concur. Much deforestation has taken place recently within its always limited range.

One race: nominate *derbyi*. The species is perhaps better regarded as monotypic (see Schuchmann 1999).

Range: Andes of s. Colombia and n. Ecuador.

Haplophaedia aureliae   Greenish Puffleg   Zamarrito Verdoso

Uncommon to fairly common but seemingly quite local in the lower and middle growth of montane forest in the subtropical zone on the east slope of the Andes south to Zamora-Chinchipe (Cordillera del Cóndor in the Chinapinza area, and along the Loja-Zamora road); at least up to now it has not been found southward into Peru. The Greenish Puffleg often seems less numerous on the slopes of the main range of the Andes than it typically is on outlying mountain ridges (e.g., Volcán Sumaco, Cordillera de Huacamayos, Cordillera de Cutucú, and Cordillera del Cóndor), but a large series (MCZ) was obtained along the Loja-Zamora road in Aug. 1965. Recorded mostly from 1500 to 2100 m, locally or in small numbers as low as 1300 m (San Rafael Falls; ANSP); also seems likely to occur somewhat higher.

The race *russata* is generally given as the subspecific designation for Ecuadorian birds (*Birds of the World*, vol. 5; J. T. Zimmer, *Am. Mus. Novitates* 1540, 1951), and we tentatively follow this, though recognizing that racial determinations in *Haplophaedia aureliae* are not yet firmly established. Fjeldså and Krabbe (1986) discuss the possibility that birds from s. Ecuador may ultimately best be separated subspecifically from those of n. Ecuador on the basis of their broader and more grayish white scaling on the lower underparts.

Range: Andes from n. Colombia to w. Bolivia. More than one species may be involved.

Haplophaedia lugens   Hoary Puffleg   Zamarrito Canoso

Uncommon in the lower and middle growth of montane forest in the subtropical zone on the west slope of the Andes in w. Pichincha (in the Mindo and Nanegal areas, and along the Chiriboga road). The scarce Hoary Puffleg seems likely to occur at appropriate elevations northward through Imbabura and Carchi, though as yet there are no records from there. Recorded mostly in a narrow zone between 1700 and 2100 m, at least occasionally down to 1500 m (near Mindo; MECN).

The Hoary Puffleg was considered to be a Near-threatened species by Collar et al. (1994), an assessment with which we concur.

Monotypic. Some authors (e.g., J. T. Zimmer, *Am. Mus. Novitates* 1540,

1951) have considered *lugens* as only a west-slope subspecies of *H. aureliae* (Greenish Puffleg), and that may prove to be the preferable course.

Range: west slope of Andes in sw. Colombia and nw. Ecuador.

Ocreatus underwoodii   **Booted Racket-tail**   Colaespátula Zamarrito

Uncommon to locally (and perhaps seasonally) fairly common at the borders of montane forest and in adjacent clearings in the foothills and subtropical zone on both slopes of the Andes. The Booted Racket-tail tends to be more numerous in the west, where it has been recorded south to El Oro and w. Loja (south to Sozoranga; Best et al. 1993). Recorded mostly from 900 to 2200 m, locally (seasonally?) ranging down to 500 m on the west slope (e.g., it was seen above San Luis in nw. Azuay in Jan. 1991; J. C. Matheus).

Two distinctly different races are found in Ecuador, *peruanus* on the east slope and *melanantherus* on the west slope; Fjeldså and Krabbe (1990) err in stating that *melanantherus* occurs on both slopes. Males differ strikingly in the color of their leg puffs (buff in *peruanus*, white in *melanantherus*), females equally strikingly in the pattern of their underparts (thickly spangled with green discs in *peruanus*, nearly immaculate white with only a few small green spots in *melanantherus*; *peruanus* also has an extensively buff crissum). These may prove to be representatives of two distinct species, *O. underwoodii* (White-booted Racket-tail) and *O. addae* (Buff-booted Racket-tail).

Range: mountains of n. Venezuela, and Andes from w. Venezuela to s. Bolivia; Sierra de Perijá.

Lesbia victoriae   **Black-tailed Trainbearer**   Colacintillo Colinegro

Fairly common to common in shrubby areas, gardens, and grassland in the temperate zone and in paramo in the Andes from Carchi south to se. Azuay (Bestión region near the border with Zamora-Chinchipe), and on both slopes of the Cordillera Las Lagunillas in extreme se. Loja and adjacent Zamora-Chinchipe, where several were seen in late Oct. and early Nov. 1992, with a male collected on 4 Nov. (ANSP). The Black-tailed Trainbearer is most numerous on the slopes above the central valley, less so on outside slopes; it is frequent in the parks and gardens of Quito. Recorded mostly from 2500 to 3800 m.

Two similar races occur in Ecuador, nominate *victoriae* in most of the species' Ecuadorian range and *juliae* in s. Loja. We do not recognize the very similar *aequatorialis*, to which the birds of Ecuador and Nariño, Colombia, have sometimes been assigned; in *Birds of the World* (vol. 5, p. 115) it was noted as being "doubtfully distinct." *Juliae* differs from the nominate race principally in having a shorter bill.

Range: Andes of ne. Colombia, and from extreme s. Colombia to s. Peru.

Lesbia nuna   **Green-tailed Trainbearer**   Colacintillo Coliverde

Uncommon to (especially southward) common in shrubby areas, borders of montane forest and woodland, and gardens in the subtropical and temperate zones on both slopes of the Andes, and on slopes above the central and interandean valleys. Unlike the Black-tailed Trainbearer (*L. victoriae*), the Green-tailed sometimes (seasonally?) occurs in humid areas, especially where deforestation has taken place. Recorded mostly between 1900 and 3000 m, in smaller numbers a bit higher and lower (as high as 3250 m west of Molleturo in Azuay; MECN). Unlike the Black-tailed Trainbearer, the Green-tailed rarely or never occurs in actual paramo.

One race: *gracilis*.

Range: Andes from w. Venezuela to w. Bolivia.

| *Ramphomicron microrhynchum* | Purple-backed | Picoespina |
| | Thornbill | Dorsipúrpura |

Rare to uncommon and local (or seasonal?) at the borders of montane forest and in adjacent clearings in the temperate zone on both slopes of the Andes. There are, however, relatively few records of the Purple-backed Thornbill from cen. and s. Ecuador; these include a 1920 specimen (AMNH) from Taraguacocha in El Oro, four specimens from Portete in Azuay (Berlioz 1932b), and sightings from the Río Mazán area in Azuay (King 1989; N. Krabbe) and in Podocarpus National Park at the Cajanuma sector as well as at Cerro Toledo (Rasmussen et al. 1996; P. Coopmans et al.; RSR et al.). Recorded mostly from 2500 to 3400 m.

One race: nominate *microrhynchum*.

Range: Andes from w. Venezuela to w. Bolivia.

| *Metallura williami* | Viridian Metaltail | Metalura Verde |

Uncommon to locally fairly common at the borders of montane forest and woodland near and not far below treeline, in adjacent shrubby paramo, and locally in patches of *Polylepis*-dominated woodland on the east slope of the Andes. On the west slope recorded only from the extreme north in Carchi on Páramo del Ángel (one seen and another mist-netted and released on 14 Nov. 1990 [Krabbe 1992], and one collected [MECN] by N. Krabbe on 21 Nov. 1991). On the east slope recorded south locally to Loja (Podocarpus National Park at Cerro Toledo, where seen in Sep. 1998 by RSR and L. Navarrete et al.), and also occurs very locally westward on slopes above interandean valleys in Azuay (e.g., Río Mazán valley near Cuenca; King 1989) and El Oro (Taraguacocha). Recorded mostly from about 3000 to 3700 m, locally somewhat lower (has been seen down to 2700–2800 m in the Podocarpus National Park area).

Two races are found in Ecuador, *primolinus* south to ne. Azuay (Palmas; Berlioz 1932b) and *atrigularis* from w. Morona-Santiago (Volcán Sangay region and along the Gualaceo-Limón road) south into El Oro, Loja, and Zamora-Chinchipe. Males differ strikingly in that *atrigularis* has a black patch on its midthroat, whereas the entire throat is glittering green in *primolinus*. Given the narrow species limits employed elsewhere in the *M. aeneocauda* (Scaled Metaltail) group (see G. R. Graves, *Wilson Bull.* 92[1]: 1–7, 1980), and the lack of known intergradation between *primolinus* and *atrigularis*, a case could perhaps be made for treating *M. atrigularis* as a species (Black-throated Metaltail) separate from *M. williami*. Examples (ANSP) from w. Nariño (Cumbal), Colombia, and possibly adjacent Carchi in Ecuador may represent an undescribed race, differing in their more coppery median breast and belly and shorter bills.

Range: Andes from n. Colombia to s. Ecuador.

Metallura baroni   **Violet-throated Metaltail**   Metalura Gorjivioleta

Uncommon to fairly common at the borders of montane woodland (including patches of *Polylepis*-dominated woodland) and in adjacent shrubby paramo in the temperate and paramo zones on the west slope of the Andes in Azuay, principally in the El Cajas area west of Cuenca; a male was also seen in a patch of secondary woodland dominated by alders (*Alnus*) above Oña in s. Azuay on 1 Mar. 1980 (RSR and D. Finch et al.). F. Ortiz-Crespo (*Bull. B. O. C.* 104[3]: 95–97, 1984) recorded details of his obtaining the first modern specimen of this Ecuadorian endemic, still known from very few examples. Details regarding the species' movements and ecological requirements in the Río Mazán valley were presented by King (1989). Recorded mostly from 3100 to 3700 m, though a few birds evidently wander lower; the puzzling Oña observation was of a male with Purple-throated Sunangels (*Heliangelus viola*) at only 1900 m, and J. Dunning (pers. comm. to RSR) mist-netted and photographed (VIREO) a male at "7000 feet near Cuenca" in Nov. 1981.

The Violet-throated Metaltail was given Vulnerable status by Collar et al. (1994). We are somewhat more optimistic concerning its prospects and thus give it a rating of Near-threatened. Despite the species' very limited range, significant portions of it do now receive some measure of protection at El Cajas National Recreation Area and Río Mazán.

Monotypic. Of note (see King 1989) at Río Mazán is the sympatric presence of two members of the supposed *M. aeneocauda* (Scaled Metaltail) superspecies, *M. baroni* and *M. williami atrigularis* (Viridian Metaltail, the latter mainly at slightly lower elevations, 3100 to 3200 m, but one was later reported seen higher, at 3400 m [R. ter Ellen]).

Range: locally in Andes of s. Ecuador.

*Metallura odomae   **Neblina Metaltail**   Metalura Neblina

Fairly common but local in patches of low woodland near treeline and in adjacent shrubby paramo on the east slope of the Andes in Zamora-Chinchipe and adjacent Loja. This recently described species (G. R. Graves, *Wilson Bull.* 92[1]: 1–7, 1980) was heretofore thought to be endemic to extreme n. Peru and was known from only about eight specimens. The Neblina Metaltail was first seen in Ecuador above the Cajanuma Station in Podocarpus National Park in 1989 (M. Kessler); here it seems to range principally in the very wet paramo around Lagunas del Compadre, where it is the most numerous hummingbird (Rasmussen et al. 1996). In recent years it has also been seen at Cerro Toledo, farther south in Podocarpus National Park (N. Krabbe; L. Navarrete and RSR et al.). Subsequent records have come from between there and the Peruvian border. In Sep. 1990 R. Williams et al. found the species in small numbers in the Río Angashcola drainage above Amaluza in s. Loja. The first Ecuadorian specimen (ANSP) of the Neblina Metaltail was obtained by RSR on the Cordillera Las Lagunillas in extreme s. Zamora-Chinchipe on 16 Dec. 1991; in Oct. 1992 it was found to be quite common there (indeed, above 3200 m being the most numerous hummingbird), and a series was obtained (ANSP, MECN). Recorded mostly from 2950 to 3400 m, in small numbers as high as 3650 m (N. Krabbe).

The Neblina Metaltail was accorded Near-threatened status by Collar et al. (1994). However, given its relative abundance in its very remote range—where habitat disturbance has, at least to date, been minimal to nonexistent—we do not believe it merits listing as even a Near-threatened species. As noted above, a population occurs in Podocarpus National Park.

Monotypic. Of note is the sympatric presence of two members of the supposed *M. aeneocauda* (Scaled Metaltail) superspecies, *M. odomae* and *M. williami atrigularis* (Viridian Metaltail), in the Cajanuma area of Podocarpus National Park, though apparently with some altitudinal segregation (*M. odomae* higher).

Range: locally in Andes of s. Ecuador and extreme n. Peru.

Metallura tyrianthina　　Tyrian Metaltail　　Metalura Tiria

Common and widespread in shrubby lower growth at borders of montane forest and woodland, clearings, and gardens in the upper subtropical and temperate zones on both slopes of the Andes, and on slopes above the central and interandean valleys. On the west slope recorded south locally to El Oro (Taraguacocha). Recorded mostly from 2300 to 3400 m, occasionally or seasonally lower (e.g., down to 2100 m on the Cordillera de Huacamayos [P. Coopmans; RSR]) or higher (e.g., to 3800 m above Yanacocha; N. Krabbe).

Two very similar races are found in Ecuador, *quitensis* in the northwest (south to Chimborazo and Cañar, and in n. Ecuador ranging as far east as the west slope of the Eastern Andes) and nominate *tyrianthina* elsewhere. *Quitensis* differs in its on average slightly longer bill.

Range: mountains of n. Venezuela, and Andes from w. Venezuela to w. Bolivia; Sierra de Perijá and Santa Marta Mountains.

Chalcostigma ruficeps　　Rufous-capped Thornbill　　Picoespina Gorrirrufa

Rare to locally uncommon in shrubby clearings and borders of montane forest in the upper subtropical and lower temperate zones on the east slope of the Andes in Zamora-Chinchipe and adjacent e. Loja. There are only a few Ecuadorian records of this scarce hummingbird, most from various sites in or near Podocarpus National Park (Loja-Zamora road, Cajanuma, Cordillera de Sabanilla, and Quebrada Honda, seeming to be somewhat more numerous at the last two localities). In Nov. 1992 a few were also seen farther south, on the east slope of the Cordillera Las Lagunillas (M. B. Robbins). The Rufous-capped Thornbill seems to favor the sparse secondary growth associated with landslides and roadcuts. Recorded only between about 2100 and 2700 m.

Monotypic. We do not recognize the very similar *aureofastigatum*, to which Ecuadorian birds have sometimes been assigned; in *Birds of the World* (vol. 5, p. 121) it was considered to be "doubtfully distinct."

Range: locally on east slope of Andes from s. Ecuador to w. Bolivia.

Chalcostigma herrani   Rainbow-bearded Thornbill   Picoespina Arcoiris

Rare to locally fairly common at the edge of montane woodland and forest (including patches at and slightly above treeline), and locally in shrubby growth on steep rocky slopes, cliffs, and landslides, in the Andes. On the west slope recorded south to Azuay (a male taken west of El Cajas at Miguir on 13 Aug. 1986; ANSP). The Rainbow-bearded Thornbill is perhaps more widespread and numerous in the south than it is in the north; in Oct.–Nov. 1992 it was found to be notably numerous on both slopes of the Cordillera Las Lagunillas (ANSP, MECN). Recorded mostly from 2800 to 3700 m.

One race: nominate *herrani*.

Range: Andes from s. Colombia to extreme n. Peru.

Fjeldså and Krabbe (1990) mention a possible sighting of the Bearded Helmetcrest (*Oxypogon guerinii*) from Páramo del Ángel in Carchi. The species is otherwise not known south of Tolima and Quindío in the Central Andes and Cundinamarca in the Eastern Andes of Colombia, and it has not been found during extensive work in adjacent Nariño, Colombia (Salaman and Mazariegos 1998). No further records have come to light, and as we have no details regarding the record—seemingly improbable so far south of the species' normal range—we regard the species as uncorroborated for Ecuador.

Chalcostigma stanleyi   Blue-mantled Thornbill   Picoespina Dorsiazul

Uncommon to fairly common in shrubby paramo and in low woodland at and near treeline (including *Polylepis*-dominated woodland groves) in the Andes from Carchi (Páramo del Ángel [Jan. 1982 sightings by RSR, PJG, and S. Greenfield] and Guandera Biological Reserve [Creswell et al. 1999]) south to Azuay (El Cajas National Recreation Area and Río Mazán; ANSP, MECN). As yet the species has not been found in adjacent Nariño, Colombia (Salaman and Mazariegos 1998). Recorded between about 3600 and 4100 m.

One race: nominate *stanleyi*.

Range: Andes of n. and cen. Ecuador, and from n. Peru to w. Bolivia.

Opisthoprora euryptera   Mountain Avocetbill   Piquiavoceta

Rare and local (and inconspicuous) in the lower growth and borders of montane forest and adjacent shrubby clearings on the east slope of the Andes. Although the species has long been known from n. Ecuador, all records from the south are relatively recent; these include a specimen (LSUMZ) and a sighting on 13–14 Oct. 1976 from Planchas in the upper Río Palora valley of Sangay National Park in nw. Morona-Santiago (RSR and J. P. O'Neill) and another sighting from near Sabanilla along the Loja-Zamora road on 25 Jul. 1978 (RSR and D. Wilcove; see Ridgely 1980). These records were overlooked by Fjeldså and Krabbe (1986), who reported on a specimen from along the Gualaceo-Limón road. An avocetbill was also seen at Cajanuma in Podocarpus National Park in Sep. 1990 (P. Coopmans) and another on the east slope of the Cordillera Las Lagunillas on 1 Nov. 1992 (T. J. Davis). Recorded mostly from 2400 to 3200 m, with one record as low as 1700 m (Sabanilla).

Monotypic.

Range: locally in Andes from s. Colombia to n. Peru.

Aglaiocercus kingi   Long-tailed Sylph   Silfo Colilargo

Uncommon to locally common in montane forest, forest borders, and adjacent clearings in the subtropical and lower temperate zones on both slopes of the Andes. On the west slope known south only to Pichincha (Chiriboga road, Tandapi). Recorded mostly from 1600 to 2600 m, in small numbers a little higher and lower.

Two distinctly different races are found in Ecuador, *mocoa* on the east slope and *emmae* on the west slope. Male *emmae* differ in lacking *mocoa*'s glittering blue throat patch, and the upperside of their elongated central rectrices is much greener (not as blue); females of the two are similar, though *emmae* are slightly more richly colored below. *A. emmae* has sometimes (e.g., Salaman and Mazariegos 1998) been considered a separate species (Green-throated Sylph), though J. T. Zimmer (*Am. Mus. Novitates* 1595, 1952) presented evidence for considering the two forms to be conspecific.

Range: mountains of n. Venezuela, and Andes from w. Venezuela to w. Bolivia; Sierra de Perijá. This excludes Venezuelan Sylph (*A. berlepschi*) of ne. Venezuela.

Aglaiocercus coelestis   Violet-tailed Sylph   Silfo Colivioleta

Uncommon to fairly common in the lower growth of montane forest and forest borders in the foothills and subtropical zone on the west slope of the Andes from Carchi south to w. Chimborazo, and again in El Oro and w. Loja (Alamor and Celica areas). This beautiful hummingbird is replaced at higher elevations in nw. Ecuador by the almost equally attractive Long-tailed Sylph (*A. kingi*). These two species are not known to be sympatric, though—especially allowing for possible seasonal elevational movements—they may occasionally be so. Recorded mostly from 800 to 1950 m, in small numbers or locally slightly higher and lower.

Two races occur in Ecuador, nominate *coelestis* south to Chimborazo and *aethereus* in El Oro and Loja; the species seems to be unrecorded from the intermediate area. *Aethereus* males have a glittering green gorget (glittering blue in the nominate race); females of the two taxa are similar.

Range: west slope of Andes in sw. Colombia and w. Ecuador.

Schistes geoffroyi   **Wedge-billed Hummingbird**   Colibrí Piquicuña

Rare to uncommon and somewhat local in the undergrowth of montane forest and forest borders in the foothills and lower subtropical zone on both slopes of the Andes. In the northwest known only from Pichincha (numerous records from the Mindo region, etc.); there is also an old record from w. Chimborazo at Pallatanga. In addition, there are several recent records from much farther south: sw. Cañar (one seen above Cochancay on 5 Aug. 1980; RSR, PJG, and R. A. Rowlett), w. Azuay (seen above San Luis in Jan. 1991 [J. C. Matheus] and near Corona de Oro on the Puerto Inca-Molleturo road in Feb. 1997 [P. Coopmans et al.]), and El Oro (Buenaventura [Robbins and Ridgely 1990; also repeated subsequent sightings]). The species' range on the west slope is presumably more continuous, and likely extends north through Imbabura and Carchi into Colombia, where it is present in small numbers in Nariño (Salaman and Mazariegos 1998). The Wedge-billed Hummingbird favors wet, lush, heavily vegetated areas, often near streams; it is often numerous around San Rafael Falls. Recorded mostly from 800 to 2000 m, lowest in the southwest and locally or in small numbers slightly higher (e.g., at 2300 m at SierrAzul; ANSP).

The species is sometimes (e.g., Schuchmann 1999) placed in the genus *Augastes*, which otherwise is known only from two species endemic to e. Brazil. Two rather different subspecies of *S. geoffroyi* occur in Ecuador, the nominate race on the east slope and *albogularis* on the west slope. Male *albogularis* differ from the nominate mainly in having a glittering green forecrown, lacking the coppery on back and uppertail-coverts, and having a more or less continuous white band across the chest. Female *albogularis* differ in having entire throat white and the small tufts at the sides of chest mainly blue (not violet). Separate species are perhaps involved.

Range: mountains of n. Venezuela, and locally in Andes from w. Venezuela to w. Bolivia; Sierra de Perijá.

Heliothryx barroti   **Purple-crowned Fairy**   Hada Coronipúrpura

Uncommon to fairly common in humid forest, secondary woodland, and borders, sometimes also out into adjacent clearings, in the more humid lowlands and foothills of w. Ecuador south to n. Manabí and through Los Ríos and n. and e. Guayas to El Oro (e.g., at Buenaventura) and w. Loja (old specimens from the Puyango area, but no recent reports). There also are a few recent reports from the slopes of the coastal Cordillera de Colonche in sw. Manabí (Cerro San Sebastián in Machalilla National Park, where one was seen on 11 Jan. 1992; J. Rasmussen et al.) and w. Guayas (Loma Alta; Becker and López-Lanús 1997). Recorded up to about 800 m.

Monotypic.

Range: s. Mexico to sw. Ecuador.

Heliothryx aurita   Black-eared Fairy   Hada Orejinegra

Uncommon to locally fairly common in the subcanopy and borders of humid forest (both terra firme and várzea) and in adjacent clearings in the lowlands and foothills of e. Ecuador. The Black-eared Fairy appears to be somewhat more numerous in the foothills and close to the Andes than it is in far e. Ecuador. Recorded up to about 1200 m.

One race: nominate *aurita*. The race *major* was described as a race of *H. aurita* from Pisagua in the lowlands of w. Ecuador in se. Los Ríos, within the range of *H. barroti* (Purple-crowned Fairy); it has been suggested (Schuchmann 1999) that what was involved was merely a young male of *H. barroti*.

Range: e. Colombia, s. and e. Venezuela, and the Guianas to n. Bolivia and Amazonian Brazil; e. Brazil.

Heliomaster longirostris   Long-billed Starthroat   Heliomaster Piquilargo

Uncommon in borders of forest and woodland (both humid and deciduous) and in shrubby areas, clearings, and gardens in the lowlands of e. and w. Ecuador. The Long-billed Starthroat is somewhat more numerous, and is certainly more widespread (occurring in both humid and arid regions), in the west; here it is recorded from w. Esmeraldas south through El Oro and w. Loja. Occurs mostly below about 700 m, but ranges higher at least locally, perhaps especially in cleared areas; it has been recorded as high as 1300 m in Zamora-Chinchipe (Palanda), and in Loja it has been seen as high as 1500 m.

Two races occur in Ecuador, nominate *longirostris* in the east and *albicrissa* in the west and also in the Río Marañón drainage of extreme s. Zamora-Chinchipe around Zumba (ANSP specimen taken on 7 Aug. 1992). *Albicrissa* may also be the race that has been recorded in the vicinity of Zamora in n. Zamora-Chinchipe, though this population could also be referable to the nominate race. We know of only one Ecuadorian specimen of the nominate race, a male (ANSP) taken near Misahuallí on 16 Dec. 1990. *Albicrissa* differs in having its crissum feathers much more broadly white-tipped (such that the entire crissum often appears white).

Range: s. Mexico to nw. Peru, n. Bolivia, and Amazonian Brazil.

[Heliomaster furcifer   Blue-tufted Starthroat   Heliomaster Barbado]

Uncertain. Known in Ecuador from only a single sighting, an apparent immature male seen from 27 to 29 Dec. 1994 in a clearing at Yuturi Lodge (250 m) in e. Napo in the lowlands of ne. Ecuador (R. Williams). This anomalous record is difficult to interpret. The Blue-tufted Starthroat is otherwise known in w. Amazonia only from an apparently undated specimen taken at Leticia in se.

Colombia (Hilty and Brown 1986). These authors suggest that it might be an austral migrant, but there does not seem to be any conclusive indication that the species does in fact vacate its normal s.-cen. South American range during the austral winter. In addition, of course, the dates of the Ecuadorian sighting, from the height of the austral summer, would not fit an austral migration pattern.

Monotypic.

Range: e. and s. Bolivia and sw. Amazonian Brazil to n. Argentina, n. Uruguay, and extreme s. Brazil; two records from w. Amazonia.

*[*Thaumastura cora*   Peruvian Sheartail   Colifina Peruana]

Uncertain. Apparently a casual (possibly seasonal) visitant to extreme s. Ecuador. There is only one Ecuadorian report, a female seen feeding with other hummingbirds (including Purple-collared Woodstars [*Myrtis fanny*]) on a shrubby hillside south of Utuana (2350 m) in s. Loja on 27 Jul. 1991 (P. Coopmans, R. Jones, and P. Scharf). The species is known in Peru as far north as Piura.

Monotypic.

Range: nw. Peru to extreme n. Chile; one report from extreme s. Ecuador.

Calliphlox mitchellii   Purple-throated Woodstar   Estrellita Gorjipúrpura

Uncommon to locally (perhaps seasonally) fairly common in the canopy and borders of montane forest and in adjacent clearings in the foothills and sub-tropical zone on the west slope of the Andes from e. Esmeraldas and Carchi south to Pichincha, and from Bolívar (a male seen above San Antonio along the road to Guaranda on 9 Aug. 1986; M. B. Robbins and G. S. Glenn) south locally through El Oro to w. Loja (2 specimens [MCZ] taken by D. Norton 10 km north of Celica in Aug. 1965). The Purple-throated Woodstar seems likely to be found in the intervening area but as yet is unrecorded. Found mostly from 800 to 1800 m.

The species was formerly placed in the genus *Philodice* (e.g., Meyer de Schauensee 1966, 1970), but now is usually merged into *Calliphlox* (e.g., Schuchmann 1999). Monotypic.

Range: mountains of e. Panama; west slope of Andes in sw. Colombia and w. Ecuador.

Calliphlox amethystina   Amethyst Woodstar   Estrellita Amatista

Rare to locally (perhaps seasonally) uncommon in the canopy and borders of humid forest, clearings, and gardens in the foothills and adjacent lowlands along the east slope of the Andes from Napo (Coca region) southward. There are relatively few records of this woodstar from Ecuador, and it has never been found any distance east of the Andes. We regard the published record (Chapman 1926) from Gima, at 3000 m in Azuay, as being from an improbably high elevation; it likely was misidentified or mislabeled. Recorded from about 300 to 1400 m.

Monotypic.

Range: extreme e. Colombia, much of Venezuela, the Guianas, e. and s. Brazil, e. Paraguay, ne. Argentina, e. and n. Bolivia, e. Peru, and e. Ecuador.

Myrtis fanny   Purple-collared Woodstar   Estrellita Gargantillada

Uncommon to locally fairly common in lighter woodland, scrub, natural grassy areas, agricultural areas, and gardens in the more arid intermontane valleys from Carchi (Río Mira valley) south into n. Pichincha (as close to Quito as the warmer and drier valleys to the north and northeast, e.g., around Tumbaco, but apparently not ranging into the city itself) and from Tungurahua south locally through Loja (scattered reports from near Loja [city], Celica, Utuana, and Amaluza). There are also a few records (of wandering individuals?) from the east slope of the Andes, mainly in s. Ecuador (e.g., at Zamora and Sabanilla in Zamora-Chinchipe), and an apparently resident population was found in 1996 in the arid upper Río Pastaza valley near Baños (L. Jost). The Purple-collared Woodstar is also known from the Río Marañón drainage of extreme s. Zamora-Chinchipe (v.o., with a male [ANSP] obtained on 10 Aug. 1992). There is even one record from the west slope of the Andes in Pichincha (a Sep. 1941 specimen [ANSP] taken by L. Ponce at Mindo). The species' Ecuadorian distribution thus seems to be very patchy and, considering the large number of early specimens that were taken, the species is now not very frequently encountered. Aside from the Baños population mentioned above, there is only one recent report of the species from the east slope of the Andes (a pattern similar to that seen for the Little Woodstar [*Chaetocercus bombus*]), a single bird seen above Zamora in Feb. 1989 (PJG et al.). Recorded mainly from 1400 to 2600 m.

One race: nominate *fanny*. J. T. Zimmer (*Am. Mus. Novitates* 1604, 1953) suggested that birds from n. Ecuador may be recognizable as a separate subspecies based on the reduced rufous found on the underparts of males, but he concluded that the limited material available did not then permit actually doing so. This is still the case.

Range: Andes from n. Ecuador to sw. Peru.

Myrmia micrura   Short-tailed Woodstar   Estrellita Colicorta

Fairly common to common in desert scrub, shrubby areas, and gardens in the arid lowlands of sw. Ecuador from sw. Manabí (Isla de la Plata and Machalilla National Park) and w. and s. Guayas south through much of El Oro and w. Loja. An old record (Oberholser 1902) from Santo Domingo de los Colorados seems decidedly unlikely. In Loja recorded up to about 800 m, but elsewhere usually below 200 m; a published report (Best et al. 1993) from much higher, at Alamor at 1450 m, is considered to be unverified.

Monotypic.
Range: sw. Ecuador and nw. Peru.

Chaetocercus mulsant   White-bellied Woodstar   Estrellita Ventriblanca

Uncommon to locally (seasonally?) fairly common in the canopy and borders of montane forest and woodland and in adjacent shrubby clearings and gardens in the subtropical zone on both slopes of the Andes, and on the slopes above the central and interandean valleys. In the west recorded south to w. Loja (Alamor and Guachanamá; AMNH, ANSP). The White-bellied is more likely than any other montane woodstar (except the Purple-collared [*Myrtis fanny*]) to occur in fairly arid regions; small numbers are seen with fair regularity in Quito. Recorded from a surprisingly wide elevational range, from about 1100 to 3500 m; at either extreme it may perhaps only be a seasonal visitant, with regular movements perhaps being involved.

This and the following three species were formerly placed in the genus *Acestrura*, but we follow Schuchmann (1999) in merging this into *Chaetocercus*. Monotypic.
Range: Andes from n. Colombia to w. Bolivia.

Chaetocercus bombus   Little Woodstar   Estrellita Chica

Rare to perhaps locally (seasonally?) uncommon at the borders of forest and woodland (both humid and deciduous) and in shrubby clearings and gardens in the lowlands of w. Ecuador, ranging locally up into the subtropical zone, from w. Esmeraldas and Pichincha (a few recent sightings from the Mindo area, none of them confirmed) south locally through much of Guayas and w. Loja. Occurs mainly in humid regions, though locally in more arid areas as well. There are even a few old records from the east slope of the Andes, several early-20th-century specimens having been taken "below San José" in w. Napo and also at Zamora. We can only assume that these were correctly labeled and that they represent seasonal wanderers. Although a rather large number of Little Woodstar specimens were taken at various sites in the 19th and early 20th centuries, there are surprisingly few recent reports of the species. Most recent reports come from the more humid lowlands of sw. Ecuador north to extreme s. Pichincha at Río Palenque (v.o.); the species may be most numerous (at least seasonally) in w. Guayas (e.g., at Loma Alta; Becker and López-Lanús 1997). The only recent Esmeraldas report involves several birds seen at Finca Paraíso de Papagaios north of Quinindé in Sep. 1995 (J. R. and V. Fletcher and P. Coopmans). Recorded up to at least 2200 m, though mostly found below 1200 m.

The Little Woodstar seems to have declined substantially in recent decades, for reasons that remain unexplained. It was given Endangered status by Collar et al. (1994), an assessment that in our view somewhat exaggerates the threats it faces (it is surely substantially less at risk than the Esmeraldas Woodstar [*C. berlepschii*], which Collar et al. [1994] also gave Endangered status). We therefore accord it only Vulnerable status. In no area is the species now known to be present regularly or in large numbers, but there are recent reports from several different sites, among them the Esmeraldas locality mentioned above, near Mindo, Río Palenque, the hills south of the lower Río Ayampe in nw. Guayas, and Buenaventura in El Oro.

Formerly placed in the genus *Acestrura*. Monotypic.

Range: locally in w. Ecuador and nw. Peru; one recent sighting from w. Nariño, sw. Colombia (Salamann and Mazariegos 1998).

Chaetocercus heliodor   Gorgeted Woodstar   Estrellita de Gorguera

Rare and local (probably overlooked; certainly difficult to identify or collect) in the canopy and borders of montane forest in the foothills and subtropical zone on both slopes of the Andes. On the east slope recorded only from w. Napo (specimens and/or recent sightings from Baeza, below Oyacachi, Cuyuja, Cordillera de Huacamayos, and Loreto road/lower Sumaco region) and nw. Morona-Santiago (3 specimens [ANSP] taken by L. Gómez at Macas in 1938–1939). On the west slope recorded only from Pichincha (a few recent sightings from the Mindo region) and sw. Chimborazo (old specimens from Pallatanga and Huigra). There is also an anomalous specimen in the Moore collection at Occidental College in Los Angeles (see G. R. Graves, *Proc. Biol. Soc. Wash.* 99[2]: 218–224, 1986) that is labeled as being from Tumbaco, a locality that seems unlikely from an ecological standpoint. Recorded mainly from 1100 to 1800 m, occasionally as high as 2200 m.

Formerly placed in the genus *Acestrura*. One race: *cleavesi*. G. R. Graves (op. cit.) discussed the systematics and distribution of this species, and pointed out that the subspecific diagnosis of the population found on the west slope awaits the collection of fully adult male specimens.

Range: Andes from w. Venezuela to s. Ecuador.

Chaetocercus berlepschi   Esmeraldas Woodstar   Estrellita Esmeraldeña

Rare to seasonally uncommon and very local in the canopy and borders of semi-humid forest and woodland in the lowlands of w. Ecuador, principally if not entirely along the slopes of the coastal cordillera. This very rare Ecuadorian endemic is known from only nine specimens with locality data, all but one of them taken in the early 20th century at Esmeraldas (city) in w. Esmeraldas in Oct.–Dec. 1912 and at Chone in n. Manabí in Dec. 1912 (AMNH); three additional birds have uncertain or vague locality data. The Esmeraldas Woodstar was finally relocated on 20 Jan. 1991, when at least three were seen (likely more were present) and a male collected (ANSP) in hills south of the lower Río Ayampe in nw. Guayas. Several were seen here the following day, but despite careful searching none could be found in Aug. 1991 or Jul. 1993, either here or in Machalilla National Park immediately to the north. A male reportedly had been seen in much the same area in Mar. 1990 (R. Jones and P. Kaestner), and a few have continued to be found here in subsequent years in the Dec.–Mar.

period. Nesting in the Ayampe area is indicated by the observation of displaying males on at least two occasions (R. Behrstock et al.; RSR et al.). The species was also believed to be present at Loma Alta in w. Guayas in Dec. 1996 (Becker and López-Lanús 1997). The only additional sites where the Esmeraldas Woodstar has been found are in Esmeraldas: southwest of Súa, where a female was seen on 29 Jan. 1993 (RSR and F. Sornoza), and at Finca Paraíso de Papagaios north of Quinindé, where at least one male was seen between 15 and 23 Sep. 1995 (J. R. and V. Fletcher, fide P. Coopmans). Recorded mostly below about 200 m, but the Loma Alta birds were "above 500 m" (Becker and López-Lanús 1997, p. 167).

The deforestation that has taken place across much of the Esmeraldas Woodstar's range has surely caused a steep decline in its overall numbers, and the species certainly deserves the formal Endangered status it was accorded by Collar et al. (1994). Given that so little is known of it, however, an argument could even be made that it deserves Critical status. Of particular concern is the fact that it has not been found in any area yet receiving formal protection, and that the species' whereabouts during most of the year remain essentially unknown.

Formerly placed in the genus *Acestrura*. Monotypic.

Range: very locally in w. Ecuador.

Trogoniformes
Trogonidae   Trogons and Quetzals

Fifteen species in two genera. Trogons and quetzals are pantropical in distribution, though none occurs in Australasia.

Pharomachrus antisianus   Crested Quetzal   Quetzal Crestado

Uncommon and seemingly local in the mid-levels and subcanopy of montane forest and forest borders, occasionally coming out into adjacent clearings (especially to feed), in the subtropical and temperate zones on both slopes of the Andes. In the west recorded south to El Oro (pairs seen at Buenaventura on 7 Aug. 1980 and 17 Aug. 1989; RSR and PJG et al.). The Crested and Golden-headed Quetzals (*P. auriceps*) occur together at numerous sites, though usually the Golden-headed is the more numerous of the two; how they separate ecologically remains uncertain. Recorded mostly from 1500 to 2500 m, in smaller numbers down to about 1100 m (along the Loreto road north of Archidona [v.o.] and at Río Bombuscaro [D. Wolf]).

Monotypic.

Range: Andes from w. Venezuela to w. Bolivia; Sierra de Perijá.

Meyer de Schauensee (1966, 1970) states that there is a record of the White-tipped Quetzal (*P. fulgidus*) from "eastern Ecuador." We do not know the basis for this statement, but from what is now known about the distribution of the species—it occurs in the mountains of n. Venezuela and n. Colombia—it now seems clear that it could not possibly occur in Ecuador and that some error must have been made.

Pharomachrus auriceps Golden-headed Quetzal Quetzal Cabecidorado

Uncommon to locally fairly common in the mid-levels and subcanopy of montane forest and forest borders, occasionally coming out into adjacent clearings (especially to feed), in the foothills and subtropical and temperate zones on both slopes of the Andes. On the west slope recorded south to w. Loja (several sites in the Alamor region, and locally around Sozoranga [Best et al. 1993; RSR et al.]). Both this species and the Crested Quetzal (*P. antisianus*) are recorded mainly through their characteristic and far-carrying vocalizations. Recorded over a wide elevational range, mostly between 1000 and 2800 m, though most numerous between 1200 and 2400 m; regularly occurs even lower in sw. Ecuador, routinely as low as 500 m in Cañar and Azuay.

One race: nominate *auriceps*. Following *Birds of the World* (vol. 5) and J. T. Zimmer (*Am. Mus. Novitates* 1380, 1948), the population found in w. Ecuador, formerly separated as the race *heliactin*, does not appear to be valid.

Range: highlands of e. Panama; Andes from w. Venezuela to w. Bolivia; Sierra de Perijá.

Pharomachrus pavoninus Pavonine Quetzal Quetzal Pavonino

Rare to uncommon in the mid-levels and subcanopy of terra firme forest in the lowlands of e. Ecuador. Long thought very rare in Ecuador, the beautiful Pavonine Quetzal has in recent decades been found at several sites (e.g., Taisha, Zancudococha, La Selva, Sacha Lodge, along the Maxus road southeast of Pompeya, and Kapawi Lodge); like certain other terra firme birds, this quetzal was presumably under-recorded earlier because of relatively limited access into its habitat. Recorded up to about 600 m (north of Canelos in Pastaza; N. Krabbe).

One race: nominate *pavoninus*.

Range: se. Colombia and s. Venezuela to se. Peru, n. Bolivia, and w. and cen. Amazonian Brazil.

Trogon melanurus Black-tailed Trogon Trogón Colinegro

Uncommon to fairly common in the mid-levels and subcanopy of humid forest (both terra firme and várzea) in the lowlands of e. Ecuador. Recorded up only to about 400 m.

T. mesurus (Ecuadorian Trogon) of w. Ecuador is here treated as a separate species; see the discussion under that species. One race of *T. melanurus* occurs in Ecuador, *eumorphus* (see J. T. Zimmer, *Am. Mus. Novitates* 1380, 1948).

Range: Panama to n. Colombia and extreme nw. Venezuela; e. Colombia, s. Venezuela, and the Guianas to n. Bolivia and Amazonian Brazil.

Trogon mesurus   Ecuadorian Trogon   Trogón Ecuatoriano

Fairly common and relatively widespread in both humid and deciduous forest and woodland and their borders in the lowlands of w. Ecuador from w. Esmeraldas (vicinity of Esmeraldas [city]) south through El Oro and w. Loja. Recorded mostly below 800 m, but in the southwest found considerably higher, in w. Loja locally up to nearly 2000 m around Celica.

T. *mesurus* is here regarded as a monotypic species distinct from *T. melanurus* (Black-tailed Trogon), based on its plumage differences (and also its striking pale iris), consistent vocal differences, and disjunct range.

Range: w. Ecuador and nw. Peru.

Trogon comptus   Chocó Trogon   Trogón del Chocó

Uncommon to locally fairly common in the mid-levels and subcanopy of very humid forest and secondary woodland in the foothills on the west slope of the Andes in Esmeraldas and Pichincha (south to Tinalandia, where several pairs remain), also ranging down into adjacent lowlands at least in n. Esmeraldas (e.g., northwest of Alto Tambo; ANSP). Also, since Feb. 1994 recorded from the coastal Cordillera de Mache in w. Esmeraldas at Bilsa, where found to be uncommon (R. P. Clay; D. Wolf et al.); we feel that the identification of the Bilsa birds as this species—and not the very similar Ecuadorian Trogon (*T. mesurus*)—has yet to be fully confirmed, though it appears likely that the Chocó Trogon is the species present there. Chocó Trogons may also range farther south in foothills along the western base of the Andes: two female trogons seen (RSR) in Aug. 1991 in appropriate humid foothill forest above Manta Real in nw. Azuay were probably this species, but it was impossible to confirm that they were not the Ecuadorian Trogon (which was definitely present in woodland in the adjacent lowlands). Recorded mostly from 300 to 800 m, but ranging lower in n. Esmeraldas (to 70 m at Playa de Oro; O. Jahn et al.).

T. *comptus* was formerly called the Blue-tailed Trogon (e.g., Meyer de Schauensee 1966, 1970; Hilty and Brown 1986). Sibley and Monroe (1990) concluded that English name was better applied to a trogon found on Sumatra and Java in Indonesia, *Harpactes reinwardtii*; we do not know how this decision was reached. Their choice of the name of White-eyed Trogon for *T. comptus* was accurate enough, though it caused, especially in Ecuador, confusion with the equally "white-eyed" *T. mesurus* (they were presumably not aware that this would be the case). We prefer to end the confusion and change the name of *T. comptus* to Chocó Trogon.

Monotypic.

Range: w. Colombia and nw. Ecuador.

Trogon massena   Slaty-tailed Trogon   Trogón Colipizarro

Uncertain. Recorded from humid forest and forest borders in the lowlands of extreme nw. Ecuador in n. Esmeraldas. J. T. Zimmer (*Am. Mus. Novitates* 1380,

1948) recorded specimens (AMNH) taken at Pambilar and Carondelet; there is also an MECN specimen from Río Caune. There are only a few recent reports of the Slaty-tailed Trogon from Ecuador, first when seen and tape-recorded near San Lorenzo in 1997 and 1998 (originally by J. Nilsson); it has also been seen in the La Chiquita Reserve near that city (M. Lysinger). A few other sightings, some from much farther south, seem to be the result of confusion with the Ecuadorian Trogon (*T. mesurus*). Recorded only below about 200 m.
One race: *australis*.
Range: s. Mexico to extreme nw. Ecuador.

Trogon viridis   Amazonian White-tailed Trogon   Trogón Coliblanco Amazónico

Common in mid-levels of humid forest and forest borders (both terra firme and várzea) in the lowlands of e. Ecuador. Ranges up locally to about 900–1000 m (regular that high in the Río Nangaritza valley), with a few locally as high as 1200 m (west slope of the Cordillera de Cutucú [PJG] and near Volcán Sumaco [P. Coopmans]).
Trans-Andean *T. chionurus* (Western White-tailed Trogon) is here treated as a separate species; see the discussion under that species. One race occurs in Ecuador, nominate *viridis*.
Range: e. Colombia, s. and e. Venezuela, and the Guianas to n. Bolivia and Amazonian Brazil; e. Brazil.

Trogon chionurus   Western White-tailed Trogon   Trogón Coliblanco Transandino

Uncommon in the mid-levels of humid forest and forest borders in the lowlands of w. Ecuador. Ranges south in small numbers to n. Manabí, n. Los Ríos (an Oct. 1950 specimen from Quevedo [ANSP]), and nw. Azuay at Manta Real (several seen in Aug. 1991; RSR), but now quite scarce and local southward (though remaining moderately numerous at Río Palenque). Recorded mainly below 500 m, in small numbers up to 800 m.
T. chionurus is here regarded as a monotypic species separate from *T. viridis* (Amazonian White-tailed Trogon), found east of the Andes, based on several distinct plumage differences (both sexes have, inter alia, considerably more white on underside of tail) as well as their strikingly different songs.
Range: Panama to n. and w. Colombia and w. Ecuador.

Trogon collaris   Collared Trogon   Trogón Collarejo

Uncommon to fairly common in the lower and mid-levels of humid forest and secondary woodland in the lowlands and foothills of e. and w. Ecuador. In the west recorded south to nw. Guayas (hills south of Río Ayampe), El Oro (e.g., Buenaventura), and w. Loja (an old record from Las Piñas). We regard Best et al.'s (1993) record of this species from Celica in sw. Loja as being uncertain; the elevation seems excessively high for the Collared Trogon, and more than likely it was the *assimilis* race of the Masked Trogon (*T. personatus*)—which they also recorded—that was involved. In the eastern lowlands the Collared Trogon is found mainly or entirely in várzea and riparian forest, but in the foothills and lower subtropics it ranges into montane forest. In the west, though occurring primarily where conditions are humid, it is found at least locally in semihumid woodland and forest (e.g., in Machalilla National Park); from e. Guayas southward it occurs only in the Andean foothills and lower subtropics. Ranges up into the lower subtropical zone on both slopes to about 1300 m, locally as high as 1500 m (Pallatanga).

Two similar races are found in Ecuador, nominate *collaris* in the east and *virginalis* in the west; *virginalis* females have a blacker face than nominate feamles.

Range: s. Mexico to sw. Ecuador, n. Bolivia, and Amazonian Brazil (but absent from much of n. Colombia and Venezuela); locally in e. Brazil. More than one species is perhaps involved.

Trogon personatus   Masked Trogon   Trogón Enmascarado

Uncommon to fairly common in mid-levels of montane forest and forest borders on both slopes of the Andes. On the west slope recorded south to El Oro (old specimens from El Chiral and Zaruma, with a few recent reports from above Buenaventura) and w. Loja (where known only from sightings in the Tierra Colorada and Celica areas in early 1991; Best et al. 1993); it also ranges on slopes above interandean valleys, at least in s. Azuay. On the west slope found between 1200 and 3000 m, on the east slope mostly from 1500 to 3000 m (but recorded up to 3350 m at Cerro Mongus; Robbins et al. 1994).

Racial variation in this species in Ecuador is complex, with more than one species perhaps being involved. *Assimilis* is the only race occurring on the west slope. On the east slope, nominate *personatus* occurs at lower elevations, being replaced (with some overlap?) at higher elevations by *temperatus*; the elevations at which the switchover occurs remain uncertain, with *temperatus* definitely ranging down at least locally to 2500 m and nominate *personatus* up to at least 2300 m. *Temperatus* differs from nominate *personatus* in having a notably smaller bill, and in males showing a darker green (less bronzy olive) upperside to the tail and virtually no white barring on the underside of the tail (the barring is conspicuous in nominate *personatus*); females are quite similar. *Assimilis* resembles nominate *personatus* in most respects. The upper-elevation taxon seems likely to be a separate species, *T. temperatus* (Highland Trogon), ranging mainly on the east slope of the Andes in Colombia and Ecuador (in the latter as far west as the El Cajas area of Azuay; P. Coopmans). Tape-recordings of the males' vocalizations are also different enough to be suggestive of species status, but the situation also needs to be analyzed outside Ecuador before we are willing to actually separate *T. temperatus* as a full species.

Range: Andes from w. Venezuela to s. Bolivia; tepuis of s. Venezuela and adjacent Guyana and Brazil; Sierra de Perijá and Santa Marta Mountains.

Trogon rufus Black-throated Trogon Trogón Golinegro

Uncommon in lower and middle growth inside terra firme forest in the lowlands of e. Ecuador; locally fairly common inside humid forest in the lowlands (especially) and foothills of nw. Ecuador, primarily ranging in Esmeraldas, but with a few—mostly old—records from Pichincha as well; a lone female was seen at Tinalandia on 21 Jul. 1994 (P. Coopmans et al.), and there are several recent reports from several sites in the province's northwest (north of Simón Bolívar near Pedro Vicente Maldonado and 5 km south of Golondrinas; v.o.). Norton (1965) discusses two specimens he obtained at Guaticocha in Aug. 1964, suggesting they were the first from a definite locality in e. Ecuador; however, in our view there seems to be no reason to discredit the specimen from Coca referred to by J. T. Zimmer (*Am. Mus. Novitates* 1380, 1948). Recorded up to 700–750 m on both slopes.

Two races are found in Ecuador, *sulphureus* in the east and *cupreicauda* in the northwest. Male *sulphureus* differ in their bronzier—not so pure green—upperside of tail, and there are slight vocal differences. Note that both sexes of both races differ strikingly in eye-ring color (yellowish) from the blue found in most races elsewhere in the species' range.

Range: Honduras to nw. Ecuador; se. Colombia, s. Venezuela, and the Guianas to e. Peru, n. Bolivia, and Amazonian Brazil; e. Brazil, e. Paraguay, and ne. Argentina.

Trogon curucui Blue-crowned Trogon Trogón Coroniazul

Uncommon in the mid-levels and subcanopy of humid forest and forest borders (both terra firme and várzea) in the lowlands and foothills of e. Ecuador; seems to favor edge situations. Ranges locally and in small numbers up along the base of the Andes to about 1100 m.

One race: *peruvianus*. J. T. Zimmer (*Am. Mus. Novitates* 1380, 1948) demonstrated that the name *bolivianus* was antedated by the name *peruvianus*.

Range: se. Colombia, e. Peru, n. and e. Bolivia, Amazonian, ne., and sw. Brazil, w. and n. Paraguay, and nw. Argentina.

Trogon violaceus Amazonian Violaceous Trogón Violáceo
              Trogon   Amazónico

Uncommon in mid-levels and subcanopy of humid forest and forest borders in the lowlands of e. Ecuador. Ranges in both várzea and terra firme forest, and here seeming to be less tied to secondary habitats and borders than the Northern Violaceous Trogon (*T. caligatus*) is in the west. Ranges up to about 500 m.

Trans-Andean *T. caligatus* is here treated as a separate species; see the discussion under that species. We follow J. T. Zimmer (*Am. Mus. Novitates* 1380, 1948) in placing Ecuadorian birds in *crissalis*, separating it from *ramonianus* of e. Peru and w. Brazil.

Range: se. Colombia, s. Venezuela, and the Guianas to n. Bolivia and Amazonian Brazil. More than one species may be involved.

Trogon caligatus   Northern Violaceous Trogon   Trogón Violáceo Norteño

Locally fairly common in the mid-levels and subcanopy of humid forest, forest borders, and secondary woodland in the lowlands of w. Ecuador; also ranges locally into more deciduous forest and woodland. Recorded from w. Esmeraldas south locally through El Oro to w. Loja (near El Limo [ANSP] and north of Guayquichuma [P. Coopmans]). Recorded up to about 900 m.

T. *caligatus* is here regarded as a species separate from *T. violaceus* (Amazonian Violaceous Trogon), found east of the Andes; this is based on several distinct plumage differences, as well as their strikingly different songs. One race of *T. caligatus* is found in Ecuador, *concinnus*.

Range: e. Mexico to n. Colombia and nw. Venezuela; w. Ecuador and extreme nw. Peru.

Coraciiformes
Alcedinidae   Kingfishers

Six species in two genera. Kingfishers are worldwide in distribution but reach their highest diversity in tropical areas of the Old World. All the American kingfishers, together with three species found in the Old World, are sometimes (e.g., Sibley and Monroe 1990) separated as a separate family, the Cerylidae.

Megaceryle torquata   **Ringed Kingfisher**   Martín Pescador Grande

Fairly common and conspicuous along rivers, larger streams and canals, around larger lakes and ponds, and in mangroves in the lowlands and foothills of e. and w. Ecuador. Unlike other Ecuadorian kingfishers, the Ringed frequently overflies land (often high above the ground) on its way to and from feeding areas, and thus it can be seen almost anywhere. Ranges up along larger rivers to at least 1300 m, infrequently higher; a vagrant or wanderer at an exceptionally high elevation was seen near Papallacta (3000 m) on 6 Feb. 1998 (L. Navarrete et al.).

This and the following species have often been placed in the genus *Ceryle*, but we follow Sibley and Monroe (1990) in separating them in *Megaceryle*. One race: nominate *torquata*.

Range: Texas and Mexico to nw. Peru, Chile, and Argentina.

*[Megaceryle alcyon   **Belted Kingfisher**   Martín Pescador Norteño]

An accidental visitant to near the coast in w. Ecuador. Known in Ecuador from only one sighting, an apparent female seen near the mouth of the Río Ayampe on 31 Dec. 1992 (Green 1996; C. Green, pers. comm.). This is the southernmost record of the Belted Kingfisher, which regularly migrates as far south as Panama (Ridgely and Gwynne 1989), with a few scattered records along the northern coast of South America. The species has also been recorded as a vagrant on the Galápagos Islands.

Monotypic (see Wetmore 1968).

Range: breeds in North America, wintering south to coasts of n. South America.

Chloroceryle amazona   **Amazon Kingfisher**   **Martín Pescador Amazónico**

Uncommon to fairly common along rivers and larger streams and lakes in the lowlands of e. Ecuador. Ranges up in small numbers along larger rivers to about 1000 m in the Zamora area.

One race: nominate *amazona*.

Range: n. Mexico to w. Colombia, n. Argentina, and Uruguay.

Chloroceryle americana   **Green Kingfisher**   **Martín Pescador Verde**

Uncommon to fairly common along rivers and streams, around the edge of lakes and ponds, and in mangroves (where it is the commonest kingfisher) in the lowlands and foothills of e. and w. Ecuador. Ranges up locally and in small numbers to at least 1300 m (near Sozoranga; Best et al. 1993).

Two races are recorded from Ecuador, nominate *americana* in the east and *cabanisii* in the west. *Cabanisii* differs in being substantially larger and in having somewhat more white in the wing and slightly more chestnut (male) or buff (female) on the breast.

Range: extreme sw. United States to n. Chile, n. Argentina, and Uruguay.

Chloroceryle inda   **Green-and-rufous**   **Martín Pescador Verdirrufo**
                        Kingfisher

Uncommon to locally fairly common along small—usually sluggish—streams, in shady vegetation overhanging or along the edge of lakes and ponds, and in backwaters of várzea forest in the lowlands of e. and w. Ecuador. Locally uncommon in Esmeraldas (e.g., at Playa de Oro; O. Jahn et al.) but not recorded at all in recent years farther south in the west; in the past it ranged south to n. Manabí (old specimens from Chone), e. Guayas (an old specimen from Bucay), and adjacent Chimborazo (MECN specimen from Pallatanga, doubtless taken somewhat lower). The Green-and-rufous Kingfisher seems especially numerous at Imuyacocha. Ranges mostly below 400 m, but occasionally found considerably higher (as wandering birds?); on the east slope it has been recorded from as high as 750 m (Guaticocha) and even 1250 m (collected [MECN] in Apr. 1998 in the Río Huamboya valley west of Macas in nw. Morona-Santiago; F. Sornoza). There is also an old specimen from Mindo (1300 m), though whether the bird was actually obtained so high cannot be known (many other "Mindo" specimens seem actually to have been taken lower).

Two similar races are found in Ecuador, nominate *inda* in the east and *chocoensis* in the west. Some authors (e.g., C. H. Fry, K. Fry, and A. Harris,

Kingfishers, Bee-eaters, & Rollers, Princeton, N.J., Princeton University Press, 1992) consider the species to be monotypic, but the differences between these taxa appear to be consistent in recently obtained ANSP material. Female *chocoensis* differ in having the back and rump lightly spotted with white (this area is plain in the nominate race), and they have somewhat more extensively white-spotted wings.

Range: Nicaragua to w. Ecuador, n. Bolivia, n. Paraguay, and Amazonian Brazil; e. Brazil.

Chloroceryle aenea   **American Pygmy Kingfisher**   **Martín Pescador Pigmeo**

Uncommon to locally fairly common along small—usually sluggish—streams, in shady vegetation overhanging or near the edge of lakes and ponds, and in backwaters of várzea forest in the lowlands of e. and w. Ecuador. In the west recorded south mainly to cen. Manabí (Chone) and s. Pichincha (Río Palenque), with two recent undated reports of a single (wandering?) female in Guayas at Cerro Blanco in the Chongón Hills (Berg 1994); at Playa de Oro in n. Esmeraldas recent survey work has revealed the species to be "markedly rarer" than the Green-and-rufous Kingfisher (*C. inda*; O. Jahn et al.). Like the Green-and-rufous—with which it often occurs—the American Pygmy is considerably more numerous and widespread in the east than it is in the west. Ranges mostly below 400 m, but has been seen (PJG) at 700 m at Tinalandia.

This species was formerly usually (e.g., Meyer de Schauensee 1966, 1970) simply called the Pygmy Kingfisher. The awkward name American Pygmy Kingfisher was introduced (e.g., AOU 1983; Sibley and Monroe 1990) because an African species, *Ispidina picta*, also has long gone by the English name of Pygmy Kingfisher. With reluctance, we follow these authors.

One race: nominate *aenea*.

Range: s. Mexico to w. Ecuador, n. and e. Bolivia, extreme n. Paraguay, and se. Brazil.

Momotidae   Motmots

Four species in three genera. Motmots are exclusively Neotropical in distribution.

Electron platyrhynchum   **Broad-billed Motmot**   **Momoto Piquiancho**

Uncommon to fairly common (especially by voice) in the lower and middle growth of humid forest, to a lesser extent also at forest borders, in the lowlands and foothills of e. and w. Ecuador. In the east found primarily in terra firme forest. In the west recorded south to n. Manabí, n. Los Ríos (Oct. 1950 specimen [ANSP] from Quevedo), and El Oro (old specimens from El Chiral and La Chonta; many recent reports from Buenaventura). The Broad-billed Motmot is apparently not found in far ne. Ecuador, evidently ranging no farther east than

La Selva and Yuturi; however, it is found—at least locally—all the way to near the Peruvian border in the southeast (recent reports from around Kapawi Lodge). Ranges mostly below 1000 m, but at least locally it also occurs in small numbers up to 1600 m on the west slope (Tandayapa, and at El Chiral in El Oro).

Two races occur in Ecuador, nominate *platyrhynchum* in the west and *pyrrho-laemum* in the east. The latter differs from the nominate race in not developing the racquets on the central retrices and in having a smaller bill.

Range: Honduras to sw. Ecuador; se. Colombia, e. Ecuador, e. Peru, n. Bolivia, and s. Amazonian Brazil.

Baryphthengus martii   Rufous Motmot   Momoto Rufo

Uncommon to fairly common (especially by voice) in the lower growth of humid forest, less often at edge or out into adjacent clearings, in the lowlands and foothills of e. and w. Ecuador. In the east found mainly in terra firme forest. In the west recorded south to n. Manabí, n. Los Ríos (Quevedo; ANSP), El Oro (Buenaventura [Robbins and Ridgely 1990; also numerous subsequent sightings]), and adjacent nw. Loja (Guaicichuma; a specimen taken by M. Olalla on 25 Dec. 1968; MECN). Ranges mostly below 900 m, but occurs locally and in small numbers up to about 1350 m on the east slope (Guaticocha) and to 1500–1600 m on the west slope (above Mindo and at Maquipucuna).

B. martii of Central America to w. Amazonia is regarded as a species separate from *B. ruficapillus* (Rufous-capped Motmot) of se. South America because of its widely disjunct range, major plumage differences, and different vocalizations; this follows most recent authors. Two races occur in Ecuador, nominate *martii* in the east and *semirufus* in the west. The nominate race differs in not developing racquets on its central rectrices, in being more tinged with olive-yellow on the mantle (not as pure a green), and in having its lower belly less bluish green; there are also vocal differences, and the two forms may represent separate species.

Range: Honduras to sw. Ecuador; se. Colombia, e. Ecuador, e. Peru, w. Bolivia, and w. and cen. Amazonian Brazil.

Momotus momota   Blue-crowned Motmot   Momoto Coroniazul

Uncommon in the lower growth of humid forest, secondary woodland, and borders in the lowlands of e. Ecuador; in the lowlands of w. Ecuador uncommon to locally fairly common in the lower growth of semihumid and deciduous forest and woodland from w. Esmeraldas southward. In the east the Blue-crowned Motmot occurs in both terra firme and várzea; it tends to avoid humid situations in the west, where it also ranges out regularly into adjacent clearings and plantations. Likely the record of this species from Miazi in the Río Nangaritza valley (Schulenberg and Awbrey 1997) refers to the similar Highland Motmot (*M. aequatorialis*). In the east the Blue-crowned Motmot occurs mainly below 300 m but has been found at least locally to as high as

500 m (Cotapino; MCZ); in the west it regularly occurs up to at least 1000 m at various localities in El Oro and Loja, locally as high as 1600–1800 m (Catacocha and Sozoranga in Loja [RSR; Best et al. 1993]).

Does not include *M. aequatorialis* of the Andes. It also remains possibile that what is presently considered to be a polytypic *M. momota* will prove to comprise two or more species. Two races of *M. momota* are found in Ecuador, *microstephanus* in the east and the near-endemic *argenticinctus* in the west. Besides having its underparts more bronzy and olive (not so pure a green), *microstephanus* differs in having the nape band entirely deep glistening violet-blue; in *argenticinctus* the same area is at least mixed with turquoise blue. There are also vocal differences.

Range: n. Mexico to n. Colombia, n. and e. Bolivia, nw. Argentina, extreme n. Paraguay, and Amazonian and sw. Brazil; ne. Brazil; w. Ecuador and extreme nw. Peru.

Momotus aequatorialis   Highland Motmot   Momoto Montañero

Rare to locally uncommon in the lower growth and borders of montane forest and secondary woodland in the foothills and lower subtropical zone on the east slope of the Andes. Although the Highland Motmot is known from the Western Andes of Colombia, there are no records from the west slope of the Andes in Ecuador. Highland Motmots appear to be especially numerous in the Cordillera de Huacamayos region. Recorded from about 1000 to 2100 m, thus substantially above the elevational range of eastern-lowland inhabiting Blue-crowned Motmots (*M. momota*).

M. aequatorialis is regarded as a species distinct from *M. momotus*, differing in its substantially larger size, voice, and montane habitat preference. Its nape band color resembles that of the western *M. m. argenticinctus*, thus differing strikingly from that of *M. m. microstephanus* of the eastern lowlands. One race occurs in Ecuador: nominate *aequatorialis*.

Range: Andes from n. Colombia to s. Peru.

Piciformes
Galbulidae   Jacamars

Nine species in four genera. Jacamars are exclusively Neotropical in distribution.

Galbalcyrhynchus leucotis   White-eared Jacamar   Jacamar Orejiblanco

Fairly common to common but rather local at borders of humid forest (especially várzea), around the edge of lakes, and locally on river islands, in the lowlands of e. Ecuador. In Ecuador this conspicuous jacamar is known primarily from the drainages of the Ríos Napo and Aguarico; in the Río Pastaza drainage it is known only from a 19th-century record from Río Copotaza and from recent sightings around Kapawi Lodge near the Peruvian border, but it likely is more widespread. The White-eared Jacamar is particularly numerous around Limoncocha and Imuyacocha; it seems much less common in the vicinity of La Selva and Sacha Lodge. Recorded up to about 400 m.

The species is sometimes called the Chestnut Jacamar.

G. leucotis of nw. Amazonia is regarded as a monotypic species distinct from *G. purusianus* (Purús Jacamar) of sw. Amazonia, following Haffer (1974).

Range: se. Colombia, e. Ecuador, ne. Peru, and extreme w. Amazonian Brazil.

Brachygalba lugubris   Brown Jacamar   Jacamar Pardo

Uncommon and distinctly local at borders of várzea and riparian forest and around the edge of lakes in the lowlands of e. Ecuador. The Brown Jacamar is numerous and quite readily found at Limoncocha, though it is distinctly less common farther east, where it may even be absent from some areas (e.g., seemingly from the lower Río Lagarto drainage). It is also absent from Jatun Sacha (B. Bochan). Recorded below about 300 m.

One race: *caquetae*.

Range: the Guianas, s. Venezuela, e. Colombia, e. Ecuador, ne. Peru, and n. Amazonian Brazil; e. Amazonian Brazil; s. Amazonian and sw. Brazil and ne. Bolivia. More than one species may be involved.

Galbula albirostris   Yellow-billed Jacamar   Jacamar Piquiamarillo

Uncommon and somewhat local in the understory of terra firme forest and mature secondary woodland in the lowlands of e. Ecuador. The Yellow-billed Jacamar is perhaps most numerous and widespread in se. Ecuador. Recorded up only to about 400 m.

G. cyanicollis (Blue-cheeked Jacamar), found south of the Amazon River in Brazil and ne. Peru, is regarded as a separate species, following Haffer (1974). One race: *chalcocephala*.

Range: the Guianas, s. Venezuela, se. Colombia, e. Ecuador, ne. Peru, and n. Amazonian Brazil.

Galbula tombacea   White-chinned Jacamar   Jacamar Barbiblanco

Uncommon to locally fairly common in the lower growth of humid forest borders (especially várzea, seeming most numerous along the edges of lakes and streams) in the lowlands of ne. Ecuador. The White-chinned Jacamar is recorded only from the drainages of the Ríos Napo and Aguarico, and it apparently does not range any distance south of the Río Napo (though it has been found at Yuturi, fide S. Howell); thus it has not been recorded even at Jatun Sacha (B. Bochan) or anywhere along the Maxus road southeast of Pompeya (RSR et al.). Two specimens were recorded (Berlioz 1937c), however, as having been taken at "Laguna de Siguin," a site believed to be somewhere in the upper Río Pastaza drainage of se. Ecuador. Recorded up to about 400 m.

One race: nominate *tombacea*.

Range: e. Colombia, ne. Ecuador, ne. Peru, and w. Amazonian Brazil.

The Bluish-fronted Jacamar (*G. cyanescens*) of sw. Amazonia is sometimes credited to the Ecuadorian avifauna (e.g., Haffer 1974) on the basis of a 19th-century specimen from Zamora. This specimen, which Haffer apparently never examined, was identified by Salvadori and Festa (1900, p. 308) as the White-chinned Jacamar, this despite its being described as having "a golden green pileum." Because the White-chinned has a sooty brown crown, Haffer seems to have assumed that the specimen referred to the Bluish-fronted (which does have a mainly green crown). In fact, the specimen in question (which we also have not seen) is almost unquestionably an example of the Coppery-chested Jacamar (*G. pastazae*), the sole *Galbula* jacamar that has been recorded anywhere near the Zamora region and which also has a green crown. In the absence of more conclusive evidence regarding the presence of the Bluish-fronted Jacamar in Ecuador, we feel we should delete the species from the country's bird list.

Galbula pastazae   **Coppery-chested Jacamar**   Jacamar Pechicobrizo

Uncommon to locally fairly common and apparently local in the lower growth of montane forest in the foothills and lower subtropical zone on the east slope of the Andes from Sucumbíos south through Zamora-Chinchipe, with the southernmost known locality being the Río Isimanchi valley about 10 road km north of Zumba (seen in Dec. 1991; RSR) and just north of there along the road to Valladolid and Loja (ANSP). The Coppery-chested Jacamar seems to be more numerous in s. Ecuador than it is northward; there is, for example, a substantial population in the Río Bombuscaro sector of Podocarpus National Park (Rasmussen et al. 1996). Long thought to be an Ecuadorian endemic, this jacamar has now been found just north of the border in Colombia (see Hilty and Brown 1986), and in Jul. 1994 it was also located just into Peru on the east slope of the Cordillera del Cóndor (Schulenberg and Awbrey 1997). A review of the Coppery-chested Jacamar's distribution and behavior is given by M. Poulsen and D. Wege (*Cotinga* 2: 60–61, 1994), though several of the species' more important sites mentioned above were not included therein. Recorded from about 750 to 1500 m. An 1896 specimen (AMNH) supposedly obtained at Ambato (2600 m) seems surely to have been mislabeled, and must have been taken somewhere farther down the Río Pastaza valley. This may have been the source for Hilty and Brown (1986, p. 312) listing the species from "2100 m" in Ecuador.

The Coppery-chested Jacamar was considered Vulnerable by Collar et al. (1994), and was also considered to be at risk by Poulsen and Wege (op. cit.). We remain less than convinced that the species is actually declining, however, and suspect that instead it has always occurred at low densities. Indeed, the patchy habitats being created through the partial deforestation now occurring in many parts of its range may actually have the effect of increasing its numbers.

Nonetheless, given the species' limited distribution and undoubtedly small total population, we accord it Near-threatened status in Ecuador.

Monotypic.

Range: along eastern base of Andes in extreme s. Colombia, Ecuador, and extreme n. Peru.

Galbula ruficauda   **Rufous-tailed Jacamar**   Jacamar Colirrufo

Uncommon to fairly common in lower growth of humid forest and woodland borders and openings, sometimes out into adjacent clearings, in the lowlands and foothills of w. Ecuador from Esmeraldas south to n. Manabí, n. Los Rios (Oct. 1950 specimens [ANSP] from Quevedo), and at least formerly to e. Guayas (old specimens from Bucay and Naranjito). There are, however, no recent reports of the Rufous-tailed Jacamar from south of extreme s. Pichincha at Río Palenque, at which locality it remains numerous. Recorded up to about 800 m at Tinalandia.

The form found in Ecuador, *melanogenia*, has sometimes been accorded specific rank (Black-chinned Jacamar), though more usually it is regarded as a trans-Andean race of *G. ruficauda* (see Haffer 1974).

Range: s. Mexico to sw. Ecuador and n. Venezuela; locally in Guyana and extreme n. Brazil; n. and e. Bolivia, s. Amazonian and e. Brazil, ne. Argentina.

Galbula chalcothorax   **Purplish Jacamar**   Jacamar Purpúreo

Rare to locally fairly common in the lower and middle growth of humid forest (mainly terra firme), and at forest borders and in adjacent clearings, in the lowlands of e. Ecuador. Recorded mostly below 500 m, but small numbers range up as high as 1000 m (e.g., at Shaime in the Río Nangaritza valley; MECN).

G. chalcothorax of w. Amazonia is regarded as a monotypic species distinct from *G. leucogastra* (Bronzy Jacamar), found disjunctly in cen. and e. Amazonia and the Guianas; this follows the suggestion of T. A. Parker III and J. V. Remsen, Jr. (*Bull. B. O. C.* 107[3]: 98, 1987).

Range: se. Colombia, e. Ecuador, e. Peru, and extreme w. Amazonian Brazil.

Galbula dea   **Paradise Jacamar**   Jacamar Paraíso

Rare to uncommon and apparently local in the canopy and borders of humid forest (mainly terra firme) in the lowlands of e. Ecuador. The Paradise Jacamar was first recorded from Ecuador by Norton et al. (1972), based on specimens taken in 1958 at Montalvo and Morete in Morona-Santiago; these specimens were presumably the basis for the species being recorded in Ecuador by Meyer de Schauensee (1970). Since then it has been found only in the lower Río Aguarico drainage, where small numbers were seen in 1976 (e.g., Ridgely 1980) and several specimens were subsequently collected (ANSP, MECN). Recorded only below 250 m.

Ecuadorian birds have been referred to the subspecies *brunneiceps* (Norton et al. 1972), but we suspect that this race may not be valid; more specimen material is needed.

Range: se. Colombia, s. Venezuela, and the Guianas to n. Bolivia and Amazonian Brazil.

Jacamerops aureus   Great Jacamar   Jacamar Grande

Rare to locally fairly common (more often heard than seen) in the mid-levels and subcanopy of very humid and humid forest (ranging primarily in terra firme) in the lowlands and foothills of e. and nw. Ecuador. In the west known primarily from Esmeraldas, but there are several recent reports from nw. Pichincha along the road north of Simón Bolívar near Pedro Vicente Maldonado (M. Lysinger et al.; P. Coopmans et al.). In the east recorded mainly below about 700 m, but ranges in small numbers on the east slope of the Andes up to as high as 1325 m (Cordillera de Cutucú; ANSP); in the west recorded only below 450 m, and in n. Esmeraldas most numerous between 200 and 400 m (O. Jahn et al.).

Two races occur in Ecuador, *isidori* in the east, whereas northwestern birds are probably referable to trans-Andean *penardi* (the latter despite the comments of Norton et al. 1972). In any case, though there is some individual variation in the tonality of their dorsal coloration, populations on either side of the Andes resemble each other closely.

Range: Costa Rica to nw. Colombia, and in nw. Ecuador; se. Colombia, s. Venezuela, and the Guianas to n. Bolivia and Amazonian Brazil.

Bucconidae   Puffbirds

Nineteen species in nine genera. Puffbirds are exclusively Neotropical in distribution.

Notharchus macrorhynchos   White-necked Puffbird   Buco Cuelliblanco

Rare to locally uncommon in the subcanopy and borders of humid forest and secondary woodland in the lowlands of e. and w. Ecuador. In the east the White-necked Puffbird occurs in both terra firme and várzea; in the west it also occurs in more deciduous forest and woodland. In e. Ecuador recorded primarily from the Río Napo drainage northward, but recent reports from the southeast at Kapawi Lodge near the Peruvian border on the Río Pastaza lead us to suspect that the White-necked Puffbird ranges widely throughout the Río Pastaza drainage as well. In the west recorded south sparingly—perhaps only formerly—to El Oro (Santa Rosa and Piedras; ANSP); there are recent records south only to se. Guayas in the Naranjal area and at Manglares-Churute Ecological Reserve (v.o.). Recorded up only to about 400 m.

The correct spelling of the species name is *macrorhynchos*, not *macrorhynchus* (Sibley and Monroe 1990).

One race: *hyperrhynchus.*
Range: s. Mexico to sw. Ecuador, n. Bolivia, and Amazonian Brazil; se. Brazil, e. Paraguay, and ne. Argentina. More than one species is perhaps involved.

Notharchus pectoralis   Black-breasted Puffbird   Buco Pechinegro

Rare to locally common in the subcanopy and borders of humid forest and in adjacent clearings in the lowlands of extreme nw. Ecuador in n. Esmeraldas. Recent reports come from along the Río Cayapas, single birds seen near San Miguel on 5 Mar. 1980 (PJG and S. Greenfield) and at El Encanto on 9 Jun. 1995 (M. Lysinger et al.); a very few were also seen at Playa de Oro in 1996 (O. Jahn et al.). In Aug. 1998, however, the species was found to be numerous east of San Lorenzo (e.g., near the Río Mataje; v.o.). Recorded only below 200 m, though in w. Colombia known to range up to 1000 m (Hilty and Brown 1986).
Monotypic.
Range: Panama to nw. Ecuador.

Notharchus tectus   Pied Puffbird   Buco Pinto

Rare to uncommon and perhaps local in the subcanopy and borders of humid forest, secondary woodland, and adjacent clearings in the lowlands of e. and w. Ecuador. In the east found both in terra firme and várzea (perhaps mostly the former), and primarily occurring in areas away from the base of the Andes (it has, for instance, never been seen at Jatun Sacha; B. Bochan). In the west recorded south to Guayas (at least formerly it occurred in the Guayaquil area; PJG) and nw. Azuay (seen in Jul. 1991 at Manta Real; Parker and Carr 1992). Pied Puffbirds may perhaps be largely absent from the wet-forest belt of n. Esmeraldas, for they were not recorded in the Río Santiago and Río Cayapas drainages during extensive survey work in 1995–1998 (O. Jahn and P. Mena Valenzuela et al.). The Pied Puffbird seems to be more numerous in w. Ecuador than it is in the east, but in the latter it likely is often overlooked, remaining high and motionless in the canopy. Recorded up to about 500 m.

Two rather different races are found in Ecuador, *subtectus* in the west and the larger *picatus* in the east. *Subtectus* is somewhat glossier black above, and it has a markedly narrower black chest band and fewer white crown dots.
Range: Costa Rica to sw. Ecuador; se. Colombia, s. Venezuela, and the Guianas to ne. Peru, ne. Bolivia, and cen. and e. Amazonian Brazil.

Bucco macrodactylus   Chestnut-capped Puffbird   Buco Gorricastaño

Uncommon and inconspicuous in the lower growth of forest borders (both terra firme and várzea), secondary woodland, and shrubby clearings in the lowlands of e. Ecuador. Stolid and quiet, the strikingly marked Chestnut-capped Puffbird is doubtless often overlooked. Recorded up to about 600 m.

One race: nominate *macrodactylus*. The species is, however, perhaps better considered to be monotypic, as was done in *Birds of the World* (vol. 6); we are not familiar with the only other described race, *caurensis*, of s. Venezuela.

Range: se. Colombia and s. Venezuela to nw. Bolivia and w. and cen. Amazonian Brazil.

Bucco tamatia   Spotted Puffbird   Buco Moteado

Very rare and apparently local in the lower and middle growth of forest (elsewhere mainly in várzea) in the lowlands of se. Ecuador. The Spotted Puffbird is not a well-known bird in Ecuador, and specimens have been recorded only from Sarayacu (BMNH) and Montalvo (MCZ, MECN). The only recent report comes from Kapawi Lodge on the Río Pastaza near the Peruvian border, where one was tape-recorded in Jan. 1997 (D. Michael, fide J. Moore); there may have been other reports from there as well. Recorded below 300 m.

One race: *pulmentum*.

Range: se. Colombia, s. Venezuela, and the Guianas to se. Ecuador, ne. Peru, Amazonian Brazil, and ne. Bolivia.

Bucco capensis   Collared Puffbird   Buco Collarejo

Rare to locally uncommon in the lower and middle growth of humid forest (mainly terra firme) in the lowlands of e. Ecuador. Aside from its far-carrying calls, the striking Collared Puffbird is a sluggish bird that doubtless is often overlooked. Mainly ranges below 500 m, though one was collected as high as 1700 m on the Cordillera de Cutucú in Jun. 1984 (ANSP).

One race: *dugandi*. The species may, however, ultimately better be considered monotypic.

Range: se. Colombia, s. Venezuela, and the Guianas to se. Peru and Amazonian Brazil.

Nystalus radiatus   Barred Puffbird   Buco Barreteado

Uncommon to locally fairly common (especially by voice) in the subcanopy and borders of humid forest, mature secondary woodland, and adjacent clearings in the foothills and adjacent more humid lowlands of w. Ecuador south to El Oro (Buenaventura; Robbins and Ridgely 1990, also subsequent sightings). Also recorded recently from the coastal Cordillera de Colonche in sw. Manabí (Cerro San Sebastián in Machalilla National Park; ANSP) and w. Guayas (Cerro La Torre [N. Krabbe] and Loma Alta [Becker and López-Lanús 1997]). In the wet-forest belt of n. Esmeraldas the Barred Puffbird appears to be uncommon at forest borders and rare inside actual forest, then occurring mainly along ridges (O. Jahn et al.). It persists in reasonable numbers in small patches of forest—continuing to be seen regularly at Tinalandia and Río Palenque—and can apparently disperse across quite cutover terrain; it routinely perches on wires in mostly agricultural country around Santo Domingo de los Colorados. Recorded mostly below 1000 m, though ranging up to 1500 m above Mindo.

The genus *Nystalus* has occasionally been merged into *Notharchus* (e.g., in the 1983 AOU Check-list), but most authors (including Sibley and Monroe 1990) maintain it as distinct. We agree, and the 1998 AOU Check-list reverted to accepting it. Monotypic.

Range: Panama to sw. Ecuador.

Nystalus striolatus   Striolated Puffbird   Buco Estriolado

Rare to locally uncommon (perhaps overlooked) in the subcanopy and borders of montane forest in the foothills and lower subtropical zone on the east slope of the Andes. Recorded from w. Napo (north to along the Loreto road north of Archidona; there is also one report from the adjacent lowlands, a bird heard at Jatun Sacha on 3 Feb. 1994 [B. Bochan]) south to Zamora-Chinchipe (Cordillera del Cóndor, and [RSR tape-recording] in the Río Isimanchi valley north of Zumba). Unlike in s. Amazonia, in Ecuador the Striolated Puffbird is not known to occur in the lowlands any distance away from the Andes, the easternmost record being the Jatun Sacha report mentioned above; near the mouth of the Río Cuyabeno into the Río Aguarico, however, P. Coopmans has heard Lawrence's Thrushes (*Turdus lawrencii*) giving what sounded like imitations of the puffbird's song, so it is conceivable that the species may range farther east at least locally. There do not appear to be any published records from adjacent n. Peru, though the species seems almost certain to occur there. The species is sluggish and difficult to see, more so than its congener the Barred Puffbird (*N. radiatus*); as with that species, the Striolated is most often recorded through its far-carrying song. Recorded mostly from 800 to 1700 m, rarely wandering down as low as 400 m (Jatun Sacha).

One race: nominate *striolatus*.

Range: along base of Andes in e. Ecuador; se. Peru, n. Bolivia, and s. Amazonian Brazil.

Malacoptila fusca   White-chested Puffbird   Buco Pechiblanco

Uncommon to locally fairly common in undergrowth inside terra firme forest in the lowlands of e. Ecuador. Unobtrusive (notably more so than the White-whiskered Puffbird [*M. panamensis*]), and as a result only rarely seen, though regularly captured in mist-nets. Ranges up in the foothills on the east slope of the Andes to 900 m (e.g., at Shaime in the Río Nangaritza valley; MECN) and locally even to 1200 m (near Volcán Sumaco; P. Coopmans et al.).

One race: nominate *fusca*.

Range: se. Colombia, sw. Venezuela, extreme nw. Amazonian Brazil, e. Ecuador, and ne. Peru; the Guianas and n. Amazonian Brazil.

The Rufous-necked Puffbird (*M. rufa*) is known to range as close to the present eastern border of Ecuador as the mouth of the Río Curaray in ne. Peru (2 specimens in AMNH); these were the basis for the species having been recorded from "eastern Ecuador" in *Birds of the World* (vol. 6), this when that locality was still part of Ecuador. As yet, though, the Rufous-necked Puffbird has not been found within the present borders of Ecuador, though it could eventually be found to occur in the extreme east. Interestingly, no *Malacoptila* puffbird has been found in the Imuyacocha/Zancudococha regions; one of the two species seems likely to occur there.

Malacoptila fulvogularis   Black-streaked Puffbird   Buco Negrilistado

Rare to uncommon (doubtless overlooked) in undergrowth inside montane forest, to a lesser extent at borders, in the foothills and subtropical zone on the east slope of the Andes, mainly in s. Ecuador in Morona-Santiago (north to the Macas area and [ANSP] on the Cordillera de Cutucú) and Zamora-Chinchipe. Long thought likely to occur northward as well, the Black-streaked Puffbird was finally first recorded in w. Napo north of Archidona along the Loreto road on 26 Aug. 1996 (M. Lysinger and P. Coopmans) and again on 28 Aug. 1996 (P. Coopmans and N. Krabbe); probably it will be found to occur elsewhere as well, though numbers perhaps are not as great as they seem to be southward. Ranges between about 1100 and 2000 m.

Given its general rarity, limited altitudinal range, and substantial levels of habitat destruction in that range, we believe the Black-streaked Puffbird deserves Near-threatened status in Ecuador. Its numbers are probably no greater than those of the Coppery-chested Jacamar (*Galbula pastazae*), with which it sometimes occurs. The puffbird was not considered to be at risk by Collar et al. (1994).

One race: nominate *fulvogularis*.

Range: locally on east slope of Andes from s. Colombia to w. Bolivia.

Malacoptila panamensis   White-whiskered Puffbird   Buco Bigotiblanco

Uncommon to fairly common in the undergrowth of humid forest and mature secondary woodland, occasionally foraging out into adjacent plantations and partially shaded clearings, in the lowlands and foothills of w. Ecuador south to nw. and e. Guayas (in the former south of the lower Río Ayampe [RSR] and Loma Alta [Becker and López-Lanús 1997]) and El Oro (old specimens from La Chonta and Portovelo; many recent records from Buenaventura). Ranges up to 900 m.

One race: *poliopis*.

Range: s. Mexico to sw. Ecuador.

The Moustached Puffbird (*M. mystacalis*) ranges very close to the Ecuador border on the west slope of the Andes in Nariño, Colombia (at Ricaurte and Guayacana; ANSP), but as yet it has not been found across the border in Ecuadorian territory. It occurs in foothill and subtropical—not primarily lowland—forest.

Micromonacha lanceolata   Lanceolated Monklet   Monjecito Lanceolado

Rare and seemingly local (doubtless mostly just overlooked; the species is notably lethargic, often going for long periods without moving at all) in the lower and middle growth of humid forest, mature secondary woodland, and borders in the lowlands and foothills of e. and w. Ecuador. In the west ranges south to sw. Cañar (one seen above Cochancay on 17 Aug. 1991; F. Sornoza and RSR) and nw. Azuay (one seen at Manta Real by T. Parker in Jul. 1991, fide K. Berg). The first, and still only, specimens (WFVZ, MECN) from w. Ecuador were taken at Filo de Monos in n. Manabí on 8 Jul. 1988 (Marín et al. 1992). Recorded up to 1100 m (in Pichincha at Pacto along the Río Guaillabamba [MECN] and along the lower Chiriboga road [v.o.]; also in Zamora-Chinchipe in the Bombuscaro sector of Podocarpus National Park), exceptionally as high as about 1300 m (one seen and tape-recorded near Mindo on 25 Dec. 1997; J. Lyons and V. Perez).

The Lanceolated Monklet was given Near-threatened status by Collar et al. (1994). However, we do not believe the species is particularly at risk and think it merely occurs at low densities and is difficult to see.

Monotypic. We follow Wetmore (1968) in synonymizing the race *austinsmithi* of Costa Rica and w. Panama; *Birds of the World* (vol. 6, p. 18) had earlier considered it "doubtfully distinct.")

Range: locally in Costa Rica and Panama; sw. Colombia and w. Ecuador; se. Colombia, e. Ecuador, e. Peru, and w. Amazonian Brazil.

Nonnula brunnea   Brown Nunlet   Nonula Parda

Rare to uncommon—perhaps mostly just overlooked—in the lower growth of humid forest, primarily in terra firme (especially favoring tangled secondary growth around treefalls) in the lowlands of e. Ecuador. Small numbers of the Brown Nunlet occur at La Selva and at Sacha Lodge, but this retiring bird is always hard to locate. Ranges up locally to about 700 m (Eugenio; MCZ), though found mostly below 400 m.

Monotypic.

Range: se. Colombia, e. Ecuador, and ne. Peru.

*Nonnula rubecula   Rusty-breasted Nunlet   Nonula Pechirrojiza

Rare and local in the lower growth of terra firme forest in the lowlands of ne. Ecuador in Sucumbíos. Thus far the Rusty-breasted Nunlet is known primarily from Cuyabeno, where C. Canaday obtained the first Ecuadorian record, a female mist-netted and collected (MECN) on 3 Aug. 1990; there have been a few subsequent sightings from here. The only additional site from which the Rusty-breasted Nunlet is known in Ecuador is Zábalo on the lower Río Aguarico, where it was seen several times in early 1993 (A. Johnson and M. Hedemark). Recorded below about 250 m.

Presumed race: *cineracea*.

Range: ne. Ecuador, ne. Peru, extreme se. Colombia, w. and cen. Amazonian Brazil, sw. Venezuela, and locally in the Guianas; se. Brazil, e. Paraguay, and ne. Argentina.

Hapaloptila castanea   **White-faced Nunbird**   Monja Cariblanca

Rare to uncommon and very local in montane forest, forest borders, and adjacent clearings in the subtropical zone on the west slope of the Andes in nw. Ecuador in Pichincha (mainly on the Tandayapa ridge); there are also a few sightings from Carchi (one seen west of Maldonado on 27 Mar. 1984 [N. Krabbe] and a pair seen on Cerro Golondrinas between El Corazon and Gualtal on 25 Jun. 1994 [J. Hendriks]). There are also a few records from the east slope, including a 1948 specimen (ANSP) from "Cerro Imbana, Loja" (a locality that cannot be definitely located but which perhaps lies in Zamora-Chinchipe), specimens (MECN) taken along the La Bonita road in Sucumbíos on 5 Jan. 1985 and 8 Apr. 1986 (J. C. Matheus), a sighting of a lone bird on the Cordillera de Huacamayos on 30 Oct. 1993 (M. Lysinger), and solitary birds recorded at Cabañas San Isidro on 16 Sep. 1995 (P. Coopmans) and 8 Sep. 1996 (M. Lysinger). Recorded between 1300 and 2400 m.

The White-faced Nunbird occurs at very low densities under the best of circumstances—indeed it appears to be absent from large areas of seemingly suitable terrain. Given, in addition, its rather limited range, we feel that the species deserves to be accorded Data Deficient status. This nunbird is certainly a much rarer bird than, for instance, two species considered to be at risk (Collar et al. 1994) that occur with it, the Toucan Barbet (*Semnornis ramphastinus*) and the Plate-billed Mountain-Toucan (*Andigena laminirostris*).

Monotypic.

Range: very locally on west slope of Andes in Colombia and nw. Ecuador, and on east slope in Ecuador and n. Peru.

Monasa nigrifrons   **Black-fronted Nunbird**   Monja Frentinegra

Common in the mid-levels, subcanopy, and borders of várzea and riparian forest, and along the edges of lakes and rivers (frequent on river islands), in the lowlands of e. Ecuador. The Black-fronted Nunbird is recorded in Ecuador primarily from the drainages of the Ríos Napo and Aguarico, but an old record from "Laguna de Siguin" (Berlioz 1937c) and recent sightings from around Kapawi Lodge on the Río Pastaza near the Peruvian border make it seem likely that the species ranges widely in the southeast as well. To about 400 m.

One race: nominate *nigrifrons*.

Range: se. Colombia, e. Ecuador, e. Peru, n. Bolivia, and s. Amazonian and sw. Brazil; very locally in e. Brazil.

Monasa morphoeus   **White-fronted Nunbird**   Monja Frentiblanca

Uncommon to fairly common in the mid-levels and subcanopy of terra firme forest and forest borders in the lowlands of e. Ecuador. Ranges up regularly to about 1000 m along the eastern base of the Andes, rarely or locally as high as 1350 m at the head of the Río Guataracu (MCZ).

One race: *peruana.*
Range: Costa Rica to n. Colombia; se. Colombia and extreme sw. Venezuela to n. Bolivia and Amazonian Brazil; e. Brazil.

Monasa flavirostris   **Yellow-billed Nunbird**   Monja Piquiamarilla

Uncommon and local in the mid-levels and canopy of humid forest borders and secondary woodland in the lowlands of e. Ecuador; unlike se. Peru, in Ecuador seems to show no special predilection for bamboo. On the whole the Yellow-billed Nunbird is much less frequently encountered than its two larger congeners, and it seems to be found mainly in areas fairly close to the Andes. In the northeast the easternmost report seems to be a sighting of a pair at "Sacha Urco," along the north bank of the Río Aguarico west of the mouth of the Río Pacuyacu, on 7 Nov. 1992 (P. Coopmans et al.). In the far southeast there are recent reports from Kapawi Lodge near the Peruvian border on the Río Pastaza. Recorded mostly below 400 m, locally as high as 750 m (group of 4 seen on 8–9 Jun. 1999 about 17 km northwest of Archidona along the road to Baeza; L. Navarrete and J. Moore).
Monotypic.
Range: se. Colombia, e. Ecuador, and ne. Peru; se. Peru and sw. Amazonian Brazil.

Chelidoptera tenebrosa   **Swallow-winged Puffbird**   Buco Golondrina

Uncommon to fairly common and conspicuous in forest and woodland borders and in clearings in the lowlands of e. Ecuador, especially along rivers and streams. Almost all records of the Swallow-winged Puffbird come from the drainages of the Ríos Napo and Aguarico. Recent reports from around Kapawi Lodge on the Río Pastaza near the Peruvian border, however, indicate that the species likely also ranges widely along the Río Pastaza. There is also an old anomalous specimen (Salvadori and Festa 1900) from "Valle de Zamora" which we suspect to be mislabeled as there are no recent records of the Swallow-winged Puffbird from anywhere near this region. Recorded up only to about 400 m.
   This very distinctive species is often called simply the Swallow-wing (e.g., Meyer de Schauensee 1966, 1970; Sibley and Monroe 1990), but we believe it is preferable to indicate to which family of birds it belongs, as suggested previously by Hilty and Brown (1986).
One race: nominate *tenebrosa.*
Range: e. Colombia, Venezuela, and the Guianas to n. Bolivia and Amazonian Brazil; e. Brazil.

Capitonidae   New World Barbets

Seven species in three genera. The New World barbets are exclusively Neotropical in distribution. Following Sibley and Ahlquist (1990) and Gill (1994), they are considered to be a family distinct from the Old World Barbets

(Megalaimidae) and the African Barbets (Lybiidae), being more closely allied to the Toucans (Ramphastidae).

Capito aurovirens   **Scarlet-crowned Barbet**   **Barbudo Coronirrojo**

Fairly common to common in the canopy and borders of várzea and riparian forest and woodland, in semiopen areas around lake margins, and on river islands in the lowlands of e. Ecuador. We regard the undocumented report of the species from the "Loja-Zamora road" in Rasmussen et al. (1994) as being unverified; all habitats there would certainly appear to be unsuitable. The Scarlet-crowned Barbet is particularly numerous at Limoncocha and Imuyacocha. Recorded up to about 400 m.
Monotypic.
Range: se. Colombia, e. Ecuador, ne. Peru, and w. Amazonian Brazil.

Capito squamatus   **Orange-fronted Barbet**   **Barbudo Frentinaranja**

Uncommon to locally fairly common in the canopy and borders of humid forest and secondary woodland in the lowlands and foothills of w. Ecuador south to cen. Manabí (Chone), s. Los Ríos (Río San Antonio), and e. Guayas; at least formerly small numbers also occurred south along the base of the Andes as far as El Oro (a female [ANSP] was taken in Oct. 1950 at Piedras). In recent years, though, the Orange-fronted Barbet has been recorded mainly northward from Los Ríos (Jauneche) and extreme s. Pichincha (Río Palenque), though a specimen (ANSP) was obtained at San José de Tambo in s. Bolívar on 9 Aug. 1986. Recorded mainly below 800 m, but locally a few occur (or wander) up to 1300 m (e.g., at Mindo).

Overall numbers of the Orange-fronted Barbet, nearly an Ecuadorian endemic, have doubtless declined because of deforestation, but populations remain substantial in areas where patchy forest and woodland is extant. We therefore do not believe that the species should be considered at risk. Collar et al. (1994), however, concluded that it merits Near-threatened status.
Monotypic.
Range: extreme sw. Colombia and w. Ecuador.

*Capito quinticolor   **Five-colored Barbet**   **Barbudo Cinco Colores**

Locally not uncommon in the canopy and borders of very humid forest and adjacent secondary woodland in the lowlands and lower foothills of extreme nw. Ecuador in n. Esmeraldas. First recorded northwest of Alto Tambo, where four specimens were obtained by M. B. Robbins at a fruiting tree on 17 Jul. 1990 (ANSP, MECN). The Five-colored Barbet was then found again in the Playa de Oro area on 5 Jun. 1997 and on a few subsequent occasions as well (O. Jahn et al.); based on tape-recordings, it was believed not to be rare in that area. The following year a few other territories were located farther south in the Tsejpi area on the Río Zapallo, an affluent of the Río Cayapas (O. Jahn).

In 1998–1999 the species was also recorded on several occasions east of San Lorenzo in the Río Mataje area (M. Lysinger). For a summary of records, see O. Jahn, M. B. Robbins, P. M. Valenzuela, P. Coopmans, R. S. Ridgely, and K.-L. Schuchmann, *Bull. B. O. C.* 120[1]: 16–22, 2000. The Five-colored Barbet was heretofore known only from a few localities in the Pacific lowlands of w. Colombia (Hilty and Brown 1986). At all Ecuadorian localities it is syntopic with the Orange-fronted Barbet (*C. squamatus*), though more restricted to the canopy of continuous forest. Recorded between about 100 and 400 m (in adjacent Nariño, Colombia, recorded regularly as high as 550 m; P. Coopmans).

Collar et al. (1994) concluded that the Five-colored Barbet merited Vulnerable status. With such limited information from Ecuador upon which to make an assessment, we simply concur.

Monotypic.

Range: sw. Colombia and extreme nw. Ecuador.

Capito auratus   Gilded Barbet   Barbudo Filigrana

Fairly common to common in humid forest (both terra firme and várzea, but especially the former) and forest borders in the lowlands of e. Ecuador, also ranging up in smaller numbers in foothill and montane forest on the east slope of the Andes. Although occurring primarily in the canopy and mid-levels of forest, Gilded Barbets also regularly follow flocks in the understory, and they are surprisingly often captured in mist-nets. Recorded mostly below about 1200 m, ranging in small numbers as high as 1700 m.

C. auratus is considered a species separate from *C. niger* (Black-spotted Barbet), following J. Haffer (*Studies in Neotropical Ornithology Honoring Ted Parker*, Ornithol. Monogr. no. 48: 281–305, 1997). *C. brunneipectus* (Brown-throated Barbet) of lower Amazonian Brazil was also treated as a full species in this paper. One race of *C. auratus* occurs in Ecuador, *punctatus*. D. Norton (*Breviora* 230, 1965) determined that the supposed Ecuadorian race *macintyrei* was better subsumed into *punctatus*, which ranges from se. Colombia to ne. Peru.

Range: se. Colombia and s. Venezuela to nw. Bolivia and w. Amazonian Brazil.

Eubucco richardsoni   Lemon-throated Barbet   Barbudo Golilimón

Uncommon in the canopy and borders of terra firme forest in the lowlands of e. Ecuador, ranging in small numbers up into the foothills along the base of the Andes. At at least one site, on the lower slopes of the Cordillera del Cóndor at Pachicutza (WFVZ), the Lemon-throated Barbet has been found to occur sympatrically with the congeneric Red-headed Barbet (*E. bourcierii*). A specimen of a male Lemon-throated Barbet x Red-headed Barbet is known (L. Short); though this specimen was evidently obtained in Ecuador, it is from an uncertain locality. Recorded up to 1000 m, but mainly found below 600 m.

One race: nominate *richardsoni*.

Range: se. Colombia, e. Ecuador, e. Peru, nw. Bolivia, and w. Amazonian Brazil.

Eubucco bourcierii   Red-headed Barbet   Barbudo Cabecirrojo

Uncommon to fairly common in the canopy and borders of montane forest and secondary woodland in the foothills and subtropical zone on both slopes of the Andes; on the west slope recorded south to ne. Guayas (an old specimen from Naranjito). The Red-headed Barbet also ranges out locally into the more humid lowlands of the northwest (south to about Río Palenque), and recently it has also been recorded from the coastal Cordillera de Mache in w. Esmeraldas (Bilsa; R. P. Clay et al.) and the Cordillera de Colonche in sw. Manabí (Cerro San Sebastián in Machalilla National Park; ANSP) and w. Guayas (hills south of the Río Ayampe [ANSP], above Salanguilla [N. Krabbe], and Loma Alta [Becker and López-Lanús 1997]). Recorded mostly from 800 to 1900 m, but in the west locally occurs much lower, even down to near sea level (150 m) in nw. Guayas south of the lower Río Ayampe (female taken on 21 Jan. 1991; ANSP).

Two races are found in Ecuador, *orientalis* on the east slope and *aequatorialis* in the west. In *aequatorialis*, males have red extending lower on the underparts and contrasting more with yellow on the upper belly; female *orientalis* have a narrow blue band behind the black forehead that is lacking in *aequatorialis*.

Range: highlands from Costa Rica to w. Venezuela and extreme n. Peru. More than one species may be involved.

Semnornis ramphastinus   Toucan Barbet   Barbudo Tucán

Uncommon to locally fairly common in the canopy and borders of montane forest and secondary woodland in the subtropical zone on the west slope of the Andes in nw. Ecuador south to w. Cotopaxi (Caripero area; N. Krabbe and F. Sornoza). Recorded mostly from 1400 to 2400 m.

The magnificent Toucan Barbet was accorded Near-threatened status by Collar et al. (1994), presumably reflecting mainly its present scarcity in much of the Colombian portion of its range (due in part to trapping pressure for intense local demand as a cagebird). In Ecuador, however, the species appears to be less at risk, for here much of its subtropical forest habitat remains relatively intact, and the impact of trapping appears to be negligible. Nonetheless, we elect to be cautious and concur with Near-threatened status for the species.

One race: nominate *ramphastinus*.

Range: west slope of Andes in sw. Colombia and nw. Ecuador.

Ramphastidae   Toucans

Nineteen species in five genera. Toucans are exclusively Neotropical in distribution.

Aulacorhynchus prasinus   Emerald Toucanet   Tucanete Esmeralda

Uncommon to locally fairly common in the canopy and borders of montane forest and in trees of adjacent clearings in the subtropical zone on the east slope of the Andes. Recorded mostly between about 1500 and 2600 m, but at least locally it ranges much higher (e.g., to 3250 m below Oyacachi; Krabbe et al. 1997).

Two very different races occur in Ecuador, *albivitta*, with a white throat, in the north and *cyanolaemus*, with a blue throat, in the south. *Albivitta* also differs in having considerable yellow along a wide band on the culmen; *cyanoleamus*, in contrast, has an all-black bill with a yellow tip. *Albivitta* has been found south only to the Baeza/Cosanga and Sumaco areas of w. Napo, whereas *cyanolaemus* has been found north only to the Macas/Cordillera de Cutucú region; it remains unknown whether the two forms are in contact, and whether they intergrade, in the intervening area. We suspect that it may be an oversimplification to treat all the many forms presently considered as races of *A. prasinus* as conspecific.

Range: highlands from n. Mexico to Panama; Andes from w. Venezuela to w. Bolivia, also locally in lowlands of se. Peru and adjacent Brazil; Sierra de Perijá and Santa Marta Mountains.

Aulacorhynchus derbianus   Chestnut-tipped   Tucanete Filicastaño
                                 Toucanet

Rare to uncommon and seemingly local in the canopy and borders of montane forest in the foothills and lower subtropical zone on the east slope of the Andes from w. Sucumbíos (Bermejo oil-field area north of Lumbaquí; RSR) south locally into Zamora-Chinchipe. The Chestnut-tipped Toucanet almost certainly occurs in adjacent s. Colombia, where it is known from only a single specimen lacking specific locality (Hilty and Brown 1986). Small numbers of this generally scarce species are found regularly in the Río Bombuscaro sector of Podocarpus National Park. Recorded mostly from about 800 to 1800 m, locally as low as 600 m (Yaupi; WFVZ).

One race: nominate *derbianus*.

Range: east slope of Andes from n. Ecuador to w. Bolivia; tepuis of s. Venezuela, extreme n. Brazil, and Guyana and Suriname.

Aulacorhynchus haematopygus   Crimson-rumped   Tucanete
                                    Toucanet   Lomirrojo

Fairly common in the canopy and borders of humid and montane forest, secondary woodland, and adjacent clearings in the foothills and subtropical zone on the west slope of the Andes south to w. Loja (Celica and Utuana), ranging in smaller numbers out into the adjacent humid lowlands (e.g., at Río Palenque); curiously, though, it seems to be absent from the lowlands and lowermost foothills of n. Esmeraldas (none, for instance, were found in 4 years of intensive survey work at Playa de Oro; O. Jahn et al.). Also recently recorded on the coastal Cordillera de Mache in w. Esmeraldas (Bilsa; R. P. Clay et al.) and the Cordillera de Colonche in sw. Manabí (Parker and Carr 1992; ANSP) and w. Guayas (above Salanguilla [N. Krabbe] and at Loma Alta [Becker and López-Lanús 1997]). The Crimson-rumped Toucanet may also occur on the east slope in n. Ecuador, though it is not definitely known on the east slope of the Andes south of s. Colombia. Birds seemingly referable to this species were seen on several occasions at San Rafael Falls in the 1970s and early 1980s (PJG; RSR et al.), but they have not been found here subsequently. Ranges mostly from 500 to 2000 m, locally or occasionally lower or higher (once recorded as high as 2750 m above Celica; Bloch et al. 1991).

Only one race, *sexnotatus*, has definitely been recorded from Ecuador, but birds on the east slope would presumably be referable to nominate *haematopygus*, which ranges on the east slope of the Colombian Andes.

Range: Andes from sw. Venezuela to sw. Ecuador; Sierra de Perijá.

Selenidera spectabilis   Yellow-eared Toucanet   Tucancillo Orejiamarillo

Uncertain. Apparently very rare and local in the subcanopy and mid-levels of very humid forest in the lowlands of extreme nw. Ecuador. Thus far recorded only from two specimens (Escuela Politécnica collection) taken by M. Olalla on 27 Mar. 1962 at "La Boca" in Esmeraldas (Norton et al. 1972); these must have been the basis for the inclusion of the species as occurring in Ecuador by Meyer de Schauensee (1970). The only subsequent record involves two birds seen near Playa de Oro in Nov. 1997 (E. Vargas Grefa, fide O. Jahn). O. Jahn (pers. comm.) suggests that the Yellow-eared Toucanet may only occur in Ecuador during very wet El Niño years. Curiously, it has not been recorded from adjacent sw. Colombia (Hilty and Brown 1986). Elsewhere in its range (e.g., Panama) it ranges primarily in foothill forest. Recorded up to 100 m.

Given the Yellow-eared Toucanet's evident rarity and its extremely small range in Ecuador, we believe it prudent to accord it Data Deficient status in the country. Elsewhere the species is not considered to be at risk (Collar et al. 1994).

Monotypic.

Range: Honduras to nw. Colombia, with records from nw. Ecuador.

Selenidera reinwardtii   Golden-collared   Tucancillo Collaridorado
                            Toucanet

Uncommon to fairly common in the subcanopy and mid-levels of terra firme forest (less often at borders) in the lowlands and foothills of e. Ecuador. Ranges mostly below 800 m, occurring up in diminishing numbers to about 1200 m (north of Archidona along the road to Loreto; P. Coopmans).

One race: nominate *reinwardtii*.

Range: se. Colombia, e. Ecuador, e. Peru, nw. Bolivia, and w. Amazonian Brazil.

Pteroglossus erythropygius   Pale-mandibled Araçari   Arasari Piquipálido

Uncommon to locally common in the canopy and borders of humid and semi-humid forest and mature secondary woodland in the lowlands and foothills of w. Ecuador from w. Esmeraldas (vicinity of Esmeraldas [city]) south to nw. Guayas (hills south of the Río Ayampe, Cerro La Torre, and Loma Alta; v.o.) and El Oro (Buenaventura and near Balzas; Robbins and Ridgely 1990). Somewhat surprisingly, the Pale-mandibled Araçari has never been reported to range south into adjacent w. Loja, where suitable habitat would still seem to occur, at least locally. Ranges up to 1300–1500 m in w. Pichincha (above Mindo and at Maquipucuna), but elsewhere usually below 1100 m.

Overall numbers of this Ecuadorian endemic have declined considerably because of the deforestation that has occurred across the greater part of its range, but the Pale-mandibled Araçari seems to persist relatively well in fragmented and disturbed forest patches, and it cannot be considered to be immediately threatened.

Monotypic, but see the discussion under *P. sanguineus* (Stripe-billed Araçari).
Range: w. Ecuador.

Pteroglossus sanguineus   Stripe-billed Araçari   Arasari Piquirrayado

Fairly common to common in the canopy and borders of humid forest and mature secondary woodland in the lowlands and foothills of nw. Ecuador in n. Esmeraldas and adjacent Imbabura. The Stripe-billed Araçari is numerous at El Placer. Ranges up at least to 800 m.

P. sanguineus of w. Colombia and adjacent Panama and Ecuador is regarded as a monotypic species distinct from *P. torquatus* (Collared Araçari) of Middle America and *P. erythropygius* (Pale-mandibled Araçari) of w. Ecuador, in accord with most recent authors. *P. sanguineus* hybridizes at least to a limited extent with *P. torquatus* in a narrow zone in nw. Colombia and e. Panama, and hybridization may be occurring with *P. erythropygius* in nw. Ecuador. ANSP specimens show no evidence that this is the case, but the area of potential contact in Esmeraldas and extreme nw. Pichincha remains poorly explored ornithologically. In the Playa de Oro area the bill pattern of two individuals mist-netted and photographed in 1996 shows characteristics between both taxa, with a third showing the pattern of *P. sanguineus*; some other individuals observed in the area have the bill pattern of *P. erythropygius* (O. Jahn and K.-L. Schuchmann et al.).
Range: e. Panama to extreme nw. Ecuador.

Pteroglossus castanotis   Chestnut-eared Araçari   Arasari Orejicastaño

Uncommon to locally fairly common in the canopy and borders of várzea and riparian forest, along the shores of larger rivers and lakes (frequent on river islands, where it is the most common araçari), and in adjacent clearings in the lowlands and to a limited extent the foothills of e. Ecuador. The Chestnut-eared

Araçari is absent from extensive areas of terra firme forest. Ranges up to about 1000 m along the eastern base of the Andes.

One race: nominate *castanotis*.

Range: se. Colombia, e. Ecuador, e. Peru, n. and e. Bolivia, and w. and cen. Amazonian and sw. Brazil, e. Paraguay, and ne. Argentina.

Pteroglossus pluricinctus   Many-banded Araçari   Arasari Bifajeado

Uncommon to fairly common in the canopy and borders of humid forest (especially terra firme, but also ranging in smaller numbers into várzea and riparian forest) in the lowlands and to a lesser extent the foothills of e. Ecuador. The Many-banded Araçari has been recorded primarily from the drainages of the Ríos Napo and Aguarico, but there are a few old specimens from the Río Pastaza drainage (Sarayacu, Río Bobonaza, and "Laguna de Siguin"; Berlioz 1932a, 1937c) and recent reports from Kapawi Lodge on the Río Pastaza near the Peruvian border; likely the species occurs widely in this region. The Many-banded is usually the most numerous araçari at La Selva and at Sacha Lodge. Ranges mostly below 800 m, a few up in the foothills to about 1200 m.

Monotypic.

Range: s. Venezuela, e. Colombia, e. Ecuador, ne. Peru, and nw. Amazonian Brazil.

Pteroglossus azara   Ivory-billed Araçari   Arasari Piquimarfíl

Uncommon in the canopy and borders of humid forest (principally terra firme) in the lowlands and to a limited extent in the foothills of e. Ecuador. Ranges along the eastern base of the Andes up locally to 900–1000 m (north of Lumbaquí in the Bermejo oil-field area [ANSP] and at Pachicutza in the Río Nangaritza valley [WFVZ, MECN]).

The correct species name is evidently *azara*, not *flavirostris* (Sibley and Monroe 1990), contra Meyer de Schauensee (1966, 1970) and most other recent authors. *Azara* is the name applied to a taxon, found only in a limited area of nw. Amazonian Brazil, that is now considered to be conspecific with the much more wide-ranging *flavirostris* of w. Amazonia; the name *azara* has priority, being two decades older. *P. mariae* (Brown-mandibled Araçari), found south of the Amazon River in w. Amazonia, is closely related to *P. azara*, differing mainly in the mandible's color and pattern; it may well be conspecific with *P. azara*. Seemingly "intermediate" individuals have been recorded, and certain individuals are hard to place in one species or the other. One race of *P. azara* occurs in Ecuador, *flavirostris*.

Range: s. Venezuela, e. Colombia, e. Ecuador, ne. Peru, and nw. Amazonian Brazil.

Pteroglossus inscriptus   Lettered Araçari   Arasari Letreado

Fairly common in the canopy and subcanopy of humid forest and forest borders, secondary woodland, and adjacent clearings in the lowlands of e. Ecuador. The

Lettered Araçari is especially numerous in várzea and riparian forest and woodland, in terra firme ranging principally in borders; it is also more apt than the other araçaris to be found in partially deforested areas with patchy second-growth. Recorded mostly below 500 m, in small numbers up to 650 m near Archidona.

One race: *humboldti*. This taxon was formerly (e.g., in *Birds of the World*, vol. 6), considered a subspecies of *P. viridis* (Green Araçari) of ne. South America.

Range: se. Colombia, e. Ecuador, e. Peru, n. Bolivia, and Amazonian Brazil (south of Amazon River); ne. Brazil.

Andigena laminirostris	Plate-billed Mountain-Toucan	Tucán Andino Piquilaminado

Locally fairly common in the canopy and borders of montane forest in the subtropical and temperate zones on the west slope of the Andes south to Chimborazo; in recent years, however, recorded only northward from w. Cotopaxi (Caripero area; N. Krabbe and F. Sornoza). The splendid Plate-billed Mountain-Toucan, a near endemic, is readily found along both the Nono-Mindo and the Chiriboga roads in Pichincha. Ranges mainly between about 1600 and 2600 m, but has been recorded as high as 3100 m in Imbabura above Apuela (LSUMZ).

The Plate-billed Mountain-Toucan was accorded Near-threatened status by Collar et al. (1994). As substantial tracts of suitable habitat still remain, with some of it being formally protected, and as the species seems to persist quite well in areas where forest cover has been partially fragmented, we question whether this toucan can be considered to be truly at risk. But, as with the often sympatric and equally well known Toucan Barbet (*Semnornis ramphastinus*), we elect to concur with Near-threatened status.

Monotypic.

Range: west slope of Andes in sw. Colombia and nw. Ecuador.

Andigena hypoglauca	Gray-breasted Mountain-Toucan	Tucán Andino Pechigris

Rare to locally fairly common in the canopy and borders of montane forest in the temperate zone on the east slope of the Andes, and locally on slopes above interandean valleys in s. Ecuador (including valleys above Cuenca in Azuay), and extending as far west as the west slope of the Andes in Azuay (Sural and Chaucha; N. Krabbe) and e. El Oro (an old specimen from Taraguacocha). Recorded mostly between about 2500 and 3300 m; an old specimen (ANSP) was recorded as having been obtained at Baeza (1900 m), considerably lower than the species has been found in recent years, and it may actually have been been obtained higher.

The Gray-breasted Mountain-Toucan was given Near-threatened status by Collar et al. (1994). From a purely Ecuadorian perspective we concur.

Ecuadorian birds have usually been referred to the mainly Peruvian race *lateralis* (e.g., by Chapman [1926] and in *Birds of the World* [vol. 6]), though Haffer (1974) assigns them to nominate *hypoglauca*, which is found primarily in Colombia. In fact, both subspecies appear to occur in Ecuador, with more northerly birds (ranging south at least to the Papallacta/Baeza region) showing characters of nominate *hypoglauca* (e.g., a dark iris and orange on the maxilla) and more southerly populations (ranging north at least to the Gualaceo-Limón road) showing characters of *lateralis* (e.g., a pale olive or yellow iris and strongly pink on the maxilla). It is not known if or where these two taxa intergrade; a specimen (ANSP) from Hacienda La Libertad in Cañar does seem somewhat intermediate, having the brown irides of the nominate race and the pink on the mandible of *lateralis*. We do not, even in examples of *lateralis* from Peru, see any appreciable amount of yellow on the flanks (contra Haffer op. cit.).

Range: Andes from s. Colombia to s. Peru.

Andigena nigrirostris	Black-billed Mountain-Toucan	Tucán Andino Piquinegro

Rare to uncommon in the canopy and borders of montane forest in the subtropical zone on the east slope of the Andes. Recorded mainly from w. Napo in the Baeza/Sumaco region, but in recent years small numbers of the Black-billed Mountain-Toucan have also been found considerably farther south; there have been several sightings in Morona-Santiago along the Gualaceo-Limón road and a few (the first on 1–2 Dec. 1991) from Zamora-Chinchipe at Quebrada Avioneta in the Romerillos sector of Podocarpus National Park (Rasmussen et al. 1994) and as far south as the Río Isimanchi valley on the east slope of the Cordillera Las Lagunillas in Nov. 1992 (A. Whittaker). Given its occurrence so close to the Peruvian border, the species seems likely to range south into adjacent northernmost Peru, though it was not recorded from Cerro Chinguela in the early 1980s (Parker et al. 1985). Ranges mostly from 1500 to 2300 m. The two east-slope *Andigena* mountain-toucans thus replace each other altitudinally, and they may even overlap locally. ANSP has old specimens of both species from Baeza (1900 m), though they may of course have been obtained at different elevations in the vicinity of the town; in addition, AMNH has specimens of both species labeled as from "above Baeza."

The Black-billed Mountain-Toucan was given Near-threatened status by Collar et al. (1994), an assessment with which we agree. At least in Ecuador its overall population size appears to be smaller than that of the Gray-breasted Mountain-Toucan (*A. hypoleuca*).

One race: *spilorhynchus*.

Range: Andes from w. Venezuela to se. Ecuador.

Ramphastos vitellinus	Channel-billed Toucan	Tucán Piquiacanalado

Uncommon to locally common (in less disturbed areas) in the canopy and borders of humid forest (especially terra firme) in the lowlands and to a lesser degree the foothills of e. Ecuador. Ranges mostly below 700 m, though small numbers occur at least locally in the foothills along the eastern base of the Andes up to about 1100 m, locally as high as 1400 m on the Cordillera de Cutucú (Jun. 1984 sightings [RSR; M. B. Robbins]).

The form of *R. vitellinus* found in Ecuador, *culminatus*, has often (e.g., Meyer de Schauensee 1966, 1970; Hilty and Brown 1986) been considered a full species (*R. culminatus*, Yellow-ridged Toucan). Following Haffer (1974), however, it is here regarded as only a subspecies of the wide-ranging *R. vitellinus*.

Range: n. and e. Colombia, w. and s. Venezuela, and the Guianas to n. Bolivia and Amazonian and cen. Brazil; e. Brazil. More than one species may be involved.

Ramphastos brevis   Chocó Toucan   Tucán del Chocó

Uncommon to locally common in the canopy and borders of humid forest in the lowlands and foothills of w. Ecuador south to El Oro (Buenaventura; Robbins and Ridgely 1990). An apparently isolated population was found in Dec. 1996 on the Cordillera de Colonche at Loma Alta (Becker and López-Lanús 1997). In many areas the Chocó Toucan is sympatric with, but usually outnumbered by, the Chestnut-mandibled Toucan (*R. swainsonii*). Mainly ranges below 900 m, occasionally or locally up to 1550 m (e.g., above Mindo; ANSP).

Both the Chocó and the Chestnut-mandibled Toucans have declined considerably in Ecuador as a result of widespread deforestation and hunting, but neither species seems as yet truly threatened, in large part because of their relatively large populations in still mainly forested Esmeraldas.

Monotypic.

Range: w. Colombia and w. Ecuador.

Ramphastos swainsonii   Chestnut-mandibled Toucan   Tucán de Swainson

Uncommon to locally common in the canopy and borders of humid forest in the lowlands and foothills of w. Ecuador south to n. Manabí, n. and e. Guayas, and El Oro (Buenaventura and near Balzas; Robbins and Ridgely 1990). Chestnut-mandibled Toucans have also recently been found to occur locally on the slopes of the coastal Cordillera de Colonche in sw. Manabí (Cerro San Sebastián; Parker and Carr 1992) and nw. Guayas (hills south of the lower Río Ayampe, Cerro La Torre, above Salanguilla, and Loma Alta; v. o.). Recorded up to about 1000 m (Buenaventura in El Oro); formerly it may have occurred higher (there are old specimens from Gualea and Pallatanga at 1500 m).

Haffer (1974) considered *swainsonii* as a race of *R. ambiguus* (Black-mandibled Toucan), though most subsequent authors (e.g., Hilty and Brown 1986; Sibley and Monroe 1990) have maintained them as separate species. The two are indubitably closely related, with similar if not identical voices; we consider them to be allospecies. Monotypic.

Range: Honduras to sw. Ecuador.

Ramphastos ambiguus   Black-mandibled Toucan   Tucán Mandíbula Negra

Rare to uncommon and local in the canopy and borders of montane forest in the foothills and lower subtropical zone on the east slope of the Andes. To date, the Black-mandibled Toucan has only been found in n. Ecuador in w. Sucumbíos (below San Rafael Falls; RSR and PJG) and w. Napo (Oyacachi, Baeza, and Volcán Sumaco), and in s. Ecuador in Zamora-Chinchipe (locally on the Cordillera del Cóndor and in the Zamora region, and [ANSP] at Panguri). It has not been recorded from the intervening area, though there is no apparent reason why it should not occur there. The Black-mandibled Toucan and White-throated Toucan (*R. tucanus*) are sympatric along the Loreto road northeast of Archidona, though neither species is numerous and the two have never been seen to flock together. Recorded mostly from 1000 to 1600 m, perhaps in small numbers a little lower and higher (the species may exhibit seasonal altitudinal movements).

Given its patchy distribution, evident rarity, and the substantial deforestation that has taken place across much of its range, we conclude that the Black-mandibled Toucan deserves Near-threatened status in Ecuador. The species as a whole is not considered to be at risk (Collar et al. 1994), though in our view it seems clear that populations in many areas are much reduced.

Does not include *R. swainsonii* (Chestnut-mandibled Toucan), found west of the Andes. Monotypic.

Range: mountains of n. Venezuela, and Andes from w. Venezuela to cent. Peru; Sierra de Perijá.

Ramphastos tucanus   White-throated Toucan   Tucán Goliblanco

Uncommon to locally common (in less disturbed areas) in the canopy and borders of humid forest (largest numbers occurring in terra firme) in the lowlands of e. Ecuador. The White-throated Toucan often occurs with the similarly plumaged Channel-billed Toucan (*R. vitellinus*); despite its larger size, the White-throated is usually the more numerous of the two. Ranges up locally and in small numbers to about 1250 m along the eastern base of the Andes (e.g., along the Loreto road north of Archidona).

One race: *cuvieri*. We follow Haffer (1974) in considering *cuvieri* as a subspecies of *R. tucanus*. *Cuvieri* has often (e.g., Meyer de Schauensee 1966, 1970; Sibley and Monroe 1990) been regarded as a separate species (Cuvier's Toucan). The correct spelling of that taxon is *cuvieri* and not *cuvierii* (contra Hilty and Brown 1986).

Range: e. Colombia, s. and e. Venezuela, and the Guianas to n. Bolivia and Amazonian Brazil.

Picidae   Woodpeckers and Piculets

Thirty-five species in nine genera. Woodpeckers and piculets are worldwide in distribution, though piculets are found only in the tropics.

Picumnus rufiventris   Rufous-breasted Piculet   Picolete Pechirrufo

Rare to locally uncommon—but probably somewhat overlooked because of its inconspicuous behavior—in the lower growth of secondary woodland, terra firme forest (where it favors tangled and viny borders and openings), and shrubby clearings in the foothills and lower subtropical zone on the east slope of the Andes, ranging out in smaller numbers into the adjacent eastern lowlands. In the northeast the Rufous-breasted Piculet is not known to occur any great distance from the Andes, with a few sightings from La Selva and along the Maxus road southeast of Pompeya being the easternmost. In the southeast it may be more wide-ranging for there are reports from Kapawi Lodge near the Peruvian border on the Río Pastaza. Mostly below 1100 m, a few ranging up to 1500 m (Palm Peak on Volcán Sumaco; MCZ).

One race: nominate *rufiventris*.

Range: se. Colombia, e. Ecuador, e. Peru, nw. Bolivia, and w. Amazonian Brazil.

[*Picumnus castelnau*   Plain-breasted Piculet   Picolete Pechillano]

Uncertain. Recorded in Ecuador only from 19th-century specimens labeled as coming from Sarayacu and "Río Napo." Elsewhere in its range the Plain-breasted Piculet is not uncommon in the lower and middle growth of várzea and riparian forest and woodland, particularly on river islands, along the upper course of the Amazon River and a few of its major tributaries. Given the extensive recent field work in seemingly ideal habitat in ne. Ecuador along the Ríos Napo and Aguarico, the absence of reports of this species seems inexplicable—unless it is in fact not present. Further evidence that this may be the case is the fact that the species was not obtained during a lengthy collecting expedition in the 1920s by the Olalla family to the mouth of the Río Curaray into the Río Napo, a site downstream from the present Ecuador-Peru border but which then lay inside Ecuador. We are thus forced to conclude that the Plain-breasted Piculet likely does not occur within the present boundaries of Ecuador. We suspect that the "Sarayacu" specimen was in reality obtained far from that site—as indubitably were numerous other specimens recorded as supposedly having been taken there—and that the "Río Napo" specimen was taken much farther down the Napo in what is now Peruvian territory.

Monotypic.

Range: along Amazon River and some of its tributaries in extreme se. Colombia and ne. Peru; old records supposedly from e. Ecuador.

Picumnus lafresnayi   Lafresnaye's Piculet   Picolete de Lafresnaye

Uncommon to fairly common in the lower and middle growth of humid forest (both terra firme and várzea), secondary woodland, and in adjacent clearings in the lowlands and foothills of e. Ecuador. Ranges up along the east slope of the Andes to about 1200 m, locally as high as 1400 m above San Rafael Falls; in Zamora-Chinchipe it may regularly range higher, having been found at 1575 m at Panguri (ANSP) and even to 1800 m above Valladolid (L. Navarrete and RSR).

Following Short (1982), *P. lafresnayi* of nw. Amazonia is regarded as a species separate from *P. aurifrons* (Bar-breasted Piculet) of Amazonia south of the Amazon River, with which it was formerly considered conspecific. One race occurs in Ecuador, nominate *lafresnayi*.

Range: se. Colombia, e. Ecuador, and ne. Peru.

Picumnus olivaceus   Olivaceous Piculet   Picolete Oliváceo

Uncommon to fairly common in the lower and middle growth of humid forest borders, secondary woodland, and shrubby clearings in the lowlands and foothills of w. Ecuador from Esmeraldas south to n. and e. Guayas (Cerro La Torre and above Salanguilla [N. Krabbe], Loma Alta [Becker and López-Lanús 1997], and Manglares-Churute Ecological Reserve [Pople et al. 1997]), El Oro, and w. Loja (old specimens from Cebollal, but no recent reports). Recorded mainly below about 900 m.

One race: the near-endemic *harterti*.

Range: Guatemala to nw. Venezuela and w. Colombia; extreme sw. Colombia to extreme nw. Peru.

Picumnus sclateri   Ecuadorian Piculet   Picolete Ecuatoriano

Uncommon in the lower growth of deciduous woodland and in arid scrub in the lowlands and foothills of sw. Ecuador, ranging from cen. Manabí to Guayas in the Guayaquil region, and disjunctly from El Oro and Loja south into Peru, in Loja ranging locally as far east as Malacatos (MECN). In Loja (from most of which the Olivaceous Piculet [*P. olivaceus*] is absent) the Ecuadorian Piculet also occurs in more humid lower montane forest and woodland. Although recorded sympatrically with the Olivaceous Piculet at a few localities (e.g., Cerro San Sebastián in sw. Manabí), the Ecuadorian Piwlet favors more arid regions, and even where the two are sympatric they never range in precisely the same habitats. In a few areas—perhaps only in the absence of the Olivaceous Piculet—the Ecuadorian Piculet does range into more humid habitats, such as the semihumid forest and woodland near El Limo in westernmost Loja (RSR). Recorded mostly below 800 m, but ranging up to 1500–1700 m in w. Loja.

Two races are found in Ecuador, *parvistriatus* in Manabí and Guayas and nominate *sclateri* in El Oro and Loja. *Parvistriatus* is somewhat less heavily marked with blackish ventrally (both breast bars and belly streaks being narrower and not so dark).

Range: sw. Ecuador and nw. Peru.

*Colaptes rupicola   Andean Flicker   Picatierra Andino

Uncommon and local in rocky paramo and patches of woodland near treeline in the Andes of extreme s. Ecuador in se. Loja. First recorded from a sighting of a pair observed going to roost in a hole dug into a roadcut on the west slope of the Cordillera Las Lagunillas along the Jimbura-Zumba road on 16 Dec. 1991 (RSR). In Oct. 1992 several specimens (ANSP, MECN) were obtained in the same area. The Andean Flicker had previously been recorded north in Peru to Cruz Blanca in n. Piura (Parker et al. 1985). Recorded between 3100 and 3350 m.

One race: *cinereicapillus*.

Range: Andes from extreme s. Ecuador to n. Chile and nw. Argentina.

Chrysoptilus punctigula   Spot-breasted   Carpintero
                                   Woodpecker   Pechipunteado

Fairly common in trees in clearings, around the margins of lakes, in várzea and riparian forest and woodland, and on river islands in the lowlands and foothills of e. Ecuador. The Spot-breasted Woodpecker appears to be increasing as a result of deforestation, and it certainly is now following clearings up onto Andean slopes where it was formerly absent. It regularly ranges up on the east slope of the Andes to about 1200 m, locally and in small numbers as high as 1600 m (e.g., around El Chaco in w. Napo).

The species has been called the Spot-breasted Flicker (when the genus *Chrysoptilus* is merged into *Colaptes*; see below).

Short (1982) suggested merging the genus *Chrysoptilus* into the relatively uniform genus encompassing the flickers, *Colaptes*. We believe that their differences far outweigh any similarities, however, and favor maintaining *Chrysoptilus*. One race of *C. punctigula* occurs in Ecuador, *guttatus*.

Range: Panama to e. Venezuela and nw. Bolivia and Amazonian Brazil and coastal Guianas, though absent from much of Amazonia north of Amazon River itself.

Piculus rivolii   Crimson-mantled Woodpecker   Carpintero Dorsicarmesí

Uncommon to fairly common and widespread in montane forest and forest borders, persisting in partially deforested areas as well, in the upper subtropical and temperate zones on both slopes of the Andes; locally also occurs on slopes above the central valley (even in and around Quito) and interandean valleys. On the west slope recorded south to El Oro (old specimens from El Chiral, Salvias, and Taraguacocha), but there are no recent reports known to us from south of Azuay. Ranges mostly from 1900 to 3300 m (Cerro Mongus; ANSP), but locally occurs lower in smaller numbers, especially in the west.

One race: *brevirostris*.

Range: Andes from w. Venezuela to w. Bolivia; Sierra de Perijá.

Piculus rubiginosus Golden-olive Woodpecker Carpintero Olividorado

Fairly common in a notably wide variety of habitats in w. Ecuador, ranging in the lowlands, foothills, and lower subtropics on the west slope from Esmeraldas south through w. Loja; habitats range from the borders of very humid foothill forest (e.g., at El Placer in Esmeraldas) through humid and deciduous woodland and forest to arid scrub in w. Guayas, El Oro, and w. Loja. However, the Golden-olive Woodpecker appears to be absent from the lowlands and lower foothills of the wet-forest belt of n. Esmeraldas; it was not recorded in four years of intensive survey work at Playa de Oro (O. Jahn and P. Mena Valenzuela et al.). On the east slope it is less numerous and more local, ranging in the canopy and borders of montane forest in the foothills and lower subtropical zone from at least w. Napo (San Rafael Falls) south to the Peruvian border; it likely occurs north into Colombia as well. In the southwest recorded up to 2100 m (above Celica; P. Coopmans); on the east slope found between about 800 and 2300 m.

Three races are found in Ecuador: *rubripileus* in the west, *buenavistae* on most of the east slope, and what is apparently referable to *coloratus* (or very close to it) in the extreme southeast on the Cordillera del Cóndor. East-slope material is surprisingly limited. This treatment follows Short (1982), in which, however, the taxa *michaelis* and *nuchalis* are not mentioned (both are recorded as occurring in Colombia, and both conceivably also range into Ecuador; see Hilty and Brown 1986, p. 335). *Rubripileus* is somewhat variable but is characterized by having some gray in the center of the male's red crown; both sexes have a mainly black throat and dense dusky to blackish breast barring. *Buenavistae* is larger, and both sexes differ in their somewhat richer bronzy olive upperparts. *Coloratus* is also a large taxon, and differs from *buenavistae* in its more extensively olive-barred belly, and apparently in a mainly gray crown in males (with red only across the nape) and more olive (less bronzy) upperparts.

Range: s. Mexico to w. Panama; n. Venezuela and n. and w. Colombia south, mainly on Andean slopes, to nw. Argentina; locally in mountains of the Guianas and s. Venezuela.

Piculus flavigula Yellow-throated Woodpecker Carpintero Goliamarillo

Rare to uncommon in the mid-levels and subcanopy of humid forest (mainly terra firme) in the lowlands of e. Ecuador. The Yellow-throated Woodpecker appears to be most numerous in the far northeastern lowlands (e.g., in the lower Río Aguarico drainage) and apparently does not occur in areas close to the base of the Andes (e.g., it has never been found at Jatun Sacha; B. Bochan). There is an early record from Sarayacu, but otherwise the only records from the southeast are recent sightings at Kapawi Lodge on the Río Pastaza near the Peruvian border; doubtless, however, the species is more widespread in the Río Pastaza drainage. Recorded only below 300 m.

One race: *magnus*. Some references (e.g., *Birds of the World*, vol. 6) list the subspecies occurring in Ecuador as nominate *flavigula*, but this clearly is not correct, for Ecuadorian males lack the red malar stripe found in males of the nominate race (which ranges in ne. South America).

Range: se. Colombia, s. Venezuela, and the Guianas to ne. Bolivia and Amazonian Brazil; e. Brazil. More than one species is perhaps involved.

Piculus leucolaemus   White-throated Woodpecker   Carpintero Goliblanco

Rare to locally uncommon in the mid-levels and subcanopy of terra firme forest in the lowlands and foothills of e. Ecuador from Sucumbíos southward. There are old specimen records from w. Napo and Morona-Santiago, and the species was recorded in the 1970s from Limoncocha. Otherwise the only recent records are from Taisha (ANSP), along the Loreto road north of Archidona (first on 14 Apr. 1992 by G. H. Rosenberg and B. Finch), in the Bermejo oil-field area north of Lumbaquí (ANSP), Jatun Sacha (only one sighting; B. Bochan), La Selva (sightings from the south bank of the Río Napo [PJG; M. Lysinger]), along the Maxus road southeast of Pompeya (numerous sightings in 1994–1995; RSR et al.), and Miazi in the Río Nangaritza valley (Schulenberg and Awbrey 1997). Until recently the White-throated Woodpecker had not been recorded from adjacent se. Colombia (Hilty and Brown 1986), but in 1998 it was found in the Serranía de Churumbelos in e. Cauca (Salaman et al. 1999). Recorded up to about 1000 m on the eastern base of the Andes (along the Loreto road).

We consider trans-Andean *P. litae* (Lita Woodpecker), *P. simplex* (Stripe-cheeked Woodpecker), and *P. callopterus* (Rufous-winged Woodpecker), the last two of which range in Middle America, as allospecies in the *P. leucolaemus* superspecies, based mainly on their different plumages and disjunct ranges. Short (1982) treats all three taxa as subspecies of *P. leucolaemus*. *P. leucolaemus* is monotypic.

Range: se. Colombia, e. Ecuador, e. Peru, n. Bolivia, and locally in w. and cen. Amazonian Brazil.

Piculus litae   Lita Woodpecker   Carpintero de Lita

Uncommon to locally fairly common in the mid-levels and subcanopy of very humid to humid forest, forest borders, and mature secondary woodland in the lowlands and foothills of nw. Ecuador, where known primarily from Esmeraldas though also found in adjacent Imbabura (Lita area) and nw. Pichincha (north of Simón Bolívar near Pedro Vicente Maldonado, Hacienda San Francisco northwest of San Miguel de los Bancos, 5 km south of Golondrinas, and north of Salto del Tigre; v.o.). Reports from farther south in Pichincha (e.g., at Tinalandia and Río Palenque) remain uncorroborated. Recorded up to about 800 m.

Given the species' small range and the deforestation that has occurred over a substantial portion of it, we consider the Lita Woodpecker to be Near-threatened. The species was not mentioned by Collar et al. (1994), perhaps because of differences in the taxonomy they employed.

P. litae of w. Colombia and nw. Ecuador is here regarded as a monotypic species separate from *P. leucolaemus* based on its smaller size, different plumage pattern, and disjunct distribution.

Range: w. Colombia and nw. Ecuador.

Piculus chrysochloros   Golden-green Woodpecker   Carpintero Verdidorado

Rare to uncommon and apparently local in the mid-levels and subcanopy of humid forest (both terra firme and várzea) and forest borders in the lowlands of e. Ecuador. The Golden-green Woodpecker is perhaps most numerous in the far northeast (e.g., in the lower Río Aguarico drainage). Recorded mostly below 300 m, but locally found as high as 600 m in w. Pastaza (north of Canelos; N. Krabbe and F. Sornoza).

We are uncertain to which race of *P. chrysochloros* Ecuadorian examples of this species should be assigned. They have been referred to *capistratus* of n. South America (e.g., Chapman 1926) but differ strikingly from that form in lacking its red moustachial streak; an ANSP male, also assigned to *capistratus* and obtained at Puerto Umbría in adjacent Putumayo, Colombia, also lacks the red malar. Possibly an undescribed subspecies is involved; more study is needed.

Range: Panama to n. Colombia and nw. Venezuela; se. Colombia, s. Venezuela, and the Guianas to n. and e. Bolivia, nw. Argentina, w. Paraguay, and Amazonian Brazil; locally in se. Brazil.

Celeus elegans   Chestnut Woodpecker   Carpintero Castaño

Uncommon to locally fairly common at all levels in humid forest (both terra firme and várzea) and forest borders in the lowlands of e. Ecuador. Occurs mostly below 500 m, but at least once recorded up to about 700 m in the foothills on the east slope of the Andes in w. Napo (Guaticocha; MCZ).

One race: *citreopygius*.

Range: e. Colombia, s. and e. Venezuela, and the Guianas to n. Bolivia and Amazonian Brazil. More than one species is perhaps involved.

Celeus grammicus   Scale-breasted Woodpecker   Carpintero Pechiescamado

Uncommon to fairly common in the mid-levels and subcanopy of humid forest (especially terra firme but also in várzea) and forest borders in the lowlands of e. Ecuador. Ranges up to about 500 m.

The species has been called the Scaly-breasted Woodpecker.

One race: *verreauxii*.

Range: se. Colombia and s. Venezuela to n. Bolivia and w. and cen. Amazonian Brazil; locally in French Guiana.

Celeus loricatus   Cinnamon Woodpecker   Carpintero Canelo

Uncommon to fairly common in the mid-levels and subcanopy of humid forest and forest borders in the lowlands and foothills of w. Ecuador south locally to n. Manabí, s. Los Ríos (old record from Babahoyo), and se. Guayas (Manglares-Churute Ecological Reserve [RSR and F. Sornoza; Pople et al. 1997]). Recorded up to 800 m.

One race: nominate *loricatus*.

Range: Nicaragua to sw. Ecuador.

Celeus flavus   Cream-colored Woodpecker   Carpintero Flavo

Uncommon in the mid-levels and subcanopy of humid forest and forest borders in the lowlands of e. Ecuador, favoring várzea and riparian forest and wood-land but also occurring regularly in terra firme. The Cream-colored Woodpecker has been recorded primarily from the drainages of the Ríos Napo and Aguarico, but recent reports from Kapawi Lodge on the Río Pastaza near the Peruvian border make it seem likely that the species ranges widely through that drainage as well; it had earlier been recorded from "Laguna de Siguin" (Berlioz 1937c). Recorded mostly below about 400 m, rarely or in small numbers up to 650 m (Archidona area; P. Coopmans).

Short (1982) summarizes information on this "very variable" species, with variation even being described in terms of "color phases," though many birds are intermediates between the bright yellow and duller buff extremes. This variation involves not only overall body color—which can be either brighter yellow or duller buff—but also the extent of cinnamon or buff on the wings and the extent of yellow on the inner secondaries and tertials. We suggest that the best taxonomic arrangement would be to unite all Amazonian populations in the nominate race (synonymizing *peruvianus*, *inornatus*, and *tetricialis*). Short (op. cit., p. 405), despite noting that "it seems futile to delimit subspecies," maintains all three as distinct; in addition there is the very distinct, and geographically isolated, *subflavus* of e. Brazil. Ecuadorian birds were formerly placed either in nominate *flavus* or *peruvianus*, or were considered intermediate between these two taxa.

Range: e. Colombia, s. and e. Venezuela, and the Guianas to n. Bolivia and Amazonian Brazil; locally in e. Brazil.

Celeus spectabilis   Rufous-headed Woodpecker   Carpintero Cabecirrufo

Apparently rare and local in the lowlands of e. Ecuador, where largely over-looked until recently. Sarayacu is the type locality for the species. Only a few additional specimens have been taken in Ecuador (see Norton et al. 1972); these include a male from Sarayacu (Mejía collection), a female from Montalvo (formerly in the Escuela Politécnica collection, now at MECN), a female from Río Conambo (MCZ), and an unpublished female taken by M. Olalla at Cotapino on 20 Oct. 1965 (MECN). In Apr. 1994 this splendid woodpecker was relocated in Ecuador by L. Jost and P. Coopmans at Sacha Lodge, where it was found to be associated with stands of *Cecropia* trees that have an understory of *Gynerium* cane and *Heliconia* in riparian forest on either side of the Río Napo. Subsequent to the realization that this was the habitat of the species here, the Rufous-headed Woodpecker has also been found at La Selva (P. Coopmans et al.), near Pompeya (RSR and F. Sornoza), at Yuturi (S. Howell), and at Kapawi Lodge (first by J. Moore and G. Rivadeneira); we suspect it will be found elsewhere as well. Of interest is the striking difference in the Rufous-headed Woodpecker's preferred habitats between Ecuador and se. Peru, where it has been found almost exclusively in association with bamboo stands. Recorded below 300 m.

One race: nominate *spectabilis*.

Range: e. Ecuador, e. Peru, nw. Bolivia, and extreme w. Brazil; one record from ne. Brazil (same species?).

Celeus torquatus   **Ringed Woodpecker**   **Carpintero Fajeado**

Rare to uncommon in the mid-levels and subcanopy of humid forest (both várzea and terra firme) in the lowlands of ne. Ecuador in the drainages of the Ríos Napo and Aguarico. The handsome Ringed Woodpecker appears to be most numerous in the lowlands well to the east of the Andes (e.g., in the lower Río Aguarico drainage and along the Maxus road southeast of Pompeya; RSR et al.). Two specimens (ANSP) from Imuyacocha taken in Mar. 1991 may be the only ones from Ecuador, though since the mid-1980s there have been sightings from several sites. Ringed Woodpeckers apparently do not occur at all in areas close to the Andes, having, for instance, never been found at Jatun Sacha (B. Bochan). Somewhat surprisingly, there still are no reports from the Río Pastaza drainage. Recorded below about 300 m.

One race: *occidentalis*.

Range: se. Colombia, s. Venezuela, and the Guianas to n. Bolivia and Amazonian Brazil; locally in e. Brazil.

Dryocopus lineatus   **Lineated Woodpecker**   **Carpintero Lineado**

Uncommon to fairly common and widespread in most forested and wooded habitats (persisting well in mainly deforested terrain) in the lowlands of both e. and w. Ecuador. In the west ranges south through El Oro and w. Loja; in the east found primarily in várzea and riparian forest and woodland as well as clearings with scattered trees, but rare or absent in extensive terra firme. Ranges up in the foothills on both slopes of the Andes to at least 1200 m.

Two races are found in Ecuador, nominate *lineatus* in the east and *fuscipennis* in the west. *Fuscipennis* differs quite strikingly in its markedly smaller size, in being browner (less black) generally and with more blurred scaling on the lower underparts, and in having pale brownish flight feathers.

Range: n. Mexico to nw. Peru, n. and e. Bolivia, e. Paraguay, ne. Argentina, and se. Brazil.

Melanerpes cruentatus   Yellow-tufted   Carpintero
                               Woodpecker   Penachiamarillo

Common and conspicuous in the subcanopy and borders of humid forest (both terra firme and várzea) and in clearings, especially favoring dead snags and sometimes around houses or even in towns, in the lowlands of e. Ecuador. The Yellow-tufted is an exceptionally noisy and social woodpecker, and in many parts of the Oriente it is the most frequently recorded member of its family; it is less numerous in extensive terra firme, where it favors snags in treefalls. Found mostly below 900 m, ranging up in decreasing numbers to about 1200 m (e.g., north of Archidona along the road to Loreto), locally as high as 1500 m (e.g., on the south slope of the Cordillera de Huacamayos).

The species has sometimes (e.g., Short 1982) been called the Red-fronted Woodpecker.

This and the next species are sometimes placed in the genus *Centurus*, but we follow most recent authors in retaining them in *Melanerpes*. Following Short (1982), we consider the race *extensus*—in which Ecuadorian birds have by some been placed—to be unrecognizable and consider the species to be monotypic.

Range: e. Colombia, s. Venezuela, and the Guianas to n. Bolivia and Amazonian Brazil.

Melanerpes pucherani   Black-cheeked Woodpecker   Carpintero Carinegro

Fairly common to common and conspicuous in the subcanopy and borders of forest and secondary woodland, in clearings with scattered tall trees (favoring dead snags), and sometimes even in gardens in the more humid lowlands of w. Ecuador south into much of Guayas and to El Oro (south to the Cordillera Larga along the Arenillas-Puyango road, where seen on 18 Sep. 1990; P. Coopmans); not recorded from adjacent w. Loja, but conceivably could occur there. Found mostly below 800 m, but recorded locally up to 1300–1500 m in mainly cleared areas above Mindo (J. Lyons and V. Perez).

Following Short (1982), we consider the characters of the supposed race of n. Middle America, *perileucus*, as being too variable to permit its recognition and consider the species monotypic.

Range: s. Mexico to sw. Ecuador.

Veniliornis fumigatus   Smoky-brown Woodpecker   Carpintero Pardo

Uncommon to fairly common in the lower and middle growth of montane forest and secondary woodland in the foothills and subtropical zone on both slopes of the Andes. In the west ranges south through w. Loja (Alamor/Celica area, Sozoranga, and Utuana). Somewhat surprisingly, the Smoky-brown Woodpecker does not seem to occur on the coastal cordillera. Recorded mostly from 600 to 2400 m, but has been found up to 2700 m above Celica in Loja (C. Rahbek et al.); most numerous between 1000 and 1800 m.

Two similar races are found in Ecuador, nominate *fumigatus* in most of the species' Ecuadorian range (we include *aureus* in this, following J. T. Zimmer, *Am. Mos. Novitates* 1159, 1942) and the dark *obscuratus* in El Oro and w. Loja. J. T. Zimmer (op. cit.) considered the birds of sw. Ecuador to be intermediate between the nominate race and *obscuratus*.

Range: n. Mexico to Panama; mountains of n. Venezuela, and Andes from w. Venezuela to nw. Argentina; Sierra de Perijá and Santa Marta Mountains.

Veniliornis passerinus   Little Woodpecker   Carpintero Chico

Uncommon to locally fairly common in várzea and riparian forest, clearings (sometimes even gardens) with scattered trees, and river islands in the lowlands and foothills of e. Ecuador; in the Zamora region found in patches of semihumid forest and secondary woodland. The Little Woodpecker is basically a nonforest *Veniliornis*, unlike the Red-stained (*V. affinis*). Found mostly below 700 m, ranging up in smaller numbers to 1200–1300 m on the lower slopes of the Andes, principally in association with clearings.

One race: *agilis*.

Range: e. Colombia, s. Venezuela, and the Guianas to n. Argentina, e. Paraguay, and se. Brazil.

Veniliornis affinis   Red-stained Woodpecker   Carpintero Rojoteñido

Uncommon to locally fairly common in the canopy and borders of humid forest (primarily terra firme), in smaller numbers also out into trees in adjacent clearings and in riparian forest and woodland, in the lowlands of e. Ecuador. Recorded mostly below about 600 m, locally as high as 850 m (Miazi in the Río Nangaritza valley; Schulenberg and Awbrey 1997).

Does not include trans-Andean *V. chocoensis* (Chocó Woodpecker); see the discussion under that species. One race of *V. affinis* occurs in Ecuador, *hilaris*.

Range: se. Colombia and sw. Venezuela to n. Bolivia and Amazonian Brazil; locally in e. Brazil. More than one species may be involved.

Veniliornis chocoensis   Chocó Woodpecker   Carpintero del Chocó

Rare to uncommon and apparently local in the canopy and borders of very humid to humid forest in the foothills of nw. Ecuador, where known primarily from Esmeraldas. The first Ecuadorian record involves a single specimen (MCZ) taken by M. Olalla on 27 Mar. 1962 at "La Boca" (Norton et al. 1972); this presumably was the basis for the species being included as occurring in Ecuador by Meyer de Schauensee (1970). The Chocó Woodpecker has since been found at El Placer (ANSP, MECN), above Alto Tambo (one collected on 24 Nov. 1991,

MECN; N. Krabbe), Playa de Oro (where not recorded from the lowlands below about 200 m; O. Jahn et al.), and Bilsa (J. P. Clay et al.). There also has been at least one recent sighting from as far south as nw. Pichincha near Simón Bolívar along the road to Pedro Vicente Maldonado (J. Nilsson, fide M. Lysinger). Recorded mostly between 200 and 700 m.

The Chocó Woodpecker was given Near-threatened status by Collar et al. (1994). Given its very limited overall range, we agree; protected populations do occur in Cotocachi-Cayapas Ecological Reserve and at Bilsa.

The trans-Andean taxon *chocoensis* of w. Colombia and nw. Ecuador was formerly considered (e.g., Meyer de Schauensee 1966, 1970) a subspecies of *V. cassini* (Golden-collared Woodpecker) of ne. South America. L. L. Short (*Auk* 91[3]: 631–634, 1974) transferred *chocoensis* to *V. affinis* (Red-stained Woodpecker), but we conclude that it is better regarded as a separate monotypic species on the basis of its distinct plumage differences from either, as well as its smaller size and disjunct range.

Range: w. Colombia and nw. Ecuador.

Veniliornis kirkii   Red-rumped Woodpecker   Carpintero Lomirrojo

Uncommon to locally fairly common in the lower and middle growth of humid and deciduous forest and woodland and their borders in the lowlands and foothills of w. Ecuador from n. Esmeraldas (north to the Playa de Oro area, where it is widespread in lowlands; O. Jahn et al.) south locally to Guayas, El Oro, and adjacent w. Loja (La Puente, Punta Santa Ana, Puyango). There seemingly are no recent records from Loja. The Red-rumped Woodpecker is quite numerous at Río Palenque and Jauneche. Recorded up to about 1200 m, but only below 300 m in n. Esmeraldas (Playa de Oro).

One race: *cecilii*.

Range: Costa Rica to n. Venezuela and extreme nw. Peru; very locally in s. Venezuela and adjacent Brazil.

Veniliornis dignus   Yellow-vented Woodpecker   Carpintero Ventriamarillo

Uncommon in montane forest and borders in the subtropical and lower temperate zones on the entire east slope of the Andes. Recently also found on the west slope of the Andes in n. Ecuador, first in Carchi along the Maldonado road (Jul.–Aug. 1988 specimens; ANSP, MECN) with another taken at nearby Lomo Laurel on 21 Nov. 1990 (N. Krabbe; MECN). Even more recently, the Yellow-vented Woodpecker has also been seen farther south on the west slope, above Mindo in Pichincha (PJG and D. Wolf; P. Coopmans), and in w. Cotopaxi (Caripero area; N. Krabbe and F. Sornoza). Recorded from 1400 to 2600 m.

Two similar races are found in Ecuador, nominate *dignus* on the west slope and *baezae* on the east slope. *Baezae* is slightly shorter-billed and more extensively barred below.

Range: Andes from extreme sw. Venezuela to s. Peru.

Veniliornis nigriceps   Bar-bellied Woodpecker   Carpintero Ventribarrado

Rare to uncommon in montane forest, forest borders, and woodland in the temperate zone on both slopes of the Andes, also ranging up locally into patches of woodland above treeline (including groves of *Polylepis*) and on slopes above the central and interandean valleys. Recorded mostly from 2800 to 3500 m.

One race: *equifasciatus*.

Range: Andes from n. Colombia to w. Bolivia.

Veniliornis callonotus   Scarlet-backed   Carpintero Dorsiescarlata
                             Woodpecker

Fairly common in deciduous and semihumid forest and woodland and their borders, and in arid scrub, in the lowlands and foothills of w. Ecuador from n. Esmeraldas (lowlands around Playa de Oro; O. Jahn et al.) south through El Oro and Loja. In the past several decades—apparently in response to the deforestation that has taken place—the Scarlet-backed Woodpecker has begun to spread into more humid regions; small numbers, for example, now occur regularly at Río Palenque, and the species was seen for the first time at Tinalandia in 1992 (P. Coopmans). Recorded mostly below about 1000 m, locally up to 1300–1400 m around Mindo (to which it also appears to have spread only recently) and to 1800 m in the Sozoranga area in Loja (Best et al. 1993) and south of Portovelo in s. El Oro (RSR et al.); only below 200 m in n. Esmeraldas (Playa de Oro).

Two races are found in Ecuador, nominate *callonotus* northward (south into Guayas and coastal El Oro) and *major* southward (in El Oro and Loja); there are numerous intergrades between them in an extensive zone of sw. Ecuador, particularly in El Oro. *Major* differs in having a white or whitish postocular stripe that broadens on the side of the neck, separating the brownish auriculars from the red (male) or black (female) crown. *Major* also has, on average, more extensive dusky scaling on the whitish underparts, but it is equalled in this regard by certain examples from well within the range of nominate *callonotus* (e.g., several recent ANSP specimens from sw. Manabí).

Range: w. Ecuador and nw. Peru; very locally in sw. Colombia.

Campephilus melanoleucos   Crimson-crested   Carpintero
                               Woodpecker   Crestirrojo

Fairly common in humid forest (both terra firme and várzea), forest borders, and adjacent clearings (especially where there are large dead trees) in the lowlands and foothills of e. Ecuador. In extensive terra firme forest the Crimson-crested Woodpecker forages mostly at middle heights and in the subcanopy (also often at edge), with the Red-necked (*C. rubricollis*) mainly foraging closer to the ground. Found mostly below 900 m, ranging in smaller numbers on the east slope of the Andes up to at least 1350 m (Guaticocha; MCZ); in Colombia it occurs higher into the subtropical zone (Hilty and Brown 1986).

This and the following four species were formerly placed in the genus *Phloeoceastes*. One race: nominate *melanoleucos*.
Range: Panama to nw. Argentina, n. Paraguay, and s. Brazil.

Campephilus gayaquilensis Guayaquil Carpintero
Woodpecker Guayaquileño

Uncommon to locally fairly common in humid and deciduous forest, forest borders, and adjacent clearings in the lowlands and foothills of w. Ecuador. Ranges up in small numbers to about 1400 m in Pichincha in the Mindo area and at Maquipucuna, and in w. Loja (Alamor region); there are old records from around 1500 m in Chimborazo (Cayandeled and Pallatanga).
The Guayaquil Woodpecker, nearly an endemic of w. Ecuador, has declined considerably because of the conversion of so much forest in its range to agricultural usage, but it appears to persist quite well in decidedly patchy and disturbed habitat, and it does not appear to be immediately threatened. A substantial population exists in Machalilla National Park.
Monotypic.
Range: sw. Colombia to nw. Peru.

Campephilus rubricollis Red-necked Woodpecker Carpintero Cuellirrojo

Rare to uncommon and apparently somewhat local in terra firme forest in the lowlands of e. Ecuador. Recorded mostly below 500 m, ranging locally or in very small numbers along the eastern base of the Andes to about 900 m (old records [Salvadori and Festa 1900] from "Valle de Zamora" and "Valle del Río Santiago").
One race: nominate *rubricollis*.
Range: se. Colombia, s. Venezuela, and the Guianas to n. Bolivia and Amazonian Brazil.

Campephilus pollens Powerful Woodpecker Carpintero Poderoso

Rare to uncommon and perhaps local in montane forest and forest borders in the subtropical and temperate zones on both slopes of the Andes. On the west slope ranges south to El Oro (old records from El Chiral and Salvias; no recent reports). Recorded mainly from 1700 to 2600 m, occasionally somewhat lower or as high as 3000–3500 m (N. Krabbe and others have found it with some regularity at Yanacocha, 3500 m).
One race: nominate *pollens*.
Range: Andes from extreme sw. Venezuela to cen. Peru.

Campephilus haematogaster Crimson-bellied Carpintero
Woodpecker Carminoso

Rare to locally uncommon in the lower growth of montane forest in the foothills and subtropical zone on the east slope of the Andes, and in the lower growth of humid forest in the lowlands and foothills of nw. Ecuador where recorded from Esmeraldas (including Bilsa, where seen in Oct. 1996; D. and M. Wolf), adjacent Imbabura, and Pichincha (south to Tinalandia, where a pair persisted into at least the early 1990s; v.o.). On the east slope recorded north to w. Sucumbíos (San Rafael Falls; v.o.); the species has not been recorded on the east slope of the Andes in adjacent Colombia (Hilty and Brown 1986), though it almost surely occurs there as one record exists from much farther north, near the Venezuelan border. On the east slope recorded between about 1000 and 1700 m; in the west found up to about 800 m (an old specimen from as high as "near Gualea" is regarded as of uncertain provenance).

Two distinctly different races are found in Ecuador, nominate *haematogaster* in the east and *splendens* in the west. The nominate race differs in being redder and less barred with dusky brown below, and in having the black throat patch extending down over the upper chest. These two forms have been considered separate, monotypic species (with *C. splendens* being called the Splendid Woodpecker), and as they also appear to differ vocally, this treatment may well be correct.

Range: Panama, n. and w. Colombia, and nw. Ecuador; east slope of Andes in n. Colombia, and from n. Ecuador to s. Peru.

Passeriformes
Furnariidae   Ovenbirds

Seventy-eight species in 31 genera. Ovenbirds are exclusively Neotropical in distribution. We follow Gill (1994) in considering the Woodcreepers (Dendrocolaptidae) as a separate family; in recent years they have often been included in the Furnariidae.

Geositta tenuirostris   Slender-billed Miner   Minero Piquitenue

Apparently a rare and local resident in open arid terrain in the upper temperate and paramo zones of cen. Ecuador. Discovered in Ecuador by N. Krabbe, who on 7 Aug. 1990 collected two and saw several others on the Planados de Guintza in a high arid valley on the slopes of Volcán Iliniza in cen. Cotopaxi (ZMCOP, MECN). A pair was subsequently obtained in disturbed paramo adjacent to small agricultural fields west of Pujilí, also in Cotopaxi, on 28 Jul. 1991 (ANSP, MECN). N. Krabbe (pers. comm.) has since made additional sightings from a few other sites in the highlands of Cotopaxi and Chimborazo (Loma Cholla Pungo, Loma Totol, and near Guamote). The Slender-billed Miner is not otherwise known from north of Cajamarca in Peru, so the discovery of this highly disjunct population came as a distinct surprise. Recorded from 3350 to 4000 m.

Ecuadorian birds were described as the race *kalimayae* (N. Krabbe, *Bull. B. O. C.* 112[3]: 166–169, 1992).

Range: Andes of cen. Ecuador, and from n. Peru to extreme n. Chile and nw. Argentina.

Cinclodes fuscus   Bar-winged Cinclodes   Cinclodes Alifranjeado

Fairly common to common in paramo and grassy areas, usually near water, in the Andes from Carchi south through Loja and adjacent Zamora-Chinchipe. Recorded mostly between 3200 and 4300 m, a few up to 4500 m.

One race: *albidiventris*. *Paramo*, the very similar race that occurs in Nariño, Colombia, was noted by Fjeldså and Krabbe (1990) as occurring in extreme n. Ecuador; we have seen no specimens of *C. fuscus* from the area in question.

Range: Andes from w. Venezuela to Tierra del Fuego, and in lowlands of Argentina and Chile.

Cinclodes excelsior   Stout-billed Cinclodes   Cinclodes Piquigrueso

Uncommon to fairly common in paramo, low shrubbery, and patches of *Polylepis* woodland, most often near water, in the Andes from Carchi south to n. Azuay (El Cajas National Recreation Area). Fjeldså and Krabbe (1990) state that the species ranges south into Loja, but we are not aware of any specific record from that province. The Stout-billed Cinclodes appears to be more numerous than the Bar-winged (*C. fuscus*) at very high elevations, though the two species occur together in many places (e.g., at Papallacta pass and in Cotopaxi National Park). Recorded between about 3300 and 4500 m.

Vaurie (1980) placed *excelsior* in the genus *Geositta*, but from its behavior, morphology, and voice this generic alignment is surely incorrect. One race occurs in Ecuador, nominate *excelsior*.

Range: Andes of Colombia and Ecuador. Excludes Royal Cinclodes (*C. aricomae*) of se. Peruvian Andes as a separate species.

Furnarius cinnamomeus   Pacific Hornero   Hornero del Pacífico

Common to very common on or near the ground in a variety of open and semi-open habitats, often but by no means always near water (though mud is required for nesting), in the lowlands of w. Ecuador from w. Esmeraldas, sw. Imbabura, and n. Pichincha southward; in Loja and Azuay also ranges in semiarid agricultural terrain well up into the subtropical zone. The Pacific Hornero is one of the more numerous and conspicuous birds of arid sw. Ecuador, where it struts about boldly on roads and even in towns. In recent decades it has taken advantage of the widespread clearing of humid forest to spread into formerly unoccupied terrain, though it never is as numerous in humid regions as where conditions are more arid. It continues to spread northward in nw. Ecuador, and its range may eventually extend as far as sw. Colombia. In recent decades the Pacific Hornero has also been found in the upper Río Marañón valley of nw. Peru, and it seems possible that it may in due course spread north as far as the Zumba region in Ecuador. Recorded widely up to 1500–2000 m in the southwest, and occurs regularly around the city of Loja; locally it is found up to 2300–2500 m in s. Loja.

F. cinnamomeus is here treated as a species separate from *F. leucopus* (Pale-legged Hornero) of Amazonia and e. Brazil, as it was earlier by Parker and Carr (1992). This split is based on *cinnamomeus*'s disjunct distribution, morphological differences from *F. leucopus*, and various behavioral and vocal differences. Monotypic.

Range: w. Ecuador and nw. Peru.

[Furnarius torridus   Bay Hornero   Hornero Castaño]

Very rare (wanderer?) on or near the ground in várzea forest and woodland and their borders along the lower Río Lagartococha in the lowlands of far ne. Ecuador. The Bay Hornero was first reported in Ecuador in Mar. 1991, when birds were seen on two occasions near the military camp of Lagartococha (RSR and M. B. Robbins); no specimens could be taken. The species also seems likely to occur on river islands in the Río Aguarico downstream from this point, but as yet it has not been found there.

The species has usually been called the Pale-billed Hornero, despite the fact that its bill is no paler than that of the Pale-legged Hornero (*F. leucopus*) of Amazonia and e. Brazil. *F. torridus* is, however, a much darker ("bay") bird generally; see Ridgely and Tudor (1994).

Monotypic. *Torridus* has sometimes been considered as only a dark morph of *F. leucopus* (e.g., Vaurie 1980), but we follow J. T. Zimmer (*Am. Mus. Novitates* 860, 1936) in considering it a full species. Recent data from ne. Peru supports this; see, e.g., G. H. Rosenberg (*Condor* 92[2]: 427–443, 1990).

Range: along Amazon River and several major tributaries in w. Brazil, ne. Peru, and extreme ne. Ecuador.

Furnarius minor   Lesser Hornero   Hornero Menor

Rare to uncommon on or near the ground in relatively early-succession growth on islands in the Río Napo in the lowlands of ne. Ecuador. First recorded in Ecuador only in 1975–1976 (Tallman and Tallman 1977), this small and relatively inconspicuous hornero has since been found on a number of islands in the Río Napo upriver to the vicinity of Hacienda Primavera and Limoncocha (including several near La Selva and Sacha Lodge). At least as yet it has not been found along other rivers such as the lower Río Aguarico; there are also no records from along the Río Pastaza. Recorded below 300 m. Monotypic.

Range: along Amazon River and several major tributaries in Brazil, ne. Peru, and ne. Ecuador.

Leptasthenura andicola   Andean Tit-Spinetail   Tijeral Andino

Uncommon to fairly common in shrubbery and low woodland near treeline, patches of *Polylepis* woodland, and paramo with scattered bushes in the Andes from Carchi south to n. Azuay (El Cajas National Recreation Area). In Feb. 1996 an isolated population was also found on the Cordillera Las Lagunillas of s. Zamora-Chinchipe (Krabbe et al. 1997); here, curiously, the species had gone undetected during earlier intensive surveys. The absence of the Andean Tit-Spinetail from the Andes of most of s. Ecuador, where seemingly suitable terrain is widespread, is puzzling. Recorded mostly from 3200 to 4000 m.

One race: nominate *andicola*. It is possible that the isolated Cordillera Las Lagunillas population may represent an undescribed subspecies (fide N. Krabbe).

Range: Andes from w. Venezuela to s. Ecuador, and locally from n. Peru to w. Bolivia; Santa Marta Mountains.

Synallaxis azarae   Azara's Spinetail   Colaespina de Azara

Fairly common to common in shrubby forest and woodland borders, regenerating clearings, and hedgerows and patches of woodland in the upper tropical, subtropical, and temperate zones on both slopes of the Andes, and on slopes above and even locally in the central and interandean valleys; however, the species is strikingly absent from outlying ridges such as the Cordilleras de Cutucú and del Cóndor. The Azara's Spinetail is one of the more widespread and ecologically tolerant montane birds in Ecuador; it occurs in both humid and fairly dry situations. Recorded mostly from 1500 to 3000 m, but ranging much lower in the south; it occurs regularly down to about 900 m at Buenaventura in El Oro (occasionally down as low as 650 m there; P. Coopmans) and is also found as low as 900–1000 m in the Zumba region of the extreme southeast.

Birds found from w. Venezuela to n. Peru were formerly sometimes specifically separated as *S. elegantior* (Elegant Spinetail), despite their being very similar both vocally and morphologically to *S. azarae* (aside from *elegantior*'s usually having 10 instead of 8 rectrices). Two races of *S. azarae* occur in Ecuador, *media* in most of the species' Ecuadorian range and *ochracea* in the southwest (north to s. Chimborazo). *Ochracea* averages longer-tailed, has 8 (not 10) rectrices, and is somewhat paler (less gray) below; some individuals show a pale superciliary.

Range: Andes from w. Venezuela to nw. Argentina. Includes what was formerly (e.g., Meyer de Schauensee 1966, 1970) a separate species, Buff-browed Spinetail (*S. superciliosa*) of s. Bolivia and nw. Argentina.

Synallaxis moesta   Dusky Spinetail   Colaespina Oscura

Uncommon to fairly common but seemingly local in the dense undergrowth of humid forest borders and secondary woodland in the foothills along the east slope of the Andes and adjacent lowlands of e. Ecuador. The localities farthest away from the Andes where the Dusky Spinetail has definitely been recorded are La Selva (repeated sightings and tape-recordings since 1993 from thickets around Garzacocha) and Kapawi Lodge on the Río Pastaza near the Peruvian border. It seems especially numerous along the westernmost part of the Loreto road north of Archidona. Recorded from about 250 to 1350 m.

One race: *brunneicaudalis*.

Range: foothills and adjacent lowlands along eastern base of Andes in Colombia, Ecuador, and n. Peru. Excludes Cabanis's Spinetail (*S. cabanisi*) of Peru and Bolivia, and Macconnell's Spinetail (*S. macconnelli*) of s. Venezuela, the Guianas, and adjacent Brazil, both of which are sometimes considered conspecific.

Synallaxis brachyura   Slaty Spinetail   Colaespina Pizarrosa

Fairly common to common in shrubby forest and woodland borders, clearings, grassy pastures, and gardens in the more humid lowlands and foothills of w. Ecuador south to nw. Guayas and through El Oro (various localities); there appear to be no records from adjacent w. Loja, where the species seems likely to occur as it has been found in adjacent Peru. Recorded mainly below 1400 m, up in small numbers in cutover terrain to around 1750 m.

Two races are found in Ecuador, *chapmani* in the northwest and *griseonucha* in the southwest, distinguished in agreement with characters given by Wiedenfeld et al. (1985, p. 309); based on ANSP material, these taxa appear to intergrade in e. Guayas and nw. Azuay. Wiedenfeld et al. (op. cit.) applied the name of *chapmani* to the birds of sw. Ecuador and adjacent Peru, but the type locality of this dark taxon is actually Jiménez in Valle, Colombia. We thus follow Wetmore (1972) in considering *chapmani* to be the name applicable to the nw. Ecuadorian population, restricting *nigrofumosa* to the area from Honduras to w. Panama. We agree with Chapman (1926) that the name *griseonucha* is applicable to the relatively pale southwestern birds. *Griseonucha* differs from *chapmani* in being somewhat paler generally, especially on its underparts where there is more silvery on the throat and whitish on the midbelly.

Range: Honduras to extreme nw. Peru.

Synallaxis albigularis   Dark-breasted Spinetail   Colaespina Pechioscura

Fairly common to common in grassy pastures, overgrown clearings, woodland borders, and early-succession riparian growth on river islands in the lowlands and foothills of e. Ecuador, now following clearings up the east slope of the Andes. Recorded commonly up to about 1500 m and found in smaller numbers up to at least 1800 m.

Two races are found in Ecuador: *rodolphei* (see J. Bond, *Proc. Acad. Nat. Sci. Phil.* 108: 244–245, 1956) in the north and nominate *albigularis* southward. Some intergradation apparently occurs along a zone that roughly parallels the upper Río Napo. *Rodolphei* is darker gray below than the nominate race, with its midbelly usually contrastingly whitish.

Range: se. Colombia to extreme nw. Bolivia and w. Amazonian Brazil.

The existence of the Pale-breasted Spinetail (*S. albescens*) on some Ecuadorian bird lists is only due to its having once been considered a subspecies of the

Dark-breasted Spinetail (e.g., Chapman 1926). The Pale-breasted Spinetail does occur as close as the arid Río Patía valley in sw. Colombia (Hilty and Brown 1986), but it seems unlikely to range as far south as Ecuador.

Synallaxis gujanensis   Plain-crowned   Colaespina Coroniparda
                           Spinetail

Fairly common in the undergrowth of riparian woodland and early-succession scrub, favoring areas with *Gynerium* cane under stands of *Cecropia* trees, in the lowlands of ne. Ecuador on islands in the Río Napo (upriver to near Jatun Sacha; B. Bochan) and the lower Río Aguarico (upriver to near the mouth of the Río Lagarto). It has not, however, been found along the Río Pastaza. The Plain-crowned Spinetail was first found in Ecuador on 21 Aug. 1986, when at least one pair was encountered on an island near San Roque (RSR et al.); since then the species has proven to be widespread in appropriate habitat along the Río Napo. A short series was obtained (ANSP, MECN) on a Río Napo island near the mouth of the Río Aguarico on 19–21 Sep. 1992, the only specimens available from Ecuador. Recorded up to about 400 m.

Ecuadorian birds are probably referable to the race *huallagae* of ne. Peru; several of the described races of *S. gujanensis* resemble each other very closely, and there is insufficient material at ANSP to make a definitive diagnosis.

Range: se. Colombia to n. Bolivia, Amazonian Brazil, the Guianas, and s. Venezuela. Excludes White-lored (or Pantanal) Spinetail (*S. albilora*) of sw. Brazil and adjacent Bolivia and Paraguay, Bananal Spinetail (*S. simoni*) of s.-cen. Brazil, and Marañón Spinetail (*S. maranonica*).

Synallaxis maranonica   Marañón Spinetail   Colaespina de Marañón

Uncommon in the undergrowth of deciduous and semihumid forest and woodland in the Río Marañón drainage of s. Zamora-Chinchipe in the vicinity of Zumba, and north to Palanda near Valladolid. The Marañón Spinetail was first recorded on 7 Sep. 1991, when one was mist-netted and photographed at Palanda (R. Williams et al.). On 12–14 Dec. 1991, at least one pair was found near the Río Mayo east of La Chonta, and a fairly large population was located in the drainage of the lower Río Ishimanchi; a small series was obtained then (ANSP and MECN), the only Ecuadorian specimens. Several observers have found the species in this region on subsequent occasions. Recorded from 650 to 1200 m.

S. maranonica is regarded as a species distinct from *S. gujanensis* (Plain-crowned Spinetail) of Amazonia on the basis of its disjunct distribution, different plumage, and different song. Monotypic.

Range: upper Río Marañón valley of nw. Peru and extreme s. Ecuador.

Synallaxis propinqua   White-bellied Spinetail   Colaespina Ventriblanca

Uncommon to fairly common in low early-succession growth on islands in the larger rivers of the lowlands of e. Ecuador. To date recorded primarily from islands in the Río Napo upriver to the Primavera vicinity, with a single (wandering?) individual having been present as far upriver as near Jatun Sacha in Mar.–Apr. 1994 (B. Bochan), and from islands in the lower Río Aguarico upriver to just below the mouth of the Río Cuyabeno (P. Coopmans et al.). The White-bellied Spinetail also occurs locally along the Río Pastaza; there is a single specimen from Sarayacu (Orcés 1974) and there are recent reports from Kapawi Lodge near the Peruvian border. The species likely occurs all along the Río Pastaza in appropriate habitat. Recorded mainly below 300 m, rarely as high as 400 m (Jatun Sacha).

Monotypic.

Range: locally on islands in Amazon River and some of its tributaries from e. Ecuador and e. Peru to e. Amazonian Brazil.

Synallaxis tithys   Blackish-headed   Colaespina Cabecinegruzca
                           Spinetail

Rare to uncommon and local in the undergrowth of deciduous forest and woodland in the more arid lowlands of sw. Ecuador in sw. Manabí (north to the road to San Clemente, south of Bahía de Caráquez; ANSP) and Guayas, and in coastal El Oro and w. Loja (east to the Macará/Sabiango region). Recorded up to 1100 m.

Overall numbers of the Blackish-headed Spinetail have declined in the past several decades because of widespread habitat destruction in many areas and the elimination by overgrazing of undergrowth in much of what wooded habitat remains. The species was given Vulnerable status by Collar et al. (1994), and we agree with this assessment. A substantial population of the species is found in Machalilla National Park in Manabí, it occurs in the Chongón Hills (including the Cerro Blanco property), and it remains relatively widespread in various parts of sw. Loja.

Monotypic.

Range: sw. Ecuador and extreme nw. Peru.

Synallaxis unirufa   Rufous Spinetail   Colaespina Rufa

Uncommon to locally fairly common in the undergrowth of montane forest and forest borders in the upper subtropical and temperate zones on both slopes of the Andes. On the west slope recorded south to w. Cotopaxi (west slope of Volcán Iliniza [N. Krabbe] and above Pilaló [RSR and G. H. Rosenberg]). Favors thickets of *Chusquea* bamboo. Recorded mostly from 2200 to 3200 m, locally as low as 2000–2100 m on the Cordillera de Huacamayos.

One race: nominate *unirufa*.

Range: Andes from w. Venezuela to s. Peru; Sierra de Perijá. Excludes Black-throated Spinetail (*S. castanea*) of mountains of n. Venezuela.

*Synallaxis rutilans   Ruddy Spinetail   Colaespina Rojiza

Rare to uncommon and evidently local in the undergrowth of terra firme forest in the lowlands of e. Ecuador; there are still only a few Ecuadorian records.

The first Ecuadorian record is a female (AMNH) taken at the mouth of the Río Lagarto Cocha (J. T. Zimmer, *Am. Mus. Novitates* 861, 1936). ANSP has two specimens that were obtained by R. Olalla at Río Conambo in Mar. 1952. Despite these earlier records, however, Ecuador was inadvertently omitted from the species' range as given by Meyer de Schauensee (1966, 1970). Another specimen (MECN) was taken by M. Olalla on 1 Feb. 1963 at Puerto Libre in Sucumbíos (Orcés 1974). The only recent confirmed records of the Ruddy Spinetail are from La Selva (where first seen and tape-recorded by M. Hedemark and A. Johnson), Cuyabeno (Aug. 1992 sightings and tape-recordings; B. Whitney et al.), Zancudococha (ANSP), along the Maxus road southeast of Pompeya (a few 1994 and 1995 sightings [N. Krabbe, C. Canaday, and RSR]), and Kapawi Lodge along the Río Pastaza near the Peruvian border. The Ruddy Spinetail apparently does not occur in areas close to the base of the Andes (it has never, for example, been found at Jatun Sacha; B. Bochan). Recorded only below 250 m.

One race: *caquetensis*.

Range: se. Colombia, s. Venezuela, and the Guianas to n. Bolivia and Amazonian Brazil.

Synallaxis cherriei   Chestnut-throated Spinetail   Colaespina Golicastaña

Very rare and local in the undergrowth of secondary woodland and forest borders in the foothills and adjacent lowlands along the eastern base of the Andes in w. Napo and w. Sucumbíos. Known in Ecuador from only three specimens: a female taken by "Olalla y Hijos" in Mar. 1921 labeled only as from "near Río Napo," the type of *napoensis* (N. Gyldenstolpe, *Ark. Zool.* 21A, no. 25, 1930); a male (UKMNH) obtained at Santa Cecilia (340 m) on 5 Jul. 1971; and a male (ANSP) obtained by F. Sornoza and RSR along the Inchillaqui road (650 m) near Archidona on 11 Apr. 1993. P. Coopmans discovered the small population near Archidona in Sep. 1989; several pairs were present in the early 1990s. In the similar Ruddy Spinetail (*S. rutilans*), the black throat patch is often difficult to discern in the field, and this has resulted in a few difficulties in distinguishing the two species; indeed for a period they were confused taxonomically as well. The birds occurring at La Selva were, for instance, first reported as Chestnut-throated Spinetails; however, the Ruddy is now definitely known to be the only species involved. Also, RSR now believes that his 1976 sightings from Zancudococha, believed at the time to refer to the Chestnut-throated Spinetail (see Hilty and Brown 1986), also more likely refer to the Ruddy (which was recently collected there). Although it remains possible, we feel that additional documentation is needed to confirm the presence of the Chestnut-throated Spinetail in Ecuador at any distance east of the Andes. Recorded from about 300 to 900 m.

The Chestnut-throated Spinetail was given Near-threatened status by Collar et al. (1994), after having earlier been given a rating of Insufficiently Known (Collar et al. 1992). We question whether the species should be considered at risk in Ecuador, mainly because all or virtually all of the (admittedly very few) records come from secondary habitats, and it seems safe to assume that such habitats will only increase in the years to come. Nonetheless, given the species' evident rarity, we will concur with Near-threatened status for the country. Obviously a high priority for future investigators will be to determine what factors are governing, and presumably limiting, its distribution. No population presently is known to occur in a formally protected area.

Few examples of this scarce spinetail have ever been obtained, and subspecific determinations have therefore remained conjectural. Ecuadorian birds are referable to *napoensis*, for which "near Río Napo" represents the type locality. The recently obtained Santa Cecilia and Archidona specimens are considerably darker than a series (ANSP) that M. A. Carriker, Jr., obtained in Oct. 1933 at Moyobamba in n. Peru. These latter have been assigned to *napoensis* (see *Birds of the World* vol. 7), but in fact they appear quite different from the one example we have at hand from Ecuador: they are markedly paler generally, being browner above, paler and more orange-rufous on the throat and breast, and notably paler gray on the belly. We conclude that the name that was originally given to them, *saturata* (M. A. Carriker, Jr., *Proc. Acad. Nat. Sci. Phil.* 86: 321–322, 1934), is indeed applicable to the Moyobamba series (and probably to birds found elsewhere in Peru as well), and that they are separable from Ecuadorian Birds. The name *saturata* had been synonymized into *napoensis* (see *Birds of the World*, vol. 7).

Range: very locally in extreme se. Colombia, e. Ecuador, e. Peru, and s. Amazonian Brazil.

Synallaxis stictothorax   Necklaced Spinetail   Colaespina Collareja

Fairly common in arid scrub (even in desert-like areas) and borders of deciduous woodland in the arid lowlands of sw. Ecuador in sw. Manabí (north along the coast to the Bahía de Caráquez area; AMNH, ANSP) and w. Guayas, extreme s. coastal El Oro (Huaquillas region), and extreme s. Loja (Zapotillo region; ANSP). The Necklaced Spinetail is numerous and widespread on the Santa Elena Peninsula in w. Guayas and also at lower levels in Machalilla National Park in Manabí. Recorded only below about 200 m.

Although traditional, the generic allocation of *stictothorax* (and the closely allied *chinchipensis* of the Río Marañón valley of nw. Peru) to *Synallaxis* seems likely to be incorrect, as first noted in *Birds of the World* (vol. 7). The general behavior, voice, nest, and (fide L. Kiff) eggs of *stictothorax* (and presumably also of *chinchipensis*) are more like that of a *Cranioleuca* spinetail, whereas the plumage and vocalizations (fide P. Coopmans) of both taxa resemble those of the monotypic genus *Siptornopsis* of n. Peru. In default of definitive information, however, we still retain them in *Synallaxis*. Two races are found in Ecuador: nominate *stictothorax* in most of the species' Ecuadorian range, with *maculata* having recently been found in s. Loja in the Zapotillo region (ANSP). The population from s. coastal El Oro (of which there is no material) may also be referable to *maculata*. *Maculata* differs from the nominate race in having a mainly rufous tail, with only the tips of the central rectrices dusky; the central rectrices are entirely dusky in the nominate race.

Range: sw. Ecuador and nw. Peru.

The Chinchipe Spinetail (*S. chinchipensis*) of deciduous woodland and scrub in the upper Río Marañón valley in nw. Peru ranges quite close (as far north as near Chirinos, fide P. Coopmans) to the Ecuadorian border south of Zumba. It may occur north into Ecuadorian territory.

Hellmayrea gularis   White-browed Spinetail   Colaespina Cejiblanca

Uncommon to fairly common in the undergrowth of montane forest in the temperate zone on both slopes of the Andes and locally on slopes above the central and some interandean valleys (e.g., the Río Mazán valley near Cuenca; King 1989). On the west slope, though previously known south only to Pichincha, the White-browed Spinetail is now known to occur, in appropriate habitat, south through Bolívar to Azuay (Chaucha; N. Krabbe). Recorded from about 2500 to 3700 m.

Sometimes placed in the genus *Synallaxis* (e.g., Meyer de Schauensee 1966, 1970; Vaurie 1980), but M. Braun and T. A. Parker III (*Neotropical Ornithology*, Ornithol. Monogr. no. 36: 333–346, 1985) present evidence for continuing to separate the species in the monotypic genus *Hellmayrea*. One race occurs in Ecuador, nominate *gularis*.

Range: Andes from w. Venezuela to cen. Peru; Sierra de Perijá.

Cranioleuca curtata   Ash-browed Spinetail   Colaespina Cejiceniza

Uncommon to fairly common in montane forest and secondary woodland, favoring tangles in the canopy and borders, in the foothills and subtropical zone on the east slope of the Andes. Recorded mostly from 900 to 1700 m, perhaps regularly occurring lower (down to 650 m) in the Río Marañón drainage near Zumba (RSR; P. Coopmans). There are even a few reports from Jatun Sacha (400 m) in the adjacent eastern lowlands (at least 4 sightings, all from the period between Aug. and Nov.; B. Bochan et al.); these perhaps represent postbreeding birds that have wandered downslope.

Vaurie (1980) merged the genus *Cranioleuca* into *Certhiaxis*, though this course has not been followed by any subsequent author, and seems certainly incorrect. One race of *C. curtata* occurs in Ecuador, *cisandina*.

Range: Andes from n. Colombia to w. Bolivia.

The Fork-tailed Spinetail (*C. furcata*), believed by C. Vaurie (*Ibis* 113: 517–519, 1971) to represent a valid species (and thus it is on a few Ecuadorian bird lists), has been demonstrated to be the immature plumage of the Ash-browed Spinetail (G. R. Graves, *Condor* 88[1]: 120–122, 1986). A specimen (ANSP) obtained at Panguri in s. Zamora-Chinchipe on 21 Jul. 1992 (T. J. Davis) is a classic example of "*furcata*," and it indeed has an unossified skull and a large bursa.

Cranioleuca erythrops   Red-faced Spinetail   Colaespina Carirroja

Uncommon to locally common in montane forest and secondary woodland, favoring tangles in the canopy and borders, in the foothills and lower subtropical zone on the west slope of the Andes south to e. Guayas and nw. Azuay (above Manta Real; Parker and Carr 1992). Also found locally on the coastal Cordillera de Mache at Bilsa (J. Hornbuckle; D. Wolf) and on the Cordillera de Colonche in sw. Manabí and w. Guayas. Contra Fjeldså and Krabbe (1990), we can find no corroborated evidence that the Red-faced Spinetail ranges as far south as El Oro. Recorded mostly from 700 to 1500 m, in small numbers up to 1850 m at Maquipucuna (P. Coopmans) and very locally as low as 150 m in the hills south of the Río Ayampe in nw. Guayas.

One race: nominate *erythrops*.

Range: highlands of Costa Rica and Panama; west slope of Andes in Colombia and Ecuador (where also on coastal cordillera).

Cranioleuca antisiensis   Line-cheeked Spinetail   Colaespina Cachetilineada

Fairly common to common in montane forest, secondary woodland, and agricultural areas with hedgerows and scattered trees and montane scrub in the subtropical and temperate zones on the west slope of the Andes, and locally above interandean valleys, in sw. Ecuador. Ranges from n. Azuay (north to around Gualaceo; numerous recent sightings) south through El Oro and Loja; in Loja ranges east as far as slopes below the Continental Divide (e.g., at Cajanuma and above Jimbura). In Aug. 1996 the Line-cheeked Spinetail was also reported to occur near the summits of hills in Manglares-Churute Ecological Reserve in se. Guayas (Pople et al. 1997). Records from farther north (see Ridgely 1980; Kirwan et al. 1996) are now regarded as being uncertain because of the realization that confusion was possible with the juvenile and immature plumages of the Red-faced Spinetail (*C. erythrops*), which can resemble the Line-cheeked. Whether the Line-cheeked and Red-faced Spinetails ever come into contact along the west slope of the Andes (e.g., in sw. Azuay) thus remains unknown. The Line-cheeked Spinetail is particularly numerous and widespread in w. Loja. Recorded mostly from 1000 to 2500 m, in smaller numbers up to 2900 m.

The species is sometimes called the Northern Line-cheeked Spinetail.

C. antisiensis of sw. Ecuador and extreme nw. Peru is regarded as a species distinct from *C. baroni* (Baron's Spinetail) of n. and cen. Peru, on the basis of its much smaller size and differences in plumage and voice. One race occurs in Ecuador, nominate *antisiensis*.

Range: Andes of s. Ecuador and nw. Peru.

*Cranioleuca vulpecula   Parker's Spinetail   Colaespina de Parker

Fairly common in the undergrowth of riparian woodland and early-succession scrub on river islands in the lowlands of e. Ecuador. The Parker's Spinetail is best known from islands in the Río Napo upriver to near Jatun Sacha (B. Bochan), and it has also been found along the lower Río Aguarico upriver to near Lagartococha. There are also recent reports of it from islands in the Río Pastaza near Kapawi Lodge close to the Peruvian border, and the species probably ranges widely along that river as well. The first Ecuador record is based on a specimen (MECN) obtained by M. Olalla on 17 Dec. 1967 at Lagartococha (Orcés 1974), likely near the mouth of that river into the Río Aguarico. Recorded up to about 400 m.

K. Zimmer (*Studies in Neotropical Ornithology Honoring Ted Parker*, Ornithol. Monogr. no. 48: 849–864, 1997) recently separated this taxon as a species distinct from *C. vulpina* (Rusty-backed Spinetail) on the basis of its distinctly different vocalizations and plumage differences. *C. vulpecula* is confined to islands in the Amazon River and the Río Napo drainage, with *C. vulpina* ranging widely in riparian scrub and thickets both to the north and south of the Amazon. Monotypic.

Range: islands in Amazon River upriver from near mouth of Río Negro in cen. Amazonian Brazil to ne. Peru and up Ríos Napo and Pastaza to e. Ecuador, and along various Amazon tributaries in w. Amazonian Brazil and n. Bolivia.

Cranioleuca gutturata   Speckled Spinetail   Colaespina Jaspeada

Rare to uncommon and seemingly local in tangled lower growth and mid-levels of humid forest and forest borders in the lowlands of e. Ecuador, principally in terra firme, favoring the edge of forest streams. Recorded mainly below 400 m, locally as high as 500–600 m (Bermejo oil-field area north of Lumbaquí; RSR).

Monotypic.

Range: se. Colombia, s. Venezuela, and the Guianas to n. Bolivia and Amazonian Brazil.

Schizoeaca fuliginosa   White-chinned Thistletail   Colicardo Barbiblanco

Uncommon to locally fairly common in the undergrowth of low woodland near treeline and in patches of *Polylepis* woodland in paramo and the upper temperate zone in n. and cen. Ecuador on both slopes of the Andes and locally on slopes above the central valley. On the west slope recorded south to Volcán Iliniza on the Pichincha-Cotopaxi border (N. Krabbe); on the east slope recorded south to nw. Morona-Santiago (Volcán Sangay massif). Found mostly from 2800 to 3500 m, locally as high as 3900–4000 m near Papallacta pass and at Antisana Ecological Reserve.

One race: nominate *fuliginosa*.

Range: Andes from sw. Venezuela to cen. Ecuador, and in n. Peru.

Schizoeaca griseomurina   Mouse-colored   Colicardo Murino
                             Thistletail

Uncommon to fairly common in the undergrowth of low woodland near and just below treeline and in patches of *Polylepis* woodland in paramo and the upper temperate zone in s. Ecuador on both slopes of the Andes and locally on slopes above interandean valleys. The Mouse-colored Thistletail ranges as far north as n. Azuay (El Cajas and Río Mazan areas) and w. Morona-Santiago (along the Gualaceo-Limón road); it is not known to occur together with the White-chinned Thistletail (*S. fuliginosa*), though the ranges of the two species come close in Morona-Santiago. Recorded mainly from about 2800 to 4000 m, locally down to 2500 m (Quebrada Honda on the south slope of the Cordillera Sabanilla; L. Navarrete and RSR et al.). Mouse-colored Thistletails were recorded even lower, at 2150 m, on the Cordillera del Cóndor (Schulenberg and Awbrey 1997, p. 68), though apparently only in the drainage of the Río Comainas in Peru (and not on the Ecuadorian side of the cordillera, though they likely occur there as well).

Vaurie (1980) considered *griseomurina* and all the other forms in the genus *Schizoeaca* to be subspecies of *S. fuliginosa*. However, we follow J. V. Remsen, Jr. (*Proc. Biol. Soc. Wash.* 94[4]: 1068–1075, 1981) in regarding most of the various taxa in the genus as allospecies. Monotypic.

Range: Andes of s. Ecuador and n. Peru.

Asthenes wyatti   Streak-backed Canastero   Canastero Dorsilistado

Uncommon and somewhat local in shrubby paramo and montane scrub in more arid regions in the Andes from Cotopaxi south to Chimborazo, and in s. Azuay and adjacent nw. Zamora-Chinchipe and n. Loja; there is one old record from farther north, at Mojanda on the Pichincha/Imbabura border (Berlioz 1927), and recently (Nov. 1999) a small population was found in montane scrub near Calacalí in n. Pichincha (v.o.). Small numbers of the Streak-backed Canastero can be found at lower levels of Cotopaxi National Park. In most of its Ecuadorian range recorded from about 3100 to 4400 m, but in Azuay and n. Zamora-Chinchipe and Loja only known between about 2900 and 3100 m.

Vaurie (1980) merged the genus *Asthenes* into *Thripophaga*, but virtually all subsequent authors have maintained them as separate genera, which they surely are. Two races of *A. wyatti* are found in Ecuador, *aequatorialis* ranging mainly from Pichincha to Chimborazo, with *azuay* in Azuay and adjacent Zamora-Chinchipe and n. Loja. The latter taxon was for many years known only from the type specimen obtained early in the 20th century at Bestión on the Azuay/Zamora-Chinchipe border. In Nov. 1992 N. Krabbe relocated it at several sites near the type locality: Loma de Santa Rosa and Guagualoma in Azuay, Río Shingata on the Azuay/Zamora-Chinchipe border, and along the road to San Antonio de Cumbe in Loja (specimens in ZMCOP, MECN, and ANSP; see N. Krabbe, *Bull. B. O. C.* 120[3]: 149–153, 2000). *Azuay* differs in having somewhat more extensive rufous on its wings and tail; it is also buffier below (not so grayish).

Range: locally in Andes of w. Venezuela and adjacent n. Colombia, and locally in Andes from n. Ecuador to s. Peru; Sierra de Perijá and Santa Marta Mountains.

Asthenes flammulata   Many-striped Canastero   Canastero Multilistado

Fairly common in grassy paramo with scattered bushes (in both humid and rather arid regions) in the Andes from Carchi south through Loja and adjacent

Zamora-Chinchipe. The Many-striped Canastero is considerably more numerous and widespread in Ecuador than is the Streak-backed (*A. wyatti*); at only a few sites are the two species known to be sympatric. Recorded mainly between about 3200 and 4200 m, locally ranging as low as 3000 m, perhaps especially in the south (e.g., on the Cordillera Las Lagunillas on the Loja/Zamora-Chinchipe border; ANSP, MECN).

One race: nominate *flammulata*.

Range: Andes from n. Colombia to cen. Peru.

Thripophaga fusciceps   **Plain Softtail**   **Colasuave Sencillo**

Apparently rare and very local in the mid-levels and subcanopy of várzea forest and forest borders in the lowlands of e. Ecuador. Thus far the Plain Softtail is known mainly from Napo, with only a few records: a specimen (AMNH) was taken at "San José Abajo" on 12 Apr. 1924, and there are several records from Limoncocha, including a specimen (LSUMZ) obtained on 29 Mar. 1976 (Tallman and Tallman 1977) and tape-recordings (LNS) made by A. Van den Berg on 16 and 20 Feb. 1981. The only subsequent reports involve a pair seen along the Maxus road southeast of Pompeya on 9 Jan. 1994 (RSR and F. Sornoza) and birds seen in Feb. 1999 at Tiputini Biodiversity Center (M. Lysinger et al.); the species has also been listed as occurring at Kapawi Lodge along the Río Pastaza near the Peruvian border, but we have seen no details concerning this. The Plain Softtail seems likely to occur in adjacent Colombia, from which it is unrecorded (Hilty and Brown 1986). A single bird seen at Jatun Sacha and "believed to be this species" (fide B. Bochan) is regarded as uncertain. Below 400 m.

Vaurie (1980) places the species in the genus *Phacellodomus*. One race of *T. fusciceps* occurs in Ecuador, *dimorpha*.

Range: very locally in ne. Ecuador, e. Peru, n. Bolivia, and Amazonian Brazil.

*Phacellodomus rufifrons   **Rufous-fronted Thornbird**   **Espinero Frentirrufo**

Fairly common in secondary woodland and clearings with scattered trees in the Río Marañón drainage of s. Zamora-Chinchipe around Zumba, also ranging north to just south of Valladolid. The Rufous-fronted Thornbird was first seen in Ecuador at some point during 1986 (M. Pearman); the first Ecuadorian specimens (ANSP, MECN) were then obtained near Zumba on 13 Aug. 1989 and along the Río Mayo east of La Chonta on 10 Aug. 1992. Recorded from about 650 to 1500 m.

P. rufifrons has gone by a number of English names in the recent literature—in part depending upon the taxonomy employed—these including Rufous-fronted Thornbird, Plain Thornbird, and Plain-fronted Thornbird.

One race: *peruvianus*. It remains possible that this may be better separated as a distinct species (Marañón Thornbird). We consider the population of n. South America that formerly was regarded as conspecific with *P. rufifrons* (e.g., Ridgely and Tudor 1994) as better representing a separate species, *P. inornatus*

(Plain Thornbird); apart from the latter's widely disjunct range, plumage and voice differ.

Range: nw. Peru and extreme se. Ecuador; n. and e. Bolivia to nw. Argentina, sw. Brazil, and n. Paraguay; e. Brazil.

Siptornis striaticollis   Spectacled Prickletail   Colapúa Frontino

Rare to uncommon and seemingly local in the canopy and borders of montane forest in the subtropical zone on the east slope of the Andes from w. Napo (Palm Peak; MCZ specimen taken by D. Norton on 7 Aug. 1964) south to Zamora-Chinchipe (Podocarpus National Park, Cordillera del Cóndor, and Panguri; ANSP, MECN). The Spectacled Prickletail is quite readily found in the Romerillos area of Podocarpus National Park (C. Rahbek). Recorded from about 1300 to 2300 m.

One race: *nortoni* (see G. R. Graves and M. B. Robbins, *Proc. Biol. Soc. Wash.* 100[1]: 121–124, 1987).

Range: locally in Andes of Colombia, and on east slope of Andes in Ecuador and extreme n. Peru.

Xenerpestes singularis   Equatorial Graytail   Colagris Ecuatorial

Rare to locally uncommon in the canopy and borders of montane forest in the foothills and lower subtropical zone along the eastern base of the Andes from w. Napo (a few sightings from along the Loreto road above Archidona [RSR and PJG et al.; P. Coopmans]) south locally to Zamora-Chinchipe (Zamora area, especially in the Bombuscaro sector of Podocarpus National Park, Panguri [ANSP, MECN], and Coangos on the Cordillera del Cóndor [Schulenberg and Awbrey 1997]). The Equatorial Graytail is probably often overlooked because it generally forages high above the ground, usually in tall epiphyte-laden trees. Recorded from about 1000 to 1600 m.

The Equatorial Graytail was given Near-threatened status by Collar et al. (1994), and we concur with this assessment. It has a distribution pattern and general abundance level comparable to those of, for example, the Napo Sabrewing (*Campylopterus villaviscensio*), Ecuadorian Piedtail (*Phlogophilus hemileucurus*), and Coppery-chested Jacamar (*Galbula pastazae*). Foothill forest in much of the Equatorial Graytail's range continues to be affected by clearing, and very little as yet receives formal protection (a notable exception being in Podocarpus National Park).

Monotypic.

Range: locally along eastern base of Andes from n. Ecuador to n. Peru.

*Xenerpestes minlosi   Double-banded Graytail   Colagris Alibandeado

Rare and local in the canopy and borders of montane forest on the west slope of the Andes in nw. Pichincha (along the road north of Simón Bolívar near Pedro Vicente Maldonado, where a pair was first recorded on 13 Sep. and 1 Oct. 1995

[P. Coopmans]; since then seen there by many other observers) and adjacent sw. Imbabura (at least one seen near the Salto del Tigre bridge over the Río Guaillabamba on 14 Jul. 1997; P. Coopmans et al.). The Double-banded Graytail had not previously been known south of s. Chocó in w. Colombia (Hilty and Brown 1986; Ridgely and Tudor 1994). An inconspicuous and obscure bird, it may also occur at least locally northward along the west slope of the Andes in nw. Ecuador. Recorded from about 400 to 500 m.

Given the Double-banded Graytail's evident rarity in Ecuador, and the fact that the forest patches where it has been found remain unprotected and vulnerable to clearance, we believe it only prudent to consider the species as Vulnerable. It has not been considered to be at risk elsewhere in its small range (Collar et al. 1992, 1994).

Race unknown, but Ecuadorian birds are likely referable to *umbraticus* (which was described from Valle in w. Colombia; see A. Wetmore, *Smith. Misc. Coll.* 117(2): 4, 1951).

Range: e. Panama to w. Colombia; nw. Ecuador.

Metopothrix aurantiacus	Orange-fronted Plushcrown	Coronifelpa Frentidorada

Uncommon to locally fairly common in secondary woodland, forest borders, clearings with scattered trees, and gardens in the lowlands of e. Ecuador. The Orange-fronted Plushcrown is known in Ecuador mostly from Napo, where it is quite easily found at various spots around Tena (including the grounds of the Auca Hotel) and at Limoncocha. There is also a 19th-century specimen from Pastaza at Sarayacu, and the species has recently been found around Kapawi Lodge along the Río Pastaza near the Peruvian border; likely it occurs at least locally along that river as well. Recorded mainly up to about 650 m around Archidona, once reported as high as 900 m north of that town (P. Coopmans and M. Lysinger).

Monotypic.

Range: se. Colombia, e. Ecuador, e. Peru, nw. Bolivia, and w. Amazonian Brazil.

Pseudocolaptes boissonneautii	Streaked Tuftedcheek	Barbablanca Rayada

Uncommon to fairly common in the canopy and borders of montane forest in the subtropical and temperate zones on both slopes of the Andes. On the west slope recorded south locally through El Oro (old specimens from Salvias and Taraguacocha; no recent reports) and Loja (Utuana [Best et al. 1993; also subsequent sightings]). Recorded mostly from 1800 to 3100 m, but locally as low as 1600 m (Panguri; ANSP) and as high as 3400–3500 m (Cerro Mongus; Robbins et al. 1994).

Two similar subspecies are found in Ecuador, *orientalis* on the east slope and in El Oro (though the specimens on which its occurrence in this latter province rests were considered "not typical" by J. T. Zimmer [*Am. Mus. Novitates* 862: 10, 1936]) and w. Loja, and nominate *boissonneautii* on the west slope from north of El Oro to the Colombian border.

Range: mountains of n. Venezuela, and Andes from w. Venezuela to w. Bolivia; Sierra de Perijá.

Pseudocolaptes johnsoni   **Pacific Tuftedcheek**   **Barbablanca del Pacífico**

Uncommon and apparently very local in the canopy and borders of montane forest in the foothills and lower subtropical zone on the west slope of the Andes. There are several old records of the Pacific Tuftedcheek from the Mindo/Gualea area in Pichincha but, surprisingly, there are only a few recent sightings from this region, where the species must be very scarce and/or local. Old records also exist from w. Chimborazo (an old AMNH specimen from "Mt. Chimborazo") and from El Oro (AMNH specimen from El Chiral taken on 1 Aug. 1920). In recent decades the Pacific Tuftedcheek has been found mainly in the southwest, with reports from Azuay (where seen above San Luis in Jan. 1991 [J. C. Matheus]) and El Oro (numerous sightings since 1985 from Buenaventura; also several specimens [ANSP, MECN]). The Pacific Tuftedcheek occurs at elevations below those at which the Streaked Tuftedcheek (*P. boissonneautii*) is found; the Pacific is perhaps confined to very wet, mossy "cloud forest" conditions, and this may account for its very local distribution. Recorded from about 700 to 1700 m.

Given the species' still poorly understood distribution and limited range, we believe the Pacific Tuftedcheek deserves Vulnerable status in Ecuador. No population is known to occur in a formally protected area. The species was not mentioned by Collar et al. (1992, 1994), at least in part because these authors employed a differing taxonomy.

P. johnsoni of w. Colombia and w. Ecuador is regarded as a monotypic species separate from *P. lawrencii* (Buffy Tuftedcheek) of Costa Rica and w. Panama, based on its very different distribution and distinct plumage differences. J. T. Zimmer (*Am. Mus. Novitates* 862, 1936) clarified the actual range of the taxon *johnsoni* (which had earlier been ascribed to the east slope of the Andes) and allied it as a subspecies of *P. lawrencii*. Robbins and Ridgely (1990) suggested possible full species status for *P. johnsoni*, and this treatment is followed here (see also Ridgely and Tudor 1994). Robbins and Ridgely (1990) overlooked the earlier El Oro record, and were also unaware of the unpublished Chimborazo specimen. Vaurie (1980) suggested that *johnsoni* was only an immature plumage of *P. boissonneautii*, but this is certainly not correct.

Range: locally on west slope of Andes in sw. Colombia and w. Ecuador.

*Berlepschia rikeri   **Point-tailed Palmcreeper**   **Palmero**

Rare to locally uncommon in stands of *Mauritia* palms (also occasionally in other palms, e.g., *Bactris* spp.) in the lowlands of ne. Ecuador in Napo and

Sucumbíos; the palmcreeper seems possible in se. Ecuador as well but is as yet unrecorded. The spectacular Point-tailed Palmcreeper was first reported from Ecuador in Sep. 1988 when it was seen and tape-recorded at La Selva (S. Hilty et al.); a few continue to be found in this area, though here as elsewhere they are elusive and difficult to track down. A nesting pair was seen, tape-recorded, and photographed (photos by R. Clements to VIREO) on the grounds of the Hotel Auca at Tena on 15 Mar. 1989 (RSR and PJG et al.); palmcreepers have continued to be found sporadically at this site. At least one was heard along the Inchillaqui road near Archidona on 3 Dec. 1992 (P. Coopmans), and one was collected (ANSP) there on 11 Apr. 1993, the first Ecuador specimen; there are several subsequent reports from there as well. One was heard near Jatun Sacha on 23–26 Aug. 1990 (P. Coopmans), and at least two pairs were found west of Jatun Sacha (same site?) on 13 Apr. 1993, when two birds were collected (ANSP, MECN); the species continues to occur there (B. Bochan). One was heard at the Lago Agrio airport on 10 Nov. 1992 (P. Coopmans). Finally, at least one pair was located along the Maxus road southeast of Pompeya in Nov. 1994 (RSR and F. Sornoza). Palmcreepers were doubtless overlooked in the past, as they are notably inconspicuous and unlikely to be recorded except when vocalizing. Recorded up to 650 m around Archidona.

Monotypic.

Range: locally from the Guianas and s. Venezuela to Amazonian Brazil, nw. Bolivia, se. Colombia, and se. Peru; ne. Ecuador.

Margarornis squamiger   Pearled Treerunner   Subepalo Perlado

Fairly common in the subcanopy and borders of montane forest and secondary woodland (including *Polylepis*-dominated woodland near treeline) in the subtropical and temperate zones on both slopes of the Andes, and locally on slopes above the central and interandean valleys. On the west slope recorded south through Azuay (e.g., at El Cajas) to El Oro (old specimens from El Chiral, Salvias, and Taraguacocha). Recorded mostly from 1800 to 3500 m, locally and in smaller numbers as high as 3800 m (e.g., near Papallacta pass) and as low as 1500–1600 m.

One race: *perlatus*.

Range: Andes from w. Venezuela to w. Bolivia; Sierra de Perijá.

Margarornis stellatus   Star-chested Treerunner   Subepalo Pechiestrellado

Rare to uncommon and very local in the mid-levels and subcanopy of montane forest on the west slope of the Andes, where recorded only from Carchi (4 specimens taken near Maldonado in Jul. 1989; ANSP and MECN), Imbabura (an old specimen from Intag), e. Esmeraldas (one seen above Cristal on 17 Oct. 1992; M. Lysinger and T. Walla), and Chimborazo (an old specimen from Pagma). The Star-chested Treerunner is perhaps confined to very wet, mossy, cloud forest conditions, which if so might account for its unusually local distribution and apparent rarity in Ecuador; it is amazing that the species still seems not to have been found in the relatively well known province of Pichincha. Recorded between about 1200 and 1900 m.

We conclude that the very local and scarce Star-chested Treerunner deserves Vulnerable status in Ecuador. The species was not mentioned by Collar et al. (1992, 1994). No area in which it is known to occur presently receives formal protection, though it seems likely to range in remote sectors of Cotacachi-Cayapas Ecological Reserve.

The species has usually (e.g., Hilty and Brown 1986) been called the Fulvous-dotted Treerunner, but we follow Ridgely and Tudor (1994) in employing the more accurate English name of Star-chested Treerunner. The species' "dots" are actually white, and they are distinctly star-shaped.

Monotypic.

Range: locally on west slope of Andes in Colombia and Ecuador.

Premnoplex brunnescens   Spotted Barbtail   Subepalo Moteado

Fairly common in the undergrowth of montane forest and mature secondary woodland in the foothills and subtropical zone on both slopes of the Andes. On the west slope recorded south to El Oro in the Buenaventura/Zaruma area and at Salvias. The Spotted Barbtail is an inconspicuous bird, with its true numbers being best revealed through mist-netting. Recorded mostly between about 900 and 2500 m, locally ranging lower on the west slope (e.g., at El Placer in Esmeraldas and Buenaventura in El Oro; ANSP, MECN).

Vaurie (1980) merges the genus *Premnoplex* into *Margarornis*, but we follow most more recent authors in maintaining it as distinct. One race occurs in Ecuador, nominate *brunnescens*.

Range: highlands of Costa Rica and Panama; mountains of n. Venezuela, and Andes from w. Venezuela to w. Bolivia; Sierra de Perijá and Santa Marta Mountains.

Premnornis guttuligera   Rusty-winged Barbtail   Subepalo Alirrojizo

Rare to uncommon but seemingly local (perhaps just under-recorded?) in the lower growth of montane forest in the subtropical zone on the east slope of the Andes. On the west slope recorded only from Pichincha (on the west slope of Volcán Pichincha, in the Mindo/Gualea/Maquipucuna region, and also along the Chiriboga road), but it likely occurs northward to the Colombian border as well. On the east slope the Rusty-winged Barbtail also appears to be quite local, with most records coming from various sites in w. Napo; southward it is apparently known only from the Cordillera de Cutucú (ANSP) and a few sites in and near Podocarpus National Park. Recorded mostly from 1600 to 2300 m, rarely as low as 1250 m (Río Bombuscaro; P. Coopmans).

Vaurie (1980) merges the genus *Premnornis* into *Margarornis*. However, in our view the relationships of this obscure species remain more enigmatic; in many respects it is more reminiscent of a foliage-gleaner than a barbtail or treerunner, and therefore we maintain it as a separate genus. One race: nominate *guttuligera*.

Range: Andes from sw. Venezuela to w. Bolivia; Sierra de Perijá.

Syndactyla subalaris   Lineated Foliage-gleaner   Limpiafronda Lineada

Uncommon to fairly common (but inconspicuous) in the undergrowth and mid-levels of montane forest in the foothills and subtropical zone on both slopes of the Andes. On the west slope recorded south to El Oro (numerous old specimens from El Chiral and near Zaruma), though there are few recent records from south of Pichincha. Recorded mostly from 1000 to 2100 m (above El Chaco; ANSP).

Vaurie (1980) merged the genus *Syndactyla* into his much-expanded *Philydor*, but subsequent authors have generally not followed this proposal. Two somewhat different races are found in Ecuador, *mentalis* on the east slope and nominate *subalaris* on the west slope. *Mentalis* is blacker on the crown and nape with more contrasting and extensive buff streaking, and it also is more profusely streaked below.

Range: highlands of Costa Rica and Panama; Andes from w. Venezuela to s. Peru.

*Syndactyla rufosuperciliata   Buff-browed   Limpiafronda
                                     Foliage-gleaner   Cejianteada

Uncommon in the lower growth of montane forest on the east slope of the Andes in extreme s. Ecuador. The Buff-browed Foliage-gleaner is known in Ecuador only from the Cordillera del Cóndor in Zamora-Chinchipe, where it was first found in Sep. 1990 by N. Krabbe and F. Sornoza above Chinapinza (ANSP and MECN; Krabbe and Sornoza 1994). In Jun. 1993 it was again found in this area (ANSP and MECN). Remarkably, on both trips the two species of *Syndactyla* were found together in precisely the same areas, and even were captured in the same mist-nets. Recorded from 1700 to 1900 m.

One race: *cabanisi*.

Range: Andes from extreme se. Ecuador to nw. Argentina; se. Brazil, e. Paraguay, ne. Argentina, and Uruguay.

Syndactyla ruficollis   Rufous-necked   Limpiafronda Cuellirrufa
                            Foliage-gleaner

Uncommon to fairly common in the undergrowth and mid-levels of montane forest, secondary woodland, and borders in the foothills and subtropical zone of sw. Ecuador in Loja; ranges east as far as the slopes below the Continental Divide (e.g., at San Pedro above Vilcabamba [J. Rasmussen et al.] and above Jimbura [RSR and F. Sornoza]. We regard the reported sightings from Buenaventura in El Oro (Collar et al. 1992) as unverified. Recorded mostly from

1300 to 2300 m, but locally ranging as high as about 2700 m (west slope of Cordillera Las Lagunillas), and occurring in smaller numbers down to 600 m even in more deciduous forest (e.g., at Tambo Negro; Best et al. 1993).

Overall numbers of the Rufous-necked Foliage-gleaner have unquestionably declined in recent decades because of forest clearance, and the species has been considered to be Endangered (Collar et al. 1992) or Vulnerable (Collar et al. 1994). It appears to tolerate some habitat disturbance, seemingly thriving in quite degraded patches of woodland, and we therefore accord it Vulnerable status. A population occurs in the El Tundo Reserve above Sozoranga in s. Loja.

In the past *ruficollis* has usually been placed in the genus *Automolus* (e.g., Meyer de Schauensee 1966, 1970; Vaurie 1980). However, its voice clearly places it in the genus *Syndactyla* (see Parker et al. 1985; Ridgely and Tudor 1994). Its subtropical distribution and plumage pattern are also more in accord with the genus *Syndactyla*. Monotypic. Based on the good series in ANSP from throughout the species' range, we find that the characters used to diagnose *celicae*, the race to which Ecuadorian birds were formerly allocated (see J. T. Zimmer, *Am. Mus. Novitates* 785, 1935), can be found throughout the species' range, and likely are dependent on age and wear. This includes the prominence of the breast streaking, the degree of rufescence on the crown and crissum, and the color of the brow. We thus suggest synonymyzing *celicae*.

Range: Andes of sw. Ecuador and nw. Peru.

Anabacerthia variegaticeps	Scaly-throated Foliage-gleaner	Limpiafronda Goliescamosa

Fairly common in the subcanopy and borders of montane forest, mature secondary woodland, and borders in the foothills and lower subtropical zone on the west slope of the Andes south to w. Loja (Alamor region). Also found recently on the coastal Cordillera de Mache in w. Esmeraldas (Bilsa; R. P. Clay et al.) and on the Cordillera de Colonche in sw. Manabí (Cerro San Sebastián in Machalilla National Park [Parker and Carr 1992; ANSP]) and w. Guayas (Loma Alta; Becker and López-Lanús 1997). An anomalous specimen was collected in the lowlands in May 1962 at Hacienda Pacaritambo (Brosset 1964). Recorded mostly from about 700 to 1700 m, locally lower (to 450–500 m at Bilsa; once at about 100 m at Pacaritambo; perhaps formerly it regularly occurred this low?).

The species has sometimes been called the Spectacled Foliage-gleaner, but see Ridgely and Tudor (1994) for the rationale for reverting to the name Scaly-throated. The 1998 AOU Check-list concurred.

Vaurie (1980) merged the genus *Anabacerthia* into his much-expanded *Philydor*. One race of *A. variegaticeps* occurs in Ecuador, *temporalis*. This disjunct South American form is perhaps specifically distinct from Middle American populations; if separated, it would be called *A. temporalis* (Spot-breasted Foliage-gleaner).

Range: highlands from s. Mexico to w. Panama; west slope of Andes in w. Colombia and w. Ecuador.

Anabacerthia striaticollis	Montane Foliage-gleaner	Limpiafronda Montana

Fairly common in the subcanopy and borders of montane forest and mature secondary woodland in the foothills and subtropical zone on the east slope of the Andes, sometimes foraging lower and coming out into adjacent clearings. Readily seen along the Loreto road north of Archidona and around San Rafael Falls. Recorded mostly between about 1000 and 1800 m, locally as high as 2100 m (northeast of El Chaco; ANSP).

One race: *montana*.

Range: mountains of n. Venezuela, and Andes from w. Venezuela to w. Bolivia; Sierra de Perijá and Santa Marta Mountains.

Hyloctistes subulatus	Eastern Woodhaunter	Rondamusgos Oriental

Uncommon to fairly common in the lower growth and mid-levels of humid forest and mature secondary woodland in the lowlands, foothills, and lower subtropical zone of e. Ecuador. In the lowlands occurs primarily in terra firme. Recorded mostly below 1100 m, but ranging in small numbers up to 1700 m at least on outlying ridges of the southeast (Cordilleras de Cutucú and del Cóndor).

The species was formerly sometimes called the Striped Foliage-gleaner.

Vaurie (1980) merged the genus *Hyloctistes* into his much-expanded *Philydor*. Because of their very different primary vocalizations, we treat trans-Andean *H. virgatus* (Western Woodhaunter) as a species separate from birds found east of the Andes, *H. subulatus*. All taxa were formerly considered conspecific under the name of *H. subulatus* (Striped Woodhaunter), though Ridgely and Tudor (1994) registered doubts as to whether this was the best course. *H. subulatus* then is monotypic.

Range: s. Venezuela and se. Colombia to nw. Bolivia and w. and cen. Amazonian Brazil.

Hyloctistes virgatus	Western Woodhaunter	Rondamusgos Occidental

Uncommon to fairly common in the undergrowth and mid-levels of humid forest and mature secondary woodland in the lowlands and foothills of w. Ecuador south in more humid areas to n. Manabí (Río de Oro), n. Guayas (Hacienda Pacaritambo; Brosset 1964), sw. Chimborazo (Puente de Chimbo), nw. Azuay (Manta Real; ANSP, MECN), and El Oro (RSR sightings from Buenaventura in Jun. 1985; also a specimen taken there by N. Krabbe on 16 Nov. 1991 [MECN] and a few subsequent records [P. Coopmans et al.]). Recorded mostly below 1100 m.

One race: *assimilis*. This species differs from *H. subulatus* (Eastern Woodhaunter; found east of the Andes) in showing virtually no streaking on its mantle and in being somewhat less flammulated below; its song also differs notably.

Range: Nicaragua to sw. Ecuador.

Ancistrops strigilatus   Chestnut-winged Hookbill   Picogancho Alicastaño

Rare to locally fairly common in the mid-levels and subcanopy of humid forest (principally terra firme) in the lowlands of e. Ecuador, where known mainly from Napo and Sucumbíos. In the southeast the Chestnut-winged Hookbill is known from Sarayacu (Berlioz 1932a), and has also been reported from around Kapawi Lodge near the Río Pastaza close to the Peruvian border, though it likely is more widespread. The Chestnut-winged Hookbill is reasonably numerous at La Selva. Recorded mostly below about 400 m, locally up to 600 m in the Bermejo oil-field area north of Lumbaquí (RSR).

One race: nominate *strigilatus*.

Range: se. Colombia to nw. Bolivia and w. and cen. Amazonian Brazil.

Philydor erythropterus   Chestnut-winged   Limpiafronda
                            Foliage-gleaner   Alicastaña

Uncommon in the mid-levels and subcanopy of terra firme forest in the lowlands of e. Ecuador. Small numbers are seen regularly with canopy flocks at La Selva and at Sacha Lodge, sometimes together with the Chestnut-winged Hookbill (*Ancistrops strigilatus*). Recorded mostly below about 400 m, locally up to 600 m in the Bermejo oil-field area north of Lumbaquí (RSR).

One race: nominate *erythropterus*.

Range: s. Venezuela; se. Colombia to n. Bolivia and s. Amazonian Brazil.

Philydor fuscipennis   Slaty-winged   Limpiafronda Alipizarrosa
                          Foliage-gleaner

Rare to uncommon and seemingly local in the lower growth and mid-levels of humid forest and mature secondary woodland in the lowlands of w. Ecuador. The handsome Slaty-winged Foliage-gleaner is known in Ecuador only from s. Pichincha (Río Palenque) south to e. Guayas and adjacent Chimborazo (old specimens [AMNH, ANSP] from Bucay and Chimbo), and nw. Azuay (Manta Real; ANSP, MECN). Somewhat surprisingly, the species has never been found northward in the more humid forests of the northwest. A population continues to occur at Río Palenque. Recorded below about 600 m.

Because of the extensive deforestation that has occurred across much of this species' Ecuadorian range, we feel it deserves Vulnerable status here. The species was not mentioned as being at risk by Collar et al. (1992, 1994). The only population occurring in a protected area is the necessarily very small one at Río Palenque; some evidence indicates that this population may itself be declining.

P. fuscipennis of Panama to w. Ecuador is regarded as a species separate from cis-Andean *P. erythrocercus* (Rufous-rumped Foliage-gleaner), with which it formerly was considered conspecific (e.g., Meyer de Schauensee 1966, 1970), following most recent authors. The disjunct Ecuadorian population of *P. fuscipennis* has normally been assigned to the race *erythronotus*, which is found in e. Panama and n. Colombia. We suspect, however, that the Ecuadorian population actually represents an undescribed form, for it is considerably more ochraceous (less olivaceous) below than examples (ANSP) of *erythronotus* from Colombia.

Range: Panama and n. Colombia; w. Ecuador.

Philydor pyrrhodes	Cinnamon-rumped Foliage-gleaner	Limpiafronda Lomicanela

Rare to locally uncommon in the lower growth of humid forest (both terra firme and várzea) in the lowlands of e. Ecuador. The Cinnamon-rumped Foliage-gleaner seems particularly to favor the vicinity of streams and often forages amongst the fronds of palms. Recorded mostly below about 300 m, locally as high as 700 m at Eugenio and Guaticocha in w. Napo (MCZ).

Monotypic.

Range: the Guianas and s. Venezuela to se. Peru, n. Bolivia, and Amazonian Brazil.

Philydor erythrocercus	Rufous-rumped Foliage-gleaner	Limpiafronda Lomirrufa

Uncommon to locally fairly common from the lower growth up to the sub-canopy of humid forest (especially terra firme) in the lowlands of e. Ecuador, and in montane forest in the foothills (where most numerous) and lower subtropical zone along the eastern base of the Andes. Recorded up to about 1300 m.

Does not include trans-Andean *P. fuscipennis* (Slaty-winged Foliage-gleaner). One race of *P. erythrocercus* occurs in Ecuador, *subfulvus*.

Range: se. Colombia and the Guianas to n. Bolivia and Amazonian Brazil.

Philydor ruficaudatus	Rufous-tailed Foliage-gleaner	Limpiafronda Colirrufa

Rare in the mid-levels of terra firme forest in the lowlands of e. Ecuador. Some published specimens and many sightings are incorrect, and result from confusion with the very similar—and, in Ecuador, much more numerous—*subfulvus* race of the Rufous-rumped Foliage-gleaner (*P. erythrocercus*). The two species are among the most frequently confused in Ecuador. For example, it would appear that J. T. Zimmer (*Am. Mus. Novitates* 785, 1935) reidentified two specimens that had been identified as Rufous-tailed Foliage-gleaners from

Zamora (Chapman 1926) as *subfulvus* Rufous-rumpeds, and it appears unlikely that the Rufous-tailed occurs in the Zamora region, or indeed anywhere in the foothills. Chapman (1926) mentions numerous other localities for the Rufous-tailed in his synonymy; many of these are likely incorrect, and some certainly are (e.g., Zamora). Zimmer (op. cit.) notes that he examined 10 Ecuadorian specimens from 3 localities, all of them in w. Napo: "below San José de Sumaco," Río Suno, and above Archidona. ANSP has a specimen from one additional locality, Cordillera de Galeras, which is in the same general area. Specimens labeled as having been taken much farther east, at the mouth of the Río Curaray (now in ne. Peru but formerly within Ecuador), are believed likely to have been mislabeled; we are not aware of any confirmatory evidence that this species occurs so far away from the Andes in Ecuador or adjacent Peru. Another specimen (UKMNH) was obtained on 8 Aug. 1971 at Santa Cecilia in w. Sucumbíos. The failure of ANSP or any other modern workers to obtain any examples of the Rufous-tailed Foliage-gleaner indicates that the species must, at a minimum, be scarce and local in Ecuador, perhaps with a distribution restricted to the lowlands and adjacent lower foothills of Napo and Sucumbíos. The Rufous-rumped Foliage-gleaner forages much more often at lower levels in forest; as a result it is the species frequently captured in mist-nets, whereas the Rufous-tailed rarely or never is. Recorded up to about 600 m.

One race: nominate *ruficaudatus*.

Range: the Guianas and s. Venezuela to n. Bolivia and Amazonian Brazil.

Philydor rufus   **Buff-fronted Foliage-gleaner**   **Limpiafronda Frentianteada**

Uncommon to locally fairly common in the canopy and borders of montane forest in the foothills and lower subtropical zone on both slopes of the Andes. On the west slope recorded from Esmeraldas (El Placer) and Imbabura (Intag) south to Pichincha, and from El Oro (a few sightings at Buenaventura, first by RSR in Jun. 1985). On the east slope recorded from w. Napo (sightings from north of Archidona along the road to Loreto [RSR; P. Coopmans]) southward. Found regularly along the lower Chiriboga road. On the west slope recorded from about 600 to 1500 m, on the east slope from about 1000 to 1700 m.

Two races occur in Ecuador, *bolivianus* on the east slope and the smaller, darker, and duller *riveti* on the west slope.

Range: highlands of Costa Rica and w. Panama; mountains of n. and s. Venezuela, and locally in Andes from n. Colombia to w. Bolivia; se. Brazil, e. Paraguay, and ne. Argentina.

Anabazenops dorsalis   **Bamboo Foliage-gleaner**   **Rascahojas de Bambú**

Rare to locally uncommon in the undergrowth of forest borders and taller secondary woodland, in many areas favoring stands of bamboo, in the foothills along the eastern base of the Andes from Napo south to Zamora-Chinchipe (old specimens [AMNH] from Zamora; also recent sightings from Miazi in the Río Nangaritza valley [Schulenberg and Awbrey 1997]). There are only a few

Ecuadorian records in the lowlands east of the Andean foothills, where recent evidence indicates that the species is found mainly if not entirely in floodplain forest, and on islands, where there is an understory of *Gynerium* cane; up to now it has been found principally near the Río Napo. A series of five specimens (UKMNH) was obtained at Santa Cecilia in w. Sucumbíos in Jul.–Aug. 1971; a specimen (LSUMZ) was taken at Limoncocha on 14 Oct. 1976 (Tallman and Tallman 1977); there are several recent reports (including tape-recordings) from La Selva south of the Río Napo (first on 5 Aug. 1991; P. Coopmans et al.) and from Sacha Lodge (L. Jost et al.). Recorded up to 1300 m along the Loreto road north of Archidona, but found mostly below 1000 m.

This species was formerly (e.g., Meyer de Schauensee 1966, 1970; Hilty and Brown 1986) called the Crested Foliage-gleaner, despite the fact that it shows absolutely no crest. Ridgely and Tudor (1994) employed the descriptive modifier of "Dusky-cheeked," but it has been suggested—especially given that some other foliage-gleaners also have dusky cheeks—that a much better modifier would be "Bamboo." The species is indeed characteristic of bamboo stands in most of its range, even if in Ecuador it is also found in *Gynerium* cane.

A. Kratter and T. A. Parker III (*Studies in Neotropical Ornithology Honoring Ted Parker*, Ornithol. Monogr. no. 48; 383–397, 1997) suggest that this foliage-gleaner, an above-ground nester, is better transferred from the genus *Automolus*, in which it was long placed (e.g., Sibley and Monroe 1990; Ridgely and Tudor 1994), and placed in the genus *Anabazenops* (together with another species from se. Brazil). Monotypic.

Range: locally from se. Colombia to nw. Bolivia, and in Amazonian Brazil.

Automolus infuscatus   Olive-backed   Rascahojas Dorsiolivácea
                             Foliage-gleaner

Fairly common in the lower growth and mid-levels of humid forest (principally terra firme) in the lowlands of e. Ecuador. This and many of the other *Automolus* foliage-gleaners are notably skulking birds; they are recorded principally by voice or through mist-net captures. Recorded up to about 700 m.

One race: nominate *infuscatus*.

Range: se. Colombia, s. Venezuela, and the Guianas to extreme nw. Bolivia and Amazonian Brazil. More than one species is probably involved.

Automolus ochrolaemus   Buff-throated   Rascahojas Golipálida
                             Foliage-gleaner

Uncommon to locally fairly common in the undergrowth of humid forest in the lowlands of both e. and w. Ecuador, in the east mainly in swampy forest and near streams in terra firme. In the west recorded south to extreme sw. Manabí in Machalilla National Park (ANSP), w. Guayas (above Salanguilla and Cerro La Torre [N. Krabbe]; Loma Alta [Becker and López-Lanús 1997]), and Los

Ríos (Jauneche). In the wet-forest belt of n. Esmeraldas the Buff-throated Foliage-gleaner is scarce and occurs mainly in disturbed and secondary habitats (O. Jahn and P. Mena Valenzuela et al.). Recorded mostly below 800 m, locally and in small numbers up to about 1300 m.

Two readily distinguished races of occur in Ecuador, *turdinus* in the east (with a buffyish throat and pale flammulations on the breast) and *pallidigularis* in the west (with a white throat and plainer underparts).

Range: s. Mexico to sw. Ecuador; the Guianas and s. Venezuela to n. Bolivia and Amazonian Brazil. More than one species is almost certainly involved.

Automolus rufipileatus	Chestnut-crowned Foliage-gleaner	Rascahojas Coronicastaña

Uncommon to locally fairly common in the undergrowth and borders of várzea forest and woodland, favoring stands of *Gynerium* cane in floodplain areas and also occurring regularly on river islands, in the lowlands of e. Ecuador. Known primarily from the northeast in Sucumbíos and Napo. The only site in the southeast from which the Chestnut-crowned Foliage-gleaner is known is around Kapawi Lodge along the Río Pastaza near the Peruvian border, but it is probably more widespread; it may be less confined to *Gynerium* cane thickets in this area (P. Coopmans). Although it vocalizes frequently, the Chestnut-crowned Foliage-gleaner is skulking and very difficult to observe (even more so than most other *Automolus* foliage-gleaners); it probably was overlooked in the past. Found mostly below 500 m, but recorded up locally to about 750 m in association with *Gynerium* cane near Archidona (P. Coopmans).

One race: *consobrinus*.

Range: locally from the Guianas, s. Venezuela, and e. Colombia to n. Bolivia and Amazonian Brazil.

Automolus melanopezus	Brown-rumped Foliage-gleaner	Rascahojas Lomiparda

Rare to uncommon and seemingly local in dense tangled undergrowth of humid forest, principally near streams or in swampy places, in the lowlands of e. Ecuador. Most Ecuadorian records of the Brown-rumped Foliage-gleaner are from Napo and Sucumbíos, and both AMNH and ANSP have series from the lower Sumaco region; one recent report exists from as far east as e. Sucumbíos at Zancudococha (P. Coopmans). In the southeast it is known from Morona-Santiago, with specimens (ANSP) from Chichirota and "Churo Yacu"; in addition, two birds were collected (ANSP, MECN) at Santiago in Jul. 1989. Brown-rumped Foliage-gleaners have also been reported from around Kapawi Lodge along the Río Pastaza near the Peruvian border (v.o.). Recorded mostly below about 400 m, but locally found up to 600 m at Canelos in Pastaza (N. Krabbe).

Monotypic.

Range: locally from e. Ecuador and adjacent se. Colombia to nw. Bolivia and sw. Amazonia Brazil.

Automolus rubiginosus   Ruddy Foliage-gleaner   Rascahojas Rojiza

Rare to uncommon and rather local in the dense undergrowth of humid forest and forest borders in the lowlands of both e. and w. Ecuador, and uncommon to locally fairly common in the foothills along both slopes of the Andes; generally more numerous in the west than the east. In the west recorded south, at least formerly, to El Oro (an early-20th-century specimen from Santa Rosa; AMNH), there are only a few recent reports from south of Pichincha (e.g., small numbers found at Manta Real in nw. Azuay in Aug. 1991; RSR). In the wet-forest belt of n. Esmeraldas the Ruddy Foliage-gleaner is almost entirely a forest-edge species and does not occur inside continuous forest (O. Jahn and P. Mena Valenzuela et al.). In the eastern lowlands it ranges east in diminishing numbers away from the Andes, with the easternmost recent localities in the northeast being at La Selva and along the Maxus road southeast of Pompeya (RSR et al.); there is also a single 1935 specimen (ANSP) from Río Tigre in extreme e. Pastaza, and the species has also been reported from around Kapawi Lodge along the Río Pastaza near the Peruvian border. As revealed by their oft-heard vocalizations, Ruddy Foliage-gleaners occur in fair numbers at both Tinalandia and Río Palenque. Recorded up to about 1300 m on the west slope, but mainly below 1000 m in the east.

Three races are found in Ecuador, their appearance differing strikingly on either side of the Andes. The dark *nigricauda* (with a blackish tail and dark grayish brown underparts aside from its rich rufous throat) ranges in the west. In most of the east occurs *brunnescens* (with a dark chestnut tail and paler, more rufescent brown lower underparts). In 1993 the race *caquetae* was found near the Colombian border in Sucumbíos (Bermejo oil-field area north of Lumbaquí; ANSP, MECN); it resembles *brunnescens*, differing in being browner (less rufescent) below, so that its rich rufous throat stands out more.

Range: locally from s. Mexico to sw. Ecuador; Santa Marta Mountains; along eastern base of Andes from w. Venezuela to nw. Bolivia; the Guianas, s. Venezuela, and n. Brazil.

Hylocryptus erythrocephalus   Henna-hooded   Rascahojas
                                 Foliage-gleaner   Capuchirrufa

Uncommon and very local on or near the ground in deciduous and semihumid forest and woodland, including areas where substantially fragmented, in the lowlands and (mainly) foothills of sw. Ecuador. In Ecuador the Henna-hooded Foliage-gleaner is known mainly from w. Loja, whence there are recent reports from several sites below Celica (ANSP), the Sozoranga area (Best et al. 1993), near Catacocha (ANSP), south of Sabanilla (ANSP, MECN), and near Macará (RSR; P. Coopmans). There are also a few recent reports from south of Portovelo in e. El Oro (first by R. A. Rowlett and J. Arvin et al.). The species has also recently been found much farther north, in Machalilla National Park in sw. Manabí, first by T. Parker in Feb. 1991 (Parker and Carr 1992); several specimens were obtained here in Aug. 1991 (ANSP, MECN). In Jul. 1992 the species was also found on the slopes of Cerro La Torre in w. Guayas (N. Krabbe), and there are recent reports from Cerro Blanco, the first on 25 Jul. 1992 (P. Coopmans and M. Van Beirs et al.; Berg 1994). Likely the Henna-hooded Foliage-gleaner ranges locally along much of the coastal Cordillera de Colonche in Manabí and Guayas. Recorded mostly from 400 to 1800 m, locally a little lower (e.g., down to 200 m at Cerro Blanco).

Numbers of the Henna-hooded Foliage-gleaner are now substantially reduced as a result of deforestation and the trampling of undergrowth in much of what wooded habitat remains. The species was accorded Vulnerable status by Collar et al. (1994), and we agree. Recent evidence does, however, indicate that—somewhat surprisingly—the Henna-hooded appears capable of persisting in quite severely degraded habitats, and this gives some cause for optimism regarding its future. Probably its largest extant population occurs in Machalilla National Park.

This distinctive species is sometimes placed in the genus *Automolus*, but we follow Ridgely and Tudor (1994) in maintaining it in *Hylocryptus*. One race: nominate *erythrocephalus*.

Range: sw. Ecuador and nw. Peru.

Thripadectes flammulatus   Flammulated   Trepamusgos Flamulado
                               Treehunter

Uncommon in the undergrowth of montane forest (especially where there is extensive *Chusquea* bamboo) in the upper subtropical and temperate zones on both slopes of the Andes, and very locally on slopes above the central and interandean valleys south to Azuay at Río Mazán (King 1989); on the west slope recorded south to Azuay at Sural (N. Krabbe) and Portete (Berlioz 1932b). Like all the *Thripadectes*, the shy Flammulated Treehunter favors areas with very dense cover, and though infrequently seen it may be regularly mist-netted in appropriate habitat. Recorded mostly between 2200 and 3500 m, locally down to 2000 m (Cordillera de Huacamayos; WFVZ, MECN).

One race: nominate *flammulatus*.

Range: locally in Andes from w. Venezuela to extreme n. Peru; Santa Marta Mountains.

Thripadectes holostictus   Striped Treehunter   Trepamusgos Listado

Uncommon to locally fairly common and seemingly local (likely under-recorded) in the undergrowth of montane forest in the subtropical and temperate zones on both slopes of the Andes; on the west slope recorded south to Chimborazo (several old specimens from the Pallatanga and Cayandeled areas). The Striped generally occurs lower than the Flammulated Treehunter (*T. flam-*

mulatus), though there is considerable overlap; like that species, the Striped favors stands of *Chusquea* bamboo. It may be most numerous in Pichincha (e.g., in the Mindo/Tandayapa area), though even there it is mainly recorded just by voice. Recorded mostly from 1500 to 2500 m, in small numbers or locally up to 3000 m, perhaps most numerous between about 2000 and 2300 m (P. Coopmans).

Two similar races are found in Ecuador, nominate *holostictus* on the east slope and *striatidorsus* on the west.

Range: locally in Andes from w. Venezuela to w. Bolivia.

Thripadectes virgaticeps	Streak-capped Treehunter	Trepamusgos Gorrirrayado

Uncommon in the undergrowth of montane forest in the subtropical zone on both slopes of the Andes in n. Ecuador. On the east slope recorded only from w. Napo in the Baeza/Sumaco region; probably it also occurs northward through Sucumbíos to the Colombian border. On the west slope recorded from Carchi (above Chical; ANSP, MECN) south into Pichincha (south to near Tandapi; P. Coopmans). Krabbe (1992) reported on a single specimen he identified as the Streak-capped Treehunter obtained at Buenaventura in El Oro in 1991 (unfortunately the bird was skeletonized). However, until the species is fully corroborated from this site, far to the south of its known range, we prefer to treat the specimen's identity as uncertain. Recorded mostly from 1300 to 2100 m.

Two races are found in Ecuador, *sumaco* on the east slope and nominate *virgaticeps* on the west slope. They are similar in plumage, though the nominate race has a longer and heavier bill. Birds from Carchi (ANSP) show little or no indication of intergradation with the similar (but slightly smaller) *sclateri* of the west slope of Colombia's Western Andes.

Range: mountains of n. Venezuela, and locally in Andes from w. Venezuela to n. Ecuador.

Thripadectes melanorhynchus	Black-billed Treehunter	Trepamusgos Piquinegro

Uncommon to locally fairly common in the undergrowth of montane forest and secondary woodland in the foothills and lower subtropical zone on the east slope of the Andes from w. Sucumbíos (north to San Rafael Falls; ANSP, MECN) south to Zamora-Chinchipe (Podocarpus National Park, Panguri, Cordillera del Cóndor). The Black-billed Treehunter has recently been found as far north in s. Colombia as e. Cauca (Salaman et al. 1999); it was formerly known there only from one record, in the north (Hilty and Brown 1986). Recorded between about 1000 and 1700 m.

One race: nominate *melanorhynchus*.

Range: locally on east slope of Andes from n. Colombia to s. Peru.

Thripadectes ignobilis   Uniform Treehunter   Trepamusgos Uniforme

Uncommon to locally fairly common in the undergrowth of montane forest in the foothills and lower subtropical zone on the west slope of the Andes south to El Oro (Buenaventura; ANSP, MECN). Recorded between 700 and 1700 m. Monotypic.

Range: west slope of Andes in w. Colombia and w. Ecuador.

Xenops rutilans   Streaked Xenops   Xenops Rayado

Uncommon to locally fairly common in the canopy and borders of montane forest and secondary woodland in the foothills and subtropical zone on both slopes of the Andes, and in deciduous forest and woodland in the lowlands and foothills of more arid w. Ecuador from w. Esmeraldas south through El Oro and Loja. The Streaked Xenops is not found in humid lowland forest, where it is replaced by the Plain Xenops (*X. minutus*), and is entirely absent from the lowlands of e. Ecuador; otherwise it occurs in a surprising range of habitats. The species seems especially numerous around Zumba in s. Zamora-Chinchipe. In the west recorded up to about 2000 m; on the east slope found mainly between 800 and 2000 m, lower (down to 700 m) in the Zumba area.

Three similar races are found in Ecuador. *Heterurus* occurs on the east slope, but is replaced southward (in Morona-Santiago and Zamora-Chinchipe) by the similar *peruvianus*; *guayae* is the race found in the west. *Guayae* has the inner web of only the 4th rectrix black, whereas in both the eastern races the inner webs of both the 3d and 4th rectrices are black.

Range: Costa Rica to n. Venezuela, Andean slopes south to nw. Argentina, cen. and e. Brazil, e. Paraguay, and ne. Argentina.

Xenops tenuirostris   Slender-billed Xenops   Xenops Picofino

Rare to locally uncommon (overlooked?) in the canopy and borders of humid forest (both terra firme and várzea, but seems more numerous in the former) in the lowlands of e. Ecuador. There are only a few Ecuadorian records of this usually scarce xenops, including two specimens from Zamora (one of them the type, but from whence there are no recent reports), a specimen from Río Suno (J. T. Zimmer, *Am. Mus. Novitates* 862, 1936), a recent specimen taken on 6 Aug. 1987 at Tayuntza in Morona-Santiago (WFVZ), and several sightings from La Selva and Sacha Lodge. The Slender-billed Xenops has also recently proven to be not uncommon with canopy flocks along the Maxus road southeast of Pompeya (RSR and S. Howell et al.). Its ecological relationship with the more widespread, and extremely similar, Streaked Xenops (*X. rutilans*) remains to be elucidated; the two species are apparently sympatric at least in the Zamora area, but the Streaked does not occur in the eastern lowlands away from the Andes. Recorded up to about 1000 m at Zamora, but found mostly below 600 m.

One race: *acutirostris*.

Range: locally from the Guianas and s. Venezuela to n. Bolivia and Amazonian Brazil.

Xenops minutus   Plain Xenops   Xenops Dorsillano

Uncommon to fairly common in the subcanopy and borders of humid forest and mature secondary woodland in the lowlands of w. Ecuador; uncommon in the lower and middle growth of both terra firme and várzea forest (principally the former) in the lowlands of e. Ecuador. In the west recorded south to sw. Manabí (Machalilla National Park), Guayas (Chongón Hills; Manglares-Churute Ecological Reserve [Pople et al. 1997]), and El Oro (numerous sightings and a VIREO photograph from Buenaventura). Given that there are recent sightings of the Plain Xenops from adjacent Peru (Parker et al. 1995), it is curious that the species is unrecorded from w. Loja; because of near-total deforestation at lower elevations, if it did occur formerly, it may no longer do so. The Plain Xenops is considerably more numerous and conspicuous in w. Ecuador than it is in the east, where its behavior also differs and it tends to remain inside forest, infrequently or never coming out to edge situations. Recorded mostly below 900 m, in small numbers up to 1300 m at least in the northwest (Maquipucuna Reserve; WFVZ).

Two races are found in Ecuador, *littoralis* in the west and *obsoletus* in the east. *Obsoletus* has a more extensively whitish throat and more whitish mottling on the chest.

Range: se. Mexico to extreme nw. Peru, n. Bolivia, and Amazonian Brazil; e. Brazil, e. Paraguay, and ne. Argentina.

Xenops milleri   Rufous-tailed Xenops   Xenops Colirrufo

Rare and seemingly local in the canopy and borders of terra firme forest in the lowlands of e. Ecuador. The Rufous-tailed Xenops was first recorded in Ecuador by Norton et al. (1972), based on specimens taken at "Morete," Montalvo, and Río Capaguari in Pastaza; these presumably were the basis for the species being included as occurring in Ecuador by Meyer de Schauensee (1970). Since then there have been a few sightings from Napo and Sucumbíos, including one bird seen by RSR and PJG et al. at La Selva on 20 Mar. 1989, another seen by J. Rowlett and R. A. Rowlett et al. at Cuyabeno in late Jul. 1991, and a few birds noted along the Maxus road southeast of Pompeya in 1993–1994 (RSR and S. Howell et al.). In addition, a female was collected (ANSP) and a few others were seen in Aug. 1993 at the Bermejo oil-field area north of Lumbaquí in Sucumbíos. There are also a few 1995–1996 sightings from the Bombuscaro sector of Podocarpus National Park (D. Wolf et al.); recorded up to 1000 m at this locality.

Monotypic.

Range: se. Colombia, s. Venezuela, and the Guianas to nw. Bolivia and Amazonian Brazil.

Sclerurus mexicanus   Tawny-throated   Tirahojas Golianteado
                             Leaftosser

Rare to locally uncommon on or near the ground inside humid and montane forest in the lowlands, foothills, and lower subtropical zone of both e. and w. Ecuador. In the east largely restricted to terra firme forest. In the west recorded south very locally to nw. Guayas (Loma Alta; Becker and López-Lanús 1997), El Oro (El Chiral), and w. Loja (one seen near El Limo on 18 Apr. 1993; RSR), but in recent years found primarily in Esmeraldas. Ranges up locally to about 1500 m, exceptionally as high as 1650 m (Panguri; ANSP); despite the elevation overlap, the Tawny-throated is not known to be actually sympatric with the Gray-throated Leaftosser (*S. albigularis*).

All members of the genus *Sclerurus* formerly went by the group name of leaf-scraper (e.g., Meyer de Schauensee 1966, 1970), though they never "scrape" with their feet but rather toss leaves with their bills. In recent years, however, the name "leaftosser," apparently first suggested by A. F. Skutch, has come into widespread usage and seems much to be preferred.

Two similar races are found in Ecuador, *peruvianus* in the east and the slightly darker *obscurior* in the west.

Range: s. Mexico to n. Colombia and w. Ecuador; the Guianas, s. Venezuela, and se. Colombia to Amazonian Brazil and n. Bolivia; e. Brazil.

Sclerurus albigularis   Gray-throated Leaftosser   Tirahojas Goligris

Very rare and local—overlooked?—on or near the ground inside montane forest in the foothills and lower subtropical zone on the east slope of the Andes from w. Napo (specimens from Ramos Urcu, head of Río Guataraco, and Guatic-ocha) south through Zamora-Chinchipe (old specimens from Zamora and Guayaba; one recent report from Coangos [Schulenberg and Awbrey 1997]). The Gray-throated Leaftosser seems unaccountably scarce in Ecuador. Recorded mostly from 1000 to 1700 m, once as low as 700 m.

One race: *zamorae*.

Range: locally in highlands of Costa Rica and w. Panama, mountains of n. Venezuela, and Andes from w. Venezuela to w. Bolivia.

*Sclerurus rufigularis   Short-billed Leaftosser   Tirahojas Piquicorto

Rare on or near the ground inside terra firme forest in the lowlands of e. Ecuador. The first two Ecuadorian records involve heretofore unpublished specimens which had long gone unrecognized because of confusion with the *peruvianus* race of the Tawny-throated Leaftosser (*S. mexicanus*). M. B. Robbins discovered an "Olalla y Hijos" specimen (ANSP) taken at Río Suno Abajo on 7 Mar. 1924; PJG discovered another specimen (MECN), taken by M. Proaño at "Río Corrientes" (a locality that unfortunately cannot be located) on 3 Aug. 1964. Recent records include two birds mist-netted and collected (ANSP) north of Tigre Playa near the Río San Miguel in Sucumbíos on 6 and 8 Aug. 1993, and sight reports as well as tape-recordings from La Selva, Sacha Lodge, Yuturi, and Kapawi Lodge in s. Pastaza (v.o.). It now appears that the Short-billed Leaftosser is a relatively widespread, though low-density, resident in suitable terra firme habitat in Ecuador. Recorded below 300 m.

One race: *brunnescens*.

Range: the Guianas, s. Venezuela, and se. Colombia to n. Bolivia and Amazonian Brazil.

Sclerurus caudacutus Black-tailed Leaftosser Tirahojas Colinegro

Uncommon to locally fairly common on or near the ground inside terra firme forest in the lowlands of e. Ecuador. Generally the most numerous *Sclerurus* in the eastern lowlands, but even so the Black-tailed Leaftosser is so inconspicuous—like its congeners—that it is not recorded very frequently, especially if one does not know its voice. Found mostly below 500 m, locally ranging as high as 950 m (Bermejo oil-field area north of Lumbaquí; ANSP).

One race: *brunneus*.

Range: se. Colombia, s. Venezuela, and the Guianas to nw. Bolivia and Amazonian Brazil; e. Brazil. More than one species may be involved.

Sclerurus guatemalensis Scaly-throated Tirahojas Goliescamoso
                          Leaftosser

Rare to uncommon and local on or near the ground inside humid forest in the lowlands and foothills of w. Ecuador from coastal n. Esmeraldas (old specimens from Pulún and San Javier) south to sw. Manabí (Machalilla National Park; ANSP), w. Guayas (above Salanguilla [N. Krabbe] and at Loma Alta [Becker and López-Lanús 1997]), and n. Los Ríos (an Oct. 1950 specimen from Quevedo; ANSP). There is also an old specimen record from Balzar in n. Guayas, and another specimen was collected at Hacienda Pacaritambo in Jun. 1962 (Brosset 1964), though it seems very doubtful that this species could still exist in the almost entirely deforested Río Guayas basin. In recent years the Scaly-throated Leaftosser has been recorded mainly from Esmeraldas and nw. Pichincha, though, as noted above, a few do persist locally in forest patches on the slopes of the coastal Cordillera de Colonche. There are even a few recent sightings from Tinalandia (PJG), though whether the species persists at Río Palenque seems doubtful. Scaly-throated Leaftossers do not seem to occur in the wet-forest belt of n. Esmeraldas, where during four years (1995–1998) of intensive field work at Playa de Oro none were recorded (O. Jahn et al.); apparently in this area it is confined to areas closer to the coast, mostly west of the Río Cayapas (P. Mena Valenzuela and O. Jahn et al.). Recorded up to 800 m at Tinalandia.

One race: *salvini*.

Range: se. Mexico to n. Colombia; w. Ecuador.

Lochmias nematura   **Sharp-tailed Streamcreeper**   **Riachuelero**

Rare to locally uncommon and inconspicuous on or near rocky streams and rivulets in montane forest in the foothills and lower subtropical zone on the east slope of the Andes. The streamcreeper remains a poorly known bird in Ecuador, but it has probably been overlooked. Recent reports include individuals mist-netted near San Rafael Falls on 13 Apr. 1979 (N. Krabbe) and others collected along the Loreto road north of Archidona on several occasions (WFVZ, ANSP), in the Río Huamboya valley of Sangay National Park in nw. Morona-Santiago (collected [MECN] in Apr. 1998; F. Sornoza), and near Zumba in s. Zamora-Chinchipe on 13 Dec. 1991 and in Aug. 1992 (ANSP, MECN). The Sharp-tailed Streamcreeper seems comparatively numerous in the Zumba region, and it has also been seen near the Río Bombuscaro entrance to Podocarpus National Park. Recorded from about 700 to 1300 m.
  One race: *sororia*.
  Range: locally from e. Panama, Colombia, and mountains of n. Venezuela south locally on east slope of Andes to nw. Argentina; s. Venezuela; s. Brazil, e. Paraguay, ne. Argentina, Uruguay.

Dendrocolaptidae   Woodcreepers

Twenty-eight species in 11 genera. Woodcreepers are exclusively Neotropical in distribution. The woodcreepers have recently often been treated as a subfamily within the Furnariidae (Ovenbirds), but they are here maintained as a family following Gill (1994) and the 1998 AOU Check-list.

Dendrocincla tyrannina   **Tyrannine Woodcreeper**   **Trepatroncos Tiranino**

Rare to locally fairly common in the lower and middle growth of montane forest, occasionally out to forest borders and adjacent clearings (especially when following mixed flocks), in the subtropical and temperate zones on both slopes of the Andes. On the west slope found south to El Oro (El Chiral) and w. Loja (Celica; B. Whitney). Recorded mainly from 1400 to 3100 m, but most numerous between 1800 and 2700 m.
  One race: nominate *tyrannina*. *Macrorhyncha*, described from e. Ecuador on the basis of its larger size, and occasionally even treated as a possible full species (e.g., *Birds of the World*, vol. 7), is here regarded as merely having been based on aberrant individual examples of nominate *tyrannina* (see J. Fjeldså and N. Krabbe, *Bull. B. O. C.* 106[3]: 120, 1986).
  Range: Andes from sw. Venezuela to s. Peru.

Dendrocincla fuliginosa   **Plain-brown Woodcreeper**   **Trepatroncos Pardo**

Fairly common to locally common and widespread in the lower and middle growth of humid forest and secondary woodland, to some extent also at borders and in adjacent clearings, in the lowlands and foothills of both e. and w.

Ecuador; in the east occurs in both terra firme and várzea, in the west in both humid and to a lesser extent also more deciduous situations. In w. Ecuador recorded south locally through Guayas (hills south of Río Ayampe, Chongón Hills, Manglares-Churute Ecological Reserve), El Oro, and w. Loja (Alamor). The Plain-brown Woodcreeper is especially often seen in attendance at swarms of army ants. In the west recorded mainly below 1000–1100 m, a few up to about 1400 m above Mindo and in w. Loja; in the east ranges up to about 1100 m (e.g., in the Zamora area).

Two similar races occur in Ecuador, *ridgwayi* in the west and *neglecta* in the east. *Neglecta*, described by W. E. C. Todd (*Ann. Carnegie Mus.* 31: 16–18, 1948), is very similar to *phaeochroa*, and future work may demonstrate that it is better synonymized into it. *Neglecta* has a slightly more whitish throat than *ridgwayi* and is more rufescent brown generally.

Range: Honduras to extreme nw. Peru, n. Bolivia, and Amazonian Brazil; ne. Brazil. Excludes Plain-winged Woodcreeper (*D. turdina*) of se. Brazil region.

*Dendrocincla merula   White-chinned   Trepatroncos Barbiblanco
                            Woodcreeper

Rare in the lower growth of terra firme forest in the lowlands of ne. Ecuador, where known only from Napo and Sucumbíos. The White-chinned Wood-creeper was first recorded in Ecuador at Limoncocha, where a specimen (LSUMZ) was taken on 12 Oct. 1976 (Tallman and Tallman 1977). Subsequently the species was seen by D. Pearson along the lower Río Yasuní in Sep. 1978, and it was recorded in small numbers at Imuyacocha off the lower Río Lagarto in both Dec. 1990 and Mar. 1991 (ANSP). In Aug. 1993, four examples were mist-netted and collected (ANSP, MECN) north of Playa Tigre near the Río San Miguel, likely an indication that the species is more numerous in the extreme northeast than elsewhere in Ecuador. The White-chinned Wood-creeper is almost invariably seen in attendance at swarms of army ants; at Imuy-acocha it was once seen at a swarm with the Plain-brown Woodcreeper (*D. fuliginosa*). Recorded only below 300 m.

One race: *bartletti*.

Range: se. Colombia, s. Venezuela, and the Guianas to n. Bolivia and Amazonian Brazil. More than one species may be involved.

Deconychura longicauda   Long-tailed   Trepatroncos Colilargo
                             Woodcreeper

Rare and seemingly local in the lower and middle growth of montane forest in the foothills and lower subtropical zone on the east slope of the Andes from w. Napo (Palm Peak [MCZ], and a few reports from along the Loreto road north of Archidona) south to Zamora-Chinchipe (Sabanilla and Panguri). The Long-tailed Woodcreeper also occurs in very small numbers (or very locally) in the eastern lowlands, where it is so far recorded only from Napo. There are a few recent reports from Jatun Sacha (B. Bochan), a sighting of one along the Maxus road southeast of Pompeya on 11 Jan. 1994 (RSR and F. Sornoza), and one seen and tape-recorded at Zancudococha on 17–18 Jan. 1995 (P. Coopmans et al.). A report from adjacent w. Nariño in sw. Colombia (Salaman 1994) is an indication that the species could also occur on the west slope, at least in Esmeraldas. Recorded up to 1700 m.

Only one race is definitely known from Ecuador, *connectens*. It is likely, however, that the birds found in the lowlands—where the species remains uncollected—will prove to represent a different subspecies, most likely *pallida*. *Pallida* differs from *connectens* in its more spotted or chevroned (less streaked) pattern on the breast.

Range: Honduras to sw. Colombia; locally from e. Ecuador, e. Peru, and n. Bolivia to Amazonian Brazil and the Guianas. More than one species may be involved.

Deconychura stictolaema	Spot-throated Woodcreeper	Trepatroncos Golipunteado

Rare and local—but probably overlooked—in lower growth inside terra firme forest in the lowlands of ne. Ecuador; thus far known from only a limited area in w. Napo. There are only a few Ecuadorian records of the Spot-throated Woodcreeper; these include the type specimen from Coca; two birds mist-netted, photographed, and released at Taracoa on 9–10 Jan. 1982 (RSR, PJG, and S. Greenfield); a male taken 40 km southeast of Coca on 25 Nov. 1988 (WFVZ); one seen and tape-recorded along the Maxus road southeast of Pompeya on 31 Oct. 1994 (N. Krabbe and M. Lysinger); and one seen south of the Río Napo near La Selva on 27 Jan. 1995 (M. Lysinger). Recorded only between about 200 and 350 m.

One race: *secunda*.

Range: locally from se. Colombia, e. Ecuador, and e. Peru to Amazonian Brazil and the Guianas.

Glyphorynchus spirurus	Wedge-billed Woodcreeper	Trepatroncos Piquicuña

Uncommon to (usually) common in the lower and middle growth of humid forest and mature secondary woodland, to a lesser extent also out to borders, in the lowlands of both e. (where mainly in terra firme) and w. Ecuador and the foothills on both slopes of the Andes. In the west ranges south to Guayas (on slopes of the Cordillera de Colonche in the northwest; on hills in Manglares-Churute Ecological Reserve in the southeast) and El Oro (an old specimen [AMNH] from La Chonta; recent records from Buenaventura). Recorded up to about 1700 m in both nw. and e. Ecuador, exceptionally as high as nearly 2000 m (Cordillera de Cutucú; Robbins et al. 1987).

Two very similar races are found in Ecuador, *castelnaudii* in the east and *sublestus* in the west. J. T. Zimmer (*Am. Mus. Novitates* 757, 1934) placed the birds of w. Napo (Río Suno area) in his subspecies *rufigularis*, which was described in the same paper. However, A. Capparella (pers. comm.) suggests that individual variation encompasses the minor differences believed by Zimmer (op. cit.) to exist between *castelnaudii* and *rufigularis*, and therefore that the latter should be synonymized into *castelnaudii*.

Range: se. Mexico to sw. Ecuador; e. Colombia, s. Venezuela, and the Guianas to n. Bolivia and Amazonian Brazil; e. Brazil. More than one species may be involved (fide A. Capparella).

Sittasomus griseicapillus   Olivaceous Woodcreeper   Trepatroncos Oliváceo

Fairly common to locally common in the mid-levels and subcanopy of forest (mainly deciduous and semihumid, in smaller numbers also in humid forest), secondary woodland, and borders in the lowlands and foothills of w. Ecuador from w. Esmeraldas south through El Oro and w. Loja. Rare to uncommon in the lowlands and foothills of e. Ecuador, in the lowlands primarily in várzea forest, at higher elevations mainly in second-growth and borders. Recorded mostly below about 1100 m, but in w. Loja ranges locally up to about 2000 m around Sozoranga; recorded as high as 1550–1700 m in s. Zamora-Chinchipe (Podocarpus National Park, Panguri, Cordillera del Cóndor), and there is an old record (Taczanowski and Berlepsch 1885) from as high as Hacienda Mapoto in Tungurahua (2100 m), though the birds may have actually occurred lower.

Two rather different races are found in Ecuador, *amazonus* in the east and *aequatorialis* in the west. *Amazonus* is larger and considerably darker generally (its wings and tail are rufous, not cinnamon-rufous) and is more grayish (not so buffy olive) below. In addition, the primary vocalizations of these two taxa differ strikingly. They are representative forms of what are almost certainly two separate species: what is now considered to be a very wide-ranging, polytypic species seems surely to consist of several allospecies. In view of the complexities involved, however, for the present we continue to regard all the taxa in *Sittasomus* as conspecific.

Range: cen. Mexico to n. Colombia and n. Venezuela; w. Ecuador and extreme nw. Peru; se. Colombia and s. Venezuela to n. Argentina and n. Uruguay.

Nasica longirostris   Long-billed Woodcreeper   Trepatroncos Piquilargo

Uncommon to locally fairly common in the mid-levels and subcanopy of humid forest in the lowlands of e. Ecuador where it is particularly numerous in várzea and riparian forest but also regularly occurs in terra firme in the vicinity of swampy places and larger streams. The spectacular Long-billed Woodcreeper is known primarily from the drainages of the Ríos Napo and Aguarico. Southward there is also a 19th-century record from Sarayacu, one from "Laguna de Siguin" (Berlioz 1937c), and recent reports from around Kapawi Lodge near the Río Pastaza close to the Peruvian frontier; the species likely ranges widely in the Río Pastaza drainage. It seems particularly numerous around Imuyacocha. Recorded up to about 400 m at Jatun Sacha (B. Bochan).

Monotypic.
Range: se. Colombia and sw. Venezuela to n. Bolivia and Amazonian Brazil.

| *Dendrexetastes rufigula* | Cinnamon-throated Woodcreeper | Trepatroncos Golicanelo |

Uncommon to locally fairly common in the canopy and (especially) borders of humid forest in the lowlands of e. Ecuador, principally in várzea and riparian forest but also (usually in smaller numbers) in terra firme, principally at borders. The Cinnamon-throated Woodcreeper is most often recorded from its far-carrying vocalizations, which are given almost entirely at dawn and dusk; the species seems particularly numerous at Imuyacocha. Ranges mostly below 500 m, but has been recorded up to 1200 m above Archidona (P. Coopmans).
One race: *devillei*.
Range: e. Ecuador and adjacent Colombia to n. Bolivia, Amazonian Brazil, the Guianas, and se. Venezuela.

| *Xiphocolaptes promeropirhynchus* | Strong-billed Woodcreeper | Trepatroncos Piquifuerte |

Uncommon and somewhat local in montane forest in the foothill, subtropical, and temperate zones on both slopes of the Andes; on the west slope ranges south to w. Loja (Alamor and Sozoranga). Also rare to uncommon and decidedly local in humid forest (especially in várzea, a few ranging into terra firme) in the lowlands of ne. Ecuador, where at least to date recorded only from Napo and Sucumbíos. Montane birds are recorded mostly from 1100 to 3000 m (locally as high as 3200 m at Cerro Mongus; Robbins et al. 1994), whereas birds of the eastern lowlands are recorded up to about 600 m. Whether the two populations come into contact in the foothills along the east slope of the Andes is uncertain.
Three races are found in Ecuador: nominate *promeropirhynchus* in most of the montane portion of the species' Ecuadorian range, *crassirostris* in w. El Oro and Loja, and *orenocensis* in the eastern lowlands. We follow Chapman (1926) in regarding *ignotus*, to which montane Ecuadorian birds have sometimes been referred (e.g., *Birds of the Americas*, pt. 4), to be indistinguishable from the nominate race. *Crassirostris* differs from the nominate race in being smaller with a purer white throat that shows no streaks. *Orenocensis* differs from the nominate race in its larger and paler (usually greenish or grayish horn, not blackish or gray) bill, redder iris, more rufescent coloration generally (especially below), blackish crown with strong buff streaking, and in having its pale supramalar stripe more or less obliterated by buff streaking (hence it is much less conspicuous). *X. orenocensis* (Great-billed, or Rusty-breasted, Woodcreeper), ranging in the Amazonian lowlands north to s. Venezuela, may prove to be a separate species from *X. promeropirhynchus* (ranging from Mexico south to the mountains of n. Venezuela and the Andes of Bolivia).

Range: locally from cen. Mexico to w. Panama, and from Venezuela to Bolivia, Amazonian Brazil, and the Guianas.

Dendrocolaptes certhia Amazonian Barred- Trepatroncos
                          Woodcreeper   Barreteado
                                            Amazónico

Uncommon to locally fairly common in the lower and middle growth of humid forest (both várzea and terra firme) in e. Ecuador. Recorded mostly below 600 m, but recorded at least locally along the eastern base of the Andes up to about 900 m at the Bermejo oil-field area north of Lumbaquí (M. B. Robbins) and at Miazi in the Río Nangaritza valley (Schulenberg and Awbrey 1997).

We follow C. F. Marantz (*Studies in Neotropical Ornithology Honoring Ted Parker*, Ornithol. Monogr. no. 48: 399–429, 1997) in recognizing trans-Andean *D. sanctithomae* (Northern Barred-Woodcreeper) as a species separate from *D. certhia*. This treatment was earlier suggested by E. O. Willis (*Bol. Mus. Par. Emílio Goeldi* 8: 151–216, 1992) and is based especially on their different primary vocalizations. One race of *D. certhia* occurs in Ecuador, *radiolatus*.

Range: se. Colombia, s. Venezuela, and the Guianas to n. Bolivia and Amazonian Brazil; ne. Brazil. Includes what was formerly often specifically separated as Concolor Woodcreeper (*D. concolor*) of s. Amazonian Brazil.

Dendrocolaptes sanctithomae Northern Barred- Trepatroncos
                               Woodcreeper   Barreteado
                                                 Norteño

Uncommon to locally fairly common in the lower and middle growth of humid forest in nw. Ecuador. The Northern Barred-Woodcreeper is most numerous in Esmeraldas, but smaller numbers occur as far south as Pichincha (e.g., a few still are present at Tinalandia). There is also a single recent report from as far south as nw. Guayas in the hills south of the Río Ayampe, where a pair was seen and tape-recorded on 1 Feb. 1996 (P. Coopmans et al.). Recorded up to about 800 m.

As noted under the previous species, *D. sanctithomae* is here treated as a species separate from the Amazonian-ranging *D. certhia* (Amazonian Barred-Woodcreeper). One race of *D. sanctithomae* occurs in Ecuador, *colombianus*.

Range: s. Mexico to w. Venezuela and nw. Ecuador.

Dendrocolaptes picumnus Black-banded Trepatroncos
                           Woodcreeper   Ventribandeado

Rare to locally fairly common in the lower and middle growth of humid forest (both terra firme and várzea) in the lowlands of e. Ecuador, ranging up in small numbers on the east slope of the Andes into the lower subtropical zone (Baeza and Cordillera de Cutucú). There are no records of the Black-banded Wood-creeper from the west slope of the Andes other than a 19th-century specimen labeled as having been taken at Cayandeled in Chimborazo (Berlepsch and Taczanowski 1884); we suspect, however, that this specimen's locality data are incorrect. Recently found to be numerous at Río Pacuyacu (P. Coopmans). Recorded at least locally as high as 1500–1900 m.

One race: *validus*.

Range: mountains from s. Mexico to w. Panama; locally from Colombia, Venezuela, and the Guianas to nw. Argentina and Paraguay. More than one species may be involved.

Xiphorhynchus picus	Straight-billed Woodcreeper	Trepatroncos Piquirrecto

Fairly common to common in várzea forest, riparian forest and woodland, and borders and nearby clearings in the lowlands of e. Ecuador, where recorded primarily from the Río Napo drainage upriver to around Jatun Sacha, and the Río Aguarico drainage upriver to at least Cuyabeno. The only site in the southeast from which the Straight-billed Woodcreeper appears to be known is in the vicinity of Kapawi Lodge near the Río Pastaza, but it likely will prove to be widespread in the Río Pastaza drainage. Recorded up to about 500 m at Tena (P. Coopmans).

The racial affinity of Ecuadorian birds remains uncertain, but they probably can be referred to *peruvianus* (if that taxon is considered distinct from *kienerii*).

Range: Panama to n. Colombia, n. Bolivia, and Amazonian and ne. Brazil.

Zimmer's Woodcreeper (*X. necopinus*) could well occur in riparian forest and on river islands along the Río Napo as far upstream as ne. Ecuador. This species, very similar to the Straight-billed Woodcreeper, is now known to occur on the lower Río Napo near Iquitos, Peru (where it was first located by J. Alvarez A. and B. M. Whitney, pers. comm.).

Xiphorhynchus obsoletus	Striped Woodcreeper	Trepatroncos Listado

Uncommon to fairly common in the lower growth of várzea forest and woodland in the lowlands of e. Ecuador, where recorded primarily from the Río Napo drainage upriver to around Limoncocha and Taracoa, and from the Río Aguarico drainage upriver to the Cuyabeno area. The only site in the southeast from which the Striped Woodcreeper is known is the vicinity of Kapawi Lodge near the Río Pastaza close to the Peruvian border, but the species likely will prove to be more widespread in the Río Pastaza drainage. Recorded only below 300 m.

One race: *palliatus*.

Range: e. Colombia, s. and e. Venezuela, and the Guianas to n. Bolivia and Amazonian Brazil.

Xiphorhynchus ocellatus   Ocellated Woodcreeper   Trepatroncos Ocelado

Uncommon to locally fairly common in the lower growth of humid forest (mainly in terra firme, smaller numbers in várzea) in the lowlands of e. Ecuador. The Ocellated is a relatively inconspicuous woodcreeper, its true numbers being better revealed by mist-netting. Recorded mainly below about 800 m, in smaller numbers up to about 1100 m.

One race: *napensis.*

Range: se. Colombia and extreme sw. Venezuela to n. Bolivia and Amazonian Brazil.

Xiphorhynchus spixii   Spix's Woodcreeper   Trepatroncos de Spix

Rare to apparently locally uncommon and apparently very local in lower growth of várzea forest in the lowlands of e. Ecuador in Napo and Sucumbíos. There are rather few records of this species from Ecuador, including material from Río Suno (2 specimens in AMNH; J. T. Zimmer, *Am. Mus. Novitates* 756, 1934), Cotapino (6 specimens taken by D. Norton in Jul. 1964; MCZ), Santa Cecilia (7 specimens taken in Jul.–Aug. 1971; UKMNH), and the Río Payamino (one specimen in ANSP). The only recent confirmed reports are of single birds mist-netted and collected at Puerto Napo in Dec. 1990 (MECN), the Río Pacuyacu on 24 Sep. 1992 (ANSP), and along the Maxus road on 12 May 1995 (MECN); at least some of the other reports likely result from confusion with the far more numerous and widespread—and extremely similar— Ocellated Woodcreeper (*X. ocellatus*). Recorded below about 400 m.

One race: *ornatus. Ornatus* has sometimes been treated (e.g., Meyer de Schauensee 1966, 1970; Hilty and Brown 1986) as a race of a species (*X. elegans*, Elegant Woodcreeper), when that is considered a separate species from *X. spixii*. However, we are convinced that both *elegans* and *ornatus* are better considered as subspecies of *X. spixii*. This treatment was first suggested by J. T. Zimmer (*Am. Mus. Novitates* 756, 1934) and also follows Ridgely and Tudor (1994).

Range: se. Colombia to n. Bolivia and Amazonian Brazil.

Xiphorhynchus guttatus   Buff-throated   Trepatroncos
                            Woodcreeper   Golianteado

Fairly common to common in mid-levels and subcanopy of humid forest and forest borders (in both terra firme and várzea) in the lowlands of e. Ecuador, ranging up in diminishing numbers into the foothills along the base of the Andes. Recorded mainly below about 700 m, locally as high as 900–1000 m (one seen north of Archidona on 3 Mar. 1991 [P. Coopmans]; also recorded at Miazi in Río Nangaritza valley [Schulenberg and Awbrey 1997]).

One race: *guttatoides.*

Range: se. Colombia, s. Venezuela, and the Guianas to n. Bolivia and Amazonian Brazil; e. Brazil. Excludes Cocoa Woodcreeper (*X. susurrans*) of

Central America, n. Colombia, and n. Venezuela (see Ridgely and Tudor 1994; AOU 1998). Includes *eytoni* of Amazonian Brazil, formerly considered a separate species (Dusky-billed Woodcreeper).

Xiphorhynchus lachrymosus	Black-striped Woodcreeper	Trepatroncos Pinto

Fairly common in very humid to humid forest and forest borders in the lowlands and lower foothills of nw. Ecuador. Recorded primarily from Esmeraldas, but there have also been a number of recent reports (first by P. Coopmans et al. 1995) from several localities in nw. Pichincha (along the road north of Simón Bolívar near Pedro Vicente Maldonado; east of Puerto Quito; and along the Pedro Vicente Maldonado-Quinindé road); there is also a single recent record from as far south as n. Manabí (female collected at Filo de Monos on 10 Jul. 1988; WFVZ). Recorded up to 450 m.
One race: nominate *lachrymosus*.
Range: Nicaragua to nw. Ecuador.

Xiphorhynchus erythropygius	Spotted Woodcreeper	Trepatroncos Manchado

Fairly common to locally common in humid forest, mature secondary woodland, and borders in the more humid lowlands of w. Ecuador, also ranging up in foothills and the lower subtropical zone on the west slope of the Andes. Ranges south to Guayas (on slopes of the coastal Cordillera de Colonche and hills in Manglares-Churute Ecological Reserve), El Oro, and w. Loja (Alamor and Celica areas). Recorded mainly below about 1400 m, but small numbers have been recorded as high as 1700 m in the northwest (Maquipucuna Reserve; P. Coopmans) and up to about 2000 m in the southwest.
One race: *aequatorialis*.
Range: e. Mexico to sw. Ecuador.

Xiphorhynchus triangularis	Olive-backed Woodcreeper	Trepatroncos Dorsioliváceo

Fairly common in montane forest, secondary woodland, and borders in the foothills and subtropical zone on the east slope of the Andes. Recorded mostly between 1000 and 2100 m, occasionally or in smaller numbers as low as 750 m near Archidona (WFVZ).
One race: nominate *triangularis*.
Range: mountains of n. Venezuela, and Andes from w. Venezuela to w. Bolivia.

Lepidocolaptes souleyetii	Streak-headed Woodcreeper	Trepatroncos Cabecirrayado

Fairly common in deciduous and semihumid forest and woodland, partially cleared areas and plantations in more humid regions, and desert scrub with scattered larger trees and cactus in the lowlands and on lower Andean slopes of w. Ecuador; recorded from Esmeraldas south through Loja (east to around Cata- mayo). Recorded mostly below 800 m, but locally up to 1800 m in w. Loja at Sozoranga (where it overlaps with the Montane Woodcreeper [*L. lacrymiger*]) and near Amaluza, and up to 1300 m around Mindo.

Two rather similar races are found in Ecuador, *esmeraldae* south into most of El Oro (e.g., at Santa Rosa and Portovelo) and nominate *souleyetii* in s. El Oro and Loja. Although the variation is clinal, the nominate race (especially examples from nw. Peru) has a longer bill.

Range: s. Mexico to n. Colombia, n. and e. Venezuela, and adjacent Brazil and Guyana; sw. Colombia to nw. Peru.

Lepidocolaptes lacrymiger	Montane Woodcreeper	Trepatroncos Montano

Uncommon to fairly common in montane forest and forest borders in the sub- tropical and temperate zones on both slopes of the Andes and locally on slopes above the central valley. On the west slope recorded south to w. Loja in the Sozoranga/Utuana region. Recorded mostly from 1500 to 3000 m, in smaller numbers down to about 1100 m.

L. lacrymiger of South America is here considered as a species separate from *L. affinis* (Spot-crowned Woodcreeper) of Middle America, this being based especially on its very different vocalizations; see the discussion in Ridgely and Tudor (1994). J. T. Zimmer (*Am. Mus. Novitates* 753, 1934) also treated the South American birds as a species distinct from those of Middle America. They were considered conspecific by Meyer de Schauensee (1966, 1970), but we are not aware of any published rationale for their having been so treated. Two races of *L. lacrymiger* are found in Ecuador, *aequatorialis* in virtually all of the species' Ecuadorian range and the similar *warscewiczi* in extreme s. Zamora- Chinchipe on the Cordillera Las Lagunillas (ANSP).

Range: mountains of n. Venezuela, and Andes from w. Venezuela to w. Bolivia; Sierra de Perijá and Santa Marta Mountains.

Lepidocolaptes albolineatus	Lineated Woodcreeper	Trepatroncos Lineado

Rare to uncommon and apparently local in the canopy and mid-levels of humid forest (mainly terra firme), in smaller numbers at forest edge and in secondary woodland, in the lowlands of e. Ecuador from Sucumbíos (Bermejo oil-field area north of Lumbaquí [ANSP] and Zancudococha [P. Coopmans]) southward through Pastaza and Morona-Santiago. The Lineated Woodcreeper seems almost certain to occur in adjacent s. Colombia, where it is unrecorded (Hilty and Brown 1986). Recorded mostly below 600 m, but ranges sparingly up along the base of the Andes to about 1000 m (seen north of Archidona along the Loreto road [J. Rowlett and R. A. Rowlett et al.]; also recorded at Miazi in the Río Nangaritza valley [Schulenberg and Awbrey 1997]). The Lineated Woodcreeper may overlap marginally with the similar Montane Woodcreeper

(*L. lacrymiger*), though as yet the two species have never actually been found together.

One race: *fuscicapillus*.

Range: s. Venezuela and the Guianas to n. Bolivia and Amazonian Brazil.

Campylorhamphus pucherani   Greater   Picoguadaña
                               Scythebill   Grande

Very rare and apparently local in the lower and mid-levels of montane forest in the upper subtropical and temperate zones on both slopes of the Andes. On the east slope recorded from a few localities in w. Napo (Sumaco area, Cuyuja), Morona-Santiago (along the Gualaceo-Limón road), and Loja (recent records from the Cajanuma sector of Podocarpus National Park, and at nearby Cerro Toledo [Krabbe et al. 1997]). On the west slope recorded only from Pichincha (an old specimen from Guanacilla [above Santo Domingo de los Colorados] and a few recent sightings from the Tandayapa area). Recorded mostly from 2000 to 2800 m, but has been found as low as 1850 m on the Peruvian side of the Cordillera del Cóndor (T. S. Schulenberg).

The Greater Scythebill was accorded Near-threatened status by Collar et al. (1994), presumably because of its (inexplicably) low numbers found throughout its range. This is certainly no less the case in Ecuador, where the reasons for its very evident rarity remain obscure. We therefore accord it Data Deficient status. A population does exist in Podocarpus National Park.

Monotypic.

Range: very locally in Andes from s. Colombia to Peru.

Campylorhamphus trochilirostris   Red-billed   Picoguadaña
                                     Scythebill   Piquirrojo

Uncommon to locally fairly common in the lower and middle growth of humid and semihumid forest in the lowlands of w. Ecuador, ranging from Esmeraldas south to w. Loja, where it also occurs up into the subtropical zone; locally it occurs also in more deciduous forest (e.g., at Cerro Blanco in the Chongón Hills). Rare and seemingly local in humid forest and mature secondary woodland in the lowlands of e. Ecuador. In the wet-forest belt of n. Esmeraldas the Red-billed Scythebill occurs only in forest borders, secondary woodland, and plantations, being replaced inside continuous forest by the Brown-billed Scythebill (*C. pusillus*; O. Jahn and P. Mena Valenzuela et al.). In the west the Red-billed is recorded mostly below 800 m except in the south (especially in Loja), where it ranges up regularly to 1500 m and in small numbers as high as 1900 m; in the east found mainly below 400 m, with a few records from along the base of the Andes to as high as 900 m.

Two races are found in Ecuador, *thoracicus* in the west and *napensis* in the east. These are similar aside from *napensis* having a somewhat shorter and proportionately more decurved bill.

Range: Panama to n. Colombia and Venezuela; extreme sw. Colombia, w.

Ecuador, and nw. Peru; Amazonia to n. Argentina and e. Brazil. More than one species may be involved.

| *Campylorhamphus procurvoides* | Curve-billed Scythebill | Picoguadaña Piquicurvo |

Very rare and local in the lower and middle growth of terra firme forest in the lowlands of ne. Ecuador. The Curve-billed Scythebill was first recorded in Ecuador at Cuyabeno in Sucumbíos, where two were seen and one tape-recorded by B. Whitney et al. on 2 Aug. 1992, and another here was seen and tape-recorded by J. Arvin et al. on 24 Aug. 1993. The only subsequent report involves a bird seen and tape-recorded at Jatun Sacha on 26 Sep. 1996 (P. Coopmans). The Limoncocha specimens (LSUMZ) obtained in 1975–1976 and reported as this species (Tallman and Tallman 1977) are actually examples of the *napensis* race of the Red-billed Scythebill (*C. trochilirostris*; fide J. P. O'Neill). To 400 m.

The presumed race occurring in Ecuador is *sanus*.

Range: locally from se. Colombia, s. Venezuela, and the Guianas to Amazonian Brazil.

| *Campylorhamphus pusillus* | Brown-billed Scythebill | Picoguadaña Piquipardo |

Rare to uncommon (and inconspicuous) in the lower and middle growth of montane forest in the foothills and lower subtropical zone on both slopes of the Andes, ranging down in very small numbers into the adjacent lowlands in n. Esmeraldas (e.g., at Playa de Oro; O. Jahn et al.). The Brown-billed Scythebill appears to be more or less continuously distributed along the east slope of the Andes, though actual records are relatively few, this doubtless mainly being a reflection of the species' inconspicuous behavior and its occurring at low densities. On the west slope it is known primarily from Pichincha northward, but there are also recent records from El Oro at Buenaventura (see Robbins and Ridgely 1990), and the species may well occur in the intervening area as well. The Brown-billed Scythebill was recently reported as being present in w. Loja at Sozoranga (Best et al. 1993), but given the species' close similarity to the Red-billed Scythebill (*C. trochilirostris*), we regard its presence at this locality, where on ecological grounds it would not be expected to occur, as unlikely. Brown-billed Scythebills appear to be especially numerous at El Placer in e. Esmeraldas. Recorded mostly from 600 to 2100 m (above El Chaco in w. Napo [ANSP] and at Cabañas San Isidro), locally down to about 100 m at Playa de Oro in n. Esmeraldas (O. Jahn et al.).

Subspecific diagnoses in *C. pusillus* are difficult. The race *guapiensis* was recently described (H. Romero-Zambrano, *Lozania* 31: 1–4, 1980) from the Pacific slope of sw. Colombia; on biogeographic grounds this would be expected to be the taxon found on the west slope of the Andes in Ecuador. Based on ANSP's substantial series of modern Ecuadorian specimens, however, we discern little if any consistent difference between specimens from the west and east slopes of the Andes, with any variation seen being individual (and not geographic). They certainly are closely similar and, if anything, the distinctions appear to be the reverse of what was described by Romero-Zambrano (op. cit.), with east slope birds being more rufescent on average, west slope birds somewhat darker and more olivaceous. For now we consider birds from throughout the Ecuadorian range of the species as belonging to the nominate race. ANSP has only one specimen from El Oro, which does stand apart on the basis of its wider ventral streaking, and if this difference proves to be discernible in any larger series that might be obtained in the future, El Oro birds might be separable.

Range: highlands of Costa Rica and Panama, and locally in Andes from sw. Venezuela to n. Peru; Sierra de Perijá.

Thamnophilidae   Typical Antbirds

Ninety-four species in 33 genera. Typical antbirds are exclusively Neotropical in distribution. Antthrushes and antpittas are separated as a full family (Formicariidae), following Sibley and Ahlquist (1990) and Sibley and Monroe (1990).

Cymbilaimus lineatus   Fasciated Antshrike   Batará Lineado

Uncommon to fairly common in the borders of humid forest and in mature secondary woodland in the lowlands of both e. and w. Ecuador. In the west recorded south to n. Manabí and s. Pichincha (e.g., at Río Palenque). The Fasciated Antshrike favors viny tangles at low and mid-levels (east of the Andes perhaps tending to occur more often higher above the ground) and is inconspicuous unless vocalizing. Recorded up to about 1000 m.

Two similar races are found in Ecuador, *fasciatus* in the west and *intermedius* in the east. The male's crown in *intermedius* is solid black, whereas in *fasciatus* it is variably barred with white.

Range: Honduras to nw. Ecuador; se. Colombia, s. Venezuela, and the Guianas to n. Biolivia and Amazonian Brazil.

Frederickena unduligera   Undulated Antshrike   Batará Ondulado

Rare to uncommon and seemingly rather local (but perhaps mainly just overlooked) in the undergrowth of terra firme forest in the lowlands of e. Ecuador, where it favors dense shrubby or viny areas around treefalls and along streams. Recorded up to about 700 m (Guaticocha; MCZ).

One race: *fulva*.

Range: se. Colombia to nw. Bolivia and w. Amazonian Brazil.

Taraba major   Great Antshrike   Batará Mayor

Uncommon to fairly common and widespread in the dense undergrowth of shrubby overgrown clearings, younger secondary woodland, and forest borders in the lowlands of both e. and w. Ecuador; more numerous in the west. In the east found primarily in várzea and riparian woodland. Recorded mostly below 1000 m, but small numbers have been found up to 1600 m in w. Loja.

Two similar races are found in Ecuador, *melanurus* in the east and *transandeanus* in the west.

Range: s. Mexico to nw. Peru, n. Argentina, and se. Brazil.

Sakesphorus bernardi Collared Antshrike Batará Collarejo

Fairly common to common in the undergrowth of deciduous woodland, shrubby second-growth, and desert scrub in the lowlands and foothills of sw. Ecuador in sw. Manabí and w. Guayas (east to around Guayaquil and on Isla Puná; a few have also been found recently at Manglares-Churute Ecological Reserve [Pople et al. 1997]) and in El Oro and w. Loja (east to near Malacatos and east of Amaluza). Occurs only in more arid regions. Recorded mostly below 1500 m, but ranging up in small numbers to 1850 m in s. Loja.

Two races are found in Ecuador, nominate *bernardi* in Manabí and Guayas and *piurae* in El Oro and Loja. The latter differs in being slightly larger, and males are slightly more rufescent on the back with more rufous on the wings.

Range: sw. Ecuador and nw. Peru.

Thamnophilus tenuepunctatus Lined Antshrike Batará Listado

Fairly common to common in shrubby regenerating clearings, secondary woodland, forest borders, and gardens in the foothills and lower subtropical zone on the east slope of the Andes, ranging out onto the adjacent eastern lowlands as far east as around Jatun Sacha. The Lined Antshrike is particularly numerous in various areas around Tena and Archidona, and in the Zumba region. Ranges mostly from 400 to 1400 m, in small numbers slightly higher; it probably is now following clearings upward.

Following Ridgely and Tudor (1994), *T. tenuepunctatus* from the east slope of the Andes from Colombia south to n. Peru is regarded as a species distinct from *T. palliatus* (Chestnut-backed Antshrike), found locally from cen. Peru to e. Brazil, on the basis of its dramatically different male plumage. Two similar races of *T. tenuepunctatus* are found in Ecuador, *tenuifasciatus* northward (south at least to the Macas region in n. Morona-Santiago [J. T. Zimmer, *Am. Mus. Novitates* 646, 1933]) and *berlepschi* southward (north at least to Cumbaratza in Zamora-Chinchipe [MCZ]).

Range: along eastern base of Andes from n. Colombia to n. Peru.

*Thamnophilus doliatus Barred Antshrike Batará Barreteado

Rare and local in dense thickets on islands in the Río Napo and lower Río Aguarico in the lowlands of ne. Ecuador. The first Ecuadorian record of the Barred Antshrike was a male seen and tape-recorded near La Selva on 23–24 Jul. 1988 (P. Scharf, RSR and PJG et al.); a few birds continue to be found on islands in the La Selva/Sacha area (v.o.). Barred Antshrikes have also been found on certain Río Napo islands near Yuturi Lodge (J. Moore; S. Howell). At least as yet there are no reports from the Río Pastaza drainage. A male collected of a pair seen near the mouth of the Río Lagartococha into the Río Aguarico on 9 Dec. 1990 (ANSP) remains the only specimen from Ecuador. Recorded below 250 m.

One race: *subradiatus*.

Range: ne. Mexico to n. Argentina and s. Brazil (but absent from parts of Amazonia).

Thamnophilus zarumae   Chapman's Antshrike   Batará de Chapman

Fairly common in the undergrowth of secondary woodland, regenerating clearings, and the borders of humid and montane forest in the foothills and subtropical zone of sw. Ecuador in El Oro and Loja, in the latter ranging east to near the Continental Divide east of Vilcabamba at San Pedro (Rasmussen et al. 1996) and the Río Angashcola valley east of Amaluza (R. Williams et al.). Recorded primarily between 800 and 2000 m, but locally it ranges considerably · higher in e. Loja, where it has been found as high as 2550 m above Amaluza and 2500 m at Utuana.

Following Ridgely and Tudor (1994) and Parker et al. (1995), *T. zarumae* is regarded as a species distinct from *T. doliatus* (Barred Antshrike), based on its different plumage (especially in males), vocalizations, and disjunct range. One race occurs in Ecuador, nominate *zarumae*.

Range: sw. Ecuador and nw. Peru.

Thamnophilus praecox   Cocha Antshrike   Batará de Cocha

Uncommon to fairly common but apparently very local in the lower growth of várzea forest, favoring thickets along small blackwater streams, in the lowlands of ne. Ecuador; at least up to now the species seems to have been found only in blackwater drainages. The Cocha Antshrike, which long was known only from a single female collected in 1926 (and described by J. T. Zimmer, *Am. Mus. Novitates* 917: 1–2, 1937), was rediscovered in Dec. 1990 by RSR, PJG, and P. Coopmans et al. at Imuyacocha, apparently very near the type locality near the Río Lagarto. Several specimens, including the first males, were taken at that time (ANSP, MECN). Soon thereafter a substantial population was found to exist there (though, so far as known, none occurs elsewhere in the drainage of the Río Lagarto); smaller numbers were also located south of the Río Aguarico at nearby Zancudococha. Subsequent to the species' distinctive vocalizations becoming known, a few pairs have also been found at La Selva, mainly along the Mandiyacu (first by PJG and G. Budney et al. in Jan. 1991). It turns out that, both there and at nearby Sacha Lodge, Cocha Antshrikes had been tape-recorded and even seen by several observers in the late 1980s, but their identity remained unrecognized at the time. Since then the Cocha Antshrike has also been located south of the Río Aguarico near the Río Pacuyacu, first on 28

Dec. 1992 (P. Coopmans); another was heard south of the Río Napo along a blackwater stream near Pompeya on 4 Feb. 1995 (RSR and F. Sornoza). Recorded below 300 m.

The endemic Cocha Antshrike was accorded Near-threatened status by Collar et al. (1994), but we cannot see that any factor is affecting it unfavorably, and we therefore do not consider the species to be at risk.

Monotypic.

Range: ne. Ecuador; seems likely to occur in adjacent Peru and Colombia.

*Thamnophilus cryptoleucus   Castelnau's Antshrike   Batará de Castelnau

Fairly common in the lower and middle growth of riparian forest on islands in the Río Napo and the lower Río Aguarico in the lowlands of e. Ecuador. The first report of the Castelnau's Antshrike in Ecuador occurred in Feb. 1985 when several pairs were found on an island in the Río Napo near Pompeya (RSR and T. Schulenberg et al.). There have been numerous subsequent sightings on various Río Napo islands downstream from the Primavera and Pompeya area; a pair was also mist-netted as far upriver as near Jatun Sacha in Feb. 1994 (B. Bochan). In Dec. 1990 the Castelnau's Antshrike was found to be equally numerous on islands in the lower Río Aguarico downstream from Zancudococha; a male was taken on 9 Dec. (ANSP), the first specimen for Ecuador. A small series (ANSP, MECN) was also obtained on a Río Napo river island near the mouth of the Río Aguarico in Sep. 1992. In the southeast the Castelnau's Antshrike is known only from islands in the Río Pastaza near Kapawi Lodge, close to the Peruvian frontier; it likely also ranges up the Río Pastaza to some extent. Recorded mostly below about 300 m, once as high as 400 m (Jatun Sacha, fide B. Bochan).

T. cryptoleucus is regarded as a species distinct from *T. nigrocinereus* (Blackish-gray Antshrike) of ne. South America, following most recent authors. Monotypic.

Range: locally along larger rivers in ne. Ecuador, e. Peru, and w. Amazonian Brazil.

Thamnophilus aethiops   White-shouldered   Batará Hombriblanco
Antshrike

Rare to uncommon in the lower growth of terra firme forest in the lowlands, foothills, and lower subtropical zone on the east slope of the Andes, favoring dense tangled thickets and viny areas. There have been no confirmed records of the White-shouldered Antshrike in the far eastern lowlands, with the easternmost acceptable reports being from Santa Cecilia (west of Lago Agrio; UKMNH specimens), a few sightings from south of the Río Napo near La Selva (P. Coopmans; M. Lysinger et al.), and Santiago in Morona-Santiago (M. B. Robbins). Recorded mostly between about 250 and 1000 m, with smaller numbers ranging up (perhaps mostly on outlying ridges?) in the subtropical zone to 1500–1700 m at Palm Peak on Volcán Sumaco (MCZ) and on the Cordillera de Cutucú (ANSP).

One race: nominate *aethiops*.
Range: s. Venezuela and se. Colombia to n. Bolivia and Amazonian Brazil; ne. Brazil. More than one species is perhaps involved.

Thamnophilus unicolor   Uniform Antshrike   Batará Unicolor

Uncommon to locally fairly common in the lower growth of montane forest in the foothills and subtropical zone on both slopes of the Andes. On the west slope, where it seems to be more numerous and widespread, recorded from Carchi (seen and tape-recorded on Cerro Golondrina in 1996; N. Krabbe) south to El Oro and w. Loja (Las Piñas, San Bartolo, west of Celica, and Tierra Colorada near Alamor). On the east slope, somewhat surprisingly, the Uniform Antshrike is not known to occur north of the Cordillera de Cutucú in Morona-Santiago; it is known primarily from Zamora-Chinchipe, where it is widespread in forested areas at appropriate elevations. Recorded mostly from 1000 to 2000 m, but locally as low as 700 m in the southwest.

Two similar races are found in Ecuador, nominate *unicolor* on the west slope and the slightly larger *grandior* on the east slope. N. Krabbe (pers. comm.) suggests that birds from Carchi may also be referable to *grandior*.
Range: locally in Andes from n. Colombia to s. Peru.

Thamnophilus schistaceus   Plain-winged Antshrike   Batará Alillano

Fairly common to common in the lower and middle growth of terra firme forest and secondary woodland, in smaller numbers also in várzea forest and at borders, in the lowlands of e. Ecuador. Ranges up on the east slope of the Andes to about 1000 m, in smaller numbers as high as 1300 m.

The species is often called the Black-capped Antshrike (e.g., Hilty and Brown 1986), though over the greater part of its range the species' crown is not black (see Ridgely and Tudor 1994).

Two races of are found in Ecuador, *capitalis* in most of the species' Ecuadorian range and *dubius* in the extreme southeast (the latter being recorded only from Zamora in Zamora-Chinchipe). *Dubius* differs in that males have less black on the crown.
Range: se. Colombia to n. Bolivia and w. and s. Amazonian Brazil.

Thamnophilus murinus   Mouse-colored Antshrike   Batará Murino

Uncommon and somewhat local in the lower and middle growth of terra firme forest (favoring hilly terrain) in the lowlands of e. Ecuador. The Mouse-colored Antshrike appears to be mainly or entirely recorded from the Río Napo southward, with small numbers being found as far north as Jatun Sacha, along the Maxus road southeast of Pompeya, and near the south bank of the Napo opposite La Selva. Reports from north of the Napo remain unconfirmed and may be the result of confusion with the Plain-winged Antshrike (*T. schistaceus*). South of the Río Napo, Mouse-colored and Plain-winged Antshrikes regularly occur together, though the Mouse-colored is more frequent in well-drained, often hilly terra firme forest; the Plain-winged is relatively more catholic in its habitat selection. Recorded up to 450 m.

One race: *canipennis.*

Range: se. Colombia, s. Venezuela, and the Guianas to extreme nw. Bolivia and Amazonian Brazil.

Thamnophilus atrinucha	Western Slaty-Antshrike	Batará Pizarroso Occidental

Uncommon to locally common in the lower growth of humid (especially) and semihumid forest and secondary woodland in the lowlands of w. Ecuador south to nw. and e. Guayas (Cerro La Torre and above Salanguilla [N. Krabbe]; Loma Alta [Becker and López-Lanús 1997]), El Oro, and w. Loja (an old record from Las Piñas; no recent reports). In the wet-forest belt of n. Esmeraldas the Western Slaty-Antshrike mainly occurs at forest borders and is less frequent inside continuous forest, where it is mainly found around treefall gaps (O. Jahn et al.). The Western Slaty-Antshrike remains particularly numerous at Río Palenque. Found mostly below 1100 m, locally and in small numbers up to 1350 m at Maquipucuna (P. Coopmans).

Following Ridgely and Tudor (1994) and M. L. Isler, P. R. Isler, and B. M. Whitney (*Studies in Neotropical Ornithology Honoring Ted Parker*, Ornithol. Monogr. no. 48: 355–381, 1997), *T. atrinucha* of Middle America to w. Ecuador is regarded as a species distinct from taxa found locally east of the Andes, on the basis of its different behavior and very different vocalizations. Formerly (e.g., Meyer de Schauensee 1966, 1970) all were united as *T. punctatus* (Slaty Antshrike). One race occurs in Ecuador, nominate *atrinucha.*

Range: Guatemala and Belize to nw. Venezuela and w. Ecuador.

*Thamnophilus leucogaster	Marañón Slaty-Antshrike	Batará Pizarroso del Marañón

Rare in the lower growth of deciduous woodland in the Río Marañón drainage of s. Zamora-Chinchipe near Zumba. The Marañón Slaty-Antshrike was first recorded in Ecuador on 12 Dec. 1991 when a male (ANSP) was collected by RSR from a pair seen, and several others were heard, in woodland patches along the Río Mayo. The species has since been seen on a few occasions in this immediate area, but no further specimens have been obtained. Recorded at 650 m.

Given the extremely small Ecuadorian range of the Marañón Slaty-Antshrike, its apparent scarcity within that range, and the fragmented and unprotected nature of its habitat there, we believe the species should be accorded Vulnerable status within the country. It was not considered to be at risk by Collar et al. (1992, 1994), doubtless because of the differing taxonomy then employed. None of its habitat is protected.

In their comprehensive taxonomic review of *T. punctatus*, M. L. Isler, P. R. Isler, and B. M. Whitney (*Studies in Neotropical Ornithology Honoring Ted*

Parker, Ornithol. Monogr. no. 48: 355–381, 1997) treated *T. leucogaster* as a taxon "of uncertain rank." We treat it as a distinct, monotypic species based on its distinctive voice and disjunct distribution. However, the poorly known taxon *huallagae*, which is found very locally in n. Peru, may be closely allied to *T. leucogaster* (perhaps even conspecific?).

Range: nw. Peru in upper Río Marañón valley and extreme s. Ecuador.

*Thamnophilus amazonicus   Amazonian Antshrike   Batará Amazónico

Rare to uncommon and apparently local in the lower growth of blackwater várzea woodland in the drainage of the Río Lagarto in extreme ne. Ecuador. First found in Ecuador on 7 Dec. 1990 when a singing male was obtained by P. Coopmans, RSR, and PJG et al. in *Macrolobium*-dominated woodland at Garzacocha (ANSP); additional birds were found here in Mar. 1991, and a solitary female was also recorded at Imuyacocha (specimens ANSP, MECN). Recorded at about 200 m.

One race: *cinereiceps*.

Range: se. Colombia, s. Venezuela, and the Guianas to n. Bolivia and Amazonian Brazil. More than one species may be involved.

Megastictus margaritatus   Pearly Antshrike   Batará Perlado

Rare and local in lower growth inside terra firme forest in the lowlands of e. Ecuador. There are relatively few Ecuadorian records of this distinctive antshrike. Single specimens from Montalvo and Río Conambo taken in the early 1960s, discussed by Norton et al. (1972), were believed to represent the first records from Ecuador; however, these authors overlooked an earlier specimen that had been taken at the mouth of the Río Lagartococha (J. W. Zimmer, *Am, Mus, Novitates* 558, 1932). Recent records include sightings from the Cuyabeno area, Río Pacuyacu (G. DeSmet, fide P. Coopmans), and along the Maxus road southeast of Pompeya (C. Canaday and RSR et al.); the species has also recently been found around Kapawi Lodge along the Río Pastaza near the Peruvian border and at the Tiputini Biodiversity Center (fide K. Zimmer). Recorded only below about 300 m.

Monotypic.

Range: locally from s. Venezuela and se. Colombia to w. Amazonian Brazil.

Pygiptila stellaris   Spot-winged Antshrike   Batará Alimoteado

Uncommon to fairly common in the mid-levels and subcanopy of humid forest (primarily in terra firme, smaller numbers in várzea) and mature secondary woodland in the lowlands of e. Ecuador; perhaps somewhat local, as it is apparently very rare or even absent at Jatun Sacha (B. Bochan). Recorded mostly below 400 m, but locally recorded somewhat higher (e.g., one was heard at 600 m, north of Canelos in Pastaza in Sep. 1996; N. Krabbe).

One race: *maculipennis.*

Range: se. Colombia, s. Venezuela, and the Guianas to n. Bolivia and Amazonian Brazil.

Thamnistes anabatinus   Russet Antshrike   Batará Rojizo

Uncommon to locally fairly common in the subcanopy and canopy of humid forest, tall secondary woodland, and borders in the foothills and lower subtropical zone on both slopes of the Andes, also locally out into the adjacent lowlands of n. Esmeraldas at Playa de Oro (O. Jahn et al.). In the west recorded south to se. Guayas (hills of Manglares-Churute Ecological Reserve; Pople et al. 1997), El Oro (La Chonta; numerous recent records from Buenaventura), and w. Loja (an old specimen from Las Piñas; no recent records). Also recorded recently on the coastal Cordillera de Mache in w. Esmeraldas (Cabeceras de Bilsa [Parker and Carr 1992] and Bilsa [J. P. Clay et al.]) and the Cordillera de Colonche in w. Guayas (Loma Alta; Becker and López-Lanús 1997). A single Oct. 1994 sighting from along the Maxus road southeast of Pompeya (English 1998) requires confirmation. Recorded mostly between 400 and 1300 m, but small numbers have been found locally as high as 1700 m on the east slope and as low as 100 m in n. Esmeraldas.

Two races are found in Ecuador, *intermedius* on the west slope and *aequatorialis* on the east slope. *Intermedius* differs in being more rufescent brown above than *aequatorialis* (which is more olivaceous brown above) and in having a distinctly rufescent crown. In addition, a male (ANSP) obtained on 25 Jul. 1992 at Panguri (1650 m) in s. Zamora-Chinchipe is intermediate in some characters toward the Peruvian race *rufescens* (being more ochraceous below, lacking the rufescent tone on crown, etc.).

Range: s. Mexico to sw. Ecuador and w. Bolivia.

Dysithamnus mentalis   Plain Antvireo   Batarito Cabecigris

Fairly common to locally very common in the lower growth of both humid and deciduous forest and secondary woodland in the lowlands and foothills of w. Ecuador from w. Esmeraldas (vicinity of Esmeraldas [city]), n. Manabí, and Pichincha south through Guayas, El Oro, and w. Loja. Uncommon to fairly common in the lower growth of montane forest in the foothills and lower subtropical zone on the entire east slope of the Andes. The Plain Antvireo is more numerous in the west—where in some forested areas it can be one of the most common birds—than it ever is on the east slope of the Andes. It is, however, replaced in the very wet forest of the northwest by the Spot-crowned Antvireo (*D. puncticeps*). In the west recorded up to about 1500 m, on the east slope mainly between 700 and 1700 m.

Three races occur in Ecuador: *aequatorialis* in the west, and *napensis* and *tambillanus* on the east slope. *Napensis* ranges along most of the east slope in Ecuador, with *tambillanus* replacing it in far se. Ecuador in s. Zamora-

Chinchipe (Cordillera del Cóndor and Panguri; ANSP). It is possible that *napensis* may ultimately best be synonymized into *tambillanus*; the two are very similar. However, western *aequatorialis* differs strikingly from east-slope birds; both sexes of *aequatorialis* are considerably paler below than east-slope birds, are tinged yellowish on the belly, and have more contrasting dark auriculars; east-slope males are quite uniform smoky gray below.

Range: s. Mexico to extreme nw. Peru, n. Bolivia, e. Paraguay, and se. Brazil (but absent from Amazonia).

Dysithamnus puncticeps	Spot-crowned Antvireo	Batarito Coronipunteado

Uncommon to locally common in the lower growth of very humid to humid forest in the lowlands and foothills of nw. Ecuador, where mainly recorded from Esmeraldas but found in small numbers south to n. Manabí (an old specimen [AMNH] from Pata de Pájaro) and Pichincha (Santo Domingo de los Colorados and Tinalandia); an Oct. 1950 specimen (ANSP) was taken as far south as Quevedo in n. Los Ríos. The Spot-crowned Antvireo appears to be particularly numerous at Bilsa; at Playa de Oro it is one of the most numerous antbirds in the foothills between 200 and 400 m (O. Jahn et al.). Recorded up to 800 m at Tinalandia.

One race: *flemmingi*.

Range: Panama to nw. Ecuador.

Dysithamnus leucostictus	White-streaked Antvireo	Batarito Albirrayado

Uncommon and rather local in the lower growth of montane forest in the subtropical zone on the east slope of the Andes from w. Sucumbíos (Puerto Libre on Río San Miguel; MECN) and w. Napo (north to Palm Peak and Río Guataraco; MCZ) south to Zamora-Chinchipe (south to above Zamora and in Podocarpus National Park). The species has not yet been found on the west slope of the Cordillera del Cóndor, though it should occur there as it has recently been found in adjacent Peru on the Cóndor's east slope at Río Comainas (Schulenberg and Awbrey 1997). The distribution of the White-streaked Antvireo seems likely to extend northward into adjacent s. Colombia along the east slope of the Andes at appropriate elevations, though it has not been recorded from this area (Hilty and Brown 1986). Recorded mostly between 1300 and 1800 m, locally and in small numbers down to 1100 m (e.g., in the Río Bombuscaro sector of Podocarpus National Park).

The species was formerly sometimes placed in the genus *Thamnomanes*; see T. S. Schulenberg (*Wilson Bull.* 95[4]: 505–521, 1983) for its shift to *Dysithamnus*. *D. leucostictus* is here regarded as a species separate from the highly disjunct *D. plumbeus* (Plumbeous Antvireo) of the lowlands of se. Brazil, following Ridgely and Tudor (1994). One race occurs in Ecuador, nominate *leucostictus*.

When specifically separated, *D. leucostictus* was formerly often called the White-spotted Antvireo, but the pattern on the male's underparts is one of streaking, not spotting; see Ridgely and Tudor (1994).

Range: locally in mountains of n. Venezuela; east slope of Andes in Colombia, Ecuador, and extreme n. Peru. More than one species may be involved.

Dysithamnus occidentalis   Bicolored Antvireo   Batarito Bicolor

Rare and apparently local (and easily overlooked) in the lower growth of montane forest and mature secondary woodland, ranging primarily along ridges around treefalls and in the dense growth springing up after landslides, in the subtropical zone on the east slope of the Andes mainly in w. Napo, with a recent record also from Morona-Santiago; also recently recorded from the far northwest. The Ecuadorian endemic race *punctitectus* is known from only seven specimens, five of them obtained in the early 20th century from "below Oya-cachi" (AMNH), "Sumaco abajo" (AMNH), and "reportedly near Baeza" (BMNH). Having gone unobserved for more than a half-century, the Bicolored Antvireo was relocated in Jan. 1991 well above the road to Loreto north of Archidona on the lower slopes of Volcán Sumaco at elevations of 1675 to 1750 m (B. Whitney, *Auk* 109[2]: 302–308, 1992). There since have been a few sightings from the north slope of the Cordillera de Huacamayos (first by D. Wolf) and below Cabañas San Isidro near Cosanga (first by RSR in Apr. 1993); in addition, two specimens (MECN) were obtained in the Río Abanico valley of Sangay National Park in nw. Morona-Santiago in Nov. 1996 (N. Krabbe and J. Palacio, *Cotinga* 11: 48, 1999). The only record from the west slope involves a single female seen and tape-recorded near El Corazón in the Reserva Cerro Golondrina in Carchi on 28 May 1996 (Krabbe and Palacio, op. cit.). On the east slope recorded between about 1500 and 2050 m, on the west slope at 2200 m.

The still poorly known Bicolored Antvireo was believed to warrant Vulner-able status by Collar et al. (1994). We concur. The species has an extremely limited known range, but it has at least been recently found in Sangay National Park (N. Krabbe and J. Palacio, op. cit.), and it may also occur in Cayambe-Coca Ecological Reserve.

D. occidentalis has usually been called the Western Antshrike (Meyer de Schauensee 1970) or, reflecting the genus in which it was being placed, Western Antvireo. Whitney (op. cit.) coined the English name of Bicolored Antvireo, based primarily on the female plumage. Given that the species is now known to occur on the east slope of the Andes as well as the west slope, and that other *Dysithamnus* are actually more "western" in distribution, the change to Bicol-ored seems warranted.

This species was formerly sometimes placed in the genus *Thamnomanes*; the genus *Thamnophilus* has also been suggested. Whitney (op. cit.) concluded that it was best included in *Dysithamnus*, and we follow this judgement. Only one subspecies is definitely known from Ecuador, the endemic *punctitectus* on the east slope. The subspecies found in Carchi remains to be determined; birds there most likely will be referable to nominate *occidentalis* of sw. Colombia. The two taxa resemble each other, but males of the nominate race are somewhat darker and blacker generally.

Range: locally on west slope of Andes in sw. Colombia and extreme nw. Ecuador, and on east slope of Andes in e. Ecuador.

Thamnomanes ardesiacus   Dusky-throated   Batará Golioscuro
                              Antshrike

Fairly common in the lower growth of humid forest (mainly terra firme) in the lowlands of e. Ecuador. Recorded mostly below 500 m, in small numbers and locally up to 900–950 m in foothills along the eastern base of the Andes (Warientza [WFVZ]; Bermejo oil-field area north of Lumbaquí [ANSP]; Miazi in the Río Nangaritza valley of Zamora-Chinchipe [Schulenberg and Awbrey 1997]).

This species is sometimes called the Saturnine Antshrike (e.g., Sibley and Monroe 1990), but this is only when *T. ardesiacus* is expanded to include what is at present generally considered to be a separate species, *T. saturninus* (with *huallagae*; Saturnine Antshrike), of e. Peru and Amazonian Brazil (south of the Amazon River). For an explanation as to why this is likely not the best course, see Ridgely and Tudor (1994).

One race: nominate *ardesiacus*. Males of this race are individually variable in the extent of black on their throats, though the extent is always limited, and most often little or none is present.

Range: the Guianas, s. Venezuela, and se. Colombia to n. Amazonian Brazil, e. Peru, and extreme nw. Bolivia.

Thamnomanes caesius   Cinereous Antshrike   Batará Cinéreo

Fairly common to common and widespread in the lower growth of humid forest (both terra firme and várzea, though especially in the former) in the lowlands of e. Ecuador. Recorded mostly below 600 m, in small numbers up to about 800 m.

One race: *glaucus*.

Range: se. Colombia, s. Venezuela, and the Guianas to ne. Peru, ne. Bolivia, and Amazonian Brazil; e. Brazil.

Myrmotherula brachyura   Pygmy Antwren   Hormiguerito Pigmeo

Fairly common to common in the mid-levels and subcanopy, especially favoring vine tangles, of humid forest (both terra firme and várzea) and secondary woodland in the lowlands of e. Ecuador. Recorded up to about 600 m in the Archidona area, with one report from as high as 900 m at Miazi in the Río Nangaritza valley of s. Zamora-Chinchipe (Schulenberg and Awbrey 1997).

Does not include *ignota*, often regarded as a trans-Andean race of *M. brachyura*. *M. brachyura* is then monotypic.

Range: se. Colombia, s. Venezuela, and the Guianas to n. Bolivia and Amazonian Brazil.

*Myrmotherula ignota   Griscom's Antwren   Hormiguerito de Griscom

Rare to locally fairly common (probably overlooked) in the mid-levels and sub-canopy of very humid and humid forest and forest borders in the lowlands and foothills of nw. Ecuador. First found in nw. Pichincha (along the road north of Simón Bolívar near Pedro Vicente Maldonado), where a pair was seen and tape-recorded on 7 Feb. 1995 (P. Coopmans, M. Lysinger, and T. Walla). Since then the species has been found on numerous occasions at several other sites in nw. Pichincha (v.o.), and it also has been located in adjacent sw. Imbabura (near the Salto del Tigre bridge over the Río Guaillabamba in Jul. 1997; P. Coopmans et al.). At about the same time the Griscom's Antwren was independently discovered in Esmeraldas, first at Playa de Oro (O. Jahn and P. Mena Valenzuela et al.) and then some 30 km southeast of San Lorenzo (a pair found on 25 Jul. 1997; D. Lane and F. Sornoza). The Griscom's Antwren, an inconspicuous species aside from its voice, is likely more widespread than this scatter of records would seem to indicate. The first specimen for Ecuador was a female (LSUMZ) obtained at the Pedro Vicente Maldonado site on 14 Jul. 1997 (F. Sornoza); another pair was collected (MECN) southeast of San Lorenzo. Recorded up to about 500 m.

Trans-Andean *M. ignota* is regarded as a species distinct from *M. brachyura* (Pygmy Antwren) of east of the Andes, based on its different song and slightly differing morphology.

Range: Panama to nw. Ecuador.

Myrmotherula obscura   Short-billed Antwren   Hormiguerito Piquicorto

Fairly common in the subcanopy and borders of humid forest (especially terra firme) in the lowlands of e. Ecuador. The ecological relationships of this species and the Pygmy Antwren (*M. brachyura*) are as yet only imperfectly understood, mainly because they are vocally so similar. The Short-billed Antwren seems to be the more frequent of the two in extensive terra firme forest (though even here it favors viny tangles around treefalls and other breaks in the canopy), with the Pygmy being more numerous in várzea and along streams and in other openings in terra firme. There is substantial overlap, however, and the two species are routinely found together at least at some localities. Recorded up locally to about 600 m in the Bermejo oil-field area north of Lumbaquí (ANSP) and near Archidona.

Monotypic.

Range: se. Colombia, e. Ecuador, ne. Peru, and w. Amazonian Brazil.

Myrmotherula multostriata   Amazonian   Hormiguerito
                                Streaked-Antwren   Rayado
                                                     Amazónico

Uncommon and rather local in shrubby growth around the margins of lakes and along streams in the lowlands of e. Ecuador, where recorded primarily from the drainages of the Río Napo and the lower Río Aguarico in Napo and Sucumbíos. There also are recent reports from the southeast around Kapawi Lodge along the Río Pastaza near the Peruvian border; the species will likely prove to be more widespread in this region. We suspect that the specimens from Gualaquiza that were recorded by Salvadori and Festa (1899b) as the Streaked Antwren (*M. surinamensis*), with which the Amazonian was formerly considered conspecific, actually refer to the Stripe-chested Antwren (*M. longicauda*). Recorded only below 300 m, and apparently not occurring in areas close to the base of the Andes; the westernmost known sites are Limoncocha and along rivers (e.g., the Tiputini and Yasuní) crossing the Maxus road southeast of Pompeya.

We follow Ridgely and Tudor (1994) and M. L. Isler, P. R. Isler, and B. M. Whitney (*Auk* 116[1]: 83–96, 1999) in excluding trans-Andean *M. pacifica* (Pacific Antwren) from this species. The latter authors, in their comprehensive review of the complex, further suggested separating *M. multostriata* of w. Amazonia as a species from *M. surinamensis* (Guianan Streaked-Antwren) of ne. South America. *M. multostriata* then becomes monotypic.

Range: se. Colombia to n. Bolivia and s. Amazonian Brazil.

Myrmotherula pacifica   Pacific Antwren   Hormiguerito del Pacífico

Fairly common in shrubby borders of humid forest and secondary woodland and in overgrown clearings and gardens in the lowlands of w. Ecuador south locally to n. Manabí (Río de Oro, Río Peripa), s. Pichincha (Río Palenque), n. Los Ríos (Quevedo; ANSP), e. Guayas (Bucay; ANSP), and nw. Azuay (seen by J. C. Matheus at San Luis in Jan. 1991). At least in n. Esmeraldas the species' original habitat was probably river-edge vegetation, from which it has spread into secondary habitats in upland areas, though remaining most numerous on the floodplain (O. Jahn). Recorded up to about 1300 m in the Mindo area, but mainly found below about 800 m.

M. pacifica is considered to be a species distinct from cis-Andean *M. multostriata* (Amazonian Streaked-Antwren) and *M. surinamensis* (Guianan Streaked-Antwren); see M. L. Isler, P. R. Isler, and B. M. Whitney (*Auk* 116[1]: 83–96, 1999). Monotypic.

Range: Panama to sw. Ecuador.

Myrmotherula longicauda   Stripe-chested   Hormiguerito
                              Antwren   Pechilistado

Locally fairly common in the subcanopy and mid-levels of secondary woodland, borders of humid forest, and clearings with scattered trees in the foothills and lower subtropical zone along the eastern base of the Andes. The Stripe-chested Antwren does not range very far out in the adjacent lowlands, the easternmost known site being Jatun Sacha. There is also an anomalous specimen (MECN), a male, labeled as having been obtained at "Valle Catamayo, Loja" by M. Olalla on 22 Nov. 1968; this locality is so far from the known range and habitat of the species that we must regard its provenance as dubious. The Stripe-chested Antwren seems particularly numerous in the Archidona and Tena regions. Recorded mostly between about 400 m (Jatun Sacha) and 1000 m, locally up to 1200 m.

Two similar races are found in Ecuador, *soderstromi* northward and *pseudoaustralis* southward (in Morona-Santiago and Zamora-Chinchipe).

Range: along eastern base of Andes from s. Colombia to w. Bolivia.

Myrmotherula hauxwelli   Plain-throated Antwren   Hormiguerito Golillano

Widespread but uncommon on or near the ground inside humid forest (both terra firme and várzea, but especially the former) in the lowlands of e. Ecuador. Recorded mostly below 400 m, in small numbers as high as 600 m.

One race: nominate *hauxwelli*. The race *suffusa*, to which Ecuadorian birds have been ascribed (see J. T. Zimmer, *Am. Mus. Novitates* 523: 10–14, 1932), is exceeedingly similar to the nominate race; female *suffusa* supposedly differ in being slightly more richly colored below and more rufescent (less olivaceous) above, but some recent specimens (ANSP) can be matched by recent Peruvian material of the nominate race. We thus propose that *suffusa* be synonymized.

Range: se. Colombia to n. Bolivia and s. Amazonian Brazil.

Myrmotherula fulviventris   Checker-throated   Hormiguerito
                                  Antwren   Ventrifulvo

Rare to locally common (more numerous northward) in the lower growth of very humid and humid forest and secondary woodland in the lowlands and foothills of w. Ecuador south to n. Manabí, s. Pichincha (Río Palenque), e. Guayas (old specimens from Bucay; recent sightings from the hills of Manglares-Churute Ecological Reserve [Pople et al. 1997]), and El Oro (Buenaventura [ANSP]; there also is an old specimen [AMNH] from the El Oro coastal plain at Santa Rosa). In addition, MECN has a male and a female supposedly obtained by M. Olalla at Guaiquichuma (1500 m) in Loja on 24 and 26 Dec. 1968, but this locality is so far south and at such a high elevation that—as with the specimen of the White-flanked Antwren (*M. axillaris*) purportedly taken at the same time—we strongly question the accuracy of its labeling. Recorded below 900 m.

One race: nominate *fulviventris*.

Range: Honduras to sw. Ecuador.

Myrmotherula haematonota   Stipple-throated   Hormiguerito
                                Antwren   Golipunteado

Fairly common but very local in the lower growth of terra firme forest in the lowlands of far ne. Ecuador in Sucumbíos. On 1–10 Aug. 1993 the Stipple-throated Antwren was found to be numerous north of Tigre Playa near the Río San Miguel; a series of specimens was taken (ANSP, MECN). Earlier published records of the Stipple-throated Antwren (e.g., Meyer de Schauensee 1966, 1970) from e. Ecuador involve what is now considered to be the Foothill Antwren (*M. spodionota*), then considered conspecific with the Stipple-throated Antwren. Recorded below 300 m.

Excludes *M. spodionota* of the east slope of the Andes. One race of *M. haematonota* occurs in Ecuador, *pyrrhonota*.

Range: s. Venezuela, se. Colombia, ne. Ecuador, ne. Peru, and w. Amazonian Brazil.

Myrmotherula fjeldsaai   Yasuní Antwren   Hormiguerito del Yasuní

Apparently local in dense undergrowth of terra firme forest, especially near streams, in the lowlands of e. Ecuador south of the Río Napo. This recently described species (N. Krabbe, M. L. Isler, P. R. Isler, B. M. Whitney, J. Alvarez A., and P. G. Greenfield, *Wilson Bull.* 111[2]: 157–165, 1999)—which for a time was confused with the White-eyed Antwren (*M. leucophthalma*) and Stipple-throated Antwren (*M. haematonota*)—is now known from several sites in Ecuador. The first specimen record involved a male (MECN) taken by R. Olalla at "Río Bufeo" on 3 Feb. 1963; it was first noticed in the MECN collection by PJG. This locality cannot be definitely located (Paynter 1993) but is believed (*fide* N. Krabbe) to be near the lower Río Bobonaza in Pastaza. Although apparently unpublished, this specimen may have been the basis for Meyer de Schauensee's (1970) inclusion of "Ecuador" in the range of the White-eyed Antwren. A female (LSUMZ) was taken at Tzapino in Pastaza in May 1976, and another female (WFVZ) was collected by F. Sibley at "Sunka," some 40 km southeast of Coca in Yasuní National Park on 24 Nov. 1988; both had also been provisionally considered to be White-eyed Antwrens. More recently, birds were found to occur in small numbers along the Maxus road southeast of Pompeya in 1994–1995, and a male and a female were then collected by N. Krabbe (MECN). The species has also been recorded at the Tiputini Biodiversity Center (v.o.) and Yuturi (S. Howell), and it has also been reported around Kapawi Lodge on the Río Pastaza near the Peruvian border. Recorded below about 250 m.

Krabbe et al. (op. cit.) suggested the English name of Brown-backed Antwren for this species. Although we recognize that this serves to distinguish the species from the (rufous-backed) Stipple-throated Antwren, it fails to differentiate it from numerous other antwren species that have equally brown backs. We prefer to name the species after the Ecuadorian national park in which the species was first recognized, and where the type specimen was obtained, Yasuní National Park; N. Krabbe (pers. comm.) is in accord with this change.

Monotypic. The very close relationship of this species with *M. haematonota* and *M. leucophthalma* was discussed by Krabbe et al. (op. cit.).

Range: e. Ecuador and ne. Peru.

Myrmotherula spodionota	Foothill Antwren	Hormiguerito Tropandino

Uncommon in the lower growth of montane forest in foothills along the eastern base of the Andes from Sucumbíos (Bermejo oil-field area north of Lumbaquí; M. B. Robbins) south to Zamora-Chinchipe. However, the only records from between w. Napo and Zamora-Chinchipe appear to be old specimens from Sarayacu in Pastaza, and from Morona-Santiago (Chiguaza [ANSP] and Jun. 1984 sightings from the Cordillera de Cutucú [RSR and M. B. Robbins]); in Apr. 1998 the species was also collected (MECN) in the Río Huamboya valley in Sangay National Park (F. Sornoza). Small numbers of the Foothill Antwren occur along the Loreto road north of Archidona, but the species appears to be most numerous and widespread in Zamora-Chinchipe. Recorded from 600 m (on the lower slopes of Volcán Sumaco; B. Whitney) to 1425 m (on the Cordillera del Cóndor; ANSP).

The species is sometimes called the Ecuadorian Antwren.

M. spodionota of the east-slope foothills is regarded as a species distinct from lowland-inhabiting *M. haematonota* (Stipple-throated Antwren), following Hilty and Brown (1986) and Ridgely and Tudor (1994). One race occurs in Ecuador, nominate *spodionota*.

Range: along e. base of Andes from s. Colombia (Salaman et al. 1999) to s. Peru.

Myrmotherula ornata	Ornate Antwren	Hormiguerito Adornado

Rare to uncommon and seemingly rather local in the lower growth of humid forest and secondary woodland in the lowlands of e. Ecuador and in the foothills on the east slope of the Andes. In some areas (perhaps mainly in the foothills, e.g., along the Loreto road north of Archidona) the Ornate Antwren favors stands of *Guadua* bamboo. In the lowlands, where bamboo is scarce in Ecuador, the species occurs primarily in dense tangled secondary growth (e.g., at treefalls), with numbers apparently highest in areas close to the Andes (for example, large series were obtained in the 1970s at Limoncocha (LSUMZ) and Santa Cecilia [UKMNH]). Farther east, numbers are small, though a few pairs have recently been located as far from the Andes as La Selva (P. Coopmans et al.) and around Kapawi Lodge along the Río Pastaza near the Peruvian border (v.o.). Recorded from about 250 to 1200 m.

One race: *saturata*.

Range: se. Colombia to n. Bolivia and s. Amazonian Brazil. More than one species may be involved.

Myrmotherula erythrura   Rufous-tailed   Hormiguerito Colirrufo
Antwren

Uncommon to locally fairly common in the lower and middle growth of humid forest (principally terra firme) in the lowlands of e. Ecuador. Recorded mainly below 700 m, but found locally up to 900 m in the Bermejo oil-field area north of Lumbaquí (sightings in Mar. 1993; M. B. Robbins).
One race: nominate *erythrura*.
Range: se. Colombia, e. Ecuador, e. Peru, and w. Amazonian Brazil.

Myrmotherula axillaris   White-flanked   Hormiguerito
Antwren   Flanquiblanco

Fairly common to common in the lower and middle growth of humid forest and mature secondary woodland in the lowlands of both e. and w. Ecuador; in the east found mainly in terra firme but also ranges into várzea and floodplain forest. In the west recorded south to n. Manabí, s. Pichincha (Río Palenque), e. Guayas (an old specimen from Bucay [ANSP]; recent sightings from hills of Manglares-Churute Ecological Reserve [Pople et al. 1997]), and nw. Azuay (Manta Real; ANSP). There is also a specimen (MECN) labeled as from Guaiquichuma in Loja, taken by M. Olalla on 23 Dec. 1968; as this is so much farther south than the species is known to occur, and at such a high elevation (1500 m), we strongly suspect that mislabeling is involved (as with the specimens of the Checker-throated Antwren [*M. fulviventris*] taken at the same time). Recorded up to about 900 m.

Two similar races are found in Ecuador, *albigula* in the west and *melaena* in the east. Females of these two taxa are extremely alike, but on average male *melaena* have more extensively black underparts.
Range: Honduras to sw. Ecuador; e. Colombia, s. and e. Venezuela, and the Guianas to n. Bolivia and Amazonian Brazil; e. Brazil. More than one species is likely involved.

Myrmotherula schisticolor   Slaty Antwren   Hormiguerito Pizarroso

Uncommon in the lower growth of montane forest in the foothills and lower subtropical zone on the east slope of the Andes. Uncommon to fairly common in the lower growth of humid forest and mature secondary woodland in the foothills and more humid lowlands of w. Ecuador south to n. Manabí, se. Guayas (old specimens from Bucay; recent reports from hills of Manglares-Churute Ecological Reserve [Pople et al. 1997]), El Oro, and w. Loja (Alamor area). Also found recently on the coastal Cordillera de Colonche in sw. Manabí (Cerro San Sebastián in Machalilla National Park [Parker and Carr 1992; ANSP]) and w. Guayas (Cerro La Torre and above Salanguilla [N. Krabbe]; Loma Alta [Becker and López-Lanús 1997]). In the west recorded locally and in small numbers as low as 100 m (but mainly above 400 m), and ranging up to about 1450 m at Maquipucuna and above Mindo; on the east slope found mostly between 900 and 1700 m.

Two races are found in Ecuador, *interior* in the east and nominate *schisticolor* in the west. Males of these two taxa are similar, but females differ strikingly. Nominate females are uniform grayish olive above, whereas *interior* females have a pure bluish gray mantle; *interior* females are also more richly colored below.

Range: s. Mexico to n. Venezuela and se. Peru.

Myrmotherula longipennis   Long-winged Antwren   Hormiguerito Alilargo

Rare to locally fairly common in the lower and middle growth of terra firme forest and mature secondary woodland in the lowlands of e. Ecuador. Where this species and the Río Suno Antwren (*M. sunensis*) occur together, the Long-winged appears to forage somewhat higher above the ground. Recorded up to about 500 m.

Two races are found in Ecuador, *zimmeri* in most of the species' Ecuadorian range and nominate *longipennis* only in the extreme northeast at Tigre Playa in Sucumbíos, just south of the Colombian border (where a series of specimens clearly referable to this race was obtained in Aug. 1993; ANSP). Males of the two taxa are quite similar, but females differ strikingly: *zimmeri* is dull bluish gray above and uniform ochraceous below, whereas the nominate race is brown above and ochraceous only on the throat and chest, extensively and contrastingly whitish on the median breast and belly. Whether these two taxa are in contact remains unknown.

Range: se. Colombia, s. Venezuela, and the Guianas to se. Peru, nw. Bolivia, and Amazonian Brazil.

Myrmotherula sunensis   Río Suno Antwren   Hormiguerito del Suno

Rare to locally uncommon in the undergrowth of terra firme forest and mature secondary woodland in the lowlands of e. Ecuador. Known primarily from Napo and Sucumbíos, the Río Suno Antwren has also been recently found around Kapawi Lodge near the Peruvian border on the Río Pastaza (v.o.). There are relatively few reports of this still poorly known antwren (some behavioral details were given by B. Whitney, *Auk* 111[2]: 473–474, 1994), doubtless in part because of its similarity to the more numerous Long-winged Antwren (*M. longipennis*). Recorded between about 250 and 500 m.

One race: nominate *sunensis*.

Range: locally in e. Ecuador, adjacent se. Colombia, ne. Peru, and w. Amazonian Brazil.

**Myrmotherula behni*   Plain-winged Antwren   Hormiguerito Alillano

Rare and local in the lower growth of montane forest in the foothills on the eastern slope of the Andes, perhaps mainly occurring on outlying ridges. The first record of the Plain-winged Antwren from Ecuador is an unpublished male (AMNH) that was obtained by L. Gómez at "Colimba" (1600 m) on 13 Jun.

1939; this locality cannot be precisely located, and though thought to be west of Macas in Morona-Santiago (Paynter 1993), we suspect that it may actually be on the slopes of the Cordillera de Cutucú. Another old specimen (AMNH), long considered of uncertain identity, was obtained at Río Avila (B. M. Whitney and J. F. Pacheco, *Bird Conserv. Int.* 5: 421–439, 1995). In recent years the Plain-winged Antwren has been found mainly on or near the slopes of Volcán Sumaco in w. Napo. D. Norton obtained eight specimens (MCZ) in 1963–1964 at "upper Río Pucuno" and "head of the Río Guataraco." F. Sibley took a male along the Loreto road above Archidona on 12 Mar. 1988 (WFVZ), and three additional specimens were obtained in this area on 10–11 Oct. 1992 (ANSP). The only record from elsewhere involves a single female seen near Chinapinza (1450 m) on the Cordillera del Cóndor in Zamora-Chinchipe on 13 Jun. 1993 (RSR). Recorded from about 800 to 1600 m; B. Whitney (*Auk* 111[2]: 471, 1994) found it to be most common around 1000 m, which is in accord with our (more limited) observations.

One race: nominate *behni*.

Range: tepui slopes of s. Venezuela and adjacent Brazil and Guyana; locally on east slope of Andes in se. Colombia and e. Ecuador.

Myrmotherula menetriesii   Gray Antwren   Hormiguerito Gris

Fairly common in the middle growth and subcanopy of terra firme forest in the lowlands of e. Ecuador. For the most part recorded below about 600 m, but the species has been taken as high as 1300 m on the upper Río Pucuno in w. Napo (MCZ), and there is also a male (ANSP) labeled as having been obtained at Macas (1000 m) in Morona-Santiago on 17 Aug. 1935.

One race: *pallida*.

Range: se. Colombia, s. Venezuela, and the Guianas to n. Bolivia and Amazonian Brazil.

Microrhopias quixensis   Dot-winged Antwren   Hormiguerito Alipunteado

Fairly common to common in the lower and middle growth and borders of humid forest and secondary woodland in the lowlands and lower foothills of w. Ecuador south locally to n. Manabí, Los Ríos (Quevedo [ANSP], and in the early 20th century recorded as far south as Babahoyo), se. Guayas (old specimens from Bucay [ANSP]; seen in 1996 in hills of Manglares-Churute Ecological Reserve [Pople et al. 1997]), and nw. Azuay (Manta Real, where a few were seen in Aug. 1991; RSR). Considerably less numerous in the lowlands of e. Ecuador, where local in dense thickets, sometimes of bamboo. In the west the Dot-winged Antwren locally shows a predilection for stands of bamboo; although it remains numerous at Río Palenque, there are only a few recent records from farther south. In the east the Dot-winged Antwren is apparently most numerous at Jatun Sacha (B. Bochan) and along the Maxus road southeast of Pompeya (RSR et al.). Recorded mostly below 500 m, in small numbers up to 750–800 m on both slopes (near Los Bancos [P. Coopmans] and near Archidona [WFVZ]).

Two races occur in Ecuador, *consobrina* in the west and the markedly larger nominate race in the east. Males of these taxa are similar aside from the difference in size, but females of the nominate race differ from *consobrina* in having a black throat (the underparts are all rufous-chestnut in *consobrina*).

Range: s. Mexico to sw. Ecuador; se. Colombia and the Guianas to n. Bolivia and Amazonian Brazil. More than one species may be involved.

Herpsilochmus dugandi   Dugand's Antwren   Hormiguerito de Dugand

Uncommon to locally fairly common in the canopy and borders of terra firme forest in the lowlands of e. Ecuador from Sucumbíos and Napo south into Morona-Santiago (Taisha) and Pastaza (Kapawi Lodge). This species and the other *Herpsilochmus* antwrens have been much under-recorded in the past because they usually remain high in the forest canopy, generally remaining undetected until their characteristic and oft-given vocalizations are recognized. Now that these vocalizations are more widely known to observers, these antwrens are known to be more numerous and wide-ranging than had been thought. All the Ecuadorian reports of the Dugand's Antwren are recent, the species only being described in 1945, from a specimen obtained in Caquetá in adjacent se. Colombia. Only three specimens have been taken in Ecuador: a male (MECN) obtained at Sarayacu on 7 Jul. 1968 and females (ANSP) obtained at the mouth of the Río Bobonaza in Pastaza on 25 Oct. 1937 and at Taisha in Morona-Santiago on 19 Aug. 1990. Recent field work demonstrates that the Dugand's Antwren is actually reasonably numerous at a number of localities, including Taisha, La Selva, Sacha Lodge, Cuyabeno, Jatun Sacha, along the Maxus road southeast of Pompeya, Tiputini Biodiversity Center, and Kapawi Lodge; there is also one report from Zancudococha (P. Coopmans). Recorded mostly below about 450 m, but locally up to 600 m near Miazal east of the Cordillera de Cutucú (B. M. Whitney and J. A. Alonso, *Auk* 115[3]: 559–576, 1998).

H. dugandi is regarded as a monotypic species distinct from the highly disjunct *H. sticturus* (Spot-tailed Antwren) of ne. South America, following most recent authors (e.g., Ridgely and Tudor 1994). Females differ markedly in plumage.

Range: se. Colombia, e. Ecuador, and ne. Peru.

*Herpsilochmus gentryi   Ancient Antwren   Hormiguerito Antiguo

Rare to uncommon and local in the canopy and subcanopy of terra firme forest in the lowlands of se. Ecuador. The Ancient Antwren—an only recently described species (B. M. Whitney and J. Alvarez A., *Auk* 115[3]: 559–576, 1998)—has been found in Ecuador only around Kapawi Lodge near the Río Pastaza, where it was first located by D. Stejskal in 1996. It inhabits forest mainly in well-drained upland sites where subsurface soils are quite sandy. Whitney and Alvarcz (op. cit.) suggest that the species may be somewhat more widely distributed in se. Ecuador. Recorded at about 200 m.

Given the lack of information on this range-restricted species, we accord it

Data Deficient status. None of its limited habitat receives formal protection, but none is known to be under any threat.

Monotypic.

Range: locally in ne. Peru and extreme se. Ecuador.

Herpsilochmus axillaris	Yellow-breasted Antwren	Hormiguerito Pechiamarillo

Rare to uncommon and seemingly local in the canopy and borders of montane forest in the foothills and lower subtropical zone along the eastern base of the Andes from Sucumbíos (San Rafael Falls and Reventador area) south through Zamora-Chinchipe (Cordillera del Cóndor and north of Zumba; ANSP). Small numbers of the Yellow-breasted Antwren occur with canopy flocks along the Loreto road north of Archidona and in the Zamora region. Recorded between about 800 and 1700 m.

One race: *aequatorialis*.

Range: locally in Andes of s. Colombia and on their east slope in Ecuador and Peru.

Herpsilochmus rufimarginatus	Rufous-winged Antwren	Hormiguerito Alirrufo

Uncommon and local in the canopy and borders of montane forest and secondary woodland in the foothills and lower subtropical zone along the eastern base of the Andes. In w. Ecuador the Rufous-winged Antwren is known only from recent sight reports, the first in Feb. 1985 at Río Palenque in s. Pichincha (RSR and T. Schulenberg et al.); small numbers have been found at this site on several occasions since. In Feb. 1991 the Rufous-winged Antwren was also found at Cerro San Mateo in w. Esmeraldas (Parker and Carr 1992), and on 1–3 Sep. 1991 several were heard and seen at Jauneche in Los Ríos (P. Coopmans). At certain sites on the east slope, notably along the Loreto road in w. Napo, this species and the Yellow-breasted Antwren (*H. axillaris*) occur in sympatry; in general the Rufous-winged tends to occur in more open, broken areas with patchier forest. On the east slope recorded mostly between about 600 and 1300 m, but there is one specimen (ANSP) recorded as having been taken as high as 1600 m at Colimba in Morona-Santiago; in the west found only below 200 m.

One race: *frater*. The racial affinity of the isolated western population remains unknown; it may represent an undescribed taxon.

Range: locally from e. Panama to n. Bolivia, e. and s. Brazil, e. Paraguay, and ne. Argentina (but lacking from much of Amazonia).

Drymophila caudata	Long-tailed Antbird	Hormiguero Colilargo

Uncommon to locally fairly common in the lower growth of montane forest and secondary woodland in the subtropical zone on both slopes of the Andes. On the west slope recorded south mainly to sw. Chimborazo (a number of old specimens from several localities in the Pallatanga, Cayandeled, and Chaguarpata region), but in recent years found primarily from Pichincha northward except for a pair seen and tape-recorded on 8 Feb. 1998 in nw. Azuay at Corona de Oro on the Puerto Inca-Molleturo road (P. Coopmans et al.). The Long-tailed Antbird is found almost entirely in association with stands of bamboo, both *Chusquea* spp. and *Guadua* spp. Recorded mostly from 1500 to 2600 m, at least on the east slope locally lower (down to 1100 m along the road to Loreto north of Archidona, and at an unusually low 750 m in Azuay).

One race: nominate *caudata*.

Range: mountains of n. Venezuela, and Andes from w. Venezuela to w. Bolivia; Sierra de Perijá and Santa Marta Mountains.

Drymophila devillei   Striated Antbird   Hormiguero Estriado

Uncertain. Apparently very rare and local in the lower growth of humid forest in the lowlands of ne. Ecuador, where known only from a few sites in w. Napo. Elsewhere in its range the species is strictly tied to stands of *Guadua* bamboo. The status of the Striated Antbird in Ecuador remains an enigma, and until very recently there were no recent records from the country. In the early 20th century several specimens (AMNH) were obtained in w. Napo (Río Suno, below San José, Avila); a male (ANSP) was obtained by T. Mena at "Payamino" on 7 Jan. 1950. The species was finally rediscovered in Ecuador on 8–9 Jun. 1999 when at least three pairs were seen and tape-recorded in mature secondary woodland with an extensive understory of *Guadua* bamboo at "El Para," some 17 km northwest of Archidona along the road toward Baeza (L. Navarrete and J. Moore). The extensive bamboo stands along the Loreto road north of Archidona presently appear to support only this species' congener, the Long-tailed Antbird (*D. caudata*), being perhaps at a little too high an elevation for the Striated. Extensive bamboo stands near the lower Río Payamino that in the early 1990s were specifically searched for the Striated Antbird supported neither *Drymophila*. Recorded between about 300 and 750 m.

The Striated Antbird's present numbers and distribution in Ecuador remain shrouded in mystery, and we therefore regard the species as warranting Data Deficient status in the country. Only the one population is known to exist, though it must be admitted that there does not appear to be anything all that special about it, and thus there is nothing to explain the species' very local distribution here. Elsewhere in its range this species is not considered to be at risk (Collar et al. 1992, 1994), and indeed it can be locally numerous.

One race: nominate *devillei*.

Range: se. Peru, n. Bolivia, and s. Amazonian Brazil; very locally in se. Colombia and ne. Ecuador.

Terenura callinota   Rufous-rumped Antwren   Hormiguerito Lomirrufo

Uncommon in the canopy and borders of montane forest in the foothills and subtropical zone on both slopes of the Andes; on the west slope recorded south to El Oro (Buenaventura; ANSP). This species and the other *Terenura* antwrens found in Ecuador have been much under-recorded in the past because of their habit of remaining high in the forest canopy, where they usually pass undetected until their vocalizations are known. Recorded between about 900 and 1800 m.

One race: nominate *callinota*.

Range: locally from Costa Rica to w. Venezuela and se. Peru; s. Guyana and s. Suriname.

Terenura humeralis	Chestnut-shouldered Antwren	Hormiguerito Hombricastaño

Rare to locally fairly common in the canopy of terra firme forest in the lowlands of e. Ecuador, mainly or entirely south of the Río Napo. As with the Dugand's Antwren (*Herpsilochmus dugandi*), with which the Chestnut-shouldered often occurs, recent field work by observers knowledgeable of the species' characteristic vocalizations has demonstrated that it is considerably more numerous and widespread in Ecuador than had been realized; because much of its range is relatively remote and difficult to access, however, the Chestnut-shouldered remains rather poorly known. Although the type material was obtained in the 19th century at Sarayacu, only a few Ecuadorian specimens have been taken subsequently. In Aug. 1990 the species was found to be reasonably numerous near Taisha in Morona-Santiago, where many canopy flocks contained a pair and where two females (ANSP, MECN) were obtained on 12 and 15 Aug.; MECN also has another female with no data. In 1994–1995 the Chestnut-shouldered Antwren proved to be quite numerous along the Maxus road southeast of Pompeya (RSR et al.); it also is known to occur on the southern side of the Río Napo at La Selva and at Sacha Lodge, at the Tiputini Biodiversity Center, and around Kapawi Lodge on the Río Pastaza near the Peruvian border. Likely it will prove to be widespread in terra firme forests of the southeast. Recorded up to about 600 m (north of Canelos in Pastaza; N. Krabbe); an old specimen supposedly procured from "Papallacta" (Goodfellow 1902) is surely mislabeled.

One race: nominate *humeralis*.

Range: e. Ecuador, e. Peru, nw. Bolivia, and w. Amazonian Brazil.

*[Terenura spodioptila	Ash-winged Antwren	Hormiguerito Alicinéreo]

Rare to perhaps locally uncommon in the canopy of terra firme forest in the lowlands and foothills of ne. Ecuador, where thus far known only from Sucumbíos north of the Río Aguarico. First recorded in Jul. 1991 when the species was seen at Cuyabeno (J. Rowlett and J. Arvin et al.); it has since been observed there by others. The only subsequent confirmed record involved several pairs seen and tape-recorded on 20–23 Aug. 1993 in the Bermejo oil-field area north of Lumbaquí (RSR and A. Capparella); unfortunately, a voucher specimen could not be obtained there. We regard the sighting reported by English (1998) from south of the Río Napo along the Maxus road south of Pompeya as requiring further confirmation. Additional field work will, however, likely reveal that the Ash-winged Antwren is more wide-ranging in extreme n. Ecuador. Recorded up to about 600 m (Bermejo).

Presumed race: *signata*.

Range: the Guianas, s. Venezuela, n. Amazonian Brazil, and extreme e. Colombia; locally in se. Colombia and ne. Ecuador.

Cercomacra cinerascens   Gray Antbird   Hormiguero Gris

Fairly common to locally common (especially by voice) in the canopy and borders of terra firme forest and secondary woodland in the lowlands of e. Ecuador. Favors viny tangles at mid-levels and in the subcanopy, often around treefalls. Recorded mostly below 700 m, locally and in small numbers up to 900 m in the Bermejo oil-field area north of Lumbaquí (M. B. Robbins) and at Miazi in the Río Nangaritza valley (Schulenberg and Awbrey 1997).

One race: nominate *cinerascens*.

Range: se. Colombia, s. Venezuela, and the Guianas to n. Bolivia and Amazonian Brazil. More than one species may be involved.

Cercomacra tyrannina   Dusky Antbird   Hormiguero Oscuro

Fairly common to common in the undergrowth and (especially) borders of humid and semihumid forest and secondary woodland in the lowlands of w. Ecuador south to n. Manabí, Los Ríos, e. Guayas, and coastal El Oro (ANSP female taken in 1950 at Piedras); less numerous and more local southward, in recent years not found south of se. Guayas in the Naranjal area. Around Playa de Oro in n. Esmeraldas the Dusky Antbird has been found almost exclusively in secondary and edge habitats, the few birds recorded in continuous forest being dispersing juveniles (O. Jahn et al.). The Dusky Antbird could also occur in the northeastern lowlands near the Colombian border, for it has been found not too far to the north in adjacent Colombia in w. Caquetá (Hilty and Brown 1986). Recorded mainly below about 800 m, locally as high as 1400 m near Mindo.

One race: nominate *tyrannina*. We follow G. R. Graves (*Studies in Neotropical Ornithology Honoring Ted Parker*, Ornithol. Monogr. no. 48: 21–35, 1997) in synonymizing *rufiventris*, in which Ecuadorian populations were formerly classified, with the nominate race.

Range: s. Mexico to sw. Ecuador, n. and e. Colombia, s. Venezuela, the Guianas, and n. Amazonian Brazil. Excludes the recently separated Willis's Antbird (*C. laeta*) of lower Amazonian and ne. Brazil.

Cercomacra nigrescens   Blackish Antbird   Hormiguero Negruzco

Uncommon to fairly common in the undergrowth of humid forest borders, secondary woodland, and shrubby clearings in the foothills and lower subtropical zone on the east slope of the Andes (perhaps more numerous southward); also uncommon and apparently local (though perhaps just overlooked) in secondary and riparian growth in the lowlands of e. Ecuador, so far mainly in the drainage of the Río Napo. The only lowland Ecuadorian records involve a male (LSUMZ) taken at Limoncocha on 22 Jul. 1976 (Tallman and Tallman 1977) and recent reports (including tape-recordings) from Jatun Sacha (B. Bochan), near the Hotel Auca in Tena (P. Coopmans et al.), along the Río Tiputini near the Maxus road southeast of Pompeya (RSR), and at the Tiputini Biodiversity Center (J. Moore et al.). Recorded mostly from 500 to 1800 m, locally in the lowlands below 500 m.

Aequatorialis is the race found in the montane part of this species' Ecuadorian range, but the specimen from Limoncocha is apparently referable to *fuscicauda*, which apparently is the taxon found in the lowlands; no further examples of the latter have been taken in Ecuador. Although available material of *fuscicauda* is limited, male *fuscicauda* are apparently more blackish, both above and below, than male *aequatorialis*, whereas female *fuscicauda* are reported to have a "brighter" back (J. T. Zimmer, *Am. Mus. Novitates* 500: 15, 1931). Given that the two taxa apparently replace each other altitudinally, with a gap, and that there are vocal differences (as first recognized in Ecuador by P. Coopmans), more than one species seems almost certainly to be involved in what is presently still called *C. nigrescens*.

Range: se. Colombia, e. Ecuador, e. Peru, n. Bolivia, and Amazonian Brazil; Suriname and French Guiana.

Cercomacra serva   **Black Antbird**   **Hormiguero Negro**

Uncommon to fairly common in the undergrowth of forest borders, secondary woodland, and riparian woodland in the lowlands and foothills of e. Ecuador. The ecological relationship of the Black Antbird to the very similar Blackish Antbird (*C. nigrescens*) remains to be fully elucidated, and in some places the two species occur together in very close proximity, for instance along the Loreto road north of Archidona (where the Blackish is the more numerous). Recorded up to about 1300 m; perhaps also occurs somewhat higher, but these reports are probably the result of confusion with the Blackish Antbird.

One race: nominate *serva*.

Range: se. Colombia, e. Ecuador, e. Peru, nw. Bolivia, and w. Amazonian Brazil.

Cercomacra nigricans   **Jet Antbird**   **Hormiguero Azabache**

Uncommon to locally common in tangled, viny undergrowth of semihumid and deciduous forest and woodland and in adjacent shrubby plantations and clearings in the lowlands of w. Ecuador from w. Esmeraldas (vicinity of Esmeraldas [city]) south to nw. Guayas (Río Ayampe) and coastal El Oro (numerous old specimens from Santa Rosa and Piedras, where the species formerly must have been numerous; curiously, however, there are no recent reports). In addition,

ANSP has two specimens labeled as having been taken by W. Clarke-Macintyre at "Valladolid, Loja" in 1938; being so far to the south, the locality seems suspect and in any case it cannot be located (Paynter 1993). In recent years the Jet Antbird has not been found south of se. Guayas in the Naranjal area (RSR). It is particularly numerous at Jauneche in Los Ríos (P. Coopmans). Recorded below about 500 m.

The species is considered to be monotypic, despite the disjunct range of the Ecuadorian population (and populations on the offshore Pearl Islands in Panama).

Range: Panama to n. Colombia; w. Ecuador; sw. Venezuela and along east base of Andes in Colombia; n. and e. Venezuela.

Myrmoborus leucophrys   White-browed Antbird   Hormiguero Cejiblanco

Uncommon to locally fairly common in the dense undergrowth of secondary woodland, regenerating clearings, and borders of terra firme forest in the foothills and adjacent lowlands of e. Ecuador. In the northeast the White-browed Antbird has not been recorded any distance away from the Andes, with the sites farthest to the east being Cotapino (MCZ), Jatun Sacha, and Sarayacu (MECN). The species has recently been reported from around Kapawi Lodge along the Río Pastaza near the Peruvian border, however, suggesting that it may range more widely in the Río Pastaza drainage. To 1100 m.

One race: nominate *leucophrys*.

Range: se. Colombia, s. Venezuela, and the Guianas to n. Bolivia and Amazonian Brazil.

*Myrmoborus lugubris   Ash-breasted Antbird   Hormiguero Pechicinéreo

Locally fairly common in the undergrowth of riparian forest and woodland on one Río Napo island in the lowlands of extreme ne. Ecuador. Thus far the Ash-breasted Antbird is known only from a single island situated near the mouth of the Río Aguarico into the Río Napo, where a small series (ANSP, MECN) was obtained on 19–21 Sep. 1992. It remains possible that the species may range somewhat farther upriver on islands in the Río Napo. Recorded at about 150 m.

One race: *berlepschi*.

Range: along Amazon River and some of its tributaries in Brazil, se. Colombia, ne. Peru, and adjacent Ecuador.

Myrmoborus myotherinus   Black-faced Antbird   Hormiguero Carinegro

Uncommon to locally common in the undergrowth of humid forest (primarily terra firme, smaller numbers in várzea) and secondary woodland in the lowlands of e. Ecuador. Recorded mostly below 700 m, in small numbers locally as high as 1300 m (upper part of Loreto road; RSR and PJG).

One race: *elegans*. *Napensis*, the taxon in which Ecuadorian birds were formerly placed, was synonymized into *elegans* by J. Haffer and J. W. Fitzpatrick (*Neotropical Ornithology*, Ornithol. Monogr. no. 36: 152–157, 1985).

Range: se. Colombia and s. Venezuela to nw. Bolivia and Amazonian Brazil.

Hypocnemis cantator   Warbling Antbird   Hormiguero Gorjeador

Uncommon to locally common in dense lower growth at the borders of humid forest and in forest openings (especially in terra firme, with smaller numbers in várzea and riparian forest) and secondary woodland in the lowlands of e. Ecuador. Recorded mostly below 600 m, locally up to 900 m (e.g., around Zamora).

One race: *saturata*. J. T. Zimmer (*Am. Mus. Novitates* 538, 1932) expressed reservations concerning the distinctness of this race, which is very similar to *peruviana*. It may well be best to synonymize it.

Range: se. Colombia, s. Venezuela, and the Guianas to n. Bolivia and Amazonian Brazil.

Hypocnemis hypoxantha   Yellow-browed   Hormiguero
                           Antbird   Cejiamarillo

Uncommon to locally fairly common in the lower growth of terra firme forest in the lowlands of e. Ecuador. This attractive antbird is much more strictly confined to terra firme forest than is the more numerous and widespread Warbling Antbird (*H. cantator*), which seems to replace the Yellow-browed at forest borders and in openings. Recorded mostly below about 400 m, locally up to 600 m (north of Canelos in Pastaza; N. Krabbe).

One race: nominate *hypoxantha*.

Range: se. Colombia, e. Ecuador, ne. Peru, and w. Amazonian Brazil; e. Amazonian Brazil.

*Hypocnemoides melanopogon   Black-chinned   Hormiguero
                                 Antbird   Barbinegro

Locally fairly common in the undergrowth of várzea forest (almost always near water, along streams and lakeshores) in the lowlands of ne. Ecuador. The Black-chinned Antbird was first reported from Ecuador from a sighting of a pair along the lower Río Yasuní on 19 Sep. 1976 (RSR). It was then found to be numerous in the Imuyacocha area near the Río Lagarto in Dec. 1990 and Mar. 1991 (ANSP and MECN) and at Cuyabeno in late Jul. 1991 (J. Rowlett and R. A. Rowlett et al.; H. Kasteleijn); small numbers were also found along the Río Pacuyacu in Sep. 1992 (RSR and T. J. Davis), at Zancudococha in Jan. 1995 (P. Coopmans et al.), and at Yuturi Lodge in Jul. 1996 (S. Howell). Recorded below 200 m.

One race: *occidentalis*.

Range: se. Colombia, s. Venezuela, and the Guianas to ne. Peru and Amazonian Brazil.

Hylophylax naevioides   Spotted Antbird   Hormiguero Moteado

Uncommon to common in the undergrowth of very humid and humid forest and mature secondary woodland in the lowlands of nw. Ecuador, where now found primarily in Esmeraldas. There are 19th-century records of the Spotted Antbird from as far south as ne. Guayas (Chimbo), but otherwise the only recent report from south of n. Manabí and s. Pichincha (where it seems to have disappeared in the 1980s) involves a population found in Aug. 1996 near the summits of hills in Manglares-Churute Ecological Reserve in se. Guayas (Pople et al. 1997). Recorded mostly below 300 m, small numbers up to about 500 m at Bilsa.
One race: nominate *naevioides*.
Range: Honduras to sw. Ecuador.

Hylophylax naevia   Spot-backed Antbird   Hormiguero Dorsipunteado

Uncommon to fairly common in the undergrowth of humid forest (principally terra firme, smaller numbers in várzea and floodplain forest) in the lowlands of e. Ecuador. Recorded mostly below 700 m, ranging up in small numbers along the eastern base of the Andes to 1000–1100 m, exceptionally to 1250 m (collected [MECN] in Apr. 1998 in the Río Huamboya valley of Sangay National Park in nw. Morona-Santiago; F. Sornoza).
One race: *theresae*.
Range: se. Colombia, s. Venezuela, and the Guianas to n. Bolivia and Amazonian Brazil. More than one species may be involved.

*Hylophylax punctulata   Dot-backed   Hormiguero Lomipunteado
                              Antbird

Rare to locally fairly common in the undergrowth of várzea forest in the lowlands of e. Ecuador; almost always occurs along streams or at the edges of lakes, usually in blackwater areas. The Dot-backed Antbird was first recorded from Ecuador by Orcés (1974), based on a specimen (MECN) obtained in Jan. 1963 by R. Olalla at "Río Bufeo" in s. Pastaza near the Río Bobonaza. Subsequent to its distinctive primary vocalization becoming known, it has been found at several other localities in the lowlands of ne. Ecuador, including Taracoa, La Selva, Cuyabeno, the Río Pacuyacu, and Imuyacocha; it may be particularly numerous around Sacha Lodge (L. Jost) and Yuturi Lodge (S. Howell). There are also recent reports from around Kapawi Lodge near the Peruvian border on the Río Pastaza. Only one additional specimen has been taken, a male obtained at Imuyacocha on 23 Mar. 1991 (ANSP). Recorded below about 300 m.
One race: nominate *punctulata*.

Range: locally in s. Venezuela, se. Colombia, e. Ecuador, ne. Peru, n. Bolivia, and Amazonian Brazil.

Hylophylax poecilinota	Scale-backed Antbird	Hormiguero Dorsiescamado

Uncommon to fairly common in the undergrowth of humid forest (principally terra firme) in the lowlands of e. Ecuador. Unlike the Spot-backed Antbird (*H. naevia*), the Scale-backed is frequently found in attendance at army antswarms. Recorded mostly below 700 m, in small numbers up along the eastern base of the Andes to about 1100 m.

The spelling of the species name is apparently *poecilinota* and not *poecilonota* (contra Meyer de Schauensee 1966, 1970). One race occurs in Ecuador, *lepidonota*.

Range: se. Colombia, s. Venezuela, and the Guianas to n. Bolivia and Amazonian Brazil.

Dichrozona cincta	Banded Antbird	Hormiguero Bandeado

Rare to locally uncommon on the ground inside terra firme forest (favoring well-drained ridges) in the lowlands of e. Ecuador. A few pairs can be found in forest on the south side of the Río Napo opposite La Selva and Sacha Lodge. Recorded up to about 450 m.

One race: *stellata*.

Range: se. Colombia and sw. Venezuela to n. Bolivia and w. and cen. Amazonian Brazil.

*Myrmochanes hemileucus	Black-and-white Antbird	Hormiguero Negriblanco

Uncommon to locally fairly common in younger successional growth on islands in the Río Napo in the lowlands of ne. Ecuador. The first Ecuadorian report is of a pair seen by T. Schulenberg and RSR et al. on 19 Mar. 1985 near Pompeya; the species having been recorded as occurring in "e. Ecuador" by Meyer de Schauensee (1970, p. 246) is presumed to have been based on specimens obtained in territory now in Peru but previously part of Ecuador. The Black-and-white Antbird has since been found by various observers on numerous Río Napo islands (including several near La Selva), and it probably occurs on all islands with appropriate habitat downstream to the Peruvian border. However, in Dec. 1990 it was not found in similar habitat on islands along the lower Río Aguarico, nor has it been found along the Río Pastaza near Kapawi Lodge. The first and only specimen from Ecuador is a male obtained on a Río Napo island near the mouth of the Río Aguarico on 20 Sep. 1992 (ANSP). Recorded below 300 m.

Monotypic.

Range: islands in Amazon River and some of its tributaries in Brazil, n. Bolivia, se. Colombia, e. Peru, and ne. Ecuador.

Schistocichla leucostigma   Spot-winged   Hormiguero Alimoteado
Antbird

Rare to uncommon in the undergrowth of terra firme forest in the lowlands of e. Ecuador, strongly favoring the vicinity of small forest streams and swampy places. Recorded mostly below 600 m, in small numbers as high as 1100 m. A 19th-century record from Hacienda Mapoto (situated at over 2000 m in Tungurahua) seems unlikely to have actually been procured at anywhere near so high an elevation.

This and the next species were formerly usually placed in the genus *Percnostola* (e.g., Meyer de Schauensee 1966, 1970). We prefer to resurrect the genus *Schistocichla* (which last apparently was employed by J. T. Zimmer, *Am. Mus. Novitates* 500, 1931; a rationale for its merger seems never to have been given) for the species *leucostigma*, *schistacea* (Slate-colored Antbird) and the extralimital *caurensis* (Caura Antbird). This is based on their rounder heads, different wing patterns, and different vocalizations as compared with "true" *Percnostola*; see also Ridgely and Tudor (1994). One race of *S. leucostigma* occurs in Ecuador, *subplumbea*.

Range: se. Colombia, s. Venezuela, and the Guianas to nw. Bolivia and Amazonian Brazil. More than one species is perhaps involved.

*Schistocichla schistacea   Slate-colored Antbird   Hormiguero Pizarroso

Rare to uncommon and local in the undergrowth of terra firme forest in the lowlands of ne. Ecuador, where thus far found only in Sucumbíos. In Ecuador the Slate-colored Antbird is mainly known from the Cuyabeno area, where it was first recorded in Jun. 1989; it was then observed, mist-netted, and photographed by C. Canaday (photos to VIREO). The only record from elsewhere involves a single female that was mist-netted and collected north of Tigre Playa near the Río San Miguel on 2 Aug. 1993 (ANSP); this remains the only Ecuadorian specimen. Given its occurrence so close to the Colombian border, the Slate-colored Antbird seems virtually certain to occur in adjacent Putumayo, Colombia, from which it is unrecorded (Hilty and Brown 1986). Recorded only between about 200 and 250 m.

Monotypic.

Range: ne. Ecuador to e. Peru, extreme se. Colombia, and w. Amazonian Brazil.

Sclateria naevia   Silvered Antbird   Hormiguero Plateado

Uncommon to locally fairly common in the undergrowth of várzea forest and woodland, along the borders of lakes and sluggish streams, and in larger swampy areas in terra firme forest (e.g., it is regular in stands of *Mauritia* palms along the Maxus road southeast of Pompeya; RSR et al.) in the lowlands of e. Ecuador. Recorded up to about 450 m in the Puerto Napo area (P. Coopmans).

One race: *argentata*.

Range: se. Colombia, s. Venezuela, and the Guianas to n. Bolivia and Amazonian Brazil.

Neoctantes niger   Black Bushbird   Arbustero Negro

Rare to locally uncommon in dense lower growth of humid forest (mainly terra firme) and secondary woodland in the lowlands of e. Ecuador. The Black Bushbird may especially favor areas of dense tangled growth around treefalls. Recorded up to about 600 m (near Archidona; P. Coopmans).

Monotypic.

Range: se. Colombia, e. Ecuador, ne. Peru, and w. Amazonian Brazil; e. Amazonian Brazil.

Although the Recurve-billed Bushbird (*Clytoctantes alixi*) has occasionally been credited to the avifauna of Ecuador on the basis of the type series supposedly obtained at "Río Napo," as noted by Chapman (1926, p. 386) this was in error. The specimens are in fact labeled vaguely as only being from "Equateur." The species, known only from n. Colombia, has no basis for being on the Ecuador list.

Pyriglena leuconota   White-backed Fire-eye   Ojo-de-Fuego Dorsiblanco

Uncommon to locally fairly common in the undergrowth of montane forest and secondary woodland in the foothills and subtropical zone on the east slope of the Andes, and in undergrowth of deciduous and humid forest and secondary woodland in the lowlands and foothills of w. Ecuador from w. Esmeraldas south through Guayas (slopes of Cordillera de Colonche, Cerro Blanco in the Chongón Hills, Manglares-Churute Ecological Reserve, etc.), El Oro, and w. Loja (Alamor/El Limo area). The White-backed Fire-eye is often seen at swarms of army ants but otherwise is mostly recorded from its frequently given and far-carrying vocalizations. In the east recorded between about 1000 and 1900–2050 m, in the west from sea level up to at least 1350 m (at Maquipucuna).

Two rather different races are found in Ecuador, *castanoptera* on the east slope and *pacifica* in the west. The bill of *pacifica* is longer and the tail shorter than in *castanoptera*; males of the two taxa are very similar in plumage, but females differ strikingly, *castanoptera* being entirely black on its head and underparts, whereas *pacifica* is drab grayish buff below and uniform brown above except for a blackish tail. Because of these differences, it has been suggested (e.g., Parker and Carr 1992) that *P. pacifica* might better be considered a monotypic species (Pacific Fire-eye). However, as the two Ecuadorian taxa are so similar vocally—as indeed are all the taxa presently considered to be comprised by *P. leuconota*—and as males are so similar in plumage, we continue to consider them conspecific. Other extralimital taxa currently considered to be races of *P. leuconota* are about equally distinct, and if *P. pacificus* is split, then certain other taxa should probably also be considered as full species.

Range: w. Ecuador and extreme nw. Peru; east slope of Andes from s.

Colombia to n. Bolivia, thence east into sw. Brazil; e. Amazonian Brazil; ne. Brazil.

Myrmeciza griseiceps   Gray-headed Antbird   Hormiguero Cabecigris

Rare and local in the lower growth of montane forest borders and secondary woodland, in some areas favoring areas with *Chusquea* bamboo (the main requirement seeming to be dense undergrowth), in the foothills and subtropical zone on the west slope of the Andes in El Oro and w. Loja; in a few areas perhaps locally uncommon. In El Oro the Gray-headed Antbird is recorded from a specimen taken in the early 1920s at La Chonta; a bird that was only heard near Zaruma in 1991 was considered to be only "possible" (Collar et al. 1992) but would appear to be confirmed by a single (wandering?) individual that was seen and heard at Buenaventura in Nov. 1995 (M. Lysinger). In w. Loja the Gray-headed Antbird is known from a few localities in the Alamor and Celica region, and it was also found (first by Best et al. 1993) at Tambo Negro and Utuana, and even more recently in the Sozoranga area (e.g., at the El Tundo Reserve; v.o.). Recorded between about 600 and 2500 m.

The Gray-headed Antbird seems always to have been a scarce bird (there still are rather few specimens), and it has undergone a marked decline in recent decades as a result of the destruction of so much forest and woodland in its restricted range and the elimination of forest undergrowth by livestock trampling much of what remains. The species does, however, appear to thrive in the dense bamboo-dominated second-growth that often grows back when pastures are abandoned; even woodland of short stature seems to be adequate if its understory is dense enough. The Gray-headed Antbird was accorded Endangered status by Collar et al. (1992, 1994), and this is surely deserved; we consider it to be the most threatened of Ecuador's many "southwestern endemics." Even more alarming, the only protected site where the species occurs is the El Tundo Reserve near Sozoranga (owned and managed by Fundación Arcoiris), which because of its relatively small size cannot support a very large population. It remains to be determined whether a population exists at Buenaventura, where the Fundación Jocotoco in 1999 began to establish a reserve (and where there is only one report of the species). Small numbers persist in woodland patches west of Celica toward Pindal (ANSP), around Alamor (e.g., north of El Limo; ANSP), and at Tambo Negro and Utuana (Best et al. 1993).

Griseiceps likely does not belong in the genus *Myrmeciza*, but it is still not known where its actual generic affinities lie; its voice and behavior seem distinctly different from that of other antbird species. Monotypic.

Range: Andes of sw. Ecuador and nw. Peru.

Myrmeciza hemimelaena   Chestnut-tailed   Hormiguero
                            Antbird   Colicastaño

Apparently rare and local in the undergrowth of terra firme forest in the lowlands of e. Ecuador. Although it is elsewhere locally much more numerous, there are only a few Ecuadorian records of the Chestnut-tailed Antbird, and the factors governing its distribution in the country remain uncertain; P. Coopmans (pers. comm.) suggests that it may be found primarily on ridges with sandier soil. These few records include a 19th-century specimen from Sarayucu, a female (MECN) taken by M. Olalla at Puerto Libre in Sucumbíos on 1 Feb. 1963, and a male (UKMNH) obtained on 5 Aug. 1971 at Santa Cecilia, west of Lago Agrio in Sucumbíos; sightings of at least one pair near Zancudococha in Sep. 1976 (RSR); and recent sightings and tape-recordings from around Kapawi Lodge near the Peruvian border on the Río Pastaza. In far s. Zamora-Chinchipe, a single male was mist-netted and collected (ANSP) near Chinapinza (1450 m) on the west slope of the Cordillera del Cóndor on 14 Jun. 1993 (however, no others were seen or heard there); the species was also recorded by T. Parker at Miazi in the Río Nangaritza valley in Jul. 1993 (Schulenberg and Awbrey 1997). Other than the Zamora-Chinchipe records, recorded below 350 m.

One race: nominate *hemimelaena*. Recently obtained tape-recordings of Ecuador birds have shown that their songs differ strikingly from those of birds south of the Río Marañón and Amazon River in Peru, two species apparently being involved (J. Alvarez A. et al., pers. comm.).

Range: e. Ecuador and adjacent Colombia to e. Peru, n. Bolivia, and s. Amazonian Brazil.

Myrmeciza atrothorax   **Black-throated Antbird**   **Hormiguero Golinegro**

Rare to locally uncommon in the undergrowth of forest borders, secondary woodland, and riparian woodland in the lowlands of e. Ecuador; favors swampy situations. Although often numerous elsewhere, the Black-throated Antbird is inexplicably scarce in Ecuador. The first record is of an immature male (LSUMZ) taken at Limoncocha as recently as 20 Jul. 1976 (Tallman and Tallman 1977); we are not aware of the source for the species having been listed as occurring in "e. Ecuador" by Meyer de Schauensee (1970, p. 249). There have been recent sightings from the upper Río Napo region near Coca, Taracoa, Puerto Napo, and Jatun Sacha; the only additional specimen is a male taken west of Jatun Sacha on 13 Apr. 1993 (ANSP). Several pairs are resident along the entrance boardwalk to Sacha Lodge (L. Jost et al.), and a few pairs were also located along the Maxus road southeast of Pompeya in 1994–1995 (RSR); in addition there have been recent reports from La Selva, Yuturi Lodge (where reportedly fairly common; S. Howell), and Kapawi Lodge along the Río Pastaza near the Peruvian border. Recorded below 400 m.

One race: *tenebrosa*. This taxon, apparently known from only five specimens (2 in AMNH, 2 in LSUMZ, and 1 in ANSP; all of them males), is much blacker than other races of *M. atrothorax*, and it also shows fewer and much smaller white dots on its wing-coverts. Based on birds observed in the field, females appear not to differ as strikingly from other races of *M. atrothorax*, though their wing spots are small (as in males). The song of Ecuadorian birds resembles that of other races, and indeed—despite their different appearance—Ecuadorian birds respond strongly to tape-recordings made elsewhere.

Range: e. Colombia, s. Venezuela, and the Guianas to n. Bolivia and Amazonian Brazil.

Myrmeciza hyperythra   Plumbeous Antbird   Hormiguero Plomizo

Fairly common to common in the undergrowth and borders of várzea and floodplain forest in the lowlands of ne. Ecuador. The Plumbeous Antbird has thus far been found only in the drainages of the Río Napo (upriver to around Limoncocha and Taracoa) and Río Aguarico (upriver to Cuyabeno); here, in appropriate habitat, it is one of the more frequently heard antbirds (e.g., around La Selva and Sacha Lodge). It was first recorded from Ecuador from a specimen obtained at the mouth of the Río Lagartococha (J. T. Zimmer, *Am. Mus. Novitates* 545, 1932). This record was overlooked by Meyer de Schauensee (1966, 1970), and by D. Pearson (*Condor* 77[1]: 98, 1975). The Plumbeous Antbird appears to be absent from the Río Pastaza drainage. Recorded only below about 300 m.

Monotypic.

Range: se. Colombia, ne. Ecuador, e. Peru, nw. Bolivia, and w. Amazonian Brazil.

Myrmeciza melanoceps   White-shouldered   Hormiguero
                           Antbird   Hombriblanco

Uncommon to locally common in the lower growth of várzea and riparian forest, secondary woodland, and borders of terra firme forest in the lowlands of e. Ecuador. The White-shouldered Antbird seems particularly numerous around Limoncocha. Recorded up to around 500 m in the Puerto Napo area.

Monotypic.

Range: se. Colombia, e. Ecuador, ne. Peru, and w. Amazonian Brazil.

Myrmeciza fortis   Sooty Antbird   Hormiguero Tiznado

Uncommon to fairly common in the undergrowth of terra firme forest in the lowlands of e. Ecuador. The Sooty Antbird is most often seen at swarms of army ants. Recorded up to about 750 m ("El Para," some 17 km northwest of Archidona along the road toward Baeza; seen on 5 Aug. 1999 by M. Lysinger), but mostly below 600 m.

One race: nominate *fortis*.

Range: se. Colombia, 'e. Ecuador, e. Peru, nw. Bolivia, and w. Amazonian Brazil.

Myrmeciza immaculata   Immaculate Antbird   Hormiguero Inmaculado

Uncommon to fairly common in the undergrowth of very humid and humid forest and mature secondary woodland in the lowlands, foothills, and subtropical zone of w. Ecuador south locally to nw. and e. Guayas, El Oro, and w. Loja (two old specimens from Las Piñas and a recent report [Best et al. 1993] from Tierra Colorada near Alamor). The Immaculate Antbird is now rather local as a result of forest destruction, though it seems capable of persisting in relatively patchy habitat (e.g., at and near Tinalandia). Found mostly below 1400 m, in small numbers up to about 2000 m.

One race: *macrorhyncha*. This taxon was formerly named *berlepschi*, but that name is preoccupied by M. *berlepschi* (Stub-tailed Antbird) now that that species has been transferred from the genus *Sipia* to *Myrmeciza*; see M. B. Robbins and R. S. Ridgely, *Bull. B. O. C.* 113[3]: 190, 1993).

Range: Costa Rica to w. Venezuela and sw. Ecuador.

Myrmeciza exsul   Chestnut-backed Antbird   Hormiguero Dorsicastaño

Fairly common to common in the undergrowth of very humid and humid forest and secondary woodland in the lowlands and foothills of w. Ecuador south to n. Manabí, n. Los Ríos (1950 specimens from Quevedo; ANSP), e. Guayas, and El Oro (an old specimen from La Chonta; recent reports from below Buenaventura). Recorded mostly below about 900 m, with a few occurring locally up to 1300 m near Mindo (in small numbers even as high as over 1500 m; J. Lyons).

The Chestnut-backed Antbird's overall numbers have doubtless declined substantially because of forest destruction, but the species seems to persist well in relatively small forest fragments, and it remains numerous at Río Palenque.

One race: *maculifer*.

Range: Honduras to sw. Ecuador.

Myrmeciza nigricauda   Esmeraldas Antbird   Hormiguero Esmeraldeño

Uncommon to locally fairly common in the undergrowth of very humid to humid forest and mature secondary woodland in the lowlands and (especially) foothills of w. Ecuador south along the base of the Andes to El Oro (Buenaventura; Robbins and Ridgely 1990). Recorded mostly from 400 to 1100 m; there are a few records from lower elevations in Esmeraldas, and it occurs locally and in small numbers up to 1300 m (e.g., around Mindo).

Formerly called *Sipia rosenbergi*. M. B. Robbins and R. S. Ridgely (*Bull. B. O. C.* 111[1]: 11–18, 1991) showed that the genus *Sipia* is best considered congeneric with *Myrmeciza*, and that what was formerly called *Myrmeciza laemosticta nigricauda* is actually the same taxon as *Sipia rosenbergi*. The species name *nigricauda* has priority over *rosenbergi*. M. *laemosticta* (Dull-mantled Antbird), long credited to Ecuador's avifauna on the basis of its supposed race *nigricauda*, therefore is not found south of n. Colombia. M. *nigricauda* is monotypic.

Range: w. Colombia and w. Ecuador.

Myrmeciza berlepschi   Stub-tailed Antbird   Hormiguero Colimocho

Uncommon to locally fairly common in the undergrowth of humid forest and secondary woodland borders in the lowlands and lower foothills of nw. Ecuador in Esmeraldas and extreme nw. Pichincha (east of Puerto Quito, where first recorded in Apr. 1995; M. Lysinger). Around Playa de Oro in n. Esmeraldas the Stub-tailed Antbird overlaps with the Esmeraldas Antbird (*M. nigricauda*) at about 400–500 m (O. Jahn et al.), with that species replacing the Stub-tailed at higher elevations. Recorded mostly below 400 m, sparingly as high as about 650 m.

Berlepschi was formerly placed in the genus *Sipia*; see comments under *Myrmeciza nigricauda*. Monotypic.

Range: w. Colombia and nw. Ecuador.

Pithys albifrons   White-plumed Antbird   Hormiguero Cuerniblanco

Uncommon to fairly common in the undergrowth of terra firme forest in the lowlands of e. Ecuador; apparently most numerous in the southeast. The White-plumed Antbird is most often seen at swarms of army ants. Recorded mostly below 600 m, in smaller numbers up to about 1100 m along the east slope of the Andes, especially in the south (Morona-Santiago and Zamora-Chinchipe).

One race: *peruviana*. Ecuadorian birds are sometimes assigned to the very similar race *brevibarba* on the basis of their slightly shorter (on average) head plumes, but this character appears to be individually variable. We cannot see that *brevibarba* is worthy of recognition.

Range: e. Colombia, s. Venezuela, and the Guianas to ne. Peru and n. Amazonian Brazil.

The White-masked Antbird (*P. castanea*) is known only from the type specimen which was taken near the Ecuadorian border at Andoas in adjacent Loreto, Peru. If it does not prove to be a hybrid—as seems likely the case—this species perhaps will be found in Ecuador as well.

Gymnopithys leucaspis   Bicolored Antbird   Hormiguero Bicolor

Uncommon to fairly common in undergrowth inside humid forest and taller secondary woodland in the lowlands of e. and w. Ecuador. In the east recorded principally from terra firme, and there one of the more widespread ant-following antbirds. In the west recorded south, at least formerly, to coastal El Oro (Santa Rosa); there are, however, no recent reports from south of e. Guayas and nw. Azuay. The Bicolored Antbird is now very local in the southwest, with only a few recent reports: small numbers were found above Manta Real in nw. Azuay in Aug. 1991 (ANSP), and it was recorded from hills in Manglares-Churute Ecological Reserve in Aug. 1996 (Pople et al. 1997). The species seems to have disappeared from Río Palenque, not having been seen there since the early 1980s. A few do, however, persist at Tinalandia (P. Coopmans). Recorded up to about 750 m in the east, to about 900 m in the west.

We follow the 1998 AOU Check-list and Ridgely and Tudor (1994) in considering populations from east and west of the Andes as conspecific under the name of *G. leucaspis*. They have been considered distinct species by some authors (e.g., Hilty and Brown 1986, Sibley and Monroe 1990), in which case the Amazonian population went by the name of *G. leucaspis* (White-cheeked Antbird) and the trans-Andean population by the name of *G. bicolor* (Bicolored Antbird). S. Hackett (*Wilson Bull.* 105[2]: 301–315, 1993) indicated that there was only "weak" genetic support for treating the two populations as distinct species, and that more study was needed before the situation could be satisfactorily resolved. Two races of *G. leucaspis* occur in Ecuador, *castanea* in the east and *aequatorialis* in the west. *Aequatorialis* differs from *castanea* in having the bare ocular area dusky (not pale bluish) and a gray border to the black (not white) cheeks; female *castanea* have a semiconcealed cinnamon-rufous dorsal patch lacking in *aequatorialis*.

Range: Honduras to sw. Ecuador; e. Colombia, e. Ecuador, ne. Peru, and w. Amazonian Brazil.

Gymnopithys lunulata Lunulated Antbird Hormiguero Lunado

Rare to uncommon and local in the undergrowth of várzea forest in the lowlands of e. Ecuador. The Lunulated Antbird was first recorded in Ecuador by E. O. Willis (*Condor* 70[2]: 128–148, 1968), who found the species at Putuimi in s. Morona-Santiago. There are subsequent records from a scatter of localities, including a female mist-netted, photographed, and released at Taracoa on 7 Jan. 1982 (RSR, PJG, and S. Greenfield); a male collected (ANSP) by F. Sornoza at Daime in Yasuní National Park on 17 Sep. 1989; a male mist-netted, photographed (photos to VIREO), and released at Cuyabeno on 3 Nov. 1989 (C. Canaday; the species has also been seen here on subsequent occasions); a female (MECN) collected by N. Krabbe along the Maxus road southeast of Pompeya on 6 Nov. 1994; and recent sightings from Sacha Lodge, La Selva, Yuturi Lodge (mainly or entirely occurring south of the Río Napo; at Yuturi Lodge considered to be "fairly common" in Jul.–Aug. 1996 [S. Howell]), and Kapawi Lodge near the Peruvian border along the Río Pastaza. Recorded below 300 m.

Monotypic.

Range: locally in e. Ecuador and ne. Peru.

Rhegmatorhina melanosticta Hairy-crested Hormiguero
 Antbird Cresticanoso

Uncommon and local in the undergrowth of terra firme forest and taller secondary woodland in the lowlands and foothills of e. Ecuador. The Hairy-crested Antbird is most often noted when attending swarms of army ants; small numbers can be found in the Archidona and Tena region. Recorded mainly below about 750 m, but north of Archidona along the Loreto road it has been recorded up to 1000 m (ANSP, MECN).

One race: nominate *melanosticta*.

Range: e. Ecuador and adjacent se. Colombia to nw. Bolivia and w. Amazonian Brazil.

Phlegopsis nigromaculata   Black-spotted   Carirrosa Negripunteada
                               Bare-eye

Uncommon to fairly common in the undergrowth of humid forest (mainly várzea and floodplain forest; occasionally also in terra firme, where perhaps mainly near streams or swampy places) and secondary woodland in the lowlands of ne. Ecuador, where found mainly north of the Río Napo. There are no records from the Río Pastaza drainage. The attractive Black-spotted Bare-eye is most often seen when in attendance at army antswarms, occasionally being present at the same swarm as the Reddish-winged Bare-eye (*P. erythroptera*), though for the most part they segregate by habitat, with the Reddish-winged more in terra firme. The Black-spotted seems to be particularly numerous near Imuyacocha. Recorded up to about 400 m in the Jatun Sacha area.

One race: nominate *nigromaculata*.

Range: se. Colombia, ne. Ecuador, e. Peru, n. Bolivia, and Amazonian Brazil.

Phlegopsis erythroptera   Reddish-winged Bare-eye   Carirrosa Alirrojiza

Rare to locally uncommon in the undergrowth of terra firme forest in the lowlands of e. Ecuador. The Reddish-winged Bare-eye is most often seen in attendance at swarms of army ants. Recorded up to about 750 m (near Archidona; WFVZ, MECN).

One race: nominate *erythroptera*.

Range: se. Colombia and sw. Venezuela to e. Ecuador, ne. Peru, nw. Bolivia, and w. Amazonian Brazil.

G. R. Graves (*Proc. Biol. Soc. Wash.* 105[4]: 834–840, 1992) has shown that *Phlegopsis barringeri* (Argus Bare-eye), described by R. Meyer de Schauensee (*Not. Naturae* 241: 1–3, 1951) on the basis of a single male obtained just north of the Ecuadorian border at Río Rumiyaco in Nariño, Colombia, is a hybrid between *P. erythroptera* and *P. nigromaculata* (Black-spotted Bare-eye).

Phaenostictus mcleannani   Ocellated Antbird   Hormiguero Ocelado

Rare to uncommon in the undergrowth of very humid and humid forest in the lowlands of nw. Ecuador, where recorded mainly from Esmeraldas; there have also been a few recent reports from nw. Pichincha (along the road north of Simón Bolívar near Pedro Vicente Maldonado; P. Coopmans and others). In Esmeraldas small numbers of the beautiful Ocellated Antbird have been found to range as far south as Bilsa, and the species has also been recorded along the Río Verde south of Chontaduro (ANSP). Recorded mostly below 400 m, sparingly up to 700 m.

Given its limited range in Ecuador, overall scarcity, and apparent strict dependence on blocks of extensive forest, we suggest that the Ocellated Antbird merits Vulnerable status in Ecuador. Small populations may be protected in the Awá Forest Reserve and the lower parts of Cotacachi-Cayapas Ecological Reserve, but their actual status and population sizes at these sites have not been ascertained. O. Jahn (pers. comm.) suggests that the species is at greater risk in Ecuador than the Tawny-faced Quail (*Rhynchortyx cinctus*), Scarlet-breasted Dacnis (*Dacnis berlepschi*), or Blue-whiskered Tanager (*Tangara johannae*), all of which can persist in somewhat fragmented and degraded forest. The Ocellated Antbird was not considered to be globally at risk by Collar et al. (1992, 1994).

One race: *pacificus*.

Range: Honduras to nw. Ecuador.

Myrmornis torquata   **Wing-banded Antbird**   Hormiguero Alifranjeado

Rare and apparently local on or near the ground inside terra firme forest in the lowlands of e. Ecuador, apparently primarily (perhaps only) close to the base of the Andes; perhaps locally more numerous. Ecuadorian records of this antbird are few, and only one of them is recent. There are 19th-century specimens from Sarayacu (BMNH) and a series of nine specimens (AMNH) obtained in the early 20th century at Río Suno; there also are old specimens from Loreto (ANSP). Recent intensive work at several areas that would appear to be suitable for this species (e.g., Jatun Sacha and along the Maxus road southeast of Pompeya) has turned up virtually no evidence of the species' presence; the only recent record we are aware of is a bird tape-recorded in Yasuní National Park in Jan. 1995 (G. Rivadeneira). Recorded below about 400 m.

We are at a loss to explain the rarity of the Wing-banded Antbird in Ecuador. Although deforestation has occurred in some regions, extensive areas of seemingly suitable terrain remain in many areas. As a result, we feel we should accord the species Data Deficient status in Ecuador; it is not regarded as being generally at risk (Collar et al. 1992, 1994).

One race: nominate *torquata*.

Range: Nicaragua; Panama to n. Colombia; se. Colombia, e. Ecuador, and ne. Peru; s. Venezuela and the Guianas to e. Amazonian Brazil.

Formicariidae   Antthrushes and Antpittas

Thirty species in seven genera. Antthrushes and antpittas are strictly Neotropical in distribution.

Formicarius colma   **Rufous-capped Antthrush**   Formicario Gorrirrufo

Uncommon and somewhat local on or near the ground inside terra firme forest in the lowlands of e. Ecuador. Recorded only below about 500 m.

One race: *nigrifrons*.

Range: se. Colombia, s. Venezuela, and the Guianas to n. Bolivia and Amazonian Brazil; e. Brazil.

Formicarius analis   **Black-faced Antthrush**   Formicario Carinegro

Uncommon to locally common on or near the ground inside terra firme and várzea forest and taller secondary woodland in the lowlands of e. Ecuador. The ecological relationship of this species with the generally scarcer Rufous-capped Antthrush (*F. colma*) in Ecuador remains puzzling. In some parts of Amazonia they seem to segregate (with the Black-faced more in second-growth and várzea, the Rufous-capped strictly in terra firme), but in Ecuador the Black-faced Antthrush occurs much more widely and is regular in hilly terra firme; at some sites (e.g., Taisha, where only terra firme exists) it is the sole *Formicarius* antthrush present, whereas in others (e.g., along the Maxus road southeast of Pompeya) the two species occur together in terra firme. Recorded up to about 1000 m along the eastern base of the Andes in the south (several localities in Zamora-Chinchipe), northward mainly below about 800 m.

One race: *zamorae*.

Range: Honduras to n. Colombia and n. Venezuela; se. Colombia and the Guianas to n. Bolivia and Amazonian Brazil. Excludes Mexican Antthrush (*F. monileger*) of n. Middle America, following S. Howell (*Cotinga* 1: 21–25, 1994). Even with *F. monileger* having been split off, *F. analis* still likely consists of more than one species.

Formicarius nigricapillus	Black-headed Antthrush	Formicario Cabecinegro

Uncommon to fairly common on or near the ground inside very humid and humid forest and mature secondary woodland in the lowlands and foothills of w. Ecuador south locally to Guayas (slopes of the Cordillera de Colonche and at Manglares-Churute Ecological Reserve) and nw. Azuay (above Manta Real [RSR and G. H. Rosenberg] and along the Puerto Inca-Molleturo road at Corona de Oro [P. Coopmans]). At Playa de Oro in n. Esmeraldas the Black-headed Antthrush apparently favors secondary forest and borders (O. Jahn et al.). Recorded mostly below about 900 m, in small numbers or locally up to 1300 m (Maquipucuna Reserve; P. Coopmans).

Widespread forest destruction across much of w. Ecuador has doubtless resulted in a substantial reduction of the Black-headed Antthrush's overall numbers. However, the species seems capable of persisting even in isolated and relatively small forest patches such as Río Palenque, and we therefore do not consider that it merits being given formal threatened status.

One race: *destructus*.

Range: Costa Rica to sw. Ecuador.

Formicarius rufipectus	Rufous-breasted Antthrush	Formicario Pechirrufo

Uncommon to fairly common on or near the ground inside montane forest and mature secondary woodland in the foothills and subtropical zone on both slopes of the Andes, in the west found south to El Oro and w. Loja (recorded at Tierra Colorada near Alamor in Feb. 1991; Best et al. 1993). The Rufous-breasted Antthrush favors steep hillsides and ravines and is even more difficult to see than the other *Formicarius* antthrushes. Recorded mostly from 1100 to 2000 m, but occurs lower (regularly down to 800 m) on the west slope.

Two distinctly different races are found in Ecuador, *carrikeri* on the west slope and *thoracicus* on the east slope. *Carrikeri* has a chestnut crown and nape, whereas *thoracicus* has an entirely blackish head.

Range: highlands of Costa Rica and w. Panama; Andes from extreme sw. Venezuela to s. Peru; Sierra de Perijá.

Chamaeza campanisona   Short-tailed Antthrush   Chamaeza Colicorto

Uncommon and somewhat local on or near the ground inside montane forest in the foothills and subtropical zone on the east slope of the Andes. Recorded between about 950 m (Bermejo oil-field area north of Lumbaquí; ANSP) and 1700 m.

One race: *punctigula*.

Range: mountains of n. Venezuela, and Andes from w. Venezuela to w. Bolivia; tepuis and mountains of s. Venezuela and Guyana; e. Brazil to e. Paraguay and ne. Argentina.

Chamaeza nobilis   Striated Antthrush   Chamaeza Noble

Rare to locally fairly common (at least by voice) on or near the ground inside terra firme forest in the lowlands of e. Ecuador. Recorded up to 700 m at Eugenio and Guaticocha in w. Napo (MCZ), but found mostly below 500 m.

The species has sometimes (e.g., Ridgely and Tudor 1994) been called the Noble Antthrush.

One race: *rubida*.

Range: se. Colombia, e. Ecuador, e. Peru, nw. Bolivia, and w. and cen. Amazonian Brazil.

Chamaeza mollissima   Barred Antthrush   Chamaeza Barreteado

Very rare to rare and seemingly local (possibly mainly just overlooked) on or near the ground inside montane forest on the east slope of the Andes. Still poorly known in Ecuador, with most specimen records coming from the Baeza/Sumaco region; in all likelihood, the species will eventually be found to range more or less continuously along the east slope where suitable habitat exists. The Barred Antthrush has recently been found on the Cordillera de Huacamayos and around the nearby Cabañas San Isidro, and one was collected (MECN) in Sucumbíos along the La Bonita road (2070 m) on 19 Sep. 1991 (N. Krabbe). There are old specimens from the Baños region but apparently no recent reports from this area. Farther south, in Zamora-Chinchipe at least one was heard near the Río Isimanchi (2200 m) in early Nov. 1992 (M. B. Robbins), and in 1997–1998 the species was found to be uncommon in and near Quebrada Honda on the south slope of the Cordillera de Sabanilla (with a specimen [MECN] taken in Dec. 1997). Recorded mostly between about 1850 m (north of Valladolid in Zamora-Chinchipe; one heard on 8 Apr. 1994 [P. Coopmans and M. Lysinger]) and 2600 m, but has been recorded as high as 3075 m at Oyacachi (Krabbe et al. 1997).

One race: nominate *mollissima.*

Range: locally in Andes from s. Colombia to w. Bolivia.

Pittasoma rufopileatum	Rufous-crowned Antpitta	Pitasoma Coronirrufa

Rare to uncommon and seemingly somewhat local on or near the ground inside very humid and humid forest and mature secondary woodland in the foothills of nw. Ecuador, where it is most widespread in Esmeraldas; there are also records from Pichincha, where it formerly occurred as far south as the southern edge of that province at Río Palenque (the last reports from this site, now isolated, date from the mid-1980s). At Playa de Oro in n. Esmeraldas the Rufous-crowned Antpitta is most numerous in the foothills between 200 and 400 m, with small numbers occurring down into the lowermost foothills to just under 100 m (O. Jahn et al.). Recorded mostly from about 100 to 700 m.

We regard the Rufous-crowned Antpitta as warranting Vulnerable status in Ecuador; no area where it is known to occur presently receives adequate protection, though small numbers must range at lower levels of Cotocachi-Cayapas Ecological Reserve and in the Awa Forest Reserve. The species was—in our view, surprisingly—not regarded as being at risk by Collar et al. (1992, 1994).

One race: nominate *rufopileatum.* The series from El Placer (ANSP, MECN) shows no approach to *harterti* of sw. Colombia, despite that site's being situated close to the Colombian border.

Range: w. Colombia and nw. Ecuador.

Grallaria gigantea	Giant Antpitta	Gralaria Gigante

Apparently rare and local (doubtless overlooked to some extent) on or near the ground inside montane forest and secondary woodland, sometimes foraging out into adjacent overgrown soggy pastures in the subtropical zone on both slopes of the Andes; recorded mainly or perhaps entirely from n. Ecuador.

On the west slope the Giant Antpitta is known mainly from Pichincha, where a substantial number of specimens were taken in the 19th and early 20th centuries at various localities on the northwestern flank of Volcán Pichincha in the Mindo, Gualea, and Nanegal region; the species must not have been uncommon then. In addition, ANSP has two specimens of the west-slope taxon *hylodroma* (see below) that were taken in 1938 by W. Clark-Macintyre, supposedly at "El Tambo, Loja." This controversial site appears, as Paynter (1993) suggests, not to be situated in Loja at all but rather along the Guayaquil-Quito railway line in Cañar. The species has also been recorded recently in the Caripero area of nw. Cotopaxi (F. Sornoza). After its having long gone undetected, in the 1990s there have been numerous reports of this elusive antpitta in the Mindo/Tandayapa region, the first being a tape-recorded bird near Tandayapa in Nov. 1991 (PJG) that for several years went unidentified; one was subsequently collected there (MECN). Since then the species has been found fairly regularly in several areas around Mindo itself, and also up to around the Bellavista Lodge. On the west slope recorded between about 1400 and 2300 m; old specimens from as high as 3350 m are regarded as of uncertain provenance.

On the east slope the Giant Antpitta is known from Carchi (El Pun), w. Napo, and Tungurahua (an ANSP specimen from "Runtun Hills"). The only definite recent east-slope records are from SierrAzul in the upper Río Cosanga valley of w. Napo, where F. Sornoza obtained one specimen on 16 Jun. 1992 (ANSP), M. B. Robbins another on 8 Mar. 1993 (ANSP), and N. Krabbe a third on 13 Oct. 1993 (MECN); there are also recent sightings from nearby Cabañas San Isidro (M. Lysinger et al.). On the east slope recorded only between about 2000 and 2400 m.

Details of some of the recent records of the Giant Antpitta have been given by N. Krabbe, G. DeSmet, P. Greenfield, M. Jácome, J. C. Matheus, and F. Sornoza M. (*Cotinga* 2: 32–34, 1994); see also Y. de Soye, K.-L. Schuchmann, and J. C. Matheus (*Cotinga* 7: 35–36, 1997).

The Giant Antpitta was accorded Vulnerable status by Collar et al. (1994), a status with which we concur. Krabbe et al. (op. cit.) suggest that one possible explanation for the apparent rarity of this species at least on the east slope of the Andes may be its preference for wet forest on relatively level terrain, a microhabitat that not only is naturally rare but is also cleared preferentially for pasture; however, both ANSP specimens were in fact taken on rather steep slopes. Nonetheless, for whatever reason, the Giant Antpitta does not appear to be a numerous bird anywhere.

Two quite different races occur in Ecuador, *hylodroma* in Pichincha and nominate *gigantea* on the east slope; the two specimens from "El Tambo, Loja" (which almost certainly were actually taken in Cañar; see above) are referable to *hylodroma*. The nominate race is much paler rufous below, with markedly broader and sparser black barring on its sides and flanks (giving more the effect of spotting). Despite the similarity in their voices, the two forms may prove to be separate species; if this is done, G. *hylodroma* of the west slope would be called the Pichincha Antpitta.

Range: locally in Andes of s. Colombia and Ecuador.

Grallaria squamigera   **Undulated Antpitta**   Gralaria Ondulada

Uncommon to locally fairly common on or near the ground inside montane forest and woodland of the upper subtropical and temperate zones on both slopes of the Andes, and locally also above interandean valleys. On the west slope recorded south locally to Azuay (one taken west of Molleturo near Paredones on 1 Mar. 1991 [N. Krabbe]) and Loja (seen and tape-recorded north of Celica in 1989 [Bloch et al. 1991] and 1991 [N. Krabbe], and mist-netted near Utuana on 12 Mar. 1991 [N. Krabbe], with others having been found here since). The Undulated Antpitta favors damp places such as seepage zones and the vicinity of streams. Recorded from about 2200 to 3700 m, occurring regularly just below treeline.

Two races are found in Ecuador, nominate *squamigera* in most of the species' Ecuadorian range and *canicauda* in the southeast (from the Cordillera de Cutucú [ANSP] in Morona-Santiago southward). The latter differs in being uniform gray above (the nominate being more olivaceous above except for its gray crown and nape) with a whitish (not buffyish) loral spot.

Range: Andes from w. Venezuela to w. Bolivia.

Grallaria guatimalensis   Scaled Antpitta   Gralaria Escamada

Rare to locally fairly common on or near the ground, especially near streams or other damp places, inside humid forest and secondary woodland in the lowlands and foothills in both e. and w. Ecuador, ranging up into the subtropical zone at least in the south. In the west recorded south to w. Loja in the Alamor/Celica region and around Sozoranga (we regard the Utuana record of Best et al. 1993 as unverified because of the high elevation involved). In the west found mainly along the base of the Andes, though also recently found on the coastal Cordillera de Mache at Bilsa and known from an early-20th-century specimen (AMNH) from Río de Oro in n. Manabí; as well there are recent reports from the Cordillera de Colonche in sw. Manabí (Cerro San Sebastián; Parker and Carr 1992) and w. Guayas ("Cerro Manglar" [AMNH]; Cerro La Torre [heard by N. Krabbe in Jul. 1992]). On the east slope decidedly local, with scattered records along the base of the Andes from w. Napo (Río Suno, below San José) south to Zamora-Chinchipe (Zamora area), though fairly numerous in the Zumba region. Until recently the Scaled Antpitta was not known at all from the eastern lowlands away from the base of the Andes. The first report from there involved a bird tape-recorded at Limoncocha on 23 Feb. 1981 (A. B. van den Berg and C. A. W. Bosman, *Bull. B. O. C.* 104[4]: 152, 1984). The species has since been found at several additional localities, including Jatun Sacha, La Selva, the Maxus road southeast of Pompeya, Taisha, and Zancudococha (where a specimen was taken on 13 Apr. 1991; ANSP). Ranges mainly below 1300 m, but at least in w. Loja recorded considerably higher (e.g., to about 2000 m around Celica [P. Coopmans; RSR]).

One race: *regulus*.

Range: locally from cen. Mexico to w. Venezuela, extreme nw. Peru, and w. Bolivia; tepuis of s. Venezuela and adjacent Brazil.

**Grallaria alleni*   Moustached Antpitta   Gralaria Bigotuda

Rare to locally fairly common (at Maquipucuna Reserve) on or near the ground inside montane forest and mature secondary woodland in the subtropical zone on both slopes of the Andes in n. Ecuador. The distribution of the Moustached Antpitta in Ecuador remains imperfectly known, and indeed only recently was the species' presence in the country confirmed by N. Krabbe and P. Coopmans (*Ibis* 142: 183–187, 2000). Evidently the Moustached Antpitta replaces the Scaled (*G. guatemalensis*) at higher elevations. The Moustached Antpitta is known in Ecuador from two specimens (MECN), both taken by N. Krabbe, one obtained on the south slope of the Cordillera de Hua-camayos (2100 m) in w. Napo on 9 Oct. 1992, the other at Maquipucuna Reserve (1850 m) in w. Pichincha on 12 Feb. 1994. The species has also been heard and seen above Mindo along the road to Tandayapa, and it likely will be found at additional sites with appropriate habitat northward toward the Colombian border; it has also been found south into w. Cotopaxi in the Caripero area (N. Krabbe and F. Sornoza). Moustached Antpittas have also been heard below Cabañas San Isidro (M. Lysinger). Recorded from about 1850 to 2200 m.

The Moustached Antpitta was accorded Endangered status by Collar et al. (1992), on account of its extremely small known range (in Colombia) at the time. Now that the species' range is known to be considerably more extensive, its status can be viewed as less critical. We therefore suggest that Vulnerable status is more appropriate.

One race: *andaquiensis*.

Range: locally in Andes of cen. Colombia and n. Ecuador.

Grallaria haplonota   **Plain-backed Antpitta**   Gralaria Dorsillana

Uncommon to fairly common but local on or near the ground inside montane forest and secondary woodland in the foothills and lower subtropical zone on both slopes of the Andes. On the east slope known from w. Sucumbíos (San Rafael Falls vicinity) and w. Napo (Loreto road north of Archidona; head of Río Guataraco [MCZ]) south locally through Morona-Santiago (Cordillera de Cutucú; also collected [MECN] in Apr. 1998 in the Río Huamboya valley west of Macas; F. Sornoza) into Zamora-Chinchipe (Cordillera del Cóndor [Krabbe and Sornoza 1994] and Río Bombuscaro sector of Podocarpus National Park). On the west slope known only from Imbabura (Hacienda Paramba), Pichincha (Mindo area), and El Oro (La Chonta; Buenaventura [ANSP]). The exception-ally secretive Plain-backed Antpitta is heard much more often than seen, and it seems likely to range along the eastern base of the Andes north into adjacent Colombia, whence not recorded (Hilty and Brown 1986). Recorded from about 1100 to 1700 m on the east slope, between about 700 and 1300 m on the west slope.

Two races are found in Ecuador, *parambae* in the west and *chaplinae* (M. B. Robbins and R. S. Ridgely, *Bull. B. O. C.* 106[3]: 101–104, 1986) in the east. *Chaplinae* differs in having its crown and back feathers faintly scaled black and somewhat less rich ochraceous underparts.

Range: mountains of n. Venezuela; locally on Pacific slope of Andes in sw. Colombia (F. G. Stiles and H. Alvarez-López, *Caldasia* 17: 607–610, 1995) and Ecuador, and on east slope in Ecuador and adjacent n. Peru (Schulenberg and Awbrey 1997).

Grallaria ruficapilla   Chestnut-crowned   Gralaria Coronicastaña
                            Antpitta

Fairly common and widespread on or near the ground in montane forest, secondary woodland, and borders in the subtropical and temperate zones on both
slopes of the Andes, and locally also above the central and interandean valleys.
The Chestnut-crowned is somewhat less difficult to see than are most other
forest-based antpittas, and it occurs more often at edge and even—especially
soon after dawn—in the semiopen. It persists well where forest and woodland
are patchy and fragmented. Recorded mostly from 2000 to 3100 m, locally
down to about 1600 m in Morona-Santiago (Río Upano valley west of Macas;
RSR) and w. Loja (e.g., the Sozoranga area).

   Does not include *G. watkinsi* (Watkins's Antpitta). Two races are found in
Ecuador, nominate *ruficapilla* in most of the species' Ecuadorian range and *connectens* in El Oro and Loja. The latter differs in its somewhat more whitish
lores and sparser streaking below. We can find no evidence to indicate that
the range of the race *albiloris*, known from nw. Peru, extends as far north as
Ecuador—despite its having sometimes (e.g., *Birds of the World*, vol. 7) been
listed as occurring here. J. T. Zimmer (*Am. Mus. Novitates* 703: 15, 1934) does
record a specimen from an unspecified site in Loja as being "exactly intermediate between *albiloris* and *connectens.*"

   Range: mountains of n. Venezuela, and Andes from w. Venezuela to n. Peru;
Sierra de Perijá.

Grallaria watkinsi   Watkins's Antpitta   Gralaria de Watkins

Fairly common on or near the ground in deciduous and montane forest and
woodland and dense regenerating scrub in the lowlands and lower subtropical
zone of sw. Ecuador in El Oro and w. Loja. Also recently found on the slopes
of the Cordillera de Colonche in sw. Manabí (north to above Puerto Cayo;
ANSP specimens from Cerro San Sebastián in Machalilla National Park) and
w. Guayas (hills south of the Río Ayampe [P. Coopmans] and on Cerro La Torre
and above Salanguilla [N. Krabbe]). Recorded from near sea level near Arenillas in El Oro to about 1800 m (e.g., west of Chaguarpamba in Loja); on the
Cordillera de Colonche found above 400 m.

   The species has sometimes been called the Scrub Antpitta.

   G. watkinsi is considered a species separate from *G. ruficapilla* (Chestnut-
crowned Antpitta), based on its strikingly different voice, pink (not bluish gray)
legs, and different habitat and elevational range, as first noted (at least in
Ecuador) by PJG, S. Greenfield, and P. Scharf in 1987; see also Parker et al.
(1995) for information from adjacent Peru. The two species replace each other
altitudinally in the Celica and Sozoranga regions, where occasionally both can
be heard from the same spot (the Chestnut-crowned from the slopes above, the
Watkins's from the valleys below), rarely even from the same forest or woodland patch. *G. watkinsi* is monotypic.

   Range: sw. Ecuador and extreme nw. Peru.

*Grallaria ridgelyi   Jocotoco Antpitta   Gralaria Jocotoco

Uncommon and extremely local on or near the ground inside montane forest with an understory of *Chusquea* bamboo in the temperate zone on the east slope of the Andes in s. Zamora-Chinchipe. The only recently described Jocotoco Antpitta (N. Krabbe, D. J. Agro, N. H. Rice, M. Jácome, L. Navarrete, and F. Sornoza M., *Auk* 116[4]: 882–890, 1999) was discovered on 20 Nov. 1997 by RSR, L. Navarrete, J. and R. Moore, and M. Ribadeneira; a pair was seen and tape-recorded (recordings to LNS) on the south side of the Cordillera de Sabanilla along the Quebrada Honda trail (2550 m). One of the same pair was relocated and photographed by RSR (photos to VIREO) the following day. The first specimen (MECN) was taken here on 28 Dec. 1997 by F. Sornoza and L. Navarrete, and a short series (ANSP, MECN) was subsequently obtained in Jan. 1998 by D. Agro, N. Krabbe, and others; an individual was also mist-netted, photographed by D. Wechsler, and released (photos to VIREO). A last specimen (MECN) was obtained in Mar. 1998 some 10 km northeast of Quebrada Honda on the east slope of Cerro Toledo (just within the boundaries of Podocarpus National Park) by F. Sornoza and M. Jácome. Intensive searches utilizing tape-playback have yet to reveal the presence of this strikingly distinctive new species anywhere else nearby. Recorded from about 2300 to 2650 m.

As noted above, present evidence indicates that this antpitta has an extremely circumscribed range, being limited by factors not yet understood. The vast majority of the terrain where the Jocotoco Antpitta occurs being privately owned and under some threat from clearing for cattle pasturage and timber removal, a group of concerned individuals decided to form an Ecuadorian-based foundation to purchase and manage land for the species' benefit. Late in 1998, with financial support from N. Simpson, this foundation—now incorporated as the Fundación Jocotoco—made initial purchases of some 700 ha of land in the heart of the species' known range, including the precise spot of initial discovery; as of late 2000, some 1500 additional hectares have been purchased. The protected land, now called the Tapichalaca Biological Reserve, is being guarded by a local resident; further expansion is anticipated. Given the Jocotoco Antpitta's extremely small known range and presumed miniscule population—in the range of 15–20 pairs have been estimated for the immediate Quebrada Honda and Tapichalaca region—we suggest that the species merits Endangered status. We continue to expect that the species will ultimately be located at least some distance to the north and south of the type locality.

Monotypic.

Range: Andes of se. Ecuador.

Grallaria nuchalis   Chestnut-naped Antpitta   Gralaria Nuquicastaña

Uncommon to fairly common and apparently local on or near the ground inside montane forest and forest borders, often where there are dense stands of *Chusquea* bamboo, in the upper subtropical and temperate zones on the east slope of the Andes; on the west slope recorded only from Imbabura and Pichincha, and apparently less numerous. The Chestnut-naped Antpitta seems to be most numerous in se. Ecuador (e.g., along the Gualaceo-Limón road, in the Cajanuma sector of Podocarpus National Park, and at the newly created Tapichalaca Reserve at Quebrada Honda north of Valladolid in s. Zamora-Chinchipe). Recorded between about 2000 and 3000 m.

Two races are found in Ecuador, nominate *nuchalis* on the east slope and *obsoleta* in the northwest. *Obsoleta* differs in having the chestnut-rufous of the head restricted to the nape and in being darker and sootier below. Its voice differs from that of the nominate race (L. Navarrete recording), and possibly separate species are involved.

Range: locally in Andes from s. Colombia to extreme n. Peru.

Grallaria rufocinerea   Bicolored Antpitta   Gralaria Bicolor

Apparently rare on or near the ground in montane forest and forest borders in the temperate zone on the east slope of the Andes in nw. Sucumbíos. The Bicolored Antpitta was first recorded in Ecuador on 11 Nov. 1999 when a pair was heard and tape-recorded, and one individual seen, along the La Bonita road (J. Nilsson, R. Jonsson, M. E. Mulhollam, and J. Sipiora). The species was formerly thought to be a Colombian endemic. Recorded at 2550 m.

G. *rufocinerea* was considered to be Endangered in its Colombian range by Collar et al. (1994). As forest habitat along the La Bonita road remains unprotected, and is very much at risk from clearance, we judge the species to be Endangered in Ecuador as well.

Presumed race: *romeroana* (see J. Hernández C. and J. V. Rodriguez M., *Caldasia* 12: 573–580, 1979).

Range: Andes of Colombia and extreme n. Ecuador.

Grallaria hypoleuca   White-bellied Antpitta   Gralaria Ventriblanca

Uncommon to fairly common on or near the ground in montane forest borders and secondary woodland in the subtropical zone on the east slope of the Andes from Sucumbíos (above San Rafael Falls) south somewhat locally into Zamora-Chinchipe (Romerillos sector of Podocarpus National Park; Chinapinza area on west slope of Cordillera del Cóndor [Krabbe and Sornoza 1994; RSR]; Panguri [ANSP]; Río Isimanchi valley below Cordillera Las Lagunillas [heard in Nov. 1992; M. B. Robbins]; and south slope of Cordillera de Sabanilla above Valladolid). The White-bellied Antpitta is numerous around Cabañas San Isidro near Cosanga in w. Napo. Recorded from about 1400 to 2200 m.

Does not include G. *flavotincta* (Yellow-breasted Antpitta). One race occurs in Ecuador, *castanea*.

Range: Andes from Colombia to extreme n. Peru. Excludes other members of G. *hypoleuca* superspecies.

Grallaria flavotincta   Yellow-breasted Antpitta   Gralaria Pechiamarillenta

Rare to locally uncommon (doubtless overlooked) on or near the ground inside montane forest in the subtropical zone on the west slope of the Andes in nw.

Ecuador. First recorded on 26 Mar. 1984 when N. Krabbe heard no fewer than five birds (and tape-recorded one) in Carchi along the Maldonado road. In w. Pichincha first recorded on 25 Mar. 1989 when one was heard above Tandayapa (RSR and PJG et al.); three days later another was heard, tape-recorded, and seen along the lower Chiriboga road (RSR and PJG et al.). There have been subsequent reports from there and several other sites in Pichincha, including Maquipucuna Reserve (P. Coopmans). Recorded from about 1500 to 2350 m.

Following Hilty and Brown (1986) and Ridgely and Tudor (1994), *G. flavotincta* of the Western Andes is treated as a monotypic species separate from *G. hypoleuca* (White-bellied Antpitta) of the east slope, based on its different plumage and slightly different song.

Range: locally on west slope of Andes in w. Colombia and nw. Ecuador.

Grallaria rufula   Rufous Antpitta   Gralaria Rufa

Uncommon to fairly common on or near the ground inside montane forest and forest borders, favoring stands of *Chusquea* bamboo, in the upper subtropical and temperate zones on both slopes of the Andes, and locally also above the central and interandean valleys. On the west slope recorded south to Azuay (west of Molleturo; N. Krabbe) and locally to El Oro (Taraguacocha; AMNH). The Rufous Antpitta is especially numerous in the Cajanuma sector of Podocarpus National Park, likely because of the prevalence of *Chusquea* there. Recorded mostly from 2200 to 3300 m, locally as low as 2000 m and as high as 3700 m (regularly up to treeline).

One race: nominate *rufula*.

Range: Andes from sw. Venezuela to w. Bolivia; Sierra de Perijá and Santa Marta Mountains. More than one species is likely involved.

Grallaria quitensis   Tawny Antpitta   Gralaria Leonada

Fairly common to common on or near the ground in agricultural areas and paramo with scattered low bushes, hedgerows, and patches of woodland (including *Polylepis*) in the temperate and paramo zones from Carchi south through e. El Oro (Taraguacocha; AMNH) and e. Loja; somewhat less numerous and widespread southward. Unlike the other Ecuadorian antpittas, the Tawny is comparatively easy to see, and often hops boldly in the open along roads or uses prominent perches; it is numerous in various localities above Quito. Recorded mostly from 3000 to 4500 m.

One race: nominate *quitensis*.

Range: Andes from n. Colombia to n. Peru.

Grallaria dignissima   Ochre-striped Antpitta   Gralaria Ocrelistada

Rare to uncommon on or near the ground inside terra firme forest (at least in some areas most often in the vicinity of small forest streams) in the lowlands of e. Ecuador. The Ochre-striped Antpitta is elusive, shy, and difficult to see unless it is vocalizing; doubtless as a result it was until recently believed to be very scarce and local. Since the late 1980s, with knowledge of its distinctive song, pairs have been found in the La Selva/Sacha Lodge area, mainly south of the Río Napo, and the species has also proven to be not uncommon along the Maxus road southeast of Pompeya (RSR et al.). Recorded below about 450 m.

This species was formerly placed in the genus *Thamnocharis*, but see G. H. Lowery, Jr., and J. P. O'Neill (*Auk* 86[1]: 1–12, 1969) for its inclusion in *Grallaria*. Monotypic.

Range: e. Ecuador and adjacent se. Colombia to ne. Peru.

Hylopezus perspicillatus   **Streak-chested Antpitta**   Tororoi Pechirrayado

Rare to locally fairly common on or near the ground inside very humid and humid forest in the lowlands and foothills of nw. Ecuador south to extreme s. Pichincha at Río Palenque (where it seems not to have been reported since the 1980s). In Ecuador the Streak-chested Antpitta is now most numerous and widespread in Esmeraldas; at Playa de Oro in the northern part of that province, it favors mature forest on flat terrain and is much less numerous in hilly areas (O. Jahn et al.). Recorded up to about 800 m.

This species has sometimes been called the Spectacled Antpitta, but see Ridgely and Tudor (1994) for the rationale for returning to the older name of Streak-chested Antpitta. The latter name was also used in the 1998 AOU Check-list.

This and the following species were formerly placed in the genus *Grallaria*, but see G. H. Lowery, Jr., and J. P. O'Neill (*Auk* 86[1]: 1–12, 1969). One race: *periophthalmicus*.

Range: Honduras to nw. Ecuador.

Hylopezus fulviventris   **White-lored Antpitta**   Tororoi Loriblanco

Uncommon on or near the ground in very dense undergrowth at the borders of terra firme and várzea forest, riparian forest, and regenerating clearings in the lowlands of e. Ecuador. The White-lored Antpitta is known primarily from the northeast, but this exceptionally secretive species (recorded almost exclusively from its distinctive song) seems almost certain to have been overlooked. In the southeast there are old records from Sarayacu and Río Bobonaza in Pastaza, and the species has recently been reported from Kapawi Lodge along the Río Pastaza near the Peruvian border. Recorded up to 750 m (near Archidona; WFVZ).

Following Ridgely and Tudor (1994), *H. fulviventris* of w. Amazonia is regarded as a species separate from trans-Andean *H. dives* (the true Fulvous-bellied Antpitta, of Honduras to w. Colombia; called the Thicket Antpitta in the 1998 AOU Check-list), on the basis of its totally different song, marked plumage differences, and disjunct range. One race is found in Ecuador, nominate *fulviventris*.

Range: se. Colombia, e. Ecuador, and ne. Peru.

Although the Fulvous-bellied Antpitta was listed by Meyer de Schauensee (1966, 1970) as occurring in w. Ecuador, there does not seem to be any actual evidence that the species has been recorded from here. Being so secretive, however, the species may have been overlooked, and it could occur in Esmeraldas; it is known to range very close to the Ecuadorian border in w. Nariño, Colombia.

Myrmothera campanisona   Thrush-like Antpitta   Tororoi Campanero

Fairly common and widespread in thickets on or near the ground inside humid forest, especially in terra firme near damp places or streams, in the lowlands of e. Ecuador. Recorded mostly below 700 m, in small numbers up along the eastern base of the Andes to about 1000 m, locally as high as 1200 m in the Río Bombuscaro sector of Podocarpus National Park.

One race: *signata*.

Range: se. Colombia, s. Venezuela, and the Guianas to nw. Bolivia and Amazonian Brazil.

Grallaricula flavirostris   Ochre-breasted Antpitta   Gralarita Ocrácea

Uncommon to locally fairly common (but easily overlooked) in undergrowth inside montane forest in the foothills and subtropical zone on both slopes of the Andes. On the west slope recorded locally from Carchi south to Pichincha (Mindo/Nanegal area), recurring again in w. Azuay (San Luis; ANSP) and El Oro (old specimens from above Zaruma, La Chonta, El Chiral, and Salvias [Chapman 1926], but recent records only from Buenaventura [ANSP, MECN]). On the east slope known from Sucumbíos south to nw. Pastaza (Sarayacu) and from several recent records from Zamora-Chinchipe (specimens [ANSP, MECN] from Panguri and an Aug. 1990 sighting from the Río Bombuscaro sector of Podocarpus National Park [E. P. Toyne]). Recorded mostly from 800 to 2000 m, but locally lower in sw. Ecuador (e.g., at San Luis in w. Azuay, where a specimen was taken at only 300 m by J. C. Matheus and F. Sornoza on 28 Jan. 1991).

The racial allocation of Ecuadorian birds presents a complex problem; see the discussion in Robbins and Ridgely (1990). Birds from the west slope of the Andes are variable individually, both with regard to bill color (either all yellow, all dark, or with maxilla dark and mandible yellow) and in the extent of their breast streaking. Typically, birds of the northwest (*mindoensis*) are boldly streaked on the breast and have the bill bicolored (maxilla dark, mandible yellow, the later sometimes mixed with brown). Birds from the southwest (*zarumae*) usually have plain ochraceous underparts (though a few show coarse breast streaking) and usually have an all-yellow bill (but in a few it is bicolored). Birds of the east slope (nominate *flavirostris*) resemble *mindoensis*; the bills of most are bicolored (though all dark in a few) and their breasts are (variably) streaked.

Range: locally in highlands of Costa Rica and Panama, and Andes from w. Colombia to w. Bolivia. More than one species is perhaps involved.

Grallaricula nana   Slate-crowned Antpitta   Gralarita Coronipizarrosa

Uncommon in undergrowth inside montane forest in the upper subtropical and lower temperate zones on the east slope of the Andes, perhaps especially favoring stands of *Chusquea* bamboo (at least often in them). Regularly heard—but not easy to see—in the Cordillera de Huacamayos and in the Cajanuma sector of Podocarpus National Park. Recorded from about 2000 to 2900 m.

One race: nominate *nana*.

Range: locally in mountains of n. Venezuela, tepuis of s. Venezuela, and Andes from w. Venezuela to n. Peru.

*Grallaricula peruviana   Peruvian Antpitta   Gralarita Peruana

Rare and local in undergrowth inside montane forest in the subtropical zone on the east slope of the Andes in s. Ecuador. The first Ecuadorian record of the Peruvian Antpitta involved a single bird collected by N. Krabbe on the Cordillera de Cutucú on 21 Apr. 1984 (Fjeldså and Krabbe 1986). The only subsequent reports are a bird seen along the Gualaceo-Limón road in Morona-Santiago on 21 Mar. 1990 (B. Whitney) and two birds mist-netted and collected (ANSP, MECN) at Quebrada Avioneta (1950 m) in the Romerillos sector of Podocarpus National Park in Zamora-Chinchipe on 5 and 7 Jan. 1992. Recorded from about 1750 to 2100 m.

The Peruvian Antpitta was given Near-threatened status by Collar et al. (1994). Given the paucity of information regarding its numbers in Ecuador, we can only agree, though noting that most of its range is remote and seems unlikely to be adversely affected by human activities in the near future. A population of unknown size does occur in Podocarpus National Park.

Monotypic.

Range: locally in Andes of s. Ecuador and n. Peru.

Grallaricula lineifrons   Crescent-faced Antpitta   Gralarita Carilunada

Rare to fairly common and perhaps local (doubtless much overlooked) in undergrowth inside montane forest and adjacent secondary woodland in the temperate zone on the east slope of the Andes. Long known in Ecuador only from the type specimen (AMNH), taken in the early 1920s at Oyacachi in w. Napo. After having been found in the 1970s at a single site in Cauca, s. Colombia (F. C. Lehmann V., J. R. Silliman, and E. Eisenmann, *Condor* 79[3]: 387–388, 1977), the Crescent-faced Antpitta finally was relocated in Ecuador in 1991. On 19 Aug. of that year a single female was obtained by RSR and F. Sornoza (with P. Stafford and R. Marsi) in undisturbed *Podocarpus*-dominated temperate forest (2900 m) at Hacienda La Libertad in Cañar. In early Mar. 1992, M. B. Robbins and G. H. Rosenberg et al. found a population in se. Carchi on the west slope of Cerro Mongus, and a small series (ANSP, MECN) was taken; in late Mar., another (perhaps smaller?) population was discovered by the same workers in n. Loja on the Cordillera de Cordoncillo south of Saraguro. These records, together with some natural-history observations, were summarized by M. B. Robbins, N. Krabbe, G. H. Rosenberg, R. S. Ridgely, and F. Sornoza M. (*Wilson Bull.* 106[1]: 169–173, 1994). The only additional localities for the species that have come to light are the Guandera Biological Reserve in se. Carchi (Creswell et al. 1999), Páramos de Matanga in sw. Morona-Santiago (Krabbe et al. 1997), and the Gualaceo-Limón road (L. Navarrete and J. Moore). This attractive antpitta seems likely to occur at least locally along the entire east slope of the Andes. Given the numbers that have been found recently, and the major southward range extension that has been involved, we remain perplexed as to why the species went so long without being rediscovered. Recorded from 2900 to 3400 m.

The Crescent-faced Antpitta was accorded Near-threatened status by Collar et al. (1994), but this was apparently without those authors having been aware of the recent spate of Ecuadorian records. As the species appears to be capable of existing in patchy and somewhat degraded habitat, we conclude that it likely is not at risk.

Monotypic.

Range: locally in Andes of s. Colombia and Ecuador.

Conopophagidae   Gnateaters

Three species in one genus. Gnateaters are exclusively South American in distribution. They have occasionally (e.g., Meyer de Schauensee 1970) been subsumed into the Formicariidae but are now generally regarded as being of family rank (e.g., Sibley and Ahlquist 1990; Sibley and Monroe 1990).

| *Conopophaga castaneiceps* | Chestnut-crowned Gnateater | Jejenero Coronicastaño |

Uncommon and inconspicuous in the undergrowth of montane forest, borders, and secondary woodland in the foothills and subtropical zone on the east slope of the Andes. The Chestnut-crowned Gnateater is known in Ecuador mainly from the south, where it has been found at various sites in Morona-Santiago and Zamora-Chinchipe south to Panguri (ANSP). Northward it is apparently recorded only from w. Napo, principally from specimens (MCZ) obtained in the Volcán Sumaco region. Recorded mostly from about 800 to 2000 m, but found locally down to 600 m (Tayuntza [WFVZ], where recorded in sympatry with the Ash-throated Gnateater, *C. peruviana*).

Two races are found in Ecuador, nominate *castaneiceps* northward (south to the Sumaco area in w. Napo) and *chapmani* southward (north to the Cordillera de Cutucú in Morona-Santiago; ANSP). Females of the nominate race differ in being darker and more olivaceous brown above with back feathers faintly edged blackish. It is unknown if the two taxa come into contact, or if they intergrade in the intervening area (where the species is presumed to occur).

Range: locally in Andes of Colombia and on east slope of Andes in Ecuador and Peru.

The Slaty Gnateater (*Conopophaga ardesiaca*) was recorded from e. Ecuador by Chapman (1926)—albeit with uncertainty—based on a female specimen that seems almost certain to be referable to the Ash-throated Gnateater.

Conopophaga peruviana   Ash-throated Gnateater   Jejenero Golicinéreo

Uncommon and inconspicuous in the undergrowth of terra firme forest in the lowlands of e. Ecuador. The inconspicuous Ash-throated Gnateater has apparently not been recorded north of the Río Napo, though it has been seen just south of that river (e.g., opposite La Selva, along the Maxus road southeast of Pompeya, and at Jatun Sacha). Recorded up to about 600 m (Tayuntza; WFVZ).

Monotypic.

Range: e. Ecuador, e. Peru, nw. Bolivia, and w. Amazonian Brazil.

Conopophaga aurita   Chestnut-belted Gnateater   Jejenero Fajicastaño

Uncommon and inconspicuous in the undergrowth of terra firme forest in the lowlands of ne. Ecuador. There are apparently no specific Ecuadorian locality records of the Chestnut-belted Gnateater from south of the Río Napo; the type locality of *occidentalis* is merely "Río Napo," which could refer to either side of the river. A specimen from Sarayacu (Berlioz 1932a) seems likely not to have been taken precisely at that locality. The species is known definitely only from more northerly sites, including Río Suno, "below San José," Limoncocha, La Selva, Cuyabeno, Tigre Playa, and Zancudococha. Recorded below 300 m.

One race: *occidentalis*.

Range: se. Colombia, ne. Ecuador, and ne. Peru through Amazonian Brazil and the Guianas.

Rhinocryptidae   Tapaculos

Fourteen species in five genera. Tapaculos are exclusively Neotropical in distribution.

Liosceles thoracicus   Rusty-belted Tapaculo   Tapaculo Fajirrojizo

Uncommon to locally common (though hard to see, and recorded almost entirely through its far-carrying and often-heard vocalizations) on or near the ground inside terra firme forest in the lowlands of e. Ecuador, ranging up in

smaller numbers into the foothills. Recorded mostly below 600 m, locally up to 900 m in the Bermejo oil-field area north of Lumbaquí (M. B. Robbins).
One race: *erithacus*.
Range: se. Colombia, e. Ecuador, e. Peru, and w. and cen. Amazonian Brazil.

Melanopareia elegans   Elegant Crescentchest   Pecholuna Elegante

Uncommon to locally fairly common in dense scrub and thick undergrowth of low woodland in the lowlands of sw. Ecuador from cen. Manabí (north to around Bahía de Caráquez and Chone) and extreme s. Pichincha (Río Palenque, where first seen on 23–24 Jul. 1992 [P. Coopmans and M. Van Beirs et al.]; the species has since become more numerous at this site) south through much of Los Ríos, Guayas, El Oro, and Loja. Although favoring more arid regions, the Elegant Crescentchest also is found locally in fairly humid areas; it seems readily to adapt to very disturbed conditions and may be spreading northward with deforestation. Recorded up locally to 2300–2400 m in s. Loja (e.g., above La Toma and near Utuana; RSR and L. Navarrete et al.).

The genus *Melanopareia* may not belong with the Rhinocryptidae. One race occurs in Ecuador, nominate *elegans*. In s. Loja there may be some approach to *paucalensis* of nw. Peru.
Range: sw. Ecuador and nw. Peru.

[*Melanopareia maranonica*   Marañón   Pecholuna del
                              Crescentchest   Marañón]

Uncertain. A crescentchest was seen by T. J. Davis on 10 Aug. 1992 in secondary scrub near the Río Mayo in the drainage of the Río Marañón near Zumba in extreme s. Zamora-Chinchipe; in addition, a song very similar to that of the Elegant Crescentchest (*M. elegans*) was heard in the area several times (F. Sornoza). One was also heard just west of La Chonta on 6 Apr. 1994 (P. Coopmans and M. Lysinger). It would appear virtually certain that a small population of the Marañón Crescentchest is resident in this region. The species was formerly considered endemic to a limited area in the upper Río Marañón valley of nw. Peru. 650–1050 m.

M. maranonica is here considered as a monotypic species separate from *M. elegans*, following most recent authors (though not Sibley and Monroe 1990).
Range: upper Río Marañón valley of nw. Peru and extreme s. Ecuador.

Myornis senilis   Ash-colored Tapaculo   Tapaculo Cinéreo

Uncommon to fairly common (recorded mainly by voice) in the undergrowth of montane forest and forest borders in the upper subtropical and (mainly) temperate zones on both slopes of the Andes and also locally on slopes above the central valley (e.g., at Pasochoa). On the west slope recorded mainly south to Pichincha, but there are also recent records from sites in w. Bolívar (Salinas) and w. Azuay (Sural and Chaucha; N. Krabbe). Recorded between 2000 m (around Cabañas San Isidro; M. Lysinger) and 3500 m, but mostly above about 2300 m.

The monotypic genus *Myornis* is sometimes merged into *Scytalopus* (e.g., Hilty and Brown 1986), but we follow Fjeldså and Krabbe (1990), who suggest that *Myornis* deserves recognition based on its long tail and very different juvenal plumage. Krabbe and Schulenberg (1997) agree. Monotypic.

Range: locally in Andes from n. Colombia to n. Peru.

Scytalopus unicolor   Unicolored Tapaculo   Tapaculo Unicolor

Fairly common to common (recorded mainly by voice) in the undergrowth of montane forest, forest borders, and secondary woodland (in less humid regions, e.g., in the southwest, even in montane scrub) in the subtropical and temperate zones up to treeline on the west slope of the Andes and on interandean slopes, also locally spilling over onto the east slope in Sucumbíos and Napo. Lower down on the east slope, uncommon to fairly common but somewhat local in montane forest undergrowth in the subtropical zone from w. Napo southward, and also on most outlying ridges. Recorded mostly from 2000 to 3500 m (locally higher, up to 4000 m, e.g., on Volcán Pichincha); lower on the east slope (see above), where recorded between about 1900 and 2450 m; in the southwest recorded mainly from 1700 to 4000 m, locally down to 1500 m.

Two races occur in Ecuador, *latrans* in most of the species' Ecuadorian range (on the west slope south to n. Cañar) and *subcinereus* in the southwest (north to Azuay at El Cajas and Bestión). *Subcinereus* differs from *latrans* principally in the females, which are considerably paler gray than in *latrans* and often have extensively brown flanks; the all-blackish males are similar. This arrangement follows that presented by Krabbe and Schulenberg (1997), but—as those authors acknowledge—it is possible that two separate species may be involved, there being subtle vocal differences; the entire complex may also be specifically distinct from isolated nominate *unicolor* of Peru.

Range: Andes from w. Venezuela to n. Peru.

Scytalopus micropterus   Equatorial Rufous-   Tapaculo Ventrirrufo
　　　　　　　　　　　　　　vented Tapaculo   Equatorial

Fairly common (recorded mainly by voice) in the undergrowth of montane forest, forest borders, and secondary woodland (often along streams) in the subtropical zone on the east slope of the Andes, ranging at least locally up into the lower temperate zone as well. The Equatorial Rufous-vented Tapaculo is quite readily found near and above San Rafael Falls. Recorded mainly between about 1250 and 2300 m (SierrAzul; ANSP).

S. micropterus is considered a monotypic species distinct from *S. femoralis* (Peruvian Rufous-vented Tapaculo) of Peru, following Krabbe and Schulenberg (1997) and based mainly on biochemical data (Arctander and Fjeldså 1994) and differences in their songs. Formerly (e.g., Meyer de Schauensee 1966, 1970) they were considered conspecific under the name of *S. femoralis* (Rufous-vented Tapaculo), with *atratus*, *bolivianus*, and *sanctaemartae* also being included (see the next account).

Range: east slope of Andes from s. Colombia to extreme n. Peru.

Scytalopus atratus	Northern White-crowned Tapaculo	Tapaculo Coroniblanco Norteño

Uncommon to fairly common (recorded mainly by voice) in the undergrowth of montane forest and at forest borders in the foothills and subtropical zone on the east slope of the Andes. The Northern White-crowned Tapaculo is quite readily found along the Loreto road north of Archidona and also in the Río Bombuscaro sector of Podocarpus National Park. Recorded between about 850 and 1650 m.

Following Krabbe and Schulenberg (1997), *S. atratus* is considered as a species distinct from *S. bolivianus* (Bolivian White-crowned Tapaculo) of s. Peru and Bolivia, and *S. sanctaemartae* (Santa Marta Tapaculo) of Colombia. *Atratus* was treated as a subspecies of *S. bolivianus* (White-crowned Tapaculo) by Ridgely and Tudor (1994). All three were formerly (e.g., Meyer de Schauensee 1966, 1970) considered subspecies of *S. femoralis* (which then was called the Rufous-vented Tapaculo). These three taxa occur at somewhat lower elevations than the two Rufous-vented Tapaculos, and usually show a small white crown spot; all three forms differ vocally. One race occurs in Ecuador, nominate *atratus*.

Range: Andes from sw. Venezuela to s. Peru; Sierra de Perijá.

Scytalopus vicinior	Nariño Tapaculo	Tapaculo de Nariño

Fairly common (recorded mainly by voice) in the undergrowth of montane forest and to a lesser extent in forest borders in the subtropical zone of nw. Ecuador south to w. Cotopaxi. The first Ecuadorian records involved the series obtained by M. B. Robbins et al. in Carchi near the Maldonado road in Jul.–Aug. 1988 (ANSP, MECN) and birds tape-recorded—first by P. Coopmans—in Pichincha and collected (ZMCOP, MECN) since 1990 along the Nono-Mindo road, at Maquipucuna, and along the upper Chiriboga road. The Nariño Tapaculo is now known to occur in fair numbers at various sites on the slopes above Mindo. An early specimen from Hacienda Paramba in Imbabura (J. T. Zimmer, *Am. Mus. Novitates* 1044, 1939) recorded under the name of *S. panamensis vicinior* is probably referable to *S. chocoensis* (Chocó Tapaculo; Krabbe and Schulenberg, op. cit.). Recorded mostly between 1250 and 2000 m, locally (in Carchi) as high as 2350 m.

Monotypic.

Range: west slope of Andes in sw. Colombia and nw. Ecuador.

*Scytalopus chocoensis	Chocó Tapaculo	Tapaculo del Chocó

Fairly common (recorded mainly by voice) in undergrowth of montane forest in the foothills of nw. Ecuador in e. Esmeraldas and adjacent sw. Imbabura (one seen and tape-recorded near the Salto del Tigre bridge over the Río Guaillabamba on 15 Jul. 1997; P. Coopmans et al.). This newly described species (Krabbe and Schulenberg 1997) is known in Ecuador mainly from Esmeraldas, where it was first found at El Placer in 1979 (O. Jakobsen, fide N. Krabbe) and first collected on 8 Jul. 1984 (N. Krabbe); a series was obtained in Jul.–Aug. 1987 (ANSP, MECN). In Jul. 1990 the species was also found to occur in very small numbers at the base of the western foothills near Alto Tambo (ANSP). Lastly, in Jun. 1997 small numbers were tape-recorded in the foothills above Playa de Oro (O. Jahn et al.). An early specimen from Hacienda Paramba in Imbabura (J. T. Zimmer, *Am. Mus. Novitates* 1044, 1939) that was recorded under the name of *S. panamensis vicinior* is probably referable to this species (Krabbe and Schulenberg, op. cit.). Recorded between about 350 and 950 m.

Monotypic.

Range: e. Panama (Cerro Pirre), and on west slope of Andes in w. Colombia and nw. Ecuador.

*Scytalopus robbinsi   El Oro Tapaculo   Tapaculo de El Oro

Uncommon (recorded mainly by voice) and local in undergrowth of montane forest in the foothills and lower subtropical zone of sw. Ecuador in El Oro and Azuay. This newly described species (Krabbe and Schulenberg 1997) is known only from Buenaventura in El Oro (ANSP) and from near Molleturo along the new Cuenca road in Azuay (ANSP, MECN). Recorded between about 700 and 1250 m.

The species has also been called the Ecuadorian Tapaculo. Given that so many other tapaculo species occur in Ecuador, we prefer to give it the name of the province in which it was discovered, El Oro (as in the El Oro Parakeet, *Pyrrhura orcesi*).

Monotypic.

Range: west slope of Andes in sw. Ecuador.

Scytalopus spillmanni   Spillmann's Tapaculo   Tapaculo de Spillmann

Fairly common (recorded mainly by voice) in the undergrowth of montane forest and forest borders in the subtropical and lower temperate zones on both slopes of the Andes in n. Ecuador. On the east slope recorded south to the Azuay/Morona-Santiago border (south bank of the Río Paute; N. Krabbe), whereas on the west slope it is known south at least to w. Cotopaxi (west of Sigchos; N. Krabbe). Spillmann's is the most numerous *Scytalopus* tapaculo along the upper portions of the Nono-Mindo and Chiriboga roads in Pichincha. It tends to favor *Chusquea* bamboo thickets. Recorded mostly between about 1900 and 3200 m, locally up to 3700 m in w. Napo.

Following Krabbe and Schulenberg (1997), *S. spillmanni* is considered a monotypic species distinct from *S. latebricola* (Brown-rumped Tapaculo) of the Santa Marta Mountains of Colombia, with *S. meridanus* (Mérida Tapaculo)

and *S. caracae* (Caracas Tapaculo) of Venezuela also being considered distinct species.
Range: Andes of Colombia and n. Ecuador.

*Scytalopus parkeri   Chusquea Tapaculo   Tapaculo de Chusquea

Uncommon to locally fairly common (recorded mainly by voice) in the undergrowth of montane forest and forest borders, principally associated with stands of *Chusquea* bamboo, in the upper subtropical and temperate zones on the east slope of the Andes in se. Ecuador. Recorded from Morona-Santiago (north to along the Gualaceo-Limón road), Zamora-Chinchipe, and e. Loja; also known from the highest ridges in the Cordillera del Cóndor. This newly described species (Krabbe and Schulenberg 1997) is numerous in the Cajanuma sector of Podocarpus National Park and at Tapichalaca Biological Reserve above Valladolid. Recorded from about 2250 to 3150 m.
Monotypic.
Range: east slope of Andes in s. Ecuador and extreme n. Peru.

Scytalopus canus   Paramo Tapaculo   Tapaculo Paramero

Uncommon and local in the undergrowth of montane forest borders and woodland, especially near treeline and in patches of *Polylepis* woodland, in the temperate and paramo zones on the east slope of the Andes. On the west slope known only from the extreme north, in Carchi on the Páramo del Ángel. Reports from elsewhere on the west slope, on slopes above the central valley, and in Azuay are (fide N. Krabbe) apparently the result of confusion with other *Scytalopus* species. Small numbers of the Paramo Tapaculo can be found in woodland patches near the pass above Papallacta. Recorded from about 3050 to 4000 m.
Following Krabbe and Schulenberg (1997), *S. canus* is considered as one of numerous separate species that formerly (e.g., Meyer de Schauensee 1966, 1970) constituted *S. magellanicus* (Andean Tapaculo). Birds found in the Andes from w. Venezuela to nw. Argentina were united under the name of *S. griseicollis* by Ridgely and Tudor (1994), with two austral species having been separated out. One race of *S. canus* occurs in Ecuador, *opacus*.
Range: Andes from w. Colombia to extreme n. Peru.

Acropternis orthonyx   Ocellated Tapaculo   Tapaculo Ocelado

Uncommon to fairly common in the undergrowth of montane forest in the upper subtropical and temperate zones on both slopes of the Andes and locally on slopes above the central valley (e.g., at Pasochoa). On the west slope recorded south to w. Cotopaxi (heard on 28 Jul. 1991 above Pilaló; RSR et al.). Until recently a poorly known bird and only rarely recorded, the spectacular Ocellated Tapaculo has since the late 1980s—when its distinctive song became known—been found much more frequently and widely. It remains, however, a difficult bird to see, with prime sites being Pasochoa, the northwestern flank of

Volcán Pichincha around Yanacocha, and near Bellavista Lodge. Recorded from about 1900 to 3500 m, but mainly found above 2500 m.
One race: *infuscata*.
Range: Andes from w. Venezuela to extreme n. Peru.

Tyrannidae	Tyrant Flycatchers

Two hundred eight species in 78 genera. Tyrant flycatchers are exclusively American in distribution, with species diversity being by far greatest in the tropics. We follow the sequence of genera and, for the most part, species in this complex and difficult family as they were set out in *Birds of the World* (vol. 8).

Phyllomyias zeledoni	White-fronted Tyrannulet	Tiranolete Frentiblanco

Rare and local in the canopy and borders of montane forest in the foothill and lower subtropical zones on both slopes of the Andes; doubtless overlooked because of the difficulty in identifying this obscure tyrannulet at the heights above the ground that it favors. There are only three specimen records from Ecuador: Chimbo in e. Guayas (BMNH), Chonta Urcu (which cannot be precisely located but apparently is in the vicinity of—perhaps between—Archidona and Puyo; Royal Ontario Museum, fide D. Agro), and Zamora (AMNH). The only recent reports involve several sightings and tape-recordings north of Archidona along the road to Loreto (first on 9 Sep. 1990; P. Coopmans et al.), one bird seen on the south slope of Volcán Sumaco on 22 Jan. 1991 (B. Whitney), a pair seen and tape-recorded at San Rafael Falls on 31 Aug. 1994 (P. Coopmans et al.), one seen above Manta Real on 20 Aug. 1991 (RSR), one heard along the lower Gualaceo-Limón road on 19 Feb. 1998 (P. Coopmans), one seen in the Río Bombuscaro sector of Podocarpus National Park on 26 Jul. 1992 (D. Wolf), and at least one seen north of Zumba on 13 Dec. 1991 (RSR). Recorded from about 600 to 1500 m.

Formerly placed in the genus *Acrochordopus*. We follow *Birds of the Americas* (vol. 13, part 5), J. T. Zimmer (*Am. Mus. Novitates* 1126, 1941), and Ridgely and Tudor (1994) in considering *P. zeledoni* as a species separate from *P. burmeisteri* (Rough-legged Tyrannulet), the latter found farther south in the Andes and in the se. Brazil region. Stiles and Skutch (1989) split the complex even further, considering *P. zeledoni* of Costa Rica and w. Panama as a third species (Zeledon's Tyrannulet), leaving the birds of the n. Andes as *P. leucogonys* (White-fronted Tyrannulet). Meyer de Schauensee (1966, 1970) and *Birds of the World* (vol. 8) treated all members of the complex as conspecific; the 1998 AOU Check-list does likewise. Voices of all forms appear to be more or less similar, but *burmeisteri* differs distinctly in plumage from the others. One race of *P. zeledoni* occurs in Ecuador, *leucogonys*.

Range: mountains of Costa Rica and w. Panama; very locally in mountains of n. Venezuela and Andes from Colombia to s. Peru; Sierra de Perijá; tepuis of s. Venezuela.

Phyllomyias griseiceps   Sooty-headed   Tiranolete Coronitiznado
                                 Tyrannulet

Uncommon to locally fairly common at the borders of humid forest and secondary woodland, and in adjacent clearings with scattered trees in the foothills and more humid lowlands of both e. and w. Ecuador. In the east found locally along and near the eastern base of the Andes from w. Napo (Jatun Sacha [ANSP] and north of Archidona along the road to Loreto [WFVZ]) south locally into Morona-Santiago and Zamora-Chinchipe (south to Zamora); the farthest east report is a single bird tape-recorded and seen along the Maxus road southeast of Pompeya on 13 Jul. 1994 (RSR and S. Howell; no other birds were ever recorded here). In the west found from Esmeraldas south to w. Guayas (hills south of the Río Ayampe; RSR), El Oro, and w. Loja (Cebollal, near El Limo, and east of Pindal). The Sooty-headed Tyrannulet is apt to be overlooked until its distinctive and oft-given song is recognized; the species is proving to be more widespread than had been realized, and it is quite numerous at both Tinalandia and Río Palenque. Recorded mostly below about 1100m, locally and in small numbers up to 1350m (e.g., at Maquipucuna Reserve; P. Coopmans).

One race: nominate *griseiceps*.

Range: e. Panama to sw. Ecuador, e. Peru, Guyana, and e. Amazonian Brazil.

Phyllomyias plumbeiceps   Plumbeous-crowned   Tiranolete
                                  Tyrannulet   Coroniplomizo

Rare to locally fairly common in the mid-levels and subcanopy of montane forest and forest borders in the subtropical zone on the east slope of the Andes. The Plumbeous-crowned Tyrannulet is doubtless overlooked because of difficulties in identification; in fact, B. Whitney (pers. comm.) considers that it is not all that rare once its vocalizations are recognized. There are only a few Ecuadorian specimens, all of them old, including two collected at Oyacachi (AMNH) and El Chaco (Paris Museum) in w. Napo, and one at Machay in Tungurahua (BMNH). More recently the species has been recorded on the south slope of Volcán Sumaco in w. Napo and above Zamora in Zamora-Chinchipe (B. Whitney), north of Archidona along the Loreto road (one seen and tape-recorded in Aug. 1998; P. Coopmans), near the Río Cosanga below Cabañas San Isidro in Apr. 1993 (RSR), and at Panguri in s. Zamora-Chinchipe in Jul. 1992 (T. J. Davis). The specimen listed for Baeza (Chapman 1926) is a misidentified Ashy-headed Tyrannulet (*P. cinereiceps*; J. T. Zimmer, *Am. Mus. Novitates* 1126, 1941). Recorded from about 1200 to 2200m.

Formerly placed in the genus *Oreotriccus*. Monotypic.

Range: locally in Andes of Colombia and on their east slope in Ecuador and Peru.

Phyllomyias nigrocapillus Black-capped Tiranolete Gorrinegro
Tyrannulet

Uncommon and apparently local in the borders of montane forest and secondary woodland in the upper subtropical and temperate zones on both slopes of the Andes, often most numerous near treeline. On the west slope until recently recorded south only to Pichincha, but there are 1996–1997 reports from Salinas in Bolívar and Chaucha in Azuay (Krabbe et al. 1997). The Black-capped Tyrannulet seems particularly numerous along the upper part of the Loja-Zamora road and is generally more numerous and widespread in s. Ecuador than it is northward. Recorded mostly from 2300 to 3300 m.

This and the next two species were formerly placed in the genus *Tyranniscus*. One race: nominate *nigrocapillus*.

Range: Andes from w. Venezuela to s. Peru; Santa Marta Mountains.

Phyllomyias cinereiceps Ashy-headed Tyrannulet Tiranolete Cabecicinéreo

Uncommon to locally fairly common in the mid-levels, subcanopy, and borders of montane forest and secondary woodland in the subtropical zone on both slopes of the Andes. On the west slope recorded south to Chimborazo (old specimens from Pallatanga and Chimbo) and El Oro (Buenaventura, where single birds were recorded on 13 Mar. 1994 and 6 Mar. 1995; P. Coopmans et al.); in recent years mostly found from Pichincha (Mindo/Nanegal region and along lower Chiriboga road) northward. The Ashy-headed Tyrannulet is perhaps especially numerous on the slopes of the Cordillera de Huacamayos in w. Napo. Recorded mostly between about 1350 and 2500 m, but in El Oro found at 900 m.

Monotypic.

Range: Andes from extreme sw. Venezuela to s. Peru.

Phyllomyias uropygialis Tawny-rumped Tiranolete Lomileonado
Tyrannulet

Uncommon in borders of montane forest and in secondary woodland and shrubby clearings in the subtropical and temperate zones on both slopes of the Andes, and locally (where patches of forest or at least woodland remain) also on slopes above the central and interandean valleys. Recorded mostly from 2100 to 3100 m, in small numbers down to 1500–1600 m (e.g., above Mindo and [Best et al. 1993] at Sozoranga) and up to 3400 m (e.g., at Yanacocha).

Monotypic.

Range: Andes from w. Venezuela to s. Bolivia.

Zimmerius chrysops Golden-faced Tyrannulet Tiranolete Caridorado

Common in the canopy and borders of montane and foothill forest and secondary woodland, and in adjacent clearings with trees and dense shrubbery, on both slopes of the Andes, also ranging out into adjacent humid lowlands on both slopes. In the western lowlands ranges south to Guayas (near the coast south to the hills south of the Río Ayampe, and along the base of the Andes south to Chimbo and Naranjito). In the east recorded out at least as far as along the Maxus road southeast of Pompeya (where it is quite numerous; RSR et al.), Yuturi Lodge (S. Howell), and Kapawi Lodge along the Río Pastaza near the Peruvian border. At none of the last three sites has the Slender-footed Tyrannulet (*Z. gracilipes*) been reported. The ecological and distributional relationships between these two lowland *Zimmerius* tyrannulets has still not been entirely resolved. Especially in the west the Golden-faced is a relatively frequently observed tyrannulet, but like many others it is recorded even more often once its voice is known. In e. Ecuador recorded from about 2200 m down into the lowlands, mainly (but not only) near the base of the Andes; in the west recorded from near sea level up locally to about 1600 m (e.g., at Maquipucuna and above Mindo).

This and the following three species were formerly placed in the genus *Tyranniscus*; see M. A. Traylor, Jr. (*Bull. Mus. Comp. Zool.* 148[4]: 129–184, 1977) for the erection of *Zimmerius*. *Z. chrysops* of Venezuela to n. Peru is regarded as a species separate from *Z. viridiflavus* (Peruvian Tyrannulet), the latter being found on the east slope of the Andes in cen. Peru; their voices and plumages are distinctly different (see Ridgely and Tudor 1994), and they were so treated by J. T. Zimmer (*Am. Mus. Novitates* 1109, 1941). We now also treat *Z. flavidifrons* (Loja Tyrannulet) as a separate species (see the next account). Two races occur in Ecuador: nominate *chrysops* on the east slope and out into the eastern lowlands and *albigularis* in w. Ecuador. Compared to nominate *chrysops*, *albigularis* is somewhat whiter (less yellow) on the throat, slightly darker and grayer on the breast, and whiter on the belly. Based on their differing voices, these two taxa may themselves represent separate species, but more work is needed to establish this.

Range: mountains of ne. Venezuela; Andes and adjacent lowlands from w. Venezuela to ne. Peru; Sierra de Perijá and Santa Marta Mountains.

Zimmerius flavidifrons   Loja Tyrannulet   Tiranolete de Loja

Fairly common to common in the canopy and borders of montane forest and woodland in the foothills and subtropical zone on the west slope of the Andes in sw. Chimborazo (the type specimen was obtained at Pallatanga, and J. T. Zimmer [*Am. Mus. Novitates* 1109, 1941] records others from there and Chimbo), Azuay, El Oro, and Loja (ranging east at least to Utuana). Precise details concerning the nature of the replacement zone between this species and the Golden-faced Tyrannulet (*Z. chrysops*) remain to be determined; Golden-faced Tyrannulets have been recorded south to se. Guayas (old specimens from Naranjo) and have also been recorded in the past from se. Chimborazo (though some specimens may have been confused with *flavidifrons?*). On the east slope known only from the Río Marañón drainage of s. Zamora-Chinchipe, where one calling bird (tape-recorded by RSR) was collected north of Zumba on 13 Dec. 1991 (ANSP) and another was heard and seen above Valladolid on 18 Jan.

1998 (RSR et al.). Recorded from about 900 m (Buenaventura in El Oro) to 2400 m (Utuana).

Z. *flavidifrons* is here regarded as a species separate from Z. *chrysops*, based primarily on its strikingly different voice; it also differs from Z. *chrysops* in having markedly less facial yellow. The two species nearly come into contact on the slopes above Valladolid in Zamora-Chinchipe (they may even overlap here). Monotypic.

Range: Andes of sw. Ecuador and nw. Peru (ANSP specimens from Palambla in Piura).

Zimmerius gracilipes   Slender-footed   Tiranolete Patidelgado
                                     Tyrannulet

Uncommon to locally fairly common (doubtless having been overlooked in the past) in the canopy and borders of humid forest (both terra firme and várzea) in the lowlands of ne. Ecuador. Definite reports of the Slender-footed Tyrannulet are rather few and are confined to areas near the Río Napo (Limoncocha, Taracoa, Sacha Lodge, La Selva) and the Río Aguarico (upriver to Cuyabeno). It may be more widespread, but so far there is no definite evidence that it occurs in the southeastern lowlands, nor are there any confirmed reports from any distance south of the Río Napo. Although their voices differ, there continues to be considerable confusion between this species and the similar Golden-faced Tyrannulet (Z. *chrysops*), a species that also occurs widely in the eastern lowlands. Only one specimen of the Slender-footed Tyrannulet has been taken in Ecuador, a female obtained at Zancudococha on 7 Apr. 1991 (ANSP). Recorded mostly below 250–300 m, with a few unconfirmed reports from Jatun Sacha (400 m).

One race: nominate *gracilipes*.

Range: se. Colombia, s. Venezuela, and the Guianas to n. Bolivia and Amazonian Brazil. More than one species may be involved.

Zimmerius cinereicapillus   Red-billed   Tiranolete Piquirrojo
                                     Tyrannulet

Rare to uncommon and local (though doubtless overlooked) in the canopy and borders of montane forest and woodland in foothills along the eastern base of the Andes. To date there is only one Ecuadorian specimen (AMNH) of the Red-billed Tyrannulet, taken at "Río Suno Arriba" in w. Napo on 4 Feb. 1923, but the species has been seen in the same general area on a few occasions in recent years along the road to Loreto north of Archidona. As the species is found in Peru, it seems likely to occur also in s. Ecuador, but as yet there appears to be only one confirmed report, a bird seen along the lower Loja-Zamora road in Zamora-Chinchipe on 28 Oct. 1993 (D. Wolf and PJG). Recorded from about 900 to 1350 m.

Monotypic.

Range: locally on east slope of Andes from n. Ecuador to s. Peru.

Ornithion inerme   White-lored Tyrannulet   Tiranolete Alipunteado

Uncommon to locally fairly common in the canopy and borders of humid forest and secondary woodland in the lowlands of e. Ecuador. As it tends to remain high in trees, the White-lored Tyrannulet is likely to be overlooked until its vocalizations are recognized; in recent years it has proven to be considerably more numerous and widespread than had been thought. Mainly found below about 600 m (e.g., near Archidona [P. Coopmans] and at the Bermejo oil-field area north of Lumbaquí [RSR]); also recorded at Miazi (900 m) in s. Zamora-Chinchipe (Schulenberg and Awbrey 1997).

Monotypic.

Range: se. Colombia, s. Venezuela, and the Guianas to n. Bolivia and Amazonian Brazil; e. Brazil.

Ornithion brunneicapillum   Brown-capped   Tiranolete Gorripardo
                                Tyrannulet

Uncommon to locally fairly common in the canopy and borders of forest and secondary woodland, and in adjacent clearings with scattered trees, in the more humid lowlands of w. Ecuador; recorded south in very small numbers to El Oro (well below Buenaventura, where seen and heard in Jun. 1985; RSR), but more numerous from e. Guayas and nw. Azuay northward. A tiny and inconspicuous tyrannulet, the Brown-capped is apt to be overlooked until its distinctive and frequently given calls are recognized. Recorded mainly below about 400 m, but locally found as high as 750–800 m (at Hacienda Paramba in Imbabura; and in Pichincha at Tinalandia and near San Miguel de los Bancos [P. Coopmans]).

One race: nominate *brunneicapillum*. However, examples from the southern part of the species' Ecuadorian range may be separable on the basis of their duskier crowns, as alluded to by Chapman (1926); they remain unnamed. ANSP has two recently obtained specimens from Ecuador: one from Azuay has the crown quite blackish, but one from Esmeraldas appears assignable to the nominate race (and has a brownish crown).

Range: Costa Rica to nw. Venezuela and sw. Ecuador.

Camptostoma obsoletum   Southern Beardless-   Tiranolete Silbador
                            Tyrannulet   Sureño

Fairly common to common in a variety of semiopen to partially wooded habitats (locally even at the edge of humid cloud forest) in the lowlands and subtropics of w. Ecuador, where found in both arid and humid regions (though more numerous in the former). Much less numerous in e. Ecuador, where local and confined to second-growth mainly near water (e.g., at Limoncocha) and also the canopy of várzea forest (e.g., at Imuyacocha on the Río Lagarto; v.o.), but perhaps now beginning to follow major valleys up to the base of the Andes (e.g., around Zamora). Also fairly common in second-growth in the Río

Marañón drainage around Zumba in extreme s. Zamora-Chinchipe, with smaller numbers ranging up in cleared areas to the Valladolid area (1600 m). Ranges up to at least 2800 m in the central valley (e.g., in Quito); in the east recorded primarily below 300 m, but locally up to 1000 m (Zamora).

At least two races occur in Ecuador, *sclateri* in the west and *olivaceum* in the east. *Sclateri* is dull and grayish overall, whereas the rather different *olivaceum* is more olive above, a paler clear yellow below, and has whitish (not buffyish) wing-bars. The latter is known in Ecuador from only a few specimens. The two taxa may be representative forms of separate species, as in addition to the plumage differences their vocalizations also differ markedly. *Maranonicum*, which presumably is the race occurring in the Río Marañón drainage, resembles western *sclateri*; these two taxa sound much alike.

Range: Costa Rica to nw. Peru, n. Argentina, and Uruguay. As noted above, more than one species is likely involved.

**Phaeomyias murina*   Mouse-colored Tyrannulet   Tiranolete Murino

Uncommon and local in clearings with scattered trees and around towns in lowlands of the northeast. The Mouse-colored Tyrannulet was only recently first found in Ecuador—though it perhaps was present earlier—and to date is known only from Sucumbíos and Napo. The first record occurred on 12 Dec. 1990, when several pairs were seen at Lago Agrio (P. Coopmans, RSR, and PJG et al.); they were there again in Mar.–Apr. 1991 (RSR et al.) and have been subsequently. Additional records involve birds seen and tape-recorded on Isla Anaconda near La Selva on 14 Jul. 1992 (P. Coopmans and M. Van Beirs et al.) and at Tena since Dec. 1993 and at Archidona since Apr. 1994 (P. Coopmans et al.). With continued forest clearance, the Mouse-colored Tyrannulet seems likely to increase and spread in the eastern lowlands. Recorded to about 600 m.

Does not include *P. tumbezana* (Tumbesian Tyrannulet). The subspecific determination of *P. murina* in Ecuador must await the collection of specimens. Based on the decidedly yellowish bellies of Ecuadorian birds, we believe it likely that they will prove assignable to the race *wagae*.

Range: Costa Rica to n. Argentina, Paraguay, and s. and e. Brazil.

Phaeomyias tumbezana   Tumbesian Tyrannulet   Tiranolete de Tumbes

Uncommon to locally fairly common in desert scrub, arid light woodland, and gardens in intermontane valleys of n. Ecuador in Imbabura (Paramba and north of Salinas; MECN) and n. Pichincha (Cumbayá; MECN), and in more arid lowlands of sw. Ecuador from cen. Manabí (Bahía de Caráquez) south through sw. Azuay, El Oro, and Loja; in Loja it ranges east to near Loja (city), Vilcabamba, and Amaluza. As it is numerous in adjacent Peru, it seems possible that the Tumbesian Tyrannulet may also range north in the Río Marañón drainage to the Zumba area in se. Zamora-Chinchipe. In Loja small numbers have been recorded as high as 2300 m at Gonzanamá (MCZ) and 2400 m at Cumbayá in Pichincha, but mostly this species is found below 1900 m.

Because of striking differences in their vocalizations and various plumage differences, we treat *P. tumbezana* as a species distinct from the widespread *P. murina* (Mouse-colored Tyrannulet). *P. tumbezana* (with races *inflava* and *maranonica*) has usually been considered conspecific with *P. murina*, though Ridgely and Tudor (1994) expressed reservations as to whether this was the correct course. One race is known to occur, nominate *tumbezana*; if the species ranges to the Zumba area (see above), *maranonica* would be the likely race to occur there.

Range: w. Ecuador and nw. Peru.

| *Pseudelaenia leucospodia* | Gray-and-white Tyrannulet | Tiranolete Grisiblanco |

Uncommon to locally fairly common in desert scrub in the lowlands of sw. Ecuador, where apparently recorded only from w. Guayas (east to near the Guayaquil area, and on Isla Puná) and on Isla de la Plata off s. Manabí; curiously, the Gray-and-white Tyrannulet has not been found on the adjacent Manabí mainland (a distribution pattern shared by the Short-tailed Field-Tyrant [*Muscigralla brevicauda*]). As yet the Gray-and-white Tyrannulet has not been found in coastal El Oro, and there is only one report from Loja, of a bird seen in the far south near Lalamor on 16 Sep. 1998 (RSR et al.). The Gray-and-white Tyrannulet is quite readily found in the dry washes near the town of Santa Elena on the Santa Elena Peninsula, and it is particularly numerous on Isla de la Plata, despite somehow not having been recorded there by F. Ortiz-Crespo and P. Agnew (*Bull. B. O. C.* 112[2]: 66–73, 1992), these authors having also overlooked Chapman's (1926) early published record from the island. Recorded only below about 100 m.

Formerly placed in the genus *Phaeomyias* (e.g., Meyer de Schauensee 1966, 1970) or more recently in *Myiopagis* (*Birds of the World*, vol. 8), in which it definitely does not belong. W. E. Lanyon (*Am. Mus. Novitates* 2914, 1988) erected the monotypic genus *Pseudelaenia* for it; we follow this course. Only one race is definitely known from Ecuador, *cinereifrons*. However, as the nominate race is known from Tumbes and Alamor in extreme n. Piura, Peru (J. T. Zimmer, *Am. Mus. Novitates* 1109, 1941), it is likely that the bird seen just across the nearby border in Ecuador (see above) is also referable to that form. The two are very similar, but *cinereifrons* is slightly grayer above and whiter below; it has been suggested (*Birds of the Americas*, pt. 5) that the two forms may not be separable.

Range: locally in sw. Ecuador and nw. Peru.

| *Tyrannulus elatus* | Yellow-crowned Tyrannulet | Tiranolete Coroniamarillo |

Uncommon to locally common in clearings, gardens, lighter woodland, forest borders, and the canopy of humid forest and woodland (both terra firme and várzea) in the lowlands of both e. and w. Ecuador; in the east in both terra firme and várzea. In the west found only in more humid regions, with small numbers ranging south to n. and e. Guayas (Balzar and in the Naranjal area) and nw. Azuay (Manta Real; RSR and G. H. Rosenberg et al.). Recorded mostly below about 600 m, but a few have been noted up to about 950 m near Zamora (P. Coopmans).

Monotypic.

Range: Costa Rica to sw. Ecuador, n. Bolivia, and Amazonian Brazil.

Myiopagis caniceps   Gray Elaenia   Elenita Gris

Rare to locally uncommon (doubtless under-recorded) in the canopy and borders of humid forest in the lowlands and lower foothills of both e. and nw. Ecuador. In the east occurs in both terra firme and várzea though mainly the former. In the west the Gray Elaenia has been recorded south locally to s. Pichincha at Río Palenque (where it has apparently not been found since about 1990); it is most numerous in n. Esmeraldas where, for instance, it was found to be a regular member of canopy flocks at Playa de Oro (O. Jahn and P. Mena Valenzuela et al.). Gray Elaenias are presumed to be of widespread occurrence in the east, though there are relatively few actual reports, with records scattered from Sucumbíos south to Morona-Santiago (Taisha) and Pastaza (Kapawi Lodge). Recorded up to about 600 m in the east, but below 400 m in the northwest.

Two rather similar races occur, *cinerea* in the east and the smaller *parambae* in the west. Aside from size, males of these taxa closely resemble each other (though they differ vocally). Females apparently differ in the color of the crown patch, white in *parambae* and pale yellowish in *cinerea*.

Range: Panama to nw. Ecuador; se. Colombia, s. Venezuela, and Guyana to n. Argentina, e. Paraguay, and s. and e. Brazil. More than one species is likely involved.

Myiopagis gaimardii   Forest Elaenia   Elenita Selvática

Fairly common in the canopy and borders of humid forest (both terra firme and várzea, but primarily the latter) and secondary woodland in the low-lands of e. Ecuador. Although it is widespread and reasonably numerous, there are rather few Ecuadorian specimens of the Forest Elaenia: an obscure, small tyrannid found mainly in the high canopy, it is heard much more often than seen. Recorded mostly below about 1000 m along the eastern base of the Andes, exceptionally as high as 1200 m along the Loreto road (P. Coopmans).

One race: nominate *gaimardii*.

Range: Panama to n. Bolivia and cen. and Amazonian Brazil; ne. Brazil.

A new species of elaenia, discovered in the early 1990s, was described by P. Coopmans and N. Krabbe (*Wilson Bull.* 112[3]: 305–312, 2000) as the Foothill Elaenia (*Myiopagis olallai*)—too late to be included here. It ranges locally in the foothills along the e. base of the Andes.

Myiopagis flavivertex   Yellow-crowned Elaenia   Elenita Coroniamarilla

Uncommon to locally fairly common (doubtless formerly overlooked) in the lower and middle growth of várzea forest in the lowlands of ne. Ecuador. There are several recent reports of the Yellow-crowned Elaenia, all from various sites along the Ríos Napo and Aguarico. The first Ecuador record was of a pair seen and tape-recorded at Taracoa on 22 Aug. 1986 (RSR et al.). Since 1988 birds have also been found at La Selva along the Mandiyacu and the shores of Mandicocha. The first specimens were taken at Imuyacocha on the Río Lagarto on 6 Dec. 1990 and in Mar. 1991 (ANSP and MECN). In Jul. 1991 the species was also found to be quite numerous at Cuyabeno (J. Rowlett and R. A. Rowlett et al.), and it has been found along the Río Pacuyacu since Sep. 1992 (RSR and T. J. Davis; P. Coopmans et al.) and at Yuturi Lodge (S. Howell). This easily overlooked species is probably widespread in appropriate habitat; as yet, however, it has not been found in the se. lowlands. Recorded below 300 m.
Monotypic.
Range: locally in e. and s. Venezuela, the Guianas, Amazonian Brazil, ne. Peru, and ne. Ecuador.

Myiopagis subplacens   Pacific Elaenia   Elenita del Pacífico

Uncommon to fairly common in the lower growth and borders of deciduous and semihumid woodland, sometimes ranging out into adjacent second-growth, in the lowlands and foothills of w. Ecuador from w. Esmeraldas, Manabí, extreme s. Pichincha (a few records from Río Palenque), and sw. Chimborazo (Pallatanga) south through El Oro and w. Loja. The Pacific Elaenia is numerous in the Chongón Hills west of Guayaquil and in appropriate habitat in Machalilla National Park in sw. Manabí, but overall numbers have doubtless declined in many areas as a result of woodland destruction. Recorded up to about 1700 m.
Monotypic.
Range: w. Ecuador and nw. Peru.

Myiopagis viridicata   Greenish Elaenia   Elenita Verdosa

Uncommon to fairly common in the lower growth and mid-levels of deciduous and semihumid forest and woodland and their borders in the lowlands of w. Ecuador from Esmeraldas (Esmeraldas [city] area; also recorded as a non-breeding wanderer as far north as Playa de Oro [O. Jahn et al.]), w. Imbabura (an old specimen from Intag), and w. Pichincha south to El Oro and w. Loja (an old specimen from Punta Santa Ana and an Apr. 1993 specimen [ANSP] from near El Limo). Given its occurrence so close to the border, the Greenish Elaenia seems likely to range south into adjacent nw. Peru. A specimen (ANSP) taken by T. Mena on 20 Nov. 1950 along the Río San Miguel just across the border in w. Putumayo, Colombia, suggests that the species may also occur in adjacent ne. Ecuador, where it should be watched for; the specimen is considered as probably belonging to an undescribed race (*Birds of the World*, vol. 8). Recorded up locally to about 1000 m in Loja, but mostly found below 500 m.

One race: *implacens*.

Range: n. Mexico to n. Colombia and Venezuela; sw. Colombia and w. Ecuador; se. Peru to n. Argentina, e. Paraguay, and s. and e. Brazil.

Elaenia flavogaster   Yellow-bellied Elaenia   Elenia Penachuda

Uncommon to fairly common in shrubby clearings and second-growth, gardens, and forest borders in the more humid lowlands and foothills of w. Ecuador from Esmeraldas, w. Carchi (La Concepción), and nw. Imbabura (Lita) south locally to w. Loja (Puyango area; RSR). Also fairly common to common in the Río Marañón drainage of extreme s. Zamora-Chinchipe around Zumba and in valleys up to near Valladolid, but absent from the eastern lowlands. Recorded mostly below about 1000 m, locally up in small numbers as high as 1200–1300 m in the Nanegal and Mindo regions, and in small numbers up to 1600 m around Valladolid in Zamora-Chinchipe (RSR et al.) and to 1800 m at Cariamanga in Loja (P. Coopmans).

One race: *semipagana*.

Range: s. Mexico to sw. Ecuador, n. Argentina, e. Paraguay, and se. Brazil (but absent from much of Amazonia); Lesser Antilles.

**Elaenia spectabilis*   Large Elaenia   Elenia Tribandeada

A rare austral winter visitor to shrubby clearings, forest borders, and riparian areas in the lowlands of e. Ecuador. There are only a few Ecuadorian reports; the Large Elaenia appears to be more numerous as a visitant to ne. Peru. The first record is a specimen (LSUMZ) taken at Limoncocha in Sep. 1976 (Tallman and Tallman 1977); a second Ecuadorian specimen was obtained at Tayuntza on 5 Aug. 1987 (WFVZ). Large Elaenias have been seen at a few localities in Napo and Morona-Santiago in Jul. and Aug., with a few arriving as early as late Mar. (e.g., one seen at Imuyacocha on the Río Lagarto on 30 Mar. 1991; RSR and F. Sornoza). Recorded up to 600 m (Tayuntza; WFVZ).

We follow Ridgely and Tudor (1994) in considering *ridleyana*, formerly sometimes considered a subspecies of *E. spectabilis*, to be a separate species endemic to Ilha Fernando de Noronha off ne. Brazil (Noronha Elaenia). *E. spectabilis* then becomes monotypic.

Range: breeds in s.-cen. Brazil, e. Bolivia, Paraguay, and n. Argentina, wintering north to e. Ecuador, se. Colombia, and Amazonian and ne. Brazil.

Elaenia gigas   Mottle-backed Elaenia   Elenia Cachudita

Uncommon and somewhat local in *Tessaria*/willow (*Salix*) scrub on river islands and in shrubby clearings with scattered trees and bushes in the lowlands and foothills of e. Ecuador; ranges away from the Andes as far east as the vicinity of Yuturi Lodge on the Río Napo and as far as Kapawi Lodge on the Río

Pastaza. The numbers and range of the striking Mottle-backed Elaenia may be slowly increasing as a result of forest clearance. Ranges up in some valleys (e.g., around Zamora and near Zumba) on the east slope of the Andes to 1000–1250 m, in smaller numbers even higher (once as high as 1800 m below Baeza; P. Coopmans et al.).

Monotypic.

Range: near eastern base of Andes from Colombia to w. Bolivia.

The Brownish Elaenia (*E. pelzelni*) ranges as close to Ecuador as the mouth of the Río Curaray into the Río Napo (Ridgely and Tudor 1994). This species is one of the few river-island specialists ranging that far upriver on the Río Napo that has not been found within present-day Ecuadorian territory. We continue to suspect that it may occur in far e. Ecuador and feel it should particularly be watched for close to the Peruvian frontier around the mouth of the Río Aguarico into the Napo.

Elaenia obscura   Highland Elaenia   Elenia Oscura

Rare and local in the lower growth and borders of montane forest and secondary woodland, sometimes where surprisingly disturbed, in the Andes of s. Ecuador in Azuay and Loja. First recorded in Ecuador from three specimens (MCZ) taken by D. Norton at Loja (city) on 27–28 Jul. 1965. Subsequently several birds were seen and tape-recorded several kilometers east of Gualaceo in Azuay on 20–21 Mar. 1990 (B. Whitney), and one was collected (MECN) 1 km east of Cariamanga, Loja, on 9 Apr. 1991 (see Krabbe 1992). The Highland Elaenia is presumed to be a scarce resident in Ecuador, as the species is not known to be migratory. Recorded from 2150 to 3000 m.

One race: nominate *obscura*.

Range: Andes from s. Ecuador to nw. Argentina; se. Brazil, e. Paraguay, and ne. Argentina.

Elaenia chiriquensis   Lesser Elaenia   Elenia Menor

Rare to uncommon and local in semiopen areas and clearings with scattered bushes and trees in foothills and the subtropical zone of nw. Ecuador in n. and e. Esmeraldas, Imbabura (Hacienda Paramba; Ibarra), and n. Pichincha (south perhaps to the Mindo region, but see below). Also recently found in the Río Marañón drainage of extreme s. Zamora-Chinchipe near Zumba (seen in Dec. 1991; RSR). Confirmed records of the Lesser Elaenia from Ecuador are few, at least in part because of its close similarity to several congeners, especially the Sierran Elaenia (*E. pallatangae*), and because the voice of the form found in Ecuador bears little resemblance to that given by Lesser Elaenias elsewhere (as discussed below, it may represent a separate species). Even specimens have been confused; the old specimens from Mindo are noted as being uncertain by Chapman (1926, p. 507), and indeed there is also no modern confirmation that the species occurs there; the species may occur mostly or entirely from Imbabura northward. Recorded between about 700 and 2800 m.

One race is known from Ecuador, *brachyptera*. However, another race of *E. chiriquensis*, the relatively widespread *albivertex*, is known from a single ANSP specimen taken by T. Mena on 21 Dec. 1950 at Cordillera (= Cerro) Pax in adjacent Nariño, Colombia, very close to the Ecuadorian border; this may be the race of *E. chiriquensis* that occurs near Zumba, where it remains uncollected. *Albivertex* is similar to *brachyptera* but is slightly paler generally and not quite so yellow on the belly. *Brachyptera*, which ranges only in sw. Colombia and nw. Ecuador, may merit full species status (Nariño Elaenia), as its voice is notably different from that of other races of *E. chiriquensis* (fide P. Coopmans).

Range: Costa Rica to n. Bolivia, e. Paraguay, and s. and e. Brazil (but absent from much of Amazonia). As noted above, more than one species may be involved.

Elaenia albiceps   White-crested Elaenia   Elenia Crestiblanca

Uncommon to locally common (more numerous southward) in secondary woodland and borders, shrubby clearings, and scrub in the subtropical and temperate zones on both slopes of the Andes and on slopes above the central and interandean valleys. The White-crested Elaenia overlaps broadly with the similar Sierran Elaenia (*E. pallatangae*), though the White-crested tends to occur more often in disturbed areas. Recorded mostly from 1900 to 3400 m, locally down to 1700 m near Sozoranga at El Tundo Reserve (Sep. 1998 sighting; RSR et al.).

One race: *griseigularis*. It is possible that the austral migrant race *chilensis* also occurs in Ecuador as it is known to range as close as Peru (and there are uncertain records of it from Colombia; Hilty and Brown 1986).

Range: Andes from s. Colombia to w. Bolivia; Pacific lowlands of w. Peru and n. Chile; breeds from cen. Chile and cen. Argentina south to Tierra del Fuego, wintering north to Peru and Amazonian and e. Brazil, possibly farther. More than one species is perhaps involved.

The Slaty Elaenia (*E. strepera*) almost surely occurs at least as a transient in Ecuador in the eastern lowlands and possibly the foothills as well, en route to and from its breeding grounds in the Andes of Bolivia and nw. Argentina and its wintering grounds in n. South America. It has been recorded as a transient (Apr.–May) as close as Putumayo in s. Colombia. See C. A. Marantz and J. V. Remsen, Jr. (*J. Field Ornithol.* 62[2]: 162–172, 1991).

Elaenia pallatangae   Sierran Elaenia   Elenia Serrana

Uncommon to common (usually more numerous southward) in borders of montane forest and secondary woodland, and in adjacent shrubby clearings, in the subtropical and lower temperate zones on both slopes of the Andes and above the central and interandean valleys. The Sierran Elaenia favors more humid areas than the White-crested Elaenia (*E. albiceps*), though it and the Sierran occur sympatrically in some places (e.g., around Nono and below

Cajanuma); possible hybrids between the two species have been recorded (fide *Birds of the World*, vol. 8). Recorded mostly between 1500 and 2800 m, locally down to 1000 m.

One race: nominate *pallatangae*.

Range: Andes from n. Colombia to w. Bolivia; tepuis of s. Venezuela and adjacent Guyana and Brazil. More than one species may be involved.

Elaenia parvirostris   **Small-billed Elaenia**   **Elenia Piquichica**

An uncommon austral winter resident in borders of humid forest and secondary woodland, shrubby clearings and gardens, and riparian areas in the lowlands of e. Ecuador. In addition, a female believed to be this species was obtained in deciduous woodland east of Mangaurco (625 m) in sw. Loja on 15 Aug. 1992 (ANSP); if its identity can be confirmed, this would be the first record of the Small-billed Elaenia from west of the Andes. E. Stresemann (*Ornith. Monatsber.* 45: 75–77, 1937) described his *Elaenia aenigma* on the basis of a specimen obtained on 13 Apr. 1931 at "Monte Iliniza" (= Illiniza; presumably actually obtained somewhere on the peak's lower slopes). *Aenigma* is now considered a synonym of *E. parvirostris* (fide *Birds of the World*, vol. 8), which would seem to indicate that there is another west-slope record of this species, but given the difficulty in identifying certain specimens of these exceedingly similar elaenias, we are reluctant to accept this without ourselves having seen the specimen. Recorded between early Apr. and Oct.; a few reports from other seasons are regarded as unverified, and based on what is known about the distribution of the species, they seem very unlikely. Ranges mostly below 400 m, but may occur higher as a transient.

Monotypic.

Range: breeds from e. Bolivia and s. Brazil south to n. Argentina, wintering north to Colombia, Venezuela, and the Guianas.

*Sublegatus obscurior   **Amazonian**   **Mosquerito Breñero**
                             Scrub-Flycatcher   **Amazónico**

Rare to uncommon and local (overlooked?) in the canopy and borders of terra firme forest and secondary woodland in the lowlands of e. Ecuador. The first definite record of the Amazonian Scrub-Flycatcher in Ecuador was a singing male taken by M. B. Robbins at Santiago in s. Morona-Santiago on 24 Jul. 1989 (ANSP); it was considered not uncommon at this locality, with at least three territorial pairs resident in forest edge situations along a ridge. Pairs have subsequently been found at the edge of humid forest at Zancudococha in Apr. 1991 (ANSP, MECN), and at La Selva (G. H. Rosenberg) and near Archidona (P. Coopmans); the species was also reported from Miazi in s. Zamora-Chinchipe in Jul. 1993 (Schulenberg and Awbrey 1997). This obscure flycatcher will likely prove to be more widespread in e. Ecuador; ANSP also has a pair of specimens taken in adjacent se. Colombia at Umbria in w. Putumayo in Dec. 1947. There is also an unsexed, undated "Ecuador" specimen in FMNH; this was the basis for the taxon's having been recorded from e. Ecuador in *Birds of the World*, vol. 8 (M. A. Traylor, Jr., pers. comm.). Recorded up to 900 m (Miazi).

S. obscurior was given full species rank in *Birds of the World* (vol. 8), but in a subsequent paper the same author, M. A. Traylor, Jr., (*Fieldiana, Zool.*, new series 13, 1982) retracted this, leaving it instead as a subspecies of *S. modestus* (Southern Scrub-Flycatcher), a species found farther south and east in South America. In part because of recently recognized differences in their primary songs, we favor treating *S. obscurior* as a species distinct from *S. modestus*, following Ridgely and Tudor (1994). Monotypic.

Range: se. Colombia, s. and e. Venezuela, and the Guianas to n. Bolivia and Amazonian Brazil. Excludes Northern Scrub-Flycatcher (*S. arenarum*) and, as noted above, Southern Scrub-Flycatcher.

| *Mecocerculus leucophrys* | White-throated Tyrannulet | Tiranillo Barbiblanco |

Uncommon to fairly common in montane forest borders, secondary woodland, and pastures and clearings with scattered shrubs and low trees in the temperate zone on both slopes of the Andes, and locally on slopes above the central and interandean valleys. The White-throated Tyrannulet also regularly occurs above treeline in patches of *Polylepis* woodland. Recorded mostly from 2800 to 3400 m, locally as high as 3600 m (e.g., at Cerro Mongus).

One race: *rufomarginatus*.

Range: mountains of n. Venezuela, and Andes from w. Venezuela to nw. Argentina; Sierra de Perijá and Santa Marta Mountains; tepuis of s. Venezuela and adjacent Brazil.

| *Mecocerculus stictopterus* | White-banded Tyrannulet | Tiranillo Albibandeado |

Fairly common to locally common in the canopy and (especially) borders of montane forest and secondary woodland in the temperate zone on both slopes of the Andes, and locally on slopes above the central and interandean valleys. On the west slope recorded south to w. Loja (various localities around Celica and at Utuana). The White-banded Tyrannulet particularly favors stands of alders (*Alnus*). Recorded mostly from 2400 to 3500 m.

One race: nominate *stictopterus*.

Range: Andes from w. Venezuela to w. Bolivia.

| *Mecocerculus poecilocercus* | White-tailed Tyrannulet | Tiranillo Coliblanco |

Fairly common to common in the canopy and (especially) borders of montane forest, sometimes foraging out into secondary woodland and shrubby clearings with scattered trees and tree ferns, in the subtropical and lower temperate zones on both slopes of the Andes. On the west slope recorded south to w. Loja in the Celica region, above Sozoranga at the El Tundo Reserve (Sep. 1998 sightings; RSR et al.), and at Utuana (Sep. 1998 sightings; RSR et al.). Recorded mostly from 1500 to 2500 m.

Monotypic.

Range: Andes from Colombia to s. Peru.

Mecocerculus calopterus   **Rufous-winged Tyrannulet**   Tiranillo Alirrufo

Rare to uncommon in the canopy and borders of montane forest and secondary woodland in the foothills and lower subtropical zone on both slopes of the Andes, mostly in s. Ecuador; perhaps increasing, though the recent upsurge in records may simply be the result of increased observer activity. The Rufous-winged Tyrannulet ranges mostly in sw. Ecuador from sw. Chimborazo and e. Guayas south through El Oro and w. Loja. In recent years it has also been recorded north to Pichincha (a number of sightings from the Mindo and Chiri-boga road/Tinalandia areas as well as 3 specimens [WFVZ, MECN] taken at Maquipucuna in Jun. 1988 [Marín et al. 1992]) and once as far north as w. Imbabura (one seen on 26 Aug. 1990 on the southeastern slope of Cordillera de Toisán; N. Krabbe). There is even one report from the lowlands away from the Andes, a bird seen at Río Palenque on 4 Dec. 1996 (P. Coopmans et al.). There are also a few sightings from the east slope of the Andes in Zamora-Chinchipe, where the species is rare along the Loja-Zamora road (see Ridgely 1980; also a few subsequent reports). More surprising was the discovery of the species on the coastal Cordillera de Colonche in sw. Manabí (Cerro San Sebastián in Machalilla National Park) in late Jan. 1991 (Parker and Carr 1992), but it is presumably rare here as none were found in Aug. 1991 (ANSP), nor have any been located subsequently. However, the species has subsequently been found at two sites on the coastal cordillera in w. Guayas (Loma Alta, where seen in Dec. 1996 [Becker and López-Lanús 1997] and in the Chongón Hills at Cerro Blanco, where seen on two occasions in Sep. 1997 [K. S. Berg]). Recorded mostly between 700 and 2000 m (in the Sozoranga region at Reserva El Tundo); at least occasionally or locally it occurs lower (to 400 m at Manta Real [ANSP] and once down to 200 m at Río Palenque).

Monotypic.

Range: Andes and adjacent lowlands of w. and s. Ecuador and n. Peru.

*Mecocerculus minor   **Sulphur-bellied**   Tiranillo Ventriazufrado
                                  Tyrannulet

Rare to locally fairly common (perhaps most numerous southward) in the canopy and borders of montane forest and secondary woodland in the sub-tropical and lower temperate zones on the east slope of the Andes. Given the number of recent records, it seems surprising that the Sulphur-bellied Tyran-nulet was only recently first recorded in Ecuador. The first specimens (MCZ) were obtained by D. Norton in Sep. 1965 along the Loja-Zamora road in Zamora-Chinchipe (Ridgely 1980). Since then Sulphur-bellied Tyrannulets have been found north through Morona-Santiago (Cordillera de Cutucú and along the Gualaceo-Limón road; ANSP, ZMCOP) to w. Napo (Cordillera de Huacamayos and SierrAzul; WFVZ and MECN; also numerous sightings) as well as south to near the Peruvian border in Zamora-Chinchipe (Panguri, Río Isimanchi valley on Cordillera Las Lagunillas, Cordillera del Cóndor, and Quebrada Honda and Valladolid area; ANSP, MECN, v.o.). The Sulphur-bellied Tyrannulet has also recently been found on the west slope of the Andes in the extreme north, where one was collected (MECN) and others seen along the Maldonado road (2500 m) in Carchi on 22 Nov. 1990 (Krabbe 1992). Recorded from about 1600 to 2800 m.

Monotypic.

Range: locally in Andes from extreme sw. Venezuela to cen. Peru.

*Serpophaga hypoleuca   River Tyrannulet   Tiranolete Ribereño

Rare to uncommon and apparently local in semiopen, early-successional growth on river islands in the lowlands of e. Ecuador, where at least so far recorded primarily along the Río Napo upriver in small numbers to near Jatun Sacha. The River Tyrannulet was first recorded for Ecuador from specimens taken along the Napo at El Edén in 1964 (Orcés 1974). As yet it has not been found along the Río Aguarico (which may be too narrow?), but there are recent reports from islands in the Río Pastaza in the vicinity of Kapawi Lodge near the Peruvian border, so it may occur upriver along that river as well. Recorded below about 400 m (Jatun Sacha).

One race: nominate *hypoleuca*.

Range: along Amazon River and some of its major tributaries in Brazil, n. Bolivia, se. Colombia, e. Peru, and e. Ecuador; along Orinoco River in Venezuela.

Serpophaga cinerea   Torrent Tyrannulet   Tiranolete Guardarríos

Fairly common to common and conspicuous along fast-flowing, usually rocky streams and rivers (in both forested and semiopen country) on both slopes of the Andes, and also in the central and interandean valleys. Recorded mostly from 700 to 3100 m, locally down to 300–400 m (e.g., at Manta Real in nw. Azuay; RSR) or even lower (to below 100 m at Playa de Oro in n. Esmeraldas; O. Jahn et al.).

One race: nominate *cinerea*.

Range: highlands of Costa Rica and w. Panama; Andes from w. Venezuela to w. Bolivia; Sierra de Perijá and Santa Marta Mountains.

*Stigmatura napensis   Lesser Wagtail-Tyrant   Rabicano Menor

Uncommon to fairly common but local on islands with early-successional growth in the Río Napo and lower Río Aguarico in the lowlands of ne. Ecuador. The first record was of at least four or five pairs seen and tape-recorded near La Providencia on a large island that had an extensive growth of *Tessaria* on 17 Aug. 1986 (RSR et al.); the species has subsequently been seen on several other islands in the La Selva and Sacha Lodge area. The first Ecuadorian specimen was taken on an island near the mouth of the Río Lagarto into the Aguarico on 10 Dec. 1990 (ANSP); more were obtained in Sep. 1992 on an island in the Río Napo near the mouth of the Aguarico (ANSP, MECN). There is as yet no evidence, however, that the wagtail-tyrant occurs on islands in the Río Pastaza. Recorded below 300 m.

Monotypic.

Range: islands in Amazon River and some of its major tributaries in Brazil, se. Colombia, ne. Peru, and ne. Ecuador; locally along Orinoco River in s. Venezuela (fide S. Hilty). Excludes Bahia Wagtail-Tyrant (*S. bahiae*) of interior ne. Brazil.

Anairetes parulus Tufted Tit-Tyrant Cachudito Torito

Fairly common in shrubby clearings and undergrowth of forest borders in the temperate zone and locally in paramo on both slopes of the Andes, and in the central and interandean valleys. Recorded mostly between about 2500 and 3500 m.

One race: *aequatorialis*.

Range: Andes from s. Colombia to Tierra del Fuego (southward also in lowlands).

**Anairetes nigrocristatus* Black-crested Tit-Tyrant Cachudito Crestinegro

Rare to uncommon and very local in humid scrub, woodland borders, and hedgerows in the temperate zone of extreme s. Loja. Known from only one site, several kilometers south of Utuana, with the first record being a pair found here on 14 Sep. 1989; two to three pairs were seen on 6 Feb. 1991 (Best et al. 1993). A single male was obtained here on 25 Apr. 1993 (ANSP), and a few pairs have been found in the same area on several subsequent occasions (v.o.). Recorded at about 2400 to 2500 m.

The Black-crested Tit-Tyrant has one of the smallest known ranges of any Ecuadorian bird, and thus while it seems to be capable of persisting in some patches of the regenerating scrub so prevalent around Utuana, its numbers are unquestionably very low and the species thus inherently at risk. We therefore accord it Near-threatened status for Ecuador. The species was not considered to be at risk by Collar et al. (1992, 1994).

The species is sometimes called the Marañón Tit-Tyrant.

A. nigrocristatus of the Andes of cen. and n. Peru and extreme s. Ecuador is considered a species separate from the considerably smaller *A. reguloides* (Pied-crested Tit-Tyrant) of w. Peru and n. Chile, following Fjeldså and Krabbe (1990) and Ridgely and Tudor (1994). *A. nigrocristatus* is considered to be monotypic, but the single Ecuadorian specimen, a male, is much blacker than any of ANSP's ample series from Peru; additional material may reveal it to be a separable subspecies.

Range: Andes of extreme s. Ecuador and nw. Peru.

Uromyias agilis   Agile Tit-Tyrant   Cachudito Agil

Uncommon to locally fairly common in the lower growth and borders of montane forest and woodland, almost always where there are stands of *Chusquea* bamboo, in the temperate zone on both slopes of the Andes, sometimes up to near treeline. On the west slope recorded south to w. Cotopaxi (seen by RSR and G. H. Rosenberg above Pilaló on 28 Jul. 1991); on the east slope south to n. Loja (Cordillera de Cordoncillo; ANSP, MECN). Recorded mostly from 2600 to 3500 m, locally down to 2350 m in Carchi (N. Krabbe) and up to 3700 m (Río Anatenorio valley in w. Napo; N. Krabbe).

Despite their behavioral and vocal differences, the genus *Uromyias* has sometimes been merged in *Anairetes* (e.g., *Birds of the World*, vol. 8). For morphological reasons for maintaining *Uromyias* as a separate genus, see W. E. Lanyon (*Am. Mus. Novitates* 2914, 1988). Monotypic.

Range: Andes from sw. Venezuela to s. Ecuador.

Pseudocolopteryx acutipennis   Subtropical   Doradito Subtropical
                                        Doradito

Rare to uncommon and very local in small sedge-(*Scirpus-*) dominated marshes, seepage zones, and adjacent shrubbery, and reedbeds fringing certain lakes, in the central valley from Imbabura (one seen by J. Blincow at Lago San Pablo in Jan. 1991, fide N. Krabbe) south to Chimborazo. Contrary to numerous suggestions that the species was an austral migrant to Ecuador (e.g., Meyer de Schauensee 1966, p. 374), recent evidence demonstates that Ecuadorian birds are in fact resident, birds having been seen throughout the year and (fide J. C. Matheus) a nest having been found. Recorded mostly from 2400 to 3300 m (at Laguna de Colta in Chimborazo), but in the early 20th century recorded as low as 850–1225 m in Chimborazo (junction of Ríos Chanchan and Chiguancay; Huigra).

The Subtropical Doradito has certainly declined in Ecuador as a result of drainage and intensive grazing, and it now occurs in small numbers and is very local; a tiny population persists near the La Ciénega north of Saquisilí in s. Cotopaxi. Given its recent decline, we believe that the species should be accorded Near-threatened status in Ecuador; it is not considered to be at risk generally (Collar et al. 1994). The fact that populations of unknown size occur in the still-extensive reedbeds surrounding certain lakes is reassuring, but as far as we are aware no population exists in a formally protected area. Monotypic.

Range: locally in Andes from n. Colombia to nw. Argentina, southward spreading into adjacent lowlands during austral winter.

Euscarthmus meloryphus   Tawny-crowned   Tirano Enano
                                     Pygmy-Tyrant   Frentileonado

Fairly common to common (but inconspicuous and mainly recorded by voice) in arid scrub, weedy and shrubby second-growth, and undergrowth of deciduous woodland in the lowlands and lower subtropical zone of w. Ecuador from w. Esmeraldas (Atacames region in Jan. 1993; RSR and F. Sornoza) and Manabí (including Isla de la Plata) southward. Also recently found to be numerous in the Río Marañón drainage around Zumba in extreme s. Zamora-Chinchipe (v.o.). Recorded up locally to about 2000 m in Azuay and Loja, but mostly found below 1500 m.

One race: *fulviceps*.

Range: n. Colombia and n. Venezuela; w. Ecuador and nw. Peru; extreme se. Peru east through n. and e. Bolivia to e. Brazil and south to n. Argentina and Uruguay.

Mionectes striaticollis	Streak-necked Flycatcher	Mosquerito Cuellilistado

Fairly common but inconspicuous in the lower and middle growth of montane forest, secondary woodland, and borders in the subtropical and lower temperate zones on both slopes of the Andes. Ranges mostly between about 1500 and 2500 m, occurring locally and/or in smaller numbers down into the foothills to about 800–900 m (near Zumba; ANSP) and farther up into the temperate zone (once as high as 3100 m on the Cordillera Las Lagunillas; ANSP).

Two rather different races occur in Ecuador, *columbianus* on the east slope and *viridiceps* on the west slope. J. T. Zimmer (*Am. Mus. Novitates* 1126, 1941) lists specimens from El Oro (Salvias) and Loja (Celica and San Bartolo) as *columbianus*, though what are almost certainly the same birds were listed as the expected *viridiceps* by Chapman (1926). They perhaps are the "intergrades" from "northeastern [sic] Loja" referred to in *Birds of the World* (vol. 8, p. 54). *Viridiceps* has the olive hood more or less concolor (not so gray) with the rest of the upperparts.

Range: Andes from n. Colombia to w. Bolivia.

Mionectes olivaceus	Olive-striped Flycatcher	Mosquerito Olivirrayado

Fairly common to common but inconspicuous in the lower and middle growth of very humid and humid forest, secondary woodland, and borders in the more humid lowlands of w. Ecuador, and in the foothills and subtropical zone on both slopes of the Andes; on the west slope recorded south to w. Loja (Las Piñas). In n. Esmeraldas the Olive-striped Flycatcher is one of the most numerous birds in the understory of forest, as attested to by mist-netting (O. Jahn et al.). The unobtrusive Olive-striped Flycatcher does not range far out into the eastern lowlands, where it appears to be only a rare, or perhaps an altitudinal, migrant or wanderer; there are, for example, only two records even from as close as Jatun Sacha (fide B. Bochan) and none from farther east. Olive-striped and Streak-necked Flycatchers (*M. striaticollis*) occur sympatrically at some sites on both slopes of the Andes (mostly in a zone between 1200 and 1700 m), and in the west the two species are not always readily separated in the field.

Recorded mostly from 100 to 2000 m; in the east scarce below about 600 m and not recorded at all below about 400 m (Jatun Sacha).

Two quite similar races occur in Ecuador, *fasciaticollis* on the east slope and *hederaceus* in the west. It is possible that *fasciaticollis* will ultimately be shown best to be merged into the extremely similar race *pallidus*. *Pallidus* is thus far known only from the lower slopes of the Eastern Andes in Colombia, considerably to the north of Ecuador, but additional work may close the apparent "gap" between the two, and we suspect they will be shown to intergrade clinally (D. Agro; RSR).

Range: Costa Rica to n. Venezuela and extreme w. Bolivia.

Mionectes oleagineus Ochre-bellied Flycatcher Mosquerito Ventriocráceo

Uncommon to fairly common and widespread in the lower growth of humid and deciduous forest and secondary woodland, borders, and adjacent clearings and gardens in the lowlands of both e. and w. Ecuador. In the west recorded south through w. Loja. In the east found in both terra firme and várzea, and— unlike in some other parts of its wide range—often not particularly associated with edge situations. The Ochre-bellied Flycatcher seems to be especially numerous at Jatun Sacha (B. Bochan). Recorded up in the foothills to about 1000 m, reaching 1600 m in w. Loja.

Formerly (e.g., Meyer de Schauensee 1966, 1970) placed in the genus *Pipromorpha*; see *Birds of the World*, vol. 8. Two similar races occur in Ecuador, *pacificus* in the west and *hauxwelli* in the east. *Hauxwelli* is somewhat more richly colored and ochraceous below, and on the edging of its wing-coverts and inner flight feathers, than *pacificus*.

Range: s. Mexico to extreme nw. Peru, n. Bolivia, and Amazonian Brazil; e. Brazil.

Leptopogon superciliaris Slaty-capped Mosquerito Gorripizarro
                              Flycatcher

Fairly common to locally common in the lower growth of montane forest, secondary woodland, and their borders in the foothills and lower subtropical zone on both slopes of the Andes, in the west also ranging down locally into more humid lowlands (e.g., at Playa de Oro [O. Jahn et al.] and Río Palenque) and on the coastal cordillera. On the west slope recorded south to El Oro and locally into w. Loja (a Sep. 1921 specimen from Las Piñas; Sep. 1998 sightings from the El Tundo Reserve near Sozoranga; RSR et al.). We suspect that the "*Phylloscartes* sp." recorded by Best et al. (1993, p. 251) from Tierra Colorada in w. Loja refers to the Slaty-capped Flycatcher; no *Phylloscartes* has ever been found in sw. Ecuador. Surprisingly, small numbers of the Slaty-capped Flycatcher have also been found recently in semihumid forests of Guayas in the Chongón Hills at Cerro Blanco (Berg 1994; R. Clay et al.) and on hills in Manglares-Churute Ecological Reserve (Pople et al. 1997). The species has also recently been found to occur on the coastal Cordillera de Mache in sw. Esmeraldas (Bilsa; D. Wolf et al.) and on the Cordillera de Colonche in sw. Manabí (Cerro San Sebastián in Machalilla National Park [Parker and Carr 1992; G. H. Rosenberg]) and w. Guayas (Cerro La Torre and above Salanguilla [N. Krabbe]; Loma Alta [Becker and López-Lanús 1997]). Recorded mostly from 200 to 1500 m, locally lower (e.g., at Cerro Blanco in Guayas and at Playa de Oro in Esmeraldas); in the east mostly from 600 to 1500 m, with a few ranging down to 400 m.

Two similar races occur in Ecuador, *transandinus* in the west and nominate *superciliaris* on the east slope. Both races have individually variable ochraceous to buffy yellowish wing-bars.

Range: Costa Rica to n. Venezuela, extreme nw. Peru, and w. Bolivia.

Leptopogon amaurocephalus Sepia-capped Mosquerito
  Flycatcher Gorrisepia

Uncommon and perhaps local in the lower and middle growth of humid forest (both terra firme and várzea, especially the former) in the lowlands of e. Ecuador; thus far recorded only from Napo, Sucumbíos, and Pastaza. Until recently the Sepia-capped Flycatcher was thought to be unrecorded from Ecuador (Meyer de Schauensee 1966; *Birds of the World*, vol. 8), though even then there were unpublished specimens of the species in the MECN from Sarayacu, Río Pucuno, and Río Arajuno; possibly this material or part of it was the source of the species being recorded from Ecuador by Meyer de Schauensee (1970). There have been a substantial number of recent records, and the species now appears to be reasonably widespread in appropriate habitat, being perhaps most numerous in a narrow zone along the eastern base of the Andes (e.g., at Santa Cecilia in Sucumbíos and at Jatun Sacha). Recorded up to about 450 m (south of Puerto Napo).

One race: *peruvianus*.

Range: s. Mexico to n. Argentina, e. Paraguay, and s. and e. Brazil (but absent from much of Amazonia).

Leptopogon rufipectus Rufous-breasted Mosquerito Pechirrufo
  Flycatcher

Uncommon and inconspicuous in the lower growth of montane forest, secondary woodland, and their borders in the subtropical zone on the east slope of the Andes. Most Ecuador records of the Rufous-breasted Flycatcher are from the north; small numbers are, for instance, found regularly in the Cuyuja and San Isidro areas. There are, however, a number of recent reports from Zamora-Chinchipe, including birds seen along the Loja-Zamora road on 7 Aug. 1990 (S. Hilty et al.), in the Romerillos area of Podocarpus National Park in Dec. 1991 and Jan. 1992 (J. Rasmussen et al.), one collected (ANSP) at Panguri on 14 Jul. 1992, and birds seen in the Río Isimanchi valley in Nov. 1992 (M. B. Robbins and T. J. Davis). In addition there is an MECN specimen taken at Cerro Imbana in adjacent n. Loja. Despite the lack of reports from the intervening area, we suspect that the species ranges all along the east slope in suitable habitat at appropriate elevations. Recorded mostly from 1600 to 2500 m.

One race: nominate *rufipectus*.

Range: locally in Andes from extreme sw. Venezuela to extreme n. Peru.

Phylloscartes gualaquizae   **Ecuadorian Tyrannulet**   **Tiranolete Ecuatoriano**

Uncommon to locally fairly common in the canopy and borders of montane forest in the foothills and lower subtropical zone on the east slope of the Andes. The Ecuadorian Tyrannulet doubtless often goes overlooked as a result of its obscure appearance and its habit of usually foraging high above the ground. Most older Ecuador records are from the south (mainly in Morona-Santiago and Zamora-Chinchipe), but there are recent sightings and one specimen (ANSP) from as far north as San Rafael Falls in w. Sucumbíos, and the species is recorded regularly north of Archidona along the Loreto road (v.o.). The Ecuadorian Tyrannulet likely also occurs in adjacent s. Colombia, from which it is unrecorded (Hilty and Brown 1986). Recorded mostly from 700 to 1400 m (early published records from considerably higher are believed likely to be in error).

Gualaquizae was formerly (e.g., Meyer de Schauensee 1966, 1970) placed in the genus *Pogonotriccus* and then was called the Ecuadorian Bristle-Tyrant. Robbins et al. (1987) showed that on behavioral grounds *gualaquizae* was a "classic" *Phylloscartes* tyrannulet and suggested calling it such.

Monotypic.

Range: east slope of Andes in Ecuador and n. Peru.

*Phylloscartes superciliaris   **Rufous-browed**   **Tiranolete Cejirrufo**
                                   Tyrannulet

Uncommon to fairly common but local in the canopy of montane forest in the lower subtropical zone of se. Ecuador, at least so far only on outlying ridges. The Rufous-browed Tyrannulet was first recorded from Ecuador in Jun. 1984 on the west slope of the Cordillera de Cutucú in Morona-Santiago at a Shuar hunting encampment named "Yapitya"; four specimens (ANSP, MECN) were taken then (Robbins et al. 1987). The species was also seen and tape-recorded near Chinapinza on the west slope of the Cordillera del Cóndor in Zamora-Chinchipe on 13–14 Jun. 1993 (RSR), and it was also recorded a month later at Coangos in the northern part of the Cordillera del Cóndor (Schulenberg and Awbrey 1997). Recorded between 1300 and 1700 m.

One race: *griseocapillus*. However, Robbins et al. (op. cit.) point out that the described races of *P. superciliaris* are all only weakly differentiated and that the species is perhaps better considered to be monotypic.

Range: locally in highlands of Costa Rica and w. Panama; very locally on east slope of Andes in Colombia, Ecuador, and extreme n. Peru; Sierra de Perijá.

Pogonotriccus ophthalmicus   Marble-faced   Orejerito
                                 Bristle-Tyrant   Carijaspeado

Uncommon to locally common in the lower and middle growth of montane forest and forest borders in the subtropical zone on both slopes of the Andes. On the west slope recorded south to Pichincha. Recorded from about 1200 to 2100 m.

This and the following two species (as well as others elsewhere is South America) were long placed in the genus *Pogonotriccus*, prior to that genus being merged into *Phylloscartes* by M. A. Traylor, Jr. (*Bull. Mus. Comp. Zool.* 148[4]: 129–184, 1977). This treatment was adopted, with reservations, by Ridgely and Tudor (1994), though Hilty and Brown (1986) had continued to maintain *Pogonotriccus* as a separate genus. Newly obtained behavioral information on certain little-known species convinces us that the behavioral and vocal differences between the two groups are distinct enough to warrant their generic separation. J. W. Fitzpatrick and D. F. Stotz (*Studies in Neotropical Ornithology Honoring Ted Parker*, Ornithol. Monogr. no 48: 40, 1997) also "strongly suggest" that two phylogenetic groups are involved. One race: nominate *ophthalmicus*.

Range: mountains of n. Venezuela; Andes from n. Colombia to w. Bolivia.

Pogonotriccus poecilotis   Variegated Bristle-Tyrant   Orejerito Variegado

Rare to uncommon in the lower and middle growth of montane forest in the subtropical zone on the east slope of the Andes. The one published record from the west slope of the Ecuadorian Andes, W. Goodfellow's listing of the species from "lower w. Pichincha" (*Ibis* ser. 8, vol. 1: 704, 1901) seems far too uncertain to credit, especially given the number of incorrectly labeled specimens that were being procured by native collectors during that era; this record has not been supported by any recent specimens or confirmed observations, and the species is not known from w. Nariño, Colombia. Recorded between about 1500 and 2000 m.

The species has been placed in the genus *Phylloscartes* (see discussion in the previous account). One race occurs in Ecuador, nominate *poecilotis*.

Range: Andes from w. Venezuela to s. Peru; Sierra de Perijá.

Pogonotriccus orbitalis   Spectacled Bristle-Tyrant   Orejerito de Anteojos

Rare to locally uncommon (perhaps often overlooked) in the lower and middle growth of montane forest in the foothills and lower subtropical zone on the east slope of the Andes. For the present the Spectacled Bristle-Tyrant continues to be a known only from the north (from the Sumaco region north into adjacent Sucumbíos near San Rafael Falls) and in Zamora-Chinchipe (where known from various sites in the Cordillera del Cóndor, and along the main slope of the Andes from around Zamora south through Podocarpus National Park to Panguri). Presumably it occurs in the intervening area as well, though as yet there seem to be a no reports. Recorded from about 700 to 1400 m.

We consider the Spectacled Bristle-Tyrant to be a Near-threatened species in Ecuador. It was not considered to be at risk by Collar et al. (1992, 1994).

The species has been placed in the genus *Phylloscartes* (see discussion in the *P. opthalmicus* [Marble-faced Bristle-Tyrant] account). Monotypic.

Range: locally on east slope of Andes from s. Colombia to w. Bolivia.

Capsiempis flaveola   Yellow Tyrannulet   Tiranolete Amarillo

Uncommon to locally fairly common in the lower and middle growth of secondary woodland and forest borders (often most numerous in dense bamboo stands) in the more humid lowlands and foothills of w. Ecuador from Pichincha (Mindo/Nanegal region) south to w. Guayas and El Oro (old specimens from Río Jubones and Santa Rosa, and recent reports from Buenaventura). The Yellow Tyrannulet also occurs very locally in ne. Ecuador in the foothills and adjacent lowlands in w. Napo and adjacent nw. Pastaza; here it is thus far known only from along the road to Loreto north of Archidona (where quite numerous), near the Río Payamino west of Coca (several pairs found and one specimen obtained in Nov. 1991; ANSP), and north of Puyo along the Tena road (1–2 birds recorded on 2 Apr. 1994; P. Coopmans et al.). There also are old specimens (AMNH) from Río Suno and "below San José." In the west recorded up to about 1500 m, in the east from about 350 to 1300 m.

This species has sometimes been placed in the genus *Phylloscartes* (e.g., *Birds of the World*, vol. 8), but W. E. Lanyon (*Condor* 86[1]: 42–47, 1984) presented convincing reasons for maintaining it in the monotypic genus *Capsiempis*; it was also so treated in the 1998 AOU Check-list. Two similar races are found in Ecuador, *magnirostris* in the west and *cerulus* in the northeast.

Range: Nicaragua to the Guianas and e. Amazonian Brazil; w. Ecuador; e. Brazil to e. Paraguay and ne. Argentina; very locally in se. Peru and n. Bolivia.

Pseudotriccus pelzelni   Bronze-olive   Tirano Enano Bronceado
                             Pygmy-Tyrant

Fairly common to common but inconspicuous in the undergrowth of montane forest and forest borders in the foothills and subtropical zone on both slopes of the Andes. On the west slope recorded south to Pichincha (lower Chiriboga road) and from Azuay (Manta Real and San Luis) south to El Oro (old specimens from El Chiral and many recent records [including ANSP and MECN specimens] from Buenaventura). Recorded mainly from 600 to 2000 m, locally lower (to 300 m) in w. Azuay.

Two races are found in Ecuador, nominate *pelzelni* on the east slope and *annectens* on the west slope. *Annectens* is larger than the nominate race and is browner above and more ochraceous below.

Range: highlands of e. Panama and adjacent nw. Colombia; Andes from n. Colombia to s. Peru.

Pseudotriccus ruficeps   Rufous-headed   Tirano Enano
                            Pygmy-Tyrant   Cabecirrufo

Uncommon to fairly common but inconspicuous in dense undergrowth of montane forest and forest borders in the upper subtropical and temperate zones on both slopes of the Andes. On the west slope recorded south to w. Cotopaxi (Caripero region; N. Krabbe and F. Sornoza). Recorded mostly from 2000 to 3300 m.
Monotypic.
Range: Andes from s. Colombia to w. Bolivia.

Corythopis torquata   Ringed Antpipit   Coritopis Fajeado

Uncommon on or near the ground inside terra firme forest in the lowlands of e. Ecuador. Found mostly below 600 m, but recorded locally and in small numbers as high as 1000 m at Pachicutza (WFVZ).
One race: *sarayacuensis*.
Range: se. Colombia, s. Venezuela, and the Guianas to n. Bolivia and Amazonian Brazil.

*Myiornis ecaudatus   Short-tailed Pygmy-Tyrant   Tirano Enano Colicorto

Apparently uncommon in borders of humid forest (both terra firme and várzea) in the lowlands of e. Ecuador. There are relatively few reports of this pygmy-tyrant in Ecuador; it doubtless has been overlooked on account of its minute size and insignificant, easily passed-over vocalizations. Meyer de Schauensee (1966) records the species from e. Ecuador, but we can find no basis for this citation. The localities from which we now have records are Taracoa (where seen on several occasions between the late 1970s and mid-1980s [RSR et al.; PJG et al.]), Taisha in Morona-Santiago (where seen in Aug. 1990; M. B. Robbins), Zancudococha (where the only known Ecuadorian specimen [ANSP] was taken on 14 Apr. 1991), Cuyabeno (where seen in Jul. 1991 by J. Rowlett and R. A. Rowlett et al.), near the Río Payamino west of Coca (P. Coopmans), Jatun Sacha (only two sightings; B. Bochan), along the Maxus road southeast of Pompeya (RSR et al.), Sacha Lodge, and Kapawi Lodge on the Río Pastaza near the Peruvian border. Recorded below about 400 m.
Excludes trans-Andean *M. atricapillus* (Black-capped Pygmy-Tyrant). One race occurs in Ecuador, nominate *ecaudatus*.
Range: e. Colombia, s. Venezuela, and the Guianas to n. Bolivia and Amazonian Brazil.

Myiornis atricapillus   Black-capped   Tirano Enano Gorrinegro
                           Pygmy-Tyrant

Uncommon to locally common in borders of humid forest and secondary woodland and adjacent plantations in the lowlands and foothills of nw. Ecuador in Esmeraldas, adjacent Imbabura, and nw. Pichincha (where known from an

MECN specimen from Pachijal; also since 1990 from numerous sightings and tape-recordings south to along the San Miguel de los Bancos-Puerto Quito highway [v.o.]). The Black-capped Pygmy-Tyrant is an inconspicuous bird, despite being quite vocal; it doubtless has been much overlooked. Recorded up to about 800 m.

The trans-Andean *M. atricapillus* is here regarded as a species separate from cis-Andean *M. ecaudatus* (Short-tailed Pygmy-Tyrant), based on its slightly different plumage and allopatric distribution, following the 1993 and 1998 AOU Check-lists and Ridgely and Tudor (1994). Monotypic.

Range: Costa Rica to nw. Ecuador.

Lophotriccus pileatus   Scale-crested   Cimerillo
                          Pygmy-Tyrant   Crestiescamado

Fairly common to common in the lower growth and borders of humid forest and secondary woodland in the lowlands and foothills of w. Ecuador, and in the foothills and lower subtropical zone on both slopes of the Andes. In w. Ecuador found south through Guayas (e.g., at Cerro Blanco in the Chongón Hills and at Manglares-Churute Ecological Reserve), El Oro, and w. Loja (Alamor region and Punta Santa Ana). In the west the Scale-crested Pygmy-Tyrant is found primarily in more humid regions, but it also ranges (usually in smaller numbers) into areas where the forest is more deciduous in character (e.g., south of Bahía de Caráquez in Manabí); in the wet-forest belt of n. Esmeraldas it is mainly a forest-edge species (O. Jahn and P. Mena Valenzuela et al.). Favors stands of bamboo in most areas, especially on the east slope. Recorded up locally to about 1700 m (e.g., on the Cordilleras de Cutucú and del Cóndor); on the east slope generally not recorded below about 700 m.

Two rather similar races occur in Ecuador, nominate *pileatus* on the east slope and *squamaecrista* in the west.

Range: Costa Rica to n. Venezuela, extreme nw. Peru, and s. Peru.

Lophotriccus vitiosus   Double-banded   Cimerillo Doblebandeado
                          Pygmy-Tyrant

Uncommon to locally fairly common in the lower growth and borders of humid forest (both terra firme and várzea) and secondary woodland in the lowlands of e. Ecuador. The Double-banded Pygmy-Tyrant is frequently very vocal, but otherwise it is a generally inconspicuous flycatcher. Recorded up to about 600 m near Archidona (P. Coopmans).

One race: *affinis*.

Range: se. Colombia, e. Ecuador, ne. Peru, n. Amazonian Brazil, and the Guianas.

Hemitriccus zosterops   White-eyed Tody-Tyrant   Tirano Todi Ojiblanco

Uncommon to locally fairly common in the lower and middle growth of terra firme forest (especially in areas with relatively hilly terrain) in the lowlands and foothills in e. Ecuador, but apparently only occurring south of the Río Napo. Thus far the White-eyed Tody-Tyrant has been recorded primarily from the southeast in Pastaza, Morona-Santiago, and Zamora-Chinchipe, but it has also recently been reported from several localities in Napo, all of them south of the Río Napo. Recorded by Meyer de Schauensee (1966, p. 367) as only "probably" occurring in Ecuador, this despite a 1948 specimen (ANSP) from "Churo Yaco." There have been a number of recent records of this small and rather inconspicuous tyrannid, including numerous specimens (MCZ, WFVZ, ANSP, and MECN) taken. Recorded up to 950–1000 m (at various localities in Zamora-Chinchipe).

    H. zosterops is considered to be a species separate from *H. griseipectus* (White-bellied Tody-Tyrant), following M. Cohn-Haft, A. Whittaker, and P. C. Stouffer (*Studies in Neotropical Ornithology Honoring Ted Parker*, Ornithol. Monogr. no. 48: 205–235, 1997). The latter species is found in Amazonia south of the Amazon River and in ne. Brazil and differs from *H. zosterops* in voice as well as plumage. *H. zosterops* and the following three species were formerly (e.g., Meyer de Schauensee 1966) placed in the genus *Idioptilon*; see *Birds of the World*, vol. 8. We are uncertain whether Ecuadorian examples of *H. zosterops* are better placed with the nominate race or referred to the very similar *flaviviridis*; M. Cohn-Haft (pers. comm.) suggests that *flaviviridis* may better be synonymized into nominate *zosterops*.

    Range: se. Colombia and sw. Venezuela to ne. Peru, n. Amazonian Brazil, and the Guianas.

Hemitriccus iohannis   Johannes's Tody-Tyrant   Tirano Todi de Johannes

Rare and local in riparian woodland at the edge of oxbow lakes in the lowlands of se. Ecuador. Thus far the Johannes's Tody-Tyrant is known only from the vicinity of Kapawi Lodge on the Río Pastaza near the Peruvian frontier. It was first recorded in Ecuador from a bird seen and tape-recorded on 2–3 Oct. 1996 (J. Moore); other observers have found the species in this area subsequently. This obscure tyrannid may eventually also be found in the lowlands of the northeast, for it has been recorded close to the Ecuadorian border in Putumayo (San Antonio), Colombia (M. E. Traylor, *Fieldiana, Zool.*, new series 13: 1–22, 1982). Recorded at about 200 m.

    H. iohannis is regarded as a monotypic species separate from *H. striaticollis* (Stripe-necked Tody-Tyrant), found mainly in s. Amazonia; this treatment follows *Birds of the World* (vol. 8) and Ridgely and Tudor (1994).

    Range: locally in se. Colombia and se. Ecuador; e. Peru, n. Bolivia, and w. Amazonian Brazil.

Hemitriccus granadensis   Black-throated Tody-Tyrant   Tirano Todi Golinegro

Uncommon to locally fairly common in the borders of montane forest and secondary woodland in the subtropical and temperate zones on the east slope of the Andes. The Black-throated Tody-Tyrant is found in the extreme north in e.

Carchi (El Pun), w. Sucumbíos (Santa Bárbara [Orcés 1974] and a sighting from northwest of El Calvario [N. Krabbe]), and w. Napo (an old specimen from Baeza, though there have been no recent reports from here), and again in the south (nw. Morona-Santiago on the Cordillera del Cutucú [ANSP] and along the Gualaceo-Limón road southward). On the west slope known only from the extreme north in Carchi (a Jan. 1980 sighting above Maldonado; RSR, PJG, and S. Greenfield). Recorded from about 1700 to 3000 m.

Two distinctly different races are found in Ecuador. The nominate race, with whitish lores and ocular area, ranges in the extreme north on the west slope in Carchi and on the east slope south to w. Napo. *Pyrrhops*, with buff lores and ocular area, is found on the east slope of the Andes in s. Ecuador. There appears to be a gap between the ranges of these two forms on the east slope. More than one species may be involved; if split, *H. pyrrhops* should be called the Buff-lored Tody-Tyrant.

Range: mountains of n. Venezuela; Andes from extreme sw. Venezuela to s. Peru; Sierra de Perijá and Santa Marta Mountains.

*Hemitriccus rufigularis	Buff-throated Tody-Tyrant	Tirano Todi Golianteado

Rare to uncommon and very local in the lower and middle growth of montane forest in a narrow elevational zone in the lower subtropical zone on the east slope of the Andes. The Buff-throated Tody-Tyrant is known from only four sites, primarily on ridges or mountains east of the actual Andes. The first Ecuadorian record involved several birds seen and tape-recorded above Guadeloupe on the west slope of the Cordillera de Cutucú in Morona-Santiago on 4 Aug. 1979 (RSR and R. A. Rowlett). Subsequent records involve a male (WFVZ) collected above Pachicutza in Zamora-Chinchipe on the west slope of the Cordillera del Cóndor on 27 Jul. 1989 (Marín et al. 1992); a male (MECN) collected along the road to Loreto above Archidona in w. Napo on 2 Sep. 1990 (Krabbe 1992; also numerous sightings from here); and at least four birds seen and heard on the south slope of Volcán Sumaco in Jan. 1991 (B. Whitney). Recorded only between 1300 and 1500 m.

The Buff-throated Tody-Tyrant was considered Near-threatened by Collar et al. (1994). Those authors were not aware that the species' range extended north into Ecuador, and this combined with the relatively remote nature of the vast majority of that range suggests to us that the species does not need to be considered at risk.

Monotypic.

Range: locally on east slope of Andes from n. Ecuador to w. Bolivia.

*Hemitriccus cinnamomeipectus	Cinnamon-breasted Tody-Tyrant	Tirano Todi Pechicanelo

Rare to uncommon in the lower growth of montane forest and woodland above Chinapinza on the west slope of the Cordillera del Cóndor in Zamora-Chinchipe, where discovered in Sep. 1990 by N. Krabbe and F. Sornoza (Krabbe and Sornoza 1994), with specimens to MECN and ANSP. Another bird was seen and collected at the same locality but slightly higher on 9 Jun. 1993 (ANSP); at this site the Cinnamon-breasted Tody-Tyrant was syntopic with (and outnumbered by) the *pyrrhops* form of the Black-throated Tody-Tyrant (*H. granadensis*). The only recently described Cinnamon-breasted Tody-Tyrant (J. W. Fitzpatrick and J. P. O'Neill, *Auk* 96[3]: 443–447, 1986) was heretofore known only from n. Peru. Recorded at 1700–1900 m.

The Cinnamon-breasted Tody-Tyrant was considered to be Near-threatened by Collar et al. (1994). Given the species' minute range, and the fact that all or significant portions of it are now being affected by human activities (on the Ecuadorian side of the Cordillera del Cóndor by extensive gold-mining), we feel the species merits Vulnerable status.

Monotypic.

Range: locally on east slope of Andes in extreme s. Ecuador and n. Peru.

Poecilotriccus ruficeps	Rufous-crowned Tody-Flycatcher	Tirano Todi Coronirrufo

Uncommon to locally fairly common but inconspicuous in the shrubby lower growth of montane forest and woodland borders and in adjacent overgrown clearings in the subtropical zone on both slopes of the Andes. On the west slope recorded south to Chimborazo (Pallatanga and Cayandeled) and Azuay (recent sightings in the Río Yunguilla valley). There are also several recent reports of single birds found at Utuana in s. Loja, first on 23 Sep. 1989 (Best et al. 1993), and an MECN specimen is labeled as being from "Cordillera de Tinajillas, Loja Province" (a locality that cannot be located; Paynter 1993). We regard the Salvadori and Festa (1900) specimens supposedly from the western lowlands (Vinces in Los Ríos and Balazar in Guayas) as surely having been mislabeled. Ranges mostly from 1500 to 2500 m.

The species is sometimes called the Rufous-crowned Tody-Tyrant.

Subspecific determinations in *P. ruficeps* are difficult as the species seems to be subject to considerable individual variation. Birds on Ecuador's east slope have traditionally (e.g., *Birds of the World*, vol. 8) been assigned to the nominate race, and birds on the west slope to *rufigenis*, but we are not entirely convinced that *rufigenis* can be separated from the nominate race. West-slope birds show individual variation in the extent of buff or tawny on the throat (in some it is quite whitish), but the color of the crown is the same as in *ruficeps*. Recent specimens and observations indicate that birds from the west slope and the east slope south at least to w. Napo have the black malar stripe vestigial or absent. It is, however, prominent in birds from the far southeast, as shown in specimens (ANSP, MECN) recently obtained in the Río Isimanchi valley on the eastern slope of the Cordillera Las Lagunillas in s. Zamora-Chinchipe. These birds from the far southeast appear to be referable to *peruvianus*, heretofore known only from extreme n. Peru. A specimen (MECN) from the Cordillera Tinajillas in Loja has also been assigned to *peruvianus*. The bird reported seen by Best et al. (1993) at Utuana in s. Loja seems to have been similar to these latter specimens.

Range: Andes from w. Venezula to n. Peru.

Poecilotriccus capitalis	Black-and-white Tody-Flycatcher	Tirano Todi Negriblanco

Rare to uncommon and local (perhaps just overlooked) in dense tangled viny thickets in the lower growth and borders of terra firme forest in the lowlands of e. Ecuador, and in bamboo thickets and viny tangles in the foothills and lower subtropical zone on the east slope of the Andes. The Black-and-white Tody-Tyrant may be especially numerous around the Tiputini Biodiversity Center (M. Lysinger). Recorded mostly from 400 to 1350 m, ranging very locally out into the lowlands as low as 200–250 m (north of Playa Tigre in Sucumbíos [ANSP], Tiputini Biodiversity Center, and around Kapawi Lodge on the Río Pastaza near the Peruvian border). It has also been recorded from Andoas in adjacent Peru (Berlioz 1937c).

The species is sometimes called the Black-and-white Tody-Tyrant.

Formerly (e.g., Meyer de Schauensee 1966, 1970) placed in the genus *Todirostrum*, with the species name then being spelled *capitale*. One race occurs in Ecuador, nominate *capitalis*. We consider *tricolor* of w. Brazil to be a subspecies of *P. capitalis*, but it is possible that *tricolor* may not even be worthy of subspecific recognition, in which case *P. capitalis* would become monotypic (see Ridgely and Tudor 1994).

Range: locally in se. Colombia, e. Ecuador, ne. Peru, and sw. Amazonian Brazil.

Poecilotriccus calopterus	Golden-winged Tody-Flycatcher	Tirano Todi Alidorado

Uncommon to locally fairly common in undergrowth at borders of humid forest and secondary woodland in the lowlands and foothills of e. Ecuador, mainly recorded fairly close to the base of the Andes. Inconspicuous but vocal, the attractive Golden-winged Tody-Flycatcher is reasonably numerous around Tena and Jatun Sacha. Recorded mostly from 400 to 1300 m, with a few reports from as far east as about 250 m at La Selva and Kapawi Lodge on the Río Pastaza near the Peruvian border; on the Cordillera del Cóndor near Chinapinza it ranges locally as high as 1450 m (RSR).

Although contra many earlier references, we follow W. E. Lanyon (*Am. Mus. Novitates* 2923, 1988) and the 1998 AOU Check-list in accepting the transfer of several undergrowth-inhabiting tody-flycatchers formerly usually placed in the genus *Todirostrum* to the genus *Poecilotriccus*; in Ecuador this involves *calopterus* and the next species. The endings of the specific names change as a result of the generic shift. Monotypic.

Range: extreme se. Colombia, e. Ecuador, and ne. Peru. Excludes Black-backed Tody-Flycatcher (*P. pulchellus*) of se. Peru.

Poecilotriccus latirostris   Rusty-fronted   Tirano Todi
                                  Tody-Flycatcher   Frentirrojizo

Uncommon to locally fairly common in the dense undergrowth of shrubby clearings, forest and woodland borders, and river islands in the lowlands of e. Ecuador. Although quite vocal, the Rusty-fronted Tody-Flycatcher is always a very inconspicuous bird; it seems most numerous near the base of the Andes. Recorded mostly below 700 m, in small numbers up in the foothills to about 1000 m.

As noted under the previous species, *P. latirostris* was formerly often placed in the genus *Todirostrum*. One race occurs in Ecuador, *caniceps*.

Range: se. Colombia, e. Ecuador, e. Peru, n. Bolivia, and w. and cen. Amazonian Brazil.

Todirostrum nigriceps   Black-headed   Espatulilla Cabecinegra
                             Tody-Flycatcher

Fairly common in the canopy and borders of humid forest, secondary woodland, and adjacent clearings in the more humid lowlands and foothills of w. Ecuador south to n. Manabí, s. Pichincha (Río Palenque), ne. Guayas (Naranjito), and nw. Azuay (Manta Real; ANSP). Ranges up to about 900 m.

Monotypic.

Range: Nicaragua to nw. Venezuela and sw. Ecuador.

*Todirostrum chrysocrotaphum   Yellow-browed   Espatulilla
                                     Tody-Flycatcher   Cejiamarilla

Uncommon to fairly common in the canopy and borders of humid forest (both terra firme and várzea), secondary woodland, and adjacent clearings with scattered tall trees in the lowlands of e. Ecuador. First recorded for Ecuador only recently; Norton et al. (1972) reported on specimens obtained at Loreto, Cotapino, Río Conambo, and Montalvo. The species had earlier been recorded from "Río Napo" (Sclater 1855 *in* Chapman 1926), though of course this specimen could have been taken outside the present limits of Ecuador. More recently the Yellow-browed Tody-Flycatcher has been found to be widespread in appropriate habitat, though it is always most often recorded by voice. Recorded up to about 600 m.

T. chrysocrotaphum of w. Amazonia is regarded as a species distinct from *T. pictum* (Painted Tody-Flycatcher) of ne. South America, following most recent authors. One race: *guttatum*.

Range: se. Colombia to n. Bolivia and w. and s. Amazonian Brazil.

Todirostrum cinereum   Common Tody-Flycatcher   Espatulilla Común

Fairly common to common in shrubby clearings, lighter woodland, and gardens in the lowlands of both e. and w. Ecuador, and ranging up in the foothills and lower subtropical zone in cleared areas on both slopes of the Andes; most numerous in w. Ecuador. At least in very wet areas such as n. Esmeraldas, the Common Tody-Flycatcher's original habitat seems to have been river-edge vegetation, from which—subsequent to partial deforestation—it has spread into clearings and plantations, where it is now numerous (O. Jahn and P. Mena Valenzuela et al.). In the east the species does not occur any distance east of the Andes, the farthest east localities being near Jatun Sacha (B. Bochan) and Santiago in Morona-Santiago (M. B. Robbins). Recorded up usually to about 1500 m, locally and in small numbers as high as 1900 m in Napo and Zamora-Chinchipe (and perhaps elsewhere); in the east not recorded below about 400 m.

Two races occur in Ecuador, *peruanum* in the east and *sclateri* in the west. *Sclateri* differs in having a whitish (not concolor yellow) throat.

Range: s. Mexico to nw. Peru, n. and e. Bolivia, extreme n. Paraguay, and s. and e. Brazil (but absent from much of Amazonia).

Todirostrum maculatum   Spotted Tody-Flycatcher   Espatulilla Moteada

Locally fairly common in riparian woodland and shrubby areas on river islands in the lower Río Aguarico and the Río Napo in the lowlands of e. Ecuador. The first report of the Spotted Tody-Flycatcher from Ecuador consisted of a single bird seen on a Río Napo island near La Selva on 24 and 26 Jul. 1989 (RSR and PJG et al.). There have been numerous subsequent reports from this area and farther downstream, but the species seems not to range above the La Selva/Sacha Lodge area. In se. Ecuador the only site where the Spotted Tody-Flycatcher has been found is on islands in the Río Pastaza near Kapawi Lodge close to the Peruvian border, but it may prove also to range farther upriver along the Pastaza. The first and still only Ecuadorian specimen (ANSP) was taken near the mouth of the Río Lagartococha into the Río Aguarico on 9 Dec. 1990. Recorded below about 250 m.

One race: *signatum*.

Range: se. Colombia, ne. Ecuador, e. Peru, nw. Bolivia, Amazonian Brazil, the Guianas, and e. Venezuela.

Cnipodectes subbrunneus   Brownish Twistwing   Alitorcido Pardo

Rare to uncommon and local in the tangled lower growth of both humid and deciduous forest and secondary woodland in the lowlands of w. Ecuador from w. Esmeraldas (Esmeraldas [city]) south to n. Manabí (Río de Oro, Río Peripa), Los Ríos (Babahoyo, Jauneche), se. Guayas (Manglares-Churute Ecological Reserve), and El Oro (an old record from La Chonta; no recent reports). Uncommon to locally fairly common in the eastern lowlands, where apparently confined to terra firme forest though occurring regularly along forest streams. There are only a few recent reports of the twistwing from w. Ecuador, though it has been found in fair numbers at Manglares-Churute Ecological Reserve in se. Guayas (ANSP; Pople et al. 1997) and at Manta Real in nw. Azuay (ANSP), and it has also been found to be reasonably numerous at Jauneche in Los Ríos

(P. Coopmans). Twistwing numbers have doubtless declined substantially in the west as a result of forest destruction, but there is some evidence that the species may be capable of persisting in relatively isolated forest patches. In e. Ecuador, though there apparently are no older records (*Birds of the World* vol. 8), a fair number of recent reports of the twistwing are available (e.g., from La Selva, Sacha Lodge, Zancudococha, Jatun Sacha, along the Maxus road southeast of Pompeya, and Kapawi Lodge); the species appears to be reasonably widespread. Recorded up to about 600 m. We view the record from the exceptionally high elevation of 1900 m at Cabañas San Isidro (B. López-Lanús, *Cotinga* 12, 74, 1999) as unverified.

This species was formerly usually called the Brownish Flycatcher, but we prefer to highlight its uniquely twisted primaries in its English name, following Ridgely and Tudor (1994).

Two similar races are known from Ecuador, *minor* in the east and nominate *subbrunneus* in the west.

Range: Panama to n. Colombia; w. Ecuador; se. Colombia, e. Ecuador, e. Peru, nw. Bolivia, and w. Amazonian Brazil.

Ramphotrigon ruficauda   Rufous-tailed Flatbill   Picoplano Colirrufo

Rare and apparently local in the lower growth of terra firme forest in the lowlands of e. Ecuador, where known primarily from the northeast in the drainages of the Ríos Napo and Aguarico. First recorded from a single individual seen by RSR at Zancudococha on 23 Sep. 1976 (see Ridgely 1980); several were collected here in Apr. 1991 (ANSP and MECN). The only additional confirmed records involve two birds collected by F. Sornoza at Daime in e. Napo on 16 Sep. 1989 (ANSP), one taken north of Playa Tigre in Sucumbíos on 8 Aug. 1993 (ANSP), and a pair seen and tape-recorded along the Maxus road southeast of Pompeya on 3 Feb. 1995 (RSR and F. Sornoza); there also are a number of reports from Sacha Lodge (S. Howell et al.) but seemingly none from the La Selva area. The only reports from the southeast come from Kapawi Lodge on the Río Pastaza near the Peruvian border, but the species likely is more widespread in the Río Pastaza drainage. Recorded only below about 300 m.

Monotypic.

Range: se. Colombia, s. and e. Venezuela, and the Guianas to n. Bolivia and Amazonian Brazil.

Ramphotrigon fuscicauda   Dusky-tailed Flatbill   Picoplano Colinegruzco

Very rare and apparently local in the lower growth of terra firme forest and secondary woodland in the lowlands of ne. Ecuador. There are only a few Ecuadorian records of this scarce species, all of them from a restricted area in w. Napo not far from the base of the Andes. These include the type specimen (AMNH) from "Río Suno Abajo" taken on 24 Mar. 1924; one collected at Cotapino on 13 Jul. 1964 (D. Norton; MCZ); one mist-netted, photographed, and released at Taracoa on 9 Jan. 1982 (RSR, PJG, and S. Greenfield); a pair seen along the road between Puerto Napo and Jatun Sacha on 26 Apr. 1989 (P. Coopmans et al.); a pair seen west of Coca on 30 Jul. 1991 (J. Rowlett and R. A. Rowlett et al.); and a pair seen and tape-recorded on 8–9 Jun. 1999 some 17 km northwest of Archidona at "El Para" (L. Navarrete and J. Moore). Recorded between 250 and 750 m.

Monotypic.

Range: locally in se. Colombia and ne. Ecuador, and in se. Peru, nw. Bolivia, and sw. Amazonian Brazil.

Ramphotrigon megacephala	Large-headed Flatbill	Picoplano de Bambú

Uncommon and very local in stands of bamboo in humid forest and forest borders in the foothills and adjacent lowlands along the eastern base of the Andes. There are relatively few Ecuadorian records of this unobtrusive small flatbill, which is usually noted through its distinctive vocalizations. The first involved a single bird seen and another heard by RSR and D. Wilcove north of Paquisha along the Río Nangaritza in Zamora-Chinchipe on 24 Jul. 1978 (see Ridgely 1980). The first Ecuadorian specimens (WFVZ, MECN) were obtained by F. C. Sibley in w. Napo southeast of Archidona in Oct. 1988 (Marín et al. 1992). Since 1987 the Large-headed Flatbill has also been found regularly in small numbers north of Archidona along the road to Loreto. One was also taken in a bamboo stand west of Coca near the Río Payamino on 11 Nov. 1991 (ANSP). Recorded between about 300 and 1300 m.

One race: *pectoralis.*

Range: locally in w. and s. Venezuela, n. and e. Colombia, and e. Ecuador; se. Peru, n. Bolivia, and sw. Amazonian Brazil; se. Brazil, e. Paraguay, and ne. Argentina.

Rhynchocyclus pacificus	Pacific Flatbill	Picoplano del Pacífico

Uncommon to fairly common (but easily overlooked) in the lower growth of very humid to humid forest in the lowlands and foothills of nw. Ecuador south to sw. Esmeraldas (Bilsa) and s. Pichincha (in small numbers at Río Palenque). Recorded up to about 800 m above Tinalandia.

R. *pacificus* of w. Colombia and nw. Ecuador is regarded as a species distinct from R. *brevirostris* (Eye-ringed Flatbill) of Middle America to nw. Colombia, following J. T. Zimmer (*Am. Mus. Novitates* 1045, 1939); see Ridgely and Tudor (1994). Monotypic.

Range: w. Colombia and nw. Ecuador.

Rhynchocyclus fulvipectus	Fulvous-breasted Flatbill	Picoplano Pechifulvo

Uncommon in lower growth of montane forest in the foothills and lower subtropical zone on both slopes of the Andes; on the west slope not found south of Pichincha (lower Chiriboga road). Recorded between about 900 and 1800 m, but on the east slope for the most part not found below about 1200 m.

Monotypic.

Range: Andes from extreme sw. Venezuela to w. Bolivia.

Rhynchocyclus olivaceus   Olivaceous Flatbill   Picoplano Oliváceo

Rare to locally uncommon in the lower and middle growth of humid forest (especially terra firme, but also ranging into várzea) in the lowlands of e. Ecuador. The Olivaceous Flatbill appears to be most common in lowlands relatively close to the Andes, perhaps especially so in Morona-Santiago (east as far as the vicinity of Taisha). Farther east and north it becomes decidedly scarce or even appears to be locally absent (e.g., there appear to be no verified reports from La Selva, fide P. Coopmans, though small numbers are found in the Cuyabeno area and in the lower Río Aguarico drainage). Recorded up to 700 m.

One race: *aequinoctialis*.

Range: Panama to nw. Bolivia, Amazonian Brazil, and the Guianas (but absent from most of e. Colombia, sw. Venezuela, and nw. Brazil); e. Brazil.

Tolmomyias sulphurescens   Yellow-olive Flatbill   Picoancho Azufrado

Fairly common in the lower and middle growth of deciduous forest, woodland, and borders in the lowlands and foothills of w. Ecuador from w. Esmeraldas (Esmeraldas [city]) south through El Oro and Loja. Uncommon to fairly common and seemingly somewhat local (perhaps more numerous southward) in the mid-levels and subcanopy of montane forest, secondary woodland, and borders in the foothills and subtropical zone on the east slope of the Andes. Occurs mostly below 500 m in the west, ranging up to 1800–2000 m at least in Loja; on the east slope found only between about 900 and 1700 m.

We have reverted to the group name of "flatbill" for all members of the genus *Tolmomyias*. This name was used long ago in *Birds of the Americas* (pt. 5) and is surely more useful than considering this group as yet another tyrannid genus bearing the group name of "flycatcher."

Three described races of *T. sulphurescens* are recorded from Ecuador: *aequatorialis* in the west, and *confusus* and *peruvianus* on the east slope of the Andes (*confusus* in Sucumbíos and Napo, *peruvianus* southward). All are quite similar in overall appearance, and *confusus* is in fact perhaps not separable from *peruvianus*. East- and west-slope birds may prove to be representatives of separate species (on either side of the Andes); the "species" *T. sulphurescens* is made up of a complex of related forms, most of them with differing voices, and a complex of separate species is certainly involved (T. Schulenberg and K. Zimmer, pers. comm.).

Range: s. Mexico to nw. Peru, n. Argentina, Paraguay, and s. Brazil.

Tolmomyias traylori   Orange-eyed Flatbill   Picoancho Ojinaranja

Apparently rare and local—though doubtless overlooked—in the canopy and borders of várzea and riparian forest and woodland in the lowlands of e. Ecuador. This only recently described species (T. S. Schulenberg and T. A. Parker III, *Studies in Neotropical Ornithology Honoring Ted Parker*, Ornithol. Monogr. no. 48: 723–731, 1997) is known in Ecuador mainly from sites near the Río Napo, having been found at Taracoa, Sacha Lodge, and La Selva. It has also recently been recorded around Kapawi Lodge near the Río Pastaza close to the Peruvian frontier (P. Coopmans; D. Michaels) and likely occurs elsewhere in the Río Pastaza drainage as well. It should also occur in the drainage of the Río Aguarico. The first Ecuadorian record involved a bird seen and tape-recorded at Taracoa on 9 Jan. 1985 (PJG). There still is no Ecuadorian specimen. Recorded only below 300 m.
    Monotypic.
    Range: se. Colombia, e. Ecuador, and ne. Peru.

Tolmomyias assimilis   Zimmer's Flatbill   Picoancho de Zimmer

Uncommon to locally fairly common in the subcanopy and mid-levels of humid forest and forest borders (in both terra firme and várzea) in the lowlands of e. Ecuador. Recorded up to about 750 m.
    Does not include *T. flavotectus* (Yellow-margined Flatbill), found west of the Andes. One race of *T. assimilis* occurs in Ecuador, *obscuriceps*. More than one species is almost surely involved in what is presently considered to be *T. assimilis* (T. Schulenberg and K. Zimmer, pers. comm.).
    English names pose a problem when, as is anticipated, the "species" *T. assimilis* is broken up into allospecies; the first step has been taken here, with the separation of trans-Andean *T. flavotectus*. It is to that form that the modifying name "Yellow-margined" was first attached (in *Birds of the Americas*, pt. 5), and because of its wing pattern that name is most appropriate for it. At the suggestion of T. Schulenberg (pers. comm.), we adopt the name of Zimmer's Flatbill for the Amazonian members of the "species" *T. assimilis* which, as noted above, seems itself likely to be split up in due course.
    Range: se. Colombia, s. Venezuela, and the Guianas to n. Bolivia and Amazonian Brazil.

Tolmomyias flavotectus   Yellow-margined Flatbill   Picoancho Alimarginado

Locally fairly common in the subcanopy and mid-levels of very humid and humid forest and forest borders in the lowlands of nw. Ecuador. Old specimens (AMNH) were taken as far south as Guayas (Guayaquil), but in recent years the Yellow-margined Flatbill has not been found south of w. Esmeraldas (Muisne area) and n. Pichincha (e.g., along and north of the San Miguel de los Bancos-Puerto Quito highway; v.o.). In the wet-forest belt of n. Esmeraldas the

Yellow-margined Flatbill also ranges into plantations and secondary growth, perhaps because the Yellow-olive Flatbill (*T. sulphurescens*) does not occur in this region (O. Jahn et al.). The Yellow-margined has likely been overlooked at least to some extent. Recorded up to about 500 m.

T. flavotectus is here considered as a monotypic species distinct from *T. assimilis* (Zimmer's Flatbill) of Amazonia, based on its radically different voice.

Range: Costa Rica to w. Ecuador.

Tolmomyias poliocephalus   Gray-crowned   Picoancho
                              Flatbill   Coroniplomizo

Uncommon to fairly common in the subcanopy and mid-levels of humid forest (both terra firme and várzea), forest borders, and at least sometimes out into adjacent clearings in the lowlands of e. Ecuador. Like the Zimmer's Flatbill (*T. assimilis*), the equally obscure Gray-crowned often remains overlooked until its distinctive voice is recognized. Recorded up locally to about 500–600 m near Archidona (P. Coopmans) and in the Bermejo oil-field area north of Lumbaquí (RSR).

One race: nominate *poliocephalus*.

Range: se. Colombia, s. Venezuela, and the Guianas to nw. Bolivia and Amazonian Brazil; e. Brazil.

Tolmomyias viridiceps   Olive-faced Flatbill   Picoancho Cabecioliváceo

Fairly common to common in clearings and gardens with scattered trees, lighter woodland, river-edge woodland, and subcanopy and borders of humid forest (especially in várzea, but smaller numbers also in terra firme) in the lowlands of e. Ecuador. Ranges mostly below 800 m, but occurs locally (primarily following clearings) up to 1100 m along the eastern base of the Andes.

We recognize the three duller, west-Amazonian subspecies (*viridiceps*, *subsimilis*, and *zimmeri*), formerly considered as races of *T. flaviventris*, as comprising a separate species, *T. viridiceps*. This follows a suggestion put forward by J. M. Bates, T. A. Parker III, A. P. Caparella, and T. J. Davis (*Bull. B. O. C.* 112[2]: 90–91, 1992). One race occurs in Ecuador, nominate *viridiceps*.

In consultation with T. Schulenberg, we suggest calling *T. viridiceps* the Olive-faced Flatbill (its face being duller and less patterned than that of other *Tolmomyias*) and *T. flaviventris* (now including *aurulentus*, *collingwoodi*, and *dissors* as races) of e. South America the Ochre-lored Flatbill. Before being split, *T. flaviventris* was called the Yellow-breasted Flycatcher.

Range: se. Colombia, e. Ecuador, e. Peru, nw. Bolivia, and w. Amazonian Brazil (east to the lower Rio Negro and the Rio Madeira).

Platyrinchus mystaceus   White-throated Spadebill   Picochato Goliblanco

Uncommon to fairly common but inconspicuous in the undergrowth of montane forest and secondary woodland in the foothills and subtropical zone on both slopes of the Andes, on the west slope ranging as far south as w. Loja (seen near El Limo in Apr. 1993; RSR). In the west—where it appears to be more numerous than it is in the east—the White-throated Spadebill also ranges out into adjacent humid lowlands (e.g., at Esmeraldas [city] and Río Palenque), and it has also recently been found on the slopes of the coastal Cordillera de Mache at Bilsa in sw. Esmeraldas (J. Hornbuckle et al.) and the Cordillera de Colonche in sw. Manabí (Machalilla National Park) and w. Guayas (Río Ayampe [RSR and D. Wilcove], Cerro La Torre and above Salanguilla [N. Krabbe]). Recorded mostly from 600 to 2000 m on the west slope, but locally ranging down to near sea level (e.g., along the Río Ayampe in Machalilla National Park); on the east slope found mostly between 1000 and 2000 m.

Two very similar races are found in Ecuador, *albogularis* on the west slope and *zamorae* on the east slope. We suspect that ultimately it may be preferable to synonymize *zamorae* into *albogularis*.

Range: Costa Rica to extreme nw. Peru, much of Venezuela, and south along east slope of Andes to w. Bolivia; French Guiana; e. and s. Brazil and ne. Bolivia, e. Paraguay, and ne. Argentina.

Platyrinchus coronatus	Golden-crowned Spadebill	Picochato Coronidorado

Uncommon to locally fairly common (though always inconspicuous) in the undergrowth of terra firme forest in the lowlands of both e. and w. Ecuador. In the west recorded south to n. Manabí and s. Pichincha (at least formerly occurring at Río Palenque); now most numerous in Esmeraldas (e.g., in the Playa de Oro region; O. Jahn et al.). Recorded mainly below 700 m, exceptionally up to 1650 m (a displaying male seen on the south slope of Cordillera de Huacamayos on 10 Oct. 1992; N. Krabbe).

Two similar races are found in Ecuador, *superciliaris* in the west and nominate *coronatus* in the east.

Range: Honduras to nw. Ecuador; se. Colombia, s. Venezuela, and the Guianas to nw. Bolivia and Amazonian Brazil.

*Platyrinchus saturatus	Cinnamon-crested Spadebill	Picochato Cresticanelo

Very rare—perhaps just overlooked?—in the undergrowth of terra firme forest in the lowlands of extreme ne. Ecuador. The Cinnamon-crested Spadebill is known from only a single record in Ecuador, a male that was mist-netted and collected (ANSP) north of Playa Tigre on the Río San Miguel in Sucumbíos on 9 Aug. 1993. 200 m.

One race: nominate *saturatus*.

Range: extreme e. Colombia, s. Venezuela, the Guianas, and n. Amazonian Brazil; very locally in ne. Ecuador, ne. Peru, and w. Amazonian Brazil.

Platyrinchus flavigularis   Yellow-throated   Picochato Goliamarillo
                              Spadebill

Rare to uncommon and local in the undergrowth of montane forest in the foothills and the lower subtropical zone on the east slope of the Andes from w. Napo (Río Guataraco and Palm Peak; MCZ) south through Zamora-Chinchipe. There are only a few Ecuadorian records of this apparently genuinely scarce spadebill, which still is known from only a few specimens; see the summary in Robbins et al. (1987). The first Ecuadorian record, overlooked in that paper, is a specimen from "Ramos Urcu," a site believed to be in the Volcán Sumaco area (W. Meise, *Auk* 71[3]: 287, 1954). Since the mid-1980s the Yellow-throated Spadebill has been found at only one other locality, Panguri in s. Zamora-Chinchipe (ANSP, MECN). Recorded from about 750 to 1700 m.

We consider the Yellow-throated Spadebill to be Near-threatened in Ecuador. It was not considered to be at risk by Collar et al. (1992, 1994).

One race: nominate *flavigularis*.

Range: locally on east slope of Andes from w. Venezuela to s. Peru; Sierra de Perijá.

Platyrinchus platyrhynchos   White-crested   Picochato Crestiblanco
                               Spadebill

Apparently rare and local in the undergrowth of terra firme forest in the lowlands of e. Ecuador. The White-crested Spadebill has perhaps been overlooked in Ecuador; in any case, records are very few. In the 19th century it was recorded from Sarayacu, and more recently birds were seen and tape-recorded on the south side of the Río Napo near La Selva on 28 Nov. 1988 and 26 Dec. 1989 (P. Coopmans) and along the Maxus road southeast of Pompeya in 1994 (P. English); several were also noted at Zancudococha in Dec. 1990 and Apr. 1991 (RSR and M. B. Robbins et al.). Other than the old Sarayacu specimen, the only records from the southeast come from around Kapawi Lodge on the Río Pastaza near the Peruvian frontier. Recorded only below about 300 m.

One race: *senex*.

Range: e. Colombia, s. Venezuela, and the Guianas to n. Bolivia and Amazonian Brazil.

Onychorhynchus coronatus   Amazonian Royal-   Mosquero Real
                             Flycatcher   Amazónico

Rare to uncommon and local in the lower growth of humid forest (mainly várzea forest and in swampy places along streams) and secondary woodland in the lowlands of e. Ecuador. Recorded mainly below 300 m, locally as high as 400 m (Santiago; ANSP).

Does not include trans-Andean *O. occidentalis* (Pacific Royal-Flycatcher) or two other geographically distant allospecies in the *O. coronatus* superspecies. One race of *O. coronatus* occurs in Ecuador, *castelnaui*.

Range: se. Colombia, s. and e. Venezuela, and the Guianas to n. Bolivia and Amazonian Brazil.

| *Onychorhynchus occidentalis* | Pacific Royal-Flycatcher | Mosquero Real del Pacífico |

Rare and local in the lower growth of deciduous and semihumid forest and secondary woodland and their borders (usually near watercourses) in the lowlands of w. Ecuador from w. Esmeraldas (Esmeraldas [city]) and s. Pichincha (Río Palenque) south through Guayas and El Oro (Santa Rosa, Piedras, and La Avanzada). Recorded mostly below 600 m (e.g., below Buenaventura; RSR and M. B. Robbins), exceptionally as high as 900 m (a pair seen at their partially constructed nest at San Miguel del Azuay in Azuay in Jan. 1992; M. Whittingham, *Bull. B. O. C.* 114[2]: 131–132, 1994).

Recent decades have seen a substantial decline in the numbers of the Pacific Royal-Flycatcher as a result of the near-total habitat destruction across most of its former range, much of which is prime agricultural land. The Pacific Royal-Flycatcher was accorded Endangered status by Collar et al. 1992 but was downgraded to Vulnerable status in Collar et al. 1994. The species, which appears nowhere to be numerous and is virtually an Ecuadorian endemic, does seem capable of persisting at least locally in somewhat degraded habitat—though it likely will not be able to do so indefinitely. Small numbers persist along the Río Ayampe at the edge of Machalilla National Park, at Jauneche in Los Ríos, at Cerro Blanco in the Chongón Hills, and at Manglares-Churute Ecological Reserve. We give the species Vulnerable status.

O. occidentalis is regarded as a species separate from the cis-Andean *O. coronatus* (Amazonian Royal-Flycatcher), following the lead of Collar et al. (1992, 1994). *O. mexicanus* (Northern Royal-Flycatcher) and *O. swainsoni* (Atlantic Royal-Flycatcher) are also considered allospecies in the *O. coronatus* superspecies. Compared to *O. coronatus*, *O. occidentalis* is somewhat larger with a longer and much brighter cinnamon tail, paler upperparts, and lacking the narrow dusky brown barring on the breast. *O. occidentalis* is monotypic.

Range: w. Ecuador and extreme nw. Peru.

| *Myiotriccus ornatus* | Ornate Flycatcher | Mosquerito Adornado |

Common at borders of montane forest and woodland and around treefalls in the foothills and subtropical zone on both slopes of the Andes. On the west slope found south to El Oro (old specimens from La Chonta and El Chiral; numerous recent records from Buenaventura [including ANSP and MECN]). Also recorded recently on the coastal Cordillera de Mache at Bilsa in sw. Esmeraldas, where numerous and first found in Feb. 1994 (R. P. Clay et al.). In addition there is a single specimen (AMNH) from the coastal Cordillera de Colonche in w. Guayas taken at "Cerro Manglar Alto" on 2 Jun. 1923; surprisingly, however, the Ornate Flycatcher has not been found during recent surveys by ANSP teams at nearby Cerro San Sebastián in sw. Manabí, or in w. Guayas by

N. Krabbe or the teams of C. D. Becker (Becker and López-Lanús 1997). Possibly the population in the southern coastal cordillera has become extinct; as this is such a conspicuous bird, it seems unlikely to have been missed if it were still present. Recorded mostly between about 800 and 2000 m. On the east slope a few range locally down to 500 m (Bermejo oil-field area north of Lumbaquí; ANSP, MECN). On the west slope the species ranges down to about 350 m in the foothills of n. Esmeraldas, and it is also known to have occurred in the humid lowlands south to n. Los Ríos; there are old specimens from Rio de Oro in Manabí (AMNH) and from Quevedo in Los Ríos (ANSP), and a few sightings from Río Palenque.

Two quite distinct races are found in Ecuador, *phoenicurus* on the east slope and *stellatus* on the west slope. *Phoenicurus* has an all-rufous tail, whereas in *stellatus* the tail is dusky, showing rufous only at its base.

Range: Andes from n. Colombia to s. Peru.

Neopipo cinnamomea   Cinnamon Neopipo   Neopipo Canelo

Very rare and seemingly local (probably overlooked) in the lower and middle growth of humid forest (apparently mainly or perhaps entirely in terra firme) in the lowlands and foothills of e. Ecuador. There are only a few records of this poorly known species from Ecuador. Besides old specimens from Sarayacu and Río Suno, a female (WFVZ) was obtained on 23 Jul. 1989 at Pachicutza (1000 m) on the lower slopes of the Cordillera del Cóndor in Zamora-Chinchipe (see Marín et al. 1992). Subsequent records that appear to be valid include a sighting and tape-recording from Cuyabeno in Jul. 1993 (P. English), two records from Jatun Sacha (one seen and tape-recorded on 28 Aug. 1991[P. Coopmans and M. Catsis]; 2 birds mist-netted in May 1994 [M. Guerrero, fide B. Bochan]), one seen along the Maxus road southeast of Pompeya on 11 Jan. 1994 (RSR and F. Sornoza), and one seen and tape-recorded near Campo Cocha between the Ríos Sotano and Nushiño on 2 Oct. 1995 (J. Nilsson). The likelihood of confusion with the similar and far more common Ruddy-tailed Flycatcher (*Terenotriccus erythrurus*) has probably precluded obtaining a more complete picture of the Cinnamon Neopipo's abundance in Ecuador; it certainly is not numerous in the country. Recorded between about 300 and 1000 m, but mostly below 400 m.

The relationships of the enigmatic genus *Neopipo* were discussed by J. A. Mobley and R. O. Prum (*Condor* 97[3]: 650–662, 1995). They unequivocally concluded that *Neopipo*, long considered a manakin genus, is not a member of the Pipridae but rather is a tyrant flycatcher in the *Myiophobus* group. One race occurs in Ecuador, nominate *cinnamomea*.

This species was formerly called the Cinnamon Manakin (e.g., Meyer de Schauensee 1966, 1970), but given the doubts concerning its familial affinities, its English name was modified to Cinnamon Tyrant-Manakin by Sibley and Monroe (1990) and Ridgely and Tudor (1994). Mobley and Prum (op. cit.) suggested shortening this to Cinnamon Tyrant, but we fear doing so causes possible confusion with *Pyrrhomyias cinnamomea* (Cinnamon Flycatcher). We prefer to give this distinctive species the unambiguous name of Cinnamon Neopipo, first suggested by S. Hilty.

Range: locally in the Guianas, Amazonian Brazil, extreme sw. Venezuela, extreme e. Colombia, e. Ecuador, e. Peru, and extreme nw. Bolivia.

Terenotriccus erythrurus   **Ruddy-tailed Flycatcher**   **Mosquerito Colirrojizo**

Uncommon to locally common in lower and middle growth of humid forest in the lowlands and foothills of both e. and w. Ecuador, in the east mainly in terra firme. In the west recorded south locally and in small numbers to n. Manabí (Río de Oro, Río Peripa), s. Pichincha (Río Palenque), Los Ríos (at least formerly at Quevedo; ANSP), and nw. Azuay (Manta Real; at least one seen by RSR in Aug. 1991). Ranges up locally to about 1000 m along the east slope of the Andes.

It has been suggested (W. E. Lanyon, *Am. Mus. Novitates* 2915, 1988) that the monotypic genus *Terenotriccus* be merged into the genus *Myiobius*, and although the anatomical evidence appears to be supportive (J. A. Mobley and R. O. Prum, *Condor* 97[3]: 650–662, 1995), their behavior and indeed overall appearance are so utterly different that we still hesitate to do so, and thus agree with the 1998 AOU Check-list in maintaining *Terenotriccus* as a separate genus. Two only slightly differentiated races of *T. erythrurus* are found in Ecuador, *signatus* in the east and *fulvigularis* in the west.

Range: s. Mexico to sw. Ecuador, n. Bolivia, and Amazonian Brazil.

Myiobius atricaudus   **Black-tailed Flycatcher**   **Mosquerito Colinegro**

Uncommon to locally fairly common in the lower growth of secondary woodland and humid forest borders in the lowlands of both e. and w. Ecuador. In the west recorded from Esmeraldas south through El Oro and w. Loja (Alamor region). In the east decidedly less numerous, here found in várzea and riparian forest (and rather strictly avoiding terra firme). In the west overall numbers of the Black-tailed Flycatcher have doubtless been reduced as a result of forest and woodland destruction. Ranges up to about 1000 m on both slopes, but has been recorded to 1300 m in Loja at Alamor.

Two quite similar races (both having the breast dull buff) are found in Ecuador, *adjacens* in the east and *portovelae* in the west.

Range: Costa Rica to extreme nw. Peru; e. Venezuela; extreme se. Colombia, e. Ecuador, and e. Peru east through s. Amazonian Brazil to ne. Brazil; se. Brazil.

Myiobius barbatus   **Whiskered Flycatcher**   **Mosquerito Bigotillo**

Uncommon and seemingly rather local in the lower growth of terra firme forest in the lowlands of e. Ecuador. Found mostly below about 600 m, but ranging locally up in larger river valleys to around 1000 m (e.g., at Zamora).

Does not include trans-Andean *M. sulphureipygius* (Sulphur-rumped Flycatcher), often considered conspecific with *M. barbatus* (e.g., by Sibley and Monroe 1990, Ridgely and Tudor 1994). One race occurs in Ecuador, nominate *barbatus*.

Range: se. Colombia, s. Venezuela, and the Guianas to e. Peru and Amazonian and cen. Brazil; e. Brazil.

Myiobius sulphureipygius	Sulphur-rumped Flycatcher	Mosquerito Lomiazufrado

Uncommon to fairly common in the lower and middle growth of very humid and humid forest and woodland in the lowlands and foothills of w. Ecuador, where recorded south to El Oro (Buenaventura; ANSP and numerous sightings) and w. Loja (AMNH specimen from Alamor [J. T. Zimmer, *Am. Mus. Novitates* 1042, 1939] and a Guayquichuma specimen in MECN). Ranges mostly below about 1000 m, but in w. Loja ranges as high as 1300 m at Alamor.

M. sulphureipygius is considered a species separate from cis-Andean *M. barbatus* (Whiskered Flycatcher), in this agreeing with the 1998 AOU Check-list. One race of *M. sulphureipygius* occurs in Ecuador, *aureatus*.

Range: s. Mexico to sw. Ecuador.

Myiobius villosus	Tawny-breasted Flycatcher	Mosquerito Pechileonado

Uncommon to locally fairly common in the lower and middle growth of montane forest and mature secondary woodland in the foothills and lower subtropical zone on both slopes of the Andes. On the west slope recorded in the past south to sw. Chimborazo (Chimbo), though in recent years not found south of Pichincha along the lower Chiriboga road. Note that, especially in w. Ecuador, there has been much confusion regarding the separation of this species from the bright *aureatus* race of Sulphur-rumped Flycatcher (*M. sulphureipygius*); this has resulted in a number of erroneous reports of the Tawny-breasted Flycatcher from outside the species' confirmed range, especially from the southwest. The Tawny-breasted Flycatcher occurs at higher elevations than its congeners, though some overlap occurs, particularly with the Sulphur-rumped. Ranges mostly from 1000 to 1700 m, but locally recorded lower (to 600 m), even in substantial numbers (e.g., at El Placer in Esmeraldas); one was also seen at Cabañas San Isidro (2000 m), unusually high, on 26 Jul. 1996 (M. Lysinger).

Two similar races occur in Ecuador, *clarus* on the east slope and nominate *villosus* on the west slope.

Range: highlands of e. Panama; locally in Andes from extreme sw. Venezuela to w. Bolivia; Sierra de Perijá.

Myiophobus flavicans	Flavescent Flycatcher	Mosquerito Flavecente

Uncommon to fairly common in the lower and middle growth of montane forest and to a lesser extent at borders in the subtropical and lower temperate zones on both slopes of the Andes. On the west slope found south to w. Loja in the Alamor/Celica region. Recorded from about 1300 to 2500 m, occurring lowest in w. Loja; in areas of potential sympatry with the Orange-crested Flycatcher (*M. phoenicomitra*), the Flavescent appears to range at somewhat higher elevations, though perhaps with very local overlap.

One race: nominate *flavicans*.

Range: mountains of n. Venezuela, and Andes from w. Venezuela to s. Peru; Sierra de Perijá.

Myiophobus phoenicomitra	Orange-crested Flycatcher	Mosquerito Crestinaranja

Uncommon to fairly common (but inconspicuous) in the lower growth of montane forest in the foothills and lower subtropical zone on both slopes of the Andes. On the west slope recorded mainly from e. Esmeraldas (El Placer) and Imbabura (Lita) south to Pichincha (Tinalandia and lower Chiriboga road areas). There are also two more southern records: a specimen (MECN) that was formerly identified as a Flavescent Flycatcher (*M. flavicans*), taken in w. Chimborazo at Pallatanga on 26 Feb. 1967 (M. Olalla), and another bird that was mist-netted and skeletonized at Buenaventura in El Oro on 15 Apr. 1991 (Krabbe 1992). On the east slope recorded from w. Napo in the Archidona and lower Volcán Sumaco areas, Tungurahua at Hacienda Mapoto (the type locality), n. Morona-Santiago (Cordillera de Cutucú [Robbins et al. 1987] and Río Huamboya valley of Sangay National Park in Apr. 1998 [MECN]), and in Zamora-Chinchipe (early-20th-century specimens from Zamora, with more recent records from the Río Bombuscaro sector of Podocarpus National Park). The species likely has a more continuous distribution on the east slope. Recorded from about 600 to 1550 m.

Two similar races are found, *litae* on the west slope and nominate *phoenicomitra* on the east slope.

Range: locally in Andes from sw. Colombia to n. Peru.

*Myiophobus roraimae	Roraiman Flycatcher	Mosquerito Roraimeño

Rare and local (perhaps overlooked?) in the lower and middle growth of montane forest in the lower subtropical zone in se. Ecuador, where at least thus far recorded only from outlying ridges (and not on the slopes of the Andes themselves). The Roraiman Flycatcher was first recorded in Ecuador in Jun. 1984 on the west slope of the Cordillera de Cutucú in Morona-Santiago at a Shuar hunting encampment named "Yapitya"; several pairs were found here, and a female (ANSP) was obtained by T. S. Schulenberg on 23 Jun. (Robbins et al. 1987). The only subsequent Ecuadorian record is of a single male mist-netted and collected (ANSP) near Chinapinza on the west slope of the Cordillera del Cóndor in Zamora-Chinchipe on 14 Jun. 1993. Recorded only between 1400 and 1700 m.

One race: *rufipennis*.

Range: very locally on east slope of Andes in s. Ecuador and Peru; tepuis of s. Venezuela and adjacent Guyana and Brazil.

Myiophobus fasciatus   Bran-colored Flycatcher   Mosquerito Pechirrayado

Fairly common in shrubby clearings and pastures, gardens, and lighter woodland in the lowlands and foothills of w. Ecuador from Esmeraldas and locally in the arid Río Mira valley of Imbabura (MECN specimen from north of Salinas taken on 19 Oct. 1983; C. G. Schmitt) south through Loja. Rare and local in similar habitat along the base of the Andes in e. Ecuador, where there are a few records from w. Napo (including AMNH specimens from below San José and Río Suno, LSUMZ specimen material from Limoncocha, and sightings from along the Loreto road), w. Pastaza (including an MECN specimen from Puyo), Morona-Santiago (ANSP specimen from Santiago), and Zamora-Chinchipe (a few sightings from Zamora area); the species has also been recorded from Kapawi Lodge in the Río Pastaza drainage near the Peruvian border. The Bran-colored Flycatcher is now known to occur sympatrically with larger numbers of the similar but duller Olive-chested Flycatcher (*M. cryptoxanthus*) in a few areas, and how—or if—they segregate ecologically remains uncertain. The Bran-colored Flycatcher does not, however, seem to be increasing as rapidly as is the Olive-chested. In the west ranges up to 1500–2000 m in sw. Azuay (Río Yunguilla valley) and Loja; in the east found between about 300 and 1100 m.

Two quite different races occur in Ecuador, *crypterythrus* (with dull grayish brown upperparts) in the west and nominate *fasciatus* (with upperparts reddish brown) in the east. Southward the latter may perhaps intergrade with the similar *saturatus* (which is found in e. Peru).

Range: Costa Rica to cen. Argentina and Uruguay (though absent from much of Amazonia, s. Venezuela, and the Guianas); sw. Colombia, w. Ecuador, and w. Peru. More than one species may be involved.

Myiophobus cryptoxanthus   Olive-chested   Mosquerito
                             Flycatcher   Pechioliváceo

Fairly common to locally common in shrubby clearings and pastures in the foothills and lower subtropical zone on the east slope of the Andes, ranging out in somewhat smaller numbers into the adjacent lowlands of e. Ecuador, with the easternmost area from which it is known being at Jatun Sacha. Recorded north to w. Sucumbíos in the Bermejo oil-field area north of Lumbaquí; the species thus seems likely to occur in adjacent s. Colombia, where it is unrecorded (Hilty and Brown 1986). As a result of continuing forest clearance, numbers of the inconspicuous (but often very vocal) Olive-chested Flycatcher seem destined to increase in Ecuador; it now is numerous along the Loreto road north of Archidona, and also in the Zumba region. In Jul. 1989 it was found to occur sympatrically with the nominate race of the Bran-colored Flycatcher (*M. fasciatus*) at Santiago in Morona-Santiago (ANSP specimens of both species), and the two have also since been found together along the Loreto road north of Archidona. Recorded mostly from 400 to 1400 m, in small numbers as high as 1800 m.

Monotypic.

Range: along eastern base of Andes in Ecuador and n. Peru.

Myiophobus pulcher   Handsome Flycatcher   Mosquerito Hermoso

Uncommon to fairly common in the canopy and borders of montane forest in the subtropical zone on both slopes of the Andes, mainly in n. Ecuador. On the west slope found south to Pilaló in sw. Cotopaxi (E. Stresemann, *Ornith. Monatsber.* 45: 75–77, 1937). On the east slope known definitely only from w. Sucumbíos (along the La Bonita road; MECN and several sightings) and w. Napo (Baeza and Cordillera de Huacamayos areas); there is also a sighting of a single bird from below Sabanilla (1650 m) in Zamora-Chinchipe on 24 Jul. 1992 (D. Wolf). Recorded from about 1500 to 2400 m.

Two races are found in Ecuador, nominate *pulcher* on the west slope and *bellus* on the east. The latter is slightly larger and has more deeply colored wing-bars, throat, and breast.

Range: locally in Andes of Colombia and Ecuador, and disjunctly in se. Peru.

Myiophobus lintoni   Orange-banded Flycatcher   Mosquerito Franjinaranja

Rare to fairly common but rather local in the mid-levels and canopy of montane forest and ridgetop elfin woodland in the temperate zone on the east slope of the Andes in s. Ecuador where it ranges north to nw. Morona-Santiago along the Gualaceo-Limón road (Fjeldså and Krabbe 1986; also subsequent sightings), in Azuay also spreading west locally onto slopes above interandean valleys (e.g., at Portete). The range-restricted Orange-banded Flycatcher has been found to be relatively numerous on the Cordillera de Cordoncillo south of Saraguro in extreme n. Loja (ANSP) and also at the Tapichalaca Biological Reserve above Valladolid; it also is known from the Cajanuma sector of Podocarpus National Park (Rasmussen et al. 1996; MECN). Curiously, however, it was not found during extensive field work in Oct.–Nov. 1992 on the east slope of the Cordillera Las Lagunillas near the Peruvian border in extreme s. Zamora-Chinchipe. Recorded mostly from 2250 to 3200 m.

The Orange-banded Flycatcher was accorded Near-threatened status by Collar et al. (1994) because of widespread deforestation across much of its limited range. We agree; the species may even merit Vulnerable status.

Monotypic.

Range: locally in Andes of s. Ecuador and extreme n. Peru.

Pyrrhomyias cinnamomea   Cinnamon Flycatcher   Mosquerito Canelo

Fairly common to common and conspicuous at borders of montane forest and woodland in the foothill, subtropical, and temperate zones on both slopes of the Andes, and locally on slopes above the central and interandean valleys. On the west slope recorded south to El Oro (Salvias). The Cinnamon Flycatcher persists well in partially deforested regions. Recorded from an unusually wide elevational range, from about 1200 to 3000 m.

One race: *pyrrhoptera*.

Range: mountains of n. Venezuela, and Andes from w. Venezuela to nw. Argentina; Sierra de Perijá and Santa Marta Mountains.

| *Mitrephanes phaeocercus* | Northern Tufted-Flycatcher | Mosquerito Moñudo Norteño |

Uncommon and local at borders of humid forest and around treefalls and in adjacent clearings in the lower foothills of nw. Ecuador in n. and e. Esmeraldas and extreme sw. Imbabura (fairly common near the Salto del Tigre bridge over the Río Guaillabamba in Jul. 1997 [P. Coopmans et al.], with one specimen [MECN] taken). Recent reports of the Northern Tufted-Flycatcher in Esmeraldas come mainly from the Alto Tambo region, but the species was also found to be uncommon at Playa de Oro during survey work from 1995 to 1998 (O. Jahn). Recorded between about 100 and 600 m.

M. *olivaceus* (Olive Tufted-Flycatcher) of the east slope of the Andes in Peru and Bolivia is regarded as a species distinct from M. *phaeocercus*, following J. D. Webster (*Auk* 85[2]: 287–303, 1968), *Birds of the World* (vol. 8), and Ridgely and Tudor (1994). The 1998 AOU Check-list, however, continued to regard them as conspecific. One race: *berlepschi*.

We follow Ridgely and Tudor (1994) in employing the group-name of "tufted-flycatcher" for the allospecies in the genus *Mitrephanes*. The species has been called the Common Tufted-Flycatcher.

Range: n. Mexico to nw. Ecuador. More than one species may be involved.

The Olive Tufted-Flycatcher has been recorded in adjacent Peru on Cerro Chinguela (Parker et al. 1985), and it may perhaps occur north into extreme s. Ecuador, though—despite the comment of Hilty and Brown (1986)—it has not been confirmed as occurring in the country. One individual was, however, believed seen (RSR) in s. Zamora-Chinchipe north of Zumba in the Río Isimanchi valley (800 m) on 13 Dec. 1991, but the observation was too brief to preclude the possibility that another species was involved, and the bird could not be relocated.

| *Contopus virens* | Eastern Wood-Pewee | Pibí Oriental |

An uncommon to locally common transient and boreal winter resident at borders of humid forest and in shrubby clearings and gardens in the lowlands and foothills of both e. and w. Ecuador. The Eastern Wood-Pewee is substantially more numerous in the eastern lowlands than elsewhere, and there it seems to favor wooded or forested stream borders. In the west it has been recorded primarily from Pichincha northward, though several were seen and heard above

Manta Real in nw. Azuay on 26–27 Jan. 1991 (RSR and F. Sornoza). At higher elevations it seems to occur principally as a migrant. Recorded mostly from Oct. to mid-Apr., a few individuals arriving as early as late Sep. or lingering into early May. Recorded mainly below about 1200 m, but transients are at least occasionally recorded much higher (once even as high as 2800 m at Quito, where 2 birds were found on 21 Oct. 1992; RSR and A. Whittaker).

Monotypic.

Range: breeds in e. North America, wintering south to w. Amazonia as far as n. Bolivia and w. Amazonian Brazil.

Contopus sordidulus   Western Wood-Pewee   Pibí Occidental

An uncommon to fairly common transient and boreal winter resident at borders of humid and montane forest and in shrubby clearings in the foothills and lower subtropical zone on both slopes of the Andes, and in the lowlands of w. Ecuador south to El Oro. The species also seems possible, particularly as a transient, in the eastern lowlands away from the Andes, but the farthest east site from which it has been recorded is only at Jatun Sacha, where it is rare, probably occurring only as a migrant (RSR). Recorded mostly from Sep. to mid-Apr., with a few arriving as early as late Aug. (e.g., one was collected [ANSP] in the Bermejo oil-field area north of Lumbaquí on 21 Aug. 1993); on the whole the Western Wood-Pewee appears both to arrive and to depart earlier than does the Eastern (*C. virens*). Ranges up at least to 1700 m; in the east not definitely recorded below about 400 m.

The nominate race of *C. sordidulus* certainly occurs in Ecuador, as perhaps do other subspecies as well; all of the described races resemble each other closely.

Range: breeds in w. North America, wintering south to e. Peru and w. Bolivia.

Contopus punensis   Tumbes Pewee   Pibí de Tumbes

Uncommon to fairly common in lighter woodland, clearings with scattered trees, forest borders, and locally even in mangroves in the lowlands and foothills of sw. Ecuador north to cen. Manabí (Bahía de Caráquez area and Chone; also Isla de la Plata) and Los Ríos. It is possible that the Tumbes Pewee may also occur in the Río Marañón drainage of se. Ecuador in the vicinity of Zumba, as it is known in adjacent Peru from habitats comparable to those found around Zumba. Recorded mostly below about 1500 m, though in sw. Azuay and w. Loja it has been found as high as 1900 m.

Based on its distinctly different vocalizations, we consider *C. punensis* to be a species separate from *C. cinereus* (Tropical Pewee) of s. Mexico to n. Argentina, of which it has long been treated as a race. Ridgely and Tudor (1994) noted this striking vocal difference but did not separate *C. punensis* as a species. One race occurs in Ecuador, nominate *punensis*.

Range: sw. Ecuador and w. Peru.

Contopus nigrescens   Blackish Pewee   Pibí Negruzco

Rare to fairly common but very local in the canopy and borders of montane and terra firme forest in the foothills and adjacent lowlands of e. Ecuador; several sightings from higher elevations in the subtropical zone are here considered suspect because of the likelihood of confusion with the far more common Smoke-colored Pewee (*C. fumigatus*). There are only a few records of the scarce Blackish Pewee in Ecuador, doubtless because in part it is being overlooked in the high forest canopy to which it is confined. Ecuadorian records include old specimens from Sarayacu and "Valle del Río Santiago," a sighting from the west slope of the Cordillera de Cutucú on 9 Jun. 1984 (RSR), and sightings near the road to Loreto above Archidona in Mar. 1990 (B. Whitney); the species also was not uncommon in the Campococha vicinity in late 1995 (J. Nilsson). Blackish Pewees proved to be fairly common near Taisha in Morona-Santiago in Aug. 1990, with four specimens being obtained (ANSP, MECN). The northernmost record involves a single specimen (ANSP) obtained by A. Capparella in the Bermejo oil-field area north of Lumbaquí on 21 Aug. 1993. The Blackish Pewee likely also occurs in adjacent s. Colombia, whence it is unrecorded (Hilty and Brown 1986). Recorded only between 400 and 900 m, but as the species has been found up to 1200 m in Peru, it may occur as high as that in Ecuador as well.

One race: nominate *nigrescens*.

Range: very locally along eastern base of Andes in Ecuador and Peru; very locally in s. Guyana and e. Amazonian Brazil.

Contopus fumigatus   Smoke-colored Pewee   Pibí Ahumado

Fairly common and conspicuous in the borders of montane forest and in clearings with scattered tall trees in the foothill, subtropical, and lower temperate zones on both slopes of the Andes, and locally on slopes above the central and interandean valleys (e.g., in Azuay). On the west slope found south through w. Loja (e.g., in the Sozoranga area). Also recorded recently on the coastal Cordillera de Mache in sw. Esmeraldas at Bilsa (J. Hornbuckle; D. Wolf) and the Cordillera de Colonche in sw. Manabí (Cerro San Sebastián in Machalilla National Park [Parker and Carr 1992; ANSP]) and w. Guayas (Cerro La Torre and above Salanguilla; N. Krabbe); a few were also found in Aug. 1996 on hills in Manglares-Churute Ecological Reserve in se. Guayas (Pople et al. 1997). Recorded mainly from 800 to 2600 m. Small numbers, however, occur down to 400 m on the coastal cordillera, occasionally even lower (an apparently resident pair was seen at only 175 m south of the Río Ayampe in nw. Guayas on 18 Jul. 1993; RSR); locally it also occurs higher (e.g., to 3000 m at Oyacachi; Krabbe et al. 1997).

C. *pertinax* (Greater Pewee) and C. *lugubris* (Dark Pewee) of Middle America are regarded as species distinct from C. *fumigatus* of South America, following most recent authors and the 1998 AOU Check-list. If all are considered conspecific under the name of C. *fumigatus*, then the species' English name becomes the Greater Pewee. Two very similar races of *C. fumigatus* are found in Ecuador, *ardosiacus* on the east slope and *zarumae* in the west.

Range: mountains of n. Venezuela, and Andes from w. Venezuela to nw. Argentina; Sierra de Perijá; tepuis of s. Venezuela.

Contopus cooperi   Olive-sided Flycatcher   Pibí Boreal

An uncommon but conspicuous transient and boreal winter resident in forest borders and adjacent clearings with scattered tall trees in both e. and w. Ecuador. Recorded mostly from Oct. to Apr., a few birds arriving as early as Sep. and lingering into early May. Found mainly in the foothills and lower subtropical zone on both slopes of the Andes (400–1500 m), in the lowlands away from the Andes occurring mainly as a transient; however, one was collected at Manglares-Churute Ecological Reserve in se. Guayas on 26 Jan. 1991 (MECN).

Although this species has traditionally been called the Olive-sided Flycatcher, the name of Boreal Pewee was adopted by Sibley and Monroe (1990).

Formerly placed in the genus *Nuttallornis*. The specific name for this species has long been considered to be *borealis*. However, R. Banks and R. Browning (*Auk* 112: 633–648, 1995), in their review of the status of old bird names for various North American species, determined that the oldest specific name for the species was *cooperi*; the 1998 AOU Check-list followed this. Monotypic.

Range: breeds in n. and w. North America, wintering south to n. and w. South America.

Empidonax virescens   Acadian Flycatcher   Mosquerito Verdoso

A rare to locally fairly common (but inconspicuous and easily overlooked except by call) boreal winter resident in the lower growth of humid forest, secondary woodland, borders, and plantations (especially of cacao) in the lowlands and foothills of w. Ecuador. Most recent reports are from Pichincha northward, but there is an old record from as far south as e. Guayas (Chimbo), and recent sightings from Manta Real in nw. Azuay on 26 Jan. 1991 (RSR), and two birds were noted in the Río Ayampe area of sw. Manabí and nw. Guayas on 30 Jan. and 1 Feb. 1996 (P. Coopmans). There also is a Dec. 1968 specimen (MECN) from Guayquichuma in extreme n. Loja, the southernmost report on record. Recorded from Oct. to Mar. The Acadian Flycatcher reaches the southern limit of its wintering range in Ecuador. Recorded up to 1500 m at Guayquichuma; we presume that a specimen (MECN) labeled as having been taken at Yanacocha (3400 m) must be incorrect.

Monotypic.

Range: breeds in e. North America, wintering south to w. Ecuador.

Empidonax alnorum   Alder Flycatcher   Mosquerito de Alisos

Uncertain because of confusion with the following species; definite identification in the field requires hearing a bird giving its song or call. The Alder

Flycatcher is probably an uncommon transient and boreal winter resident in the lowlands of e. Ecuador, occurring in the same habitats as the Willow Flycatcher (*E. traillii*). Definite reports known to us are few: four birds were mist-netted and collected (ANSP, MECN) in young riparian growth on an island in the Río Napo near the mouth of the Río Aguarico on 20–21 Sep. 1992; one was heard and seen at Zancudococha on 25 Sep. 1976 (RSR); one was heard and seen near La Selva on 9 Jan. 1991 (R. Behrstock et al.); and individuals were seen and heard on various dates at and near Jatun Sacha (RSR; B. Bochan et al.). Recorded from Sep. to mid-Apr. So far recorded only below about 450 m, though likely occurring higher as a transient.

    E. alnorum is considered a species separate from the very similar *E. traillii*, following all recent authorities. The English name of Traill's Flycatcher was formerly used for the expanded species, and still often is when the two species cannot be distinguished. Monotypic.

    Range: breeds in n. North America, wintering south to w. South America.

Empidonax traillii   Willow Flycatcher   Mosquerito de Sauces

A rare boreal winter resident in shrubby clearings, overgrown pastures, and lighter woodland in the lowlands of ne. Ecuador near the base of the Andes. Despite the doubts expressed by the 1998 AOU Check-list, there are a number of reports of singing ("fitz-bew") Willow Flycatchers from various localities in w. Napo; these include several in the Lago Agrio/Coca area on 18–19 Feb. 1985 (RSR and T. Schulenberg); single birds heard near Tena on 14 Mar. 1989 and 17 Mar. 1990 (RSR and PJG); and single birds at Jatun Sacha on 12 Apr. 1993 (RSR) and 26 Oct. 1993 (B. Bochan). Recorded from Oct. to Apr. An MECN specimen from Cotapino is recorded as having been taken on 8 Sep., but this specimen cannot be definitely identified as either a Willow Flycatcher or an Alder Flycatcher (*E. alnorum*). Recorded below about 500 m, though possibly occurring higher as a transient; a nonvocalizing "Traill's" Flycatcher (either a Willow or Alder) was seen on Páramo del Ángel, Imbabura (3500 m), on 13 Nov. 1990 (N. Krabbe and J. Sterling).

    E. traillii has several described subspecies, all of them extremely similar and not likely identifiable on their wintering grounds; the eastern-breeding nominate race and *campestris* would seem to be the most likely to occur in Ecuador.

    Range: breeds in North America, wintering south to nw. South America.

Lathrotriccus euleri   Euler's Flycatcher   Mosquerito de Euler

Rare to uncommon and local (probably overlooked) in the lower growth of humid forest (both várzea and terra firme, but primarily the latter), secondary woodland (locally in stands of bamboo), and borders in the lowlands of e. Ecuador, ranging up at least locally onto the lower slopes of the Andes. The Euler's Flycatcher does not seem to be found in the lowlands well away from the Andes, the farthest east site from which it is known being La Selva. Recorded up to 1300 m (a pair seen north of Archidona along the road to Loreto on 7 Apr. 1991; RSR and PJG).

Formerly placed in the genus *Empidonax*; see W. E. Lanyon and S. Lanyon (*Auk* 103[2]: 341–350, 1986) for the erection of the genus *Lathrotriccus*. We follow J. T. Zimmer (*Am. Mus. Novitates* 1042, 1939) in placing Ecuadorian examples of *L. euleri* in the race *bolivianus*.

Range: Venezuela and e. Colombia south to n. Argentina, Uruguay, and s. Brazil (but absent from a broad area from the Guianas west through n. Amazonia); during austral winter withdraws from southern part of breeding range, but this population apparently does not reach Ecuador.

Lathrotriccus griseipectus	Gray-breasted Flycatcher	Mosquerito Pechigris

Rare to locally fairly common in the lower growth of humid and deciduous forest and woodland in the lowlands and subtropical zone on the west slope of the Andes in w. Ecuador from w. Esmeraldas (Río Verde; MCZ) and Pichincha (including recent reports from Río Palenque; there are also some recent reports [P. Coopmans and others] from Tinalandia, perhaps all of the same territorial individual that no longer appears to be present) south locally through El Oro and Loja. In addition, a single specimen (BMNH) is labeled as having been taken in Jan. 1914 at "Mindo, 1830 m," though whether it was actually obtained at so high an elevation in such a wet region is unknown (Collar et al. 1992; see also Kirwan et al. 1996). The Gray-breasted Flycatcher is apparently also rare in the Río Marañón drainage of s. Zamora-Chinchipe near Zumba, where a few were found in the drainage of the lower Río Isimanchi in Dec. 1991 (RSR). Recorded mainly below about 1700 m, but ANSP has a 1948 specimen recorded as having been taken at Loja (city; 2200 m); most numerous below about 800 m.

From the number of specimens that were collected, the nearly endemic Gray-breasted Flycatcher seems once to have been a reasonably numerous bird in w. Ecuador. In recent years it has been encountered at relatively few sites, however, and it clearly is threatened by the widespread deforestation that has taken place over the past few decades. Collar et al. (1992, 1994) accorded it Vulnerable status, and we agree. The largest extant populations may be found in Machalilla National Park in sw. Manabí, where the species was reasonably common in Aug. 1991 (ANSP), especially around Cerro San Sebastián, and at Cerro Blanco in the Chongón Hills of Guayas; it is also numerous in the small Jauneche Reserve in Los Ríos (P. Coopmans).

Formerly placed in the genus *Empidonax*, but given this species' close morphological, vocal, and behavioral similarity to *Lathrotriccus euleri* (Euler's Flycatcher), it seems clear that *griseipectus* should also be placed in that genus; see Ridgely and Tudor (1994) and Parker et al. (1995). Monotypic.

Range: w. Ecuador and nw. Peru.

Cnemotriccus fuscatus   Fuscous Flycatcher   Mosquerito Fusco

Uncommon to locally fairly common (but always inconspicuous) inside the lower growth of riparian woodland and shrubby areas on river islands in the lowlands of e. Ecuador. First seen by J. A. Rowlett on an island in the Río Napo near Primavera in Jul. 1984, the Fuscous Flycatcher has subsequently been found by numerous observers, both in that area and at various points down-stream in the Río Napo; small numbers have also been noted as far up the Río Napo as the vicinity of Jatun Sacha (B. Bochan). The first Ecuadorian speci-mens (ANSP, MECN) were obtained in Sep. 1992 on an island in the Río Napo near the mouth of the Río Aguarico. The Fuscous Flycatcher has also recently been found on islands in the Río Pastaza in the vicinity of Kapawi Lodge near the Peruvian border, and it may range widely along that river as well. Recorded below about 400 m.

One race: *fuscatior*.

Range: n. and e. Colombia, Venezuela, and the Guianas south to n. Argentina, Paraguay, and se. Brazil. More than one species may be involved.

Sayornis nigricans   Black Phoebe   Febe Guardarríos

Fairly common and conspicuous in semiopen areas, often near human habita-tion, along streams and rivers in the foothills and subtropical zone on both slopes of the Andes, and in the central and interandean valleys. Recorded mostly from 500 to 2800 m, locally and in small numbers descending lower where suit-able conditions exist near the base of the Andes, both in the west (as low as just below 100 m at Playa de Oro in Esmeraldas [O. Jahn et al.] and down to 200 m east of La Avanzada in El Oro and east of Macará in sw. Loja [RSR et al.]) and the east (locally as low as 400 m near Jatun Sacha; B. Bochan). A spec-imen recorded from Esmeraldas (city) by Chapman (1926) was presumably taken in the hills inland from that city and not actually at sea level.

One race: *angustirostris*.

Range: sw. United States south through Middle America to mountains of n. Venezuela and Andes from w. Venezuela to nw. Argentina.

Pyrocephalus rubinus   Vermilion Flycatcher   Mosquero Bermellón

Fairly common to common and conspicuous in open areas with scattered trees and bushes, mostly in more arid regions, in the lowlands of w. Ecuador from w. Esmeraldas southward (including Isla de la Plata), and in the central and interandean valleys from Carchi (Río Mira valley) southward. In e. Ecuador known only as a rare austral migrant, with only three known records, the first two from s. Morona-Santiago: a male collected by T. Maxwell at Santiago on 2 Aug. 1989 (ANSP); a female seen sallying from the forest canopy along a small stream near Taisha on 9 Aug. 1990 (RSR); and a male seen at Cuyabeno in late Jul. 1991 (J. Rowlett and R. A. Rowlett et al.). The species was recorded from "eastern Ecuador" in *Birds of the World* (vol. 8, p. 152) though no details were given. The Vermilion Flycatcher is numerous and widespread in the Río Marañón drainage of n. Peru (including the Jaen/San Ignacio area), and it thus seems surprising that it has not (yet?) been found near Zumba. Ranges up to at least 3000 m in the highlands.

Two races occur in Ecuador, *piurae* being a widespread resident in the west and nominate *rubinus* a rare austral migrant to the east. Males are very similar, but females of the nominate race tend to have no pink on the belly (the pink is often extensive in *piurae*), or they may even have the lower belly tinged with pale yellow; they are also more extensively streaked on their underparts (extending down over the flanks, in some birds over the entire belly).

Range: sw. United States south, especially in more arid regions, through Middle America and mostly w. South America to cen. Argentina and Uruguay; during austral winter southern breeders migrate north into Amazonia; Galápagos Islands (probably not conspecific).

Ochthoeca fumicolor   Brown-backed Chat-Tyrant   Pitajo Dorsipardo

Fairly common to common and usually conspicuous in semiopen shrubby areas and borders of woodland patches (including groves of *Polylepis*) from the upper temperate zone on both slopes of the Andes up into shrubby paramo above treeline, and also on slopes above the central and interandean valleys. On the west slope recorded south to e. El Oro (Taraguacocha), but not recorded from w. Loja. Recorded mostly from 2800 to 4200 m.

One race: *brunneifrons*.

Range: Andes from w. Venezuela to w. Bolivia. More than one species may be involved.

**Ochthoeca leucophrys*   White-browed Chat-Tyrant   Pitajo Cejiblanco

Apparently rare and local in montane scrub and woodland in Azuay and adjacent n. Loja. Only recently found in Ecuador, the White-browed Chat-Tyrant was first reported from a single bird observed south of Oña in s. Azuay on 23 Jan. 1988 (R. A. Rowlett et al.). Single birds have been seen and heard in this area on a few subsequent occasions, and one was collected (MECN) by N. Krabbe and P. Coopmans on 24 Oct. 1994 just across the border into Loja. In addition, a pair of chat-tyrants seen by several observers (first by P. Coopmans in Nov. 1988; also by C. Green on 27 Dec. 1992) farther north in Azuay around the Hostería Uzhupud near Gualaceo are now believed to have referred to the White-browed Chat-Tyrant, though there was some debate concerning the possibility that they could have been Jelski's Chat-Tyrants (*O. jelskii*); in any case, in more recent years (since 1994) the birds have not been found here. The White-browed Chat-Tyrant had heretofore not been definitely recorded north of n. Peru in Cajamarca, though possible sightings were also apparently made in the Celica region (Fjeldså and Krabbe 1990); we wonder whether these too may have actually been Jelski's Chat-Tyrants. Recorded from about 2200 to 2800 m.

The Ecuadorian specimen appears to be referable to the race *dissors* of n. Peru (fide N. Krabbe).

Range: Andes of s. Ecuador, and from n. Peru to n. Chile and nw. Argentina.

Ochthoeca rufipectoralis	Rufous-breasted Chat-Tyrant	Pitajo Pechirrufo

Uncommon to fairly common at borders of montane forest, secondary woodland, and in adjacent clearings in the temperate zone on both slopes of the Andes, in some areas ranging up to near treeline; also locally on slopes above central and interandean valleys (e.g., at Pasochoa and in Azuay). On the west slope recorded south to e. El Oro (Taraguacocha); not recorded from w. Loja. Less numerous overall on the west slope. Recorded mostly from 2500 to 3300 m, locally a little higher.

One race: *obfuscata*.

Range: Andes from n. Colombia to w. Bolivia; Sierra de Perijá and Santa Marta Mountains.

Ochthoeca cinnamomeiventris	Slaty-backed Chat-Tyrant	Pitajo Dorsipizarro

Fairly common but inconspicuous in the lower growth of montane forest and mature secondary woodland, almost invariably near water, in the upper subtropical and temperate zones on both slopes of the Andes. On the west slope recorded south to Chimborazo (an old specimen from Pagma), but in recent years recorded south only into w. Cotopaxi. Recorded mostly from 1700 to 2800 m, locally up in small numbers to about 3000 m, exceptionally as high as 3300 m (Cordillera Las Lagunillas; Krabbe et al. 1997).

One race: nominate *cinnamomeiventris*.

Range: Andes from n. Colombia to extreme n. Peru. Excludes Blackish Chat-Tyrant (*O. nigrita*) of Venezuelan Andes and Maroon-belted Chat-Tyrant (*O. thoracica*, with *angustifasciata*) of east slope of Andes from n. Peru to Bolivia; see J. García-Moreno, P. Arctander, and J. Fjeldså (*Condor* 100[4]: 629–640, 1998).

Ochthoeca frontalis	Crowned Chat-Tyrant	Pitajo Coronado

Uncommon and inconspicuous in the undergrowth of montane forest borders and stunted woodland, especially just below and at treeline, in the temperate zone on both slopes of the Andes, and also locally on slopes above interandean valleys (perhaps especially in Azuay; at least known from El Cajas and Río Mazan). On the west slope recorded south to w. Bolívar (Salinas) and w. Azuay (Chaucha; N. Krabbe et al.). Recorded from 2800 to 4000 m, locally somewhat lower.

This and the next two species were for a time separated in the genus *Silvicultrix*, following W. E. Lanyon (*Am. Mus. Novitates* 2846, 1986), but we follow J. García-Moreno, P. Arctander, and J. Fjeldså (*Condor* 100[4]: 629–640, 1998) in subsuming that genus into *Ochthoeca*. One race occurs in Ecuador, nominate *frontalis*. We follow M. A. Traylor, Jr. (*Neotropical Ornithology*, Ornithol. Monogr. no. 36: 430–442, 1985) in considering the extremely similar race *orientalis*, to which Ecuadorian birds were formerly ascribed, as a synonym of the nominate race.

Range: Andes from n. Colombia to cen. Peru. Excludes Kalinowski's Chat-Tyrant (*O. spodionota*) of east slope of Andes from cen. Peru to w. Bolivia; see García-Moreno et al. (op. cit.).

Ochthoeca jelskii   Jelski's Chat-Tyrant   Pitajo de Jelski

Rare to uncommon and local in the undergrowth of montane forest borders and woodland in the subtropical and temperate zones on the west slope of the Andes in s. Ecuador, where thus far known only from Loja; an inconspicuous bird, the Jelski's Chat-Tyrant has perhaps to some extent been overlooked. Sites from which it is known include the Celica/Guachanamá area (though, surprisingly, there seem to be no recent reports from here); Cajanuma Divide (MCZ specimen taken by D. Norton on 2 Oct. 1965; also found on two occasions in Nov.–Dec. 1991 along the entrance road to Podocarpus National Park [Rasmussen et al. 1996]); the Sozoranga, Cariamanga, and Utuana regions (Best et al. 1993; also others), where it seems to be more numerous and widespread than elsewhere in Ecuador; and above Jimbura on the west side of the Cordillera Las Lagunillas (seen in Dec. 1991; RSR). Recorded mostly from 2200 to 2800 m.

O. *jelskii* is regarded as a species distinct from *O. pulchella* (Golden-browed Chat-Tyrant) of Peru and Bolivia, following M. E. Traylor, Jr. (*Neotropical Ornithology*, Ornithol. Monogr. no. 36: 430–442, 1985). Alternatively it has been treated as a subspecies of *O. frontalis* (Crowned Chat-Tyrant; Fjeldså and Krabbe 1990). Monotypic.

Range: Andes of s. Ecuador and nw. Peru.

Ochthoeca diadema   Yellow-bellied   Pitajo Ventriamarillo
                        Chat-Tyrant

Uncommon and inconspicuous in the undergrowth of montane forest and secondary woodland and their borders in the upper subtropical and temperate zones on both slopes of the Andes. On the west slope recorded south to w. Cotopaxi and locally as far south as w. Azuay (Sural; N. Krabbe). The Yellow-bellied Chat-Tyrant was found on the upper slopes of the Cordillera de Cutucú (Robbins et al. 1987), though it appears—somewhat surprisingly—to be absent from the Cordillera del Cóndor. Ranges mostly from 2200 to 3100 m, thus mainly below the elevational range of the behaviorally similar Crowned Chat-Tyrant (*O. frontalis*).

One race: *gratiosa*.

Range: mountains of n. Venezuela, and Andes from w. Venezuela to n. Peru; Sierra de Perijá and Santa Marta Mountains.

Ochthornis littoralis   Drab Water-Tyrant   Guardarríos Arenisco

Fairly common and conspicuous along the larger rivers in the lowlands of e. Ecuador. The Drab Water-Tyrant favors places where eroding banks have exposed branches and roots, but it also occurs on sandbars, especially near piles of accumulated driftwood. Recorded only below about 400 m.

Sometimes placed in the genus *Ochthoeca* (e.g., *Birds of the World*, vol. 8), but this species' very different behavior and its lowland range argue otherwise, and we therefore maintain it as a monotypic genus (see Ridgely and Tudor 1994). Monotypic.

Range: se. Colombia, s. Venezuela, and the Guianas to n. Bolivia and w. and cen. Amazonian Brazil.

Cnemarchus erythropygius   Red-rumped   Alinaranja
                              Bush-Tyrant   Lomirrojiza

Rare to uncommon but conspicuous in open, usually grassy, areas with scattered shrubs and small trees, and around patches of woodland (including *Polylepis* groves), in the temperate and paramo zones on both slopes of the Andes and on interandean ridges. On the west slope recorded south to Azuay at El Cajas National Recreation Area. Recorded from about 2850 to 4100 m.

Formerly placed in the genus *Myiotheretes*; see W. E. Lanyon (*Am. Mus. Novitates* 2846, 1986) for the resurrection of the monotypic genus *Cnemarchus* for this species. One race occurs in Ecuador, nominate *erythropygius*.

Range: locally in Andes from n. Colombia to w. Bolivia; Santa Marta Mountains.

Myiotheretes striaticollis   Streak-throated   Alinaranja Golilistada
                               Bush-Tyrant

Uncommon but conspicuous in semiopen shrubby or grassy areas, and at borders of montane forest and woodland, in the temperate zone on both slopes of the Andes, and in the central and interandean valleys. Recorded mostly from about 2400 to 3200 m, exceptionally as low as 1700–1800 m near Alamor (Best et al. 1993).

One race: nominate *striaticollis*.

Range: Andes from w. Venezuela to nw. Argentina; Sierra de Perijá and Santa Marta Mountains.

Myiotheretes fumigatus   Smoky Bush-Tyrant   Alinaranja Ahumada

Uncommon in the middle growth and subcanopy of montane forest and borders in the upper subtropical and temperate zones on both slopes of the Andes, and locally on slopes above interandean valleys. On the west slope not recorded south of the Cañar/Bolívar border area. Recorded between about 2000 and 3200 m.

Two similar races are recorded from Ecuador, nominate *fumigatus* in the north and *cajamarcae* in the southeast (north to Cañar).

Range: Andes from w. Venezuela to s. Peru; Sierra de Perijá.

**Hirundinea ferruginea*   Cliff Flycatcher   Tirano de Riscos

Uncommon and local around cliffs and rocky roadcuts in the foothills and subtropical zone on the east slope of the Andes. First recorded in Ecuador only in 1978, from sightings near San Rafael Falls in w. Napo and along the Loja-Zamora road in Zamora-Chinchipe (see Ridgely 1980). The Cliff Flycatcher has since been recognizably photographed, and several specimens have been collected. The first specimen (WFVZ) was obtained north of Archidona along the Loreto road on 17 Oct. 1987, with others (WFVZ, MECN) having been taken near there on 22 Nov. 1991 (see Marín 1992); the species continues to be seen along that road. Additional specimens (ANSP, MECN) were obtained above Zamora in Jan. 1992 and near Las Palmas in w. Napo on 29 Aug. 1993. The species was also seen near Zumba on 6 Apr. 1994 (P. Coopmans and M. Lysinger) and at Miazi in the Río Nangaritza valley (Schulenberg and Awbrey 1997). The Cliff Flycatcher is probably increasing in Ecuador as a result of newly constructed roads, which create extensive "artificial" habitat. It remains puzzling that the species has yet to be located between w. Napo and s. Zamora-Chinchipe, but we suspect that it will eventually be found to range more or less continuously where suitable habitat exists. Recorded from about 900 to 1700 m.

It has been suggested (e.g., Sibley and Monroe 1990) that it may be more appropriate to recognize two species of cliff-flycatcher, *H. ferruginea* (Northern Cliff-Flycatcher) and *H. bellicosa* (Southern Cliff-Flycatcher), but we conclude that the evidence to do so is not persuasive. Ecuadorian birds would belong with the former. One race of *H. ferruginea* occurs in Ecuador, *sclateri*.

Range: Andes from w. Venezuela to w. Argentina (rather local northward); tepuis of e. Colombia, s. Venezuela, and the Guianas; e. Bolivia and s. and e. Brazil south to e. Paraguay, ne. Argentina, and Uruguay.

Agriornis montana   Black-billed Shrike-Tyrant   Arriero Piquinegro

Uncommon but conspicuous in paramo and in the upper temperate zone in open grassy and agricultural areas with a few scattered bushes and boulders, sometimes around buildings, in the Andes. Recorded mostly from 3000 to 4000 m, but ranges lower (regularly down to 2500 m, locally down to 2000 m) in Azuay and Loja.

One race: *solitaria*.

Range: Andes from s. Colombia to cen. Chile and s. Argentina (in Argentina also in lowlands).

Agriornis andicola   White-tailed Shrike-Tyrant   Arriero Coliblanco

Very rare and local in paramo and in grassy, sparsely vegetated shrubby areas in the temperate zone of the Andes from Imbabura (an early-20th-century record near Laguna Yaguarcocha; Lönnberg and Rendahl 1922) and Pichincha (a few old specimens; only one recent report, a pair seen and tape-recorded on several occasions from Nov. 1999 onward in montane scrub west of Calacalí (discovered by F. Sornoza) southward. Recorded from 2400 m (Cajanuma Divide in Loja; MCZ) to 3100 m.

The White-tailed Shrike-Tyrant has declined in recent decades, in Ecuador as elsewhere in its Andean range; the reasons for its apparently precipitous drop in numbers remain unknown. Most of the few recent Ecuadorian reports come from s. Azuay and n. Loja, whence there is a specimen (MCZ) taken at Cajanuma Divide by D. Norton on 13 Aug. 1965, a sighting of a single bird south of Saraguro on 21 Jul. 1978 (RSR and D. Wilcove), a specimen obtained by N. Krabbe at Bestión on 2 Nov. 1992 (ANSP), and apparently nesting birds seen by N. Krabbe in s. Azuay along the Cuenca-Loja road near La Paz on 12 Feb. 1995 and in n. Loja near Saraguro on the Selva Alegre road on 13 Feb. 1995. In the mid 19th century the White-tailed Shrike-Tyrant was described as "common on the paramo" near Quito, but we (and N. Krabbe) suspect that comments such as this may in part be the result of confusion with the similar Black-billed Shrike-Tyrant (*A. montana*). N. Krabbe (*Cotinga* 1: 33–34, 1994) discussed some of the parameters that may be governing the White-tailed Shrike-Tyrant's present distribution and declining numbers. None of its habitat is formally protected in Ecuador. The species was accorded Vulnerable status by Collar et al. (1992, 1994), and that seems appropriate in Ecuador as well.

The species was formerly called *Agriornis albicauda*. One race occurs in Ecuador, nominate *andicola*.

Range: very locally in Andes from n. Ecuador to n. Chile and nw. Argentina.

Muscisaxicola alpina   Paramo   Dormilona del Páramo
                                Ground-Tyrant

Fairly common to common in paramo (perhaps most numerous where it is relatively arid) in the Andes of n. and cen. Ecuador, apparently ranging no farther south than Azuay at El Cajas National Recreation Area. In Oct.–Nov. 1992 the Paramo Ground-Tyrant was found to be absent from seemingly ideal habitat in the Cordillera Las Lagunillas (3300–3600 m) in extreme se. Loja and adjacent Zamora-Chinchipe, this despite its having been found just to the south at Cerro Chinguela in adjacent Peru (Parker et al. 1985). The Paramo Ground-Tyrant occurs mainly above the elevational range of its congeners, and indeed it is one of the highest-ranging Ecuadorian birds. It is especially numerous at higher elevations in Cotopaxi National Park, where pairs and small groups are readily found along the road to the mountaineering refugio. Recorded mostly from 3800 to 4600 m, but seemingly absent from paramo at slightly lower elevations (this appears especially to be the case if elevations at that site do not reach at least 4000 m); occasionally bad weather forces birds somewhat lower (to about 3500 m).

The birds of the Colombian and Ecuadorian Andes differ from the population (*grisea*) found in the Peruvian and Bolivian Andes to about the same degree as *grisea* differs from the recently specifically separated *M. cinerea* (Cinereous Ground-Tyrant) of the Andes from s. Peru to Chile and Argentina (*Birds of the World*, vol. 8; Ridgely and Tudor 1994). R. T. Chesser (*Molec. Phylogen. and Evol.* 15[3]: 369–380, 2000) presented data demonstrating that *M. grisea* (Plain-capped Ground-Tyrant) should be treated as a separate species. One race occurs in Ecuador, nominate *alpina*.

Range: Andes from n. Colombia to extreme n. Peru.

Muscisaxicola albilora	White-browed Ground-Tyrant	Dormilona Cejiblanca

An uncommon austral winter visitant to paramo, pastures, and fields in the temperate zone in the Andes north to Pichincha on Volcán Pichincha. To date there are records mainly for the Jun.–Aug. period, but the species may be present in Ecuador somewhat longer; one anomalous record involved a sighting of a single bird near Alto Tambo in e. Esmeraldas (650 m) on 24 Nov. 1994 (N. Krabbe). Pichincha is at the northern limit of the species' normal wintering range, but there is also a record of a vagrant from Gorgona Island off sw. Colombia (B. Ortiz-Von Halle, *Caldasia* 77: 209–214, 1990). Recorded mainly from 2400 to 3700 m, occasionally as high as 4000 m.

Monotypic.

Range: breeds in Andes of cen. and s. Chile and Argentina, wintering north in Andes to n. Ecuador, straggling to Gorgona Island off sw. Colombia.

Muscisaxicola maculirostris	Spot-billed Ground-Tyrant	Dormilona Piquipinta

Uncommon to locally fairly common in open, usually arid, regions with sparse vegetation, and in predominantly agricultural terrain in the temperate zone of the central and interandean valleys and on the slopes above it (but not on the outside slopes of the Andes) from Pichincha south to Azuay (an old specimen from Hacienda El Paso near Nabón; Chapman 1926); there are no recent reports from south of Chimborazo. The absence of the Spot-billed Ground-Tyrant from Loja, where much terrain appears to be ideal for the species, seems inexplicable—but is apparently real, for the species is also not known in nw. Peru north of s. Cajamarca (the nominate race). The species' absence from northernmost Ecuador (Imbabura, etc.), where much terrain likewise would appear to be suitable, is almost equally puzzling, but it is also absent from adjacent Colombia (Hilty and Brown 1986). Recorded mostly from 2400 to 3500 m, occasionally up to about 3800 m.

One race: the endemic *rufescens*.

Range: Andes in n. Colombia, and from n. Ecuador south to s. Chile and Argentina.

Muscisaxicola fluviatilis Little Ground-Tyrant Dormilona Chica

Uncertain. Apparently a casual visitant to e. Ecuador. There are only three Ecuadorian records of the Little Ground-Tyrant, all of them recent. One was seen on a sandbar in the Río Napo upstream from La Selva on 1 Aug. 1990 (G. DeSmet, R. M. La Fontaine, and L. Ratty, fide P. Coopmans), and presumably the same bird was found independently there on 20 Aug. 1990 (S. Hilty et al.). Another was reported seen at the Hotel Auca near Tena on 24 Jul. 1990 (C. Green). A male (ANSP) was taken by F. Sornoza 5 km south of Palanda in Zamora-Chinchipe on 10 Jul. 1992. The Little Ground-Tyrant was heretofore not known from any closer to Ecuador than the Iquitos area of ne. Peru, and the Ecuadorian birds must be presumed to be wanderers from that population. Recorded up to 1150 m.

Monotypic.

Range: e. Peru, n. Bolivia, and sw. Amazonian Brazil; wanders (?) to e. Ecuador.

[Muscisaxicola macloviana Dark-faced Dormilona
        Ground-Tyrant Carinegruzca]

Presumably an accidental austral migrant. The Dark-faced Ground-Tyrant is known in Ecuador only from a single sighting of two birds 2 km west of Huaquillas in coastal s. El Oro on 2 Jan. 1989 (J. Sterling). Presumably these were unusually early overshoots from their normal wintering range.

Presumed race: *mentalis*. It is virtually certain that this is the subspecies of *M. macloviana* that was involved in the Ecuador sighting, for the apparently nonmigratory nominate race of the species is resident on the Falkland Islands.

Range: breeds in Andes of s. Chile and Argentina and on Falkland Islands, wintering north to Uruguay, cen. Argentina, and coastal w. Peru, straggling as far north as s. Ecuador.

Muscigralla brevicauda Short-tailed Field-Tyrant Tiranito Colicorto

Uncommon to locally fairly common in open barren areas, sometimes with a scattering of bushes or low trees, and in predominantly agricultural terrain in the more arid lowlands and foothills of sw. Ecuador from w. and n. Guayas (Santa Elena Peninsula and north to Balzar) and s. Los Ríos (Babahoyo) south mainly near the coast to coastal El Oro; also ranges in some arid interior valleys from sw. Azuay (Río Rircay; Best et al. 1993) south into w. Loja (as far east as around Catamayo and Vilcabamba). The Short-tailed Field-Tyrant also occurs on Isla de la Plata off s. Manabí, but curiously it appears to be absent from the adjacent Manabí mainland. It is especially numerous on the Santa Elena Peninsula. Ranges up to about 1500 m in arid intermontane valleys of sw. Azuay (Río Rircay) and Loja, but elsewhere it occurs much lower, with largest numbers along the coast.

This species is sometimes placed in the genus *Muscisaxicola*, but several considerations argue strongly for retaining it in its own genus (see W. E. Lanyon, *Am. Mus. Novitates* 2846, 1986; Ridgely and Tudor 1994). Monotypic.

Range: sw. Ecuador, w. Peru, and extreme n. Chile.

Knipolegus poecilurus   **Rufous-tailed Tyrant**   **Viudita Colicolorada**

Rare to uncommon and local at borders of montane forest and in adjacent shrubby clearings in the foothills and subtropical zone on the east slope of the Andes. Thus far recorded only from the north in w. Sucumbíos (San Rafael Falls and Bermejo oil-field area) and w. Napo (Baeza), and from the south in Zamora-Chinchipe (Zamora, Sabanilla, and above Chinapinza in the Cordillera del Cóndor), but presumably the species occurs in the intervening area as well. As the Rufous-tailed Tyrant is reported to be "common" at La Planada Reserve on the west slope of the Andes in adjacent Nariño, Colombia (Salaman 1994; also fide P. Coopmans), it seems surprising that it has yet to be found on the Andes' west slope in extreme n. Ecuador. We remain uncertain why this species remains so scarce in Ecuador, in particular as one might have expected that the second-growth habitat now so prevalent would by now have resulted in an increase in numbers. Recorded mostly from 1000 to 2000 m, locally down to 650 m in the Bermejo oil-field area north of Lumbaquí (ANSP).

One race: *peruanus*. A recent specimen (ANSP) from near the Colombian border (Bermejo) shows no approach to the nominate race of Colombia, which has a paler rufous belly.

Range: locally in mountains of n. Venezuela, and in Andes from w. Venezuela to s. Peru; Sierra de Perijá; tepuis of s. Venezuela and adjacent Brazil.

It seems virtually certain that the Andean Tyrant (*K. signatus*) occurs in Ecuador on the Cordillera del Cóndor, though up to now the species has only been recorded from the Peruvian side of the range (e.g., in the drainage of the Río Comainas; Schulenberg and Awbrey 1997). See Ridgely and Tudor (1994) for the complex systematic and nomenclatural history of this species.

* *Knipolegus orenocensis*   **Riverside Tyrant**   **Viudita Ribereña**

Rare and apparently local in early-succession growth on river islands (especially in stands of *Tessaria*) in the lowlands of ne. Ecuador. The Riverside Tyrant was first recorded in Ecuador on an island in the Río Napo near the mouth of the Río Aguarico, where a family group was found on 20–22 Sep. 1992 (ANSP). There are a few subsequent records, including displaying males tape-recorded on an island in the Río Napo near La Selva in Sep. 1994 (G. Rivadeneira, fide J. Moore) and near Añangu on 30 Jan. 1996 (N. Krabbe). The species has not been found along the Río Pastaza. Recorded below 250 m.

One race: *sclateri*.

Range: locally along Amazon River and some of its major tributaries in Brazil, se. Colombia, ne. Peru, and ne. Ecuador; along Orinoco River in Venezuela and ne. Colombia. More than one species is probably involved.

Knipolegus poecilocercus	Amazonian Black-Tyrant	Viudita Negra Amazónica

Rare, local, and inconspicuous (easily overlooked) in the tangled lower growth of blackwater várzea woodland and thickets in extreme ne. Ecuador. The Amazonian Black-Tyrant is known from only a single site in Ecuador, Imuyacocha near the Río Lagarto, where it was first found on 8 Dec. 1990 (ANSP and MECN); small numbers have been seen subsequently. It may occur westward in similar habitat (e.g., at Cuyabeno), but if so it has yet to be found. 200 m.

The species was formerly placed in the genus *Phaeotriccus*; see *Birds of the World*, vol. 8, and Ridgely and Tudor (1994). Monotypic.

Range: locally in e. Colombia, s. Venezuela, Guyana, ne. Ecuador, ne. Peru, and Amazonian Brazil.

Colonia colonus	Long-tailed Tyrant	Tirano Colilargo

Uncommon to locally common (but always conspicuous) in the borders of humid forest and in adjacent clearings, especially where there are numerous dead snags, in the lowlands of both e. and w. Ecuador, in the east favoring the edge of hilly terra firme forest and secondary woodland. In the west found only in more humid regions and ranging south to n. Manabí, n. Guayas (Pacaritambo), and n. Los Ríos (two 1950 specimens from Quevedo; ANSP). There also are two specimens (MECN) labeled as having been taken on 23–24 Dec. 1969 by M. Olalla much farther south, at Guaiquichuma in extreme n. Loja (southeast of Portoviejo in El Oro); this locality seems so anomalous that we are inclined to think that it likely is not accurate. In the east recorded mainly close to the Andes, though a few individuals have been seen as far east as La Selva (especially south of the Río Napo), and the species is quite numerous along the Maxus road southeast of Pompeya (RSR et al.); it also has been seen around Kapawi Lodge on the Río Pastaza near the Peruvian border. Long-tailed Tyrants seem most numerous in the southeast (e.g., around Zamora and in the Río Nangaritza valley). Recorded mostly below about 1100 m, locally as high as 1575 m at Panguri in s. Zamora-Chinchipe (ANSP); in the eastern lowlands scarce and local below 400 m.

Three races are found in Ecuador, *leuconotus* in the west and *fuscicapillus* in most of the east (south to the Cordillera del Cóndor; ANSP), with the similar *niveiceps* having recently been recorded in the extreme southeast at Panguri in s. Zamora-Chinchipe (ANSP). Trans-Andean *leuconotus* differs from the two eastern races in having much more mottled white on the back and rump.

Range: Honduras to nw. Ecuador; s. Venezuela and the Guianas; e.

Colombia, e. Ecuador, e. Peru, n. Bolivia, s. Amazonian and se. Brazil, e. Paraguay, and ne. Argentina.

Fluvicola nengeta   Masked Water-Tyrant   Tirano de Agua Enmascarado

Fairly common in open and semiopen shrubby areas, especially around marshes and ponds and along streams and rivers, in the lowlands of w. Ecuador from w. Esmeraldas (Atacames and Muisne areas, and around Quinindé), to which it seems only recently to have spread, and Pichincha south to coastal El Oro. Ranges up locally to 800 m (e.g., at Tinalandia), and has even recently (Oct. 1998) been reported (fide J. Lyons) as high as 1300 m at Mindo, but found mostly below 300 m.

One race: *atripennis*. It has been suggested that the two widely allopatric (but similar) populations presently treated as subspecies of *F. nengeta* represent separate species; if split, the Ecuadorian populations would be known as the Ecuadorian Water-Tyrant (*F. atripennis*).

Range: w. Ecuador and extreme nw. Peru; e. Brazil.

*[*Fluvicola pica*   Pied Water-Tyrant   Tirano de Agua Pinto]

An accidental wanderer to river shores in the lowlands of ne. Ecuador. There is only a single report, of a bird seen along the Río Napo near Misahuallí on 17 Jul. 1974 (C. Leck, *Am. Birds* 34[3]: 313, 1980). The observer (pers. comm. to RSR) believes the bird pertained to the form *pica*, which has never been recorded in e. Colombia south of s. Meta (Hilty and Brown 1986). 400 m.

F. albiventer (Black-backed Water-Tyrant) of e. and s. South America is regarded as a species separate from *F. pica* of n. South America because of its different plumage, vocalizations, and disjunct distribution; see Ridgely and Tudor (1994). *F. albiventer* is known to be an austral migrant as far north as ne. Peru.

Range: e. Panama, n. Colombia, Venezuela, the Guianas, and n. Brazil; accidental in e. Ecuador.

*[*Arundinicola leucocephala*   White-headed   Tirano de Ciénega
                                   Marsh-Tyrant   Cabecialbo]

Apparently a casual wanderer to grassy areas and early-successional growth on river islands in the lowlands of ne. Ecuador; thus far recorded only along the Río Napo. The first Ecuadorian record involved a male seen on 1 Aug. 1990 on an open Río Napo island covered with short grass upstream from La Selva (G. DeSmet, R. M. La Fontaine, and L. Ratty, fide P. Coopmans). A female was reported downstream from Añangu on 29 Dec. 1995 (N. Krabbe). Presumably these represent wandering, nonterritorial individuals from ne. Peru where, somewhat surprisingly, the species seems not to have been recorded along the Río Napo upstream from near its mouth into the Amazon River. Recorded below 300 m.

The genus *Arundinicola* is sometimes merged into *Fluvicola* (e.g., *Birds of the World*, vol. 8), but behaviorally and in plumage it is quite different (being, inter alia, sexually dimorphic); see W. E. Lanyon (*Am. Mus. Novitates* 2846, 1986). Monotypic.

Range: n. and e. Colombia, Venezuela, and the Guianas to n. Argentina, Paraguay, and s. Brazil (but absent from much of w. and cen. Amazonia).

Attila spadiceus   Bright-rumped Attila   Atila Polimorfo

Uncommon to locally fairly common in the mid-levels, subcanopy, and borders of terra firme forest in the lowlands of e. Ecuador, and in humid forest, borders, and adjacent secondary woodland and plantations in the lowlands and foothills of nw. Ecuador south to n. Manabí and s. Pichincha (Río Palenque), with one record from s. Manabí (Isla Silva; E. Stresemann, *Ornith. Monatsber.* 46: 115–118, 1938). The Bright-rumped Attila is more numerous in the west, at least where forest remains, than it ever seems to be in the east. Recorded mostly below about 1300 m, rarely as high as 2000 m in Pichincha.

Two similar races are found in Ecuador, nominate *spadiceus* in the east and *parambae* in the west. Substantial individual plumage variation occurs in both races but is especially frequent in the nominate race; in addition to the usual olive morph, comparatively scarce gray and rufous morphs are known. The rufous morph has been reported as the most frequent at Jatun Sacha (B. Bochan), which based on specimen collections is most unusual.

Range: w. Mexico to nw. Ecuador, n. Bolivia, and Amazonian Brazil; e. Brazil.

Attila citriniventris   Citron-bellied Attila   Atila Ventricitrino

Rare to uncommon and local in the subcanopy of terra firme forest in the lowlands of e. Ecuador; perhaps overlooked, as except for its loud voice an inconspicuous bird. There are only a few confirmed Ecuadorian records, including specimens taken by R. Olalla at Chichirota in Pastaza on 12 Dec. 1948 (ANSP) and two others taken at Zancudococha in e. Napo on 2 Dec. 1990 and 7 Apr. 1991 (ANSP and MECN). The species has also been found recently at Cuyabeno, Pañacocha, and near the Río Pacuyacu; other records remain, in our view, unconfirmed. The ecological relationship of the Citron-bellied Attila with the generally more numerous and widespread Bright-rumped Attila (*A. spadiceus*) remains to be clarified. Recorded below about 300 m.

Monotypic.

Range: locally in sw. Venezuela, extreme e. Colombia, nw. Amazonian Brazil, ne. Peru, and e. Ecuador.

*[Attila bolivianus   White-eyed Attila   Atila Ojiblanco]

Uncertain. Known in Ecuador from a single sighting, a nonvocalizing bird seen in várzea forest at Sacha Lodge on 7 Jun. 1995 (S. Howell and S. Webb). Presumably a wandering bird; the closest area where the White-eyed Attila normally is found is in the lower Río Napo region of ne. Peru. This puzzling record represents the only report of the species from anywhere north of the Río Marañón. Recorded at about 250 m.

The species was formerly called the Dull-capped Attila; see Ridgely and Tudor (1994).

Presumed race: *nattereri*.

Range: ne. Ecuador (one report), e. Peru, extreme se. Colombia, n. and e. Bolivia, and s. Amazonian Brazil.

Attila cinnamomeus　　Cinnamon Attila　　Atila Canelo

Locally fairly common in the subcanopy and mid-levels of várzea forest and in forest and woodland along the margins of lakes and rivers in the lowlands of e. Ecuador, but not ranging close to the base of the Andes; it has, for instance, never been recorded from Jatun Sacha (B. Bochan). Aside from its loud voice, the Cinnamon Attila is often rather inconspicuous. Recorded below about 300 m.

Monotypic.

Range: e. Colombia, s. and e. Venezuela, and the Guianas to n. Bolivia and Amazonian Brazil.

Attila torridus　　Ochraceous Attila　　Atila Ocráceo

Uncommon in the subcanopy and mid-levels of humid forest, forest borders, secondary woodland, and adjacent plantations and small clearings in the lowlands and foothills of w. Ecuador from Esmeraldas and Pichincha south to w. Guayas, El Oro, and w. Loja (south to Alamor, Celica, Sozoranga, and Utuana). The Ochraceous Attila appears, however, to be absent from the wet-forest belt of n. Esmeraldas, where it was not found during four years of field work in the Playa de Oro region (O. Jahn and P. Mena Valenzuela et al.). Ranges up to about 1500 m in w. Loja in the Alamor/Celica region, smaller numbers reaching as high as 2400 m at Utuana in Loja (seen on 13 Sep. 1998; RSR et al.); in Pichincha and Esmeraldas, however, the species does not appear to range above about 300 m.

Numbers of the nearly endemic Ochraceous Attila have declined seriously in recent decades; there are, for instance, no fewer than six specimens in ANSP taken in Oct. 1950 at Piedras and Santa Rosa in coastal El Oro, an area in which all suitable forest habitat has long since been converted to banana cultivation and other agricultural pursuits. However—and fortunately—this species seems to be somewhat more resilient than would have been expected had it been a true forest species, and it appears to be capable of persisting in patchy forest and woodland fragments (far more likely to do so than the Bright-rumped Attila [*A. spadiceus*]). A few pairs persist at both Río Palenque and Tinalandia, though there are only a few recent sightings from the latter. The Ochraceous Attila is now most numerous in the foothills, especially southward, as this is where comparatively more forest remains; a good population is found, for instance, at the higher levels of Machalilla National Park in sw. Manabí. The species was accorded Endangered status by Collar et al. 1992 but was downgraded to Vulnerable in Collar et al. 1994. We concur with the latter assessment.

Monotypic.

Range: extreme sw. Colombia to extreme nw. Peru.

Rhytipterna simplex   **Grayish Mourner**   Copetón Plañidero Grisáceo

Uncommon to fairly common in the subcanopy and mid-levels of humid forest, less often at borders, in the lowlands of e. Ecuador; occurs in both terra firme and várzea, though mainly the former. Recorded mostly below 700 m, though ranging in very small numbers up locally into the foothills to about 1100 m.

One race: *frederici*.

Range: se. Colombia, s. Venezuela, and the Guianas to n. Bolivia and Amazonian Brazil; e. Brazil.

Rhytipterna holerythra   **Rufous Mourner**   Copetón Plañidero Rufo

Uncommon to locally fairly common in the subcanopy and mid-levels of very humid and humid forest, less often at borders, in the lowlands and foothills of nw. Ecuador, where most numerous and widespread in Esmeraldas. Southward the Rufous Mourner has been recorded from sw. Imbabura (2 seen and tape-recorded near the Salto del Tigre bridge over the Río Guaillabamba in Jul. 1997; P. Coopmans et al.) and Pichincha. The only Pichincha records are from near the San Miguel de Los Bancos–Puerto Quito highway and in the Pedro Vicente Maldonado area (v.o.), Tinalandia (where it has not been seen since 1980; PJG), and Río Palenque (where there have been only a few recent sightings, all of single, perhaps only wandering, birds). Recorded up to about 700 m.

One race: *rosenbergi*.

Range: s. Mexico to nw. Ecuador.

Sirystes sibilator   **Eastern Sirystes**   Siristes Oriental

Uncommon and local in the canopy and borders of humid forest in the lowlands of e. Ecuador, occurring primarily in várzea and floodplain forest, less frequently in terra firme. To some extent the Eastern Sirystes may have been overlooked as it tends to remain high in the canopy, largely unnoticed unless its voice is known. Recorded below 400 m.

Does not include trans-Andean *S. albogriseus* (Western Sirystes). One race occurs in Ecuador, *albocinereus*.

Range: locally from sw. Venezuela, e. Colombia, and the Guianas to n. Bolivia and Amazonian and se. Brazil, e. Paraguay, and ne. Argentina. More than one species could be involved.

*Sirystes albogriseus   **Western Sirystes**   Siristes Occidental

Uncommon and decidedly local (overlooked?) in the canopy and borders of humid forest in the lowlands of nw. Ecuador. The Western Sirystes is unaccountably scarce and local in its Ecuadorian range, even allowing for ongoing deforestation, and was only first recorded from the country in the 1970s from sight reports at Río Palenque. It has since been recorded only from w. Esmeraldas (southeast of Muisne [ANSP] and at Playa de Oro [O. Jahn et al.]), sw. Imbabura (one heard and tape-recorded near the Salto del Tigre bridge over the Río Guaillabamba on 15 Jul. 1997; P. Coopmans), Pichincha (in the northwest from a number of sightings south to along the San Miguel de los Bancos–Puerto Quito highway [P. Coopmans and others]; also from Río Palenque, though it seems to have declined there of late [P. Coopmans]), and n. Manabí (a specimen, the first from Ecuador, taken on 12 Jul. 1988 at Filo de Monos; WFVZ). Recorded below 500 m.

S. *albogriseus* is regarded as a species separate from cis-Andean S. *sibilator* (Eastern Sirystes) in recognition of its strikingly different vocalizations and somewhat different plumage pattern. They have usually been regarded as conspecific (Sirystes), though Ridgely and Tudor (1994) expressed reservations as to whether this was the correct course. Monotypic.

Range: Panama to nw. Ecuador.

Myiarchus tuberculifer   Dusky-capped Flycatcher   Copetón Crestioscuro

Uncommon to locally common in the canopy and borders of forest and woodland and in adjacent clearings in the lowlands, foothills, and (in s. Ecuador) subtropical and even lower temperate zones in both e. and w. Ecuador. The Dusky-capped is the most widespread *Myiarchus* flycatcher in Ecuador, and it occurs in both humid and (in the west) rather arid regions; in the eastern lowlands it ranges in the canopy of both várzea and terra firme forest. Ranges mostly below 1500 m, but up to about 2500–2700 m in the south, ranging especially high in Loja.

Three rather distinct races occur in Ecuador: nominate *tuberculifer* in the eastern lowlands, *nigriceps* in the western lowlands and foothills, and *atriceps* in the foothills and subtropical zone of s. Ecuador (El Oro, Loja, and Zamora-Chinchipe). *Atriceps* is larger than the other two; both *atriceps* and *nigriceps* have blackish crowns, whereas the crown of the nominate race is sepia brown.

Range: extreme sw. United States to w. Peru, nw. Argentina, and Amazonian Brazil; se. Brazil.

**Myiarchus swainsoni*   Swainson's Flycatcher   Copetón de Swainson

A rare to locally fairly common austral winter resident in shrubby clearings with scattered trees, along margins of lakes and rivers, and on river islands in the lowlands of e. Ecuador. This *Myiarchus* flycatcher was until recently overlooked; there still are relatively few Ecuadorian records, with the first being made as late as 1976 (a specimen from Limoncocha in LSUMZ). Recorded from Apr. to Sep. Recorded below 400 m.

Only one race, *ferocior*, has been collected in Ecuador (W. E. Lanyon, *Bull AMNH*. 161: 427–628, 1978), but it seems virtually certain that nominate *swainsoni* occurs as well, for specimens of it are known from as close as Caquetá in se. Colombia (Lanyon, op. cit.). Both are known to be austral migrants. *Ferocior* is very pale overall with distinctive contrasting dark ear-coverts; it is even paler than the nominate race, which is relatively pale for a *Myiarchus*. The two subspecies have equally pale lower mandibles.

Range: s. Venezuela and the Guianas to cen. Argentina and Uruguay, southern breeders migrating north as far as w. Amazonia in austral winter.

Myiarchus ferox   **Short-crested Flycatcher**   Copetón Cresticorto

Fairly common at the borders of humid forest and secondary woodland and in shrubby clearings in the lowlands and foothills of e. Ecuador. The Short-crested Flycatcher seems to be increasing as a result of deforestation. Recorded mainly below about 1000 m, in small numbers as high as 1350 m (perhaps following clearings upslope?); we regard the provenance of specimens mentioned by W. E. Lanyon (*Bull. AMNH 161:* 574, 1978) from much higher elevations on the east slope of the Andes (e.g., at Cuyuja) as conjectural.

One race: nominate *ferox*.

Range: e. Colombia, s. Venezuela, and the Guianas to n. and e. Bolivia, e. Paraguay, ne. Argentina, and se. Brazil.

Myiarchus cephalotes   **Pale-edged Flycatcher**   Copetón Filipálido

Uncommon in the canopy and borders of montane forest and secondary woodland, and in adjacent clearings in the subtropical zone on the east slope of the Andes. Recorded from about 1000 to 2275 m.

One race: nominate *cephalotes*.

Range: mountains of n. Venezuela, and Andes from w. Venezuela to w. Bolivia.

Myiarchus phaeocephalus   **Sooty-crowned**   Copetón
                                Flycatcher   Coronitiznado

Fairly common in deciduous woodland, arid scrub, and sometimes at the edge of mangroves in the lowlands and foothills of w. Ecuador from w. Esmeraldas (vicinity of Esmeraldas [city]) and w. Pichincha south through El Oro and w. Loja. Also recently found in the Río Marañón drainage around Zumba in extreme s. Zamora-Chinchipe (specimen taken on 13 Aug. 1989; ANSP). Ranges up to about 1100 m.

Both of this species' very similar races are recorded from Ecuador: nominate *phaeonotus* occurs in most of the species' range, with *interior* (having a slightly browner crown) ranging in the Río Marañón drainage.

Range: w. Ecuador and nw. Peru.

*Myiarchus crinitus   **Great Crested Flycatcher**   Copetón Viajero

Apparently a casual boreal winter visitant to the canopy and borders of humid forest in the lowlands of e. Ecuador in Napo; the species may have been overlooked and could occur elsewhere. There are only three reports of the Great Crested Flycatcher from Ecuador, all of them recent. Two involved birds that were tape-recorded but remained unseen (one near Coca at San Carlos on 19 Feb. 1987 [P. Coopmans] and one along the Mandiyacu at La Selva on 18 Mar. 1989 [RSR and PJG et al.]), and there was a sighting of two birds near Tena on 27 Feb. 1992 [PJG and D. Wolf et al.]). These are the southernmost reports known. Recorded below 500 m.

Monotypic.

Range: breeds in e. North America, wintering south to nw. South America.

The Brown-crested Flycatcher (*M. tyrannulus*) occurs in deciduous woodland and borders in the Río Marañón drainage of n. Peru around Jaen, not too far south of Zumba. As habitats around both towns are comparable, it would seem possible that the species could occur north into Ecuadorian territory.

Pitangus sulphuratus   **Great Kiskadee**   **Bienteveo Grande**

Fairly common along the margins of lakes and rivers, and in clearings and gardens, in the lowlands of e. Ecuador. The Great Kiskadee is less "dominant" in Ecuador than it is in much of tropical and subtropical South America, and it is entirely absent from the seemingly ideal lowlands of the west; it is not particularly numerous even in e. Ecuador. Numbers may be increasing, however, as a result of deforestation. Recorded up locally and in small numbers to about 1000–1100 m along the eastern base of the Andes, and with clearing may continue to spread upward.

One race: nominate *sulphuratus*.

Range: extreme sw. United States to cen. Argentina.

Philohydor lictor   **Lesser Kiskadee**   **Bienteveo Menor**

Uncommon to locally fairly common along the margins of lakes and sluggish rivers, and in marshy shrubby clearings in the lowlands of e. Ecuador. The Lesser Kiskadee is particularly numerous along the shores of Limoncocha. Recorded mostly below about 500 m, locally as high as 850 m in river valleys along the eastern base of the Andes (e.g., along the Río Upano near Macas; P. Coopmans).

The species was formerly placed in the genus *Pitangus*; see W. E. Lanyon (*Am. Mus. Novitates* 797, 1984) for the erection of the monotypic genus *Philohydor*, in part based on its very different cup-shaped nest. One race occurs in Ecuador, nominate *lictor*.

Range: Panama to n. and e. Bolivia and Amazonian Brazil; e. Brazil.

Megarynchus pitangua   **Boat-billed Flycatcher**   **Mosquero Picudo**

Uncommon to locally common in the canopy and borders of forest and woodland (both humid and deciduous) and in adjacent clearings in the lowlands and foothills of both e. and w. Ecuador. In the east the Boat-billed Flycatcher is less numerous in the canopy of continuous terra firme forest. In the wet-forest zone of n. Esmeraldas it seemingly is restricted to disturbed and secondary habitats (O. Jahn and P. Mena Valenzuela et al.). Recorded mostly below about 1300 m, but ranges up locally and in small numbers as high as 1900 m in e. Loja (e.g., above Amaluza; R. Williams et al.).

Two similar races occur in Ecuador, nominate *pitangua* in the east and the nearly endemic *chrysogaster* in the west. The latter differs in its more tawny-orange (not orange-yellow) coronal patch. The correct spelling of the generic name is *Megarynchus*, not *Megarhynchus*.

Range: n. Mexico to n. and e. Bolivia, Paraguay, ne. Argentina, and se. Brazil; w. Ecuador and extreme nw. Peru.

Myiozetetes similis   Social Flycatcher   Mosquero Social

Common to very common in shrubby clearings, gardens and residential areas, and humid forest and woodland borders in the lowlands and foothills of both e. and w. Ecuador. In the west ranges north into w. Esmeraldas to the vicinity of Borbón, where it is scarce (much outnumbered by the Rusty-margined [*M. cayanensis*]); it occurs south through El Oro and Loja. The Social is generally a conspicuous and noisy flycatcher, though in more forested regions it is relatively local; it spreads fairly quickly subsequent to clearing, and appears to be moving north through coastal Esmeraldas. Ranges up on the east slope of the Andes to about 1400 m, in small numbers as high as 1750–1850 m; now apparently spreading upslope in cleared areas. In the west ranges mainly below 1300 m, but has been recorded locally as high as 1900 m in e. Loja (above Amaluza; R. Williams et al.) and exceptionally as high as 2400 m at Utuana (Best et al. 1993).

Two similar races are found in Ecuador, nominate *similis* in the east and the nearly endemic *grandis* in the west. *Grandis* is slightly brighter yellow below, and the pale edging on its wing-coverts averages slightly wider.

Range: n. Mexico to n. Bolivia, e. Paraguay, and se. Brazil; w. Ecuador and extreme nw. Peru. More than one species may be involved.

Myiozetetes cayanensis   Rusty-margined   Mosquero Alicastaño
                            Flycatcher

Fairly common to common in clearings, forest and woodland borders (to a lesser extent also canopy), gardens, and agricultural regions in the more humid lowlands and foothills of w. Ecuador south to El Oro (e.g., Santa Rosa; ANSP) and adjacent w. Loja (an old specimen from Cebollal; also seen in Feb. 1991 at Tierra Colorada near Alamor [Best et al. 1993]). Despite a few reported sightings, there are still no confirmed records of the Rusty-margined Flycatcher from e. Ecuador; we have been unable to determine the basis for its being mentioned as occurring in "eastern Ecuador" in *Birds of the World*, vol. 8 (p. 211); older specimens seem likely to have been misidentified Social Flycatchers (*M. similis*). Recorded mostly below 1000 m, locally to 1400 m in the Mindo region.

One race: *hellmayri*.
Range: Panama to sw. Ecuador, n. Bolivia, and Amazonian and se. Brazil (but absent from much of w. and cen. Amazonia).

Myiozetetes granadensis   Gray-capped Flycatcher   Mosquero Cabecigris

Fairly common to common in borders of humid forest (to some extent also forest canopy), shrubby clearings, and gardens and residential areas in the lowlands and foothills of e. and nw. Ecuador. In the west recorded south to n. Manabí and s. Pichincha (Río Palenque). There is also a 19th-century record from Tumbes, Peru (Taczanowski 1877), but there have been no subsequent records from anywhere in sw. Ecuador (or from nw. Peru), and it appears likely that the specimen was confused with the Social Flycatcher (*M. similis*). In the eastern lowlands the Gray-capped is more likely than the Social Flycatcher to occur in mainly forested regions away from the vicinity of water. Mostly below 1000 m, though locally ranging up to about 1300 m in the Mindo area.

Two very similar races are found in Ecuador, *obscurior* in the east and *occidentalis* in the west. *Obscurior* is slightly larger.
Range: Honduras to w. Ecuador and (perhaps) extreme nw. Peru; se. Colombia and s. Venezuela to n. Bolivia and w. Amazonian Brazil.

Myiozetetes luteiventris   Dusky-chested Flycatcher   Mosquero Pechioscuro

Rare to uncommon in the canopy and borders of terra firme forest in the lowlands of e. Ecuador. Ecuadorian records of this obscure tyrannid were until recently relatively few, but with knowledge of its distinctive vocalizations it has proven to be considerably more widespread than had been heretofore realized. Ecuadorian specimens exist from "Río Napo" and Sarayacu (BMNH), Río Suno (AMNH), Río Bufeo and Montalvo (MECN), Tayuntza (WFVZ), and east of Puerto Napo (ANSP). There are recent sightings from numerous other sites, and the species is now proving to be a widespread, low-density resident in appropriate habitat. Recorded up to about 600 m.

Formerly usually placed in the genus *Tyrannopsis* (e.g., Meyer de Schauensee 1966, 1970), but *luteiventris* clearly belongs in the genus *Myiozetetes*; see W. E. Lanyon (*Am. Mus. Novitates* 2797, 1984). One race occurs in Ecuador, nominate *luteiventris*.
Range: locally from se. Colombia, s. Venezuela, and the Guianas to se. Peru, extreme nw. Bolivia, and Amazonian Brazil.

Conopias cinchoneti   Lemon-browed Flycatcher   Mosquero Cejilimón

Uncommon in borders of montane forest and in adjacent clearings with scattered tall trees in the foothills and subtropical zone on the east slope of the Andes. On the west slope recorded mainly from the extreme north in Carchi, where several pairs were seen and tape-recorded below Maldonado in Aug. 1988 (M. B. Robbins); considerably to the south, an evident wandering bird was heard and seen in Jun. 1997 above Mindo (M. Lysinger). Recorded mostly from 1000 to 2000 m, locally down to 750 m (near Tayuza in Morona-Santiago; RSR et al.).

Only the nominate race of *C. cinchoneti* is definitely known to occur in Ecuador, though the Carchi birds (see above) probably will prove to be referable to the similar *icterophrys* (a race otherwise found on the west slope of the Andes in sw. Colombia).

Range: locally in Andes from w. Venezuela to s. Peru; Sierra de Perijá.

Conopias albovittata   White-ringed Flycatcher   Mosquero Aureola

Uncommon to fairly common but apparently local in the canopy and borders of very humid and humid forest in the lowlands of nw. Ecuador, where known mainly from Esmeraldas but also recently found in nw. Pichincha (e.g., along the road north of Simón Bolívar near Pedro Vicente Maldonado, where first found on 12 Sep. 1995 [P. Coopmans]; also in other areas nearby) and adjacent sw. Imbabura (seen and tape-recorded near the Salto del Tigre bridge over the Río Guaillabamba in Jul. 1997; P. Coopmans et al.). The published report (Berg 1994) from much farther south, at Cerro Blanco in the Chongón Hills of Guayas, remains unsubstantiated and seems very unlikely. There are relatively few Ecuadorian records of the White-ringed Flycatcher, though in recent years it has proven to be widespread in Esmeraldas: in Jul. 1990 it was numerous northwest of Alto Tambo (ANSP, MECN), and it has also proven not uncommon east of Muisne (ANSP), at Bilsa, and at Playa de Oro (O. Jahn et al.). Recorded up to at least 500 m.

This and the following species have sometimes been placed in the genus *Coryphotriccus*, but we follow W. E. Lanyon (*Am. Mus. Novitates* 2797, 1984) in subsuming *Coryphotriccus* into *Conopias*. Trans-Andean *C. albovittata* is here regarded as a species separate from *C. parva* (Yellow-throated Flycatcher) of Amazonia; see Ridgely and Tudor (1994). One race occurs in Ecuador, nominate *albovittata*.

Range: Honduras to nw. Ecuador.

**Conopias parva*   Yellow-throated Flycatcher   Mosquero Goliamarillo

Uncertain. Known only from two recent records from the lowlands of e. Ecuador. The first involves a single bird seen and tape-recorded on 10 Aug. 1992 in the canopy of terra firme forest at Cuyabeno in Sucumbíos (B. Whitney et al.; tape-recording to LNS). More recently a pair was seen and tape-recorded at Kapawi Lodge on the Río Pastaza near the Peruvian border in Aug. 1999 (P. Coopmans et al.). The Yellow-throated Flycatcher was not previously recorded any closer to Ecuador than south of the Amazon River east of Iquitos in ne. Peru (M. B. Robbins, A. P. Capparella, R. S. Ridgely, and S. W. Cardiff, *Proc. Acad. Nat. Sci. Phil.* 143: 152, 1991). 200 m.

We regard *C. parva* of Amazonia as a species separate from trans-Andean *C. albovittata* (White-ringed Flycatcher) based on its different voice, plumage, and highly disjunct distribution; this follows Sibley and Monroe (1990) and Ridgely and Tudor (1994). Monotypic.

Range: the Guianas, s. Venezuela, extreme e. Colombia, and n. Brazil; very locally in ne. Ecuador and ne. Peru.

*Conopias trivirgata   **Three-striped Flycatcher**   **Mosquero Trirrayado**

Rare and very local in the canopy and borders of várzea forest in the lowlands of far ne. Ecuador in e. Napo and e. Sucumbíos. First found along the Río Pacuyacu, when on 8 Nov. 1992 an obviously territorial pair was seen and tape-recorded (P. Coopmans et al.; tape-recording to LNS); two pairs were found here, one of them along the boardwalk, on 20 Nov. 1993 (P. Coopmans et al.). On 21 Jan. 1994 a calling pair was seen and tape-recorded at Imuyacocha (P. Coopmans et al.). The Three-striped Flycatcher was not previously recorded any closer to Ecuador than e.-cen. Peru; the fact that both this species and the Yellow-throated Flycatcher (*C. parva*) were recorded for the first time in Ecuador within a few months of each other, and at sites only a short distance apart, is remarkable. Recorded at about 200 m.

Race unknown, but Ecuadorian birds would presumably be referable to *berlepschi*.

Range: very locally in Amazonia; se. Brazil, e. Paraguay, and ne. Argentina.

Myiodynastes maculatus   **Streaked Flycatcher**   **Mosquero Rayado**

Uncommon to fairly common in the canopy and borders of forest and secondary woodland in the lowlands and foothills of both e. and w. Ecuador. In the wet-forest belt of n. Esmeraldas it appears to be only a scarce nonbreeding visitor (O. Jahn and P. Mena Valenzuela et al.). In e. Ecuador the Streaked Flycatcher is generally less numerous and more local than it is in the west, despite the presence there of two overlapping races, a resident and an austral migrant form, the latter being recorded from about Apr. through Aug. (specific records including an AMNH specimen from Río Suno, an MECN specimen from Montalvo, ANSP specimens taken at Zancudocone on 2 and 9 Apr. 1991, and recent sightings from La Selva, Jatun Sacha, and along the Maxus road southeast of Pompeya). In the east, resident birds are essentially restricted to riparian growth and the edge of várzea forest, with austral migrants ranging somewhat more widely, regularly in the larger trees left standing in clearings. Recorded up to about 1000 m, but *solitarius* (see below) has not been found above about 400 m.

Three races have been recorded in Ecuador. Nominate *maculatus* is resident in the east, *chapmani* is resident in the west, and *solitarius* is an uncommon austral migrant to the eastern lowlands. *Chapmani* and the nominate race are quite similar, but *solitarius* differs strikingly, being markedly darker above with blacker streaks, more coarsely streaked below, and with a mainly blackish tail that shows only narrow rufous-chestnut edging.

Range: e. Mexico to nw. Peru, n. Argentina, and Uruguay; southern breeders (*solitarius*) migrate north into Amazonia during austral winter, and northern breeders (*insolens*) migrate south to n. South America during boreal winter.

| *Myiodynastes luteiventris* | Sulphur-bellied Flycatcher | Mosquero Ventriazufrado |

An uncommon to fairly common transient and rare to uncommon boreal winter resident in the canopy and borders of humid forest (both terra firme and várzea/riparian areas), secondary woodland, and adjacent shrubby clearings in the lowlands of e. Ecuador. The Sulphur-bellied Flycatcher has not been recorded at all from w. Ecuador, nor has it ever been found on either slope of the Andes; it would seem possible in such areas as a transient, however. Recorded mostly from Oct. to Apr. but is especially numerous during northward passage in Mar. and Apr., with a few birds lingering into May and once as late as Jun. (one seen at Jatun Sacha on 6 Jun. 1991; B. Bochan). Although this species is often thought to winter only in Peru and Bolivia (e.g., *Birds of the World*, vol. 8), there are in fact numerous recent records from various sites in e. Ecuador during the midwinter months (Dec.–Feb.), so it is not merely a transient in the country. Recorded only below about 400 m.

Monotypic.

Range: breeds from extreme sw. United States to Costa Rica, wintering in w. Amazonia.

| *Myiodynastes bairdii* | Baird's Flycatcher | Mosquero de Baird |

Uncommon to fairly common in deciduous woodland borders, arid scrub with scattered taller trees, and towns in the lowlands and foothills of sw. Ecuador from cen. Manabí (north to the northern side of the Bahía de Caráquez) south through El Oro and w. Loja. The Baird's Flycatcher is especially numerous around Zapotillo in extreme sw. Loja. Ranges up to about 1000 m in Loja.

Monotypic.

Range: sw. Ecuador and nw. Peru.

| *Myiodynastes chrysocephalus* | Golden-crowned Flycatcher | Mosquero Coronidorado |

Fairly common at borders of montane forest and adjacent clearings, especially along streams, in the foothills and subtropical zone on both slopes of the Andes. In the west recorded south to w. Loja in the Alamor area. Recorded from about 800 to 2500 m (the latter in Carchi; ANSP); mostly 1000–2200 m.

One race: *minor*.

Range: highlands of e. Panama; mountains of n. Venezuela, and Andes from w. Venezuela to extreme nw. Argentina; Sierra de Perijá and Santa Marta Mountains.

Legatus leucophaius   Piratic Flycatcher   Mosquero Pirata

Uncommon to locally or seasonally common at borders of humid forest and secondary woodland and in clearings with scattered tall trees in the lowlands and to a lesser extent the foothills of both e. and w. Ecuador. In the west ranges south mainly to e. Guayas, exceptionally (only wandering birds?) to sw. Manabí (at least 2 calling birds at Cerro San Sebastián in Aug. 1991; RSR), nw. Azuay (one seen on 27 Jan. 1991 at Manta Real; RSR) and El Oro (west of La Avanzada at Saracay; P. Coopmans). Singing Piratic Flycatchers are conspicuous, but the species otherwise is only infrequently noted. It seems particularly numerous in the Puerto Napo/Jatun Sacha area, but even there it is difficult to find when not vocalizing, approximately May–Sep. (B. Bochan). The Piratic Flycatcher likely engages in long- or short-distance migratory movements, but details remain uncertain. Recorded mostly below 800 m, rarely to 1300 m (e.g., around Mindo); seen once as high as 1600 m (a single silent bird at Vilcabamba on 6 Sep. 1998; RSR and PJG et al.).

Only the nominate race has definitely been recorded from Ecuador. The similar *variegatus*, which is migratory from its breeding grounds in n. Middle America, could occur during the Sep.–Jan. period.

Range: s. Mexico to sw. Ecuador, n. Argentina, and se. Brazil.

Empidonomus varius   Variegated Flycatcher   Mosquero Variegado

An uncommon austral winter resident in riparian growth, clearings, and humid forest and woodland borders in the lowlands of e. Ecuador; very small numbers also occur in the canopy of terra firme forest, perhaps primarily as transients. In Ecuador the Variegated is much less of a forest-based bird than the Crowned Slaty Flycatcher (*Griseotyrannus aurantioatrocristatus*). Recorded from late Mar. to late Sep. Recorded up to 500 m.

One race: nominate *varius*.

Range: breeds from e. Venezuela and the Guianas south to cen. Argentina, southern breeders migrating north into w. Amazonia during austral winter.

Griseotyrannus aurantioatrocristatus   Crowned Slaty   Mosquero
                                           Flycatcher   Coronado

A fairly common austral winter resident to the canopy and borders of humid forest (both terra firme and várzea) and clearings with scattered trees in the lowlands and foothills of e. Ecuador. Recorded mostly from late Mar. to Sep., exceptionally later (e.g., solitary lingerers along the Maxus road southeast of Pompeya on 9 and 12 Nov. 1994; RSR et al.). Ranges mostly below about 1100 m. Seen once at 1600 m (a solitary bird found at Vilcabamba on 6–7 Sep. 1998 by PJG and RSR et al., the only record from west of the east slope of the Andes); another was seen at the remarkably high elevation of 2500 m at Cuyuja on 23 May 1996 (L. Navarrete et al.).

Formerly placed in the genus *Empidonomus*. For the erection of the monotypic genus *Griseotyrannus*, see W. E. Lanyon (*Am. Mus. Novitates* 2797, 1984). One race occurs in Ecuador, nominate *aurantioatrocristatus*.

Range: breeds from n. and e. Bolivia and ne. Brazil south to cen. Argentina, southern breeders migrating north into w. Amazonia during austral winter, a few reaching Venezuela.

Tyrannopsis sulphurea　　　**Sulphury Flycatcher**　　　**Mosquero Azufrado**

Rare to uncommon and local primarily in and near groves of *Mauritia* palms in the lowlands of e. Ecuador, where recorded primarily from the drainages of the Ríos Napo and Aguarico. The Sulphury Flycatcher is perhaps more numerous than the few Ecuador records would indicate, for much of its favored habitat is difficult to access; in 1994–1995 it proved to be widespread in appropriate habitat along the Maxus road southeast of Pompeya. There are few Ecuadorian specimens with specific locality information, with one in the MECN labeled as being from "Río Due." Recently acquired knowledge of its vocalizations has enabled observers to find the Sulphury Flycatcher at a number of scattered localities, including Taracoa, La Selva, Zancudococha, east of the mouth of the Río Cuyabeno (P. Coopmans), and Jatun Sacha (where it seems to be decidedly scarce). The only record from the Río Pastaza drainage is of a pair tape-recorded near Kapawi Lodge on 13 Dec. 1996 (P. Coopmans). Recorded below about 400 m.

Monotypic.

Range: locally from e. Colombia, s. and e. Venezuela, and the Guianas to nw. Bolivia and Amazonian Brazil.

Tyrannus melancholicus　　　**Tropical Kingbird**　　　**Tirano Tropical**

Common to very common and conspicuous in virtually all semiopen and agricultural habitats, and in woodland and forest borders, in the lowlands and on lower slopes of the Andes in both e. and w. Ecuador. Although extremely widespread in Ecuador, the Tropical Kingbird does tend to avoid very arid, desertlike areas such as are found on the Santa Elena Peninsula in w. Guayas, and it is also absent from areas still supporting extensive continuous forest. Numbers have doubtless increased substantially as a result of the clearing of formerly forested areas, and the species rapidly colonizes such places soon after they are opened up. The Tropical Kingbird may also occur as an austral winter visitant; at Jatun Sacha numbers have been observed to increase "greatly" in Apr.–May, subsequently dropping off again (B. Bochan). Ranges mostly below 1500 m, smaller numbers occurring higher (e.g., to about 2500 m around Cuyuja), with at least one report from as high as Quito (2800 m).

One race: nominate *melancholicus*.

Range: extreme sw. United States to cen. Argentina.

Tyrannus albogularis   White-throated Kingbird   Tirano Goliblanco

A very rare austral winter visitant to the borders of humid forest and clearings in the lowlands of e. Ecuador; has probably been overlooked amongst the omnipresent Tropical Kingbird (*T. melancholicus*). Three sightings are known to us, all of single birds: near Limoncocha on 4 Aug. 1981 (S. Hilty and D. Finch et al.), along the Río Napo at La Selva on 26 Jul. 1988 (RSR and PJG et al.), and on Isla Anaconda in the Río Napo upriver from La Selva on 14 Aug. 1994 (P. Coopmans and D. Pitman et al.). Recorded below 300 m.

Monotypic.

Range: breeds locally from the Guianas and se. Venezuela to n. and e. Bolivia and s. Brazil; southern breeders migrate into w. Amazonia during austral winter.

Tyrannus niveigularis   Snowy-throated Kingbird   Tirano Goliníveo

Fairly common as a breeder (ca. Jan.-Jul.) in desert scrub, shrubby areas, and deciduous woodland borders in the more arid lowlands of w. Ecuador from cen. Manabí (including Isla de la Plata) south through El Oro and in extreme sw. Loja (only in the Zapotillo/La Ceiba area; ANSP). During the nonbreeding season (especially Jun.–Nov.), the Snowy-throated Kingbird disperses into more humid areas, and it then also engages in some northward migration. Apparently all records north of n. Manabí and s. Pichincha involve nonbreeding birds; at that season the species is fairly common in partially cleared areas at Playa de Oro in n. Esmeraldas, even defending feeding territories there (O. Jahn et al.). Ranges mostly below 500 m, occasionally up to 700 m (e.g., at Tinalandia; PJG) and rarely even to 1350 m at Mindo (one seen on 29 Sep. 1998; J. Lyons).

Monotypic.

Range: extreme sw. Colombia to nw. Peru.

[Tyrannus dominicensis   Gray Kingbird   Tirano Gris]

Accidental. Known from only a single sighting: one was seen at the edge of mangrove forest near Puerto Pitahaya in El Oro on 16 Apr. 1993 (RSR). This represents the southernmost report, though wanderers have also been seen as far south as Buenaventura on the Pacific coast of Colombia (Hilty and Brown 1986).

Likely race: nominate *dominicensis*.

Range: breeds in s. United States, West Indies, and Venezuela, wintering south to n. Colombia, extreme n. Brazil, and the Guianas.

Tyrannus tyrannus   Eastern Kingbird   Tirano Norteño

A fairly common to briefly common transient and rare boreal winter resident in the canopy and borders of humid forest and woodland and in clearings in the lowlands and (rarely) foothills of e. Ecuador. There are also a few records of transient Eastern Kingbirds in nw. Ecuador south to Pichincha in the Mindo area and at Tinalandia (one seen on 21 Sep. 1993; P. Coopmans). Recorded mostly in Oct. and Nov., and again in Mar. and Apr., with a few arriving as early as late Sep. and lingering into early May. Numbers are generally low—or the species may be entirely absent—from Dec. to early Mar., though MECN does have a series of six specimens obtained by R. Olalla in Jan. 1962 at Conambo on the Río Conambo. Recorded mainly below 500 m, smaller numbers up to 1300 m; at least occasionally it occurs higher, likely mainly as a transient (e.g., one was seen in Quito on 28–29 Nov. 1994 [S. Hansson] and four juveniles were seen at the remarkably high elevation of 3700 m at Antisana Ecological Reserve on 20 Oct. 2000 [RSR et al.]).

Monotypic.

Range: breeds in North America, wintering primarily in w. Amazonia (stragglers farther south and east).

Tyrannus savana   **Fork-tailed Flycatcher**   **Tijereta Sabanera**

A rare to occasionally fairly common and conspicuous transient and less numerous austral winter visitant to borders of forest and woodland and clearings in the lowlands of e. Ecuador. There is no evidence of breeding in Ecuador. Recorded from Mar. (in some years as early as Feb.) to Sep. (rarely Oct.) During both its northward passage in Mar. and Apr. and its return southward passage in Sep., the Fork-tailed Flycatcher is most often seen migrating with larger numbers of Eastern Kingbirds (*T. tyrannus*), especially along rivers and lake margins. There are comparatively few reports from May to Aug., though an MECN specimen from Sarayacu was taken on 3 Jul. 1971. Elsewhere in Ecuador there are only a few reports, though a few individuals may straggle on a more or less annual basis into the highlands and perhaps also to the northwestern lowlands. Recent highland reports include one bird seen near Cuenca in late Oct. 1987 (R. Jones); two seen at Yahuarcocha in Imbabura on 14 Jul. 1991 (R. Williams et al.); one seen at Parque La Carolina in Quito on 21 Oct. 1992 (A. Whittaker and RSR); and one seen at Tumbaco on 4 May 1999 (C. Bustamente, M. Lysinger, and P. Coopmans). There are only two modern reports from the western lowlands: a loose group of about 12 birds seen south of Muisne in w. Esmeraldas on 7 Sep. 1992 (T. J. Davis and F. Sornoza) and a group of about six seen north of Santa Rosa in coastal El Oro on 28 Apr. 1993 (RSR and F. Sornoza et al.). Mostly below 400 m, stragglers higher.

Formerly (e.g., Meyer de Schauensee 1966, 1970) placed in the genus *Muscivora* and called *M. tyrannus*; see *Birds of the World*, vol. 8. Only one race has been recorded in Ecuador, austral migrant nominate *savana*.

Range: breeds in open and semiopen areas from Mexico to cen. Argentina (largely absent as a breeder across Amazonia); southern breeders migrate north to n. South America during austral winter.

Pachyramphus xanthogenys   Yellow-cheeked   Cabezón
                               Becard   Cachetiamarillo

Locally uncommon to fairly common in clearings with scattered tall trees and
at borders of humid forest in the foothills and lower subtropical zone along the
eastern base of the Andes from w. Sucumbíos (sightings near El Reventador;
RSR) south through Zamora-Chinchipe (including the Zumba area, where it
seems especially numerous). The Yellow-cheeked Becard seems likely also to
occur in adjacent Colombia, whence it has not yet been recorded (Hilty and
Brown 1986). Recorded from about 650 to 1700 m.

This and the following eight members of the genus *Pachyramphus*, as well as
the three members of the genus *Platypsaris*, were formerly (e.g., Meyer de
Schauensee 1966, 1970; Hilty and Brown 1986) considered members of the w.
Cotingidae (Cotingas). They are, however, apparently most closely allied to the
Tyrannidae (Tyrant Flycatchers; see the 1998 AOU Check-list), but their exact
taxonomic placement requires additional study. The becards and the genus
Tityra have also been considered a subfamily in the Tyrannidae (*Birds of the
World*, vol. 8). *P. xanthogenys* of Ecuador and Peru is here regarded as a species
distinct from *P. viridis* (Green-backed Becard) of e. South America, based on its
different plumage and highly disjunct distribution, following Ridgely and Tudor
(1994). One race occurs in Ecuador, nominate *xanthogenys*.

Range: along eastern base of Andes from n. Ecuador to cen. Peru.

Pachyramphus versicolor   **Barred Becard**   **Cabezón Barreteado**

Uncommon to locally fairly common in the canopy and borders of montane
forest and mature secondary woodland in the subtropical and lower temperate
zones on both slopes of the Andes. On the west slope recorded south in the
past to Chimborazo (Cayandeled), but in recent years found south only to
w. Cotopaxi. Recorded mostly from 1500 to 2600 m, in small numbers down
to 1200 m (e.g., along the Loreto road) and locally as high as 2850 m (at
Huashapamba in n. Loja; R. Williams et al.) and even 3000 m (in w. Cotopaxi
and at Oyacachi) and 3050 m (at Cerro Toledo; Krabbe et al. 1997); exceptionally, one was mist-netted at 3500 m at Yanacocha (B. O. Poulsen and T.
Laessoe, fide N. Krabbe).

Two very similar races are found in Ecuador, nominate *versicolor* in most of
the species' Ecuadorian range and *meridionalis* in Zamora-Chinchipe. The latter
may prove not to be a recognizable taxon.

Range: highlands of Costa Rica and w. Panama; Andes from w. Venezuela to
w. Bolivia; Sierra de Perijá.

Pachyramphus spodiurus   **Slaty Becard**   **Cabezón Pizarroso**

Rare to locally uncommon in semihumid and deciduous woodland, clearings
and plantations with at least scattered taller trees, and along dry washes in arid
scrub in the lowlands of w. Ecuador from w. Esmeraldas (old specimens from

Esmeraldas [city]; no recent reports) and Pichincha (north to near Pedro Vicente Maldonado, where one was seen on 19 Nov. 1996; P. Coopmans et al.) south through El Oro and w. Loja (Mangaurco, north of Zapotillo, and east of Macará; ANSP, MECN, and P. Coopmans). Although not typically a bird of humid regions, the Slaty Becard may move into such areas at least locally when these are deforested. Many reports of this species are considered doubtful because of frequent confusion with the similar—and far more numerous and widespread—One-colored Becard (*Platypsaris homochrous*). Recorded mostly below 600 m, but locally as high as 1100 m (a pair seen and tape-recorded below Cruzpamba on 2 Mar. 1999; L. Navarrete et al.).

The Slaty Becard, a virtual Ecuadorian endemic, seems never to have been really numerous and has been thought to be declining as a result of the widespread forest and woodland destruction that has occurred across most of its range. Collar et al. (1992) therefore considered it to warrant Endangered status. However, recent evidence indicates that the species is relatively inconspicuous—and thus likely often overlooked—and that it actually can tolerate a substantial amount of habitat disturbance, persisting readily in many such areas. Collar et al. (1994) downgraded its status to only Near-threatened. At least for the present we concur with the latter assessment, but we suspect that ultimately the species may prove not to be at risk at all.

Monotypic.

Range: w. Ecuador and extreme nw. Peru.

[*Pachyramphus rufus*   Cinereous Becard   Cabezón Cinéreo]

Uncertain. The only records of the Cinereous Becard from Ecuador are a few old specimens, none of them recent and none entirely confirmed. The species was recorded (Salvadori and Festa 1899b) from late-19th-century specimens taken in the foothills along the eastern base of the Andes in Zamora-Chinchipe ("Valle del Zamora" and "Valle del Río Santiago"). The identity of these specimens needs to be rechecked, particularly in light of the fact that at that time the separation of female Cinereous from Chestnut-crowned (*P. castaneus*) and other becards was not well understood (see J. T. Zimmer, *Am. Mus. Novitates* 894: 2–4, 1936). We have not been able to examine the supposed Ecuadorian material. The closest confirmed records of the Cinereous Becard come from ne. Peru near the mouth of the Río Napo east of Iquitos.

Race uncertain, but Ecuador birds—if confirmed as being of this species (see above)—would probably be referable to *juruanus*.

Range: e. Panama to w. Colombia, much of Venezuela, the Guianas, Amazonian Brazil, and ne. Peru.

Pachyramphus castaneus   Chestnut-crowned Becard   Cabezón Nuquigris

Rare to locally fairly common at the borders of humid forest (both terra firme and várzea) and in adjacent clearings with tall trees in the lowlands and (to a lesser extent) foothills of e. Ecuador. The Chestnut-crowned Becard seems less numerous and more local in Ecuador than it is in at least some other parts of its range; numbers seem greatest along the Maxus road southeast of Pompeya (RSR et al.). Recorded mostly below 1000 m, but locally ranging as high as 1500 m at Coangos on the slopes of the Cordillera del Cóndor (Schulenberg and Awbrey 1997).

One race: *saturatus*.

Range: n. Venezuela and locally on tepui slopes of s. Venezuela; se. Colombia to n. Bolivia and Amazonian Brazil; se. Brazil, e. Paraguay, and ne. Argentina.

Pachyramphus cinnamomeus   Cinnamon Becard   Cabezón Canelo

Fairly common to common at borders of humid forest and secondary woodland (at least locally also in forest canopy) and in clearings with tall trees in the more humid lowlands and foothills of w. Ecuador south to El Oro (old specimen records from Portovelo and Santa Rosa); the southernmost recent sites seem to be in nw. Azuay (Manta Real [ANSP] and Corona de Oro along the Molletura-Cuenca road [P. Coopmans et al.]). In addition, there is a single specimen (MECN) of the Cinnamon Becard from well to the south and much farther inland than the species is otherwise known, a male obtained by M. Olalla at Valle Catamayo in Loja on 28 Apr. 1974; whether this individual was correctly labeled cannot be known, but it seems unlikely. We also regard the record from Cerro Blanco near Guayaquil (Berg 1994) as uncertain. Recorded mostly below 800 m, in smaller numbers as high as 1500 m (e.g., above Mindo and at Pallatanga in Chimborazo).

One race: nominate *cinnamomeus*.

Range: s. Mexico to nw. Venezuela and sw. Ecuador.

Pachyramphus polychopterus   White-winged Becard   Cabezón Aliblanco

Uncommon to fairly common in secondary woodland, forest borders, clearings with scattered trees, and river islands in the lowlands of e. Ecuador; uncommon in secondary woodland, montane forest borders, and clearings in the foothills and lower subtropical zone of the northwest in e. Esmeraldas, w. Imbabura, and n. Pichincha (south to around Mindo). Apparently the White-winged Becard does not occur at all in the northwestern lowlands, reports from there being the result of confusion with the Black-and-white Becard (*P. albogriseus*). The White-winged Becard is considerably more numerous in e. Ecuador than it is in the northwest. The absence of early specimen records from various heavily worked collecting localities in n. Pichincha is puzzling but is perhaps correlated with the local increase of secondary growth, this species' favored habitat. In the east recorded up to about 900 m; in the northwest recorded between about 600 and 1500 m.

Two rather different (especially in males) races are known from Ecuador, *tenebrosus* in the east and *dorsalis* in the northwest. Male *tenebrosus* are basically black with white wing and tail markings, with females more rufescent above; male *dorsalis* are clear gray below and on the upper back and rump, with females more olivaceous above.

Range: Guatemala to nw. Ecuador, n. and e. Bolivia, n. Argentina, and Uruguay. More than one species may be involved.

Pachyramphus marginatus   Black-capped Becard   Cabezón Gorrinegro

Uncommon to locally fairly common in the canopy and borders of humid forest (primarily terra firme) in the lowlands of e. Ecuador. The Black-capped Becard is probably often overlooked because it almost always remains high above the ground; it is most frequently recorded through its oft-given vocalizations. Recorded up to about 700 m.

One race: *nanus*.

Range: se. Colombia, s. Venezuela, and the Guianas to n. Bolivia and Amazonian Brazil; e. Brazil.

Pachyramphus albogriseus   Black-and-white   Cabezón
                              Becard   Blanquinegro

Uncommon in the canopy and borders of montane forest and secondary woodland in the foothills and subtropical zone on the east slope of the Andes. In the west uncommon and rather local in the canopy and borders of forest and woodland (deciduous as well as humid) in the lowlands as well as in the subtropical zone north to Esmeraldas (an old record from San Javier; recent reports from the Atacames/Súa area, with a few records from Playa de Oro [O. Jahn et al.]) and Carchi (southwest of Chical; ANSP). Recorded mainly below about 2000 m, rarely up to 2500 m, and exceptionally as high as 2850 m (near the Cajanuma station of Podocarpus National Park; R. Williams et al.) or even 3200 m (Yanacocha; M. Lysinger); on the east slope not recorded below about 900 m.

Two similar races are found in Ecuador, *guayaquilensis* in the west and *salvini* on the east slope and apparently in the subtropics of the northwest. The subspecific affinity of birds from nw. Ecuador is difficult to determine. J. T. Zimmer (*Am. Mus. Novitates* 894, 1936) described *guayaquilensis* on the basis of its smaller size and very slightly more gray in the tail of males. He ascribed a female from San Javier in Esmeraldas to *guayaquilensis*, a male from Mindo to *salvini*. Recently obtained material (ANSP) from the subtropical zone of Carchi and Pichincha appears to us, at least on the basis of size, to be referable to *salvini*; perhaps *guayaquilensis* occurs north as far as n. Esmeraldas only in the coastal lowlands.

Range: highlands of Costa Rica and w. Panama; mountains of n. Venezuela and Andes of Venezuela, Ecuador, and n. and cen. Peru (singularly lacking from Colombian Andes), also lowlands of w. Ecuador and nw. Peru; Sierra de Perijá and Santa Marta Mountains.

Platypsaris homochrous   One-colored Becard   Cabezón Unicolor

Locally fairly common in the canopy and borders of forest (both humid and deciduous), secondary woodland, and even rather low scrub with a few taller trees in the lowlands and foothills of w. Ecuador from Esmeraldas south through El Oro and w. Loja (east mainly to the Macará region). Found mostly below 1000 m, but recorded locally up to about 1500 m at Pallatanga in Chimborazo.

This and the following two species have sometimes (e.g., *Birds of the World*, vol. 8; 1998 AOU Check-list) been placed in the genus *Pachyramphus*, but they appear to form a cohesive group rather different from *Pachyramphus*, being larger and heavier-billed, less vocal, and with different nest shape and placement. We favor treating the long-recognized genus *Platypsaris* as distinct. One race: nominate *homochrous*.

Range: e. Panama to nw. Venezuela and extreme nw. Peru.

Platypsaris minor   Pink-throated Becard   Cabezón Golirrosado

Rare to locally uncommon in the canopy and borders of humid forest (mainly terra firme) in the lowlands of e. Ecuador. Recorded up to about 600 m (north of Canelos in Pastaza; N. Krabbe).

Monotypic.

Range: se. Colombia, s. Venezuela, and the Guianas to n. Bolivia and Amazonian Brazil.

*Platypsaris validus   Crested Becard   Cabezón Crestado

Apparently rare in the canopy and borders of montane forest on the east slope of the Andes in s. Zamora-Chinchipe, where known only from Quebrada Honda. The Crested Becard was first seen at this site on 13 Jan. 1998 (RSR et al.), and an immature male was collected there by D. Agro on 23 Jan. (MECN). The species had not previously been recorded from north of Ayacucho in s. Peru (Ridgely and Tudor 1994), more than 1000 km to the south. Recorded at about 2550–2600 m.

Presumed race: *audax*. With the acquisition of additional specimen material, the presumably isolated Ecuadorian population could prove to be subspecifically distinct.

Range: Andean slopes from s. Peru to nw. Argentina, thence eastward through s. and e. Brazil; one record from s. Ecuador.

Tityra cayana   Black-tailed Tityra   Titira Colinegra

Fairly common and conspicuous in the canopy and borders of humid forest (both terra firme and várzea), secondary woodland, and clearings with scattered trees (particularly where there are dead snags) in the lowlands of e. Ecuador. Recorded up only to about 500 m.

This species and the following two members of the genus *Tityra* were formerly (e.g., Meyer de Schauensee 1966, 1970; Hilty and Brown 1986) placed with the Cotingidae (Cotingas). One race is found in Ecuador, nominate *cayana*.

Range: e. Colombia, much of Venezuela, and the Guianas to Paraguay, ne. Argentina, and se. Brazil.

Tityra semifasciata   Masked Tityra   Titira Enmascarada

Fairly common and conspicuous in the canopy and borders of humid forest, secondary woodland, and clearings with scattered trees in the lowlands adjacent to the Andes and foothills of e. Ecuador, and in the lowlands and foothills of w. Ecuador south locally to Guayas (Cerro La Torre, Loma Alta, Pacaritambo, and Manglares-Churute Ecological Reserve) and nw. Azuay (Manta Real; ANSP); there are also a few sightings (wandering birds?) from as far south as Buenaventura in El Oro (P. Coopmans et al.; R. Williams et al.). The Masked Tityra is not usually found any distance well east of the Andes, and thus although it is frequent at Jatun Sacha (fide B. Bochan), in more than two years of observations there were none seen along the Maxus road southeast of Pompeya. A recent report (fide J. Moore) of the species from Kapawi Lodge on the Río Pastaza in the far southeast thus comes as a distinct surprise; perhaps it is not of regular occurrence this far east. Mostly below about 1100 m, smaller numbers up to 1500 m.

Two races are found in Ecuador, *fortis* in the east and *nigriceps* in the west. Although the races are similar, male *nigriceps* differ in being slightly whiter.

Range: n. Mexico to n. and w. Venezuela and w. Ecuador; e. Colombia to n. and e. Bolivia and Amazonian Brazil (though largely absent from n. Amazonia).

Tityra inquisitor   Black-crowned Tityra   Titira Coroninegra

Uncommon in the canopy and borders of humid forest and secondary woodland and adjacent clearings in the lowlands of both e. and w. Ecuador, in the west also in semihumid forest. In the west recorded south locally to n. and e. Guayas (old records from as far south as Guayaquil and Bucay; also a few recent reports from Manglares-Churute Ecological Reserve [Pople et al. 1997; RSR et al.]), with the southernmost report being a pair seen in nw. Azuay at Puerto Inca along the lower Molleturo-Cuenca road on 8 Feb. 1998 (P. Coopmans et al.). Recorded up to about 700 m.

Two rather different races are found in Ecuador, *buckleyi* in the east and *albitorques* in the west. Male *albitorques* have the tail mainly white, with black reduced to a broad subterminal band; in male *buckleyi* the tail is all black.

Range: s. Mexico to w. Ecuador, n. and e. Bolivia, e. Paraguay, ne. Argentina, and se. Brazil.

Cotingidae   Cotingas

Thirty-three species in 19 genera. Cotingas are exclusively Neotropical in distribution. Included are the Sharpbill (*Oxyruncus cristatus*) and the Andean Cock-of-the-rock (*Rupicola peruviana*), both formerly considered as being in distinct families (Oxyruncidae and Rupicolidae, respectively); this follows Sibley and Ahlquist (1990) and Sibley and Monroe (1990) as well as most other recent authors. The taxonomic affinity of two genera here placed with the cotingas (*Lipaugus* and *Laniocera*) remains uncertain.

*[Oxyruncus cristatus   Sharpbill   Picoagudo]

Uncertain. Known in Ecuador only from a report by T. Parker and A. Luna in the canopy and borders of humid foothill forest at Miazi in the upper Río Nangaritza valley of s. Zamora-Chinchipe in late Jul. 1993 (Schulenberg and Awbrey 1997). Tragically, Parker died in a plane crash shortly thereafter, and details of this record perished with him. The Sharpbill is known from comparable habitats and elevations in adjacent Amazonas, Peru, so its discovery in s. Ecuador had been anticipated. About 900 m.

The racial affinity of Ecuadorian birds is unknown.

Range: locally in Costa Rica, Panama, extreme s. Ecuador, Peru, w. Bolivia, s. Venezuela and the Guianas, e. Amazonian Brazil, and e. Brazil to e. Paraguay and ne. Argentina.

Ampelion rubrocristatus   Red-crested Cotinga   Cotinga Crestirroja

Uncommon to fairly common at borders of montane forest and woodland (including patches of woodland in paramo such as groves of *Polylepis*), secondary woodland, and locally in shrubby areas in mostly agricultural terrain in the temperate zone on both slopes of the Andes, and on slopes above the central and interandean valleys. The Red-crested Cotinga is generally a quite conspicuous bird, often perching in the tops of low trees and shrubs. Recorded mostly from 2500 to 3500 m, locally as high as 3900 m; an exceptional record involves a (wandering?) bird seen at 2200 m above Cabañas San Isidro on 5 Dec. 1998 (M. Lysinger).

The genus *Ampelion* has sometimes been placed with the Phytotomidae (Plantcutters), when the latter are maintained as a family separate from the Cotingidae, but recent evidence (R. O. Prum, *Occas. Pap. Mus. Zool. Univ. Mich.* 723, 1990) suggests that *Ampelion*'s traditional placement with the Cotingidae is the preferable course. Monotypic.

Range: Andes from w. Venezuela to w. Bolivia.

Ampelion rufaxilla   Chestnut-crested Cotinga   Cotinga Cresticastaña

Rare and apparently very local in the canopy and borders of montane forest in the subtropical zone on the east slope of the Andes, where thus far found only in the extreme north and extreme south. The Chestnut-crested Cotinga was first reported in Ecuador by J. C. Matheus, who in Nov. 1984 observed several along the La Bonita road near La Alegria in extreme w. Sucumbíos. There have been several sightings from the Quebrada Honda/Tapichalaca Reserve area on the south side of the Cordillera Sabanilla in s. Zamora-Chinchipe, first on 28 Nov. 1987 (D. Platt and M. Vestergaard). The first Ecuadorian specimen (ANSP) was obtained west of Zumba near San Andrés in extreme s. Zamora-Chinchipe on 11 Aug. 1992. These are the sole records; the Chestnut-crested Cotinga thus seems to be inexplicably absent from much of the east slope of the Andes in Ecuador. Recorded from about 1800 to 2700 m.

Two similar races occur in Ecuador, *antioquiae* along the La Bonita road (fide N. Krabbe) and nominate *rufaxilla* in the far south.

Range: locally in Andes from w. Colombia to w. Bolivia.

Doliornis remseni   Chestnut-bellied Cotinga   Cotinga Ventricastaña

Rare and local in montane forest and low windswept woodland near treeline on the east slope of the Andes. The only newly described Chestnut-bellied Cotinga has now been found at five localities. *Doliornis* cotingas were first seen in Ecuador on 7 Mar. 1989 in the Cajanuma sector of Podocarpus National Park (H. Bloch et al. 1991); there have also been a few subsequent sightings of birds here (see Rasmussen et al. 1996), but unfortunately the observers' field descriptions were apparently not precise enough to preclude confusion with the Bay-vented Cotinga (*D. sclateri*) of Peru. Subsequently, a small series (ANSP, MECN) of the Chestnut-bellied Cotinga was obtained by M. B. Robbins and G. H. Rosenberg on the west slope of Cerro Mongus in se. Carchi in Mar. 1992, and two more were taken here in Jun. 1992 by F. Sornoza. One pair was subsequently found (with the male collected; ANSP) in extreme s. Zamora-Chinchipe on the east slope of the Cordillera Las Lagunillas on 29 Oct. 1992 (contra Collar et al. 1994, p. 137, this lies in extreme s. Ecuador near the Peruvian border, not in extreme n. Peru). The specimens show no geographical variation, and differ in several characters from the Bay-vented Cotinga of the Peruvian Andes north to San Martín. They were thus described as a new species by M. B. Robbins, G. H. Rosenberg, and F. Sornoza M. (*Auk* 111[1]: 1–7, 1994). Additional localities found in the past few years are near the crest of the Gualaceo-Limón road, where one bird was seen and photographed (photo to VIREO) on 2 Jan. 1999 (G. H. Rosenberg), and at Guandera Biological Reserve in se. Carchi, where the species was found in "very small" numbers (Creswell et al. 1999, p. 60). The Chestnut-bellied Cotinga will probably prove to be locally distributed wherever there is appropriate habitat along the entire east slope of the Andes. Recorded from 2900 to 3500 m.

The Chestnut-bellied Cotinga is still known from only the five Ecuadorian sites mentioned above, one of which (Cerro Mongus) is at considerable risk because of habitat clearance and two of which (Cordillera Las Lagunillas and the Gualaceo-Limón road) remain totally unprotected, though because of their high elevation they will hopefully not be too greatly disturbed in the near future. Collar et al. (1994) accorded the species Vulnerable status, and we agree; its status may need to be upgraded in the near future.

The genus *Doliornis* has sometimes been merged into *Ampelion* (e.g., Snow 1975, 1982), but recent evidence (S. M. Lanyon and W. E. Lanyon, *Auk* 106[3]: 422–432, 1989) suggests that it is best maintained, and this conclusion was supported by M. B. Robbins et al. (op. cit.). Monotypic.

Range: locally in Andes of e. Ecuador. Presumably this species is also the *Doliornis* recently found in the Central Andes of Colombia (L. M. Renjifo M., *Bull. B. O. C.* 114[2]: 101–103, 1994), though it was there published as the Bayvented Cotinga.

Pipreola arcuata　　　Barred Fruiteater　　　Frutero Barreteado

Uncommon to locally fairly common (but inconspicuous; perhaps more numerous southward) in the lower and middle growth of montane forest and forest borders in the temperate zone on both slopes of the Andes, and at least formerly also locally on slopes above the central valley in n. Ecuador. On the west slope recorded south to Bolívar at Salinas (Krabbe et al. 1997). Recorded mostly from 2500 to 3300 m, locally and in small numbers as low as 2250 m near the crest of the Cordillera de Huacamayos (M. Lysinger).

One race: nominate *arcuata*.

Range: Andes from w. Venezuela to w. Bolivia.

Pipreola riefferii　　　Green-and-black Fruiteater　　　Frutero Verdinegro

Uncommon to fairly common in the lower and middle growth of montane forest and forest borders in the subtropical and lower temperate zones on both slopes of the Andes. On the west slope recorded south locally to Bolívar (an old specimen from Chillanes) and El Oro (an old specimen from Salvias), but in recent years found only from w. Cotopaxi northward. Recorded mostly from 1700 to 2900 m, exceptionally as high as 3300 m (Cerro Mongus; MECN).

Two races, differing slightly in males, are found in Ecuador: *occidentalis* on the west slope and *confusa* on the east slope. *Confusa*, described by J. T. Zimmer (*Am. Mus. Novitates* 893, 1936), is very similar to *chachapoyas* of n. Peru, and some of the characters that Zimmer used to diagnose *confusa* seem barely to hold up in the larger series now available; further study may demonstrate that *confusa* is better synonymized into *chachapoyas*. Male *occidentalis* have the midbelly clear yellow, whereas male *confusa* have the belly rather heavily mottled and streaked with green.

Range: mountains of n. Venezuela, and Andes from w. Venezuela to cen. Peru.

Pipreola lubomirskii　　　Black-chested Fruiteater　　　Frutero Pechinegro

Uncommon in the lower and middle growth of montane forest and forest borders in the subtropical zone on the east slope of the Andes. The Black-chested Fruiteater does occur in small numbers on both slopes of the Cordillera de Huacamayos, but in general it is a scarce and infrequently encountered bird in Ecuador. Recorded only from a narrow elevational zone between about 1500 and 2100 m.

The Black-chested Fruiteater was considered to be Near-threatened by Collar et al. (1994), and despite the remoteness of a significant proportion of its habitat (e.g., on the Cordilleras de Cutucú and del Cóndor), we agree with this assessment.

Lubomirskii has been considered (e.g., *Birds of the World*, vol. 8) a race of *P. aureopectus* (Golden-breasted Fruiteater) of n. South America, though most subsequent authors (e.g., Snow 1982; Sibley and Monroe 1990) have treated it as a separate species. Monotypic.

Range: Andes (mainly on east slope) from s. Colombia to n. Peru.

Pipreola jucunda   Orange-breasted Fruiteater   Frutero Pechinaranja

Rare to locally fairly common in the lower and middle growth of montane forest and forest borders in the foothills and lower subtropical zone on the west slope of the Andes. In recent years the Orange-breasted Fruiteater has been found south only to Pichincha (e.g., along the Chiriboga road), but in the 19th century it was recorded south to Chimborazo (Cayandeled). Recorded from 600 to 1700 m, tending to be more numerous toward the upper limit of its elevational range.

Jucunda has been considered (e.g., *Birds of the World*, vol. 8), like *P. lubomirskii* (Black-chested Fruiteater), to be a race of *P. aureopectus* (Golden-breasted Fruiteater) of n. South America, though most subsequent authors (e.g., Snow 1982; Sibley and Monroe 1990) have treated it as a separate species. Monotypic.

Range: west slope of Andes in sw. Colombia and w. Ecuador.

Pipreola frontalis   Scarlet-breasted Fruiteater   Frutero Pechiescarlata

Uncommon to locally fairly common in the mid-levels and subcanopy of montane forest in the foothills and lower subtropical zone on the east slope of the Andes from w. Napo (Volcán Sumaco region, e.g., along the Loreto road north of Archidona) south locally to Zamora-Chinchipe (Cordillera del Cóndor and Panguri; ANSP). Recorded from about 1000 to 1700 m; Snow (1982) records the species down to 700 m, but we do not know the basis for this, and no recent reports from so low are known to us.

The Scarlet-breasted Fruiteater was accorded Near-threatened status by Collar et al. (1994), but given its relative abundance in much of its Ecuadorian range, we do not consider it to be at risk. A substantial portion of its range remains remote and very little threatened by human activities, and a significant portion is encompassed within Podocarpus National Park.

One race: *squamipectus*. This form may deserve specific separation from nominate *frontalis*, which is found in the Andes from cen. Peru to Bolivia. If split, *P. squamipectus* would be called the Bluish-fronted Fruiteater.

Range: east slope of Andes from n. Ecuador to w. Bolivia.

Pipreola chlorolepidota   Fiery-throated Fruiteater   Frutero Golifuego

Rare to uncommon (but probably overlooked) in the lower and middle growth of montane forest in the foothills on the east slope of the Andes and out locally into the adjacent eastern lowlands. Although generally occurring at lower elevations than the Scarlet-breasted Fruiteater (*P. frontalis*), the two species occur syntopically at certain localities (e.g., north of Archidona along the road to Loreto). Recorded mainly from a rather narrow elevational zone between about 600 and 1250 m (Río Huamboya valley in Sangay National Park; MECN), but also found in small numbers as low as 300 m, apparently especially in Pastaza (e.g., at Montalvo; MECN) and Morona-Santiago (e.g., at Taisha; ANSP).

The Fiery-throated Fruiteater was given Near-threatened status by Collar et al. (1994). It has seemingly never been a numerous bird, and we agree with this assessment. In our view it remains possible that the species will ultimately be found to be more numerous and widespread than is presently recognized.

Monotypic.

Range: locally in foothills and adjacent lowlands along east slope of Andes from s. Colombia to s. Peru.

Ampelioides tschudii   Scaled Fruiteater   Frutero Escamado

Rare to uncommon in the subcanopy and mid-levels of montane forest in the foothills and subtropical zone on both slopes of the Andes; on the west slope recorded south to El Oro (an old specimen from Salvias, and specimens [ANSP, MECN] and numerous recent sightings from Buenaventura) and w. Loja (specimens from Las Piñas [AMNH] and north of Celica [MCZ]; also seen at Tierra Colorada in Feb. 1991 [Best et al. 1993]). The Scaled Fruiteater was also found recently on the coastal Cordillera de Mache in w. Esmeraldas at Bilsa, first on 26 Feb. 1994 (R. P. Clay et al.), and on the Cordillera de Colonche at Loma Alta (Becker and López-Lanús 1997). Recorded mostly from 900 to 1900 m, in the west occurring in smaller numbers down to 650 m.

The Scaled Fruiteater was given Near-threatened status by Collar et al. (1992, 1994). However, we consider the species to be too numerous and wide-ranging to deserve any threatened listing, at least in Ecuador.

Monotypic.

Range: locally in Andes from w. Venezuela to w. Bolivia.

Iodopleura isabellae   White-browed Purpletuft   Yodopleura Cejiblanca

Uncommon in the canopy and borders of humid forest (both terra firme and várzea), forest borders, and adjacent clearings in the lowlands of e. Ecuador. Although often conspicuous, regularly perching on high exposed branches, the White-browed Purpletuft never seems to be a particularly numerous bird. Recorded mostly below about 500 m, in smaller numbers up to 600–700 m; once reported as high as 900 m at Miazi in s. Zamora-Chinchipe (Schulenberg and Awbrey 1997).

One race: nominate *isabellae*.

Range: sw. Venezuela and e. Colombia to n. Bolivia and Amazonian Brazil.

Laniisoma buckleyi   Andean Laniisoma   Laniisoma Andino

Very rare to rare and apparently local in the lower and middle growth inside terra firme forest and (less often) at forest borders in the foothills along the eastern base of the Andes and out into the adjacent lowlands of e. Ecuador. A relatively inactive bird which seems not to vocalize very often (and seems often not to be responsive to tape playback), the Andean Laniisoma has almost certainly been underrecorded. From north to south, there are records from the head of the Río Guataraco (MCZ), the Loreto road (B. Whitney), Archidona and Quijos (Meyer de Schauensee 1966, 1970), Jatun Sacha (only one report, fide B. Bochan), Conambo on the Río Conambo in Pastaza (MECN), Santiago and Warientza in Morona-Santiago (ANSP and WFVZ, respectively), and Zamora-Chinchipe at Miazi in the Río Nangaritza valley (Schulenberg and Awbrey 1997) and in the Río Bombuscaro sector of Podocarpus National Park (one seen in Jun. 1992; P. Coopmans). There is also a 19th-century specimen, the type, from "Pindo" (which cannot be located but doubtless is somewhere near Sarayacu in Pastaza). Found exclusively in hilly areas, at elevations ranging from about 400 to 1350 m.

We assess the status of the scarce and little-recorded Andean Laniisoma in Ecuador as Near-threatened. Its close relative the Brazilian Laniisoma (*L. elegans*) was accorded Vulnerable status by Collar et al. (1992, 1994); those authors seem not to have addressed the status of Andean forms.

We follow the suggestion of Collar et al. (1992) that *buckleyi*, together with *venezuelensis* and *cadwaladeri*, should be considered an Andean species (Andean Laniisoma) distinct from the monotypic *L. elegans* of se. Brazil.

L. elegans has traditionally (e.g., Meyer de Schauensee 1966, 1970; Hilty and Brown 1986) been called the Shrike-like Cotinga, a notably inappropriate English name as it is not in the slightest shrike-like. The suggestion of R. Prum and W. E. Lanyon (*Condor* 91[2]: 459, 1989) to employ the modifier Elegant was followed by Ridgely and Tudor (1994), but now that the species has been split, this name is better no longer used for either of its constituent semispecies. Rather than use—as Prum and Lanyon (op. cit.) suggest—the group name of "mourner" for the genus, we prefer simply to employ the generic name of Laniisoma.

Prum and Lanyon (op. cit.) propose that the genus *Laniisoma* belongs in what they term the "*Schiffornis* assemblage"; it may be most closely related to *Laniocera*. One race occurs in Ecuador, nominate *buckleyi*.

Range: locally in foothills along east slope of Andes from w. Venezuela to w. Bolivia.

Laniocera hypopyrra   Cinereous Mourner   Plañidera Cinérea

Rare to locally uncommon in the lower and middle growth of terra firme forest in the lowlands of e. Ecuador. The explanation for the Cinereous Mourner's apparent scarcity in Ecuador remains uncertain; there are recent records from Taisha and Zancudococha, and a few sightings from Cuyabeno, the south side of the Río Napo near La Selva, along the Maxus road southeast of Pompeya, and Kapawi Lodge. Recorded mostly below 400 m, but ranges at least occasionally or locally as high as 850 m (Miazi in the Río Nangaritza valley of Zamora-Chinchipe; Schulenberg and Awbrey 1997).

The systematic position of the genus *Laniocera* remains uncertain. It is sometimes placed in the Tyrannidae, but recent evidence (R. O. Prum and W. E. Lanyon, *Condor* 91[4]: 444–461, 1989) indicates that it belongs in what they term the "*Schiffornis* assemblage." For now we retain it with the cotingas. Monotypic.

Range: the Guianas and s. Venezuela to n. Bolivia and Amazonian Brazil.

Laniocera rufescens   Speckled Mourner   Plañidera Moteada

Rare to locally uncommon in the lower and middle growth of very humid and humid forest in the lowlands and lower foothills of nw. Ecuador in Esmeraldas (south to the Río Verde area [ANSP] and Bilsa) and nw. Pichincha (one seen and tape-recorded 5 km south of Golondrinas along the Pedro Vicente Maldonado–Quinindé road on 12 Jul. 1997; P. Coopmans et al.). Reports from farther south in Pichincha are less certain, though PJG observed a single bird on at least one occasion in the early 1980s at Río Palenque. Recorded up to about 500 m at Bilsa.

Given its limited range in Ecuador, overall scarcity, and apparent strict dependence on large blocks of continuous forest (as revealed by extensive field work in n. Esmeraldas; O. Jahn and P. Mena Valenzuela et al.), we feel that the Speckled Mourner merits Near-threatened status in Ecuador. Small populations are perhaps protected in the Awá Forest Reserve and the lower parts of Cotacachi-Cayapas Ecological Reserve, but the species' status and population size at both sites remain unknown. The Speckled Mourner was not considered to be at risk by Collar et al. (1992, 1994).

One race: *tertia*.

Range: s. Mexico to nw. Ecuador.

Lathria cryptolophus   Olivaceous Piha   Piha Olivácea

Uncommon and seemingly local in the lower and middle growth of montane forest in the subtropical zone on both slopes of the Andes. On the west slope not definitely recorded south of Pichincha (various localities in the Mindo/Nanegal region), but ANSP has a male of the west-slope race labeled as having been taken at "El Tambo, Loja," a locality that cannot be located with certainty but that Paynter (1993) believes likely to be in Cañar (which would be reasonable biogeographically). Recorded from about 1000 to 1800 m.

We place this and the following species in the genus *Lathria*; they long were classified in *Lipaugus*. This follows genetic evidence presented by R. O. Prum, N. H. Rice, J. A. Mobley, and W. W. Dimmick (*Auk* 117[1]: 236–241, 2000). Two similar races are found in Ecuador, *mindoensis* on the west slope and nominate *cryptolophus* on the east slope.

Range: locally in Andes from s. Colombia to cen. Peru.

Lathria subalaris   **Gray-tailed Piha**   Piha Coligris

Rare to uncommon and seemingly local (perhaps locally more numerous) in the lower and middle growth of montane forest in the foothills on the east slope of the Andes from Sucumbíos (Bermejo oil-field area north of Lumbaquí; ANSP, MECN) south to Morona-Santiago and Zamora-Chinchipe (Pachicutza and Miazi [WFVZ and MECN; Schulenberg and Awbrey 1997]). The Gray-tailed Piha seems to occur mainly at elevations lower than those of the similar (but much less vocal) Olivaceous Piha (*L. cryptolophus*). Small numbers occur in the Bermejo oil-field area and along the Loreto road north of Archidona. Recorded from about 500 to 1400m.

We consider the Gray-tailed Piha to merit Near-threatened status, at least in Ecuador. Somewhat surprisingly, it was not mentioned at all by Collar et al. (1992, 1994).

Monotypic.

Range: locally in foothills along east slope of Andes from s. Colombia to s. Peru.

Lipaugus vociferans   **Screaming Piha**   Piha Gritona

Uncommon to locally common in lower and middle growth of humid forest (both terra firme and várzea) in the lowlands of e. Ecuador. Recorded mostly from its far-carrying and oft-repeated vocalizations, the Screaming Piha is actually not seen all that often. Inexplicably, its numbers seem to have declined at Jatun Sacha (B. Bochan). Recorded mainly below 500m, but locally and in small numbers as high as 900m (Bermejo oil-field area north of Lumbaquí [M. B. Robbins; RSR]).

Monotypic.

Range: the Guianas and s. Venezuela to n. Bolivia and Amazonian Brazil; e. Brazil.

Lipaugus fuscocinereus   **Dusky Piha**   Piha Oscura

Rare to uncommon and apparently local in the subcanopy and borders of montane forest and in adjacent clearings and second-growth woodland (including alder [*Alnus*] groves) in the subtropical and lower temperate zones on the east slope of the Andes. Small numbers of the Dusky Piha still occur in the remaining areas of forest near Cuyuja along the road to Lago Agrio; even more are found at SierrAzul and Cabañas San Isidro near Cosanga. Recorded from about 1700 to 2600m.

Monotypic.

Range: locally in Andes from n. Colombia to extreme n. Peru.

Lipaugus unirufus   **Rufous Piha**   Piha Rojiza

Uncommon to locally common in lower and middle growth inside very humid and humid forest in the lowlands and lower foothills of nw. Ecuador. The

Rufous Piha apparently is now found only in Esmeraldas, where it ranges as far south as east of Muisne and Bilsa. Formerly it occurred as far south as s. Pichincha (e.g., at Río Palenque, where there have been no reports since the 1970s). Recorded up to about 700 m at El Placer in e. Esmeraldas (RSR and M. B. Robbins et al.).

One race: *castaneotinctus*.

Range: s. Mexico to nw. Ecuador.

Porphyrolaema porphyrolaema   Purple-throated   Cotinga
                                 Cotinga   Golipúrpura

Rare to uncommon and apparently local in the mid-levels, subcanopy, and borders of humid forest (both terra firme and várzea) and adjacent clearings in the lowlands of e. Ecuador. Recorded up only to about 400 m (Taisha; ANSP).

The Purple-throated Cotinga was given Near-threatened status by Collar et al. (1992, 1994). Given the relatively immense size of its range, and the vast expanses of still-unaltered lowland forest habitat suitable for it, we find its listing inexplicable; if it is considered Near-threatened, then so too must be nearly every other uncommon species restricted to w. Amazonia.

Monotypic.

Range: se. Colombia to se. Peru and w. Amazonian Brazil.

Cotinga nattererii   Blue Cotinga   Cotinga Azul

Rare to locally uncommon in the canopy and borders of humid forest and in trees in adjacent clearings in the lowlands of nw. Ecuador, where recorded mainly from Esmeraldas though there is one recent report from nw. Pichincha (a male seen west of San Miguel de los Bancos on 23 Oct. 1995; J. R. and V. Fletcher, fide P. Coopmans). Although the Blue Cotinga appears to be a genuinely scarce bird in Ecuador, small numbers have recently been seen in the Atacames and Muisne area of w. Esmeraldas. Recorded in Ecuador mostly below 300 m, but with one sighting at about 900 m at San Miguel de los Bancos.

Monotypic.

Range: Panama to w. Venezuela and nw. Ecuador.

Cotinga maynana   Plum-throated Cotinga   Cotinga Golimorada

Uncommon to locally fairly common in the canopy and borders of humid forest (both terra firme and várzea) and adjacent clearings in the lowlands of e. Ecuador. Inexplicably, the Plum-throated Cotinga's numbers seem to have declined precipitously at Jatun Sacha (B. Bochan). Recorded mainly below about 700 m, with one exceptional sighting of a male at 1200 m north of Archidona along the road to Loreto on 17 Jul. 1992 (P. Coopmans and M. Van Beirs et al.).

Monotypic.

Range: se. Colombia to n. Bolivia and w. Amazonian Brazil.

Cotinga cayana   **Spangled Cotinga**   Cotinga Lentejuelada

Uncommon to locally fairly common in the canopy and borders of humid forest (mainly terra firme) and adjacent clearings in the lowlands of e. Ecuador. The Spangled Cotinga is routinely found in the same areas as the Plum-throated Cotinga (*C. maynana*), and sometimes the two even can be found feeding simultaneously in the same fruiting trees, rarely with Purple-throated Cotingas (*Porphyrolaema porphyrolaema*) present as well. Recorded mostly below about 400 m, locally to 600 m (seen near Canelos in Pastaza; N. Krabbe).
Monotypic.
Range: the Guianas and s. Venezuela to n. Bolivia and Amazonian Brazil.

Xipholena punicea   **Pompadour Cotinga**   Cotinga Púrpura

Uncertain. Apparently very rare and local in the canopy of terra firme forest in the lowlands of se. Ecuador. The Pompadour Cotinga is known in Ecuador from only a single specimen, a female taken by A. Proaño at "Río Corrientes" on 23 Apr. 1964 (ANSP). This locality cannot be precisely located but is believed to be in se. Pastaza near where the Río Bobonaza empties into the Río Pastaza. The Pompadour Cotinga is not otherwise known from any closer than ne. Peru in sandy-soil forests west of Iquitos (fide J. Alvarez A.). There have been no subsequent Ecuadorian records of this spectacular cotinga, though we expect that it will eventually be found in the area surrounding Kapawi Lodge. Recorded at about 200 m.
Monotypic.
Range: the Guianas and s. Venezuela to cen. Amazonian Brazil and ne. Bolivia; isolated records from se. Ecuador and ne. Peru.

Carpodectes hopkei   **Black-tipped Cotinga**   Cotinga Blanca

Rare to locally fairly common in the canopy and borders of very humid and humid forest and in tall trees of adjacent clearings in the lowlands of nw. Ecuador south to Pichincha. The Black-tipped Cotinga is now found mainly in Esmeraldas, and the species was, for instance, found to be quite common northwest of Alto Tambo in Jul. 1990 (M. B. Robbins et al.); it also is still reasonably numerous in remaining forests southeast of Muisne (ANSP). It likewise seems to be persisting quite well in the forest patches remaining in nw. Pichincha and adjacent sw. Imbabura (v.o.). There is a 1977 specimen (MECN) from as far south as Santo Domingo de los Colorados, and presumed wandering individuals are still seen occasionally even farther south (e.g., at Río Palenque and Tinalandia). Most numerous below 500 m, ranging up in smaller numbers to 700 m at least seasonally.
The Black-tipped Cotinga was given Near-threatened status by Collar et al. (1992, 1994). We concur.
The species is sometimes called the White Cotinga.
Monotypic.
Range: e. Panama to nw. Ecuador.

Gymnoderus foetidus   Bare-necked   Cuervo Higuero
                         Fruitcrow   Cuellopelado

Fairly common and generally conspicuous along the edge of lakes and rivers, on islands, and in the canopy of humid forest (both várzea and terra firme) in the lowlands of e. Ecuador. In extensively forested areas the Bare-necked Fruitcrow is mainly seen flying overhead (sometimes high above the ground, unlike other cotingas), and the species is apparently prone to wandering and may occur only seasonally at some localities. Recorded mainly below about 300 m, occasionally wandering as high as 400 m at Jatun Sacha (B. Bochan).
Monotypic.
Range: the Guianas and s. Venezuela to n. Bolivia and Amazonian Brazil.

Querula purpurata   Purple-throated Fruitcrow   Querula Golipúrpura

Uncommon to locally common in the mid-levels and subcanopy of humid forest (mainly terra firme) and mature secondary woodland in the lowlands of both e. and w. Ecuador. In the west there are old records of the Purple-throated Fruitcrow from as far south as e. Guayas (Bucay), but no recent reports exist from south of s. Pichincha at Río Palenque (where small numbers persist). In the wet-forest belt of n. Esmeraldas it occurs predominantly in forest-edge situations and in secondary woodland, avoiding continuous forest (O. Jahn et al.). Purple-throated Fruitcrows seem to be particularly numerous in the patchy forests of nw. Pichincha. Recorded mostly below about 500 m, a few occurring locally as high as 700 m (e.g., at Tinalandia; P. Coopmans).
Monotypic.
Range: Costa Rica to w. Ecuador; the Guianas and s. Venezuela to n. Bolivia and Amazonian Brazil.

Pyroderus scutatus   Red-ruffed Fruitcrow   Cuervo Higuero Golirrojo

Very rare and local in humid forest of the lowlands and foothills of nw. Ecuador. There are only a few Ecuadorian records of this spectacular fruitcrow, including specimens from Imbabura (Hacienda Paramba), Carchi (Las Tablas), and Pichincha (Santo Domingo de los Colorados). The only recent report is of a single bird of unknown sex seen on 2 Aug. 1997 in the valley southeast of Maldonado in Carchi, just south of Hacienda Puente Palo (D. Lane and F. Sornoza). Prior to that, the most recent report had come from 1972. We are at a loss to explain this species' evident rarity in Ecuador, for there is still a considerable expanse of little-modified habitat in some of its Ecuadorian range, and there has also been considerable recent field work in the region; the species is more numerous in the Andes of Colombia. Recorded up to about 1500 m.
Given the paucity of Ecuadorian data on the inexplicably rare Red-ruffed Fruitcrow, we give the species Data Deficient status for Ecuador. It was not considered to be generally at risk by Collar et al. (1992, 1994).
One race: *occidentalis*.

Range: locally in n. Guyana and se. Venezuela; mountains of n. Venezuela, and Andes of Venezuela, Colombia, nw. Ecuador, and n. Peru; se. Brazil, e. Paraguay, and ne. Argentina.

Cephalopterus ornatus	Amazonian	Pájaro Paraguas
	Umbrellabird	Amazónico

Rare to locally fairly common in two distinctly different habitats, in the canopy and borders of riverine and várzea forest and on river islands in the eastern lowlands, and in the canopy and borders of montane forest in the foothills and lower subtropical zone along the eastern base of the Andes. No contact seems to exist between the two populations. The lowland population mainly occurs below 300 m (occasionally wandering as high as 400 m at Jatun Sacha, fide B. Bochan), whereas the foothill population occurs mainly between about 900 and 1300 m (though an early specimen was recorded as having been taken at Hacienda Mapoto in Tungurahua, at 2100 m—though it may actually have been obtained lower down).

The Amazonian Umbrellabird appears to be declining in both parts of its disjunct distribution, in the foothills mainly from habitat destruction, along lowland rivers presumably from hunting and disturbance. It may deserve formal threatened status in Ecuador (and elsewhere?) but was not considered to be at risk by Collar et al. (1992, 1994).

Monotypic. Despite the elevational split in its range, no subspecies are recognized.

Range: e. Colombia, s. Venezuela, and s. Guyana to n. Bolivia and Amazonian Brazil.

Cephalopterus penduliger	Long-wattled	Pájaro Paraguas
	Umbrellabird	Longuipéndulo

Rare to locally uncommon in the canopy and borders of humid forest in the foothills and more humid lowlands of w. Ecuador south to El Oro (La Chonta; a few have also been seen in recent years at Buenaventura). It has long been thought (e.g., Ridgely and Tudor 1994) that the Long-wattled Umbrellabird might engage in seasonal altitudinal movements, though the details of such movements were never established. In the wet-forest belt of n. Esmeraldas at Playa de Oro, however, the species is now known to range down regularly as low as about 200 m, and more or less permanently occupied leks have been found at that elevation (O. Jahn, E. E. Vargas Grefa, and K.-L. Schuchmann, *Bird Conserv. Int.* 9[1]: 81–94, 1999). In the past there reportedly were leks down to 100 m (O. Jahn, pers. comm.). Thus whether altitudinal movements in fact take place still remains unknown. The spectacular Long-wattled Umbrellabird is often encountered at El Placer in Esmeraldas; an apparently resident population was also found in 1994 at Bilsa in the coastal Cordillera de Mache in w. Esmeraldas (first by R. P. Clay et al.). Recorded up occasionally to about 1500 m, but usually between about 150 and 1100 m.

With the massive deforestation that has taken place across much of w. Ecuador in the past half-century, the Long-wattled Umbrellabird has, not unexpectedly, become much rarer and more local, and lowland occurrences have become unusual (though the odd individual still occasionally turns up even at now-isolated Río Palenque). The species was accorded Vulnerable status by Collar et al. (1994), and we agree with this assessment. At least locally (e.g., in the Playa de Oro region of n. Esmeraldas) this umbrellabird is still hunted for food (O. Jahn et al.).

Monotypic.

Range: sw. Colombia and w. Ecuador.

Phoenicircus nigricollis   Black-necked   Cotinga Roja Cuellinegra
Red-Cotinga

Rare to uncommon and decidedly local in the lower and (especially) middle growth of terra firme forest and (less often) at forest borders in the lowlands of e. Ecuador. Recorded up only to about 400 m (Taisha [ANSP]; there is only a single observation at Jatun Sacha, fide B. Bochan).

The beautiful Black-necked Red-Cotinga appears to have been reduced in overall numbers and is now found only locally in Ecuador; habitat destruction has doubtless played a role in this decline, as perhaps also has persecution by some indigenous peoples, some of whom covet this species' red feathers more than those of any other bird. The Black-necked Red-Cotinga remains tolerably numerous in forests around Zancudococha, and there is a lek near the tower at Yuturi Lodge, but inexplicably it appears to be extremely scarce and local along the nearly uninhabited Maxus road southeast of Pompeya (RSR), a road that traverses vast expanses of undisturbed and presumably suitable terra firme forest. The species may have always been relatively scarce and local, so we therefore consider it to be only Near-threatened; it was not considered to be at risk by Collar et al. (1992, 1994).

Monotypic.

Range: se. Colombia and extreme sw. Venezuela to ne. Peru and w. and cen. Amazonian Brazil.

Rupicola peruviana   Andean Cock-of-   Gallo de la Peña Andino
the-rock

Uncommon to locally fairly common in and near forested gorges and ravines in the foothills and subtropical zone on both slopes of the Andes. On the west slope not definitely recorded south of Cotopaxi (male seen east of Quevedo on 28 Jul. 1991 [RSR and G. H. Rosenberg] and in the Caripero area [N. Krabbe and F. Sornoza]), but reports from local residents indicate that cocks-of-the-rock may occur south to n. Azuay (e.g., above Manta Real). Substantial numbers occur along the Nono-Mindo road (especially near Tandayapa), at certain sites around Mindo, and at San Rafael Falls, but they can often be frustratingly hard to see well. The Andean Cock-of-the-rock has a wide elevational range, occurring as low as 600 m where there is appropriate habitat (e.g., at El Placer in Esmeraldas), with small numbers ranging as high as about 2500 m; most numerous from 900 to 2100 m.

Andean Cocks-of-the-rock are capable of persisting in rather deforested areas, mainly because the steep gorges they favor are usually the last places to be cut, and because their nest sites are usually so inaccessible. Although overall numbers have doubtless declined to an unknown extent, we do not consider the species to be at risk, nor do Collar et al. (1992, 1994).

Two quite different races occur in Ecuador, *sanguinolenta* on the west slope and *aequatorialis* on the east slope. Males of the former are brilliant blood red, whereas males of the latter are reddish orange; female *sanguinolenta* are also notably redder than their eastern counterparts.

Range: Andes from w. Venezuela to w. Bolivia.

Pipridae   Manakins

Twenty species in 13 genera. Manakins are exclusively Neotropical in distribution. The taxonomic affinity of four genera here placed with the manakins (*Tyranneutes*, *Piprites*, *Schiffornis*, and *Sapayoa*) remains uncertain.

Pipra erythrocephala   Golden-headed   Saltarín Capuchidorado
                          Manakin

Fairly common in the lower growth of humid forest (principally terra firme) and mature secondary woodland, also occasionally out to forest borders, in the lowlands of e. Ecuador. Ranges mostly below about 600 m, in smaller numbers up into the foothills to about 1100 m, rarely (wanderers?) a bit higher.

One race: *berlepschi*.

Range: e. Panama to n. and e. Colombia, ne. Peru, n. Amazonian Brazil, and the Guianas.

Pipra mentalis   Red-capped Manakin   Saltarín Cabecirrojo

Fairly common to locally common in the lower and middle growth of very humid and humid forest and mature secondary woodland, also occasionally out to borders, in the lowlands of w. Ecuador. Now found mainly in Esmeraldas and nw. Pichincha, the Red-capped Manakin formerly occurred south at least as far as n. Manabí (Río de Oro, Río Peripa), n. Guayas (Pacaritambo), and n. Los Ríos (a series of specimens [ANSP] taken in Oct. 1950 by T. Mena). In addition, along the base of the Andes a male was mist-netted in late Jan. 1991 as far south as San Luis in nw. Azuay (J. C. Matheus and F. Sornoza), and a female (ANSP) was mist-netted on 17 Aug. 1991 at Manta Real just north of that site; the species is obviously rare this far south, and there were no other records. Recorded below about 500 m.

Most of lowland w. Ecuador has been severely deforested in recent decades, and overall numbers of the Red-capped Manakin have undergone a massive decline as a result; for instance, the species has apparently gone extinct at Río Palenque, having persisted there into the 1970s (R. Webster, RSR). Nonetheless, as the species remains so numerous in the far northwest, we do not believe that it yet warrants formal status listing in Ecuador.

One race: *minor*.

Range: s. Mexico to w. Ecuador.

Pipra filicauda   **Wire-tailed Manakin**   Saltarín Cola de Alambre

Fairly common to locally common in undergrowth inside humid forest (both terra firme and várzea, in the former primarily along streams) in the lowlands of e. Ecuador. Ranges mostly below 500 m, but recorded in small numbers as high as 700–750 m.

Formerly placed in the monotypic genus *Teleonema* (e.g., Meyer de Schauensee 1966, 1970), but that is now generally subsumed into *Pipra*. One race occurs in Ecuador, nominate *filicauda*, which apparently intergrades with the similar *subpallida* near the Colombian border.

Range: n. and w. Venezuela to ne. Peru and w. Amazonian Brazil.

Dixiphia pipra   **White-crowned Manakin**   Saltarín Coroniblanco

Rare to locally fairly common in lower growth of montane forest in the foothills and lower subtropical zone on the east slope of the Andes, in smaller numbers also out locally into hilly terra firme forest in the eastern lowlands (e.g., at "Laguna de Siguin," Montalvo, and Kapawi Lodge in the Río Pastaza drainage; and on the south side of the Río Napo near La Selva and along the Maxus road southeast of Pompeya). Recorded mainly from 500 to 1500 m, in small numbers as low as 250 m.

This genus *Dixiphia* was separated from *Pipra* by R. O. Prum (*Condor* 96[4]: 692–702, 1994; *Evolution* 48: 1657–1675, 1994). Based on their different displays and vocalizations, the birds of Andean slopes and w. Amazonia, *D. coracina* (Sclater's Manakin), may deserve to be specifically separated from birds of the e. Amazonian lowlands, true *D. pipra* (White-crowned Manakin). One race of *D. pipra* occurs in Ecuador, *coracina*.

Range: locally from Costa Rica to sw. Colombia and e. Peru, more widely east through Amazonian Brazil and s. Venezuela to the Guianas.

Lepidothrix coronata   **Blue-crowned Manakin**   Saltarín Coroniazul

Fairly common to common in undergrowth inside very humid and humid forest as well as taller secondary woodland in the lowlands of both e. and nw. Ecuador; in the east occurs mostly in terra firme. In the west the Blue-crowned Manakin is now found mainly in Esmeraldas, with small numbers occurring locally south into n. Manabí and n. Pichincha (though it seems to have disappeared from Río Palenque); there is a specimen in ANSP taken in Oct. 1950 by T. Mena from as far south as Quevedo in Los Ríos. Recorded up to about 900 m in the Bermejo oil-field area north of Lumbaquí (M. B. Robbins).

This and the following species, formerly placed in the genus *Pipra*, were separated in the genus *Lepidothrix* by R. O. Prum (*Condor* 96[4]: 692–702, 1994; *Evolution* 48: 1657–1675, 1994). Two similar races occur in Ecuador, nominate *coronata* in the east and *minuscula* in the west. Males of the latter are slightly deeper black, and the blue of their crowns is slightly darker, a more ultramarine (not azure) blue. Females of the two are very similar.

Range: Costa Rica to w. Ecuador; s. Venezuela and se. Colombia to nw. Bolivia and w. Amazonian Brazil.

Lepidothrix isidorei   **Blue-rumped Manakin**   **Saltarín Lomiazul**

Uncommon to locally fairly common but easily overlooked in undergrowth inside montane forest and forest borders in the foothills and lower subtropical zone on the east slope of the Andes from w. Napo (San Rafael Falls; ANSP) south to Zamora-Chinchipe (Zumba area and Podocarpus National Park). The lovely Blue-rumped Manakin is perhaps most numerous near Zamora, especially in the Río Bombuscaro sector of Podocarpus National Park. Whether this species and the Blue-crowned Manakin (*L. coronata*) are ever sympatric remains unknown; as yet they have not been found to occur together, though their elevational ranges do come very close in some places. Recorded mostly from 1000 to 1700 m, in small numbers down to about 800 m (Warientza; WFVZ).

One race: nominate *isidorei*. Birds of n. Peru, *leucopygia*, may represent a separate species (Milky-rumped Manakin), in which case *L. isidorei* would become monotypic.

Range: locally on east slope of Andes in s. Colombia, Ecuador, and n. Peru.

Chiroxiphia pareola   **Blue-backed Manakin**   **Saltarín Dorsiazul**

Uncommon to locally fairly common in undergrowth inside humid forest (mainly terra firme) in the lowlands of e. Ecuador south through Pastaza and Morona-Santiago. A single report (Schulenberg and Awbrey 1997) from the far southeast in the Río Nangaritza valley of Zamora-Chinchipe at Miazi (900 m) is unique. Ranges mostly below 500 m, but recorded in small numbers up to 700–750 m.

One race: *napensis*.

Range: se. Colombia, e. Ecuador, most of e. Peru, extreme nw. and ne. Bolivia, most of Amazonian Brazil, the Guianas, and se. Venezuela; e. Brazil. Excludes Yungas Manakin (*C. boliviana*) of Andean slopes in s. Peru and Bolivia.

Masius chrysopterus   **Golden-winged Manakin**   **Saltarín Alidorado**

Uncommon to fairly common (but easily overlooked) in the lower and middle growth of montane forest and forest borders in the foothills and subtropical zone on both slopes of the Andes. On the west slope recorded south to El Oro (e.g., Buenaventura, where it remains quite numerous) and w. Loja (old specimens from Alamor and Las Piñas; no recent reports). Also recently recorded from the coastal Cordillera de Mache at Bilsa in w. Esmeraldas, where first found in Feb. 1994 (R. P. Clay et al.). Recorded mostly from 800 to 2000 m, but locally ranges as low as 400 m in the west (e.g., above Manta Real in nw. Azuay; ANSP).

Three races have been recorded from Ecuador. *Coronulatus* occurs on the entire west slope, *pax* on most of the east slope, and the very similar *peruvianus* was recently found to be the form occurring in s. Zamora-Chinchipe at Quebrada Avioneta in Podocarpus National Park and Panguri (ANSP). Male *coronulatus* have flame orange on the center of the nape, whereas male *pax* have their blunter, shiny, and scale-like mid-nape feathers much browner; male *peruvianus* resemble male *pax*.

Range: Andes from w. Venezuela to n. Peru.

Manacus manacus   **White-bearded Manakin**   Saltarín Barbiblanco

Fairly common in undergrowth (often dense) of secondary woodland and forest borders in the lowlands of both e. and w. Ecuador. In the west ranges south locally through El Oro and w. Loja (old specimens from Alamor; recent reports only from near El Limo [RSR]). In most areas occurs only up to about 800 m, but in the Alamor region recorded as high as about 1300 m.

Four described races, all of them quite similar (even in males; females are identical), are recorded from Ecuador. *Interior* ranges in the east, whereas in the west the following are found: *bangsi* (n. Esmeraldas and Imbabura), *leucochlamys* (w. Esmeraldas, Manabí, and Pichincha south through Guayas and adjacent Azuay), and *maximus* (e. Guayas and adjacent Chimborazo south through El Oro into w. Loja). Some of these should very likely be synonymized, in particular the western taxa.

Range: n. and e. Colombia, w. and s. Venezuela, and the Guianas to ne. Bolivia and Amazonian Brazil; sw. Colombia to extreme nw. Peru; e. Brazil, e. Paraguay, and ne. Argentina.

Birds of the World (vol. 8, p. 261) states that the Golden-collared Manakin (*M. vitellinus viridiventris*) occurs in n. Esmeraldas. In fact, however, it ranges south only to sw. Cauca in Colombia (Hilty and Brown 1986; Ridgely and Tudor 1994), being replaced there by the White-bearded.

Machaeropterus regulus   **Striped Manakin**   Saltarín Rayado

Uncommon to fairly common (but easily overlooked except through its vocalizations) and seemingly local in the lower and middle growth inside terra firme forest in the lowlands and foothills of e. Ecuador. In the southeast recorded up to about 1000–1100 m on the lower slopes of the Cordillera de Cutucú and in the Zamora area and Río Nangaritza valley, but northward usually not found above 600–700 m.

One race: *striolatus*.

Range: n. Colombia; w. and s. Venezuela, e. Colombia, e. Ecuador, ne. Peru, and w. Amazonian Brazil; e. Brazil. More than one species may be involved.

Machaeropterus deliciosus   Club-winged Manakin   Saltarín Alitorcido

Uncommon to fairly common, but seemingly local (and usually inconspicuous), in lower and middle growth inside montane forest in the foothills and lower subtropical zone on the west slope of the Andes, also ranging at least locally out into more humid lowlands (perhaps especially in Esmeraldas, e.g., at Playa de Oro [O. Jahn]). Found mostly from Pichincha northward, but since Jan. 1988 Club-winged Manakins have also been seen regularly at Buenaventura in El Oro; curiously, intensive work there in the mid-1980s did not reveal the presence of the species, and we suspect that it may actually have been absent. In addition, a single male (MECN) was collected by M. Olalla at Guaiquichuma (1500 m) in w. Loja on 23 Dec. 1968. Ranges mostly from 600 to 1500 m, but apparently undertakes a distinct seasonal elevational migration, with nonbreeding birds being found regularly down to below 100 m at Playa de Oro during the dry season (O. Jahn et al.). In addition, ANSP has two specimens taken in Oct. 1950 by T. Mena near Quevedo (100 m) in Los Ríos, and in the 1970s the Club-winged Manakin was also seen on a few occasions at Río Palenque.

The species was formerly (e.g., Meyer de Schauensee 1966, 1970) placed in the monotypic genus *Allocotopterus*, but that is now generally subsumed into *Machaeropterus* (e.g., Sibley and Monroe 1990; Ridgely and Tudor 1994). Monotypic.

Range: west slope of Andes in sw. Colombia and w. Ecuador.

Chloropipo holochlora   Green Manakin   Saltarín Verde

Uncommon to fairly common (but easily overlooked) in undergrowth inside humid forest in the foothills and adjacent lowlands of both e. and w. Ecuador. In the east not found any great distance east of the Andes (and there largely restricted to hilly terra firme forest), with the farthest east reports being a few records from along the Maxus road southeast of Pompeya (MECN; P. English et al.). In the west the Green Manakin is now found mainly in Esmeraldas and w. Imbabura, though there is one recent report from as far south as nw. Azuay (a female mist-netted and collected above Manta Real on 21 Aug. 1991; ANSP). There are also a few sightings from Río Palenque in s. Pichincha (e.g., in Aug. 1986; RSR and PJG et al.), and a specimen (ANSP) was taken by T. Mena in Oct. 1950 near Quevedo in Los Ríos. Prior to extensive deforestation, Green Manakins likely were more numerous and widespread through the humid western lowlands. In the west recorded mostly from 400 to 1100 m, but down to 100 m in n. Esmeraldas at Playa de Oro (O. Jahn and P. Mena Valenzuela); southward, small numbers occur lower (to 200 m) at least locally. On the east slope recorded mostly between 300 and 1200 m, locally as high as 1500 m (Palm Peak; MCZ), there in sympatry with the Jet Manakin (*C. unicolor*).

Two rather different races are found in Ecuador, nominate *holochlora* in the east and *litae* in the west. *Litae* is duller, more olive above, and darker on the throat and breast; the nominate race is a brighter, more mossy green above. It has been suggested (R. O. Prum, pers. comm.) that *litae* may represent a separate species (Chocó Manakin).

Range: e. Panama to w. Ecuador; se. Colombia, e. Ecuador, and e. Peru.

*Chloropipo unicolor   Jet Manakin   Saltarín Azabache

Uncommon and apparently local (but easily overlooked) in undergrowth inside montane forest in the subtropical zone on the east slope of the Andes, with numbers perhaps being highest on outlying ridges. Although only first recorded from Ecuador in Aug. 1979 (see Ridgely 1980), since then there have been several records of the Jet Manakin from w. Napo (Volcán Sumaco region) southward. The localities from which it has been recorded are Palm Peak on Volcán Sumaco (MCZ), the Río Upano valley in ne. Morona-Santiago (AMNH; Ridgely 1980), the Cordillera de Cutucú (ANSP and MECN), the Cordillera del Cóndor (ANSP; Krabbe and Sornoza 1994), and Panguri in s. Zamora-Chinchipe (ANSP, MECN). The Jet Manakin has so far been recorded only from within a narrow elevational zone, 1450–1700 m, but it may also occur slightly higher and lower.

Monotypic.

Range: east slope of Andes in Ecuador and Peru.

Chloropipo flavicapilla   Yellow-headed Manakin   Saltarín Cabeciamarillo

Very rare (but probably overlooked) in undergrowth inside montane forest in the subtropical zone on the east slope of the Andes, where as yet recorded only from w. Napo and Tungurahua. There are only three Ecuadorian records of this beautiful manakin: a 19th-century specimen (BMNH) from Hacienda Mapoto in Tungurahua; a female seen on the south slope of the Cordillera de Huacamayos above Archidona on 10 Nov. 1990 (PJG); and a male seen on the south slope of Volcán Sumaco on 22 Jan. 1991 (B. Whitney). Recorded from about 1500 to 2100 m.

The Yellow-headed Manakin was given Near-threatened status by Collar et al. (1994). Given its rarity and very limited range in Ecuador, we favor according it Vulnerable status here, and we suggest that it may deserve upgrading in Colombia as well.

Monotypic.

Range: locally in Andes of Colombia and n. Ecuador.

Heterocercus aurantiivertex   Orange-crested Manakin   Saltarín Crestinaranja

Rare to locally uncommon in the lower growth of várzea forest and woodland in the lowlands of e. Ecuador, mainly or perhaps entirely confined to blackwater regions. Recorded mostly from the drainages of the Ríos Napo and

Aguarico; there is also a 19th-century record from Sarayacu, and a series of specimens was obtained at "Laguna de Siguin" (Berlioz 1937c), but there have been no recent reports from anywhere in the Río Pastaza drainage. The Orange-crested Manakin—a virtual Ecuadorian endemic—has recently been reported at a scatter of localities near the Río Napo (Taracoa, Sacha Lodge, La Selva, and Yuturi Lodge) and the Río Aguarico (Cuyabeno, Zancudococha, and Imuyacocha and Garzacocha on the Río Lagarto). Recorded only below 300 m.

This species was formerly usually called the Orange-crowned Manakin, though it is only the usually inconspicuous crest—and not the entire crown—that is colored; see Ridgely and Tudor (1994).

Monotypic.

Range: e. Ecuador and (fide J. Alvarez J.) locally in ne. Peru; probably also in adjacent se. Colombia.

Tyranneutes stolzmanni   Dwarf Tyrant-Manakin   Saltarincillo Enano

Fairly common (but easily overlooked, except through its vocalizations) in the lower and middle growth of humid forest (both terra firme and várzea, but especially the former) in the lowlands of e. Ecuador. Ranges mostly below 500 m, locally up to 900 m at the Bermejo oil-field area north of Lumbaquí (M. B. Robbins) and to 900–1000 m at Zamora and in the Río Nangaritza valley at Miazi (Schulenberg and Awbrey 1997) and Shaime (Balchin and Toyne 1998).

The relationships of the genus *Tyranneutes* remain uncertain, though it does not seem to be a "true" manakin (R. O. Prum, pers. comm.). For now we retain it with the manakins. Monotypic.

Range: se. Colombia and sw. Venezuela to n. Bolivia and Amazonian Brazil.

Piprites chloris   Wing-barred Piprites   Piprites Alibandeado

Uncommon and seemingly local in the subcanopy and mid-levels of terra firme forest, forest borders, and mature secondary woodland in the lowlands and foothills of e. Ecuador. Meyer de Schauensee (*Not. Naturae* no. 234, 1951, p. 7) discusses a specimen (ANSP) of the Wing-barred Piprites that was procured by L. Gómez supposedly at "La Palma," a site in w. Pichincha; it was identified by Meyer de Schauensee as referring to the taxon *antioquiae*, which is otherwise known only from w. Colombia. The specimen is actually typical of *tschudii* (found east of the Andes), however; indeed, Meyer de Schauensee (op. cit.) states that it "does not differ appreciably from the Amazonian birds." We suspect that mislabeling was involved and that the specimen was actually procured somewhere in e. Ecuador; the Wing-barred Piprites is not otherwise recorded from west of the Andes in Ecuador. This is the source for the citation of *antioquiae* from Pichincha (*Birds of the World*, vol. 8, p. 250). Recorded mostly below about 1100 m, with small numbers recorded as high as 1500 m on the slopes of Volcán Sumaco (B. Whitney) and in the Río Upano valley in w. Morona-Santiago (RSR).

The species was formerly called the Wing-barred Manakin. See Ridgely and

Tudor (1994) for the rationale for no longer calling members of the genus "manakins," followed in the 1998 AOU Check-list.

Although still uncertain, there is evidence indicating that the affinities of the genus *Piprites* do not lie with the Pipridae (R. O. Prum, pers. comm.); its arboreal behavior and overall comportment certainly are not manakin-like. As noted above, only one race, *tschudii*, occurs in Ecuador.

Range: n. and e. Colombia, much of Venezuela (but not in llanos), and the Guianas to n. Bolivia and Amazonian Brazil; se. Brazil, e. Paraguay, and ne. Argentina.

Schiffornis turdinus   **Thrush-like Schiffornis**   **Chifornis Pardo**

Uncommon to locally fairly common (recorded almost entirely by voice) in undergrowth inside humid forest and mature secondary woodland in the lowlands, foothills, and lower subtropical zone of both e. and w. Ecuador. In the west recorded south to n. Manabí and s. Pichincha, thence south along the base of the Andes to se. Guayas (hills of Manglares-Churute Ecological Reserve; Pople et al. 1997), El Oro, and w. Loja (old specimens from Alamor and La Puente; no recent reports); the species also has been found on the coastal Cordillera de Colonche at Loma Alta (Becker and López-Lanús 1997). In the east found primarily in the foothills and lower subtropical zone, occurring only very locally in the lowlands, where it favors terra firme forest. Most of the lowland sites from which the species is known are in the far northeast: an AMNH specimen from Río Lagarto, a specimen (ANSP) from north of Tigre Playa taken on 7 Aug. 1993, and a few reports from Zancudococha, Cuyabeno, and Yuturi. In addition, in 1994–1995 the species was found to be not uncommon along the Maxus road southeast of Pompeya (RSR et al.); it has also recently been found in the far southeast around Kapawi Lodge on the Río Pastaza near the Peruvian border. In the west recorded up to about 1300 m, on the east slope recorded from 900 m up at least locally to as high as 1700 m (Cordillera de Cutucú in Morona-Santiago; ANSP). Birds of the eastern lowlands are recorded only from 200 to 300 m.

R. O. Prum and W. E. Lanyon (*Condor* 91[2]: 444–461, 1989) presented evidence showing that the members of the genus *Schiffornis* are not true manakins but rather should be considered part of a separate "*Schiffornis* assemblage"; these authors suggested that the English group name be switched to "mourner" to reflect this relationship, and this usage was followed by Ridgely and Tudor (1994). However, in view of the continuing uncertainty as to the affinities of the genus (and other apparently unrelated genera), we opt to follow the 1998 AOU Check-list in employing the group name of "Schiffornis," in accord with the English group names for the genera *Sapayoa* and *Piprites*. Formerly the species was called the Thrush-like Manakin.

Three races are found in Ecuador: *rosenbergi* in the west, *aeneus* on the east slope of the Andes, and *amazonus* very locally in the northeastern lowlands. The first two are quite similar in appearance, though *aeneus* has a browner crown. *Amazonus*, however, differs strikingly in being paler generally (less brown and more olive above, except for its rufescent brown crown) and it is decidedly more grayish below. More than one species is almost surely involved.

Range: s. Mexico to extreme nw. Peru, n. Bolivia, and Amazonian Brazil; e. Brazil.

Schifformis major   Várzea Schiffornis   Chifornis de Várzea

Uncommon and local (recorded mainly by voice) in the undergrowth of várzea forest in the lowlands of e. Ecuador; largely avoids blackwater areas. Although first recorded from Ecuador only in the 1970s (see Ridgely 1980), the Várzea Schiffornis has since been found at a number of localities in the drainages of the Río Napo (upriver to the Taracoa and Limoncocha areas) and Río Aguarico (upriver to Cuyabeno). In the southeast it is known only from recent reports from Kapawi Lodge on the Río Pastaza near the Peruvian border. There still is no Ecuadorian specimen of the species. Recorded only below about 300 m.

The species was formerly called the Greater Manakin (e.g., Meyer de Schauensee 1966, 1970). R. O. Prum and W. E. Lanyon (*Condor* 91[2]: 444–461, 1989) point out that this inaccurate English name was simply the result of the larger *S. turdinus* (Thrush-like Schiffornis) being placed in a separate genus when *S. major* was described. They suggested calling *S. major* either the Cinnamon or the Várzea Mourner. In view of its strict association with várzea forest, we opt to employ the latter modifier, following Ridgely and Tudor (1994); we follow the 1998 AOU Check-list in using Schiffornis as the group name for the genus.

One race: nominate *major*.

Range: sw. Venezuela and se. Colombia to n. Bolivia and w. Amazonian Brazil.

Sapayoa aenigma   Broad-billed Sapayoa   Sapayoa

Rare to locally uncommon (but often overlooked) in the lower and middle growth of very humid forest in the lowlands of nw. Ecuador in Esmeraldas and nw. Pichincha (along the road north of Simón Bolívar near Pedro Vicente Maldonado, where one was seen on 24 Sep. 1995 by J. Nilsson, fide PJG). Favors areas near streams. There are relatively few Ecuadorian records of the unobtrusive Broad-billed Sapayoa. That it may locally not be all that rare is borne out by M. Olalla's having collected no fewer than six specimens (MECN) in a fairly short period of time in 1963 and 1964; it also is not at all rare at Playa de Oro (O. Jahn et al.). Recorded up to about 500 m.

The Broad-billed Sapayoa is scarce and highly range-limited in Ecuador; we thus favor giving it Near-threatened status here. Elsewhere it is not considered to be at risk (Collar et al. 1992, 1994), though in our view this assessment should perhaps be revisited.

The species was formerly called the Broad-billed Manakin, but we opt to follow Ridgely and Tudor (1994) in calling it the Broad-billed Sapayoa, in recognition of the fact that the species is certainly not a typical manakin. The 1998 AOU Check-list opted to shorten the species' name to simply Sapayoa, but as the bill is indeed quite broad, we conclude that it is worth retaining the modifier.

As with *Piprites*, there is recent evidence indicating that the affinities of *Sapayoa* lie not with the Pipridae but more likely with the Tyrannidae. We retain it with the manakins; the 1998 AOU Check-list left it as incertae sedis (of "uncertain placement") along with several other genera (*Schiffornis*, *Piprites*, *Lipaugus*, *Laniocera*, *Pachyramphus*, *Platypsaris*, and *Tityra*). Monotypic.

Range: Panama to nw. Ecuador.

Corvidae   Crows, Jays, and Magpies

Six species in two genera. The family is worldwide in distribution, though only jays occur in most parts of the Neotropics.

Cyanolyca armillata   Black-collared Jay   Urraca Negricollareja

Very rare in montane forest in the upper subtropical and lower temperate zones on the east slope of the Andes in n. Ecuador. Recorded from extreme e. Carchi at El Pun (19th century specimens; Salvadori and Festa 1899a), w. Sucumbíos (along the La Bonita road), and w. Napo at Oyacachi (an ANSP specimen taken by A. Proaño at Oyacachi on 16 Oct. 1950; also recent sightings [Krabbe et al. 1997]). The only recent confirmed records come from Oyacachi and from along the La Bonita road; at the latter, a specimen (MECN) was taken at La Alegria (2320 m) on 17 Nov. 1984 by J. C. Matheus, and birds were heard near Quebrada Las Ollas in Sep. 1991 by N. Krabbe. It is curious that there are no reports of the Black-collared Jay from the well-watched Papallacta/Cuyuja area just south of Oyacachi; the few recent reports remain unsubstantiated and are believed to be the result of confusion with the far more numerous Turquoise Jay (*C. turcosa*); however, in the Río Oyacachi valley just to the north, only the Black-collared Jay occurs (Krabbe et al. 1997). Recorded between about 2100 and 3150 m.

Given the Black-collared Jay's apparent rarity and limited range in Ecuador, and the deforestation that is taking place in much of its range, we assess its status as Data Deficient. Virtually nothing is known about the factors governing its numbers and distribution in Ecuador. The species was not regarded as being generally at risk by Collar et al. (1992, 1994).

C. armillata of Venezuela to Ecuador is regarded as a species distinct from the White-collared Jay (*C. viridicyana*) of Peru and Bolivia, following Ridgely and Tudor (1989) and most other recent authors. The expanded species was formerly called the Collared Jay. One race occurs in Ecuador, *quindiuna*. We treat *angelae* as a synonym of *quindiuna*, following most recent authors (e.g., Fjeldså and Krabbe 1990).

Range: Andes from w. Venezuela to n. Ecuador.

Cyanolyca turcosa   Turquoise Jay   Urraca Turquesa

Fairly common and widespread in montane forest, forest borders, and adjacent clearings (favoring alders [*Alnus*]) in the upper subtropical and temperate zones on both slopes of the Andes. On the west slope recorded south to El Oro (Taraguacocha). A few Turquoise Jays persist in remnant forest patches above the central valley, such as at Pasochoa. Most numerous from 2000 to 3000 m, occasionally wandering as high as 3500 m (e.g., at Yanacocha; N. Krabbe).

Monotypic.

Range: Andes from s. Colombia to extreme n. Peru.

Cyanolyca pulchra   Beautiful Jay   Urraca Hermosa

Rare to locally uncommon in montane forest and forest borders in the subtropical zone on the west slope of the Andes from Carchi south to Pichincha (where recorded south to along the Chiriboga road). The Beautiful Jay is much less conspicuous than the far more numerous Turquoise Jay (*C. turcosa*), which occurs almost entirely at higher elevations. Small numbers continue to be found in the Tandayapa area and above Mindo. Recorded mostly between 1300 and 2000 m.

The Beautiful Jay was given Near-threatened status by Collar et al. (1992, 1994). We agree, though registering the concern that continuing deforestation in its very limited range may have caused the species to have become more seriously at risk. At least in Pichincha it seems to have undergone a substantial decline in recent years, even in areas where seemingly suitable forest cover remains.

Monotypic.

Range: west slope of Andes in w. Colombia and nw. Ecuador.

Cyanocorax violaceus   Violaceous Jay   Urraca Violácea

Fairly common to common in humid forest borders (both terra firme and várzea), secondary woodland, and on river islands in the lowlands of e. Ecuador. Occurs mostly below 500 m, but smaller numbers range up locally along major river valleys into the foothills on the east slope of the Andes to as high as 900 m, locally even to 1050 m (along the Puyo-Tena road; P. Coopmans et al.).

One race: nominate *violaceus*.

Range: e. Colombia, s. Venezuela, e. Ecuador, e. Peru, nw. Bolivia, and w. Amazonian Brazil; s. Guyana.

Cyanocorax mystacalis   White-tailed Jay   Urraca Coliblanca

Uncommon to fairly common in deciduous woodland and desert scrub in the lowlands of sw. Ecuador, where restricted to arid and semiarid regions; recorded only from extreme s. Los Ríos (a 19th-century record from Vinces; no recent reports) and sw. Guayas (from the Chongón Hills area westward), and in s. El Oro and Loja (east to above Catamayo). In Loja locally recorded up to about 2000 m in the La Toma/Catamayo region (RSR and F. Sornoza), but mostly occurs no higher than about 1200–1500 m.

The White-tailed Jay seems to have declined in recent years across much of its Ecuadorian range, perhaps especially so in Guayas, but even in El Oro and w. Loja numbers now seem relatively low. We are uncertain what has caused this drop in numbers, but habitat degradation seems likely to be a primary factor (e.g., in the western part of the Santa Elena Peninsula). Direct persecution, and even trapping for the local cagebird market, may also be important, at least in some areas. For the present we do not believe the species need be considered formally at risk, but its status bears monitoring.

Monotypic.

Range: sw. Ecuador and nw. Peru.

Cyanocorax yncas   Inca Jay   Urraca Inca

Fairly common at borders of montane forest and secondary woodland, sometimes ranging out into mostly cleared areas in hedgerows and small woodlots, in the subtropical zone on the east slope of the Andes. The Inca Jay occurs mainly below the range of the Turquoise Jay (*Cyanolyca turcosa*); the two never seem to flock together. The Inca Jay ranges mostly from 1300 to 2200 m, locally down to 700 m in the Zumba area of Zamora-Chinchipe (where it occurs in more deciduous woodland); it seems surprisingly scarce on the Cordillera del Cóndor.

The disjunct population of birds found in n. Middle America (*C. luxuosus*, Green Jay), which inhabits the lowlands, is here specifically separated from birds found in the Andes. There are distinct differences in habitat and social behavior, as well as plumage and vocal distinctions. Only one race is definitely recorded from Ecuador, nominate *yncas*. However, it is probable that the population found in the Río Marañón drainage in the Zumba region is referable to *longirostris* of the upper Río Marañón valley in nw. Peru; as yet no specimens have been taken in this area. *Longirostris* becomes seasonally more blue-backed as its plumage wears (N. K. Johnson and R. E. Jones, *Wilson Bull.* 105[3]: 389–398, 1993).

Range: mountains of n. Venezuela, and Andes from w. Venezuela to w. Bolivia.

Vireonidae   Vireos, Peppershrikes, and Shrike-Vireos

Twelve species in four genera. The family is purely American in distribution. Despite their dissimilar external appearance, the vireos are now classified as part of the corvoid assemblage (Sibley and Ahlquist 1990; Gill 1994), hence the radical shift in family sequencing.

Cyclarhis gujanensis   Rufous-browed Peppershrike   Vireón Cejirrufo

Fairly common to common in a variety of wooded and scrubby habitats in the lowlands of w. Ecuador, occuring north in small numbers to w. Esmeraldas and Pichincha but more numerous from Manabí and Los Ríos southward. Recently found also to be locally uncommon to fairly common in the canopy and borders of montane forest and mature secondary woodland on the east slope of the Andes in se. Ecuador, primarily in Zamora-Chinchipe but also recorded as far north as nw. Morona-Santiago along the Gualaceo-Limón road (P. Coopmans et al.). Ranges inland through s. Loja and Azuay up at least locally to about 2500 m, occasionally even a bit higher; on the east slope recorded mostly from about 900 to 1900 m, with one report from as high as 3100 m on the Páramos de Matanga in Morona-Santiago (Krabbe et al. 1997).

The race occurring in most of the species' Ecuadorian range, *virenticeps*, has a striking bright olive crown, long deep chestnut superciliary, and bright lemon yellow breast (sometimes extending up over the throat). In the southeast, the first specimen (ANSP) of the race *contrerasi*—previously known only from the drainage of the upper Río Marañón valley in n. Peru—was obtained in humid montane forest on the east slope of the Andes north of Zumba in Zamora-Chinchipe on 15 Aug. 1989. Another was taken on 15 Sep. 1990 above Chinapinza on the Cordillera del Cóndor (ANSP) and a third at Panguri in s. Zamora-Chinchipe on 18 Jul. 1992 (ANSP). All three have the crown essentially chestnut though variably intermixed with olive, especially in its center, and the throat and breast are mainly olive and the belly gray (thus with underparts rather resembling those of the Black-billed Peppershrike [*C. nigrirostris*]). In addition, ANSP has an unusual (juvenile) specimen of *C. gujanensis* from adjacent sw. Nariño, Colombia (La Guayacana) which has a solid rufous crown and nape; it may represent an undescribed form. Presumably it was this form that was seen by Salaman et al. (1994) at their nearby Río Ñambí site in Nariño.

Range: e. Mexico to n. Argentina and Uruguay (but absent from parts of w. Amazonia); extreme sw. Colombia to nw. Peru.

Cyclarhis nigrirostris   Black-billed Peppershrike   Vireón Piquinegro

Uncommon in the canopy and borders of montane forest and mature secondary woodland in the foothills and subtropical zone on both slopes of the Andes. On the west slope recorded south mainly to Pichincha, but there is also at least one sighting from El Oro (above Buenaventura, where seen in Aug. 1980; PJG, RSR, and R. A. Rowlett). On the east slope recorded definitely south only to w. Napo, with several sightings from farther south being regarded as uncertain because of likely confusion with the surprisingly similar *contrerasi* form of the Rufous-browed Peppershrike (*C. gujanensis*), which until recently was not known to occur in Ecuador. A. Brosset (*Oiseau* 34: 122–135, 1964) obtained what was published as specimens of the Black-billed Peppershrike from the lowlands of n. Guayas at Hacienda Pacaritambo, but we strongly suspect that this must have been in error for the *virenticeps* race of the Rufous-browed Peppershrike. In the west recorded from 650 to 1800 m or more, on the east slope mostly between about 1200 and 2400 m.

Two races occur in Ecuador, nominate *nigrirostris* on the east slope and *atrirostris* on the west slope. *Atrirostris* differs in being somewhat darker gray below.

Range: Andes of Colombia and Ecuador.

Vireolanius leucotis   Slaty-capped Shrike-Vireo   Vireón Coroniplomizo

Uncommon to fairly common but seemingly local (and heard much more often than seen) in the canopy and borders of humid forest and mature secondary woodland in the lowlands and foothills of both e. and w. Ecuador. In the west recorded south to w. Esmeraldas in the Muisne area, Pichincha, and along the base of the Andes to nw. Azuay (Manta Real; RSR and G. H. Rosenberg). In the eastern lowlands not found any distance away from the Andes, the easternmost site being along the Maxus road southeast of Pompeya (RSR et al.). Recorded mostly from 200 to 1100 m, locally ranging as high as 1300 m in the Río Bombuscaro sector of Podocarpus National Park (P. Coopmans); in the eastern lowlands not found below about 250 m.

Formerly placed in the genus *Smaragdolanius*. Two races occur in Ecuador, nominate *leucotis* in the east and *mikettae* in the west. The nominate race has a white cheek stripe lacking in *mikettae*; *mikettae* also has pink, not bluish gray, legs.

Range: sw. Colombia and nw. Ecuador; extreme se. Colombia, e. Ecuador, e. Peru, n. Bolivia, s. Amazonian Brazil, the Guianas, and se. Venezuela. More than one species is possibly involved.

Vireo olivaceus   Red-eyed Vireo   Vireo Ojirrojo

Fairly common to common in a variety of wooded and forested habitats, also often ranging out into groves of trees and in clearings, in the lowlands of both e. and w. Ecuador. Recorded mostly below 1300 m.

The status of the Red-eyed Vireo in Ecuador is complex, perhaps more so than that of any other bird. Three subspecies are known to occur in the country, and an additional two are likely to do so. More research on this species remains to be done, and in our view considering all forms as conspecific is likely an oversimplification. For clarity's sake we segregate the component forms into separate paragraphs.

1. Nominate *olivaceus* occurs as a fairly common to common transient and uncommon boreal winter resident mainly in the eastern lowlands, where it seems to occur primarily in the canopy and borders of humid forest; an exceptional record involves an obviously migrating male collected (ANSP) in

woodland near treeline (3250 m) in s. Loja on the Cordillera Las Lagunillas on 1 Nov. 1992. Nominate *olivaceus* may also occur in w. Ecuador, but its occurrence there could easily be "masked" by the presence of resident Red-eyed Vireos, and there appear to be no specimen records. Recorded mostly from Sep. to Apr.

2. *Chivi* occurs as an austral migrant (its period of occurrence is uncertain, probably about May to Aug.), apparently mainly or entirely in the eastern lowlands; like nominate *olivaceus*, it seems to occur primarily in and near humid forest. *Chivi* and *olivaceus* would be very difficult to distinguish in the field, though their wing formulae differ markedly. Neither of these two taxa seems to sing in Ecuador.

3. Resident in w. Ecuador is the race *griseobarbatus*, which ranges from w. Esmeraldas, Carchi (La Concepción; Salvadori and Festa 1899), and w. Imbabura (Intag, Hacienda Paramba, and Lita) south to El Oro and Loja; it appears, however, to be absent from extensive areas of humid forest and has not definitely been recorded during extensive work at Playa de Oro in n. Esmeraldas (O. Jahn). The Red-eyed Vireo is found, at least at times, on Isla de la Plata (occasionally it can even be one of the most common landbirds on the island, and it then may sing vigorously; P. Coopmans), and it apparently breeds here at least in some years; we suspect that *griseobarbatus* engages in local movements, but the details of these remain little understood. Recorded up to about 1800 m in the Celica region. This race shows considerable yellow on its underparts (with a pattern and coloration not unlike those of the Yellow-green Vireo [*V. flavoviridis*]), though its facial pattern is typical of a Red-eyed), and it sings regularly during its breeding season, from about Feb. to May.

4. Red-eyed Vireos of an uncertain race are apparently resident in woodland in the Río Marañón drainage of s. Zamora-Chinchipe, where they have been seen throughout the year, with frequent singing noted in Apr. 1994 (P. Coopmans and M. Lysinger). These probably represent *pectoralis*, the subspecies recorded from the upper Río Marañón valley of nw. Peru in Cajamarca; no specimens have been taken in Ecuador. *Pectoralis* resembles *griseobarbatus* but is somewhat duller, with the throat and breast more grayish buff, not so white.

5. Finally, what seems likely to be the race *solimoensis* was found to be resident and singing in riparian woodland on a Río Napo island near the mouth of the Río Aguarico in Sep. 1992 (RSR and T. J. Davis); unfortunately, no specimens could be obtained. This subspecies has been recorded as close to Ecuador as the mouth of the Río Curaray into the Napo in ne. Peru (J. T. Zimmer, *Am. Mus. Novitates* 1127, 1941); its being recorded from e. Ecuador in *Birds of the World* (vol. 14) is doubtless the result of that locality's having formerly been in Ecuador. It may occur elsewhere in similar habitat in the eastern lowlands, and what was perhaps this race has also been found at Jatun Sacha, where a nest was even located in Feb. 1988 (B. Bochan).

Range: breeds in North America, wintering mainly in Amazonia; Colombia,

Venezuela, and the Guianas to n. Argentina and Uruguay, southern breeders migrating north to Amazonia during austral winter.

Vireo flavoviridis   Yellow-green Vireo   Vireo Verdiamarillo

An uncommon to fairly common boreal winter resident and fairly common transient in the canopy and borders of humid forest, secondary woodland, and clearings in the lowlands of e. Ecuador. The Yellow-green Vireo seems possible elsewhere, especially as a transient, but as yet it is unrecorded. In the west it could easily be mistaken for the common resident *griseobarbatus* race of the Red-eyed Vireo (*V. olivaceus*), as we suspect is the explanation for the record from Santo Domingo de los Colorados (Ménégaux 1911). Recorded mainly from Oct. to Apr., with one report from as early as late Sep. The Yellow-green Vireo regularly gives brief snatches of song, and it sings much more often in Ecuador than do long-distance migrant Red-eyeds. To date recorded only below about 400 m, but seems likely to occur higher at least as a transient.

V. *flavoviridis* was formerly sometimes treated as conspecific with V. *olivaceus* (e.g., *Birds of the World*, vol. 14), but see N. Johnson and R. Zink (*Wilson Bull.* 97[4]: 421–435, 1985) for its resurrection as a full species; this treatment has been followed by all recent authors, including the 1998 AOU Check-list. Two races are known from Ecuador: nominate *flavoviridis* is by far the more common, but there is also at least one specimen (ANSP) of *forreri* (which breeds on the Tres Marías Islands off w. Mexico), a male taken by T. Mena at San José on 24 Mar. 1949. *Forreri* is a large subspecies, and its head markings are faint. *Insulanus* (which breeds on the Pearl Islands off Panama) may also occur; the validity of this race has been questioned.

Range: breeds in Middle America, wintering in w. Amazonia.

The Black-whiskered Vireo (*V. altiloquus*) breeds in the West Indies and surrounding areas and winters primarily in Amazonia. It has been recorded as close to Ecuador as ne. Peru and could occur in small numbers in Ecuador's far-eastern lowlands.

Vireo leucophrys   Brown-capped Vireo   Vireo Gorripardo

Fairly common to common in montane forest, forest borders, secondary woodland, and adjacent clearings with scattered trees in the subtropical zone on both slopes of the Andes. On the west slope recorded south to El Oro and w. Loja (numerous localities in the Alamor and Celica region, and at Sozoranga and Utuana). On the east slope recorded mostly from 1300 to 2600 m, but occurs substantially lower on the west slope (regularly down to 600 m, sometimes even lower), especially in El Oro; there is also one report from substantially higher, 3150 m at Oyacachi (Krabbe et al. 1997).

V. *leucophrys* of Middle America and South America is regarded as a species separate from V. *gilvus* (Warbling Vireo) of North America and Mexico, following most recent authors, including the 1998 AOU Check-list. Two slightly differentiated races occur in Ecuador, nominate *leucophrys* on the east slope and *josephae* on the west slope. *Josephae* has a somewhat darker and browner crown than nominate *leucophrys*.

Range: highlands of e. Mexico to Panama; mountains of n. Venezuela, and Andes from w. Venezuela to w. Bolivia; Sierra de Perijá and Santa Marta Mountains.

The Chocó Vireo (*V. masteri*) was recently described (P. G. W. Salaman and F. G. Stiles, *Ibis* 138[4]: 610–619, 1996) from montane forests between 1200 and 1600 m on the west slope of the Andes in Colombia, where it ranges as far south as w. Nariño at Río Ñambi. There seems to be no biogeographic reason why this easily overlooked vireo should not also occur in adjacent n. Ecuador, though to date there have been no records of it.

Hylophilus semibrunneus   Rufous-naped Greenlet   Verdillo Nuquirrufo

Uncommon and somewhat local in the middle growth and subcanopy of montane forest in the foothills and lower subtropical zone on the east slope of the Andes in n. Ecuador, where recorded only from w. Napo. Until recently the only specimens from Ecuador (AMNH) were birds taken in the 1920s at San José de Sumaco on the lower slopes of Volcán Sumaco. Since the late 1980s small numbers have been seen north of Archidona along the road to Loreto, where they were first recorded by J. Rowlett and R. A. Rowlett et al. in 1990; a specimen was taken here on 13 Nov. 1991 (ANSP). The Rufous-naped Green-let was also seen in Jan. 1991 on the lower slopes of Volcán Sumaco up to about 1400 m (B. Whitney), and there are subsequent sightings from here (v.o.) and one specimen (MECN) taken on 28 Aug. 1996. Recorded between about 900 and 1400 m.

Monotypic. We follow *Birds of the Americas* (pt. 8) and *Birds of the World* (vol. 14) in not recognizing *leucogastra*, described as a subspecies of *H. semibrunneus* from the Sumaco region.

Range: locally in Andes of Colombia and n. Ecuador; Sierra de Perijá.

Hylophilus hypoxanthus   Dusky-capped Greenlet   Verdillo Ventriamarillo

Fairly common to common (but mostly recorded by voice) in the canopy and subcanopy of humid forest (mainly terra firme) in the lowlands of e. Ecuador. Recorded mostly below about 400 m, locally up to about 600 m along the eastern base of the Andes (Bermejo oil-field area north of Lumbaquí; ANSP) and north of Canelos in Pastaza (N. Krabbe).

One race: *fuscicapillus*.

Range: se. Colombia and sw. Venezuela to n. Bolivia and w. and s. Amazonian Brazil.

Hylophilus decurtatus   Lesser Greenlet   Verdillo Menor

Fairly common to common in the canopy and borders of humid forest and sec-ondary woodland in the lowlands of w. Ecuador, occurring at least locally and in smaller numbers in more deciduous situations as well; ranges south to Guayas (hills south of the Río Ayampe; Chongón Hills), El Oro, and w. Loja (various localities in the Alamor region). Recorded mostly below about 1100 m, but a few are found as high as 1400 m in Loja.

One race: *minor. H. minor* (Lesser Greenlet) of e. Panama and nw. South America was formerly sometimes considered a distinct species from *H. decurtatus* (Gray-headed Greenlet) of Middle America, but all recent authors have treated *minor* as a subspecies of *H. decurtatus*.

Range: e. Mexico to extreme nw. Peru.

Hylophilus olivaceus Olivaceous Greenlet Verdillo Oliváceo

Locally fairly common in overgrown clearings and lower growth at the borders of humid and montane forest and secondary woodland in the foothills along the eastern base of the Andes. The Olivaceous Greenlet ranges north to Napo in the Sumaco/Archidona region, with a few occurring as far north as the El Chaco area, but the species seemingly is absent from the Bermejo/Reventador area slightly farther north. It ranges south into s. Zamora-Chinchipe in the Zumba region, doubtless also occurring southward into adjacent Peru. Recorded mostly from 600 to 1450 m.

Monotypic.

Range: along eastern base of Andes in Ecuador and n. Peru.

*Hylophilus thoracicus Lemon-chested Greenlet Verdillo Pechilimón

Rare and seemingly local—though doubtless overlooked—in the canopy and borders of terra firme forest in the lowlands of e. Ecuador, where now recorded from several well-separated sites in Morona-Santiago, Pastaza, and Napo. The Lemon-chested Greenlet was first found in Ecuador on 5 Aug. 1989 when a singing male (ANSP) was obtained by M. B. Robbins at Santiago in Morona-Santiago. Subsequent records have involved birds seen and tape-recorded at Taisha, Morona-Santiago, in Aug. 1990 (RSR and M. B. Robbins), a few seen and tape-recorded along the Maxus road southeast of Pompeya in 1994–1995 (RSR et al.), one seen and tape-recorded at Zancudococha on 17 and 20 Jan. 1995 (P. Coopmans et al.), and birds seen and tape-recorded in Sep.–Oct. 1996 at Kapawi Lodge on the Río Pastaza near the Peruvian border (J. Moore and G. Rivadeneira). Recorded up to about 400 m.

One race: *aemulus*.

Range: e. Ecuador, e. Peru, n. Bolivia, locally in Amazonian Brazil, the Guianas, and se. Venezuela; se. Brazil. More than one species may be involved.

Hylophilus ochraceiceps Tawny-crowned Verdillo Coronileonado
           Greenlet

Rare to uncommon and local in the lower growth of wet and humid forest in the lowlands and foothills of e. and nw. Ecuador. In the northwest the Tawny-crowned Greenlet is known primarily from Esmeraldas, mainly from the north where it is recorded from old specimens taken at Pulún and from recent material (ANSP, MECN) obtained northwest of Alto Tambo, and the Playa de Oro area (where it is considered to be fairly common; O. Jahn et al.). In addition, a single individual was mist-netted as far south as the Río Verde south of Chontaduro on 3 Aug. 1992 (it was skeletonized; UKMNH); in n. Pichincha the species has been found along the road north of Simón Bolívar near Pedro Vicente Maldonado, first on 1 Oct. 1995 (P. Coopmans). In the west recorded up only to about 450 m, but in the east found as high as 700 m.

Two races are found in Ecuador, *bulunensis* in the northwest and *ferrugineifrons* in the east. *Ferrugineifrons* differs in having more grayish (not so olive) underparts.

Range: s. Mexico to nw. Ecuador; se. Colombia, s. Venezuela, and the Guianas to nw. Bolivia and Amazonian Brazil.

Turdidae	Thrushes

Twenty-two species in six genera. Thrushes are worldwide in distribution. The family is sometimes subsumed into a much larger Muscicapidae.

Myadestes ralloides   Andean Solitaire   Solitario Andino

Locally fairly common (though generally inconspicuous) in the lower growth and borders of montane forest, especially in ravines or near water, on both slopes of the Andes; on the west slope recorded south to El Oro and w. Loja (small numbers in the Alamor/Celica region). Recorded mostly from 1000 to 2500 m on the east slope, but occurs lower in the west (locally down to 600 m, routinely as low as 800–900 m).

Two similar races occur in Ecuador, *plumbeiceps* on the west slope and *venezuelensis* on the east slope. Despite their similar appearance, their songs differ markedly.

Range: mountains of n. Venezuela, and Andes from w. Venezuela to w. Bolivia; Sierra de Perijá. Excludes Black-faced Solitaire (*M. melanops*) of Costa Rica and w. Panama and Varied Solitaire (*M. coloratus*) of e. Panama and adjacent Colombia.

Cichlopsis leucogenys   Rufous-brown Solitaire   Solitario Rufimoreno

Rare to uncommon and very local in the lower and middle growth of montane forest in the foothills of nw. Ecuador, where recorded only from e. Esmeraldas and Pichincha. Until recently known only from two specimens obtained early in the 20th century around Mindo in Pichincha ("Mindo Huila"); the species has not subsequently been found so far south. In 1979 N. Krabbe and O. Jakobsen discovered a population of the Rufous-brown Solitaire at El Placer in Esmeraldas, and it has since proved to be not uncommon at that locality. The only other Ecuadorian site known is Playa de Oro, where during intensive surveys from 1995 to 1998 the species was considered a very rare nonbreeding visitant, with the only record being one seen on 25 Feb. 1997 (O. Jahn). Recorded mostly between about 500 and 1200 m, occasionally (seasonally?) down to 100 m in n. Esmeraldas.

The Rufous-brown Solitaire was given Near-threatened status by Collar et al. (1992, 1994); from a purely Ecuadorian standpoint we consider it to warrant Vulnerable status. The species may eventually prove to be more widespread in Ecuador than the current paucity of records indicates, so much of its range being difficult to access. Up to the present, no population is known to occur in any formally protected area.

Cichlopsis is regarded as a monotypic genus, separated from *Myadestes* because of its strikingly different song and plumage, following Ridgely and Tudor (1989) and Sibley and Monroe (1990). One race of *C. leucogenys* occurs in Ecuador, *chubbi*.

Range: very locally in sw. Colombia and nw. Ecuador, east slope of Andes in Peru, tepuis of Venezuela and Guyana, and se. Brazil.

Entomodestes coracinus   **Black Solitaire**   **Solitario Negro**

Rare to locally uncommon and inconspicuous in the lower growth of montane forest and forest borders in nw. Ecuador, where recorded south to Pichincha in the Chiriboga area. The often overlooked Black Solitaire does not appear to be regular at any reasonably accessible locality in Ecuador, but small numbers do occur above Mindo; it also is recorded at Maquipucuna. Recorded mainly between about 1100 and 1600 m.

We consider the scarce, local, and range-restricted Black Solitaire to warrant Near-threatened status in Ecuador, where it never appears to be very numerous and seems quite sensitive to habitat disturbance and fragmentation. The species was not considered to be at risk by Collar et al. (1992, 1994).

Monotypic.

Range: west slope of Andes in sw. Colombia and nw. Ecuador.

Catharus fuscater   **Slaty-backed Nightingale-Thrush**   **Zorzal Sombrío**

Uncommon to fairly common (recorded most often by voice) in the undergrowth of montane forest on the west slope of the Andes, where recorded locally from Carchi (near the Maldonado road in Jul. 1988; ANSP) south to El Oro and w. Loja. There are also a few records from the east slope, where the species is evidently rare or local (or both) and is known only from w. Sucumbíos, w. Napo, and Zamora-Chinchipe. These east-slope records include several birds seen and tape-recorded along the La Bonita road on 18 Sep. 1991 (N. Krabbe), an old Olalla specimen (AMNH) from Puente del Río Quijos, an immature female (WFVZ) taken 5 km east of Cosanga on 16 Oct. 1988 (there are also a few reports from the nearby Cordillera de Huacamayos), a few records from Podocarpus National Park (including an ANSP specimen from Quebrada Avioneta [Rasmussen et al. 1996]; P. Coopmans), and specimens (ANSP and

MECN) taken along the Río Isimanchi on the east slope of the Cordillera Las Lagunillas in Nov. 1992. The Slaty-backed occurs mainly at elevations above those of the Spotted Nightingale-Thrush (*C. dryas*), ranging principally from 1200 to 2600 m, occasionally somewhat lower on the west slope.

Only one race, nominate *fuscater*, is definitely known from Ecuador, but M. B. Robbins believes that birds from extreme s. Zamora-Chinchipe are probably referable to *caniceps* (a taxon found primarily in Peru); the insufficiency of specimen material precludes a more definite statement. *Caniceps* is similar to nominate *fuscater*, differing primarily in being not quite so dark above.

Range: highlands of Costa Rica and Panama; locally in Andes from w. Venezuela to w. Bolivia; Sierra de Perijá and Santa Marta Mountains.

Catharus dryas   Spotted Nightingale-Thrush   Zorzal Moteado

Uncommon to fairly common (recorded most often by voice) in the undergrowth of montane forest in foothills and the subtropical zone on both slopes of the Andes, but generally more numerous and widespread on the west slope. On the east slope recorded south primarily to Morona-Santiago, with only a few recent records from Zamora-Chinchipe, including a specimen (ANSP) from Panguri taken on 15 Jul. 1992 (Rasmussen et al. 1996) and sightings from the Cordillera del Cóndor (Coangos; Schulenberg and Awbrey 1997). On the west slope known from Pichincha (Maquipucuna) south through w. Loja (Alamor and Celica areas). Also found locally on the coastal Cordillera de Mache in sw. Esmeraldas (Bilsa [J. Hornbuckle; D. Wolf]) and the Cordillera de Colonche in sw. Manabí (Cerro San Sebastián in Machalilla National Park [Parker and Carr 1992; ANSP]) and w. Guayas (an old specimen [AMNH] from Cerro de Manglaralto, and recent sightings from Cerro La Torre and above Salanguilla [N. Krabbe], south of the Río Ayampe [RSR], and at Loma Alta [Becker and López-Lanús 1997]). Recorded mostly from 650 to 1800 m, locally occurring lower in the west (e.g., down to 400 m at Manta Real in nw. Azuay); exceptionally (?), this species occurs as low as 175 m south of the Río Ayampe, where two birds were found on 18 Jul. 1993 (RSR).

One race: *maculatus*. *Ecuadoreanus* of w. Ecuador is regarded as not meriting recognition (see Robbins and Ridgely 1990).

Range: s. Mexico to Honduras; Andes from w. Venezuela to nw. Argentina; Sierra de Perijá.

Catharus minimus   Gray-cheeked Thrush   Zorzal Carigris

An uncommon transient and rare boreal winter resident in the lower growth of humid forest (especially terra firme) and forest borders in the lowlands and foothills of e. Ecuador. What presumably were transients have also been seen on very rare occasions in the Quito region (J. C. Matheus), and a single bird was a window-kill in Mindo on 19 Dec. 1998 (J. Lyons, specimen to MECN). The species also seems possible in the western lowlands, particularly on migration, but remains unrecorded from there. The Gray-cheeked Thrush is a generally unobtrusive bird, much less conspicuous than the Swainson's Thrush (*C. ustulatus*) and less apt to come out to forest borders; numbers have often been best revealed through mist-netting. Recorded from Oct. to Apr. Found mostly below 1300 m, very rarely in the central valley at 2700–2800 m.

Two very similar races have been recorded in Ecuador, nominate *minimus* (e.g., one was taken at Zancudococha on 5 Apr. 1991; ANSP) and *aliciae* (Chapman 1926). On a global level, populations of the latter taxon are far greater. These subspecies would be distinguishable in the field only with difficulty (see I. A. McLaren, *Birding* 27[5]: 358–364, 1995). H. Ouellet (*Wilson Bull.* 105[4]: 545–572, 1993) has shown that *bicknelli*, formerly considered a subspecies of *C. minimus*, is better considered as a full species (Bicknell's Thrush), and this treatment was adopted in the 1998 AOU Check-list. Although there has been some confusion in the past, Ouellet (op. cit.) concludes that there is no evidence that *C. bicknelli* occurs in South America; it apparently winters exclusively in the West Indies.

Range: breeds in n. North America, wintering mainly in n. and w. Amazonia.

Catharus ustulatus   Swainson's Thrush   Zorzal de Swainson

A locally common to very common transient and uncommon boreal winter resident in the lower growth of forest, borders, secondary woodland, and clearings in the lowlands and (especially) on lower Andean slopes in both e. and w. Ecuador. The Swainson's Thrush is considerably more numerous in e. Ecuador than it is in the west; in the latter it ranges mainly on the lower slopes of the Andes, where it has been found south to w. Loja (Alamor region and Sozoranga; Best et al. 1993), but it also ranges west to the coast (e.g., in nw. Guayas in the hills south of the Río Ayampe). Recorded from Oct. to Apr. At least for brief periods the Swainson's Thrush can be astonishingly numerous along the eastern base of the Andes, mainly during passage periods in late Oct. and again in Mar. It is also regularly recorded, though always in smaller numbers, up to 3000 m and occasionally even higher in the Andes and the central valley (e.g., around Quito; the highest report is of a single bird seen at 3800 m on Volcán Pichincha on 15 Nov. 1990 [R. ter Ellen]); at higher elevations it likely occurs mainly as a transient.

Only two passerine birds banded elsewhere have ever been recovered in Ecuador. Both, remarkably, were Swainson's Thrushes. One of these was banded in Costa Rica in Oct. 1979, only to be found a few months later in the Oriente, in Jan. 1980.

The sole race definitely known to occur in Ecuador is the eastern-breeding *swainsoni*; western-breeding races apparently winter mainly in Middle America.

Range: breeds in n. and w. North America, wintering mainly from Middle America to w. South America.

The Veery (*C. fuscescens*) also seems likely to occur in the eastern low-lands of Ecuador, at least as a rare transient. It breeds in North America and has been reported as close to Ecuador as Meta in e. Colombia (Hilty and Brown 1986).

Platycichla leucops   Pale-eyed Thrush   Mirlo Ojipálido

Rare to uncommon and seemingly local and/or erratic in montane forest and forest borders in the foothills and lower subtropical zone on both slopes of the Andes. On the west slope recorded south mainly to Pichincha, with one sighting from El Oro (2 males seen above Buenaventura on 8 Aug. 1980; RSR, PJG, and R. A. Rowlett). Recorded mostly from 1000 to 2000 m, occasionally wandering somewhat higher.
Monotypic.
Range: locally in mountains of n. Venezuela, and Andes from w. Venezuela to w. Bolivia; tepuis of Venezuela and adjacent Guyana and Brazil.

Turdus chiguanco   Chiguanco Thrush   Mirlo Chiguanco

Uncommon to common in semiopen, often agricultural, country and gardens in arid sectors of the central valley and interandean slopes north to w. Cotopaxi (slopes of Volcán Iliniza, vicinity of La Ciénega, etc.), a few rarely wandering as far north as Pichincha (east of Quito in the Río Guaillabamba valley); most numerous from Chimborazo southward. The Chiguanco Thrush is perhaps now spreading northward in the central valley; all records from Cotopaxi and Pichincha date from the 1980s and 1990s. Although there is some overlap, in general the Great Thrush (*T. fuscater*) favors more humid situations than does the Chiguanco. Recorded mostly from 1500 to 3200 m.
One race: nominate *chiguanco*. We follow *Birds of the Americas* (pt. 7) in subsuming *conradi*, to which Ecuadorian birds have been assigned, into the nominate race. If *anthracinus* of Bolivia and Argentina is separated as a full species (Coal-black Thrush), as may be the most appropriate course, *T. chiguanco* will become monotypic.
Range: Andes from cen. Ecuador to n. Chile and w. Argentina.

Turdus fuscater   Great Thrush   Mirlo Grande

Very common to common and conspicuous at borders of montane forest and secondary woodland and in clearings, agricultural areas with hedgerows, and gardens in the temperate zone and up to treeline on both slopes of the Andes, and (in n. Ecuador) in the central valley. The Great Thrush is one of the most frequently seen birds in the city of Quito. Recorded mostly from 2500 to 4000 m, but in w. Loja routinely occurs lower, a few down to about 1800 m.
Two races occur in Ecuador, the dark *quindio* in most of the species' Ecuado-

rian range and *gigantodes* in the south from Azuay south through El Oro, Loja, and adjacent Zamora-Chinchipe. *Gigantodes* is slightly paler.

Range: Andes from w. Venezuela to w. Bolivia; Sierra de Perijá and Santa Marta Mountains.

Turdus serranus   Glossy-black Thrush   Mirlo Negribrilloso

Fairly common in the canopy and mid-levels of montane forest and forest borders in the subtropical and temperate zones on both slopes of the Andes; on the west slope recorded south to w. Loja in the Celica area (MCZ) and at Sozoranga and Utuana (first by Best et al. 1993). The Glossy-black is much more of a forest bird than the Great Thrush (*T. fuscater*), and it rarely remains long in the open except when singing. Recorded mostly from 1500 to 2800 m, locally (perhaps only seasonally?) up to 3100–3350 m on the east slope. Exceptionally, several singing birds were recorded at the very high elevation of 3750 m at Loma Yanayacay on the northwest slope of Volcán Pichincha in mid-Mar. 1996, but not before or since (Krabbe et al. 1997).

One race: *fuscobrunneus*.

Range: mountains of n. Venezuela, and Andes from w. Venezuela to nw. Argentina; Sierra de Perijá.

Turdus fulviventris   Chestnut-bellied Thrush   Mirlo Ventricastaño

Uncommon to fairly common and rather local in the lower growth of montane forest and forest borders in the subtropical zone on the east slope of the Andes. Perhaps most numerous on outlying ridges such as the Cordilleras de Huacamayos, de Cutucú, and del Cóndor—and less so, even at appropriate elevations, on the main east-Andean slope. The Chestnut-bellied Thrush occasionally hops on the ground at roadsides, but despite this, in most areas it is not a very frequently reported bird; it appears to be more numerous in s. Ecuador than it is northward. Recorded mainly from 1500 to 2500 m. In 1996–1997 small numbers, perhaps postbreeding wanderers, were observed on several occasions between Jul. and Nov. in the Bombuscaro sector of Podocarpus National Park at 1000–1100 m (D. and M. Wolf; RSR et al.); whether they regularly occur this low remains unknown.

Monotypic.

Range: locally in Andes from w. Venezuela to n. Peru; Sierra de Perijá.

Turdus reevei   Plumbeous-backed Thrush   Mirlo Dorsiplomizo

Uncommon to locally (and apparently seasonally) common in deciduous and semihumid forest and woodland in the lowlands and lower Andean slopes of sw. Ecuador. Found mainly from cen. Manabí (Bahía de Caráquez and Chone) and s. Los Ríos (Jauneche) south in more arid regions through El Oro and Loja; in Loja found mainly in the west and south, but there is at least one record from as far east as near Malacatos (MCZ), and one bird was seen at Vilcabamba on 6 Sep. 1998 (RSR and PJG et al.). The northernmost record is a Jul. 1977 report from Río Palenque in extreme s. Pichincha. The Plumbeous-backed Thrush seems to be seasonal and erratic in its occurrence in many areas, but detailed information on its presumed local movements remains unavailable. There is even a single report from the east slope of the Andes, an immature seen along the Loja-Zamora road (1600 m) on 27 Jul. 1992 (D. Wolf et al.). Recorded mainly below about 1600 m, but a few nonbreeding birds have been found as high as 2500 m in Loja.

Monotypic.

Range: sw. Ecuador and nw. Peru.

Turdus maranonicus   Marañón Thrush   Mirlo del Marañón

Locally fairly common to common in secondary woodland, clearings, and gardens in the Río Marañón drainage of s. Zamora-Chinchipe in the Zumba region, in smaller numbers ranging north as far as Valladolid (several seen in Jan. 1998; RSR et al.). At least in recent years the species has also been seen at the borders of relatively humid montane forest and woodland (P. Coopmans; RSR). The Marañón Thrush was first seen in Ecuador in 1986 (M. Pearman), with specimens obtained near Zumba in Aug. 1989 and Aug. 1992 (ANSP, MECN). Recorded from about 650 to 1650 m.

Monotypic.

Range: upper Río Marañón valley of nw. Peru and extreme s. Ecuador.

Turdus ignobilis   Black-billed Thrush   Mirlo Piquinegro

Uncommon to fairly common in clearings, lighter secondary woodland, and humid forest borders in the lowlands of e. Ecuador; absent from the Río Marañón drainage. Except along larger rivers, the Black-billed Thrush is generally absent from extensively forested regions; it is, however, capable of rather quickly colonizing newly cleared areas. Ranges up to at least 1200 m along the eastern base of the Andes.

One race: *debilis*.

Range: w.-cen. Colombia; the Guianas and Venezuela to n. Bolivia and w. and cen. Amazonian Brazil.

Turdus lawrencii   Lawrence's Thrush   Mirlo Mímico

Uncommon in the mid-levels and canopy of humid forest (both terra firme and várzea, but especially the former) in the lowlands of e. Ecuador. There are relatively few published Ecuadorian records of the Lawrence's Thrush, and very few specimens have ever been taken in the country. However, subsequent to the recent discovery of its remarkable song, which is replete with superb mimicry, the species has been found to range quite widely though it seems to occur mainly in areas well to the east of the base of the Andes (numbers at Jatun Sacha are, for example, very small, fide B. Bochan). Recorded mostly below about 400 m, locally up to 600 m (north of Canelos in Pastaza; N. Krabbe).

Monotypic.

Range: s. Venezuela and se. Colombia to n. Bolivia and w. and cen. Amazonian Brazil.

Turdus obsoletus   Pale-vented Thrush   Mirlo Ventripálido

Generally rare and very local in humid forest in the foothills on the west slope of the Andes, but comparatively numerous at El Placer in Esmeraldas; we suspect that the Pale-vented Thrush's apparent scarcity elsewhere in its Ecuadorian range may more reflect the inaccessible nature of most of its habitat than actual rarity. It ranges primarily in the northwest, being recorded locally south to Pichincha, but there is also an old record from sw. Chimborazo (Chimbo), and small numbers are also present at Buenaventura in El Oro (specimen [MECN] taken in Jun. 1985, with photos in VIREO; also a few subsequent sightings and tape-recordings). A Dec. 1996 record (Becker and López-Lanús 1997) from Loma Alta in nw. Guayas remains to be confirmed. Recorded mostly from 600 to 1100 m, but has also been found as high as 1500 m above Mindo and at Maquipucuna in Pichincha, and singing birds have been tape-recorded at 450 m near Playa de Oro in n. Esmeraldas (O. Jahn); at least seasonally (only formerly?) a few range as low as 200 m at Río Palenque (PJG).

Trans-Andean *T. obsoletus* is regarded as a species separate from cis-Andean *T. fumigatus* (Cocoa Thrush) and *T. hauxwelli* (Hauxwell's Thrush) of the lowlands east of the Andes, mainly because of its different habitat, elevation requirements, and song (see Ridgely and Tudor 1989). One race of *T. obsoletus* occurs in Ecuador, *parambanus*.

Range: locally from Costa Rica to sw. Ecuador.

Turdus hauxwelli   Hauxwell's Thrush   Mirlo de Hauxwell

Rare to locally uncommon in the lower and middle growth of humid forest (primarily in várzea, occasionally also wandering into terra firme) and on river islands in the lowlands of e. Ecuador. Mainly recorded below about 300 m, but at least once believed seen at 400 m (Jatun Sacha; B. Bochan).

T. hauxwelli of w. Amazonia is regarded as a species separate from *T. fumigatus* (Cocoa Thrush) of e. South America, and from trans-Andean *T. obsoletus* (Pale-vented Thrush), following Ridgely and Tudor (1989) and Sibley and Monroe (1990). Monotypic.

Range: se. Colombia to n. Bolivia and w. Amazonian Brazil.

Turdus maculirostris   Ecuadorian Thrush   Mirlo Ecuatoriano

Uncommon to locally fairly common in humid and deciduous forest, forest borders, secondary woodland, and adjacent clearings in the lowlands of w. Ecuador and on the west slope of the Andes from w. Esmeraldas and Imbabura (Río Chota valley and around Ibarra) south to Guayas (hills south of the lower

Río Ayampe, and Chongón Hills), El Oro, and w. Loja. Since the early 1980s small numbers of Ecuadorian Thrushes have begun to spread into the central valley in the Ibarra region, where the species apparently was first seen in 1983 (T. Southerland et al.). Ranges up locally to about 1900 m, with a few as high as about 2200 m in Imbabura around Ibarra.

T. maculirostris is regarded as a species distinct from *T. nudigenis* (Bare-eyed Thrush) of ne. South America because of its differing morphology and highly disjunct range, following Ridgely and Tudor (1989). Monotypic.

Range: w. Ecuador and extreme nw. Peru.

Turdus albicollis   White-necked Thrush   Mirlo Cuelliblanco

Fairly common (but reclusive and most often recorded when vocalizing, or by mist-net captures) in the lower growth of humid forest (mainly terra firme) and mature secondary woodland in the lowlands of e. Ecuador. Ranges up in foothills along the base of the Andes to about 1100 m.

Excludes *T. assimilis* (White-throated Thrush) of Middle America and trans-Andean *T. daguae* (Dagua Thrush). One race of *T. albicollis* occurs in Ecuador, *spodiolaemus*.

Range: e. Colombia, Venezuela, and the Guianas to Amazonian Brazil, Bolivia, and nw. Argentina; e. Brazil, e. Paraguay, and ne. Argentina.

Turdus daguae   Dagua Thrush   Mirlo Dagua

Uncommon to fairly common but decidedly local in humid forest in the lowlands of nw. Ecuador; recorded mainly from Esmeraldas, with 19th-century specimens from Cachaví, recent specimens from northwest of Alto Tambo and east of Muisne (ANSP, MECN), and sightings and tape-recordings from Playa de Oro (where considered to be one of the most numerous birds in lower foothill forest; O. Jahn et al.). There are also some reports of the Dagua Thrush—most of them recent—from farther south. These include single birds seen and tape-recorded at Río Palenque in s. Pichincha on 10 Mar. 1981 (A. B. van den Berg and C. A. W. Bosman, *Bull. B. O. C.* 104[4]: 153, 1984), several sites in nw. Pichincha (northwest of Pedro Vicente Maldonado; v.o.), and a specimen (ANSP) taken by T. Mena on 6 Oct. 1950 at Quevedo in n. Los Ríos. In addition, Dagua Thrushes were found in Aug. 1996 near the summit of hills in Manglares-Churute Ecological Reserve in se. Guayas (Pople et al. 1997), and in Dec. 1996 the species was found to be "common" in cloud forest on the coastal Cordillera de Colonche at Loma Alta in w. Guayas (Becker and López-Lanús 1997). The status of the Dagua Thrush south of Esmeraldas requires further clarification. Recorded up to about 600 m (but in Colombia known to occur as high as 900 m; Hilty and Brown 1986).

Daguae has usually been considered a subspecies of *T. assimilis* (White-throated Thrush), with that species recently usually (e.g., Ridgely and Tudor 1989; Sibley and Monroe 1990) having been considered distinct from cis-Andean *T. albicollis* (White-necked Thrush). Now that more information is available regarding the voice of *daguae*—it is distinctly different from that of *T. assimilis* in Middle America—we consider it more appropriate to treat *T. daguae* as a separate monotypic species, differing not only in voice but also in several morphological features. *Daguae*'s voice actually more closely resembles that of cis-Andean *T. albicollis*, suggesting that *daguae* may be more closely related to that species.

Range: e. Panama to sw. Ecuador.

Turdus nigriceps   Andean Slaty-Thrush   Mirlo Pizarroso Andino

Uncommon and local (and apparently only seasonal) in the canopy and borders of montane forest and secondary woodland in the subtropical zone of s. Ecuador, where thus far known primarily from s. Loja; in addition, an immature female (ANSP) of the Andean Slaty-Thrush was mist-netted and collected at Panguri (1575 m) in s. Zamora-Chinchipe on 17 Jul. 1992 (Rasmussen et al. 1996), and one was heard singing near Chinapinza (1450 m) on the Cordillera del Cóndor on 14 Jun. 1993 (RSR). Recorded only between about 1400 and 1800 m.

Until the late 1980s the Andean Slaty-Thrush was known in Ecuador only from two 19th-century specimens (BMNH) obtained by Buckley at "Monji," a locality that cannot definitely be located; in *Birds of the Americas* (pt. 7) it was believed it to be situated in se. Ecuador, though Chapman (1926) and Paynter (1993) thought the site more likely situated on the west slope. In Feb. 1989 two independent observers, R. A. Rowlett and P. Coopmans, discovered substantial numbers of singing birds in sw. Loja east of Celica; despite much searching, however, none could be relocated in the same area during an ANSP expedition in Aug. 1989, nor has the species subsequently been found here at this time of year. Andean Slaty-Thrushes were again present here in Jan. 1990 (J. Arvin et al.) and in Apr. 1992 (ANSP). Additional breeding populations have been found near Cariamanga, where a juvenile (MECN) was taken by N. Krabbe on 8 Apr. 1991; near Catacocha, where found by Best et al. (1993) in Mar. 1991, with a male (ANSP) taken on 2 Apr. 1992; and at Sozoranga in Feb.–Mar. 1991 (Best et al. 1993). The Andean Slaty-Thrush is thus now known to breed locally in s. Loja, but the evidence suggests that it vacates this region during the second half of the year (at least there are no reports from that period); where it goes at this time remains conjectural, but the pattern is reminiscent of that shown by the Black-and-white Tanager (*Conothraupis speculigera*). The Panguri specimen of the Andean Slaty-Thrush, however, seems likely to represent an austral migrant; T. S. Schulenberg (*Bull. B. O. C.* 107[4]: 186–187, 1987) has shown that on the east slope of the Andes in Peru the species appears to occur only as an austral migrant.

We regard the breeding population of the Andean Slaty-Thrush in sw. Loja as a Near-threatened species because of the extensive deforestation that has taken place in this region in recent decades. The species was not regarded as generally at risk by Collar et al. (1992, 1994).

T. nigriceps is regarded as a monotypic species separate from birds of se.

South America, *T. subalaris* (Eastern Slaty-Thrush); see Ridgely and Tudor (1989). The expanded species had formerly been called the Slaty Thrush.

Range: breeds on Andean slopes of s. Bolivia and nw. Argentina, and locally in nw. Peru and sw. Ecuador; s. breeders migrate north to east slope of Andes of Peru and s. Ecuador during austral winter.

Mimidae   Mockingbirds and Thrashers

Two species in one genus. The family is strictly American in distribution, with only mockingbirds occurring in South America.

Mimus longicaudatus   Long-tailed Mockingbird   Sinsonte Colilargo

Fairly common to common and conspicuous in arid scrub, undergrowth of light woodland, and even barren, desert-like areas with very sparse vegetation in the lowlands of sw. Ecuador north to cen. Manabí (Bahía de Caráquez area; also on Isla de la Plata); ranges inland through Loja east to near Loja (city) and around Malacatos. The Long-tailed Mockingbird strictly avoids more humid regions and is generally most numerous near the coast. Recorded locally up to about 1900 m in Loja.

Two slightly differentiated races occur in Ecuador: *albogriseus* occurs on the mainland, and the slightly larger (especially in bill measurements) *platensis* is endemic to Isla de la Plata. The latter is the only endemic bird taxon found on that island.

Range: sw. Ecuador and w. Peru.

**Mimus gilvus*   Tropical Mockingbird   Sinsonte Tropical

Uncertain. Known in Ecuador only from several recent sightings of one or two birds seen northwest of Otavalo from Sep. 1996 onwards; first located by C. Vogt (fide J. Nilsson) on 20 Sep. and subsequently tape-recorded by J. Nilsson. In Colombia, where the species appears to be increasing and spreading (it was only first found in Nariño in 1996; see R. Strewe, *Bull. B. O. C.* 120[3]: 194, 2000), the Tropical Mockingbird favors semiopen agricultural or even suburban areas. Although the possibility of these being escaped cagebirds cannot be ruled out, our presumption is that the Otavalo birds represent genuine wanderers, and that they may even presage the species' becoming established as a resident in n. Ecuador, where suitable habitat would appear to be widespread. As a further indication that this may indeed be happening, a single (wandering?) Tropical Mockingbird was seen near Cosanga on 5 Sep. 1998 (M. Lysinger). Recorded at about 2600 m and at 1900 m.

Presumed race: *tolimensis*.

Range: s. Mexico to Honduras; Panama (introduced, but now spreading); w. and n. Colombia, much of Venezuela, the Guianas, and coastal e. Brazil; Lesser Antilles.

Cinclidae   Dippers

One species. Dippers are essentially Holarctic in distribution, but two species occur in South America.

Cinclus leucocephalus   **White-capped Dipper**   Cinclo Gorriblanco

Fairly common along fast-flowing rocky streams and rivers on both slopes of the Andes, with a few also persisting in the central valley north and east of Quito (e.g., along the Río Guaillabamba). Elevation seems not to matter particularly as long as the species' ecological preconditions are met; the dipper is recorded from at least 700 to 3800 m, at least occasionally somewhat higher and lower (e.g., to 400 m in Azuay at Manta Real; RSR).

One race: *leuconotus*.

Range: Andes from w. Venezuela to w. Bolivia; Siera de Perijá and Santa Marta Mountains.

Hirundinidae   Swallows and Martins

Seventeen species in nine genera. The family is worldwide in distribution.

Progne tapera   **Brown-chested Martin**   Martín Pechipardo

Uncommon in open and semiopen areas in the lowlands of both e. and w. Ecuador; in the west known only from s. Los Ríos and Guayas south into coastal El Oro. Brown-chested Martins have been observed breeding in old hornero (*Furnarius*) nests in sw. Ecuador (e.g., in El Oro in Apr. 1989), but in e. Ecuador they nest (sometimes in small colonies) in burrows dug into banks, most often along rivers. Ranges mainly below 600 m.

The species is often separated in the monotypic genus *Phaeoprogne*, but we follow F. H. Sheldon and D. W. Winkler (*Auk* 110[4]: 798–824, 1993) in merging it into *Progne*; this was also done in the 1998 AOU Check-list. Two races occur in Ecuador, austral migrant *fusca* (which so far is recorded only from the eastern lowlands but is possible elsewhere) and the resident nominate *tapera*. The latter has a less well-defined breast band and fewer spots on its median breast.

Range: n. and e. Colombia, Venezuela, and the Guianas to n. Argentina, southern breeders wintering in n. South America and Panama (casually north to Costa Rica); sw. Ecuador and extreme nw. Peru.

Progne chalybea   **Gray-breasted Martin**   Martín Pechigris

Fairly common to locally very common in semiopen, agricultural, and (especially) urban areas in the lowlands of both e. and w. Ecuador, in Loja occurring as far east as e. Loja just west of the Continental Divide (e.g., at Vilcabamba and Amaluza). Particularly numerous in the southwest, where large groups assemble to roost on electrical wires in various towns, notably in Arenillas, El Oro, where the thousands gathering during some nights can at times present an impressive spectacle. Gray-breasted Martins are relatively scarce in mainly forested regions, and thus they remain quite local in the east, where they are present mainly around towns, along river and lake shores, and in larger clearings; they can quickly colonize even small clearings in extensive terra firme, such as the oil rigs along the Maxus road southeast of Pompeya (RSR). Found primarily in the lowlands, but in s. Ecuador ranges up to about 2000 m in semiarid valleys. A few have also occurred, at least casually (perhaps involving long-distance migrants), in the central valley (e.g., at Lago San Pablo [2570 m] in Imbabura, where 3 were seen on 30 Aug. 1989; RSR and M. Weinberger).

One race: nominate *chalybea*.

Range: n. Mexico to nw. Peru and n. Argentina; northern breeders winter southward and southern breeders winter northward, but their precise seasonal distributions remain uncertain.

*[*Progne subis*   **Purple Martin**   Martín Purpúreo]

Apparently a very rare transient, but perhaps overlooked. Few Ecuadorian reports of the Purple Martin are available, all of them sight reports, most or all from the highlands from Imbabura south to Azuay; it has been reported from Dec. to Apr. The species may also occur in e. Ecuador, as it has been collected at Iquitos, Peru (Ridgely and Tudor 1989).

The subspecies of *P. subis* involved is not known.

Range: breeds in North America south to n. Mexico, wintering in South America.

*[*Progne elegans*   **Southern Martin**   Martín Sureño]

Apparently a casual austral winter visitant to semiopen areas in the eastern lowlands. There are only two Ecuadorian sightings of the Southern Martin. Three individuals (a male and 2 females) were seen at Imuyacocha, together with other migratory swallows, on 1–2 Apr. 1991 (RSR and J. Gerwin); and two male martins, which RSR believes were almost certainly this species and not the very similar Purple Martin (*P. subis*), were seen at Jatun Sacha on 13 Apr. 1992 (B. Bochan). Large numbers of the Southern Martin occur regularly as close as ne. Peru around Iquitos (D. Oren, *Condor* 82[3]: 344–345, 1980), so the species has perhaps been overlooked in Ecuador and may occur more regularly. Recorded below 400 m.

P. elegans is here regarded as a monotypic species separate from the other forms in the *P. modesta* complex, following the 1983 and 1998 AOU Checklists. South American birds are sometimes united under the name of *P. modesta* (e.g., Ridgely and Tudor 1989). Birds occurring on the Ecuador mainland surely are austral migrant *elegans* and not *P. modesta* (Galápagos Martin; endemic to the Galápagos Islands) or *P. murphyi* (Peruvian Martin; resident in coastal w. Peru and extreme n. Chile).

Range: breeds in s. South America, wintering in w. Amazonia.

Tachycineta albiventer   **White-winged Swallow**   **Golondrina Aliblanca**

Fairly common to common over and near water, and in adjacent clearings, in the lowlands of e. Ecuador. The White-winged Swallow tends to favor more open areas than the White-banded Swallow (*Atticora fasciata*), though the two species are often found together. Mostly below 400 m, thus not penetrating the Andes as much as the White-banded Swallow does.

Monotypic.

Range: n. and e. Colombia, Venezuela, and the Guianas to n. Argentina and s. Brazil.

*Tachycineta stolzmanni   **Tumbes Swallow**   **Golondrina de Tumbes**

Uncommon and very local in arid, severely overgrazed scrub far from any water in the lowlands of extreme sw. Ecuador in sw. Loja around Zapotillo. First recorded from Ecuador when a pair (ANSP, MECN) was taken 3 km north of Zapotillo on 6 Apr. 1992. Despite much searching, no others were seen in the area at that time. The Tumbes Swallow was then found to be somewhat more numerous up to about 20 km southwest of Zapotillo on 21–22 Apr. 1993, when a series of six specimens (ANSP, MECN) was obtained. The species may not be resident in this area, however, as none were found during an extended visit to the Zapotillo region in Sep. 1998 (RSR et al.). 150 m.

T. stolzmanni of nw. Peru and adjacent Ecuador is regarded as a species separate from *T. albilinea* (Mangrove Swallow) of Middle America, based on its widely disjunct distribution, smaller size, various plumage differences, and habitat differences (see M. B. Robbins, G. H. Rosenberg, F. Sornoza M., and M. A. Jácome, *Studies in Neotropical Ornithology Honoring Ted Parker*, Ornithol. Monogr. no. 48: 609–612, 1997). Monotypic.

Range: extreme sw. Ecuador and nw. Peru.

Unidentified swallows seen in and around mangrove forests near Machala, El Oro, on 19 Jul. 1991 (J. F. Rasmussen et al.) were believed at the time to possibly have been Tumbes Swallows, but RSR considers that austral migrant White-rumped Swallows (*T. leucorrhoa*) or Chilean Swallows (*T. meyeni*) are just as likely to have been involved, for these two species are known to be long-distance migrants with some potential for vagrancy. Unfortunately, they were not seen well enough to permit positive identification, but observers should be alert to the possibility of the two migratory species occurring in Ecuador.

*[Tachycineta bicolor   **Tree Swallow**   **Golondrina Bicolor]**

An accidental boreal winter visitant to the lowlands of nw. Ecuador. The Tree Swallow is known in Ecuador from only one sighting, a loose group of about six individuals seen on 2 Oct. 1996 as they foraged over secondary growth along the Río Santiago near Playa de Oro in n. Esmeraldas (O. Jahn, pers. comm.). This represents the southernmost record for this North American swallow; it is a casual vagrant to Colombia, with one record from as far south as Laguna La Cocha in Nariño (Hilty and Brown 1986). 50 m.

Monotypic.

Range: breeds in North America, wintering primarily in s. United States and Middle America, smaller numbers southward and casually as far south as n. South America.

Notiochelidon murina   **Brown-bellied Swallow**   Golondrina Ventricafé

Fairly common to common in paramo (especially near roadcuts and cliffs) and open areas in more forested regions in the temperate zone and the central and interandean valleys. The Brown-bellied Swallow is regular around Quito, though here outnumbered by the Blue-and-white (*N. cyanoleuca*). Recorded mostly from 2500 to 4000 m, occasionally higher or lower (perhaps lowest on the crest of the Cordillera de Huacamayos, at 2200 m).

One race: nominate *murina*.

Range: Andes from w. Venezuela to w. Bolivia; Sierra de Perijá and Santa Marta Mountains.

Notiochelidon cyanoleuca   **Blue-and-white**   Golondrina
                              Swallow   Azuliblanca

A common and widespread resident in open and semiopen areas at higher elevations, mainly in towns and around habitations, in primarily forested regions favoring the vicinity of roadcuts; also an uncommon transient and austral winter resident in semiopen areas in the lowlands and foothills of the east, occurring from about late Mar. to Sep. Recorded mostly between 500 and 3000 m, small numbers occasionally ranging somewhat higher; in the west occurs locally down to sea level, breeding in at least a few coastal areas (e.g., sw. Manabí).

Two races occur in Ecuador, nominate *cyanoleuca* being the widespread breeding form and southern-breeding *patagonica* occurring as an austral migrant. *Patagonica* differs in having less black on its crissum (essentially restricted to the sides) and in having a grayer underwing.

Range: highlands of Costa Rica and w. Panama; Colombia and Venezuela to Tierra del Fuego, in the north and west mainly in Andes and other highland areas; southern breeders are strongly migratory, occurring north to n. South America and Panama, casually even farther north.

*Notiochelidon flavipes   **Pale-footed Swallow**   Golondrina Nuboselvática

Rare and seemingly local over montane forest and in adjacent clearings in the temperate zone on the east slope of the Andes. Recorded first from sightings by

RSR in Oct. 1976 in Sangay National Park (Ridgely 1980). The first Ecuador specimens (WFVZ, MECN) were obtained by F. Sibley at 2900 m on the slopes of Cerro Pan de Azucar from 29 Sep. to 11 Oct. 1989 (Marín et al. 1992). The only additional specimen obtained is a female (ANSP) taken by F. Sornoza at Quebrada Honda in Zamora-Chinchipe on 31 Dec. 1991 (Rasmussen et al. 1996). Small numbers have in recent years been seen in Podocarpus National Park at Cajanuma (Rasmussen et al. 1996) and Cerro Toledo (N. Krabbe), and in Aug. 1991 a few were seen at Hacienda La Libertad in Cañar (RSR et al.); small numbers were also noted regularly on the east slope of Cordillera Las Lagunillas in Oct.–Nov. 1992 (RSR and M. B. Robbins et al.). The Pale-footed Swallow has likely been under-recorded as it is easily overlooked among the far more numerous Blue-and-white Swallow (*N. cyanoleuca*), with which it regularly consorts; the Pale-footed may prove to be more or less continuously distributed in appropriate habitat. Recorded mostly from 2650 to 3300 m, once as low as 2500 m in Podocarpus National Park and noted as high as 3500 m on Cerro Pan de Azucar.

Monotypic.

Range: locally in Andes from w. Venezuela to w. Bolivia.

Atticora fasciata   **White-banded Swallow**   Golondrina Fajiblanca

Fairly common to common along forest-bordered rivers and larger streams in the lowlands of e. Ecuador. The pretty White-banded Swallow often perches in small groups on branches protruding from the water. Ranges up in small numbers to about 1100 m along the Río Zamora near Zamora, but found mostly below about 900 m.

Monotypic.

Range: e. Colombia, s. Venezuela, and the Guianas to n. Bolivia and Amazonian Brazil.

Neochelidon tibialis   **White-thighed Swallow**   Golondrina Musliblanca

Uncommon to fairly common but somewhat local in borders of humid forest and adjacent clearings in the lowlands and (especially) foothills of nw. Ecuador; in e. Ecuador occurs in comparable habitats but is decidedly less numerous and more local. In the west recorded south to e. Guayas (old specimens from Chimbo) and nw. Azuay (seen by J. C. Matheus at San Luis in Jan. 1991), though most numerous in Esmeraldas. In the east found mostly near the base of the Andes (it is especially numerous above Archidona), though there are reports from as far east as the Maxus road southeast of Pompeya (RSR et al.) as well as recent reports from Kapawi Lodge on the Río Pastaza near the Peruvian border. In the west recorded up to about 800 m; in the east found between about 250 and 1250 m (collected [MECN] in Apr. 1998 in the Río Huamboya valley west of Macas in nw. Morona-Santiago; F. Sornoza).

Two rather distinct races are found in Ecuador, the small and dark *minima* in the west and the larger and somewhat paler *griseiventris* in the east; *griseiventris* also shows a pale rump.

Range: Panama to nw. Ecuador; locally from se. Colombia, s. Venezuela, and the Guianas to n. Bolivia and Amazonian Brazil; e. Brazil.

Stelgidopteryx ruficollis	Southern Rough-winged Swallow	Golondrina Alirrasposa Sureña

Uncommon to locally common in semiopen areas, especially near water, in the lowlands of both e. and w. Ecuador; considerably more numerous in w. Ecuador than in the east, and in the latter found primarily and in largest numbers in areas fairly close to the Andes (farther east mainly along larger rivers). In the west occurs primarily in more humid regions. Some austral migrants may also occur; flocks of Southern Rough-wingeds in association with Blue-and-white Swallows (*Notiochelidon cyanoleuca*) in the Jatun Sacha area have been seen briefly in Feb.–Mar. and were believed likely to have involved austral migrants (B. Bochan). Ranges up to at least 1500 m.

S. *serripennis* (Northern Rough-winged Swallow) of North and Middle America (breeding south to Costa Rica) is regarded as a separate species, following F. G. Stiles (*Auk* 98[2]: 282–293, 1981) and the 1983 and 1998 AOU Check-lists. Two races of S. *ruficollis* are found in Ecuador, *uropygialis* in the west and nominate *ruficollis* in the east. *Uropygialis* has a conspicuously whitish rump, whereas the nominate race has the rump barely contrasting.

Range: Costa Rica to nw. Peru, n. Argentina, and Uruguay.

Riparia riparia	Sand Martin	Martín Arenero

An uncommon transient in open and semiopen areas, mostly in the lowlands of both e. and w. Ecuador; there are also a few reports from the central valley, where the species has been seen principally near various lakes. Recorded mainly from Sep. to Nov. and again from Mar. to Apr., with two reports from as early as late Aug. (Laguna de Colta in Chimborazo, Puerto Viejo in Manabí). There is, however, only one report from during the northern winter: several dozen were seen along the Río Napo near Primavera in Feb. 1985 (RSR and T. Schulenberg et al.). To about 2500 m.

In most American literature this species is called the Bank Swallow, but we follow Ridgely and Tudor (1989) and Sibley and Monroe (1990) in employing the English name of Sand Martin, which is used universally in Old World.

One race: nominate *riparia*.

Range: breeds in North America, wintering mainly in South America; also in Old World.

Hirundo rustica	Barn Swallow	Golondrina Tijereta

A fairly common to common transient in open and semiopen areas throughout, with numbers generally greatest in the west. Recorded mainly from Sep. to Nov.

(a few arriving as early as Aug., once even in late Jul.) and again from Feb. to Apr. Smaller numbers occur during the northern winter months as well, when the species is most numerous in agricultural terrain of the southwestern lowlands. Passage birds can occur almost anywhere, with small numbers routinely seen overflying continuous forest or even paramo. The Barn Swallow is most numerous in the lowlands, but fairly large numbers also occur in the central valley and the Andes, especially during migration. To 4000 m.

One race: *erythrogaster*.

Range: breeds in North America south to n. Mexico (a few have also recently been found nesting in Argentina), wintering in Middle and South America; also in Old World.

Petrochelidon pyrrhonota   Cliff Swallow   Golondrina de Riscos

A rare to uncommon transient in open and semiopen areas throughout. Recorded mainly from Sep. to Oct. and from Mar. to Apr. Cliff Swallows have been recorded mainly in the lowlands, but very small numbers also occur in the central valley, probably mainly during migration. Several were seen near Otavalo in Sep. 1980 (PJG), and two were noted in Quito on 21 Oct. 1992 (RSR and A. Whittaker); at least four were seen with migrating Barn Swallows (*Hirundo rustica*) as high as 4000 m at Antisana Ecological Reserve on 20 Oct. 2000 (RSR and PJG et al.).

The species is sometimes called the American Cliff Swallow.

Pyrrhonota has sometimes been placed in the genus *Hirundo*, as has the following species, but we follow F. H. Sheldon and D. W. Winkler (*Auk* 104[1]: 97–108, 1993) and the 1998 AOU Check-list in recognizing the genus *Petrochelidon*. The racial affinity of *P. pyrrhonota* recorded from Ecuador remains uncertain; very few specimens have been taken. No examples of birds appearing to be the mainly Mexican-breeding *melanogaster* (with a dark chestnut forehead) have been seen.

Range: breeds in North America south to Mexico, wintering mainly in South America.

Petrochelidon rufocollaris   Chestnut-collared   Golondrina
                                     Swallow   Ruficollareja

Uncommon to locally common in semiopen and agricultural areas, and in towns and cities (where it usually nests), in sw. Ecuador; recorded from Manabí (north to the Bahía de Caráquez area), Guayas, El Oro, and w. Loja. The Chestnut-collared Swallow nests in colonies during the rainy season (Jan.–Aug.), dispersing at other seasons; it then may be distinctly local, sometimes gathering in very large flocks (as have been seen in w. Manabí and El Oro). The largest colonies presently known are found in Loja at Celica and Sozoranga and in El Oro at Portovelo; Chestnut-collareds also nest on certain buildings in the city of Guayaquil (though after breeding, birds are mainly seen outside the city limits; large numbers sometimes roost under the bridge over the Río Daule).

Numbers of the Chestnut-collared Swallow may be on the rise; for example, it now nests in substantial numbers in various towns and cities on the Santa Elena Peninsula of w. Guayas, though it apparently was not present there in the 1950s (at least it was not mentioned by Marchant [1958] in his comprehensive review of the avifauna of that region). Recorded as high as 2000–2100 m in Loja.

P. rufocollaris of Ecuador and Peru is regarded as a species distinct from *P. fulva* (Cave Swallow) of sw. United States, Mexico, and West Indies because of its different plumage, nest structure, and wide range disjunction, following Ridgely and Tudor (1989) and Sibley and Monroe (1990). *P. fulva* may itself comprise more than a single species, as suggested by O. H. Garrido, A. T. Peterson, and O. Komar (*B. O. C. Bull.* 119[2]: 80–91, 1999). One race of *P. rufocollaris* occurs in Ecuador, *aequatorialis*. See K. C. Parkes (*Bull. B. O. C.* 113[2]: 119–120, 1993) for the suppression of the name *chapmani*, formerly the subspecific name applied to Ecuadorian birds.

Range: sw. Ecuador and w. Peru.

Troglodytidae   Wrens

Twenty-six species in 10 genera. Wrens are essentially American in distribution, with only one species reaching the Old World.

Donacobius atricapillus   **Black-capped Donacobius**   **Donacobio**

Fairly common to common in grassy and marshy vegetation around lakes, locally along rivers, and in marshes and damp pastures in the lowlands of e. Ecuador; a few also sometimes range out into crops such as in cornfields. Ranges mostly below 700 m, but small numbers occur higher where suitable damp habitat with rank growth exists (locally as high as 1400 m, e.g., in the Reventador area); perhaps increasing at higher elevations with the rapid spread of pasturage.

The species was formerly often called the Black-capped Mockingthrush.

The Black-capped Donacobius is now considered a wren and not a member of the Mimidae (R. Kiltie and J. W. Fitzpatrick, *Auk* 101[4]: 804–811, 1984). One race occurs in Ecuador, *nigrodorsalis*.

Range: e. Panama to n. and e. Bolivia, e. Paraguay, ne. Argentina, and e. Brazil.

Campylorhynchus turdinus   **Thrush-like Wren**   **Soterrey Mirlo**

Fairly common to common in the canopy and borders of humid forest and secondary woodland (even in areas where quite disturbed) and in clearings with scattered large trees in the lowlands and foothills of e. Ecuador. Recorded mostly below 700 m, in smaller numbers up along the base of the Andes to around 1100 m, a few occurring locally as high as 1300 m in Zamora-Chinchipe (Río Bombuscaro sector of Podocarpus National Park).

One race: *hypostictus*.

Range: se. Colombia, e. Ecuador, e. Peru, n. and e. Bolivia, w. and cen. Amazonian Brazil, and n. Paraguay; e. Brazil.

Campylorhynchus zonatus   **Band-backed Wren**   Soterrey Dorsibandeado

Uncommon to locally fairly common in the canopy and borders of humid and wet forest in the lowlands and foothills of nw. Ecuador south to cen. Manabí (Chone) and extreme s. Pichincha (Río Palenque). In n. Esmeraldas the Band-backed Wren ranges mainly in areas in and close to human settlement (O. Jahn). Ranges up to about 800 m above Tinalandia.

One race: *brevirostris*. Examples that approach "*aenigmaticus*" in appearance were collected at El Placer, Esmeraldas, in Aug. 1987 (ANSP). *Aenigmaticus* was described as a race of the superficially somewhat similar cis-Andean *C. turdinus* (Thrush-like Wren), despite its being found west of the Andes in sw. Nariño, Colombia. What it certainly represents, however, is an intermediate or hybrid population between *C. albobrunneus* (White-headed Wren) of w. Colombia and Panama and *C. zonatus brevirostris* of Ecuador. Despite their dramatically different appearance, *C. albobrunneus* and *C. zonatus* may perhaps best be considered conspecific; vocally they are essentially identical, and their ranges are parapatric.

Range: s. Mexico to w. Panama; n. Colombia; nw. Ecuador.

Campylorhynchus fasciatus   **Fasciated Wren**   Soterrey Ondeado

Fairly common to common in scrub, deciduous woodland (most often at borders), and agricultural areas and gardens with thorny hedgerows in the more arid lowlands of sw. Ecuador north to cen. Manabí (around Rocafuerte and Portoviejo) and n. Los Ríos (Jauneche), locally ranging into more humid terrain along the base of the Andes at least in se. Guayas and nw. Azuay. Also occurs inland through much of the more arid parts of e. El Oro and Loja. Curiously, though, and despite being found in the Río Marañón valley in nw. Peru, the Fasciated Wren does not seem to occur in the Marañón drainage around Zumba in extreme se. Ecuador. Ranges up locally to about 2500 m (e.g., east of Gonzanamá in Loja).

One race: *pallescens*. *Pallescens* appears to intergrade with nw. Peruvian nominate *fasciatus* in s. Loja, though individual variation in all populations is substantial.

Range: sw. Ecuador and nw. Peru.

Odontorchilus branickii   **Gray-mantled Wren**   Soterrey Dorsigris

Uncommon and perhaps local in the canopy and borders of montane forest in the foothill and lower subtropical zones on the east slope of the Andes. On the west slope uncommon and seemingly local (but easily overlooked) in the foothills of the northwest south to nw. Pichincha. In the west recorded from Esmeraldas (sightings northwest of Alto Tambo in Jul. 1990 [M. B. Robbins] and numerous sightings and tape-recordings at Playa de Oro from 1995 to 1998, where the species was considered to be widespread and fairly common [O. Jahn et al.]), w. Imbabura (a 19th-century specimen taken at "Hacienda Paramba"; also seen in Jul. 1997 near the Salto del Tigre bridge over the Río Guaillabamba [P. Coopmans et al.]), and Pichincha (a 1967 specimen [MECN] from Santo Domingo de los Colorados, and repeated sightings since Sep. 1995 along the road north of Simón Bolívar near Pedro Vicente Maldonado [v.o.]). On the east slope recorded from about 1100 to 1900 m, but in the northwest thus far recorded much lower, mainly between 200 and 500 m, locally down to under 100 m at Playa de Oro.

Two races occur in Ecuador, nominate *branickii* on the east slope and *minor* in the northwest. The latter differs in its plain gray, unbarred central rectrices but is (despite its name) at most only marginally smaller. The two differ vocally and perhaps represent distinct species.

Range: locally on east slope of Andes from s. Colombia to w. Bolivia; west slope of Andes in sw. Colombia and nw. Ecuador.

Cinnycerthia unirufa   Rufous Wren   Soterrey Rufo

Uncommon to locally common in the undergrowth of montane forest in the temperate zone on both slopes of the Andes. On the west slope recorded south to Bolívar (Salinas; Krabbe et al. 1997) and Azuay (Portete; Berlioz 1932b). Rufous Wrens are most numerous where there is an understory of *Chusquea* bamboo. Recorded mainly from 2200 to 3400 m.

One race: *unibrunnea*.

Range: Andes from sw. Venezuela to extreme n. Peru; Sierra de Perijá.

Cinnycerthia olivascens   Sepia-brown Wren   Soterrey Caferrojizo

Uncommon to locally fairly common in the undergrowth of montane forest in the subtropical zone on the entire east slope of the Andes. There appear to be no previous published records from the west slope, though since the late 1970s small numbers of Sepia-brown Wrens have been seen by numerous observers at various sites in Pichincha (e.g., along the Chiriboga and Mindo-Tandayapa roads). The species has also been seen and tape-recorded in nw. Cotopaxi (N. Krabbe). The only west-slope specimen of which we are aware is a female (ANSP) obtained south-southwest of Chical (off the Maldonado road) in Carchi on 17 Aug. 1988. The Sepia-brown Wren occurs mainly at elevations below the range of its congener the Rufous Wren (*C. unirufa*), though overlap of the two species occurs locally. Recorded mainly from 1500 to 2500 m.

We follow R. T. Brumfield and J. V. Remsen, Jr., (*Wilson Bull.* 108[2]: 205–227, 1996) in considering *C. olivascens* of Colombia to n. Peru and *C. fulva* (Fulvous Wren) of s. Peru and w. Bolivia as species separate from

C. peruana (Peruvian Wren) of Peru (Amazonas to Ayacucho). Brumfield and Remsen (op. cit.) called *C. olivascens* the Sharpe's Wren (however, we favor continuing to call this best known of the complex the Sepia-brown Wren) and *C. fulva* the Superciliated Wren (despite that name's having long been in use for *Thryothorus superciliaris*). *C. olivascens* is monotypic.

Range: Andes from n. Colombia to n. Peru.

Cistothorus platensis   Grass Wren   Soterrey Sabanero

Fairly common in damp grassy and shrubby situations and in paramo throughout the highlands (and locally in fields in the central valley). Recorded mostly between 3000 and 4000 m, smaller numbers occurring locally lower (down to 2200 m along the Loja-Zamora road) and probably also a bit higher.

The complex musical song of the comparatively long-tailed form found in Ecuador (*aequatorialis*) bears no resemblance to the simple harsh song of North American birds (*stellaris* group; Sedge Wren). Although the systematics of the group are still not fully worked out (for the most recent revision, see M. A. Traylor, Jr., *Fieldiana Zool.* new series 48, publ. 1392, 1988), more than one species must surely be involved. Given the continuing uncertainty regarding species limits, however, we feel we must continue to regard all forms as conspecific.

Range: North America south locally in highlands to w. Panama; mountains of n. Venezuela, and Andes from w. Venezuela to Tierra del Fuego, also occurring in lowlands of Chile and Argentina and north locally into s.-cen. Brazil; tepuis of se. Venezuela and adjacent Guyana; Sierra de Perijá and Santa Marta Mountains.

Thryothorus nigricapillus   Bay Wren   Soterrey Cabecipinto

Common in dense undergrowth at borders of humid and wet forest and secondary woodland, often near streams and in *Heliconia* thickets, in the lowlands and foothills of w. Ecuador south to cen. Manabí, n. Guayas (Hacienda Pacaritambo; A. Brosset, *Oiseau* 34: 122–135, 1964), and along the base of the Andes to El Oro (La Chonta and Buenaventura). Ranges up to about 1400 m in the Mindo area, but mostly below 900–1000 m.

By far the most widespread race of *T. nigricapillus* occurring in Ecuador is nominate *nigricapillus*. However, in recent years *connectens* (which has somewhat more extensive black barring on its underparts) has been shown to be the subspecies present at El Placer and northwest of Alto Tambo in n. Esmeraldas (ANSP, MECN).

Range: Nicaragua to sw. Ecuador.

Thryothorus euophrys   Plain-tailed Wren   Soterrey Colillano

Uncommon to common in *Chusquea* bamboo-dominated understory of montane forest in the temperate zone on both slopes of the Andes and locally above the central valley (e.g., at Pasochoa). On the west slope ranges south to w. Chimborazo (old records from Ceche, Cayandeled, and Llagos). The Plain-tailed Wren becomes more numerous southward on the east slope of the Andes. Recorded mostly from 2200 to 3200 m, locally occurring somewhat lower (e.g., it is regular at ca. 2000 m, below the crest of the Cordillera de Huacamayos) and considerably higher (e.g., above 3500 m on Volcán Pichincha; N. Krabbe).

Three rather well-marked races are found in Ecuador. Nominate *euophrys* of the west slope has more heavily spotted underparts than does *longipes*, which is found on almost the entire east slope. The spotting below seen in *longipes* appears to decrease clinally from north to south (though there is also considerable individual variation); the crown of *longipes* is intermixed with gray, not so purely brown as seen in the nominate race. Birds of extreme se. Ecuador (e.g., on the east slope of Cordillera Las Lagunillas in Zamora-Chinchipe; ANSP) are referable to *atriceps* (which is found mainly in n. Peru, north of the Río Marañón); these are unspotted and more grayish below and have a decidedly grayer (not so brown) crown. Contra Fjeldså and Krabbe (1990, p. 543), recently obtained examples (ANSP, MECN) of the population found in Podocarpus National Park appear to be referable to *longipes* and not to *atriceps*, the break between the two forms apparently occurring just south of the park.

Range: Andes from extreme s. Colombia to cen. Peru.

Thryothorus mystacalis   Whiskered Wren   Soterrey Bigotillo

Uncommon to locally fairly common in dense undergrowth and viny tangles at lower and middle levels of forest borders and secondary woodland in the more humid lowlands, foothills, and lower subtropical zone of w. Ecuador from w. Esmeraldas south to sw. Manabí (Machalilla National Park), n. Guayas (slopes of Cordillera de Colonche; Hacienda Pacaritambo), El Oro, and w. Loja (Alamor region). Recorded up locally to about 1500 m.

T. mystacalis of Venezuela to Ecuador is regarded as a species separate from *T. genibarbis* (Moustached Wren), found in the lowlands of s.-cen. South America. This follows Ridgely and Tudor (1989) and most other recent authors and is based especially on their very different songs. One race occurs in Ecuador, nominate *mystacalis*.

Range: mountains of n. Venezuela; Andes from w. Venezuela to w. Ecuador (where also in lowlands); Sierra de Perijá.

Contra Fjeldså and Krabbe (1990), the *juruanus* race of Moustached Wren does not occur in the lowlands of e. Ecuador; it extends north only to e.-cen. Peru (Ridgely and Tudor 1989). In a lapsus, Robbins et al. (1987) listed the Moustached Wren from the Cordillera de Cutucú.

Thryothorus coraya   Coraya Wren   Soterrey Coraya

Fairly common in lower (especially) and middle growth at the borders of terra firme and várzea forest and secondary woodland, favoring dense viny tangles, in the lowlands of e. Ecuador. The Coraya seems less tied to the vicinity of water than the Buff-breasted Wren (*T. leucotis*), though (fide P. Coopmans) at Kapawi Lodge—where the Buff-breasted is much less numerous—the Coraya is found in shrubby streamside vegetation. Found mostly below 700 m, but ranges locally (e.g., along the Loreto road north of Archidona) up along the base of the Andes to about 1200 m.

One race: *griseipectus*.

Range: se. Colombia, s. Venezuela, and the Guianas to e. Peru and parts of Amazonian Brazil.

Thryothorus leucotis	Buff-breasted Wren	Soterrey Pechianteado

Uncommon to locally fairly common in the undergrowth of várzea forest and woodland, and in tangled thickets around the margins of lakes and along rivers, in the lowlands of e. Ecuador. In the Río Pastaza drainage of se. Ecuador known only from recent sightings at Kapawi Lodge near the Peruvian border. The Buff-breasted Wren does not occur on river islands in Ecuador—where usually no *Thryothorus* wren is present—favoring instead their banks and associated backwaters; it is considerably less associated with forest than the Coraya Wren (*T. coraya*). Recorded only below about 300 m.

One race: *peruanus*.

Range: Panama to e. Peru, nw. Bolivia, and Amazonian and s.-cen. Brazil.

Thryothorus superciliaris	Superciliated Wren	Soterrey Cejón

Uncommon to fairly common in arid scrub, undergrowth of deciduous woodland, and thickets and hedgerows in agricultural regions in the more arid lowlands of sw. Ecuador north to cen. Manabí (Bahía de Caráquez area) and s. Los Ríos (Babahoyo); smaller numbers range locally into somewhat more humid regions such as the woodland south of the Río Ayampe in nw. Guayas. The Superciliated Wren is most numerous in the desert scrub of w. Guayas and sw. Manabí. Ranges up to about 1500 m in the upper Río Catamayo valley of w. Loja, in small numbers as high as 1850 m above La Toma (P. Coopmans; RSR et al.).

Two only slightly differentiated races occur in Ecuador, nominate *superciliaris* in Manabí and Guayas and *baroni* in El Oro and w. Loja.

Range: sw. Ecuador and nw. Peru.

Thryothorus sclateri	Speckle-breasted Wren	Soterrey Pechijaspeado

Fairly common to locally common in dense undergrowth and thickets in deciduous and semihumid forest and woodland and in their borders in the lowlands and lower subtropical zone of sw. Ecuador from cen. Manabí (Bahía de Caráquez area) and Los Ríos (Jauneche) south, mainly in more arid regions, through El Oro and w. Loja. With ongoing clearing of humid forest, the Speckle-breasted Wren may slowly be spreading northward; one was recorded at Río

Palenque in s. Pichincha, where the species had not previously been noted, on 23 Oct. 1997 (P. Coopmans et al.). Also recently found in the Río Marañón drainage of s. Zamora-Chinchipe, mostly around Zumba though small numbers have been recorded north to near Valladolid. The Speckle-breasted Wren is numerous in the Chongón Hills and in Machalilla National Park; it also is one of the most common birds at Jauneche in Los Ríos (P. Coopmans). To about 1600 m.

   T. sclateri of Colombia to n. Peru is regarded as a species separate from *T. maculipectus* (Spot-breasted Wren) of Mexico to Nicaragua and *T. rutilus* (Rufous-breasted Wren) of Costa Rica to Venezuela, following Ridgely and Tudor (1989) and most other recent authors; these three taxa are best treated as allospecies. Two races of *T. sclateri* are found in Ecuador, *paucimaculatus* in most of the species' Ecuadorian range and nominate *sclateri* in the Río Marañón drainage. The slightly larger nominate race differs from *paucimaculatus* especially in its much bolder and denser black speckling and barring below. Parker and Carr (1992) go so far as to list these two taxa as separate species, though they do not give any explanation why they adopted that course; a monotypic *T. sclateri* would be called the Marañón Wren.

   Range: very locally on Andean slopes in Colombia; sw. Ecuador and nw. Peru.

Thryothorus leucopogon   **Stripe-throated Wren**   **Soterrey Golirrayado**

Rare and apparently local to locally fairly common (apparently mainly northward) in viny tangles and thickets in the lower and mid-levels of very humid to humid forest and secondary woodland in the lowlands and foothills of w. Ecuador, ranging south to n. Manabí and n. Pichincha, and south along the base of the Andes locally to nw. Azuay (Manta Real, where found in Aug. 1991; ANSP). There are relatively few Ecuadorian records of this generally scarce wren from south of Esmeraldas; however, surveys from 1995 to 1998 revealed it, with the Southern Nightingale-Wren (*Microcerculus marginatus*), to be the most numerous wren inside forest in the lowlands (below 250 m) at Playa de Oro in n. Esmeraldas (O. Jahn et al.). Recorded up locally to about 750 m.

   The species has been considered (e.g., Meyer de Schauensee 1966, 1970) to be conspecific with *T. thoracicus* (Stripe-breasted Wren) of s. Middle America. However, we agree with the 1983 and 1998 AOU Check-lists that *T. leucopogon* is better regarded as distinct. One race: nominate *leucopogon*.

   Range: e. Panama to sw. Ecuador.

   The Sooty-headed Wren (*T. spadix*) could range into the foothills of extreme nw. Ecuador as it is known from immediately adjacent Colombia in sw. Nariño (Hilty and Brown 1986).

Troglodytes aedon   **House Wren**   **Soterrey Criollo**

Common and widespread in clearings, scrub, gardens, and woodland borders in the lowlands of both e. and w. Ecuador in both arid and humid regions, ranging well up into the highlands; also occurs on Isla de la Plata. The House Wren is considerably less numerous and more local in the eastern lowlands, however, especially away from the Andes, though it does seem to be capable of colonizing newly created clearings in generally forested zones; it quickly did so at the major installations along the Maxus road southeast of Pompeya during 1994–1995 (RSR et al.). Occurs at least locally as high as about 3300 m (e.g., below the Papallacta pass).

R. T. Brumfield and A. P. Capparella (*Condor* 98[3]: 547–556, 1996) suggested treating *T. musculus* (Southern House-Wren) of the Lesser Antilles and s. Mexico southward as a full species separate from *T. aedon* (Northern House-Wren) of North America and Mexico. The situation in Mexico appears to be somewhat unclear, however, and we opt to continue to regard all mainland members of the group as conspecific, as was done in the 1998 AOU Check-list. One race of *T. aedon* occurs in Ecuador, *albicans*.

Range: North America to Tierra del Fuego; Lesser Antilles. More than one species is likely involved.

Troglodytes solstitialis	Mountain Wren	Soterrey Montañés

Uncommon to fairly common in montane forest and woodland, ranging mostly in dense growth at borders and in openings, in the subtropical and temperate zones on both slopes of the Andes; on the west slope recorded south to w. Loja (several recent reports from the Alamor/Celica region [Bloch et al. 1991; Best et al. 1993]). Also found recently on the coastal Cordillera de Colonche of sw. Manabí (Cerro San Sebastián in Machalilla National Park [Parker and Carr 1992; ANSP]) and nw. Guayas (Loma Alta; Becker and López-Lanús 1997). Recorded mostly from 1500 to 3200 m, but locally ranges a little higher (e.g., regularly to at least 3400 m on the east side of the Papallacta pass); on the coastal cordillera found between about 700 and 850 m.

One race: nominate *solstitialis*.

Range: Andes from w. Venezuela to nw. Argentina; Sierra de Perijá.

Henicorhina leucosticta	White-breasted Wood-Wren	Soterrey Montés Pechiblanco

Fairly common to common in the undergrowth of humid forest and forest borders in the lowlands and foothills of e. and nw. Ecuador. In the west found mainly in Esmeraldas, but also recorded south in small numbers to sw. Imbabura and n. Pichincha (as far south as the Mindo region). In e. Ecuador the White-breasted Wood-Wren is most numerous in terra firme forest, with smaller numbers ranging into secondary growth and floodplain habitats. Ranges mainly below about 1000 m, locally occurring as high as 1500 m in w. Napo on the slopes of Volcán Sumaco (B. Whitney); recorded only up to 900 m on the west slope (near Lita; N. Krabbe), with old records from higher elevations (e.g., Mindo) probably involving birds actually taken lower down.

Two rather different races are found in Ecuador, *hauxwelli* in the east and *inornata* in the northwest. *Hauxwelli* has a black crown and a clear white throat and breast, whereas *inornata* has a rufous brown crown and a dingy whitish throat and breast.

Range: s. Mexico to nw. Ecuador; the Guianas, s. Venezuela, n. Amazonian Brazil, se. Colombia, e. Ecuador, and ne. Peru.

Henicorhina leucophrys	Gray-breasted Wood-Wren	Soterrey Montés Pechigris

Common in the undergrowth of montane forest borders and in overgrown clearings in the subtropical and temperate zones on both slopes of the Andes. On the west slope ranges south to El Oro and w. Loja (Alamor/Celica region); also found on the coastal Cordillera de Colonche in sw. Manabí (Cerro San Sebastián in Machalilla National Park and east of Puerto Cayo [ANSP; v.o.]) and w. Guayas (Cerro La Torre and above Salanguilla; N. Krabbe). Ranges mostly from 1500 to 3000 m, but occurs locally down to 750 m in Pichincha (e.g., in the San Miguel de los Bancos area, where it occurs sympatrically with the White-breasted Wood-Wren [*H. leucosticta*]; P. Coopmans), and regularly as low as 500 m on the west slope from Azuay southward and on the coastal cordillera.

Two rather different races occur in Ecuador, nominate *leucophrys* in most of the species' Ecuadorian range, and *hilaris* in the southwest on Andean slopes from sw. Chimborazo south through El Oro to w. Loja, and also on the coastal cordillera. *Hilaris* is a markedly paler grayish white below than the nominate race, which is decidedly gray on its throat and breast. Birds ascribed to the race *brunneiceps* have been reported from the west slope in extreme n. Ecuador in Imbabura (*Birds of the World*, vol. 9; Fjeldså and Krabbe 1990). We have not seen these specimens; *brunneiceps* is, in any case, very similar in plumage to the nominate *leucophrys*.

Range: highlands of cen. Mexico to Panama; mountains of n. Venezuela, Andes from w. Venezuela to w. Bolivia; Sierra de Perijá and Santa Marta Mountains. More than one species is apparently involved (in Colombia).

*Henicorhina leucoptera	Bar-winged Wood-Wren	Soterrey-Montés Alibandeado

Fairly common in the lower growth and borders of montane forest and borders near Chinapinza on the west slope of the Cordillera del Cóndor in Zamora-Chinchipe. Here a series of specimens (MECN, ANSP) was obtained in Sep. 1990 by N. Krabbe and F. Sornoza (*Bull. B. O. C.* 114[1]: 55–61, 1994); the species was again found in the same general region in Jun. 1993 (ANSP, MECN). This only recently described species (J. W. Fitzpatrick, J. W. Terborgh, and D. E. Willard, *Auk* 94[2]: 195–210, 1977) was heretofore known only from n. Peru. It is recorded only between about 1700 and 1950 m, replacing the Gray-breasted Wood-Wren (*H. leucophrys*) at higher elevations on the Cordillera del Cóndor but with some overlap, the Gray-breasted there occurring up to only about 1700–1800 m.

The Bar-winged Wood-Wren was given Near-threatened status by Collar et al. (1992, 1994), doubtless primarily in consideration of the species' very limited range. Given its relative abundance in that range and its tolerance for somewhat disturbed situations, however, we do not consider it to be truly at risk.

Monotypic.

Range: outlying ridges on east slope of Andes in extreme s. Ecuador and n. Peru.

Cyphorhinus thoracicus   **Chestnut-breasted Wren**   **Soterrey Pechicastaño**

Uncommon and apparently somewhat local in the undergrowth of montane forest in the foothills and subtropical zone on the entire east slope of the Andes from Sucumbíos (San Rafael Falls; v.o.) south to Zamora-Chinchipe (several sites in and near Podocarpus National Park, and at Panguri [T. J. Davis]). The Chestnut-breasted Wren seems likely to range north on the east slope of the Andes into adjacent Colombia, though it has apparently not been found in this area (Hilty and Brown 1986). Recorded mostly from 1100 to 2200 m.

One race: *dichrous*.

Range: Andes from w. Colombia to w. Bolivia.

Cyphorhinus arada   **Musician Wren**   **Soterrey Virtuoso**

Rare to uncommon and somewhat local in the undergrowth of humid forest (mainly terra firme, with smaller numbers in várzea) in the lowlands and foothills of e. Ecuador; seems less numerous in Ecuador than it is in some other parts of Amazonia. Ranges up along the base of the Andes to about 1000 m, with a few recorded locally as high as about 1300 m in the Volcán Sumaco/Loreto road area in w. Napo (MCZ; also recent sight records) and in the Río Bombuscaro sector of Podocarpus National Park (P. Coopmans and others). At its upper elevational limit the Musician Wren thus overlaps, at least to a limited extent, with the Chestnut-breasted Wren (*C. thoracicus*).

Although some references (e.g., *Birds of the World*, vol. 9; Sibley and Monroe 1990) spell the species name as *aradus*, it is our understanding that "arada" was the Cayenne Indian name for the bird and that it is thus a noun in apposition (and not an adjective that must agree in gender). One race of *C. arada* occurs in Ecuador, *salvini*.

Range: se. Colombia, se. Venezuela, and the Guianas to n. Bolivia and Amazonian Brazil.

Cyphorhinus phaeocephalus   **Song Wren**   **Soterrey Canoro**

Locally fairly common in the undergrowth of very humid to humid forest in the lowlands and foothills of w. Ecuador south to nw. and se. Guayas (an old specimen [AMNH] from Cerro de Manglaralto, a Dec. 1996 report from Loma Alta [Becker and López-Lanús 1997], and Aug. 1996 records from the hills of Manglares-Churute Ecological Reserve [Pople et al. 1997]) and El Oro (an old record from La Chonta, many recent reports from Buenaventura). Recorded up to about 900 m.

Trans-Andean *C. phaeocephalus* is regarded as a species separate from *C. arada* (Musician Wren; found east of the Andes), following most recent authors. One race occurs in Ecuador, nominate *phaeocephalus*.
Range: Honduras to sw. Ecuador.

Microcerculus marginatus	Southern Nightingale-Wren	Soterrey Ruiseñor Sureño

Fairly common to common in the undergrowth of wet and humid forest and mature secondary woodland in the lowlands of e. and w. Ecuador, in the east occurring mainly in terra firme. In the west recorded south mainly to n. Manabí and s. Pichincha, but occurring in small numbers and locally as far south as sw. Manabí (Cerro San Sebastián in Machalilla National Park; ANSP), Guayas in the northwest (hills south of Río Ayampe [RSR; P. Coopmans] and Loma Alta [Becker and López-Lanús 1997]) and southeast (hills of Manglares-Churute Ecological Reserve; Pople et al. 1997) and nw. Azuay (Manta Real [Parker and Carr 1992; RSR et al.]). Ranges mostly below about 700 m, a few as high as 1300 m in Pichincha in the Mindo area and at Maquipucuna Reserve.

Subspecific diagnoses in *M. marginatus* present difficulties, and more material is needed, especially from w. Ecuador, in order to better define the ranges of various taxa. In the east is found nominate *marginatus*, which has mostly white underparts. In the northwest the similar *occidentalis* occurs; this form ranges at least in n. Esmeraldas, adjacent Imbabura, and nw. Pichincha (5 km south of Golondrinas; one collected [MECN] on 16 Jul. 1997 by D. Lane). Southward in w. Ecuador is found the rather different *taeniatus*, which is boldly scaled with dusky on all of its underparts aside from the white throat. *Taeniatus* is known north to s. Esmeraldas on the west bank of the Río Verde [ANSP] and at Bilsa (fide J. Hornbuckle) and to s. Pichincha (Tinalandia), and it is also recorded from Manabí (north to Filo de Monos; WFVZ), e. and extreme w. Guayas, and nw. Azuay. We suspect that the AMNH specimen from Naranjo ascribed by Chapman (1926, p. 575) to *occidentalis* is actually a juvenile *taeniatus*, which is more in accord with what is now understood about the pattern of geographic variation in w. Ecuador. The nature of the contact zone—if there is any—or even possible overlap of white-breasted *occidentalis* and scaly-breasted *taeniatus* remains to be determined; their respective ranges come very close in, for instance, Pichincha.

Despite their diverse appearance, all Ecuadorian birds have similar songs, and they thus are here considered conspecific. Further study may demonstrate that this treatment is an oversimplification. *M. marginatus* of Panama and South America is, however, regarded as a species distinct from *M. philomela* (Northern Nightingale-Wren) of Mexico to Costa Rica, following data presented by F. G. Stiles (*Wilson Bull.* 95[2]: 169–183, 1983), as well as all subsequent authors, including the 1998 AOU Check-list. *M. philomela* has a very different song, though in plumage it resembles nominate *marginatus*. Trans-Andean birds have sometimes been considered specifically distinct from cis-Andean *M.*

marginatus (as *M. luscinia*, Scaly-breasted Wren, or Whistling Nightingale-Wren), but more study is needed in order to determine whether this is the appropriate course. In s. Amazonia the song of what is still classified as nominate *marginatus* again changes very dramatically; a separate species is clearly involved here, but its nomenclatural status remains unclear (see Ridgely and Tudor 1989).

Range: Costa Rica to sw. Ecuador, n. Bolivia, and Amazonian Brazil (but absent from ne. South America).

Microcerculus bambla   **Wing-banded Wren**   **Soterrey Alifranjeado**

Uncommon to fairly common but local in the undergrowth of humid forest in the foothills along the east slope of the Andes, and in the lowlands adjacent to the Andes; recorded mainly from w. Napo, in the drainage of the upper Río Napo. The Wing-banded Wren occurs primarily at elevations above those at which the Southern Nightingale-Wren (*M. marginatus*) is found, though the two species occur sympatrically at sites such as Jatun Sacha (B. Bochan) and west of Coca in terra firme near the lower Río Payamino (ANSP). The Wing-banded's easternmost locality is along the far end of the Maxus road southeast of Pompeya, where at least one bird was mist-netted in Nov. 1994 (C. Canaday et al.) and another heard on 28 Jan. 1995 (RSR and F. Sornoza). The only record from outside Napo is a 19th-century specimen taken at Sarayacu in Pastaza, but this bird, like numerous other "Sarayacu" specimens, may have actually been taken at a substantial distance from that settlement. As with the much more widespread and generally more numerous Southern Nightingale-Wren, the Wing-banded Wren is recorded mainly by voice; it is regularly heard along the western end of the road to Loreto north of Archidona. Recorded mainly between 400 and 1300 m, but small numbers occur up to about 1500 m on the slopes of Volcán Sumaco (Palm Peak, MCZ; B. Whitney) and on the south slope of the Cordillera de Huacamayos, and locally as low as 250 m.

One race: *albigularis*.

Range: the Guianas, s. Venezuela, and n. Amazonian Brazil; locally near base of Andes in e. Ecuador and e. Peru.

Polioptilidae   Gnatcatchers and Gnatwrens

Five species in three genera. The family is exclusively American in distribution. All five species were formerly usually considered to be part of one of several mainly Old World families, either the Sylviidae or the Muscicapidae.

Microbates cinereiventris   **Tawny-faced Gnatwren**   **Soterillo Carileonado**

Uncommon to locally fairly common but inconspicuous in the lower growth of humid forest in the lowlands and foothills of both e. and w. Ecuador. In the west recorded south to n. Manabí (Pata de Pájaro, Río de Oro), s. Pichincha

(Río Palenque), e. Guayas (old specimens from Naranjito; recently found in the hills of Manglares-Churute Ecological Reserve [Pople et al. 1997]), El Oro (Buenaventura), and w. Loja (old records from Alamor). The Tawny-faced Gnatwren seems less numerous and more local in e. than in w. Ecuador, and in the east it is not recorded in areas well away from the Andes (the easternmost locality being La Selva); in the west it is now found primarily in foothill areas— so much forest having been destroyed in the lowlands, where it presumably was once widespread—and it is particularly numerous at Tinalandia. Ranges mostly up to about 1000 m, locally a bit higher in w. Loja.

The species was formerly (e.g., Meyer de Schauensee 1966, 1970) called the Half-collared Gnatwren; virtually all more recent references have named it the Tawny-faced Gnatwren.

Two races are found in Ecuador, nominate *cinereiventris* in the west and *hormotus* (see S. L. Olson, *Proc. Biol. Soc. Wash.* 93[1]: 72–73, 1980) in the east. The nominate race differs in having a blackish postocular stripe lacking in *hormotus*.

Range: Nicaragua to sw. Ecuador and e. Peru.

Microbates collaris   Collared Gnatwren   Soterillo Collarejo

Rare and local in the lower growth of terra firme forest in the lowlands of ne. Ecuador in Sucumbíos. The Collared Gnatwren was first recorded in Ecuador from a single individual that was mist-netted, photographed, and released at Cuyabeno on 9 Sep. 1989 (C. Canaday; photos to VIREO); there have been a few subsequent sightings from there. The only additional locality is north of Tigre Playa near the Río San Miguel, where two specimens (ANSP, MECN) were obtained in Aug. 1993. Recorded below about 250 m.

One race: *colombianus* (see K. C. Parkes, *Proc. Biol. Soc. Wash.* 93[1]: 66, 1980).

Range: the Guianas, s. Venezuela, se. Colombia, ne. Ecuador, ne. Peru, and n. Amazonian Brazil.

Ramphocaenus melanurus   Long-billed Gnatwren   Soterillo Piquilargo

Fairly common in viny tangled lower and middle growth of humid and semi-humid forest borders and secondary woodland in the lowlands and to a lesser extent the foothills of both e. and w. Ecuador. In the west recorded south through most of Guayas, El Oro, and w. Loja; the Long-billed Gnatwren avoids, however, the wet-forest belt in much of n. Esmeraldas (O. Jahn and P. Mena Valenzuela et al.). Ranges up to about 1300 m in the west; in the east, however, not known above 600 m (north of Canelos in Pastaza; N. Krabbe).

Three races are found in Ecuador: *rufiventris* occurs in the west, whereas in the east *duidae* occurs northward (especially the upper Río Napo region northward) and *badius* ranges elsewhere (Pastaza and Morona-Santiago southward, also in the far east). *Duidae* and *badius* are similar, but *rufiventris* differs from both in having a grayer back, more uniform cinnamon-buff underparts and cheeks, and wider white tail-tips. Separate species on either side of the Andes are perhaps involved; the trans-Andean species would be called *R. rufiventris* (Long-billed Gnatwren) and the cis-Andean species *R. melanurus* (Straight-billed Gnatwren).

Range: s. Mexico to extreme nw. Peru, n. Bolivia, and Amazonian Brazil; e. Brazil.

Polioptila plumbea   **Tropical Gnatcatcher**   **Perlita Tropical**

Fairly common to common in the canopy and borders of woodland, forest borders, and arid scrub in lowlands and foothills of w. Ecuador, ranging inland through Loja east to the Malacatos, Vilcabamba, and Amaluza areas. Uncommon and seemingly local (perhaps overlooked to some extent) in the lowlands of e. Ecuador, where only found in the canopy and borders of terra firme and várzea forest and occurring little or not at all in secondary habitats; so far recorded primarily from the northeast in Napo and Sucumbíos, but recent reports from Kapawi Lodge on the Río Pastaza near the Peruvian border make it seem likely that the species occurs widely through the far southeast as well. In the west recorded up to about 1900 m in Loja (though in most areas not found above about 1500 m); in the east not found at all in areas close to the base of the Andes, and thus not recorded from above about 300 m.

Two distinctly different—especially in males—races are found in Ecuador. In the west ranges *bilineata*, in which males have white extending up over their cheeks and face to above the eye; in the east the subspecies found is uncertain (probably *parvirostris*, but few or no specimens have been taken; we have not examined any). In any case, eastern males have a black crown extending straight back from below the eye. In addition to the plumage differences, their songs are markedly dissimilar. Two species are almost certainly involved, but complications in cis-Andean populations preclude us from splitting them at this time. If split, trans-Andean *P. bilineata* is perhaps best called the White-faced Gnatcatcher—it has been called the White-browed Gnatcatcher in some literature, though the effect is not really that of a white brow.

Range: s. Mexico to nw. and se. Peru and Amazonian and ne. Brazil.

The Marañón Gnatcatcher (*P. maior*) of the upper Río Marañón drainage in nw. Peru (north to the Jaen/San Ignacio area), here regarded as a species separate from *P. plumbea*, seems likely to occur in the Marañón drainage of far se. Ecuador around Zumba. However, as yet it has—somewhat surprisingly—not been found here.

Polioptila schistaceigula   **Slate-throated Gnatcatcher**   **Perlita Pechipizarrosa**

Rare to locally uncommon in the canopy and borders of very humid to humid forest and forest borders in the lowlands and foothills of nw. Ecuador from the Colombian border south to w. Esmeraldas (southeast of Muisne; ANSP) and s.

Pichincha (where found, at least formerly, at Río Palenque though there have been no reports from there since the late 1980s). There have also been numerous sightings since early 1994 of Slate-throated Gnatcatchers from patchy foothill forest and secondary woodland west of San Miguel de los Bancos in nw. Pichincha, along the road north of Simón Bolívar near Pedro Vicente Maldonado (v.o.); the species is also known from adjacent sw. Imbabura (a pair seen and tape-recorded near Salto del Tigre bridge over the Río Guaillabamba on 14 Jul. 1997; P. Coopmans et al.). Intensive surveys from 1995 to 1998 also revealed it to be not uncommon at Playa de Oro in n. Esmeraldas (O. Jahn). Recorded up to about 750 m in nw. Pichincha.

Given the very extensive deforestation that has taken place across much of the Slate-throated Gnatcatcher's Ecuadorian range, and the species' general rarity and apparent reliance on relatively undisturbed forest, we believe it merits Near-threatened status in Ecuador. It was not considered to be generally at risk by Collar et al. (1992, 1994).

Monotypic. Two specimens (ANSP), both males, recently obtained in Esmeraldas do not differ appreciably from ANSP's single example, also a male, from nw. Colombia (Santander). Hilty and Brown's (1986) and Ridgely and Gwynne's (1989) comments concerning possible subspecific variation in the species thus seem unlikely to be warranted; at least the variation seen (if any) is not striking. The Ecuador examples are not generally blacker, but they do show some white flecking on the upper throat not seen in the Santander individual.

Range: e. Panama to nw. Ecuador.

Motacillidae   Pipits and Wagtails

One species. The family is worldwide in distribution; only pipits occur in most of the New World.

Anthus bogotensis   **Paramo Pipit**   **Bisbita del Páramo**

Uncommon to locally fairly common in grasslands, paramo, and adjacent small fields on both slopes of the Andes and on slopes above the central and interandean valleys. In the west recorded south to El Oro at Taraguacocha. Ranges mostly between 3000 and 4000 m, occasionally somewhat lower or higher.

One race: nominate *bogotensis*.

Range: Andes from w. Venezuela to nw. Argentina (more local southward).

Parulidae   New World Warblers

Thirty-one species in 12 genera. New World Warblers are exclusively American in distribution. They were formerly considered a subfamily in a broadly inclusive Emberizidae; we follow Gill (1994) in giving them family rank.

*Vermivora chrysoptera   Golden-winged Warbler   Reinita Alidorada

A very rare boreal winter visitant to the borders of montane forest and secondary woodland on both slopes of the Andes in n. Ecuador. There are only a few Ecuadorian records, all of them recent sightings. These include males seen above Mindo on 20 Feb. 1981 (J. Egbert, *Am. Birds* 35[3]: 346, 1981) and near Mindo on 3 Mar. 1995 (M. Lysinger et al.); evidently there are several more recent sightings as well (v.o.). In addition, one was seen near San Rafael Falls in w. Napo on 20 Jan. 1985 (J. Terborgh). The Ecuador reports are the southernmost known for the Golden-winged Warbler, small numbers of which occur regularly as far south as w. Colombia (Hilty and Brown 1986). So far only recorded between about 1300 and 1500 m.

Populations of the Golden-winged Warbler have declined substantially on their breeding grounds, mostly as a result of competition with the closely related Blue-winged Warbler (*V. pinus*).

Monotypic.

Range: breeds in e. North America, wintering in s. Middle America and nw. South America.

*Vermivora peregrina   Tennessee Warbler   Reinita Verdilla

A very rare to rare and perhaps erratic (not annual) boreal winter visitant to montane forest borders, secondary woodland, and trees in gardens in the subtropical zone on either slope of the Andes in n. Ecuador (also found once in the central valley); known from Pichincha, Imbabura, and w. Napo. The Tennessee Warbler was first recorded in Feb. 1980 when several were seen—perhaps a "mini-invasion" occurred—at a total of four localities: above Tandayapa, along the lower Chiriboga road, at Lago San Pablo, and at San Rafael Falls (see Ridgely 1980). Since then the species has to our knowledge been reported only at and above Mindo, where a few are seen fairly regularly from late Oct. to early Mar. (fide J. Lyons). These Ecuador reports represent the southernmost known for the Tennessee Warbler, which occurs regularly, locally in substantial numbers, in Colombia and Venezuela (Ridgely and Tudor 1989). Recorded from about 1300 to 2400 m.

Monotypic.

Range: breeds in n. North America, wintering mainly in s. Middle America and nw. South America.

Parula pitiayumi   Tropical Parula   Parula Tropical

Common in a variety of forested and wooded habitats in w. Ecuador, where it occurs in both humid and fairly arid regions, ranging from deciduous woodland in the coastal lowlands to montane forest and forest borders in the subtropical zone; the Tropical Parula avoids, however, the very wet forests of the northwestern lowlands and lower foothills. In e. Ecuador recorded only from the foothills and subtropical zone on the east slope of the Andes. In the west ranges from the lowlands up to at least 2000 m, whereas in the east it is recorded primarily between 900 and 1800 m.

Two similar races occur in Ecuador, *pacifica* in the west and *alarum* on the east slope.

Range: Texas and Mexico to nw. Peru, w. Colombia, much of Venezuela, and mountains of the Guianas and extreme n. Brazil; east slope of Andes from Ecuador south to n. Argentina, in Bolivia and Argentina spreading east across lowlands to e. Brazil and Uruguay.

Dendroica aestiva   Yellow Warbler   Reinita Amarilla

A rare to uncommon boreal winter resident (recorded from Sep. to Apr.) in shrubby areas, clearings, gardens, and on river islands in the lowlands of both e. and w. Ecuador. There is also a single sighting of a presumed transient female at Parque Carolina in Quito on 21 Oct. 1992 (RSR and A. Whittaker). Occurs primarily below about 700 m.

We consider migratory populations of the Yellow Warbler complex as a species separate from populations that are resident along the coasts of Middle America and n. South America and on the West Indies; see the following species account. This follows the line of evidence suggested by data presented by N. Klein and W. M. Brown (*Evolution* 48: 1914–1932, 1994); behavioral and plumage differences are also quite marked. Most examples of boreal migrants wintering in Ecuador are referable to nominate *aestiva*, but *amnicola*, *morcomi*, and *sonorana* have also been recorded (J. T. Zimmer, *Am. Mus. Novitates* 1428: 5–6, 1949). "The identification of wintering examples of the North American forms of this species is extremely difficult" (Zimmer, op. cit.). In summary, compared to nominate *aestiva*, breeding plumage male *morcomi* are slightly greener above and paler yellow below, whereas male *sonorana* are yellower (less green) above and paler yellow below, and also show finer and sparser chestnut streaking on the underparts. Females of both these races are relatively pale compared with females of nominate *aestiva*. The northern breeding race *amnicola* is darker and duller than nominate *aestiva*. Whether any of these could be identified in the field seems doubtful.

Range: breeds in North America and n. Mexico, wintering south to n. South America.

Dendroica petechia   Mangrove Warbler   Reinita Manglera

Locally fairly common along the Ecuadorian coast wherever there are mangroves; on a few islands (notably Isla Puná) also occurs in scrubby woodland near the coast. As a result of the clearance of mangrove forests, the Mangrove Warbler is in many areas now quite local in Ecuador, with its largest extant population occurring in numerous areas around the Río Guayas estuary.

As noted in the previous account, *D. petechia* is here treated as a species separate from the highly migratory *D. aestiva* (Yellow Warbler). Populations resident in mainland Middle and South America (and the Galápagos) have sometimes been separated into a third species, *D. erithachorides* (Mangrove Warbler), leaving resident West Indian birds as *D. petechia* (Golden Warbler). One race of *D. petechia* occurs in Ecuador, *peruviana*.

Range: locally along coast from n. Mexico to nw. Peru and ne. Venezuela; West Indies; Galápagos Islands.

Dendroica cerulea   Cerulean Warbler   Reinita Cerúlea

A rare to uncommon transient and boreal winter visitant in montane forest and forest borders in e. Ecuador, mainly in the foothills and lower subtropical zone of the Andes, but there are also a few reports from the lowlands adjacent to the Andes, where the species likely occurs only as a transient and has not been found any farther east than Limoncocha. Somewhat surprisingly, there are only a few reports of the Cerulean Warbler from w. Ecuador, all from the west slope of the Andes in Pichincha, and all of them recent (though likely this is merely the result of increased observer coverage of late). One was seen above Tandayapa in Pichincha on 26 Nov. 1993 (P. Coopmans and A. Ball), and there have been several subsequent sightings from the Mindo region, almost always of single individuals accompanying mixed flocks. Ecuadorian records fall mainly in the period from late Oct. to early Apr., but Chapman (1926, p. 594) mentions two Aug. specimens (one as early as 10 Aug.); whether the Cerulean regularly arrives so early remains to be determined, but it is known to be among the first of the North American breeding parulids to depart its nesting grounds. Recorded mainly in a narrow zone between about 500 and 1400 m, with a few reports (probably mainly of transients) from the lowlands as low as 300 m, and as high as 2000 m on Andean slopes (e.g., one was seen at Cabañas San Isidro on 12 Sep. 1996; M. Lysinger).

Based on recent data from its breeding grounds, the Cerulean Warbler has been shown to have undergone a steep decline (C. S. Robbins, J. W. Fitzpatrick, and P. B. Hamel *in* J. M. Hagan III and D. W. Johnston, eds., *Ecology and Conservation of Neotropical Migratory Landbirds*, Washington, D.C., Smithsonian Institution Press, 1992, pp. 549–562), and it would be useful to monitor its status in Ecuador as closely as possible.

Monotypic.

Range: breeds in e. North America, wintering primarily in n. South America and along eastern base of Andes south to Bolivia.

Dendroica striata   Blackpoll Warbler   Reinita Estriada

A fairly common transient and less numerous boreal winter resident in the canopy and borders of humid forest (both terra firme and várzea), secondary woodland, clearings, and gardens in the lowlands of e. Ecuador, ranging in smaller numbers up into the foothills and lower subtropical zone on the east slope of the Andes (where it likely occurs primarily as a transient). The only report of the Blackpoll Warbler from w. Ecuador is of a transient female seen on 14 Apr. 1997 at Playa de Oro in n. Esmeraldas (O. Jahn). Recorded mostly from Oct. to Apr., with at least one report from as early as Sep. and one bird lingering as late as early May. Occurs mostly below about 800 m, with smaller numbers being found up to 1400 m; there is also a single sighting from as high as 2800 m (a bird seen at Quito in Nov. 1992; PJG).

Monotypic.

Range: breeds in n. North America, wintering mainly in Amazonia.

Dendroica castanea   **Bay-breasted Warbler**   **Reinita Pechicastaña**

A casual boreal winter visitor to the canopy and borders of humid and montane forest and woodland. There are only three Ecuadorian records, the first being a heretofore unpublished specimen (AMNH) collected by R. Miketta on 5 Feb. 1900 at Hacienda Paramba in Imbabura. More recently, an individual in nonbreeding plumage was seen by PJG above San Rafael Falls in Napo on 21 Feb. 1980 (Ridgely 1980); and there is a highly unusual report of "at least two singing" individuals seen at Cabeceras de Bilsa southeast of Muisne in w. Esmeraldas in Feb. 1991 (Parker and Carr 1992, p. 28). The Bay-breasted Warbler occurs regularly as far south as w. Colombia (Hilty and Brown 1986); the Ecuador records are the southernmost for the species. Recorded up to about 1500 m.

Monotypic.

Range: breeds in n. North America, wintering in s. Middle America and nw. South America.

Dendroica fusca   **Blackburnian Warbler**   **Reinita Pechinaranja**

A fairly common to common transient and boreal winter resident in the canopy and borders of montane forest and secondary woodland the length of both slopes of the Andes, but much more numerous on the east slope. Recorded mainly from Oct. to Apr., with one surprisingly late specimen (ANSP) taken on 20 May at San Gabriel in Carchi. Recorded mainly between about 900 and 2800 m, but also occasionally elsewhere, with a few reports of likely transients in the lowlands adjacent to the Andes (there are, for example, a few recent sightings from Playa de Oro, Jatun Sacha, and Kapawi Lodge), the central valley (e.g., around Quito, where quite regular), and up to near treeline; presumed transients have also been seen in the coastal Cordillera de Mache in sw. Esmeraldas at Bilsa.

Monotypic.

Range: breeds in e. North America, wintering primarily in montane n. and w. South America.

[Dendroica virens   **Black-throated Green Warbler**   **Reinita Cariamarilla]**

An accidental boreal winter visitor to borders of montane forest and secondary woodland on the west slope of the Andes. The first record was a breeding plumage male seen at Mindo on 4 Mar. 1998 (M. Lysinger); another was seen at the Cajanuma sector of Podocarpus National Park in Feb. 1999 (M. Lysinger et al.). The Black-throated Green Warbler is rare even in Colombia's Andes (Hilty and Brown 1986; Ridgely and Tudor 1989); the Ecuadorian records represent the southernmost reports of the species. Recorded at 1400 and 2500 m.

Presumed race: nominate *virens*.

Range: breeds in n. and e. South America, wintering mainly in Middle America and West Indies, sparingly to n. South America.

*[*Dendroica pensylvanica*   Chestnut-sided   Reinita
                            Warbler   Flanquicastaña]

An accidental boreal winter visitor to the borders of humid forest and secondary woodland. The only Ecuadorian records involve breeding-plumage males seen at Tinalandia on the surprisingly late date of 3 May 1981 (PJG) and at Playa de Oro in n. Esmeraldas on 24 Apr. 1997 (O. Jahn). The Chestnut-sided Warbler is rare even in Colombia (Hilty and Brown 1986; Ridgely and Tudor 1989); the Ecuadorian reports are the southernmost known for the species. Recorded at 70 and 700 m.

Monotypic.

Range: breeds in e. North America, wintering mainly in s. Middle America, a few as far south as Colombia.

Mniotilta varia   Black-and-white Warbler   Reinita Blanquinegra

A rare to uncommon boreal winter resident in humid forest and woodland, mainly in the foothills and subtropical zone on both slopes of the Andes; also very rare (presumably only as a transient) in the central valley and the western lowlands. Recorded from Sep. to Apr. Most reports of the Black-and-white Warbler are from the west slope of the Andes south through Pichincha, but there is also a single sighting records records from the lowlands of s. Pichincha at Río Palenque, and there are a few records from the east slope of the Andes as far south as the Cordillera de Cutucú (Robbins et al. 1987). The southernmost report is of a bird seen about 25 km south of Portovelo in El Oro on 28 Jan. 1990 (R. A. Rowlett and J. Arvin et al.). Recorded mostly from 700 to 1900 m, transients exceptionally as high as 3000 m (one seen near Papallacta on 15 Sep. 1996; L. Navarrete et al.).

Monotypic. It has been suggested (e.g., Howell and Webb 1995) that the monotypic genus *Mniotilta* should perhaps be merged with the genus *Dendroica*; the name *Mniotilta* would have priority.

Range: breeds in e. North America, wintering south to nw. South America.

Setophaga ruticilla   American Redstart   Candelita Norteña

A rare to uncommon boreal winter resident in humid and deciduous forest borders, secondary woodland, plantations, and mangroves in the lowlands of w. Ecuador south to se. Guayas (an Oct. 1976 sighting from Naranjal; RSR); in the east recorded along the base of the Andes south to Zamora-Chinchipe (Zamora area and the Río Bombuscaro sector of Podocarpus National Park), with a single report from the eastern lowlands (a female seen on a Río Napo island near La Selva on 18 Nov. 1995; P. Coopmans et al.). There are also several records of likely transients from the central valley in the Quito region. Recorded mostly from Oct. to Mar., but one exceptionally early bird was seen on 29 Aug. 1991 at Tinalandia (D. Agro et al.), and a specimen was taken as late as 24 Apr. There also are two old anomalous Jun. specimens from Ibarra (Goodfellow 1901). Ranges up to about 1300 m, rarely as high as 2800 m at Quito.

Monotypic. It has been suggested (e.g., Howell and Webb 1995) that the monotypic genus *Setophaga* should perhaps be merged with the genus *Dendroica/Mniotilta*.

Range: breeds in North America, wintering south to n. and w. South America.

Protonotaria citrea   Prothonotary Warbler   Reinita Protonotaria

A casual boreal winter visitor to secondary woodland and adjacent clearings and plantations, most often near water, in the lowlands and on the lower Andean slopes of n. Ecuador, where known only from Esmeraldas (an old specimen from Esmeraldas [city], and a sighting from San Lorenzo in late Mar. 1979 by N. Krabbe and O. Jakobsen), Pichincha (several sightings from the Mindo area), and w. Napo (two Oct. sightings from Jatun Sacha; B. Bochan). These are the southernmost records for the species. The beautiful Prothonotary Warbler should be looked for in mangroves along the Pacific coast in Esmeraldas; in its normal wintering range it is most numerous in this habitat. Recorded from Oct. to Mar. Ranges up to about 1300 m.

Monotypic.

Range: breeds in e. North America, wintering south to n. South America.

Seiurus noveboracensis   Northern Waterthrush   Reinita-Aquática Norteña

A rare boreal winter resident in mangroves and in shrubbery and undergrowth near water (both around lakes and along streams) in the lowlands of e. and w. Ecuador. Recorded mainly from n. Ecuador in Esmeraldas, Pichincha, and Napo. Southward there are few reports of the Northern Waterthrush, including a specimen taken at Mera in w. Pastaza and sightings (perhaps of the same individual) from along the lower Río Ayampe on the Manabí/Guayas border on 27 Dec. 1990 (N. Krabbe) and in Jan. 1991 (Parker and Carr 1992), several records (including mist-netted birds) in Dec. 1996 from Loma Alta on the coastal Cordillera de Colonche in nw. Guayas (Becker and López-Lanús 1997), and a few sightings from Cerro Blanco (Berg 1994). These represent the southernmost reports for the species. Recorded from Oct. to Apr. Ranges mostly below about 400 m, but on rare occasions higher (at 1100 m along the Loreto road north of Archidona on 15 Mar. 1990 [RSR and PJG et al.] and at 2000 m at Cabañas San Isidro on 23 Feb. 1999 [T. Brock and J. Hendriks]).

Monotypic.

Range: breeds in n. North America, wintering south to n. South America.

*Seiurus aurocapillus   Ovenbird   Reinita Hornera

An accidental boreal migrant to humid forest and mature secondary woodland in the foothills on both slopes near the base of the Andes. There are only two Ecuadorian records, both involving birds that were mist-netted, photographed, and released. The first was captured at Tinalandia in Mar. 1975 (J. Dunning; photo to VIREO), the second in the Río Bombuscaro sector of Podocarpus National Park in Zamora-Chinchipe on 27 Nov. 1991 (R. Williams et al.). There are only a few records of the Ovenbird even from n. Colombia and n. Venezuela (Ridgely and Tudor 1989); these Ecuador reports are by far the southernmost records for the species. Recorded at 700–1000 m.

Presumed race: nominate *aurocapillus*.

Range: breeds in North America, wintering mainly in Middle America and West Indies, a few reaching n. South America.

Geothlypis semiflava   Olive-crowned Yellowthroat   Antifacito Coronioliva

Uncommon to fairly common in rank tall grass and adjacent shrubbery in the more humid lowlands and foothills of w. Ecuador south to cen. Manabí, Los Ríos, and El Oro (Portovelo and Buenaventura). The Olive-crowned Yellowthroat appears to be spreading and increasing as a result of deforestation. Ranges mostly below 1200 m, in small numbers up to 1500 m and exceptionally as high as 2300 m (a singing male tape-recorded above Tandayapa on 1 Oct. 1995; P. Coopmans).

One race: nominate *semiflava*.

Range: Honduras to w. Panama; w. Colombia and w. Ecuador.

Geothlypis auricularis   Black-lored Yellowthroat   Antifacito Lorinegro

Uncommon to fairly common in shrubbery and woodland borders in the lowlands and foothills of w. Ecuador from n. Esmeraldas (recent reports from Playa de Oro; O. Jahn), Manabí, and s. Pichincha (Río Palenque) south somewhat locally into El Oro and Loja. Also recently found in the Río Marañón drainage of extreme se. Ecuador in s. Zamora-Chinchipe in the Zumba region (first in Aug. 1989; RSR), a few ranging north to the Valladolid area. The Black-lored Yellowthroat usually favors somewhat less humid areas than the Olive-crowned (*G. semiflava*), but the two species are widely sympatric; the Black-lored may be spreading northward in n. Esmeraldas, and it is possible that it may in due course spread north into adjacent Nariño, Colombia. Recorded mainly below about 1100 m, locally as high as 1500 m in Loja; in the Río Marañón drainage found between about 850 and 1650 m.

Formerly called *G. aequinoctialis* (Masked Yellowthroat), but we follow B. P. Escalante-Pliego (*Acta XX Congr. Int. Ornithol.*: 333–341, 1991) in considering the taxon occurring in w. Ecuador, *auricularis*, as a species separate from *G. aequinoctialis* of ne. South America, *G. velata* (Southern Yellowthroat) of s. South America, and *G. chiriquensis* (Chiriquí Yellowthroat) of w. Panama and adjacent Costa Rica. *Peruviana* is retained as a subspecies of *G.*

auricularis. Peruviana is formally recorded only from the upper Río Marañón valley of nw. Peru, though this almost surely is the taxon occurring in the Zumba/Valladolid area; there still, however, are no Ecuadorian specimens from there. *Peruviana* is larger than *auricularis* but similar in plumage. The song of Zumba birds (presumed to be *peruviana*) is "quite different" from that of *auricularis* (fide P. Coopmans); *peruviana* could itself prove to be a separate monotypic species (Marañón Yellowthroat).

Range: w. Ecuador and nw. Peru.

Oporornis philadelphia   Mourning Warbler   Reinita Plañidera

A very rare to rare boreal winter resident in dense shrubby forest and woodland borders, especially near water, and river islands mainly in ne. Ecuador south to Tungurahua (an old record from Mapoto). There is also one recent record from the lowlands of the northwest in Esmeraldas (a male mist-netted and photographed near Playa de Oro on 11 Mar. 1996; O. Jahn and K.-L. Schuchmann et al.). The Mourning Warbler seems to be most numerous in the Limoncocha and Jatun Sacha areas of w. Napo; at Jatun Sacha it especially favors damp floodplain thickets dominated by *Gynerium* cane (B. Bochan). The only recent report away from the lowlands is of a bird seen on the Cordillera de Huacamayos on 12 Nov. 1991 (C. Rahbek). These are the southernmost records for the species. Recorded from Nov. to Mar. Ranges from the eastern lowlands up on the east slope of the Andes to about 2500 m.

Monotypic.

Range: breeds in n. North America, wintering in s. Middle America and nw. South America.

**Oporornis agilis*   Connecticut Warbler   Reinita Ojianillada

An accidental boreal migrant to undergrowth of secondary woodland and humid forest as well as thickets at borders. The only Ecuadorian record of the Connecticut Warbler is of an individual mist-netted, photographed, and released in the lowlands of coastal n. Esmeraldas at Playa de Oro on 21 Nov. 1996 (O. Jahn, M. E. Jara Viteri, and K.-L. Schuchmann, *Wilson Bull.* 111[2]: 281–282, 1999). Presumably this individual was an off-course south-bound migrant; it represents the first record for the species from west of the Andes. The species, a notorious skulker, might also occur as a transient in the lowlands of e. Ecuador; there are a few records from se. Colombia (Hilty and Brown 1986). 50 m.

Monotypic.

Range: breeds in n. North America, wintering in cen. South America.

Wilsonia canadensis   Canada Warbler   Reinita Collareja

A fairly common transient and boreal winter resident in the lower growth of montane forest, secondary woodland, and shrubby borders on the east slope of the Andes; rare in similar habitats on the west slope. The Canada Warbler is also recorded as a transient in the eastern lowlands, with substantial numbers having been noted on passage at Limoncocha and Jatun Sacha in Oct. and Nov. and again in Mar. and Apr.; however, only a few have been noted just a little farther east, along the Maxus road southeast of Pompeya (RSR et al.). There are also a few reports of transients from the western lowlands (e.g., at Río Palenque). Recorded mainly from Oct. to Apr., but there are a few specimens and sightings from as early as Aug. Ranges especially between about 500 and 2000 m, but occurring lower as a transient.

Monotypic.

Range: breeds in e. North America, wintering mainly in n. and w. South America.

Myioborus miniatus   Slate-throated Whitestart   Candelita Goliplomiza

Fairly common to common in montane forest, forest borders, secondary woodland, and adjacent clearings in upper tropical and subtropical zones mainly on both slopes of the Andes. On the west slope found south through El Oro and w. Loja, and also recently found on the coastal Cordillera de Colonche in sw. Manabí (Machalilla National Park; Parker and Carr 1992, ANSP) and w. Guayas (Cerro La Torre and above Salanguilla; N. Krabbe). Recorded mostly from about 800 to 2300 m, in small numbers up locally as high as 2600 m (e.g., in s. Loja); in the west occurs much lower locally, in small numbers or occasionally even down to near sea level in the Río Ayampe valley on the Manabí/Guayas border.

After much consideration, and following various recent references (e.g., Curson et al. 1994), we opt to employ the more accurate group name of "whitestart" for all members of the genus *Myioborus*. This replaces the long-used but misleading group name of "redstart." All members of the genus of course have white in the tail, not red.

The validity of some of the races of *M. miniatus* recorded as occurring in Ecuador is unclear, and in any case all resemble each other closely. Basically, *ballux* is the race to which birds from both slopes of the Andes have usually been assigned. On the east slope *ballux* ranges south to Tungurahua at Mocha (J. T. Zimmer, *Am. Mus. Novitates* 1428, 1949). Birds found southward on the east slope (in Morona-Santiago and Zamora-Chinchipe) have been assigned to *verticalis*. Those found southward on the west slope (El Oro and Loja) have been assigned to *subsimilis* (Zimmer, op. cit.). *Subsimilis* in particular is very similar to *ballux*, and it may be best to synonymize it. Indeed *ballux* itself may not deserve separation from *verticalis* (fide M. B. Robbins), though it averages a bit more orange below and has slightly more extensive white in its tail.

Range: mountains from n. Mexico to Panama; mountains of n. Venezuela, and Andes from w. Venezuela to w. Bolivia; Sierra de Perijá and Santa Marta Mountains; tepuis of Venezuela and adjacent Guyana and Brazil.

Myioborus melanocephalus   Spectacled Whitestart   Candelita de Anteojos

Common in montane forest, borders, secondary woodland, and adjacent shrubby clearings in the upper subtropical and temperate zones on both slopes of the Andes, and locally on slopes above the central and interandean valleys (e.g., numerous at Pasochoa). On the west slope recorded south to Azuay (Chaucha; N. Krabbe et al.) and El Oro (Taraguacocha). Recorded from about 2300 to 4000 m, occurring regularly in patches of *Polylepis*-dominated woodland at and above treeline. The Spectacled Whitestart occurs mainly at elevations above those where the Slate-throated (*M. miniatus*) is found, though there is some overlap and occasionally the two species even forage together in the same flock.

Formerly usually called the Spectacled Redstart; see the discussion under the Slate-throated Whitestart.

Some individuals of *M. melanocephalus* in n. Ecuador (e.g., on the Páramo del Ángel, around Santa Bárbara along the road to La Bonita, and on the west slope of Cerro Mongus) have (as alluded to by Robbins et al. 1994) yellow on their forecrown and lores, indicating some level of intergradation with the Golden-fronted Whitestart (*M. ornatus*) of the Colombian Andes. Birds that look much like *M. ornatus* have been seen south as far as Oyacachi in w. Napo, where a pair of seemingly phenotypically pure *ornatus* was seen on 25 Jan. 1991 (B. Whitney). In populations of *M. ornatus* in s. Colombia, some individuals also show characters of *M. melanocephalus* (P. Coopmans). It may ultimately prove to be preferable to consider these species as conspecific under the name of *M. ornatus* (Variable Whitestart). One race of *M. melanocephalus* occurs in Ecuador: *ruficoronatus*.

Range: Andes from s. Colombia to w. Bolivia.

Basileuterus nigrocristatus   **Black-crested Warbler**   **Reinita Crestinegra**

Fairly common to common in dense shrubby growth at the edge of montane forest and secondary woodland, sometimes even in patches of regenerating scrub and hedgerows in agricultural regions, in the temperate zone on both slopes of the Andes, and on slopes above the central and interandean valleys. The Black-crested Warbler avoids the interior of continuous, closed-canopy forest, but it is regular in areas dominated by *Chusquea* bamboo. Recorded mostly from 2500 to 3500 m, ranging up into woodland patches near treeline, occurring locally and in smaller numbers down to about 2000 m (e.g., on outlying ridges such as the Cordillera de Huacamayos), in w. Napo exceptionally as low as 1750 m (this perhaps especially in response to disturbed conditions).

Monotypic.

Range: mountains of n. Venezuela, and Andes from w. Venezuela to n.-cen. Peru; Sierra de Perijá.

Basileuterus luteoviridis   **Citrine Warbler**   **Reinita Citrina**

Uncommon to fairly common in the lower growth of montane forest and forest borders in the temperate zone on the east slope of the Andes from

w. Sucumbíos and w. Napo south through Zamora-Chinchipe (Podocarpus National Park, Cordillera Las Lagunillas). It is also possible that the Citrine Warbler occurs on the west slope of the Andes in the extreme north, for it was recently recorded as "rare" in immediately adjacent w. Nariño, this presumably representing the distinctive *richardsoni* race of Colombia's Western Andes (Salaman 1994). Despite the apparent lack of earlier records from s. Ecuador (see *Birds of the World*, vol. 14), the Citrine Warbler actually appears to be more numerous there (particularly in Zamora-Chinchipe) than it is northward. Recorded mostly from 2500 to 3200 m, but ranges lower (a few regularly at around 2200 m) on the Cordillera de Huacamayos, where exceptionally it occurs as low as 2000 m (at which elevation a specimen was taken on 11 Oct. 1988; WFVZ); it also is found slightly higher at some sites (e.g., as high as 3400 m at Cerro Mongus in se. Carchi; Robbins et al. 1994).

One race: nominate *luteoviridis*. Contra J. Curson (*Cotinga* 1: 16, 1994) and Curson et al. (1994), there is no evidence that *striaticeps*, the race of *B. luteoviridis* found in most of the species' Peruvian range, ranges into Ecuador. Specimens (ANSP) from within a few kilometers of the Peruvian border in the Río Isimanchi valley on the east slope of the Cordillera Las Lagunillas do not, in fact, show any indication of intergradation with *striaticeps*.

Range: Andes from w. Venezuela to w. Bolivia.

Basileuterus chlorophrys　　Chocó Warbler　　Reinita del Chocó

Fairly common to locally common in the lower and middle growth of montane forest and secondary woodland in the foothills and lower subtropical zone on the west slope of the Andes from Esmeraldas south to e. Guayas (Chimbo) and Chimborazo (Pallatanga). Also recorded recently on the coastal Cordillera de Mache at Bilsa in w. Esmeraldas, where found to be uncommon in Feb. 1994 (R. P. Clay et al.). An old record from Babahoyo would seem best disregarded as the species has never been recorded from the hot tropical lowlands. Recorded mostly between about 400 m (at Playa de Oro in Esmeraldas; O. Jahn et al.) and 1200 m.

B. chlorophrys is considered a species separate from *B. chrysogaster* (Cuzco Warbler) of the east slope of the Andes in cen. and s. Peru, based in particular on its utterly different song. Monotypic.

Range: west slope of Andes of sw. Colombia and w. Ecuador.

Basileuterus tristriatus　　Three-striped Warbler　　Reinita Cabecilistada

Fairly common to common in the lower growth of montane forest and (to a lesser extent) forest borders and secondary woodland in foothills and the subtropical zone on both slopes of the Andes. On the west slope ranges south to nw. Azuay (seen above Manta Real in Aug. 1991; RSR). Ranges mostly between about 1000 and 2000 m, locally occurring somewhat lower on the west slope (e.g., to 700–800 m at and above Tinalandia).

Three races are found in Ecuador: *daedalus* on the west slope, and the very similar *baezae* and nominate *tristriatus* on the east slope. *Baezae* ranges south to nw. Morona-Santiago, where it intergrades with the nominate race, which is found from about there southward. *Daedalus* is dull buffy yellowish below, whereas both of the two intergrading east-slope races, especially the nominate, are much yellower below.

Range: mountains of Costa Rica and Panama and in n. Venezuela, and Andes from w. Venezuela to w. Bolivia; Sierra de Perijá.

Basileuterus trifasciatus   **Three-banded Warbler**   **Reinita Tribandeada**

Uncommon to locally fairly common in the lower growth of montane forest, secondary woodland, and borders on the west slope of the Andes in El Oro and Loja, also ranging east on interandean slopes to various localities on the west slope of the Eastern Andes in Loja (north to near Loja [city]). Small numbers are found regularly at Buenaventura in El Oro, but the Three-banded Warbler is especially numerous around Celica and Sozoranga in w. Loja. Ranges mostly between about 800 and 2400 m, locally occurring as high as almost to 3000 m (e.g., near the summit of Cerro Guacha Urco near Celica in w. Loja, where seen in Aug. 1989; RSR et al.).

One race: *nitidior*. Vocally the species resembles *B. tristriatus* (Three-striped Warbler).

Range: Andes of sw. Ecuador and nw. Peru.

Basileuterus coronatus   **Russet-crowned Warbler**   **Reinita Coronirrojiza**

Fairly common in the lower growth of montane forest, secondary woodland, and (to a lesser extent) borders in the subtropical and temperate zones on both slopes of the Andes, and locally on slopes above various interandean valleys. On the west slope found south locally through Loja. Recorded mostly between about 1500 and 3000 m, locally higher on west slope.

Strong racial variation within this species occurs in Ecuador. On the west slope south to Chimborazo is found *elatus*, which has a bright yellow breast and belly. On the west slope from Azuay south to El Oro and Loja, and also ranging well eastward onto slopes above intermontane valleys, is found *castaneiceps*; it lacks all yellow below, being pale grayish across the breast and whitish on its belly. On virtually the entire east slope is found *orientalis*, which resembles *castaneiceps* though *orientalis* has a tinge of pale yellow on its midbelly. In the extreme southeast (recent specimens [ANSP, MECN] from s. Zamora-Chinchipe on the Cordillera del Cóndor and on the east slope of the Cordillera Las Lagunillas) birds intermediate between *orientalis* and *chapmani* (which otherwise is known from nw. Peru) are found; these birds resemble *castaneiceps* but are grayer on the breast and slightly darker above.

Range: Andes from w. Venezuela to w. Bolivia. More than one species is perhaps involved.

Basileuterus fraseri   Gray-and-gold Warbler   Reinita Grisidorada

Fairly common to locally common in the lower growth of humid and deciduous forest and woodland in the lowlands and lower Andean slopes in sw. Ecuador north sparingly to w. Esmeraldas and s. Pichincha (e.g., a few reports from Río Palenque where, fide P. Coopmans, the species may be increasing), but more numerous from cen. Manabí and Los Ríos southward. We suspect that the Gray-and-gold Warbler may engage in local seasonal movements, moving into more mesic habitats during the dry season and then breeding in more deciduous habitats during the rainy season. Ranges mostly below 1700 m (northward only below 300 m), but in s. Loja it occurs locally up to about 2100 m.

Two readily distinguished races occur in Ecuador. *Ochraceicrista* (with an orange-ochraceous coronal streak) is found in Manabí and all but extreme se. Guayas where, in the Río Chimbo valley area, it appears to intergrade with nominate *fraseri* (whose coronal streak is mainly yellow); the Chimbo valley birds show a range of variation in coronal streak color. Nominate *fraseri* ranges from there south through at least most of El Oro and Loja. A series of nine specimens (ANSP) taken in 1950 in coastal El Oro at Santa Rosa and Piedras is clearly referable to the seemingly "out-of-range" *ochraceicrista*; Chapman (1926), however, assigns a long AMNH series from Santa Rosa (only one of which remained in the AMNH in Mar. 1995) to the nominate race. This puzzling situation merits further investigation.

Range: w. Ecuador and nw. Peru.

Basileuterus fulvicauda   Buff-rumped Warbler   Reinita Lomianteada

Uncommon to fairly common along forested streams and smaller rivers in more humid lowlands and foothills in both e. and w. Ecuador, but overall less numerous in the eastern lowlands. In the west the Buff-rumped Warbler ranges south in foothills and more humid adjacent lowlands through El Oro and w. Loja (Cebollal). Recorded up to about 1000 m on both slopes of the Andes.

The species is sometimes placed in the genus *Phaeothlypis*, but we follow Ridgely and Tudor (1989) and Sibley and Monroe (1990) in considering that genus as better included within *Basileuterus*, mainly because certain South American species seem intermediate in various respects between the two. *B. fulvicauda* of Middle America and w. South America is regarded as a species distinct from *B. rivularis* (Riverside Warbler) of e. South America, based on its distinct plumage differences and disjunct range. Two similar races are found in Ecuador, *semicervinus* in the west and nominate *fulvicauda* in the east; *semicervinus* is somewhat buffier below and has slightly more extensive buff on the tail.

Range: Honduras to extreme nw. Peru; se. Colombia to se. Peru, nw. Bolivia, and w. Amazonian Brazil. More than one species may be involved.

Thraupidae   Tanagers, Honeycreepers, Bananaquit, and
                             Plushcap

One hundred forty-three species in 47 genera. The family is American in distribution, reaching by far its highest diversity in the Neotropics. The tanagers
were formerly considered as a subfamily in a broadly inclusive Emberizidae;
we follow Gill (1994) in giving the group full family rank. The Swallow
Tanager (*Tersina viridis*) and the Plushcap (*Catamblyrhynchus diadema*) were
formerly (e.g., Meyer de Schauensee 1966, 1970) placed in monotypic families,
Tersinidae and Catamblyrhynchidae, respectively. The Coerebidae (Honeycreepers) were then also so recognized, though all are now usually thought
to be tanagers; however, the 1998 AOU Check-list recognized *Coereba*
(Bananaquit) as a monotypic family. Sibley and Ahlquist (1990) and Sibley and
Monroe (1990) suggest that certain genera (e.g., *Phrygilus*, *Sicalis*, *Sporophila*,
and *Poospiza*) now placed with the Emberizine Finches may be better considered as members of the Thraupidae, but we retain them within their traditional
family.

Coereba flaveola   **Bananaquit**   **Mielero Flavo**

Uncommon to locally common in secondary woodland, forest borders,
clearings, and gardens in the more humid lowlands and foothills of both e. and
w. Ecuador. The Bananaquit becomes, however, much less numerous in the
eastern lowlands away from the Andes, and it is entirely absent (or nearly so)
from most areas with extensive forest cover; it is especially numerous in w.
Esmeraldas, and here it does range in the canopy of humid forest (e.g., at Bilsa).
Ranges up in the foothills to about 1100 m, in smaller numbers locally to at
least 1800 m (the latter perhaps especially in cleared areas on the east slope
of the Andes in s. Ecuador).
   One race is found in most of the species' Ecuadorian range, *intermedia*; in
the Río Marañón drainage of s. Zamora-Chinchipe around Zumba, where
Bananaquits are numerous, the larger-billed *magnirostris* (ANSP) occurs.
   Range: s. Mexico to nw. Peru, n. and e. Bolivia, e. Paraguay, ne. Argentina,
and se. Brazil (but absent or nearly so from w. Amazonia); West Indies except
Cuba.

*Cyanerpes nitidus   **Short-billed Honeycreeper**   **Mielero Piquicorto**

Rare and local in the canopy and borders of humid forest, evidently mainly in
terra firme, in the lowlands of e. Ecuador, mainly in the better-known northeast. There are only a few definite reports of the Short-billed Honeycreeper from
Ecuador, and no specimens have been taken; it may have been overlooked to
some extent. The only Ecuadorian record prior to the few recent sight reports
consists of the species being listed as occurring in "eastern Ecuador" in *Birds
of the Americas* (pt. 8, p. 266) and Meyer de Schauensee (1966, p. 458). No
details concerning any Ecuador record were mentioned in either reference, and

it is possible that the record refers to a specimen obtained outside Ecuador's present borders. Recent sightings include a male seen near Lago Agrio in Sucumbíos on 11 Aug. 1980 (RSR and T. Butler et al.), several reports from Jatun Sacha (fide B. Bochan), a few reports from La Selva (RSR and PJG et al.; P. Coopmans), and sightings from Kapawi Lodge on the Río Pastaza near the Peruvian border. The Short-billed Honeycreeper is not recorded from Peru any closer than the Iquitos region and near the lower Río Napo, but it is known from w. Caquetá in s. Colombia (whence there are 5 specimens from Belén; ANSP). Recorded below about 400 m.

Monotypic.

Range: se. Colombia, s. Venezuela, and Guyana to ne. Peru and w. and cen. Amazonian Brazil.

Cyanerpes caeruleus   **Purple Honeycreeper**   Mielero Purpúreo

Fairly common to common in the canopy and borders of humid forest and secondary woodland, sometimes out into adjacent clearings, in the lowlands and to a lesser extent the foothills in both e. and w. Ecuador. In the west recorded south at least occasionally to El Oro at Buenaventura (seen in Sep. 1991 by R. Williams et al.), but mainly recorded north of s. Pichincha (Río Palenque), se. Guayas (hills of Manglares-Churute Ecological Reserve; Pople et al. 1997), and nw. Azuay (Manta Real and on the Puerto Inca-Molleturo road). Found mainly below 1200 m, a few occasionally wandering higher.

Two only slightly differentiated races occur in Ecuador, *chocoanus* in the west and *microrhynchus* in the east.

Range: e. Panama to sw. Ecuador, n. Bolivia, and Amazonian Brazil.

Cyanerpes cyaneus   **Red-legged Honeycreeper**   Mielero Patirrojo

Rare to uncommon and local in the canopy and borders of humid and semi-humid woodland and adjacent clearings in the lowlands of nw. Ecuador from Esmeraldas south to n. Manabí and Pichincha (rarely as far south as Río Palenque, from whence there is at least one sighting). Perhaps most numerous in the lowlands of n. Esmeraldas (e.g., around Borbón; PJG and S. Greenfield); in the Playa de Oro area of n. Esmeraldas it occurs primarily in river-edge vegetation (O. Jahn et al.). The Red-legged Honeycreeper is rare and local in the lowlands of e. Ecuador, where there are only a few scattered records. The sole specimen from the east seems to be a male taken by T. Mena at Montalvo on 27 Jun. 1959 (Orcés 1974); a specimen is listed by J. T. Zimmer (*Am. Mus. Novitates* 1203, 1942) as from "Napo," but whether it was actually taken within the present borders of Ecuador cannot be known. In addition, a male was seen in *Macrolobium*-dominated woodland at Garzacocha off the Río Lagarto on 27 Mar. 1991 (RSR), and another male was reported seen at Sacha Lodge on 29 Dec. 1992 (M. Catsis, fide P. Coopmans); the species has also been reported from Kapawi Lodge on the Río Pastaza near the Peruvian border. Apparently only recorded below about 300 m.

Two only slightly differentiated races occur in Ecuador, *pacificus* in the west and presumably *dispar* in the east.

Range: e. Mexico to nw. Ecuador, n. Bolivia, and Amazonian Brazil; e. Brazil; Cuba.

Chlorophanes spiza   Green Honeycreeper   Mielero Verde

Fairly common in the canopy and borders of humid forest and secondary woodland in lowlands and foothills of both e. and w. Ecuador; in the west recorded south to nw. (hills south of the Río Ayampe) and se. (Manglares-Churute Ecological Reserve) Guayas, El Oro, and w. Loja (an old record from Las Piñas and [RSR] an Apr. 1993 sighting from near El Limo). Recorded up in small numbers to about 1100 m, but mainly occurs below 800 m.

Two races are found in Ecuador, *exsul* in the west and *caerulescens* in the east. Males of the latter are bluer, especially below, but females are very similar.

Range: s. Mexico to extreme nw. Peru, n. Bolivia, and Amazonian Brazil; e. Brazil.

Iridophanes pulcherrima   Golden-collared   Mielero Collarejo
                             Honeycreeper

Uncommon to locally fairly common in the canopy and borders of montane forest and in adjacent clearings in the foothills and lower subtropical zone on the east slope of the Andes. Rare on the west slope, where known from a few sightings from Carchi (a male seen near Chical in Aug. 1988; D. Wechsler) and e. Esmeraldas (El Placer [N. Krabbe; RSR et al.]) and old specimens from n. Pichincha (Gualea, Guanacilla, and Santo Domingo de los Colorados). Curiously, and despite the early specimens, there are no recent reports of this species from Pichincha, where it must surely still occur, perhaps especially in the northwest. The Golden-collared Honeycreeper is relatively numerous in the vicinity of San Rafael Falls. Recorded mainly between about 1100 and 2000 m (e.g., at Cabañas San Isidro; M. Lysinger); in the northwest recorded between 650 and 1500 m.

This species has sometimes (e.g., *Birds of the World*, vol. 13), been assigned to the genus *Tangara*, though more often it has been placed in its own monotypic genus, *Iridophanes*. Based on bill shape and behavior, we suspect that the species is closer to various "honeycreeper" genera (e.g., *Chlorophanes*) than it is to *Tangara* (despite its remarkable plumage similarity to *T. cyanoptera*, the Black-headed Tanager of n. South America), and so we maintain the genus. Two races occur in Ecuador, nominate *pulcherrima* on the east slope and *aureinucha* in the northwest. The latter differs principally in having a longer bill and, in the male, a sootier (less black) head.

Range: Andes from s. Colombia to s. Peru.

Dacnis cayana   Blue Dacnis   Dacnis Azul

Fairly common to common in the canopy and borders of humid forest, secondary woodland, and in clearings and gardens with scattered trees in the lowlands and foothills of nw. Ecuador; apparently less numerous in n. Esmeraldas. Uncommon to locally fairly common in comparable habitats in the eastern lowlands. In the west recorded south in small numbers (and perhaps only formerly) to e. Guayas (an early-20th-century specimen from Naranjito; AMNH) and Los Ríos (2 taken on 16 Oct. 1950 at Quevedo; ANSP), but in recent years not found south of s. Pichincha at Río Palenque. Recorded mainly below 1000 m, in small numbers up to 1400–1600 m in the southeast.

Two races are found in Ecuador, *glaucogularis* in the east and *baudoana* in the west. Male *baudoana* are much darker, more ultramarine blue as compared to the paler turquoise blue of male *glaucogularis*; female *baudoana* show more extensive and brighter blue on the head than female *glaucogularis*.

Range: Honduras to w. Ecuador, n. and e. Bolivia, e. Paraguay, ne. Paraguay, and se. Brazil.

Dacnis lineata   Black-faced Dacnis   Dacnis Carinegro

Fairly common in the canopy and borders of humid forest, secondary woodland, and adjacent clearings in the lowlands and—often especially—foothills of e. Ecuador. Recorded mostly below about 1200 m, rarely as high as 1500–1700 m in Zamora-Chinchipe.

Trans-Andean *D. egregia* (Yellow-tufted Dacnis) is here considered to be a separate species. *D. lineata* then becomes monotypic.

Range: e. Colombia, s. Venezuela, and the Guianas to n. Bolivia and Amazonian Brazil.

Dacnis egregia   Yellow-tufted Dacnis   Dacnis Pechiamarillo

Uncommon to fairly common in the canopy and borders of humid and deciduous forest, secondary woodland, and adjacent clearings and gardens in the lowlands of w. Ecuador from n. Esmeraldas, e. Imbabura (Intag), and Pichincha south to El Oro (above La Avanzada; ANSP, MECN). Recorded up to about 900 m.

Trans-Andean *D. egregia* is regarded as a species distinct from cis-Andean *D. lineata* (Black-faced Dacnis), based on its striking plumage differences and disjunct range. One race of *D. egregia* occurs in Ecuador, the endemic *aequatorialis*.

Range: w. Colombia; w. Ecuador.

Dacnis flaviventer   Yellow-bellied Dacnis   Dacnis Ventriamarillo

Uncommon to fairly common in the canopy and borders of humid forest (both terra firme and várzea), secondary woodland, and adjacent clearings in the lowlands of e. Ecuador. The Yellow-bellied Dacnis seems especially to favor areas near water, but smaller numbers are also regularly found in terra firme. Recorded mostly below about 500 m; in the north a few have been found as high as 700 m (MCZ specimen from Guaticocha), whereas in Zamora-

Chinchipe small numbers range up the Río Nangaritza valley as high as 900–1000 m (Balchin and Toyne 1998), with one exceptional report from as high as 1050 m (2 females seen above Zamora on 26 Oct. 1993; D. Wolf and PJG).

Monotypic.

Range: se. Colombia and s. Venezuela to n. Bolivia and w. and cen. Amazonian Brazil.

Dacnis venusta   **Scarlet-thighed Dacnis**   **Dacnis Musliescarlata**

Rare to locally uncommon in the canopy and borders of very humid and humid forest, secondary woodland, and adjacent clearings in the lowlands and foothills of nw. Ecuador, where recorded mainly from Pichincha northward; most records are from Esmeraldas and nw. Pichincha. Two undated specimens were apparently taken as far south as n. Guayas at El Empalme (Orcés 1974), but these specimens seem to have been lost. The Scarlet-thighed Dacnis occurs in small numbers at both Río Palenque and Tinalandia (perhaps only seasonally, though it is known to breed at the former), and since 1995 it has also been found to occur in substantial numbers along the road north of Simón Bolívar near Pedro Vicente Maldonado and in nearby areas (v.o.); it also has been seen regularly at Playa de Oro in n. Esmeraldas (O. Jahn et al.). Recorded up to about 800 m.

One race: *fuliginata*.

Range: Costa Rica to w. Ecuador.

Dacnis berlepschi   **Scarlet-breasted Dacnis**   **Dacnis Pechiescarlata**

Rare to locally uncommon in the canopy and borders of very humid and humid forest in the lowlands and foothills of nw. Ecuador south to s. Pichincha (Río Palenque, where first recorded on 7 Aug. 1979 [PJG]; however, there seem to be no reports from the 1990s). There are only a few known Ecuadorian sites for this colorful dacnis. It was reported seen in 1986 at Ventanas (500 m) along the railroad line to San Lorenzo in Esmeraldas (R. J. Evans et al.), and a pair was seen southeast of Muisne on 29 Jan. 1993 (RSR). An area where the Scarlet-breasted Dacnis is now known to be numerous and regular is in nw. Pichincha along the road north of Simón Bolívar near Pedro Vicente Maldonado, where it was discovered on 6 Feb. 1995 (P. Coopmans, M. Lysinger, and T. Walla) and has since been seen on numerous occasions (v.o.); it has also been seen south of Golondrinas and in adjacent sw. Imbabura (near the Salto del Tigre bridge over the Río Guaillabamba; P. Coopmans et al.). Intensive surveys from 1995 to 1998 also revealed it to be not uncommon at Playa de Oro in n. Esmeraldas (O. Jahn et al.). Recorded up to about 600 m, but at Playa de Oro most numerous below 250 m.

The Scarlet-breasted Dacnis was given Vulnerable status by Collar et al. (1994), and we agree with this assessment. It probably occurs in the lower parts of Cotocachi-Cayapas Ecological Reserve and in the Awá Forest Reserve in Esmeraldas.

Monotypic.
Range: sw. Colombia and nw. Ecuador.

Dacnis albiventris   White-bellied Dacnis   Dacnis Ventriblanco

Very rare in the canopy and borders of humid forest and secondary woodland (both terra firme and várzea) in the lowlands of e. Ecuador. The explanation for the rarity of this distinctive dacnis—over its entire range, not just in Ecuador—remains uncertain. Specimens have been taken at Río Copotaza (Orcés 1944) and Sarayacu. Otherwise there are sightings from Limoncocha (Jul. 1976; RSR), Cuyabeno (J. Arvin), along the Maxus road southeast of Pompeya (Jul. and 9 Nov. 1994; S. Howell, RSR and F. Sornoza), and Tiputini Biodiversity Center (a male seen in Feb. 1999; M. Lysinger). Recorded only below about 300 m.

The White-bellied Dacnis was given Near-threatened status by Collar et al. (1992, 1994). We are not, however, aware that it is under any particular threat—certainly an abundance of suitable habitat exists, and seems likely to do so for a long time to come—and thus, despite its very evident rarity, we do not consider the species to be at risk in Ecuador.
Monotypic.
Range: locally in se. Colombia, sw. Venezuela, e. Ecuador, and ne. Peru; limited area in cen. Amazonian Brazil.

Conirostrum speciosum   Chestnut-vented Conebill   Picocono Culicastaño

Apparently rare and local in the canopy and borders of humid forest and secondary woodland along the eastern base of the Andes, where thus far known only from a few sites in w. Napo and Zamora-Chinchipe. The sole Ecuadorian specimen is a male (AMNH) obtained by the Olallas at "Río Suno Abajo" in w. Napo on 8 Mar. 1924. Recent reports involve a pair seen west of Loreto on 29 Mar. 1990 (B. Whitney) and numerous sightings from the Zamora area, the first being a male seen on 29 Jul. 1991 (PJG). The Zamora sightings have mainly involved birds attracted to flowering *Inga* trees; the largest number reported from there is six, seen on 27 Jul. 1992 (D. Wolf et al.). Recorded between about 400 and 1200 m.

This and the following species should perhaps be separated from *Conirostrum* in the genus *Ateleodacnis*. One race of *C. speciosum* occurs in Ecuador, *amazonum*.
Range: locally in cen. Venezuela, and in the Guianas and extreme n. Brazil; Amazonian Brazil, nw. Bolivia, and e. Peru extending north along base of Andes in e. Ecuador; e. Bolivia, Paraguay, n. Argentina, and widely in s. and e. Brazil.

*[Conirostrum bicolor   Bicolored Conebill   Picocono Bicolor]

Uncertain. Known in Ecuador only from a sighting of a single bird, presumably a wandering individual, on an island in the Río Napo near Primavera on 4 Jul.

1985 (J. Rowlett et al.). Despite extensive field work by various observers on many river islands in subsequent years, and despite an abundance of seemingly appropriate habitat for it, there have been no additional reports of the species; Bicolored Conebills were, for instance, clearly absent from a Río Napo island situated near the Peruvian border near the mouth of the Río Aguarico where the species was specifically looked for in Sep. 1992 (RSR et al.). Specimens (AMNH) of the Bicolored Conebill have been obtained along the Río Napo in ne. Peru as close to Ecuador as the mouth of the Río Curaray (*Birds of the Americas*, pt. 8), and these are presumably the basis for the species having been recorded from "eastern Ecuador in vicinity of Río Napo" in *Birds of the World* (vol. 14, p. 84). Why the species fails to regularly range upstream into Ecuadorian territory remains unknown. Recorded at about 300 m.

Presumed race: *minor*.

Range: locally in mangroves along coast from n. Colombia to se. Brazil, and along Amazon River upriver to ne. Peru; a single report from e. Ecuador.

Conirostrum cinereum   Cinereous Conebill   Picocono Cinéreo

Uncommon to fairly common in shrubby areas, lighter woodland, patches of *Polylepis* woodland, and gardens in the Andes from Carchi south to Loja (above Jimbura on west slope of Cordillera Las Lagunillas; RSR), favoring more arid regions. Most numerous on slopes above the central valley, in smaller numbers on the outside slopes of the Andes (perhaps especially in cleared areas); regularly occurs up to the edge of paramo, sparingly even onto shrubby paramo itself. The Cinereous Conebill is frequent in the parks and residential areas of Quito and numerous other highland cities and towns; it seems less numerous in s. Ecuador. Recorded mostly from 2500 to 3500 m, locally down to 2300 m and in small numbers as high as 4000 m.

One race: *fraseri*. J. T. Zimmer (*Am. Mus. Novitates* 1193, 1942) notes that there is an old undated specimen (AMNH) of the mainly Peruvian race *littorale* from Caleta Grau in Tumbes, Peru. This record, recently overlooked, extends the range of this distinctive form northward to within a short distance of the Ecuadorian border; there have also been a few recent reports of it from along the Pan-American Highway south of Tumbes (fide P. Coopmans). *Littorale* should therefore be watched for in coastal sw. El Oro near Huaquillas; it differs from *fraseri* of the highlands in being slightly smaller, distinctly grayer above with a white (not buff) superciliary, and more drab buffy grayish (not uniform ochraceous buff) underparts. Separate species may be involved, in which case a monotypic *C. fraseri* would be called Fraser's Conebill.

Range: Andes from s. Colombia to n. Chile and w. Bolivia, also in lowlands of w. Peru and Chile.

Conirostrum sitticolor   Blue-backed Conebill   Picocono Dorsiazul

Fairly common in montane forest, secondary woodland, and borders in the temperate zone on both slopes of the Andes, and locally on slopes above the central and interandean valleys (e.g., at Pasochoa). On the west slope recorded south locally into El Oro (Taraguacocha) and Loja (Utuana; Best et al. 1993). Found mostly from 2500 to 3500 m, but has been recorded to 3800 m on Volcán Pichincha (N. Krabbe); usually most numerous not too far below treeline.

One race: nominate *sitticolor*.

Range: Andes from w. Venezuela to w. Bolivia; Sierra de Perijá.

Conirostrum albifrons   Capped Conebill   Picocono Coronado

Uncommon to locally fairly common in the canopy and borders of montane forest, secondary woodland, and borders in the upper subtropical and temperate zones on both slopes of the Andes. On the west slope recorded south at least formerly to "Río Chanchan" in e. Guayas (though these specimens were probably actually taken at higher elevations, and likely came from somewhere in w. Chimborazo) and w. Azuay (Portete; Berlioz 1932b). In recent years the species has been found mainly from Pichincha northward on the west slope, though it has also recently been seen in the Caripero region of w. Cotopaxi (N. Krabbe and F. Sornoza). The Capped Conebill is considerably more numerous and widespread on the east slope than it is in the west. Recorded mostly between 2000 and 2800 m, locally to 3000 m at Oyacachi in w. Napo (N. Krabbe et al.).

One race, *atrocyaneum*, occurs in most of the species' Ecuadorian range. In Nov. 1999, however, the mainly Colombian race *centralandium* was seen along the La Bonita road in w. Sucumbíos (J. Nilsson et al.). Males of this race differ from mate *atrocyaneum* in their snowy white crown; females are similar.

Range: mountains of n. Venezuela, and Andes from sw. Venezuela to w. Bolivia.

Oreomanes fraseri   Giant Conebill   Picocono Gigante

Rare to locally uncommon in patches of *Polylepis*-dominated woodland, mainly above the upper limit of continuous forest, on both slopes of the Andes. Recorded in Ecuador from Carchi (seen by N. Krabbe on Páramo del Ángel on 18 Nov. 1991) south locally to Azuay (El Cajas National Recreation Area) and n. Loja (specimen from El Parotillo, San José, in the Moore collection [Occidental College, Los Angeles], fide F. Vuilleumier; the actual site cannot be located). The Giant Conebill is found in small numbers in woodland patches along the road crossing Papallacta pass, and high on the north slope of Volcán Pichincha (above Yanacocha). Recorded mostly between 3500 and 4200 m.

The Giant Conebill was given Near-threatened status by Collar et al. (1992, 1994), and we concur with their assessment. The species always occurs at relatively low densities, and its habitat in many areas continues to be reduced by human activities (removal of wood, grazing, and burning).

One race: nominate *fraseri*.

Range: locally in Andes of extreme s. Colombia and Ecuador, and from cen. Peru to extreme n. Chile and w. Bolivia.

Xenodacnis parina   Tit-like Dacnis   Xenodacnis

Fairly common but very local in patches of low shrubby woodland dominated by *Gynoxys* and in patches of *Polylepis* woodland in w. Azuay, where first found in and near El Cajas National Recreation Area west of Cuenca. The Tit-like Dacnis was first recorded from Ecuador in Jul. 1978 (Ridgely 1980). Two specimens (MECN) were obtained at El Cajas National Recreation Area by F. Ortiz-Crespo in Dec. 1981, and three more (ANSP) were taken on 13 Aug. 1986 just west of the recreation area (M. B. Robbins and G. S. Glenn). Farther north, the Tit-like Dacnis was also seen in nw. Morona-Santiago on the "Collanes Plain" in Sangay National Park on 25–26 Jun. 1992 (T. R. Mark), and it was also seen in 1990 on nearby Volcán Altar (C. A. Vogt, fide N. Krabbe). A sighting reported from Páramo del Ángel in Carchi (Fjeldså and Krabbe 1990) must be regarded as unverified; recent extensive field work in this region has failed to reveal the species' presence there. Recorded only from about 3700 to 4000 m.

The Tit-like Dacnis was given Near-threatened status by Collar et al. (1992, 1994), and we concur with their assessment. Despite the fact that in Ecuador both of the species' known sites are located within formally protected areas, its population here must be very small, and thus it must be viewed as inherently at risk.

The racial affinity of the Ecuador population remains uncertain: the one ANSP male from near Cajas is unfortunately not fully adult, but it seems similar to the pair of races found in n. Peru, *bella* and *petersi* (these two closely resemble each other, and *bella* should perhaps be synonymized into *petersi*).

Range: locally in Andes from s. Ecuador to s. Peru.

Diglossopis caerulescens   Bluish Flowerpiercer   Pinchaflor Azulado

Locally uncommon to fairly common in montane forest and forest borders in the subtropical and lower temperate zones on the east slope of the Andes; recorded from the La Bonita road in w. Sucumbíos, the Cordillera de Huacamayos in w. Napo (where the species is especially numerous, consistently found at least near the ridgetop), east of Gualaceo and on the Cordillera de Cutucú in Morona-Santiago, and at various localities in e. Azuay and Zamora-Chinchipe, locally spilling over into adjacent Loja on the west slope of the Cordillera Las Lagunillas (seen once in Oct. 1992; M. B. Robbins). The Bluish Flowerpiercer was not recorded during two expeditions to the mid-elevation slopes of the Cordillera del Cóndor (Sep. 1990 and Jun. 1993), but it was found to be present at somewhat higher elevations at Achupallas (Schulenberg and Awbrey 1997). On the west slope it is known only from a few sightings in the northwest in Carchi (along the Maldonado road) and Pichincha (Nono-Mindo road, Yanacocha, and Chiriboga road); there is also a old record from w. Azuay (two Jan. 1948 specimens from Portete; ANSP). Until recently the Bluish Flowerpiercer was considered a rare bird in Ecuador (see Norton and Orcés 1972), but in recent years there has been a long series of records demonstrating that its distribution is much more extensive than had been thought; presumably in the past it was overlooked. Recorded mostly from 1700 to 2700 m, very rarely as high as 3200 m.

This and the following three species are now usually separated from *Diglossa* in the genus *Diglossopis*; see W. J. Bock (*Neotropical Ornithology*, Ornithol. Monogr. no. 36: 319–332, 1986) and Sibley and Monroe (1990). Despite the comments of Norton et al. (1972), we find that most recently obtained specimens (including one in ANSP from as far north as SierrAzul in w. Napo) are best referred to the comparatively gray and pale race *media*, but that one taken northeast of El Chaco (also in ANSP) appears to be intermediate toward the darker and bluer *saturata* of Colombia. Likely there is a south-north cline toward darker and bluer coloration, though Fjeldså and Krabbe (1990) suggest that the population on the Cordillera de Cutucú could represent an undescribed subspecies. The population from the northwest also seems likely to be referable to *saturata*, but no specimens we are aware of have been taken there; furthermore, Fjeldså and Krabbe (1990) suggest that they too may belong to an unnamed subspecies (which would range also in the Central and Western Andes of Colombia).

Range: mountains of n. Venezuela, and Andes from w. Venezuela to w. Bolivia; Sierra de Perijá.

Diglossopis cyanea   Masked Flowerpiercer   Pinchaflor Enmascarado

Fairly common to common in the borders of montane forest and woodland, and in shrubby clearings and gardens, in the temperate zone on both slopes of the Andes; also found in woodland patches and scrub above the central and interandean valleys (e.g., at Pasochoa). Recorded mainly from 2400 to 3500 m, occasionally occurring higher (e.g., in patches of *Polylepis*), and in the southwest (El Oro and w. Loja) regularly found lower (to 2000 m).

Two races occur in Ecuador, nominate *cyanea* in most of the species' Ecuadorian range and the only slightly differentiated *dispar* in El Oro and w. Loja.

Range: mountains of n. Venezuela, and Andes from w. Venezuela to w. Bolivia; Sierra de Perijá.

Diglossopis glauca   Golden-eyed Flowerpiercer   Pinchaflor Ojidorado

Uncommon to fairly common in the canopy and borders of montane forest in the foothills and subtropical zone on the east slope of the Andes from Sucumbíos (San Rafael Falls vicinity) south to cen. Zamora-Chinchipe (Podocarpus National Park, Cordillera del Cóndor; ANSP, v.o.); the species does not, however, seem to occur in s. Zamora-Chinchipe (e.g., it was conspicuously absent at Panguri; T. J. Davis). The Golden-eyed Flowerpiercer is especially numerous north of Archidona along the Loreto road. Recorded mainly from 1000 to 1800 m, occasionally a bit higher.

The species was formerly called the Deep-blue Flowerpiercer, but it stands out among the flowerpiercers not for its deep-blue coloration (many others are equally blue) but for its conspicuous golden yellow iris.

One race: *tyrianthina*. However, examination of material in ANSP leads M. B. Robbins and RSR to question the validity of *tyrianthina*, which is the taxon found in the Ecuadorian and Colombian portion of the species' range. *D. glauca* may perhaps be better regarded as monotypic.

Range: east slope of Andes from s. Colombia to w. Bolivia.

Diglossopis indigotica   Indigo Flowerpiercer   Pinchaflor Indigo

Very rare to rare and local in the canopy and borders of mossy forest on the west slope of the Andes in nw. Ecuador from Carchi (where seen in Aug. 1988 south of Chical; M. B. Robbins) south to Pichincha (old records from the Nanagel area and Canzacoto). A sight report from much farther south in El Oro (a single bird seen briefly in appropriate habitat at Buenaventura [1000 m] on 1 Apr. 1989; RSR and PJG et al.) requires further confirmation. There are extremely few Ecuadorian records of this scarce flowerpiercer, and it has not been found to occur regularly at any accessible locality in the country; despite much recent field work in the Nanegal/Mindo region, no one has found it there. In Ecuador recorded mostly from 1600 to 2000 m, but in Colombia found considerably lower, to 700 m (Hilty and Brown 1986).

Given its very limited range, overall scarcity, and apparently very strict habitat requirements, we believe the Indigo Flowerpiercer merits Vulnerable status in Ecuador. The species was not considered at risk by Collar et al. (1992, 1994), perhaps only in oversight.

Monotypic.

Range: west slope of Andes in sw. Colombia and nw. Ecuador.

Diglossa lafresnayii   Glossy Flowerpiercer   Pinchaflor Satinado

Locally uncommon to locally common in montane woodland, forest borders, and shrubby paramo near treeline on both slopes of the Andes. On the west slope recorded south to Azuay (El Cajas and Portete areas), but seemingly absent from w. Loja and El Oro. The Glossy is much more a forest-based bird than the Black Flowerpiercer (*D. humeralis*), and unlike that species the Glossy seems rarely to range up above treeline into woodland patches found at higher elevations (where Blacks are regularly to be found). Recorded mainly from 2700 to 3500 m.

D. lafresnayii of w. Venezuela to extreme n. Peru is regarded as a species separate from *D. mystacalis* (Moustached Flowerpiercer) of the Peruvian and Bolivian Andes and from *D. gloriosissima* (Chestnut-bellied Flower-piercer) of the Western Andes of Colombia, following F. Vuilleumier (*Am. Mus. Novitates* 2831, 1969) and most subsequent authors. *D. lafresnayii* then becomes monotypic.

Range: Andes from w. Venezuela to n. Peru.

Diglossa humeralis   Black Flowerpiercer   Pinchaflor Negro

Fairly common to common in montane forest borders, woodland (including *Polylepis* patches and even *Eucalyptus* plantations), shrubby areas and clearings, and gardens in the temperate zone and paramo on both slopes of the Andes, including slopes above the central and interandean valleys. Recorded mostly from 2500 to 4000 m, occasionally somewhat lower (this usually in cleared areas).

  D. *humeralis* of Colombia to nw. Peru is regarded as a species distinct from *D. carbonaria* (Gray-bellied Flowerpiercer) of Bolivia, *D. brunneiventris* (Black-throated Flowerpiercer) of n. Peru to w. Bolivia and extreme n. Chile (and also nw. Colombia), and *D. gloriosa* (Mérida Flowerpiercer) of w. Venezuela, following G. R. Graves (*Condor* 84[1]: 1–14, 1982). The expanded species was formerly usually called the Carbonated Flowerpiercer. One race of *D. humeralis* occurs in Ecuador, *aterrima*.

  Range: Andes from n. Colombia to n. Peru; Sierra de Perijá and Santa Marta Mountains.

Diglossa albilatera   White-sided Flowerpiercer   Pinchaflor Flanquiblanco

Fairly common in montane forest, forest borders, and secondary woodland in the upper subtropical and temperate zones on both slopes of the Andes. On the west slope recorded south mainly to Chimborazo and w. Azuay (Portete; Berlioz 1932b); the species has also recently been found at Utuana in s. Loja (first by Best et al. 1993). Ranges mostly from 1900 to 3100 m, but recorded down to 1700 m on the Cordillera del Cóndor (ANSP).

  Two only slightly differentiated races occur in Ecuador, nominate *albilatera* in most of the species' Ecuadorian range and *schistacea* in much of Loja.

  Range: mountains of n. Venezuela, and Andes from w. Venezuela to cen. Peru; Sierra de Perijá and Santa Marta Mountains.

Diglossa sittoides   Rusty Flowerpiercer   Pinchaflor Pechicanelo

Uncommon to locally fairly common in shrubby clearings, gardens, and borders of montane woodland and forest in the subtropical zone on both slopes of the Andes, and locally also in and above the central and interandean valleys. The Rusty Flowerpiercer appears to be more numerous and widespread in Loja than it is elsewhere in Ecuador. Recorded mostly from 1700 to 2800 m, locally and in small numbers as low as 1500 m in Loja.

  D. *sittoides* of South America is regarded as a species distinct from *D. baritula* (Cinnamon-bellied Flowerpiercer) of n. Middle America and *D. plumbea* (Slaty Flowerpiercer) of s. Middle America, following most recent authors (see, most recently, S. J. Hackett, *Auk* 112[1]: 156–170, 1995); this is also the treatment adopted in the 1998 AOU Check-list. One race of *D. sittoides* occurs in Ecuador, *decorata*.

Range: mountains of n. Venezuela, and Andes from w. Venezuela to nw. Argentina; Sierra de Perijá and Santa Marta Mountains.

Hemithraupis guira   Guira Tanager   Tangara Guira

Fairly common in the canopy and borders of forest and secondary woodland, and in adjacent clearings, in the lowlands and foothills of both e. and w. Ecuador. In the west recorded from w. Esmeraldas (Muisne area; ANSP) south at least to Guayas (hills south of Río Ayampe, Cerro Blanco in the Chongón Hills, and Manglares-Churute Ecological Reserve) and nw. Azuay (Manta Real [ANSP] and San Miguel del Azuay [M. Whittingham]). As there is a single 19th-century record of the Guira Tanager from Santa Lucia in Tumbes, Peru, it is possible that the Guira Tanager could also occur in El Oro; however, recent field workers have failed to find it anywhere in the far southwestern region of Ecuador. In the east found mainly in secondary growth and forest borders and recorded primarily near the base of the Andes, with records extending from Napo south locally to Zamora-Chinchipe (e.g., it is numerous in the Zumba area; ANSP); there are only a few reports (of wandering birds?) from as far east as La Selva and Cuyabeno. Recorded up to about 1100 m.

Two races are found in Ecuador, *guirina* in the west and *huambina* in the east. Male *guirina* differ in having a more ochraceous, not so yellow, superciliary; females of the two are similar.

Range: w. and n. Colombia, n. and e. Venezuela, and the Guianas to nw. and ne. Argentina and se. Brazil; w. Ecuador and extreme nw. Peru.

Hemithraupis flavicollis   Yellow-backed Tanager   Tangara Lomiamarilla

Uncommon to fairly common in the canopy and borders of humid forest (especially terra firme) in the lowlands and foothills of e. Ecuador. Ranges up at least locally to 900–950 m along the eastern base of the Andes in the Río Nangaritza valley around Miazi in Zamora-Chinchipe (Schulenberg and Awbrey 1997); elsewhere usually below 700 m.

One race: *peruana*.

Range: e. Panama and n. Colombia; se. Colombia, s. Venezuela, and the Guianas to n. Bolivia and w. and cen. Amazonian Brazil; e. Brazil.

Erythrothlypis salmoni   Scarlet-and-white   Tangara Escarlatiblanca
                              Tanager

Uncommon to locally fairly common in the canopy and borders of very humid and humid forest and secondary woodland in the lowlands and foothills of nw. Ecuador, where recorded primarily from Esmeraldas (southwest to along the Río Verde south of Chontaduro; T. J. Davis). There are also numerous recent (1995 onward) sightings of this spectacular tanager from nw. Pichincha, along the road north of Simon Bolívar near Pedro Vicente Maldonado (v.o.) and from Hacienda San Francisco near San Miguel de los Bancos (male seen on 12 Sep.

1995; P. Coopmans); it has also been seen in adjacent sw. Imbabura (male seen near Salto del Tigre bridge over the Río Guaillabamba on 15 Jul. 1997; P. Coopmans et al.). In Ecuador the Scarlet-and-white Tanager is perhaps most numerous in n. Esmeraldas in the El Placer area and also at Playa de Oro; at the latter it is most numerous in the foothill zone between 300 and 400 m (O. Jahn et al.). Recorded up to about 700 m.

The monotypic genus *Erythrothlypis* has often (e.g., *Birds of the World*, vol. 13) been merged into *Chrysothlypis* (which is also monotypic), but the two species are so different that their merger seems improbable; were this to be done, then likely both should be merged into *Hemithraupis*. Monotypic.

Range: w. Colombia and nw. Ecuador.

Thlypopsis sordida   Orange-headed Tanager   Tangara Cabecinaranja

Uncommon to fairly common in riparian woodland and shrubbery on islands in the lowlands of e. Ecuador, with smaller numbers beginning to spread into clearings and second-growth away from riparian areas (e.g., since 1989 a few have been seen in such areas around Lago Agrio and Tena). The Orange-headed Tanager is known mainly from recent reports along the Río Napo and the lower Río Aguarico in the northeast; the first specimen from this region was obtained at the mouth of the Río Lagartococha (J. T. Zimmer, *Am. Mus. Novitates* 1345, 1947). In the southeast it is known from 19th-century specimen material from Sarayacu, and there are also recent reports from Kapawi Lodge on the Río Pastaza near the Peruvian border. Recorded up to about 500 m.

One race: *chrysopis*.

Range: se. Venezuela; se. Colombia and e. Ecuador to n. Argentina, Paraguay, and s. and e. Brazil.

Thlypopsis ornata   Rufous-chested Tanager   Tangara Pechicanela

Uncommon and somewhat local at borders of montane forest and (especially) secondary woodland and adjacent clearings in the upper subtropical and temperate zones on both slopes of the Andes and locally above the central and interandean valleys. On the east slope in recent times the species has apparently been recorded only from s. Ecuador, in Morona-Santiago and Zamora-Chinchipe, though there is also an old specimen from El Pun in e. Carchi (Salvadori and Festa 1899a). The Rufous-chested Tanager seems more numerous and widespread in s. Ecuador than northward. Recorded mostly from 1800 to 3000 m, locally or in small numbers as low as 1500 m, with an old specimen labeled as having been taken as low as 1200 m at Huigra in Chimborazo.

Two similar races are found in Ecuador, nominate *ornata* in most of the species' Ecuadorian range and *media* in El Oro (Taraguacocha), Loja, and Zamora-Chinchipe. The subspecific affinity of the El Pun specimen remains uncertain.

Range: Andes from s. Colombia to s. Peru.

Thlypopsis inornata   Buff-bellied Tanager   Tangara Ventrianteada

Uncommon to locally fairly common in shrubby secondary woodland, clearings, and forest borders in the Río Marañón drainage of se. Ecuador in extreme s. Zamora-Chinchipe around Zumba. The Buff-bellied Tanager was discovered here in Aug. 1989, when three specimens (ANSP and MECN) were obtained by RSR and others; additional specimens (ANSP) were taken here on 12 Dec. 1991 and 8 Aug. 1992. The species, which formerly was known only from a limited area in adjacent Peru, has since been seen here by several other observers. Recorded from about 650 to 1200 m.
Monotypic.
Range: upper Río Marañón valley of extreme s. Ecuador and nw. Peru.

Pipraeidea melanonota   Fawn-breasted Tanager   Tangara Pechianteada

Uncommon to locally fairly common in montane forest borders, secondary woodland, and adjacent clearings in the subtropical and temperate zones on both slopes of the Andes; also found locally in the central valley (including at least seasonally or locally in Quito itself) and southward in interandean valleys. On the west slope recorded south through w. Loja. Basically a nonforest tanager, the Fawn-breasted occurs in only very small numbers in extensively forested regions. Recorded mostly from 1000 to 2800 m, but it regularly occurs lower on the west slope (e.g., at Tinalandia), and has been found as low as 400 m above Cochancay in Cañar and at Manta Real in nw. Azuay (RSR); there is also one record of two singing birds at 3100 m at Río Anatenorio in w. Napo (N. Krabbe et al.).
One race: *venezuelensis*.
Range: mountains of n. Venezuela, and Andes from w. Venezuela to nw. Argentina; locally on tepuis of Venezuela; se. Brazil, e. Paraguay, ne. Argentina, and Uruguay.

Chlorophonia cyanea   Blue-naped Chlorophonia   Clorofonia Nuquiazul

Rare and local (but inconspicuous and probably overlooked at least to some extent) in the subcanopy of montane forest and forest borders, sometimes ranging out into tall trees in adjacent clearings, in the foothills and lower subtropical zone on the entire east slope of the Andes; occasionally (or temporarily?) somewhat more numerous (e.g., on the Cordillera de Cutucú in Jun.–Jul. 1984; Robbins et al. 1987). There is also at least one sighting from the eastern lowlands, at Limoncocha on 2 Sep. 1971 (Tallman and Tallman 1977); this would indicate the possibility of some altitudinal movements or wandering down into the lowlands, but the fact that the species has not been reported since, or closer to the Andes at sites such as Jatun Sacha (B. Bochan), indicates that such movements must be infrequent at best. Reports of the Blue-naped Chlorophonia from the west slope of the Andes must be considered as unverified; the species is also not known from Colombia's Western Andes south of Valle (Hilty and Brown 1986). Recorded mainly between about 800 and 1700 m.

One race: *longipennis.*

Range: mountains of n. Venezuela, and Andes from w. Venezuela to w. Bolivia; Sierra de Perijá and Santa Marta Mountains; tepuis of Venezuela and Guyana; se. Brazil, e. Paraguay, and ne. Argentina.

Chlorophonia flavirostris	Yellow-collared Chlorophonia	Clorofonia Cuellidorada

Rare to locally fairly common in the canopy and borders of montane forest and in large trees of adjacent clearings in the foothills and lower subtropical zone on the west slope of the Andes south to Pichincha (along lower Chiriboga road, rarely at Tinalandia). The lovely Yellow-collared Chlorophonia is quite numerous around El Placer in Esmeraldas, but it is less common (and perhaps irregular) southward, where small numbers can sometimes be found above Mindo and along the lower Chiriboga road; it may occur in the latter areas primarily between Dec. and Apr. (M. Lysinger). Recorded mostly between about 450 m (above Playa de Oro in Esmeraldas; O. Jahn) and 1500 m.

Monotypic.

Range: west slope of Andes in sw. Colombia and nw. Ecuador; a few records from e. Panama.

Chlorophonia pyrrhophrys	Chestnut-breasted Chlorophonia	Clorofonia Pechicastaña

Rare to occasionally (or locally) uncommon in the canopy and borders of montane forest and mature secondary woodland in the subtropical and temperate zones on both slopes of the Andes. On the west slope known only from the northwest, where recorded from Carchi and Pichincha (e.g., above Mindo). Especially on the east slope the Chestnut-breasted Chlorophonia is probably more widespread at appropriate elevations than the relatively few Ecuadorian records would appear to indicate; it is regular around Cabañas San Isidro (M. Lysinger). Recorded mostly from 1500 to 2750 m, locally (or seasonally?) up to 3000 m at Angashcola in e. Loja (R. Williams et al.).

Monotypic.

Range: locally in Andes from w. Venezuela to cen. Peru; Sierra de Perijá.

Euphonia laniirostris	**Thick-billed Euphonia**	**Eufonia Piquigruesa**

Uncommon to common in the canopy and borders of humid forest, deciduous woodland, and in clearings and gardens in the lowlands and (to a lesser extent) foothills of both e. and w. Ecuador; more numerous in the west. In the west recorded from n. Esmeraldas and n. Pichincha south through El Oro and w. Loja, though avoiding drier regions. Generally more local in the east, though apparently relatively numerous at Jatun Sacha (B. Bochan). Recorded mostly below about 1200–1500 m, but at least on the west slope ranging locally as high as 1800 m (e.g., above Tandayapa and Mindo, and at Sozoranga in s. Loja); it may be spreading upward in response to clearing.

Two races occur in Ecuador, *melanura* in the east and *hypoxantha* in the west. Male *melanura* have an all-black tail (with no white tail spots as in *hypoxantha*), and they have a smaller yellow crown patch than is seen in *hypoxantha*. Females of the two races are very similar.

Range: Costa Rica to n. and e. Bolivia and w. and s. Amazonian Brazil; w. Ecuador and nw. Peru.

Euphonia cyanocephala   Golden-rumped Euphonia   Eufonia Lomidorada

Uncommon and local (and perhaps seasonal) in lighter woodland, clearings and gardens, and borders of montane forest on the entire west slope of the Andes and in the central valley; occurs both in humid and quite arid regions. Definite records from the actual east slope of the Andes are relatively few but include an old specimen from Machay in Tungurahua, many recent sightings from Cabañas San Isidro in w. Napo, and one near Valladolid in Zamora-Chinchipe on 4 Apr. 1994 (P. Coopmans and M. Lysinger). The Golden-rumped Euphonia is quite regular in the Quito region, where it can sometimes be seen in city parks. Occurs mostly between about 1200 and 2800 m, occasionally wandering down as low as 700 m (Tinalandia).

E. *cyanocephala* of South America is regarded as a separate species from *E. musica* (Antillean Euphonia) of the West Indies and *E. elegantissima* (Blue-hooded Euphonia) of Middle America, following most recent authors; the expanded species was called the Blue-hooded Euphonia, and the 1998 AOU Check-list suggests that *E. elegantissima* be called the Elegant Euphonia. The name *cyanocephala* has priority over *aureata*, which has sometimes been used for the South American species (e.g., Isler and Isler 1987), as noted by Sibley and Monroe (1990). Only two races of *E. cyanocephala* have been definitely recorded from Ecuador, *pelzelni* south to Chimborazo and *insignis* in Azuay and Loja. Male *insignis* have a yellow, not black, frontlet; females are similar. Various references (e.g., *Birds of the World*, vol. 13), state or imply that nominate *cyanocephala* (sometimes under the name of *aureata*) also occurs in "eastern Ecuador." However, we are not aware of any specific records of this taxon from the country, though it is possible that the Zamora-Chinchipe sighting (above) may represent a population that would belong there. Nominate *cyanocephala* resembles *pelzelni*, even in males (also having a black frontlet), though it perhaps averages somewhat deeper ochraceous below.

Range: mountains of n. Venezuela, and Andes from w. Venezuela to nw. Argentina; Sierra de Perijá; locally on tepuis of s. Venezuela and in the Guianas; se. Brazil, e. Paraguay, and ne. Argentina.

Euphonia xanthogaster   Orange-bellied Euphonia   Eufonia Ventrinaranja

Fairly common to common and widespread in humid forest, forest borders, and mature secondary woodland in the lowlands of both e. and w. Ecuador, becoming common to very common in the foothills and subtropical zone on both slopes of the Andes. On the west slope recorded south through w. Loja

(Alamor/Celica region and near Sozoranga [Sep. 1998; RSR et al.]). The Orange-bellied is generally the most numerous euphonia in montane forest and woodland throughout Ecuador. Recorded up to about 2000 m, exceptionally as high as 2750 m in w. Loja above Celica (Bloch et al. 1991).

The races occurring in Ecuador differ only slightly from each other. The marginally smaller *chocoensis* is found in the lowlands and foothills of n. Esmeraldas and adjacent Imbabura (Lita and Hacineda Paramba; J. T. Zimmer, *Am. Mus. Novitates* 1225, 1943), *quitensis* in the remainder of the west, and *brevirostris* in the east. Male *brevirostris* have the crown patch and underparts slightly more ochraceous than in the other two Ecuadorian races. S. L. Olson (*Proc. Biol. Soc. Wash.* 94[1]: 102–104, 1981) suggested that the race *oressinoma* (of the Western and Central Andes of Colombia) may occur south into Ecuador, but there are no definite records of it; it is, in any case, extremely similar to both *quitensis* and *chocoensis*.

Range: e. Panama to extreme nw. Peru, nw. Bolivia, w. and cen. Amazonian Brazil, nw. and s. Venezuela, and Guyana; se. Brazil. More than one species may be involved.

Euphonia minuta   White-vented Euphonia   Eufonia Ventriblanca

Rare to locally uncommon in the canopy and borders of humid forest and secondary woodland in the lowlands and lower foothills of both e. and w. Ecuador; in the west not known from south of w. Esmeraldas (Muisne area) and s. Pichincha (a few sightings from Río Palenque). The White-vented Euphonia is perhaps more numerous at Jatun Sacha than elsewhere in Ecuador (B. Bochan). Recorded up to about 700 m at Tinalandia.

Two only slightly different races occur in Ecuador, *humilis* in the northwest and nominate *minuta* in the east.

Range: Guatemala and s. Belize to nw. Ecuador; e. Colombia, s. and e. Venezuela, and the Guianas to n. Bolivia and Amazonian Brazil.

*Euphonia chlorotica   Purple-throated Euphonia   Eufonia Golipúrpura

Uncommon to fairly common in the canopy and borders of deciduous forest, secondary woodland, and partially cleared areas in the Río Marañón drainage of s. Zamora-Chinchipe around Zumba. The Purple-throated Euphonia was first recorded on 13 Aug. 1989 near Zumba, when a male (ANSP) was collected by T. C. Maxwell. A calling male tape-recorded near Zumba on 3 May 1989 was then identified subsequently (P. Coopmans). A second male (also ANSP) was obtained by RSR in the nearby Río Mayo valley on 12 Dec. 1991, and the species has since proven to be widespread and tolerably numerous in this region. Recorded between about 650 and 1100 m.

Ecuadorian birds are assignable to the race *taczanowskii*, following the diagnosis presented by J. T. Zimmer (*Am. Mus. Novitates* 1225, 1943), but we note that many races of *E. chlorotica* are exceedingly similar to each other.

Range: ne. Colombia, s. Venezuela, Guyana, and s. Suriname; extreme s.

Ecuador, e. Peru, n. and e. Bolivia, much of Brazil (but not in some parts of Amazonia), Paraguay, and n. Argentina.

Euphonia saturata   Orange-crowned Euphonia   Eufonia Coroninaranja

Uncommon and somewhat local in deciduous woodland, forest borders, and clearings with scattered trees in the lowlands and foothills of w. Ecuador from w. Esmeraldas (Esmeraldas [city] region) and n. Pichincha (Nanegal and Mindo region) south through El Oro and w. Loja. Recorded at least locally as high as 1500 m around Mindo and in sw. Loja.
    Monotypic.
    Range: sw. Colombia to extreme nw. Peru.

Euphonia fulvicrissa   Fulvous-vented Euphonia   Eufonia Ventrileonada

Uncommon to locally fairly common in very humid to humid forest, forest borders, and secondary woodland in the lowlands of nw. Ecuador, where known primarily from Esmeraldas; there is also a single recent record from n. Manabí (a female [WFVZ] taken on 12 Jul. 1988 at Filo de Monos; see Marín et al. 1992), and since 1995 there have also been numerous records from nw. Pichincha (along the road north of Simón Bolívar near Pedro Vicente Maldonado and 5 km south of Golondrinas; v.o.). In Ecuador the Fulvous-vented Euphonia has not been recorded above about 500 m, though elsewhere (e.g., Panama) it often ranges farther up into the foothills (Ridgely and Gwynne 1989).
    One race: *purpurascens*.
    Range: Panama to nw. Ecuador.

Euphonia rufiventris   Rufous-bellied Euphonia   Eufonia Ventrirrufa

Uncommon to fairly common in the canopy, subcanopy, and borders of humid forest (both terra firme and várzea) in the lowlands of e. Ecuador. Ranges mostly below 500 m, with small numbers up into foothills along the base of the Andes to 700–900 m (including birds reported in the Río Nangaritza valley of Zamora-Chinchipe at Miazi; Schulenberg and Awbrey 1997).
    Monotypic.
    Range: se. Colombia and s. Venezuela to n. Bolivia and w. and cen. Amazonian Brazil.

Euphonia mesochrysa   Bronze-green Euphonia   Eufonia Verdibronceada

Uncommon to locally fairly common in the canopy and borders of montane forest in the foothills and lower subtropical zone along the entire east slope of the Andes. The Bronze-green Euphonia is regularly found around San Rafael Falls, and above Archidona along the road to Loreto. Recorded mostly between about 1100 and 1800 m, locally down to 900 m (Miazi; Schulenberg and Awbrey 1997).

One race: nominate *mesochrysa*.
Range: east slope of Andes from s. Colombia to w. Bolivia.

Euphonia chrysopasta   White-lored Euphonia   Eufonia Loriblanca

Uncommon to (usually) fairly common in the canopy and borders of humid forest (both terra firme and várzea), and to a lesser extent in secondary woodland and adjacent clearings, in the lowlands of e. Ecuador. Recorded mostly below about 600 m, locally ranging as high as 900 m (Bermejo oil-field area north of Lumbaquí [M. B. Robbins] and Miazi in the Río Nangaritza valley [Schulenberg and Awbrey 1997]).

The species was formerly called the Golden-bellied Euphonia, but we follow Isler and Isler (1987), Ridgely and Tudor (1989), and Sibley and Monroe (1990) in employing the English name of White-lored Euphonia.

One race: nominate *chrysopasta*.
Range: e. Colombia, s. Venezuela, and the Guianas to n. Bolivia and Amazonian Brazil.

Chlorochrysa calliparaea   Orange-eared Tanager   Tangara Orejinaranja

Fairly common to locally common in the canopy and borders of montane forest in the foothills and subtropical zone on the east slope of the Andes. The Orange-eared Tanager is numerous north of Archidona along the road to Loreto and around San Rafael Falls. Recorded mostly between 1000 and 1700 m, in small numbers as high as 1975 m (on the Cordillera de Cutucú; Robbins et al. 1987).

One race: *bourcieri*.
Range: east slope of Andes from w. Venezuela to w. Bolivia.

Chlorochrysa phoenicotis   Glistening-green   Tangara Verde
                             Tanager   Reluciente

Uncommon in the canopy and borders of montane forest in the foothills and lower subtropical zone on the west slope of the Andes, where found from Carchi south locally to El Oro (a sighting from Buenaventura on 17 Sep. 1990 by P. Coopmans et al.). The beautiful Glistening-green Tanager seems to favor very wet, often mossy conditions. Recorded from about 600 to 1700 m, occasionally up to 2000 m above Mindo.
Monotypic.
Range: west slope of Andes in w. Colombia and w. Ecuador.

Tangara rufigula   Rufous-throated Tanager   Tangara Golirrufa

Locally fairly common in the canopy and borders of montane forest and secondary woodland in the foothills on the west slope of the Andes in the northwest from Carchi south to Pichincha, and then again in the southwest in El Oro (La Chonta and Buenaventura); it perhaps occurs in the intervening area but as yet is unrecorded. The Rufous-throated Tanager is the most numerous *Tangara* at El Placer in Esmeraldas, but it appears to be absent from many areas even at appropriate elevations, apparently because it favors mossy "cloud forest" conditions (which themselves are quite local). Recorded mostly from 600 to 1400 m, locally somewhat lower (e.g., down to 450 m at Playa de Oro in Esmeraldas; O. Jahn) and occasionally slightly higher.

Monotypic.

Range: west slope of Andes in w. Colombia and w. Ecuador.

Tangara palmeri   **Gray-and-gold Tanager**   **Tangara Doradigris**

Uncommon to locally fairly common in the canopy and borders of humid forest and in adjacent clearings in the foothills on the west slope of the Andes south to Pichincha (Tinalandia). Also recently recorded from the coastal Cordillera de Mache in sw. Esmeraldas (southeast of Muisne [Parker and Carr 1992] and at Bilsa [first by R. P. Clay et al. in 1994]) and the Cordillera de Chindul in n. Manabí (2 specimens [WFVZ] taken at Filo de Monos on 15 Jul. 1988; see Marín et al. 1992). Although most numerous in Esmeraldas, small numbers of the Gray-and-gold Tanager continue to be found at Tinalandia. Recorded mainly from 400 to 1000 m, but has been found lower (down to 100 m) in n. Esmeraldas (e.g., at Playa de Oro; O. Jahn et al.) and on the coastal cordillera.

Monotypic.

Range: e. Panama to nw. Ecuador.

Tangara arthus   **Golden Tanager**   **Tangara Dorada**

Common and widespread in the canopy and borders of montane forest, secondary woodland, and in adjacent clearings in the foothills and subtropical zone on both slopes of the Andes. On the west slope ranges south to w. Loja in the Celica area. Recorded mainly between about 900 and 2000 m, but on the west slope locally lower (as low as 550 m, e.g., above Manta Real in nw. Azuay; RSR) and also higher (up to 2200 m above Mindo).

Two races occur in Ecuador, *goodsoni* on the west slope and *aequatorialis* on the east slope. *Goodsoni* has uniform golden yellow underparts, whereas the larger *aequatorialis* has a variable but often strong suffusion of rufous on the throat and chest.

Range: mountains of n. Venezuela, and Andes from w. Venezuela to w. Bolivia; Sierra de Perijá.

*Tangara florida   **Emerald Tanager**   **Tangara Esmeralda**

Locally uncommon to fairly common in the canopy and borders of montane forest and secondary woodland in the foothills of nw. Ecuador. First recorded from a Nov. 1978 sight report (PJG) from near San Miguel de los Bancos in nw. Pichincha (Ridgely 1980); more recently there have been numerous sightings from 1995 onward along the road north of Simón Bolívar near Pedro

Vicente Maldonado (though deforestation has resulted in the species becoming quite local in this area); it has also been seen in adjacent sw. Imbabura near the Salto del Tigre bridge over the Río Guaillabamba (v.o.). The Emerald Tanager has also been found to be fairly common at El Placer in w. Esmeraldas, where four specimens (ZMCOP) were taken by N. Krabbe on 4–12 Jul. 1984 (Fjeldså and Krabbe 1986) and another in Aug. 1987 (ANSP); the species has also been seen above nearby Lita in adjacent w. Imbabura (P. Coopmans and P. Boesman). Emerald Tanagers were also found recently on the coastal Cordillera de Mache in sw. Esmeraldas at Bilsa (first by J. P. Clay et al. in 1994). Like the Rufous-throated Tanager (*T. rufigula*)—with which it often occurs—the Emerald Tanager seems to be quite local in Ecuador, apparently because of its preference for very wet, mossy "cloud forest." Recorded between about 400 and 1200 m, in small numbers and locally down to 200 m above Playa de Oro in n. Esmeraldas (O. Jahn et al.).

We follow Wetmore et al. (1984) in synonymizing the race *auriceps*, to which South American birds had formerly been ascribed. *T. florida* then becomes monotypic.

Range: Costa Rica to nw. Ecuador.

Tangara icterocephala   **Silver-throated Tanager**   **Tangara Goliplata**

Uncommon to locally fairly common in the canopy and borders of montane forest and secondary woodland in the foothills and lower subtropical zone on the west slope of the Andes south to El Oro and w. Loja (Las Piñas and Alamor). Also recently recorded on the coastal Cordillera de Mache in w. Esmeraldas at Bilsa (first by J. P. Clay et al. in Feb. 1994) and the Cordillera de Colonche in sw. Manabí (Machalilla National Park; ANSP) and w. Guayas (Cerro La Torre and near Salanguilla [N. Krabbe] and Loma Alta [Becker and López-Lanús 1997]; also in hills south of Río Ayampe [ANSP]). Recorded mainly from 500 to 1350 m, but in nw. Guayas has been found down to about 150 m near the Río Ayampe.

One race: nominate *icterocephala*.

Range: Costa Rica to sw. Ecuador.

Tangara xanthocephala   Saffron-crowned Tanager   **Tangara Coroniazafrán**

Fairly common in the canopy and borders of montane forest and in adjacent clearings in the subtropical zone on the east slope of the Andes. Rare to uncommon on the west slope of the Andes, where in the past recorded south to Chimborazo (Cayandeled), though there are no recent records from south of Pichincha (and the species is far from numerous even there). Recorded mostly between 1500 and 2300 m, locally and in small numbers as low as 1100 m (w. Napo, Zamora-Chinchipe).

One race: *venusta*.

Range: Andes from w. Venezuela to w. Bolivia; Sierra de Perijá.

Tangara chrysotis   Golden-eared Tanager   Tangara Orejidorada

Uncommon to locally fairly common in the canopy and borders of montane forest, secondary woodland, and adjacent clearings in the lower subtropical zone on the east slope of the Andes. The beautiful Golden-eared Tanager can be found regularly with the tanager flocks in the vicinity of San Rafael Falls. Recorded mostly in a narrow elevational range between about 1100 and 1700 m, but at least occasionally (or locally) it occurs lower (e.g., one was seen at 850 m in the Río Isimanchi valley north of Zumba on 13 Dec. 1991; RSR).

One race: nominate *chrysotis*. In *Birds of the World* (vol. 13) doubt is expressed as to the validity of the race *cochabambae* from Peru and Bolivia. Although he did not deal with the issue directly, J. V. Remsen, Jr. (*Gerfaut* 74: 163–179, 1984), in his review of the distribution of the species, implicitly appeared to accept it.

Range: east slope of Andes from s. Colombia to w. Bolivia.

Tangara parzudakii   Flame-faced Tanager   Tangara Cariflama

Fairly common in the canopy and borders of montane forest and secondary woodland and in adjacent clearings in the subtropical zone on both slopes of the Andes. On the west slope recorded in the northwest from Carchi to Pichincha (lower Chiriboga road), in w. Chimborazo (an old specimen from Cayandeled), and then again in the southwest in El Oro (an old specimen from El Chiral; many recent records from Buenaventura); whether the species' distribution is more continuous on the west slope remains unknown. Recorded mainly from 1500 to 2400 m, but locally much lower on the west slope (down to 700 m at Buenaventura in El Oro).

Two races occur in Ecuador, nominate *parzudakii* on the east slope and *lunigera* on the west slope. The nominate race has a fiery red forecrown and lower cheeks, whereas *lunigera* is somewhat smaller with yellow-orange forecrown and cheeks, less extensive yellow on the hindneck, and duller and less opalescent underparts.

Range: Andes from sw. Venezuela to s. Peru.

Tangara ruficervix   Golden-naped Tanager   Tangara Nuquidorada

Uncommon to fairly common in the canopy and borders of montane forest and secondary woodland in the subtropical zone on both slopes of the Andes. On the west slope found south to El Oro (e.g., at Buenaventura) and recently discovered in small numbers as far as w. Loja (Alamor and Tierra Colorada; Best et al. 1993). Recorded mainly from 1400 to 2400 m, but a few range locally as low as about 900 m at Buenaventura in El Oro.

Two similar races occur in Ecuador, *leucotis* on the west slope and *taylori* on the east slope.

Range: Andes from n. Colombia to w. Bolivia.

Tangara cyanotis   Blue-browed Tanager   Tangara Cejiazul

Uncommon to locally fairly common in the canopy and borders of montane forest in the foothills and lower subtropical zone on the east slope of the Andes. The Blue-browed Tanager can regularly be found with tanager flocks just above San Rafael Falls. Recorded mainly from 1400 to 1900 m, occasionally a little higher or lower.

One race: *lutleyi.*
Range: east slope of Andes from s. Colombia to w. Bolivia.

Tangara labradorides   Metallic-green Tanager   Tangara Verdimetálica

Rare to locally fairly common in the canopy and borders of montane forest and secondary woodland in the subtropical zone on the west slope of the Andes south to Pichincha (lower Chiriboga road). Since the late 1970s small numbers of the Metallic-green Tanager have also been found on the east slope of the Andes in Morona-Santiago (sightings from along the Gualaceo-Limón road) and Zamora-Chinchipe (sightings from along the Loja-Zamora road [Ridgely 1980], at Romerillos in adjacent Podocarpus National Park (Rasmussen et al. 1994), and near Valladolid [RSR et al.]). To date the only specimen from se. Ecuador is one collected (MECN) at Chinapinza on the west slope of the Cordillera del Cóndor on 17 Sep. 1990 (Krabbe and Sornoza 1994); however, the species apparently is quite scarce there. Recorded mainly from 1300 to 2000 m.

Two races occur in Ecuador, nominate *labradorides* on the west slope and *chaupensis* in the southeast. *Chaupensis* is markedly more opalescent green than the opalescent blue of the nominate race, especially on the foreneck and lesser wing-coverts.
Range: Andes from n. Colombia to n. Peru.

Tangara nigroviridis   Beryl-spangled Tanager   Tangara Lentejuelada

Fairly common to common and widespread in the canopy and borders of montane forest, secondary woodland, and adjacent clearings in the subtropical zone on both slopes of the Andes. On the west slope recorded from Carchi to Pichincha and in El Oro and w. Loja (Alamor and Celica region); it seems likely to occur in the intervening area but as yet is unrecorded. Recorded mainly from 1400 to 2500 m, locally occurring somewhat lower on the west slope (down to 1100 m).

Two races occur in Ecuador, *cyanescens* on the west slope and nominate *nigroviridis* on the east slope; recent ANSP specimens from the Cordilleras de Cutucú and del Cóndor appear to be intermediate toward the Peruvian race *berlepschi.* The nominate race has somewhat greener (less blue) opalescent spangles on the underparts than does *cyanescens.* We follow *Birds of the World* (vol. 13) in synonymizing *consobrina*, the race in which west-slope birds were formerly often placed, into *cyanescens.*

Range: mountains of n. Venezuela, and Andes from w. Venezuela to w. Bolivia; Sierra de Perijá.

Tangara vassorii   Blue-and-black Tanager   Tangara Azulinegra

Uncommon to fairly common in canopy and borders of montane forest and woodland in the upper subtropical and temperate zones on both slopes of the Andes, and locally on slopes above the central and interandean valleys (e.g., at Pasochoa, and on Cerro Colambo above Gonzanamá [Sep. 1998; RSR et al.]). On the west slope found south to w. Loja. Recorded mostly between 2000 and 3300 m, ranging in small numbers as high as 3600 m, and regularly found up to near treeline.

One race: nominate *vassorii*.

Range: Andes from w. Venezuela to w. Bolivia.

Tangara heinei   Black-capped Tanager   Tangara Gorrinegra

Uncommon in the borders of montane forest, secondary woodland, and clearings in the subtropical zone on both slopes of the Andes in n. Ecuador. The Black-capped Tanager was first recorded from the west slope of the Andes only in Feb. 1980, when pairs were seen above Tandayapa in Pichincha (see Ridgely 1980); since then there have been numerous sightings from Pichincha, principally if not entirely in the Tandayapa/Nanegal region, and there is also one sighting from Imbabura (by P. Coopmans at Hacienda La Florida, east of Apuela, on 9 Mar. 1991). Given that extensive collecting took place in w. Pichincha in the 19th and early 20th centuries, the fact that the species was never taken during that period makes us suspect that it has only recently colonized this region, presumably from Colombia; this may have been in response to increased clearing (the species being basically a nonforest *Tangara*). On the east slope known mainly from w. Sucumbíos and w. Napo, though there are also records from Tungurahua (specimen [MCZ] taken by D. Norton at Río Negro in the Río Pastaza valley in Sep. 1964) and even Zamora-Chinchipe (males seen near Palanda in Sep. 1991 [R. Williams et al.] and south of Valladolid on 16 Feb. 1998 [P. Coopmans and H. Batjes]). Recorded between about 1100 and 1900 m.

Monotypic.

Range: mountains of n. Venezuela, and Andes from w. Venezuela to n. Ecuador; Sierra de Perijá and Santa Marta Mountains.

Tangara viridicollis   Silver-backed Tanager   Tangara Dorsiplateada

Uncommon to locally fairly common in the canopy and borders of montane forest and woodland and in adjacent clearings in the subtropical zone on the west slope of the Andes in s. Ecuador, where recorded mainly from Loja (east to the lower edge of Podocarpus National Park near the Cajanuma Divide and above Yangana; v.o.) though also occurring in adjacent s. El Oro; the northernmost report is an old specimen from Gima in s. Azuay (Chapman 1926). The Silver-backed Tanager has also recently been found to occur on the east slope of the Andes in s. Zamora-Chinchipe, where it was uncommon at Panguri in Jul. 1992 (ANSP), here in sympatry with the Straw-backed Tanager (*T. argyrofenges*); a few Silver-backed Tanagers have also been seen recently in the Valladolid area (as first noted by Rasmussen et al. 1996) and east of Zumba (P. Coopmans and M. Lysinger). The species is widespread and reasonably numerous in the remaining forest patches around Alamor and west of Celica in w. Loja. Recorded mainly between about 1300 and 2300 m, rarely as high as 2750 m (perhaps higher; Gima is at ca. 2950 m); on the east slope found from 1400 to 1600 m.

The species was formerly called the Silvery Tanager (Meyer de Schuauensee 1966, 1970), but see Isler and Isler (1987), Ridgely and Tudor (1989), and Sibley and Monroe (1990).

One race: *fulvigula*.

Range: Andes from s. Ecuador to s. Peru.

Tangara argyrofenges   Straw-backed Tanager   Tangara Dorsipajiza

Rare in the canopy of montane forest in the subtropical zone on the east slope of the Andes in extreme se. Ecuador in s. Zamora-Chinchipe. The Straw-backed Tanager was first found on 22 Jul. 1992, when a male was collected (ANSP) at Panguri by F. Sornoza; he and T. J. Davis saw a few others there in the following week, but no additional specimens could be obtained (Rasmussen et al. 1996). Of particular interest was the sympatry at Panguri of this species and its close relative the Silver-backed Tanager (*T. viridicollis*); often the two were noted foraging in the same flock. The Straw-backed Tanager was previously known north only to south and east of the Río Marañón in n. Peru. Recorded between about 1350 and 1600 m.

The Straw-backed Tanager appears to be very local throughout its limited range, and it never seems to be numerous. As so little is known about it in Ecuador, we give it Data Deficient status in the country. It was not considered to be at risk by Collar et al. (1992, 1994).

The species was formerly (e.g., Meyer de Schauensee 1966, 1970) called the Green-throated Tanager, but see Isler and Isler (1987), Ridgely and Tudor (1989), and Sibley and Monroe (1990).

One race: *caeruleigularis*.

Range: locally on east slope of Andes from extreme s. Ecuador to w. Bolivia.

Tangara vitriolina   Scrub Tanager   Tangara Matorralera

Fairly common in patches of woodland, brushy areas, and agricultural regions with hedgerows and scattered trees in arid intermontane valleys of nw. Ecuador from Carchi and Imbabura south to n. Pichincha in the Río Tumbaco valley east of Quito; has occasionally even been seen in Quito itself. The Scrub Tanager has also apparently begun to spread into cleared areas in more humid regions

(e.g., it is now regular at Hacienda La Florida, east of Apuela in Imbabura, fide P. Coopmans). Recorded from about 800 to 2500 m.
Monotypic.
Range: lower Andean slopes and valleys in w. Colombia and n. Ecuador.

Tangara cyanicollis   Blue-necked Tanager   Tangara Capuchiazul

Common in clearings, secondary woodland, and forest borders (both humid and deciduous, but essentially a nonforest tanager) in the lowlands and foothills of w. Ecuador, though absent from arid regions and mostly absent from very humid areas. Also common in the foothills and subtropical zone on the east slope of the Andes, but here not found at all in the lowlands, even at sites immediately adjacent to the Andes (there are no records, for example, from Jatun Sacha, fide B. Bochan). In the west recorded from n. Esmeraldas (Ventanas and Alto Tambo) and adjacent Imbabura (Lita area) south to w. Guayas (Cerro La Torre and above Salanguilla; N. Krabbe), and ranging in very small numbers to El Oro (Santa Rosa [AMNH] and near Piñas [RSR and PJG]); the species is not known to range south into adjacent Loja. In the west found primarily in the lowlands, locally ranging as high as 1400 m, especially in partially deforested regions (e.g., in the Mindo area). On the east slope recorded between about 500 and 1800 m.

Two well-marked races are found in Ecuador, *caeruleocephala* in the east and the endemic *cyanopygia* in the west. *Caeruleocephala* has a glistening straw rump, a more purplish throat, and glistening coppery wing-coverts; *cyanopygia* has its rump and entire hood glistening turquoise, and its wing-coverts are greener.
Range: Andes from w. Venezuela to w. Bolivia, and in more humid lowlands of w. Ecuador; Sierra de Perijá; also locally in s. Amazonian Brazil.

Tangara nigrocincta   Masked Tanager   Tangara Enmascarada

Uncommon to locally fairly common in the canopy and borders of humid forest (both terra firme and várzea) and secondary woodland in the lowlands of e. Ecuador. Recorded mostly below 600 m, though found in small numbers as high as 1000 m along their eastern base (e.g., at Shaime on the slopes of the Cordillera del Cóndor; WFVZ, MECN).
Monotypic.
Range: se. Colombia, s. Venezuela, and Guyana to n. Bolivia and w. and s. Amazonian Brazil.

Tangara larvata   Golden-hooded Tanager   Tangara Capuchidorada

Uncommon to fairly common in the borders and canopy of humid forest and secondary woodland, and in adjacent clearings and gardens, in the lowlands of nw. Ecuador south to n. Manabí and n. Los Ríos (an Oct. 1950 specimen from Quevedo; ANSP). In recent years the Golden-hooded Tanager has not been recorded south of s. Pichincha at Río Palenque. In the wet-forest zone of n. Esmeraldas the Golden-hooded Tanager ranges mainly in clearings and forest borders, not regularly entering the canopy of more continuous forest (O. Jahn et al.). Recorded up to 800 m (above Tinalandia, and above Lita in w. Imbabura), but mainly occurs below 500 m.

The species was formerly (e.g., Meyer de Schauensee 1966, 1970) called the Golden-masked Tanager, but see Isler and Isler (1987), Ridgely and Tudor (1989), and Sibley and Monroe (1990).

Trans-Andean *T. larvata* is here treated as a a species separate from *T. nigrocincta* (Masked Tanager) found east of the Andes, following most recent authors. One race of *T. larvata* occurs in Ecuador: *fanny*.

Range: s. Mexico to nw. Ecuador.

Tangara mexicana   Turquoise Tanager   Tangara Turquesa

Fairly common to common in clearings, secondary woodland, and canopy and borders of humid forest (both várzea and terra firme) in the lowlands of e. Ecuador. Although generally considered a bird of secondary habitats (e.g., Hilty and Brown 1986; Ridgely and Tudor 1989), recent field work along the Maxus road southeast of Pompeya has revealed that substantial numbers of Turquoise Tanagers occur in the canopy of extensive terra firme forest as well. Its numbers do nonetheless seem highest in disturbed areas, and the species therefore seems destined to increase in Ecuador. Recorded mostly below about 600 m, ranging in smaller numbers as high as 1000–1100 m along the eastern base of the Andes (e.g., along the Loreto road and in the Zamora/Río Nangaritza area).

One race: *boliviana*.

Range: se. Colombia and s. and e. Venezuela south to n. Bolivia and Amazonian Brazil; se. Brazil. More than one species may be involved.

Tangara velia   Opal-rumped Tanager   Tangara Lomiopalina

Uncommon in the canopy and borders of humid forest (both terra firme and várzea) and secondary woodland in the lowlands and (in reduced numbers) foothills of e. Ecuador. Recorded as high as 1075 m (Cordillera de Cutucú; Robbins et al. 1987), but occurs mostly below 600 m.

One race: *iridina*.

Range: se. Colombia, s. Venezuela, and the Guianas to n. Boliva and Amazonian Brazil; e. Brazil. More than one species may be involved.

Tangara callophrys   Opal-crowned Tanager   Tangara Cejiopalina

Uncommon to fairly common in the canopy and borders of humid forest (both terra firme and várzea) and secondary woodland in the lowlands of e. Ecuador. The Opal-crowned often occurs with the Opal-rumped Tanager (*T. velia*), even in the same flock, but in Ecuador the Opal-crowned is usually more numerous. Recorded mostly below 600 m, locally up to 750 m (seen on 8–9 Jun. 1999 northwest of Archidona along the road to Baeza; L. Navarrete and J. Moore), but more numerous away from the Andes.

Monotypic.

Range: se. Colombia, e. Ecuador, e. Peru, nw. Bolivia, and w. Amazonian Brazil.

Tangara chilensis   Paradise Tanager   Tangara Paraíso

Fairly common to common in the canopy (especially) and borders of humid forest (both terra firme and várzea), secondary woodland, and clearings with scattered trees in the lowlands and foothills of e. Ecuador. The gaudy Paradise Tanager is generally most numerous along the base of the Andes. Occurs mostly below 1200 m, ranging in small numbers up to about 1700 m (Cordillera de Cutucú; Robbins et al. 1987).

One race: nominate *chilensis*.

Range: se. Colombia, s. Venezuela, and the Guianas to n. Bolivia and Amazonian Brazil (but absent from its eastern and part of its central sections).

Tangara schrankii   Green-and-gold Tanager   Tangara Verdidorada

Fairly common to common in humid forest (both terra firme and várzea), secondary woodland, and clearings with scattered trees in the lowlands and foothills of e. Ecuador. Unlike most other *Tangara*, the Green-and-gold Tanager not only ranges in the canopy and borders of forest but also regularly accompanies understory flocks inside forest and woodland. Ranges mostly below 1100 m, in small numbers as high as 1500–1600 m (e.g., at Palm Peak [MCZ] and Panguri [T. J. Davis]).

One race: nominate *schrankii*.

Range: se. Colombia and s. Venezuela to nw. Bolivia and w. Amazonian Brazil.

Tangara johannae   Blue-whiskered Tanager   Tangara Bigotiazul

Rare to locally fairly common in the borders and canopy of very humid and humid forest, secondary woodland, and adjacent clearings in the lowlands of nw. Ecuador, where recorded primarily from Esmeraldas and adjacent Imbabura (an old record from Hacienda Paramba); the species is frequently encountered in the Playa de Oro area (O. Jahn et al.). Farther south the Blue-whiskered Tanager is known only from a few sites in Pichincha: there were a few sightings in the 1980s at Río Palenque (PJG et al.); a male was taken (date?) at Santo Domingo de los Colorados (Orcés 1944); and since 1995 there have been numerous sightings along the road north of Simón Bolívar near Pedro Vicente Maldonado and 5 km south of Golondrinas (v.o.). The species has also been seen and collected (MECN) in adjacent sw. Imbabura (near the Salto del Tigre bridge over the Río Guaillabamba in Jul. 1997). The southernmost record involves a female (ANSP) taken by T. Mena on 16 Oct. 1950 at Quevedo in n.

Los Ríos. The present status of the Blue-whiskered Tanager south of nw. Pichincha is uncertain, and it may now be nearly extirpated from this region by the near-total deforestation that has taken place there. Recorded mostly below about 500 m, but once found as high as 900 m at Lita.

The Blue-whiskered Tanager was given Near-threatened status by Collar et al. (1992, 1994). The species seems relatively scarce throughout its limited range, and although it may be capable of persisting in partially deforested terrain, its precise ecological requirements remain poorly understood. Given the extensive deforestation that has occurred in much of its Ecuadorian range, and that it is not known to occur in any protected area, we accord the species Vulnerable status in Ecuador.

Monotypic.

Range: w. Colombia and nw. Ecuador.

Tangara punctata   Spotted Tanager   Tangara Punteada

Fairly common in the canopy and borders of montane forest and adjacent clearings in the foothills and lower subtropical zone on the east slope of the Andes from w. Sucumbíos (San Rafael Falls; ANSP and many sightings) south to Zamora-Chinchipe (Zamora area and Podocarpus National Park). The Spotted Tanager seems likely to occur northward into adjacent Colombia, though as yet it is unrecorded from there (Hilty and Brown 1986). It is one of the more numerous *Tangara* north of Archidona along the first part of the road to Loreto. Recorded mostly from 900 to 1500 m, in smaller numbers up to 1700 m or even a little higher.

One race: *zamorae*.

Range: east slope of Andes from n. Ecuador to w. Bolivia; s. Venezuela, the Guianas, and n. Amazonian Brazil.

Tangara xanthogastra   Yellow-bellied Tanager   Tangara Ventriamarilla

Uncommon to locally fairly common in the canopy and borders of humid forest (both terra firme and várzea), secondary woodland, and adjacent shrubby clearings in the lowlands and foothills of e. Ecuador. Recorded mainly below about 1100 m, rarely or locally as high as 1350 m at the head of the Río Guataracu (MCZ).

One race: nominate *xanthogastra*.

Range: se. Colombia and s. Venezuela to w. Bolivia and w. Amazonian Brazil.

Tangara gyrola   Bay-headed Tanager   Tangara Cabecibaya

Uncommon to fairly common in the canopy and borders of humid forest (in the east in both terra firme and várzea, more numerous in the former), secondary woodland, and adjacent clearings in the lowlands and foothills of both e. and w. Ecuador. In the west ranges south to nw. Guayas, and in the foothills south through e. Guayas, El Oro, and w. Loja (Alamor and El Limo area); in the wet-forest belt of n. Esmeraldas largely replaced by the Rufous-winged Tanager (*T. lavinia*; O. Jahn et al.). The Bay-headed Tanager is rather uncommon in the eastern lowlands, and it becomes decidedly scarce in the lowlands well to the east of the Andes where a few have been recorded as far east as along the Maxus road southeast of Pompeya and at La Selva. There are also three specimens (ANSP) from the far southeast that were taken in Jul.–Aug. 1949 by R. Olalla at Montalvo and "Churo Yaco" as well as recent sightings from Kapawi Lodge. On the west slope recorded up to about 1500 m, on the east slope up to 1700 m.

Apparently three races occur in Ecuador: *nupera* occurs throughout the species' western range; *catharinae* is found along the east slope of the Andes and out into the adjacent lowlands; and birds that appear to be referable to the very similar (but slightly smaller) *parva* occur in the far southeast. The two eastern races differ from *nupera* in having a conspicuous golden nuchal band and shoulders.

Range: Nicaragua to extreme nw. Peru, n. Bolivia, and Amazonian Brazil (but local in Amazonia).

Tangara lavinia   **Rufous-winged Tanager**   Tangara Alirrufa

Uncommon to locally common in the canopy and borders of humid forest and secondary woodland in the lowlands of nw. Ecuador where known primarily from n. Esmeraldas. In nw. Pichincha this species appears to be distinctly scarcer than farther north, being known only from a few recent sightings, the first being a bird seen near San Miguel de los Bancos on 31 Aug. 1994 (P. Coopmans et al.). The Rufous-winged Tanager can be locally numerous: in Jul. 1990 it was found to be the most common *Tangara* at a camp about 20 km northwest of Alto Tambo (ANSP, MECN), and it also is the most numerous *Tangara* in the Playa de Oro region between about 50 and 200 m (O. Jahn et al.). Recorded up to about 750 m in Pichincha.

One race: nominate *lavinia*.
Range: Honduras to nw. Ecuador.

Iridosornis porphyrocephala   **Purplish-mantled**   Tangara
                                        Tanager   Dorsipurpurina

Very rare in the lower growth of montane forest in the subtropical zone on the west slope of the Andes in n. Ecuador in Carchi (where seen by N. Krabbe along the Maldonado road in Mar. 1983) and Imbabura (19th-century specimen from Intag [1200 m, but probably taken somewhat higher]). Somewhat surprisingly, the species was not encountered during a lengthy expedition in Jul.–Aug. 1988 by ANSP personnel to this region. There also have been a few recent sightings in the Tandayapa area of nw. Pichincha, first on 7 Mar. 1996 (M. Lysinger). The Purplish-mantled Tanager is also known from a single specimen that supposedly was taken in Loja at "El Tambo" (Orcés 1944); as Paynter (1993) suggests, this controversial locality seems more likely actually to be along the Guayaquil-Quito railway in Cañar, but it has also been thought (N. Krabbe) perhaps to be situated in Loja on the slopes of Urutisinga above the town itself. Ridgely and Tudor (1989, p. 289) considered the specimen "uncertain" because of its improbable derivation, seemingly so far away from the west slope of the Andes; we now suspect that Paynter (1993) is right and that it was actually obtained in Cañar. We regard the reported sighting from the Loja-Zamora road (Rasmussen et al. 1994) to be uncorroborated. In Colombia recorded between about 1500 and 2400 m.

The Purplish-mantled Tanager was given Near-threatened status by Collar et al. (1992, 1994). Given its obvious rarity in Ecuador—it appears to be more numerous in Colombia—and that it is not known to occur within any protected area, we accord it Vulnerable status in Ecuador.

Monotypic.

Range: locally in Andes of w. Colombia and nw. Ecuador.

Iridosornis analis   Yellow-throated Tanager   Tangara Goliamarilla

Rare to fairly common but apparently local in the lower growth of montane forest and forest borders in the subtropical zone on the east slope of the Andes from Sucumbíos (San Rafael Falls area) south through Zamora-Chinchipe (Podocarpus National Park; Cordillera del Cóndor [ANSP]). The Yellow-throated Tanager seems somewhat more numerous and widespread in s. Ecuador than it is northward, but even there numbers may be highest on outlying ridges such as the Cordilleras de Cutucú and del Cóndor. Recorded between about 1400 and 2300 m.

Monotypic.

Range: Andes from s. Colombia to s. Peru.

Iridosornis rufivertex   Golden-crowned   Tangara Coronidorada
                             Tanager

Uncommon to locally fairly common in the lower growth of montane forest, forest borders, and shrubbery at or near treeline and down into the temperate zone on the east slope of the Andes. On the west slope not recorded south of Pichincha (upper Chiriboga road). Recorded mainly from 2500 to 3300 m, but locally occurs down to 2250 m on the Cordillera de Huacamayos (N. Krabbe) and up to 3550 m (at Cerro Mongus; Robbins et al. 1994).

Two very similar subspecies occur in Ecuador, *subsimilis* in the northwest and nominate *rufivertex* on the east slope. Fjeldså and Krabbe (1990) suggest that *subsimilis* may not be worthy of recognition; we agree.

Range: Andes from sw. Venezuela to extreme n. Peru.

The Yellow-scarfed Tanager (*I. reinhardti*) of the Andes of Peru and w. Bolivia was reported seen on two occasions in 1989 in humid montane forest in the Cajanuma sector of Podocarpus National Park (Bloch et al. 1991; Rasmussen et al. 1994). None have been seen there since, however, and the observers no longer consider (pers. comm. to RSR) their sightings to be unequivocal, and that aberrant Golden-crowned Tanagers possibly were involved. That species occurs commonly in the same area; even farther south along the main ridge of the Andes (e.g., on the Cordillera Las Lagunillas [ANSP] and at Cerro Chinguela in adjacent Peru [Parker et al. 1985]) only the Golden-crowned Tanager has ever been found, with confirmed records of the Yellow-scarfed only from south and east of the Río Marañón in n. Peru. At least on the basis of presently available evidence, the Yellow-scarfed Tanager cannot be considered a member of Ecuador's avifauna.

| *Anisognathus igniventris* | Scarlet-bellied Mountain-Tanager | Tangara Montana Ventriescarlata |

Fairly common to common in montane woodland and forest borders, and in hedgerows and patches of woodland in partially agricultural terrain, in the temperate zone on both slopes of the Andes and on slopes above the central and interandean valleys, locally ranging up into patches of *Polylepis*-dominated woodland near or even slightly above treeline. On the west slope recorded south to El Oro (Taraguacocha). Recorded mainly from 2500 to 3500 m, but small numbers occur locally as high as 3800–3900 m (e.g., near Papallacta pass).
One race: *erythrotus*.
Range: Andes from sw. Venezuela to w. Bolivia.

| *Anisognathus lacrymosus* | Lacrimose Mountain-Tanager | Tangara Montana Lagrimosa |

Fairly common to common in montane forest, forest borders, and secondary woodland in the upper subtropical and temperate zones on the east slope of the Andes. In s. Ecuador the Lacrimose Mountain-Tanager also occurs on the west slope in Azuay (sightings from above Girón in Jul. 1978 [RSR and D. Wilcove] and from Sural and Chaucha [N. Krabbe]) and at Taraguacocha in El Oro (AMNH). There are also a few specimens labeled as having been taken on the west slope in n. Ecuador (e.g., from Tambillo and "Monji"; J. T. Zimmer, *Am. Mus. Novitates* 1262, 1944), but as there have been no recent reports of this conspicuous tanager from any west-slope localities in this region, we consider these specimens likely to have been mislabeled. Recorded mainly from 2300 to 3200 m, in small numbers somewhat higher and lower (e.g., near the crest of the Cordillera de Huacamayos down to 2100 m).
Two very similar subspecies occur in Ecuador, *palpebrosus* in most of the species' Ecuadorian range and *caerulescens* in the far south in El Oro, Loja, and adjacent Zamora-Chinchipe.
Range: Andes from w. Venezuela to cen. Peru; Sierra de Perijá.

Anisognathus somptuosus   Blue-winged   Tangara Montana
                                Mountain-Tanager   Aliazul

Fairly common to common in montane forest, forest borders, secondary woodland, and adjacent clearings in the subtropical and lower temperate zones on
both slopes of the Andes. On the west slope found south to w. Loja in the
Alamor and Celica region, where at least now the species is comparatively
uncommon because of habitat loss. Recorded mostly from 1200 to 2500 m,
locally higher.

Four races are recorded from Ecuador, of which the most divergent is
baezae, which occurs on the east slope as far south as Morona-Santiago, and
which has a moss green back. The back is black in all other Ecuadorian races,
and these three are all generally similar: nominate *somptuosus* occurs on the
east slope in Zamora-Chinchipe, *cyanopterus* on the west slope as far south
as Chimborazo and ne. Azuay (an old record from Yerbebuena; Salvadori
and Festa 1899a), and *alamoris* in s. Azuay, El Oro, and w. Loja. *Cyanopterus*
differs in having its flight-feather edging a cobalt blue similar to the shoulder
patch; in the nominate race and *alamoris*, this flight-feather edging is paler,
more turquoise blue. *Alamoris* may not deserve recognition, and should
probably be synonomized into nominate *somptuosus*. Contra Meyer de
Schauensee (1966, 1970) and many other authors, on the basis of priority the
correct species name is *somptuosus* and not *flavinuchus* (Sibley and Monroe
1990).

    Range: mountains of n. Venezuela; Andes from sw. Venezuela to s.
Bolivia.

Anisognathus notabilis   Black-chinned   Tangara Montana
                                Mountain-Tanager   Barbinegra

Uncommon to locally fairly common in the borders and canopy of montane
forest in the foothills and subtropical zone on the west slope of the Andes
south to Pichincha (lower Chiriboga road), and also found recently in El Oro
(e.g., at Buenaventura; Robbins and Ridgely 1990); it is not known whether
the species ranges in the intervening area. There is also a single record of a
specimen supposedly taken in Loja at "El Tambo" (Orcés 1944); this site
remains controversial, but Paynter (1993) considers that it refers to a town of
that name along the Guayaquil-Quito railway in Cañar. Although it has
also been suggested (N. Krabbe) that the site might be situated in Loja on the
slopes of Urutisinga above the town itself, this seems biogeographically improbable given this mountain-tanager's known distribution. We believe that mislabeling was involved and that the collecting site was likely in Cañar. Recorded
mostly from 1400 to 2200 m, but in El Oro found between about 800 and
1100 m.

    Monotypic.

    Range: Andes of w. Colombia and w. Ecuador.

Buthraupis montana   Hooded Mountain-   Tangara Montana
                         Tanager   Encapuchada

Fairly common in the canopy and borders of montane forest, forest borders, and adjacent clearings in the upper subtropical and temperate zones on both slopes of the Andes. On the west slope recorded in the past south to Chimborazo (Matus), though there are no recent reports from south of w. Bolívar (Salinas; N. Krabbe et al.). Recorded mostly from 1900 to 3200 m.

One race: *cucullata*.

Range: Andes from sw. Venezuela to w. Bolivia; Sierra de Perijá.

Buthraupis wetmorei   Masked Mountain-   Tangara Montana
                          Tanager   Enmascarada

Rare to uncommon and seemingly local (but perhaps under-recorded, much of its range being remote and difficult to access) in low woodland at and just below treeline on the east slope of the Andes. Until recently thought to be a very rare bird, known only from the type locality of Sangay National Park in sw. Morona-Santiago, the Masked Mountain-Tanager in recent years has been found at a number of additional sites, and it now appears to occur wherever suitable treeline habitat exists. From north to south, these recent sites are: Guandera Biological Reserve in e. Carchi (Creswell et al. 1999), the west slope of Cerro Mongus in e. Carchi (ANSP, MECN; Robbins et al. 1994); the east side of Papallacta pass and in the Cordillera de los Llanganates in w. Napo (MECN); along the Gualaceo-Limón road and on the Páramos de Matanga in w. Morona-Santiago; in Podocarpus National Park (Cajanuma sector) in e. Loja (Rasmussen et al. 1994, 1996) and at nearby Cerro Toledo; and on the east slope of Cordillera Las Lagunillas in s. Zamora-Chinchipe (ANSP, MECN). Recorded between about 2950 and 3600 m.

The Masked Mountain-Tanager was given Vulnerable status by Collar et al. (1992, 1994), and we agree with this assessment. The primary threat to the species is the burning of paramo grasslands, which gradually reduces the extent of the shrubby ecotonal treeline habitat it requires.

Monotypic.

Range: locally in Andes from s. Colombia to extreme n. Peru.

Buthraupis eximia   Black-chested   Tangara Montana
                        Mountain-Tanager Pechinegra

Rare to locally uncommon in montane forest and forest borders and in low woodland near treeline in the temperate zone on both slopes of the Andes. On the west slope known south to Chimborazo (19th-century record from Matus), but there are recent records south only to Pichincha. Recorded mostly from about 2750 to 3300 m, locally as high as 3700 m (on Volcán Pichincha).

Two very similar races occur in Ecuador, *chloronota* in most of the species' Ecuadorian range (both slopes) and *cyanocalyptra* in Morona-Santiago,

Zamora-Chinchipe, and Loja. *Cyanocalyptra* should perhaps be synonymized into *chloronota*, as noted by Fjeldså and Krabbe (1990).
Range: Andes from sw. Venezuela to extreme n. Peru.

Bangsia rothschildi   Golden-chested   Tangara Pechidorada
                          Tanager

Rare to uncommon in the mid-levels and subcanopy of very humid forest in a narrow elevational zone in the lower foothills of nw. Ecuador, where recorded only from n. Esmeraldas and sw. Imbabura (seen in Jul. 1997 near the Salto del Tigre bridge over the Río Guaillabamba; P. Coopmans et al.). The Golden-chested Tanager occurs at elevations somewhat below those of the Moss-backed Tanager (*B. edwardsi*); in recent years small numbers of the Golden-chested Tanager have been found northwest of Alto Tambo (ANSP; N. Krabbe), but the species has not been found at the slightly higher elevations (600–700 m) of El Placer. At Playa de Oro the Golden-chested Tanager is rare at 100 m, most numerous between 250 and 350 m, and has already been replaced by the Moss-backed Tanager at 450 m (O. Jahn et al.). Recorded between 100 and 600 m in Ecuador, though in Colombia known from as high as 1100 m (Hilty and Brown 1986).

Given its limited range (in much of which deforestation is proceeding rapidly) and its evident rarity, we accord the Golden-chested Tanager Near-threatened status in Ecuador. It was not considered to be at risk by Collar et al. (1992, 1994).

This and the following species are sometimes (e.g., *Birds of the World*, vol. 13) placed in the genus *Buthraupis*, but we follow Ridgely and Tudor (1989) and Sibley and Monroe (1990) in maintaining them in the genus *Bangsia*. Monotypic.
Range: west slope of Andes in w. Colombia and nw. Ecuador.

Bangsia edwardsi   Moss-backed Tanager   Tangara Dorsimusgosa

Locally uncommon to common in montane forest, forest borders, and secondary woodland in foothills on the west slope of the Andes in nw. Ecuador south to nw. Pichincha in the San Miguel de los Bancos area. Although decidedly less numerous southward, at El Placer in e. Esmeraldas the Moss-backed Tanager is one of the most frequently seen birds. Recorded mostly from 500 m (Playa de Oro in n. Esmeraldas; O. Jahn et al.) to 1100 m, locally as high as 1700 m in Carchi (ANSP).
Monotypic.
Range: west slope of Andes in sw. Colombia and nw. Ecuador.

*Wetmorethraupis sterrhopteron   Orange-throated   Tangara
                                       Tanager   Golinaranja

Rare to uncommon and very local in the canopy and borders of montane forest on the west slope of the southern Cordillera del Cóndor in se. Zamora-

Chinchipe. The striking Orange-throated Tanager was until recently known only from adjacent Peru. In Ecuador it is known only from a series of four specimens (WFVZ, MECN) which were obtained on 27 Jul. and 3 Aug. 1990 at Shaime in the upper Río Nangaritza valley (see Marín et al. 1992). In Jun. 1993 the species could not, however, be located at comparable elevations slightly to the north on the Cordillera del Cóndor between Paquisha and Chinapinza (RSR and F. Sornoza). The species' absence from here may have been the result of deforestation at appropriate elevations; in 1978 this area still supported extensive and very luxuriant forest (RSR and D. Wilcove). The species was also seen in Jul. 1993 at Miazi, just to the north (Schulenberg and Awbrey 1997). Recorded at about 900–1000 m.

The Orange-throated Tanager was accorded Vulnerable status by Collar et al. (1992) but was upgraded to Endangered by Collar et al. (1994). Given the widespread deforestation that has evidently occurred at elevations suitable for the species, we are in accord with the latter assessment and rank the species as Endangered. None of its potential range as yet receives any protection.

Monotypic.

Range: locally near base of Andes in extreme se. Ecuador and extreme ne. Peru.

| *Dubusia taeniata* | Buff-breasted | Tangara-Montana |
| | Mountain-Tanager | Pechianteada |

Uncommon to locally fairly common (recorded especially by voice) at borders and in lower growth of montane forest and secondary woodland and in adjacent clearings, sometimes even montane scrub, in the temperate zone on both slopes of the Andes and locally on slopes above the central and interandean valleys. On the west slope recorded south to El Oro at Taraguacocha. Recorded mostly from 2500 to 3500 m.

One race: nominate *taeniata*.

Range: Andes from w. Venezuela to s. Peru; Sierra de Perijá and Santa Marta Mountains.

| *Chlorornis riefferii* | Grass-green Tanager | Tangara Carirroja |

Fairly common in the subcanopy and borders of montane forest and forest borders in the upper subtropical and temperate zones on both slopes of the Andes. On the west slope recorded south at least formerly to se. Chimborazo (old specimens from Chaguarpata and Llagos), but there are no recent records of the Grass-green Tanager from south of w. Cotopaxi. Recorded mostly between 2000 and 2900 m, locally in the northwest as high as 3150 m at Intag in Imbabura and 3500 on Volcán Corazón in Pichincha (Krabbe et al. 1997).

One race: nominate *riefferii*.

Range: Andes from n. Colombia to w. Bolivia.

Creurgops verticalis   Rufous-crested Tanager   Tangara Crestirrufa

Uncommon to locally fairly common in the canopy and borders of montane forest in the subtropical and lower temperate zones on the east slope of the Andes. Recently found also on the west slope of the Andes in the extreme north, where known only from a male taken and another seen along the Maldonado road in Carchi on 28–29 Mar. 1984 (Fjeldså and Krabbe 1986). Recorded mostly from 1500 to 2500 m.
Monotypic.
Range: locally in Andes from sw. Venezuela to cen. Peru.

Tersina viridis   Swallow Tanager   Tersina

Locally and irregularly fairly common to common in humid forest borders, secondary woodland, and (especially) clearings in the foothills and lower subtropical zone on both slopes of the Andes, and in the lowlands adjacent to the Andes on both slopes; however, it seems unaccountably scarce in the southeast (e.g., despite several old specimen records, there are virtually no recent reports from the seemingly suitable Zamora region). In w. Ecuador recorded south in substantial numbers to El Oro (e.g., around Buenaventura) and at least occasionally (or formerly?) as far as w. Loja (old records from Punta Santa Ana, Cebollal, and Alamor [Chapman 1926]; no recent reports). The Swallow Tanager's numbers seem to vary seasonally in many or all areas, and it apparently has a propensity to wander; it may even be truly migratory. Numbers at Jatun Sacha, for example, appear to be highest from Oct. to May, though the species is scarce or absent during other months (B. Bochan); this also is the basic pattern along the Maxus road southeast of Pompeya (RSR et al.). A few even wander (?) to the far east, such as the six seen along the Río Lagarto near Garzacocha on 7 Dec. 1990 (RSR and PJG et al.) and others reported at Kapawi Lodge in Sep.–Oct. 1996 (J. Moore and G. Ribadeneira). Recorded up to about 1400 m.
Formerly (e.g., Meyer de Schauensee 1966, 1970; Hilty and Brown 1986) usually treated as constituting a monotypic family (Tersinidae), the Swallow Tanager is now considered as a typical tanager (Thraupidae), apparently closest to the genus *Thraupis*. One race occurs in Ecuador, *occidentalis*.
Range: e. Panama to sw. Ecuador, n. and e. Bolivia, Paraguay, ne. Argentina, and se. Brazil (but absent from much of e. Amazonia and ne. Brazil).

Thraupis episcopus   Blue-gray Tanager   Tangara Azuleja

Uncommon to very common in a variety of habitats ranging from forest borders (even the canopy of várzea forest) and secondary woodland to clearings and gardens and even mostly urban areas in the lowlands of both e. and w. Ecuador, but considerably more numerous and widespread in w. Ecuador than in most parts of the east. Ranges up in reduced numbers on both slopes of the Andes to the lower subtropical zone (in humid regions found mainly in cleared areas).

Small numbers of Blue-gray Tanagers, probably escaped cagebirds, have also been found locally in the central valley (e.g., in the Quito area, near Salcedo, and Cuenca). Found mostly below 1500 m, ranging higher in smaller numbers (locally up to about 2500 m), probably spreading upward as a result of deforestation.

Three races are known from Ecuador: *quaesita* west of the Andes and *coelestis* and *caerulea* east of the Andes, *caerulea* occurring only in Zamora-Chinchipe. *Quaesita* has essentially plain bluish wings, whereas both of the similar eastern races have conspicuous whitish lesser wing-coverts (though the wings are plainer in juvenile birds). In the Río Marañón drainage (where the species ranges up to around Valladolid) the situation seems to be more complex: here most birds show white on the wing-coverts, but others do not, and still others appear to be intermediate in this respect. More study is needed, but presumably west-slope (plain-winged) birds are spreading up from the dry upper Río Marañón valley in nw. Peru and coming into contact with east-slope (white-winged) birds.

Range: e. Mexico to w. Peru, nw. Bolivia, and Amazonian Brazil.

Thraupis palmarum   **Palm Tanager**   **Tangara Palmera**

Common in gardens, clearings, and borders and (in smaller numbers) the canopy of humid forest and woodland, especially where palms are prevalent, in the lowlands and foothills of both e. and w. Ecuador. In the west found south to w. Loja in the Alamor region. Recorded mostly below 1300 m, but occasional (wandering?) birds have been seen higher on Andean slopes (once as high as 1900 m; P. Coopmans).

Two races are found in Ecuador, *melanoptera* in the east and *violilavata* in the west. The latter is glossier generally and has less contrasting black on the wings.

Range: Honduras to sw. Ecuador, n. and e. Bolivia, n. Paraguay, and se. Brazil.

Thraupis cyanocephala   **Blue-capped Tanager**   **Tangara Gorriazul**

Uncommon to fairly common in the borders of montane forest, secondary woodland, and clearings in the subtropical and temperate zones on both slopes of the Andes, and locally on slopes above the central and interandean valleys. On the west slope recorded south through w. Loja in the Alamor/Celica region, and at Sozoranga and Utuana (Best et al. 1993; RSR). In general the Blue-capped Tanager does not seem to be as numerous in Ecuador as it is in many other parts of its range. Recorded mostly from 1800 to 2900 m.

One race: nominate *cyanocephala*.

Range: mountains of n. Venezuela, and Andes from w. Venezuela to w. Bolivia; Sierra de Perijá and Santa Marta Mountains.

Thraupis bonariensis   Blue-and-yellow   Tangara Azuliamarilla
                           Tanager

Uncommon to locally fairly common in lighter woodland, scrub, gardens, and hedgerows in agricultural regions in more arid portions of the central valley from Carchi (a pair seen near Mira on 15 Dec. 1995; P. Coopmans) and Imbabura (near Tumbabiro; ANSP) south to s. Azuay (Oña area; RSR) and extreme n. Loja (near Papaya in the Río Paquishapa valley; N. Krabbe). There are very few reports of the Blue-and-yellow Tanager from the outside slopes of the Andes, though it would seem that the species could spread into such areas as these become increasingly deforested; a pair was, however, seen near Papallacta in Napo on 1 Feb. 1990 (M. B. Robbins, J. C. Matheus, and F. Sornoza). The Blue-and-yellow Tanager remains unrecorded from Colombia (Hilty and Brown 1986), but as the species occurs so close to the border, it might be expected to at least occasionally wander north into Colombian territory. Recorded mostly from 1800 to 3000 m.

One race: *darwinii*.

Range: Andes from n. Ecuador to n. Chile and w. Bolivia, also east in lowlands across Argentina to Uruguay and se. Brazil, spreading north into Paraguay during austral winter. More than one species may be involved.

Ramphocelus carbo   Silver-beaked Tanager   Tangara Concha de Vino

Common to very common in shrubby clearings, gardens, secondary woodland, and forest borders (especially in várzea) in the lowlands and—in smaller numbers—up into the foothills of e. Ecuador. Overall numbers of this numerous tanager are doubtless increasing with the gradual spread of settled areas. Recorded mostly below about 1100 m, but ranging locally as high as 1300 m in w. Napo, where it is following recently created clearings upward.

One race: nominate *carbo*.

Range: e. Colombia, much of Venezuela, and the Guianas to n. and e. Bolivia and s. Brazil.

Ramphocelus nigrogularis   Masked Crimson   Tangara
                               Tanager   Enmascarada

Fairly common in humid forest, secondary woodland, and forest borders in the lowlands of e. Ecuador. The Masked Crimson Tanager is generally found near water and is always most numerous in such areas; in regions with extensive terra firme forest, it is found almost exclusively along rivers and streams. Ranges up to about 600 m (e.g., around Archidona, and at Tayuntza in Morona-Santiago [WFVZ]).

Monotypic.

Range: se. Colombia to nw. Bolivia and w. Amazonian Brazil.

Ramphocelus icteronotus Lemon-rumped Tangara Lomilimón
           Tanager

Common to very common and conspicuous in shrubby clearings, gardens, and forest and woodland borders in the more humid lowlands and foothills of w. Ecuador south to nw. Guayas (hills south of the Río Ayampe [RSR] and at Loma Alta [Becker and López-lanús 1997]), El Oro, and w. Loja (various localities in the Alamor region and below Celica). Recorded mostly below about 1600 m, locally and in small numbers a little higher (e.g., in w. Loja).

We follow Hilty and Brown (1986) in calling this species the Lemon-rumped Tanager; in most recent literature it is called the Yellow-rumped Tanager.

R. icteronotus of Panama to sw. Ecuador is sometimes regarded as conspecific with *R. flammigerus* (Flame-rumped Tanager) of Colombia's Río Cauca valley (e.g., *Birds of the World*, vol. 13; Ridgely and Tudor 1989; Sibley and Monroe 1990). Meyer de Schauensee (1966, 1970) and Hilty and Brown (1986) considered them separate species. The differences between *R. icteronotus* and *R. flammigerus* are comparable to those between two Middle American *Ramphocelus* taxa (*R. passerinii* [Passerini's Tanager] and *R. costaricensis* [Cherrie's Tanager]) presently considered to be distinct species (S. J. Hackett, *Mol. Gen. Evol.* 5: 368–382, 1996; AOU 1998), and we thus believe it appropriate also to recognize *icteronotus* as a full species. Monotypic.

Range: Panama to sw. Ecuador.

Calochaetes coccineus Vermilion Tanager Tangara Bermellón

Uncommon to locally fairly common in the canopy and borders of montane forest in the subtropical zone on the east slope of the Andes. The Vermilion Tanager is regularly found near San Rafael Falls and in the Romerillos area of Podocarpus National Park. Recorded mostly between about 1100 and 1800 m.

Monotypic.

Range: east slope of Andes from s. Colombia to cen. Peru.

Piranga lutea Highland Hepatic-Tanager Piranga Bermeja Montañero

Uncommon to fairly common and somewhat local in deciduous and semihumid woodland, forest, and adjacent clearings mainly—but not entirely—in more arid regions in the lowlands of w. Ecuador, and up on the west slope of the Andes well into the subtropical zone. Recorded from w. Esmeraldas (Esmeraldas [city]) and Manabí (Bahía de Caráquez area) south through El Oro and w. Loja. Also recently found to occur—apparently only locally or in very small numbers—in the canopy and borders of montane forest and in clearings in the foothills along the eastern base of the Andes in s. Ecuador, where there are reports from Morona-Santiago (a pair seen at Indanza on 3 Aug. 1991; PJG et al.) and Zamora-Chinchipe (2 males seen near Mayacu on the west slope of the Cordillera del Cóndor on 14 Jun. 1993 [RSR] and several reports from the

Zamora region along the lower Loja-Zamora road and in the Bombuscaro sector of Podocarpus National Park [D. Wolf et al.]). It is possible that the Highland Hepatic-Tanager may be increasing in se. Ecuador as a result of deforestation. In w. Ecuador recorded up to about 1900 m (especially southward), locally as high as 2400 m at Utuana in s. Loja (Sep. 1998; RSR et al.); in the southeast recorded between about 1000 and 1350 m.

We follow K. J. Burns (*Auk* 115[3]: 621–634, 1998) in considering birds found from Costa Rica south through n. and w. South America, primarily in the highlands, as a species, *P. lutea* (Highland Hepatic-Tanager), separate from lowland-inhabiting *P. flava* (Lowland Hepatic-Tanager) of e. and s. South America. *P. hepatica* of the Middle American highlands is also regarded as a separate species (Northern Hepatic-Tanager). One race of *P. lutea* occurs in Ecuador, nominate *lutea*.

Range: mountains of n. Venezuela, and Andes from w. Venezuela to w. Bolivia, also in lowlands of w. Ecuador and nw. Peru; tepuis of Venezuela and mountains of the Guianas.

Piranga rubra   Summer Tanager   Piranga Roja

An uncommon to fairly common boreal winter resident in humid forest borders, secondary woodland, clearings with scattered trees, and gardens in the lowlands of both e. and w. Ecuador, ranging in smaller numbers up into the subtropical zone on both slopes of the Andes; there are also several records from the central valley in the Quito region. In the eastern lowlands Summer Tanagers have been recorded primarily in areas fairly close to the base of the Andes, with numbers becoming much smaller in the far east, where they may occur primarily as transients. Recorded from early Oct. to early Apr. Recorded mostly below 1500 m, but ranges at least occasionally as high as 2700–2800 m in the Quito area (where it is perhaps mainly a transient).

One race: nominate *rubra*.

Range: breeds in United States and n. Mexico, wintering south to n. South America.

Piranga olivacea   Scarlet Tanager   Piranga Escarlata

An uncommon to occasionally fairly common transient and less numerous boreal winter resident in the canopy and borders of humid forest, secondary woodland, plantations, and clearings in the lowlands of e. and w. Ecuador, ranging up in smaller numbers into the lower subtropical zone on both slopes of the Andes. The Scarlet Tanager is much more numerous in e. Ecuador (especially along the eastern base of the Andes) than it ever is in the west, in the latter—where perhaps primarily occurring as a transient—being known from only a scatter of records south to e. Guayas and Chimborazo in the Chimbo/Pallatanga area. Recorded from Oct. to early Apr. Recorded mostly below 1500 m, occasionally ranging higher (to 2500 m or more), especially on migration.

Monotypic.
Range: breeds in e. North America, wintering mainly in w. Amazonia.

Piranga leucoptera   **White-winged Tanager**   **Piranga Aliblanca**

Uncommon to locally fairly common in the canopy and borders of montane forest and in adjacent clearings in the foothills and lower subtropical zone on both slopes of the Andes. On the west slope recorded south to El Oro (an old Zaruma specimen; also many recent reports from Buenaventura) and w. Loja (a Feb. 1991 sighting from Tierra Colorada near Alamor; Best et al. 1993). On the east slope recorded from w. Napo (a few sightings from along the Loreto road north of Archidona; v.o.) and Tungurahua (Mapoto) southward. Recorded mostly from 800 to 1800 m, in the southwest locally down to 600 m.

The species is sometimes placed in the genus *Spermagra* (Howell and Webb 1995). One race occurs in Ecuador, *ardens*.

Range: highlands from e. Mexico to w. Panama; mountains of n. Venezuela, tepui slopes of s. Venezuela and adjacent Brazil, and Andes from w. Venezuela to w. Bolivia; Sierra de Perijá.

Piranga rubriceps   **Red-hooded Tanager**   **Piranga Capuchirroja**

Rare to locally uncommon in the canopy and borders of montane forest in the upper subtropical and temperate zones on the east slope of the Andes, and on the west slope in the north (south to Pichincha). The Red-hooded Tanager is scarce on the west slope, where it was long known only from two old specimens taken at Mindo (Lönnberg and Rendahl 1922); more recently it has been recorded from Carchi (specimens [ANSP, MECN] taken along the Maldonado road in Jul. 1988) as well as Pichincha (a few sightings from the Tandayapa region [R. Johnson, P. Coopmans et al.]). Recorded mostly from 2200 to 3000 m.

Monotypic.
Range: Andes from n. Colombia to n. Peru.

Chlorothraupis olivacea   **Lemon-spectacled**   **Tangara Ojeralimón**
                                  Tanager

Fairly common in the lower and middle growth and borders of very humid forest in the lowlands and lower foothills of nw. Ecuador in Esmeraldas, where it reaches its southwesternmost limit along the Río Verde south of Chontaduro (ANSP). Although it has a limited range in Ecuador, the Lemon-spectacled Tanager can be quite numerous northward toward the Colombian border: in Jul. 1989 it was, for instance, found to be common northwest of Alto Tambo (ANSP, MECN), and it is also fairly common at Playa de Oro (O. Jahn and P. Mena Valenzuela et al.). It appears, however, to be decidedly less numerous at the southern edge of its range: in Jun. 1992, along the Río Verde south of Chontaduro, only one was seen with another bird mist-netted (T. J. Davis; ANSP). In Ecuador recorded only below about 450 m, but in Colombia has been found as high as 1500 m (Hilty and Brown 1986); at Playa de Oro most numerous between 50 and 200 m (O. Jahn et al.).

The species was formerly called the Lemon-browed Tanager, but see Isler and Isler (1987), Ridgely and Gwynne (1989), and Ridgely and Tudor (1989). Monotypic.

Range: e. Panama to nw. Ecuador.

*Chlorothraupis frenata   Olive Tanager   Tangara Oliva

Locally fairly common in the lower and middle growth of montane forest in the foothills along the east slope of the Andes; thus far known only from one site in the extreme north and a single sighting in the south. The Olive Tanager was first recorded in Ecuador from a pair seen and heard along the Loja-Zamora road just above Zamora on 7 Aug. 1990 (S. Hilty et al.); despite much recent field work at various sites in the region, this remains the sole report of the species from s. Ecuador. In Mar. and Aug. 1993, Olive Tanagers were discovered to be quite numerous in the Bermejo oil-field area north of Lumbaquí in Sucumbíos, with a series of specimens (ANSP, MECN) taken. The species should be watched for elsewhere at appropriate elevations along the eastern base of the Andes. Recorded between about 600 and 1100 m.

C. *frenata* is here regarded as a species distinct from C. *carmioli* (Carmiol's Tanager) of s. Middle America, this based on its widely disjunct distribution and differing behavior and vocalizations. Monotypic.

Range: locally along eastern base of Andes from s. Colombia to w. Bolivia.

Chlorothraupis stolzmanni Ochre-breasted Tangara Pechiocrácea
              Tanager

Uncommon to locally common in the lower growth and borders of montane forest in the foothills and lower subtropical zone on the west slope of the Andes south to El Oro (old records from La Chonta; many recent reports from Buenaventura). Also recently recorded on the coastal Cordillera de Mache in sw. Esmeraldas (Cabeceras de Bilsa [Parker and Carr 1992] and Bilsa [first by J. P. Clay et al.]). The Ochre-breasted Tanager is particularly numerous at El Placer in Esmeraldas (where it and the Moss-backed Tanager [*Bangsia edwardsi*] are the two most common tanagers), and also in El Oro, but it seems markedly less common in between (e.g., in Pichincha). Recorded mostly from 600 to 1500 m (above Mindo), locally lower (e.g., ranging down to 400 m above Manta Real in nw. Azuay [RSR] and above Playa de Oro in n. Esmeraldas [O. Jahn]).

Two races occur in Ecuador, nominate *stolzmanni* in most of the species' Ecuadorian range and *dugandi* in n. Esmeraldas at El Placer (where first recorded by J. Fjeldså and N. Krabbe, *Bull. B. O. C.* 106[3]: 122–123, 1986). *Dugandi* differs in having a grayer crown and somewhat paler underparts.

Range: west slope of Andes in w. Colombia and w. Ecuador.

Habia rubica   Red-crowned   Tangara Hormiguera Coronirroja
                      Ant-Tanager

Rare to locally fairly common in the lower growth of humid forest (especially terra firme), and to a lesser extent at forest borders, in the lowlands of e. Ecuador. Not an especially numerous bird in Ecuador, the Red-crowned Ant-Tanager appears to be most common near the base of the Andes, perhaps especially in Morona-Santiago, though a series of six specimens (UKMNH) was obtained in Jul.–Aug. 1971 at Santa Cecilia, west of Lago Agrio in Sucumbíos. Ranges up to about 600–700 m along the eastern base of the Andes.

One race: *rhodinolaema*.

Range: Mexico to Panama; very locally in n. Colombia and n. Venezuela; se. Colombia south to n. Bolivia and Amazonian Brazil; e. Brazil, e. Paraguay, and ne. Argentina.

Eucometis penicillata   Gray-headed Tanager   Tangara Cabecigris

Rare to uncommon and local in the lower growth of várzea forest, *Mauritia* palm swamps, and humid forest borders in the lowlands of e. Ecuador, primarily in the drainages of the Ríos Napo and Aguarico. In the southeast the Gray-headed Tanager is known only from sightings from Kapawi Lodge on the Río Pastaza near the Peruvian border. Recorded up to only about 400 m (Jatun Sacha).

One race: nominate *penicillata*.

Range: s. Mexico to n. Colombia and n. Venezuela; sw. Venezuela, se. Colombia, ne. Ecuador, e. Peru, n. and e. Bolivia, n. Paraguay, parts of Amazonian and sw. Brazil, and the Guianas.

Mitrospingus cassinii   Dusky-faced Tanager   Tangara Carinegruzca

Fairly common in dense lower growth at borders of humid forest and secondary woodland, often near rivers and streams, in the lowlands and foothills of w. Ecuador south to w. Esmeraldas (Muisne and Bilsa area), extreme s. Pichincha (Río Palenque), and nw. Azuay (Manta Real; ANSP). Recorded up to about 800 m near Tinalandia.

One race: nominate *cassinii*.

Range: Costa Rica to sw. Ecuador.

Tachyphonus rufus   White-lined Tanager   Tangara Filiblanca

Rare to locally fairly common in shrubby clearings, gardens, and lighter woodland in the lowlands and foothills of w. Ecuador from Esmeraldas (El Placer/Lita area, Playa de Oro) and Carchi (Maldonado road) south to Bolívar (an old record from Balzapamba; Orcés 1944); in recent years, however, recorded mainly from Pichincha northward, with only one report from farther south, at Loma Alta in nw. Guayas (Becker and López-Lanús 1997). In e. Ecuador recorded mainly from along the base of the Andes in w. Napo (Tena and Archidona regions; P. Coopmans et al.), Morona-Santiago (Macas region), and Zamora-Chinchipe (Zamora area and in the Río Marañón drainage from around Zumba, where the species is particularly numerous and ranges north to Valladolid). There are also 1996 reports from Yuturi Lodge (S. Howell). Although still scarce and local in Ecuador, the White-lined Tanager may be increasing, likely in response to the continued clearing of forest. Recorded up to 1700 m.

Monotypic.

Range: Costa Rica and Panama; locally in nw. Ecuador and n. and w. Colombia, extending east through n. Venezuela and the Guianas to e. and interior Brazil, e. Paraguay, and ne. Argentina; locally in e. Ecuador and e. Peru.

Tachyphonus cristatus   **Flame-crested Tanager**   **Tangara Crestiflama**

Uncommon to locally fairly common in the canopy and mid-levels of humid forest (especially terra firme) and forest borders in the lowlands of e. Ecuador. Ranges mainly below about 600 m, but occurring locally and in small numbers as high as 900–1000 m along the eastern base of the Andes (perhaps especially in Zamora-Chinchipe).

One race: *fallax*.

Range: se. Colombia, s. Venezuela, and the Guianas to ne. and extreme se. Peru, n. Bolivia, and Amazonian Brazil; e. Brazil.

Tachyphonus surinamus   **Fulvous-crested Tanager**   **Tangara Crestifulva**

Uncommon to locally fairly common in the lower and middle growth of humid forest (especially terra firme) and forest borders in the lowlands of e. Ecuador. The Fulvous-crested Tanager seems more numerous south of the Río Napo than it does farther north. Ranges up to about 900 m in foothills along the eastern base of the Andes (e.g., in the Bermejo oil-field area north of Lumbaquí; ANSP).

One race: *brevipes*.

Range: se. Colombia, s. Venezuela, and the Guianas to ne. Peru and Amazonian Brazil.

Tachyphonus luctuosus   **White-shouldered**   **Tangara Hombriblanca**
                          Tanager

Uncommon to fairly common in the canopy and borders of humid and deciduous forest and secondary woodland in the lowlands of w. Ecuador; generally rare in the canopy and borders of humid forest and secondary woodland in the lowlands of e. Ecuador, though in Aug. 1989 the species was found to be fairly common around Santiago in Morona-Santiago (M. B. Robbins; ANSP). In the west recorded from n. Esmeraldas south to El Oro (Santa Rosa, Río Jubones, La Chonta, Palmales, and Buenaventura). Recorded mostly below 800 m, locally and in small numbers as high as 1300 m around Mindo.

Two quite similar races are found in Ecuador, *panamensis* in the west and nominate *luctuosus* in the east.

Range: Honduras to n. Colombia and nw. Venezuela; w. Ecuador and extreme nw. Peru; the Guianas, sw. and se. Venezuela, and se. Colombia to n. Bolivia and Amazonian Brazil. More than one species may be involved.

Tachyphonus delatrii   Tawny-crested Tanager   Tangara Crestinaranja

Rare to locally common in the lower and middle growth of very humid to humid forest and forest borders in the lowlands and (especially) the foothills of nw. Ecuador; most numerous in Esmeraldas, and one of the more numerous forest birds between 50 and 250 m at Playa de Oro (O. Jahn et al.). In the 19th century, specimens were taken as far south as Pallatanga in w. Chimborazo (in all likli-hood these birds were actually obtained on the slopes well below the town of Pallatanga, perhaps even as low as adjacent e. Guayas). In recent years, however, the Tawny-crested Tanager has not been found south of Pichincha (e.g., along the lower Chiriboga road); Tawny-cresteds are locally common in nw. Pichin-cha (e.g., along the road north of Simón Bolívar near Pedro Vicente Maldon-ado; v.o.). Recorded mostly below 800 m.

Monotypic.

Range: Honduras to nw. Ecuador.

Lanio fulvus   Fulvous Shrike-Tanager   Tangara Fulva

Uncommon to locally fairly common in the subcanopy and mid-levels of humid forest (mainly terra firme) and forest borders in the lowlands and foothills of e. Ecuador. Ranges regularly up in the foothills of the Andes to about 1100 m, in small numbers even higher (recorded as high as 1500 m at Palm Peak [MCZ] and exceptionally to 1750 m at Sabanilla [AMNH; Chapman 1926]).

One race: *peruvianus*.

Range: the Guianas, extreme sw. and s. Venezuela, se. Colombia, e. Ecuador, ne. Peru, and n. Amazonian Brazil.

Heterospingus xanthopygius   Scarlet-browed   Tangara Cejiescarlata
                                Tanager

Uncommon to locally fairly common in the canopy and borders of humid forest and mature secondary woodland in the lowlands and foothills of w. Ecuador. In the 19th century the Scarlet-browed Tanager was recorded south to e. Guayas (Chimbo), but in recent years it has not been found south of s. Pichincha (e.g., at Tinalandia and Río Palenque, at both of which small numbers persisted as of the late 1990s). The Scarlet-browed Tanager is most numerous in Esmeraldas, but substantial numbers are also found in nw. Pichincha (e.g., along the road north of Simón Bolívar near Pedro Vicente Maldonado; v.o.). Ranges up to about 800 m (above Tinalandia, and near Lita in w. Imbabura); most numerous in the foothills between 200 and 500 m.

One race: *berliozi*.

Range: e. Panama to nw. Ecuador.

Chlorospingus ophthalmicus	Common Bush-Tanager	Clorospingo Común

Fairly common but local in the subcanopy and borders of montane forest and adjacent clearings in the subtropical zone on the east slope of the Andes. On the west slope the Common Bush-Tanager is known primarily from sw. Ecuador in El Oro (recently from the Buenaventura and Zaruma areas) and adjacent w. Loja (Vicentino; Best et al. 1993); there are also old records (Chapman 1926) from as far north as se. Azuay (Gima) and even s. Bolívar (Chillanes). We regard the specimen recorded as taken at Gualea (J. T. Zimmer, *Am. Mus. Novitates* 1367, 1947) as having surely been mislabeled; there are no recent records of this bush-tanager, conspicuous where it occurs, from nw. Ecuador. On the east slope recorded mostly from 1500 to 2500 m, but locally ranging down to 1100 m in Zamora-Chinchipe; however, in El Oro and Loja it occurs much lower, only between about 700 and 1450 m.

Somewhat surprisingly, given the species' Ecuadorian distribution, only one race is known from Ecuador, *phaeocephalus*.

Range: highlands from s. Mexico to w. Panama; mountains of n. Venezuela, and Andes from w. Venezuela to nw. Argentina; Sierra de Perijá. More than one species may be involved.

Chlorospingus canigularis	Ashy-throated Bush-Tanager	Clorospingo Golicinéreo

Uncommon to fairly common in the canopy and borders of montane forest and adjacent clearings in the foothills and lower subtropical zone on the east slope of the Andes. On the west slope known only from Pichincha (Mindo/Nanegal area, Chiriboga road, and Tinalandia) and from Chimborazo (old specimens from Pallatanga and Chimbo) and Azuay (sightings in Aug. 1991 at Manta Real; RSR et al.) south to w. Loja (Las Piñas and Alamor). Recently also found on the coastal Cordillera de Colonche in sw. Manabí (Cerro San Sebastián in Machalilla National Park [Parker and Carr 1992; ANSP]) and w. Guayas (Cerro La Torre and above Salanguilla [N. Krabbe]; Loma Alta [Becker and López-Lanús 1997]). On the west slope recorded mostly from 700 to 1300 m, but locally ranging down to 400 m (e.g., at Manta Real in nw. Azuay); on the east slope mostly from 1000 to 1900 m.

Three races are found in Ecuador: *signatus* ranges along the entire east slope; *conspicillatus* is presumably the race found in Pichincha (but apparently no specimens have ever been taken in nw. Ecuador); and *paulus* is found in the southwest. *Signatus* has a narrow white postocular streak and dusky auriculars; *paulus* has a uniform gray head, with *conspicillatus* being similar though slightly larger.

Range: highlands of Costa Rica and w. Panama; Andes from sw. Venezuela to s. Peru. More than one species may be involved.

Chlorospingus semifuscus   Dusky Bush-Tanager   Clorospingo Oscuro

Fairly common to common in montane forest and forest borders in the subtropical zone on the west slope of the Andes south to w. Cotopaxi (Caripero area; N. Krabbe and F. Sornoza). The Dusky Bush-Tanager has also recently been reported from much farther south, at Buenaventura in El Oro, this based on sightings in Feb. and Mar. 1991 (Best et al. 1993). However, as no other observers have found this obscure species here, and as juvenile Common Bush-Tanagers (*C. opthalmicus*) can look almost equally nondescript, we prefer to await specimens—or at least additional sightings—before regarding the Dusky Bush-Tanager as being confirmed as occurring at this locality. Recorded mostly from 1200 to 2300 m.

The species was formerly called the Dusky-bellied Bush-Tanager, but see Isler and Isler (1987), Ridgely and Tudor (1989), and Sibley and Monroe (1990).

One race: nominate *semifuscus*.

Range: west slope of Andes in w. Colombia and nw. Ecuador.

Chlorospingus flavovirens   Yellow-green   Clorospingo
                               Bush-Tanager   Verdiamarillo

Very rare and local in mossy forest and forest borders in the foothills on the west slope of the Andes in nw. Ecuador. Recorded only from Pichincha (a single specimen in BMNH that is labeled as being from Santo Domingo de los Colorados [L. Griscom, *Auk* 52(1): 94–95, 1935] though it probably was actually taken on a higher ridge to the east) and e. Esmeraldas (a few recent sightings at El Placer; here too the species may be more numerous at slightly higher elevations on ridges just to the east). In Colombia mainly recorded between about 700 and 1050 m (Collar et al. 1992), but recently seen as low as 500 m in Nariño (P. Coopmans; P. Salaman).

The Yellow-green Bush-Tanager was given Vulnerable status by Collar et al. (1994). We concur. This extremely local species is not known to occur in any protected area in Ecuador.

Monotypic.

Range: locally in foothills on west slope of Andes in sw. Colombia and nw. Ecuador.

Chlorospingus flavigularis   Yellow-throated   Clorospingo
                                 Bush-Tanager   Goliamarillo

Common in the lower growth of montane forest and forest borders in the foothills and subtropical zone on both slopes of the Andes. On the west slope known south to El Oro (La Chonta and Buenaventura). Also recorded recently on the coastal Cordillera de Mache in sw. Esmeraldas at Bilsa (first by J. P. Clay et al. in Feb. 1994) and on the Cordillera de Colonche in w. Guayas at Loma Alta (several birds mist-netted and photographed in Dec. 1996; Becker and López-Lanús 1997). Recorded mostly from 700 to 1800 m, in smaller numbers as high as 2100 m; on the west slope occurs locally down to 300–400 m (e.g., above Manta Real in nw. Azuay; also old specimens from the ne. Guayas area).

Two distinctly different races are found in Ecuador, nominate *flavigularis* on the east slope and *marginatus* on the west slope. The latter is notably dingier and less pure gray below, and it has the yellow on its throat more restricted to the sides; its dawn song is notably different (P. Coopmans).

Range: highlands of w. Panama; Andes from n. Colombia to s. Peru. More than one species is probably involved.

Chlorospingus parvirostris   Yellow-whiskered   Clorospingo
                                 Bush-Tanager   Bigotudo

Uncommon to fairly common and seemingly local (perhaps more numerous and widespread southward) in the lower and middle growth of montane forest and forest borders in the subtropical zone on the east slope of the Andes from w. Sucumbíos south to Zamora-Chinchipe. Almost all of the Ecuadorian records of this species are recent. Specimens were evidently obtained at Machay in Tungurahua (J. T. Zimmer, *Am Mus. Novitates* 1367, 1947), though in *Birds of the World* (vol. 13, p. 260) it was specifically stated that the species was "not recorded from Ecuador, where it probably occurs." Recent specimens from the Cordillera de Cutucú in Morona-Santiago (ANSP) and from along the Loja-Zamora road in Zamora-Chinchipe (MCZ) are discussed by Robbins et al. (1987); there also are numerous recent sight reports and an additional specimen (ANSP) from various sites in or near Podocarpus National Park (Bloch et al. 1991; Rasmussen et al. 1996). In Nov. 1992 the Yellow-whiskered Bush-Tanager was also found farther south in Zamora-Chinchipe in the Río Isimanchi valley on the east slope of the Cordillera Las Lagunillas (ANSP), and substantial numbers were noted in Sep. 1998 above Valladolid (RSR et al.). Northward, one was taken by N. Krabbe along the Gualaceo-Limón road on 20 Jun. 1987 (Fjeldså 1987, p. 62), and two were seen west of Mera in Tungurahua on 21 Aug. 1987 (RSR et al.); the species is also known from San Rafael Falls, where two specimens (ANSP) were obtained on 26 Aug. 1993 (there are also sightings from here), and one was taken (MECN) at Quebrada Las Ollas along the La Bonita road in Sucumbíos on 18 Sep. 1991 (N. Krabbe). Recorded in Ecuador from about 1200 to 2250 m, thus mainly occurring at elevations above the range of the Yellow-throated Bush-Tanager (*C. flavigularis*), though there is some overlap, with the two species occasionally even occurring in the same flock.

The species was formerly called the Short-billed Bush-Tanager, but see Isler and Isler (1987), Ridgely and Tudor (1989), and Sibley and Monroe (1990).

One race: *huallagae*.

Range: east slope of Andes from cen. Colombia to w. Bolivia.

Cnemoscopus rubrirostris	Gray-hooded Bush-Tanager	Tangara Montés Capuchigris

Uncommon to fairly common in the canopy and borders of montane forest and (especially) secondary woodland in the temperate zone on the east slope of the Andes. Uncommon and very local on the west slope of the Andes in n. Ecuador, where thus far recorded only from Pichincha (Nono-Mindo and Chiriboga roads). Recorded mostly between about 2200 and 3000 m, locally down to 1900 m on the Cordillera de Huacamayos (N. Krabbe).

One race: nominate *rubrirostris*.

Range: Andes from sw. Venezuela to s. Peru.

Urothraupis stolzmanni	Black-backed Bush-Tanager	Quinuero Dorsinegro

Uncommon to locally fairly common in humid shrubby woodland near treeline and in patches of *Polylepis*-dominated woodland on the east slope of the Andes from se. Carchi (Cerro Mongus; ANSP, MECN) and w. Napo (Oyacachi and Papallacta pass area) south to s. Morona-Santiago (Páramos de Matanga; Krabbe et al. 1997). Small groups of the Black-backed Bush-Tanager can be found in woodland patches on both sides of the Continental Divide near Papallacata pass. Recorded between about 3200 and 4000 m.

Monotypic.

Range: locally in Andes of s. Colombia and on east slope of Andes in Ecuador.

Hemispingus atropileus	Black-capped Hemispingus	Hemispingo Coroninegro

Fairly common in the lower growth (especially where there is abundant *Chusquea* bamboo) of montane forest and forest borders and in secondary woodland in the temperate zone on both slopes of the Andes, locally also on slopes above the central valley (e.g., at Pasochoa). On the west slope recorded south locally to w. Bolívar (Salinas) and w. Azuay (Sural; N. Krabbe). Found mostly between about 2250 and 3200 m, locally as low as 2000–2100 m on the Cordillera de Huacamayos and as high as 3500 m at Cerro Mongus.

One race: nominate *atropileus*. Birds found on the Cordillera Las Lagunillas in extreme s. Ecuador in Oct.–Nov. 1992 did not appear to show any approach to the distinctive Peruvian form *auricularis*, but unfortunately the one specimen that was obtained was skeletonized. This racial diagnosis agrees with that presented by Ridgely and Tudor (1989) and Fjeldså and Krabbe (1990), but Isler and Isler (1987) indicate that *auricularis*, which is known primarily from south of the Río Marañón in Peru, has also been recorded north of the Marañón in Piura.

Range: Andes from sw. Venezuela to s. Peru.

Hemispingus superciliaris	Superciliaried Hemispingus	Hemispingo Superciliado

Uncommon to locally fairly common in montane forest and forest borders and secondary woodland in the temperate zone on both slopes of the Andes, also locally on slopes above the central and interandean valleys. On the west slope recorded south to Pichincha and again in Azuay (specimen from Sural; MECN), El Oro (old specimens from Salvias and Taraguacocha), and e. Loja (specimens [MCZ] and sightings from Cajanuma (v.o.) and the west slope of Cordillera Las Lagunillas [M. B. Robbins]). Recorded mostly between about 2400 and 3200 m, but locally ranges higher in forest and woodland up to near treeline (e.g., on the north slope of Volcán Pichincha where it is regular up to 3500 m at Yanacocha) and has been seen (N. Krabbe) as high as 3700 m at Loma Yanayacu.

Two races are found in Ecuador, *nigrifrons* in the northern and central part of the species' Ecuadorian range (south on the east slope to Morona-Santiago on the Gualaceo-Limón road), and *maculifrons* in w. Azuay, El Oro, and e. Loja. *Nigrifrons* has the forecrown and auricular area blackish to dark gray, whereas in *maculifrons* this area is considerably paler, a more grayish olive.

Range: Andes from w. Venezuela to w. Bolivia. More than one species may be involved.

Hemispingus frontalis	Oleaginous Hemispingus	Hemispingo Oleaginoso

Uncommon and seemingly rather local in the lower growth of montane forest in the subtropical zone on the east slope of the Andes. Recent records from the west slope of the Andes in Imbabura (see Fjeldså 1987) and Carchi are now believed to be incorrect and to have involved juveniles of some other *Hemispingus* species (N. Krabbe, pers. com.). Recorded mostly between about 1500 and 2500 m.

One race: nominate *frontalis*.

Range: mountains of n. Venezuela, and Andes from w. Venezuela to s. Peru; Sierra de Perijá.

Hemispingus melanotis   Black-eared   Hemispingo Orejinegro
                            Hemispingus

Rare to uncommon and local (and generally inconspicuous) in the lower growth of montane forest, especially where there is a dense understory of *Chusquea* bamboo, in the upper subtropical and lower temperate zones on the east slope of the Andes. The Black-eared Hemispingus is known primarily from n. Ecuador south to Tungurahua (Baños and "Ambato"), but there are also several recent reports from Zamora-Chinchipe in the Podocarpus National Park area (including a Jan. 1998 specimen [MECN] from Quebrada Honda); we suspect that the species ranges all along the east slope where conditions are favorable. Recorded mostly between about 1800 and 2700 m.

We treat the forms *ochraceus* and *piurae*, often regarded as subspecies in a highly polytypic *H. melanotis*, as allospecies. One race occurs in Ecuador, nominate *melanotis*.

Range: Andes from sw. Venezuela to w. Bolivia.

Hemispingus ochraceus   Western Hemispingus   Hemispingo Occidental

Rare and apparently local in the lower growth of montane forest and secondary woodland on the west slope of the Andes. The still poorly known Western Hemispingus has been recorded only from Pichincha (2 specimens [WFVZ] taken in Jun. 1989 at Las Palmeras by M. Marín et al.; also several recent sightings by various observers from along the Nono-Mindo and Chiriboga roads, and above Selva Alegre), Chimborazo (old specimens from Cayandeled and Chaguarpata), and Azuay (3 specimens [MECN, Universidad de Azuay] taken at Sural on 3 Mar. 1991 [N. Krabbe and F. Toral]). Unlike the Black-eared Hemispingus (*H. melanotis*) and Piura Hemispingus (*H. piurae*), the Western Hemispingus does not seem to show any particular predilection for an understory of *Chusquea* bamboo. Recorded from 1600 to 2200 m.

H. ochraceus is here considered as a monotypic species separate from *H. melanotis* of the east slope of the Andes. It differs notably in plumage.

Range: west slope of Andes in sw. Colombia and w. Ecuador.

*Hemispingus piurae   Piura Hemispingus   Hemispingo de Piura

Uncommon and local in lower growth of montane woodland and forest, primarily where there is a dense understory of *Chusquea* bamboo, in s. Loja. The Piura Hemispingus was first recorded in Ecuador at Utuana in Sep.–Oct. 1989 (Best et al. 1993); the species has since also been reported from the El Tundo Reserve near Sozoranga. There have been a number of recent sightings from the Utuana area, though it is thought that forest burning and clearing may have adversely affected the species' numbers there. One female, the first and only specimen (MECN) for Ecuador, was obtained on 13 Sep. 1998 (F. Sornoza and RSR et al.). Recorded between about 2000 and 2500 m.

The Piura Hemispingus has an extremely limited range in Ecuador, and much of its forest/woodland habitat there has been either degraded or destroyed. Although it seems capable of persisting in secondary habitats, we nonetheless believe it prudent to accord the species Near-threatened status. The species was not considered to be at risk by Collar et al. (1992, 1994), doubtless because of the differing taxonomy they employed.

H. piurae is here considered as a species distinct from *H. melanotis* (Black-eared Hemispingus) of the east slope of the Andes. *Piurae* differs strikingly in plumage, and though similar, their vocalizations do differ. One race: nominate *piurae*.

Range: locally in Andes of extreme sw. Ecuador and nw. Peru.

Hemispingus verticalis	Black-headed Hemispingus	Hemispingo Cabecinegro

Rare to locally fairly common in the borders and canopy of montane forest and shrubby areas near treeline in the temperate zone on the east slope of the Andes, ranging west on interandean slopes in Azuay to the Río Mazan valley (King 1989). The Black-headed Hemispingus appears to be more numerous in Zamora-Chinchipe (e.g., in the Cajanuma sector of Podocarpus National Park [Rasmussen et al. 1996] and on the Cordillera Las Lagunillas [ANSP]) than it is farther north in Ecuador. Ranges mostly from 2700 to 3400 m.

Monotypic.

Range: locally in Andes from sw. Venezuela to extreme n. Peru.

Conothraupis speculigera	Black-and-white Tanager	Tangara Negriblanca

At least in certain years a locally fairly common breeder during the rainy season (depending on the year, approximately Feb.–May) in scrub, clearings, and lower growth of deciduous woodland in w. Loja from east of Zambi and east of Catamayo south to around Macará and Sozoranga and west to the Pindal and Sabanilla region southwest of Celica; it probably also nests in interior s. El Oro. A number of specimens (WFVZ, ANSP, MECN) of the Black-and-white Tanager have been obtained in recent years from various sites in Loja, and there are numerous additional sightings. There are also a few records of Black-and-white Tanagers from elsewhere in w. Ecuador, these presumably relating to over-shooting migratory individuals: there has been at least one sighting of a male from Río Palenque in s. Pichincha (on 8 Feb. 1980 by RSR and D. Finch et al.; Ridgely 1980), a male was taken at Valle de Yunguillas in Azuay (Orcés 1944), a singing male was seen on Isla de la Plata on 19 Feb. 1991 (P. Coopmans and A. Mayer), and there have been a few reports (e.g., Berg 1994) from Cerro Blanco in the Chongón Hills west of Guayaquil in Guayas. Occurs mostly between 500 and 1700 m when breeding, occasionally or locally as high as 1950 m, at least in the Catamayo/La Toma area.

For the remainder of the year (which corresponds to the dry season), the

Black-and-white Tanager is largely if not entirely absent from the region where it breeds. The evidence now available indicates that the species is a long-distance and apparently a cross-Andean migrant, though details of these movements remain sketchy. The only Ecuador record from the nonbreeding season and from east of the Andes is an immature male taken at Tayuntza (600 m) in Morona-Santiago on 3 Aug. 1987 (WFVZ; see Marín et al. 1992). The species' nonbreeding habitat preferences remain poorly known, but in se. Peru a few individuals have been seen in the understory of riparian and floodplain forest and woodland, and in adjacent semiopen areas.

The Black-and-white Tanager was given Near-threatened status by Collar et al. (1992, 1994). However, given what is now known concerning its preference for secondary and disturbed areas, at least in the breeding season, it is clear that the species cannot be viewed as at risk, at least not in Ecuador.

Monotypic.

Range: breeds in sw. Ecuador and nw. Peru; during nonbreeding season recorded mainly from e. Peru, with a few records also from se. Ecuador and w. Bolivia.

Cissopis leveriana   **Magpie Tanager**   **Tangara Urraca**

Fairly common to common in the borders of humid forest, secondary woodland, and trees in adjacent clearings in the lowlands of e. Ecuador. Ranges up in partially cleared areas in the foothills and lower subtropical zone to about 1200 m, in small numbers as high as 1500–1800 m.

One race: nominate *leveriana*.

Range: e. Colombia, s. Venezuela, and Guyana to n. Bolivia and w. and s. Amazonian Brazil; e. Brazil, e. Paraguay, and ne. Argentina.

*Schistochlamys melanopis   **Black-faced Tanager**   **Tangara Carinegra**

Recently found to be uncommon to fairly common in clearings and scrubby secondary woodland in the Río Marañón drainage of s. Zamora-Chinchipe in the Zumba area and north to around Valladolid. First found in Ecuador only in 1986 (M. Pearman); the first specimens (ANSP, MECN) were obtained near Zumba on 8–9 Aug. 1992. The Black-faced Tanager appears—likely in response to deforestation—to be increasing in Ecuador; it may have only recently spread into the country from adjacent Peru. There is some evidence that its range is continuing to expand northward: the species now is regular north to the Valladolid area, and a pair was even seen on the Pacific slope near Yangana in extreme se. Loja on 8 Apr. 1994 (P. Coopmans and M. Lysinger). As the species is recorded from adjacent Colombia (Hilty and Brown 1986), it seems possible that the Black-faced Tanager will soon spread south into ne. Ecuador as well. Recorded mainly from 650 to 1600–1700 m, once as high as 1800 m (Yangana).

One race: *grisea*.

Range: n. and e. Colombia and Venezuela to n. and e. Bolivia and s. Brazil, but absent from much of cen. and w. Amazonia.

Sericossypha albocristata   White-capped Tanager   Tangara Caretiblanca

Rare to locally uncommon in the canopy and borders of montane forest in the subtropical and temperate zones on the east slope of the Andes. Many of the recent Ecuadorian reports of this spectacular tanager come from the south (e.g., along the Gualaceo-Limón road in Morona-Santiago and at various points in or along the edge of Podocarpus National Park in Zamora-Chinchipe), but it has also been seen regularly on the Cordillera de Huacamayos and at nearby SierrAzul and Cabañas San Isidro. Recorded between about 1750 and 3000 m, but most numerous from 1900 to 2700 m; has been seen as low as 1400 m along the Loja-Zamora road (D. Wolf).
Monotypic.
Range: locally in Andes from sw. Venezuela to cen. Peru.

Catamblyrhynchus diadema   Plushcap   Gorradiadema

Uncommon to locally fairly common in the lower growth of montane forest, forest borders, and secondary woodland (favoring areas where there is a dense growth of *Chusquea* bamboo) in the subtropical and temperate zones on both slopes of the Andes, and also occurring locally on slopes above the central and interandean valleys (e.g., at Pasochoa). On the west slope recorded south to w. Loja (e.g., in the Celica area and at Utuana). Recorded mostly from 2000 to 3500 m, occasionally somewhat higher or lower.

Formerly (e.g., Meyer de Schauensee 1966, 1970; Hilty and Brown 1986) usually treated as a monotypic family (Catamblyrhynchidae); the Pluschcap's taxonomic affinities remain uncertain but probably are closest to the tanagers.

C. *diadema* was formerly usually called the Plush-capped Finch, but as it appears not to be a true "finch," it seems preferable to shorten its English name to simply the Plushcap; see Ridgely and Tudor (1989) and Sibley and Monroe (1990).

One race is known from Ecuador, nominate *diadema*. However, the possibility that birds from the southwest may be referable to *pallida* should not be ruled out. This race was described by M. A. Carriker, Jr. (*Proc. Acad. Nat. Sci, Phil.* 86:330, 1934) on the basis of a single specimen from Abra de Porculla in nw. Peru. It was synonymized by C. E. Hellmayr in *Birds of the Americas* (pt. 11, p. 5), but the specimen—contra Hellmayr (op. cit.)—is not "young" but in fact is an adult *Catamblyrhynchus* though quite pale above and below. The only specimens from sw. Ecuador we are aware of are two (MCZ) taken by D. Norton north of Celica on 23 Aug. 1965. Unfortunately both are juveniles, thus not useful in determining their subspecific affinity; additional material is therefore required.
Range: mountains of n. Venezuela, and Andes from w. Venezuela to nw. Argentina; Sierra de Perijá and Santa Marta Mountains.

Cardinalidae   Saltators, Grosbeaks, and Cardinals

Fifteen species in seven genera. The family is American in distribution. The saltators and grosbeaks have formerly often been considered a subfamily in a broadly inclusive Emberizidae; we follow Gill 1994 in giving them full family rank. The genus *Paroaria* perhaps is better placed with the Emberizine finches (Emberizidae; e.g., Sibley and Ahlquist 1990; Sibley and Monroe 1990), but we retain it within its traditional family.

Saltator maximus   **Buff-throated Saltator**   **Saltador Golianteado**

Fairly common to common in borders of humid forest (to a lesser extent also in canopy, especially in várzea), secondary woodland, and shrubby clearings and gardens in the lowlands and foothills of both e. and w. Ecuador. In the west found only in more humid areas, and recorded south (especially in the foothills) through El Oro and w. Loja (Alamor region). Mostly below 900 m, but ranges up in smaller numbers into the lower subtropical zone to 1300–1500 m (especially on the west slope, e.g., around Alamor and Mindo).
One race: nominate *maximus*.
Range: s. Mexico to extreme nw. Peru, n. Bolivia, and Amazonian and cen. Brazil; e. Brazil.

Saltator atripennis   **Black-winged Saltator**   **Saltador Alinegro**

Uncommon to locally fairly common in the canopy and borders of montane and humid forest, secondary woodland, and shrubby clearings and plantations in the more humid lowlands, foothills, and lower subtropical zone on the west slope of the Andes south to El Oro and w. Loja (east of Alamor at Tierra Colorada; Best et al. 1993), ranging down in smaller numbers into adjacent humid lowlands. Also recorded recently on the coastal Cordillera de Mache at Bilsa in w. Esmeraldas (first by R. P. Clay et al. in 1994). The Black-winged Saltator remains numerous at both Tinalandia and Río Palenque. Recorded mostly between 500 and 1500 m, ranging locally down to about 200 m in the adjacent lowlands (e.g., at Río Palenque) and as low as 50 m at Playa de Oro in Esmeraldas (O. Jahn et al.); very small numbers occur up to 1700 m in the Mindo/Tandayapa area.
Monotypic. We regard the race *caniceps*, to which Ecuador birds have usually been assigned, as best being synonymized; the variation in crown color, involving a greater or lesser extent of black, seems better ascribed to individual rather than to geographic variation, as does variation in bill size.
Range: Andes from n. Colombia to sw. Ecuador.

Saltator coerulescens   **Grayish Saltator**   **Saltador Grisáceo**

Fairly common to common in borders of várzea forest, riparian areas (e.g., river islands), clearings with scattered trees, and gardens in the lowlands and foothills of e. Ecuador, including the Río Marañón drainage of s. Zamora-Chinchipe around Zumba. Regularly ranges up on the lower slopes of the Andes to 1200–1300 m, locally and in small numbers as high as 1600 m in the El Chaco area of w. Napo.

One race: *azarae*.

Range: Mexico to Costa Rica; n. and e. Colombia, Venezuela, and the Guianas to n. Argentina and Uruguay. More than one species is probably involved.

Saltator nigriceps   **Black-cowled Saltator**   **Saltador Capuchinegro**

Uncommon to locally fairly common in montane woodland, forest borders, and dense regenerating scrub on the west slope of the Andes in sw. Ecuador; recorded mainly from Loja. The Black-cowled Saltator is known from the Loja (city) vicinity south and west through the lower Cajanuma area to above Amaluza and Jimbura (on the west slope of the Cordillera Las Lagunillas; ANSP), the Sozoranga/Utuana region, and around Celica; it also ranges into adjacent El Oro, where one bird was seen above Balzas on 6 Feb. 1988 (PJG and P. Scharf). Recorded mostly from 1700 to 2900 m, occasionally as low as 1550 m at Catacocha (Best et al. 1993).

S. *nigriceps* has occasionally (e.g., *Birds of the World*, vol. 13) been regarded as a subspecies of *S. aurantiirostris* (Golden-billed Saltator) of the Andes and s.-cen. South America from Peru to Argentina, but we follow most recent authors in treating it as a distinct species. Its voice differs, and clinal variation between the two species does not occur. Monotypic.

Range: Andes of sw. Ecuador and nw. Peru.

Saltator striatipectus   **Streaked Saltator**   **Saltador Listado**

Fairly common to common in deciduous woodland, arid scrub, agricultural areas, and gardens in the lowlands of w. Ecuador from w. Esmeraldas (Atacames area; RSR) and w. Pichincha south through El Oro and Loja. Also locally common in arid intermontane valleys in the highlands of n. Ecuador in Carchi and Imbabura (Río Mira and Río Chota valleys) and Pichincha (south to the Río Guaillabamba valley and Tumbaco region, east of Quito). Recently also found in extreme se. Ecuador in the Río Marañón drainage around Zumba in s. Zamora-Chinchipe, ranging north to around Valladolid. In n. Ecuador recorded mostly from 1300 to 2500 m; elsewhere found below about 1500 m, in small numbers a little higher.

We follow G. Seutin, J. Brawn, R. E. Ricklefs, and E. Bermingham (*Auk* 110[1]: 117–126, 1993) in treating birds of the Lesser Antilles (*S. albicollis*, Lesser Antillean Saltator) as a separate species from the populations of Central and South America (*S. striatipectus*, Streaked Saltator); these authors suggest that more than one species may be involved in the mainland Neotropics as well. The racial allocation of Ecuadorian birds has been the source of confusion (see Chapman 1926), but recently obtained material (ANSP, MECN) has helped to clarify the situation. Birds from the highlands of n. Ecuador are best referred to nominate *striatipectus* (which also ranges in adjacent Colombia, where *flavidicollis* does not occur, contra many references). This taxon is rather gray above and is white below with grayish streaking especially across its breast; its bill is black. The birds of lowland w. Ecuador are *flavidicollis*. Adults of this taxon have a longer white superciliary, olive upperparts, uniform whitish underparts (often tinged yellowish), and usually a yellow-tipped bill; immatures show a variable amount of dusky-olive streaking on the breast. Birds from the Río Marañón drainage are *peruvianus*, another "streaked" taxon which is generally similar to the disjunct *striatipectus* though *peruvianus* has more olive upperparts and breast streaking, a less prominent brow, and more black on the bill. Birds from Isla Puná, in the heart of the range of *flavidicollis*, seem to retain streaking on their underparts even as adults and are grayish above like *striatipectus*; they likely deserve to be named as a separate taxon.

Range: Costa Rica and Panama; n. and w. Venezuela, w. Colombia, w. Ecuador, and w. Peru.

Saltator cinctus   **Masked Saltator**   Saltador Enmascarado

Rare and seemingly local and perhaps nomadic—but probably mainly just overlooked—in the lower and middle growth of montane forest in the subtropical and lower temperate zones on the east slope of the Andes. The still poorly known Masked Saltator has been recorded at only a few Ecuadorian localities. The first was the Cordillera de Cutucú in Morona-Santiago, which is the type locality for the species (see Robbins et al. 1987). Additional recent localities include (from north to south) the Cordillera de Huacamayos in w. Napo (a few sightings since 1989; P. Coopmans and others), the Gualaceo-Limón road in Morona-Santiago (one seen by D. Ward at 2100m on 4 Feb. 1990, fide J. Arvin), Podocarpus National Park and the Cordillera de Sabanilla in Zamora-Chinchipe (v.o.), the Río Angashcola valley east of Amaluza in s. Loja (R. Williams et al.), and the Río Isimanchi valley on the east slope of the Cordillera Las Lagunillas (T. J. Davis). Contra earlier indications, recent observers have not noted any particular association with *Chusquea* bamboo, but the species has frequently been observed to eat the fruits of the tree *Podocarpus oleifolius* (J. Tobias and R. Williams). Recorded from about 2000 to 2700m.

The Masked Saltator was given Near-threatened status by Collar et al. (1992, 1994). This assessment seems appropriate from a strictly Ecuadorian perspective as well.

Monotypic.

Range: locally in Central Andes of Colombia and on east slope of Andes in Ecuador and Peru.

Saltator grossus   **Slate-colored Grosbeak**   Picogrueso Piquirrojo

Fairly common (especially by voice) in the mid-levels and subcanopy of humid forest and forest borders and mature secondary woodland in the lowlands of both e. and w. Ecuador. In the west recorded south in more humid regions, especially in the foothills, to n. Manabí (Filos de Mono [WFVZ] and Río de Oro [AMNH]), n. Los Ríos (an Oct. 1950 specimen from Quevedo; ANSP), and El Oro (La Chonta and Buenaventura). Ranges up in smaller numbers into the lower subtropical zone on both slopes of the Andes to about 1200 m, on the east slope occasionally as high as 1700 m (Cordillera de Cutucú; Robbins et al. 1987).

This species was formerly placed in the genus *Pitylus*, but we follow J. W. Tamplin, J. W. Demastes, and J. V. Remsen, Jr. (*Wilson Bull.* 105[1]: 93–113, 1993) in merging *Pitylus* into *Saltator*; this change was also adopted by the 1998 AOU Check-list. Two weakly differentiated races occur in Ecuador, *saturatus* in the west and nominate *grossus* in the east.

Range: Honduras to sw. Ecuador; se. Colombia, s. Ecuador, and the Guianas to n. Bolivia and Amazonian Brazil.

Paroaria gularis   Red-capped Cardinal   Cardenal Gorrirrojo

Fairly common to common but always conspicuous in shrubbery along shores of lakes, ponds, and more sluggish rivers and streams in the lowlands of e. Ecuador. The Red-capped Cardinal has been recorded primarily from the drainages of the Ríos Napo and Aguarico, but specimens were also taken at "Laguna de Siguin" somewhere in Pastaza (Berlioz 1937c), and the species has recently been found around Kapawi Lodge on the Río Pastaza near the Peruvian border; it likely occurs widely in the Río Pastaza drainage as well. Most numerous below 300 m, with a few ranging up to about 400 m.

One race: nominate *gularis*.

Range: e. Colombia, much of Venezuela, and the Guianas to n. Bolivia and Amazonian Brazil (though largely absent from much of s. Venezuela and n. Amazonian Brazil).

Pheucticus chrysogaster   Southern Yellow-   Picogrueso Amarillo
                              Grosbeak   Sureño

Uncommon to locally common in lighter woodland, deciduous forest, desert scrub, and agricultural areas and gardens, mainly but not entirely in arid regions, in the lowlands of w. Ecuador from cen. Manabí (Manta and Bahía de Caráquez area) southward. Also fairly common in intermontane valleys and slopes from Carchi south through El Oro and Loja, and occasional or perhaps local on outside slopes of the Andes as well, perhaps primarily subsequent to an area's being partially deforested. On the west slope there are many recent records from the Tandayapa area in Pichincha; on the east slope there are old specimens from Sabanilla and Zamora (Chapman 1926), and in Aug. 1991 the species was also seen at Hacienda La Libertad in Cañar (RSR et al.). Southern Yellow-Grosbeaks are particularly numerous and widespread in s. Ecuador, especially in Loja, but they are also common in the arid Río Mira valley of Imbabura. Ranges locally from sea level up in arid regions to about 3500 m (on rare occasions wandering even higher).

P. chrysogaster of South America is regarded as a species distinct from *P. chrysopeplus* (Mexican Yellow-Grosbeak) of Mexico and *P. tibialis* (Black-thighed Grosbeak) of Costa Rica and Panama, following Ridgely and Tudor (1989), Sibley and Monroe (1990), and the 1998 AOU Check-list. One race occurs in Ecuador, nominate *chrysogaster*.

    P. chrysogaster (when split) was formerly called the Golden-bellied Grosbeak (e.g., Meyer de Schauensee 1966, 1970). When all forms were considered conspecific under the name of *C. chrysopeplus*, Yellow Grosbeak was usually its English name.

    Range: locally in mountains of n. Venezuela, and in Sierra de Perijá and Santa Marta Mountains; Andes from extreme s. Colombia to Peru, also in arid lowlands of sw. Ecuador and nw. Peru.

Pheucticus aureoventris	Black-backed Grosbeak	Picogrueso Dorsinegro

Uncommon and local in scrub, agricultural regions, and gardens in arid intermontane valleys from Carchi (Río Mira valley) and Imbabura south to Chimborazo in the Guamote area; seems most numerous in Chimborazo. The Black-backed Grosbeak is apparently not recorded from farther south in Ecuador, though why it should be absent from this region remains unknown, numerous areas appearing suitable. It is essentially absent from the outside slopes of the Andes but does range down the Río Pastaza valley as far as around Baños. Although it tends to occur in more xeric habitats than the more widespread Southern Yellow-Grosbeak (*P. chrysogaster*), the two species do occur sympatrically at a few locations (e.g., around the western base of Volcán Cotopaxi). Ranges mainly from 1500 to 3200 m, rarely a bit higher.

    One race: *crissalis*. Fjeldså and Krabbe (1990) suggest that hybridization between this species and *P. chrysogaster* may be occurring in Ecuador, and indeed there does seem to be some variation in the amount of yellow and black mottling seen on certain individuals of both species.

    Range: locally in Andes from w. Venezuela to nw. Argentina, also spreading out into lowlands in Bolivia and Argentina, extending as far east as extreme sw. Brazil.

Pheucticus ludovicianus	Rose-breasted Grosbeak	Picogrueso Pechirrosado

A rare to uncommon boreal winter visitant to the canopy and borders of montane, humid, and deciduous forest and clearings with scattered trees; seems most numerous in the foothills and subtropical zone, and occasionally in the central valley. Recorded south mainly to Guayas and Tungurahua, but there are single sightings from Loja (one seen at Vilcabamba on 30 Nov. 1991; J. Rasmussen et al.) and Zamora-Chinchipe (a male seen near Zamora on 2 Jan. 1995; D. Wolf and PJG). Most Rose-breasted Grosbeak records come from w. Ecuador, but there are also a few reports from the east slope of the Andes and adjacent eastern lowlands (especially at Jatun Sacha); in addition, a single bird was seen as far east as the Río Pacuyacu, south of the Río Aguarico in e. Napo, on 8 Nov. 1992 (P. Coopmans et al.). Recorded from Oct. to Mar. Recorded up to about 3300 m, but above 2000 m probably occurring mainly as a transient.

Monotypic.

Range: breeds in North America, wintering south to nw. South America.

Cyanocompsa cyanoides   Blue-black   Picogrueso Negriazulado
                            Grosbeak

Uncommon to fairly common in the lower growth of humid forest, mature secondary woodland, and borders in the lowlands and foothills of both e. and w. Ecuador. In the west recorded south to nw. Guayas and through El Oro and w. Loja (an old record from the Puyango area at La Puente; no recent reports); however, it may be largely absent from the wet-forest belt of n. Esmeraldas, at least where extensively forested (O. Jahn et al.). Recorded mostly below about 1000 m, but in recent years a few Blue-black Grosbeaks have been seen regularly in the Mindo region at around 1300 m, and one apparently territorial male was found as high as 1700 m on 6 Jun. 1998 (J. Lyons and V. Perez).

The genus *Cyanocompsa* has been merged into the genus *Passerina* in a few references (e.g., *Birds of the World*, vol. 13), but most authors (e.g., Ridgely and Tudor 1989; J. W. Tamplin, J. W. Demastes, and J. V. Remsen, Jr., *Wilson Bull.* 105[1]: 93–113, 1993) doubt that the two genera are all that closely related. The 1998 AOU Check-list also maintained the two genera. Two races occur in Ecuador, nominate *cyanoides* in the west and *rothschildii* in the east. Males of the latter are somewhat brighter blue generally and have a paler and brighter blue brow; females are similar.

Range: s. Mexico to extreme nw. Peru, n. Bolivia, and Amazonian Brazil.

*Guiraca caerulea   Blue Grosbeak   Picogrueso Azul

An accidental boreal winter visitant to shrubby areas and clearings in the lowlands of ne. Ecuador. There is only a single Ecuadorian record, a female taken by one of the Olallas at Edén on the Río Napo in e. Napo on 1 Dec. 1964 (specimen donated to AMNH by G. Orcés; see Ridgely 1980). This is the southernmost record of the Blue Grosbeak, and it represents only the second South American specimen. 225 m.

The species is by some authors (e.g., in *Birds of the World*, vol. 13), placed in the genus *Passerina*, but we maintain the genus *Guiraca*, following most recent authors, including J. W. Tamplin, J. W. Demastes, and J. V. Remsen, Jr. (*Wilson Bull.* 105[1]: 93–113, 1993) and the 1998 AOU Check-list. The specimen was identified as belonging to nominate *caerulea* by E. Eisenmann.

Range: breeds from s. North America to Costa Rica, wintering south to Panama, accidentally to nw. South America.

Parkerthraustes humeralis   Yellow-shouldered   Picogrueso
                                Grosbeak   Hombriamarillo

Rare to locally uncommon (but likely often overlooked) in the canopy and borders of terra firme forest in the lowlands of e. Ecuador from Sucumbíos south into Morona-Santiago. The Yellow-shouldered Grosbeak does not seem to occur any distance east of the Andes in Ecuador; the easternmost sites known are along the Maxus road southeast of Pompeya, where pairs were regularly found with canopy flocks in 1994–1995 (RSR et al.), La Selva (a single Sep. 1993 sighting; P. Coopmans et al.), and Yuturi (S. Howell). Recorded mainly below about 600 m, with a few ranging locally as high as 900 m at the Bermejo oil-field area north of Lumbaquí (M. B. Robbins).

As has been noted by numerous authors, this distinctive species does not appear to belong in the genus *Caryothraustes*, in which it was long placed; see especially J. W. Tamplin, J. W. Demastes, and J. V. Remsen, Jr. (*Wilson Bull.* 105[1]: 93–113, 1993) and J. W. Demastes and J. V. Remsen, Jr. (*Wilson Bull.* 106[4]: 733–738, 1994). More recently, J. V. Remsen (*Studies in Neotropical Ornithology Honoring Ted Parker*, Ornithol. Monogr. 48: 89–90, 1997) erected the monotypic genus *Parkerthraustes* for the species. Monotypic.

Range: se. Colombia, e. Ecuador, e. Peru, n. Bolivia, and s. Amazonian Brazil.

*[Spiza americana   Dickcissel   Llanero]

An accidental boreal winter visitant to grassy areas and clearings in the lowlands of e. Ecuador. There is only a single Ecuadorian record, a sighting of what was apparently an immature bird seen by R. Clay, S. Jack, and J. Vincent with a mixed seedeater (*Sporophila*) flock near Jatun Sacha on 27 Jan. 1994 (R. Clay, *Cotinga* 11: 49, 1999). This represents the southernmost record of the species. 400 m.

The relationships of the genus *Spiza* remain uncertain (see J. W. Tamplin, J. W. Demastes, and J. V. Remsen, Jr., *Wilson Bull.* 105[1]: 93–113, 1993). Some authors consider that the species may be better placed with the Icteridae; we follow the 1998 AOU Check-list in considering it a cardinalid. Monotypic.

Range: breeds in cen. North America, wintering primarily in llanos of Venezuela and ne. Colombia.

Emberizidae   Emberizine Finches

Fifty-three species in 22 genera. The family is worldwide in distribution. The placement of certain genera that here have been retained in this family remains uncertain.

Rhodospingus cruentus   Crimson-breasted   Pinzón Pechicarmesí
                            Finch

Uncommon to seasonally common in shrubby and grassy areas, lighter woodland, and forest borders in the lowlands of w. Ecuador north to w. Pichincha and w. Esmeraldas, occurring inland into w. Loja as far as west of Sabiango (MCZ); it also occurs (at least seasonally) on Isla de la Plata off s. Manabí. The Crimson-breasted Finch is most numerous in the more arid parts of sw. Ecuador, especially in Guayas and El Oro, and this is apparently where most (all?) birds breed in the Jan.–May period; singing males are then conspicuous there. In many areas, though, the species appears to be erratic or seasonal, with non-breeding birds moving about in small flocks, often accompanying *Sporophila* seedeaters. The species then appears—like the Snowy-throated Kingbird (*Tyrannus niveigularis*)—to move northward and into more humid regions; an immature male was mist-netted as far north as Playa de Oro in n. Esmeraldas on 27 Nov. 1996 (O. Jahn and K.-L. Schuchmann), an indication that a few may even occur north into adjacent Nariño, Colombia. Recorded up to 900 m below Alamor in w. Loja (R. Williams), but mostly below 500 m.

The species was long called the Crimson Finch (e.g., Meyer de Schauensee 1966, 1970; Ridgely and Tudor 1989), but Sibley and Monroe (1990) altered this to "Crimson Finch-Tanager" in order to avoid duplicating the name of the Australian finch *Neochmia phaeton*, which also went by the name of Crimson Finch. However, we feel that employing the group name "finch-tanager" causes potential confusion with the Tanager Finch (*Oreothraupis arremonops*). We conclude that simply adding "breasted" to the modifier is the best course.

R. A. Paynter, Jr. (*Bull. B. O. C.* 91: 79–81, 1971) suggested that the affinities of *Rhodospingus* lie with the tanagers rather than with the emberizine finches, with which *Rhodospingus* has usually been associated. Based on vocalizations and behavior, however, we are not entirely convinced that this is the case for *Rhodospingus* any more than it is for various other genera. We thus retain it within the emberizine "finch" group. Monotypic.

Range: w. Ecuador and nw. Peru; one record from Gorgona Island off Colombia (B. Ortiz-Von Halle, *Caldasia* 16: 209–214, 1990).

Volatinia jacarina   Blue-black Grassquit   Semillerito Negriazulado

Common to very common in agricultural regions, grassy areas, and around habitations in the lowlands and lower montane areas of w. Ecuador from Esmeraldas south through El Oro and Loja. Less numerous and more local in the eastern lowlands, though spreading and increasing as a result of deforestation; at least seasonally fairly common in the Río Marañón drainage around Zumba in extreme s. Zamora-Chinchipe. Blue-black Grassquits also wander up into the central valley on increasingly frequent occasions, for example in the Río Tumbaco valley east of Quito (N. Krabbe), and others have been seen at times in the Ibarra region. Most numerous below about 1000 m, but locally occurs up to 1300–1400 m in the Mindo and Maquipucuna regions, to 1500 m (occasionally higher) in southwestern intermontane valleys, and locally as high as 2300–2500 m in northern intermontane valleys.

Two races occur in Ecuador, *peruviensis* in the west and presumably *splen-*

dens in the east. In fact, however, we have not seen any specimens from e. Ecuador and suspect that none have actually been taken. Male *splendens* differ in having black flight feathers (these being distinctly brownish in *peruviensis*); females are similar.

Range: n. Mexico to extreme n. Chile, n. Argentina, and Uruguay.

*Tiaris olivacea   **Yellow-faced Grassquit**   **Semillerito Cariamarillo**

A recent arrival in Ecuador, the Yellow-faced Grassquit was first recorded from a specimen (LSUMZ) taken at Limoncocha on 30 Mar. 1976 (Tallman and Tallman 1977). Yellow-faced Grassquits have not been since recorded from this site, and there is only a single additional report from anywhere in the eastern lowlands, that of an immature male seen at Jatun Sacha on 20 Nov. 1994 (A. Jaramillo). In the early 1980s small numbers of Yellow-faced Grassquits began to be found in cleared areas and pastures in subtropical and upper tropical zones on the west slope of the Andes from Carchi (in the Río Mira valley and near Maldonado) and Imbabura (near Lita) south into n. Pichincha. The species is now fairly common in a region encompassing the area between San Miguel de los Bancos, Nanegalito, and Tandayapa; a female (WFVZ) was collected at Maquipucuna on 12 Jul. 1988. We anticipate that the Yellow-faced Grassquit will continue to increase and spread in Ecuador, and this prediction is supported by one having been seen as far south as Tinalandia in Jul. 1995 (P. Coopmans). In the west found mostly between 600 and 1800 m; in the east recorded only between 300 and 400 m.

One race: *pusilla*.

Range: e. Mexico to Panama, and mainly on lower Andean slopes of sw. Venezuela, Colombia, and nw. Ecuador (wandering to lowlands of ne. Ecuador); Greater Antilles.

Tiaris obscura   **Dull-colored Grassquit**   **Semillerito Oscuro**

Uncommon to locally (and perhaps seasonally) common but somewhat local in shrubby clearings, woodland and forest borders, and gardens in the lowlands and foothills of w. Ecuador mainly from Imbabura and w. and s. Esmeraldas (Atacames/Súa area, and near Quinindé on 8 Sep. 1994; P. Coopmans) south through El Oro and Loja; also considered to be a "nonbreeding visitor" to Playa de Oro in n. Esmeraldas (O. Jahn and K.-L. Schuchmann). The Dull-colored Grassquit is probably increasing as a result of ongoing deforestation. It was also recently found in the Río Marañón drainage of s. Zamora-Chinchipe in the vicinity of Zumba (P. Coopmans; RSR). Recorded mostly below 1400 m, but ranges up in arid intermontane valleys (particularly in e. Loja) as high as 1750 m.

The species was formerly (e.g., Meyer de Schauensee 1966, 1970) placed in the genus *Sporophila* and called the Dull-colored Seedeater, but more recently it has been placed in *Tiaris* (and called a grassquit) on account of its domed nest and buzzy song. One race of *T. obscura* occurs in Ecuador, *pauper*. See

J. M. Bates (*Studies in Neotropical Ornithology Honoring Ted Parker*, Ornithol. Monogr. no. 48: 91–110, 1997) for a detailed review of subspecific variation in the species.

Range: locally from n. and w. Venezuela to w. Peru, nw. Argentina, and w. Paraguay.

Oryzoborus angolensis   **Lesser Seed-Finch**   **Semillero Menor**

Uncommon to locally common in shrubby clearings, woodland and forest borders, and grassy areas in the lowlands of both e. and w. Ecuador, including the Río Marañón drainage of s. Zamora-Chinchipe in the Zumba region. Unlike some *Sporophila* seedeaters, the Lesser Seed-Finch tends to avoid areas with intensive agriculture. In the west recorded south to n. Manabí, Los Ríos, e. Guayas, and El Oro, occurring only in more humid regions. Found mostly below 900 m, but ranges in smaller numbers up in the foothills to about 1200 m, exceptionally as high as 1700 m (Cordillera del Cóndor; ANSP) where perhaps only occurring as a wanderer.

Two rather different (in males) races are found in Ecuador, *funereus* in the west and *torridus* in the east. Male *funereus* are essentially all black, whereas male *torridus* have the breast and belly chestnut; females are similar. These are still sometimes (e.g., in the 1983 and 1998 AOU Check-lists) considered as representatives of separate species, trans-Andean O. *funereus* (Thick-billed Seed-Finch) and cis-Andean O. *angolensis* (Chestnut-bellied Seed-Finch), but widespread intergradation between the two putative species has been reported in w. Colombia (see S. L. Olson, *Auk* 98[2]: 379–381, 1981); in addition, a male with all-black underparts was seen near Macas in Morona-Santiago in Aug. 1979 (RSR).

Range: s. Mexico to sw. Ecuador, n. and e. Bolivia, e. Paraguay, ne. Argentina, and se. Brazil.

Oryzoborus crassirostris   **Large-billed**   **Semillero Piquigrande**
                            Seed-Finch

Rare to uncommon and very local in damp grassy fields and shrubby areas near water in the lowlands of w. Ecuador, and in shrubbery around the marshy margins of certain lakes and damp grassy areas in the lowlands of e. Ecuador. In the west known only from Esmeraldas (El Placer; ZMCOP, ANSP), s. Los Ríos (19th-century record from Babahoyo), and El Oro (two ANSP specimens taken at Santa Rosa in Oct. 1950). In the northeast, where the species has only recently been recorded and may be increasing and spreading, known primarily from the lower Río Aguarico drainage; it remains possible, however, that the recent spate of observations is due more to observers being aware of the species than to an actual increase in numbers. Here the Large-billed Seed-Finch was first reported from a male, apparently this species, seen at Limoncocha on 5 Apr. and 3 May 1987, the significance of which was only realized much later (P. Coopmans). At Imuyacocha first reported in Mar. 1990 (F. Ortiz-Crespo); in Mar. 1991, ANSP workers also found small numbers of this species here, and a female was collected on 30 Mar.; a few were also located subsequently at nearby Zancudococha, where another female was taken on 5 Apr. There are also sightings of a male at Jatun Sacha on 9 Apr. 1992 (B. Bochan) and of a subadult male along the Maxus road southeast of Pompeya on 6 Nov. 1994 (RSR and F. Sornoza). The only site in the southeast from which the species has been reported is Kapawi Lodge on the Río Pastaza near the Peruvian border. In the west recorded as high as 700 m (El Placer), but in the east only below about 400 m.

The Large-billed Seed-Finch was given Near-threatened status by Collar et al. (1992, 1994). We are not certain what evidence this assessment was based on, but suspect that the species' frequent capture for the cagebird market, at least in some countries, was a major factor. Fortunately, small finches are not often captured in Ecuador. Nonetheless, because of the species' evident rarity in Ecuador, we also accord it Near-threatened status.

Occidentalis is the race ranging in w. Ecuador, with nominate *crassirostris* being found in the northeast; these taxa are very similar in appearance, though they differ in biometrics. *Occidentalis* (which is found also in w. Colombia) is here considered a subspecies of *O. crassirostris* and not (as in Meyer de Schauensee 1970, followed by Ridgely and Tudor 1989) a subspecies of *O. maximiliani* (Great-billed Seed-Finch), this being based on distributional considerations and its short tail.

Range: much of lowland Colombia, Venezuela, and the Guianas south locally to sw. Ecuador, ne. Peru, and n. Amazonian Brazil. Excludes Nicaraguan Seed-Finch (*Oryzoborus nuttingi*) of s. Middle America.

Oryzoborus atrirostris   Black-billed Seed-Finch   Semillero Piquinegro

Rare and local in damp grassy areas, shrubbery around marshy margins of oxbow lakes, and regenerating clearings in the lowlands of e. Ecuador. So far recorded mainly from Napo (near Archidona, Limoncocha, La Selva, and Jatun Sacha) and w. Morona-Santiago (Logroño [seen in Jun. 1984 by RSR, PJG, and S. Greenfield]; Tayuntza [female (WFVZ) obtained on 5 Aug. 1987]; and Santiago [uncommon in Jul.–Aug. 1989; ANSP, MECN]). There are also recent sightings from Kapawi Lodge in the far southeastern lowlands near the Peruvian border. The Black-billed Seed-Finch was first recorded in Ecuador only in the mid-1970s (Tallman and Tallman 1977; LSUMZ), and it may yet prove to be somewhat more widespread. Nonetheless in recent years the numbers of this species being reported have seemed to decline, while at the same time the number of Large-billeds (*O. crassirostris*) being found seems to be on the rise. Whether these two trends are correlated remains unknown. Recorded mostly below about 600 m (once to 850 m near Macas; P. Coopmans).

Collar et al. (1992, 1994) considered the Great-billed Seed-Finch (*O. maximiliani*), of which the Black-billed was then treated as a subspecies, to warrant Near-threatened status. As with the Large-billed Seed-Finch, we suspect that this assessment was based primarily on the species being frequently trapped for the cagebird market in the eastern part of its range. Fortunately this activity is not prevalent in Ecuador. Nonetheless, based on its evident rarity here, we consider the species to be Near-threatened in Ecuador.

O. atrirostris is treated as a species distinct from *O. maximiliani* (which is now considered to be found only in e. South America), following Sibley and Monroe (1990); this is based on geographical considerations and on males having a very different black (not white) bill. One race of *O. atrirostris* occurs in Ecuador, nominate *atrirostris*. However, the relationship between nominate *atrirostris* and the still poorly known *gigantirostris*, supposedly found disjunctly at the northern and southern ends of the species' range, remains uncertain. In our view *O. atrirostris* could turn out to be better considered as a monotypic species showing individual variation in bill size.

Range: locally in extreme se. Colombia, e. Ecuador, e. Peru, and n. Bolivia.

Sporophila schistacea   Slate-colored Seedeater   Espiguero Pizarroso

Rare and local (seemingly erratic in its appearances at any locality, and perhaps nomadic) in the lower growth of humid forest borders and in adjacent clearings in the foothills and the lower subtropical zone on the west slope of the Andes south to Pichincha. Also recorded very locally from the eastern lowlands, where thus far reported only from Napo. There are relatively few records of this scarce seedeater from Ecuador; it is not known to occur anywhere on a regular basis, and it does not normally associate with other seedeaters. In some areas a predilection for bamboo has been noted. Recent reports include small numbers seen at a rice field cut out of várzea forest near Jatuncocha in far e. Napo on 19 Sep. 1976 (see Ridgely 1980); two males (MCZ) taken at Cotapino in w. Napo on 3 Jul. 1964 (D. Norton); two singing males seen above Mindo in Pichincha on 17 Feb. 1980 (RSR and D. Finch et al.); males seen above El Placer in e. Esmeraldas on 10 Aug. 1987 (RSR et al.) and on 16 Dec. 1993 (J. Sterling et al.); and birds seen and tape-recorded near Quinindé in Aug.–Sep. 1995 (J. R. Fletcher, fide P. Coopmans). In the northwest recorded between 200 and 1400 m, in the east below 400 m.

Despite its evident rarity in Ecuador, there is no indication that the Slate-colored Seedeater is actually declining, and indeed continuing deforestation may ultimately result in an increase in some regions.

Two rather similar races are recorded from Ecuador, *incerta* in the northwest and *longipennis* in the east.

Range: locally from Belize and Honduras to nw. Ecuador, se. Peru, n. Bolivia, s. Venezuela, the Guianas, and e. Amazonian Brazil.

The similar Gray Seedeater (*S. intermedia*) has been recorded close to the Ecuadorian border at La Planada in Nariño, Colombia (Salaman 1994), and it might range, at least as a nonbreeding visitant or wanderer, south into adjacent n. Ecuador.

Sporophila murallae   Caquetá Seedeater   Espiguero de Caquetá

Rare to uncommon and somewhat local in damp grassy clearings and shrubbery along the margins of lakes and rivers, and on river islands, in the lowlands of e. Ecuador south through at least e. Morona-Santiago. One might have expected that the Caquetá Seedeater would have increased as a result of deforestation during the past several decades, but at least as yet it has shown little indication of doing so. Recorded up to about 400 m.

We follow F. G. Stiles (*Ornitol. Neotrop.* 7[2]: 75–107, 1996), who considered *S. murallae* as a species distinct from *S. americana* (Wing-barred Seedeater) of ne. South America and trans-Andean *S. corvina* (formerly *S. aurita*; Variable Seedeater). Monotypic.

Range: se. Colombia, e. Ecuador, ne. Peru, and w. Amazonian Brazil.

Sporophila corvina   Variable Seedeater   Espiguero Variable

Common and widespread in grassy and shrubby areas, agricultural regions, gardens, and around habitations in the more humid lowlands of w. Ecuador from Esmeraldas south through Guayas El Oro, and w. Loja. Ranges up in smaller numbers in cleared or semiopen areas to about 1300–1500 m, especially in the southwest.

F. G. Stiles (*Ornitol. Neotrop.* 7[2]: 75–107, 1996) considered the trans-Andean population of the *S. americana* complex to represent a species distinct from *S. corvina*. Stiles (op. cit.) also demonstrated that the correct specific name is *corvina*, not *aurita*. One race of *S. corvina* occurs in Ecuador, *ophthalmica*.

Range: s. Mexico to nw. Peru.

Sporophila lineola   Lined Seedeater   Espiguero Lineado

Apparently a casual austral migrant to grassy areas along rivers and in cleared areas in the lowlands of e. Ecuador, but status still relatively unclear; recorded from Aug. to Dec. There are only only a few Ecuadorian records of the Lined Seedeater, all of them recent and all from the northeast; it remains possible that the species is being overlooked to some extent. There is no evidence of nesting or even of song, and birds seen (only adult males can be separated from the more numerous Lesson's Seedeater [*S. bouvronides*], with which the Lined often consorts) have usually been with mixed-species *Sporophila* flocks. Apparently the first Ecuadorian record involved a male (MECN) taken by M. Olalla at Pañacocha on 28 Dec. 1969; this possibly is the source for the species having been recorded from Ecuador by Meyer de Schauensee (1970). More recent records include a male seen at Jatun Sacha on 27–29 Aug. and (same individual?) on 21 Sep. 1992 (B. Bochan); at least three males seen in an extensive grassy area on a large Río Napo island near the mouth of the Río Aguarico on 21–22 Sep. 1992 (T. J. Davis and RSR); and a male seen along the Maxus road southeast of Pompeya on 6 Nov. 1994 (RSR and F. Sornoza). Recorded below 400 m.

Does not include *S. bouvronides*. Monotypic.

Range: breeds in n. Argentina, Paraguay, and s. and (a separate population) ne. Brazil, migrating northward during austral winter (southern population) or northwestward during second half of year (caatinga population); birds found in Ecuador should belong to southern-breeding population (see J. M. Cardoso da Silva, *Bull. B. O. C.* 115[1]: 14–21, 1995).

Sporophila bouvronides   **Lesson's Seedeater**   **Espiguero de Lesson**

Apparently a rare visitor to the lowlands of e. Ecuador, but status still poorly understood. Only a few Ecuadorian specimens of the Lesson's Seedeater are known, including a male (ANSP) taken by R. Olalla at Montalvo in Pastaza on 31 Aug. 1949; other males (MECN) were taken by M. Olalla at Edén on 4 Dec. 1964, and three were obtained at Lagartococha on 12–16 Dec. 1967. Chapman (1926) also mentions an old record from Machay in Tungurahua. A female *S. lineola/S. bouvronides* "pair" (females of the two species are indistinguishable to species) was taken by T. Mena at Avila in Napo on 5 Feb. 1950 (ANSP). Sightings include a male seen with other seedeaters near Tena on 16 Feb. 1990 (J. Arvin and R. A. Rowlett et al.); a male seen at Imuyacocha on 6 Dec. 1990 (J. Sterling) and one again on 30 Mar. 1991 (RSR and F. Sornoza); at least four males with a large mixed seedeater flock at Lago Agrio on 5 Apr. 1991 (RSR); a male with other seedeaters at Jatun Sacha on 27–28 Nov. 1992 (A. Long and J. Tobias); two males seen at Zancudococha on 9 Mar. 1992 (D. Wolf); at least 20 of both sexes seen on a Río Napo island near La Selva on 3 Dec. 1992 (G. H. Rosenberg et al.); six or more males seen along the Maxus road southeast of Pompeya on 6 Nov. 1994, with one still there on 30 Jan. 1995. There is no evidence of breeding in Ecuador, where the species has never been heard to sing. Recorded mostly from Nov. to Apr., with one anomalous Aug. record. Recorded below 400 m.

S. bouvronides is regarded as a species distinct from *S. lineola* (Lined Seedeater), following most recent authors (see especially P. Schwartz, *Ann. Carnegie Mus.* 45: 277–285, 1975). One race of *S. bouvronides* occurs in Ecuador, *restricta*.

Range: breeds locally in n. Colombia, n. Venezuela, and the Guianas, ranging south and west into w. Amazonia when not nesting.

Sporophila luctuosa   **Black-and-white**   **Espiguero Negriblanco**
                          Seedeater

An uncommon to locally fairly common breeder (mainly during first half of calendar year) in grassy areas, shrubby clearings, and roadsides in the sub-tropical and temperate zones on both slopes of the Andes and in the central and interandean valleys. During the nonbreeding season (at least Aug.–Jan.) irregular numbers range down into the eastern lowlands, where they apparently mostly remain quite close to the Andes. Black-and-white Seedeaters do not seem to hold territories in the lowlands, but here occur in groups, primarily among mixed flocks of other *Sporophila*. At least occasionally they range farther east: 10 or more adult males (and probably at least as many female-plumaged birds) were among a large mixed-species *Sporophila* flock on a Río Napo island on the Peruvian border near the mouth of the Río Aguarico on 21–22 Sep. 1992 (ANSP), and six or more adult males were seen with a seedeater flock on another Napo island near La Selva on 3 Dec. 1992 (G. H. Rosenberg et al.; others have been seen here since); no fewer than 75 were with a large mixed-species seedeater flock along the Maxus road southeast of Pompeya on 6 Nov. 1994 (RSR and F. Sornoza). However, the species appears to be distinctly rare in the western lowlands, where the only report appears to be a (wandering?) male seen at Playa de Oro in n. Esmeraldas on 25 Nov. 1996 (O. Jahn et al.). Breeding Black-and-white Seedeaters appear to be most numerous in far n. Ecuador, and again in the south (e.g., in Loja and Zamora-Chinchipe); more information is, however, needed, both as to the species' breeding status in many areas and as to the timing of its movements into the eastern lowlands. Recorded up to at least 2400 m in Pichincha and Loja.

Monotypic.

Range: locally in Andes and adjacent lowlands from w. Venezuela to w. Bolivia; Santa Marta Mountains.

Sporophila nigricollis	**Yellow-bellied Seedeater**	**Espiguero Ventriamarillo**

Fairly common to common in grassy areas, shrubby clearings, and roadsides in the more humid lowlands of w. Ecuador south through El Oro and Loja, also ranging in substantial numbers up into the foothills and subtropical zone on the west slope of the Andes, especially in arid intermontane valleys of the north. Less numerous and more local in the east, where found principally on lower Andean slopes and in adjacent lowlands; common, however, in the Río Marañón drainage of s. Zamora-Chinchipe from Zumba north to the Valladolid area. The Yellow-bellied Seedeater is not found at all any great distance east of the Andes, with the easternmost report being a male seen with Chestnut-bellied Seedeaters (*S. castaneiventris*) along the Maxus road southeast of Pompeya on 11 Nov. 1995 (RSR). Ranges up locally in cleared areas on both slopes of the Andes to over 2000 m, and also regular and apparently breeding (fide N. Krabbe) in the Tumbaco valley east of Quito (2400 m).

Two races are formally recorded from Ecuador, nominate *nigricollis* in the east and *vivida* in the west. The population found in the Río Marañón drainage—where no specimens have been taken—is probably referable to *inconspicua*, which otherwise is known from Peru. Male *vivida* are brighter yellow below than the nominate race; male *inconspicua* have less black on the face and bib than the nominate race; females of all races are identical.

Range: Costa Rica to Colombia, Venezuela, Guyana, and nw. and e. Peru; e. and s. Brazil.

Sporophila peruviana   Parrot-billed   Espiguero Pico de Loro
                               Seedeater

Uncommon to (seasonally) fairly common to common in grassy areas, shrubby clearings, agricultural regions (especially in hedgerows and on fallow or regenerating fields), and desert scrub in the more arid lowlands of sw. Ecuador from cen. Manabí (around Bahía de Caráquez; also on Isla de la Plata, at least at times) south locally through El Oro and Loja, in Loja ranging east at least seasonally as far as the Catamayo region. The Parrot-billed Seedeater is rather erratic in its occurrences in much of Ecuador, and during the nonbreeding season (Jun.–Jan.) it ranges widely in flocks, sometimes occurring with larger numbers of Chestnut-throated Seedeaters (*S. telasco*). Generally numbers are highest on the Santa Elena Peninsula. Ranges up in the valleys of interior El Oro and Loja to about 1200–1400 m, especially in scrub and sugarcane fields around Catamayo, but mainly occurs below about 800 m.

One race: *devronis*.
Range: sw. Ecuador and w. Peru.

*Sporophila simplex   Drab Seedeater   Espiguero Simple

Rare to uncommon and apparently local (though perhaps mostly just overlooked, and can be at least seasonally numerous) in arid scrub and adjacent agricultural fields in certain intermontane valleys in sw. Ecuador in Loja and s. Azuay; presumably it also occurs in the valleys of interior e. El Oro. The Drab Seedeater was first found in Ecuador by D. Norton, who collected three specimens (MCZ) at Hacienda Yamaná on 2–8 Jul. 1965, and subsequently by R. A. Paynter, Jr., and D. Norton who together obtained two more specimens (MCZ) 5 km west of Sabiango on 13 Oct. 1965; they later took an additional three birds (also in MCZ) 10 km northwest of Oña in Azuay on 2–3 Nov. 1965. Since the late 1980s the Drab Seedeater has been seen by various observers at several sites in s. Azuay (principally near Oña, also in the Ríos Yunguilla and Rircay valleys) and near Catamayo and Gonzanamá in e. Loja. Although unrecorded, it is possible that the species may also range in the Río Marañón drainage as far north as around Zumba. Recorded from about 650 to 1900 m.

Monotypic.
Range: locally on Andean slopes and valleys of sw. Ecuador and w. Peru.

Sporophila telasco   Chestnut-throated   Espiguero Gorjicastaño
                             Seedeater

Fairly common to seasonally common in grassy and shrubby areas, agricultural regions (where often particularly numerous at the edge of sugarcane fields), and around habitations in the lowlands of w. Ecuador from w. Esmeraldas south through El Oro and w. Loja. The Chestnut-throated Seedeater occurs in both arid and quite humid regions, but in the former it is found almost exclusively in irrigated situations. Although unrecorded, it is possible that the species may also range in the Río Marañón drainage as far north as around Zumba. Recorded mainly in the lowlands below about 500 m, only rarely (or perhaps only seasonally) occurring up into the foothills. In Apr. 1993, however, possibly in response to the heavy rains of that season, large numbers were found as high as around Alamor (1200 m) in w. Loja (RSR); in addition, a population appears to be resident in the wide valleys near and south of Catamayo in e. Loja, at 1200–1400 m.

Monotypic.

Range: sw. Colombia to extreme n. Chile.

The probably closely related Tumaco Seedeater (*S. insulata*), which is known only from around Tumaco Island in adjacent coastal w. Nariño in sw. Colombia, might also occur in coastal n. Esmeraldas. Having gone unrecorded since its discovery in 1912, the species was rediscovered in Jul. 1994 in grassy seaside scrub on an island near the type locality (P. Salaman, *Cotinga* 4: 33–35, 1995). Collar et al. (1992) considered the Tumaco Seedeater to be Endangered, and at that time it was even feared possibly already Extinct.

| *Sporophila minuta* | Ruddy-breasted Seedeater | Espiguero Pechirrojizo |

Rare to uncommon and local in grassy fields and adjacent secondary scrub in the lowlands of nw. Ecuador, also ranging up on the west slope into the lower subtropical zone (principally in arid intermontane valleys but also following clearing at least to some extent). Recorded from Esmeraldas (including several recent reports from the Borbón area), Carchi, Imbabura (mainly from the Río Mira valley, and also seen near Lita), and Pichincha (a few recent sightings from the San Miguel de los Bancos area; PJG et al.). Recorded up to about 1500 m.

One race: nominate *minuta*.

Range: w. Mexico to nw. Ecuador, e. Colombia, much of Venezuela, the Guianas, and e. Amazonian Brazil. Excludes Tawny-bellied Seedeater (*S. hypoxantha*) of interior s. South America.

| *Sporophila castaneiventris* | Chestnut-bellied Seedeater | Espiguero Ventricastaño |

Common and widespread in agricultural regions, grassy areas, shrubby margins of lakes and rivers, and river islands in the lowlands of e. Ecuador. The pretty Chestnut-bellied Seedeater has obviously benefited from the deforestation that has swept so much of e. Ecuador, and it doubtless will continue to spread and increase in numbers. Ranges in cleared areas on the east slope of the Andes up to about 1300 m, locally and in small numbers as high as at least 1800 m (below Baeza; P. Coopmans and M. Lysinger).

Monotypic.

Range: e. Colombia and sw. Venezuela to n. Bolivia, locally in Amazonian Brazil, and the Guianas.

Amaurospiza concolor   **Blue Seedeater**   **Semillero Azul**

Rare and local (probably overlooked to some extent) in the lower growth and borders of montane forest and mature secondary woodland in the foothills and subtropical zone on the west slope of the Andes south to Loja (female taken by D. Norton 5 km southeast of Gonzanamá on 7 Oct. 1965; MCZ). Recorded from Pichincha (in the Mindo/Chiriboga area and above Tinalandia), Chimborazo (old specimens from the Pallatanga/Chimbo area), and El Oro (near Portovelo and near Zaruma). Also recently recorded from the coastal Cordillera de Colonche in sw. Manabí (Cerro San Sebastián in Machalilla National Park; Parker and Carr 1992) and w. Guayas (above Salanguilla; N. Krabbe). The association of the Blue Seedeater with bamboo that has been widely noted in Middle America (Ridgely and Gwynne 1989; Stiles and Skutch 1989; Howell and Webb 1995) is less strict in Ecuador, where the species seems at least as often to be found away from it. Recorded mostly from 1100 to 2300 m, locally down to about 800 m on the coastal cordillera and at Tinalandia.

One race: *aequatorialis*.

Range: highlands from s. Mexico to Panama; west slope of Andes in sw. Colombia and w. Ecuador.

Catamenia inornata   **Plain-colored Seedeater**   **Semillero Sencillo**

Fairly common and widespread in paramo and grassy areas with scattered shrubbery in the temperate zone of the Andes from Carchi south through El Oro and Loja. Recorded mainly from 2600 to 3800 m.

One race: *minor*.

Range: Andes from w. Venezuela to sw. Peru and nw. Argentina.

Catamenia homochroa   **Paramo Seedeater**   **Semillero Paramero**

Rare to uncommon and local (and perhaps erratic or seasonal) in shrubbery at the edge of montane forest in the temperate zone and near treeline on both slopes of the Andes. On the west slope recorded from Pichincha (Volcán Pichincha area) south locally to Chimborazo, but probably also occurs north to the Colombian border. On the east slope known only from w. Sucumbíos and w. Napo, the road east of Salcedo (N. Krabbe), Tungurahua (an old record from San Rafael), along the Gualaceo-Limón road near the Azuay/Morona-Santiago border (P. Coopmans), and a few localities in Loja (Cordillera de Cordoncillo [ANSP] and Uritusinga, and at several sites in Podocarpus National Park [v.o.]) and adjacent Zamora-Chinchipe (Cordillera Las Lagunillas; T. J. Davis). Recorded mainly from 2500 to 3500 m.

One race: nominate *homochroa*.

Range: Andes from w. Venezuela to w. Bolivia; Sierra de Perijá and Santa Marta Mountains; tepuis of s. Venezuela.

Catamenia analis   Band-tailed Seedeater   Semillero Colifajeado

Uncommon to locally fairly common in low shrubby areas, hedgerows in agricultural regions, and arid scrub in the subtropical and temperate zones of the Andes, mainly in drier interandean valleys and on the slopes above the central valley. The Band-tailed Seedeater is entirely lacking from the east slope of the Andes, and also is not found on their west slope in the north. Ranges from Carchi south through Loja; though the Band-tailed Seedeater was not recorded from Loja until recently (see Krabbe 1992), there have been recent reports of it from several localities, mainly in the province's extreme north. Recorded mostly from 1500 to 3000 m.

One race: *soederstromi*.

Range: Andes from n. Colombia to n. Chile and nw. Argentina; Santa Marta Mountains.

Phrygilus unicolor   Plumbeous Sierra-Finch   Frigilo Plomizo

Fairly common to common in paramo and shrubbery near or slightly below treeline in the Andes. The Plumbeous Sierra-Finch occurs higher than any other bird on Volcán Chimborazo, there ranging up at least as high as the mountaineering refugio at 4800 m. Recorded mostly between 3000 and 4300 m.

One race: *geospizopsis*.

Range: Andes from w. Venezuela to Chile and Argentina.

Phrygilus plebejus   Ash-breasted Sierra-Finch   Frigilo Pechicinéreo

Fairly common to common in shrubby and (to a lesser extent) grassy areas in the arid highlands and interior valleys from s. Carchi (La Concepción; Salvadori and Festa 1899a) and Imbabura (Río Mira valley near Salinas and Tumbabiro; M. B. Robbins) south through much of Loja; generally more numerous southward. The Ash-breasted Sierra-Finch is also recorded locally in small numbers in desert scrub near the coast of s. El Oro in the Huaquillas region; a female (MECN) was obtained here on 11 Oct. 1980 (F. Ortiz-Crespo). Recorded mainly between about 1500 and 3500 m, but locally occurs down to 200–400 m in s. Loja and even as low as sea level in El Oro.

One race: *ocularis*.

Range: Andes from n. Ecuador to n. Chile and nw. Argentina, also in arid lowlands of sw. Ecuador and nw. Peru.

Phrygilus alaudinus   Band-tailed Sierra-Finch   Frigilo Colifajeado

Uncommon and somewhat local in open, barren, stony or sandy areas with sparse grass cover and scattered low bushes in more arid highlands and the central and interandean valleys from Carchi (Río Mira valley at La Concepción and [M. B. Robbins] near Salinas and Tumbabiro) south locally through Azuay and Loja. The Band-tailed Sierra-Finch is also recorded locally in the arid coastal lowlands of w. Guayas (Santa Elena Peninsula) and extreme sw. El Oro

(Huaquillas region). Recorded mainly from 1200 to 3000 m, in the southwest also locally at or near sea level.

Two races occur in Ecuador, *bipartitus* in the highlands and the paler and somewhat smaller *humboldti* near the coast.

Range: Andes from n. Ecuador to nw. Argentina and cen. Chile; arid lowlands of sw. Ecuador and nw. Peru.

*[*Piezorhina cinerea*   Cinereous Finch   Pinzón Cinéreo]

Uncertain. There is only one report of this distinctive Peruvian endemic, a single bird that was seen 20 km north of El Empalme in the Río Catamayo valley (1100 m) on 15 Mar. 1990 (B. Whitney). Whether the Cinereous Finch is resident there or this bird was simply a vagrant from its normal range in nw. Peru remains unknown. That area, as well as seemingly even more suitable habitat around Zapotillo in extreme s. Loja, has been searched on several subsequent occasions, but always without success; it seems possible that relatively wet recent years may have caused the species to disappear, or at least have discouraged it from prospecting northward from Peru, where it inhabits open, desert-like plains with scattered small shrubs. It has also been suggested (N. Krabbe, pers. comm.) that, as the species is not infrequent as a cagebird in Peru, the Ecuador bird could merely represent an escape from captivity. The Cinereous Finch also occurs just south of the Ecuador border in Tumbes, Peru, and thus might wander to the Huaquillas region of extreme s. El Oro.

Monotypic.

Range: nw. Peru, with one report from sw. Ecuador.

Sicalis flaveola   Saffron Finch   Pinzón Sabanero Azafranado

Locally fairly common to common around habitations and in agricultural regions in semiarid lowlands of sw. Ecuador in coastal El Oro and in Loja, where recorded inland into e. Loja as far as Loja (city), Vilcabamba, and Amaluza. In addition, small numbers of Saffron Finches have long occurred around Guayaquil and more recently at Cerro Blanco in Guayas, though for some reason they have shown little or no sign of increasing there. The species is conspicuously absent, however, from seemingly ideal terrain on the Santa Elena Peninsula. Also recorded from the Río Marañón drainage of extreme se. Ecuador in s. Zamora-Chinchipe, where known only from a single bird seen near Zumba on 7 Apr. 1994 (P. Coopmans and M. Lysinger); it seems possible that the Saffron Finch, which ranges in the upper Río Marañón valley of adjacent Peru, may increase in this region. Saffron Finches are particularly numerous in the desert-like areas of extreme s. coastal El Oro (e.g., around Huaquillas) and s. Loja (e.g., around Macará and Zapotillo). Ranges locally as high as about 2000 m around Loja (city).

One race: *valida*.

Range: cen. Panama (introduced); much of n. Colombia and n. Venezuela;

sw. Ecuador and nw. Peru; n. and e. Bolivia, Paraguay, n. Argentina, Uruguay, and s. and e. Brazil. More than one species is perhaps involved.

| *Sicalis luteola* | Grassland Yellow-Finch | Pinzón Sabanero Común |

Uncommon to fairly common but rather local in damp grassy areas with *Scirpus* sedges and *Juncus* rushes around the edge of marshes and lakes and in pastures and fields with tall grass in the highlands from Carchi south into Azuay. Somewhat surprisingly, there are only a few records from farther south, where in Loja especially the status of this yellow-finch requires clarification: one was collected (ANSP) by F. A. Colwell at Loja on 10 May 1948, another was obtained (WFVZ) by L. Kiff at San Pedro on 5 Mar. 1989, about 10 were seen at Gualedal on 5 Nov. 1992 (N. Krabbe), and a flock of about 10 nonbreeding birds was seen near Catamayo on 13 Jan. 1998 (RSR et al.). Recorded mostly between 2200 and 3200 m, but Loja birds have been seen as low as 1300 m.

Recent evidence points to a decline in overall numbers of the Grassland Yellow-Finch through much of the central valley, though not yet to the point where the species would be considered threatened. The species is certainly not generally at risk across its vast range (Collar et al. 1994), though some other populations (e.g., in Panama; RSR) are also believed to be decreasing.

One race: *bogotensis*.

Range: locally from s. Mexico to w.-cen. Panama; locally in Andes from n. Colombia to s. Peru; ne. Colombia, n. and e. Venezuela, and Guyana; lower Amazonian Brazil; cen. Chile, cen. and n. Argentina, Uruguay, and s. Brazil, migrating north to e. Bolivia and ne. Brazil. More than one species may be involved.

| *Sicalis taczanowskii* | Sulphur-throated Finch | Pinzón Sabanero Golisurfureo |

Rare to uncommon and apparently local and erratic in dry desert-like areas with extensive bare ground and at most sparse low shrubbery in the lowlands of sw. Ecuador. Recorded mainly from w. Guayas on the Santa Elena Peninsula, and there also is an old record from Guayaquil (Chapman 1926). A few flocks of Sulphur-throated Finches were also seen in the early 1980s in coastal s. El Oro in the Huaquillas region (RSR et al.). Marchant (1958, pp. 385–386) recorded "vast flocks" on the Santa Elena Peninsula in the 1950s, but he specifically noted that there was "no indication of breeding." He therefore suggested that the species might only be a "post-breeding immigrant," presumably moving north from Peru during the dry season; he appears to have found it primarily between the months of Aug. and Feb. In subsequent years the numbers of Sulphurthroated Finches have not been nearly so high, and the species has seemed especially rare since the late 1980s, perhaps at least in part because in this period there have been years with unusually heavy rains. When seen at all, the species continues to be found during the same general time period. Recorded only below 50 m.

Overall numbers of this species appear to have decreased in recent years and, as noted above, this certainly seems to be the case in Ecuador; information from Peru, however, is scanty. The species has not been considered to be generally at risk (Collar et al. 1994), but we accord it Data Deficient status for Ecuador. Its decline on the Santa Elena Peninsula would appear to echo the decline of two other species formerly found there, the Least Seedsnipe (*Thinocorus rumicivorus*) and the Tawny-throated Dotterel (*Oreopholus ruficollis*).

Formerly placed in the monotypic genus *Gnathospiza*, but with some reluctance we follow *Birds of the World* (vol. 13) in merging that into *Sicalis*. Monotypic.

Range: sw. Ecuador and nw. Peru.

Haplospiza rustica   Slaty Finch   Pinzón Pizarroso

Uncommon, local, and seemingly somewhat erratic or nomadic (but occasionally or temporarily more numerous, as it was in Jul. 1988 above Maldonado in Carchi; M. B. Robbins et al.) in the lower growth and borders of montane forest in the subtropical and temperate zones on both slopes of the Andes. On the west slope recorded in the past south to the border area between Chimborazo and e. Guayas (Llagos, Río Chanchán), though in recent years found south mainly to Pichincha; it was mist-netted at Salinas in Bolívar in Feb. 1995 (N. Krabbe et al.). The Slaty Finch favors stands of *Chusquea* bamboo and is especially numerous when these are seeding. Recorded from about 1500 to 3300 m.

One race: nominate *rustica*.

Range: highlands from s. Mexico to w. Panama; mountains of n. Venezuela, and Andes from w. Venezuela to w. Bolivia; locally on tepuis of Venezuela; Sierra de Perijá and Santa Marta Mountains.

**Coryphospingus cucullatus*   Red Pileated-Finch   Brasita de Fuego Rojo

Uncommon in secondary scrub and gardens in the Río Marañón drainage of extreme s. Zamora-Chinchipe in the Zumba area. The Red Pileated-Finch was first recorded from a singing male that was collected there on 15 Aug. 1989 (ANSP). Subsequently small numbers have been found to range north to near Valladolid (RSR). Recorded from about 1100 to 1400 m.

The species is often called the Red-crested Finch, but see Ridgely and Tudor (1989).

One race: *fargoi*.

Range: locally on east slope of Andes in extreme s. Ecuador and n. Peru; n. and e. Bolivia and s. Brazil to n. Argentina.

Atlapetes pallidinucha   Pale-naped   Matorralero
Brush-Finch   Nuquipálido

Fairly common in shrubbery near treeline and at edge of montane forest and woodland in the temperate zone on the east slope of the Andes. We consider the several old records of the Pale-naped Brush-Finch from the west slope of the

Andes (e.g., from around Nono) to be erroneous or at least doubtful; all reliable modern records come from the east slope. Recorded mostly from 2700 to 3700 m, occasionally or in small numbers as low as 2400–2500 m (e.g., in the vicinity of Quebrada Honda) and locally as high as 3900 m at Papallacta pass.

One race: *papallactae*.

Range: Andes from sw. Venezuela to extreme n. Peru.

Atlapetes latinuchus	Rufous-naped Brush-Finch	Matorralero Nuquirrufo

Common but somewhat local in shrubby borders of montane forest, secondary woodland and scrub, and hedgerows in mainly agricultural regions on both slopes of the Andes and on slopes above the central and interandean valleys. The Rufous-naped Brush-Finch occurs along the entire west slope of the Andes, but on the east slope it is found mainly from sw. Morona-Santiago (Gualaceo-Limón road) southward; it also spills over the Continental Divide onto the east slope in the extreme north, along the upper part of the La Bonita road in Sucumbíos (RSR). The general complementarity of this species' range with that of the Slaty Brush-Finch (*A. schistaceus*)—as pointed out by J. V. Remsen, Jr., and W. S. Graves IV (*Auk* 112[1]: 210–224, 1995)—is indeed striking. In Ecuador the Slaty occurs only on the east slope, where it ranges south to the Gualaceo-Limón road; from about there southward only the Rufous-naped is found, with overlap between the two species occurring along parts of that road. The two are syntopic at one site in far n. Ecuador (La Bonita road, there even ranging in the same flocks), and both species have also been recorded from Cerro Mongus (Robbins et al. 1994; RSR). Recorded mainly between 1500 and 3200 m, but occurring higher in nw. Ecuador (regularly to 3400–3500 m, locally as high as 3600–3750 m in forest on Volcán Pichincha).

We follow J. García-Moreno and J. Fjeldså (*Ibis* 141[2]: 199–207, 1999) in separating forms found in the northern Andes, long considered conspecific with *A. rufinucha*, as a species distinct from *A. rufinucha* (Bolivian Brush-Finch) of the Bolivian Andes and other species found in the Andes of s. Peru. Three rather well-marked races of *A. latinuchus* occur in Ecuador. *Spodionotus* occurs on the west slope from the Colombian border south to Chimborazo and on interandean slopes south at least to the west side of Papallacta pass above Pifo. *Comptus* occurs on the west slope from Chimborazo (where it intergrades with *spodionotus*) south through w. Loja. Nominate *latinuchus* occurs on the east slope from n. Azuay (Palmas; Berlioz 1932b) and sw. Morona-Santiago (Gualaceo-Limón road) south to the Peruvian border. *Comptus* resembles *spodionotus* but differs in having a strong black submalar stripe, yellow on the lores, and usually a paler crown color (especially to the rear). The nominate race resembles *spodionotus* aside from having a prominent white wing speculum. As García-Moreno and Fjeldså (op. cit.) acknowledge, some of these northern taxa will also likely prove to be separate species; they did not specifically address the issue.

Range: Andes from n. Colombia to n. Peru; Sierra de Perijá.

Atlapetes tricolor   Tricolored Brush-Finch   Matorralero Tricolor

Locally fairly common in the lower growth of montane forest borders and secondary woodland in the foothills and subtropical zone on the west slope of the Andes from e. Esmeraldas (El Placer) south to Pichincha and w. Cotopaxi (Caripero area; N. Krabbe and F. Sornoza), and in Azuay (Manta Real; ANSP) and El Oro (La Chonta and Buenaventura). There is no apparent reason why the Tricolored Brush-Finch should not also occur in the intervening region, but as yet it remains unrecorded from this area; the absence of records seems particularly noteworthy given the amount of collecting at appropriate elevations that occurred in Chimborazo and e. Guayas in the late 19th and early 20th centuries. Recorded mostly between about 600 and 1800 m, but at Caripero in w. Cotopaxi found much higher (2200–2400 m).

One race: *crassus*. In *Birds of the World* (vol. 13) it is suggested that this taxon may ultimately best be subdivided.

Range: west slope of Andes in sw. Colombia and w. Ecuador; east slope of Andes in Peru. More than one species is perhaps involved.

Atlapetes schistaceus   Slaty Brush-Finch   Matorralero Pizarroso

Uncommon in shrubby borders of montane forest and secondary woodland (including woodland at or near treeline) on the east slope of the Andes south to nw. Morona-Santiago (Gualaceo-Limón road). Contra Ridgely and Tudor (1989), the Slaty Brush-Finch does not occur in s. Ecuador, where its southern limit was long considered uncertain (see *Birds of the World*, vol. 13). It has been suggested (J. V. Remsen, Jr., and W. S. Graves IV, *Auk* 112[1]: 210–224, 1995) that the absence of the Slaty Brush-Finch from s. Ecuador results from its being replaced there by the Rufous-naped Brush-Finch (*A. latinuchus*). We consider the several old specimens of the Slaty Brush-Finch supposedly obtained on the west slope of the Andes to refer only to mislabeled birds; despite a tremendous amount of recent field work in this region, there are no modern west-slope reports of the species. Recorded mostly between about 2500 and 3400 m.

One race: nominate *schistaceus*.

Range: Andes of w. Venezuela and Colombia and on east slope of Andes in n. and cen. Ecuador; east slope of Andes in cen. Peru; Sierra de Perijá.

Atlapetes leucopterus   White-winged   Matorralero Aliblanco
                          Brush-Finch

Fairly common but somewhat local in shrubby growth and lighter woodland in more arid regions on slopes above the central and interandean valleys from Imbabura south through El Oro and Loja. Small numbers of White-winged Brush-Finches have also apparently begun to move into cleared areas on the west slope, where the species is now found locally in Imbabura (Hacienda La Florida; P. Coopmans), Pichincha (the Nanegal, Maquipucuna, and Tandayapa areas), Cotopaxi (above Pilaló; RSR and G. H. Rosenberg), and Azuay (Sural; N. Krabbe). Recorded mostly from 1000 to 2600 m, northward tending to occur higher, southward lower.

Two races occur in Ecuador, nominate *leucopterus* south to Chimborazo and w. Azuay (Sural) and *dresseri* in Loja and s. El Oro. *Dresseri* is similar to the nominate race but has a larger area of black on the forecrown and a variable amount of white on the face. This variation is readily seen in recently obtained ANSP material: in some individuals the white is limited to the ocular area, whereas in others it is much more extensive, encorporating much of the sides of the head. For a discussion of this situation, interpreted as an unusually frequent tendency toward partial albinism, see J. W. Fitzpatrick (*Auk* 97[4]: 883–887, 1980).

Range: Andes of w. Ecuador and nw. Peru.

Atlapetes seebohmi	Bay-crowned Brush-Finch	Matorralero Coronicastaño

Rare to locally fairly common in the lower growth of montane woodland and scrub, and in regenerating cleared areas in the subtropical zone of sw. Ecuador in Loja (perhaps ranging also into adjacent e. El Oro), generally becoming more numerous southward (e.g., in the Utuana, Sozoranga, and Sabiango regions). The Bay-crowned Brush-Finch ranges north and east to near the city of Loja (MCZ), from there occurring south and west locally to the Peruvian frontier. Recorded mostly from 1300 to 2300 m, locally as high as 2600 m.

One race: *simonsi*. Material (ANSP) recently obtained at and near the type locality (Celica) of *celicae* demonstrates that that taxon is not worthy of recognition, a suggestion made earlier by R. A. Paynter, Jr. (in *Birds of the World*, vol. 13). *Celicae*, then known only from the type specimen, was distinguished from *simonsi* by its supposedly slightly smaller size and narrow black frontlet. Recent specimens show that these characters are too variable to support continued recognition of the form. *Seebohmi* (of nw. Peru in La Libertad and Ancash) and *simonsi* have sometimes been treated as subspecies of *A. nationi* (Rusty-bellied Brush-Finch) of cen. and s. Peru, but we favor regarding *A. seebohmi* and *A. nationi* as separate species (see Ridgely and Tudor 1989).

Range: Andes of sw. Ecuador and nw. Peru.

Atlapetes albiceps	White-headed Brush-Finch	Matorralero Cabeciblanco

Uncommon to locally fairly common in the lower growth of deciduous woodland and scrub in the lowlands and foothills of sw. Ecuador in s. Loja, where found north only to the El Empalme region (ANSP, MECN) in the Río Catamayo valley east of Celica, and at Río Casanga (AMNH). The White-headed occurs at lower elevations than the other Ecuadorian brush-finches, being recorded between about 200 and 1100 m.

Monotypic.

Range: sw. Ecuador and nw. Peru.

Atlapetes pallidiceps   Pale-headed   Matorralero Cabecipálido
                                 Brush-Finch

Very rare and local in low woodland and scrub in a few arid intermontane
valleys of s. Azuay, where known from the Río Jubones drainage at Girón, north
of Oña, and Valle de Yunguilla. In 1965 the Pale-headed Brush-Finch was considered "fairly common" at a site 10 km northwest of Oña (R. A. Paynter, Jr.,
Bull. Mus. Comp. Zool. 143[4]: 312, 1972), but even then this and the nearby
Río Léon were the only localities where the species could be found. A few years
later M. Olalla collected a series of this species between 31 Dec. 1968 and 6
Jan. 1969; these are labeled as having been taken considerably farther south,
at "Valle de Cazanga" in sw. Loja in the Río Catamayo valley west of Catacocha (probably between 700 and 1000 m). Four of these specimens are in the
MECN, and two were donated to the AMNH. This Loja locality is totally unexpected given what had been understood about the distribution of this very rare
and localized bird, and recent surveys in the Río Catamayo valley have revealed
the presence of only a single brush-finch species, the White-headed (*A. albiceps*;
White-wingeds [*A. leucopterus*] occur at somewhat higher elevations). We thus
suspect that the "Valle de Cazanga" specimens may have been mislabeled, as
suggested too by Collar et al. (1992). Known with reasonable certainty from
between about 1500 and 2100 m.

For several decades there were no reports of the Pale-headed Brush-
Finch, this despite several concerted searches for it, including one (in Mar. 1992;
M. B. Robbins and G. H. Rosenberg) that reached what was apparently
Paynter's 1965 site, now quite degraded. As a result the species has been treated
as warranting Endangered status (Collar et al. 1992) or even Critical status
(Collar et al. 1994). Present evidence supports its having Critical status, and we
so designate it here. In Nov. 1998, N. Krabbe et al. finally succeeded in locat-
ing a small population in a 25-ha patch of degraded scrub woodland in the
Río Yunguilla valley near Girón (A. Agreda, N. Krabbe, and O. Rodriguez,
Cotinga 11: 50–54, 1999). Some 12 pairs are estimated to be present in this
area, and there are thought to be a very few other individuals and pairs in even
smaller patches of habitat nearby. The welcome discovery that the species still
exists has led to a series of research and conservation efforts on its behalf, and
the area where the species was found has now been purchased by the Fundación
Jocotoco, which will manage this and adjacent areas with an eye to increasing
the species' population size.

Monotypic.

Range: very locally in Andes of sw. Ecuador.

Atlapetes leucopis   White-rimmed   Matorralero de Anteojos
                              Brush-Finch

Rare, local, and inconspicuous in the undergrowth of montane forest in the subtropical and temperate zones on the east slope of the Andes south to Azuay
(Palmas), nw. Morona-Santiago (Gualaceo-Limón road, where first found in
1984 by N. Krabbe [Fjeldså and Krabbe 1986]), and Loja (Loma Angashcola;

R. Williams). The White-rimmed Brush-Finch was also recorded recently from the west slope of the Andes in Imbabura, first from specimens (LSUMZ, MECN) taken along the road from Laguna Cuicocha to Apuela and Selva Alegre on 9 Sep. 1983 (C. Gregory and D. Schmitt); there have also been subsequent records, including two MECN specimens, from the Cordillera de Toisán at Loma Taminanga (N. Krabbe). Recorded between about 2200 m (along the crest of the Cordillera de Huacamayos; v.o.) and 3100 m.

The White-rimmed Brush-Finch was accorded Near-threatened status by Collar et al. (1992, 1994). Based on its evident rarity and limited range in Ecuador, we agree, though it remains possible that the species is more inconspicuous than rare. Populations should occur in both Cayambe-Coca Ecological Reserve and Sangay National Park, though the species remains undocumented in either; the area from which it is known in nw. Ecuador lies close to (within?) Cotocachi-Cayapas Ecological Reserve.

Monotypic.

Range: locally in Andes of s. Colombia and Ecuador.

Buarremon brunneinucha	Chestnut-capped Brush-Finch	Matorralero Gorricastaño

Fairly common but inconspicuous in the undergrowth of montane forest and secondary woodland in the upper tropical and subtropical zones on both slopes of the Andes. On the west slope the Chestnut-capped Brush-Finch is recorded south to El Oro (early-20th-century records from El Chiral, Zaruma, and Salvias, but—surprisingly—no recent reports). Also recently found on the coastal Cordillera de Mache in w. Esmeraldas (Bilsa; first by R. P. Clay et al. in 1994) and the Cordillera de Colonche in sw. Manabí (Cerro San Sebastián in Machalilla National Park [Parker and Carr 1992; ANSP]) and w. Guayas (Cerro La Torre and above Salanguilla [N. Krabbe]; Loma Alta [Becker and López-Lanús 1997]). Recorded mostly from 700 to 2500 m, but locally ranging as high as 2750–3150 m in Podocarpus National Park at Cerro Toledo (Krabbe et al. 1997), and recorded down to 550 m in the west at Manta Real.

We follow S. J. Hackett (Ph.D. diss., Louisiana State Univ., Baton Rouge, La., 1992) and J. V. Remsen, Jr., and W. S. Graves IV (*Auk* 112[1]: 225–236, 1995) in removing this and the following species from the genus *Atlapetes* and placing them in the genus *Buarremon*. Two races of *B. brunneinucha* are found in Ecuador, *frontalis* in most of the species' Ecuadorian range and *inornata* on the west slope of the Andes from sw. Chimborazo south to nw. Azuay (Manta Real) and on at least the southern end of the coastal cordillera. *Inornata* differs strikingly in lacking the black chest band. The distribution of the two taxa in sw. Ecuador is an unusual one, with the El Oro population of *frontalis* being isolated from the remainder of that taxon's range by unsuitable terrain to the east and by *inornata* to the north; the single specimen (ANSP) from the geographically intermediate area, taken at Manta Real in Azuay, is closer to *inornata*.

Range: highlands from e. Mexico to Panama; mountains of n. Venezuela, and Andes from w. Venezuela to s. Peru; Sierra de Perijá.

Buarremon torquatus	Stripe-headed Brush-Finch	Matorralero Cabecilistado

Fairly common but inconspicuous in the undergrowth of montane forest, secondary woodland, and borders in the upper subtropical and temperate zones on both slopes of the Andes, and locally on slopes above the central valley and interandean valleys. On the west slope the Stripe-headed Brush-Finch occurs from Carchi south to w. Cotopaxi (above Pilaló; RSR and G. H. Rosenberg) and again in El Oro and w. Loja. There is only a single old record from the intervening area, a 19th-century specimen from Ceche in Chimborazo; however, we surmise that this apparent absence is due to a lack of much field work at appropriate localities, especially as the subspecies is the same both to the north and south. Recorded mostly between 1900 and 3500 m in the north, but southward the species does not seem to range as high (typically only to ca. 3000 m), and it also occurs much lower in El Oro and w. Loja (routinely down to 800–900 m). A pair was even seen at 200 m in deciduous woodland near La Avanzada on 1 Apr. 1989 (RSR and PJG et al.), but whether the species is regularly found this low and in this habitat remains unknown.

Two very similar races occur in Ecuador, *assimilis* on the entire east slope and on the west slope south at least to Pichincha, and *nigrifrons* southward on the west slope and inland through much of Loja. *Nigrifrons* has slightly less gray in the crown (therefore there is more black); birds from Pichincha appear to be more or less intermediate though averaging closer to *assimilis*.

Range: mountains of n. Venezuela, and Andes from w. Venezuela to nw. Argentina; Sierra de Perijá and Santa Marta Mountains. Excludes Black-headed Brush-Finch (*Buarremon atricapillus*) of Costa Rica to Colombia.

Lysurus castaneiceps	Olive Finch	Pinzón Oliváceo

Rare to uncommon but inconspicuous (and doubtless under-recorded) in the undergrowth of montane forest, especially in ravines and along rivers, in the foothills and subtropical zone on the east slope of the Andes from Sucumbíos (San Rafael Falls; ANSP) south to Zamora-Chinchipe (Panguri [T. J. Davis] and in the Río Isimanchi valley near Zumba [RSR]). On the west slope the Olive Finch is seemingly even less numerous, and is recorded south only to Pichincha (an old specimen from Nanegal and a few recent reports from the Mindo area); one was collected above Maldonado in Carchi on 12 Aug. 1988 (ANSP). Small numbers of the scarce Olive Finch occur in the Río Bombuscaro sector of Podocarpus National Park (Rasmussen et al. 1996) and along the Loreto road north of Archidona. Recorded from about 800 to 1800 m.

Following most recent authors (e.g., AOU 1983, 1998; Ridgely and Tudor 1989), *L. crassirostris* (Sooty-faced Finch) of Costa Rica to nw. Colombia is considered a species separate from *L. castaneiceps* of the Andes. *L. castaneiceps* is monotypic.

Range: locally in Andes from w. Colombia to s. Peru.

Oreothraupis arremonops   **Tanager Finch**   **Pinzón Tangara**

Very rare and local in the undergrowth of montane forest and (perhaps especially) forest borders in the subtropical zone on the west slope of the Andes in Imbabura and Pichincha. In Imbabura the Tanager Finch is known only from two specimens (BMNH) taken in Dec. 1877 at Intag (Collar et al. 1992). In Pichincha there are a few old records from the Nanegal/Mindo area, an MCZ specimen from "Tandayapa Cordillera" taken by M. Olalla in Jul. 1968, and recent sightings in the vicinity of Bellavista on the Tandayapa ridge (first, fide N. Krabbe, by T. Lassoe in Jul. 1987). The handsome Tanager Finch seems unaccountably scarce in Ecuador, but probably it has merely been overlooked. An account of its behavior, together with a description of the first nest to be found, is presented by H. F. Greeney, M. Lysinger, T. Walla, and J. Clark (*Ornithol. Neotrop.* 9[2]: 205–207, 1998). Recorded between about 1300 and 2300 m.

The Tanager Finch was accorded Vulnerable status by Collar et al. (1992, 1994), and this seems appropriate from a purely Ecuadorian standpoint. The species' rarity throughout its small range remains unexplained, but whether human activities have actually had any impact on its numbers remains unknown.

Monotypic.

Range: locally on west slope of Andes in sw. Colombia and nw. Ecuador.

Arremon aurantiirostris   **Orange-billed**   **Saltón Piquinaranja**
                             Sparrow

Uncommon to fairly common but generally inconspicuous in the undergrowth of humid forest and secondary woodland in the lowlands and foothills of both e. and w. Ecuador. In the west recorded in more humid regions (southward especially in the foothills) south to Guayas, El Oro, and w. Loja (Puyango area and near El Limo); in the wet-forest belt of n. Esmeraldas, however, the species seems to occur only along streams, rivers, and at forest borders and does not range inside continuous forest (O. Jahn et al.). In the east, where generally less numerous, the Orange-billed Sparrow is found mostly near the base of the Andes; it is reasonably numerous as far east as Limoncocha and Jatun Sacha but is rare at La Selva, where there have been only a few sightings. An Aug. 1950 specimen (ANSP) from Montalvo, as well as reports from Kapawi Lodge on the Río Pastaza near the Peruvian border, indicate that the Orange-billed Sparrow ranges farther east in se. Ecuador than it does farther north. There are also specimens from the mouth of the Río Curaray (AMNH, ANSP) in what is now ne. Peru; possibly, however, these were mislabeled. Recorded mainly below about 1100 m, in small numbers as high as 1300–1350 m at Maquipucuna Reserve and 1250 m north of the town of Sumaco (P. Coopmans).

Three races occur in Ecuador. *Spectabilis* is found in the east; in the west *occidentalis* ranges south to w. Guayas (woodland south of Río Ayampe, Loma Alta, and Cerro Manglaralto [AMNH]) and Los Ríos, with *santarosae* occurring from e. Guayas and nw. Azuay southward. *Spectabilis* differs in having a flame-orange shoulder in males (and a different song); the shoulders are yellow in both western races, which resemble each other closely.

Range: s. Mexico to extreme nw. and ne. Peru. More than one species may be involved.

Arremon abeillei   Black-capped Sparrow   Saltón Gorrinegro

Fairly common in the undergrowth of deciduous forest and woodland and their borders, in smaller numbers out into adjacent dense secondary scrub, in the more arid lowlands of sw. Ecuador from cen. Manabí (north to just south of Bahía de Caráquez) and n. Los Ríos (north to Jauneche) south through El Oro and w. Loja, locally ranging also into undergrowth of more humid forest (e.g., at Cerro San Sebastián in Machalilla National Park). Also recently found in the drainage of the Río Marañón in extreme s. Zamora-Chinchipe near Zumba, with a single male taken (ANSP); more were obtained in this area in Dec. 1991 and Aug. 1992. In w. Loja found up to about 1600 m (e.g., below Celica), but ranges mostly below 700 m; in the Zumba area occurs between about 700 and 1100 m.

Two races occur in Ecuador, nominate *abeillei* in the southwest and the distinctly different *nigriceps* in the Río Marañón drainage. *Nigriceps* differs from nominate *abeillei* in having a clear olive back and rump and a white superciliary extending to the lores (not stopping above the eyes). It may prove to be a separate species (*A. nigriceps*, Marañón Sparrow), which would leave *A. abeillei* as monotypic.

Range: sw. Ecuador and nw. Peru.

Arremonops conirostris   Black-striped   Saltón Negrilistado
                             Sparrow

Uncommon to fairly common and rather local in shrubby clearings, pastures with tall rank grass, and agricultural areas with hedgerows in lowlands and foothills of w. Ecuador from Esmeraldas south to nw. Guayas, El Oro, and w. Loja (south to the Puyango and Alamor areas). Recorded up to about 1400 m around Mindo.

The Black-striped Sparrow would have seemed likely to increase in Ecuador in response to the widespread deforestation that has occurred across its range, but at least to date it has shown little indication of actually doing so.

One race: *striaticeps*.

Range: Honduras to n. Colombia, much of Venezuela, and extreme n. Brazil; sw. Colombia to extreme nw. Peru. More than one species is perhaps involved.

In 1996 a form of *Arremonops* sparrow, in appearance similar to the Black-striped Sparrow, was found in the lowlands of se. Ecuador on river islands in the Río Pastaza near Kapawi Lodge (D. Stejskal, fide J. Moore). Its song differs from those of trans-Andean populations of the species, more resembling that of the nominate race of n. Colombia and Venezuela.

Ammodramus aurifrons	Yellow-browed Sparrow	Sabanero Cejiamarillo

Common to very common in grassy and shrubby areas, pastures, roadsides, and on islands in larger rivers in the lowlands of e. Ecuador. The Yellow-browed Sparrow is less numerous in mainly forested regions, but even in such areas it seems to colonize openings and roadcuts relatively rapidly; as a result of deforestation it is becoming more numerous and widespread. Recorded mainly below 1100 m, but smaller numbers are found regularly in suitable habitat up to about 1600 m, rarely even occurring as high as 2500 m (wanderers near Cuyuja in w. Napo).

Formerly placed in the genus *Myospiza* (e.g., Meyer de Schauensee 1966, 1970), but most or all recent authors have subsumed this into *Ammodramus*. One race occurs in Ecuador, nominate *aurifrons*.

Range: cen. Venezuela, e. Colombia, e. Ecuador, e. Peru, n. Bolivia, and Amazonian Brazil.

Ammodramus savannarum	Grasshopper Sparrow	Sabanero Saltamontes

Very rare—if indeed still extant—in pastures and fields with tall grass in the highlands of Pichincha. The Grasshopper Sparrow is known in Ecuador only from a pair of century-old records. Ménégaux (1911) recorded a specimen from "Quito"; C. E. Hellmayr (*Birds of the Americas*, pt. 11) stated that he examined the specimen in question, an adult. *Birds of the Americas* (pt. 11, p. 501) also refers to a specimen of a "young" bird from Cayambe. Recorded at about 2800–2900 m.

There have been no confirmed records of the Grasshopper Sparrow in Ecuador for almost a century. Given the recent expansion of the city of Quito and the much more intensive agricultural practices now employed throughout the central valley, it seems all too likely that the species has been extirpated in Ecuador. Preferring to be marginally more optimistic, we give it Critical status. Its rediscovery on some remote and relatively undisturbed pasture would come as a welcome surprise. The species as a whole was not considered to be globally at risk (Collar et al. 1992, 1994), though the taxon found in nw. South America clearly is.

One race: *caucae* (also ranging in Colombia).

Range: North America and locally from Mexico to Panama; very locally in sw. Colombia and n. Ecuador; Greater Antilles.

Aimophila stolzmanni   Tumbes Sparrow   Sabanero de Tumbes

Uncommon to locally common in dense, tangled shrubby areas with scattered stands of tall grass and arid scrub in Loja, where it ranges from the interior valleys in the Catamayo/San Pedro de la Benedita region (MCZ; also many recent sightings) and east to around Vilcabamba (MECN) southward. The Tumbes Sparrow appears to be especially numerous in interior Loja between Catamayo and Gonzanamá. Found mostly below about 1300 m, and ranging down to about 150 m near Zapotillo in extreme s. Loja; in the Catamayo region, however, recorded locally as high as 1800–1900 m (D. Wolf; RSR).

This species was formerly placed in the monotypic genus *Rhynchospiza*, but we follow R. A. Paynter, Jr. (*Breviora* 278, 1967) and other recent authors in merging this into *Aimophila*. Monotypic.

Range: sw. Ecuador and nw. Peru.

Zonotrichia capensis   Rufous-collared Sparrow   Chingolo

Common to very common, widespread, and conspicuous in virtually any open grassy or shrubby area, and around habitations, throughout the temperate and subtropical zones of the Andes, locally down into the foothills. Recorded mostly between about 1500 and 3500 m, but locally found lower on both slopes (e.g., down to about 900 m at Buenaventura in El Oro), and at least occasionally also somewhat higher.

One race: *costaricensis*.

Range: highlands from s. Mexico to w. Panama; mountains of n. Venezuela, and Andes from Venezuela to Chile and Argentina, spreading to coast in Peru and Chile and widely into lowlands across s. South America to e. Brazil; Sierra de Perijá and Santa Marta Mountains; tepuis of Venezuela and adjacent Guyana and n. Brazil; locally on savannas in Guianas and e. Amazonian Brazil; Netherlands Antilles.

Poospiza hispaniolensis   Collared Warbling-   Pinzón Gorgeador
                            Finch   Collarejo

Uncommon to locally common in arid scrub and woodland in the coastal lowlands of sw. Manabí (north in small numbers, at least seasonally, to the Portoviejo area), w. Guayas (especially on the Santa Elena Peninsula, also east to the Playas region), and Isla Puná. The species is also known from a few sightings from coastal far s. El Oro; in interior s. Loja, though unrecorded previously, there are numerous recent reports of singing and presumably breeding birds in Mar. and Apr. in the Celica/Catamayo/Gonzanamá region (v.o.). The Collared Warbling-Finch is much less numerous in El Oro and Loja, where it may occur only seasonally, than it is in w. Guayas. The species is particularly common on Isla de la Plata off s. Manabí, where it is the most numerous landbird (and is also notably tame). Recorded mostly below 300 m, but the Loja population occurs at 950–1450 m.

Monotypic.
Range: sw. Ecuador and w. Peru.

Icteridae   American Orioles and Blackbirds

Thirty species in 16 genera. The family is strictly American in distribution. They were formerly considered a subfamily in a broadly inclusive Emberizidae; we follow Gill (1994) in giving them full family rank.

Cacicus cela   Yellow-rumped Cacique   Cacique Lomiamarillo

Fairly common to common in the canopy and borders of humid forest, secondary woodland, and trees in clearings and around habitations in the lowlands of e. Ecuador, especially favoring várzea and the vicinity of water. Uncommon to locally fairly common in the borders of humid and deciduous forest and woodland, and in clearings, in the more humid lowlands of w. Ecuador from w. Esmeraldas (vicinity of Esmeraldas [city]) south into coastal El Oro (south to the Arenillas region) and sw. Loja at Mangaurco (seen in Aug. 1992; T. J. Davis and F. Sornoza) and Tambo Negro (Best et al. 1993). In the east the Yellow-rumped Cacique tends to avoid extensive terra firme forest, or at least it is much less numerous in such areas; however, in partly cleared areas caciques can be very numerous and conspicuous, more so than they ever are in the west. Recorded up to about 650 m in the west (near Mangaurco in sw. Loja; T. J. Davis and F. Sornoza); locally as high as 1000 m in the east (around Zamora).

Two races are found in Ecuador, the nearly endemic *flavicrissus* in the west (also ranging into immediately adjacent Peru) and nominate *cela* in the east. *Flavicrissus* differs in being smaller with a dark (mainly dusky) bill and in having less extensive yellow on the tail. As the two races' vocalizations also differ notably, they are perhaps better regarded as separate species. If split, trans-Andean birds (*C. vitellinus*, including *flavicrissus*) could be known as the Western Yellow-rumped Cacique and cis-Andean birds as the Eastern (or Amazonian) Yellow-rumped Cacique. *Flavicrissus* also appears to stand apart from *vitellinus* and may also deserve species status; more study is needed.

Range: Panama to extreme nw. Peru, n. and e. Bolivia, and Amazonian Brazil; e. Brazil.

Cacicus leucoramphus   Northern Mountain-   Cacique Montañes
                           Cacique   Norteño

Uncommon to locally fairly common in the canopy and borders of montane forest, forest borders, and adjacent clearings in the upper subtropical and temperate zones on the east slope of the Andes, extending west locally into forests above interandean valleys at least in s. Azuay (old specimens from Portete [Berlioz 1932b], sightings in this area in Jul. 1978 [RSR and D.

Wilcove] and in the Río Yunguilla valley in Nov. 1998 [A. Agreda, N. Krabbe, and O. Rodriguez, *Cotinga* 11: 50–54, 1999]). Recorded mostly from 2000 to 3100 m.

Northern forms (nominate *leucoramphus* and *peruvianus*) are here regarded as a species separate from those found from s. Peru into Bolivia (Southern Mountain-Cacique, *C. chrysonotus*), based on newly acquired information regarding their quite different vocalizations. These two groups have often been considered conspecific under the name of *C. chrysonotus* (Mountain Cacique). Note that in many recent references (e.g., Meyer de Schauensee 1966, 1970; Ridgely and Tudor 1989; Fjeldså and Krabbe 1990), the name for the (expanded) Mountain Cacique is given in error as *C. leucoramphus*; actually the name *chrysonotus* has priority. One race occurs in Ecuador, nominate *leucoramphus*.

Range: Andes from sw. Venezuela to cen. Peru.

Cacicus haemorrhous   **Red-rumped Cacique**   Cacique Lomirrojo

Very rare and local in the canopy and borders of terra firme forest in the lowlands of e. Ecuador, principally or entirely in areas some distance east of the Andes. First recorded in Ecuador from 19th-century specimen(s?) taken at Sarayacu in Pastaza. Since then there have been only a few reports, including small numbers seen at La Selva (P. Coopmans et al.), Sacha Lodge (L. Jost et al.), and along the Maxus road southeast of Pompeya (RSR et al.); these have principally come from hilly forest south of the Río Napo. The only subsequent Ecuadorian specimens are two females from Montalvo (MECN) and another female from north of Tigre Playa in Sucumbíos taken on 5 Aug. 1993 (ANSP). We believe that the birds reported as Red-rumped Caciques from the Río Bombuscaro sector of Podocarpus National Park (see Bloch et al. 1991; Rasmussen et al. 1994) actually refer to the similar Subtropical Cacique (*C. uropygialis*). At the very least, the Subtropical Cacique is numerous there, and despite several searches the Red-rumped Cacique has not subsequently been found by various observers. Recorded only below about 300 m.

One race: nominate *haemorrhous*.

Range: se. Colombia, s. Venezuela, and the Guianas to n. and e. Bolivia and Amazonian Brazil; e. Brazil, e. Paraguay, and ne. Argentina.

Cacicus uropygialis   **Subtropical Cacique**   Cacique Subtropical

Uncommon to locally fairly common in the canopy and borders of montane forest and secondary woodland in the subtropical zone on the east slope of the Andes. Subtropical Caciques are regularly seen, often together with nominate Russet-backed Oropendolas (*Psarocolius angustifrons*), on the north slope of the Cordillera de Huacamayos along the Baeza-Tena road; the caciques are also numerous in the Bombuscaro sector of Podocarpus National Park. Recorded between about 1000 and 2100 m.

C. uropygialis of the subtropical zone in the Andes is regarded as a monotypic species separate from birds of the Pacific-slope lowlands, C. *microrhynchus* (Scarlet-rumped Cacique), on the basis of its much larger size and different vocalizations; formerly (e.g., Meyer de Schauensee 1966, 1970; Ridgely and Tudor 1989) they were considered conspecific as C. *uropygialis* (Scarlet-rumped Cacique). Jaramillo and Burke (1999) also so treated it.

Range: Andes from sw. Venezuela to cen. Peru; Sierra de Perijá.

Cacicus microrhynchus   Scarlet-rumped   Cacique Lomiescarlata
                               Cacique

Fairly common to locally common in the canopy and borders of humid forest and secondary woodland in the more humid lowlands and foothills of w. Ecuador, where found south to n. Manabí and s. Pichincha (Río Palenque), and along the base of the Andes to El Oro (La Chonta and Buenaventura). Ranges mostly below 900 m, occasionally as high as 1100–1300 m (e.g., in the Mindo region).

C. *microrhynchus* is regarded as a species separate from birds of the subtropical zone in the Andes (in Ecuador on their east slope only), C. *uropygialis* (Subtropical Cacique). One race of C. *microrhynchus* occurs in Ecuador, *pacificus*.

Range: Honduras to sw. Ecuador.

Cacicus sclateri   Ecuadorian Cacique   Cacique Ecuatoriano

Rare and local in the canopy and borders of humid forest and mature secondary woodland (mainly in várzea and along rivers) in the lowlands of e. Ecuador. Although long recorded only from Napo and Sucumbíos, in 1996 the Ecuadorian Cacique was found to be present at Kapawi Lodge on the Río Pastaza near the Peruvian border; indeed it seems to be more numerous here than it is in the northeast (J. Moore; P. Coopmans). Recorded mostly below 400 m, locally as high as 750 m (a pair seen on 8–9 Jun. 1999 about 17 km northwest of Archidona along the road to Baeza; L. Navarrete and J. Moore).

The species was formerly called the Ecuadorian Black Cacique. We follow Ridgely and Tudor (1989), Sibley and Monroe (1990), and Jaramillo and Burke (1999) in shortening its English name.

Monotypic.

Range: e. Ecuador and ne. Peru.

Cacicus solitarius   Solitary Cacique   Cacique Solitario

Uncommon to fairly common but local in dense undergrowth of regenerating clearings, grassy areas, and early-succession woodland especially near water in the lowlands of e. Ecuador. The Solitary Cacique is known principally from w. Napo and w. Sucumbíos, but there are also recent reports from the far southeast at Kapawi Lodge on the Río Pastaza near the Peruvian border, indicating that the species likely is widespread in appropriate habitat throughout the southeast as well. It does, however, seem to avoid blackwater drainages. Recorded mainly below about 400 m, but found in small numbers as high as 750 m north of Archidona.

The species was formerly called the Solitary Black Cacique. We follow Ridgely and Tudor (1989), Sibley and Monroe (1990), and Jaramillo and Burke (1999) in shortening its English name.

Monotypic.

Range: sw. Venezuela, se. Colombia, e. Ecuador, e. Peru, n. and e. Bolivia, Paraguay, n. Argentina, and Amazonian, cen., and ne. Brazil.

Amblycercus holosericeus	Yellow-billed Cacique	Cacique Piquiamarillo

Uncommon, local, and inconspicuous (except when vocalizing) in dense lower growth of deciduous forest and woodland and borders in the lowlands of w. Ecuador from n. Esmeraldas (an old record from Cachabí, and recently at Playa de Oro [O. Jahn and P. Mena Valenzuela et al.]) south through Guayas, El Oro, and Loja, ranging up in the mountains of s. Loja in the Alamor/Cebollal area and at Sozoranga. Also uncommon and seemingly local (underrecorded?) in the undergrowth of montane forest and forest borders in the upper subtropical and temperate zones on the east slope of the Andes from w. Napo southward, spreading west above interandean valleys in Azuay (Río Mazán; N. Krabbe). Yellow-billed Caciques appear to be spreading and increasing in n. Esmeraldas as a result of ongoing deforestation, and they there occur in a much more humid region than is usually the case in w. Ecuador. Especially in the Andes the distribution of the Yellow-billed Cacique appears to be governed by the presence of stands of *Chusquea* bamboo; even in the western lowlands it seems most frequent where (a different) bamboo is present. In the west recorded up to about 1700 m (in Loja). On the east slope recorded between about 1900 and 3100 m; there is also one record from much lower, 1500–1600 m, on the Cordillera del Cóndor at Coangos (Schulenberg and Awbrey 1997).

The monotypic genus *Amblycercus* is sometimes merged into *Cacicus*, though usually it is kept as distinct, largely because of its very different cup-shaped nest. Two races of *A. holosericeus* occur in Ecuador, *flavirostris* in the west and *australis* on the east slope of the Andes. Despite their disjunct ranges, the two taxa are quite similar, both in size, soft-part coloration, and voice. A. W. Kratter (*Condor* 95[4]: 641–651, 1993) also showed that they were quite close from a genetic standpoint.

Range: e. Mexico to extreme nw. Peru; mountains of n. Venezuela, and locally in Andes from w. Venezuela to w. Bolivia; Sierra de Perijá.

Ocyalus latirostris	Band-tailed Oropendola	Oropéndola Colifajeada

Uncertain. Very rare and apparently local in the lowlands of e. Ecuador; the species should perhaps be regarded as only hypothetical. There is only one specimen from Ecuador, that taken in the 19th century at Sarayacu (Chapman 1926), though it may actually have been otbained some distance from that site. The specimen recorded from Archidona mentioned by Lönnberg and Rendahl (1922) was reidentified as an example of *Zarhynchus wagleri* (N. Gyldenstolpe, *Kungl. Svenska Vet.-Akad. Handl. Band* 22[3]: 293, 1945), and obviously was mislabeled as to locality. Otherwise the Band-tailed Oropendola is known only from a single uncorroborated sighting of a pair at La Selva in 1987 and from another sighting of a single bird seen along the lower Río Lagarto on 11 Jan. 1993 (RSR et al.). Elsewhere the species is found mainly or entirely in the canopy and borders of várzea forest and on river islands (Hilty and Brown 1986; Ridgely and Tudor 1989). Recorded below 300 m.

Monotypic.

Range: e. Ecuador, ne. Peru, extreme se. Colombia, and w. Amazonian Brazil.

Zarhynchus wagleri	Chestnut-headed Oropendola	Oropéndola Cabecicastaña

Rare to uncommon and local in the canopy and borders of humid forest and secondary woodland in the lowlands of nw. Ecuador south to Manabí (an old record from Río Peripa) and sw. Pichincha (Santo Domingo de los Colorados area; surprisingly, the species has never been recorded not far to the south at Río Palenque). In addition, very far to the south, since 1988 a few have also been seen between Buenaventura and La Avanzada in El Oro (PJG et al.; P. Coopmans et al.), and the species was also reported seen at San Luis in Azuay in Jan. 1991 (J. C. Matheus). In Ecuador the handsome Chestnut-headed Oropendola seems most numerous in w. Esmeraldas; even so, during four years (1995–1998) of field work in the Playa de Oro region of n. Esmeraldas no colonies were ever located (O. Jahn and P. Mena Valenzuela et al.). Recorded up to about 700 m in Esmeraldas.

The monotypic genus *Zarhynchus* has often recently been subsumed into *Psarocolius* (e.g., Ridgely and Tudor 1989; Sibley and Monroe 1990), but we prefer to emphasize its distinct characters—in particular its proportionately large bill—by maintaining it. In addition, the display of males differs markedly from other orpendolas'. One race occurs in Ecuador, *ridgwayi*.

Range: s. Mexico to sw. Ecuador.

Clypicterus oseryi	Casqued Oropendola	Oropéndola de Casco

Rare to uncommon and apparently local in the canopy and borders of humid forest (mainly terra firme) in the lowlands of e. Ecuador north to Sucumbíos and Napo; apparently it does not occur—or at most is very rare—close to the Andes (e.g., it has never been seen at Jatun Sacha; B. Bochan). There are AMNH specimens from Río Suno and MECN specimens from Río San Miguel at Puerto Libre, Río Cotapino, Río Copotaza, Sarayacu, and Montalvo. Recent sightings exist from various localities, including Cuyabeno, La Selva, Sacha Lodge, along the Maxus road southeast of Pompeya, Imuyacocha, and Kapawi Lodge. Although unrecorded in Colombia (Hilty and Brown 1986), given the records in Ecuador from very close to the Colombian border, the Casqued Oropendola seems almost certain to range north into that country. Recorded below about 300 m.

The monotypic genus *Clypicterus* has often recently been merged into an expanded *Psarocolius* (e.g., Ridgely and Tudor 1989; Sibley and Monroe 1990), but—as with *Zarhynchus*—we now favor emphasizing its distinct characters by maintaining it. Monotypic.

Range: e. Ecuador, e. Peru, extreme w. Amazonian Brazil, and extreme nw. Bolivia.

Psarocolius decumanus	Crested Oropendola	Oropéndolo Crestada

Fairly common in the canopy and borders of humid forest (both terra firme and várzea), secondary woodland, adjacent clearings, and along rivers and on islands (perhaps occurring on islands principally to roost) in the lowlands of e. Ecuador. Recorded mostly below 500 m, ranging up in smaller numbers into the foothills to about 1000 m.

One race: nominate *decumanus*.

Range: Panama to nw. and ne. Argentina, e. Paraguay, and se. Brazil.

Psarocolius angustifrons	Russet-backed Oropendola	Oropéndola Dorsirrojiza

Common and widespread in the canopy and borders of humid forest, trees of adjacent clearings, and (especially) along river shorelines and on river islands in the lowlands of e. Ecuador. Uncommon to fairly common but somewhat local in the canopy and borders of montane forest and in trees in adjacent clearings in the subtropical zone on both slopes of the Andes; on the west slope found south, at least formerly, to El Oro (specimens from El Chiral, Zaruma, and Salvia) and adjacent w. Loja (Las Piñas); however, and inexplicably, there appear to be no recent reports from south of w. Cotopaxi. Reports from the lowlands of Guayas at Cerro Blanco in the Chongón Hills (Berg 1994) remain uncorroborated (and seem unlikely); also puzzling is the report (Best et al. 1993) of this species from Sozoranga in s. Loja in Aug.–Sep. 1989, far south of its present range (though conceivably a wandering bird could have been involved). In the east occurs up locally to over 2000 m on the east slope of the Andes; on the west slope ranges mainly between about 1000 and 2100 m.

Three rather different races are found in Ecuador: *atrocastaneus* on the west slope of the Andes; nominate *angustifrons* on the east slope south to n. Morona-Santiago and east through the lowlands; and *alfredi* of se. Ecuador in s. Morona-Santiago and Zamora-Chinchipe. *Atrocastaneus* has an orange-yellow bill, a contrasting yellow forehead, and overall rufescent plumage. The nominate race has a black bill and duller, somewhat more olivaceous plumage. *Alfredi* has a yellowish ivory bill, a yellowish facial area and throat, and a quite rich rufescent back and olivaceous underparts. It has been suggested (e.g., Hilty and Brown 1986, Jaramillo and Burke 1999) that more than one species might be involved, with the contact zone between the nominate race and *alfredi* in particular needing to be studied in greater detail. A specimen (ANSP) showing characters intermediate between those two taxa was taken on 23 Apr. 1990 at Valladolid in Zamora-Chinchipe.

Range: n. and w. Venezuela, w. and se. Colombia, Ecuador, e. Peru, n. Bolivia, and w. Amazonian Brazil.

Psarocolius viridis   Green Oropendola   Oropéndola Verde

Rare to locally fairly common in the canopy and borders of terra firme forest (only occasionally ranging into várzea or riparian forest) and adjacent clearings in the lowlands of e. Ecuador. Until the 1990s there were rather few Ecuadorian records of the Green Oropendola, and it still has mainly been recorded from the northeast in the drainages of the Ríos Napo and Aguarico. Notably, this is considered to be the most numerous oropendola at Jatun Sacha (B. Bochan), and it is also not uncommon along the Maxus road southeast of Pompeya (RSR et al.). In the southeast the Green Oropendola is known only from recent sightings at Kapawi Lodge on the Río Pastaza near the Peruvian border, but it likely will be found to range more widely in the Río Pastaza drainage. Ranges mostly below 600 m, but locally recorded up to about 900 m (Bermejo oil-field area north of Lumbaquí; ANSP).

Monotypic.

Range: e. Colombia, s. Venezuela, and the Guianas to ne. Peru and Amazonian Brazil.

Psarocolius yuracares   Olive Oropendola   Oropéndola Oliva

Rare to uncommon and local in the canopy and borders of humid forest (ranging principally in terra firme) and in large trees of adjacent clearings in the lowlands of e. Ecuador. Although essentially a terra firme bird, the Olive Oropendola is also sometimes found roosting with other oropendola species on river islands. Recorded mostly below 300 m, a few up to about 400 m at Jatun Sacha (where very rare; B. Bochan) and locally to 600 m (north of Canelos in Pastaza; N. Krabbe and F. Sornoza). An old specimen from "Valle del Río Santiago" (Salvadori and Festa 1899a) may have been taken higher than that, but as there are no recent reports from that region we prefer to regard its provenance as uncertain.

This species and several others (*montezuma, guatimozinus,* and *cassini*) were formerly often separated in the genus *Gymnostinops* (e.g., Meyer de Schauensee 1966, 1970), mainly on the basis of their unfeathered cheeks, but most recent authors have merged it into *Psarocolius* (e.g., *Birds of the World*, vol. 14; Ridgely and Tudor 1989; Sibley and Monroe 1990). *P. viridis* (Green Oropendola) does seem to provide a link between the two groups, having a limited amount of bare skin around the eyes and at the base of the bill. *Yurucares* has sometimes (e.g., Ridgely and Tudor 1989) been treated as being conspecific with *P. bifasciatus* (Pará Oropendola) of lower Amazonian Brazil, but the nature of their contact zone remains little studied, and we follow Jaramillo and Burke (1999) in considering them to be separate species. One race of *P. yuracares* occurs in Ecuador, the nominate race.

Range: s. Venezuela and se. Colombia to n. Bolivia and Amazonian Brazil.

Molothrus bonariensis   Shiny Cowbird   Vaquero Brilloso

Fairly common and widespread in semiopen and agricultural regions in the more humid lowlands of w. Ecuador; seemingly most numerous in s. Loja (e.g., around Macará, where in Sep. 1998 thousands could be seen going to roost after feeding in rice fields; RSR et al.). Less numerous and more local in similar habitat as well as along the shores of rivers and lakes in the eastern lowlands. Recorded mostly below about 900–1000 m, but regularly higher (to 1400–1600 m) where open conditions prevail. The Shiny Cowbird has also recently been seen in small numbers up into the highlands at least as high as 2700 m (e.g., in the Río Guaillabamba area and north of Latacunga, occasionally even in the Tumbaco area [N. Krabbe]), though whether it is a permanent breeding resident in this region remains to be determined; postbreeding wanderers from the western lowlands are perhaps involved.

Three races occur in Ecuador: *aequatorialis* in most of w. Ecuador, *occidentalis* in the far southwest in Loja, and *riparius* in the eastern lowlands (where only a few specimens have been taken, all males, one [AMNH] at Río Suno and two [MECN] in 1965 at Río Cuyabeno and Río Arajuno). *Occidentalis* females are much paler generally than female *aequatorialis*, being dull pale brownish above and dull grayish white below with vague brownish mottling (not more or less uniform brownish gray). Males of all forms, however, are similar.

Range: Panama to cen. Chile and cen. Argentina, but absent from large parts of Amazonia (though increasing and spreading); also spreading north through West Indies, with a few having now reached se. United States.

Molothrus oryzivorus   Giant Cowbird   Vaquero Gigante

Uncommon to fairly common at the borders of humid forest and secondary woodland (also often along riverbanks) and in adjacent fields and agricultural areas in the lowlands of e. Ecuador and the more humid lowlands of the west. In the west ranges south locally through El Oro and w. Loja. The Giant Cowbird is more numerous in the east than it is in the west, this perhaps simply being a reflection of the relative numbers of its various host oropendolas. Ranges regularly well up into the foothills and subtropical zone on both slopes of the Andes (perhaps especially when breeding), in small numbers as high as about 2000 m.

Formerly placed in the monotypic genus *Scaphidura*; see S. M. Lanyon and K. E. Omland (*Auk* 116[3]: 629–639, 1999) and *Auk* 117[3]: 853, 2000. One race occurs in Ecuador, nominate *oryzivorus*.

Range: s. Mexico to extreme nw. Peru, n. and e. Bolivia, e. Paraguay, ne. Argentina, and se. Brazil.

Dives warszewiczi   Scrub Blackbird   Negro Matorralero

Fairly common to common, noisy, and conspicuous in agricultural regions, shrubby areas and groves of trees, lighter woodland, and gardens in the lowlands and foothills of w. Ecuador from sw. Esmeraldas (Quinindé area) and nw. Pichincha (Santo Domingo/Tinalandia area, with a few in Mindo and around Pedro Vicente Maldonado and Puerto Quito) southward. Scrub Blackbirds are more numerous southward, being especially common in e. Guayas, El Oro, and Loja. They favor moderately arid to semihumid regions; the species avoids truly desert-like conditions and is, for instance, decidedly scarce on the Santa Elena Peninsula. The Scrub Blackbird is now, however, beginning to spread into more humid, formerly forested regions such as around Santo Domingo and Mindo. Curiously it has not (yet?) moved into the seemingly ideal terrain now found so widely in coastal and w. Esmeraldas, though we expect that eventually it will do so (and may ultimately even spread as far north as Colombia). As the species occurs in the upper Río Marañón valley of nw. Peru, it may also spread north into the Zumba region as well. Recorded mostly below about 1000 m, but in the southwest (Azuay, El Oro, and Loja) also ranging up in agricultural terrain of intermontane valleys and on slopes to about 2100 m; wandering (?) birds have even been seen as high as 2800 m on the Cordillera de Cordoncillo in Loja (P. Coopmans).

D. *warszewiczi* of w. Ecuador and w. Peru is considered a distinct species from *Dives dives* (Melodious Blackbird) of Middle America, following most recent authors. One race occurs in Ecuador, nominate *warszewiczi*.

Range: nw. Ecuador to sw. Peru.

Quiscalus mexicanus   Great-tailed Grackle   Clarinero Coligrande

Fairly common but quite local along beaches and shorelines, coastal rivers, and (especially) in and around mangroves along the coast of w. Ecuador from Esmeraldas south to cen. Manabí in the Bahía de Caráquez area, and from around Guayaquil in Guayas south through El Oro. The Great-tailed Grackle's distribution in Ecuador seems to be closely tied to the presence of at least remnant patches of mangroves, and it forages away from them only to a limited extent; in areas without them, such as in s. Manabí and on the Santa Elena Peninsula, it does not occur at all. The species does not occur at all inland, though whether it will eventually spread into settled and agricultural areas—as it has in the United States and Middle America—bears watching. In the 1990s there was evidence that it may be beginning to spread inland along larger rivers in n. Esmeraldas, where it is now resident around Borbón (some 20 km from the coast) with stragglers also occurring upriver (O. Jahn et al.).

The species was formerly placed in the genus *Cassidix*. One race of *Q. mexicanus* occurs in Ecuador, *peruvianus*.

Range: sw. United States to Panama; along Pacific coast of Colombia, Ecuador, and extreme nw. Peru, and along Caribbean coast of Colombia and nw. Venezuela.

Lampropsar tanagrinus   Velvet-fronted   Clarinero Frentiafelpado
                             Grackle

Uncommon and seemingly local in lower and middle growth of várzea forest and along the shores of oxbow lakes in the lowlands of ne. Ecuador in Napo and Sucumbíos; seems most numerous in blackwater drainages. The Velvet-fronted Grackle is known only from near the Río Napo upriver to the area of Limoncocha and Taracoa, and from the lower Río Aguarico area upriver as far as Cuyabeno and the Río Pacuyacu. To date there are no records from the southeast. Two specimens (ANSP) were obtained at Imuyacocha and Zancudococha on 6 Dec. 1990 and 21 Mar. 1991, respectively (an additional example remained at MECN and one was skeletonized). These represent the first Ecuadorian specimens with precise locality data; as J. V. Remsen, Jr., C. G. Schmitt, and D. C. Schmitt (*Gerfaut* 78: 376, 1988) noted, the scanty previously available material may have actually been obtained in adjacent Peru. Recorded below 300 m.

One race: nominate *tanagrinus*.

Range: se. Colombia, ne. Ecuador, ne. Peru, n. Bolivia, and w. and cen. Amazonian Brazil; locally in e. and s. Venezuela, Guyana, and extreme n. Brazil.

*Agelaius xanthophthalmus   Pale-eyed Blackbird   Negro Ojipálido

Rare and very local in the grassy and marshy borders of oxbow lakes and marshes in the lowlands of ne. Ecuador in Napo. The Pale-eyed Blackbird is known principally from Limoncocha, where it was discovered in 1975 with a few specimens (LSUMZ) obtained (Tallman and Tallman 1977); several pairs continue to be present there, though not being very conspicuous, they are not always easy to find. This only recently described species of blackbird (L. L. Short, *Occas. Pap. Mus. Zool. LSU* 36: 1–8, 1969) is also known from a few sightings at La Selva, primarily around the shores of Garzacocha (but also once along the edge of the Río Napo; S. Hilty et al.); however, there is no evidence of a resident population there (fide M. Lysinger and P. Coopmans). More surprisingly, a pair was also seen in a small roadside marsh north of Coca in Jul. 1996 (M. Lysinger), but the species has not subsequently been found at this site either. Despite careful searching, the Pale-eyed Blackbird has not been found at other sites that would seem to provide suitable habitat (e.g., Imuyacocha); it appears not to be found in blackwater drainages. Recorded at about 200–300 m.

Given its extremely small numbers and highly restricted distribution in Ecuador, we believe the Pale-eyed Blackbird warrants Near-threatened status. It was not considered to be at risk by Collar et al. (1992, 1994). There appears to be no particular threat to its marshy habitat at present, but the increasingly large human population around Limoncocha—an area supposedly under reserve status—could easily result in changes there.

Monotypic.

Range: very locally in e. Ecuador and e. Peru.

Icterus chrysocephalus Moriche Oriole Bolsero de Morete

Uncommon and somewhat local in the canopy and borders of humid forest (both terra firme and várzea) and adjacent clearings in the lowlands of e. Ecuador. Despite its English name, the Moriche Oriole seems to show no particular predilection for *Mauritia* palms. Recorded mostly below about 500 m, exceptionally as high as 900 m (Bermejo oil-field area north of Lumbaquí; M. B. Robbins).

Chrysocephalus is sometimes (e.g., *Birds of the World*, vol. 14) considered conspecific with *I. cayanensis* (Epaulet Oriole)—and it may prove to be so, intergradation and even mixed pairs having been noted in the Guianas—but we follow most recent authors in continuing to regard the two as separate species. When split, *I. chrysocephalus* is monotypic.

Range: se. Colombia, s. Venezuela, the Guianas, e. Ecuador, ne. Peru, and n. Amazonian Brazil.

Icterus croconotus Orange-backed Troupial Turpial Dorsinaranja

Uncommon to fairly common in shrubby growth along rivers and around the margins of oxbow lakes, and in adjacent regenerating or at least heavily vegetated clearings, in the lowlands of e. Ecuador. Recorded mainly below about 400 m, but small numbers occur up to about 750 m in the Archidona region.

Croconotus (with *stictifrons*) of Amazonia to sw. Brazil is here considered a species separate from *I. icterus* (Venezuelan Troupial) and *I. jamacaii* (Campo Troupial), in accord with Jaramillo and Burke (1999). Despite their distinct plumage differences, in recent years the three have most often been considered conspecific (e.g., Ridgely and Tudor 1989; Sibley and Monroe 1990), though *I. croconotus* was regarded as a full species by Hilty and Brown (1986). One race occurs in Ecuador, nominate *croconotus*.

The species has also been called the Orange-backed Oriole, but it seems preferable to segregate the three allospecies as "troupials."

Range: se. Colombia, e. Ecuador, e. Peru, n. and e. Bolivia, w. Paraguay, n. Argentina, locally in Amazonian and sw. Brazil, and s. Guyana.

**Icterus galbula* Baltimore Oriole Balsero de Baltimore

A casual boreal winter visitor to the borders of humid forest and woodland and adjacent clearings in the lowlands and foothills of n. Ecuador on both the west and east slopes. There are only a few sight reports of the Baltimore Oriole from Ecuador, most of them from Pichincha at the well-watched localities of Tinalandia and Río Palenque. In addition, a male was seen along the Río Napo above Primavera on 18 Feb. 1985 (RSR and T. Schulenberg et al.). Together these represent the southernmost reports of the species, which has been recorded regularly only as far south as n. Colombia (Hilty and Brown 1986). Recorded from Nov. to early Apr. Recorded below 800 m.

The species was formerly sometimes called the Northern Oriole, this when eastern-breeding birds in North America were considered conspecific with western-breeding birds, as for several decades was the case (e.g., in the 1983 AOU Check-list). Recent evidence, however, appears to demonstrate that their separation as full species is preferable, and this course was adopted in the 1998 AOU Check-list. *I. galbula* is monotypic when it is specifically separated from *I. bullocki* (Bullock's Oriole).

Range: breeds in e. North America, wintering south to nw. South America.

Icterus graceannae   White-edged Oriole   Bolsero Filiblanco

Uncommon to fairly common in desert scrub and lighter deciduous woodland in the more arid lowlands of sw. Ecuador from cen. Manabí (Manta) south into w. Guayas (Santa Elena Peninsula east to the western fringes of the Guayaquil area), and again in s. El Oro and through much of Loja where it ranges inland east to near Malacatos (MCZ). In general the White-edged is replaced in more humid regions by the Yellow-tailed Oriole (*I. mesomelas*), though in some places (e.g., the Chongón Hills west of Guayaquil and near Macará) they are sympatric. Recorded mostly below about 400 m, but in Loja ranging as high as 1500–1700 m.

Monotypic.

Range: sw. Ecuador and nw. Peru.

Icterus mesomelas   Yellow-tailed Oriole   Bolsero Coliamarillo

Fairly common in woodland and forest borders, adjacent clearings and gardens, and banana plantations in the more humid lowlands and foothills of w. Ecuador from Esmeraldas south to El Oro and Loja (where it ranges east as far as the Vilcabamba area; RSR); in n. Esmeraldas it seems to be largely restricted to the coastal lowlands (O. Jahn et al.). As the Yellow-tailed Oriole is known from the Río Marañón drainage of nw. Peru, it seems possible that it may also extend northward to the Zumba area, though as yet it is unrecorded from there. Recorded mostly below 900 m, but in Loja occurs regularly to about 1500 m and locally even as high as 1700–1750 m (near Amaluza and at Sozoranga).

One race: *taczanowskii*.

Range: s. Mexico to nw. Venezuela and nw. Peru.

The Yellow-backed Oriole (*I. chrysater*) has been recorded in Colombia south to Ricaurte and Yananchá (ANSP) just north of the Ecuadorian boundary in Nariño. It would seem possible that the species could at least wander across the border into adjacent Ecuador, and it should be watched for.

Gymnomystax mexicanus   Oriole Blackbird   Negro Bolsero

Uncommon on river islands and sandbars along rivers in the lowlands of e. Ecuador, where known primarily from along the lower Río Aguarico and the Río Napo upriver to around Coca. There are also recent sightings from the far southeast at Kapawi Lodge along the Río Pastaza near the Peruvian border, likely an indication that Oriole Blackbirds also range farther upriver along the Río Pastaza. Recorded mainly below about 300m, with occasional wandering birds as high as 400m at Jatun Sacha (B. Bochan).

Monotypic.

Range: ne. Colombia, much of Venezuela, and locally in Guyana; extreme se. Colombia, e. Ecuador, ne. Peru, and Amazonian Brazil.

** Sturnella militaris*   Red-breasted Blackbird   Pastorero Pechirrojo

Recently found to be locally fairly common in pastures and fields with lush tall grass in the lowlands of e. Ecuador, where so far known mainly from w. Sucumbíos and Napo in the Lago Agrio, Coca, and Limoncocha region south to around Tena (e.g., a male was taken on 17 Oct. 1987; WFVZ), with a few occurring south into n. Pastaza in the Puyo region. The Red-breasted Blackbird was also seen at Taisha, Morona-Santiago, in Aug. 1990 (D. Wechsler), and since Dec. 1990 a few have been present at Zancudococha (ANSP). The species was first found in Ecuador in 1964 (H. M. Stevenson, *Wilson Bull.* 84[1]: 99, 1972; also see Norton et al. 1972), but this conspicuous blackbird appears to be capable of colonizing grassy clearings relatively quickly, and in recent decades it has been extending its range across much of Amazonia. Recorded up only to about 400m.

S. militaris is, together with *S. superciliaris* (White-browed Blackbird), sometimes separated in the genus *Leistes*. T. A. Parker III and J. V. Remsen, Jr. (*Bull. B. O. C.* 107[3]: 105, 1987) suggested that perhaps they still should be, but we agree with L. L. Short (*Am. Mus. Novitates* 2349, 1968) who concluded that *Leistes* is best subsumed into *Sturnella*. Further, a meadowlark of s. South America (*S. defilippi*, Pampas Meadowlark) more or less bridges the two groups. *S. militaris* is monotypic; *S. superciliaris* of s. South America is now usually regarded as a separate species (e.g., Ridgely and Tudor 1989; Sibley and Monroe 1990).

Range: Costa Rica to ne. Peru, extreme n. Bolivia, and Amazonian Brazil.

Sturnella bellicosa   Peruvian Meadowlark   Pastorero Peruano

Common and conspicuous in pastures and shrubby areas in agricultural regions, and in desert scrub, in the more arid lowlands of w. Ecuador from coastal n. Esmeraldas southward (including Isla de la Plata and Isla Puná). The Peruvian Meadowlark has recently also begun to spread in small numbers north in the central valley as far as Cotopaxi (having long occurred in the Azuay and Chimborazo highlands) and even onto the east slope of the Andes at least in cleared areas of Cañar (in the Río Mazar drainage below Hacienda La Libertad: 6 or more singing males seen in Aug. 1991; RSR and F. Sornoza). Ranges up locally to 2500–3000m in the intermontane valleys of s. Ecuador.

Formerly (e.g., Meyer de Schauensee 1970) this species was called the Peruvian Red-breasted Meadowlark. We follow Ridgely and Tudor (1989), Sibley and Monroe (1990), and Jaramillo and Burke (1999) in shortening its English name.

One race: nominate *bellicosa*.

Range: nw. Ecuador to n. Chile. We suspect that it was this species of "red-breasted meadowlark" (and not *S. militaris*) that was seen in 1976 in sw. Colombia (see Hilty and Brown 1986, p. 573), but evidently there have been no reports of it subsequently.

Dolichonyx oryzivorus   Bobolink   Tordo Arrocero

Apparently a rare to uncommon transient in grassy areas in the lowlands of e. Ecuador, passing through rapidly and perhaps somewhat overlooked; there are, however, only a few records from w. Ecuador. The Bobolink has so far been recorded mainly on southward passage (Oct.–Nov.), when it was considered "common" at Limoncocha (Tallman and Tallman 1977). It seems less frequent on northward passage, when it appears to occur only within a narrow time window extending from early Apr. to early May. Alternate-plumaged males were seen at Limoncocha on 5 Apr. 1987 (P. Coopmans) and at Zancudococha on 4 Apr. 1991 (RSR), and six were noted with a seedeater (*Sporophila*) flock at Lago Agrio on 6 Apr. 1991 (RSR); in addition, a male (MECN) was taken by M. Olalla at Río Due in Sucumbíos on 4 May 1976. In w. Ecuador the Bobolink is known from one specimen record, an immature male taken in Nov. on the Santa Elena Peninsula (Marchant 1958, p. 385), and from sightings at Playa de Oro in n. Esmeraldas on 14 Oct. 1996 (O. Jahn et al.) and at Mindo on 22 Nov. 1998 (J. Lyons and V. Perez); however, it may be more common than this paucity of reports would seem to indicate as there are numerous records from the Galápagos Islands. Recorded mainly below about 300 m, though the Mindo record was at 1300 m.

Monotypic.

Range: breeds in North America, wintering primarily in interior s.-cen. South America.

Fringillidae   Cardueline Finches

Six species, all in the same genus. The family is worldwide in distribution.

Carduelis magellanica   Hooded Siskin   Jilguero Encapuchado

Common and widespread in semiopen, scrubby, and agricultural areas, around habitations, and in gardens and city parks in the highlands. The Hooded Siskin has been recorded primarily from the west slope of the Andes and the central and interandean valleys (as well as the slopes above these valleys) and is found only locally on the east slope (often just spilling over the Continental Divide a short distance, e.g., down to the Papallacta area). Ranges mostly from about 1000 to 3500 m, but recorded locally down to 650 m in Loja.

This and the following five species were formerly (e.g., Meyer de Schauensee 1966, 1970) placed in the genus *Spinus*; this generic change has forced changes in the endings of some of the species names. Two races of *C. magellanica* are found in Ecuador, with their respective ranges being somewhat unclear. *Capitalis* occurs in the highlands south at least to Azuay (and perhaps farther south at high elevations), and *paula* is found in the south (primarily in El Oro and Loja, locally in Zamora-Chinchipe at Zamora). Male *capitalis* have the rump more or less concolor olive with the back, whereas male *paula* are somewhat brighter generally and have the rump contrastingly yellow; females are very similar.

Range: Andes from s. Colombia to n. Chile, also lowlands of w. Peru; tepuis of Venezuela and adjacent Guyana and Brazil; e. Bolivia and cen. Brazil to n. and e. Argentina.

Carduelis siemiradzkii   Saffron Siskin   Jilguero Azafranado

Uncommon to locally fairly common in deciduous woodland and scrub in the lowlands of sw. Ecuador from sw. Manabí south to w. Loja. Until recently the Saffron Siskin was thought to be quite a rare bird; it was then known to occur only in Guayas, mainly in the Guayaquil region (e.g., the Chongón Hills) but also as far north as Balzar, and on Isla Puná. It is now known to be fairly common at least locally and seasonally, however, and to have a considerably wider range than had been realized; surprisingly, and despite its being recorded in adjacent Peru, there continue to be no records from El Oro. Toward the northwestern edge of its range, the Saffron Siskin was first found along the lower Río Ayampe on the Manabí/Guayas border in Dec. 1990, and a male was taken in the hills just south of the Río Ayampe in nw. Guayas on 18 Jan. 1991 (ANSP; also see Krabbe 1992). Others were subsequently found nearby on the slopes of Cerro San Sebastián in Machalilla National Park in Jan. and Aug. 1991 (Parker and Carr 1992; ANSP), with the northernmost report being a few sightings since 1991 in the Cordillera de Colonche east of Puerto Cayo (P. Coopmans; RSR); the species was also found on Cerro La Torre in w. Guayas in Jul. 1992 (N. Krabbe). Saffron Siskins have also been found on several occasions since 1994 in Manglares-Churute Ecological Reserve (K. Berg and N. Benavides; Pople et al. 1997). In Apr. 1992 and 1993, fairly large numbers (ANSP, MECN) were located about 10 km south of Sabanilla in Loja, and in 1993 the species also was found east of Pozul along the road to Pindal and in the hills west of Macará along the road to Zapotillo (ANSP, MECN); however, very few were present west of Macará in Sep. 1998 (RSR et al.). The Macará region appears to represent the species' easternmost limit, and the species seems absent from the Río Sabiango valley east of Macará (Best et al. 1993; RSR). Recorded mainly below 600 m, but has been found at least seasonally up to 1300 m in s. Loja.

The near-endemic Saffron Siskin was accorded Vulnerable status by Collar et al. (1992, 1994), being presumed to have declined as a result of deforestation.

Although we certainly do not dispute that deforestation has affected huge areas of sw. Ecuador, our recent experience is that the Saffron Siskin is by no means restricted to undisturbed areas, and in fact appears quite tolerant of habitat disturbance (not differing in this way from the closely related Hooded Siskin [*C. magellanica*]). We therefore accord the species only Near-threatened status. Populations are protected in Machalilla National Park and at Cerro Blanco in the Chongón Hills (Berg 1994).

Monotypic. It has been suggested that *siemiradzkii* might better be considered as only a subspecies of *C. magellanica*, but we do not think this likely as the two species seem to occur in near sympatry in Loja (e.g., there is a recent ANSP specimen of *C. magellanica paula* from near Mangaurco in sw. Loja, close to where *C. siemiradzkii* has been found).

Range: sw. Ecuador and extreme nw. Peru.

Carduelis olivacea   Olivaceous Siskin   Jilguero Oliváceo

Uncommon to locally fairly common in the canopy and borders of montane forest and adjacent clearings in the foothills and subtropical zone on the east slope of the Andes from Sucumbíos (Bermejo oil-field area, where seen in Mar. 1993; M. B. Robbins) south through Zamora-Chinchipe. The Olivaceous Siskin seems likely to occur north into adjacent Colombia on the east slope of the Andes, where it is unrecorded (Hilty and Brown 1986). Recorded mainly between about 900 and 1700 m, with a few ranging at least occasionally as low as 750 m above Archidona (P. Coopmans).

Monotypic. Further work is needed to determine the relationship of *C. olivacea* to *C. magellanica* (Hooded Siskin), of which *C. olivacea* is essentially an olive "version," though there are consistent and distinct differences, at least in female plumages. The supposed sympatry of *C. olivacea* and *C. magellanica paula* at Zamora is noteworthy. Zamora is the type locality of *paula*, and several specimens of both taxa were taken here in the early 20th century; however, in recent years only what are apparently *C. olivacea* have been found in this region. Males of the two species are only marginally distinguishable (especially northward in *C. olivacea*'s range), though at least in Ecuador females are readily separated on the basis of *olivacea*'s being olive (not grayish) below.

Range: east slope of Andes from n. Ecuador to w. Bolivia.

*Carduelis spinescens   Andean Siskin   Jilguero Andino

Rare to locally fairly common (but apparently somewhat erratic) in paramo with *Espeletia*, woodland near treeline, and agricultural areas in the Andes of n. Ecuador, where recorded only from Carchi and Pichincha. The first Ecuadorian record is from Páramo del Ángel, where a small flock was seen on 2 Jan. 1982 (RSR, PJG, and S. Greenfield, photos by the last); two specimens (MECN) were taken here in Nov. 1990 and Nov. 1991 (Krabbe 1992). In Mar. 1992 both this species and the Hooded Siskin (*C. magellanica*) were found to be fairly common on the west slope of Cerro Mongus, where they often fed in mixed flocks at and above the paramo/treeline ecotone, with several specimens being obtained (ANSP, MECN; see M. B. Robbins, N. Krabbe, G. H. Rosenberg, and F. Sornoza M., *Ornitol. Neotrop.* 5: 61–63, 1994). In June 1993 almost no siskins were found here, however, and there were no Andeans at all (RSR et al.). Farther south in Pichincha the Andean Siskin appears to be only an occasional wanderer, with just one sighting, a single male seen with a flock of Hooded Siskins above Nono on 23 Mar. 1989 (RSR and PJG et al.). Recorded from about 2800 to 3600 m.

One race: *nigricauda*. Robbins et al. (op. cit.) discussed racial variation in *C. spinescens* and demonstrated that *nigricauda* differs from the nominate subspecies in being nearly monomorphic in plumage. Both sexes have black crowns and show some yellow at base of their rectrices; they may prove to be separable from *nigricauda*.

Range: mountains of n. Venezuela, and Andes from w. Venezuela to n. Ecuador; Sierra de Perijá and Santa Marta Mountains.

Carduelis xanthogastra   Yellow-bellied Siskin   Jilguero Ventriamarillo

Rare to uncommon and local in the canopy and borders of montane forest and in trees in adjacent clearings in the foothills and subtropical zone on the west slope of the Andes. The Yellow-bellied Siskin is recorded only from Pichincha (sightings—the first apparently in 1989—at Tinalandia, the Nanegalito/Mindo area, the Otonga Reserve west of San Francisco de las Pampas, and north of Simón Bolívar), nw. Azuay (above Manta Real in Aug. 1991; ANSP), and El Oro (La Chonta and Buenaventura). It also recently was found on the coastal Cordillera de Mache in w. Esmeraldas at Bilsa (first by J. R. Clay et al. in Feb. 1994). Recorded from about 500 to 2200 m.

The lack of older records from the heavily collected Pichincha area makes us suspect that the Yellow-bellied Siskin is increasing at least in that province, for reasons unknown.

One race: nominate *xanthogastra*.

Range: mountains of Costa Rica and w. Panama; locally in mountains of n. Venezuela, and Andes from w. Venezuela to w. Bolivia; Sierra de Perijá.

Carduelis psaltria   Lesser Goldfinch   Jilguero Menor

Very local and usually rare (and erratic?) in montane and deciduous woodland and shrubby clearings in the foothills and subtropical zone on both slopes of the Andes. Although mainly recorded from a few well-separated west-slope localities, Lesser Goldfinches have also recently (1996) been found to be numerous—more so than they are in any other part of Ecuador—on natural grassy slopes of the arid upper Río Pastaza valley near Baños (L. Jost). On the west slope, they have been recorded from the northwest in Esmeraldas (an Oct. 1983 sighting at Cachaco; N. Krabbe), Imbabura (an Aug. 1986 sighting at Cristal [Durham University Expedition] and seen near Tumbabiro in Mar. 1992 [M. B.

Robbins et al.] and around Intag [N. Krabbe]), Pichincha (a few old specimens but only one recent report, of at least 2 males seen with a flock of Hooded Siskins [*C. magellanica*] near Nono on 22 Aug. 1995 [P. Coopmans et al.]) and in the south in El Oro (sightings from below Buenaventura in Jun. 1985; RSR and K. Berlin) and Loja at Utuana (2 specimens [ANSP] taken from several small flocks seen on 25 Apr. 1993), near Catamayo (seen on 12 Oct. 1995; D. Wolf et al.), and on the entrance road into the Cajanuma sector of Podocarpus National Park (seen on 29 Oct. 1993; D. Wolf and PJG). The Lesser Goldfinch is unaccountably scarce and local in Ecuador; it remains unclear what factors are limiting its distribution. Recorded up to 2700 m in Pichincha.

The species was formerly (e.g., Meyer de Schauensee 1966, 1970) often called the Dark-backed Goldfinch.

One race: *columbiana*.

Range: w. United States to Panama; locally from n. Venezuela and w. Colombia to n. Peru.

Passeridae   Old World Sparrows

One species. Members of the family occur naturally only in the Old World.

Passer domesticus   House Sparrow   Gorrión Europeo

Locally common in certain towns and cities, with numbers in many places seemingly fluctuating though the general trend is undeniably upward. The House Sparrow apparently spread into sw. Ecuador in the 1970s, presumably from populations in w. Peru (F. Ortiz-Crespo, *Rev. Univ. Católica* 5[16]: 193–197, 1977). As of the mid-1990s, populations were found in many towns and cities in the lowlands of w. Ecuador from n. and w. Esmeraldas and w. Pichincha southward, and they now also extend up into the highlands as far north as s. Pichincha (seen in Machachi in Feb. 1993; P. Coopmans). A few are even found on the east slope of the Andes in the towns of Macas, Zamora, and Zumba. House Sparrows are particularly numerous on the Santa Elena Peninsula, but (so far?) have not spread into the northern highlands, though we fear they may be poised to do so. Populations in some towns seem to wax and wane (e.g., at Zumba in Aug. 1989 the species was numerous and conspicuous, whereas in Dec. 1991 it was quite uncommon). The House Sparrow was introduced into s. South America from the Old World in the 19th century. Recorded up to about 2500 m.

One race: nominate *domesticus*.

Range: native to Eurasia but widely introduced in the Americas.

Bibliography

Amadon, D., and J. Bull. 1988. Hawks and owls of the world: a distributional and taxonomic list. *Proc. West. Found. Vert. Zool.* 3(4): 295–357.

American Ornithologists' Union. 1983. *Check-list of North American Birds.* 6th ed. Washington, D. C.: American Ornithologists' Union.

———. 1998. *Check-list of North American Birds.* 7th ed. Washington, D. C.: American Ornithologists' Union.

Arctander, P., and J. Fjeldså. 1994. Andean tapaculos of the genus *Scytalopus* (Aves, Rhinocryptidae): a study of modes of differentiation, using DNA sequence data. Pp. 205–255 *in* V. Loeschcke, J. Tomink, and S. K. Jain, eds. *Conservation Genetics.* Basel, Switzerland: Birkhauser Verlag.

Balchin, C. S., and E. P. Toyne. 1998. The avifauna and conservation status of the Río Nangaritza valley, southern Ecuador. *Bird Conserv. Int.* 8(3): 237–253.

Beaman, M. 1995. *Palearctic Birds: A Checklist of the Birds of Europe, North Africa, and Asia North of the Foothills of the Himalayas.* Stonyhurst, U.K.: Harrier Publications.

Becker, C. D., and B. López-Lanús. 1997. Conservation value of a garua forest in the dry season: a bird survey in Reserva Ecológica de Loma Alta, Ecuador. *Cotinga* 8: 66–74.

Berg, K. S. 1994. New and interesting records of birds from a dry forest reserve in Southwest Ecuador. *Cotinga* 2: 14–19.

Berlepsch, H., and L. Taczanowski. 1883. Liste des oiseaux recueillis par MM. Stolzmann et Siemiradzki dans l'Ecuadeur occidental. *Proc. Zool. Soc. London* 1883: 536–577.

———. 1884. Deuxieme liste des oiseaux recueillis dans l'Ecuadeur occidental par MM. Stolzmann et Siemiradzki. *Proc. Zool. Soc. London* 1884: 281–313.

Berlioz, J. 1927. Étude d'une collection d'oiseaux de l'Equateur donnée au Muséum par M. Clavery. *Bull. Mus. Nat. Hist. Nat. Paris* 33: 353–357, 486–493.

———. 1928a. Étude d'une collection d'oiseaux de l'Equateur donnée au Museum par M. Clavery. *Bull. Mus. Nat. Hist. Nat. Paris* 34: 71–78.

———. 1928b. Notes sur quelques especes rares d'oiseaux de l'Ecuador. *Bull. Mus. Nat. Hist. Nat. Paris* 34: 437–442.

———. 1932a. Contribution a l'étude des oiseaux de l'Ecuador. *Bull. Mus. Nat. Hist. Nat. Paris* ser. 2, 4: 228–242.

———. 1932b. Nouvelle contribution a l'étude des oiseaux de l'Ecuador. *Bull. Mus. Nat. Hist. Nat. Paris* ser. 2, 4: 620–628.

———. 1937a. Notes ornithologiques au cours d'un voyage en Ecuador. *Oiseau* 7: 389–416.

———. 1937b. Notes sur quelques oiseaux rares ou peu connus de l'Equateur. *Bull. Mus. Nat. Hist. Nat. Paris* ser. 2, 9: 114–118.

———. 1937c. Étude d'une collection d'oiseaux de l'Ecuador oriental (Mission Flornoy). *Bull. Mus. Nat. Hist. Nat. Paris* ser. 2, 9: 354–361.

Best, B. J., ed. 1992. *The Threatened Forests of South-west Ecuador.* Leeds, U.K.: Biosphere Publications.

Best, B. J., C. T. Clarke, M. Checker, A. L. Broom, R. M. Thewlis, W. Duckworth, and A. McNab. 1993. Distributional records, natural history notes, and conservation of some poorly known birds from southwestern Ecuador and northwestern Peru. *Bull. B. O. C.* 113(2): 108–119 and 113(4): 234–255.

Best, B. J., and M. Kessler. 1995. *Biodiversity and Conservation in Tumbesian Ecuador and Peru.* Cambridge, U.K.: BirdLife International.

Best, B. J., R. S. R. Williams, and T. Heijnen. 1997. *A Guide to Birdwatching in Ecuador and the Galápagos Islands.* Otley, West Yorkshire, U.K.: Biosphere Publications.

Bierregaard, R. O. 1994. Family Accipitridae (hawks and eagles), neotropical species accounts. Pp. 52–275 *in* J. del Hoyo, A. Elliott, and J. Sargatal, eds. *Handbook of the Birds of the World.* Vol. 2. Barcelona, Spain: Lynx Edicions.

Blake, E. R. 1977. *Manual of Neotropical Birds.* Vol. 1. Chicago: University of Chicago Press.

Bloch, H., M. K. Poulsen, C. Rahbek, and J. F. Rasmussen. 1991. A Survey of the Motane [sic] Forest Avifauna of the [sic] Loja Province, Southern Ecuador. Cambridge, U.K.: International Council for Bird Preservation.

Bochan, B. In prep. "The birds of Jatun Sacha Biological Station."

Brosset, A. 1964. Les oiseaux de Pacaritambo (ouest de l'Ecuador). *Oiseau* 34: 1–24.

Brown, L., and D. Amadon. 1968. *Hawks, Eagles, and Falcons of the World.* Feltham, Middlesex, U.K.: Country Life Books.

Burger, J., and M. Gochfeld. 1996. Family Laridae (gulls). Pp. 572–623 *in* J. del Hoyo, A. Elliott, and J. Sargatal, eds. *Handbook of the Birds of the World.* Vol 3. Barcelona, Spain: Lynx Edicions.

Carboneras, C. 1992. Family Hydrobatidae (storm-petrels). Pp. 258–271 *in* J. del Hoyo, A. Elliott, and J. Sargatal, eds. *Handbook of the Birds of the World.* Vol 1. Barcelona, Spain: Lynx Edicions.

Chantler, P. 1999. Family Apodidae (swifts). Pp. 388–466 *in* J. del Hoyo, A. Elliott, and J. Sargatal, eds. *Handbook of the Birds of the World.* Vol. 5. Barcelona, Spain: Lynx Edicions.

Chantler, P., and G. Driessens. 1995. *Swifts: A Guide to the Swifts and Treeswifts of the World.* Sussex, U.K.: Pica Press.

Chapman, F. M. 1926. The distribution of bird life in Ecuador. *Bull. AMNH* 55: 1–784.

Cleere, N., and D. Nurney. 1998. *Nightjars: A Guide to the Nightjars, Nighthawks, and Their Relatives.* New London, Conn.: Yale University Press.

Cohn-Haft, M. 1999. Family Nyctibiidae (potoos). Pp. 288–301 *in* J. del Hoyo, A. Elliott, and J. Sargatal, eds. *Handbook of the Birds of the World.* Vol. 5. Barcelona, Spain: Lynx Edicions.

Collar, N. J., and P. Andrew. 1988. *Birds to Watch: The ICBP World Checklist of Threatened Birds.* Cambridge, U.K.: International Council for Bird Preservation.

Collar, N. J., M. J. Crosby, and A. J. Stattersfield. 1994. *Birds to Watch 2: The World List of Threatened Birds.* Cambridge, U.K.: BirdLife International.

Collar, N. J., L. P. Gonzaga, N. Krabbe, A. Madroño N., L. G. Naranjo, T. A. Parker III, and D. C. Wege. 1992. *Threatened Birds of the Americas: The ICBP/IUCN Red Data Book.* Cambridge, U.K.: International Council for Bird Preservation.

Creswell, W., R. Mellanby, S. Bright, P. Catry, J. Chaves, J. Freile, A. Gabela, M. Hughes, H. Martineau, R. MacLeod, F. McPhee, N. Anderson, S. Holt, S. Barabas, C. Chapel, and T. Sanchez. 1999. Birds of the Guandera Biological Reserve, Carchi province, north-east Ecuador. *Cotinga* 11: 55–63.

Curson, J., D. Quinn, and D. Beadle. 1994. *Warblers of the Americas: An Identification Guide.* Boston: Houghton Mifflin.

del Hoyo, J. 1994. Family Cracidae (chachalacas, guans, and currasows). Pp. 310–363 *in* J. del Hoyo, A. Elliott, and J. Sargatal, eds. *Handbook of the Birds of the World.* Vol. 2. Barcelona, Spain: Lynx Edicions.

English, P. H. 1998. "Ecology of mixed-species understory flocks in Amazonian Ecuador." Ph.D. diss., University of Texas, Austin, Tex.

Enticott, J., and D. Tipling. 1997. *A Photographic Handbook of the Seabirds of the World*. London: New Holland Limited.

Fjeldså, J. 1987. *Birds of Relict Forests of the High Andes*. Copenhagen: Zoological Museum of the University of Copenhagen.

Fjeldså, J., and N. Krabbe. 1986. Some range extensions and other unusual records of Andean birds. *Bull. B. O. C.* 106(3): 115–124.

———. 1990. *Birds of the High Andes*. Copenhagen: Zoological Museum of the University of Copenhagen.

Forshaw, J. 1989. *Parrots of the World*. 3d ed. Willoughby, Australia: Lansdowne Editions.

Gill, F. 1994. *Ornithology*. Rev. ed. New York: W. H. Freeman.

Gochfeld, M., and J. Burger. 1996. Family Sternidae (terns). Pp. 624–667 *in* J. del Hoyo, A. Elliott, and J. Sargatal, eds. *Handbook of the Birds of the World*. Vol. 3. Barcelona, Spain: Lynx Edicions.

Goodfellow, W. 1901. Results of an ornithological journey through Colombia and Ecuador. *Ibis* 1901: 300–319, 458–480, 699–715.

———. 1902. Results of an ornithological journey through Colombia and Ecuador. *Ibis* 1902: 59–67, 207–233.

Green, C. 1996. Birding Ecuador. 2d ed. Tucson, Ariz.: n. p.

Haffer, J. 1974. *Avian speciation in tropical South America*. Publ. Nuttall Ornithol. Club no. 14.

Hardy, J. W., B. B. Coffey, Jr., and G. B. Reynard. 1999. Voices of the New World Owls. Rev. ed. (by T. Taylor). Gainesville, Fla.: ARA Records.

Hayman, P., J. Marchant, and T. Prater. 1986. *Shorebirds: An Identification Guide to the Waders of the World*. London: Croom Helm.

Hellmayr, C. E. (in part with C. B. Cory or B. Conover). 1924–1949. Catalogue of birds of the Americas. *Field Mus. Nat. Hist., Zool. Ser.*, vol. 13, pts. 1–11.

Hilty, S. L., and W. L. Brown. 1986. *A Guide to the Birds of Colombia*. Princeton, N.J.: Princeton University Press.

Hinkelmann, C., and K.-L. Schuchmann. 1997. Phylogeny of the hermit hummingbirds Trochilidae: Phaethornithinae]. *Stud. Neotrop. Faun. and Environ.* 32: 142–163.

Howell, S. N. G., and S. Webb. 1995. *A Guide to the Birds of Mexico and Northern Central America*. Oxford, U.K.: Oxford University Press.

Isler, M. L., and P. R. Isler. 1987. *The Tanagers: Natural History, Distribution, and Identification*. Washington, D. C.: Smithsonian Institution Press.

Jaramillo, A., and P. Burke. 1999. *New World Blackbirds: The Icterids*. Princeton, N.J.: Princeton University Press.

Juniper, T., and M. Parr. 1998. *Parrots: A Guide to the Parrots of the World*. New Haven, Conn.: Yale University Press.

King, J. R. 1989. Notes on the birds of the Rio Mazan valley, Azuay province, Ecuador, with special reference to *Leptosittaca branickii, Hapalopsittaca amazonina pyrrhops*, and *Metallura baroni*. *Bull. B. O. C.* 109(3): 140–147.

Kirwan, G., T. Marlow, and P. Coopmans. 1996. A review of avifaunal records from Mindo, Pichincha prov., north-western Ecuador. *Cotinga* 6: 47–57

Krabbe, N. 1992. Notes on distribution and natural history of some poorly known Ecuadorean birds. *Bull. B. O. C.* 112(3): 169–174.

Krabbe, N., B. O. Poulsen, A. Frolander, and O. R. Barahona. 1997. Range extensions of cloud forest birds from the high Andes of Ecuador: new sites for rare or little-recorded species. *Bull. B. O. C.* 117(4): 248–256.

Krabbe, N., and T. S. Schulenberg. 1997. Species limits and natural history of *Scytalopus* tapaculos (Rhinocryptidae), with descriptions of the Ecuadorian taxa, including three new species. *Studies in Neotropical Ornithology Honoring Ted Parker*, Ornithol. Monogr. no. 48: 47–88.

Krabbe, N., and F. Sornoza M. 1994. Avifaunistic results of a subtropical camp in the Cordillera del Condor, southeastern Ecuador. *Bull. B. O. C.* 114(1): 55–61.

Lévêque, R. 1964. Notes on Ecuadorian Birds. *Ibis* 106(1): 52–62.

Livezey, B. C. 1997. A phylogenetic classification of waterfowl (Aves: Anseriformes), including selected fossil species. *Ann. Carnegie Mus.* 66[4]: 457–496.

Lönnberg, E., and H. Rendahl. 1922. A contribution to the ornithology of Ecuador. *Arkiv. Zool.* 14(25): 1–87.

Madge, S., and H. Burn. 1988. *Waterfowl: An Identification Guide to the Ducks, Geese, and Swans of the World*. Boston: Houghton Mifflin.

Marchant, S. 1958. The birds of the Santa Elena Peninsula, S. W. Ecuador. *Ibis* 100: 349–387.

Marín, M. A., J. M. Carrión B., and F. C. Sibley. 1992. New distributional records for Ecuadorian birds. *Ornitol. Neotrop.* 3: 27–34.

Marks, J. S., R. J. Cannings, and H. Mikkola. 1999. Family Strigidae (typical owls). Pp. 76–242 *in* J. del Hoyo, A. Elliott, and J. Sargatal, eds. *Handbook of the Birds of the World*. Vol. 5. Barcelona, Spain: Lynx Edicions.

Martínez-Vilalta, A., and A. Motis. 1992. Family Ardeidae (herons). Pp. 376–429 *in* J. del Hoyo, A. Elliott, and J. Sargatal, eds. *Handbook of the Birds of the World*. Vol. 1. Barcelona, Spain: Lynx Edicions.

Ménégaux, A. 1908. Étude d'une collection d'oiseaux de l'Equateur donneé au Muséum d'Histoire Naturelle. *Bull. Soc. Philom. Paris* ser. 9, 10: 84–100.

——. 1911. Étude des oiseaux de l'Equateur, rapportés par le Dr. Rivet. *Arc. de Méridien Equatorial* 9[1]: 1–128.

Meyer de Schauensee, R. 1964. *The Birds of Colombia*. Narberth, Penn.: Livingston Publishing.

——. 1966. *The Species of Birds of South America with Their Distribution*. Narberth, Penn.: Livingston Publishing.

——. 1970. *A Guide to the Birds of South America*. Wynnewood, Penn.: Livingston Publishing.

Mills, E. L. 1967. Bird records from southwestern Ecuador. 1967. *Ibis* 109: 534–538.

Monroe, B. L., Jr. 1968. *A Distributional Survey of the Birds of Honduras*, Ornithol. Monogr. no. 7.

Murphy, R. C. 1936. *Oceanic Birds of South America: A Study of Species of the Related Coasts and Seas, Including the American Quadrant of Antarctica, Based upon the Brewster-Sanford Collection in the American Museum of Natural History*. 2 vols. New York: MacMillan.

Norton, D. W. 1965. Notes on some non-passerine birds from eastern Ecuador. *Breviora* 230: 1–11.

Norton, D. W., G. Orcés V., and E. Sutter. 1972. Notes on rare and previously unreported birds from Ecuador. *Auk* 89(4): 889–894.

Oberholser, H. C. 1902. Catalogue of a collection of hummingbirds from Ecuador and Colombia. *Proc. U.S. Nat. Mus.* 24: 309–342.

Orcés V., G. 1944. Notas sobre la distribución geográfica de algunas aves neotropicas (del Ecuador y noroeste del Peru). *Flora* 4: 103–123.

——. 1974. Notas acerca de la distribución geográfica de algunas aves del Ecuador. *Ciencia y Naturaleza* 15(1): 8–11.

Ortiz-Crespo, F., P. J. Greenfield, and J. C. Matheus. 1990. *Aves del Ecuador, Continente y Archipiélago de Galápagos*. Quito: Fundación Ecuatoriana de Promoción Turística and Fundación Ornitológica del Ecuador (CECIA).

Parker, T. A., III, and J. L. Carr, eds. 1992. Status of Forest Remnants in the Cordillera de la Costa and Adjacent Areas of Southwestern Ecuador. Washington, D. C.: Conservation International.

Parker, T. A., III, S. A. Parker, and M. A. Plenge. 1982. *An Annotated Checklist of the Birds of Peru*. Vermillion, S.D.: Buteo Books.

Parker, T. A., III, T. S. Schulenberg, G. R. Graves, and M. J. Braun. 1985. The avifauna of the Huancabamba region, northern Peru. *Neotropical Ornithology*, Ornithol. Monogr. no. 36: 169–197.

Parker, T. A., III, T. S. Schulenberg, M. Kessler, and W. H. Wust. 1995. Natural history and conservation of the endemic avifauna in north-west Peru. *Bird Conserv. Int.* 5(2/3): 201–231.

Paynter, R. A., Jr. 1993. *Ornithological Gazetteer of Ecuador.* 2d ed. Cambridge, Mass.: Museum of Comparative Zoology.

Peters, J. L. 1934–1986. *Check-list of Birds of the World.* Vols. 1–15. Cambridge, Mass.: Museum of Comparative Zoology.

Pinto, O. M. de O. 1978. *Novo Catálogo das Aves do Brazil.* Pt. 1. São Paulo, Brazil: n. p.

Pople, R. G., I. J. Burfield, R. P. Clay, D. R. Cope, C. P. Kennedy, B. López L., J. Reyes, B. Warren, and E. Yagual. 1997. Bird Surveys and Conservation Status of Three Sites in Western Ecuador: Final Report of Project Ortalis '96. Cambridge, U.K.: CSB Conservation Publications.

Rasmussen, J. F., C. Rahbek, E. Horstman, M. K. Poulsen, and H. Bloch. 1994. *Aves del Parque Nacional Podocarpus: Una Lista Anotada.* Quito: Fundación Ornitológica del Ecuador (CECIA).

Rasmussen, J. F., C. Rahbek, B. O. Poulsen, M. K. Poulsen, and H. Bloch. 1996. Distributional records and natural history notes on threatened and little known birds of southern Ecuador. *Bull. B. O. C.* 116(1): 26–46.

Ridgely, R. S. 1980. Notes on some rare or previously unrecorded birds in Ecuador. *Am. Birds* 34(3): 242–248.

Ridgely, R. S., P. J. Greenfield, and M. Guerrero G. 1998. *Una Lista Anotada de las Aves del Ecuador Continental.* Quito: Fundación Ornitológica del Ecuador (CECIA).

Ridgely, R. S., and J. A. Gwynne. 1989. *A Guide to the Birds of Panama.* Rev. ed. Princeton, N.J.: Princeton University Press.

Ridgely, R. S., and G. Tudor. 1989. *The Birds of South America.* Vol. 1. Austin, Tex.: University of Texas Press.

———. 1994. *The Birds of South America.* Vol. 2. Austin, Tex.: University of Texas Press.

Robbins, M. B., and N. Krabbe, G. H. Rosenberg, and F. Sornoza M. 1994. The tree line avifauna at Cerro Mongus, prov. Carchi, northeastern Ecuador. *Proc. Acad. Nat. Sci. Phil.* 145: 209–216.

Robbins, M. B., and R. S. Ridgely. 1990. The avifauna of an upper tropical cloud forest in southwestern Ecuador. *Proc. Acad. Nat. Sci. Phil.* 142: 59–71.

Robbins, M. B., R. S. Ridgely, T. S. Schulenberg, and F. B. Gill. 1987. The avifauna of the Cordillera de Cutucú, Ecuador, with comparisons to other Andean localities. *Proc. Acad. Nat. Sci. Phil.* 139: 242–259.

Salaman, P., ed. 1994. Surveys and Conservation of Biodiversity in the Chocó, South-west Colombia. Cambridge, U.K.: BirdLife International.

Salaman, P., T. M. Donegan, and A. M. Cuervo. 1999. Ornithological surveys in Serranía de los Churumbelos, southern Colombia. *Cotinga* 12: 29–39.

Salaman, P., and L. A. Mazariegos H. 1998. The hummingbirds of Nariño, Colombia. *Cotinga* 10: 30–36.

Salvadori, T., and E. Festa. 1899a. Viaggo del Dr. Enrico Festa nell' Ecuador. Parte prima—passeres oscines. *Boll. Mus. Zool. Anat. Comp. Torino* 15, no. 357.

———. 1899b. Viaggo del Dr. Enrico Festa nell' Ecuador. Parte seconda—passeres clamatores. *Boll. Mus. Zool. Anat. Comp. Torino* 15, no. 362.

———. 1900. Viaggo del Dr. Enrico Festa nell' Ecuador. Parte terza—Trochili-Tinami. *Boll. Mus. Zool. Anat. Comp. Torino* 15, no. 368.

Schuchmann, K.-L. 1999. Family Trochilidae (hummingbirds). Pp. 468–680 *in* J. del Hoyo, A. Elliott, and J. Sargatal, eds. *Handbook of the Birds of the World.* Vol. 5. Barcelona, Spain: Lynx Edicions.

Schulenberg, T. S., and K. Awbrey. 1997. The Cordillera del Cóndor region of Ecuador and Peru: a biological assessment. RAP Working Paper no. 7. Washington, D. C.: Conservation International.

Short, L. 1982. *Woodpeckers of the World*. Greenville, Del.: Delaware Museum of Natural History.

Sibley, C. G. 1996. *Birds of the World*. CD-ROM, ver. 2.0. N. p.

Sibley, C. G., and J. E. Ahlquist. 1990. *Phylogeny and Classification of Birds: A Study in Molecular Evolution*. New Haven, Conn.: Yale University Press.

Sibley, C. G., and B. L. Monroe, Jr. 1990. *Distribution and Taxonomy of Birds of the World*. New Haven, Conn.: Yale University Press.

Sick, H. 1993. *Birds In Brazil: A Natural History*. Princeton, N.J.: Princeton University Press.

Snow, D. 1975. The classification of the manakins. *Bull. B. O. C.* 95[1]: 20–27.

———. 1982. *The Cotingas: Bellbirds, Umbrellabirds, and Other Species*. Ithaca, N.Y.: Cornell University Press.

Stattersfield, A. J., M. J. Crosby, A. J. Long, and D. C. Wege. 1997. *A Global Directory of Endemic Bird Areas*. Cambridge, U.K.: BirdLife International.

Stiles, F. G., and A. Skutch. 1989. *A Guide to the Birds of Costa Rica*. Ithaca, N.Y.: Cornell University Press.

Stotz, D. F., R. O. Bierregaard, M. Cohn-Haft, P. Petermann, J. Smith, A. Whittaker, and S. V. Wilson. 1992. The status of North American migrants in central Amazonian Brazil. *Condor* 94[3]: 608–621.

Stotz, D. F., J. W. Fitzpatrick, T. A. Parker III, and D. K. Moskovits. *Neotropical Birds: Ecology and Conservation*. 1996. Chigago: University of Chicago Press.

Taczanowski, L. 1877. Liste des oiseaux recueillis en 1876 au nord de Pérou occidental par MM. Jelski et Stolzmann. *Proc. Zool. Soc. London* 1877: 319–333, 744–754.

Taczanowski, L., and H. Berlepsch. 1885. Troisieme liste des oiseaux recueillis par M. Stolzmann dans L'Ecuadeur. *Proc. Zool. Soc. London* 1885: 67–124.

Tallman, D. A., and E. J. Tallman. 1977. Adiciones y revisiones a la lista de la avifauna de Limóncocha, Provincia de Napo, Ecuador. *Rev. Univ. Católica* 5: 217–224.

Taylor, P. B. 1996. Family Rallidae (rails, gallinules, and coots). Pp. 108–209 *in* J. del Hoyo, A. Elliott, and J. Sargatal, eds. *Handbook of the Birds of the World*. Vol. 3. Barcelona, Spain: Lynx Edicions.

Taylor, P. B., and B. van Perlo. 1998. *Rails: A Guide to the Rails, Crakes, Gallinules, and Coots of the World*. New Haven, Conn.: Yale University Press.

Vaurie, C. 1980. Taxonomy and geographical distribution of the Furnariidae (Aves, Passeriformes). *Bull. AMNH* 166(1): 1–357.

Wege, D. C., and A. J. Long. 1995. *Key Areas for Threatened Birds in the Neotropics*. Cambridge, U.K.: BirdLife International.

Wetmore, A. 1965. The birds of the Republic of Panamá. Pt. 1. *Smith. Misc. Coll.* vol. 150.

———. 1968. The birds of the Republic of Panamá. Pt. 2. *Smith. Misc. Coll.* vol. 150.

———. 1972. The birds of the Republic of Panamá. Pt. 3. *Smith. Misc. Coll.* vol. 150.

Wetmore, A., R. A. Pasquier, and S. L. Olson. 1984. The birds of the Republic of Panamá. Pt. 4. *Smith. Misc. Coll.* vol. 150.

Wiedenfeld, D. A., T. S. Schulenberg, and M. B. Robbins. 1985. Birds of a tropical deciduous forest in extreme northwestern Peru. *Neotropical Ornithology*, Ornithol. Monogr. no. 36: 305–315.

Wiersma, P. 1996. Family Charadriidae (plovers). Pp. 384–442 *in* J. del Hoyo, A. Elliott, and J. Sargatal, eds. *Handbook of the Birds of the World*. Vol. 3. Barcelona, Spain: Lynx Edicions.

Winkler, H., D. A. Christie, and D. Nurney. 1995. *Woodpeckers: A Guide to the Woodpeckers, Piculets, and Wrynecks of the World*. Sussez, U.K.: Pica Press.

Zusi, R. L. 1996. Family Rynchopidae (skimmers). Pp. 668–677 *in* J. del Hoyo, A. Elliott, and J. Sargatal, eds. *Handbook of the Birds of the World*. Vol. 3. Barcelona, Spain: Lynx Edicions.

Index of English Names

Index of Scientific Names